DATE DUE

GAYLORD			PRINTED IN U.S.A.

HANDBOOK OF
Muscle Foods Analysis

HANDBOOK OF
Muscle Foods Analysis

Edited by
LEO M.L. NOLLET
FIDEL TOLDRÁ

CRC Press
Taylor & Francis Group
Boca Raton London New York

CRC Press is an imprint of the
Taylor & Francis Group, an **Informa** business

CRC Press
Taylor & Francis Group
6000 Broken Sound Parkway NW, Suite 300
Boca Raton, FL 33487-2742

© 2009 by Taylor & Francis Group, LLC
CRC Press is an imprint of Taylor & Francis Group, an Informa business

---

Library of Congress Cataloging-in-Publication Data

Handbook of muscle foods analysis / editors, Leo M.L. Nollet, Fidel Toldra.
    p. ; cm.
    "A CRC title."
    Includes bibliographical references and index.
    ISBN 978-1-4200-4529-1 (hardcover : alk. paper)
    1. Meat--Analysis. I. Nollet, Leo M. L., 1948- II. Toldrá, Fidel. III. Title.
    [DNLM: 1. Meat--analysis. 2. Chemistry, Analytical--methods. 3. Food Contamination--analysis. 4. Nutritive Value. QU 50 H2357 2009]

TX556.M4H38 2009
664'.907--dc22
                            2008025420

Visit the Taylor & Francis Web site at
http://www.taylorandfrancis.com

and the CRC Press Web site at
http://www.crcpress.com

Contents

v

PART IV: SENSORY QUALITY

PART V: SAFETY

Foreword

What started with a contribution in one of my former books has resulted in a series of books and a great and solid friendship.

What is most important? The answer is obvious.

I very warmly thank Fidel, my colleague editor of these handbooks on analysis techniques and methodologies of animal products.

I greatly appreciate his input, support, and help in bringing this project to a successful completion.

Leo Nollet

Preface

Muscle foods include a vast number of foods such as meat, poultry, and seafood. These muscle foods represent some of the most important foods in Western societies.

Handbook of Muscle Foods Analysis is the first handbook of a series of books related to analysis techniques and methodologies for foods of animal origin. This book constitutes a reference volume in the analysis of muscle foods with descriptions of the main analytical techniques and methodologies, and their application to the compounds involved in sensory, nutritional, and technological quality as well as in safety.

This book contains 42 chapters. Chapter 1 "Importance of Analysis and Some Basic Concepts" introduces readers to the main topic of this book.

Part I *Chemistry and Biochemistry*, Chapters 2 through 16, focuses on the analysis of the main chemical and biochemical compounds of muscle foods. These compounds are amino acids, peptides, proteins, proteomics, proteinases, exopeptidases, lipases, glucohydrolases, fatty acids, triacylglycerols, phospholipids, cholesterol products, oxidation products, aroma compounds, carbohydrates, and nucleotides and nucleosides. Each chapter first gives some generalities about the different components in meat; next they discuss sample preparation, cleanup, separation, and detection methods. Recent literature on the topic is also reviewed.

In this part no chapter on antioxidants can be found. The reader is directed to the chapter on antioxidants included in the second of this series of books, *Handbook of Processed Meats and Poultry Analysis.*

Part II, Chapters 17 through 20, *Processing Control*, describe the analysis of technological quality and the use of some nondestructive techniques. These techniques include basic methodologies, water and moisture contents, microstructure, and physical sensors.

In Part III *Nutritional Quality*, Chapters 21 through 25 detail the analysis of nutrients in muscle foods. Meat scientists and consumers who are concerned about calories, essential amino acids, fatty acids, vitamins, and minerals, and trace elements will find all necessary information on their analysis and detection in meat.

Part IV, Chapters 26 through 31, *Sensory Quality,* deal with sensory quality and the main analytical tools used to determine color, texture, flavor, etc. Different parameters for the sensory evaluation of meat are discussed in depth.

Finally, in Part V *Safety*, the 11 chapters from 32 through 42 are devoted to safety, especially of analytical tools for the detection of pathogens, parasites, allergens, adulterations, and chemical toxic compounds, environmental, generated, or intentionally added, which can be found in muscle foods. There is a great concern about genetically modified organisms, adulterations, and authenticity of meat products; therefore, the reader will find ample information on these topics in this part.

This handbook will provide readers with a full overview of the analytical tools available for the analysis of muscle foods, and the role of these techniques and methodologies in the analysis of technological, nutritional, and sensory quality as well as safety aspects. In summary, readers will find the main types of available worldwide techniques and methodologies for the analysis of muscle foods.

The editors of this handbook would like to very cordially thank all the contributing authors. This book is the result of their enthusiastic cooperation and help. They are appreciated for their scientific and in-depth knowledge of the different and diverse topics.

Fidel Toldrá

Department of Food Science, Instituto de Agroquimica y Tecnologia de Alimentos (CSIC), Burjassot, (Valencia), Spain

Leo M.L. Nollet

University College Ghent, Member of Ghent University Association, Faculty of Applied Engineering Sciences, Gent, Belgium

Editors

Leo M.L. Nollet received an MS (1973) and PhD (1978) in biology from the Katolieke Universiteit Leuven, Belgium. Dr. Nollet is the editor and associate editor of numerous books. He edited for Marcel Dekker, New York—now CRC Press of Taylor & Francis Group—the first and second editions of *Food Analysis by HPLC* and *Handbook of Food Analysis*. The last edition is a three-volume book. He also edited *Handbook of Water Analysis* (first and second editions) and *Chromatographic Analysis of the Environment 3rd Edition* (CRC Press).

Along with F. Toldrá he has coedited two books—*Advanced Technologies for Meat Processing* (CRC Press) and *Advances in Food Diagnostics* (Blackwell Publishing) published in 2006 and 2007, respectively. With M. Poschl he has coedited *Radionuclide Concentrations in Foods and the Environment*, also published in 2006 (CRC Press).

Dr. Nollet has coedited with Y.H. Hui and other colleagues several books: *Handbook of Food Product Manufacturing* (Wiley, 2007), *Handbook of Food Science, Technology and Engineering* (CRC Press, 2005), and *Food Biochemistry and Food Processing* (Blackwell Publishing, 2006). Finally, he edited the *Handbook of Meat, Poultry and Seafood Quality* (Blackwell Publishing, 2007).

Dr. Nollet is member of the Editorial Advisory Board of the Food Science & Technology Series, CRC Press (Taylor & Francis, United States). He is head of the task group Task Team Affiliated Publications of The European Federation of Food Science and Technology.

Fidel Toldrá, PhD, is a research professor at the Department of Food Science, Instituto de Agroquímica y Tecnología de Alimentos (CSIC), Spain and serves as an European editor of *Trends in Food Science and Technology*, editor-in-chief of *Current Nutrition & Food Science* and as member of the Food Additives Panel of the European Food Safety Authority. Professor Toldrá has served as author, editor, or associate editor of 14 books on food chemistry and biochemistry and meat processing. Some of the recent ones are *Advanced Technologies for Meat Processing* (CRC Press, 2006), *Handbook of Food Science, Technology and Engineering* (CRC Press, 2006), *Advances in Food Diagnostics* (Blackwell Publishing, 2007), *Handbook of Food Product Manufacturing* (Wiley, 2007), and *Handbook of Fermented Meat and Poultry* (Blackwell Publishing, 2007). He was awarded the 2002 International Prize for Meat Science and Technology, given by the International Meat Secretariat during the 14th World Meat Congress held in Berlin. He has been elected in 2008 as Fellow of the International Academy of Food Science and Technology (IAFOST). His research is focused on the (bio)chemistry and analysis of foods of animal origin.

Contributors

Maria Luisa Marina Alegre
Analytical Chemistry Department
Faculty of Chemistry
University of Alcalá
Madrid, Spain

A. Alegría
Nutrition and Food Chemistry
Faculty of Pharmacy
University of Valencia
Valencia, Spain

Teresa Antequera
Food Science, School of Veterinary Sciences
University of Extremadura
Cáceres, Spain

María-Concepción Aristoy
Instituto de Agroquímica y Tecnología de
  Alimentos
Valencia, Spain

Jean-Denis Bailly
Mycotoxicology Research Unit
National Veterinary School of Toulouse
Toulouse, France

R. Barberá
Nutrition and Food Chemistry
Faculty of Pharmacy
University of Valencia
Valencia, Spain

Hakan Benli
Department of Animal Science
Texas A&M University
College Station, Texas, U.S.A.

Rui J. Branquinho Bessa
Estação Zootécnica Nacional
Instituto Nacional de Investigação Agrária e
  das Pescas
Vale de Santarém, Portugal

Brian C. Bowker
United States Department of Agriculture
Agriculture Research Service
Animal and Natural Resources Institute
Beltsville, Maryland, U.S.A.

Neura Bragagnolo
Department of Food Science
Faculty of Food Engineering
State University of Campinas
Campinas San Paulo, Brazil

Gianfranco Brambilla
Istituto Superiore di Sanitá
Rome, Italy

Catherine M. Burgess
Ashtown Food Research Centre
Dublin, Republic of Ireland

Paulo Cézar B. Campagnol
Public University of Santa Maria
Santa Maria, Brazil

Amy E. Claflin
Department of Animal Science
Texas A&M University
College Station, Texas, U.S.A.

Rogério Manoel Lemes de Campos
Public University of Santa Maria
Santa Maria, Brazil

Rosires Deliza
Embrapa Agroindustria de Alimentos
Rio de Janeiro, Brazil

Stefaan de Smet
Laboratory for Animal Nutrition and
 Animal Product Quality
Department of Animal Production
Ghent University
Melle, Belgium

Alessandro di Domenico
Istituto Superiore di Sanitá
Rome, Italy

Anthony Dolan
Ashtown Food Research Centre
Dublin, Republic of Ireland

Judy A. Driskell
University of Nebraska
Lincoln, Nebraska, U.S.A.

Geraldine Duffy
Ashtown Food Research Centre
Dublin, Republic of Ireland

Janet S. Eastridge
United States Department of Agriculture
Agriculture Research Service
Animal and Natural Resources Institute
Beltsville, Maryland, U.S.A.

J. Stephen Elmore
Department of Food Biosciences
University of Reading
Reading, United Kingdom

Mario Estévez
Applied Chemistry and Microbiology
University of Helsinki
Helsinki, Finland

Ellen Mosleth Færgestad
Department of Chemistry, Biology and
 Food Science
Norwegian University of Life Sciences
Matforsk/Nofima Food
Ås, Norway

R. Farré
Nutrition and Food Chemistry
Faculty of Pharmacy, University of Valencia
Valencia, Spain

Juana Fernández-López
Departamento de Tecnologia Agroalimentaria
Escuela Politécnica Superior de Orihuela
Miguel Hernández University
Orihuela (Alicante), Spain

Mónica Flores
Instituto de Agroquímica y Tecnología de
 Alimentos
Valencia, Spain

Igor Fochi
Istituto Superiore di Sanitá
Rome, Italy

Daniel Y.C. Fung
Department of Animal Sciences and Industry
Kansas State University
Manhattan, Kansas, U.S.A.

Geert H. Geesink
Co-operative Research Centre for Beef
Genetic Technologies
School of Rural Science and Agriculture
 of New England
Armidale, Australia

Andrea Germini
Faculty of Agriculture
Universita degli Studi di Parma
Parma, Italy

David W. Giraud
University of Nebraska
Lincoln, Nebraska, U.S.A.

M. Beatriz A. Gloria
Laboratorio de Bioquimica
  de Alimentos
Universidade Federal
  de Minas Gerais
Belo Horizonte, Minas Gerais, Brazil

Marion L. Greaser
Department of Animal Sciences
Muscle Biology Laboratory
University of Wisconsin
Madison, Wisconsin, U.S.A.

Philippe Guerre
Mycotoxicology Research Unit
National Veterinary School of Toulouse
Toulouse, France

Kristin Hollung
Department of Chemistry, Biology
  and Food Science
Norwegian University of Life Sciences
Matforsk/Nofima Food
Ås, Norway

Karl O. Honikel
Federal Centre for Nutrition and Food
Kulmbach, Germany

Anna Laura Iamiceli
Istituto Superiore di Sanitá
Rome, Italy

Xiaohong Jia
Department of Chemistry, Biology
  and Food Science
Norwegian University of Life Sciences
Matforsk/Nofima Food
Ås, Norway

Dafni Kagli
Department of Food Science &
  Technology
Laboratory of Food Microbiology and
  Biotechnology
Agricultural University of Athens
Athens, Greece

Jimmy T. Keeton
Department of Animal Science
Texas A&M University
College Station, Texas, U.S.A.

Young-Nam Kim
University of Nebraska
Lincoln, Nebraska, U.S.A.

M.J. Lagarda
Nutrition and Food Chemistry
Faculty of Pharmacy
University of Valencia
Valencia, Spain

V. Larrea
Department of Food Technology
Polytechnical University of Valencia
Valencia, Spain

D.A. Ledward
Department of Food Biosciences
University of Reading
Reading, United Kingdom

M.A. Lluch
Department of Food Technology
Polytechnical University of Valencia
Valencia, Spain

Maria Concepción García López
Analytical Chemistry Department
Faculty of Chemistry
University of Alcalá
Madrid, Spain

Marios Mataragas
Department of Food Science & Technology
Laboratory of Food Quality Control
  and Hygiene
Agricultural University of Athens
Athens, Greece

David Morcuende
Food Technology
University of Extremadura
Cáceres, Spain

Elena Muriel
Food Science
School of Veterinary Sciences
University of Extremadura
Cáceres, Spain

Anu Näreaho
Witold Stefanski Institute of Parasitology
Polish Academy of Sciences
Warsaw, Poland

Geoffrey R. Nute
Division of Farm Animal Science
School of Veterinary Science
University of Bristol
Langford, United Kingdom

George-John E. Nychas
Department of Food Science & Technology
Laboratory of Food Microbiology and
  Biotechnology
Agricultural University of Athens
Athens, Greece

Ernie W. Paroczay
United States Department of Agriculture
Agriculture Research Service
Animal and Natural Resources Institute
Beltsville, Maryland, U.S.A.

José Angel Pérez-Alvarez
Departamento de Tecnologia Agroalimentaria
Escuela Politécnica Superior de Orihuela
Miguel Hernández University
Orihuela, Spain

I. Pérez-Munuera
Department of Food Technology
Polytechnical University of Valencia
Valencia, Spain

Trinidad Perez-Palacios
Food Science, School of Veterinary Sciences
University of Extremadura
Cáceres, Spain

José A. Mestre Prates
Faculdade de Medicina Veterinária
Universidade Tecnica de Lisboa
Pólo Universitário do Alto da Ajuda
Lisbon, Portugal

A. Quiles
Department of Food Technology
Polytechnical University of Valencia
Valencia, Spain

Katleen Raes
Research Group EnBiChem
Department of Industrial Engineering and
  Technology
University College West-Flanders
Kortrijk, Belgium

Milagro Reig
Institute of Food Engineering for Development
Polytechnical University of Valencia
Valencia, Spain

Jean-Pierre Renou
Research Unit 370 QuaPA
INRA Theix
Champanelle, France

Stefano Rossi
Faculty of Agriculture
Universita degli Studi di Parma
Parma, Italy

Andreas Rossmann
Isolab GmbH
Schweitenkirchen, Germany

Jorge Ruiz
Food Science
School of Veterinary Sciences
University of Extremadura
Cáceres, Spain

Susanne Rummel
Bayerische Staatssammlung für Paäontologie
und Geologie
München, Germany

Hanns-Ludwig Schmidt
Isolab
Schweitenkirchen, Germany

Miguel A. Sentandreu
Instituto de Agroquímica y Tecnología de
Alimentos
Valencia, Spain

Stephen B. Smith
Department of Animal Science
Texas A&M University
College Station, Texas, U.S.A.

Morse B. Solomon
United States Department of Agriculture
Agriculture Research Service
Animal and Natural Resources Institute
Beltsville, Maryland, U.S.A.

Nicole Tanz
Isolab GmbH
Schweitenkirchen, Germany

Alfredo Teixeira
Escola Superior Agraria de Bragança
Instituto Politécnico de Bragança
Bragança, Portugal

Nelcindo Nascimento Terra
Public University of Santa Maria
Santa Maria, Brazil

Fidel Toldrá
Department of Food Science
Instituto de Agroquimica y Tecnologia de
Alimentos
Valencia, Spain

Alessandro Tonelli
Faculty of Agriculture
Universita degli Studi di Parma
Parma, Italy

Ron K. Tume
Commonwealth Scientific and Industrial
Research Organisation
Food Science Australia
Queensland, Australia

Saskia M. van Ruth
Instituutvoor Voedselveiligheid
Wageningen Universiteit en Research
Centrum
Wageningen, The Netherlands

Eva Veiseth
Department of Chemistry, Biology and
Food Science
Norwegian University of Life Sciences
Matforsk/Nofima Food
Ås, Norway

Sonia Ventanas
Food Technology
University of Extremadura
Cáceres, Spain

Eric Verdon
E.V. Community Reference Laboratory for
Antimicrobial and Dye Residues in Food
from Animal Origin
Fougéres, France

Véronique Verrez-Bagnis
Department of Marine Food Sciences
and Techniques
Ifremer, France

Chapter 1

Importance of Analysis and Some Basic Concepts

D.A. Ledward

Contents

1.1   Introduction

Muscle foods have been consumed for generations and form the focal point of many dishes around the world. In the West the main meat-producing animals are cattle, pigs, sheep, and to a lesser extent goats, deer, and rabbits. Within the classification of muscle foods one also needs to consider poultry, of which chicken, turkey, geese, and ducks are the most widely consumed, as well as a whole range of fish and shellfish. Although these are the main species consumed in the West, many more species are consumed elsewhere in the world and many of these are increasingly being seen in the West, such as crocodile, ostrich, and a range of game meats.

It is readily apparent that the composition of these species varies considerably, and even within one species wide variations occur. Although meat is defined as the edible portion of the flesh of these species, this definition is open to very wide interpretation, and given the cost of meat compared to many other foods, accurate, reliable, and rapid means of analysis are of paramount importance for obvious economic reasons: Fat is considerably cheaper than striated muscle, which

1

itself can vary quite markedly in eating quality and thus cost. The price differential between a high-quality grilling meat such as beef fillet (psoas major) or sirloin (longissimus dorsi) and the shin and neck muscles of cattle can be as much as a factor of 3 or 4. Although the definition normally refers to the musculature of the species, it is often widened to include organs such as the liver, kidneys, brains, and other edible tissues.

Even for compositional analysis, simple chemical analysis is often suspect because, although fat and protein content can be easily determined by such methods, their determination in meat can be problematical. For example, the protein content, the most valuable component, is traditionally determined by estimation of the nitrogen content of the meat or meat product using either the Kjeldahl or Dumas method. Other nitrogenous compounds will, of course, complicate the analysis; even if these are correctly estimated, the various meat proteins themselves have different nitrogen contents, so that reliable conversion factors are difficult to determine. Thus, in muscle foods, a factor of 6.25 is traditionally used, which equates to the protein having a nitrogen content of 16%. However, collagen, a protein that is an essential part of muscle, has a higher nitrogen content (about 17.5%) since it contains approximately 33% of the lowest molecular weight amino acid, glycine, in its structure; thus, a lower conversion factor has to be used, approximately 5.7.

In addition to chemical analysis of the content of the major components, with regard to the quality of meat, subtle variations that can occur in the nature of these components are responsible for organoleptic/sensory differences that are reflected in the flavor, tenderness, color, or nutritional quality of the meat. A few such examples are discussed later.

Another important area of interest is the presence or lack of numerous trace compounds. These trace compounds may be components of the muscle or fat, or they may be contaminants or additives from the environment or diet. We are all aware of the problems that bovine spongiform encephalopathy (BSE) presented to the meat industry, and much analytical effort has been devoted in recent years to estimate or determine the presence or lack of these prions in meat and other animal tissues. In addition, specific bacteria that may be harmful to health, in addition to toxic contaminants that may inadvertently become part of the muscle due to diet or the environment in which the animal has been raised are of major and increasing concern, posing challenging analytical problems. Of equal importance to these contaminants in affecting the wholesomeness of the food are those trace or minor compounds that form an integral part of the muscle and have major impacts on the eating quality of the meat, including the texture (toughness, juiciness), flavor, color, or nutritional quality. Many flavor compounds are present at parts per billion or even less, and thus analytical procedures that are both sensitive and specific are needed.

1.2 Inherent Differences and Those Due to Management

In addition to the obvious differences due to species, there are very wide differences in the composition of muscles due to such factors as breed, sex, age, anatomical location, training and exercise, and, of course, the plane of nutrition. Even when all these factors are controlled, there still occurs interanimal variability and this presumably genetic effect is far from well understood. The effect of all these factors on muscle composition has been thoroughly discussed in many established texts, including those of Lawrie and Ledward[1] and Warriss.[2] Perhaps the most interesting is the differences observed in the eating quality and to some extent the composition of muscles from different parts of the animal. These differences are manifest in many ways, and even quite closely related muscles on the carcass may have very different quality attributes. For example, the muscles of a T-bone beefsteak, that is, psoas major (fillet), and the longissimus dorsi (sirloin) are both high-quality, tender muscles, yet they vary quite markedly in their color stability. The psoas major muscle of a beef carcass is very color-unstable, while the longissimus dorsi muscle is relatively color-stable; thus, when displayed in air the bright red oxymyoglobin initially present at the surface of the muscle will very rapidly convert to the brown, undesirable metmyoglobin in the psoas major muscle while the longissimus muscle will retain its red color for a longer period. Another well-established difference is in the texture of muscles; to some extent this reflects the use to which the muscle was put during life. Thus, in most animals and birds, the well-used muscles are usually darker in color and tougher than those that are relatively inactive. For example, in broiler chickens the breast muscles are invariably pale in color and very tender while the leg muscles are darker in color and less tender. This is because the leg muscle is used extensively, so that higher concentrations of the oxygen storage protein, myoglobin, accumulate, and this iron-containing protein is mainly responsible for the color of muscles. In addition, collagen (the structural protein contained in the sheet of connective tissue surrounding each muscle fiber, each muscle bundle, and the intact muscle) tends to become far more heavily cross-linked in the leg. These cross-links become very stable on maturation of the animal, so that even after prolonged, dry cooking the connective tissue is not broken down and the muscle is tough. To break down this connective tissue protein, it is essential that the meat be cooked in the presence of water for a considerable period of time so that hydrolysis of the peptide chains of the collagen can occur. This difference due to collagen quality between the leg and breast muscles of poultry is of little practical importance since the birds are usually slaughtered at only a few weeks of age. With advancing age, though, collagen cross-linking continues, so that laying hens at the end of their lives must be boiled to make them edible. This phenomenon is of more importance in red-meat animals, which are slaughtered when much older, and explains why on a beef carcass some muscles must be stewed or braised to make them acceptably tender while others can be grilled or roasted.

As with the protein of muscle, the fat component (both quantity and quality) also varies with such factors as species, plane of nutrition, and sex. It is readily apparent from their consistency that the fat of ruminants (sheep and cattle) tends to be more saturated and thus harder than that of nonruminants (poultry and pigs), while that of most fish is so unsaturated that it is liquid at room temperature. The degree of unsaturation is of paramount importance in dictating the keeping quality of muscle foods (as well as the flavor), since unsaturated fats, especially polyunsaturated ones, are far more susceptible to oxidation, a process that can give rise to off-flavors and even potentially toxic compounds. However, in recent years, for nutritional reasons emphasis has been placed not just on the ratio of polyunsaturated (P) to saturated (S) fat (where P:S ratios greater than 0.4 are deemed to be desirable) but also on the n-3 to n-6 levels, where n-6 to n-3 ratios of less than 4 are deemed to be optimal for health.[3] Although species is the major determinant of these ratios, the ratios can be manipulated by diet—relatively easily in the case of nonruminants, where the fatty acid composition of the flesh tends to reflect the diet, but also to some extent in ruminants. For example, grass-fed animals have better n-6 to n-3 ratios than concentrate/grain-fed ones,[4] and also the flavor of the meats differ.[5] Because these highly unsaturated fats are so labile, their determination poses analytical problems.

In addition to differences due to many preslaughter factors, preslaughter handling, the slaughtering process itself, and postslaughter handling of the meat can have major impacts on eating quality and thus analytical measurements. Again, there are numerous texts dealing with basic meat science and the conversion of muscle to meat[1,2] and thus the remainder of this chapter will be devoted to a very brief overview of how some of these factors, pre- and postslaughter, may effect the analytical and quality aspects of meat.

1.3 Preslaughter Considerations

It is readily apparent that for both welfare and quality reasons animals should be handled as gently as possible in environments that are relatively stress-free. This treatment not only prevents damage to the carcass (bruising, for example) but also ensures that the glycogen reserves within the muscles are maintained at their normal values. After slaughter, during postmortem glycolysis, glycogen is converted anaerobically to lactic acid so that in a normal carcass the pH decreases from a little over 7 to around 5.5, the extent and rate depending on what is often referred to as the glycolytic rate. However, if an animal is stressed over a period of time before slaughter, then the glycogen is used up in the muscle and after death the pH drops less than normal. In extreme cases the pH may remain well above 6 following slaughter, and this is detrimental to meat on many grounds. Although the water-holding capacity is increased,[1] the higher pH encourages bacterial growth and hence spoilage; also, instead of becoming a bright red on exposure to air, the meat remains purplish in color, which most consumers find unattractive; such meat is referred to as dark, firm, and dry (DFD). In other situations in some breeds of animals, especially pigs, the pH may drop very rapidly from around 7 to 5.5 or less, so that an acidic pH is achieved while the muscle temperature is still high, above 30°C. Under these conditions proteins in the muscle, especially myosin, may denature or partially denature[6] and thus lose their ability to hold water. Thus, the meat becomes very moist-looking and is referred to as pale, soft, and exudative (PSE). The time in which the muscles of a normal, well-rested animal reach their ultimate pH value is very dependent on species. In cattle and sheep, under normal chilling conditions, the ultimate pH is not normally realized for at least 24 h, while in pigs the ultimate pH is normally achieved within 6–8 h. In broiler chickens, postmortem glycolysis is very rapid, and the ultimate pH value is achieved within an hour. Although pH itself is relatively easily determined, the determination of related properties, such as water- and fat-holding capacity, color, glycolytic potential, enzymic activity, and cooked quality are less straightforward and are often empirical in nature.

1.4 Postslaughter Handling

It is obvious from the previous section that before slaughter animals must be treated with respect, and several slaughtering techniques have been developed.[2] These will not be discussed here.

Following slaughter, as postmortem glycolysis proceeds, creatine phosphate (CP) in the muscles becomes depleted so that adenosine triphosphate (ATP) cannot be regenerated, as would be the case in living muscle. As the ATP levels decrease at a certain low level, rigor is established within the muscle. In this process the actin and myosin of the myofibrillar become locked together and the muscle becomes rigid. It is critical for good meat quality, especially tenderness, that when this process occurs the muscle is as stretched as possible. If excessive shortening takes place, then the muscle can become very tough and water is expressed from the shrunken fibers. If the meat is allowed to go into rigor at a high temperature, then the depletion of ATP occurs very rapidly and the muscle is "warm shortened" and becomes tough. However, if the meat is chilled or frozen rapidly while the pH of the muscle is still high, then because of damage induced there is a large efflux of calcium ions and again ATP rapidly becomes depleted and the meat again becomes very tough—that is, it "cold shortens." It is obvious, therefore, that treatment of the carcass immediately after slaughter is very important. For hygienic reasons it is desirable that the temperature be lowered as quickly as possible. However, this can induce cold shortening in the muscles, which is undesirable. It is recommended that the temperature of the muscle should remain between 10 and 15°C until the pH has dropped to at least 6. This cold shortening phenomena is primarily a problem for those meat carcasses that go into rigor very slowly, that is, the ultimate pH is reached very slowly. Most research in this area has been carried out on cattle and sheep since, because of their more rapid rate of glycolysis, pork carcasses do not pose the same problems regarding cold shortening, although the problem is not totally unknown. In poultry, because the pH drops to its ultimate value so rapidly, cold shortening is virtually impossible to induce. It should be noted that rigor in chickens often takes place when the temperature is still quite high, and thus one might anticipate warm shortening and thus toughness occurring. Although this undoubtedly can occur, because broiler chickens are normally slaughtered at 6 weeks of age or less, it is very difficult to make the meat unacceptably tough and thus this phenomena is only of academic interest. However, the biochemistry of postmortem glycolysis is an ongoing area of interest and, because of its relationship to so many quality aspects, is continuing to present challenges to research/analytical scientists.

1.5 Postmortem Considerations

Although the animal dies, after slaughter the muscle (or the meat as it ultimately becomes) is still in many respects a living organism, and changes continue to take place during both chilled and frozen storage. Many of these changes are desirable, such as the breakdown of protein due to the action of endogenous enzymes, primarily the calpains and perhaps the cathepsins,[1] which bring about a marked improvement in tenderness. For this reason, high-quality grilling/roasting meats from red-meat animals such as beef and sheep are often aged or matured for several weeks at chill temperatures to encourage this process. In addition, many other changes take place within the muscle that give rise to flavor changes. Many of these important flavor compounds and precursors often increase in concentration with aging or conditioning but are still only present in the muscle in parts per billion or less. In addition, these important molecules are only a few of the thousand or more volatiles present in the cooked product.[7] The analytical problems thus speak for themselves. Although many of the changes occurring on aging/conditioning meats are beneficial to quality, there are other changes that are undesirable.

The most obvious detrimental change is the growth of bacteria on the meat, which can lead to both spoilage and increased risk of food poisoning. During storage other undesirable quality changes may occur, perhaps the most important of which involves the color of the meat. At its normal ultimate pH, when fresh meat is first exposed to air it is purple in color due to the predominant haem pigment being reduced myoglobin, in which the iron, in the ferrous state, is in an octahedral environment with only five of the six potential ligand positions occupied. On exposure to air, any small molecule of the correct electronic configuration will rapidly attach itself to the sixth position, which will affect the color of the meat. On exposure to air, oxygen is this ligand; this gives rise to the bright red oxymyoglobin compound that many consumers associate with freshness and quality in meat. However, on prolonged exposure to air, oxidation of the ferrous ion in the haem environment to the ferric form occurs; the oxygen molecule attached to the sixth coordination side is replaced by a water molecule to form metmyoglobin. Metmyoglobin is brown in color and is deemed to be undesirable, although it should be stressed that this compound does not affect the eating quality of the meat with regard to cooked color, flavor, or texture. However, many sophisticated packaging systems have been devised with the intention of keeping the meat of good microbial quality while still maintaining the bright red color desired by consumers. The simplest of such techniques is vacuum packing; unfortunately, in this case no oxygen remains in the pack and thus the meat retains its purple color. A more successful system is to use a mixture of 80% oxygen and 20% carbon dioxide. The carbon dioxide is inhibitory to *Pseudomonas* species that are the main spoilage organisms on many red meats, and the high oxygen concentration minimizes alteration of the oxymyoglobin to the metmyoglobin, so that an extended shelf life is possible. Again, the chemistry involved in these color changes, and that involved in many meat and fish products (cured, fermented, cooked, dried, pressure treated, etc.) is complex and can only be understood using sensitive and selective analytical procedures.

Although many of the flavor changes associated with prolonged storage of fresh meat, but not fish, at chill temperatures are deemed to be desirable, undesirable flavors can occur, of which perhaps the most important is the production of rancid flavors due to oxidation of the polyunsaturated fatty acids present in the meat or fish. These oxidative reactions can take place in the presence of oxygen, even at sub-zero temperatures, and it is this development of rancid odors and flavors that probably limits the shelf life of frozen meat and fish in the domestic refrigerator. This is the major reason why different muscle foods have very different suggested shelf lives during frozen storage—far less for poultry and pork, which have relatively high concentrations of polyunsaturated fatty acids, than for beef and lamb, whose fats are more saturated. Fish, especially fatty fish that have very high concentrations of polyunsaturated fats, have even shorter suggested storage lives when frozen in air, although textural changes are also important in fish. Since the oxidation of fats can give rise to many trace volatiles, the challenge to the analyst is clear.

In many instances muscle products are further processed, either to prolong their shelf life (including curing, fermentation, canning, pressure treatment, drying, and smoking) or to impart desirable organoleptic properties or add value.[8–10] The chemical and physical changes taking place during these processes are complex and have tested the skills of analysts for generations.

From the above-mentioned brief considerations, it is readily apparent that meat analysis is very complex. One of the major problems is that it is impossible to institute good quality "ring" testing, because the meat will invariably change in composition and quality during storage, irrespective of what storage conditions are used. Thus, it is not easy to do inter-laboratory comparisons, as can be done with more stable foods, and thus it is very important that all meat analysis carried out in a laboratory have well proven and established quality assurance procedures. In the scientific literature, authors will compare their results with those of other groups, and although they often agree, sometimes they disagree, which can give rise to the development of some weird and wonderful theories. Another possible factor in some of these comparisons is the analytical skills and procedures of the groups involved. It is indeed unfortunate that for most research publications there is very little support offered to confirm the quality of the results reported; the reader simply has to take them at their face value.

I believe the situation is becoming more acute as more and more pressure is put on authors to publish, to the extent that the quality of the data may not be of the required standard or the number of replications may be insufficient to yield robust data. Even if many replicates are taken, if the procedure itself is flawed, the results will be valueless, since they will clutter and confuse rather than clarify our knowledge.

A further complication is that with the development of so many automated and semi-automated techniques, some analysts do not fully understand the principles behind the techniques they are using to measure the concentration of a particular compound or compounds, or to determine the intrinsic properties of a product. The printout comes with a number that they quote uncritically (often to too many significant places) as an indisputable fact. As the editor of a scientific journal (*Meat Science*), I am well aware that even with the most rigorous of review processes one invariably has to take an author's data at face value, even though one may have some reservations about the quality. I hope that this situation is changing as more funding bodies insist on good quality assurance processes for all data generated, not just that which may be used in a court of law.

I believe that this book will go at least some way to addressing these problems and that future generations of analysts will be better armed to carry out robust, reliable analysis of meat and muscle foods.

References

1. Lawrie, R.A. and Ledward, D.A., *Lawrie's Meat Science*, 7th edition, Woodhouse Publishers, Cambridge, UK, 2006.
2. Warriss, P.D., *Meat Science, An Introductory Text*, CABI Publishing, Reading, UK, 2000.
3. Wood, J.D., Richardson, R.I., Nute, G.R., Fisher, A.V., Campo, M.M., Kasapidou, E., Sheard, P.R., and Enser, M., Effects of fatty acids on meat quality: A review. *Meat Sci.*, 66, 21–32, 2003.
4. Fisher, A.V., Enser, M., Richardson, R.I., Wood, J.D., Nute, G.R., Kurt, E., Sinclair, L.A., and Wilkinson, R.G., Fatty acid composition and eating quality of lamb types derived from four diverse breed × production systems. *Meat Sci.*, 55, 141–147, 2000.
5. Sañudo, C., Enser, M., Campo, M.M., Nute, G.R., Maria, G., Sierra, I., and Wood, J.D., Fatty acid composition and fatty acid characteristics of lamb carcasses from Britain and Spain. *Meat Sci.*, 54, 339–346, 2000.
6. Hector, D.A., Brew-Graves, C., Nassan, N., and Ledward, D.A., Relationship between myosin denaturation and the colour of low-voltage-electrically-stimulated beef. *Meat Sci.*, 31, 279–286, 1992.
7. Farmer, L.J., Meat flavour, in *The Chemistry of Muscle-Based Foods*, D.A. Ledward, D.E. Johnston, and M.K. Knight, Eds., The Royal Society of Chemistry, London, 169–182, 1992.
8. Rankin, M.D., *Handbook of Meat Product Technology*, Blackwell Science, UK, 2000.
9. Toldrá, F., *Research Advances in the Quality of Meat and Meat Products*, Research Signpost, Trivandrum, India, 2002.
10. Feiner, G., *Meat Products Handbook*, Woodhead Publishing Ltd., Cambridge, UK, 2006.

CHEMISTRY AND BIOCHEMISTRY

Chapter 2

Amino Acids

María-Concepción Aristoy and Fidel Toldrá

Contents

2.1 Introduction

Meat amino acids are the basic components of the muscle protein structure. Only a small part of amino acids are in free form constituting part of the soluble nitrogen material in the muscle sarcoplasm.

When meat is ingested as food, free amino acids are directly absorbed into the organism while meat proteins are easily hydrolyzed into peptides and their amino acids by enzymatic digestion, and, in the same way, absorbed into the body. Not all proteins have the same nutritional value, since protein quality strongly depends on its amino acid composition and digestibility.[1] Meat proteins are considered high-quality proteins because of their balanced content in amino acids, especially in all the essential amino acids necessary for physical and mental well-being. Nevertheless, not all meat proteins are of such high quality, since some of them, like the connective tissue proteins (collagen, elastin, etc.), are poor in essential amino acids. Meat proteins are very good from the point of view of nutritive value but are also high in price. This is the reason for the practice of replacing, in a fraudulent way, meat proteins with other, cheaper proteins from vegetables such as soy or from animal connective tissue such as collagen. On the other hand, free amino acids also contribute to meat taste and indirectly to aroma by generation of volatile compounds through Maillard reactions and Strecker degradations.[2]

Thus, the analysis of amino acids, which has been widely reviewed elsewhere,[3] is important for several reasons, including the evaluation of the nutritive value and/or sensory quality of a given meat. In this chapter, methods for the analysis of amino acids in meat are described.

2.2 Analysis of Meat Amino Acids

The strategy to follow for the analysis of amino acids depends on the final goal, which may be summarized as follows: (i) the analysis of the amino acids profile: free amino acids or total amino acids (free plus hydrolyzed) profile, (ii) the analysis of the whole amino acid content (amino nitrogen), or (iii) the analysis of a single amino acid or a group of amino acids. The structure of non-essential amino acids is shown in Table 2.1. In addition, the analysis of essential amino acids may be included under the last-named objective, but this will be the subject of Chapter 22.

2.2.1 Analysis of the Amino Acids Profile

2.2.1.1 Sample Preparation

Sample preparation will depend on whether free amino acids or total amino acids have to be analyzed.

2.2.1.1.1 Sample Preparation for Free Amino Acids Profile

Sample preparation for the free amino acids analysis includes their extraction and the cleanup or deproteinization of the extract.

The extraction consists in the separation of the free amino acids fraction from the insoluble portion of the matrix, in this case from the muscle. It is usually achieved by homogenization of the ground sample in an appropriate solvent by using a Stomacher™ or Polytron™, or may be done by means of a simple stirring in warm solvent. The extraction solvent can be hot water, 0.01–0.1N hydrochloric acid solution, or diluted phosphate buffers. In some cases, concentrated strong acid solutions such as 4% of 5-sulfosalicylic acid,[4,5] 5% of trichloroacetic acid (TCA),[6] or rich alcohol-containing solution (>75%) such as ethanol[7–9] or methanol[10] have been successfully used as extraction solvents, with the additional advantage that proteins are not extracted, meaning that there is no need for further cleaning of the sample. Once extracted and homogenized, the solution is centrifuged at more than $10,000 \times g$ under refrigeration (4°C) to separate the supernatant from the nonextracted materials (pellet) and filtered through glass-wool to remove any fat material remaining on the surface of the supernatant.

Table 2.1 Molecular Mass and Chemical Structure of Non-essential Amino Acids. Data for Essential Amino Acids are Shown in Chapter 22 (Table 22.1)

Name	Abbreviations	Molecular mass	Structure
Glycine	Gly, G	75.07	$H-\underset{\underset{NH_2}{\mid}}{CH}-COOH$
Alanine	Ala, A	89.09	$H_3C-\underset{\underset{NH_2}{\mid}}{CH}-COOH$
Arginine	Arg, R	174.20	$H_2N-\underset{\overset{NH}{\|\|}}{C}-NH-CH_2-CH_2-CH_2-\underset{\underset{NH_2}{\mid}}{CH}-COOH$
Aspartic acid	Asp, N	133.10	$HO-\underset{\overset{O}{\|\|}}{C}-CH_2-\underset{\underset{NH_2}{\mid}}{CH}-COOH$
Asparagine	Asn, D	132.12	$H_2N-\underset{\overset{O}{\|\|}}{C}-CH_2-\underset{\underset{NH_2}{\mid}}{CH}-COOH$
Glutamic acid	Glu, E	141.13	$HO-\underset{\overset{O}{\|\|}}{C}-CH_2-CH_2-\underset{\underset{NH_2}{\mid}}{CH}-COOH$
Glutamine	Gln, Q	146.15	$H_2N-\underset{\overset{O}{\|\|}}{C}-CH_2-CH_2-\underset{\underset{NH_2}{\mid}}{CH}-COOH$
Tyrosine	Tyr, Y	181.19	$HO-\langle\bigcirc\rangle-CH_2-\underset{\underset{NH_2}{\mid}}{CH}-COOH$
Cysteine	Cys, C	121.16	$HS-CH_2-\underset{\underset{NH_2}{\mid}}{CH}-COOH$
Serine	Ser, S	105.09	$HO-CH_2-\underset{\underset{NH_2}{\mid}}{CH}-COOH$
Proline	Pro, P	115.13	(pyrrolidine ring with N–H and COOH)

Source: Aristoy, M.C. and Toldrá, F. in *Handbook of Food Analysis*, Marcel Dekker, New York, 2004, 5-83 to 5-123. With permission.

Sample cleanup, in the case of meat samples, is necessary for the elimination of proteins and polypeptides by means of the deproteinization process, which can be achieved through different chemical or physical procedures. Chemical methods include the use of concentrated strong acids such as phosphotungstic (PTA), sulfosalicylic (SSA), perchloric (PCA), trichloroacetic (TCA), and picric (PA) acids or organic solvents such as methanol, ethanol, or acetonitrile.[11] Under these conditions, proteins precipitate by denaturation while free amino acids remain in solution. Some physical methods consist in the centrifugation through cut-off membrane filters (1,000, 5,000, 10,000, 30,000 Da) that allow free amino acids through while retaining large compounds.[9,12,13] All these methods give a sample solution rich in free amino acids but free of proteins.

Differences among these chemical and physical methods are related to differences in the cut-off molecular weight, recovery of amino acids, compatibility with derivatization (pH, presence of salts, etc.), or separation method (interferences in the chromatogram, etc.), and so forth. Thus, some of them give low recoveries of some amino acids, as in the case of PTA, which is the most efficient (cut-off is around 700 Da) but causes losses of acidic and basic amino acids, especially lysine. The same problem, although less severe, was observed using the cut-off membrane filters.[11] Sample pH and the membrane used can affect amino acid recoveries,[14] and prewashing of filters is recommended to improve those recoveries.[13] It is important to consider that strong acids will leave a very low pH medium, interfering with the precolumn derivatization[13] where, in general, high pH is necessary to accomplish the majority of the derivatization reactions. Thus, it is essential to completely eliminate the acid by evaporation or neutralization (adjusting the pH of the sample solution). This is not a problem when the amino acids have to be analyzed by ion-exchange chromatography and postcolumn derivatization; indeed, SSA was commonly used before ion exchange amino acid analysis because it gives an appropriate pH.[4] 3–4% SSA has been extensively used as a deproteinizing agent for the analysis of physiological fluids such as plasma and urine[15] but also in foods.[5,9,16] Nevertheless, low recoveries of some amino acids have been reported.[11] Using 0.6N PCA, which is easily neutralized by the addition of KOH or potassium bicarbonate, the deproteinization procedure can be very simple and no interferences have been described.[7,11] 12% TCA is normally used to fractionate cheese or sausage extracts to study the proteolysis course during ripening, with free amino acids analyzed in the soluble fraction;[17–19] PA has not been extensively used.[20–22]

The use of organic solvents, mixing 2 or 3 volumes of organic solvent with one volume of extract, has given very good results,[14,23,24] with amino acid recoveries around 100% for all them,[11] with the additional advantage of easy evaporation to concentrate the sample. Some comparative studies on these deproteinization techniques have been published.[11,25]

2.2.1.1.2 Sample Preparation for Total or Hydrolyzed Amino Acids Profile

Sometimes the total amino acid profile is requested because it gives information on the nutritional value of meat. Total amino acids include free amino acids plus those from muscle proteins. Before the analysis, proteins must be hydrolyzed into their constituent amino acids. Main hydrolysis methods are described in the following text.

The most common method for hydrolyzing meat proteins is by means of an acid digestion. Typically, samples are treated with constant boiling 6N hydrochloric acid in an oven at around 110°C for 20–96 h. Digestion at 145°C for 4 h has also been proposed.[26–28] These temperatures in such an acidic and oxidative medium may degrade some amino acids. Nitrogen atmosphere and sealed vials are required during the hydrolysis to minimize the degradation. The hydrolysis may be accomplished using either liquid-phase or vapor-phase methods. Liquid-phase, where the hydrochloric acid contacts directly with sample, is well suited to hydrolyze large amounts or complex samples. When limited amounts of sample are available, the vapor-phase hydrolysis method is preferred. Data obtained by Knecht and Chang[29] demonstrated that the most likely source of contaminants is aqueous 6N hydrochloric acid. In this case, gas-phase hydrolysis can reduce to about 25% the background generated by normal liquid-phase hydrolysis. In the vapor-phase hydrolysis method, the tubes containing the samples are located inside large vessels containing the acid. Upon heating, only the acid vapor comes into contact with sample, thus excluding nonvolatile contaminants. In both cases, liquid-phase and vapor-phase, oxygen is removed and substituted by nitrogen or another inert gas, creating an appropriate atmosphere inside the vessels to assure low amino acid degradation. Thus, a system capable of alternative air evacuating/inert gas purging to get a correct deaeration inside is valuable.[30] Some commercial systems such as the Pico-Tag Workstation® are available. This system includes special vessels (flat-bottom glass tubes) fitted with a heat-resistant plastic screw cap equipped with a Teflon valve which permits the alternative air evacuating/inert gas purging, and also disposes of an oven to accomplish the hydrolysis.[31,32]

The use of microwave technology for the hydrolysis has been assayed by some authors.[30,33–38] Sample manipulation (sample evaporation to dryness, addition of constant boiling hydrochloric acid and additives, and performance under vacuum) is similar to that of a conventional oven but the duration of the treatment is shorter (less than 20 min).

Hydrolysis may be improved by optimizing the temperature and time of incubation[31] or with the addition of amino acid oxidation protective compounds. The presence of appropriate anti-oxidants/scavengers during hydrolysis can prevent losses of the most labile amino acids, all of them essential amino acids, such as tyrosine, serine, threonine, methionine, and tryptophan. Thus, protective agents currently in use, up to 1% phenol or 0.1% sodium sulfite, improve the recovery of nearly all of these except tryptophan and cysteine.

Tryptophan is often completely destroyed by hydrochloric acid hydrolysis, although considerable recoveries have been found if no oxygen is present. Some additives have been proposed to protect tryptophan against oxidation, as described in Chapter 22. Alkaline instead of acid hydrolysis is also proposed (see the following text).

Cyst(e)ine is partially oxidized during acid hydrolysis, yielding several adducts: cystine, cysteine, cystein sulfinic acid, and cysteic acid, making its analysis rather difficult. The previous performic acid oxidation of cysteine to cysteic acid, in which methionine is also oxidized to methionine sulfone,[28,33–40] improves cysteine (and methionine) recoveries, making the posterior analysis easier. The use of alkylating agents to stabilize the previous hydrolysis of cysteine constitutes a valid alternative. Good recoveries have been achieved using 3-bromopropionic acid,[41] 3-bromo-propylamine,[42] 4-vinyl pyridine,[43,44] or 3,3′-dithiodipropionic acid.[13,45,46] Finally, glutamine and asparagine are deamidated during hydrolysis and are codetermined with glutamic and aspartic acids as glx and asx, respectively. Some methods have been proposed to analyze them based on the blockage of the amide residue with [bis(trifluoroacetoxy)iodo] benzene (BTI) to acid stable L-2,4-diaminobutyric acid (DABA) and L-2,4-diaminopropionic acid (DAPA), respectively, before hydrolysis. Owing to the lack of reproducibility,[47] this methodology was optimized for the analysis of glutamine in peptides and proteins by a carefully control of temperature, duration, and BTI/sample ratio on the conversion of bound glutamine to the stable DABA.[48]

As can be observed in this section, no single set of conditions will yield the accurate determination of all amino acids. In fact, it is a compromise of conditions that offers the best overall estimation for the largest number of amino acids. In general, the 22–24 h acid hydrolysis at 110°C (vapor-phase or liquid-phase hydrolysis), with the addition of a protective agent such as 1% phenol, yields acceptable results for the majority of amino acids, enough for the requirements of any food industry. Additionally, when the analysis of cyst(e)ine is necessary, a hydrolysis procedure using the performic acid oxidation previous to the hydrolysis is a good alternative.

When high sensitivity is required, the pyrolysis from 500°C for 3 h[29] to 600°C overnight[49] of all glass material in contact with the sample is advisable, as well as the analysis of some blank samples to control the level of background present. The optimization of conditions for hydrolysis based on the study of hydrolysis time and temperature, acid-to-protein ratio, presence and concentration of oxidation protective agents, importance of a correct deaeration, and so on has been extensively reported in papers[26,27,31,50,51] and books.[52,53]

An alternative to acid hydrolysis is the alkaline hydrolysis with 4.2 M of NaOH, KOH, LiOH, or BaOH, with or without the addition of 1% (w/v) thiodiglycol for 18 h at 110°C, which is recommended by many authors[37,49,54–57] for a better tryptophan determination.

A third way to hydrolyze proteins consists in the enzymatic hydrolysis by proteolytic enzymes such as trypsin, chymotrypsin, carboxypeptidase, papain, thermolysin, or pronase. This option is chosen to analyze specific amino acid sequences or single amino acids because of their specific and well-defined activity.[58,59]

2.2.1.2   Derivatization

Once free amino acids are extracted, or freed amino acids are obtained, they are separated from each other by high-performance liquid chromatography (HPLC), capillary zone electrophoresis (CZE), or gas liquid chromatography (GLC) for individual analysis. Before or after this separation, amino acids may be derivatized to allow their separation or to enhance their detection.

Derivatization is a standard practice in the amino acid analysis. The quality of a derivatizing agent is evaluated based on the following: It must be able to (1) react with both primary and secondary amino acids; (2) give a quantitative and reproducible reaction; (3) yield a single derivative of each amino acid under mild and simple reaction conditions; (4) have the possibility of automation with good stability of the derivatization products; and (5) no interference due to by-products or excess reagent. It is worth noting that the use of sufficient reagent is of special importance when dealing with biological samples since reagent-consuming amines, although unidentified, are always present.[4]

Two types of derivatives are obtained depending on the chosen separation and/or detection technique. The first type are those derivatives that enhance amino acid detection; these include derivatives for spectroscopic or electrochemical detection (ECD). The derivatives formed will be separated by HPLC or CZE; the choice of derivative is important because their spectral (high ultraviolet absorbing or fluorescence properties) or electrochemical characteristics will affect the sensitivity and selectivity of detection. The second type are those derivatives that allow GLC amino acid separation by increasing their volatility and temperature stability or the efficient separation of racemic mixtures of the amino acids.

2.2.1.2.1   Derivatives for Spectroscopic Detection

By labeling the amino acids with reagents that enable ultraviolet or visible absorption or add fluorescent properties to the molecule, the detection is improved not only in sensitivity but also in selectivity. The derivatization reaction can be performed after separation of the amino acids (postcolumn derivatization) or before separating them (precolumn derivatization). While postcolumn techniques should be run online for maximum accuracy, precolumn techniques can be run either offline or online.

2.2.1.2.1.1   Postcolumn Derivatization

Postcolumn derivatization involves the separation of the free amino acids themselves through the liquid chromatographic column, the introduction of a suitable derivatizing reagent into the effluent system from the column, the flow of both combined liquids through a mixing manifold followed by a reaction coil, and finally the pumping of the derivatized amino acids through an online detector system. This method has been employed in the classical Moore and Stein-type commercial amino acid analyzers. Obviously, the main drawback of this derivatization method is the required additional equipment, namely, another pump to introduce the reagent as well as mixing and sometimes heating devices. Another disadvantage is the peak broadening produced by the dead volume introduced behind the column. Although this broadening may not affect the process significantly when using standard-bore columns with flow rates above 1 mL/min, postcolumn derivatization is not suitable for narrow-bore HPLC.

The reagents that have been usually employed for postcolumn derivatization are ninhydrin, fluorescamin, and o-phthalaldehyde (OPT). These are described in the following text.

Ninhydrin: This was the original Moore and Stein reagent. It reacts with all 20 amino acids at acid pH (between 3 and 4), giving colored derivatives without interfering by-products nor multiple derivatization products. Those with a primary amino group give a blue reaction product with a maximum of absorbance at 570 nm, while secondary amines yield a brownish reaction product that absorbs around 440 nm. For a full amino acid analysis, either a wavelength switching facility or a continuous dual wavelength detection is necessary. The sensitivity is below 1 nmol but rarely reproducible below 100 pmol. An improved methodology consists of the use of sodium borohydride as ninhydrin reducing agent, which yields a highly stable derivative and forms no precipitate in the flow-through system of the analyzer.[60]

Fluorescamine: This was the first reagent introduced[60–64] for potential improvement of the sensitivity beyond that achieved with ninhydrin. It forms a fluorescent derivative (fluorescamine itself does not fluoresce) with primary amino acids, but not with secondary amino acids. The fluorescence is recorded at a wavelength emission of 475 nm after excitation at 390 nm. The reaction takes place in a very short time (seconds). A major drawback is the fact that the reaction takes place only under alkaline conditions, though the separation through the ion exchange column takes place under acidic conditions. This necessitates the addition of a second postcolumn pump to introduce an alkaline buffer before fluorescamine. This reagent has been used in precolumn derivatization for special applications, including the analysis of 3-methylhistidine (3-MH) as a marker of the content of lean meat in meat products.[65]

o-Phthalaldehyde (OPT): This can be detected either spectrophotometrically (UV at 338 nm) or with fluorescence (λ_{ex} = 230 or 330 nm, λ_{em} = 455 nm) for higher sensitivity at the femtomole range.[66] When OPT was introduced for amino acid analysis, it was used exclusively in the postcolumn mode[66–68] but most applications are currently performed with the precolumn technique, as described in Section 2.2.1.2.1.2.

2.2.1.2.1.2 Precolumn Derivatization

Under this option, the formed molecule improves sensitivity and selectivity at the detection; also in addition, the derivatizing agent confers hydrophobicity to the amino acid molecule, making it suitable for separation by partition chromatography in a reversed-phase (RP) column. The most common derivatizing agents for meat amino acids are described in the following text.

Phenylisothiocyanate (PITC): This reagent was initially used in protein sequencing (Edman's reagent) and has been found suitable for amino acid analysis. The methodology involves the conversion of primary and secondary amino acids to their phenylthiocarbamyl (PTC) derivatives, which are detectable at UV (254 nm), with detection limits around 5–50 pmol. All PTC-amino acids have similar response factors, which constitutes an advantage. The PTC-amino acids are moderately stable at room temperature for 1 day, and much longer when kept under frozen (especially dry) storage conditions. The methodology is well described in the literature.[12,68–71] Sample preparation is quite tedious, requiring a basic medium (pH = 10.5), which is achieved by the addition of triethylamine, and including several drying steps, the last of which is necessary to eliminate the excess of reagent, which may cause some damage to the chromatographic column. It is important to ensure a basic pH in order to get adequate derivatization recovery, which is more critical when amino acids from acid hydrolysis are being analyzed, since no buffer is used during the reaction. The reaction time is less than 10 min, though 20 min is recommended for a complete reaction.[70,71] The composition and quality of the derivatization reagents seem to be essential for correct aspartic and glutamic acid recovery.[12,72] Also, the presence of some salts, divalent cations, and metals may cause insolubility of aspartic and glutamic acids, but recovery of almost 100% can be achieved for the other amino acids when analyzing hydrolyzed samples with NaCl to protein ratios as high as approximately 100:1.[73] The chromatographic separation takes around 20 min for hydrolyzed amino acids and 60 min for physiological amino acids. Figure 2.1 shows a separation of 21 physiological PTC-amino acids.

Sarwar et al.[74] reported a modification of this method in which the analysis of 27 physiological amino acids could be performed in 22 min (30 min including equilibration). PTC-cystine shows a poor linearity that makes the quantitation of free cystine infeasible with this method.[75] The selection of the column is critical to get a good resolved separation, especially when the analysis of physiological amino acids is involved. Some trademark columns are less susceptible to the damage caused by the reagent than others, including that recommended by Waters Corporation (Milford, MA) which is a Nova-Pack C18® column (in 15 cm for hydrolyzed or in 30 cm for physiological amino acids). This method is available as a commercially prepackaged system called "Pico-Tag" (Waters Associates), which includes the analytical column, standards, and solvents. Currently,

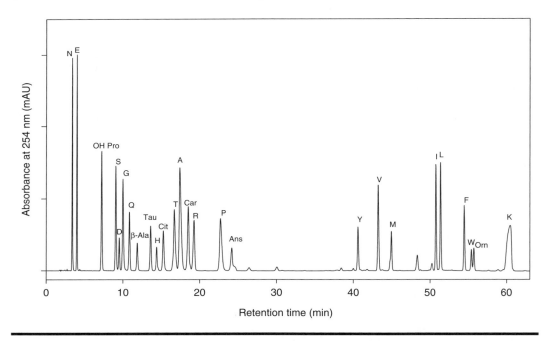

Figure 2.1 Reversed-phase HPLC chromatogram of a model solution with standards after PITC derivatization. N: asp, E: glu, S: ser, D: asn, G: gly, Q: gln, H: his, T: thr, A: ala, Car: carnosine, R: arg, P: pro, Ans: anserine, Y: tyr, V: val, M: met, I: ile, L: leu, F: phe, W: try, L: lys. (Reproduced from Aristoy, M.C. and Toldrá, F., *Handbook of Food Analysis*, Marcel Dekker, New York, 2004. With permission.)

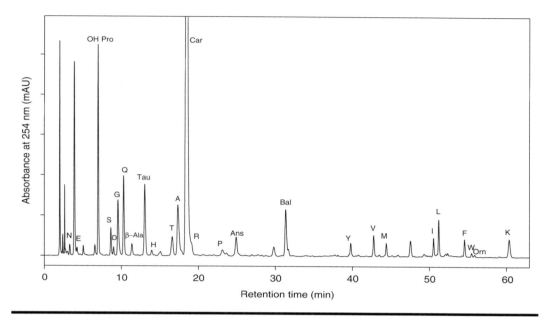

Figure 2.2 Reversed-phase HPLC chromatogram of a deproteinized pork meat extract after PITC derivatization. Bal: balenine. Nomenclature as in Figure 2.1. (Reproduced from Aristoy, M.C. and Toldrá, F., *Handbook of Food Analysis*, Marcel Dekker, New York, 2004. With permission.)

PITC is one of the preferred precolumn derivatizing agents for analysis of physiological amino acids from foods and feeds by HPLC, especially for the analysis of meat amino acids due to the presence in such samples of the dipeptides carnosine, anserine, and balenine, which are well separated, as shown in Figure 2.2.

Butylisothiocyanate (BITC): This reagent has been used in Woo et al.[76] and Woo[77] for the analysis of amino acids in foods. It forms butylthiocarbamyl amino acids at 40°C for 30 min, and yields a similar sensitivity to PTC amino acids. The advantage of BITC reagent over PITC consists in its high volatility. The excess of reagent and by-products are removed in about 10 min, significantly shorter than 60 min required using the PITC reagent.

4-Dimethyl-aminoazobenzene-4'-sulfonyl chloride (Dabsyl-Cl): This reagent was first described in 1975 for use in amino acid analysis.[78,79] Detection is by absorption in the visible range, presenting a maximum from 448 to 468 nm. The high wavelength of absorption makes the baseline chromatogram very stable with a large variety of solvents and gradient systems. Detection limits are in the low pmol range.[49] Derivatives are very stable (weeks) and can be formed from both primary and secondary amino acids. The reaction time is around 15 min at 70°C and takes place in a basic medium with an excess of reagent. Reaction efficiency is highly matrix dependent and variable for different amino acids, being especially affected by the presence of high levels of some chloride salts.[24] To overcome this problem and obtain an accurate calibration, standard amino acid solutions should be derivatized under similar conditions. By-products originating from an excess of reagent absorb at the same wavelength and thus appear in the chromatogram. Nevertheless, Stocchi et al.[49] obtained a good separation of 35 dabsyl-amino acids and by-products in a 15 cm C18 column packed with 3 μm particle size. Commercial System Gold/Dabsylation Kit™ uses this technique (Beckman Instruments, Palo Alto, CA).

1-Dimethylamino-naphtalene-5-sulfonyl chloride (Dansyl-Cl): This has been commonly used for N-terminus analysis of peptides and proteins. The use of this reagent for typical amino acid analysis was first described by Tapuhi et al.[80] Dansyl-Cl reacts with both primary and secondary amines to give a highly fluorescent derivative. It can be used for fluorescence (λ_{ex} = 350, λ_{em} = 510 nm) or UV (λ = 250 nm) detection. The dansylated amino acids are stable for 1 day,[81] or up to 7 days when kept at $-4°C$[82] and protected from light. The sample derivatization is rather simple, needing only a basic pH (around 9.5) and a reaction time of 1 h at room temperature (in the dark), 15 min at 60°C,[80] or even 2 min at 100°C. However, the reaction conditions (pH, temperature, and excess of reagent) must be carefully calibrated to optimize the product yield and to minimize secondary reactions.[82,83] Even so, this will commonly form multiple derivatives with histidine, lysine, and tyrosine. Histidine gives a very poor fluorescence response (10% of the other amino acids), reinforcing the poor reproducibility of its results.[75] Another problem is the large excess of reagent needed to ensure a quantitative reaction. This excess is hydrolyzed to dansyl sulfonic acid, which is present in excess and is highly fluorescent, probably appearing in the chromatogram as a huge peak. However, this methodology reveals excellent linearity for cystine and also cystine-containing short chain peptides.[75,84]

9-Fluorenylmethyl chloroformate (FMOC): This reagent has been used for many years as a blocking reagent in peptide synthesis. It yields stable derivatives (days) with primary and secondary amines. The derivative is fluorescent (λ_{ex} = 265 nm, λ_{em} = 315 nm), being detected at the femtomole range. The major disadvantage is due to the reagent, by itself or hydrolyzed, which is highly fluorescent and the excess may interfere in the chromatogram. It must be extracted (with pentane or diethyl ether) or converted into noninterfering adduct before injection. The first option was included in the automated AminoTag™ method[85] developed by Varian Associates Limited. In the second option, the reaction of the excess of reagent with a very hydrophobic amine such as 1-adamantylamine (ADAM) gives a late-eluting noninterfering peak.[86] This method is preferred because the addition of ADAM is more easily automated. The reaction time is fast (45–90 s) and does not require heating. To obtain reliable and precise results, reaction conditions, such as FMOC/amino acid ratio, as well as reaction time must be optimized very carefully. An automated precolumn derivatization routine, which includes the addition of ADAM, is of great advantage because it guarantees the repeatability of parameters. Tryptophan adducts do not fluoresce, and histidine and cyst(e)ine adducts fluoresce weakly.

o-Phthaldialdehyde (OPA): This reagent reacts with primary amino acids in the presence of a mercaptan cofactor to give highly fluorescent 1-alkylthio-2-alkyl-substituted isoindols.[87,88] The fluorescence is recorded at 455 or 470 nm after excitation at 230 or 330 nm and the reagent itself is not fluorescent. OPA derivatives can be detected by UV absorption (338 nm) as well. The choice of the mercaptan can affect derivative stability, chromatographic selectivity, and fluorescent intensity;[67,89–91] 2-mercaptoethanol, ethanethiol, and 3-mercaptopropionic acid are the most frequently used. The derivatization is fast (1–3 min) and is performed at room temperature in an alkaline medium, pH 9.5. OPA amino acids are not stable; this problem is overcome by standardizing the time between sample derivatization and column injection by automation. This is relatively easy because the reaction is fast and no heating is necessary. Currently, many automatic injectors are programmable and are able to achieve automatic derivatizations. Some reports have been published proposing several methods of automation,[92,93] and some of them have been patented and commercially marketed (AutoTag OPA™ from Waters Associates). One of the main disadvantages of this procedure is the inability of OPA to react with secondary amines. This drawback can be overcome in two ways. In postcolumn derivatization it is usually the case that oxidation converts secondary to primary amines with hypochlorite or Cloramine T before OPA derivatization,[67,68,94] but in the precolumn technique it is normal to combine the OPA with other derivatization methods such as FMOC, which should take place sequentially. This is the basis of the AminoQuant™ system developed and marketed by Agilent Technologies and described by Schuster[23] and Godel et al.[89]

The yield with lysine and cysteine is low and variable. The addition of detergents such as Brij 35 to the derivatization reagent seems to increase the fluorescence response of lysine.[95] In the case of cysteine, several methods have been proposed before derivatization. These methods include the conversion of cysteine and cystine to cysteic acid by oxidation with performic acid, or carboxymethylation[96] of the sulfhydryl residues with iodoacetic,[97] or the formation of the mixed disulfide S-2-carboxyethylthiocysteine (Cys-MPA) from cysteine and cystine using 3,3′-dithiodipropionic acid,[45] as incorporated by Godel et al.[89] into the automatic sample preparation protocol described by Schuster.[23] In these methods, cysteine and cystine are quantified together. Another proposal[98] consists of a slight modification in the OPA derivatization method, using 2-aminoethanol as a nucleophilic agent and altering the order of the addition of reagents in the automated derivatization procedure.[23]

6-Aminoquinolyl-N-hydroxysuccinimidyl carbamate (AQC): This reacts with primary and secondary amines from amino acids, peptides, and proteins, yielding very stable derivatives (1 week at room temperature) with fluorescent properties ($\lambda_{ex} = 250$ nm, $\lambda_{em} = 395$ nm), which are separated by reversed-phase (RP)-HPLC. Ultraviolet detection (254 nm) may also be used. Sensitivity is in the femtomole range, making them quite adequate for biochemical research.[99] The main advantage of this reagent is that the yield and reproducibility of the derivatization reaction is scarcely interfered with by the presence of salts, detergents, lipids, and other compounds naturally occurring in biological samples and foods. Furthermore, the optimum pH for the reaction is in a broad range, from 8.2 to 10. Both facts facilitate sample preparation. The excess of reagent is consumed during the reaction to form aminoquinoline (AMQ), which is only weakly fluorescent under the amino acid derivative detection conditions and does not interfere in the chromatogram. Reaction time is short, 1 min, but 10 min at 55°C would be necessary if tyrosine mono-derivative is required, because both mono- and di-derivatives are the initial adducts from tyrosine. Fluorescence of tryptophan derivative is very poor, and UV detection at 254 nm may be used to analyze it. In this case, the AMQ peak is very large at the beginning of the chromatogram, and may interfere with the first eluting peaks.[40] The chromatographic separation of these derivatives has been optimized for the amino acids from hydrolyzed proteins, but the resolution of physiological amino acids is still incomplete and needs to be improved,[15] which is the main drawback of this method. Cysteic acid and methionine sulfone, which are the adducts after performic acid oxidation of cistine/cysteine and methionine, respectively, are well separated in the chromatogram (see the corresponding chromatogram in Chapter 22). The methodology has been commercialized as a prepackaged AccQ Tag™ kit by Waters Corporation. Figure 2.3 shows a chromatogram from an AQC-amino acids model solution.

2.2.1.2.2 Derivatives for Electrochemical Detection (ECD)

These derivatives are molecules with electroactive functional groups. All of them also have spectroscopic properties, specifically, OPA/mercaptoethanol, or OPA/sulfite,[100,101] and naphtalene-2,3-dicarboxaldehyde[102] (NDA); in addition to fluorescent properties, the derivatives generated possess electroactivity (750 mV). Both reagents react with primary amines, and the sensitivity

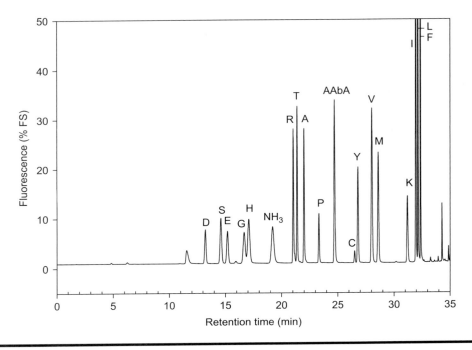

Figure 2.3   Reversed-phase HPLC chromatogram of an amino acids model solution after ACQ-Tag derivatization. Nomenclature as in Figure 2.1. AABA: alpha aminobutyric acid, C: Cysteine.

reached when connecting with capillary electrophoresis (CE) is at the femtomole level. 6-AQC[103] and PITC[104] again have the advantage of reacting with secondary amines.

2.2.1.2.3   Derivatives to Allow Separation

2.2.1.2.3.1   Volatile Derivatives

The aim here is to obtain a volatile and thermostable molecule suitable for analysis by GLC. Reactions consist of two stages: an esterification with an acidified alcohol followed by *N*-acylation with an acid anhydride in an anhydrous medium.

2.2.1.2.3.2   Chiral Derivatives

One of the techniques used to resolve racemic mixtures of D- and L-amino acids is the use of amino acid derivatives formed with fluorescent chiral reagents that permit both an effective separation of the isomers in an RP-HPLC column or in CE followed by an enhanced detection.[104–109]

2.2.1.3   Separation

The analysis of individual amino acids needs prior separation unless a very selective method of detection is used. The separation of the individual amino acids in a mixture requires very efficient separation techniques such as chromatographic (liquid or gas chromatography) or CE methods. The choice mainly depends on the equipment available or personal preferences, because each possible methodology has advantages and drawbacks.

2.2.1.3.1 Liquid Chromatographic (LC) Methods

2.2.1.3.1.1 Cation Exchange Chromatography (CEC)

This methodology is based on the amino acids' charge; the underivatized amino acids are separated using sulfonated polystyrene beads as the stationary phase and aqueous sodium citrate buffers as the mobile phase. The elution involves a stepwise increase in both pH and sodium or lithium ion concentration. Under these conditions, the more acidic amino acids elute first; those with more than one primary amino group or possessing a guanidyl residue elute at the end of the chromatogram. The original method required two separate columns and needed about 4 h to achieve a complete analysis. After separation, amino acids were converted into colored ninhydrin derivatives for spectrophotometric (colorimetric) detection.

The classical procedure has been improved with a new polystyrene matrix that offers better resolution power due to its smaller particle size, speed, pellicular packaging, and better detection systems. The latest generation of Moore and Stein amino acid analyzers uses OPA or 4-fluoro-7-nitrobenzo-2,1,3-oxadiazole postcolumn derivatization to obtain highly fluorescent derivatives with enhanced sensitivity, permitting 5–10 pmol sensitivity as standard. Nevertheless, the detection levels reached with ninhydrin, over 100 pmol injected, are sufficient for food analysis. ECD with enzyme reactor coupled with CEC is also applied to the analysis of D-,L-amino acids. This analysis combines the efficiency of HPLC with the specificity of enzymes and the sensitivity of ECD.[110]

Currently, the separation times for the 20 amino acids naturally occurring in meat proteins take around 1 h and somewhat longer for physiological amino acids. Buetikofer and Ardo[111] present the separation of the 21 amino acids (and internal standard, norvaline) from cheese extract in 110 min; Grunau and Swiader[112] report the separation of 99 amine-containing compounds in only 2 h. There are other reports applying this technique to amino acid analysis in foods.[67,113]

There are many manufacturers (Beckman, Biotronik, Dionex, LKB, Pickering, etc.) that offer integrated commercial systems, including the column, buffer system, and an optimized methodology with the advantage of the ease of use and reliability.

The advantage of this method is its accurate results for all known sample types (food, biological fluids, feed, plants), which makes it a reference method for amino acid analysis. Thus, each new methodology must compare its results with those obtained by CEC. The main drawbacks of this methodology are the high cost of the ion exchange amino acid analyzer and its maintenance, the very complex mobile phase composition, and the long time for analysis.

2.2.1.3.1.2 Reversed-Phase High-Performance Liquid Chromatography (RP-HPLC)

RP-HPLC is widely used and has the advantage of requiring standard equipment that can be shared by different types of analysis. This fact and the proliferation of precolumn derivatizing agents (see Section 2.2.1.2.1) have stimulated the development of RP-HPLC methods to analyze amino acids in all kind of matrices (food, plants, biological fluids, and tissues). In this case, prior amino acid derivatization is necessary to confer hydrophobicity to the amino acid molecule, making it suitable for partition based on chromatography. This derivatization allows the spectroscopic (ultraviolet or fluorescent) detection of amino acids.

If the choice of the derivative reaction is a challenge, the choice of the RP-column is likewise not an easy subject because of the great variability of commercially available RP-columns. The most widely used column packaging consists of alkyl-bonded silica particles, mainly octadecyl-silane. However, the selectivity obtained with each trademark column varies due to the particular chemistry employed in their manufacture, resulting in different densities of bonded-phase coverage on the silica particle and hydrophobic behavior and, as a consequence, different selectivity. The presence of residual uncapped silanol groups on the silica surface, accessible to sample molecules, can cause unwanted tailing of peaks (especially for the basic amino acids). In such cases, the addition of a strong cation (e.g., triethylamine) to the mobile phase can overcome the problem. Currently, columns are more carefully manufactured, with these silanol groups blocked or inaccessible by steric impediment, avoiding the tailing. Owing to these variables, different selectivity may be found among same columns, even those made by the same manufacturer. Only columns manufactured in the same batch are guaranteed to give the same selectivity, assuming the rest of parameters are fixed. This means that when transferring a published method to a particular set of samples, it will be necessary to readjust the chromatographic conditions to get good separation of all amino acids.

Typical analytical column dimensions are 15 cm in length (for hydrolyzed amino acids) or 25–30 cm (for physiological amino acids), packed with 5 μm particles, or shorter columns, 10 or 15 cm long, packed with 3 μm particles. Mobile phase requirements consist in the ability to dissolve the sample while remaining transparent to the detection system. Mobile phase composition combines an aqueous buffered phase with an organic phase consisting of acetonitrile and/or methanol and/or tetrahydrofuran. The buffer may be constituted by less than 100 mM concentration of acetate or phosphate. A finely adjusted binary (most often used) or ternary gradient elution is often necessary when the overall amino acid profile from hydrolyzed and, especially, physiological amino acids have to be analyzed.

2.2.1.3.2 Gas Liquid Chromatographic (GLC) Methods

The very high resolution capacity is the main advantage of gas chromatography as compared with liquid chromatographic techniques, especially since the capillary columns appeared. The determination of meat amino acids by GLC is not often used, and the few examples described only involve hydrolyzed samples.[58,114,115] These authors compared GLC with cation exchange chromatography, reporting different conclusions when analyzing some hydrolyzed food samples, including meat. Nevertheless, the technique is very efficient, and it is worthwhile to mention the separation of 32 nonprotein amino acids from edible seeds, nuts, and beans,[21] or other results obtained in honey,[116] milk,[117] and cheese.[118] In many cases, GLC has been combined with mass spectrometry for detection and identification, especially in the analysis of D-isomers,[117,119–121] where the separation was achieved by using chiral-GC stationary phases.

Recently, a very fast GC analysis of physiological amino acids has been developed, capable of separating 50 compounds including amino acids, dipeptides, and amines. This methodology has been patented as EZ:faast® and commercialized by Phenomenex (Torrance, CA). The method yields a full amino acid profile (33 amino acids) in 15 min, including a 7-min extraction-derivatization step plus 8 min for the gas chromatographic separation. Protein removal is not required, and the derivatives are stable and ready for GC/FID, GC/NPD, GC/MS, and LC/MS. Described applications are available for the analysis of physiological amino acids in blood, plasma, and urine matrices, but a meat application has not been described yet, and the ability of this method to analyze the natural meat dipeptides carnosine, anserine, and balenine is unknown.

GLC is, in summary, a very highly efficient technique adequate to amino acid analysis, although application to meat samples is scarcely described. GLC is not very expensive because no solvent is used, and the equipment is very versatile and usually available in any analytical laboratory.

2.2.1.3.3 Capillary Electrophoretic (CE) Methods

The technique of Capillary Zone Electrophoresis (CZE) is extremely efficient for separation of charged solutes.[122,123] The high efficiency, speed, and the low requirements in terms of sample amount make this technique very interesting when compared with classical electrophoresis and chromatographic techniques. The difficulty of separating amino acids by this technique derives from their structure. Amino acids constitute a mix of basic, neutral, and acidic constituents, and even though a particular pH can significantly improve the resolution of one kind, it is likely to cause overlap with the others. Under the conditions of electroosmotic flow in CE, a species with different charge can be simultaneously analyzed, but with serious doubt as to its adequate resolution. CZE shows poor ability for the separation of neutral compounds, which constitutes an important limitation of this technique. Terabe et al.[124,125] introduced a modified version of CZE in which surfactant-formed micelles were included in the running buffer to provide a two-phase chromatographic system for separating neutral compounds together with charged ones in a CE system. This technique has also been termed "Micellar electrokinetic capillary chromatography" (MECC).[126] Basic theoretical considerations of this technique are described.[127]

With few exceptions,[128–132] derivatization is used to improve separation, to enhance ultraviolet detection, or to allow fluorescence detection of amino acids. Good separations have been reported for precapillary derivatized amino acids with Dansyl-Cl,[127,133–136] PITC,[137,138] phenylthiohydantoin,[139–141] and OPA.[107,142] Liu et al.[142] compared the separation of OPA-amino acid derivatives by CZE with normal and micellar solutions, showing that higher efficiency is obtained by the MECC methods with sodium dodecyl sulfate (SDS) as the micelle-forming substance. SDS is indeed the most used additive to form micelles in this kind of analysis, though other additives have been assayed, including dodecyltrimethylammonium bromide,[139] Tween 20,[127] or octylglucoside.[107] Other additives commonly used in this analysis are organic modifiers (acetonitrile, isobutanol, methanol, tetrahydrofurane, etc.). The effect of these additives on the electroosmotic mobility and electrophoretic mobility of the micelle has been studied.[143,144] Urea has also been proposed as a useful additive.[145]

The CE coupled to electrospray ionization mass spectrometry (CE-ESI-MS) allows the direct amino acid analysis without derivatization,[132] using 1 M formic acid as the electrolyte. Protonated amino acids are separated by CE and detected selectively by a quadrupole mass spectrometer with a sheath flow electrospray ionization (ESI) interface.

When sensitivity is the objective, it is relatively easy to analyze low picomole levels of OPA derivatives in micellar solutions using a conventional fluorometric detector,[142] which is usually sufficient for food analysis. Several reviews covering high sensitivity detection following CE, including theory and practice, are available in the literature.[146–151]

2.2.1.4 Detection

The choice of detection method depends on several factors, such as the technique used for the separation or the requirements for sensitivity and selectivity. There are some detectors for GLC systems, such as flame ionization detector (FID) or flame photometric detector (FPD), and others for liquid systems (Flow Injection Analysis [FIA], HPLC, CZE), such as spectroscopic (colorimetric, ultraviolet, or fluorescence) or electrochemical detectors; mass spectrometry (MS) may be used with either gas or liquid systems.

2.2.1.4.1 Detectors Specific for GLC

The FID is a universal detector for gas chromatography and is the most widely used; thermionic-N-P or FPD are selective for organic compounds containing phosphorus and nitrogen, being much more sensitive than FID for such compounds. A thermionic-N-P detector was used by Buser and Erbersdobler[152] and FPD by Kataoka et al.[153] to analyze phosphoserine, phosphothreonine, and phosphotyrosine. The main advantage of these detectors is their high sensitivity and wide linear range.

2.2.1.4.2 Spectroscopic Detectors

2.2.1.4.2.1 Colorimetric Detection

The colorimetric detection of amino acids is achieved after derivatization with nonselective colorimetric reagents such as ninhydrin (490 nm) or dabsyl chloride (448–468 nm), which were described in Section 2.2.1.2.1

2.2.1.4.2.2 Ultraviolet and Fluorescence Detection

Amino acids, in their native form, absorb at 210 nm and thus cannot be used for spectroscopic detection, since this is a very unspecific detection wavelength. Only three amino acids (phenylalanine, tyrosine, and tryptophan) possess a chromophore moiety that confers a suitable maximum absorbance for more specific ultraviolet detection (280 nm for tyrosine and tryptophan and 254 nm for phenylalanine). Tryptophan also possesses native fluorescence (λ_{ex} = 295 nm, λ_{em} = 345 nm) that facilitates a more selective detection.[59] Thus, the spectroscopic detection of amino acids requires their prior derivatization to obtain an ultraviolet-absorbing or fluorescent molecule. An explanation of the different possibilities in pre- or postcolumn derivatization has been given above (Section 2.2.1.2.1). Ultraviolet and fluorescent spectroscopic detection may be used after HPLC, CE, FIA, or thin layer chromatographic runs.

2.2.1.4.2.3 Laser-Induced Fluorescence Detection

Laser-induced fluorescence (LIF) detection was introduced as an alternative to fluorescence detection when looking for more selective and sensitive detectors with a wide linear dynamic range (3 orders of magnitude) to cover new high-sensitivity applications (chiral analysis, o-tyrosine analysis, biomedical, or pharmaceutical research) and instrumentation (CE or microcolumn liquid chromatography). Indeed, it has found widespread use as a detector for CZE,[108,138,154] though it has also been used with HPLC.[155–157] Amino acids must be derivatized in the same way as for fluorescence detection. Some reviews covering high-sensitivity detection following CE have been published.[147,149]

2.2.1.4.2.4 Mass Spectrometry

MS is based on the conversion of components of a sample into rapidly moving gaseous ions that can be resolved on the basis on their mass-to-charge ratios, which are characteristic of each ion, allowing its identification. The identification of the 22 protein amino acids may not be a problem, although this detector may be used for more complex identifications, as in D- and L-isomer mixtures, nonprotein amino acids, etc. Unfortunately, the high cost of purchase and maintenance of mass spectrometers has inhibited their widespread use in the food industry and/or food control. Nevertheless, applications in the literature are more and more frequently found.

Mass spectrometer detectors were first connected to GC equipment. Good compatibility between the two techniques, in particular when capillary columns were available, allowed rapid development in the use of these complementary techniques. Applications in foods, including the identification of nonprotein amino acids,[21] chiral amino acids,[117] *o*-tyrosine in chicken[158] or pork[156] tissues, and others,[120,121] have been reported.

MS has also been used as a spectroscopic detector after HPLC or CZE, offering the additional advantage of analyzing the amino acids without derivatization, which means minor sample manipulation and, due to its high specificity, reduced problems related to matrix interferences or poor resolution between peaks. The combination HPLC-MS detector is much more problematic than the GLC application because of the incompatibility between the techniques (solvents from chromatography, high mobile phase flow rate vs. vacuum). However, these difficulties have recently been overcome with the development of new interfaces, and the technique is widespread, though still expensive. One of the main requirements for samples to be analyzed by MS is that analytes, amino acids in this case, must be ionized. Three types of ionization modes (atmospheric pressure microwave induced plasma ionization [AP-MIPI], atmospheric pressure chemical ionization [APCI], and Electrospray Ionization [ESI]) were compared by Kwon and Moini[159] in regard to sensitivity. The best results were obtained by using AP-MIPI in conjunction with a dual oscillating capillary nebulizer. An application to food analysis has recently been reviewed.[160] In regard to the use of MS as a detector after CE amino acid separation, the analysis of 19 amino acids by CE-ESI-MS in only 17 min with minimal sample preparation and no matrix interference has been reported.[132] This included the optimization of important parameters such as the choice of a volatile electrolyte for the electrophoresis (1 M formic acid compatible with MS) and the composition and flow rate of the sheath liquid to obtain the best sensitivity.

2.2.1.4.3 Electrochemical Detection (ECD)

ECD is based upon the electrical properties of a solution of analyte when it forms part of an electrochemical cell. It basically consists in one electrode or an array of electrodes mounted in a cell with an applied potential difference. Any electrical measure such as current, potential, conductance, or charge is related to the analyte concentration. Only amino acids with aromatic rings or sulfur-containing side chains are sufficiently electrochemically active to be detected by this method.[161] There are many applications in clinical or biomedical research.[162] Typical applications in foods involve the determination of glutathione, and cysteine in vegetables.[163]

Several modes of detecting other amino acids with non–electrochemically active properties have been developed: (i) the derivatization of amino acids by attaching to them an amine or carboxylic acid that is electrochemically active such as NDA,[102,164] OPA/mercaptoethanol or OPA/sulfite,[100,101] 6-AQC,[165] or PITC;[103] (ii) the generation of chemical reactions at the electrode surfaces to produce electrochemically active products.[166,167] This method is inexpensive and may have useful applications in analyzing amino acids in foods or in process monitoring and control during the production of amino acids. (iii) The use of immobilized enzymes (amino acid oxidases together with peroxidases) to react with amino acids, again yielding electrochemically active products (hydrogen peroxide, ammonia, etc.) for detection.[165,168–173] Owing to the high degree of selectivity of this detection system, very simple pretreatment of samples (often a dilution step is sufficient) is required and, in some cases, the analysis could even be accomplished in the sample solution or after FIA technique.[162,165–167,173]

2.2.2 Analysis of the Whole Amino Acid Content

The simplest objective consists in the analysis of the whole amino acid amount without discriminating among them. This analysis does not discriminate between free amino acids and small peptides and is based on the reaction of the α-amino group with reagents such as OPA,[174] cadmium-ninhydrin,[175–178] or trinitro-benzene-sulfonic acid (TNBS),[16,179–182] which are the most frequently used. These reagents render chromophores that increase the amino acids' ultraviolet response at a higher wavelength or confer visible or fluorescent characteristics on them.

Methods for this analysis have been extensively described and compared.[16,183–186] They generally include the precipitation of proteins, reagent addition, and colorimetric, UV-absorption, or fluorescent determination of the amine nitrogen in the supernatant. Applications to meat or meat products have been reported.[182,187]

2.2.3 Special Applications

Sometimes the main goal involves the analysis of one amino acid or a group of amino acids in meat for various purposes, such as an index of adulteration or fraud, detection of contamination, processing control, detection of bad processing practices, and so on. Some specific applications are given the following sections.

2.2.3.1 Determination of 3-Methylhistidine (3-MH)

The widespread use of nonmeat protein in meat products has prompted the development of methods for quantification of meat proteins. Protein-bound 3-MH has been detected only in the contractile meat proteins myosin and actin, and is absent from collagen and from all nonmeat proteins and, therefore, has been used as an index of meat content.[188,189] Analysis of this amino acid may be achieved using any of the described methods, but the RP-HPLC separation of the fluorescamine precolumn derivatives and fluorescence detection has been the most frequently used.[190–192] Owing to the cationic nature of this amino acid, cation exchange HPLC may be a good alternative,[67] as shown in Figure 2.4, where good separation of underivatized basic amino acids and naturally occurring dipeptides present in meat, is achieved with postcolumn OPA detection.

A GC method for the *N*-heptafluorobutyryl isobutyl esters of meat amino acids, including 3-MH, has also been published.[193] Careful removal of the sarcoplasmic (soluble) proteins is necessary before hydrolysis of the myofibrillar proteins to avoid the interference of 3-MH resulting from the hydrolysis of the naturally occurring dipeptide balenine (β-alaline-3-methylhistidine), which is present in the sarcoplasmic fraction of some animal skeletal muscles, including pork.[194]

2.2.3.2 Determination of OH-Proline and OH-Lysine

4-Hydroxyproline and 5-hydroxylysine are specific amino acids located in the primary structure of collagen. Collagen is a low-quality protein, since it is poor in essential amino acid content, and thus its presence in foods decreases their value. These amino acids have been used as an index of the collagen content in meats and meat products.[8,189,195–198]

The analysis of these amino acids is achieved through any of the described methods, RP-HPLC being the most used, but taking into account the fact that 4-OH-proline is a secondary amino acid that does not react with some derivative reagents such as OPA (see Section 2.2.1.2.1).

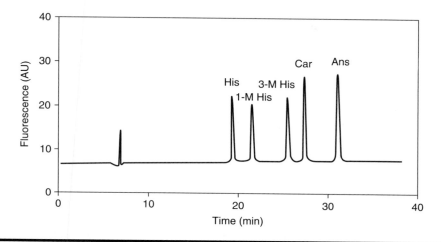

Figure 2.4 **Cation-exchange chromatogram of OPA postcolumn derivatized histidine-related compounds. (Reproduced from Aristoy, M.C. and Toldrá, F., *Handbook of Food Analysis*, Marcel Dekker, New York, 2004. With permission.)**

In addition to HPLC methods, specific colorimetric methods[67] and GC-MS of the *N(O)*-trifluoroacetyl *n*-propyl ester derivatives[196] have also been proposed.

2.2.3.3   Determination of o-Tyrosine

o-Tyrosine has been proposed as an indicator of food irradiation.[199] When phenylalanine is irradiated with gamma rays, it is oxidized, yielding *o*- and *m*-tyrosine isomers. The conversion yield is proportional to the absorbed dose and temperature during irradiation. As many foods contain constant levels of phenylalanine in proteins, the level of *o*-tyrosine may be a good indicator of food irradiation. Nevertheless, it may be taken into account that only low levels of naturally occurring *o*-tyrosine are present in some foods. The application is especially common in pork, chicken, fish, and shrimp. Methods of determining *o*-tyrosine should be highly sensitive and selective, as are GC-MS,[158] HPLC with fluorescence detector,[200] HPLC-LIF,[156,157] or HPLC-ED.[201,202]

2.2.3.4   Determination of Sulfur Amino Acids

The importance of sulfur amino acids is due to their high reactivity, reducing power, and their influence in meat flavor. Furthermore, methionine is an essential amino acid, while cyst(e)ine is essential in premature infants. Special sample preparation treatment and care are necessary for the determination of such compounds. The hydrolysis of samples for cyst(e)ine determination requires specific procedures, already described in Section 2.2.1.1.2. Their analysis is dealt with in more detail in the essential amino acids chapter (see Chapter 22).

2.2.3.5   Determination of Essential Amino Acids and Taurine

The analysis of taurine and essential amino acids is also dealt in the essential amino acids chapter (see Chapter 22).

2.3 Conclusions

To obtain the total amino acid profile of a given meat, the most important factors to take into account are resolution power and selectivity. The highest resolution is obtained by gas chromatography with capillary column techniques, but tedious and time-consuming sample derivatization is required. In general, cation-exchange and postcolumn derivatization or RP-HPLC precolumn derivatization techniques are the preferred methods. Very careful control of the derivatization reactions and chromatographic conditions is necessary for a consistent and reproducible analysis. Since many peaks corresponding to protein and nonprotein amino acids, nucleosides, small peptides, and so on may appear in the chromatogram, a complete resolution of the whole peaks is very difficult. As a result, the analytical technique for a determined sample must be carefully chosen based on the literature. Among RP-HPLC precolumn derivatization methods, the combination OPA/FMOC derivatization or the PITC derivatives are the most frequently used. The main advantage of the first is the possibility of automation, and the main advantage of the latter is the high resolution power. The convenience of purchasing commercial available kits must be evaluated.

When amino acids from meat proteins are to be analyzed, the first decision is the choice of the hydrolysis method. In general, acid hydrolysis with 6N HCl (110°C for 22 h or 145°C for 4 h) with an oxidation protective agent, such as phenol, and taking care to avoid the presence of oxygen by means of vacuum and nitrogen purging is sufficient for the majority of food analytical purposes. Particular hydrolysis problems related to certain amino acids are described in Section 2.2.1.1.2. Requirements in resolution are not so exigent as for physiological amino acids, because fewer amino acids are involved in the analysis and no peptides will appear in the chromatogram. Any separation strategy may give good results and, once again, the convenience of purchasing commercially available prepackaged kits should be evaluated.

References

1. Swaisgood, H.E., Catignani, G.L. Protein digestibility: in vitro methods of assessment. *Adv. Food Nutr. Res.*, 35, 185–236, 1991.
2. Toldrá, F. Meat: chemistry and biochemistry. In: *Handbook of Food Science, Technology and Engineering.* Y.H. Hui, J.D. Culbertson, S. Duncan, I. Guerrero-Legarreta, E.C.Y. Li-Chan, C.Y. Ma, C.H. Manley, T.A. McMeekin, W.K. Nip, L.M.L. Nollet, M.S. Rahman, F. Toldrá, Y.L. Xiong, Eds., Vol. 1, CRC Press, Boca Raton, FL, pp. 28-1 to 28-18, 2006.
3. Aristoy, M.C., Toldrá, F. Amino acids. In: *Handbook of Food Analysis.* L.M.L. Nollet, Ed., 2nd ed., Vol. 1, Marcel Dekker, New York, NY, pp. 5-83 to 5-123, 2004.
4. Godel, H. et al. Measurement of free amino acids in human biological fluids by high-performance liquid chromatography. *J. Chromatogr.*, 297, 49–61, 1984.
5. Arnold, U. et al. Analysis of free amino acids in green coffee beans. I. Determination of amino acids after precolumn derivatisation using 9-fluorenylmethylchloroformate. *Z. Lebensm. Unters. Forsch.*, 199, 22–25, 1994.
6. Shibata, K., Onodera, M., Aihara, S. High-performance liquid chromatographic measurement of tryptophan in blood, tissues, urine and foodstuffs with electrochemical and fluorometric detections. *Agric. Biol. Chem.*, 55, 1475–1481, 1991.
7. Ali Qureshi, G., Fohlin, L., Bergström, J. Application of high-performance liquid chromatography to the determination of free amino acids in physiological fluids. *J. Chromatogr.*, 297, 91–100, 1984.
8. Nguyen, Q., Zarkadas, C.G. Comparison of the amino acid composition and connective tissue protein contents of selected bovine skeletal muscles. *J. Agric. Food Chem.*, 37, 1279–1286, 1989.

9. Hagen, S.R. et al. Precolumn phenylisothiocyanate derivatization and liquid chromatography of free amino acids in biological samples. *Food Chem.*, 16, 319–323, 1993.

10. Antoine, F.R. et al. HPLC method for analysis of free amino acids in fish using *o*-phthaldialdehyde precolumn derivatization. *J. Agric. Food Chem.*, 47, 5100–5107, 1999.

11. Aristoy, M.C., Toldrá, F. Deproteinization techniques for HPLC amino acid analysis in fresh pork muscle and dry-cured ham. *J. Agric. Food Chem.*, 39, 1792–1795, 1991.

12. Cohen, S.A., Strydon, J. Amino acid analysis utilizing phenylisothiocyanate derivatives. *Anal. Biochem.*, 174, 1–16, 1988.

13. Krause, I. et al. Simultaneous determination of amino acids and biogenic amines by reversed-phase high-performance liquid chromatography of the dabsyl derivatives. *J. Chromatogr. A*, 715, 67–79, 1995.

14. Sarwar, G., Botting, H.G. Rapid analysis of nutritionally important free amino acids in serum and organs (liver, brain, and heart) by liquid chromatography of precolumn phenylisothiocyanate. *J. Assoc. Off. Anal. Chem.*, 73, 470–475, 1990.

15. Reverter, M., Lundh, T., Lindberg, J.E. Determination of free amino acids in pig plasma by precolumn derivatization with 6-aminoquinolyl-*N*-hydroxysuccinimidyl carbamate and high-performance liquid chromatography. *J. Chromatogr. B*, 696, 1–8, 1997.

16. Izco, J.M., Torre, P., Barcina, Y. Ripening of Ossau-Iratzy cheese: determination of free amino acids by RP-HPLC and of total free amino acids by the TNBS method. *Food Control*, 11, 7–11, 2000.

17. Büetikofer, U., Ardö, Y. Quantitative determination of free amino acids in cheese. *Bull. Int. Dairy Fed.*, 337(Part 2), 24–32, 1999.

18. Ordóñez, A.I. et al. Characterization of the casein hydrolysis of Idiazabal cheese manufactured from ovine milk. *J. Dairy Sci.*, 81, 2089–2095, 1998.

19. Sanz, Y. et al. Effect of pre-ripening on microbial and chemical changes in dry fermented sausages. *Food Microbiol.*, 14, 575–582, 1997.

20. Brückner, H., Hausch, M. D-Amino acids in dairy products: detection, origin and nutritional aspects. I. Milk, fermented milk, fresh cheese and acid curd cheese. *Milchwissenschaft*, 45, 357–360, 1990.

21. Oh, C.H. et al. Rapid gas chromatographic screening of edible seeds, nuts and beans for non-protein and protein amino acids. *J. Chromatogr. A*, 708, 131–141, 1995.

22. Sugawara, H., Itoh, T., Adachi, S. Assignment of an unknown peak on the high performance liquid chromatogram of free amino acids isolated from fresh chicken egg albumen using picric acid as the deproteinizing agent. *Jpn. J. Zootech. Sci.*, 55, 892–893, 1984.

23. Schuster, R. Determination of amino acids in biological, pharmaceutical, plant and food samples by automated precolumn derivatization and high-performance liquid chromatography. *J. Chromatogr.*, 431, 271–284, 1988.

24. Jansen, E.H.J.M. et al. Advantages and limitations of precolumn derivatization of amino acids with dabsyl chloride. *J. Chromatogr.*, 553, 123–133, 1991.

25. Blanchard, J. Evaluation of the relative efficacy of various techniques for deproteinizing plasma samples prior to high-performance liquid chromatographic analysis. *J. Chromatogr. Biomed. Appl.*, 226, 455–460, 1981.

26. Lucas, B., Sotelo, A. Amino acid determination in pure proteins, foods, and feeds using two different acid hydrolysis methods. *Anal. Biochem.*, 123, 349–356, 1982.

27. Gehrke, C.W. et al. Sample preparation for chromatography of amino acids: acid hydrolysis of proteins. *J. Assoc. Off. Anal. Chem.*, 68, 811–821, 1985.

28. Gehrke, C.W. et al. Quantitative analysis of cystine, methionine, lysine, and nine other amino acids by a single oxidation–4 hours hydrolysis method. *J Assoc. Off. Anal. Chem.*, 70, 171–174, 1987.

29. Knecht, R., Chang, J.Y. Liquid chromatographic determination of amino acids after gas-phase hydrolysis and derivatization with (dimethylamino)azobenzenesulfonyl chloride. *Anal. Chem.*, 58, 2375–2379, 1986.

30. Woodward, C., Gilman, L.B., Engelhart, W.G. An evaluation of microwave heating for the vapor phase hydrolysis of proteins. *Int. Lab.*, 9, 40–45, 1990.

31. Molnár-Perl, I., Khalifa, M. Tryptophan analysis simultaneously with other amino acids in gas phase hydrochloric acid hydrolyzates using the Pico-Tag™ Work Station. *Chromatographia*, 36, 43–46, 1993.

32. Molnár-Perl, I., Khalifa, M. Analysis of foodstuff amino acids using vapour-phase hydrolysis. *LC–GC Int.*, 7, 395–398, 1994.

33. Moore, S. On the determination of cystine and cysteic acid. *J. Biol. Chem.*, 243, 235–237, 1963.

34. Hirs, C.H.W. Performic acid oxidation. *Methods Enzymol.*, 11, 197–199, 1967.

35. MacDonald, J.L., Krueger, M.W., Keller, J.H. Oxidation and hydrolysis determination of sulfur amino acids in food and feed ingredients: collaborative study. *J. Assoc. Off. Anal. Chem.*, 68, 826–829, 1985.

36. Elkin, R.G., Griffith, J.E. Hydrolysate preparation for analysis of amino acids in sorghum grains: effect of oxidative pre-treatment. *J. Assoc. Off. Anal. Chem.*, 36, 1117–1121, 1985.

37. Meredith, F.I., McCarthy, M.A., Leffler, R. Amino acid concentrations and comparison of different hydrolysis procedures for American and foreign chestnuts. *J. Agric. Food Chem.*, 36, 1172–1175, 1988.

38. Alegría, A. et al. HPLC method for cyst(e)ine and methionine in infant formulas. *J. Food Sci.*, 61, 1132–1135, 1170, 1996.

39. Akinyele, A.F., Okogun, J.I., Faboya, O.P. Use of 7-chloro-4-nitrobenzo-2-oxa-1,3-diazole for determining cysteine and cystine in cereal and legume seeds. *J. Agric. Food Chem.*, 47, 2303–2307, 1999.

40. Bosch, L., Alegría, A., Farré, R. Application of the 6-aminoquinolyl-*N*-hydroxysuccinimidyl carbamate (AQC) reagent to the RP-HPLC determination of amino acids in infant formula. *J. Chromatogr. B*, 831, 176–183, 2006.

41. Bradbury, A.F., Smith, D.G. The use of 3-bromopropionic acid for the determination of protein thiol groups. *J. Biochem.*, 131, 637–642, 1973.

42. Hale, J.E., Beidler, D.E., Jue, R.A. Quantitation of cysteine residues alkylated with 3-bromopropylamine by amino acid analysis. *Anal. Biochem.*, 216, 61–66, 1994.

43. Fullmer, C.S. Identification of cysteine-containing peptides in protein digests by high-performance liquid chromatography. *Anal. Biochem.*, 142, 336–339, 1984.

44. Morel, M.H., Bonicel, J. Determination of the number of cysteine residues in high molecular weight subunits of wheat glutenin. *Electrophoresis*, 17, 493–496, 1996.

45. Barkholt, V., Jensen, A.L. Amino acid analysis: determination of cysteine plus half-cystine in proteins after hydrochloric acid hydrolysis with a disulfide compound as additive. *Anal. Biochem.*, 177, 318–322, 1989.

46. Tuan, Y.H., Phillips, R.D. Optimized determination of cystine/cysteine and acid-stable amino acids from a single hydrolysate of casein- and sorghum-based diet and digesta samples. *J. Agric. Food Chem.*, 45, 3535–3540, 1997.

47. Fouques, D., Landry, J. Study of the conversion of asparagines and glutamine of proteins into diaminopropionic and diaminobutyric acid using with [bis(trifluoroacetoxy)iodo] benzene prior to amino acid determination. *Analyst*, 116, 529–531, 1991.

48. Khun, K.S., Stehle, P., Fürst, P. Quantitative analyses of glutamine in peptides and proteins. *J. Agric. Food Chem.*, 44, 1808–1811, 1996.

49. Stocchi, V. et al. Reversed-phase high-performance liquid chromatography separation of dymethylaminoazobenzne sulfonyl- and dimethylaminoazobenzne thiohydantoin-amino acid derivatives for amino acid analysis and microsequencing studies at the picomole level. *Anal. Biochem.*, 178, 107–117, 1989.

50. Zumwalt, R.W. et al. Acid hydrolysis of proteins for chromatographic analysis of amino acids. *J. Assoc. Off. Anal. Chem.*, 70, 147–151, 1987.

51. Albin, D.M., Wubben, J.E., Gabert, V.M. Effect of hydrolysis time on the determination of amino acids in samples of soybean products with ion-exchange chromatography or precolumn derivatization with phenyl isothiocyanate. *J. Agric. Food Chem.*, 48, 1684–1691, 2000.

52. Ambler, R.P. Standards and accuracy in amino acid analysis. In: *Amino Acid Analysis*. J.M. Rattenbury, Ed., Ellis Horwood, Chichester, UK, pp. 119–137, 1981.
53. Williams, A.P. Determination of amino acids and peptides. In: *HPLC in Food Analysis*. R. McCrae, Ed., Academic Press, New York, NY, pp. 285–311, 1982.
54. Hugli, T.E., Moore, S. Determination of the tryptophan content of proteins by ion-exchange chromatography of alkaline hydrolysates. *J. Biol. Chem.*, 247, 2828–2834, 1972.
55. Zarkadas, C.G. et al. Assessment of the protein quality of beefstock bone isolates for use as an ingredient in meat and poultry products. *J. Agric. Food Chem.*, 43, 77–83, 1995.
56. Yazzie, D.Y. et al. The amino acid and mineral content of baobab (*Adansonia digitata* L.) leaves. *J. Food Comp. Anal.*, 7, 189–193, 1994.
57. Viadel, B. et al. Amino acid profile of milk-based infant formulas. *Int. J. Food Sci. Nutr.*, 51, 367–372, 2000.
58. Ihekoronye, A.I. Quantitative gas-liquid chromatography of amino acids in enzymic hydrolysates of food proteins. *J. Sci. Food Agric.*, 36, 1004–1012, 1985.
59. García, S.E., Baxter, J.H. Determination of tryptophan content in infant formulas and medical nutritionals. *J. AOAC Int.*, 75, 1112–1119, 1992.
60. Standara, S., Drdak, M., Vesela, M. Amino acid analysis: reduction of ninhydrin by sodium borohydride. *Nahrung*, 43, 410–413, 1999.
61. Udenfried, S. et al. Fluorescamine: a reagent for assay of amino acids, peptides, proteins and primary amines in the picomole range. *Science*, 178, 871–872, 1972.
62. Weigele, M. et al. Fluorogenic ninhydrin reaction. Structure of the fluorescent principle. *J. Am. Chem. Soc.*, 94, 4052–4054, 1972.
63. Weigele, M. et al. Novel reagent for the fluorometric assay of primary amines. *J. Am. Chem. Soc.*, 94, 5927, 1972.
64. Castell, J.V., Cervera, M., Marco, R. A convenient micro-method for the assay of primary amines and proteins with fluorescamine. A re-examination of the conditions of reaction. *Anal. Biochem.*, 99, 379–391, 1979.
65. White, W.J.P., Lawrie, R.A. Practical observations on the methodology for determining of 3-methyl-L-histidine using fluorescamine derivatives. *Meat Sci.*, 12, 117–123, 1984.
66. Dong, M.W., Gant, J.R. High-speed liquid chromatographic analysis of amino acids by post-column sodium hypochlorite-*o*-phthalaldehyde reaction. *J. Chromatogr.*, 327, 17–25, 1985.
67. Ashworth, R.B. Amino acids analysis for meat protein evaluation. *J. Assoc. Off. Anal. Chem.*, 70, 80–85, 1987.
68. Ashworth, R.B. Ion-exchange separation of amino acids with postcolumn orthophthalaldehyde detection. *J. Assoc. Off. Anal. Chem.*, 70, 248–252, 1987.
69. Heinrikson, R.L., Meredith, S.C. Amino acid analysis by reverse-phase high-performance liquid chromatography, precolumn derivatization with phenylisothiocyanate. *Anal. Biochem.*, 136, 65–74, 1984.
70. Bidlingmeyer, B.A., Cohen, S.A., Tarvin, T.L. Rapid analysis of amino acids using pre-column derivatization. *J. Chromatogr.*, 336, 93–104, 1984.
71. Bidlingmeyer, B.A. et al. New rapid high sensitivity analysis of amino acids in food type samples. *J. Assoc. Off. Anal. Chem.*, 70, 241–247, 1987.
72. Mora, E. et al. Quantitation of aspartate and glutamate in HPLC analysis of phenylthiocarbamyl amino acids. *Anal. Biochem.*, 172, 368–376, 1988.
73. Khalifa, M., Molnar-Perl, I. The analysis of phenylthiocarbamyl amino acids in the presence of sodium chloride. *LC–GC Int.*, 9, 143–147, 1996.
74. Sarwar, G., Botting, H.G., Peace, R.W. Complete amino acid analysis in hydrolysates of foods and feces by liquid chromatography of precolumn phenylisothiocyanate derivatives. *J. Assoc. Off. Anal. Chem.*, 71, 1172–1175, 1988.
75. Fürst, P. et al. HPLC analysis of free amino acids in biological material—an appraisal of four pre-column derivatization methods. *J. Liq. Chromatogr.*, 12, 2733–2760, 1989.

76. Woo, K.L., Hwang, Q.C., Kim, H.S. Determination of amino acids in the foods by reversed-phase high-performance liquid chromatography with new precolumn derivative, butylthiocarbamyl amino acid, compared to the conventional phenylthiocarbamyl derivatives and ion exchange chromatography. *J. Chromatogr. A*, 740, 31–40, 1996.

77. Woo, K.L. Determination of amino acids in foods by reversed-phase high-performance liquid chromatography with new precolumn derivatives, butylthiocarbamyl, and benzylthiocarbamyl derivatives compared to the phenylthiocarbamyl derivative and ion exchange chromatography. *Methods Mol. Biol.*, 159, 141–167, 2000.

78. Liu, J.K., Chang, J.Y. Chromophore labelling of amino acids with 4-dimethyl-aminoazobenzene-4′-sulfonyl chloride. *Anal. Chem.*, 47, 1634–1638, 1975.

79. Chang, J.Y., Knecht, R., Braun, D.G. Amino acid analysis in the picomole range by precolumn derivatization and high-performance liquid chromatography. *Methods Enzymol.*, 91, 41–48, 1983.

80. Tapuhi, Y. et al. Dansylation of amino acids for high-performance liquid chromatography analysis. *Anal. Biochem.*, 115, 123–129, 1981.

81. de Jong, C. et al. Amino acid analysis by high-performance liquid chromatography. An evaluation of the usefulness of pre-column Dns derivatization. *J. Chromatogr.*, 241, 345–350, 1982.

82. Martín, P. et al. Dansyl amino acids behavior on a Radial Pak C18 column: derivatization of grape wine musts, wines and wine vinegar. *J. Liq. Chromatogr.*, 7, 539–558, 1984.

83. Prieto, J.A., Collar, C., Benedito de Barber, C. Reversed-phase high-performance liquid chromatographic determination of biochemical changes in free amino acids during wheat flour mixing and bread baking. *J. Chromatogr. Sci.*, 28, 572–577, 1990.

84. Stehle, P. et al. In vivo utilization of cystine-containing synthetic short chain peptides after intravenous bolus injection in the rat. *J. Nutr.*, 118, 1470–1474, 1988.

85. Burton, C. Fully automated amino acid analysis using precolumn derivatization. *Int. Lab.*, 30, 32–34, 36, 38, 1986.

86. Betner, I., Földi, P. The FMOC-ADAM approach to amino acid analysis. *LC–GC Int.*, 6, 832–840, 1988.

87. Simons, S.S., Johnson, D.F. Reaction of *o*-phthalaldehyde and thiols with primary amines; formation of 1-alkyl(and aryl)thio-2-alkylisoindoles. *J. Org. Chem.*, 43, 2886–2891, 1978.

88. Alvarez-Coque, M.C.G. et al. Formation and instability of *o*-phthalaldehyde derivatives of amino acids. *Anal. BioChem.*, 178, 1–7, 1989.

89. Godel, H., Seitz, P., Verhoef, M. Automated amino acid analysis using combined OPA and FMOC-Cl precolumn derivatization. *LC–GC Int.*, 5, 44–49, 1992.

90. Lookhart, G.L., Jones, B.L. High-performance liquid chromatography analysis of amino acids at the picomole level. *Cereal Chem.*, 62, 97–102, 1985.

91. Euerby, M.R. Effect of differing thiols on the reversed-phase high-performance liquid chromatographic behaviour of *o*-phthaldialdehyde-thiol-amino acids. *J. Chromatogr.*, 454, 398–405, 1988.

92. Winspear, M.J., Oaks, A. Automated pre-column amino acid analyses by reversed-phase high-performance liquid chromatography. *J. Chromatogr.*, 270, 378–382, 1983.

93. Willis, D.E. Automated pre-column derivatization of amino acids with *o*-phthalaldehyde by a reagent sandwich technique. *J. Chromatogr.*, 408, 217–225, 1987.

94. Ishida, Y., Fujita, T., Asai, K. New detection and separation method for amino acids by high-performance liquid chromatography. *J. Chromatogr.*, 204, 143–148, 1981.

95. Jarret, H.W. et al. The separation of *o*-phthalaldehyde derivatives of amino acids by reversed-phase chromatography on octylsilica columns. *Anal. Biochem.*, 153, 189–198, 1986.

96. Gurd, F.R.N. Carboxymethylation. *Methods Enzymol.*, 25, 424–438, 1972.

97. Pripis-Nicolau, L. et al. Automated HPLC method for the measurement of free amino acids including cysteine in musts and wines; first applications. *J. Sci. Food Agric.*, 81, 731–738, 2001.

98. Park, S.K., Boulton, R.B., Noble, A.C. Automated HPLC analysis of glutathion and thiol-containing compounds in grape juice and wine using pre-column derivatization with fluorescence detection. *Food Chem.*, 68, 475–480, 2000.

99. Liu, H. et al. Determination of submicromolar concentrations of neurotransmitter amino acids by fluorescence detection using a modification of the 6-aminoquinolyl-*N*-hydroxysuccinimidyl carbamate method for amino acid analysis. *J. Chromatogr. A*, 828, 383–395, 1998.

100. Wang, J., Chatrathi, M.P., Tian, B. Micromachined separation chips with a precolumn reactor and end-column electrochemical detector. *Anal. Chem.*, 72, 5774–5778, 2000.

101. Tcherkas, Y.V., Kartsova, L.A., Krasnova, I.M. Analysis of amino acids in human serum by isocratic reversed-phase high-performance liquid chromatography with electrochemical detection. *J. Chromatogr. A*, 913, 303–308, 2001.

102. Weng, Q., Jin, W. Determination of free intracellular amino acids in single mouse peritoneal macrophages after naphthalene-2,3-dicarboxaldehyde derivatization by capillary zone electrophoresis with electrochemical detection. *Electrophoresis*, 22, 2797–2803, 2001.

103. Sherwood, R.A., Titheradge, A.C., Richards, D.A. Measurement of plasma and urine amino acids by high-performance liquid chromatography with electrochemical detection using phenylisothiocyanate derivatization. *J. Chromatogr.*, 528, 293–303, 1990.

104. Kuneman, D.W., Braddock, J.K., McChesney, L.L. HPLC profile of amino acids in fruit juices as their (1-fluoro-2,4dinitrophenyl)-5-L-alanine amide (FDAA) derivatives. *J. Agric. Food Chem.*, 36, 6–9, 1988.

105. Brückner, H., Westhauser, T. Chromatographic determination of D-amino acids as native constituents of vegetables and fruits. *Chromatographia*, 39, 419–426 1994.

106. Calabrese, M., Stancher, B., Riccobon, P. High-performance liquid chromatography determination of proline isomers in Italian wines. *J. Sci. Food Agric.*, 69, 361–366, 1995.

107. Tivesten, A., Lundqvist, A., Folestad, S. Selective chiral determination of aspartic and glutamic acid in biological samples by capillary electrophoresis. *Chromatographia*, 44, 623–633, 1997.

108. Dongri-Jin, T. et al. Determination of D-amino acids labelled with fluorescent chiral reagents, *R*(–) and *S*(+)-a-(3-isothiocyanatopyrrolidin-1-yl)-7-(*N*,*N*-dimethylaminosulfonyl)-2,1,3-benzoxadiazoles, in biological and food samples by liquid chromatography. *Anal. Biochem.*, 269, 124–132, 1999.

109. Toyo'oka, T. et al. *R*(–)-4-(3-Isothiocyanatopyrrolidin-1-yl)-7-(*N*,*N*-dimethylaminosulfonyl)-2,1,3-benzoxadiazole, a fluorescent chiral tagging reagent, sensitive resolution of chiral amines and amino acids by reversed-phase liquid chromatography. *Biomed. Chromatogr.*, 15, 56–67, 2001.

110. Voss, K., Galensa, R. Determination of L- and D-amino acids in foodstuffs by coupling of high-performance liquid chromatography with enzyme reactors. *Amino Acids*, 18, 339–352, 2000.

111. Buetikofer, U., Ardo, Y. Quantitative determination of free amino acids in cheese. *Bull. Int. Dairy Fed.*, 337(Part 2), 24–32, 1999.

112. Grunau, J.A., Swiader, J.M. Chromatography of 99 amino acids and other ninhydrin-reactive compounds in the Pickering lithium gradient system. *J. Chromatogr.*, 594, 165–171, 1992.

113. Ardo, Y., Gripon, J.C. Comparative study of peptidolysis in some semi-hard round-eyed cheese varieties with different fat contents. *J. Dairy Res.*, 62, 543–547, 1995.

114. de Schrijver, R. et al. Hydrolysate preparation and comparative amino acid determination by cation-exchange and gas-liquid chromatography in diet ingredients. *Cerevisia Biotechnol.*, 16, 26–37, 1991.

115. Ogunsua, A.O. Amino acid determination in conophor nut by gas-liquid chromatography. *Food Chem.*, 28, 287–298, 1988.

116. Paetzold, R., Brückner, H. Gas chromatographic detection of D-amino acids in natural and thermally treated bee honeys and studies on the mechanism of their formation as result of the Maillard reaction. *Eur. Food Res. Technol.*, 223, 347–354, 2006.

117. Brückner, H., Schieber, A. Determination of free D-amino acids in Mammalia by gas chromatography-mass spectrometry. *J. High Resolut. Chromatogr.*, 23, 576–582, 2000.

118. Bertacco, G., Boschelle, O., Lercker, G. Gas chromatographic determination of free amino acids in cheese. *Milchwissenschaft*, 47, 348–350, 1992.

119. Erbe, T., Brückner, H. Chiral amino acid analysis of vinegars using gas chromatography-selected ion monitoring mass spectrometry. *Z. Lebensm. Unters. Forsch.*, 207, 400–409, 1998.

120. Katona, Z.F., Sass, P., Molnar-Perl, I. Simultaneous determination of sugars, sugar alcohols, acids and amino acids in apricots by gas chromatography-mass spectrometry. *J. Chromatogr. A*, 847, 91–102, 1999.

121. Starke, I., Kleinpeter, E., Kamm, B. Separation, identification, and quantification of amino acids in L-lysine fermentation potato juices by gas chromatography-mass spectrometry. *Fresenius J. Anal. Chem.*, 371, 380–384, 2001.

122. Jorgenson, J.W., Lukacs, K.D. Free-zone electrophoresis in glass capillaries. *Clin. Chim.*, 27, 1551–1553, 1981.

123. Jorgenson, J.W., Lukacs, K.D. Capillary zone electrophoresis. *Science*, 222(4621), 266–72, 1983.

124. Terabe, S. et al. Electrokinetic separations with micellar solutions and open-tubular capillaries. *Anal. Chem.*, 56, 111–113, 1984.

125. Terabe, S., Otsuka, K., Ando, T. Electrokinetic chromatography with micellar solution and open-tubular capillary. *Anal. Chem.*, 57, 834–841, 1985.

126. Burton, D.E., Sepaniak, M.J., Maskarinec, M.P. Analysis of B6 vitamers by micellar electrokinetic capillary chromatography with laser-excited fluorescence detection. *J. Chromatogr. Sci.*, 24, 347–351, 1986.

127. Matsubara, N., Terabe, S. Separation of 24 dansylamino acids by capillary electrophoresis with a non-ionic surfactant. *J. Chromatogr. A*, 680, 311–315, 1994.

128. Wu, J. et al. Ultrasensitive detection for capillary zone electrophoresis using laser-induced capillary vibration. *Anal. Chem.*, 63, 2216–2218, 1991.

129. Ye, J., Baldwin, R.P. Determination of amino acids and peptides by capillary electrophoresis and electrochemical detection at a copper electrode. *Anal. Chem.*, 66, 2669–2674, 1994.

130. Klampfl, C.W. Analysis of organic acids and inorganic anions in different types of beer using capillary electrophoresis. *J. Agric. Food Chem.*, 47, 987–990, 1999.

131. Klampfl, C.W. et al. Determination of underivatized amino acids in beverage samples by capillary electrophoresis. *J. Chromatogr. A*, 804, 347–355, 1998.

132. Soga, T., Heiger, D.N. Amino acid analysis by capillary electrophoresis electrospray ionisation mass spectrometry. *Anal. Chem.*, 72, 1236–1241, 2000.

133. Skočir, E., Prosek, M. Determination of amino acid ratios in natural products by micellar electrokinetic chromatography. *Chromatographia*, 41, 638–644, 1995.

134. Skočir, E., Vindevogel, J., Sandra, P. Separation of 23 dansylated amino acids by micellar electrokinetic chromatography at low temperature. *Chromatographia*, 39, 7–10, 1994.

135. Tsai, C.F. et al. Enantiomeric separation of dansyl-derivatized DL-amino acids by beta-cyclodextrin-modified micellar electrokinetic chromatography. *J. Agric. Food Chem.*, 46, 979–985, 1998.

136. Cavazza, A. et al. Rapid analysis of essential and branched-chain amino acids in nutraceutical products by micellar electrokinetic capillary chromatography. *J. Agric. Food Chem.*, 48, 3324–3329, 2000.

137. Castagnola, D.V. et al. Optimization of phenylthiohydantoinamino acid separation by micellar electrokinetic capillary chromatography. *J. Chromatogr.*, 638, 327–333, 1993.

138. Arellano, M. et al. Several applications of capillary electrophoresis for wines analysis. Quantitation of organic and inorganic acids, inorganic cations, amino acids and biogenic amines. *J. Int. Sci. Vigne Vin*, 31, 213–218, 1997.

139. Otsuka, K., Terabe, S., Ando, T. Electrokinetic chromatography with micellar solutions separation of phenylthiohydantoin-amino acids. *J. Chromatogr.*, 332, 219–226, 1985.

140. Otsuka, K. et al. Optical resolution of amino acid derivatives by micellar electrokinetic chromatography with *N*-dodecanoyl-L-serine. *J. Chromatogr. A*, 680, 317–320, 1994.

141. Waldron, K.C. et al. Miniaturized protein microsequencer with PTH amino acid identification by capillary electrophoresis. I. An argon pressurized delivery system for adsorptive and covalent sequencing. *Talanta*, 44, 383–399, 1997.

142. Liu, J., Cobb, K.A., Novotny, M. Separation of pre-column *ortho*-phthalaldehyde-derivatized amino acids by capillary zone electrophoresis with normal and micellar solutions in the presence of organic modifiers. *J. Chromatogr.*, 468, 55–65, 1989.

143. Chen, N., Terabe, S. A quantitative study on the effect of organic modifiers in micellar electrokinetic chromatography. *Electrophoresis*, 16, 2100–2103, 1995.

144. Chen, N., Terabe, S., Nakagawa, T. Effect of organic modifier concentrations in micellar electrokinetic chromatography. *Electrophoresis*, 16, 1457–1462, 1995.

145. Otsuka, K., Terabe, S. Effect of methanol and urea on optical resolution of phenylthiohydantoin-DL-amino acids by micellar electrokinetic chromatography with sodium *N*-dodecanoyl-L-valinate. *Electrophoresis*, 11, 982–984, 1990.

146. Yeung, E.S., Kuhr, W.G. Indirect detection methods for capillary electrophoresis. *Anal. Chem.*, 63, 275A–282A, 1991.

147. Novotny, M.V., Cobb, K.A., Liu, J. Recent advances in capillary electrophoresis of proteins, peptides and amino acids. *Electrophoresis*, 11, 735–749, 1990.

148. Monnig, C.A., Kennedy, R.T. Capillary electrophoresis. *Anal. Chem.*, 66, 280R–314R, 1994.

149. Issaq, H.J., Chan, K.C. Separation and detection of amino acids and their enantiomers by capillary electrophoresis: a review. *Electrophoresis*, 16, 467–480, 1995.

150. Matsubara, N., Terabe, S. Micellar electrokinetic chromatography. *Methods Enzymol.*, 270(Part A), 319–341, 1996.

151. Corradini, C., Cavaza, A. Application of capillary zone electrophoresis (CZE) and micellar electrokinetic chromatography (MEKC) in food analysis. *It. J. Food Sci.*, 10, 299–316, 1998.

152. Buser, W., Erbersdobler, H.F. Determination of amino acids by gas-liquid chromatography and nitrogen selective detection. *Z. Lebensm. Unters. Forsch.*, 186, 509–513, 1988.

153. Kataoka, H., Sakiyama, N., Makita, M. Distribution and contents of free *o*-phosphoamino acids in animal-tissues. *J. Biochem.*, 109, 577–580, 1991.

154. Novatchev, N., Ulrike, H. Evaluation of amino sugar, low molecular peptide and amino acid impurities of biotechnologically produced amino acids by means of CE. *J. Pharm. Biomed. Anal.*, 28, 475–486, 2002.

155. Beale, S.C. et al. 3-Benzoyl-2-quinolinecarboxaldehyde: a novel fluorogenic reagent for the high-sensitivity chromatographic analysis of primary amines. *Talanta*, 36, 321–325, 1989.

156. Miyahara, M. et al. New LASER fluorometric detection for *ortho*-tyrosine in gamma-irradiated phenylalanine solution and pork. *Food Irrad. Jpn.*, 34, 3–8, 1999.

157. Miyahara, M. et al. Detection of irradiation of meats by HPLC determination for *o*-tyrosine using novel laser fluorometric detection with automatic pre-column reaction. *J. Health Sci.*, 46, 304–309, 2000.

158. Karam, L.R., Simic, M.G. Formation of *o*-tyrosine by radiation and organic solvents in chicken tissue. *J. Biol. Chem.*, 265, 11581–11585, 1990.

159. Kwon, J.Y., Moini, M. Analysis of underivatized amino acid mixtures using high performance liquid chromatography/dual oscillating nebulizer atmospheric pressure microwave induced plasma ionisation-mass spectrometry. *J. Am. Soc. Mass Spectrom.*, 12, 117–122, 2001.

160. Qu, J. et al. Rapid determination of underivatized pyroglutamic acid, glutamic acid, glutamine and other relevant amino acids in fermentation media by LC-MS-MS. *Analyst*, 127, 66–69, 2002.

161. Dou, L., Krull, I.S. Determination of aromatic and sulfur-containing amino acids, peptides, and proteins using high-performance liquid chromatography with photolytic electrochemical detection. *Anal. Chem.*, 62, 2599–2606, 1990.

162. Zhao, C., Zhang, J., Song, J. Determination of L-cysteine in amino acid mixture and human urine by flow-injection analysis with a bioamperometric detector. *Anal. Biochem.*, 297, 170–176, 2001.

163. Mills, B.J. et al. Glutathione and cyst(e)ine profiles in vegetables using high performance liquid chromatography with dual electrochemical detection. *J. Food Comp. Anal.*, 10, 90–101, 1997.

164. Lunte, S.M., Wong, O.S. Naphthalenedialdehyde-cyanide: a versatile fluorogenic reagent for the LC analysis of peptides and other primary amines. *LC–GC Int.*, 2, 20–22, 24, 26–27, 1989.

165. Li, G.D., Krull, I.S., Cohen, S.A. Electrochemical activity of 6-aminiquinolyl urea derivatives of amino acids and peptides: application to high-performance liquid chromatography with electrochemical detection. *J. Chromatogr. A*, 724, 147–157, 1996.

166. Sarkar, P., Turner, A.P.F. Application of dual-step potential on a single screen-printed modified carbon paste electrodes for detection of amino acids and proteins. *Fresenius J. Anal. Chem.*, 364, 154–159, 1999.

167. Fung, Y.S., Mo, S.Y. Determination of amino acids and proteins by dual-electrode in a flow system. *Anal. Chem.*, 67, 1121–1124, 1995.

168. Villarta, R.L., Cunningham, D.D., Guilbault, G.G. Amperometric enzyme electrodes for the determination of l-glutamate. *Talanta*, 38, 49–55, 1991.

169. Hale, P.D. et al. Glutamate biosensors based on electrical communication between l-glutamate oxidase and a flexible redox polymer. *Anal. Lett.*, 24, 345–356, 1991.

170. Wang, A.J., Arnold, M.A. Dual-enzyme fiber-optic biosensor for glutamate based on reduced nicotinamide adenine dinucleotide luminescence. *Anal. Chem.*, 64, 1051–1055, 1992.

171. Olson-Cosford, R.J., Kuhr, W.G. Capillary biosensor for glutamate. *Anal. Chem.*, 68, 2164–2169, 1996.

172. Sarkar, P. et al. Screen-amperometric biosensors for the rapid measurement of l- and d-amino acids. *Analyst*, 124, 865–870, 1999.

173. Domínguez, B. et al. Chiral analysis of amino acids using electrochemical composite bienzyme biosensors. *Anal. Biochem.*, 298, 275–282, 2001.

174. Nielsen, P.M., Petersen, D., Dambmann, C. Improved method for determining food protein degree of hydrolysis. *J. Food Sci.*, 66, 642–646, 2001.

175. Folkertsma, B., Fox, P.F. Use of the Cd-ninhydrin reagent to assess proteolysis in cheese during ripening. *J. Dairy Res.*, 59, 217–224, 1992.

176. Baer, A. et al. Microplate assay of free amino acids in Swiss cheeses. *Lebensm.-Wiss. Technol.*, 29, 58–62, 1996.

177. Pavia, M. et al. Proteolysis in Manchego-type cheese salted by brine vacuum impregnation. *J. Dairy Sci.*, 83, 1441–1447, 2000.

178. Katsiari, M.C. et al. Proteolysis in reduced sodium Kefalograviera cheese made by partial replacement of NaCl with KCl. *Food Chem.*, 73, 31–43, 2001.

179. Adler-Nissen, J. Determination of the degree of hydrolysis of food protein hydrolysates by trinitrobenzenesulfonic acid. *J. Agric. Food Chem.*, 27, 1256–1261, 1979.

180. Crowell, E.A., Ough, C.S., Bakalinsky, A. Determination of alpha amino nitrogen in musts and wines by TNBS method. *Am. J. Enol. Vit.*, 36, 175–177, 1985.

181. Bouton, Y., Grappin, R. Measurement of proteolysis in cheese: relationship between phosphotungstic acid-soluble N fraction by Kjeldahl and 2,4,6-trinitrobenzenesulphonic acid-reactive groups in water-soluble N. *J. Dairy Res.*, 61, 437–440, 1994.

182. Hernández-Herrero, M.M. et al. Protein hydrolysis and proteinase activity during the ripening of salted anchovy (*Engraulis encrasicholus* L.): a microassay method for determining the protein hydrolysis. *J. Agric. Food Chem.*, 47, 3319–3324, 1999.

183. Fields, R. The rapid determination of amino groups with TNBS. In: *Methods in Enzymology.* C.H.W. Hirs, S.N. Timasheff, Eds., Vol. 25, Academic Press, New York, NY, pp. 464–468, 1972.

184. Puchades, R., Lemieux, L., Simard, R.E. Determination of free amino acids in cheese by flow injection analysis with an enzymic reactor and chemiluminescence detector. *J. Food Sci.*, 55, 1555–1558, 1989.

185. Lemieux, L., Puchades, R., Simard, R.E. Free amino acids in Cheddar cheese: comparison of quantitation methods. *J. Food Sci.*, 55, 1552–1554, 1990.

186. Panasiuk, R. et al. Determination of alpha-amino nitrogen in pea protein hydrolysates: a comparison of three analytical methods. *Food Chem.*, 62, 363–367, 1998.

187. Lizasso, G., Chasco, J., Beriain, M.J. Microbiological and biochemical changes during ripening of salchichón, a Spanish dry cured sausage. *Food Microbiol.*, 16, 219–228, 1999.

188. Lawrie, R.A. 3-Methylhistidine as an index of meat content. *Food Sci. Technol. Today*, 2, 208–210, 1988.

189. Khalili, A.D., Zarkadas, C.G. Determination of myofibrillar and connective tissue protein contents of young and adult avian (*Gallus domesticus*) skeletal muscle and the *N*-tau-methylhistidine content of avian actins. *Poultry Sci.*, 67, 1593–1614, 1989.

190. Fuchs, G., Kuivinen, J. Methods for determination of the content of pure meat extract in mixed meat products. *Var. Foeda.*, 41, 103–118, 1989.

191. Jones, A.D., Shorley, D., Hitchcock, C.H.S. The determination of 3-methylhistidine and its application as an index for calculating meat content. *J. Assoc. Public. Anal.*, 20, 89–94, 1982.

192. Johnson, S.K., Lawrie, R.A. Actin-bound 3 methylhistidine as an index of myofibrillar protein in food. *Meat Sci.*, 22, 303–311, 1988.

193. Larsen, T.W., Thornton, R.F. Analysis of the amino acid 3-methylhistidine by gas-liquid chromatography. *Anal. Biochem.*, 109, 137–141, 1980.

194. Carnegie, P.R., Hee, K.P., Bell, A.W. Ophidine (beta-alanyl-L-3-methylhistidine, "Balenine") and other histidine dipeptides in pig muscles and tinned hams. *J. Sci. Food Agric.*, 33, 795–801, 1982.

195. Kumar, S., Pedersen, J.W. Methods of improving mechanically deboned poultry meat quality—a review. *Avian Res.*, 67, 108–115, 1983.

196. Johnson, S.K. The determination of hydroxyproline in meat using gas chromatography-mass spectrometry. *Meat Sci.*, 22, 221–227, 1988.

197. Zarkadas, C.G. Assessment of the protein quality of selected meat products based on their amino acid profiles and their myofibrillar and connective tissue protein contents. *J. Agric. Food Chem.*, 40, 790–800, 1992.

198. Zarkadas, G.C., Karatzas, C.D., Zarkadas, C.G. Assessing the myofibrillar and connective tissue protein contents and protein quality of beef tripe. *J. Agric. Food Chem.*, 44, 2563–2572, 1996.

199. Offermanns, N.C., McDougall, T.E., Guerrero, A.M. Validation of *o*-tyrosine as a marker for detection and dosimetry of irradiated chicken meat. *J. Food Protect.*, 56, 47–50, 1993.

200. Ibe, F.I. et al. Detection of *o*-tyrosine in irradiated chicken by reverse-phase HPLC and fluorescence detection. *Food Addit. Contam.*, 8, 787–792, 1991.

201. Bernwieser, I. et al. Identification of irradiated chicken meat: determination of *o*- and *m*-tyrosine by HPLC with an electrode array detector. *Ernährung*, 19, 159–162, 1995.

202. Krach, C. et al. HPLC with coulometric electrode array detection. Determination of *o*- and *m*-tyrosine for identification of irradiated shrimps. *Z. Lebensm. Unters. Forsch.*, 204, 417–419, 1997.

Chapter 3

Peptides

María-Concepción Aristoy, Miguel A. Sentandreu, and Fidel Toldrá

Contents

3.1   Introduction

There are several peptides naturally present in muscle. Carnosine (β-alanyl-L-histidine), anserine (β-alanyl-L-1-methylhistidine), and balenine (β-alanyl-L-3-methylhistidine) are dipeptides that express some physiological properties in muscle.[1,2] The content of these dipeptides is especially high in muscles with glycolytic metabolism, though this also depends on the animal species, age, and diet.[3–5] Beef and pork contain more carnosine and less anserine, lamb has similar amounts of carnosine and anserine, while poultry is very rich in anserine. Balenine is present in minor amounts in pork muscle.[6] Glutathione (GSH) is a cysteine-containing tripeptide (glutamine, glycine, and cysteine) naturally present in fresh meats. This tripeptide plays an essential role in the antioxidant system, as well as in the intercellular redox state, by the blockage of reactive oxygen species and free radicals. This thiol compound exists in two forms, the reduced (GSH) and the oxidized form (GSSG), and the ratio of the two forms is crucial for the characterization of the oxidative stress in cells. The interrelation among tissue GSH, nutrition, and oxidative stress is widely discussed elsewhere.[7] Most meats have been found to contain approximately twice the GSH found in poultry and 2–10 times more GSH than fish products.[8]

Other peptides have been reported to be generated during the aging of beef, pork, chicken, and rabbit.[9] Some of these peptides have been characterized, and some of them are related to meat tenderness or flavor: for instance, the 30-kDa peptide originated from troponin T through the action of calpain,[10] which is related to meat tenderization,[11,12] and more recently a 32-kDa peptide,[13] a 110-kDa polypeptide from the degradation of C protein,[14] and some peptides of 1282.8 Da resulting from the sarcoplasmic protein glyceraldehyde 3-phosphate dehydrogenase, 1734.8 Da from troponin T, and 5712.9 Da from creatine kinase.[14]

The study of peptides in meat has focused on several goals: (i) The detection of natural muscle dipeptides such as carnosine, anserine, and balenine, or the tripeptide GSH, which have some functional properties in postmortem muscle;[1] (ii) the detection of muscle dipeptides as biochemical markers for the detection of meat in feeds;[5,6] (iii) the prediction of tenderness based on the appearance of a 30-kDa fragment from the degradation of troponin T,[11] or the generation of other small peptides from the degradation of sarcoplasmic proteins;[14] (iv) the contribution of small peptides to flavor;[15,16] (v) the assessment of muscle metabolism pattern based on peptide levels;[3] and (vi) changes in the peptide fractions during the aging of meat as potential markers of meat quality.[14,17–19] Thus, the isolation and analysis of muscle peptides may have a broad range of relevant applications. The most important techniques for the extraction and analysis of muscle peptides are described in this chapter.

3.2   Sample Preparation and Extraction

The procedure for sample preparation is quite detailed in the literature and is quite similar to those followed for other meat-soluble compounds such as free amino acids[3,20] or nucleotides.[21] Muscle tissue must be excised from fat and other visible connective tissues. The muscle has to be finely ground and a representative sample, at least 5–10 g, taken for the analysis. The weighed tissue is homogenized with sufficient (typically in the range of 1/2, 1/5, or 1/10 p/v) of redistilled water, dilute saline solutions, acidic solutions (0.1N hydrochloric acid), or neutral phosphate buffer by vortex-mixing, or by using several disposals, including Polytron™, ultra-turrax, Stomacher™, and so on. The homogenate is then centrifuged (typically at $10,000 \times g$ for 20 min) in cold and the supernatant is filtered through glass-wool or cheesecloth. The supernatant is collected and usually deproteinized by means of protein precipitation in 2.5–3-fold volume of acetonitrile, methanol, or ethanol or by lowering the pH by adding perchloric acid (PCA) up to 0.5 M or trichloroacetic acid (TCA) up to 4–5%. Nevertheless, different acid concentrations have been used to fractionate the peptidic extract from cheese,[22,23] making use of the different solubilities of the peptides as a function of their size or amino acid composition.

A comparison of six methods for the extraction of meat peptides in the molecular weight range of 3–17 kDa has been reported.[24] In many cases, extraction and deproteinization stages are achieved in only one step by extracting the sample with a deproteinizing solvent such as 0.6N PCA[25–27] or 5% TCA.[28] Precipitated proteins are separated by centrifugation; the supernatant can be dried in a speed-vacuum or by lyophilization or directly stored in the freezer until the analysis.

3.3 Fractionation

While dipeptides and tripeptides can be readily analyzed in the obtained extract (methods for the analysis of these compounds are described in the next section), larger peptides may usually need further fractionation previous to the analysis.

Several methods of fractionation have been described based upon their differential properties, including size, charge, or polarity. In this section, methods based on size are described; methods based on charge and polarity are described in Sections 3.4.3.1 and 3.4.3.2.

3.3.1 Ultrafiltration

Ultrafiltration is a preparative technique with several applications in peptide analysis. With this technique it is possible to isolate the peptide fraction of interest based on size,[26,29–33] but it has also been used to fractionate[34] or to concentrate peptide extracts. A wide variety of membranes of several different materials, filtration surfaces, and especially different cut-off sizes are available.

3.3.2 Gel Filtration Chromatography

This technique is usually performed for preparative purposes. The sample containing the extracted peptides can be applied on a gel filtration column for further fractionation. Peptides are separated by size, eluting first the biggest peptides. In this chromatography, neither mobile nor stationary phases must interact with the sample peptides. The range of size for fractionation depends on the type of gel. For example, Sephadex G-25 gel (Pharmacia) is adequate to separate peptides within the range 500–5.000 Da,[16,30,35,36] while G-10 (Pharmacia) is adequate for much smaller peptides (below 700 Da), as in meat extracts[37] or in cheese extracts.[38] Cheese peptides smaller than 1000 Da were also fractionated[31] using a Toyopearl HW-40S (Tosoh Corp., Tokyo, Japan). This type of chromatography requires the gel be packaged into long and narrow columns (typically 2.6 × 60 cm) for maximal separation efficiency.

The elution is made with water, 0.01N HCl, or diluted phosphate buffers at low flow rates (i.e., 30–120 mL/h) under refrigeration conditions. The eluent is monitored by ultraviolet (UV) absorption at 214, 254, or 280 nm to track the elution of the compounds of interest. A typical chromatogram of a pork meat sample is shown in Figure 3.1. The column can be calibrated with standards of known molecular mass, and within the desired molecular weight range. Typical

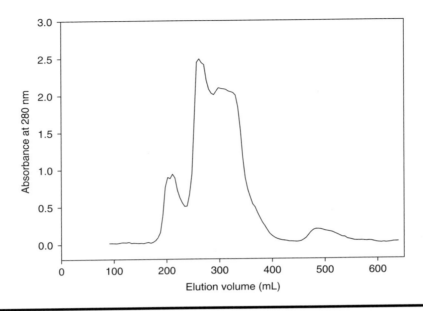

Figure 3.1 **G-25 gel filtration profile from a deproteinized pork meat extract, monitored at 280 nm.**

standards for the G-25 gel column are bovine serum albumin (68 kDa), egg albumin (45 kDa), chymotrypsinogen A (25 kDa), myoglobin (18 kDa), cytochrome *c* (12.5 kDa), aprotinin (6.5 kDa), ristocetin A sulfate (2.5 kDa), pepstatin (686 Da), and glycin (75 Da).

Gel filtration can also be performed by HPLC, in which the injection volumes used to be substantially lower. In this process, columns such as TSK-2000 SW (Tosoh Corp.) or Superdex peptide (Amersham Biosciences, Uppsala, Sweden) among others can be used. The eluent should be neutral to acid diluted phosphate or acetate buffers.[36,39] The column can also be calibrated with standards of known molecular mass as described above. Trademark catalogs are usually a good guide for these applications.

Fractions are usually collected for further analysis (sensory, further peptides separation, etc.). The size of the fractions depends on the specific purpose.

3.4 Analysis

Small peptides such as di- or tripeptides can be analyzed directly in the deproteinized extract (Section 3.2), while bigger peptides may be analyzed after fractionation of such deproteinized extract (Section 3.3).

3.4.1 Analysis of Dipeptides

The cationic nature of meat dipeptides (carnosine, anserine, and balenine) makes cation exchange HPLC the method of choice for its analysis. Postcolumn UV detection at 214 nm is enough for most of the applications in meat, but when higher sensitivity is required, the limit of detection in complex samples can be substantially enhanced with *o*-phthaldialdehyde (OPA) postcolumn

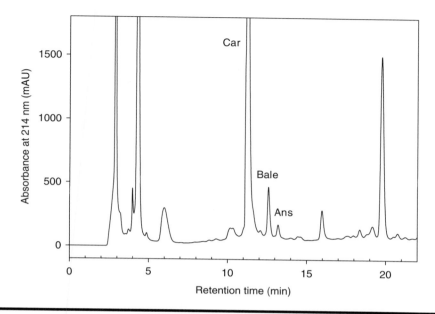

Figure 3.2 **Cation-exchange HPLC chromatogram from a deproteinized pork meat extract where separation of carnosine (Car), anserine (Ans), and balenine (Bale) is shown.**

derivatization.[5,6] In this case, fluorescence detection is possible ($\lambda_{ex} = 340$, $\lambda_{em} = 450$ nm). A cation exchange chromatographic separation of muscle dipeptides is shown in Figure 3.2. The silica-based SCX column (spherisorb) employed in this analysis renders a very robust analysis. For the postcolumn derivatization with OPA, a T-zero death volume piece connects the outlet from the column with the OPA-derivatization-reagent inlet and to a 200 cm long and 0.01 in. i.d. stainless steel reaction coil, kept at room temperature, which takes the derivatizing column effluent to the fluorescence detector set at 340 and 445 nm for excitation and emission wavelengths, respectively. The preparation of the OPA reactive is reported elsewhere.[40] Precolumn OPA[25,41] or PITC[3] derivatization and reversed-phase (RP) HPLC separation may also be used.

The quantitation of balenine has an added difficulty due to the absence of a commercially available balenine standard. To overcome this problem, Aristoy et al.[5] obtained a solution of balenine from pork loin which was further valorized by hydrolysis and analysis of the constituent amino acids (β-alanine and 3-methylhistidine) as described in that manuscript. Successful separation of carnosine, anserine, and balenine through hydrophilic interaction chromatography (HILIC) has been recently reported.[42]

3.4.2 Analysis of Glutathione

Some methods have been published on the analysis of GSH, including direct spectral,[43] separation (HPLC or capillary electrophoresis [CE]), and enzymatic methods. Separation HPLC methods are linked to different types of detection such as electrochemical,[39] UV/Vis,[44] or fluorimetric detection. When spectrophotometric detection (direct or combined with chromatographic separation) is used, the derivatization of the GSH molecule may be required, where some reagents are proposed, including iodoacetic acid and 1-fluoro-2,4-dinitrobenzene for UV/Vis

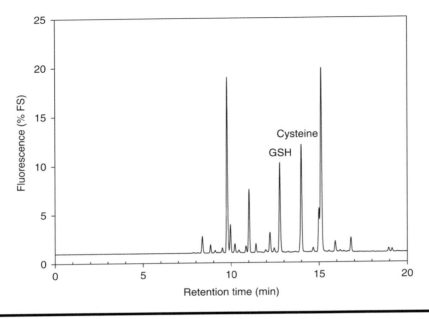

Figure 3.3 **Reversed-phase HPLC chromatogram of OPA precolumn derivatized pork loin thiol compounds. GSH, glutathione.**

detection[45] or OPA,[43] 9-fluorenyl-methoxy carbonyl succinimide,[46] or *N*-hydroxysuccinimidyl-alpha-naphthylacetate[47] for fluorimetric detection. An example of OPA precolumn derivatization and reversed phase separation is shown in Figure 3.3. CE coupled to laser-induced fluorescence detection has been developed in which fluorophore 5-iodoacetamidofluorescein (5-IAF) was chosen for GSH derivatization, which forms a fluorescent adduct that is excited at 488 nm for detection. The reaction conditions (temperature, time, and 5-IAF/GSH molar ratio) were optimized.[48] Enzymatic assays for the determination of both GSH and GSSG have been also developed. In the enzymatic assay, GSH reductase reduces GSSG with simultaneous oxidation of the specific substrate (NADPH), which is sequentially photometrically detected.[49]

3.4.3 Analysis of Other Peptides

The whole or fractionated meat peptidic extract needs further analysis through high-performance techniques such as HPLC (reversed-phase and cation exchange), CE, or polyacrylamide gel electrophoresis.

3.4.3.1 Reversed-Phase HPLC

Owing to its high resolutive power, this is the most common HPLC methodology for analyzing peptidic extracts. Indeed, RP-HPLC is widely used to generate peptide maps from digested proteins or peptidic extracts. With this technique, peptides are separated as a function of their polarity, which is directly related to their amino acid composition. There are many types of available reverse-phase columns, including those based on silica support with octadecylsilane (C-18)

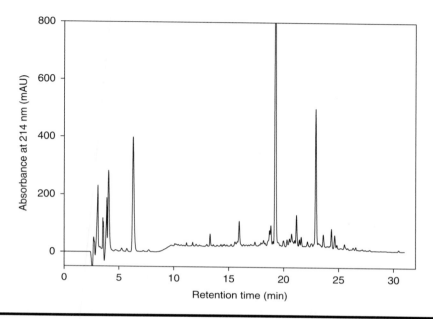

Figure 3.4 **Reversed-phase HPLC chromatogram of a deproteinized pork loin extract.**

covalently bonded and packaged in 250×4.6 mm columns, the most often used. The typical silica pore size for small-peptide analysis ranges from 60 to 100 Å, reserving 300 Å for polypeptide analysis.[29,50] A mobile phase gradient using acetonitrile as organic modifier is typically used, while 0.1% trifluoroacetic acid is the preferable volatile buffer, although 0.1% formic acid can also be used.[29] The eluent can be monitored at different wavelengths (214, 254, and 280 nm) or spectra can be obtained if using a diode array detector. Hydrophilic peptides elute first while hydrophobic peptides are retained in the column and take longer to elute. A chromatogram of pork muscle peptides is shown in Figure 3.4.

3.4.3.2   Ion-Exchange Chromatography

This type of chromatography also offers good separation of peptides and may be complementary to the RP-HPLC technique (Section 3.4.3.1) for peptide analysis. An example is shown in Figure 3.2. Acid peptides are separated better in anion exchange columns,[34] while neutral or basic peptides are separated better in cation exchange columns. In general, silica-based columns offer better performance than polymeric ones. The best results are obtained using a nonvolatile salt such as NaCl to achieve the elution of the retained peptides. This is a drawback when peptide identification by mass spectrometry is required, because in this case volatile additives must be used. To overcome this problem, the separation used to be followed by an RP-HPLC or normal-phase HPLC[34] in which salt is separated from the retained peptides.

3.4.3.3   Capillary Electrophoresis

CE has been applied to obtain peptide mapping based on its high efficiency and speed. Its application to the characterization of proteins by mean of their peptide mapping hydrolyzates has been demonstrated in the literature.[51] Its application to the analysis of peptides from crude extracts may be useful in the case of dairy foods such as milk or cheese[52,53] because of the simplicity of the milk protein composition. However, in the case of meat or meat products,[16,36,54] it is very complicated due to the more complex protein and peptide composition, the presence of interference (mainly amino acids and nucleotides), the scant information giving by this technique, and the difficulty of collecting fractions that permit further purification and characterization of the separated peptides. The widespread use of the combination of CE and mass spectrometry may enhance its applicability in the meat peptides area. However, this technique has shown very good utility for peptidase kinetic studies, where peptide hydrolysis can be controlled in very short time spans.[55–58]

3.4.3.4 Gel Electrophoresis

Sodium dodecyl sulfate polyacrylamide gel electrophoresis (SDS-PAGE) is useful for the separation of peptides in the range of 1–30 kDa present in meat. However, some cautions are necessary, including the use of low acrylamide concentration (below 10%) and tricine-SDS-PAGE. The separation is related to the p*K* values of the functional groups of tricine that define the electrophoretic mobility of peptides.[59] This system is also useful for the isolation of extremely hydrophobic peptides that will be identified by mass spectrometry. Applications to meat peptides ranging from 3 to 17 kDa have recently been reported.[24] 2 D electrophoresis is being used for proteomics as described in Chapter 5.

3.5 Peptide Sequencing

The main step for the structural characterization of proteins and peptides is the determination of their amino acid sequences. This is of crucial importance when the main functional properties are investigated since it can help to deduce some peptide/protein characteristics. Currently, there is an enormous and increasing interest in the characterization of the primary sequence of many food proteins as a basis for better knowledge of food properties. The interest in proteomics (analysis of the protein sequence) is common to other scientific fields, such as cell biology. This phenomenon is in close relation to the important development of modern mass spectrometry detectors applied to proteomics, experienced during the last decades. In fact, this discipline constitutes one of the key areas for biological research, together with genomics (gene analysis), transcriptomics (analysis of gene expression), and metabolomics (analysis of metabolite profiles).[60] The determination of the amino acid sequence of peptides and proteins is today much more feasible than just 15 years ago. In addition, the remarkable developments in cDNA sequencing allow the deduction of the corresponding protein chain, making it unnecessary to determine the entire amino acid sequence of a given protein for its identification, especially in the case of large structures.[61,62] In the case of free peptides, however, sequencing of the amino acid chain continues to be the method of choice to get the primary structure. This is particularly true in the food science field since, for example, many foods contain a large number of peptides resulting from hydrolysis of proteins during processing. This is the case for products obtained from milk or fish hydrolyzates,[63] or for the case of dry-cured meat products such as dry-fermented sausages or dry-cured ham.[64] The different alternatives that have been used for peptide/protein sequencing are described, with special emphasis on current trends.

Before the advent of the modern proteomic technology, Edman degradation has been extensively used to determine the amino acid sequence of peptides and proteins. This technique consists in the sequential cleavage of the N-terminal amino acid through the formation of a phenylthio-hydantoin (PTH)-amino acid derivative that is further identified by its specific retention time in HPLC. The automation of this process, which constitutes the basis of protein sequencers, made it more affordable to perform many cycles of this reaction, identifying at each step the corresponding N-terminal amino acid of the peptide/protein of interest.[65] Despite its utility, now decades old, this technique has some important limitations such as the time required for each cycle, the absolute requirement for a free amino terminus to perform the cleavage, and the necessity of working with highly purified samples. Researchers having much expertise in this technique can deduce the sequencing of a mixture of two peptides, but only if these peptides are in such concentrations in the mixture sufficiently different as to allow for clear different HPLC detection responses at each cycle. All these features make this process quite tedious and time consuming.

Another approach to peptide sequencing was developed just 3 years after Edman degradation, involving the derivatization of peptides into more volatile compounds suitable to analysis by GC-MS in order to obtain a mass spectrum, from which the peptide sequence could be deduced.[66] The technique was somewhat complementary to Edman sequencing in those cases where the latter showed important limitations, such as N-terminal blocked, highly hydrophobic peptides or peptides containing certain posttranslational modifications.[61]

These strategies were later displaced by the development of the so-called soft ionization techniques for peptide identification, which made possible the ionization of intact peptides without the need for derivatization. The first step was the invention of "fast atom bombardment" (FAB) ionization,[67] followed more recently by matrix-assisted laser desorption ionization (MALDI)[68] and electrospray ionization (ESI).[69] These two techniques are the principal ionization methods used for the actual mass spectrometers applied to proteomics. Peptides are ionized and their molecular masses are determined with high accuracy by following their trajectories in a vacuum system. Peptide ions can subsequently be fragmented in order to obtain the necessary structural information to determine their amino acid sequence. The main advantages compared to the previous methods are the high sensitivity, short time to collect data, and relatively easy sample preparation. Other important advantages are the possibility of analyzing complex peptide mixtures together with N-terminal blocked or other modified peptides.

Sequencing using mass spectrometry mainly uses peptides instead of proteins as a starting material for getting the sequence. As a general rule, mass spectrometers appear to be more efficient in obtaining sequence information from peptides no longer than 20 amino acids.[70] Furthermore, peptides are easier to keep in solution. For that reason, the first step after protein purification is to digest the target protein into peptides by using a specific endopeptidase such as trypsin. This enzyme specifically cleaves proteins at the C-terminus of arginine and lysine, allowing a predictable generation of peptides and an easier interpretation of mass spectra.

In view of the complexity of actual proteomic technology using mass spectrometry, only the most common situations and strategies that can be followed for obtaining the sequence of the peptides/polypeptides of interest are discussed in this chapter.

3.5.1 Peptides from Proteins

3.5.1.1 Polypeptide Digestion

The polypeptides of interest could have been purified previously by different chromatographic procedures, but the last purification step usually consists in SDS-PAGE. The portion of gel corresponding to the polypeptide is excised to perform an in-gel protein digestion using trypsin or any other suitable enzyme. If SDS-PAGE is not the last purification step, then trypsin will be added to the purified sample (in-solution digestion).

3.5.1.2 Sequencing of Tryptic Peptides

From this point, the obtained peptides can be identified in two main ways:

1. *Peptide mass fingerprinting using MALDI-TOF MS*: The peptide mixture is spotted onto a special metal plate and then combined with a matrix, normally a low-molecular-weight aromatic acid such as α-cyano-hydroxycinnamic acid or 2,5-dihydroxybenzoic acid, in order to promote the ionization of peptides when receiving the impact of a laser beam. Single-charged intact peptides are generated in this way, and then a time-of-flight (TOF) mass spectrometer will generate a mass spectrum showing the molecular masses (*m/z* values) of the peptides present in the mixture. The set of the whole tryptic peptide masses constitutes the "mass fingerprint" that will be specific to the analyzed protein. Thus, matching the obtained peptide masses to the theoretical tryptic peptide masses calculated for each protein in the database will allow the identification of the protein.

2. *Capillary HPLC coupled to the mass spectrometer*: The peptide fragments are previously separated in a capillary HPLC column that is directly coupled to the electrospray (ESI) ion source of a mass spectrometer. The column outlet is directly connected to the needle of the ESI ion source, so that the peptide flows through the needle and is ionized by the application of high voltage near the entrance of the mass spectrometer. The ions enter the vacuum of the mass spectrometer for analysis. There are different types of mass spectrometers that can be coupled to this system, including the quadrupole-ion trap, quadrupole-TOF, or triple quadrupole. The main difference with respect to the mass fingerprinting approach is that these systems allow tandem mass spectrometry, also known as MS/MS. In this mode, a first mass spectrum is carried out to obtain the *m/z* values of the intact peptides. Contrary to MALDI, ESI tends to generate double or triple-charged intact peptide ions, and this must be taken into consideration at the time to obtain the molecular mass of the peptides from the *m/z* values obtained in this first spectrum. In a second step, one particular peptide ion (called "precursor ion") is isolated and fragmented in order to obtain a second mass spectrum of the resulting fragments (called "product ions"), from which the sequence of the original peptide can be deduced. This is possible because the fragmentation of peptides in the mass spectrometer is adjusted to occur mainly through the cleavage of the amide bonds, which generates a b-type ion series when the charge is retained by the N-terminal fragment or a y-type ion series when it is retained by the C-terminal fragment. Apart from these, other kind of fragments can occur if other type of bonds are cleaved into the peptide, as illustrated in Figure 3.5. The mass difference between neighboring peaks in either the y- or b-ion series is directly related to the amino acid composition of the precursor peptide.[70] Interpretation of the spectrum can be made by using some specific computer algorithms, such as Sequest[71] or Mascot,[72] but one must keep in mind that this approach can fail under some circumstances. For such situations, an expert in the interpretation of tandem-MS spectra would probably solve the problem if data are of sufficient quality and with enough information.

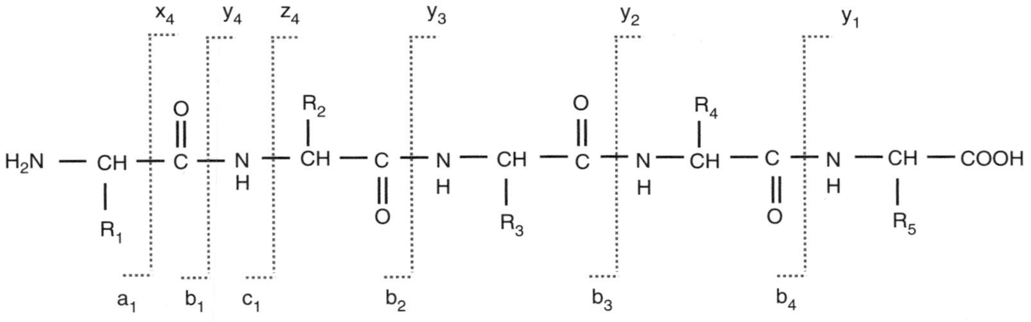

Figure 3.5 Fragmentation pattern of peptides observed in tandem mass spectrometry (MS/MS). The selected peptide ion is fragmented by collision with molecules of an inert gas. Breakdown preferentially occurs at the amide bonds because it is the level it requires the lowest energy to split the peptide. The fragment ions generated by this way are named according to a specific nomenclature. Thus, these fragments generated by cleavage of the amide bond are known as b-ions when the charge is retained on the N-terminal fragment, whereas y-ions are formed when the charge is retained on the C-terminal part. Apart from these, other types of ions can be formed, including a- or c-ions (N-terminal side), or x- and z-ions (C-terminal side).

3.5.2 Free Peptides

In many cases the research is focused on the identification of peptides present in complex matrices, which have been generated by the action of proteolytical processes that are not known in detail. This is the case of many processed foods such as fermented milk products[73] or dry-cured meat products.[74] In these cases there is no need to generate peptides by trypsin digestion or other methods because peptides are already present in the sample. However, since they have not been generated by digestion with a known enzyme, the identification of the peptide sequence will be more complicated. In such cases, the "peptide mass fingerprinting" approach will not be useful, and the only alternative is the development of different purification procedures able to fractionate the peptides as much as possible but without the need to purify to homogeneity. The first step normally implies a clean-up procedure to obtain a peptide extract devoid of the contaminants (proteins, lipids, sugars) that can be present in the initial crude extract. The obtained peptide extract is normally fractionated by different types of chromatography, including size exclusion, ion exchange, or reverse-phase. The way the sample elutes from the last purification step, which can be either online or off-line coupled to the mass spectrometer, must be compatible with MS analysis. This can be done performing an HPLC reverse-phase chromatography, for example, in which samples are eluted by increasing the organic content of the mobile phase and so avoiding the use of salt, which can suppress the further ionization of peptides. As very hydrophilic peptides are not retained in reverse-phase columns, a good alternative in those cases would be the use of hydrophilic interaction chromatography, which is also compatible with MS analysis.[42] The sequence of the eluted peptides can be determined by tandem mass spectrometry, as explained in Section 3.5.1.2. For the obtained peptide sequences, the protein of origin can be identified by searching for amino acid sequence similarity against online protein databases. The BLAST search engine (http://www.bork.embl-heidelberg.de) can be used for this purpose.

References

1. Chan, K.M. and Decker, E.A. Endogenous skeletal muscle antioxidants. *Crit. Rev. Food Sci. Nutr.* 34: 403–426, 1994.
2. Gianelli, M.P. et al. Effect of carnosine, anserine and other endogenous skeletal peptides on the activity of porcine muscle alanyl and arginyl aminopeptidases. *J. Food Biochem.* 24: 69–78, 2000.
3. Aristoy, M.C. and Toldrá, F. Concentration of free amino acids and dipeptides in porcine skeletal muscles with different oxidative patterns. *Meat Sci.* 50: 327–332, 1998.
4. Cornet, M. and Bousset, J. Free amino acids and dipeptides in porcine muscles: differences between red and white muscles. *Meat Sci.* 51: 215–219, 1999.
5. Aristoy, M.C., Soler, C., and Toldrá, F. A simple, fast and reliable methodology for the analysis of histidine dipeptides as markers of the presence of animal origin proteins in feeds for ruminants. *Food Chem.* 84: 485–491, 2004.
6. Aristoy, M.C. and Toldrá, F. Histidine dipeptides HPLC-based test for the detection of mammalian origin proteins in feeds for ruminants. *Meat Sci.* 67: 211–217, 2004.
8. Jones, D.P. et al. Glutathion in foods listed in the national Cancer Institute's health habits and history food frequency questionnaire. *Nutr. Cancer* 17: 57–75, 1992.
9. Nishimura, T. Influence of peptides produced during postmortem conditioning on improvement of meat flavor. In: *Research Advances in the Quality of Meat and Meat Products* (F. Toldrá, ed.), Research Signpost, Trivandrum, India, pp. 65–78, 2002.
10. Negishi, H., Yamamoto, T., and Kuwata, T. The origin of the 30 kDa component appearing during post-mortem ageing of bovine muscle. *Meat Sci.* 42: 289–303, 1996.
11. Olson, D.G. and Parrish, F.C., Jr. Relationship of myofibril fragmentation index to measures of beef steak tenderness. *J. Food Sci.* 42: 506–511, 1977.
12. McBride, M.A. and Parrish, F.C. The 30,000 Dalton component of tender bovine longissium muscle. *J. Food Sci.* 42: 1627–1629, 1977.
13. Okumura, T., Yamada, R., and Nishimura, T. Survey of conditioning indicators for pork loins: changes in myofibrils, proteins and peptides during postmortem conditioning of vacuum-packed pork loins for 30 days. *Meat Sci.* 64: 467–473, 2003.
14. Stoeva, S., Byrne, C.E., Mullen, A.M., Troy, D.J., and Voelter, W. Isolation and identification of proteolytic fragments from TCA soluble extracts of bovine M. Longissimus dorsi. *Food Chem.* 69: 365–370, 2000.
15. Spanier, A.M. et al. Meat flavor: contribution of proteins and peptides to the flavor of beef. In: *Quality of Fresh and Processed Foods* (F. Shahidi, A.M. Spanier, C.-T. Ho, and T. Braggins, eds.), Advances in Experimental Medicine and Biology, Vol. 542, Kluwer Academic/Plenum, New York, pp. 33–49, 2004.
16. Aristoy, M.-C. and Toldrá, F. Isolation of flavor peptides from raw pork meat and dry-cured ham. In: *Food Flavors: Generation, Analysis and Process Influence* (G. Charalambous, ed.), Elsevier Science, Amsterdam, pp. 1323–1344, 1995.
17. Moya, V.J. et al. Evolution of hydrophobic polypeptides during the ageing of exudative and non-exudative pork meat. *Meat Sci.* 57: 395–401, 2001.
18. Moya, V.J. et al. Pork meat quality affects peptide and amino acid profiles during the ageing process. *Meat Sci.* 58: 197–206, 2001.
19. Flores, M. et al. Nitrogen compounds as potential biochemical markers of pork meat quality. *Food Chem.* 69: 371–377, 2000.
20. Flores, M. et al. Non-volatile components effects on quality of serrano dry-cured ham as related to processing time. *J. Food Sci.* 62: 1235–1239, 1997.
21. Batlle, N., Aristoy, M.-C., and Toldrá, F. Early postmortem detection of exudative pork meat based on nucleotide content. *J. Food Sci.* 65: 413–416, 2000.
22. Moughan, P.J. et al. Perchloric and trichloroacetic acids as precipitants of protein in endogenous ileal digesta from rat. *J. Sci. Food Agric.* 52: 13–21, 1990.

23. Rohm, H., Benedikt, J., and Jaros, D. Comparison of 2 precipitation methods for determination of phosphotungstic acid-soluble nitrogen in cheese. *Food Sci. Technol. Lebensm.-Wiss. Technol.* 27: 392–393, 1994.

24. Claeys, E. et al. Quantification of fresh meat peptides by SDS-PAGE in relation to ageing time and taste intensity. *Meat Sci.* 67: 281–288, 2004.

25. Dunnett, M. and Harris, R.C. High-performance liquid chromatographic determination of imidazole dipeptides, histidine, 1-methylhistidine and 3-methylhistidine in equine and camel muscle and individual muscle fibres. *J. Chromatogr. B* 688: 47–55, 1997.

26. Bauchart, C. et al. Small peptides (<5 kDa) found in ready-to-eat beef meat. *Meat Sci.* 74: 658–666, 2006.

27. Liu, Y., Xu, X., and Zhou, G. Changes in taste compounds of duck during processing. *Food Chem.* 102: 22–26, 2007.

28. Nishimura, T. et al. Components contributing to the improvement of meat taste during storage. *Agric. Biol. Chem.* 52: 2323–2330, 1988.

29. Sforza, S. et al. Effect of extended aging of Parma dry-cured ham on the content of oligopeptides and free amino acids. *J. Agric. Food Chem.* 54: 9422–9429, 2006.

30. Jang, A. and Lee, M. Purification and identification of angiotensin converting enzyme inhibitory peptides from beef hydrolysates. *Meat Sci.* 69: 653–661, 2005.

31. Salles, C. et al. Evaluation of taste compounds in water-soluble extract of goat cheeses. *Food Chem.* 68: 429–435, 2000.

32. Sommerer, N. et al. Isolation of a peptidic fraction from the goat cheese water-soluble extract by nanofiltration for sensory evaluation studies. *Dev. Food Sci.* 40: 207–217, 1998.

33. Sommerer, N. et al. A liquid chromatographic purification method to isolate small peptides from goat cheese for their mass spectrometry analysis. *Sci. Aliment.* 18: 537–551, 1998.

34. Kim, S., Kim, S., and Song, K.B. Purification of an ACE inhibitory peptide from hydrolysates of duck meat protein. *Nutraceuts. Food* 8: 66–69, 2003.

35. Spanier, A.M., Edwards, J.W. and Dupuy, H.P. The warmed-overflavor process in beef: a study of meat proteins and peptides. *Food Technol.* 42: 110, 112–118, 1988.

36. Rodríguez-Núñez, E., Aristoy, M.C., and Toldrá, F. Peptide generation in the processing of dry-cured ham. *Food Chem.* 53: 187–190, 1995.

37. Cambero, M.I., Seuss, I., and Honikel, K.O. Flavor compounds of beef broth as affected by cooking temperature. *J. Food Sci.* 57: 1285–1290.

38. Gonzalez de Llano, D., Polo, M.C., and Ramos, M. Production, isolation and identification of low molecular mass peptides from blue cheese by high performance liquid chromatography. *J. Dairy Res.* 58: 363–372, 1991.

39. Vignaud, C. et al. Separation and identification by gel filtration and high-performance liquid chromatography with UV or electrochemical detection of the disulphides produced from cysteine and glutathione oxidation. *J. Chromatogr. A* 1031: 125–133, 2004.

40. Hernández-Jover, T. et al. Ion-pair high-performance liquid chromatographic determination of biogenic amines in meat and meat products. *J. Agric. Food Chem.* 44: 2710–2715, 1996.

41. Maynard, L.M. et al. High levels of dietary carnosine are associated with increased concentrations of carnosine and histidine in rat soleus muscle. *J. Nutr.* 131: 287–290, 2001.

42. Mora, L., Sentandreu, M.A., and Toldrá, F. Hydrophilic chromatographic determination of carnosine, anserine, balenine, creatine, and creatinine. *J. Agric. Food Chem.* 55: 4664–4669, 2007.

43. Hissin, P.J. and Hilf, R. Fluorometric method for determination of oxidized and reduced glutathione in tissues. *Anal. Biochem.* 74: 214–226, 1976.

44. Davey, M.W., Dekempeneer, E., and Keulemans, J. Rocket-powered high-performance liquid chromatographic analysis of plant ascorbate and glutathione. *Anal. Biochem.* 31: 74–81, 2003.

45. Reed, D.J. et al. High-performance liquid chromatography analysis of nanomole levels of glutathione, glutathione disulfide, and related thiols and disulfides. *Anal. Biochem.* 106: 55–62, 1980.

46. You, J.M. et al. High-performance liquid chromatographic determination of amino acids and oligo-peptides by pre-column fluorescence derivatization with 9-fluorenyl-methoxy carbonyl succinimide. *J. Liq. Chromatogr. Rel. Technol.* 2: 2103–2115, 1998.

47. Zhao, Y.Y. et al. *N*-Hydroxysuccinimidyl-alpha-naphthyl acetate as pre-column derivating reagent to separate and determine peptides and their hydrolysates by reversed phase HPLC. *Chem. J. Chin. Universities-Chinese* 17: 1044–1047, 1996.

48. Musenga, A. et al. Sensitive and selective determination of glutathione in probiotic bacteria by capillary electrophoresis-laser induced fluorescence. *Anal. Bioanal. Chem.* 387: 917–924, 2007.

49. Hadley, W.M., Bousquet, W.F., and Miya, T.S. Modified fluorescence assay for oxidized and reduced glutathione in tissue. *J. Pharm. Sci.* 63: 57–59, 1974.

50. Kitamura, S. et al. Mechanism of production of troponin T fragments during postmortem aging of porcine muscle. *J. Agric. Food Chem.* 53: 4178–4181, 2005.

51. Saz, J.M. and Marina, M.L. High performance liquid chromatography and capillary electrophoresis in the analysis of soybean proteins and peptides in foodstuffs. *J. Sep. Sci.* 30: 431–451, 2007.

52. Albillos, S.M. et al. Prediction of the ripening times of ewe's milk cheese by multivariate regression analysis of capillary electrophoresis casein fractions. *J. Agric. Food Chem.* 54: 8281–8287, 2006.

53. Steffan, W. et al. Characterization of casein lactosylation by capillary electrophoresis and mass spectrometry. *Eur. Food Res. Technol.* 222: 467–471, 2006.

54. Toldrá, F. and Aristoy, M.C. Availability of essential amino acids in dry-cured ham. *Int. J. Food Sci. Nutr.* 44: 215–216, 1993.

55. Sentandreu, M.A. and Toldrá, F. Purification and biochemical properties of dipeptidylpeptidase I from porcine skeletal muscle. *J. Agric. Food Chem.* 48: 5014–5022, 2000.

56. Sentandreu, M.A. and Toldrá, F. Partial purification and characterization of dipeptidyl peptidase II from porcine skeletal muscle. *Meat Sci.* 57: 93–103, 2001.

57. Sentandreu, M.A. and Toldrá, F. Biochemical properties of dipeptidylpeptidase III purified from porcine skeletal muscle. *J. Agric. Food Chem.* 46: 3977–3984, 1998.

58. Sentandreu, M.A. and Toldrá, F. Dipeptidyl peptidase IV from porcine skeletal muscle: purification and biochemical properties. *Food Chem.* 75: 159–168, 2001.

59. Schägger, H. Trcine-SDS-PAGE. *Nat. Prot.* 1: 16–22, 2006.

60. Kussmann, M. and Affolter, M. Proteomic methods in nutrition. *Curr. Opin. Clin. Nutr. Metab. Care* 9: 575–583, 2006.

61. Biemann, K. Laying the groundwork for proteomics—mass spectrometry from 1958 to 1988. *Int. J. Mass Spectrom.* 259: 1–7, 2007.

62. Johnson, R.S. and Walsh, K.A. Sequence analysis of peptide mixtures by automated integration of Edman and mass spectrometric data. *Protein Sci.* 9: 1083–1091, 1992.

63. Yamamoto, N., Ejiri, M., and Mizuno, S. Biogenic peptides and their potential use. *Curr. Pharm. Des.* 9: 1345–1355, 2003.

64. Toldrá, F. *Dry Cured Meat Products.* Food and Nutrition Press, Trumbull, CN, 2002.

65. Edman, P. and Begg, G. A protein sequenator. *Eur. J. Biochem.* 1: 80–91, 1967.

66. Nau, H. and Biemann, K. Amino-acid sequencing by gas chromatography mass spectrometry using perfluoro-dideuteroalkylated peptide derivatives. 1. Gas-chromatographic retention indexes. *Anal. Biochem.* 73: 139–153, 1976.

67. Barber, M. et al. Fast atom bombardment of solids (FAB)—a new ion-source for mass-spectrometry. *J. Chem. Soc. Chem. Commun.* 7: 325–327, 1981.

68. Karas, M. and Hillenkamp, F. Laser desorption ionization of proteins with molecular masses exceeding 10000 Daltons. *Anal. Chem.* 60: 2299–2301, 1988.

69. Fenn, J.B. et al. Electrospray ionization for mass-spectrometry of large biomolecules. *Science* 246: 64–71, 1989.

70. Steen, H. and Mann, M. The ABC's (and XYZ's) of peptide sequencing. *Nat. Rev. Mol. Cell Biol.* 5(9): 699–711, 2004.

71. Eng, J.K., Mccormack, A.L., and Yates, J.R. An approach to correlate tandem mass-spectral data of peptides with amino-acid-sequences in a protein database. *J. Am. Soc. Mass Spectrom.* 5: 976–989, 1994.
72. Perkins, D.N. et al. Probability-based protein identification by searching sequence databases using mass spectrometry data. *Electrophoresis* 20: 3551–3567, 1999.
73. Jauhiainen, T. and Korpela, R. Milk peptides and blood pressure. *J. Nutr.* 137: 825S–829S, 2007.
74. Sentandreu, M.A. et al. Proteomic identification of actin-derived oligopeptides in dry-cured ham. *J. Agric. Food Chem.* 55: 3613–3619, 2007.

Chapter 4

Proteins

Marion L. Greaser

Contents

4.1   Introduction

Proteins are the major nonwater ingredients of muscle foods and constitute almost 20% of the weight of lean muscle tissue. Proteins are polymers of amino acids that perform diverse functions. They make up the enzymes that catalyze biochemical reactions to sustain life. Muscle tissue also contains large quantities of proteins involved in the process of muscle contraction. Since the amino acid composition of proteins from food animals is similar to human muscle and muscle makes up almost 50% of our body's weight, muscle foods are an excellent source of the amino acids needed for growth, repair, and maintenance. Protein content determinations are required for measuring muscle food composition to meet human nutritional requirements,[1] for product formulation and to meet legal requirements for product identity in meat industry, and for research on the role of specific proteins in foods and biological processes.

This chapter reviews the major methodology for analysis of muscle proteins and discusses some of the advantages and disadvantages of the different approaches. Analysis of total protein, classes of proteins, and methods to measure quantities of specific proteins are reviewed in this chapter. The topic is too broad to allow for an in-depth coverage; thus the goal will rather be to provide a general overview and a few examples. For more details on topics such as amino acids, peptides, proteomics, and muscle enzymes, the reader should also refer to other related chapters in this volume.

4.2 Major Methods for Measuring Protein Nitrogen

4.2.1 Kjeldahl Method

The first widely used method for analysis of proteins was initiated by Johan Gustav Christoffer Thorsager Kjeldahl, a Danish chemist. He was employed by Carlsbad Laboratories (associated with Carlsbad Brewery) and was given the task of determining the amount of protein in various grain sources. The yield of beer had been found to be inversely related to the grain's protein content. In 1883, Kjeldahl published a method to measure the nitrogen content of a sample.[2] The method involves three steps—digestion, distillation, and titration. The material to be tested was first digested in sulfuric acid at high heat, which converted the nitrogen-containing substances into ammonium sulfate. Sulfuric acid alone proved to be too slow to achieve complete digestion (it boils at 338°C), but addition of potassium sulfate plus a divalent metal catalyst (Kjeldahl used mercury, but copper or selenium also work well) raised the boiling point to 373°C and accelerated the digestion. After cooling the digest, diluting it with water, and adding an excess of alkali to convert ammonium ions into ammonia, the ammonia was distilled and trapped in boric acid. The amount of ammonia that had been generated was then determined by back titration with acid. The method proved to be extremely robust and was adopted for the analysis of a wide variety of substances. The procedure came to be known as the "Kjeldahl method," and its flexibility, resistance to problems with interfering substances, and high reproducibility led to its early adoption as an official method for protein determination in meat and meat products.[3] This method has been the "gold standard" with which all the other methods have been compared for more than 100 years. Figure 4.1 shows a painting of Johan Kjeldahl in his laboratory along with the apparatus for digestion and distillation.

4.2.2 Dumas Method

Another method for determining nitrogen content of organic materials was developed by Jean-Baptiste Dumas, a French chemist. He found that nitrogen could be analyzed by first combusting the material at high heat in the presence of oxygen and subsequently measuring the volume of gaseous nitrogen released. Although his method was originally described in 1833, the results were somewhat variable and the Kjeldahl method proved to be much more reliable. In recent years, the Dumas combustion method has made a comeback. Its main advantage is that results may be

Figure 4.1   Copy of a painting of Johan Kjeldahl in his laboratory. Digestion and distillation apparatus are visible on the bench. (From The Carlsberg Laboratory 1876/1976. With permission of the Carlsberg Foundation.)

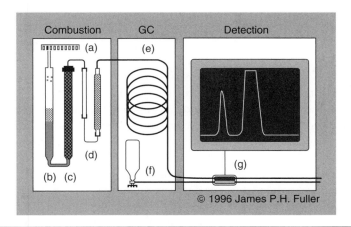

Figure 4.2   Diagram of instrumentation and steps in a Dumas protein combustion. (From http://www.uga.edu/~sisbl/udumas.html, University of Georgia. With permission.)

obtained more quickly (about 5 min for a single sample versus 2–3 h for the Kjeldahl method). Disadvantages still remain: the equipment required is more expensive than the Kjeldahl method, and the Dumas method analyzes single samples prepared sequentially. Thus, using any of the methods, the same total time is needed to obtain results from a group of 15–20 samples.

Figure 4.2 shows the various steps in the combustion and analysis process. A small sample (~0.2–1 g) is wrapped in a piece of foil and placed in a multiplace sample holder (Figure 4.2a).

The sample is dropped into a combustion tube (Figure 4.2b) heated to ~1000°C where the meat sample is mixed with pure oxygen, and the organic components are totally converted into CO_2, N_2, H_2O, and nitrogen-containing oxides. Helium gas then flushes these products through a copper-containing column (Figure 4.2c) where the nitrogen oxides are fully converted into N_2. Water and CO_2 are removed by a trap (Figure 4.2d) and the remaining volatiles pass through a gas chromatograph (Figure 4.2e). The nitrogen gas then flows through a thermal conductivity meter where the signal is compared to a separate parallel stream of pure helium. The area under the nitrogen peak is determined (Figure 4.2g), and the nitrogen content calculated by comparing with results from known standards (such as ethylenediaminetetraacetic acid [EDTA] or lysine) run at the same time as the unknown. As with the Kjeldahl, the protein content is calculated using the conversion factor of 6.25 X N.

A comparison of the results using the Kjeldahl digestion and Dumas combustion methods has been made.[4] Using both methods, 12 laboratories analyzed the same set of 15 meat samples. The samples included both raw and cooked products with varying fat content. The nitrogen values obtained by both methods were in remarkably good agreement. Only 2 of 360 determinations by the Dumas combustion method were considered outliers. Combustion instruments from three different commercial manufacturers gave similar precision, repeatability, and reproducibility. Thus, both the AOAC Kjeldahl method 928.08 and the AOAC combustion method 992.15 are now accepted for total nitrogen and estimated protein in whole meat samples. It should be mentioned that accurate analysis by both methods requires adequate mixing and careful sampling. The U.S. Department of Agriculture's *Chemistry Laboratory Guidebook* (1986) provides directions for sample preparation. Emulsified meat products need to be ground through a 1/8 in. plate twice, and coarse ground products passed through the grinder three times to ensure adequate mixing. Since only 200–900 mg of sample is typically used for combustion, replicates should be used because sampling errors can be easily made.

4.3 Muscle Protein Analysis

An early paper that systematically examined the composition of muscle tissue was conducted by von Furth.[5] He used a variety of salt solutions to prepare protein-containing extracts. Nitrogen was determined by the Kjeldahl method. The nitrogen composition of various food items, including meat, was reported as early as 1906.[6] Grossfeld[7] studied the water and nitrogen content of minced meat and sausage products, and he concluded the "nitrogen X 6.25" factor, originally proposed by Kjeldahl, was appropriate for protein estimation in such products. However, to more accurately estimate actual protein content in a meat or muscle sample, several corrections need to be applied. First, not all nitrogen in a meat sample comes from protein. Nitrogen is present in peptides, free amino acids, DNA, RNA, nucleotides, some phospholipids, some complex carbohydrates, and some vitamins. Typically these nonprotein materials are separated by treating a meat homogenate with trichloroacetic acid that causes the proteins to precipitate. The soluble fraction, when analyzed by the Kjeldahl procedure, is referred to as "nonprotein nitrogen."[8] The nitrogen content of protein is approximately 16% by weight, thus multiplying nitrogen values by 6.25 should yield protein content of meat. However, meat samples are very heterogeneous in regard to their content of fat and connective tissue, and the protein conversion factor for the latter is not 16%. This was recognized very early in protein analyses with reported values of 17.9% nitrogen in collagen (gelatin) and 16.9% in elastin.[9] Thus meat or muscle samples containing larger proportions of collagen will underestimate the total protein content.[10] Apparently not everyone has got this message, since the review by Benedict[11] has an extensive discussion on the errors in protein estimation with high collagen–containing meat products. Benedict suggested that separate analyses for collagen be conducted, and that the nitrogen conversion factor be altered using the formula 6.25 (−) 0.0085A, where A is the percent collagen. The amount of collagen is usually determined by hydroxyproline content. The conversion factor for hydroxyproline to collagen is 8, so the correct formula for meats with significant collagen content would be 6.25 (−) 0.068B, where B is the percent hydroxyproline.

4.4   Protein Solubility Classes

Proteins in meat have been divided into several classes based on their cell location and solubility. Bate-Smith[8,12] recognized there were differences between intracellular and extracellular proteins. The sarcoplasmic proteins are readily soluble in water or dilute salt-containing solutions and consist primarily of the glycolytic enzymes. Extraction of muscle with strong salt solutions dissolves the major portion of the sarcoplasmic plus myofibril proteins.[8,12] Most of the strong salt soluble proteins can be brought out of solution by diluting the ionic strength to ~0.05. Robinson[13] extended these studies by using a higher pH extracting solution (pH 8.5) and a further extraction with 0.1 M NaOH. The residue after these sequential extractions was referred to as the connective tissue or stroma fraction and was believed to consist primarily of the proteins, collagen and elastin.[13] Perry[14] found that preparations of myofibrils had very low quantities of proteins soluble in 0.08 M borate buffer, pH 7.1. He also found that the washed myofibrils could be almost completely dissolved in high (>0.5 M) concentrations of salts.

A fractionation scheme for the separation of the various solubility classes of muscle proteins is shown in Figure 4.3. The estimated quantities of the major proteins that constitute these classes are shown in Tables 4.1 and 4.2.[14] The values shown for the amounts of the various proteins are based on the assumption that one is starting with lean meat (i.e., low fat content). Homogenization of the tissue in 3–8 volumes of dilute salt such as 0.08 M borate buffer (pH 7.1),[15] 0.03 M potassium phosphate (pH 7.4),[16] or 0.1 M KCl–0.066 M potassium phosphate (pH 7.1)[17] followed by centrifugation yields a supernatant soluble fraction (containing the sarcoplasmic proteins) and an insoluble precipitate. The proteins obtained would be soluble in pure water, but the latter's pH may be lower than neutrality and result in some of the proteins becoming insoluble due to isoelectric precipitation. If the step 1 precipitate (Figure 4.3) is resuspended in a high salt-containing solution such as 0.5–1.0 NaCl, the myofibril protein is solubilized and can be separated from the insoluble proteins by centrifugation. This residual insoluble protein, referred to a stroma or connective tissue

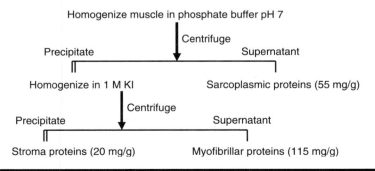

Figure 4.3   Fractionation of muscle proteins. The method shown is a sequential extraction protocol to separate proteins into three different solubility classes.

Table 4.1   Major Sarcoplasmic Proteins in Skeletal Muscle

Enzyme	Content (mg/g)
Phosphorylase	2.0
Amylo-1,6-glucosidase	0.1
Phosphoglucomutase	0.6
Phosphoglucose isomerase	0.8
Phosphofructokinase	0.35
Aldolase	6.5
Triose phosphate isomerase	2.0
Glyceraldehyde 3-phosphate dehydrogenase	0.3
Glyceraldehyde phosphate dehydrogenase	11.0
Phosphoglycerate kinase	1.4
Phosphoglycerate mutase	0.8
Enolase	2.4
Pyruvate kinase	3.2
Lactic dehydrogenase	3.2
Creatine kinase	5.0
AMP kinase	0.4
Total protein	~40

Source: Adapted from Scopes, R.K., *Biochem. J.* 134, 197, 1973. With Permission.

Table 4.2   Major Myofibril Proteins in Muscle

Protein	Percent of Myofibrillar Protein
Myosin (heavy + light chains)	43
Actin	22
Titin	8
Tropomyosin	5
Troponin (TnI + TnT + TnC)	5
Nebulin	3
Myosin binding protein C	2
M-protein	2
Alpha-actinin	2
Total	92

Note: Compiled from various sources.

protein, is composed primarily of collagen and elastin. The sarcoplasmic and myofibrillar fraction proteins are intracellular whereas the stroma proteins are found outside the muscle cell. The relative content of the different classes of proteins has been estimated as 55 mg/g for sarcoplasmic, 115 mg/g of myofibrillar, and 20 mg/g for stroma.[18]

The diagram in Figure 4.3 is a simplification of procedures necessary to accurately estimate the true content of each of these protein classes. Part of the sarcoplasmic proteins will be trapped in the soluble phase of the pellets after the first centrifugation. Thus to completely remove this fraction, one would need to extract and centrifuge more than once. Also the extent of soluble protein extraction will depend on the degree of particle size reduction and time of extraction. Mixing ground meat with a low salt buffer will result in most of the sarcoplasmic proteins being extracted, but homogenization in a Waring, Polytron, or similar type of blender will improve the percent of protein extracted. Even mincing tissue to pieces as small as a half-grain of rice did not allow as efficient protein extraction as using 20 μm microtome sections.[16] Similarly the amount of myofibril protein extracted depends on the number of extractions, the volume of extracting solution, and the type of extracting salt. During the first 10–30 min of high salt treatment, the primary protein dissolved is myosin. Longer extraction times resulted in increasing the actin content. Bate-Smith[8,12] found that at least six sequential extractions with high salt were necessary to remove 95% of the salt soluble protein. Bate-Smith[19] also showed that LiCl was more efficient than NaCl or KCl in extracting the salt soluble proteins. An extensive study by Helander[16] compared a number of different salts, and his results showed that the most efficient extractant was KI–0.1 M potassium phosphate, pH 7.4. This is probably due to the fact that the iodide ion causes F-actin to depolymerize. A major factor affecting the solubilization of the myofibrillar proteins is the high viscosity of the actin–myosin complexes that remain associated even after dissolving these proteins from the tissues. Although many early workers used an overnight extraction with high salt to dissolve the myofibrillar protein fraction, Helander[16] found that three successive extractions with 10 volumes each (solvent relative to tissue) over a total period of 8 h were sufficient to quantitatively solubilize the myofibrillar proteins. Robinson[13] followed his pH 8.5 high salt soluble extraction with an additional treatment with 0.1 M NaOH to remove the final salt soluble protein. It is unclear whether this latter alkali soluble fraction is really a salt soluble protein or a partially soluble stroma protein.

A group of proteins that do not really fit in one of the mentioned classes are the membrane proteins. These are found in the sarcolemma, nuclei, sarcoplasmic reticulum, T tubules, mitochondria, and Golgi apparatus of the muscle cells. Membrane proteins are also found in other cell types in muscle tissues such as fibroblasts, blood cells, and fat cells. Normally these proteins will be distributed between different fractions depending on the fractionation protocol. The centrifugation to separate the sarcoplasmic proteins is typically conducted using speeds of 5,000–10,000 X G. Such speeds would probably sediment large cell membrane fragments, nuclei, and mitochondria but leave most sarcoplasmic reticulum fragments in the supernatant. The latter fragments would also end up in the myofibrillar fraction supernatant after high salt extraction. Thus part of the membranes would end up in the "sarcoplasmic" fraction and the remaining part in the "myofibrillar."

Another problem with the protein fractionation scheme occurs when analyzing early postmortem tissue. If muscle is homogenized while ATP is still remaining, some of the myofibril protein gets released in the low ionic strength buffer.[20] This can be observed by the extracted appearance of many myofibrils and the presence of filaments that can be pelleted by centrifugation at 30,000 or 60,000 X G.[19] Therefore, a small part of the myofibrillar protein would end up in the "sarcoplasmic" fraction when the usual lower speed centrifugation is used.

4.5 Methods for Analysis of Soluble Proteins

In addition to the Kjeldahl and Dumas procedure, a number of methods have been used to assay muscle protein content. Most of these methods require a soluble protein extract. Thus protein fractions that are merely suspensions may be difficult or impossible to accurately pipet into the assay system. Whole muscle homogenates often contain connective tissue fragments that clog pipet tips and may thus give large variations of total protein values between replicates. Added to this, two more problems are common. First, pipeting myofibrillar or high salt extracts may also be difficult because of their high viscosity. Second, whole muscle homogenates or protein fractions may be high in fat content that results in significant turbidity. Anyone attempting to assay protein concentrations of solutions needs to be aware of these potential problems and use appropriate means to reduce assay variability.

4.5.1 Biuret

The biuret assay reagent is a mixture of potassium hydroxide, copper sulfate, and potassium sodium tartrate. Copper (Cu^{+2}) forms a complex with the peptide bonds under alkaline conditions when this reagent is added to a protein solution. Copper is converted into Cu^{+1}, and the color of the solution changes from blue to purple. Sample absorbance is measured at 540 nm in a spectrophotometer. The word "biuret" refers to a compound formed from the condensation of two urea molecules. When mixed with copper, biuret gives a similar type of color change as do proteins. The procedure that is still widely used was described by Gornall et al.[21] Protein concentrations are determined from a calibration curve prepared using bovine serum albumin or immunoglobulins as standards. The assay is not as sensitive to interfering substances as some other more recently introduced methods described in this chapter, but it requires higher protein concentrations to produce an adequate signal. The strongly alkaline solution reduces problems from turbidity due to insoluble proteins, but large amounts of lipids cause cloudiness that still can be a problem. Known interfering compounds include Tris buffer, ammonium ions, sucrose, primary amines, and glycerol.[22] The effect of Tris on the protein assay includes effects on both blank values as well as the slope of the standard curve.[23] In all protein assays, it is a good practice to prepare standard curves using the same buffer as used in the proteins to be assayed.

4.5.2 Lowry

This assay was developed by Lowry et al.[24] The method combines the biuret copper reaction with the addition of another solution (the Folin-Ciocalteau phenol reagent—a mixture of phosphotungstic acid, phosphomolybdic acid, and phenol) to enhance the sensitivity of the assay. This second reagent reacts with the tyrosine, tryptophan, and cysteine residues in the protein. The color response varies more widely between different proteins than with biuret alone, since there is a wide range in proportions of these amino acids in different proteins. The assay is sensitive to interference from strong acids or bases, chelating agents such as EDTA that tie up the copper, or reducing agents (such as 2-mercaptoethanol, dithiotreitol, or phenols) that alter the copper oxidation state.[25] The absorbance of the mixtures can be read at 500 nm if the protein solutions are at a higher concentration or at 750 nm for more dilute samples. A disadvantage of this assay system is the requirement of adding two separate reagents. A further problem is that the standard curves are typically nonlinear and that the timing of the reagent addition and the mixing of the assay samples are critical for reproducible results.[25]

4.5.3 Bicinchoninic Acid

The bicinchoninic acid (BCA) assay is a different modification of the biuret reaction.[26] In this case, BCA replaces the Folin reagent of the Lowry assay. BCA forms a 2:1 complex with the Cu^{+1}, and this complex has a purple color and an absorbance maximum at 562 nm (Figure 4.4). Since

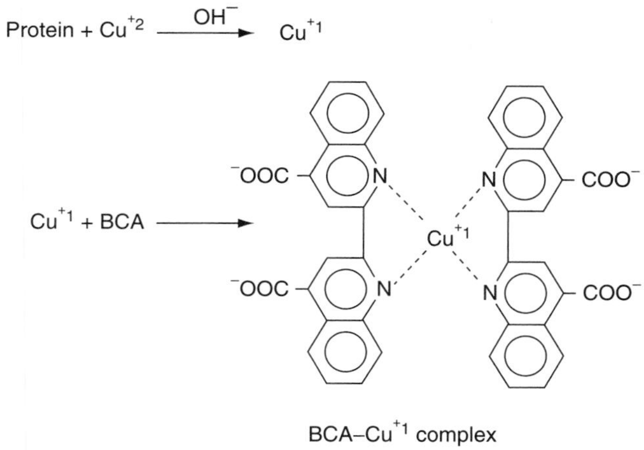

Figure 4.4 Color reaction steps using the bicinchoninic reagent. (From Smith, P.K. et al., *Anal. Biochem.,* **150, 76, 1985. With permission.)**

the BCA reagent is stable under alkaline conditions, whereas the Folin reagent used in the Lowry procedure is not, a one-step reagent addition can be used in the BCA assay. An additional advantage of this method is that it tolerates higher concentrations of detergents (such as 1% sodium dodecyl sulfate (SDS) or 1% Triton X-100) than the Lowry.

4.5.4 Coomassie Blue Dye Binding (Bradford)

The Bradford assay[27] uses Coomassie Blue G-250 binding to proteins to determine their concentration. The dye is dissolved in a strong acid solution (8.5% phosphoric acid). Unbound Coomassie Blue G-250 exists in both red and green forms, and incubation with proteins results in the conversion of the red form into blue (bound form). The reaction is rapid, with the color forming within 2 min. Protein binding causes a shift in the absorbance maximum from 465 to 595 nm. The dye binds primarily to arginine side chains with lesser interaction with lysine, histidine, tyrosine, tryptophan, and phenylalanine residues. Bradford assays give fewer problems with interfering substances (Tris, EDTA, or sulfhydryl-reducing agents) than either the Lowry or the BCA methods. However, detergents such as SDS or Triton X-100 give strong interference with the Bradford assay. A further difficulty with the Bradford method is that the dye or dye–protein complexes tend to stick to cuvettes. An acetone and acid wash is necessary to remove the dye from the cuvette optical surfaces.[27]

4.5.5 Problems with the Colorimetric Protein Assays

A major problem with these assays, and one that is not universally recognized, is the extreme variability of the values from different individual proteins obtained versus the true concentration. Results are quite different when different methods are compared. Each of these assays depends on a standard curve for calibration with serum albumin and immunoglobulins most commonly used. Variations in aromatic amino acid content have been alluded to the foregoing as a partial cause of variation. However, not all the variation can be logically explained. An example of the

Table 4.3 Protein Concentration Estimates by Different Assay Methods

Protein	Biuret	Lowry	Bradford
Alcohol dehydrogenase	5.8	5.0	7.8
α-Amylase	6.8	6.0	8.3
Bovine serum albumin	9.7	8.4	21.1
Carbonic anhydrase	8.8	8.9	13.0
Catalase	7.6	6.3	9.7
α-Chymotrypsin	9.4	11.6	7.8
Cytochrome c	25.7	11.3	25.3
Ovalbumin	10.2	10.1	9.4
Gamma globulin (bovine)	10.0	10.0	10.0
Hemoglobin	16.2	8.3	19.9
Histones	9.7	9.2	15.8
Lysozyme	10.4	12.6	9.9
Myoglobin	13.7	7.9	20.7
Pepsin	9.8	12.4	4.1
Ribonuclease	11.8	15.9	5.3
Transferrin	8.5	9.0	12.6
Trypsin inhibitor (soybean)	9.1	10.3	6.1
Trypsin	11.4	15.5	4.9

Note: Proteins were prepared at 10 mg/mL and assayed by the three methods using bovine gamma globulin as a standard.

Source: Taken from Anonymous, *Bio-rad Protein Assay*, Bulletin 1069, Bio-rad Laboratories, 1984. With permission.

problems this variability may produce is that the color yield using serum albumin is 20–50% higher than that of immunoglobulin in the Lowry, BCA, and Bradford methods.[25] Thus the choice of standard has a marked effect on the protein concentration estimates. Proteins that have strongly divergent amino acid compositions (basic and acidic proteins, gelatin, etc.) are particularly inaccurate. A large number of proteins in side-to-side comparisons of the different assays are shown in Table 4.3.[28] Note that proteins that normally have color in the visible range (cytochrome c, hemoglobin, myoglobin) are excessively divergent. However, even a conventional protein such as trypsin gives a threefold greater concentration estimate with the Lowry versus the Bradford assay (Table 4.3). In many cases, the absolute protein concentration is not needed and the value from one of these colorimetric assays will suffice. However, when an accurate concentration is required, several approaches may be used. First, if you have a pure protein and the ultraviolet extinction coefficient is known, you may assay by both UV absorption and one of the colorimetric methods and obtain a correction factor for subsequent use. Second, you can use the Kjeldahl or Dumas method and a nitrogen conversion factor based on the known amino acid sequence. Third, you can determine the protein concentration by subjecting an aliquot of the protein solution to amino acid analysis. This involves digestion of the protein in strong acid and then determining the amounts of each amino acid using the ninhydrin procedure on an amino acid analyzer. The latter approach requires corrections for destruction during hydrolysis of part of the serine, threonine, and tyrosine; for the destruction of tryptophan, cysteine, and cystine; and for incomplete hydrolysis of leucine and isoleucine. Although these corrections may seem overwhelming, in practice the fact that the stoichiometry of the amino acids is known for proteins that have been sequenced means that the amounts of stable, fully hydrolyzed amino acids can be used as a subset to estimate total protein quite accurately.

4.5.6   *Ultraviolet Protein Absorbance*

The biuret, Lowry, BCA, and Bradford assays are all destructive in that part of the protein solution for which the concentration is being determined, and must be discarded. A nondestructive method for protein assay is the direct measurement of the absorbance in the ultraviolet range, typically at 275–280 nm. This approach is only possible if a number of conditions are met. First, an extinction coefficient must be known to convert the absorbance values to concentration. Second, the method is only valid with essentially pure proteins. Third, the protein solutions should have minimal visible turbidity. Fourth, the protein needs to be dissolved in a solution that contains minimal amounts of compounds that absorb at the same wavelength. Published extinction coefficients exist for most of the major proteins, so this condition is easily met. Whether one has a pure protein to measure concentration must be determined by other methods. Regarding turbidity, corrections can be made if the protein does not normally absorb light in the visible range by determining the absorbance at 350 nm, raising the ratio of wavelengths 350 over 280 to the fourth power, multiplying this factor times the 350 nm absorbance, and subtracting the resulting number from the 280 absorbance. Absorption by other solution components is not much of a problem with typical salts (NaCl, KCl) or buffers, but sulfhydryl reducing compounds such as mercaptoethanol and dithiotreitol (DTT) give significant interference. For example, a 1 mM DTT solution in its oxidized form may contribute nearly 0.3 optical density units to the absorbance readings. Contamination by nucleotides can also be particularly troublesome, because, although their absorbance maximum is approximately 260 nm, they still absorb light significantly in the 280 nm region.

4.6   **Electrophoretic Methods**

Polyacrylamide gel electrophoresis has been used extensively for the analysis and quantitation of muscle proteins. In most cases the proteins have been dissolved in SDS prior to electrophoresis. SDS dissociates proteins and any protein–protein complexes to the polypeptide subunit level. When a protein extract is placed at the top of an acrylamide gel and a current is applied, the proteins migrate through the gel with the speed of migration being related to the subunit size (small polypetides move faster). The most widely used procedure was described by Laemmli.[29] The procedure uses two gel layers—a separating gel and a stacking gel. The purpose of the latter is to concentrate the proteins into a tight zone before they are separated in the lower gel region. After the electrophoretic run, the gel is usually stained with Coomassie Blue (R-250 or G-250) or with silver (for more sensitive detection). The quantity of specific proteins can be determined by scanning the gel lanes. The area under the peaks is then compared with a standard curve prepared using known concentration of the same pure protein applied in parallel lanes on the same gel. Simply determining the total staining area for a lane and then calculating the percentage in a specified band will give inaccurate results. This is true for the same reason as was described earlier for the colorimetric assays of different individual proteins—each protein has a unique dye binding capacity. Caution must be exercised since there may be more than one protein in a single scanned gel band. One may require a two-dimensional gel system where the proteins are separated first by isoelectric focusing or in urea and the strip placed on an SDS gel for the second dimension.[30] The proteins are thus separated by both charge and size. A further caution is that the protein concentrations need to be in a relatively linear portion of the scanned absorbance versus concentration-loaded range. This may require that different loads be applied for the analysis of different proteins. For example, determining the amount of myosin heavy chain and the amount of one of the myosin light chains in the same gel lane may not be possible, since the ratio of these two proteins is greater than 10:1. An advantage of the SDS method for protein analysis over that using the ELISA techniques (described later) is that SDS can solubilize cooked meat samples and proteins that are insoluble even in high salt solutions. This ability has been utilized to detect soy protein in meat.[31]

The typical Laemmli gel system is unsatisfactory for dealing with titin (Mr 3,000,000–3,700,000 Da) and nebulin (Mr ~800,000 Da), two unusually large molecular weight proteins in muscle. An improved electrophoretic system has been described to resolve these proteins better.[32] The critical difference from the conventional Laemmli system was the use of DATD (N-N′ diallyltartardiamide) as the cross-linking agent in the stacking gel instead of bis-acrylamide. This results in larger pores but in a mechanically stable gel. Apparently bis-acrylamide cross-linked gels cause some retardation of the very large proteins such that proper stacking does not occur. Fritz et al.[32] also showed that inclusion of a sulfhydryl reducing agent in the upper reservoir buffer improved the migration of the high molecular weight proteins. Although the sample buffer contains high concentrations of sulfhydryl reducing agents, these agents move through the gel at the same speed as the tracking dye and the sulfhydryl groups of the proteins then begin to form intermolecular disulfide bonds in the mildly alkaline running buffer environment. The problem of disulfide bond formation during the electrophoretic run is more evident with large format gels and only with the largest sized proteins. A new system using agarose and SDS has since been developed to separate isoforms of titin.[33]

Another method for quantitation of proteins separated by SDS gels was described by Yates and Greaser.[34] They separated myofibrillar proteins on polyacrylamide gels and then determined the quantity of specific proteins by cutting out the respective protein band, hydrolyzing the protein in strong acid, and determining the protein content by amino acid analysis. The results indicated that the myosin heavy chain constituted about 43% and actin 22% of the myofibrillar protein.[34] A similar approach was used to demonstrate that the molar stoichiometry of the troponin subunits TnI:TnT:TnC was 1:1:1.[35] This evidence refuted a published report that suggested there were two moles of TnI per one each of the other two subunits.

A widely used method for measuring protein concentrations in protein mixtures is western blotting.[36,37] Figure 4.5 is a diagram showing the different steps in the blotting procedure. In this method, the proteins are first separated on an SDS polyacrylamide gel. The gel is then placed on top of a piece of nitrocellulose or polyvinylidene difluoride (PVDF) membrane and the proteins are moved onto the membranes by passing current through the sandwich transversely (the original method transferred the proteins passively, thus the term "blotting"). The remaining sites on the membrane are then blocked with an inert protein such as serum albumin or nonfat dry milk. Next the membrane is incubated with an antibody (usually either a polyclonal or a monoclonal IgG) that recognizes one of the proteins in the original mixture. Unbound antibody is washed away, and the blot is incubated with a second species-specific antibody that recognizes the first IgG. If the antibody was a polyclonal produced in a rabbit, the second antibody would be an antirabbit IgG. The second antibody has a covalently attached enzyme (usually horseradish peroxidase or alkaline phosphatase). After incubation and washing away unbound secondary antibody, the membrane is incubated with a substrate. Either a colored precipitate may be formed at the site of the secondary enzyme or the substrate may luminesce and the light signal captured on film. Commercial products are available for enhanced chemiluminescent (ECL) detection that is highly sensitive with detection levels in the nanogram range.

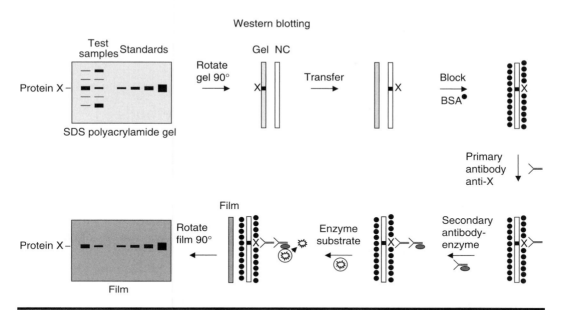

Figure 4.5   Diagram of steps used in western blotting. After an SDS polyacrylamide gel is run, the gel is placed alongside a sheet of nitrocellulose or PVDF and the protein electrophoretically moved transversely to the membrane. After the remaining membrane sites are blocked with bovine serum albumin or nonfat dry milk, the membrane is incubated with a primary antibody against protein X. Following washing, a secondary antibody (directed against the primary antibody and containing a covalently attached enzyme) is then bound to the primary. A final wash then precedes incubation with a substrate. The light emitted is finally recorded on film.

The determination of protein concentrations by western blotting requires a number of conditions. First, the antibody must be specific for the protein you are assaying. Second, the percent protein transferred should be approximately 100% and consistent between experiments. Third, you must apply a range of concentrations of a pure standard to the same gel as the unknowns to be able to determine concentrations. The large numbers of steps involved with blotting mean that protocols must be carefully followed.

Primary antibodies for many proteins may be obtained commercially, but their quality may vary widely. In many cases an antibody versus a human protein may be prepared but the application is with a different food animal species. Polyclonal antibodies can usually be used for related species, but monoclonal antibodies are more risky, since many may fail to recognize proteins from different animals. In some instances antibodies may recognize more than one isoform of a protein. For example, there are four major myosin heavy chain isoforms in skeletal muscle, but isoform-specific antibodies produced against rat myosins do not react with the same isoforms in a pig.[38]

Transfer percentage during blotting may vary considerably. Transfer of high molecular weights is more difficult. In fact if one stains a polyacrylamide gel after a typical transfer, the amount of staining for titin appears essentially the same as a companion gel that did not undergo transfer although all the proteins smaller than myosin heavy chain have been transferred from the blotted gel. One can also transfer too long and find that smaller proteins are moved through the transfer membrane. These problems, in addition to the variable primary antibody binding with time and temperature, make it imperative that standards be run on the same blot as the unknowns to give accurate protein estimations.

4.7 Enzyme-Linked Immunosorbent Assay Methods

The enzyme-linked immunosorbent assay (ELISA) procedure has been extensively used in the analysis of proteins. In this procedure, an antibody that reacts specifically with a protein in a tissue extract can be used to determine the concentration. Soluble proteins are required for such an assay. The protein solution is typically placed in a polystyrene microtiter plate where some bind tightly to the plastic surface (the steps that follow are quite similar to those used with western blotting; see Figure 4.5). After unbound protein is washed away, the microtiter wells are treated with an inert protein to block the rest of the sites on the polystyrene. The plates are next incubated with an antibody solution. The antibody forms a tight complex with the specific protein that is bound on the plastic surface. Excess antibody is washed away and the wells are then exposed to an enzyme-linked antibody that recognizes the bound antibody. Thus if the first antibody is a mouse monoclonal, the second antibody would be an antimouse immunoglobulin. The second antibody typically has a covalently linked enzyme such as horseradish peroxidase or alkaline phosphatase. After washing the plates to remove unbound secondary antibody, the wells are treated with a substrate appropriate for the enzyme. Depending on the microtiter plate reader available, either colorimetric or fluorescent substrates (the latter being more sensitive) may be used. The absorbance of the solution is obtained and compared with standards prepared from known concentrations of the pure protein being assayed. The procedure can be used to measure protein or antibody concentrations.

ELISA methods have been used in a variety of applications for muscle protein or meat analysis.[39] Examples include detection of foreign species in meat products;[40–43] identification of non-meat protein ingredients such as milk, egg, or soy protein;[44] measurement of calpastatin (a muscle calpain inhibitor) content in muscle tissue;[45] detection of central nervous tissue components in meat products;[46] and identification of various microbial contaminants in meat products.[47,48] A requirement in each of these applications is that the protein in question must be soluble and not be complexed to other proteins. Therefore, ELISAs have limited use for measuring myofibrillar proteins in mixtures because many of them typically associate into large multiprotein aggregates. A further problem occurs with cooked meat products because most proteins become insoluble with heating. A few proteins, however, survive heat treatment in the native state—serum albumin is an example.[39]

4.8 Special Problems with Collagen

In Section 4.3, the effects of collagen on protein estimations with Kjeldahl and combustion methods have been described. The unusual amino acid composition of the protein and its heat-derived product gelatin result in a higher nitrogen percentage than most of the rest of the muscle proteins. Both Gornall and Lowry noted that the biuret and Lowry methods, respectively, gave values in error in determining gelatin concentration. Gelatin should give 84 and 63% of the expected true protein concentration using the biuret and Lowry procedures, respectively.[49] Hydroxyproline analysis appears to be the best approach for collagen estimation.[50,51] This amino acid is only found in collagen and elastin. However, since the elastin content in muscle tissue is nearly 20-fold lower in concentration than collagen, the assumption that hydroxyproline predicts collagen is reasonable.

Collagen causes potential sampling problems when assaying total protein in meat samples since it is so unevenly distributed spatially in the tissue. Grinding meat may result in part of the large collagen pieces being retained behind the grinder plate. Homogenization in Waring blender or Polytron spinning blade devices also often leaves strands of stringy connective tissue that are not fully broken down and mixed with the slurry. One solution, to effectively mix and sample meat products, has been to pulverize tissue frozen in liquid nitrogen. However, this approach has not been routinely adopted because it is expensive and labor-intensive.

4.9   Summary

The accurate analysis of proteins from muscle or meat products provides many problems that are similar to those with assay in other biological materials. However, the chemical and physical properties of collagen plus its widely varying content in meat products provide a special challenge. In addition, the presence of a significant content of unusually large molecular weight proteins in muscle tissue and the very high degree of protein–protein interaction among the proteins from the myofibril require unique and specialized methodology for protein analysis.

References

1. Phillips, K.M. et al. Quality-control materials in the USDA National Food and Nutrient Analysis Program (NFNAP). *Anal. Bioanal. Chem.* 384, 1341, 2006.
2. Kjeldahl, J. Neue methode zur Besttimmung des Stickstoffs in organischen Korpern. *Fresenius Zeit. Anal. Chem.* 22, 366, 1883.
3. Anonymous AOAC Report of the Committee on Editing Tentative and Official Methods of Analysis. Meat and meat products. Vol. 1, Williams and Wilkins, Baltimore, 1916, p. 271.
4. King-Brink, M. and Sebranek, J.G. Combustion method for determination of crude protein in meat and meat products: collaborative study. *J. AOAC. Int.* 76, 787, 1993.
5. Von Furth, O. Uber die Erweisskorper des Muskelplasmas. *Arch. Exp. Path. Pharmak.* 36, 231, 1895.
6. Atwater, W.O. and Bryant, A.P. The chemical composition of American food materials. *US Off. Expt. Stas. Bull.* 28, 1, 1906.
7. Grossfeld, J. Die Ermittelung des Wasserzusatzes in Hacfleisch und Fleischwursten. *Zeitschr. Fur Untersuchung d. Nahr. U Genubmittel.* 42, 173, 1921.
8. Bate-Smith, E.C. A scheme for the approximate determination of the proteins of muscle. *J. Soc. Chem. Ind. (London)* 53, 351, 1934.
9. Richards, A.N. and Gies, W.J. Chemical studies of elastin, mucoid, and other proteids in elastic tissue, with some notes on ligament extractives. *Am. J. Physiol.* 7, 93, 1902.
10. Mitchell, H.H., Hamilton, T.S., and Haines, W.T. Some factors affecting the connective tissue content of beef muscle. *J. Nutrition* 1, 165, 1928.
11. Benedict, R.C. Determination of nitrogen and protein content of meat and meat products. *J. Assoc. Off. Anal. Chem.* 70, 69, 1987.
12. Bate-Smith, E.C. The proteins of meat. *J. Soc. Chem. Ind. (London)* 54, 152, 1935.
13. Robinson, D.S. Changes in the protein composition of chick muscle during development. *Biochem. J.* 52, 621, 1952.

14. Scopes, R.K. Studies with a reconstituted muscle glycolytic system. The rate and extent of creatine phosphorylation by anaerobic glycolysis. *Biochem. J.* 134, 197, 1973.
15. Perry, S.V. The bound nucleotide of the isolated myofibril. *Biochem. J.* 51, 495, 1952.
16. Helander, E. On quantitative muscle protein determination. *Acta Physiol. Scand.* 41 (141), 1, 1957.
17. Dickerson, J.W.T. The effect of growth on the composition of avian muscle. *Biochem. J.* 75, 33, 1960.
18. Scopes, R.K. Characterization and study of sarcoplasmic proteins. In *The Physiology and Biochemistry of Muscle as a Food, 2*, Briskey, E.J., Cassens, R.G., and Marsh, B.B. (Eds.), University of Wisconsin Press, Madison, p. 471, 1970.
19. Bate-Smith, E.C. Native and denatured muscle proteins. *Proc. Royal Soc. London. Ser. B,* 124, 136, 1937.
20. Greaser, M.L. et al. Postmortem changes in subcellular fractions from normal and pale, soft, and exudative porcine muscle. 2. Electron microscopy. *J. Food Sci.* 34, 125, 1969.
21. Gornall, A.G., Bardawill, C.J., and David, M.M. Determination of serum proteins by means of the biuret reaction. *J. Biol. Chem.* 177, 751, 1949.
22. Sapan, C.V., Lundblad, R.L., and Price, N.C. Colorimetric protein assay techniques. *Biotechnol. Appl. Biochem.* 29 (Pt 2), 99, 1999.
23. Robson, R.M., Goll, D.E., and Temple, M.J. Determination of proteins in "Tris" buffer by the biuret reaction. *Anal. Biochem.* 24, 339, 1968.
24. Lowry, O.H. et al. Protein measurement with the folin phenol reagent. *J. Biol. Chem.* 193, 265, 1951.
25. Stoscheck, C.M. Quantitation of protein. *Methods Enzymol.* 182, 50, 1990.
26. Smith, P.K. et al. Measurement of protein using bicinchoninic acid. *Anal. Biochem.* 150, 76, 1985.
27. Bradford, M. A rapid and sensitive method for the quantitation of microgram quantities of protein utilizing the principle of protein dye-binding. *Anal. Biochem.* 72, 248, 1976.
28. Anonymous. *Bio-rad Protein Assay*, Bulletin 1069, Bio-rad Laboratories, Richmond, California 28, 1984.
29. Laemmli, U.K. Cleavage of structural proteins during the assembly of the head of bacteriophage T4. *Nature* 227, 680, 1970.
30. O'Farrell, P.H. High resolution two-dimensional electrophoresis of proteins. *J. Biol. Chem.* 250, 4007, 1975.
31. Lee, Y.B. et al. Quantitative determination of soybean protein in fresh and cooked meat–soy blends. *J. Food Sci.* 40, 380, 1975.
32. Fritz, J.D., Swartz, D.R., and Greaser, M.L. Factors affecting polyacrylamide gel electrophoresis and electroblotting of high-molecular-weight myofibrillar proteins. *Anal. Biochem.* 180, 205, 1989.
33. Warren, C.M., Krzesinski, P.R., and Greaser, M.L. Vertical agarose gel electrophoresis and electroblotting of high molecular weight proteins. *Electrophoresis* 24, 1695, 2003.
34. Yates, L.D. and Greaser, M.L. Quantitative determination of myosin and actin in rabbit skeletal muscle. *J. Mol. Biol.* 168, 123, 1983.
35. Yates, L.D. and Greaser, M.L. Troponin subunit stoichiometry and content in rabbit skeletal muscle and myofibrils. *J. Biol. Chem.* 258, 5770, 1983.
36. Burnette, W.N. Western blotting: electrophoretic transfer of proteins from sodium dodecyl sulfate—polyacrylamide gels to unmodified nitrocellulose and radiographic detection with antibody and radioiodinated protein A. *Anal. Biochem.* 112, 195, 1981.
37. Towbin, H., Staehelin, T., and Gordon, J. Electrophoretic transfer of proteins from polyacrylamide gels to nitrocellulose sheets: procedure and some applications. *Proc. Natl. Acad. Sci. USA* 76, 4350, 1979.
38. Greaser, M.L., Okochi, H., and Sosnicki, A.A. Role of fiber types in meat quality. *Proc. 47th Int. Cong. Meat Sci. Technol., Krakow* 34, 2001.
39. Fukal, L. Modern immunoassays in meat-product analysis. *Nahrung.* 35, 431, 1991.
40. Dincer, B. et al. The effect of curing and cooking on the detection of species origin of meat products by competitive and indirect ELISA techniques. *Meat Sci.* 20, 253, 1987.

41. Martin, D.R., Chan, J., and Chiu, J.Y. Quantitative evaluation of pork adulteration in raw ground beef by radial immunodiffusion and enzyme-linked immunosorbent assay. *J. Food Prot.* 61, 1686, 1998.

42. Patterson, R.L. and Jones, S.J. Review of current techniques for the verification of the species origin of meat. *Analyst* 115, 501, 1990.

43. Chen, F.C. and Hsieh, Y.H. Detection of pork in heat-processed meat products by monoclonal antibody-based ELISA. *J. AOAC. Int.* 83, 79, 2000.

44. Leduc, V. et al. Immunochemical detection of egg-white antigens and allergens in meat products. *Allergy* 54, 464, 1999.

45. Doumit, M.E. et al. Development of an enzyme-linked immunosorbent assay (ELISA) for quantification of skeletal muscle calpastatin. *J. Anim. Sci.* 74, 2679, 1996.

46. Hossner, K.L. et al. Comparison of immunochemical (enzyme-linked immunosorbent assay) and immunohistochemical methods for the detection of central nervous system tissue in meat products. *J. Food Prot.* 69, 644, 2006.

47. Mattingly, J.A. et al. Rapid monoclonal antibody-based enzyme-linked immunosorbent assay for detection of Listeria in food products. *J. Assoc. Off. Anal. Chem.* 71, 679, 1988.

48. Wilhelm, E. et al. Salmonella diagnosis in pig production: methodological problems in monitoring the prevalence in pigs and pork. *J. Food Prot.* 70, 1246, 2007.

49. Zhou, P. and Regenstein, J.M. Determination of total protein content in gelatin solutions with the Lowry and biuret assay. *J. Food Sci.* 71, C474, 2006.

50. Woessner, J.F., Jr. The determination of hydroxyproline in tissue and protein samples containing small proportions of this imino acid. *Arch. Biochem. Biophys.* 93, 440, 1961.

51. Bergman, I. and Loxley, R. New spectrophotometric method for the determination of proline in tissue hydrolyzates. *Anal. Chem.* 42, 702, 1970.

Chapter 5

Proteomics

Kristin Hollung, Eva Veiseth, Xiaohong Jia,
and Ellen Mosleth Færgestad

Contents

5.1 Introduction

The quality traits of muscle foods are influenced by genetics, environmental factors, and processing conditions. However, the underlying molecular mechanisms are far from understood. Although the genes remain constant during the lifetime of the animal, the expression of the genes to mRNA and proteins is very dynamic and is regulated by a large number of factors such as environmental and processing conditions. The proteome is the protein complement of the genome and consists of the total amount of proteins expressed at a certain point of time, and may thus be viewed as the mirror image of the gene activity.[1] The genome contains the information on which genes are available, whereas the proteome contains information on which genes are actually being expressed. In contrast to the genome, the proteome is continuously changing based on the factors influencing either protein synthesis or degradation. Thus, analyzing the proteome can be viewed as analyzing snapshots into a system undergoing constant change. In this regard, the proteome can be seen as the molecular link between the genome and the functional quality of the muscle or meat. Thus understanding the variations and different components of the proteome with regard to certain quality or processing parameters will provide the knowledge that can be used in optimizing the conversion of muscles into meat.[2–4]

5.2 Analytical Tools

Proteomics are the tools used to analyze the proteomes. Over the past decade there have been significant improvements of methods in this field, largely driven by medical science, as a demand for understanding the functions of the genomes following the large-scale genome sequencing projects.

Basically, there are two analytical strategies used in proteomics. One is based on two-dimensional gel electrophoresis (2-DE) for separation of proteins, followed by identification of separated proteins by mass spectrometry (MS). The other is based on liquid chromatography (LC) in one or more dimensions, coupled with MS. Both strategies work well for qualitative comparisons of samples, but quantitative comparisons are more challenging.

The following sections give an overview of the two strategies. Despite several attempts there is no single method spanning the whole proteome in one experiment.[5–9]

5.2.1 Muscle Protein Fractionation

Several authors have tried to estimate the total number of different proteins expressed from a genome including splice variants and posttranslational modifications in a eukaryotic cell, with numbers ranging from 100,000 to 500,000.[10,11] These proteins are localized in different compartments: some are part of large protein aggregates, like the myofibrillar proteins, some are localized in membranes, whereas others are enzymes localized in the cytoplasm. With the help of prefractionation of the proteins based on solubilization and extraction procedures, the enrichment of soluble sarcoplasmic or myofibrillar proteins can be achieved. These proteins will have very different chemical properties that can be used to isolate the proteins in different fractions. Different strategies for fractionation of proteins are reviewed in Righetti et al.[10] Several issues should be considered before developing a sample preparation strategy. It is advisable to keep sample preparation as simple as possible to avoid protein losses. However, if only a subset of the proteins in the tissue or cells is of interest, prefractionation can be employed during sample preparation. Depending on the project and hypothesis, different strategies for extraction and prefractionation of muscle proteins should be considered. Approximately 1000 different proteins, or 10% of the total number of proteins, can be analyzed on one 2-DE image.

The presence of high-abundance proteins in a tissue or cell often masks low-abundance proteins and thus generally prevents their detection and identification in proteome studies. The use of prefractionation methods can assist in the identification and detection of low-abundance proteins that may ultimately prove to be informative biomarkers. For comprehensive proteome analysis by 2-DE, prefractionation is essential for the following reasons. First, by partitioning the proteome into compartments, the complexity of each compartment is dramatically reduced, facilitating spot identification and quantitative analysis. Second, there is a pronounced bias inherent in 2-DE toward abundant proteins. It has the effect of masking low-abundance proteins. Prefractionation enriches low-abundance proteins. Third, the amount of any specific protein that can be resolved on a 2-DE is limited. Prefractionation allows the proteins present in a particular fraction to be loaded at high levels, further increasing the representation of low-abundance proteins. Finally, relative to whole cell preparations, the number of proteins that are solubilized during the differential extraction procedures is greatly increased, yielding a more comprehensive representation of the proteome.

Several established protein and peptide fractionation techniques include stepwise extractions of proteins, immunodepletion, reverse-phase or ion exchange chromatography, and gel filtration.[10] The choice of technique depends on the subset of proteins that is of interest. In muscle cells, structure proteins such as actin and tubulin are high-abundance proteins and hence dominate

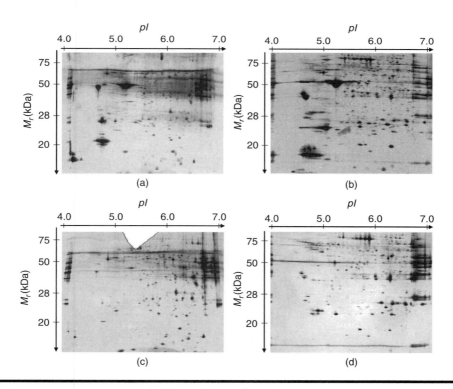

Figure 5.1 **2-DE images of proteins from salmon muscle and bovine longissimus muscle separated by pH 4–7 and 12.5% SDS-PAGE. Fifty micrograms of protein was loaded on each gel. (a) Total protein extract of salmon muscle in urea-buffer (7 M urea, 2 M thiourea, 2% CHAPS, 1% DTT, 0.5% IPG buffer, pH 3–10); (b) total protein extract of bovine longissimus muscle in urea-buffer; (c) salmon muscle proteins soluble in TES-buffer (10 mM Tris, pH 7.6, 1 mM EDTA, 0.25 M sucrose); (d) bovine longissimus muscle proteins soluble in TES-buffer.**

in 2-DE. Figure 5.1 shows representative 2-DE gels of a total protein extract and water-soluble proteins extracted from a salmon or bovine muscle sample, respectively.

5.2.2 Separation by 2-DE

Separation of proteins by 2-DE has been performed for decades.[12] However, major technical improvements such as the introduction of immobilized pH gradients have been important for the reproducibility of the method; for a review see Refs 13 and 14. The technique provides lot of information on both the proteins present as well as the possible modifications. A schematic diagram of the workflow of a proteome 2-DE analysis is shown in Figure 5.2. In short, the proteins are first separated by charge using isoelectric focusing (IEF), and then the focused proteins are separated by mass using sodium dodecyl sulfate polyacrylamide gel electrophoresis (SDS-PAGE).

IEF is usually performed on immobilized pH gradients. These may vary in length, usually from 7 to 24 cm, and in pH range, including wide gradients (e.g., 3–10), medium gradients (e.g., 4–7), or narrow gradients (e.g., 4.5–5.5).[13] The complexity of the sample determines which gradient will provide maximum information. Proteomes of low complexity can usually be separated on a wide gradient, whereas a narrow gradient is needed to separate more complex samples. A narrow gradient will reduce the number of overlapping protein spots. Following IEF, the proteins are separated by SDS-PAGE in the second dimension.

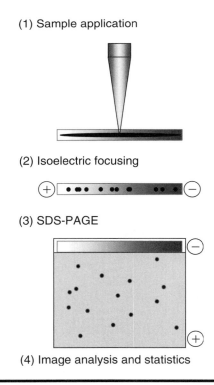

Figure 5.2   Schematic drawing of the workflow in two-dimensional electrophoresis. (1) Sample is applied onto the gel strip. (2) Proteins are separated by charge, isoelectric focussing. (3) Proteins are separated by SDS-PAGE based on MW. (4) Gel images are aligned and statistically analyzed for variations between samples.

5.2.3 Protein Detection and Quantification

After separation by 2-DE, the proteins must be visualized. This can be challenging because of the great differences in protein concentration in each protein spot on the gel. Thus the visualization method needs to be highly sensitive to detect the low-abundance proteins, it must have a high and linear dynamic range, and finally it should be compatible with protein identification methods such as MS. Unfortunately, there are no methods meeting all of these demands.[15–17]

Basically two different strategies can be used to visualize the proteins on a 2-DE gel. Either the proteins are labeled with radioactivity or fluorescent tags before separation or the proteins are stained in the gel after electrophoresis. Fluorescent dyes have the advantage of a high linear dynamic range, but they are relatively expensive compared to traditional stains such as silver or coomassie brilliant blue (CBB).

Incorporation of radiolabeled amino acids is a very sensitive method, which is applicable to studies with bacteria or cell cultures.[16–18] However, this is not useful for tissues and biopsies. Another approach can be to label the proteins with fluorescent dyes after extraction and before electrophoresis using the difference gel electrophoresis (DIGE) method.[19] A small fraction of the total proteins in the samples are labeled with the fluorescent dyes; these proteins will have a molecular weight (MW) slightly higher than the unlabelled counterparts. The use of DIGE facilitates the separation of up to three samples on the same gel using different fluorescent dyes for each sample. Co-migration of the same proteins from the different samples occurs, thus simplifying the analysis between the samples. However, there are a few challenges with the DIGE technique when proteins are to be picked for MS-identification due to the fact that only a small portion of the proteins in the sample are labeled and since the unlabeled proteins have a slightly lower MW, only limited parts of the particular protein will be excised.

To date, the most common staining methods for visualization of proteins on 2-DE gels are CBB and silver staining methods. Although CBB is not as sensitive as silver, it is compatible with downstream MS methods and is cheap and easy to perform. Silver staining is more sensitive than CBB stains, but several silver staining protocols are not compatible with MS identification of the protein spots. Fluorescent staining with SYPRO Ruby is also widely used because of the increased dynamic range. Several efforts have been made both to improve the sensitivity and to lower the costs.[20,21]

5.2.4 Handling of 2-DE Images and Statistical Analysis of Proteome Data from 2-DE

Although very informative, the gel images are complex consisting of a very high number of more or less overlapping protein spots, where the position of the protein spots may vary from one gel image to another. Furthermore, the staining intensity and background may be variable throughout the gels and from one gel to another. Thus, the process of analyzing the gel images to search for the information revealed by the proteins is a critical and complex step of the process.

Several commercially available softwares are designed to align and analyze the images from 2-DE experiments.[22] However, due to the noisy appearance of the images this is not an easy task, and improvements are still needed to get an optimal analysis. In principle, there are two different approaches for matching data from one gel to those from another. One approach is to detect spots in each gel and match them with one another. This approach is time-consuming and faces a number of challenges resulting in a significant number of missing values in the data analysis, which do not reflect the true absence or presence of a protein in the sample.[23] An alternative approach is based on alignment of the images rather than on matching of the protein spots. When the images are aligned, spots can be detected across all gels simultaneously using common boundaries around the spots for all gel images.[24] Thereafter protein spot tables can be generated without missing values. However, overlapping protein spots and saturated protein spots are still major challenges for the data analysis. A useful approach to overcome these challenges is to analyze the aligned gel images pixel by pixel.[25] Variation from one gel to another even in strongly overlapping protein spots and saturated protein spots can then be detected.

Proteomics data need special attention in analysis and statistical validation of the outcome, which is in contrast to traditional experiments with few measurements/variables in many samples/animals. Here the number of samples is usually limiting because of workload, but the amount of data that is collected from each animal exceeds several hundred observations. This is also a challenge in other -omics data, such as transcriptomics, and must be handled with care during the statistical analysis.

Several statistical approaches have been used to analyze proteomics data. Multivariate analyses such as principal component analysis (PCA)[26,27] are now included in several softwares for analysis of 2-DE experiments. Multivariate approaches have also been used for selection of significant changes in the 2-DE data based on the design parameters.[28–30] Assessment of hierarchical clustering methodologies, commonly used in transcriptomics studies, has also been performed on proteomics data.[31] The different statistical methods will shed light on different aspects of the proteomics data as has been discussed in several papers.[32–33]

5.2.5 Mass Spectrometry

An alternative technique to separate and identify peptides is high-performance liquid chromatography (HPLC) coupled with MS. A schematic workflow of a typical LC-MS experiment is shown in Figure 5.3. HPLC includes ion-exchange (IEX), size-exclusion (SEC), hydrophobic/hydrophilic-interaction (HIC/HILIC), and reversed-phase chromatography (RPC). Several comprehensive systems combining two, three, or more chromatography steps coupled with MS for protein identification have been developed. For a review, see Ref. 34. Usually the total protein lysate is digested to generate a mixture of peptide fragments and loaded first onto a strong cation-exchange chromatography column (SCX).[35] Separated fractions of the peptides may then be displaced onto a reverse-phase (RP) chromatography column using a salt step gradient. The peptides are eluted from the RP column into the mass spectrometer. The process is repeated; each fraction generates a peptide mass list in the MS and specific peptides may be selected for MS/MS fragmentation. Based on this principle of HPLC and MS/MS, multidimensional protein identification technology (MudPIT) was developed for rapid large-scale proteome analysis.[36] The use of MudPIT helped in the identification of over 1450 proteins of *Saccharomyces cerevisiae*, including low-abundance proteins and membrane proteins. MS is an analytical tool used for measuring the molecular mass of a sample. Mass spectrometers can be divided into three fundamental parts, namely, the ionization source, the mass analyzer, and the detector. The ionization methods used for the majority of biochemical analyses are electrospray ionization (ESI)[37] and matrix-assisted laser desorption ionization (MALDI).[38] Once the sample molecules are vaporized and ionized they are transferred electrostatically into the analyzer where they are separated from the matrix ions, and individually detected, based on their mass-to-charge (*m/z*) ratios.

Several separation methods have been developed for enrichment of specific proteins and peptides in LC-MS/MS experiments. An example is the use of immobilized metal-affinity chromatography

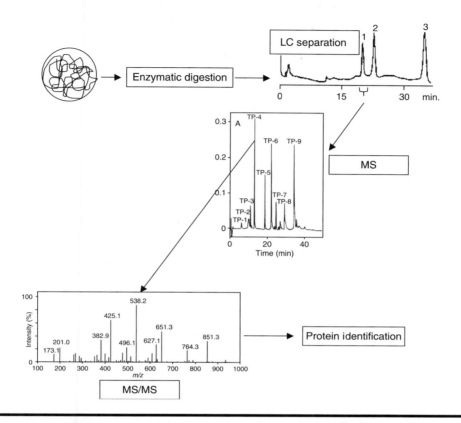

Figure 5.3 Schematic workflow of an LC-MS experiment. First, the proteins in the sample are digested to generate peptides, which are separated by liquid chromatography (LC). Peaks are eluted and applied directly into the mass spectrometer. Peaks from the MS spectra are selected for MS/MS fragmentation, which in turn can be used in database searches for protein identification or sequencing of peptides.

(IMAC) using Fe^{3+} or Ca^{3+} to enrich for phosphorylated peptides.[39,40] More developed methods based on IMAC technology have been applied to study the phosphoproteome of different organisms.[41–44] For enrichment of glycosylated peptides lectin affinity chromatography may be used.[45,46] Another method aimed to reduce the complexity of proteome is called surface-enhanced laser desorption ionization (SELDI) technology.[47] This technique utilizes specially engineered chromatography chips to capture proteins based on their intrinsic properties.

For identification of isolated proteins, for example, from 2-DE experiments, the most common strategy is peptide mass fingerprinting (PMF) where the unknown protein is cleaved into peptides by a residue-specific enzyme such as trypsin. A schematic workflow of a protein identification experiment is shown in Figure 5.4. The absolute masses of the peptides are accurately measured with a mass spectrometer such as MALDI-TOF or ESI-TOF, and matched against theoretical peptide libraries generated from protein sequence databases to create a list of likely protein identifications. The quality of PMF data depends mainly on the accuracy of the MS data, the number of the matching peptides and their MW, the control over the digest chemistry, and the availability of comprehensive sequence databases. Hence PMF-based protein identification is greatly facilitated in organisms for which the genomes have been fully sequenced.[48]

(1) Spot picking

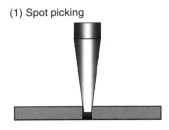

(2) In-gel digestion

(3) Desalting and concentration

(4) MS–peptide mass fingerprint

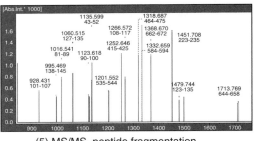

(5) MS/MS–peptide fragmentation

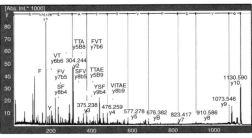

(6) Database search

Figure 5.4   Schematic drawing of the workflow in identification of proteins by MS and MS/MS. (1) Protein spot of interest is picked from the 2-DE gel. (2) Proteins are trypsinated in-gel. (3) Peptides are desalted and applied to the mass spectrometer. (4) A peptide mass fingerprint is obtained. (5) Selected peptide peaks are fragmented, and the resulting MS/MS fragments are analyzed. (6) Both the MS and MS/MS fragments are used in database searches to identify the protein in the gel.

A major disadvantage of MALDI-TOF is that protein sequence information cannot be obtained. However, tandem MS (MS/MS) with multiple steps of mass analysis can give true protein sequence data. The MS/MS contains two analyzers separated by a collision cell that has the function to fragment and activate the precursor ion from the first analyzer. The fragmentation spectrum can be recorded in the second analyzer and true sequence data can be achieved. Often a combination of MS and MS/MS is used to confirm the identity of a protein.

5.2.6   Quantification

Several methods have been developed for protein quantification by MS based on chemical isotope labeling, such as ICAT (isotope-coded affinity tagging),[49] SILAC (stable isotope labeling with amino acids in cell culture),[50] and iTRAQ (isotope tags for relative and absolute quantitation). The isotopic tags used in ICAT bind to cysteine residues within the protein. These tags are chemically identical, but exist in light and heavy isotopic forms. When bound to the same peptides in different samples, a concrete mass change will be evident when analyzed by MS.[49] SILAC is a simple and accurate approach for mass spectrometric–based quantitative proteomics. The method relies on the incorporation of amino acids with substituted stable isotopic nuclei (^{13}C, ^{15}N, ^{2}H) and on metabolic incorporation of a given "light" or "heavy" form of the amino acid into the proteins. Pairs of chemically identical peptides of different stable-isotope composition can be differentiated in a mass spectrometer based on their mass difference. The ratio of peak intensities in the mass spectrum for such peptide pairs accurately reflects the abundance ratio for the two proteins. An improved approach to ICAT and SILAC named iTRAQ has been developed over the past years. By this method, up to four mass tags that label the N terminus of every peptide are incorporated. Each tag has the same mass, and the same peptides from each sample appear at the same mass in the MS spectrum. The relative abundances of the ions generated by fragmentation of each tag provide relative quantification.[51]

5.2.7   Bioinformatics and Data Handling of LC-MS Data

The huge number of peaks generated in an LC-MS run needs special attention in the interpretation. There is an immense need for filtering out noise and extract the relevant information. Several softwares are available to analyze these spectra.[22] Since the *m/z* values generated in the MS depend on mass accuracy and resolution of the mass spectrometer, this is usually a reliable parameter. However, the retention times from the LC are dependent on the analytical method and may vary significantly from one experiment to the other. This can be a challenge for the comparison and alignment of spectra of different samples. Owing to the lack of common standards across laboratories it is often difficult to get access to essential information to verify and utilize the results. However, there are initiatives aiming at making common standards for reporting of MS and LC-MS data.[52]

The next step after identification of proteins and differences is the coupling of the proteomics data to genomics or transcriptomics data. This involves coupling of complex data matrices, which is not a straightforward task.

5.2.8   Protein Arrays

Another approach for analysis of proteomes is through the use of protein arrays. Antibody arrays, where samples are incubated onto arrays of immobilized antibodies, can be used for protein expression profiling. Haab et al.[53] performed a proof-of-concept study where 115 different antibodies were spotted onto a slide array, which was subsequently incubated with a mixture of two protein samples labeled with different fluorescent dyes. The relative concentration of each protein in the two samples was determined by comparing the fluorescence intensities of the two dyes at each spot. A different approach has been used by Schulz et al.,[54] where sarcoplasmic reticulum samples from 24-porcine-longissimus muscles were printed onto glass slides and probed with monoclonal antibodies for seven target proteins (one antibody per slide). Significant changes in several of these proteins were found in relation to different halothane-genotyped animals. In addition to antibody arrays, functional protein arrays can be used for investigation of, for example, protein–protein, protein–lipid, and enzyme–substrate interactions.[55,56]

5.2.9   Limitations of Proteomics

Similar to all advanced methods it is necessary to ensure that the experimental design is made such that it is possible to analyze the results. The choice of extraction method for the proteins determination of the which proteins that can be studied; proteins that are not extracted will thus not be considered. Very hydrophobic proteins, membrane proteins, and high-MW proteins are often difficult to solubilize and analyze by 2-DE.[13–57] It is also important to keep in mind that whatever method is chosen for proteome analysis, there is no protocol yet providing an analysis of the complete proteome in one run. Several groups have tried to compare different strategies on the same samples ending up with partly overlapping results.[5–9] Usually, a few hundred to several thousand proteins can be analyzed in one experimental setup, but this is only a small part of the entire proteome. Thus it is important to draw conclusions based on the proteins that are actually under investigation and not extrapolate the results to the proteins that failed to be analyzed. As discussed in Sections 5.2.4 and 5.2.7, analysis of the data is not straightforward, and careful consideration should be made while choosing the most optimal strategy.

5.3 Proteomics of Muscle Foods

In the field of muscle food science, proteomics is a fairly new tool. During the past years several studies have been published where proteomics sheds light on different aspects of muscle foods. The applications include studies of genetic variation,[58–60] variation between muscles,[59,61] response to feeding,[62,63] preslaugter conditions,[64,65] processing conditions,[66–68] aging,[29,69–75] storage conditions,[30,76,77] and quality traits.[60,72,78] Several species have been studied, such as cattle,[29,58,73,79–81] pork,[60,62,64,66,69,72,78] lamb,[59,61] chicken,[63] sea bass,[74] cod,[30,75] trout,[65,76,77,82] and salmon.[67,83] In most of these studies, 2-DE has been used in combination with MS identification of specific proteins as the proteomics tool. However, in the studies of peptides gel-free techniques are used.[79,82]

5.3.1 Postmortem Changes

Studying postmortem changes in the proteome will lead to an increased understanding of the biochemical mechanisms accounting for meat quality traits such as tenderness. Lately, proteomics has been used to study changes occurring in muscles during postmortem storage in pork,[69,70] cattle,[29,73] sea bass,[74] rainbow trout,[82] and cod.[75]

Several metabolic enzymes involved in energy metabolism were changed in abundance in pork longissimers dorsi (LD) postmortem.[69,70] This is further confirmed to also be the case in cattle LD muscles,[29,73] and supports the observed shift in energy metabolism toward the glycolytic pathway. Furthermore, an increase in abundance of enzymes involved in aerobic energy metabolism during the first hours after slaughter supports that aerobic metabolism may proceed for several hours postmortem.

In a proteome study performed in pork LD muscle, several of the myofibrillar substrates for μ-calpain were identified.[71] Among the substrates that were degraded by *in vitro* incubation with μ-calpain were desmin, actin, myosin heavy chain, myosin light chain I, troponin T, tropomyosin α1, tropomyosin α4, thioredoxin, and CapZ. Earlier studies using one-dimensional SDS-PAGE have concluded that actin is not degraded postmortem.[84–86] However, using 2-DE, the resolution allows for a better separation of the different actin fragments and demonstrates the potential of 2-DE. Some of the actin fragments observed in postmortem pork muscle were also significantly correlated with tenderness.

5.3.2 Proteome Changes Associated with Meat Quality Traits

Six proteins underwent changes related to tenderness (Warner Bratzler shear force) in a proteome study of pork LD muscle.[72] A great increase in fragments of actin during postmortem storage was observed. Three actin fragments as well as a myosin heavy chain fragment were correlated to shear force. Moreover, myosin light chain II and the glycolytic enzyme triose phosphate isomerase were correlated to tenderness. These enzymes and proteins have been confirmed in another study of pork LD muscle[78] where additional proteins related to tenderness, drip loss, and Hunter L^* value were also identified.

Recently, 2-DE and MS have also been used to investigate the biochemical mechanisms behind variation in meat color. Comparison of the sarcoplasmic proteome in pig SM semimembranosus muscle from two groups of animals having a light or dark meat color revealed that 22 proteins were changed in abundance.[60] The animals, 12 animals in each group, were selected from samples of 1000 pigs based on extreme L^*-values. Although the dark muscles had an increased abundance of mitochondrial proteins, indicating a more oxidative metabolism, the light muscles had an increased abundance of cytosolic proteins involved in glycolysis.

Proteome analyses have also been used to study protein changes in dry-cured hams. In one study, the myofibrillar proteins of raw meat and dry-cured hams after 6, 10, and 14 months of ripening were analyzed by 2-DE.[66] Actin, tropomyosin, and myosin light chains disappeared during the ripening period, and were almost completely hydrolyzed after 12 months. In a pilot study of Norwegian dry-cured hams, we have observed earlier a variation in the protein degradation pattern between hams ripened for 6 months from different producers.[68]

5.3.3 Analysis of Peptides Related to Muscle Quality

The low-MW peptides in bovine pectoralis profundus muscle[79] and rainbow trout muscle,[82] which appeared during postmortem storage and cooking, were analyzed directly by MS. A combination of MALDI-TOF MS and nano-LC-ESI MS/MS analyses was used to determine peptide composition and identification. The peak patterns from the MALDI-TOF MS analysis were very reproducible and several peptides were identified as specific for the treatment. In the bovine study, three of the identified proteins were known targets of postmortem proteolysis (troponin T, nebulin, and cypher protein), whereas the other two proteins were the connective tissue proteins (procollagen types I and IV). These are known to be very stable during meat aging.

Nano-LC MS/MS has also been used to identify diagnostic peptides for adulteration of meat products with soybean proteins.[88] Several subunits of glycinin A, one of the major seed proteins in soybean, were easily identified. This demonstrates that the proteomics approach is extremely very useful to identify reliable markers for the content and amount of added soybean proteins in meat products, and hence of major importance to both the industry and the food authorities.

5.4 Conclusions

Proteomics has turned out to be a promising and powerful tool in muscle food science over the past years. As a result an increasing number of studies are emerging in the literature using proteomics as the key approach to unleash the molecular mechanisms behind different genetic backgrounds or processing techniques of muscle foods. Although the genetic background is important, the quality of muscle foods is largely influenced by processing and environmental conditions. Thus proteomics seems to be an effective tool to reflect the important mechanisms that might contribute to influence on the development of a satisfactory quality for the industry and the consumer. In contrast to traditional methods where only one or a few proteins are studied at a time, proteomics can help in the study of several hundred proteins simultaneously. It has the potential to significantly enhance the understanding of molecular mechanisms underlying meat quality.

Acknowledgement

This work has been supported by the Fund for the Research Levy on Agricultural Products in Norway.

References

1. Wilkins, M.R. et al., From proteins to proteomes: large scale protein identification by two-dimensional electrophoresis and amino acid analysis, *Biotechnology (NY)*, 14, 61, 1996.
2. Bendixen, E. et al., Proteomics, an approach towards understanding the biology of meat quality, in *Indicators of Milk and Beef Quality*, vol. 112, J.F. Hocquette and S. Gigli (Eds.), Wageningen Academic Publishers, Wageningen, 2005, p. 81.
3. Bendixen, E., The use of proteomics in meat science, *Meat Sci,* 71, 138, 2005.
4. Hollung, K. et al., Application of proteomics to understand the molecular mechanisms behind meat quality, *Meat Sci,* 77, 97, 2007.
5. Frohlich, T. et al., Analysis of the HUPO Brain Proteome reference samples using 2-D DIGE and 2-D LC-MS/MS, *Proteomics,* 6, 4950, 2006.
6. Bodenmiller, B. et al., Reproducible isolation of distinct, overlapping segments of the phosphoproteome, *Nat Methods,* 4, 231, 2007.
7. McDonald, T. et al., Expanding the subproteome of the inner mitochondria using protein separation technologies: one- and two-dimensional liquid chromatography and two-dimensional gel electrophoresis, *Mol Cell Proteomics,* 5, 2392, 2006.
8. Bodnar, W.M. et al., Exploiting the complementary nature of LC/MALDI/MS/MS and LC/ESI/MS/MS for increased proteome coverage, *J Am Soc Mass Spectrom,* 14, 971, 2003.
9. Wu, W.W. et al., Comparative study of three proteomic quantitative methods, DIGE, cICAT, and iTRAQ, using 2D gel- or LC-MALDI TOF/TOF, *J Proteome Res,* 5, 651, 2006.
10. Righetti, P.G. et al., Prefractionation techniques in proteome analysis: the mining tools of the third millennium, *Electrophoresis,* 26, 297, 2005.
11. Stasyk, T., and Huber, L.A., Zooming in: fractionation strategies in proteomics, *Proteomics,* 4, 3704, 2004.
12. O'Farrell, P.H., High resolution two-dimensional electrophoresis of proteins, *J Biol Chem,* 250, 4007, 1975.
13. Gorg, A., Weiss, W., and Dunn, M.J., Current two-dimensional electrophoresis technology for proteomics, *Proteomics,* 4, 3665, 2004.
14. Gorg, A. et al., The current state of two-dimensional electrophoresis with immobilized pH gradients, *Electrophoresis,* 21, 1037, 2000.
15. Lopez, J.L., Two-dimensional electrophoresis in proteome expression analysis, *J Chromatogr B,* 849, 190, 2007.
16. Patton, W.F., A thousand points of light: the application of fluorescence detection technologies to two-dimensional gel electrophoresis and proteomics, *Electrophoresis,* 21, 1123, 2000.
17. Patton, W.F., Detection technologies in proteome analysis, *J Chromatogr B,* 771, 3, 2002.

18. McLaren, J., Argo, E., and Cash, P., Evolution of coxsackie B virus during in vitro persistent infection: detection of protein mutations using two-dimensional polyacrylamide gel electrophoresis, *Electrophoresis,* 14, 137, 1993.

19. Unlu, M., Morgan, M.E., and Minden, J.S., Difference gel electrophoresis: a single gel method for detecting changes in protein extracts, *Electrophoresis,* 18, 2071, 1997.

20. Lamanda, A. et al., Improved Ruthenium II tris (bathophenantroline disulfonate) staining and destaining protocol for a better signal-to-background ratio and improved baseline resolution, *Proteomics,* 4, 599, 2004.

21. Rabilloud, T. et al., A comparison between Sypro Ruby and ruthenium II tris (bathophenanthroline disulfonate) as fluorescent stains for protein detection in gels, *Proteomics,* 1, 699, 2001.

22. Palagi, P.M. et al., Proteome informatics I: bioinformatics tools for processing experimental data, *Proteomics,* 6, 5435, 2006.

23. Grove, H. et al., Challenges related to analysis of protein spot volumes from two-dimensional gel electrophoresis as revealed by replicate gels, *J Proteome Res,* 5, 3399, 2006.

24. Luhn, S. et al., Using standard positions and image fusion to create proteome maps from collections of two-dimensional gel electrophoresis images, *Proteomics,* 3, 1117, 2003.

25. Færgestad, E.M. et al., Analysing two-dimensional electrophoresis gel images on the level of pixels; a novel approach for analysing the proteome pattern, Pixel-based analysis of multiple images for the identification of charges: a noval approach applied to unravel proteome patterns of 2-D electrophoresis gel images, *Proteomics,* (in press).

26. Martens, H., and Martens, M., Modified Jack-knife estimation of parameter uncertainty in bilinear modelling by partial least squares regression (PLSR), *Food Qual Prefer,* 11, 5, 2000.

27. Næs, T. et al., *A User-Friendly Guide to Multivariate Calibration and Classification,* NIR Publications, Chichester, 2002.

28. Jessen, F. et al., Extracting information from two-dimensional electrophoresis gels by partial least squares regression, *Proteomics,* 2, 32, 2002.

29. Jia, X. et al., Changes in enzymes associated with energy metabolism during the early post mortem period in longissimus thoracis bovine muscle analyzed by proteomics, *J Proteome Res,* 5, 1763, 2006.

30. Kjaersgard, I.V., Norrelykke, M.R., and Jessen, F., Changes in cod muscle proteins during frozen storage revealed by proteome analysis and multivariate data analysis, *Proteomics,* 6, 1606, 2006.

31. Meunier, B. et al., Assessment of hierarchical clustering methodologies for proteomic data mining, *J Proteome Res,* 6, 358, 2007.

32. Jacobsen, S. et al., Multivariate analysis of two-dimensional gel electrophoresis protein patterns—practical approaches, *Electrophoresis,* 28, 1289, 2007.

33. Maurer, M.H. et al., Comparison of statistical approaches for the analysis of proteome expression data of differentiating neural stem cells, *J Proteome Res,* 4, 96, 2005.

34. Link, A.J., Multidimensional peptide separations in proteomics, *Trends Biotechnol,* 20, S8, 2002.

35. Link, A.J. et al., Direct analysis of protein complexes using mass spectrometry, *Nat Biotechnol,* 17, 676, 1999.

36. Washburn, M.P., Wolters, D., and Yates, J.R., Large-scale analysis of the yeast proteome by multidimensional protein identification technology, *Nat Biotechnol,* 19, 242, 2001.

37. Fenn, J.B. et al., Electrospray ionization for mass spectrometry of large biomolecules, *Science,* 246, 64, 1989.

38. Hillenkamp, F. et al., Matrix-assisted laser desorption ionization mass-spectrometry of biopolymers, *Anal Chem,* 63, A1193, 1991.

39. Ficarro, S.B. et al., Phosphoproteome analysis by mass spectrometry and its application to *Saccharomyces cerevisiae, Nat Biotechnol,* 20, 301, 2002.

40. Peters, E.C., Brock, A., and Ficarro, S.B., Exploring the phosphoproteome with mass spectrometry, *Mini-Rev Med Chem,* 4, 313, 2004.

41. Ndassa, Y.M. et al., Improved immobilized metal affinity chromatography for large-scale phosphoproteomics applications, *J Proteome Res,* 5, 2789, 2006.

42. Kinoshita-Kikuta, E. et al., Enrichment of phosphorylated proteins from cell lysate using a novel phosphate-affinity chromatography at physiological pH, *Proteomics,* 6, 5088, 2006.

43. Imanishi, S.Y., Kochin, V., and Eriksson, J.E., Optimization of phosphopeptide elution conditions in immobilized Fe(III) affinity chromatography, *Proteomics,* 7, 174, 2007.

44. Ishihama, Y. et al., Enhancement of the efficiency of phosphoproteomic identification by removing phosphates after phosphopeptide enrichment, *J Proteome Res,* 6, 1139, 2007.

45. Durham, M., and Regnier, F.E., Targeted glycoproteomics: serial lectin affinity chromatography in the selection of *O*-glycosylation sites on proteins from the human blood proteome, *J Chromatogr A,* 1132, 165, 2006.

46. Qiu, R.Q., and Regnier, F.E., Use of multidimensional lectin affinity chromatography in differential glycoproteomics, *Anal Chem,* 77, 2802, 2005.

47. Tang, N., Tornatore, P., and Weinberger, S.R., Current developments in SELDI affinity technology, *Mass Spectrom Rev,* 23, 34, 2004.

48. Suckau, D. et al., A novel MALDI LIFT-TOF/TOF mass spectrometer for proteomics, *Anal Bioanal Chem,* 376, 952, 2003.

49. Gygi, S.P. et al., Quantitative analysis of complex protein mixtures using isotope-coded affinity tags, *Nat Biotechnol,* 17, 994, 1999.

50. Ong, S.E. et al., Stable isotope labeling by amino acids in cell culture, SILAC, as a simple and accurate approach to expression proteomics, *Mol Cell Proteomics,* 1, 376, 2002.

51. Unwin, R.D., Evans, C.A., and Whetton, A.D., Relative quantification in proteomics: new approaches for biochemistry, *Trends Biochem Sci,* 31, 473, 2006.

52. Orchard, S. et al., Common interchange standards for proteomics data: public availability of tools and schema, *Proteomics,* 4, 490, 2004.

53. Haab, B.B., Dunham, M.J., and Brown, P.O., Protein microarrays for highly parallel detection and quantitation of specific proteins and antibodies in complex solutions, *Genome Biol,* 2, 2001.

54. Schulz, J.S. et al., Microarray profiling of skeletal muscle sarcoplasmic reticulum proteins, *Biochim Biophys Acta,* 1764, 1429, 2006.

55. Zhu, H., Bilgin, M., and Snyder, M., Proteomics, *Annu Rev Biochem,* 72, 783, 2003.

56. Wilson, D.S., and Nock, S., Functional protein microarrays, *Curr Opin Chem Biol,* 6, 81, 2002.

57. Fey, S.J., and Larsen, P.M., 2D or not 2D. Two-dimensional gel electrophoresis, *Curr Opin Chem Biol,* 5, 26, 2001.

58. Bouley, J. et al., Proteomic analysis of bovine skeletal muscle hypertrophy, *Proteomics,* 5, 490, 2005.

59. Hamelin, M. et al., Proteomic analysis of ovine muscle hypertrophy, *J Anim Sci,* 84, 3266, 2006.

60. Sayd, T. et al., Proteome analysis of the sarcoplasmic fraction of pig semimembranosus muscle: implications on meat color development, *J Agric Food Chem,* 54, 2732, 2006.

61. Hamelin, M. et al., Differential expression of sarcoplasmic proteins in four heterogeneous ovine skeletal muscles, *Proteomics,* 7, 271, 2007.

62. Lametsch, R. et al., Changes in the muscle proteome after compensatory growth in pigs, *J Anim Sci,* 84, 918, 2006.

63. Corzo, A. et al., Protein expression of pectoralis major muscle in chickens in response to dietary methionine status, *Br J Nutr,* 95, 703, 2006.

64. Morzel, M. et al., Proteome changes during pork meat ageing following use of two different pre-slaughter handling procedures, *Meat Sci,* 67, 689, 2004.

65. Morzel, M. et al., Modifications of trout (*Oncorhynchus mykiss*) muscle proteins by preslaughter activity, *J Agric Food Chem,* 54, 2997, 2006.

66. Di Luccia, A. et al., Proteomic analysis of water soluble and myofibrillar protein changes occurring in dry-cured hams, *Meat Sci,* 69, 479, 2005.

67. Morzel, M. et al., Use of two-dimensional electrophoresis to evaluate proteolysis in salmon (*Salmo salar*) muscle as affected by a lactic fermentation, *J Agr Food Chem,* 48, 239, 2000.

68. Sidhu, M.S., Hollung, K., and Berg, P., Proteolysis in Norwegian dry-cured hams; preliminary results, 51st ICOMST—The International Congress of Meat Science and Technology, Baltimore, MD. 2005.

69. Lametsch, R., and Bendixen, E., Proteome analysis applied to meat science: characterizing postmortem changes in porcine muscle, *J Agric Food Chem,* 49, 4531, 2001.

70. Lametsch, R., Roepstorff, P., and Bendixen, E., Identification of protein degradation during postmortem storage of pig meat, *J Agric Food Chem,* 50, 5508, 2002.

71. Lametsch, R. et al., Identification of myofibrillar substrates for mu-calpain, *Meat Sci,* 68, 515, 2004.

72. Lametsch, R. et al., Postmortem proteome changes of porcine muscle related to tenderness, *J Agr Food Chem,* 51, 6992, 2003.

73. Jia, X. et al., Proteome analysis of early post-mortem changes in two bovine muscle types: M. longissimus dorsi and M. semitendinosis, *Proteomics,* 6, 936, 2006.

74. Verrez-Bagnis, V. et al., Protein changes in post mortem sea bass (*Dicentrarchus labrax*) muscle monitored by one- and two-dimensional gel electrophoresis, *Electrophoresis,* 22, 1539, 2001.

75. Kjaersgard, I.V., and Jessen, F., Proteome analysis elucidating post-mortem changes in cod (*Gadus morhua*) muscle proteins, *J Agric Food Chem,* 51, 3985, 2003.

76. Kjaersgard, I.V., and Jessen, F., Two-dimensional gel electrophoresis detection of protein oxidation in fresh and tainted rainbow trout muscle, *J Agric Food Chem,* 52, 7101, 2004.

77. Kjaersgard, I.V. et al., Identification of carbonylated protein in frozen rainbow trout (*Oncorhynchus mykiss*) fillets and development of protein oxidation during frozen storage, *J Agric Food Chem,* 54, 9437, 2006.

78. Hwang, I.H. et al., Assessment of postmortem proteolysis by gel-based proteome analysis and its relationship to meat quality traits in pig longissimus, *Meat Sci,* 69, 79, 2005.

79. Bauchart et al., Small peptides (< 5 kDa) found in ready-to-eat beef meat, *Meat Sci,* 74, 658, 2006.

80. Bouley, J., Chambon, C., and Picard, B., Proteome analysis applied to the study of muscle development and sensorial qualities of bovine meat, *Sci Aliment,* 23, 75, 2003.

81. Bouley, J., Chambon, C., and Picard, B., Mapping of bovine skeletal muscle proteins using two-dimensional gel electrophoresis and mass spectrometry, *Proteomics,* 4, 1811, 2004.

82. Bauchart, C. et al., Peptides in rainbow trout (*Oncorhynchus mykiss*) muscle subjected to ice storage and cooking, *Food Chem,* 100, 1566, 2007.

83. Jessen, F. et al., Effect of pre slaughter stress on muscle protein expression in Atlantic salmon, TAFT Conference, Reykjavik, Iceland, June 11–14, 2003.

84. Bandman, E., and Zdanis, D., An immunological method to assess protein-degradation in postmortem muscle, *Meat Sci,* 22, 1, 1988.

85. Huff-Lonergan, E., Parrish, F.C., and Robson, R.M., Effects of postmortem aging time, animal age, and sex on degradation of titin and nebulin in bovine longissimus muscle, *J Anim Sci,* 73, 1064, 1995.

86. Koohmaraie, M., Muscle proteinases and meat aging, *Meat Sci,* 36, 93, 1994.

87. Geesink, G.H., and Koohmaraie, M., Postmortem proteolysis and calpain/calpastatin activity in callipyge and normal lamb biceps femoris during extended postmortem storage, *J Anim Sci,* 77, 1490, 1999.

88. Leitner, A. et al., Identification of marker proteins for the adulteration of meat products with soybean proteins by multidimensional liquid chromatography-tandem mass spectrometry, *J Proteome Res,* 5, 2424, 2006.

Chapter 6

Muscle Enzymes: Proteinases

Geert H. Geesink and Eva Veiseth

Contents

6.1 General Introduction

Multiple factors including palatability, water-holding capacity, color, nutritional value, and safety determine meat quality. The importance of these traits varies depending on both the end product and the consumer profile. Flavor, juiciness, and tenderness influence the palatability of meat. Among these traits, tenderness is the most important.[1] Furthermore, surveys of beef packers, purveyors, restaurateurs, and retailers indicate that tenderness is also one of the top quality concerns.[2] This concern is warranted because a number of studies have shown that a significant proportion of retail meat cuts can be considered tough.[3–5]

The three factors that determine meat tenderness are background toughness, the toughening phase, and the tenderization phase. Although the toughening and tenderization phases take place during the postmortem storage, or aging period, background toughness exists at the time of slaughter and does not change during the storage period. The effect of postmortem storage on tenderness is illustrated in Figure 6.1.

The background toughness of meat is defined as *the resistance to shearing of the unshortened muscle*,[6] and variation in the background toughness is due to the connective tissue component of muscle. In particular, the organization of the perimysium appears to affect the background toughness, because a general correlation between characteristics of the perimysium and tenderness of muscles has been found for both beef and chicken.[7]

The toughening phase is caused by sarcomere shortening during rigor development.[8,9] For beef, this process usually occurs within the first 24 h postmortem.[10] The relationship between sarcomere shortening and meat toughness was first reported by Locker.[11] Later, it was shown that

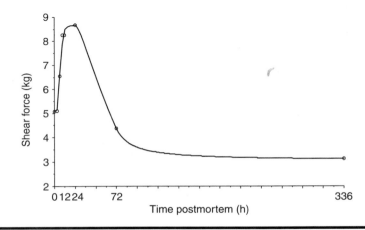

Figure 6.1 Tenderness of ovine *longissimus*, measured by Warner-Bratzler shear force, at various times postmortem. (From Wheeler, T.L. and Koohmaraie, M., *J. Anim. Sci.*, 72, 1232, 1994. With permission.)

there is a strong negative relationship between sarcomere length and meat toughness when sarcomeres are shorter than 2 µm, and that the relationship is poorer at longer sarcomere lengths.[12–14] The known effect of the interaction of the rate of postmortem glycolysis and temperature decline on shortening has led to industry recommendations on optimal processing conditions and procedures aimed at minimizing shortening.[15]

Although the toughening phase is similar for carcasses under similar processing conditions, the tenderization phase is highly variable. There is a large variation in both the rate and extent of tenderization, and this results in inconsistency of meat tenderness. It has been known for a long time that meat tenderness improves during refrigerated storage, and it was suggested almost a century ago that this is due to enzymatic activity.[16] It is now well established that postmortem proteolysis of myofibrillar and associated proteins is responsible for this process. The question that proteinase(s) are primarily responsible for postmortem tenderization has been a matter of debate for decades.[17,18]

The proteinase systems that have been studied in relation to postmortem tenderization include the calpain system, cathepsins, the multicatalytic proteinase complex, and, to a lesser extent, matrix metalloproteinases (MMP) and caspases. This chapter discusses the involvement of these proteinase systems in meat tenderization and refers the reader to key publications concerning the methods to study these proteinase systems with regard to meat tenderization.

6.2 The Calpain System

6.2.1 Introduction

Calpains are calcium-activated proteinases with an optimum activity at neutral pH. In mammalian skeletal muscle, the calpain system consists of at least three proteinases—µ-calpain, m-calpain, and calpain 3, also referred to as skeletal muscle-specific calpain, or p94, and an inhibitor of µ- and m-calpain, calpastatin.

Both µ- and m-calpain are composed of two subunits with molecular weights of 28 and 80 kDa.[19–22] An important characteristic of µ- and m-calpain is that they undergo autolysis in the presence of calcium. Autolysis reduces the Ca^{2+}-requirement for half maximal activity of µ- and m-calpain.[23–26] Initial autolysis of the large subunit of µ-calpain produces a 78 kDa fragment, followed by a 76 kDa fragment.[27] Initial autolysis of the large subunit of m-calpain produces a 78 kDa fragment only.[28] Further autolysis of µ- and m-calpain leads to lower molecular weight fragments of the large subunit and loss of activity.

Calpain 3 is a protein of 94 kDa with sequence homology to the large subunits of µ- and m-calpain.[29] Purification and characterization of calpain 3 has been extremely difficult for several reasons. Unlike µ- and m-calpain, calpain 3 cannot be easily extracted from skeletal muscle due to its association with the myofibrillar protein, titin.[30] Expression of calpain 3 *in vitro* is hampered by rapid autolysis of the enzyme at physiological levels of calcium, and furthermore, the autolysis is not affected by calpain inhibitors.[31]

Calpastatin is the endogenous specific inhibitor of µ- and m-calpain.[32] Several isoforms of this protein exist, but the predominant form in skeletal muscle contains four calpain-inhibiting domains.[33] Calpastatin requires calcium to bind and inhibit calpains.[34,35] Calpastatin is also a substrate for the calpains and can be degraded in the presence of calcium.[36,37] Degradation of calpastatin does not lead to complete loss of inhibitory activity, and even after extensive proteolysis some inhibitory activity remains.[38,39]

For further details regarding the calpain system, see Ref. 40.

The evidence for the involvement of the calpain system in postmortem proteolysis and tenderization comes from a variety of studies:

1. Incubation of myofibrils with calpains produces the same proteolytic pattern as observed in postmortem muscle.[41–43]
2. Infusion or injection of muscles with calcium accelerates postmortem proteolysis and tenderization, whereas infusion or injection of muscles with calpain inhibitors inhibit postmortem proteolysis and tenderization.[44–48]
3. Differences in the rate of proteolysis and tenderization between species can be explained by the variation in calpastatin activity.[49,50]
4. Differences in the rate of postmortem proteolysis and tenderization between *Bos taurus* and *Bos indicus* cattle can be explained by the variation in calpastatin activity.[51,52]
5. The toughening effect of treatment with ß-agonists can be explained by an increase in calpastatin activity.[53,54]
6. The greatly reduced rate and extent of postmortem proteolysis and tenderization in callipyge lamb can be attributed to elevated levels of calpastatin in these animals.[55,56]
7. Overexpression of calpastatin in transgenic mice results in a large reduction in postmortem proteolysis of muscle proteins.[57]

From these cited studies and others, it is clear that the calpain system plays an important role in postmortem proteolysis and tenderization. However, which of the calpains is responsible for postmortem proteolysis and tenderization remains an important question.

An important characteristic of the calpains is that they autolyze once activated, ultimately leading to a loss of activity.[23–25] In bovine and ovine postmortem muscle, the extractable activity of μ-calpain declines but the activity of m-calpain is remarkably stable.[55,58–60] This observation led Koohmaraie et al.[61] to conclude that μ-calpain, but not m-calpain, is responsible for postmortem tenderization.

Using western blotting, it has been established that calpain 3 does autolyze in postmortem muscle, indicating that it is activated in postmortem muscle.[62–64] However, in contrast to μ- and m-calpain, calpain 3 is not inhibited by calpastatin.[31] This observation excludes a major role of calpain 3 on postmortem proteolysis and tenderization, given the great influence of calpastatin activity on these events. This conclusion was further corroborated by results of a recent study showing that postmortem proteolysis is not affected in calpain 3 knockout mice.[65]

The conclusion of Koohmaraie et al.[61] that μ-calpain is responsible for postmortem proteolysis was recently confirmed using μ-calpain knockout mice.[66] The results of this study clearly showed that postmortem proteolysis was largely inhibited in μ-calpain knockout mice (Figure 6.2). The limited proteolysis that did occur could be attributed to m-calpain, which is activated to some extent in postmortem murine skeletal muscle, contrary to what is observed for m-calpain in muscles of meat-producing animals.[57,66]

6.2.2 Analytical Methods

6.2.2.1 Chromatography and Calpain/Calpastatin Assays

The components of the calpain system need to be separated to quantify their activity. Chromatography is a means to obtain this objective. Studies on the role of the calpain system in tenderization have used two approaches. The first one is separation of the calpains from calpastatin using hydrophobic interaction chromatography, followed by anion exchange chromatography to

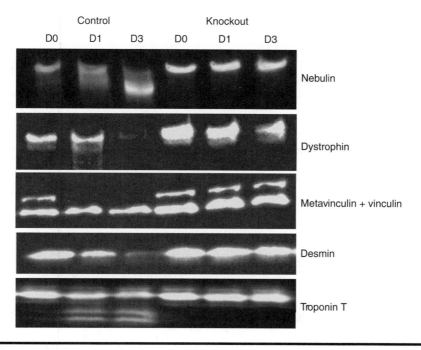

Figure 6.2 **Western blot analysis of myofibrillar proteins in muscle extracts of control and μ-calpain knockout mice at death (D0) and after 1 (D1) and 3 days (D3) storage at 4°C. (From Koohmaraie, M. and Geesink, G.H.,** *Meat Sci.,* **74, 34, 2006. With permission.)**

separate μ- and m-calpain.[67] This method has been largely abandoned because the recovery of calpain activity after hydrophobic interaction chromatography was only about 50% as compared to the recovery after anion exchange chromatography.[68] The second method is separation of calpastatin, μ- and m-calpain using anion exchange chromatography. Using this method, care must be taken to elute with a shallow gradient to completely separate calpastatin and μ-calpain.[68] Another important aspect in the quantification of the different components of the calpain system is the buffering capacity of the extraction buffer. Extraction of postrigor muscle using a buffer with insufficient buffering capacity can result in severely reduced extraction efficiency.[69] Given the difference in pH between pre- and postrigor muscle, different buffers should be used to extract these tissues and care should be taken that the pH of the extract is above 6.2.[69]

Whether frozen storage of muscle tissue results in loss of calpastatin activity or not has been a matter of debate. Koohmaraie[68] reported a progressive drop in extractable calpastatin activity during frozen storage up to 6 weeks at −70°C. Similar results on calpastatin activity were reported after frozen storage of beef at −30°C, and lamb at −20°C.[70,71] In contrast, Kristensen et al.[72] reported that calpastatin, μ-calpain, and m-calpain activities were stable in porcine longissimus muscle during 22 weeks of frozen storage at both −20 and −80°C. Recent results from our lab using beef muscle (Table 6.1) seem to confirm the results of Kristensen et al.[72] Although we did observe an effect of extracting frozen muscle instead of fresh muscle on calpastatin and μ-calpain activities, this seemed to be due to the extraction efficiency from fresh versus frozen tissue rather than to the period of frozen storage. Clearly, the ability to freeze and store samples without significant loss of calpastatin and calpain activity facilitates quantification of calpastatin and calpains in experiments involving large numbers of samples. Therefore, additional research is needed to overcome the debate

Table 6.1 Calpain and Calpastatin Activities after Extraction from Fresh or Frozen Bovine Muscle (Means ± S.E.; *n* = 8/Treatment)

	Storage Time/Storage Temperature			
Item	*Fresh*	*1 Week/−80°C*	*5 Weeks/−80°C*	*8 Weeks/−20°C*
Calpastatin	3.11 ± 0.11	2.78 ± 0.11	2.63 ± 0.11	2.96 ± 0.11
μ-Calpain	1.16 ± 0.11	0.79 ± 0.11	0.76 ± 0.11	0.77 ± 0.11
m-Calpain	0.90 ± 0.11	0.93 ± 0.11	0.91 ± 0.11	1.01 ± 0.11

Note: Means, within a row, not containing a common superscript letter are significantly different ($p < 0.05$).

on whether samples for determination of calpain and calpastatin can be stored frozen without significant loss of activity or not.

As per the preceding description, variation in calpastatin content rather than μ-calpain activity is the main determinant of the speed and extent of postmortem proteolysis. Because of the heat stability of calpastatin, its quantification can be achieved without chromatography by simply heating a muscle extract, and thus inactivating the calpains. After heating, the clarified extract can be used to assay for calpastatin activity using partially purified m-calpain to determine its inhibitory activity. Shackelford et al.[73] provide a detailed description of this procedure.

A possible source of erroneous results for calpastatin assays is the stability of partially purified m-calpain against which the inhibitory activity of calpastatin is assayed. During chilled storage, m-calpain preparations gradually lose activity. However, proteolytically inactive m-calpain apparently retains its activity to bind calpastatin.[72] As a result, the inhibitory activity of calpastatin may be severely underestimated when a significant amount of m-calpain activity has been lost by the standard against which calpastatin is assayed. It is thus advisable to freeze partially purified m-calpain in liquid nitrogen and store samples at −80°C to retain constant m-calpain activity against which to assess calpastatin activity over a period longer than a few weeks.

The most commonly used substrate for calpain assays is casein. In this assay, the increase in trichloroacetic acid (TCA) soluble degradation products is used as a measure of calpain activity. Koohmaraie[68] has described the details of this assay method. More sensitive assay methods employing [^{14}C]-, or Bodipy-labeled casein have been described by others.[74,75] A comparison of these assay methods and the effect of sample size on the accuracy of the measurements have been described by Kent et al.[76]

6.2.2.2 Zymography

An alternative to chromatography to separate the different components of the calpain system and assay for μ- and m-calpain is casein zymography. This technique consists of nondenaturing electrophoresis of samples on polyacrylamide gels containing casein. After electrophoresis, the gels are incubated in a Ca^{2+}-containing buffer to activate the calpains. After staining of the gels, activity of the calpains is assessed by measuring the area of the clear bands, indicative of degraded casein. For a detailed description of this technique, see Ref. 60. An advantage of this technique over chromatography is that it allows for separate quantification of native and partially autolyzed calpains (Figure 6.3).

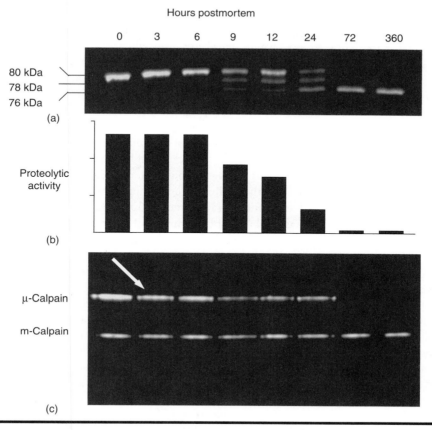

Figure 6.3 Assessment of autolysis and activity of μ-calpain during postmortem storage of ovine muscle using western blotting (a), caseinolytic activity after anion exchange chromatography (b), and zymography (c). The arrow in the zymogram indicates autolyzed μ-calpain. (From Veiseth, E. et al., *J. Anim. Sci.*, 70, 3035, 2001. With permission.)

6.2.3   Concluding Remarks

From the foregoing data it is evident that μ-calpain and calpastatin are the main determinants of postmortem tenderization. Discovering the mechanisms of μ-calpain activity regulation in early postmortem muscle, and as a result, development of methods to maximize the tenderizing capacity of μ-calpain should aid the meat industry in producing a product that is consistently tender.

6.3   Cathepsins

6.3.1   Introduction

Cathepsins are composed of a complex family of proteinases that are localized within lysosomes in the cytoplasm of cells. This proteinase family includes both cysteine and aspartic proteinases, and although most of them are characterized as endopeptidases, some have been identified as exopeptidases.[77–80] Cathepsins are synthesized as proenzymes, and are converted into active proteinases by autocatalytic cleavage of the N-terminal propeptide at acidic pH.[79] The cysteine proteinase cathepsins are, in addition, regulated by endogenous protein inhibitors, that is, cystatins. Cystatins are divided into three groups, where stefins are intracellular inhibitors and cystatins and kininogens are extracellular inhibitors.[81] Cathepsins are mostly active in the slightly acidic and reducing milieu of the lysosomes, and once activated they have an enormous disruptive potential. This is illustrated by the fact that several pathological conditions (e.g., cancer, apoptosis, and muscular dystrophy) are associated with secretion of cathepsins from lysosomes.

The most abundant of these proteinases are the cysteine proteinases cathepsins B and L, and the aspartic proteinase cathepsin D; these are also the most frequently studied cathepsins in muscle. Cathepsin B can act both as an endo- and exopeptidase (carboxydipeptidase), and its optimum pH range for proteolytic activity is 5.5–6.5; however, cathepsin B is unstable at neutral pH. Cathepsin L is an endopeptidase with a pH optimum of approximately 5, and in contrast to cathepsin B, this proteinase is stable at neutral pH and retains 30–40% of its maximal proteolytic activity at pH 7.[82] Cathepsin D has the lowest pH optimum of these proteinases (pH 3–5), and shows no activity at pH 7 or above due to conformational changes that alter the substrate binding site. These three cathepsins have been found to play roles in muscle development, pathological conditions in muscle, and different types of muscle atrophy (for a review on cathepsins in skeletal muscle, see Ref. 83).

The role of cathepsins in muscle foods pertaining to their involvement in postmortem proteolysis in meat has been much disputed. Two of the main arguments used against the potential role of cathepsins in meat tenderization have been that (1) cathepsins are contained within lysosomes during postmortem storage of meat, and thus have no access to the myofibrils, and (2) little degradation of actin and myosin occur in postmortem muscle although these proteins are very good substrates for cathepsins. Owing to the lysosomal localization of cathepsins, several groups have investigated treatments that could induce liberation of cathepsins from lysosomes. Early indications showed that a combination of high temperature and low muscle pH increased the breakdown of lysosomes, leading to a release of cathepsins.[84]

Ertbjerg et al.[85] found an increased activity of cathepsin B + L in the soluble fraction by injecting lactic acid into muscles shortly after slaughter. Moreover, the injected muscles had a reduced shear force compared to the control samples. Although some of the tenderizing effect was caused by pH-induced swelling of the muscle structure, an increased proteolysis of myofibrillar proteins was observed, suggesting that cathepsins contribute to textural changes in meat. Comparison of bovine *longissimus dorsi* muscles with fast and slow pH declines have also revealed an enhanced release of cathepsin B and L from lysosomes in the fast glycolytic muscles.[86] Similarly, electrical stimulation of carcasses is known to accelerate the pH decline in muscles, and results of several studies suggest that this treatment induces a release of cathepsins from the lysosomes.[87–90] Additionally, an elevated ultimate pH, induced by a combined epinephrine and exercise treatment of pigs prior to slaughter, caused a reduced release of cathepsins from lysosomes.[91]

The role of cathepsins has also been investigated in muscle from several fish species. Aoki and Ueno[92] studied the role of cathepsin B and L in mackerel muscle during postmortem storage, and found that their activities decreased in the lysosomal fraction and consequently increased in the supernatant fraction during the storage period. Additionally, myofibrillar incubations of purified cathepsin B and L from mackerel muscle showed that cathepsin L readily hydrolyzed myosin, troponin T and I, and tropomyosin, whereas cathepsin B had no effect on these myofibrillar proteins. Thus, Aoki and Ueno[92] concluded that the main cause of postmortem softening of mackerel muscle was the proteolytic activity of cathepsin L. Cathepsin L has also been identified as a likely candidate for postmortem tenderization of carp muscle.[93] However, in studies of sea bass muscle, cathepsin D was found to be the most likely candidate contributing to postmortem degradation, because the changes induced during myofibrillar incubation of cathepsin D were similar to that observed during postmortem storage.[94] As in mammalian muscle, the lysosomal localization of cathepsins is a barrier for their involvement in postmortem degradation of fish muscle. High-pressure treatment may be a means to release cathepsins from the lysosomes. It induces the release of cathepsins from the lysosomes in sea bass muscle,[95] similar to its reported effects on mammalian muscle.[96,97]

6.3.2 Analytical Methods

Protocols for purification of cathepsins are available and should be used for characterization studies of these proteinases. In short, these protocols involve a series of steps including autolysis, acetone fractionation, ion-exchange chromatography, and gel filtration.[98,99] For partial purification, hydrophobic interaction chromatography is an efficient method for separating cathepsin B, H, and L from other proteins, but this method cannot separate the individual proteinases. An alternative method for cathepsin D purification is affinity chromatography with immobilized pepstatin.[78] Nevertheless, in muscle research, the activities of cathepsins are usually determined on relatively crude muscle extracts; primarily because these methods are cheaper and less time-consuming than the more elaborate purification methods.

6.3.2.1 Extraction and Fractionation

The most commonly used extraction method for determining cathepsin activities in muscle is to homogenize muscle samples in a sodium acetate buffer containing EDTA and a detergent (e.g., Triton X-100 or Brij 35), followed by centrifugation of the homogenate. The resulting supernatant is retained and assayed for cathepsin activity.[100]

Since much of the debate regarding the role of cathepsins in postmortem muscle relates to their lysosomal localization, subcellular fractionation methods have also been used to study their localization in muscle during postmortem storage.[101] To study subcellular localization, muscle samples are finely minced and treated with a bacterial proteinase to improve the isolation of intact lysosomes, and homogenized using a Potter-Elvehjem-type homogenizer. The homogenate is subsequently separated into different subcellular fractions by serial centrifugation: $1,100 \times g$ (myofibrillar fraction), $3,000 \times g$ (heavy mitochondrial fraction), $27,000 \times g$ (lysosomal fraction), and $100,000 \times g$ (microsomal [pellet] and soluble fractions).

For subcellular fractionation it is essential that the analysis is performed on fresh samples, since ice crystal formation in the tissue during freezing may cause membrane rupture and thus affect the localization of the cathepsins. Whole muscle homogenates, however, can be prepared from both fresh and frozen samples. Since cathepsins are stable during frozen storage, all sample extracts, regardless of whether they result from a whole muscle homogenate or subcellular fractionation, can be frozen prior to activity measurements; however, multiple freeze-thaw cycles should be avoided.

6.3.2.2 Removal and Determination of Cystatin Activity

The mentioned extraction and fractionation methods are widely used in the analysis of cathepsins from muscle; however, it is important to keep in mind that these crude extracts also contain the endogenous inhibitors of cathepsins, which would lead to an underestimation of cathepsin activities. Cystatins are inhibitors of the cysteine cathepsins, and can be removed by affinity chromatography with immobilized papain before measurement of cathepsin activities.[102] Moreover, the cystatin activity can be estimated based on measurements of cathepsin activities before and after cystatin removal. Alternatively, the activity of the cystatins can be determined by heating the supernatant from a whole muscle extract at 100°C for 3 min to remove all proteinase activity, before an assay of inhibitory activity against a standard solution of papain.[103]

6.3.2.3 Activity Measurements

The best and most commonly used assay methods for the cysteine cathepsins (e.g., cathepsin B, L, and H) are the fluorometric procedures based on aminomethylcoumarylamide (-NMec) substrates.[103,104] In these assays, samples are preincubated in an activation buffer for 5 min at 40°C before the substrate solution is added. The following incubation is stopped after 10 min by the addition of 100 mM monochloroacetic acid in acetate buffer. The fluorescence is measured in a spectrophotofluorimeter with excitation wavelength set at 360 nm and emission wavelength set at 460 nm. A standard curve is made with aminomethylcoumarin in the stop buffer, and 1 unit is defined as the release of 1 µmol of product per minute. The stopped assays are convenient for studies where large numbers of samples need to be analyzed. However, fluorometric assays can also be performed as continuous rate assays. For a continuous assay the reaction mixture is the same as in the stopped assay, but the incubation takes place in a thermostatically controlled (40°C) cell in a spectrophotofluorimeter. These continuous assay procedures are particularly suitable for kinetic studies of cathepsins.

There are several -NMec-conjugated substrates, and some of these are highly specific for the different cysteine cathepsins. For the determination of cathepsin B activity, Z-Arg-Arg-NMec can be used, whereas Arg-NMec is specific for cathepsin H. Cathepsin L, on the contrary, has no specific substrates, but the activity of cathepsin B + L can be assayed with Z-Phe-Arg-Nmec; and by assaying the same sample with this and the cathepsin B–specific substrate (Z-Arg-Arg-NMec), the activity of cathepsin L can be calculated. Often, however, only the combined cathepsin B + L activity is reported.

The most frequently used assay method for cathepsin D activity is that of Anson.[105] In this assay, cathepsin D is incubated with hemoglobin as a substrate, leading to liberation of peptides that are soluble in the presence of TCA. The concentrations of peptides are then determined by, for example, measuring the A_{280} of the supernatants or performing protein assays. For the detection of small amounts of proteolytic activity, the hemoglobin can also be labeled with a radioactive isotope, and, in those cases, the radioactivity of the TCA-soluble supernatant is measured.

6.3.3 Concluding Remarks

Undoubtedly, the cathepsins have a great potential to cause proteolysis in postmortem muscle if released from the lysosomes, because they readily degrade many of the myofibrillar proteins. Nevertheless, no direct links between tenderness of fresh meat and cathepsin activities have so far been proven. During production of dry-cured hams, García-Garrido et al.[106] found the residual activity of cathepsin B + L to be a reliable indicator of textural defects associated with strong proteolysis. In conclusion, it seems as though cathepsins may play a more significant role in processed meat quality than in fresh meat.

6.4   Proteasome (Multicatalytic Proteinase Complex)

6.4.1   Introduction

Proteasomes are large proteinase complexes involved in many cellular processes, including the cell cycle, responses to oxidative stress, and gene expression.[107,108] The core of the proteasomes (20S proteasome) is a barrel-like complex composed of 28 proteins, having three substrate specificities: chymotrypsin-like, trypsin-like, and peptidyl-glutamyl peptide-hydrolyzing.[109] Proteolytic activity is regulated by a number of activators and inhibitors.[108]

The first study on the possible role of proteasome in postmortem proteolysis and tenderization was published 15 years ago.[74] The results of this study indicated that myofibrils are poor substrates for purified ovine proteasome, and that the limited proteolysis does not match degradation patterns in postmortem muscle. Moreover, substances that inhibit postmortem proteolysis, like EDTA, did not inhibit the activity of proteasome *in vitro*. In later studies, more extensive degradation of bovine myofibrillar proteins was observed during longer incubation times.[110,111] However, there was little similarity in protein substrate specificity or ultrastructural changes to that observed during postmortem aging of muscle. Studies on purified rabbit proteasome corroborated the previously mentioned studies in the sense that inhibitors of postmortem proteolysis did not inhibit proteasome *in vitro*, and that the proteolysis pattern of myofibrillar proteins did not match those observed during the aging of muscle.[112,113] The possibility that proteasome might play an indirect role in postmortem tenderization, that is, by degrading calpastatin, was addressed in a study by Doumit and Koohmaraie.[36] Similar to the degradation patterns of myofibrillar proteins, the degradation pattern of calpastatin by proteasome *in vitro* did not mimic that observed during aging of muscle. Despite the evidence that proteasome plays a minor, if any, role in postmortem proteolysis, the issue was revisited in a recent study by Dutaud et al.[114] Based on similarities between ultrastructural changes observed during aging and after incubation of myofibrils with purified proteasome, the authors concluded that the effect of proteasome on myofibrils mimics structural changes observed in type I muscles and meat with high pH.

6.4.2   Analytical Methods

The analytical methods employed in the aforementioned studies involve a series of purification steps to facilitate the study of characteristics of the purified enzyme. Typically, these involve a number of steps: extraction, ammonium sulfate precipitation, and chromatography (ion exchange, hydrophobic interaction, size exclusion). In addition, dialysis and concentration steps are used to prepare the samples for the following purification step. Activity measurements, protein determinations, and SDS-PAGE are employed to monitor the purification process. For activity measurements, a number of different substrates can be employed.[74,112,113] To quantify proteasome content in crude tissue extracts in studies involving large amounts of samples, the techniques described earlier are too labor intensive. Aubry et al.[115] reported on two immunological methods (enzyme-linked immunosorbent assay [ELISA] and radial immunodiffusion) to quantify proteasome content efficiently in crude tissue extracts.

6.4.3   Concluding Remarks

Given the results of the studies mentioned earlier, it appears unlikely that proteasome plays an important role, if any, in postmortem tenderization. Nevertheless, unlike studies on the involvement of the calpain system in postmortem tenderization, studies on the involvement of proteasome in postmortem tenderization have not progressed beyond the stage of comparing *in vitro* and *in situ* effects on myofibrillar proteins. Thus, a further logical step would be to explore whether animal models known to affect postmortem tenderization (the callipyge phenotype, ß-agonist treatment, etc.) affect the activity of proteasome too.

6.5 Caspases

6.5.1 Introduction

Caspases are a group of at least 14 cysteine peptidases (some of which are species specific) that are involved in apoptosis, or programmed cell death.[116] Caspases are produced as inactive zymogens, possessing a large and a small subunit preceded by an N-terminal prodomain. Upstream caspases, known as initiators, can autolytically activate themselves. Downstream effector caspases can sequentially be activated by the initiator caspases.[117] Caspases can be distinguished into three different classes, based on their role in apoptosis: caspases involved in inflammatory processes (caspases 1, 4, 5), caspases involved in initiating apoptosis (caspases 8, 9, 10), and effector caspases (caspases 3, 6, 7).[116]

The first description of a caspase (caspase 1) involved in programmed cell death in *Caenorhabditis elegans* was published in 1993.[118] At the time of writing of this chapter, a search on PubMed revealed over 27,000 publications with the word "caspase." In meat science, however, the study of a possible role of the caspases in postmortem proteolysis and tenderization was not addressed until 2006. Ouali et al.[18] hypothesized that apoptotic processes may occur in the early postmortem period in muscle tissue. However, this hypothesis was not supported by actual data from targeted experiments. Kemp et al.[119] presented evidence that the activated form of caspase 3 can be detected in early postmortem porcine muscle, and also showed that caspase activity changes early postmortem in porcine muscle.[120] Moreover, they observed a negative relationship between shear force and the degradation of a putative caspase substrate, alpha II spectrin.[120]

Given the limited amount of information, it is too early to determine the role, if any, of the caspases in postmortem proteolysis and tenderization. Koohmaraie[121] states that a good guide for future research is the criteria for proteinases to be involved in meat tenderization. First, they must be endogenous to muscle cells; second, they must be able to replicate, *in vitro*, degradation patterns of myofibrillar proteins during aging; and third, they must have access to the myofibrils in the tissue. The first criterion is well established for the caspases, but the second and third criteria still need to be addressed for postmortem muscle tissue. The criteria that have been mentioned apply to proteinases directly involved in postmortem proteolysis of structural muscle proteins. However, we may also speculate on an indirect role of the caspases. It has been shown in a number of studies that calpastatin is a caspase substrate.[122–126] In addition, it has been shown that calpastatin activity at 1 day postmortem, but not at death, is a reasonable predictor of tenderness after aging.[51,52] Calpastatin activities at 1 day postmortem are generally considerably lower than at death. For instance, in a study by Whipple et al.,[52] activities at 1 day postmortem were approximately 50% of at-death activities. It is, therefore, conceivable that caspases degrade calpastatin to a certain extent during the early postmortem period, thereby determining the efficacy of μ-calpain in proteolysis of structural proteins during further aging, and as a result, tenderization.

6.5.2 Analytical Methods

Owing to the limited attention the proteinase system has received in meat science, there is little information available on tried and tested assays for postmortem muscle. Some information on extraction methods, assay kits, and antibodies that work for porcine muscle can be found in Refs 119, 120. Nevertheless, given the large interest in research on caspases, there are a large amount of caspase research tools commercially available.

6.5.3 Concluding Remarks

As mentioned earlier, the number of studies on the involvement of caspases in postmortem proteolysis is too few to evaluate their role in postmortem tenderization. However, a paper has been published on the possible roles of caspases in meat tenderization.[18] This work could serve as a source of testable hypotheses. In addition, some recently published studies indicate that caspases can be detected in their active form in early postmortem muscle.[119,120] Given the overwhelming evidence that μ-calpain and calpastatin are the major agents determining postmortem proteolysis, and therefore, tenderization, the authors do not anticipate a direct role for the caspases in proteolysis of structural muscle proteins. It is anticipated, however, that the role of the caspases may be limited to early postmortem events determining the efficacy of μ-calpain during the actual aging period to affect tenderization.

6.6 Matrix Metalloproteinases

6.6.1 Introduction

MMPs are a family of more than 10 different zinc-dependent endopeptidases.[127–129] The MMPs are involved in degradation of the extracellular matrix (ECM), and play an important role in both normal and pathological tissue remodeling processes such as embryonic development, tissue repair, and inflammation. These proteinases are often divided into four groups: collagenases (e.g., MMP-1 and -13), gelatinases (e.g., MMP-2 and MMP-9), stromelysins (e.g., MMP-3 and MMP-7), and membrane-type (e.g., MT-MMP). Most of these proteinases are secreted into the extracellular space as inactive zymogens. To be active, it is necessary for the MMPs to disrupt the Cys-Zn^{2+} (cysteine switch) interaction, followed by removal of the propeptide. *In vivo*, most of these MMPs are activated by tissue, plasma, or bacterial proteinases; however, *in vitro* they can be activated by both proteinases and nonproteolytic agents such as denaturants and SH-reactive agents. The proteolytic activities of MMPs are controlled during activation and also by tissue inhibitors of metalloproteinases (TIMPs). TIMPs are the major endogenous inhibitors of MMP activities in tissues. TIMP-1 binds to the active forms of MMPs by forming noncovalent complexes, whereas TIMP-2 stabilizes the inactive forms of MMPs and thus inhibits the activation of the proteinases.

In muscle, MMPs are known to play regulatory roles during growth and development.[130] In a study of human and murine muscle cell cultures, MMP-9 (i.e., gelatinase B) activity was secreted by single cell and prefusion cultures, whereas MMP-2 (i.e., gelatinase A) activity was secreted at all stages including myotubes.[131] Moreover, expression of TIMP-1 mRNA increased as the cells progressed through the different stages, and reached a maximum in myotubes. Thus, Lewis et al.[131] concluded that the ratio of MMPs to TIMPs may be important in determining myoblast migration and differentiation. MMPs are also involved in muscle atrophy conditions. Reznick et al.[132] found that MMP-2 and MMP-9 were responsible for a large part of the degradation of ECM following immobilization-induced atrophy in the hind limb muscles of a rat.

In muscle food research, relatively few investigations on MMPs have been performed, although these proteinases could potentially affect meat texture due to their ability to degrade proteins in the connective tissue. However, during the past few years, the interest toward this proteolytic system in relation to muscle foods has increased, and studies of MMPs and their inhibitors have now been performed on beef, lamb, and fish muscles.[133–136] Balcerzak et al.[133] characterized the MMPs, including their activators and inhibitors, present in bovine skeletal muscles, and detected activities of MMP-2 and MMP-9 in addition to expression of mRNA encoding multiple MMPs and TIMPs. Based on their findings, Balcerzak et al.[133] concluded that skeletal muscle cells, in addition to the intramuscular fibroblasts, express an extensive system of MMPs and related proteins. However, no investigations of MMP activity levels in relation to meat quality were performed in this study. For fish, studies of muscle MMPs have focused on gelatinase activities and TIMPs. MMPs with similarities to MMP-2 and MMP-9 have been identified in cod, salmon, and wolffish muscle, whereas a TIMP-2-like protein has been identified in cod muscle.[135,136]

In a study using ovine skeletal muscle, Sylvestre et al.[134] investigated the effect of growth rate on the level of pro- and active MMP-2 in muscle and indicators of collagen alterations. The level of active MMP-2 was found to be 90% higher in the fast-growing lambs compared to slow-growing lambs. Moreover, the level of free OH-Pro (an indicator of collagen degradation) was increased in the fast-growing lambs at the time of slaughter, and the level of free OH-Pro further increased during the 21 days postmortem storage period in muscles from these animals. In contrast, the level of free OH-Pro in muscles of slow-growing lambs did not change during the same storage period. Based on these results, it was concluded that the growth rate at slaughter not only influenced the *in vivo* turnover of collagen, but also affected the postmortem collagen degradation.[134] However, the impact of this degradation on meat tenderness was not evaluated, and further investigations are necessary to determine the potential role of MMPs in meat quality.

6.6.2 Analytical Methods

One of the most commonly used methods for analysis of MMP activities is zymography with casein or gelatine.[137] Zymography can also be used to analyze TIMP activities, using a method referred to as reverse zymography.[138] By incorporating MMPs into the substrate-containing gels, inhibitory activity can be visualized as stained areas on the gel following incubation and staining. Recently, methods for real-time zymography and real-time reverse zymography have been developed by incorporating FITC-labeled substrates.[139]

In addition to zymography, activities of MMPs can also be analyzed employing more classical activity assays, where MMPs are incubated with suitable substrates under defined time–temperature conditions. The substrates used in such assays are either native proteins or peptides, which in some cases can be labeled with, for example, fluorescence or radioactivity. For a review of MMP activity assay methods, see Ref. 140.

6.6.3 Concluding Remarks

To date, no links between MMP activities and muscle food quality have been proven. Yet, this proteolytic system has the potential to cause extensive degradation of the muscular connective tissue. Compared to the myofibrillar and cytoskeletal proteins, relatively few changes have been documented for proteins in the connective tissue during postmortem storage of meat. In fish muscles, on the contrary, more extensive changes of the connective tissue proteins during storage have been documented, and a significant role for the MMPs would thus be more likely.

References

1. Miller, M.F. et al., Consumer thresholds for establishing the value of beef tenderness, *J. Anim. Sci.*, 79, 3062, 2001.
2. Smith, G.C. et al., Improving the quality, consistency, competitiveness and marketshare of beef: the final report of the second blueprint for total quality management in the fed-beef (slaughter steer/heifer) industry, National Beef Quality Audit, National Cattlemen's Beef Association, Denver CO, USA, 1995.
3. Morgan, J.B. et al., National beef tenderness survey, *J. Anim. Sci.*, 69, 3274, 1991.
4. George, M.H. et al., An audit of retail beef loin steak tenderness conducted in eight US cities, *J. Anim. Sci.*, 77, 1646, 1999.
5. Bickerstaffe, R. et al., Impact of introducing specifications on the tenderness of retail meat, *Meat Sci.*, 59, 303, 2001.
6. Marsh, B.B. and Leet, N.G., Studies in meat tenderness III. The effect of cold shortening on tenderness, *J. Food Sci.*, 31, 450, 1966.
7. Strandine, E.J., Koonz, C.H., and Ramsbottom, J.M., A study of variations in muscle of beef and chicken, *J. Anim. Sci.*, 8, 483, 1949.
8. Koohmaraie, M., Doumit, M., and Wheeler, T.L., Meat toughening does not occur when rigor shortening is prevented, *J. Anim. Sci.*, 74, 2935, 1996.
9. Wheeler, T.L. and Koohmaraie, M., Prerigor and postrigor changes in tenderness of ovine longissimus muscle, *J. Anim. Sci.*, 72, 1232, 1994.
10. Wheeler, T.L. and Koohmaraie, M., The extent of proteolysis is independent of sarcomere length in lamb longissimus and psoas major, *J. Anim. Sci.*, 77, 2444, 1999.
11. Locker, R.H., Degree of muscle contraction as a factor in tenderness of beef, *Food Res.*, 25, 304, 1960.
12. Bouton, P.E. et al., A comparison of the effects of aging, conditioning and skeletal restraint on the tenderness of mutton, *J. Food Sci.*, 38, 932, 1973.
13. Herring, H.K. et al., Tenderness and associated characteristics of stretched and contracted bovine muscles, *J. Food Sci.*, 32, 317, 1967.
14. Wheeler, T.L., Shackelford, S.D., and Koohmaraie, M., Variation in proteolysis, sarcomere length, collagen content, and tenderness among major pork muscles, *J. Anim. Sci.*, 78, 958, 2000.
15. Simmons, N.J. et al., Integrated technologies to enhance meat quality, *Meat Sci.*, 74, 172, 2006.
16. Hoagland, R., McBryde, C.N., and Powick, W.C., Changes in fresh beef during cold storage above freezing, *U.S. Dept. Agric. Bull.*, 433, 1, 1917.
17. Koohmaraie, M. and Geesink, G.H., Contribution of postmortem muscle biochemistry to the delivery of consistent meat quality with particular focus on the calpain system, *Meat Sci.*, 74, 34, 2006.
18. Ouali, A. et al., Revisiting the conversion of muscle into meat and the underlying mechanisms, *Meat Sci.*, 74, 44, 2006.
19. Dayton, W.R. et al., A Ca^{2+}-activated protease possibly involved in myofibrillar protein turnover. Purification from porcine muscle, *Biochemistry*, 15, 2150, 1976.
20. Dayton, W.R. et al., A Ca^{2+}-activated protease possibly involved in myofibrillar protein turnover. Partial characterization of the purified enzyme, *Biochemistry*, 15, 2159, 1976.
21. Dayton, W.R. et al., A calcium activated protease possibly involved in myofibrillar protein turnover. Isolation of a low-calcium requiring for the protease, *Biochim. Biophys. Acta*, 659, 48, 1981.
22. Emori, Y. et al., Isolation and sequence analysis of cDNA clones for the small subunit of rabbit calcium-dependent protease, *J. Biol. Chem.*, 261, 9472, 1986.

23. Dayton, W.R., Comparison of the low- and high-calcium-requiring forms of the calcium-activated proteinase with their autolytic breakdown products. *Biochim. Biophys. Acta*, 709, 166, 1982.

24. Nagainis, P.A. et al., Autolysis of high-Ca^{2+} and low Ca^{2+}-forms of the Ca^{2+}-dependent proteinase from bovine skeletal muscle, *Fed. Proc.*, 42, 1780, 1983.

25. Suzuki, K. et al., Autolysis of calcium-activated neutral proteinase of chicken skeletal muscle, *J. Biochem.*, 90, 1787, 1981.

26. Suzuki, K. et al., Limited autolysis of Ca^{2+}-activated neutral protease (CANP) changes its sensitivity to Ca^{2+} ions, *J. Biochem.*, 90, 275, 1981.

27. Inomata, M. et al., Activation mechanism of calcium activated neutral protease, *J. Biol. Chem.*, 263, 19783, 1988.

28. Brown, N. and Crawford, C., Structural modifications associated with the change in Ca^{2+} sensitivity on activation of m-calpain, *FEBS Lett.*, 322, 65, 1993.

29. Sorimachi, H., Ishiura, S., and Suzuki, K., Molecular cloning of a novel mammalian calcium dependent protease distinct from both m- and μ-types, *J. Biol. Chem.*, 264, 20106, 1989.

30. Sorimachi, H. et al., Muscle-specific calpain, p94, responsible for limb girdle dystrophy type 2A, associates with connectin through IS2, a p94-specific sequence, *J. Biol. Chem.*, 270, 31158, 1995.

31. Sorimachi, H. et al., Muscle specific calpain, p94, is degraded by autolysis immediately after translation, resulting in disappearance from muscle, *J. Biol. Chem.*, 268, 10593, 1993.

32. Maki, M. et al., Analysis of structure-function relationship of pig calpastatin by expression of mutated cDNAs in *Escherichia coli, J. Biol. Chem.*, 263, 10254, 1988.

33. Lee, W.J. et al., Molecular diversity in amino-terminal domains of human calpastatin by exon skipping, *J. Biol. Chem.*, 267, 8437, 1992.

34. Cottin, P., Vidalenc, P.L., and Ducastaing, A., Ca^{2+}-dependent association between Ca^{2+}-activated neutral proteinase (CaANP) and its specific inhibitor, *FEBS Lett.*, 136, 221, 1981.

35. Imajoh, S. and Suzuki, K., Reversible interaction between Ca^{2+}-activated neutral protease (CANP) and its endogenous inhibitor, *FEBS Lett.*, 187, 47, 1985.

36. Doumit, M.E. and Koohmaraie, M., Immunoblot analysis of calpastatin degradation: evidence for cleavage by calpain in postmortem muscle, *J. Anim. Sci.*, 77, 1467, 1999.

37. Mellgren, R.L., Mericle, M.T., and Lane, R.D., Proteolysis of the calcium-dependent protease inhibitor by myocardial calcium-dependent protease, *Arch. Biochem. Biophys.*, 246, 233, 1986.

38. DeMartino, G.N. et al., Proteolysis of the protein inhibitor of calcium-dependent proteinases produces lower molecular weight fragments that retain inhibitory activity, *Arch. Biochem. Biophys.*, 262, 189, 1988.

39. Nakamura, M. et al., Fragmentation of an endogenous inhibitor upon complex formation with high- and low Ca^{2+}-requiring forms of calcium activated neutral proteases, *Biochemistry*, 28, 449, 1989.

40. Goll, D.E. et al., The calpain system, *Physiol. Rev.*, 83, 731, 2003.

41. Geesink, G.H. and Koohmaraie, M., Effect of calpastatin on degradation of myofibrillar proteins by μ-calpain under postmortem conditions, *J. Anim. Sci.*, 77, 2685, 1999a.

42. Huff-Lonergan, E. et al., Proteolysis of specific muscle structural proteins by μ-calpain at low pH and temperature is similar to degradation in postmortem bovine muscle, *J. Anim. Sci.*, 74, 993, 1996.

43. Koohmaraie, M., Schollmeyer, J.E., and Dutson, T.R., Effect of low-calcium-requiring calcium activated factor on myofibrils under varying pH and temperature conditions, *J. Food Sci.*, 51, 28, 1986.

44. Koohmaraie, M. et al., Acceleration of postmortem tenderization in ovine carcasses through activation of Ca^{2+}-dependent proteases, *J. Food Sci.*, 53, 1638, 1988.

45. Koohmaraie, M., Whipple, G., and Crouse, J.D., Acceleration of postmortem tenderization in lamb and Brahman-cross beef carcasses through infusion of calcium chloride, *J. Anim. Sci.*, 68, 1278, 1990.

46. Wheeler, T.L., Crouse, J.D., and Koohmaraie, M., The effect of postmortem time of injection and freezing on the effectiveness of calcium chloride for improving beef tenderness, *J. Anim. Sci.*, 70, 3451, 1992.

47. Koohmaraie, M., Inhibition of postmortem tenderization in ovine carcasses through infusion of zinc, *J. Anim. Sci.*, 68, 1476, 1990.
48. Uytterhaegen, L., Claeys, E., and Demeyer, D., Effects of exogenous protease effectors on beef tenderness and myofibrillar degradation and solubility, *J. Anim. Sci.*, 72, 1209, 1994.
49. Koohmaraie, M. et al., Postmortem proteolysis in longissimus muscle from beef, lamb and pork carcasses, *J. Anim. Sci.*, 69, 617, 1991.
50. Ouali, A. and Talmant, A., Calpains and calpastatin distribution in bovine, porcine and ovine skeletal muscles, *Meat Sci.*, 28, 331, 1990.
51. Shackelford, S.D. et al., An evaluation of tenderness of the longissimus muscle of Angus versus Brahman crossbred heifers, *J. Anim. Sci.*, 69, 171, 1991.
52. Whipple, G. et al., Evaluation of attributes that affect longissimus muscle tenderness in Bos taurus and Bos indicus cattle, *J. Anim. Sci.*, 68, 2716, 1990.
53. Garssen, G.J. et al., Effects of dietary clenbuterol and salbutamol on meat quality in veal calves, *Meat Sci.*, 40, 337, 1995.
54. Koohmaraie, M. et al., Effect of the ß-adrenergic agonist $L_{644,969}$ on muscle growth and meat quality traits, *J. Anim. Sci.*, 69, 4823, 1991.
55. Geesink, G.H. and Koohmaraie, M., Postmortem proteolysis and calpain/calpastatin activity in callipyge and normal lamb biceps femoris during extended storage, *J. Anim. Sci.*, 77, 1490, 1999.
56. Koohmaraie, M. et al., A muscle hypertrophy condition in lamb (callipyge): characterization of effects on muscle growth and meat quality traits, *J. Anim. Sci.*, 73, 3596, 1995.
57. Kent, M.P., Spencer, M.J., and Koohmaraie, M., Postmortem proteolysis is reduced in transgenic mice overexpressing calpastatin, *J. Anim. Sci.*, 82, 794, 2004.
58. Ducastaing, A. et al., Effects of electrical stimulation on post-mortem changes in the activities of two Ca dependent neutral proteinases and their inhibitor in beef muscle, *Meat Sci.*, 15, 193, 1985.
59. Kretchmar, D.H. et al., Alterations in postmortem degradation of myofibrillar proteins in lamb fed a β-adrenergic agonist, *J. Anim. Sci.*, 68, 1760, 1990.
60. Veiseth, E. et al., Effect of post-mortem storage on μ-calpain and m-calpain in ovine skeletal muscle, *J. Anim. Sci.*, 70, 3035, 2001.
61. Koohmaraie, M. et al., Effects of postmortem storage on Ca^{++}-dependent proteases, their inhibitor and myofibril fragmentation, *Meat Sci.*, 19, 187, 1987.
62. Andersson, L.V.B. et al., Characterization of monoclonal antibodies to calpain 3 and protein expression in muscle from patients with limb-girdle muscular dystrophy type 2 A., *Am. J. Pathol.*, 153, 1169, 1998.
63. Ilian, M.A., Bekhit, A.E.D., and Bickerstaffe, R., The relationship between meat tenderization, myofibril fragmentation and autolysis of calpain 3 during post-mortem aging, *Meat Sci.*, 66, 387, 2004.
64. Parr, T. et al., Relationship between skeletal muscle-specific calpain, p94/calpain 3 and tenderness on conditioned pork longissimus muscle, *J. Anim. Sci.*, 77, 661, 1999.
65. Geesink, G.H., Taylor, R.G., and Koohmaraie, M., Calpain 3/p94 is not involved in postmortem proteolysis, *J. Anim. Sci.*, 83, 1646, 2005.
66. Geesink, G.H. et al., μ-Calpain is essential for postmortem proteolysis of muscle proteins, *J. Anim. Sci.*, 84, 2834, 2006.
67. Gopalakrischna, R. and Barsky, S.H., Hydrophobic association of calpains with subcellular organelles, *J. Biol. Chem.*, 261, 13936, 1986.
68. Koohmaraie, M., Quantification of Ca^{2+}-dependent protease activities by hydrophobic and ion-exchange chromatography, *J. Anim. Sci.*, 68, 659, 1990.
69. Veiseth, E. and Koohmaraie, M., Effect of extraction buffer on estimating calpain and calpastatin activity in postmortem ovine muscle, *Meat Sci.*, 57, 325, 2001.
70. Whipple, G. and Koohmaraie, M., Freezing and calcium chloride marination effects on beef tenderness and calpastatin activity, *J. Anim. Sci.*, 70, 3081, 1992.
71. Duckett, S.K. et al., Effect of freezing on calpastatin activity and tenderness of callipyge lamb, *J. Anim. Sci.*, 76, 1869, 1998.

72. Kristensen, L., Christensen, M., and Ertbjerg, P., Activities of calpastatin, μ-calpain and m-calpain are stable during frozen storage of meat, *Meat Sci.*, 72, 116, 2006.
73. Shackelford, S.D. et al., Heritabilities and phenotypic and genetic correlation for bovine postrigor calpastatin activity, intramuscular fat content, Warner-Bratzler shear force, retail product yield, and growth rate, *J. Anim. Sci.*, 72, 857, 1994.
74. Koohmaraie, M., Ovine skeletal muscle multicatalytic proteinase complex (proteasome): purification, characterization, and comparison of its effects on myofibrils with μ-calpain, *J. Anim. Sci.*, 70, 3697, 1992.
75. Thompson, V.F. et al., A bodipy fluorescent microplate assay for measuring activity of calpain and other proteases, *Anal. Biochem.*, 279, 170, 2000.
76. Kent, M.P. et al., An assessment of extraction and assay techniques for quantification of calpain and calpastatin from small tissue samples, *J. Anim. Sci.*, 83, 2182, 2005.
77. Barrett, A.J., Cathepsin D and other carboxyl proteinases, in *Proteinases in Mammalian Cells and Tissues*, Barrett, A.J. (Ed.), North-Holland, Amsterdam, 1977, chap. 5.
78. Kirschke, H. and Barrett, A.J., Chemistry of lysosomal proteases, in *Lysosomes: Their Role in Protein Breakdown*, Glaumann, H. and Ballard, F.J. (Eds.), Academic Press, New York, 1987, chap. 6.
79. Turk, B., Turk, D., and Turk, V., Lysosomal cysteine proteases: more than scavengers, *Biochim. Biophys. Acta*, 1477, 98, 2000.
80. Turk, V., Turk, B., and Turk, D., Lysosomal cysteine proteases: facts and opportunities, *EMBO J.*, 20, 4629, 2001.
81. Turk, V. and Bode, W., The cystatins: protein inhibitors of cysteine proteinases, *FEBS Lett.*, 285, 213, 1991.
82. Kirschke, H., Langner, J., Wiederanders, B., Ansorge, S., and Bohley, P., Cathepsin L. A new proteinase from rat-liver lysosomes, *Eur. J. Biochem.*, 74, 293, 1977.
83. Bechet, D., Tassa, A., Taillandier, D., Combaret, L., and Attaix, D., Lysosomal proteolysis in skeletal muscle, *Int. J. Biochem. Cell Biol.*, 37, 2098, 2005.
84. Moeller, P.W., Fields, P.A., Dutson, T.R., Landmann, W.A., and Carpenter, Z.L., High temperature effects on lysosomal enzyme distribution and fragmentation of bovine muscle, *J. Food Sci.*, 42, 510, 1977.
85. Ertbjerg, P. et al., Relationship between proteolytic changes and tenderness in pre-rigor lactic acid marinated beef, *J. Sci. Food Agric.*, 79, 970, 1999.
86. O'Halloran, G.R., Troy, D.J., Buckley, D.J., and Reville, W.J., The role of endogenous proteases in the tenderisation of fast glycolysing muscle, *Meat Sci.*, 47, 187, 1997.
87. Dutson, T.R., Smith, G.C., and Carpenter, Z.L., Lysosomal enzyme distribution in electrically stimulated ovine muscle, *J. Food Sci.*, 45, 1097, 1980.
88. Wu, F.Y., Dutson, T.R., Valin, C., Cross, H.R., and Smith, S.B., Aging index, lysosomal enzyme activities, and meat tenderness in muscles from electrically stimulated bull and steer carcasses, *J. Food Sci.*, 50, 1025, 1985.
89. Pommier, S.A., Vitamin A, electrical stimulation, and chilling rate effects on lysosomal enzyme activity in aging bovine muscle, *J. Food Sci.*, 57, 30, 1992.
90. Maribo, H., Ertbjerg, P., Andersson, M., Barton-Gade, P., and Møller, A.J., Electrical stimulation of pigs—effect on pH fall, meat quality and cathepsin B+L activity, *Meat Sci.*, 52, 179, 1999.
91. Ertbjerg, P., Henckel, P., Karlsson, A., Larsen, L.M., and Møller, A.J., Combined effect of epinephrine and exercise on calpain/calpastatin and cathepsin B and L activity in porcine longissimus muscle, *J. Anim. Sci.*, 77, 2428, 1999.
92. Aoki, T. and Ueno, R., Involvement of cathepsins B and L in the post-mortem autolysis of mackerel muscle, *Fd. Res. Int.*, 30, 585, 1997.
93. Ogata, H., Aranishi, F., Hara, K., Osatomi, K., and Ishihara, T., Proteolytic degradation of myofibrillar components by carp cathepsin L, *J. Sci. Food Agric.*, 76, 499, 1998.
94. Ladrat, C., Verrez-Bagnis, V., Noël, J., and Fleurence, J., In vitro proteolysis of myofibrillar and sarcoplasmic proteins of white muscle of sea bass (*Dicentrarchus labrax* L.): effects of cathepsins B, D and L, *Food Chem.*, 81, 517, 2003.

95. Chéret, R., Delbarre-Ladrat, C., de Lamballerie-Anton, M., and Verrez-Bagnis, V., High-pressure effects on the proteolytic enzymes of sea bass (*Dicentrarchus labrax* L.) fillets, *J. Agric. Food Chem.*, 53, 3969, 2005.

96. Jung, S., Ghoul, M., and de Lamballerie-Anton, M., Changes in lysosomal enzyme activities and shear values of high pressure treated meat during ageing, *Meat Sci.*, 56, 239, 2000.

97. Kubo, T., Gerelt, B., Han, G.D., Sugiyama, T., Nishiumi, T., and Suzuki, A., Changes in immuno-electron microscopic localization of cathepsin D in muscle induced by conditioning or high-pressure treatment, *Meat Sci.*, 61, 415, 2002.

98. Barrett, A.J., Cathepsin D. Purification of isoenzymes from human and chicken liver, *Biochem. J.*, 117, 601, 1970.

99. Barrett, A.J., Human cathepsin B1. Purification and some properties of the enzyme, *Biochem. J.*, 131, 809, 1973.

100. Kirschke, H., Wood, L., Roisen, F.J., and Bird, J.W.C., Activity of lysosomal cysteine proteinase during differentiation of rat skeletal muscle, *Biochem. J.*, 214, 871, 1983.

101. Ertbjerg, P., Larsen, L.M., and Møller, A.J., Effect of pre-rigor lactic acid treatment on lysosomal enzyme release in bovine muscle, *J. Sci. Food Agric.*, 79, 95, 1999.

102. Koohmaraie, M. and Kretchmar, D.H., Comparisons of four methods for quantification of lysosomal cysteine proteinase activities, *J. Anim. Sci.*, 68, 2362, 1990.

103. Kristensen, L., Therkildsen, M., Aaslyng, M.D., Oksbjerg, N., and Ertbjerg, P., Compensatory growth improves meat tenderness in gilts but not in barrows, *J. Anim. Sci.*, 82, 3617, 2004.

104. Barrett, A.J., Fluorimetric assays for cathepsin B and cathepsin H with methylcoumarylamide substrates, *Biochem. J.*, 187, 909, 1980.

105. Anson, M.L., The estimation of pepsin, trypsin, papain, and cathepsin with hemoglobin, *J. Gen. Physiol.*, 22, 79, 1938.

106. García-Garrido, J.A. et al., Activity of cathepsin B, D, H and L in Spanish dry-cured ham of normal and defective texture, *Meat Sci.*, 56, 1, 2000.

107. Chiechanover, A., Orian, A., and Schwartz, A.L., Ubiquitinin-mediated proteolysis: biological regulation via destruction, *Bioessays*, 22, 442, 2000.

108. Rechsteiner, M. and Hill, C.P., Mobilizing the proteolytic machine: cell biological roles of proteasome activators and inhibitors, *Trends Cell Biol.*, 15, 27, 2005.

109. Heinemeyer, W. et al., The active sites of the eukaryotic 20S proteasome and their involvement in subunit precursor processing, *J. Biol Chem.*, 272, 25200, 1997.

110. Taylor, R.G. et al., Proteolytic activity of proteasome on myofibrillar structures, *Mol. Biol. Reports*, 21, 71, 1995.

111. Robert, N. et al., The effect of proteasome on myofibrillar structures in bovine skeletal muscle, *Meat Sci.*, 51, 149, 1999.

112. Matsuishi, M. and Okitani, A., Proteasome from rabbit skeletal muscle: some properties and effects on muscle proteins, *Meat Sci.*, 45, 451, 1997.

113. Otsuka, Y. et al., Purification and properties of rabbit muscle proteasome, and its effect on myofibrillar structure, *Meat Sci.*, 49, 365, 1998.

114. Dutaud, D. et al., Bovine muscle 20S proteasome. II: Contribution of the 20S proteasome to meat tenderization as revealed by an ultrastructural approach, *Meat Sci.*, 74, 337, 2006.

115. Aubry, L. et al., Bovine muscle 20S proteasome. III: Quantification in tissue crude extracts using ELISA and radial immunodiffusion techniques and practical applications, *Meat Sci.*, 74, 345, 2006.

116. Fuentes-Prior, P. and Salvesen, G.S., The protein structures that shape caspase-activity, specificity, activation and inhibition, *Biochem. J.*, 384, 201, 2004.

117. Danial, N.N. and Korsmeyer, S.J., Cell death: critical control points, *Cell*, 116, 205, 2004.

118. Yuan, J. et al., The C. elegans cell death gene ced-3 encodes a protein similar to mammalian interleukin-beta-converting enzyme, *Cell*, 75, 641, 1993.

119. Kemp, C.M. et al., Comparison of the relative expression of caspase isoforms in different porcine skeletal muscles, *Meat Sci.*, 73, 426, 2006.

120. Kemp, C.M., Bardsley, R.G., and Parr, T., Changes in caspase activity during the postmortem conditioning period and its relationship to shear force in porcine longissimus muscle, *J. Anim. Sci.*, 84, 2841, 2006.

121. Koohmaraie, M., Biochemical factors regulating the toughening and tenderization processes of meat, *Meat Sci.*, 43, S193, 1996.

122. Wang, K.K.W. et al., Caspase-mediated fragmentation of calpain inhibitor protein calpastatin during apoptosis, *Arch. Biochem. Biophys.*, 356, 187, 1998.

123. Kato, M. et al., Caspases cleave the amino-terminal calpain inhibitory unit of calpastatin during apoptosis in human Jurkat T cells, *J. Biochem.*, 127, 297, 2000.

124. Shi, Y. et al., Downregulation of the calpain inhibitor protein calpastatin by caspases during renal ischemia-reperfusion, *Am. J. Physiol. Renal Physiol.*, 279, F509, 2000.

125. Barnoy, S. and Kosower, N.S., Caspase-1-induced calpastatin degradation in myoblast differentiation and fusion: cross-talk between the caspase and calpain systems, *FEBS Lett.*, 546, 213, 2003.

126. Vaisid, T., Kosower, N.S., and Barnoy, S., Caspase-1 activity is required for neuronal differentiation of PC12 cells: cross-talk between the caspase and calpain systems, *Biochim. Biophys. Acta*, 1743, 223, 2005.

127. Woessner, J.F. Jr., Matrix metalloproteinases and their inhibitors in connective tissue remodelling, *FASEB J.*, 5, 2145, 1991.

128. Birkedal-Hansen, H., Proteolytic remodeling of extracellular matrix, *Curr. Opin. Cell Biol.*, 7, 728, 1995.

129. Nagase, H. and Woessner, J.F. Jr., Matrix metalloproteinases, *J. Biol. Chem.*, 274, 21491, 1999.

130. Carmeli, E. et al., Matrix metalloproteinases and skeletal muscle: a brief review, *Muscle Nerve*, 29, 191, 2004.

131. Lewis, M.P. et al., Gelatinase-B (matrix metalloproteinase-9; MMP-9) secretion is involved in the migratory phase of human and murine muscle cell cultures, *J. Muscle Res. Cell Motil.*, 21, 223, 2000.

132. Reznick, A.Z. et al., Expression of matrix metalloproteinases, inhibitor, and acid phosphatise in muscles of immobilized hindlimbs of rats, *Muscle Nerve*, 27, 51, 2003.

133. Balcerzak, D. et al., Coordinate expression of matrix-degrading proteinases and their activators and inhibitors in bovine skeletal muscle, *J. Anim. Sci.*, 79, 94, 2001.

134. Sylvestre, M.N. et al., Elevated rate of collagen solubilization and postmortem degradation in muscles of lambs with high growth rates: possible relationship with activity of matrix metalloproteinases, *J. Anim. Sci.*, 80, 1871, 2002.

135. Lødemel, J.B. and Olsen, R.L., Gelatinolytic activities in muscle of Atlantic cod (*Gadus morhua*), spotted wolffish (*Anarhichas minor*) and Atlantic salmon (*Salmo salar*), *J. Sci. Food Agric.*, 83, 1031, 2003.

136. Lødemel, J.B. et al., Detection of TIMP-2-like protein in Atlantic cod (*Gadus morhua*) muscle using two-dimensional real-time reverse zymography, *Comp. Biochem. Physiol.*, 139B, 253, 2004.

137. Herron, G.S. et al., Secretion of metalloproteinases by stimulated capillary endothelial cells, *J. Biol. Chem.*, 261, 2814, 1986.

138. Oliver, G.W. et al., Quantitative reverse zymography: analysis of picogram amounts of metalloproteinase inhibitors using gelatinase A and B reverse zymograms, *Anal. Biochem.*, 244, 161, 1997.

139. Hattori, S. et al., Real-time zymography and reverse zymography: a method for detecting activities of matrix metalloproteinases and their inhibitors using FITC-labeled collagen and casein as substrates, *Anal. Biochem.*, 301, 27, 2002.

140. Lombard, C. et al., Assays of matrix metalloproteinases (MMPs) activities: a review, *Biochimie*, 87, 265, 2005.

Chapter 7

Muscle Enzymes: Exopeptidases and Lipases

Fidel Toldrá, Miguel A. Sentandreu, and Mónica Flores

Contents

7.1   Introduction

Exopeptidases are proteolytic enzymes that are able to cleave small peptides or free amino acids from proteins and peptides. There are two important groups of exopeptidases in muscle: dipeptidyl peptidases and aminopeptidases.

Dipeptidyl peptidases (DPP) constitute a group of enzymes capable of releasing different dipeptide sequences from the N-termini of proteins and peptides. Four different DPP activities have been described in mammalian tissues: DPP I (EC 3.4.14.1) and DPP II (EC 3.4.14.2) located in the lysosomes; DPP III (EC 3.4.14.4) located in the cytosol; and DPP IV (EC 3.4.14.5), which is present in the cell membrane.[1]

Muscle aminopeptidases constitute another group of exopeptidases, which have important roles in the proteolysis phenomena in the muscles. Aminopeptidases catalyze the cleavage of amino acids from the amino terminus of protein and peptide substrates. The classification of these enzymes is based on their efficiency against specific residues, location, susceptibility to inhibition by bestatin, metal ion content, and pH of activity.[2] Five different aminopeptidase activities have been described in mammalian tissues: leucyl aminopeptidase (LAP) (EC 3.4.11.1), aminopeptidase B or arginyl aminopeptidase (RAP) (EC 3.4.11.6), pyroglutamyl aminopeptidase (PGAP) (EC 3.4.11.8), alanyl aminopeptidase (AAP) (EC 3.4.11.14), and methionyl aminopeptidase (MAP) (EC 3.4.11.18), all of them located in the cytosol.[1,3–5]

Muscle exopeptidases have an important role in the regulation of protein turnover and in postmortem muscle where they are involved in the latest steps of the proteolytic chain. DPP, together with muscle aminopeptidases, are responsible for the complete breakdown of polypeptides and peptide fragments resulting from proteins, leading to an accumulation of free amino acids and dipeptides and contributing to flavor development in processed meats.[6,7]

Muscle lipases also play an important role in the lipolysis of the postmortem meat and during the processing of meat products. Lysosomal acid lipase is the most important lipase of muscle with an acid optimum pH. This enzyme is located in lysosomes and is responsible for the hydrolysis of triacylglycerols and the release of fatty acids at positions 1 and 3 that can contribute to flavor development and lipid oxidation.[6] It has been recently reported that the metabolism of triacylglycerols, which is regulated by acid lipases in muscle tissues, is not affected by conjugated linolenic acid (CLA) levels in the feeds.[8]

7.2   Assay of Dipeptidyl Peptidases Activity

DPP have been completely purified from porcine skeletal muscle and characterized for optimal assay conditions.[9–12] Some of these peptidases are referred to by other names in the literature. For instance, DPP I is also known as dipeptidyl aminopeptidase I or dipeptidyl arylamidase I,[13]

DPP II is also known as dipeptidyl aminopeptidase II or dipeptidyl arylamidase II,[14] DPP III is also known as dipeptidyl arylamidase III, red cell angiotensinase, neutral angiotensinase, or enkephalinase B,[15] whereas DPP IV has received different names such as dipeptidyl aminopeptidase IV, postproline dipeptidyl aminopeptidase IV, X-proline dipeptidyl aminopeptidase, or glycylproline naphthylamidase.[16]

The methods for the assay of DPP activity basically differ on the reaction conditions for each particular enzyme as described later. For all these enzymes, there are several methodologies that can be used depending on the available equipment in the laboratory and the type of substrate to be used. The use of fluorescence, based on the use of substrates with the fluorescent group 7-amino-4-methylcoumarin, gives much more sensitivity than color measurement, based on substrates with the colorimetric group paranitroanilide.

7.2.1 Extraction

Samples of muscle (between 5 and 25 g), free from visible fat and connective tissue, are homogenized 1:10 in either 100 mM citric acid buffer, pH 5 (DPP I and DPP III) or 100 mM sodium phosphate buffer, pH 7 (DPP II and DPP IV). Homogenization is achieved by using an Ultra Turrax or a polytron homogenizer, 3 strokes of 10 s each at 24,000 rpm, while cooling in ice. Homogenates are centrifuged at 17,000 × *g* for 20 min, and respective supernatants are filtered through glass wool and collected for further enzyme assays. All operations are carried out at 4°C.

7.2.2 Dipeptidyl Peptidase I (EC 3.4.14.1)

This enzyme needs reducing conditions for activity. The use of 5 mM of dithiothreitol (DTT) is recommended (see Figure 7.1). The enzyme is inhibited by thiol inhibitors such as E-64 and

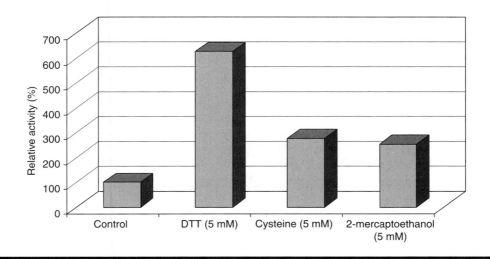

Figure 7.1 Activation of DPP I activity by addition of 5 mM of DTT, cysteine, or 2-mercaptoethanol to the reaction media. (Adapted from Sentandreu, M.A. and Toldrá, F., *J. Agric. Food Chem.*, 48, 5014–5022, 2000.)

iodoacetic acid (0.5 mM), 3,4-dichloroisocoumarin (0.5 mM), and Co^{2+} (0.5 mM). DPP I shows good stability on exposure to heat. If necessary, when DPP I activity has to be assayed in crude extracts, the extracts can be submitted to heat treatment (55°C for 30 min) before its activity assay to eliminate interferences from other enzyme activities. Examples of assay techniques, based on color, fluorescence, or peptide determination, are discussed in the following sections.

7.2.2.1   Color Determination

This type of assay can be performed using colorimetry microtiter plates. The enzyme solution (50 μL) is added to 250 μL of reaction buffer containing the substrate; the plate is shaken and immediately read in a microtiter plate colorimeter at $\lambda_{abortion} = 410$ nm when using *p*-nitroanilide substrate derivatives (Table 7.1). After incubation for 20 min at 37°C, the plate is read again under the same conditions and the activity calculated. Several determinations can be carried out along the entire incubation period for kinetics assays.

7.2.2.2   Fluorescence Determination

In this case, fluorescent microtiter plates must be used. The enzyme solution (50 μL) is added to 250 μL of reaction buffer, the plate is shaken and immediately read in a microtiter plate fluorimeter at $\lambda_{excitation} = 355$ nm and $\lambda_{emission} = 460$ nm when using 7-amino-4-methylcoumarin derivatives (Table 7.1). After incubation for 20 min at 37°C, the plate is read again in the same conditions and the activity calculated. Several determinations along the entire incubation period can be carried out for kinetics assays.

7.2.2.3   Peptide Determination

In this case, the generation of dipeptides as product of the enzyme reaction on natural peptides used as substrates (Table 7.1) is determined by free solution capillary electrophoresis. The enzyme solution (100 μL) is added to 500 μL of reaction buffer in Eppendorf tubes, and immediately shaken. The Eppendorf tubes are incubated at 37°C and aliquots (40 μL) are taken every 30 min, including a control at time zero, and the reaction stopped by adding 10 μL of 0.6 M acetic acid. Samples are injected for 4 s by vacuum in the free solution capillary electrophoresis and run at +30 kV at 35°C in a standard 72 cm capillary (50 μm internal diameter and 50 cm to detector)

Table 7.1   Types of Available Substrates (0.5 mM Final Concentration) for the Assay of DPP I and Its Relative Activity When Assayed in a Reaction Buffer Containing 50 mM Na$^+$ Acetate/Acetic Acid Buffer, pH 5.5, and 5 mM DTT

Type of Substrate	Substrate	Enzyme Activity (Relative Activity, %)
Colorimetric	Ala-Ala-pNA	11
Fluorescent	Ala-Arg-AMC	100
	Gly-Arg-AMC	81
Peptides	Val-Tyr-Ile-His-Pro-Phe	100
	Tyr-Gly-Gly-Phe-Leu	68

Source: Adapted from Sentandreu, M.A. and Toldrá, F., *J. Agric. Food Chem.*, 48, 5014–5022, 2000.

in 20 mM citric acid, pH 2.5. Eluting peptides are detected by ultraviolet (UV) absorption at 200 nm.

7.2.3 Dipeptidyl Peptidase II (EC 3.4.14.2)

DPP II is inhibited by sulfonyl fluorides like phenyl methyl sulfonyl fluoride (PMSF) or Pefabloc SC (0.5 mM), Diprotin A (Ile-Pro-Ile) (0.5 mM), and Cu^{2+} (0.5 mM). Despite being considered a serine peptidase inhibitor, 3,4-DCI does not remarkably inhibit DPP II activity. Assay techniques, based on color, fluorescence, or peptide determination, are the same as for DPP I except for the presence of 0.04 mM bestatin instead of DTT (Table 7.2). Caution must be taken against monovalent cations of salts commonly employed for buffers because these cations cause remarkable inhibition of DPP II activity (Figure 7.2). Thus, attention must be paid to the final cation concentration in the reaction buffer.

Table 7.2 Types of Available Substrates (0.5 mM Final Concentration) for the Assay of DPP II and Its Relative Activity When Assayed in a Reaction Buffer Containing 50 mM Na⁺ Acetate/Acetic Acid Buffer, pH 5.5, and 0.04 mM Bestatin

Type of Substrate	Substrate	Enzyme Activity (Relative Activity, %)
Colorimetric	Arg-Pro-pNA	100
	Gly-Pro-pNA	95
	Ala-Ala-pNA	81
Fluorescent	Gly-Pro-AMC	17
	Lys-Ala-AMC	7
Peptides	Met-Ala-Ser	100
	Gly-Pro-Ala	96
	Ala-Ala-Ala	54

Source: Adapted from Sentandreu, M.A. and Toldrá, F., *Meat Sci.*, 57, 93–103, 2001.

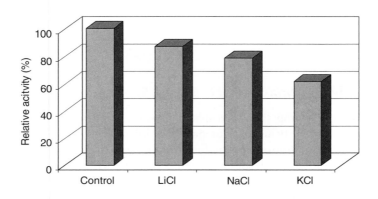

Figure 7.2 Effect of the presence of 50 mM of LiCl, NaCl, and KCl in the reaction media on porcine muscle DPP II activity. Activity without salt addition (first column) was taken as 100%. (Adapted from Sentandreu, M.A. and Toldrá, F., *Meat Sci.*, 57, 93–103, 2001.)

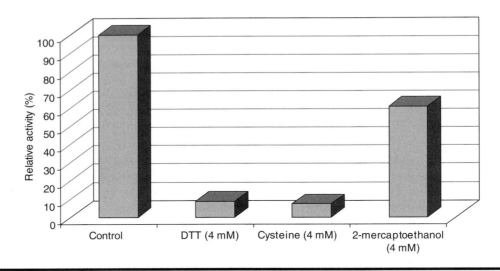

Figure 7.3 **Effect of 4 mM reducing agents DTT, cysteine, or 2-mercaptoethanol on the activity of DPP III. (Adapted from Sentandreu, M.A. and Toldrá, F., *J. Agric. Food Chem.*, 46, 3977–3984, 1998.)**

7.2.4 Dipeptidyl Peptidase III (EC 3.4.14.4)

Contrary to DPP I, reducing agents such as cysteine, DTT, and mercaptoethanol (4 mM) cause a strong inhibition of DPP III activity (Figure 7.3); thus, their use must be avoided in reaction buffers. This enzyme is also inhibited by chelating agents such as *o*-phenanthroline (0.8 mM), EDTA and EGTA (4 mM); 3 *p*-chloromercuribenzoate and DTNB (0.05 mM); 4-DCI (0.05 mM), Hg^{2+} (0.05 mM), and Cd^{2+} (0.2 mM), and the dipeptide Tyr-Tyr (0.4 mM). DPP III is activated by Co^{2+} (0.05 mM). The enzyme is sensitive to temperature and is inactivated in just 30 min when incubated at 55°C.

Assay techniques, based on color, fluorescence, or peptide determination, are the same as for DPP I, except pH 8, the presence of 0.05 mM Co^{2+}, and the absence of any reducing agent in the reaction buffer (Table 7.3).

7.2.5 Dipeptidyl Peptidase IV (EC 3.4.14.5)

DPP IV is inhibited by diisopropylfluorophosphate, puromycin (0.05 mM), diprotin A (0.5 mM), and Co^{2+} (0.5 mM). Despite being a serine peptidase inhibitor, 3,4-DCI does not remarkably inhibit DPP IV activity, as it also happens for DPP II. This enzyme is considered as a serine peptidase due to its sensitivity to diisopropylfluorophosphate, which is confirmed by numerous sequencing studies of its primary structure catalytic triad (Ser/Asp/His) by cloning and sequencing of the respective cDNA.[17–19] Assay techniques, based on color, fluorescence, or peptide determination, are the same as for DPP I except that pH is 8 (Table 7.4).

Table 7.3 Types of Available Substrates for the Assay of DPP III and Its Relative Activity When Assayed in a Reaction Buffer Containing 50 mM Na+ Tetraborate/K+ Phosphate Buffer, pH 8, and 0.05 mM Co^{2+}

Type of Substrate	Substrate	Enzyme Activity (Relative Activity, %)
Colorimetric	Ala-Ala-pNA	50
Fluorescent	Arg-Arg-AMC	100
	Ala-Arg-AMC	69
Natural peptides	Gly-Gly-Phe-Leu	100
	Phe-Gly-Gly-Phe	58
	Arg-Phe-Arg-Ser	54
	Arg-Arg-Lys-Ala-Ser-Gly-Pro	37
	Val-Gly-Ser-Glu	31
	Tyr-Gly-Gly-Phe-Leu	31
	Arg-Ser-Arg-His-Phe	29

Source: Adapted from Sentandreu, M.A. and Toldrá, F., *J. Agric. Food Chem.*, 46, 3977–3984, 1998.

Table 7.4 Types of Available Substrates for the Assay of DPP IV and Its Relative Activity When Assayed in a Reaction Buffer Containing 50 mM Tris-Base Buffer, pH 8, and 5 mM DTT

Type of Substrate	Substrate	Enzyme Activity (Relative Activity, %)
Colorimetric	Arg-Pro-pNA	100
	Gly-Pro-pNA	66
	Ala-Ala-pNA	14
Fluorescent	Gly-Pro-AMC	46
	Lys-Ala-AMC	6
Peptides	Arg-Pro-Lys-Pro	100
	Arg-Pro-Pro-Gly-Phe	43
	Met-Ala-Ser	18

Source: Adapted from Sentandreu, M.A. and Toldrá, F., *Food Chem.*, 75, 159–168, 2001.

7.3 Assay of Aminopeptidase Activity

Several aminopeptidases, AAP, MAP, and RAP, have been purified from porcine skeletal muscle and characterized,[20–22] whereas LAP and PGAP have been isolated from human muscle.[23] Aminopeptidases have received different names due to the multiple substrate specificity of many of these enzymes. In general, the term arylamidase has also been used to describe the aminopeptidase activity. LAP has been named leucine aminopeptidase or cytosol aminopeptidase. RAP or arginyl aminopeptidase is generally recognized as aminopeptidase B and also named as Cl-activated

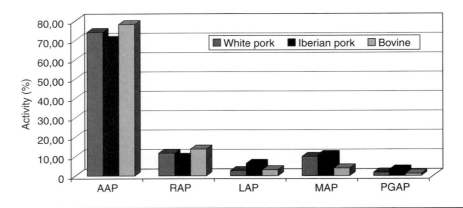

Figure 7.4 Aminopeptidase activity (%) in different mammalian species (white pork, Iberian pork, and bovine) measured using fluorescent substrates.

arginine aminopeptidase and arylamidase II. AAP has been named thiol-activated aminopeptidase, puromycin-sensitive aminopeptidase, soluble AAP, aminopeptidase C, and cytosol aminopeptidase III. AAP has been also known as major aminopeptidase due to its major activity, in relation to the other aminopeptidases, in the muscle (see Figure 7.4). MAP has been named peptidase M and aminopeptidase H. Finally, PGAP has been named pyroglutamyl peptidase I.[1] Typical procedures for the assay of muscle aminopeptidases are discussed in the following sections.

7.3.1 Extraction

Samples of ground muscle (between 5 and 25 g), free of fat and visible connective tissue, are homogenized 1:5 in 50 mM phosphate buffer, pH 7.5, containing 5 mM of EGTA. Homogenization is achieved by using an Ultra Turrax or a polytron homogenizer, 3 strokes of 10 s each at 24,000 rpm, while cooling in ice. Homogenates are centrifuged at $10,000 \times g$ for 20 min, and the supernatant is filtered through glass wool and collected for further enzyme assays. All operations are carried out at 4°C.

7.3.2 Leucyl Aminopeptidase (EC 3.4.11.1)

LAP is a zinc metalloenzyme with an alkaline optimal pH, which hydrolyzes preferably Leu but may hydrolyze other hydrophobic amino acids such as Met as well (Table 7.5).[4,23,24] LAP is activated by heavy metal ions but unaffected by the addition of NaCl (Figure 7.5). Its activity can be measured in a reaction buffer containing 50 mM borate buffer, pH 9.5, with 5 mM $MgCl_2$ and including 0.25 mM Leu-AMC or 0.5 mM Leu-pNa as substrate. The activity is measured at its optimal basic pH that avoids the interference of other aminopeptidases usually active at neutral pH.

7.3.2.1 Fluorescence Determination

This can be performed by using 7-amino-4-methyl coumarin (AMC) substrates and fluorescence multiwell plates. Different substrates are shown in Table 7.5. The enzyme solution (50 μL) is added

Table 7.5   Types of Available Fluorescent and Colorimetric Substrates for the Assay of Muscle Aminopeptidases and Its Relative Activity When Assayed at Respective Optimal Conditions

Type of Substrate	AAP (%)	RAP (%)	MAP (%)	PGAP (%)	LAP (%)
Fluorescent					
Phe-AMC	210.0	8.0	36	—	—
Lys-AMC	130.0	42.0	81	—	—
Met-AMC	124.0	2.1	100	—	50
Ala-AMC	100.0	2.5	57	—	—
Leu-AMC	98.0	0.2	48	—	100
Arg-AMC	64.0	100.0	31	—	—
Tyr-AMC	10.7	—	19	—	—
Ser-AMC	7.3	—	19	—	—
Pro-AMC	5.8	5.1	0	—	—
Gly-AMC	5.0	—	26	—	—
Val-AMC	3.3	—	3	—	—
pGlu-AMC	0.0	—	6	100	—
N-CBZ-Phe-Arg-AMC	0.0	0.0	5	—	—
Z-Arg-Arg-AMC	0.0	0.0	4	—	—
Colorimetric					
Leu-pNA			47		
Arg-pNA		150.0	43		
Ala-pNA	25.0		19		
Glu-pNA			26		
pGlu-pNA				20	

Source: Adapted from Flores, M., Aristoy, M.C. and Toldrá, F., *Z. Lebens. Unter. Forsch. A*, 205, 343–346, 1997; Flores, M., Aristoy, M.C. and Toldrá, F., *Biochimie*, 75, 861–867, 1993; Flores, M., Aristoy, M.C. and Toldrá, F., *J. Agric. Food Chem.*, 44, 2578–2583, 1996; Flores, M., Marina, M. and Toldrá, F., *Meat Sci.*, 56, 247–254, 2000.

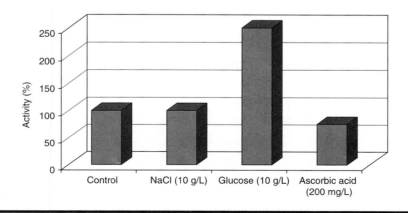

Figure 7.5   Activity of LAP measured in the optimum reaction buffer using Leu-AMC as substrate in the presence of NaCl, glucose, and ascorbic acid. (Adapted from Flores, M., Aristoy, M.C. and Toldrá, F., *Z. Lebens. Unter. Forsch. A*, 205, 343–346, 1997.)

Table 7.6   Kinetic Parameters, K_m (mM) and V_{max} (μmol/h mg), of Muscle Aminopeptidases Using Different Fluorescent and Colorimetric Substrates and Assayed under Respective Optimal Conditions

	AAP		RAP		MAP		PGAP	
	K_m	V_{max}	K_m	V_{max}	K_m	V_{max}	K_m	V_{max}
Fluorescent Substrate								
Ala-AMC	0.240	0.560			0.198	2.95		
Arg-AMC			0.188	0.071	0.259	1.58		
Leu-AMC					0.037	2.60		
Met-AMC					0.078	2.32		
pGlu-AMC							0.034	0.039
Colorimetric Substrate								
Ala-pNa	0.681	0.007	2.47	0.219	0.252	3.56		
Arg-pNa								
Leu-pNa					0.120	3.56		
pGlu-pNa							0.488	0.004

to 250 µL of reaction buffer; the plate is shaken and immediately read in a microtiter plate fluorimeter at $\lambda_{excitation}$ = 355 nm and $\lambda_{emission}$ = 460 nm. After incubation for 15 min at 37°C, the plate is read again under the same conditions and the activity calculated. Several determinations at fixed time intervals can be done for kinetics assays. Kinetic constants are shown in Table 7.6.

7.3.2.2   Color Determination

This type of assay can be performed using as substrate paranitroanilide (pNa) derivatives and colorimetric microtiter plates. The enzyme solution (50 µL) is added to 250 µL of reaction buffer; the plate is shaken and immediately read in a microtiter plate colorimeter at $\lambda_{abortion}$ = 410 nm. After incubation for 20 min at 37°C, the plate is read again under the same conditions and the activity calculated. Several determinations can be performed along the incubation time for kinetics assays. Kinetic constants are shown in Table 7.6.

7.3.2.3   Peptide Determination

In this case, the reduction of peptide as substrate of the enzyme reaction is determined by free solution capillary electrophoresis. The enzyme solution (100 µL) is added to 500 µL of reaction buffer in Eppendorf tubes and immediately shaken. The Eppendorf tubes are incubated at 37°C and aliquots (40 µL) are taken every 30 min, including a control at time zero, and the reaction stopped by adding 10 µL of 0.5 M sodium acetate/chloroacetate buffer, pH 4.3. Samples are injected for 4 s by vacuum in the free solution capillary electrophoresis and run at +30 kV at 35°C in a standard 72 cm capillary (50 µm internal diameter and 50 cm to detector) in 20 mM citric acid, pH 2.5. Eluting peptides are detected by UV absorption at 200 nm.

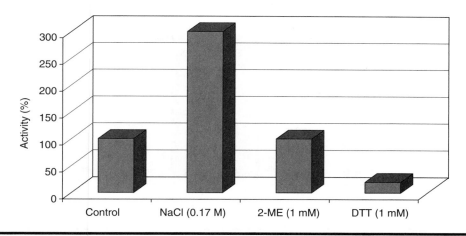

Figure 7.6   Activity of RAP measured in the optimum reaction buffer using Arg-AMC as substrate in the presence of NaCl, 2-mercaptoethanol (2-ME), and DTT. (Adapted from Flores, M., Aristoy, M.C. and Toldrá, F., *Biochimie*, 75, 861–867, 1993.)

7.3.3   Aminopeptidase B (EC 3.4.11.6)

Aminopeptidase B or arginyl aminopeptidase (RAP) is a chloride-activated enzyme hydrolyzing basic amino acids (Table 7.5).[20,25] It is inhibited by bestatin and not activated by reducing agents (Figure 7.6). Its activity can be measured in a reaction buffer containing 50 mM phosphate buffer, pH 6.5, with 0.2 M NaCl and including as substrate 0.1 mM Arg-AMC. The activity may be measured in crude extracts in the presence of 0.25 mM puromycin to inhibit the activity of the main muscle aminopeptidase, AAP, that interferes in the assay.

Assay techniques based on fluorescence and color determinations are the same as for LAP, but use 0.1 mM Arg-AMC or 2.5 mM Arg-pNa as substrate, respectively.

7.3.4   Alanyl Aminopeptidase (EC 3.4.11.14)

AAP is the main aminopeptidase in porcine, bovine, and human skeletal muscles[3,21] and has a broad substrate specificity (Table 7.5). It is inhibited by bestatin and puromycin and is also slightly inhibited by the addition of NaCl, whereas it is activated by sulfhydryl compounds and cobalt and calcium ions (Figure 7.7). Its activity can be measured in a reaction buffer containing 50 mM phosphate buffer, pH 6.5, and 2 mM 2-mercaptoethanol, including 0.1 mM Ala-AMC as substrate. The activity may be measured in crude extracts in the absence of inhibitors, as no interferences with other aminopeptidases have been described.

Assay techniques based on fluorescence and color determinations are the same as for LAP using 0.1 mM Ala-AMC or 1.0 mM Ala-pNa as substrate, respectively.

7.3.4.1   Peptide Determination

In this case, the generation of amino acids as the product of the enzyme reaction over peptides is determined by capillary electrophoresis. The enzyme solution (100 μL) is added to 500 μL of reaction buffer in Eppendorf tubes and is immediately shaken. Then, an aliquot (40 μL) is taken

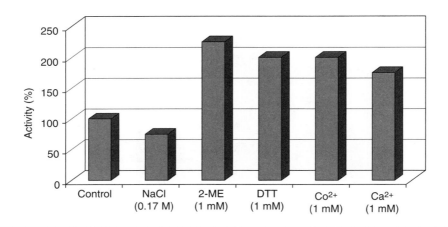

Figure 7.7 Activity of AAP measured in the optimum reaction buffer using Ala-AMC as substrate in the presence of NaCl, 2-mercaptoethanol (2-ME), DTT, cobalt, and calcium ions. (Adapted from Flores, M., Aristoy, M.C. and Toldrá, F., *J. Agric. Food Chem.*, 44, 2578–2583, 1996.)

Table 7.7   Kinetic Parameters, K_m (mM) and V_{max} (μmol/h mg), of Muscle AAP Using Different Peptide Substrates and Assayed under Optimal Conditions

Peptide	K_m	V_{max}
Ala-Tyr	0.36	0.64
Ala-Phe	0.39	0.92
Ala-Trp	0.72	1.64
Ala-Met	1.22	2.81
Ala-Leu	nh	nh
Ala-Gly	nh	nh
Leu-Phe	0.13	0.28
Leu-Met	0.61	3.06
Ala-Ala	nh	nh
TriAla	0.07	10.10
TetraAla	0.87	8.10
PentaAla	1.30	9.30

Note: nh—not hydrolyzed.

at different times of incubation (45°C) and the reaction stopped by adding 10 μL of 0.5 M sodium acetate/chloroacetate buffer. A capillary electrophoresis system is used for the analysis and the detection is done by UV absorption at 200 nm. Kinetics parameters obtained against different peptides are shown in Table 7.7.

7.3.5   Methionyl Aminopeptidase (EC 3.4.11.18)

MAP has high aminopeptidase and low endopeptidase activities (Table 7.5).[22,26,27] However, it is the only aminopeptidase that is not inhibited by bestatin although puromycin inhibits its activity. It is activated by sulfhydryl compounds and calcium ions.[28] Addition of NaCl also does not affect

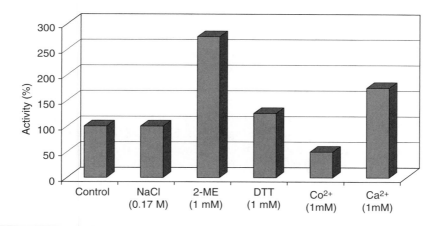

Figure 7.8 Activity of MAP measured in the optimum reaction buffer using Ala-AMC as substrate in the presence of NaCl, 2-mercaptoethanol (2-ME), DTT, cobalt, and calcium ions. (Adapted from Flores, M., Marina, M. and Toldrá, F., *Meat Sci.*, 56, 247–254, 2000.)

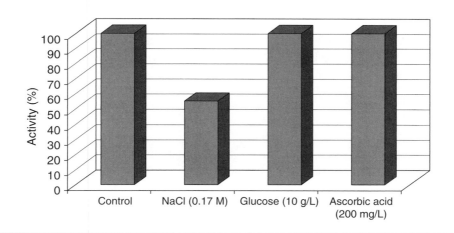

Figure 7.9 Activity of PGAP measured in the optimum reaction buffer using pGlu-AMC as substrate in the presence of NaCl, glucose, and ascorbic acid. (Adapted from Flores, M., Aristoy, M.C. and Toldrá, F., *Z. Lebens. Unter. Forsch. A*, 205, 343–346, 1997.)

its activity (Figure 7.8). The measurement of its activity can be done in a reaction buffer containing 100 mM phosphate buffer, pH 7.5, with 10 mM DTT and including 0.15 mM Ala-AMC as substrate. The activity may be measured in crude extracts but the addition of 0.05 mM bestatin is necessary to inhibit AAP activity.[29] Assay techniques based on fluorescence and color determinations are the same as for LAP using 0.15 mM Ala-AMC or 0.5 mM Leu-pNa as substrate, respectively.

7.3.6 Pyroglutamyl Aminopeptidase (EC 3.4.11.8)

PGAP specifically hydrolyzes pyroglutamic amino acid at basic pH (Table 7.5).[4,23] It is activated by sulfhydryl compounds and reducing agents, whereas NaCl addition inhibits its activity (Figure 7.9).

The measurement of its activity can be done in a reaction buffer containing 50 mM borate buffer, pH 8.5, and including 0.1 mM pGlu-AMC as substrate. The activity may be measured in crude extracts and the addition of a specific inhibitor is not necessary due to the basic pH and the specific substrate hydrolyzed by PGAP. Assay techniques based on fluorescence and color determinations are the same as for LAP using 0.1 mM pGlu-AMC or 0.5 mM pGlu-pNa as substrate, respectively.

The measurement of the DPP and aminopeptidase activities using fluorescent and colorimetric substrates has been applied to the control of pork quality. The activities of AAP, RAP, and DPP IV were assayed in the same meat samples using both substrates,[29] and the results are shown in Figure 7.10. Good linear relationships have been obtained for the three enzymes (AAP, $r^2 = 0.72$; RAP, $r^2 = 0.85$; DPP IV, $r^2 = 0.93$), meaning that there is a good relation between the fluorescent and colorimetric assays. The figure also shows that meat samples classified as exudative classes (PSE and RSE) showed lower activity values for the three enzymes than the nonexudative meats (RFN and DFD). These assays showed that colorimetric substrates, which are cheaper than the fluorescent ones, can be effectively used for the assay of muscle enzymes.

7.4 Assay of Lipase Activity

Lysosomal acid lipase catalyzes the hydrolysis of fatty acids from their glycerol esters. This enzyme is located in the lysosomes and has an acid optimal pH = 5. Lysosomal acid lipase is able to release free fatty acids from tri-, di-, and mono-acylglycerols although at different rates, TG > DG > MG.[30,31] As other lipases, lysosomal acid lipase has positional specificity for primary esters at the sn-1 and sn-3 positions of glycerides, the highest rate being in position sn-1.

7.4.1 Extraction

Samples of ground meat (between 5 and 25 g) are homogenized 1:5 or 1:10 in 50 mM Tris/HCl buffer, pH 7, containing 5 mM of EGTA and 0.2% Triton X-100.[32] Homogenization is achieved by using an Ultra Turrax or a polytron homogenizer, 3 strokes of 10 s each at 24,000 rpm, while cooling in ice. Homogenates are centrifuged at 10,000 × *g* for 20 min, and the supernatant is filtered through glass wool and collected for further enzyme assays. All operations are carried out at 4°C.

7.4.2 Lysosomal Acid Lipase (EC 3.1.1.3)

The assay of lysosomal acid lipases has been optimized for different types of substrates including triacylglycerols and luminiscent, fluorescent, and colorimetric compounds. These assays are briefly described in the following sections.

7.4.2.1 Fluorescent Assay

Lipase activity can be determined using 4-methylumbelliferyl oleate as substrate, which has been prepared by rapid homogenization in 2-methoxy ethanol.[32,33] The enzyme releases the fluorescent compound 4-methylumbelliferone that can be determined with a fluorometer. This substrate is not specific and could be hydrolyzed by other nonspecific carboxyl esterases.[34] The reaction medium consists of 0.1 M citric acid/0.2 M disodium phosphate buffer, pH 5, containing 0.8 mg/mL of BSA, 0.05 % (v/v) of Triton X-100, and 0.2 mM of substrate. Fluorescence multiwell

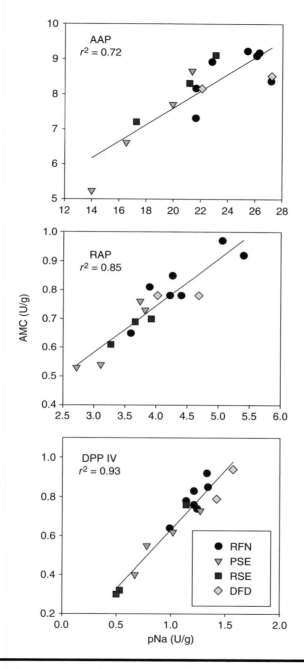

Figure 7.10 Activity of AAP, RAP, and DPP IV using fluorescent and colorimetric substrates in different quality of pork samples: normal (RFN: red, firm, and nonexudative), exudative (PSE: pale, soft, and exudative; RSE: red, soft, and exudative), and nonexudative (DFD: dark, firm, and dry) meats. (Adapted from Toldrá, F. and Flores, M., *Food Chem.*, 69, 387–395, 2000.)

plates may be used for a large number of simultaneous determinations. The reaction mixture consists of 50 μL of enzyme extract and 250 μL of reaction medium containing the substrate. The reaction mixtures are incubated at 37°C for 20 min, and fluorescence is measured at excitation and emission wavelengths of 328 nm and 470 nm, respectively. Kinetic constants K_m and V_{max} were found as 0.03 mM and 0.82 μmol/L min, respectively.[35]

7.4.2.2 Colorimetric Assay

Several chromogenic substrates have been proposed for the measurement of lipase activity. For instance, the use of 1,2-*O*-dilauryl-rac-glycero-3-glutaric acid-(6′-methylresorufin) ester (DGGR) as substrate that, after incubation with the lipase, releases an unstable dicarbonic acid ester. This reaction product is spontaneously hydrolyzed to glutaric acid and methylresorufin, the latest being a blue-purple chromophore that absorbs at 580 nm.[36] Another substrate recently proposed is 5,5′-dithiobis(2-nitrobenzoic acid), which in combination with the hydrolysis product of 2,3-dimercapto-1-propanol tributyrate after incubation at 37°C, gives an absorbance that can be read at 405 nm.[37]

7.4.2.3 Triacylglycerol Substrate Assay

Trioleoylglycerol (triolein) constitutes a good substrate because it is liquid at the reaction temperature and can be easily hydrolyzed by the lipase.[38] The release of oleic acid, as a product of the enzyme reaction after incubation at 37°C, can be determined by gas chromatography or high performance liquid chromatography (HPLC) if the substrate disappearance is also monitored.[39] In some cases, triolein labeled with ^{14}C in its fatty acids has been used for increased sensitivity.[30,33] Triolein has long-chain fatty acids (oleic acid) and is preferred to tributyrin as substrate because lysosomal acid lipase prefers to hydrolyze long-chain fatty acids from their glycerol esters. In general, these assays are more tedious to be performed. The pH stat technique that consists in the titration of the free fatty acids released to maintain a constant pH has poor sensitivity for muscle lipases.[40]

References

1. McDonald, J.K. and Barrett, A.J. (Eds.) *Mammalian Proteases: A Glossary and Bibliography. vol 2., Exopeptidases.* Academic Press, London, 1986.
2. Taylor, A. Aminopeptidases: structure and function. *FASEB J.,* 7: 290–298, 1993.
3. Lauffart, B. and Mantle, D. Rationalization of aminopeptidase activities in human skeletal muscle soluble extract. *Biochim. Biophys. Acta,* 956: 300–306, 1988.
4. Flores, M., Aristoy, M.C. and Toldrá, F. Curing agents affect aminopeptidase activity from porcine skeletal muscle. *Z. Lebens. Unter. Forsch. A,* 205: 343–346, 1997.
5. Toldrá, F. Biochemistry of processing meat and poultry. In: *Food Biochemistry and Food processing.* Y.H. Hui, W.K. Nip, L.M.L. Nollet, G. Paliyath and B.K. Simpson (Eds.), Blackwell Publishing, Ames, IA, pp. 315–335, 2006.
6. Toldrá, F. and Flores, M. The role of muscle proteases and lipases in flavor development during the processing of dry-cured ham. *Crit. Rev. Food Sci. Nutr.,* 38: 331–352, 1998.
7. Toldrá, F. Biochemical proteolysis basis for improved processing of dry-cured meats. In: *Advanced Technologies for Meat Processing.* L.M.L. Nollet and F. Toldrá (Eds.), CRC Press, Boca Raton, FL, pp. 329–351, 2006.

8. Martín, D. et al. Monounsaturated fatty acid content on pig muscle and adipose tissue lipase and esterase activity. *J. Agric. Food Chem.* 54: 9241–9247, 2006.

9. Sentandreu, M.A. and Toldrá, F. Purification and biochemical properties of dipeptidylpeptidase I from porcine skeletal muscle. *J. Agric. Food Chem.* 48: 5014–5022, 2000.

10. Sentandreu, M.A. and Toldrá, F. Partial purification and characterization of dipeptidyl peptidase II from porcine skeletal muscle. *Meat Sci.* 57: 93–103, 2001.

11. Sentandreu, M.A. and Toldrá, F. Biochemical properties of dipeptidylpeptidase III purified from porcine skeletal muscle. *J. Agric. Food Chem.* 46: 3977–3984, 1998.

12. Sentandreu, M.A. and Toldrá, F. Dipeptidyl peptidase IV from porcine skeletal muscle. *Food Chem.* 75: 159–168, 2001.

13. Turk, B. et al. Dipeptidylpeptidase I. In: *Handbook of Proteolytic Enzymes.* A.J. Barrett, N.D. Rawlings and J.F. Woessner (Eds.), 2nd ed., Elsevier, Amsterdam, The Netherlands, pp. 1192–1196, 2004.

14. McDonald, J.K. and Ohkubo, I. Dipeptidylpeptidase II. In: *Handbook of Proteolytic Enzymes.* A.J. Barrett, N.D. Rawlings and J.F. Woessner (Eds.), 2nd ed., Elsevier, Amsterdam, The Netherlands, pp. 1938–1943, 2004.

15. Chen, J.M. and Barrett, A.J. Dipeptidylpeptidase III. In: *Handbook of Proteolytic Enzymes.* A.J. Barrett, N.D. Rawlings and J.F. Woessner (Eds.), 2nd ed., Elsevier, Amsterdam, The Netherlands, pp. 809–812, 2004.

16. Misumi, Y. and Ikehara, Y. Dipeptidylpeptidase IV. In: *Handbook of Proteolytic Enzymes.* A.J. Barrett, N.D. Rawlings and J.F. Woessner (Eds.), 2nd ed., Elsevier, Amsterdam, The Netherlands, pp. 1905–1909, 2004.

17. Marguet, D. et al. cDNA cloning for mouse thymocyte-activating molecule. A multifunctional ecto-dipeptidyl peptidase IV (CD26) included in a subgroup of serine proteases. *J. Biol. Chem.* 267: 2200–2208, 1992.

18. Kiyama, M. et al. Sequence analysis of the *Porphyromonas gingivalis* dipeptidyl peptidase IV gene. *Biochim. Biphys. Acta*, 1396: 39–46, 1998.

19. Rawlings, N.D. and Barrett, A.J. Evolutionary families of peptidases. *Biochem. J.*, 290: 205–218, 1993.

20. Flores, M., Aristoy, M.C. and Toldrá, F. HPLC purification and characterization of porcine muscle aminopeptidase B. *Biochimie*, 75: 861–867, 1993.

21. Flores, M., Aristoy, M.C. and Toldrá, F. HPLC purification and characterization of soluble alanyl aminopeptidase from porcine skeletal muscle. *J. Agric. Food Chem.*, 44: 2578–2583, 1996.

22. Flores, M., Marina, M. and Toldrá, F. Purification and characterization of a soluble methionyl aminopeptidase from porcine skeletal muscle. *Meat Sci.*, 56: 247–254, 2000.

23. Mantle, D., Lauffart, B. and Gibson, A. Purification and characterization of leucyl aminopeptidase and pyroglutamyl aminopeptidase from human skeletal muscle. *Clin. Chim. Acta*, 197: 35–46, 1991.

24. Ledeme, N. et al. Human liver L-Leucine aminopeptidase: evidence for two forms compared to pig liver enzyme. *Biochimie*, 65: 397–404, 1983.

25. Toldrá, F. et al. Comparison of aminopeptidase inhibition by amino acids in human and porcine skeletal muscle tissues in vitro. *Comp. Biochem. Physiol.*, 115B: 445–450, 1996.

26. Nishimura, T. et al. Purification and properties of aminopeptidase H from bovine skeletal muscle. *J. Agric. Food Chem.*, 42: 2709–2712, 1994.

27. Rhyu, M.R. et al. Purification and properties of aminopeptidase H from chicken skeletal muscle. *Eur. J. Biochem.*, 208: 53–59, 1992.

28. Freitas, J.O., Termignoni, C. and Guimaraes, J.A. Microsomal methionine aminopeptidase: properties of the detergent-solubilized enzyme. *Int. J. Biochem.*, 17: 1285–1291, 1985.

29. Toldrá, F. and Flores, M. The use of muscle enzymes as predictors of pork meat quality. *Food Chem.*, 69: 387–395, 2000.

30. Imanaka, T. et al. Characterization of lysosomal acid lipase purified from rabbit liver. *J. Biochem.* 96: 1089–1101, 1984.

31. Imanaka, T. et al. Positional specificity of lysosomal acid lipase purified from rabbit liver. *J. Biochem.* 98: 927–931, 1985.

32. Motilva, M., Toldrá, F. and Flores, J. Assay of lipase and esterase activities in fresh pork meat and dry-cured ham. *Z. Lebensm. Unters. Forsch.* 195: 446–450, 1992.

33. Stead, D. Microbial lipases: their characteristics, role in food spoilage and industrial uses. *J. Dairy Res.* 53: 481–505, 1986.

34. Negre, A.E. et al. New fluorometric assay of lysosomal acid lipase and its application to the diagnosis of Wolman and cholesteryl ester storage diseases. *Clin. Chim. Acta* 149: 81–88, 1985.

35. Martín, D. et al. Optimisation of the assay of lipase and esterase activity in porcine muscle homogenates. Proceedings of the 4th EuroFed Lipid Congress, Madrid, 1–4 October 2006, p. 364.

36. Panteghini, M., Bonora, R. and Pagani, F. Measurement of pancreatic lipase activity in serum by a kinetic colorimetric assay using a new chromogenic substrate. *Ann. Clin. Biochem.* 38: 365–370, 2001.

37. Choi, S.J., Hwang, J.M. and Kim, S.I. A colorimetric microplate assay method for high throughput analysis of lipase activity. *J. Biochem. Mol. Biol.* 36: 417–420, 2003.

38. Jensen, R.G. Detection and determination of lipase (acylglycerol hydrolase) activity from various sources. *Lipids* 18: 650–657, 1983.

39. Veeraragavan, K. A simple and sensitive method for the estimation of microbial lipase activity. *Anal. Biochem.* 186: 301–305, 1990.

40. Taylor, F. Flow-through pH-stat method for lipase assay. *Anal. Biochem.* 148: 149–153, 1985.

Chapter 8

Muscle Enzymes: Glycohydrolases

Véronique Verrez-Bagnis

Contents

8.1 Introduction

In vivo, the muscle metabolic balance depends on glycolytic reactions and fatty acid oxidation, leading to the formation of CO_2, H_2O, and energy. Glycolysis is an ubiquitous metabolic pathway. In addition to its catabolic role, glycolysis also serves an anabolic function by providing C3 precursors for the synthesis of fatty acids and amino acids. Postmortem, in the absence of oxygen, glycolysis is the only source of energy for muscle cell to maintain its homeostasis.

Meat quality mainly depends on postmortem anaerobic metabolism of glycogen and glucose. And thus, key enzymes of glycolysis have been extensively studied to understand underlying mechanisms of variability in meat quality as well as glycolytic intermediates.

8.2 Definition

Glycohydrolases are enzymes that cleave covalent bonds of sugars. These enzymes can be coenzyme free. They can cleave either O-osidic- or N-osidic bond where one H_2O molecule is involved. These soluble enzymes are substrate specific, with pattern recognition (L or D molecule, α or β bond).

Glycohydrolases are involved in glycolytic pathways and they play an important role in muscle tissue, which in turn determines the meat quality traits. In the muscle tissue, there are varying proportions of fibers of several metabolic types that exhibit different properties and metabolic capacities. Glycohydrolases can be classified either as ATP-consumer or ATP-producer. This chapter describes the important glycohydrolases involved in glycogen, glucose 6P, and glucose degradation (see Figure 8.1).

A number of factors such as genetic, fiber type, exercise, and nutritional and hormonal state[1–11] affect the glycohydrolases content. Muscle fibers can be classified according to their contractile speed and metabolic pattern: slow-twitch red, fast-twitch red, and fast-twitch white muscle. For example, the activity of lactate dehydrogenase reflects the anaerobic metabolism of a muscle. Briand et al.[4] showed that the activity and profile of lactate dehydrogenase isoenzyme varied. The strongest lactate dehydrogenase is found in the muscle exhibiting the highest speed of contraction (myosin ATPase; see Table 8.1 and Figure 8.2).

8.3 Methods of Enzyme Activity Determination

In muscle cell, glucose and glycogen are the main reactants of Embden–Meyerhof pathway (Figure 8.1). For glycogen degradation, the first step consists in the liberation of a glucose phosphate—phosphate group is on the first carbon. This reaction requires one molecule of ATP. Glucose degradation requires phosphorylation to form glucose 6-phosphate.

The regulation of the glycolytic pathway deals with two segments: the section between glycogen and fructose 1,6-diphosphate, which involves the regulatory enzymes glycogen phosphorylase and phosphofructokinase (PFK), and the segment between triosephosphate and lactate, which is dominated by mass-action effects.

The determination of the important enzymes of Embden–Meyerhof pathway is presented in the following section.

8.3.1 The Formation of Glucose 6-Phosphate

The following sections present the specific enzymes involved in the formation of glucose 6-phosphate, a metabolic junction, depending on its source: glycogen or glucose.

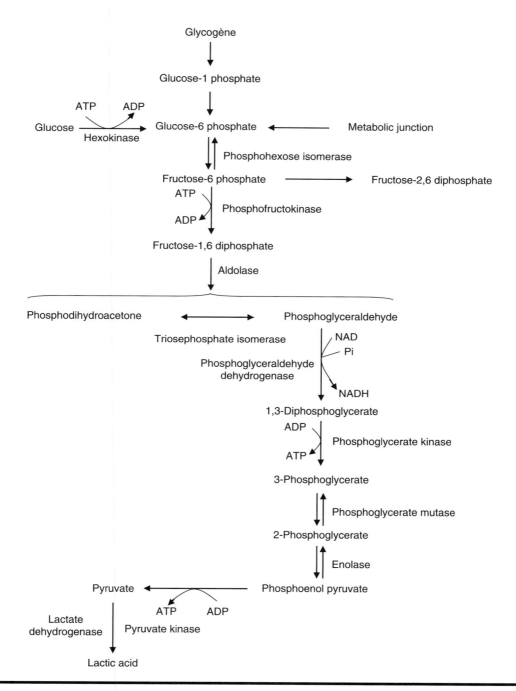

Figure 8.1   Main glycohydrolases involved in glycogen, glucose 6P, and glucose degradation.

Table 8.1 Percentage of M Chain Related to Myofibrillar ATPase in Different Sheep Muscles

	LDH Isoenzymes (%)				
	M4	M3H	M2H2	MH3	H4
Heart	0	0	11.0	20.3	68.6
Infraspinatus	13.3	4.4	27.2	17.2	37.9
Triceps brachii	23.6	6.3	28.3	14.4	28.5
Adductor	40.9	4.7	28.4	10.3	15.7
Semimembranosus	42.7	6.3	22.7	9.9	18.4
Biceps femoris	33	4.3	28.3	12.5	21.9
Tensor fascia lata	53.3	3.2	24.1	7.0	12.5

Source: Adapted from Briand M. et al., *Eur. J. Appl. Physiol.*, 46, 347–358, 1981. With permission.

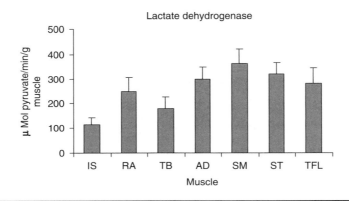

Figure 8.2 Lactic dehydrogenase activity in sheep muscles (IS—infraspinatus; RA—rectus adbominis; TB—triceps brachii; AD—adductor; SM—semimembranosus; ST—semitendinosus; TFL—tensor fascia lata). (Briand M. et al., *Eur. J. Appl. Physiol.*, 46, 347–358, 1981. With permission.)

8.3.1.1 Glycogen Phosphorylase (450 kDa, 4 Subunits)

These enzymes are found in the cytoplasm of the cells containing glycogen (muscle and liver). Glycogen phosphorylases catalyze the first reaction of glycogenolysis, phosphorylation, and hydrolytic action on the α-1,4 glycosidic linkage of the last carbon of one glycogen molecule branch with the formation of glucose 1-phosphate. The allosteric regulation of the enzyme activity depends on the adenosine mono phosphate (AMP) level (activation) and on glucose, glucose-6-phosphate and ATP levels (inhibition).

$$(\text{Glucose})_n + \text{PO}_4\text{H}_3 \xrightarrow[\text{glycogen phosphorylase}]{} (\text{glucose})_{n-1} + \text{glucose 1-phosphate}$$

A turbidimetric method has been developed for the continuous monitoring of the enzyme reaction catalyzed by glycogen phosphorylase. This method is based on the registration of the turbidity of glycogen solution at wavelengths above 300 nm. The increase in the turbidity is strictly proportional to the quantity of glucose 1-phosphate formed during the enzyme reaction. The method has the advantage of continuity, and it is suitable for determining the initial rate of catalytic synthesis or degradation of glycogen in a relatively simple and fast way. This method is based on the analysis of the dependence of the initial rate of the enzymic reaction on the proportion of the substrate of the forward reaction: $[Pi]/([Pi] + [G\text{-}1\text{-}P])$.

8.3.1.2 Phosphoglucomutase (62 kDa)

Phosphoglucomutase converts glucose 1-phosphate into glucose 6-phosphate. This enzyme contains a serine residue essential for its catalytic activity. The hydroxyl group of serine is esterified by the phosphoric acid, which changes from the first carbon position to the sixth one. The esterification of the hydroxyl group of the serine by di-isopropyl-fluorophosphate inactivates the phosphoglucomutase. The activation of this enzyme requires Mg^{2+} and an organic cofactor, glucose 1,6-diphosphate:

Phosphoenzyme + glucose 1-phosphate → enzyme + glucose 1,6-diphosphate

Enzyme + glucose 1,6-diphosphate → phosphoenzyme + glucose 6-phosphate

Assay Mixture	Concentration in Assay
2.50 mL triethanolamine buffer (0.1 M; pH 7.6)	84 mM
0.20 mL G 1-P, K salt (20 mg/mL)	3.5 mM
0.05 mL G 1,6-P_2, CHA salt (1 mg/mL)	0.02 mM
0.10 mL EDTA (10 mg/mL in buffer)	0.9 mM
0.05 mL MgCl$_2$ (0.1 M)	1.7 mM
0.05 mL NADP, Na salt (10 mg/mL)	0.19 mM
0.01 mL G6P-DH (1 mg/mL)	0.47 U/mL
0.02 mL enzyme solution in EDTA solution (1 mM)	

Note: CHA—cyclohexilammonium.

The determination of activity of phosphoglucomutase is based on the nicotinamide adenine dinucleotide phosphate (NADPH) absorbance ($\lambda = 340$ nm).

$$\text{Glucose 1-phosphate} \xrightarrow[\text{phosphoglucomutase}]{} \text{glucose 6-phosphate}$$

$$\text{Glucose 6-phosphate} + NADP^+ \xrightarrow[\text{glucose 6-phosphate dehydrogenase}]{} \text{gluconate 6-phosphate} + NADPH + H^+$$

8.3.1.3 Hexokinase (96 kDa)

Hexokinase catalyzes the phosphorylation of D-glucose on carbon 6 by ATP. This initial step converts a neutral D-glucose molecule into a negative molecule. This enzyme requires Mg^{2+} as a cofactor and is inhibited by its own reaction product: glucose 6-phosphate.

$$\text{D-glucose} + ATP \xrightarrow[\text{hexokinase}]{} \text{D-glucose 6-phosphate} + ADP$$

Assay Mixture	Concentration in Assay
1.00 mL triethanolamine buffer (50 mM; pH 7.6)	40 mM
1.00 mL glucose (100 mg/mL in buffer)	222 mM
0.20 mL MgCl$_2$ (0.1M)	8.0 mM
0.20 mL NADP, Na salt (10 mg/mL in buffer)	0.91 mM
0.10 mL ATP, Na salt (10 mg/mL)	0.64 mM
0.01 mL G6P-DH (1 mg/mL)	0.55 U/mL
0.02 mL enzyme solution in EDTA solution (1 mM)	

The determination of activity of hexokinase is based on the NADPH absorbance ($\lambda = 340$ nm).

$$\text{D-glucose} + \text{ATP} \xrightarrow{\text{hexokinase}} \text{D-glucose 6-phosphate} + \text{ADP}$$

$$\text{D-glucose 6-phosphate} + \text{NADP}^+ \xrightarrow{\text{glucose 6-phosphate dehydrogenase}} \text{D-gluconate 6-phosphate} + \text{NADPH} + \text{H}^+$$

8.3.2 Next Steps in the Embden–Meyerhof Pathway

The following enzymes are common to the glycogenolysis and glycolysis pathways.

8.3.2.1 Phosphohexose Isomerase

Phosphohexose isomerase catalyzes the first reaction of the cytoplasmic glycolysis to the formation of fructose 6P, and glucose 6P is a metabolic junction. This is the first substrate of the cytoplasmic glycolysis. The enzyme transfers the hydrogen from carbon 2 to carbon 1 (isomerization), leading to a change in the aldehyde function of carbon 1 to an alcohol function. For the carbon 2, the change leads to a ketone function instead of an alcohol one. This reaction is reversible.

$$\text{D-glucose 6-phosphate} \underset{\text{phosphohexose isomerase}}{\rightleftharpoons} \text{D-fructose 6-phosphate}$$

Assay Mixture	Concentration in Assay
2.50 mL triethanolamine buffer (0.1 M; pH 7.6)	85 mM
0.10 mL F-6-P, Na salt (13 mg/mL)	1.4 mM
0.20 mL MgCl$_2$ (0.1 M)	6.8 mM
0.10 mL NADP, Na salt (10 mg/mL in buffer)	0.39 mM
0.01 mL G6P-DH (1 mg/mL)	0.46 U/mL
0.02 mL enzyme solution in EDTA solution (1 mM)	

The determination of activity of phosphohexose isomerase is based on the NADPH absorbance ($\lambda = 340$ nm).

$$\text{Fructose 6-phosphate} \xrightarrow[\text{phosphohexose isomerase}]{} \text{glucose 6-phosphate}$$

$$\text{Glucose 6-phosphate} + \text{NADP}^+ \xrightarrow[\text{glucose 6-phosphate dehydrogenase}]{} \text{D-gluconate 6-phosphate} + \text{NADPH} + \text{H}^+$$

8.3.2.2 *Phosphofructokinase 1 (Molecular Weight 340 kDa, 4 Subunits)*

PFK plays a key role in glycolysis, and represents the limiting step because of its allosterical regulation. This enzyme catalyzes the phosphorylation of fructose-6-P on its carbon 1. ATP, in the presence of magnesium is the coenzyme, gives energy and the phosphate group. At this stage one proton is released. This reaction is exergonic and nonreversible. In the glycolysis pathway, the PFK is the slowest enzyme and its activity is inhibited by ATP, the final product of glycolysis.

$$\text{D-fructose 6-phosphate} + \text{ATP} \xrightarrow[\text{phosphofructokinase}]{} \text{D-fructose 1,6-diphosphate} + \text{ADP}$$

PFK1 contains four identical subunits and for each subunit there is one catalytical site. This enzyme is inhibited by high ATP concentration in the cell, long-chain fatty acids and citrate, and is activated by AMP and ADP levels.

Assay Mixture	Concentration in Assay
2.00 mL Tris buffer (0.1 M; pH 8.5)	70 mM
0.10 mL MgSO$_4$ 7H$_2$O (10 mg/mL; KCL 10 mg/mL)	1.4 mM
0.10 mL PEP, CHA salt (10 mg/mL)	0.71 mM
0.10 mL F-1, 6-P$_2$, Na salt (10 mg/mL)	0.64 mM
0.20 mL F-6-P, Na salt (10 mg/mL)	1.8 mM
0.20 mL ATP, Na salt (10 mg/mL)	1.1 mM
0.10 mL NADH, Na salt (10 mg/mL)	0.4 mM
0.03 mL pyruvate kinase (2 mg/mL)	4.2 U/mL
0.01 mL LDH (5 mg/mL)	9.6 U/mL
0.02 mL enzyme solution in ammonium sulfate solution (2.0 M; pH 7.5)	

The determination of activity of PFK is based on the NAD$^+$ absorbance ($\lambda = 340$ nm).

$$\text{D-fructose 6-phosphate} + \text{ATP} \xrightarrow[\text{phosphofructokinase}]{} \text{D-fructose 1,6-diphosphate} + \text{ADP}$$

$$\text{ADP} + \text{phosphoenolpyruvate} \xrightarrow[\text{pyruvate kinase}]{} \text{ATP} + \text{pyruvate}$$

$$\text{Pyruvate} + \text{NADH} + \text{H}^+ \xrightarrow[\text{pyruvate kinase}]{} \text{L-lactate} + \text{NAD}^+$$

8.3.2.3   Aldolase (150 kDa, 4 Subunits)

This enzyme cleaves a 6-carbon sugar (fructose-1,6-diphosphate) in two 3-carbon molecules (D-phosphoglyceraldehyde and phosphodihydroxyacetone).

This enzyme requires H_2O to open the hemiacetal bridge. At carbon 4, this enzyme forms a double bond between carbon and oxygen by subtracting one hydrogen and the radical containing the first three carbons. The carbons 4, 5, and 6 give D-3-phosphoglyceraldehyde and carbons 1, 2, 3 give phosphodihydroxyacetone. Although endergonic, this enzymatic reaction occurs because of the low level of triosephosphates in the cytoplasm compared to the fructose 1,6-diphosphate concentration.

$$\text{D-fructose 1,6-diphosphate} \xrightarrow{\text{aldolase}} \text{D-3-phosphoglyceraldehyde}$$
$$+ \text{ phosphodihydroxyacetone}$$

Assay Mixture	Concentration in Assay
3.00 mL triethanolamine buffer (0.1 M; pH 7.6)	94 mM
0.05 mL NADH, Na salt (10 mg/mL)	0.2 mM
0.10 mL F-1, 6-P_2, CHA salt (30 mg/mL)	1.1 mM
0.02 mL GDH/TIM (2 mg/mL) 6.7 U/mL TIM	1 U/mL GDH
0.01 mL enzyme solution in buffer	

Note: GDH/TIM—glycerol 3 phosphate dehydrogenase/triosephosphate isomerase.

$$\text{D-fructose 1,6-diphosphate} + H_2O \xrightarrow{\text{aldolase}} \text{D-3-phosphoglyceraldehyde}$$
$$+ \text{ phosphodihydroxyacetone}$$

$$\text{D-3-phosphoglyceraldehyde} \xrightarrow{\text{triosephosphate isomerase}} \text{phosphodihydroxyacetone}$$

$$2 \text{ phosphodihydroxyacetone} + 2\text{NADH} + H^+ \xrightarrow{\text{glycerol 3 phosphate dehydrogenase}}$$
$$\text{glycerol-3-phosphate} + 2\text{NAD}^+$$

The determination of activity of aldolase is based on the NAD^+ absorbance ($\lambda = 340$ nm).

8.3.2.4   Triosephosphate Isomerase (60 kDa, 2 Subunits)

Phosphodihydroxyacetone and 3-phosphoglyceraldehyde products of aldolase are isomers. Triosephosphate isomerase catalyzes an internal oxydo-reduction reaction between carbon 1 and carbon 2 of phosphodihydroxyacetone.

In the Emdben–Meyerhof pathway, only 3-phosphoglyceraldehyde is used.

$$\text{Phosphodihydroxyacetone} \xrightarrow{\text{triosephosphate isomerase}} \text{D-3-phosphoglyceraldehyde}$$

Assay Mixture	Concentration in Assay
2.50 mL triethanolamine buffer (0.3 M; pH 7.6)	243 mM
0.50 mL GAP (4 mg/mL)	3.8 mM
0.05 mL NADH, Na salt (10 mg/mL)	0.2 mM
0.01 mL GDH (10 mg/mL)	1.3 U/mL
0.01 mL enzyme solution in buffer	

$$\text{D-3-phosphoglyceraldehyde} \xrightarrow{\text{triosephosphate isomerase}} \text{phosphodihydroxyacetone}$$

$$2\,\text{phosphodihydroxyacetone} + 2\text{NADH} + \text{H}^+ \xrightarrow{\text{glycerol 3 phosphate dehydrogenase}}$$

$$\text{glycerol-3-phosphate} + 2\text{NAD}^+$$

The determination of activity of triosephosphate isomerase is based on the NAD^+ absorbance ($\lambda = 340$ nm).

8.3.2.5 Phosphoglyceraldehyde Dehydrogenase (148 kDa, 4 Subunits)

Phosphoglyceraldehyde dehydrogenase catalyzes the first oxidation of the cytoplasmic glycolysis and requires NAD coenzyme. 3-Phosphoglyceraldehyde forms a complex with the enzyme phospho-glyceraldehyde dehydrogenase by a covalent bond. The thiol (SH) group of a cystein of the enzyme reacts with the aldehyde group of the 3-phosphoglyceraldehyde (also called acyl-enzyme). This complex can react then with NAD^+ leading to a coupled oxydo-reduction reaction with NADH formation and oxidation on carbon 1 of 3-phosphoglyceraldehyde giving a 3-phosphoglycerate. A phosphate ion allows the liberation of the product while transferring energy. The final product of the phosphoglyceraldehyde dehydrogenase is the 1,3-diphosphoglycerate, an energy-rich molecule.

$$\text{D-3-phosphoglyceraldehyde} + \text{Pi} + \text{NAD} \xrightarrow{\text{phosphoglyceraldehyde dehydrogenase}}$$

$$\text{1,3-phosphoglycerate} + \text{NADH}$$

8.3.2.6 Phosphoglycerate Kinase (50 kDa)

Phosphoglycerate kinase catalyzes the direct transfer of the energy from phosphoryl radical of 1,3-diphosphoglycerate to ADP, in other words the phosphorylation of ADP and ATP, which is similar to the events in the respiratory chain. At this step, the first energy-rich molecule ATP is formed, and the product is 3-phosphoglycerate.

$$\text{1,3-Phosphoglycerate} + \text{ADP} \xrightarrow{\text{phosphoglycerate kinase}} \text{3-phosphoglycerate} + \text{ATP}$$

8.3.2.7 Phosphoglycerate Mutase (57 kDa, 2 Subunits)

Phosphoglycerate mutase converts 3-phosphoglycerate into 2-phosphoglycerate by transferring the phosphate group from carbon 3 to carbon 2. Its action mechanism is similar to the phosphoglucomutase: Mg^{2+} is essential and the enzymatic action requires 2,3-diphosphoglycerate as an organic cofactor.

$$\text{Phosphoenzyme} + \text{3-phosphoglycerate} \rightarrow \text{enzyme} + \text{2,3-diphosphoglycerate}$$

$$\text{enzyme} + \text{2,3-diphosphoglycerate} \xrightarrow[\text{phosphoglycerate mutase}]{} \text{phosphoenzyme}$$
$$+ \text{3-phosphoglycerate}$$

8.3.2.8 Enolase (85 kDa, 2 Subunits)

This enzyme catalyzes the dehydratation of 2-phosphoglycerate. H_2O is formed from a hydrogen of the esterified secondary alcohol group and from the hydroxyl group of the primary alcohol group. The secondary alcohol group is turned into an enol group, which has intermediate properties between alcohol and acid groups. Then the phosphoryl radical is linked through an ester bond, rich of energy. Enolase acts with Mg^{2+} as cofactor and is inhibited by fluor ions.

$$\text{2-Phosphoglycerate} \xrightleftharpoons[\text{phosphoglycerate kinase}]{} \text{phosphoenolpyruvate} + H_2O$$

Assay Mixture	Concentration in Assay
2.50 mL triethanolamine buffer (0.1 M; pH 7.6)	83 mM
0.10 mL $MgSO_4$ (0.1 mM)	3.3 mM
0.05 mL NADH, Na salt (10 mg/mL)	0.2 mM
0.10 mL glycerate 2P, Na salt (10 mg/mL)	0.9 mM
0.20 mL ADP, Na salt (10 mg/mL)	1.1 mM
0.02 mL LDH (5 mg/mL)	18.5 U/mL
0.02 mL pyruvate kinase (2 mg/mL)	2.7 U/mL
0.02 mL enzyme solution in triethanolamine buffer (0.1 M; pH 7.6)	

$$\text{2-Phosphoglycerate} \xrightarrow{\text{enolase}} \text{phosphoenolpyruvate} + H_2O$$

$$\text{Phosphoenolpyruvate} + \text{ADP} \xrightarrow{\text{pyruvate kinase}} \text{pyruvate} + \text{ADP}$$

$$\text{Pyruvate} + \text{NADH} + H^+ \xrightarrow{\text{lactate dehydrogenase}} \text{L-lactate} + NAD^+$$

The determination of activity of enolase, also called phosphoglycerate kinase, is based on the NAD^+ absorbance ($\lambda = 340$ nm).

8.3.2.9 Pyruvate Kinase (235 kDa, 4 Subunits)

Pyruvate kinase catalyzes the transfer of phosphate group from phosphoenolpyruvate to ADP. The reaction is irreversible due to the heat (31 kJ/mol) released by it. Ions like Mg^{2+} and Mn^{2+} are essential to form an active complex with the enzyme. Conversely, ion Ca^{2+} acts as a competitor of the previous ions, leading to a nonactive enzyme complex.

Pyruvate kinase in muscle or pyruvate kinase M is inhibited by phenylalanine and its activity is reduced when high level of ATP are found in the muscle cell as well as other energizing molecules such as long-chain fatty acids, citrate, acetyl CoA, or alanine.

$$\text{Phosphoenolpyruvate} + \text{ADP} \xrightarrow[\text{pyruvate kinase}]{} \text{pyruvate} + \text{ATP}$$

Assay Mixture	Concentration in Assay
2.50 mL triethanolamine buffer (0.1 M; pH 7.6)	85.6 mM
0.20 mL PEP, CHA salt (3.75 mg/mL)	0.54 mM PEP
Triethanolamine + PEP, CHA salt are dissolved in 0.05 M MgSO$_4$/0.2 M KCL	2.5 mM MgSO$_4$/10 mM KCL
0.20 mL ADP, neutralized with KOH (30 mg/mL)	4.7 mM
0.20 mL NADH, Na salt (10 mg/mL)	0.2 mM
0.01 mL LDH (5 mg/mL)	9.2 U/mL
0.02 mL enzyme solution in buffer	

$$\text{Phosphoenolpyruvate} + \text{ADP} \xrightarrow[\text{pyruvate kinase}]{} \text{pyruvate} + \text{ADP}$$

$$\text{Pyruvate} + \text{NDH} + \text{H}^+ \xrightarrow[\text{lactate dehydrogenase}]{} \text{L-lactate} + \text{NDA}^+$$

The determination of the activity of pyruvate kinase is based on the NAD$^+$ absorbance ($\lambda = 340$ nm).

8.3.2.10 Lactate Dehydrogenase (134 kDa, 4 Subunits)

Lactate dehydrogenase consists of a system of five isoenzymes, each one representing one of the four unit combinations (H$_4$, H$_3$M, H$_2$M$_2$, HM$_3$, and M$_4$) of two subunits (heart type) and M (muscle type), controlled by two different genes. The relative proportions of these isoenzymes is characteristic of a particular tissue. For instance, the isoenzyme H4, when inhibited by strong concentrations of pyruvate, predominates in aerobic tissue like heart, whereas the M chain form (M4), which has a strong affinity for pyruvate and a high V max, predominates in anaerobic tissue.

In the absence of oxygen, these enzymes realize the last step of the set of glycolysis reactions. The coenzyme NAD that has been reduced by oxidative phosphorylation cannot be oxidized by the respiratory chain and has to transmit its hydrogen to pyruvate. LDH has zinc as cofactor and NAD as coenzyme.

$$\text{Pyruvate} + \text{NADH} + \text{H}^+ \xrightarrow[\text{lactate dehydrogenase}]{} \text{lactic acid} + \text{NAD}^+$$

Assay Mixture
Triethanolamine buffer (50 mM; pH 7.6)
EDTA 5 mM
Pyruvate 0.33 mM
NADH, Na salt 0.234 mM

The reaction is monitored spectrophotometrically at 340 nm.

References

1. Bergmeyer H.U., *Methods of Enzymatic Analysis*, Academic Press, New York, 1974.
2. Blomstrand E., Ekblom B. and Newsholme E., Maximum activities of key glycolytic and oxidative enzymes in human muscle from differently trained individuals. *J. Physiol.* 381: 111–118, 1986.
3. Bouchard C. et al., Genetic effects in human skeletal muscle fiber type distribution and enzyme activities. *Can. J. Physiol. Pharmacol.* 64: 1245–1251, 1986.
4. Briand M. et al., Metabolic types of muscle in the sheep: I. Myosin ATPase, glycolytic, and mitochondrial enzyme activities. *Eur. J. Appl. Physiol.* 46: 347–358, 1981.
5. Connet R., The effect of exercise and restraint on pectoral muscle metabolism in pigeons. *J. Appl. Physiol.* 63: 2360–2365, 1987.
6. Chaplin S., Munson M. and Knuth S., The effect of exercise and restraint on pectoral muscle metabolism in pigeons. *J. Comp. Physiol. B.* 167: 197–203, 1997.
7. Dalle Zotte A. et al., Effect of age, diet and sex on muscle energy metabolism and on related physicochemical traits in the rabbit. *Meat Sci.* 43: 15–24, 1996.
8. Dalrymple R., Cassens R. and Kastenschmidt L., Glycolytic enzyme activity in developing red and white muscle. *J. Cell Physiol.* 83: 251–258, 1983.
9. Hue L. and Rider M., Role of fructose 2,6-bisphosphate in the control of glycolysis in mammalian tissues. *Biochem. J.* 245: 313–324, 1987.
10. Lebret B. et al., Influence of the three RN genotypes on chemical composition, enzyme activities, and myofiber characteristics of porcine skeletal muscle. *J. Anim. Sci.* 77: 1482–1489, 1999.
11. Talmant A. et al., Activities of metabolic and contractile enzymes in 18 bovine muscles. *Meat Sci.* 18: 23–40, 1986.

Chapter 9

Fatty Acids

Katleen Raes and Stefaan De Smet

Contents

9.1 Introduction

The fatty acid content and composition of muscle foods are of importance in view of their implications for human health, and for the sensory quality of meat and meat products. Muscle lipids are composed of polar lipids, mainly phospholipids (PL) located in the cell membranes, and neutral lipids consisting mainly of triacylglycerols (TAG) in the adipocytes that are located along the muscle fibers and in the interfascicular area. A small amount of TAG is also present as cytosolic droplets in the muscle fibers. The content of PL in the muscle is relatively independent of the total fat content and varies between 0.2 and 1% of muscle weight. However, the content of muscle TAG is strongly related to the total fat content and varies from 0.2% to more than 5%. Muscle PL are particularly rich in polyunsaturated fatty acids, whereas TAG contain more saturated and monounsaturated fatty acids. Hence, the muscle fatty acid profile may vary largely, depending on the fat level and other genetic and nutritional factors.

Approximately 20 fatty acids, with a chain length ranging between C12 and C22, account for more than 85% of muscle fatty acids, but the number of minor fatty acids is much larger. Species is a major source of variation in meat fatty acid composition, with meats from ruminant animals having in general lower amounts of polyunsaturated and higher amounts of saturated fatty acids, because of biohydrogenation of unsaturated fatty acids in the rumen. The rumen microbial metabolism also gives rise to a large number of minor fatty acids, for example, trans, conjugated, and odd- and branched-chain fatty acids. Above all, large changes in fatty acid composition can be brought about by the choice of the dietary fat source, particularly in monogastric animals. Indeed, fat deposition in muscle is determined by *de novo* fatty acid synthesis and uptake of preformed fatty acids from circulation.

Muscle foods generally consist of variable proportions of muscle proteins and intra- and intermuscular fat, and may also contain subcutaneous adipose tissue and added animal fat in case of meat products. It should be noted that the fatty acid composition also varies among various fat tissues. By far, the most common method of fatty acid analysis for muscle foods is solvent extraction, followed by methylation and gas chromatography (GC) identification. Considerations for the application of these methods to muscle and fat tissue from farm animals are discussed in the following sections.

9.2   Sample Preservation

Unsaturated fatty acids, especially polyunsaturated ones, are susceptible to oxidation. Hence, factors promoting oxidation should be controlled as much as possible during sampling, preservation of the muscle tissue, and extraction procedures. Maintaining a low temperature and avoiding exposure to oxygen and light are critical in this respect.

Sampling of muscle tissue after slaughter can be done during the normal cutting process. Samples of fresh meat or meat products should preferably be stored vacuum packed and may be refrigerated when analyzed within a few days, or may be deep frozen (−20°C or lower) until later analysis. However, at freezing temperatures, even at −80°C, the lipid fraction undergoes modifications: lipolysis (liberation of free fatty acids from the glycerol backbone), as well as oxidation processes continue to take place. The rate of these processes lowers significantly at lower temperatures, but the processes nevertheless go on slowly. Hence, the sooner samples are analyzed, the lower the risk of fatty acid losses due to oxidation. However, when stored properly, samples may be kept frozen for several months until further processing. It is not advisable to freeze-dry meat samples before extraction as lyophilized tissues are particularly difficult to extract, and rehydration of the sample is often necessary to ensure quantitative recovery of the lipids.[1] Endogenous water-soluble antioxidants in the meat tissue are also destroyed or become less active after lyophilization.

9.3 Lipid Extraction

During lipid extraction, attention has to be paid to proper solubilization of the lipids in organic solvents and to removal of the nonlipid compounds. As meat lipids consist of both polar and non-polar fractions, a combination of organic solvents with different polarity is necessary. The extraction solvent needs to be fairly polar, as an important (both quantitatively and nutritionally) part of the lipids in meat is associated with the cell membranes. However, solvents that are too polar can give rise to chemical side reactions or are not suitable for dissolving nonpolar simple lipids and TAG. Besides the extraction, solvents need to overcome the interactions between the lipids and the tissue matrix (e.g., protein–lipid interactions).[1]

The most frequently used solvent mixture for fresh meat and meat products consists of chloroform/methanol in a ratio of 2/1 (v/v) as described by Folch et al.[2] The tissue is homogenized with chloroform/methanol (2/1; v/v) to a final 20-fold dilution of the tissue sample. In this system, the ternary compound, which is often not taken into account, is water (or endogenous water) resulting in chloroform/methanol/water in a ratio of approximately 8/4/3 (v/v/v). The Folch procedure has been avoided by some research groups due to the toxicity of chloroform and the large volumes that are needed. The method described by Bligh and Dyer[3] is in fact an adaptation of the Folch method, using less solvent. It should be mentioned that in the original method of Bligh and Dyer, developed for fish tissues rich in PL, the recovery of nonpolar lipids is quite low. The authors, therefore, clearly stated in their paper that it is necessary to perform a reextraction of the tissue residue with pure chloroform and add this extract to the filtrate before solvent evaporation. Iverson et al.[4] have compared the original Folch method with the Bligh and Dyer method for quantitative total lipid determinations. In samples containing less than 2% lipids, the two methods gave comparable results. However, for fish samples with more than 2% lipids, the Bligh and Dyer method gave an underestimation of the total lipid content, which increased significantly with increasing lipid content.

Another extraction solvent mixture that is regularly used is hexane/2-propanol (3/2; v/v), originally proposed by Hara and Radin.[5] Different solvent extraction methods for beef total lipids were compared by Sahasrabudhe and Smallbone[6] with chloroform/methanol, *n*-hexane/isopropanol, and ethanol/diethylether for wet material, and soxhlet extraction for freeze-dried material using petroleumether, diethylether, chloroform/methanol, methylene chloride/methanol. The last two methanol-containing solvent mixtures appeared to be as effective as the wet methods in isolating neutral and polar lipids.

To date, the chloroform/methanol mixture seems to be the best extraction solvent and is still the most used one. After the extraction procedure, contaminating compounds need to be removed. The most appropriate way to do this is to add one-fourth volume of water or dilute aqueous salt solution and shake the mixture.[2] After the solvents are separated into two phases, the lower phase (chloroform/methanol/water: 86/14/1; v/v/v) contains the purified lipid.

During the extraction procedure, it is important to take into account that solvent properties can vary considerably depending on the source (e.g., stabilizing solvent for chloroform[1]), which can influence the extraction efficiency. As all solvents contain small amounts of lipids, it is necessary to distil them before use. The extraction procedure should be carried out in glass recipients, and plastic should be avoided. As meat lipids contain polyunsaturated fatty acids including the long chain ones, which are highly sensitive to oxidation, it is recommended to add antioxidants—the most used one is butylated hydroxy toluene (BHT). However, care must be taken in gas chromatography analysis as BHT elutes with the fatty acids and may interfere with the latter on the chromatogram.[1]

Concentration of extracts is often required before esterification, or separation of the lipid (classes) can be performed. Concentrating lipid extracts is mostly done by removing solvents under vacuum in a rotary film evaporator, keeping the temperature below 40°C. Extracts need to be stored at −20°C.

9.4 Fatty Acid Esterification

Although it is possible to separate underivatized fatty acids by GC, it is recommended to use ester derivatives of fatty acid mixtures. The ester derivatives are more volatile than their corresponding underivatized fatty acids and are thus more suitable for GC analysis. They are less polar and do not tend to absorb on the packaging of the column or to dimerize, avoiding peak tailing, peak asymmetry, and peak shouldering.[7,8] Most frequently, fatty acids are converted into fatty acid methyl esters (FAME), whereas butyl or other ester derivatives are less used.

The choice of saponification–esterification and of the esterification solution will influence the separation and isomerization of the investigated fatty acids. Esterification solutions can be primarily divided into the following groups: acid-catalyzed, base-catalyzed, diazomethane, and combinations of the acid- and base-catalyzed solvents.

9.4.1 Acid-Catalyzed Esterification

The most frequently used acid-catalyzed esterification solvent is a methanolic hydrogen chloride solution. It is important to note that cholesterol esters and TAG, an important fraction in meat tissues, are not soluble in methanolic hydrogen chloride solution. To make them soluble, an inert solvent should be added; often benzene has been used for this purpose.[9,10] However, due to its high toxicity, the use of benzene is not recommended and is not frequently used anymore. Methylene chloride,[11] toluene,[12] and tetrahydrofuran[13] are a few alternatives to benzene that have been cited in literature.

Similar results have been obtained with a solution of 1–2% concentrated sulfuric acid in methanol compared to a 5% methanolic hydrogen chloride solution. Free fatty acids can be rapidly esterified by heating in 10% sulfuric acid in methanol until the reflux temperature has been reached. However, this procedure is not recommended for polyunsaturated fatty acids due to the strong oxidizing nature of sulfuric acid.

A commonly used solvent is boron trifluoride in methanol, which allows esterification of most lipid classes in a very reasonable time.[14] However, high boron trifluoride concentrations in methanol produce methoxy artifacts from unsaturated fatty acids, which poses a serious concern.[15] These artifacts are not necessarily formed with less concentrated solutions.[14,15] Some evidence exists that artifact formation is most likely with aged reagents.[16] It is certainly advisable to check the solution (especially commercially bought ones) periodically before use to limit the artifact products. Several authors also found losses of the lower fatty acid esters (lower than C_{12}).[17–19] Sample size is another critical factor when using boron trifluoride. If samples do contain less than 200 mg fatty acids, substantial losses may occur.[20]

9.4.2 Base-Catalyzed Esterification

As basic catalyzing transesterification methods, potassium hydroxide in anhydrous methanol, methanolic sodium, and potassium methoxide are often used in concentrations between 0.5 and 2 M. However, potassium containing transesterification solvents have some drawbacks (artifact formations, hydrolysis of lipids in presence of trace amounts of water, etc.) and are therefore less recommended.[1,21,22] Transesterification of sodium methoxide in methanol is quite a rapid procedure; however, lipids need to be in solution before methanolysis proceeds, if necessary by adding inert solvents.

Another base-catalyzed esterification solvent is tetramethylguanidine in methanol as described by Schuchardt and Lopes.[23] However, this method seems less advisable to use on meat samples as difficulties are experienced in esterifying PL, an important lipid fraction in the meat matrix, and sometimes also of the free fatty acids.[24,25]

9.4.3 Diazomethane

Diazomethane is another esterification solvent that is less frequently used. Diazomethane has the advantage that it can esterify very rapidly unesterified fatty acids under mild conditions. For complex lipid mixtures, esterification with diazomethane should be preceded by alkaline hydrolysis of lipids.[26] However, it is highly explosive.[1,27,28] Owing to this characteristic, and the fact that it is not readily available and should thus be self-prepared, it is not frequently used. It has also been suggested that diazomethane could react with double bonds in fatty acids—an unwanted side reaction.[28,29] The reagent does not esterify fatty acid esters, and can thus only be used on hydrolyzates of fatty acids—a characteristic of diazomethane limiting its use for meat.[28]

9.4.4 Which Esterification Solvent to Choose?

The choice of the esterification solvent or solvent combinations will influence the final fatty acid profile observed by GC. Following are some of the drawbacks of esterification solvents. Acid-catalyzed reagents form FAME from both free fatty acids and O-acyl lipids,[13,14] whereas base-catalyzed reagents cause only transesterification, that is, conversion of O-acyl lipids into FAME, but not the formation of FAME from free fatty acids.[22,30] An additional feature of methylation of lipids under acidic conditions is the formation of allylic methoxy ether artifacts, observed with hydrogen chloride, boron trifluoride, and acetyl chloride.[25] When boron trifluoride–methanol is used as esterifying solution, it should be noted that it can react with BHT, producing methoxy derivatives, coeluting with methyl pentadecanoate or methyl hexadecanoate on GC.[31] Dimethylacetals can be formed from plasmalogens by reaction of acidic transesterifying reagents. These dimethylacetals tend to elute just ahead of their corresponding esters.

If the cis–trans isomerism of double bonds is of no interest, acid catalysis is a very useful technique. However, methylation under acid conditions is not adequate if the steric integrity of conjugated double bonds needs to be maintained.[25,28,32,33] Kramer et al.[25] proposed to use sodium methoxide to preserve the steric configuration of the double bonds and to reduce artifact formation. By using sodium methoxide stereochemistry was maintained but neither free fatty acids nor fatty acids on sphingomyelin were derivatized.[34]

9.4.5 In Situ *Transesterification*

Since muscle lipids are associated with proteins, efficient extractions are time-consuming and require large volumes of solvents. Formation of FAME on the matrix itself, without prior extraction, may save time and money. Park and Goins[35] compared *in situ* transesterification by adding directly to beef muscle methylene chloride, 0.5 N sodium hydroxide and 14% boron trifluoride in methanol with transesterification on a beef lipid extract (prepared by the Folch method) using the same solvent mixtures. For most fatty acids, good agreements were observed between both methods, except for arachidonic acid that was considerably lower with the *in situ* transesterification procedure. Also, other fatty acids, mostly linked with the cell membranes, for example, linoleic acid and oleic acid, were somewhat lower. However, Rule[36] found significantly higher contents of arachidonic acid in lamb muscle by using direct esterification based on 14% boron trifluoride in methanol compared to extraction and similar esterification conditions. Recently, Eras et al.[37] demonstrated that no differences were found in the total fat content and fatty acid profile between an *in situ* procedure (based on preparation of pentyl esters using chlorotrimethylsilane) or after prior extraction for beef, pork, and chicken muscle and fat. However, no long-chain polyunsaturated fatty acids were reported in this study and the extraction procedure was based on a hot extraction, using ethanol and hexane/diethylether (1/1; v/v). In general, *in situ* preparation of ester derivatives of fatty acids is not frequently used for meat samples. The yield of the obtained methyl esters is highly variable primarily due to the high water content of the samples.[1] As it is known that the yield of FAME produced by alkaline esterification is more influenced when water is present, it is advised to use acid-based esterification solvents for direct esterification procedures.[1] The high amount of bound fatty acids in the lipid fraction of meat will also negatively influence the FAME yield produced by *in situ* esterification procedures.[1]

9.5 Fatty Acid Quantification

9.5.1 Identification of Fatty Acids

The identification of the fatty acid profile and its quantification are performed preferably by GC. The availability of long capillary columns makes it possible to identify also minor fatty acids. This is important especially for ruminant products. Ruminant products are characterized by fatty acids derived from microorganisms present in the rumen (e.g., odd- and branched-chain fatty acids, conjugated linoleic acids [CLA], trans fatty acids), whereas monogastric products have a much simpler fatty acid profile (mainly linear even fatty acids). Rule et al.[38] observed the resolution of 96 peaks in ruminant muscle when a 100 m column was used. In most literature, only 30 fatty acids are reported, accounting for at least 85% of the total fatty acids present. More and more interest is given to the minor fatty acids due to their potential biological importance. It is obvious that if all fatty acids are subject of the study, no single GC separation will suffice, even when very long columns are used.

The choice of the column is of course important for the resolution of the different fatty acids. The length of the column and the packaging material influence the separation. For meat lipids, capillary columns with polar stationary phases (e.g., 100% cyanoethylsilicone oil [SP-2340, OV-275]; 100% cyanopropylsilicone [CP-Sil88] or 68% biscyanopropyl-32% dimethylsiloxane [SP-2330]; bis-cyanopropyl polysiloxane [SP-2560]; and 70% cyanopropyl polysilphenylene-siloxane [BPX-70]) are preferred, as they show a much higher resolution capacity of unsaturated FAME compared to the apolar stationary phases.[7] Also, geometrical double-bond isomers can be separated on very polar stationary phases.[7] However, the presence of dimethylacetals, originating from plasmalogen PL, can interfere with some FAME.[39] Care has to be taken when minor components are analyzed, for example, it is known that on a 100 m CP-Sil88 C18:3n-3 isomers can overlap with C20:0 or C20:1 peaks—a finding that is not observed with SP-2340 columns.[40] Even with long polar capillary columns, some isomers, for example, trans C18:1 and CLA are not completely separated and a combination of methods is then advised (e.g., Ag+-TLC (thin layer chromatography); multicolumn Ag+-HPLC (high-performance liquid chromatography); two-dimensional GC; GC-MS).[41–44] The identification of FAME occurring in biological samples is mainly based on comparing their retention times with those of commercially available purified individual standards. However, nonavailability of some fatty acid standards results in difficulties to identify some minor fatty acids. One may use well-defined (certified) natural products to overcome this problem. Christie[1] recommended a mixture of pig liver lipids, cod liver oil, and linseed, as this mixture contains all the major fatty acid classes. Alternatively, one can identify FAME based on their relative retention times (RRT) or equivalent chain length (ECL) values.[1] RRT is defined as the ratio between the retention time of the studied FAME and that of a reference FAME (C16:0 or C18:0). ECL values are based on the Kovats' retention index and result in a ready visualization of the position of a particular FAME, relative to saturated straight-chain fatty acids. Unknown FAME can then be identified by comparing actual ECL values with those tabulated in literature.

The detection of FAME is generally performed with flame ionization detector (FID). This detector is very sensitive and has a good linearity over a wide range of concentrations. The amounts of FAME are then visualized by peak areas and calculated using relative response factors, given the detector response between a FAME and a reference FAME, usually the internal standard. For higher fatty acids, response factors are approximately 1, and are often ignored in the calculations.

An internal standard is not only used to determine response factors, but mainly for the absolute quantification of the FAME in the sample (on weight basis). Therefore, a proper choice of the internal standard is necessary. Often C17:0 is used as internal standard for biological matrices. However, as C17:0 is present in relatively high amounts in ruminant products (1–1.5% of total FAME), it is advised to use other internal standards such as C19:0 or C23:0.

9.5.2 Determination of Trans Isomers

The isomeric profile of trans isomers, in particular of C18:1 isomers, is of interest for meat and other tissues of ruminants, and results from the specific hydrogenation of unsaturated fatty acids by rumen bacteria. Also, the total trans fatty acid content of food has been the subject matter of many studies due to the effect of the total or individual isomers on human health. The discovery and the health and metabolism effects of the CLA isomers has led to increasing interest in the quantification and the isomeric distribution profile of the trans C18:1 isomers. The content and isomeric distribution of trans polyenoic acids have not been studied well in meat (products). The focus here is primarily on the trans C18:1 isomers as they are quantitatively more important and because of their relevance in relation to CLA.

The trans fatty acid content in ruminant beef or lamb meat varies between 2 and 8% of the total FAME, depending on the animals diet, age, sex, and breed.[45–47] Only few papers are available that give a detailed profile of the trans C18:1 isomers.[46–48] The most important one in ruminant meat is t11 C18:1 (vaccenic acid), followed by t13 and t14.

A detailed profile of the trans monoenoic acids requires extensive and precise laboratory work. It is impossible to separate them by simple GC, even on long capillary columns (120 m) due to the high concentration of oleic acid (c9 C18:1) in meat, which overlaps with several of the trans C18:1 isomers. Therefore, a preseparation of the cis and trans monoenoic fatty acids is required. This is often performed by argentation thin-layer chromatography (Ag+-TLC) or by argentation high performance liquid chromatography (Ag+-HPLC)—a procedure based on separation of fatty acids depending on their degree of unsaturation, and more importantly also on the geometry of the double bonds. Using normal Ag⁺-TLC FAME can only be separated by the total number of cis and trans double bonds, and not on their specific locations.[49] Even with this preseparation step, the use of a long capillary column to identify the isomeric profile, such as CP-Sil88 for FAME (100 m × 0.25 mm × 0.2 μm) or SP-2340, SP-2560 (100 m × 0.25 mm × 20 μm) is advised. Wolff and Bayard[50] clearly demonstrated the increased resolution of the individual trans C18:1 isomers of beef tallow by GC on a 100 m column instead of a 50 m one (CPSil 88), after presepa-ration by Ag⁺-TLC. The higher resolution is mainly in the area of t9, t10, t11, t12 C18:1 isomers, which may be baseline resolved from other isomers. However, t13 and t14 are still coeluting as a single peak. These findings have also been confirmed by Raes et al.[46] on different ruminant tissues. The different cis C18:1 isomers can also be separated by this procedure.

The temperature program has a large effect on the resolution of the trans isomers. It is advised to work isothermally and the resolution improves with lowering the temperature. However, this leads to an increase in the time of analysis. Wolff and Bayard[50] worked at 160°C (He as carrier gas, 120 kPa) with a total analysis time of 80 min, whereas Raes et al.[46] worked at 170°C (H₂ as carrier gas, 160 kPa) with a total analysis time of 90 min.

Instead of using Ag⁺-TLC, stable silver ion column for HPLC may also be used.[51] Separations of positional and geometrical isomers of unsaturated fatty acids is possible with Ag⁺-HPLC.[52] In contrast to Ag⁺-TLC, chromatographic conditions are better controlled, analysis time is short-ened, and recoveries are high (95 versus 60% for Ag⁺-HPLC versus Ag⁺-TLC, respectively).[53] Cis and trans separation is possible by using an UV detector on the HPLC system.

9.5.3 Determination of CLA Isomers

The determination of CLA isomers is of particular interest in products of ruminants due to their formation, along with trans C18:1 isomers, during the biohydrogenation processes in the rumen. The most important trans fatty acid, t11 C18:1, is further converted into c9, t11 CLA, the quanti-tatively most important CLA, by delta-9 desaturase activity in the intermediary metabolism. The CLA concentrations reported for ruminant meat vary between 4.3 and 19 mg/g lipid for lamb meat and between 1.2 and 10 mg/g lipid for beef.[54] Nonruminant meat (pork, chicken, and horse meat) contains much lower amounts of CLA (1 mg/g lipid of lower),[54] whereas turkey meat shows unexpectedly high CLA contents (2–2.5 mg/g lipid).[41,55] The CLA content of other, less common meats in human diets, varies considerably (between 1 and 5 mg/g FAME) (elk, water buffalo, zebu-type cattle, bison)[38,56] with extremely high values reported for kangaroo (38 mg/g fatty acids in adipose tissue).[57] Large variations in the CLA content are not only reported among animal species but also in muscles of the same species. Hence, for ruminant meat types, or when CLA isomers are fed to animals, knowledge of the isomeric distribution of CLA may be required. However, analysis of CLA isomers is hampered by the lack of well-characterized reference materials. Commercially available CLA standard mixtures generally contain four major positional isomers (t8, c10; c9, t11; t10, c12; c11, t13) as well as smaller amounts of corresponding cis, cis and trans, trans isomers of each positional isomer. Only a few pure isomers are commercially available.[58]

Current methods for CLA quantification in biological samples consist of lipid extraction and fractionation of lipid classes by TLC or HPLC, before further sample workup and analysis are performed using Ag⁺-HPLC, GC, mass spectrometry, Fourier transform infrared spectroscopy, and nuclear magnetic resonance.

The most widely used method for total CLA quantification is GC-FID. CLA isomers need to be esterified before GC analysis. The most used esterification procedure is the transformation of the isomers into methyl esters. The choice of the methylation solvent is extremely important to maintain the original isomeric profile of the different CLA isomers. It is well-known that acid-based esterification solvents cause geometrical isomerization of the CLA isomers, resulting in increased proportions of trans, trans isomers.[32,59] Alkali-catalyzed methylation methods are the most reliable procedures for determining the CLA isomeric distribution. However, as mentioned earlier, these alkali-based esterification solvents are not capable of methylating free fatty acids. Therefore, combined esterification procedures, alkaline followed by an acid-based one, are proposed. Sodium methoxide is normally used as alkaline esterification solvent, whereas both methanolic hydrogen chloride[25,32] and boron trifluoride in methanol[33,60] are proposed as acid esterification solvent.

The identification of the CLA isomers by GC is mostly based on comparison with the retention time of the limited number of available standards. Highly polar cyanosilicone capillary columns of 100 m are recommended to resolve most of the closely related CLA isomers. On a 100 m CP-Sil88 column, CLA isomers elute just after α-linolenic acid methyl esters.[59] The elution order of CLA isomers with this column is first all cis, trans, followed by cis, cis and finally, all trans, trans isomers. For the same positional isomer, the cis, trans isomer elutes before the trans, cis geometric isomer. The CLA region in a GC chromatogram obtained with a CP-Sil88 column (100 m) was shown to be rather free of interfering FAME except for one peak corresponding with C21:0, which coelutes with t10, c12 or in the cis, cis area. It should be clarified that a polar capillary GC column is absolutely mandatory for the analysis of the closely related geometric and positional isomers of CLA, as nonpolar capillary columns (e.g., methylsilicone or phenylmethylsilicone phases) fail to resolve CLA isomeric profiles. Shorter polar columns (50 or 60 m) are more prone to interferences than a 100 m column.

Although the information on CLA isomeric composition provided by GC is incomplete, GC is often the only method used in the analysis of fatty acids for CLA. A further identification of CLA isomers by GC-MS lacks specificity. The mass spectra of CLA FAME isomers are indistinguishable from one another and from methylene-interrupted octadecadienoic acid methyl esters such as linoleate. However, by using GC-MS, the C18:2 methyl esters can be discriminated from other fatty acids in the same region of the chromatogram by the selective ion with a molecular mass of 294.[58] To distinguish between different CLA positional isomers, dimethyloxazolyne (DMOX) derivatives of unsaturated fatty acids are preferred as their separation is influenced by both position and geometry of double bonds of CLA.[61] All DMOX CLA derivatives contain ions at m/z 113, 126, and 333.

CLA FAME can also be selectively detected by their characteristic UV absorbance of 233 nm. The identification of isomers in HPLC chromatograms is based on coinjections of known references materials. The Ag^+-HPLC procedure was shown to separate the different trans, trans compounds, followed by a chromatographic zone where cis, trans isomers eluted. After the cis, trans area, cis, cis CLA isomers should be located. The best option to achieve a resolution of most CLA isomers in a reasonable analysis time is by using three Ag^+-HPLC columns in series.

Recently, Nuernberg et al.[33] compared four esterification procedures to quantify ten different CLA isomers in beef fat by using Ag^+-HPLC: a base-catalyzed method followed by an acid-catalyzed method at room temperature ($NaOCH_3/BF_3$, room temperature); a base-catalyzed method followed by an acid-catalyzed method at 60°C ($NaOCH_3/BF_3$, 60°C); a base-catalyzed method at 50°C ($NaOCH_3$, 50°C); a base hydrolysis and acid-catalyzed method at 60°C ($NaOH/MeOH$ and $BF_3/MeOH$, 60°C). They suggest to use a combination of a base-catalyzed ($NaOCH_3$) followed by an acid-catalyzed method (14% $BF_3/MeOH$) at 60°C for 10 min. With this procedure, the highest recovery with low variation, at the highest concentrations of all isomers, was found. The most abundant isomer was c9, t11, followed by t11, c13.

Sometimes, methyl oleate, if present in high amounts as in meat, can mask the cis, cis area. The major problem with Ag^+-HPLC is the retention volume drift that occurs over time. Ag^+-HPLC has also the disadvantage that it cannot distinguish the conjugated fatty acids with different carbon lengths. Further, the sample-load of nonCLA FAME strongly affects the resolution of CLA isomers by Ag^+-HPLC as nonconjugated FAME respond poorly at 233 nm.

9.6 Separation in Lipid Classes

Intramuscular fat of fresh meat consists, primarily, of TAG and PL. PL are amphiphilic molecules with lipophilic acyl chains and a hydrophilic head, and are components of biological membranes. The PL content in meat is nearly constant and typically ranges between 0.5 and 0.6 g/100 g of fresh muscle. In muscle, nine different PL classes—phosphatidylethanolamine, phosphatidylcholine, phosphatidylinositol, cardiolipin, sphingomyelin, phosphatidylserine, lysophosphatidylethanolamine, lysophosphatidylcholine, and phosphatidic acid—have been described. Phosphatidylethanolamine is usually the most abundant PL, followed by phosphatidylcholine in animal tissues.

HPLC is effective in the separation of selected lipid classes, but the separation of all tissue lipid classes is beyond the ability of a single HPLC column. TLC is currently the most reliable method for complete resolution of neutral and PL classes. The advantages of TLC, especially for qualitative and semiquantitative purposes, are its flexibility, simplicity, speed, and low cost. Several samples can be run simultaneously. After separation and visualization with a nondestructive reagent, spread on the plate, the individual components can be recovered by scraping them off the plate and using them for further analysis (fatty acid analysis, molecular species). Full recovery (as measured by GC) of the plates is often problematic with TLC plates. However, other systems, such as solid phase extraction (SPE) columns, do not give better recovery results. Skipski et al.[62] demonstrated good recoveries (±95%) of the different lipid classes in rat liver as measured by infrared spectrometry. Although a whole range of commercial SPE columns are available on the market, they are rarely used, especially because they lack specificity. In general, acidic PL, such as phosphatidylinositol and phosphatidic acid, are retained well on the packaging of the columns.[63]

9.7 Other Quantification Methods

Rapid methods for routine analysis of the fatty acid composition of muscle foods, have been investigated to overcome the laborious and expensive extraction, methylation and GC quantification methods: Near-infrared spectroscopy (NIRS) has been most studied and has the potential for a relatively accurate prediction of the total fatty acid content of meats and the content of classes of fatty acids (e.g., sum of saturated, cis-monounsaturated, trans-monounsaturated or polyunsaturated fatty acids).[64–66] The prediction accuracy for fatty acids strongly depends on the individual fatty acid, and is, in general, satisfactory or good for the most abundant fatty acids in meat (e.g., oleic acid and palmitic acid). Some studies have reported satisfactory results also for less abundant fatty acids, but most studies did not succeed in providing accurate predictions of minor fatty acids. The low level of fat and thus also individual fatty acids in most meats may interfere at this point. The prediction accuracy of NIRS is also considerably higher when applied to subcutaneous fat compared to intact meat or muscle-based foods. There are some indications that mid-infrared and Raman spectroscopy may yield higher prediction accuracy than NIRS.[67–69]

Appropriate data treatment procedures, for example, partial least squares regression and other multivariate statistics, are also required to develop and validate prediction models from spectroscopic data. In addition, the usefulness of prediction equations from spectroscopic data depends heavily on the calibration and validation of the models. Since fresh meats from different species and meat products vary in gross composition and physicochemical properties at several points, sufficient attention should be paid to developing models that are appropriate and specific for the foods that are to be measured. Infrared spectroscopic methods thus offer the potential for rapid fatty acid analyses, but have their limitations too, and more research is needed at this moment for widespread application. Similarly, more advanced separation and quantification methods, for example, GC-GC, need additional research.

9.8 Conclusions

GC is the primary choice for analyzing the fatty acid profile of muscle-based foods due to its high resolution and sensitivity. Fatty acids must first be converted into their nonpolar, volatile derivatives, mostly methyl esters. Attention must be paid to the choice of a suitable esterification procedure, depending on the fatty acids of interest. If a detailed profile of trans isomers or CLA isomers is aimed, preseparation by Ag^+-TLC or Ag^+-HPLC is necessary before further separation on a long polar capillary column by GC. At this point of time, alternative methods need an in-depth investigation before they can be used routinely for the fatty acid profile determination and quantification of muscle-based foods.

References

1. Christie, W.W. (Ed.), *Gas Chromatography and Lipids*, Oily Press, Ayr, 1989.
2. Folch, J., Lees, M., and Stanley, G.H.S., A simple method for the isolation and purification of total lipids from animal tissues, *J. Biol. Chem.*, 226, 497, 1957.
3. Bligh, E.G. and Dyer, W.J., A rapid method of total lipid extraction and purification, *Can. J. Biochem. Physiol.*, 37, 911, 1959.
4. Iverson, S.J., Lang, S.L.C., and Cooper, M.H., Comparison of the Bligh and Dyer and Folch methods for total lipid determination in a broad range of marine tissues, *Lipids*, 36, 1283, 2001.
5. Hara, A. and Radin, N.S., Lipid extraction of tissues with a low toxicity solvent, *Anal. Biochem.*, 90, 420, 1978.
6. Sahasrabudhe, M.R. and Smallbone, B.W., Comparative evaluation of solvent extraction methods for the determination of neutral and polar lipids in beef, *JAOCS*, 60, 801, 1983.
7. Eder, K., Gas chromatographic analysis of fatty acid methyl esters, *J. Chromatogr. B*, 671, 113, 1995.
8. Shanta, N.C. and Napolitano, G.E., Gas chromatography of fatty acids, *J. Chromatogr.*, 624, 37, 1992.
9. Sukhija, P.S. and Palmquist, D.L., Rapid method for determination of total fatty acid content and composition of feedstuffs and feces, *J. Agric. Food Chem.*, 36, 1202, 1988.
10. Sattler, W., Puhl, H., Hayn, M., Kostner, G.M., and Esterbauer, H., Determination of fatty acids in the main lipoprotein classes by capillary gas chromatography: BF3/methanol transesterification of lyophilized samples instead of Folch extraction gives higher yields, *Anal. Biochem.*, 198, 184, 1991.
11. Iverson, S.J., Sampugna, J., and Oftedal, O.T., Positional specificity of gastric hydrolysis of long-chain n-3 polyunsaturated fatty acids of seal milk triglycerides, *Lipids*, 27, 870, 1992.
12. Ulberth, F. and Henninger, M. One step extraction/methylation method for determining the fatty acid composition of processed foods, *JAOCS*, 69, 174, 1992.
13. Christie, W.W., Preparation of fatty acid methyl esters, *Inform*, 3, 1031, 1992.

14. Morrison, W.E. and Smith, L.M., Preparation of fatty acid methyl esters and dimethylacetals form lipids with boron fluoride-methanol, *J. Lipid Res.*, 5, 600, 1964.

15. Lough, A.K., The production of methoxy-substituted fatty acids as artefacts during the esterification of unsaturated fatty acids with methanol containing boron trifluoride, *Biochem. J.*, 90, 4C, 1964.

16. Fulk, W.K. and Shorb, M.S., Production of an artefact during methanolysis of lipids by boron trifluoride-methanol, *J. Lipid Res.*, 11, 276, 1970.

17. Vorbeck, M.L. et al., Preparation of methyl esters of fatty acids for gas-liquid chromatography. Quantitative comparison of methylation techniques. *Analyt. Chem.*, 33, 1512, 1961.

18. McGinnis, G.W. and Dugan, L.R., A rapid low temperature method for preparation of methyl esters of fatty acids, *JAOCS*, 42, 305, 1965.

19. Bannon, C.D. et al., Analysis of fatty acid methyl esters with accuracy and reliability, *J. Chromatogr.*, 247, 71, 1982.

20. Solomon, H.L. et al., Sample size influence on boron-methanol procedure for preparing fatty-acid methyl-esters, *JAOCS*, 51, 424, 1974.

21. Hubscher, G., Hawthorne, J.N., and Kemp, P., The analysis of tissue phospholipids: hydrolysis procedure and results with pig liver, *J. Lipid Res.*, 1, 433, 1960.

22. Glass, R.L., Alcoholysis, saponification and the preparation of fatty acid methyl esters, *Lipids*, 6, 919, 1971.

23. Schuchardt, U. and Lopes, O.C., Tetramethylguanidine catalyzed transesterification of fats and oils: a new method for rapid determination of their composition, *JAOCS*, 65, 1940, 1988.

24. Park, S.J. et al., Methylation methods for the quantitative analysis of conjugated linoleic acid (CLA) isomers in various lipid samples, *J. Agric. Food Chem.*, 50, 989, 2002.

25. Kramer, J.K. et al., Evaluating acid and base catalysts in the methylation of milk and rumen fatty acids with special emphasis on conjugated dienes and total *trans* fatty acids, *Lipids*, 32, 1219, 1997.

26. Brondz, I., Development of fatty acid analysis by high-performance liquid chromatography, gas chromatography and related techniques, *Anal. Chim. Acta*, 465, 1, 2002.

27. Agrawal, V.P. and Schulte, E., HPLC of Fatty acid isopropylidene hydrazides and its application in lipid analysis, *Anal. Biochem.*, 131, 356, 1983.

28. Rosenfeld, J.M., Application of analytical derivatizations to the quantitative and qualitative determination of fatty acids, *Anal. Chim. Acta*, 465, 93, 2002.

29. Mueller, H.W., Diazomethane as a highly selective fatty acid methylating reagent for use in gas chromatographic analysis, *J. Chromatogr. B*, 679, 208, 1996.

30. Luddy, F.E. et al., A rapid and quantitative procedure for the preparation of methyl esters of butteroil and other fats, *JAOCS*, 45, 549, 1968.

31. Moffat, C.F., McGill, A.S., and Anderson, R.S., The production of artifacts during preparation of fatty acid methyl esters from fish oils, food products and pathological samples, *J. High Resolut. Chromatogr.*, 14, 322, 1991.

32. Raes, K., De Smet, S., and Demeyer, D., Effect of double-muscling in Belgian Blue young bulls on the intramuscular fatty acid composition with emphasis on conjugated linoleic acid and polyunsaturated fatty acids, *Anim. Sci.*, 73, 253, 2001.

33. Nuernberg, K. et al., Comparison of different methylation methods for the analysis of conjugated linoleic acid isomers by silver ion HPLC in beef lipids, *J. Agric. Food. Chem.*, 55, 598, 2007.

34. Carreau, J.P. and Dubacq, J.P., Adaptation of macro-scale method to the micro-scale for fatty acid methyl transesterification of biological extracts, *J. Chromatogr.*, 151, 384, 1978.

35. Park, P.W. and Goins, R.E., In situ preparation of fatty acid methyl esters for analysis of fatty acid composition in foods, *J. Food Sci.*, 59, 1262, 1994.

36. Rule, D.C., Direct transesterification of total fatty acids of adipose tissue and freeze dried muscle and liver with boron-trifluoride in methanol, *Meat Sci.*, 46, 23, 1997.

37. Eras, J. et al., Chlorotrimethylsilane, a reagent for the direct analysis of fats and oils present in vegetable and meat samples, *J. Chromatogr. A*, 1047, 157, 2004.

38. Rule, D.C. et al., Comparison of muscle fatty acid profiles and cholesterol concentrations of bison, beef cattle, elk and chicken, *J. Anim. Sci.*, 80, 1202, 2002.

39. Eder, K., Reichlmayr-Lais, A.M., and Kirchgessner, M., Studies on the methanolysis of small amounts of purified phospholipids for GC analysis of small amounts of fatty acid methyl esters, *J. Chromatogr.*, 607, 55, 1992.

40. Ackman, R.G., The gas chromatograph in practical analyses of common and uncommon fatty acids for the 21st century, *Anal. Chim. Acta*, 465, 175, 2002.

41. Chin, S.F. et al., Dietary sources of conjugated linoleic dienoic isomers of linoleic acid, a newly recognized class of anticarcinogens, *J. Food Compos. Anal.*, 5, 185, 1992.

42. Kramer, J.K.G. and Zhou, J., Conjugated linoleic acid and octadecenoic acids: extraction and isola tion of lipids, *Eur. J. Lipid Sci. Technol.*, 103, 594, 2001.

43. Western, R.J. et al., Positional and geometric isomer separation of FAME by comprehensive 2-D GC, *Lipids*, 37, 715, 2002.

44. Precht, D. and Molkentin, J., Analysis and seasonal variation of conjugated linoleic acid and further cis-/trans-isomers of C18:1 and C18:2 in bovine milk fat, *Kieler Michwirtschaftl Forschungsberichte*, 51, 63, 1999.

45. Pfalzgraf, A., Timm, M., and Steinhart, H., Gehalte von trans Fettsäuren in Lebensmitteln, *Z. Ernährungswiss.*, 33, 24, 1994.

46. Raes, K. et al., Effect of diet and dietary fatty acids on the transformation and incorporation of C18 fatty acids in double-muscled Belgian Blue young bulls, *J. Agric. Food Chem.*, 52, 6035, 2004.

47. Dannenberger, D. et al., Effect of diet on the deposition of *n*-3 fatty acids, conjugated linoleic and C18:1 *trans* fatty acid isomers in muscle lipids of German Hollstein Bulls, *J. Agric. Food Chem.*, 52, 6607, 2004.

48. Wolff, R.L., Content and distribution of trans-18:1 acids in ruminant milk and meat fats. Their importance in European diets and their effect on human milk, *JAOCS*, 72, 259, 1995.

49. Morris, L.J., Wharry, D.M., and Hammond, E.W., Chromatographic behaviour of isomeric long-chain aliphatic compounds. II. Argentation thin-layer chromatography of isomeric fatty acids, *J. Chromatogr.*, 31, 69, 1967.

50. Wolff, R.L. and Bayard, C.C., Improvement in the resolution of individual trans C18:1 isomers by capillary gas-liquid chromatography: use of a 100-m CP-Sil88 column, *JAOCS*, 72, 1197, 1995.

51. Christie, W.W. A stable silver-loaded column for the separation of lipids by high performance liquid chromatography, *J. High Resol. Chromatogr. Commun.*, 10, 148, 1987.

52. Christie, W.W. and Breckenridge, G.H.M., Separation of cis and trans isomers of unsaturated fatty acids by HPLC in the silver ion mode, *J. Chromatogr.*, 469, 261, 1989.

53. Aveldano, M.I., van Rollins, M., and Horrocks, L.A., Separation and quantitation of free fatty acids and methyl esters by reversed phase high pressure liquid chromatography, *J. Lipid Res.*, 24, 83, 1983.

54. Schmid, A. et al., Conjugated linoleic acid in meat and meat products: a review, *Meat Sci.*, 73, 29, 2006.

55. Fritsche, J. and Steinhart, H., Analysis, occurrence and physiological properties of trans fatty acids (TFA) with particular emphasis on conjugated linoleic acid isomers (CLA)—a review, *Fett/lipid*, 100, 190, 1998.

56. De Mendoza, M.G. et al., Occurrence of conjugated linoleic acid in longissimus dorsi muscle of water buffalo (*Bubalus bubalis*) and zebu-type cattle raised under savannah conditions, *Meat Sci.*, 69, 93, 2005.

57. Engelke, C.F. et al., Kangaroo adipose tissue has higher concentrations of cis9trans11 conjugated linoleic acid than lamb adipose tissue, *J. Anim. Feed Sci.*, 13, 689, 2004.

58. De la Fuente, M.A., Luna, P., and Juarez, M., Chromatographic techniques to determine conjugated linoleic acid isomers, *Trends Analyt. Chem.*, 25, 917, 2006.

59. Kramer, J.K.G., Blackadar, C.B., and Zhou, J.Q., Evaluation of two GC columns (60-m SUPELCO-WAX 10 and 100-m CP sil 88) for analysis of milkfat with emphasis on CLA, 18:1, 18:2 and 18:3 isomers, and short- and long-chain FA, *Lipids*, 37, 823, 2002.

60. Park, Y. et al., Comparison of methylation procedures for conjugated linoleic acid and artifact formation by commercial (trimethylsilyl) diazomethane, *J. Agric. Food Chem.*, 49, 1158, 2001.
61. Roach, J.A.G. et al., Chromatographic separation and identification of conjugated linoleic acid isomers, *Anal. Chim. Acta*, 465, 207, 2002.
62. Skipski, V.P. et al., Quantitative analysis of simple lipid classes by thin layer chromatography, *Biochim. Biophys. Acta*, 151, 10, 1967.
63. Egberts, J. and Buiskool, R., Isolation of the acidic phospholipid phosphatidylglycerol from pulmonary surfactant by sorbent extraction chromatography, *Clin. Biochem.*, 34, 163, 1988.
64. Gonzalez-Martin, I. et al., Determination of fatty acids in the subcutaneous fat of Iberian breed swine by near infrared spectroscopy (NIRS) with a fibre-optic probe, *Meat Sci.*, 65, 713, 2003.
65. Berzaghi, P., Near-infrared reflectance spectroscopy as a method to predict chemical composition of breast meat and discriminate between different n-3 feeding sources, *Poult. Sci.*, 84, 128, 2005.
66. Pla, M. et al., Prediction of fatty acid content in rabbit meat and discrimination between conventional and organic production systems by NIRS methodology, *Food Chem.*, 100, 165, 2007.
67. Ripoche, A. and Guillard, A.S., Determination of fatty acid composition of pork fat by Fourier transform infrared spectroscopy, *Meat Sci.*, 58, 299, 2001.
68. Flatten, A. et al., Determination of C22:5 and C22:6 marine fatty acids in pork fat with Fourier transform mid-infrared spectroscopy, *Meat Sci.*, 69, 433, 2005.
69. Beattie, J.R. et al., Prediction of adipose tissue composition using Raman spectroscopy: average properties and individual fatty acids, *Lipids*, 41, 287, 2006.

Chapter 10

Triacylglycerols

Stephen B. Smith and Ron K. Tume

Contents

10.1 Introduction

Fatty acids in meat are located primarily in adipose tissues, and the majority of the fatty acids are stored as highly nonpolar *triacylglycerols* (Figure 10.1). Triacylglycerols coalesce into the large lipid vacuoles, which are the central features of adipocytes, and are also stored as lipid vacuoles in some type I muscle fibers. There is a renewed interest in not only the fatty acid composition but also the positional distribution of fatty acids within the triacylglycerol structure, as this influences the metabolic effects of dietary fatty acids as well as the functionality of the lipid. Stearic acid located

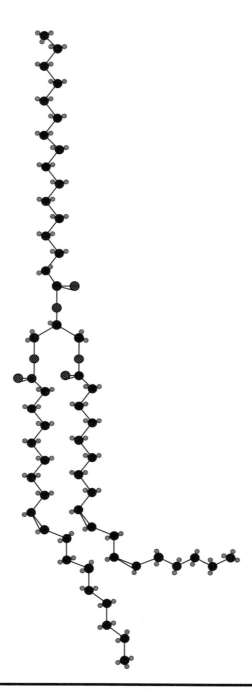

Figure 10.1 Typical structure of a triacylglycerol: sn-1 fatty acid (on left), oleic acid; sn-2 fatty acid, palmitic acid; sn-3 fatty acid, linoleic acid. This triacylglycerol would be common in porcine adipose tissue. In bovine and ovine adipose tissue, palmitic acid would occupy the sn-1 position and stearic or oleic acid would occupy the sn-2 position. Large filled circles represent carbon; large shaded circles represent oxygen; small shaded circles represent hydrogen. (From Smith, S.B., Smith, D.R., and Lunt, D.K., *Encyclopedia of Meat Science*, 2004.)

in the outer positions of triacylglycerols is poorly absorbed [1], which may in part explain its hypocholesterolemic nature. There is a greater proportion of palmitic acid in the middle position of triacylglycerols in the plasma of young piglets fed Cow's milk, which is enriched with palmitic acid in the middle position, than in plasma of pigs fed a formula with palmitic acid in the outer triacylglycerol positions [2]. This is consistent with the direct absorption of 2-monoacylglycerols from the digestive tract, which ultimately may influence the potential for palmitic acid to elevate the level of cholesterol in humans. This and similar reports suggest that positional distribution of fatty acids, in combination with their degree of unsaturation, can have a significant impact on human cholesterol metabolism. Also, placement of stearic acid in the outer position of triacylglycerols can increase the melting point by as much as 10°C relative to triacylglycerols with stearic acid solely in the middle position [3]. Thus, not only the composition of triacylglycerols, but also the positional distribution of the fatty acids within triacylglycerols, has a significant impact on the functionality of lipids from animal tissues.

10.2 Chemistry and Nomenclature

10.2.1 Triacylglycerol Structure

Relative to the other lipid classes, the structure of triacylcerols is relatively simple. Triacylglycerols consist of a glycerol (i.e., three-carbon alcohol) backbone containing three fatty acids in ester linkage. The glycerol, primarily, is derived from glycerol-3-phosphate, which in turn is produced by the metabolism of glucose or, in liver, by the phosphorylation of free glycerol catalyzed by glycerokinase. In adipose tissue and intestinal mucosal cells, 2-monoacylglycerol (produced by the partial hydrolysis of triacylglycerols) provides a portion of the carbon backbone for triacylglycerol synthesis.

The most abundant fatty acids of animal triacylglycerols are palmitic (16:0), stearic (18:0), and oleic acid (18:1n-9), typically comprising 20–25%, 10–30%, and 30–55% of the total fatty acids in muscle and adipose tissues. Smaller but significant quantities of myristic (14:0), palmitoleic (16:1n-7), linoleic (18:2n-6), and α-linoleic acid (18:3n-3) are contained in triacylglycerols, their concentrations dictated by the species and the diet. Lipids from ruminant tissues also contain measurable amounts of odd-chained fatty acids, trans fatty acids, and conjugated fatty acids, due to the absorption and metabolism of the products of ruminal fermentation. Virtually all Δ9 desaturase products of saturated fatty acids can be detected in triacylglycerols from animal tissues. However, with the exception of oleic acid, the concentrations of monounsaturated fatty acids are sparingly low in tissues in which Δ9 desaturase activity is depressed. This indicates that dietary fats are not significant sources of monounsaturated fatty acids other than oleic acid.

10.2.2 Stereospecific Numbering

A brief description of the stereospecific numbering (i.e., *sn-*) system for triacylglycerols is necessary. The fatty acids in the outer positions can be distinguished based on the specific isomerization of the ʟ-glycerol. The alcohol group of the number two (i.e., sn-2) carbon extends to the left of the carbon in the Fischer projection of glycerol, and in this representation the top carbon is designated as the sn-1 carbon, whereas the bottom carbon is the sn-3 carbon. The numbering of triacylglycerols, therefore, is stereospecific, which provides a convenient means of describing triacylglycerol composition.

10.3 Determining Triacylglycerol Composition

Lipid extracts from animal tissues can be analyzed for total fatty acid composition; the triacylglycerol classes can be separated by silver-ion chromatography; the positional distribution of fatty acids can be quantified by specific lipase digestion; or some combination of these methods can be used to provide essentially complete information about the composition of the triacylglycerols. Several texts [3–5] provide exhaustive details about the procedures that are overviewed here.

10.3.1 Lipid Extraction

Lipids frequently are extracted by the Folch [6] or Bligh and Dyer [7] procedures. Both procedures use a mixture of chloroform and methanol as solvents, and effectively extract phospholipids as well as neutral lipids. The Bligh and Dyer (1957) method is useful for extraction of lipids from muscle tissues, in that it allows for the high water content (70–80%), which may cause phase separation and thereby incomplete extraction of lipids.

Once lipids have been extracted and insoluble cellular structures have been removed by centrifugation or filtering, a polar solvent is added to separate polar solutes from lipids. Phase separation is accomplished by adding potassium chloride [6] or water [7]. Nonesterified fatty acids can also be removed by using a solution of potassium carbonate or potassium bicarbonate in place of potassium chloride, in that it raises the pH of the mixture and thereby converts the fatty acids to their potassium salts. These have sufficient solubility in the aqueous phase to allow their removal from the other lipid classes.

Lipids extracted from adipose tissues primarily consist of triacylglycerols, but also contain significant quantities of phospholipids, mono- and diacylglycerols, and cholesterol esters. Thus, any measurement of a total lipid extract of muscle or adipose tissue represents the composition of all of these components. If desired, neutral lipids can be separated from the phospholipids, and triacylglycerols can be isolated from other neutral lipids, by thin-layer chromatography. Using silica gel as the stationary phase on glass or plastic plates, solvent systems with low polarity separate neutral lipids from phospholipids and also effectively isolate triacylglycerols from cholesterol esters and the other glyceride species [8].

10.3.2 Measurement of Fatty Acid Composition

Gas/liquid chromatography of fatty acid methyl ester derivatives of tissue fatty acids is the method of choice for quantifying fatty acids, based on its high reproducibility and relatively low cost. This procedure is not restricted to the analysis of triacylglycerols; it is the most commonly used procedure for the analysis of the fatty acid composition of animal tissues. Many of the investigators, including ourselves [9], originally used packed columns that were typically 15–25 cm in length and provided only modest separation of the fatty acid methyl ester peaks. The primary shortcoming of these was their inability to separate the various 18-carbon monounsaturated fatty acids. Thus, values reported for oleic acid also included significant quantities of *trans*-vaccenic acid (18:1trans-11) and *cis*-vaccenic acid (18:1cis-11 or 18:1n-7, the elongation product of palmitoleic acid). In addition, values for palmitoleic acid may have included 17:0, and 17:1n-8 frequently was combined with other, minor peaks. Identification of the various 18-carbon monounsaturated fatty acids, in addition to other less abundant fatty acids, was made possible by the development

Table 10.1   sn-2, and sn-1/3 Fatty Acid Composition (wt%) and Slip Points (°C) of Triacylglycerol Classes in Adipose Tissue Lipids from Corn-Fed Cattle, Japanese Black Cattle, and Cottonseed-Fed Cattle

Position/Fatty Acid	Corn-Fed	Japanese Black	Cottonseed-Fed
sn-2			
Palmitic	22	16	20
Palmitoleic	5	6	3
Stearic	10	8	16
trans-Vaccenic	3	0.4	7
Oleic	52	62	43
cis-Vaccenic	3	1	0.5
Linoleic	2	3	4
α-Linolenic	0.1	0.1	0.6
sn-1/3			
Palmitic	33	42	39
Palmitoleic	4	4	1
Stearic	13	8	28
trans-Vaccenic	5	1	18
Oleic	35	35	11
cis-Vaccenic	3	6	2
Linoleic	4	6	2
α-Linolenic	0.5	0.3	0
Slip point	30.7	22.8	41.5

Note: The corn-fed and cottonseed-fed steers were produced in Australia, whereas the Japanese Black cattle were raised in Japan.

Source: From Smith, S.B., Yang, A., Larson, T.W., and Tume, R.K., *Lipids*, 33, 197, 1998. With permission.

of the capillary columns of up to 100 m in length. Fatty acid compositions such as those reported in Table 10.1 now include most of the known fatty acids present in animal tissues in reasonable abundance [10].

Methylation of the triacylglycerol fatty acids is a prerequisite to their quantification by gas/liquid chromatography. This can be accomplished by trans esterification in boron trifluoride under alkaline conditions [11], which methylates only those fatty acids that are in ester linkage. Alternatively, trans methylation can be conducted under acidic conditions, resulting in methylation of both glyceride fatty acids and nonesterified fatty acids.

10.3.3   Separation of Triacylglycerol Classes

Triacylglycerols can be separated from total lipid extracts by argentation thin-layer chromatography (TLC) or silver ion high-performance liquid chromatography (HPLC) [5,12], both of which employ the strong attraction of silver ions to the double bonds of unsaturated fatty acids. TLC plates and HPLC columns are impregnated with 5–10% silver nitrate, which separates the triacylglycerol classes based on the total number of double bonds contained within the triacylglycerol

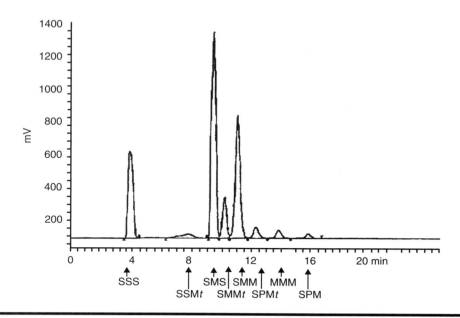

Figure 10.2 Chromatogram from high-performance lipid chromatographic separation of triacylglycerol species from bovine adipose tissue samples. Triacylglycerol species were separated on a 5 μm silica column impregnated with 5% silver nitrate. *Abbreviations*: **SSS, sn-1,2,3-saturated fatty acid triacylglycerol; SSM*t*, triacylglycerols containing two saturated and one trans monounsaturated fatty acids; SMS, triacylglycerols containing two saturated and one monounsaturated fatty acids; SMM*t*, triacylglycerols containing one saturated, one monounsaturated, and one trans monounsaturated fatty acid; SMM, triacylglycerol containing one saturated and two monounsaturated fatty acids; SPM*t*, triacylglycerol containing one saturated, one polyunsaturated, and one trans monounsaturated fatty acid; MMM, sn-1,2,3-monounsaturated fatty acid triacylglycerol; SPM, triacylglycerol containing one saturated, one monounsaturated, and one polyunsaturated fatty acid. The order of the acronym indicates the relative abundance of fatty acid classes in each position (see Tables 10.2, 10.3, and 10.5). (Adapted from Smith, S.B., Yang, A., Larson, T.W., and Tume, R.K.,** *Lipids*, **33, 197, 1998.)**

(Figure 10.2) [13]. The HPLC method has the advantage that individual triacylglycerol classes can be quantified by evaporative light scattering, yielding the relative mass of each fraction. By either method, triacylglycerol species can be collected and analyzed individually for total fatty acid composition, or they can be analyzed for the positional distribution of the fatty acids within each triacylglycerol species.

Triacylglycerols separated by silver ion chromatography can be classified only on the basis of the total number of double bonds, not based on the stereospecific position of the fatty acids in the triacylglycerol. Thus, SSS is an sn-1,2,3-saturated fatty acid triacylglycerol; SSM*t* consists of two saturated and one trans monounsaturated fatty acids; MMM is an sn-1,2,3-monounsaturated fatty acid triacylglycerol; SPM consists of one saturated, one monounsaturated, and one polyunsaturated fatty acid (Table 10.2), but positional distribution of the individual fatty acids may be unknown. In Table 10.2, the order of the acronym (e.g., SMS) indicates the relative concentration of fatty acids at each position, which was deduced by measuring the positional distribution of the fatty acids within the triacylglycerols.

Table 10.2 Fatty Acid Composition (wt%) of the Major Triacylglycerol Classes of Lipids from Adipose Tissue of Cottonseed-Fed Cattle

Position/Fatty Acid	SSS	SMS	SMMt	SMM	MMM
sn-2					
Palmitic	14	15	16	12	10
Palmitoleic	0	1	1	3	0
Stearic	82	31	18	13	29
trans-Vaccenic	0	5	21	3	0
Oleic	0	46	39	67	53
Linoleic	0	0	1	0	9
sn-1/3					
Palmitic	26	24	21	31	4
Palmitoleic	0	0	1	3	5
Stearic	75	50	22	12	37
trans-Vaccenic	0	9	28	5	17
Oleic	0	8	21	42	35
Linoleic	0	0	2	0	0

Note: SSS, sn-1,2,3-saturated fatty acid triacylglycerol; SMS, triacylglycerols containing two saturated and one monounsaturated fatty acids; SMM*t*, triacylglycerols containing one saturated, one monounsaturated, and one trans monounsaturated fatty acid; SMM, triacylglycerols containing one saturated and two monounsaturated fatty acids; MMM, sn-1,2,3-monounsaturated fatty acid triacylglycerols.

Source: From Smith, S.B., Yang, A., Larson, T.W., and Tume, R.K., *Lipids*, 33, 197, 1998. With permission.

10.3.4 Positional Distribution of Fatty Acids

Esterification of saturated fatty acids, especially stearic acid, to the outer positions of triacylglycerols can cause a disproportionate increase in the melting point of the lipids, or may increase the hypercholesterolemic nature of dietary lipids. Therefore, it is informative to measure the positional distribution of fatty acids within triacylglycerols. Fischer et al. [14] first described the use of the nonpancreatic lipase (from *Rhizopus arrhizus delemar*) for the analysis of fatty acid distributions in lipid classes. Lipase from *R. arrhizus* removes only the fatty acids in the sn-1/3 positions. Methylation under alkaline conditions methylates the fatty acid of the resultant 2-monoacylglycerol, but not nonesterified fatty acids; thus, the fatty acid composition at the sn-2 position can be determined unambiguously. The lipase-digested and undigested (total) triacylglycerols, therefore, can be used to calculate the average composition of fatty acids in the sn-1/3 positions [15]:

Average % sn-1/3 = (3 × % fatty acid in total lipids) − (2 × % fatty acid in the sn-2 position)

If more detailed information is needed, conversion of 2-monoacylglycerols to phospholipids followed by digestion with phospholipase A$_2$ can distinguish the fatty acids in the outer sn-positions [15]. Phospholipase A$_2$ hydrolyzes the fatty acids from the sn-2 positions of phospholipids and other phosphate esters. Partial hydrolysis of triacylglycerols to 1,2- or 2/3-diacylglycerols by Grignard reagent (EtMgBr) is followed by conversion to the respective phosphate esters [16]. Phospholipase A$_2$ digests only sn-3 synthetic phosphate esters, leaving the 1-*lyso*-phosphate ester and the 2,3-phosphate ester; separation and methylation of these products yield the fatty acid composition at the sn-1 carbon and the average of the sn-2,3 carbons. The fatty acid composition of the sn-3 position, therefore, can be calculated [16]:

$$\% \text{ sn-3} = (3 \times \% \text{ fatty acids in total lipids}) - (\% \text{ sn-2 fatty acids}) - (\% \text{ sn-1 fatty acids})$$

or

$$\% \text{ sn-3} = (2 \times \text{average } \% \text{ sn-2,3 fatty acids}) - (\% \text{ sn-2 fatty acids})$$

In practice, the production and isolation of the synthetic phosphate esters is exacting, whereas determination of the sn-2 and average sn-1/3 fatty acid composition, though less informative, is relatively straightforward.

10.4 Natural Variations in Triacylglycerol Structure

10.4.1 Ruminant Adipose Tissue Triacylglycerols

Until recently, it was accepted that lipids from ruminant adipose tissue and muscle were resistant to dietary modification [9,17]. In most texts, only single compositions are listed for ovine or bovine adipose tissue lipids, with little regard to differences in production conditions or differences across breed types [4,5]. However, for a period of time in Australia, cattle were produced with vast differences in the fatty acid composition of their adipose tissue lipids; under some conditions, adipose tissue lipids were remarkably enriched in stearic acid [10,18]. These could be contrasted with lipids from Japanese Black cattle raised in Japan, noted for their high concentrations of monounsaturated fatty acids [19,20]. This provided an opportunity to document changes in triacylglycerol composition in cattle that had not been accomplished previously.

Slip points (estimators of melting points; Ref. 4) were nearly 20°C higher in adipose tissue lipids from cattle fed whole cottonseed for >400 days than in lipids of Japanese Black cattle fed a grain-based diet for >500 days in Japan (Table 10.1; data derived from Ref. 10). Similarly, adipose tissue lipids from cottonseed-fed cattle were enriched in stearic and *trans*-vaccenic acid, especially at the sn-1/3 positions. Lipids from corn-fed steers were intermediate in slip points and saturated and trans fatty acids compared to those from the Japanese and cottonseed-fed steers.

Fatty acid composition was also measured for each of the triacylglycerol species. This approach provides considerable insight into the factors that influence the functionality of beef lipids (Table 10.2). The relative abundance of saturated, monounsaturated, and trans monounsaturated fatty acids differed substantially among triacylglycerol species; this is to be expected, as they were separated on the basis of the number of double bonds. It is noteworthy that silver ion HPLC can distinguish between cis and trans monounsaturated fatty acids, in that it effectively separated SMM*t* from SMM triacylglycerols (confirmed by the relative abundance of *trans*-vaccenic acid in the SMM*t* triacylglycerols).

It is apparent from the data in Table 10.2 that SSS triacylglycerols are not homogeneous, but contain a mixture of saturated fatty acids (primarily palmitic and stearic acid). Not surprisingly, the MMM triacylglycerols contain a combination of oleic, palmitoleic, and myristoleic (14:1n-5) monounsaturated fatty acids [10]. However, the MMM triacylglycerol species from the cottonseed-fed cattle contained an unusual mixture of saturated, monounsaturated, *trans*-vaccenic, and linoleic acid ([10]; Table 10. 2). This combination of fatty acids allowed it to migrate with the same retention time as authentic MMM triacylglycerols, although it clearly could have been classified as SPM*t*. This illustrates the importance of confirming the fatty acid composition of triacylglycerol species.

Table 10.3   Distribution of Triacylglycerol Classes (Relative Percentages) of Adipose Tissue Lipids from Corn-Fed Cattle, Japanese Black Cattle, and Cottonseed-Fed Cattle

Triacylglycerol Class	Corn-Fed	Japanese Black	Cottonseed-Fed
SSS	4	2	9
SSM*t*	<1	<1	5
SMS	23	19	34
SMM*t*	6	2	17
SMM	47	55	27
MMM	14	19	3
SPM	3	2	1

Note: The corn-fed and cottonseed-fed steers were produced in Australia, whereas the Japanese Black cattle were raised in Japan. Abbreviations not listed in Table 10.2: SSM*t*, triacylglycerols containing two saturated and one trans monounsaturated fatty acids; SPM*t*, triacylglycerol containing one saturated, one polyunsaturated, and one trans monounsaturated fatty acid; SPM, triacylglycerol containing one saturated, one monounsaturated, and one polyunsaturated fatty acid. The order of the acronym indicates the relative abundance of fatty acid classes in each position.

Source: From Smith, S.B., Yang, A., Larson, T.W., and Tume, R.K., *Lipids*, 33, 197, 1998. With permission.

Palmitic acid in bovine adipose tissue triacylglycerols is concentrated in the sn-1/3 positions, whereas oleic acid is concentrated in the sn-2 position. In the highly saturated lipids from the cottonseed-fed cattle, there is a remarkably high concentration of palmitic acid in the outer positions. This increases the potential for these lipids to be particularly hypercholesterolemic as the sn-1/3 fatty acids are hydrolyzed by pancreatic lipase [2].

With higher slip points, there is a profound increase in the triacylglycerol species SSS, SMS, and SSM*t*, and SMM*t*, reflecting the greater concentrations of saturated and trans monounsaturated fatty acids in these lipids. For the saturated fatty acids, the marked elevation in stearic acid (Table 10.1) accounted for the enrichment of triacylglycerols with SSS and SMS in samples from cottonseed-fed cattle (Table 10.3); lipids from lamb adipose tissue contain a similar triacylglcyerol composition (see below). These data clearly indicate that there can be substantial differences in the makeup of triacylgylycerols in lipids of ruminant species.

10.4.2   *Pig and Sheep Adipose Tissue Lipids*

The fatty acid composition of pig adipose tissue triacylglycerols closely resembles the lipid composition of their diets. Thus, diets enriched with oleic acid [9,21] or linoleic acid [22] will increase the corresponding fatty acids in their adipose tissue and muscle lipids. Inhibiting $\Delta 9$ desaturase activity by feeding conjugated linoleic acid to pigs causes an enrichment of the sn-1/3 position with stearic acid, concomitant with an increase in slip point (Table 10.4).

Unlike lipids from adipose tissues from cattle and sheep, lipids from pig adipose tissue contain substantial quantities of linoleic acid. Correspondingly, the MPP (triacylglycerol containing two monounsaturated and one polyunsaturated fatty acid) and PPP (sn-1,2,3 polyunsaturated

Table 10.4   sn-2, and sn-1/3 Fatty Acid Composition (wt%) and Slip Points (°C) of Triacylglycerol Classes in Adipose Tissue Lipids from Pigs Fed Diets Supplemented with Corn Oil, Beef Tallow, or Mixed Isomers of Conjugated Linoleic Acid

Position/Fatty Acid	Corn Oil	Beef Tallow	Conjugated Linoleic Acid
sn-2			
Palmitic	23	27	33
Stearic	9	12	13
Oleic	37	34	24
Linoleic	20	15	15
CLA isomers	0.1	0.5	3.3
sn-1/3			
Palmitic	21	29	41
Stearic	10	14	25
Oleic	38	34	16
Linoleic	22	13	9
CLA isomers	0.2	0	1.2
Slip point	28	29	38

Source: Adapted from King, D.A., Behrends, J.M., Jenschke, B.E., Rhoades, R.D., and Smith, S.B., *Meat Sci.*, 67, 675, 2004. With permission.

Table 10.5   Distribution of Triacylglycerol Classes (Relative Percentages) of Adipose Tissue Lipids from Pasture-Fed Sheep, Sorghum-Fed Pigs, and Canola Oil-Fed Pigs

Triacylglycerol Class	Pasture-Fed Sheep	Sorghum-Fed Pigs	Canola Oil-Fed Pigs
SSS	15	2	1
SSM*t*	1	ND	ND
SMS	40	20	12
SMM*t*	9	1	ND
SMM	25	31	25
MMM	3	10	12
SPM	2	16	19
MMP	ND	7	14
MPP	ND	2	4
PPP	ND	1	5

Note: ND—not detectable. Sheep grazed native pasture in Queensland, Australia. Pigs were fed either a standard sorghum-based diet or a sorghum-based diet enriched with canola oil. Abbreviations not listed in Tables 10.2 and 10.3: MMP, triacylglycerol containing two monounsaturated and one polyunsaturated fatty acid; MPP, triacylglycerol containing one monounsaturated and two polyunsaturated fatty acids; PPP, sn-1,2,3 polyunsaturated triacylglycerol.

Source: Smith, S.B. and Tume, R.K., unpublished.

triacylglycerol) triacylglycerol species, which are absent in ruminant adipose tissue lipids, are measurable in pig adipose tissue lipids (Table 10.5). Conversely, SSS triacylglycerols and triacylglycerols containing trans fatty acids are much lower in pig adipose tissue lipids than in lipids from ruminant species. Although diets of ruminants contain relatively high concentrations of linoleic acid, ruminal hydrogenation of polyunsaturated fatty acids to stearic and *trans*-vaccenic acid is essentially complete, especially in pasture-fed animals [23]. This clearly is reflected in the composition of their adipose tissue triacylglycerols.

10.5   Conclusion

It is well established that the fatty acid composition of muscle and adipose tissues of livestock species can be modified by diet, specific production practices, or breed type. We now know that triacylglcyerol species as well as the positional distribution of fatty acids within the triacylglycerols can also be modified substantially. Alterations in triacylglycerol species largely explain the differences in lipid functionality.

References

1. Monsma, C. C. and Ney, D. M. 1993. Interrelationship of stearic acid content and triacylglycerol composition of lard, beef tallow and cocoa butter in rats. *Lipids* 28:539–547.
2. Innis, S. M., Dyer, R., Quinlan, P. T., and Diersen-Schade, D. 1996. Dietary triacylglycerol structure and saturated fat alter plasma and tissue fatty acids in piglets. *Lipids* 31:497–505.
3. Gunstone, F. D., Harwood, J. L., and Padley, F. B. 1994. *The Lipid Handbook*, 2nd ed. Chapman and Hall, London.
4. Gunstone, F. D. 1996. *Fatty Acid and Lipid Chemistry*. Blackie Academic & Professional, London.
5. Christie, W. W. 2003. *Lipid Analysis*. The Oily Press, Bridgwater, England.
6. Folch, J., Lees, M., and Stanley, G. H. S. 1957. A simple method for the isolation and purification of total lipids from animal tissues. *J. Biol. Chem.* 226:497–509.
7. Bligh, E. G. and Dyer, W. J. 1959. A rapid method of total lipid extraction and purification. *Can. J. Biochem. Physiol.* 37:911–917.
8. Hammond, E. 1993. *Chromatography for the Analysis of Lipids*. CRC Press, Boca Raton, FL.
9. St. John, L. C., Young, C. R., Knabe, D. A., Schelling, G. T., Grundy, S. M., and Smith, S. B. 1987. Fatty acid profiles and sensory and carcass traits of tissues from steers and swine fed an elevated monounsaturated fat diet. *J. Anim. Sci.* 64:1441–1447.
10. Smith, S. B., Yang, A., Larson, T. W., and Tume, R. K. 1998. Positional analysis of triacylglycerols from bovine adipose tissue lipids varying in degree of unsaturation. *Lipids* 33:197–207.
11. Morrison, W. R. and Smith L. M. 1964. Preparation of fatty acid methyl esters and dimethylacetals from lipids with boron flouride-methanol. *J. Lipid Res.* 5:600–607.
12. Christie, W. W. 1987. A stable silver-loaded column for the separation of lipids by high performance liquid chromatography. *J. High Resolut. Chromatogr. Commun.* 10:148–150.
13. Smith, S. B., Smith, D. R., and Lunt, D. K. 2004. Chemical and physical characteristics of meat: Adipose tissue. In: W. Jensen, C. Devine, and M. Dikemann (Ed.) *Encyclopedia of Meat Sciences.* pp. 225–238, Elsevier Science Publishers, Oxford.
14. Fischer, W., Heinz, E., and Zeus, M. 1973. The suitability of lipase from *Rhizopus arrhizus delemar* for analysis of fatty acid distributions in dihexosyl diglycerides, phospholipids and plant sulfolipids. *Hoppe-Seyler's Z. Physiol. Chem.* 354:1115–1123.

15. Williams, J. P., Khan, M. U., and Wong, D. 1995. A simple technique for the analysis of positional distribution of fatty acids on di- and triacylglycerols using lipase and phospholipase A₂. *J. Lipid Res.* 36:1407–1412.

16. Brockerhoff, H. 1971. Stereospecific analysis of triglycerides. *Lipids* 6:942–956.

17. Chang, J. H. P., Lunt, D. K., and Smith, S. B. 1992. Fatty acid composition and fatty acid elongase and desaturase activities in tissues of steers fed high-oleate sunflower seed. *J. Nutr.* 122:2074–2080.

18. Yang, A., Larsen, T. W., Smith, S. B., and Tume, R. K. 1999. Δ⁹ desaturase activity in bovine subcutaneous adipose tissue of different fatty acid composition. *Lipids* 34:971–978.

19. Sturdivant, C. A., Lunt, D. K., Smith, G., and Smith S. B. 1992. Fatty acid composition of subcutaneous and intramuscular adipose tissues and *M. longissimus dorsi* of Wagyu cattle. *Meat Sci.* 32:449–458.

20. Zembayashi, M., Nishimura, K., Lunt, D. K., and Smith, S. B. 1995. Effect of breed type and sex on the fatty acid composition of subcutaneous and intramuscular lipids of finishing steers and heifers. *J. Anim. Sci.* 73:3325–3332.

21. Klingenberg, I. L., Knabe, D. A., and Smith, S. B. 1995. Lipid metabolism in pigs fed tallow or high oleic acid sunflower oil. *Comp. Biochem. Biophys.* 110:183–192.

22. King, D. A., Behrends, J. M., Jenschke, B. E., Rhoades, R. D., and Smith, S. B. 2004. Positional distribution of fatty acids in triacylglycerols from subcutaneous adipose tissue of pigs fed diets enriched with conjugated linoleic acid, corn oil, or beef tallow. *Meat Sci.* 67:675–681.

23. Chung, K. Y., Lunt, D. K., Choi, C. B., Chae, S. H., Rhoades, R. D., Adams, T. H., Booren, B., and Smith, S. B. 2006. Lipid characteristics of subcutaneous adipose tissue and *M. longissimus thoracis* of Angus and Wagyu steers fed to U.S. and Japanese endpoints. *Meat Sci.* 73:432–441.

Chapter 11

Analysis of Phospholipids in Muscle Foods

Jorge Ruiz, Elena Muriel, Trinidad Perez-Palacios, and Teresa Antequera

Contents

11.1 Introduction

There has been an enormous number of studies during the last 50 years devoted to studying the effect of different factors on the composition of triglycerides (TG) in muscle and muscle-based food products. However, the amount of research concerning polar lipids in muscle and muscle foods is much smaller, and most of this has been aimed at studying the fatty acid (FA) profile of total polar lipids. Nevertheless, in the last decade there has been a growing interest in more detailed analysis of this lipid fraction in muscle foods, since it has been suggested by several researchers that lipid oxidation in muscle and muscle foods initiates and mainly takes place in membrane phospholipids (PL).[1,2] The high sensitivity of PL to oxidation in meat and meat products has two primary causes: the high proportion of long chain polyunsaturated FA (PUFA), which are very susceptible to oxidation, and the close contact of PL with catalysts of lipid oxidation located in the aqueous phase of the muscle cell (such as reactive oxygen species formed during normal metabolism of mitochondria or during peroxisomal function, the iron released from heme-proteins, or these iron containing proteins themselves), since PL are the major cell and organelle membrane components.[3]

It has been suggested that PL may increase the oxidative stability of fatty foods, acting as synergists of tocopherols.[4] Moreover, PL can also bind heavy metals, which act as prooxidants, to produce inactive, undissociated salts.[5] On the other hand, the FA composition of PL has been directly linked to the oxidation of ω-3 PUFA by specific lipoxygenases[6,7] and by autooxidation reactions.[8,9] In addition, the FA composition of total PL is a subject of great interest because increasing the proportion of long chain PUFA may increase the susceptibility of muscle foods to lipid oxidation.[3] Some years ago it was believed that the FA profile of cell membranes was only slightly influenced by external factors. However, currently there is evidence in mammals, birds, and marine animals that the FA composition of muscle PL can be influenced by several factors, including the muscle fiber type,[10,11] age,[12] and especially the diet.[11,13]

The distribution of FA in the *sn*-1 and *sn*-2 positions of muscle PL may show important consequences on the quality of muscle foods: It is well known that the *sn*-2 position of glycerophospholipids is more protected against oxidative deterioration,[14] and thus the study of potential factors affecting changes in the stereospecific distribution of FA in muscle PL classes is of interest. Moreover, the occurrence and amount of alkylacyl and alkenylacyl glycerophospholipids seems to influence the susceptibility of membrane PL to lipid oxidation, protecting molecular species of PL against oxidative degradation.[14,15]

While most research concerning muscle and muscle foods PL has been devoted to studying their total FA profile, only a few works have been aimed at studying in more detail the muscle PL classes or the distribution of alkyl, alkenyl, and acyl chains in such molecular species of PL. The present chapter reviews the analytical procedures aimed at analyzing different aspects of muscle PL composition.

11.2 Muscle Phospholipids

PLs are the main structural lipid components of cell and organelle membranes in animal tissues. As a consequence of their preferential location in membranes, these compounds show strong bindings with membrane proteins through hydrogen bonds and hydrophobic interactions,[16] and thus their extraction has been difficult. There is no general agreement on the best way to classify PL, but most classifications contain a category for glycerol-containing PL (glycerophosphatides or phosphoglycerides) and one for sphingolipids (sphingosylphosphatides or phosphosphingolipids).[17] PL are usually considered complex lipids or polar lipids and are the major chemical class in this family.

Phosphoglycerides have a glycerol backbone in which the *sn*-2 position is acylated by an FA, whereas the *sn*-1 position may show either an alkyl or an alkenyl chain linked by an ether linkage or an acyl chain by an ester linkage, and the *sn*-3 position is esterified with phosphoric acid. The whole molecule is called phosphatidyl, while groups in which only one of the carbons in the glycerol is acylated are named lysophosphatidyl. The phosphate group can remain free, giving rise to phosphatidic acid, or it can be substituted with an alcohol or an amino alcohol derivative, giving rise to the major PL classes, including phosphatydilcholine (PC), phosphatidylethanolamine (PE), phosphatidylserine (PS), phosphatidylinositol (PI), and cardiolipin (CL) (Figure 11.1). Sphingomyelin (SPH) is a member of the sphingolipid family derived from sphingosine (a long chain amino alcohol) to which a FA is linked by an amide bond and a molecule of phosphoryl choline esterifies the primary hydroxyl group (Figure 11.1).

The amount and type of PL classes in muscle is rather constant as compared to triacylglycerols.[18] Depending on the neutral lipid (NL) content, the proportion of PL in muscle foods may range from 40% of total lipid extract in low NL content foods (such as marine invertebrates)[17] to 4–5% in high intramuscular fat content muscles.[18] In most muscle foods, major PL are PC and PE, usually accounting for more than 50%. The occurrence of lysophospholipids is due either to lysis during the sampling and analysis, or to the presence of such molecules due to processing conditions or their natural occurrence in some muscle foods. Processing of meat products may lead to a decrease in all PL classes, due either to the activity of endogenous or microbial phospholipases[19,20] or to the *in situ* autoxidation of membrane PL.

Principles implied in the regulation of FA distribution in the *sn*-1 and *sn*-2 positions of PL from different animal tissues have not yet been totally elucidated.[21,22] In general, the *sn*-2 position of all PL classes is mainly occupied by long chain PUFA, while in the *sn*-1 position saturated FA are the major family.[23,24]

PC and PE are neutral PL, while PS is weakly acidic and PI is strongly acidic.[25] Owing to this feature, the correct separation of PS and PI is difficult, and thus these two PL are usually isolated together.[10,26] In addition, PL are compounds with hydrophobic and hydrophilic parts, which causes high surface activity, which in turn causes problems in chromatographic separation.

11.3 Extraction and Isolation of Muscle Phospholipids

Prior to PL isolation, total lipids should be extracted from the muscle or muscle food matrix; it is necessary to choose a solvent which will not only dissolve all lipid classes promptly but will overcome the interaction between the lipids and the matrix. The solvent used is contingent upon which types of lipids are present, neutral simple lipids or polar complex lipids, as PL. NLs, such as cholesterol or TG, can be extracted using nonpolar organic solvents such as hexane or chloroform. PL are only sparingly soluble in hydrocarbon solvents, but they dissolve in more polar solvents such as methanol, ethanol, or chloroform. Chloroform:methanol mixtures dissolve most PL, but phosphoinositides or lysophospholipids are poorly soluble in this solvent mixture. Water:methanol mixtures may dissolve significant amounts of the most polar lipids.

There is a great diversity of methodologies for the extraction of total lipids in terms of the combination and proportions of solvents. The most popular method of liquid-liquid extraction for lipids was introduced by Folch et al.[27] who used a mixture of chloroform:methanol (2:1, v/v). In this method, contaminating compounds are removed by shaking the solvent mixture with 25% of its volume with a potassium chloride solution. A later variation of this method was published by Bligh and Dyer,[28] who incorporated water into the solvent system; the proportion sample:solvent

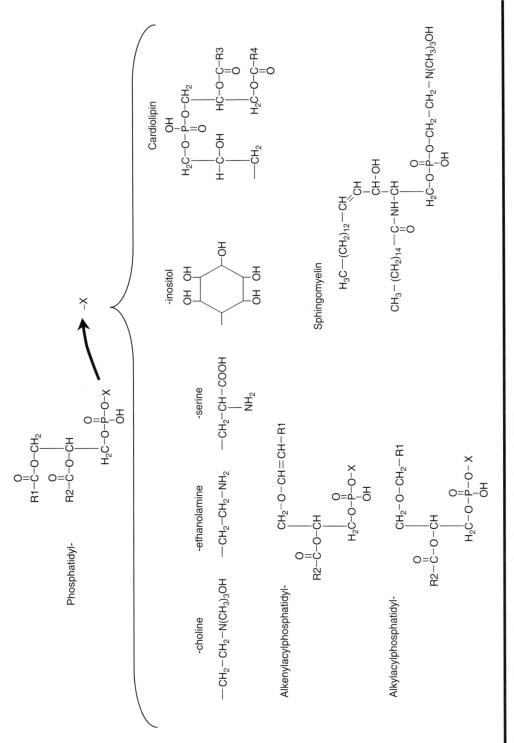

Figure 11.1 Chemical structures of main muscle phospholipids.

was also modified. Evaluation and intercomparison studies of the latter method have shown that the solvent composition alters the yields of extracted lipids, in particular PL.

After lipid extraction from muscle, the next stage in the analytical procedure involves the fractionation of the complex mixture into the various classes of lipids. Later, each lipid fraction must be isolated and analyzed. The fractionation procedure to be used will largely depend on the lipid classes present in the gross extract. Thus, the fractionation procedure must be adapted to the situation; a procedure that can work with one kind of tissue may not be effective with another source.

11.3.1 *Thin-Layer Chromatography*

Thin-layer chromatography (TLC) was the earliest chromatographic method used to isolate PL, and it is still frequently used today. However, TLC is relatively slow, produces low lipid recovery, and may result in oxidation of PUFA due to prolonged exposure to air.[29] For the separation of PL, silica gel glass plates are most commonly used, although alumina may also be used. The mobile phase consists of a solvent system that varies in polarity. Most systems incorporate varying ratios of chloroform, methanol, and water. Triethylamine, ethanol, hexane, and isopropanol are also common solvents used in the mobile phase.[30] Figure 11.2 shows a schematic TLC separation of major lipid classes in pork on a TLC plate developed using a solvent mixture of chloroform:methanol:acetic acid (100:5:2, v/v/v) following the method described by Brockerhorff.[31]

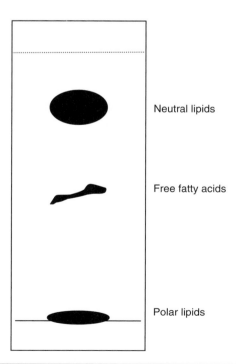

Figure 11.2   Schematic thin layer chromatography separation of major muscle lipid classes on a silica gel 60 plate developed twice in the same direction with a solvent mixture of chloroform:methanol:acetic acid (100:5:2, v/v/v).[31]

To visualize PL on the TLC plates, detection reagents are used. Spots corresponding to PL can be charred by addition of phosphomolybdic acid, sulfuric acid, or copper sulfate in phosphoric acid, with subsequent heating.

High-performance TLC (HP-TLC) constitutes an advance in TLC.[32] HP-TLC uses grades that are finer, allowing plates to be thinner and smaller. This allows for faster separation time and better separation efficiency. Lorenz et al.[33] described a method of HP-TLC for the separation of lipid classes from the muscle of bulls. To obtain the PL fraction, previously extracted lipids were separated on precoated silica gel 60 HP-TLC plates using the solvent mixture *n*-hexane/diethyl ether/ acetic acid (70:30:2, v/v). PL were viewed under UV light after spraying with 2,7-dichlorofluoresceine (0.1% in ethanol, w/v). The PL band was scraped off and eluted with chloroform:methanol (1:2, v/v), followed by chloroform:methanol (2:1, v/v). Afterwards, the extract can be used for further separation of PL classes or for quantitative analysis. Also, Dannenberger et al.[34] made use of the method of HP-TLC for extraction of PL from beef fat.

11.3.2 Column Chromatography and Solid-Phase Extraction

Marmer and Maxwell[35] described a method for muscle lipid separation that avoids prior extraction of intramuscular fat. This method presented an alternative to the traditional liquid-liquid extraction with chloroform/methanol, and allows lipid isolation by solvent elution of a dry column composed of a tissue sample, anhydrous sodium sulfate, and celite 545 diatomaceous earth ground together. To isolate total lipids, the dry column is eluted with a mixture of dichloromethane:methanol (90:10, v/v). Alternatively, lipids may be isolated and simultaneously separated into neutral and polar fractions by a sequential elution procedure: NL are eluted with dichloromethane, followed by elution of polar lipids with dichloromethane:methanol (90:10). This method has been extensively used by some researchers for isolating muscle neutral and polar lipids.[36,37]

However, traditional column chromatography is time-consuming and requires relatively large quantities of solvents. Furthermore, open column chromatography usually involves a gradient elution to separate overlapping sample peaks. A rapid, efficient, high-recovery method to separate PL involves solid-phase extraction (SPE) technology, including a prepacked SPE cartridge and vacuum assisted elution.[38] SPE utilizes a solid phase and a liquid phase for partitioning of the sample. The solid phase is a cartridge made from an adsorbent support material, usually silica or silica modified with cyanopropyl-, aminopropyl-, or dihydroxypropoxypropyl groups.[38] The liquid phase consists of organic solvents such as methanol, chloroform, or hexane that are used to elute the PL and other lipid fractions from the column. A correct separation of lipid classes is important to obtain a reliable FA composition of each fraction, since FA compositions of lipid classes are very different from each other[13] and coelution might lead to wrong conclusions about the FA profile.

Juaneda and Rocquelin[39] described an SPE method for separating PL and nonphosphorus lipids from rat heart using silica cartridges. The extracted lipids were diluted in chloroform to obtain a solution containing about 30 mg of lipids in 500 μL of solvent. The sample was then dropped at the top of a silica cartridge of 500 mg using a 500 μL syringe. The fraction containing the nonphosphorus lipids was first eluted using 20 mL of chloroform, and subsequently the fraction containing the PL was eluted with 30 mL of methanol. These authors considered it very important to avoid the formation of air bubbles between the top of the cartridges and the solvent. For samples greater than 100 mg there probably is a risk of cartridge saturation. The presence of mono-glycerides in the sample resulted in an impure PL fraction. In this case, these authors considered it necessary to use a solvent system in between, consisting of 5 mL of a mixture of chloroform: methanol (49:1, v/v), which elutes all monoglycerides from the sample. This method was modified for separation of lipid classes of muscle from Iberian pigs,[40] NL being eluted by using 30 mL of chloroform, while monoglycerides were eluted with 25 mL of acetone, and PL were subsequently collected with 30 mL of methanol.

Another extensively used method for lipid class separation among meat researchers is that described by Kaluzny et al.[41] However, this method does not seem to be accurate enough in separating lipid fractions from other biological materials.[38] Pinkart et al.[42] showed an almost 50% elution of PL with the NL fraction using this method for microbial lipids. Several researchers have also found inadequate elution of the acidic PL with methanol in the PL fraction using this method.[42–44] Different authors have proposed modifications of the Kaluzny method aimed at obtaining an adequate separation of PL and NL and to completely elute PL.[38,42] At our laboratory, this SPE method was optimized for analysis of lipid fractions in muscle foods[45] (Figure 11.3). In this optimized method, PL were eluted in two different fractions, the first with 2.5 mL methanol:chloroform (6:1, v/v) and the second with 2.5 mL of sodium acetate in methanol:chloroform (6:1, v/v). The two fractions containing PL were mixed and analyzed together. Regardless of the method, the vacuum was adjusted to generate a flow of 1 mL min^{-1}. The study of the obtained fractions using both methods revealed more efficient separation of muscle neutral and polar lipids and free fatty acids (FFA) by the optimized method.

11.3.3 High-Performance Liquid Chromatography

Separation of major muscle lipid classes has rarely been carried out by high-performance liquid chromatography (HPLC). Nevertheless, isolation of PL can be accurately achieved by this technique, usually employing normal stationary phases, among which those enriched in hydroxyl groups have shown the best results.[46,47] The introduction of the evaporative light-scattering detector (ELSD) brought a major advance in detection of lipid classes by HPLC. This detector is sensitive to the mass of the vaporized analyte and is not limited by the absorption characteristic of the individual components and/or the nature of the eluents. The only requirement is that analyzed compounds must be less volatile than the solvent. An optimized HPLC–ELSD method for the analysis of lipid classes from human plasma[47] has been recently adapted for the separation of major lipid classes in Iberian pig muscle,[48] allowing the correct separation of cholesteryl esters, TG, free cholesterol, monoglycerides, and PL. However, these authors observed that the response of the ELSD was different for various lipid classes and that the response is dependent on concentration. For accurate quantification it is imperative to use an external standard, with individual calibration curves for each analyzed compound.

11.4 Separation of Muscle Phospholipid Classes

Fractionation of muscle PL into major PL classes is becoming a common analytical procedure for meat scientists because it allows the subsequent analysis of individual PL classes and therefore yields important information for each one.[43] Usually, the separation is carried out after a successful extraction of total PL from their source, but there are methods which allow the correct separation of PL directly from the crude lipids.

The most common analytical techniques used for the isolation of muscle PL classes are TLC, SPE, and HPLC. Recently, other methods, such as capillary electrophoresis[49] or nuclear magnetic resonance (NMR) spectroscopy,[50] have also been used for the analysis of individual PL.

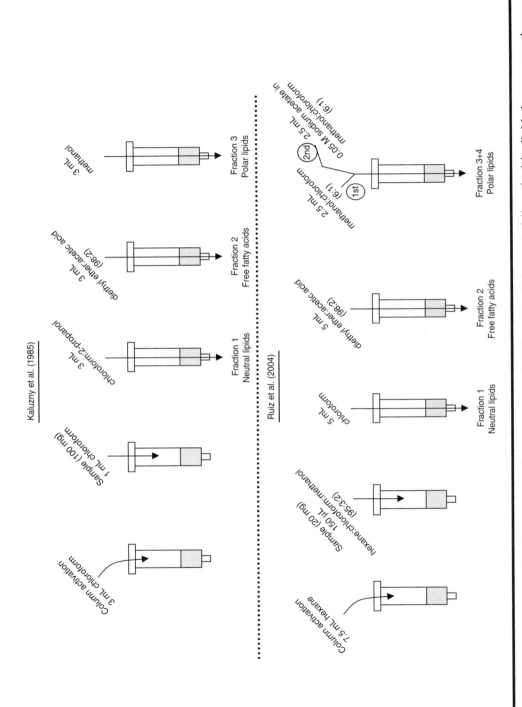

Figure 11.3 Schematic comparison between an adaptation to muscle lipids of the Kaluzny et al.[41] method for lipid classes separation and the method proposed by Ruiz et al.[45]

11.4.1 Thin-Layer Chromatography

TLC was the earliest chromatographic method used to assess PL, and it is frequently used today. One-dimensional TLC is one of the most frequently used methods for the separation of PL classes, mainly because of its simplicity, rapidity, and high resolving power,[51] considered as efficient as two-dimensional techniques by some authors.[52] HP-TLC allows faster separation of PL classes, higher efficiency, improved resolution, and lower detection limits for PL classes.[30]

As with the use of TLC for isolation of total PL, previously explained, the most common TLC coating for the isolation of PL classes is silica gel, although alumina may also be used. The mobile phase consists of a solvent system that varies in polarity. Most systems incorporate varying ratios of chloroform, methanol, acetic acid, and water. Triethylamine, ethanol, hexane, and isopropanol are also common solvents for this purpose. One of the major difficulties in the isolation of individual PL classes by this technique is the separation of PS and PI. Both the use of ammonium sulfate in the developing solvent[51,53] and the impregnation of the TLC plate with boric acid[52,54] allow a correct separation of these two PL classes.

Visualization of PL is usually carried out by UV or by spraying the plates with phosphomolybdic acid, sulfuric acid, and copper sulfate in phosphoric acid. Other authors[55] have reported a higher sensitivity using successive applications of fluorochrome 1,6-diphenyl-1,3,5-hexatriene (DPH) and molybdenum blue reagent than by using them individually.

In a recent study,[34] an HP-TLC method was used to isolate individual PL classes from pig muscle. The PL fraction was first isolated by using HP-TLC. Up to 20 mg of the total PL was applied to the HP-TLC plate in a narrow band. Earlier, the TLC plate was dipped into a solvent mixture of *n*-hexane:diethyl ether (9:1, v/v) for 1 h to remove all impurities from the silica gel phase. The solvent mixture for developing was chloroform:isopropanol:triethylamine:methanol:0.25% potassium chloride (30:25:18:9:6, v/v/v/v/v). Identification of individual PL classes was performed by spraying a solution of molybdenum blue.

11.4.2 Solid-Phase Extraction

Although SPE is a less widespread technique for separation of muscle PL classes than TLC, it is also commonly used.[56] As for isolation of total PL, the solid phase is a cartridge usually made from silica or silica modified with cyanopropyl, aminopropyl, or dihydroxypropoxypropyl groups.[38] The liquid phase consists of organic solvents than are used to elute the individual PL classes from the column. Making use of differences in polarity, pH, and solvent strength and adjusting their volumes, the separation of individual PL classes can be achieved.[57,58] Suzuki et al.[56] developed a method for fractionating five PL classes from plasma and liver of rats: PC, PE, PS, phosphatidylglycerol (PG), and CL. These authors claimed that separation of neutral PL (PC and PE) mainly depended on the polarity of the eluent, while pH and salt concentrations of the eluent had close relations with the separation of acidic PL (PS, PG, and CL). Using this method, PI was not eluted and was finally retained in the cartridge. Pietch and Lorenz[57] carried out the correct elution of PC, PE, PS, and PI from a mixture of PL standards. These authors adjusted the eluent volumes for 100 and 500 mg aminopropyl cartridges. Banni et al.[59] separated PC, PE, PS, and PI from the liver of rats following this latter method,[57] with minor modifications. However, using this method, a coelution of PC and PE and of PE with PS was found,[58] while PI was finally retained in the cartridge. Some modifications were suggested,[58] such as increased solvent volumes for the elution of PC, PE, and PS and a different solvent mixture to allow PI elution (Figure 11.4). This method was checked using both a mixture of PL standards and muscle samples, showing a correct elution of PC, PE, PS, and PI.

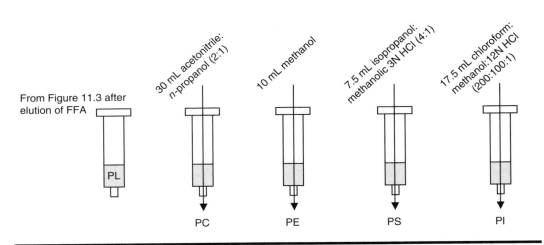

Figure 11.4 Schematic solid-phase extraction separation of major muscle phospholipid classes following the method described by Pérez-Palacios et al.[58] PC: phosphatidylcholine; PE: phosphatidylethanolamine; PS: phosphatidylserine; PI: phosphatidylinositol.

11.4.3 High-Performance Liquid Chromatography

HPLC is the most commonly used technique currently for separating PL classes. Most procedures describe the use of normal-phase HPLC rather than reverse-phase, using silica columns 100–250 mm in length. Some methods describe the use of stationary phases of silica gel modified with diol-, cyanopropyl-, or aminopropyl- to alter its polarity. Almost all methods make use of gradient elution with either a binary or ternary solvent system. Gradient elution is preferred owing to the wide range of polarity of muscle PL classes. Starting with a solvent of low polarity and ending with a solvent mixture of higher polarity allows good separation of all PL classes.[60] Most solvent systems begin with a combination of chloroform:methanol and either increase the proportion of methanol or include water for the gradient. Other commonly used solvents include hexane, 2-propanol, ethanol, water, and acetonitrile. Ammonium hydroxide or ammonia may be also added to improve the separation of PI.[61] Ammonium salt will help prevent the degradation of the column.[62] The most common detectors for the analysis of PL by HPLC are UV and ELSD. Refractive index and fluorescence detectors have also been described as detectors in methods for PL class separation, but they are less frequently used. UV detectors have high sensitivity for PL detection; however, its response is dependent on the number of double bonds. In addition, chloroform, ethyl ether, and acetone cannot be used in the mobile phase because they absorb in the same UV range as most PL (190–210 nm). Mobile phases may include acetonitrile, methanol, water, hexane, and isopropanol.[63] ELSD does not show such limitations, but the solvents of the mobile phase used with this detector must be more volatile than the sample and should not contain organic ions.[64]

The separation of individual muscle PL classes is usually performed after separation of total muscle PL from the lipid extract, but the isolation of PL classes transferring the whole crude lipid extract into the HPLC injector has also been described.[65,66]

Several different HPLC methods have been described for the separation of PL classes in a variety of biological samples that would also be useful for muscle PL class separation. The difficulty in the separation of muscle PS and PI has also been observed by using HPLC methods, and several authors

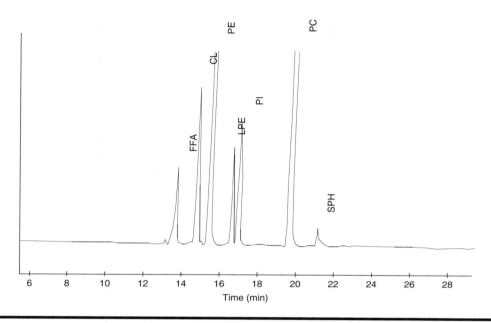

Figure 11.5 **Chromatogram of major muscle phospholipid classes separated by using HPLC–ELSD following the method described by Leseigneur-Meynier and Gandemer.[67] (Chromatogram generously provided by J.F. Tejeda). FFA: free fatty acids; CL: cardiolipin; PE: phosphatidylethanolamine; LPE: lyso-PE; PI: phosphatidylinositol; PC: phosphatidylcholine; SPH: sphingomyelin.**

provide results of these two PL classes together (Figure 11.5).[10,40,67] Nevertheless, other authors have published HPLC methods that did allow the isolation of PS and PI from animal muscle. For example, Seewald and Eichinger[65] described a sensitive HPLC method with UV detector for the separation of pork PC, PE, PS, PI, CL, SPH, lyso-PC, and lyso-PE. In this method, total lipid extracts dissolved in chloroform:methanol (2:1, v/v) were analyzed by HPLC, and PL classes were subsequently separated through a successive elution with acetonitrile, acetonitrile containing 0.2% phosphoric acid, and finally acetonitrile containing 0.2% phosphoric acid changing continuously to methanol containing 0.2% phosphoric acid.

Caboni et al.[68] described an HPLC method with ELSD detector to successfully fractionate total PL from cooked beef into PC, PE, PS, PI, PG, and SPH. These authors used two consecutive columns, a LiChrosorb 60 column followed by a Spherisorb silica column, and a binary gradient composed of chloroform:methanol:ammonium hydroxide 30% (80:19.5:0.5, v/v/v), and chloroform:methanol:water:ammonium hydroxide 30% (60:34:5.5:0.5, v/v/v/v).

11.5 Quantification of Total Phospholipids and Phospholipid Classes

The quantification of either total PL or isolated PL classes (either scraped from the TLC plate or eluted from the SPE cartridge or from the HPLC column) has been extensively performed by determination of the amount of phosphorus after digestion of PL, following the method described by Barlett.[69] The released inorganic phosphorus is thereafter analyzed by reaction with ammonium molybdate to form phosphomolybdic acid, which is subsequently reduced, forming a blue compound that is determined spectrophotometrically. There have been numerous modifications of this method,[70,71] but all of them based in the same concept. Determination of phosphorus content implies the use of free-phosphorous detergents for cleaning all the glassware. It should be remembered that certain PL classes, such as CL, comprise two phosphate groups, and thus contain 2 mol of phosphorus per mole of lipid.

Other methods for quantifying the total phosphorus in the digested sample or in the ashes include inductively coupled plasma-atomic emission spectrometry, atomic absorption spectrometry with a graphite furnace, or precipitation of phosphate as quinolinium molybdophosphate and gravimetric determination.[72] However, all these other methods have been far less frequently used with muscle foods.

Spots of separated PL classes by TLC have been directly used for quantitative purposes using a number of different procedures, such as densitometry after spraying with copper acetate,[73–75] laser-excited fluorescent detection, and subsequent integration of variable pixel intensities.[76] Some authors have even quantitatively analyzed the lipid constituents, including PC and PE, using gravimetric analysis of fractions collected from column chromatography.[77,78]

A widely used procedure for quantifying total PL or PL classes separated by either TLC or SPE (or even by HPLC) is the derivatization of the obtained PL classes to methyl esters of FAs and subsequent quantification of these derivatives using an internal standard and analysis by gas chromatography.[25,79] This procedure has been used for studying changes in total polar lipids of dry cured ham[20] and dry-cured loin.[23,80]

The TLC coupled to a flame ionization detector (Iatroscan) has been proved as a suitable method for the quantification of polar lipids. In this method, TLC is performed on quartz rods coated with silica. The rods are then passed through a flame ionization detector to quantify each lipid band.[72] This technique has been used for separation, identification and quantification of lipid classes (including PL classes) in beef,[81] duck breast,[82] and fish.[83]

As explained previously, UV detection (at 205–210 nm) is sufficiently sensitive to detect unsaturated lipids separated by HPLC, and has been extensively used to detect spots of muscle and muscle foods PL in TLC, but it is not an adequate method for quantification, since the detector response is proportional to the total number of double bonds and, in addition, saturated lipids are not detected. Moreover, solvents such as chloroform, ethyl ether, or acetone cannot be used in the mobile phase of HPLC, since they have strong absorbing properties in the UV range.

The ELSD is currently a widely accepted tool for the quantitative analysis of polar lipids. Thus, HPLC–ELSD has been used to quantify PL classes in fish muscles[84,85] or livestock muscle and meat products.[10,40] ELSD measures the intensity of light scattered by lipids that remain after the solvent has been evaporated owing to nebulization.[86]

11.6 Positional Analysis of Glycerophospholipids

11.6.1 Enzymatic Hydrolysis

Enzymatic hydrolysis procedures have been developed for determining the positional distribution of FA of a number of simple and complex glycerol-containing lipids. One of the most popular enzymatic methods for the stereospecific analysis of PL is based on the use of phospholipase A_2 (EC. 3.1.1.4).

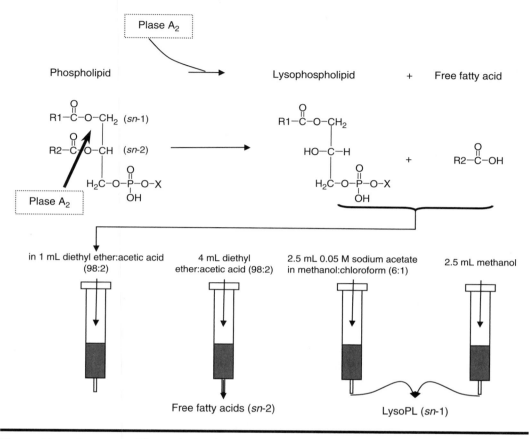

Figure 11.6 **Stereospecific analysis of muscle phospholipids by lysis of acyl chains in the *sn*-2 position of phospholipids using phospholipase A₂ (Plase A₂) and subsequent separation by solid-phase extraction of free fatty acids and lysophospholipids following the method described by Muriel et al.[23]**

This enzyme specifically hydrolyzes the ester bond in position 2 of natural glycerophosphatides, yielding an FFA released from the *sn*-2 position and a lysophosphatide containing the FA in *sn*-1 position (Figure 11.6). Thus, isolation of the two obtained fractions allows the analysis of the individual positions. The enzyme is available from different sources, mainly from snake venoms (*Crotalus ademanteus, Ophiophagus hannah*) and the honeybee (*Apis mellifera*).[25] The main advantage of these enzymes is their simplicity of use, providing the researchers a rapid and reliable method. On the other hand, it has been demonstrated that different types of phospholipase A₂ show specificity for certain FA combinations.[22] In this sense it has been pointed out that although phospholipase A₂ from *C. ademanteus* venom has been widely used, currently, the use of *O. hannah* is preferred due to its action, which appears to be independent of the FA compositions of the PL.[25] The phospholipase from *A. mellifera* venom is also being increasingly used for specific purposes. The enzymes from snake venoms are completely stereospecific for L-glycerophosphatides and will react with most natural compounds of this type except for phosphoinositides. Calcium ions are essential for the reaction, which is also stimulated by diethyl ether.

Muriel et al.[23] described the following method for the stereospecific analysis of PL from pig longissimus dorsi muscle (Figure 11.6): A stock solution of 1 mg of phospholipase A_2 from *A. mellifera* (1,225 units mg^{-1}) in 23.11 mL of 1 M Tris buffer (pH 8.9) containing calcium chloride (4 mM) was made up. Hydrolysis of PL was performed by adding 300 μL of the phospholipase A_2 stock solution to PL (2–5 mg) in 6 mL of diethyl ether with BHT (0.05%).[87] The mixture was incubated for 3 h in N_2 atmosphere with continuous stirring at 25°C. Thereafter it was washed with methanol:chloroform (2:1, v/v) and dried with anhydrous sodium sulfate.[88] The reaction mixture was separated in NH_2-aminopropyl minicolumns (500 mg),[45] obtaining two fractions: FFA and 1-lysophospholipids. FA and alkenyl chains were detected and identified after transesterification and GC-FID analysis.*

11.6.2 Nuclear Magnetic Resonance

NMR spectroscopy is a specific and nondestructive technique that has been successfully applied to the study of lipids;[89] since the late 1980s, high-resolution NMR has been increasingly applied to the analysis of food and food components. In the field of edible oils and fats, NMR provides useful information on the composition of palm, seed, and olive oils[89–91] and some terrestrial animal fats.[92] However, its application to more complicated lipid substrates, such as those extracted from marine sources, has proved to be more difficult.[93,94] Since the beginning of the 1990s, high-resolution NMR has been applied for the assessment of lipid classes and acyl stereospecific positions of FAs in marine PLs and triacylglycerols.[95] However, there are few works dealing with the application of this analytical technique in PL from muscle foods. Sacchi et al.[96] demonstrated that ^{13}C-NMR spectroscopy is a useful, direct, and nondestructive method for the study of fish lipid and oil composition. The major advantage of NMR over the classical methods is essentially the absence of enzymatic or chemical manipulation of lipid samples. This constitutes an important consideration for the positional analysis of marine lipids, considering the possibility that FA are transposed onto glycerol[21] and/or unsaturated damage during enzymatic or chemical hydrolysis.[97] ^{13}C-NMR offers the simultaneous possibility of monitoring the composition of lipid classes, FFA content and composition, and acyl positional distribution.[96]

Medina and Sacchi[98] developed a method for studying the acyl stereospecific analysis of PL in tuna muscles via high-resolution ^{13}C-NMR spectroscopy. In this paper the authors studied the PL acyl profile of skipjack tuna (*Katsuwonus pelamis*), with prior treatment of PL with phospholipase C to avoid interference of the phosphate polar head, improving the spectral resolution, and with no alteration of the acyl positional distribution. The authors obtained good quantitative results when this method is compared with chromatographic methods (and with previous hydrolysis with phospholipase C as well).

The use of ^{31}P-NMR spectroscopy appears to be a promising technique for individual identification and quantification of PL classes, since it does not require their previous extraction from the matrix and separation; it is able to differentiate between the various PL and to quantify tiny amounts of PL even in the presence of other components, since only substances containing phosphorous are detected.[25,99] Nevertheless this technique has scarcely been used for muscle or muscle foods,[100] and only a bit more for other lipid sources.[50,101]

---

* Alkyl glycerophospholipids, in which the ether linkage to the glycerol backbone is an ether bond, are not detected when using common methylation methods, since the ether bond is stable to acidic conditions 25.

11.7   Molecular Species of Muscle Phospholipids

The analysis of the molecular species allows a deeper knowledge about the complete structure of the PL molecule: The acyl, alkyl, or alkenyl residues in the *sn*-1 position, the acyl residue in the *sn*-2 position, and the polar head in the *sn*-3 will be elucidated. Separation and quantification of molecular species of a given PL class often represents an important tool for studying lipid metabolism. This is the reason why several methods for separating PL molecular species have been published recently.[102]

Olsson and Salem[103] carried out an exhaustive revision of the more frequently used methods in the analysis of molecular species during the last three decades of the twentieth century. This work revised silver-impregnated TLC and PL hydrolysis with phospholipase C, followed by derivatization of the 1,2-diacyl-*sn*-glycerol fraction obtained, and analysis through HPLC (coupled to UV or MS detector) or by gas chromatography. These authors found some works dealing with highly selective analysis with LC-MS in PL from fish, sheep, and bovine samples.[104,105] Patton et al.[106] applied HPLC-UV without derivatization of FA to the analysis of molecular species in rat liver, obtaining excellent separations.

The analysis of molecular species begins with the separation of PL classes, mainly PC and PE in muscles, followed by the enzymatic hydrolysis of the polar head with phospholipase C.

Scott et al.[107] studied the molecular species composition of PC and PE in ruminant muscle (sheep and steer longissimus dorsi) fed with fish oil supplement. In this study, after separation of PL classes, PL were hydrolyzed with phospholipase C, and the released 1,2-diradylglycerols were extracted to prepare benzoate derivatives. Once the benzoate derivatives were obtained, they were separated into three subclasses (diacyl, alkylacyl, and alkenylacyl) by TLC on silica gel G. The molecules separated were visualized under UV light. The areas of gel corresponding to these derivatives were removed and extracted for further separation by HPLC. Identification of the individual molecular species was achieved by a combination of HPLC analysis of standards and GC analysis of the FA methyl esters of the peaks corresponding to the eluted species.

The ELSD has become increasingly used in the field of lipid analysis[25] and currently is applied to detection and quantification of molecular species without previous derivatization. In this vein, Vecchini et al.[102] described a method for analyzing molecular species of glycerophospholipids based on the combination of UV detector and ELSD, which do not need derivatization of diglycerides. The pure glycerophospholipids were converted into diglycerides by hydrolysis catalyzed by phospholipase C of *Bacillus cereus* (which presents a broader specificity than phospholipase C from *Clostridium welchii* or *Clostridium perfringens*). Diacylglycerols, alkylacylglycerols, and alkenylacylglycerols were separated and quantified by HPLC on a silicic acid column coupled to an ELSD, and the three peaks detected were separately pooled using a fraction collector. Molecular species of diradylglycerol were separated by chromatography on a reverse phase octadecyl-silica column. Detection was performed by a cascade system of UV and ELSD. Identification of chromatographic peaks was performed by comparing with those of commercial or previously prepared standards. These researchers found good quantitative results using ELSD, providing that the instrumental parameters are carefully controlled.

Kuksis[108] reviewed the use of mass spectrometry (MS) in the field of molecular species of PL. This author highlighted some works dealing with this topic.[43,109–111]

Kim and Salem[43] developed a technique for rapid and detailed molecular species analysis of PL using reverse-phase HPLC with online thermospray/mass spectrometry (TS/MS). In conjunction with a hexane/methanol/0.2 M ammonium acetate mixture as mobile phase, the technique was generalized for mixtures of PC and PE. Kerwin et al.[109] described the use of positive and negative ion electrospray/mass spectra (ES/MS) and ES/MS/MS to identify polar head-groups and their alkyl, alkenyl, and acyl constituents of PC and PE. The spectra were acquired by ES/MS/MS using a flow injection of chloroform extracts of biological samples without prior chromatographic separation of the glycerophospholipid.

Other groups[110,111] reported improved ES/MS analysis of molecular species of PL by combining this instrumentation with HPLC. Myher et al.[110] combined normal-phase HPLC with ES/MS for the analysis of total lipid extracts of PL. The PL were resolved on a silica column by elution with a gradient of chloroform/methanol/water/ammonia.

Kim et al.[111] used 0.5% ammonium hydroxide in a water/methanol/hexane mixture and a C18 column to resolve complex mixtures of molecular species of PL, which were detected as protonated and sodiated molecular species.

References

1. Igene, J.O., Role of triglycerides and phospholipids on development of rancidity in model meat systems during frozen storage, *Food Chem.*, 5, 263, 1980.
2. Pikul, J., Leszczynski, D.E. and Kummerow, F.A., Relative role of phospholipids, triacylglycerols and cholesterol esters on malonaldehyde formation in fat extracted from chicken meat, *J. Food Sci.*, 49, 704, 1984.
3. Erickson, M.C., Lipid oxidation of muscle foods, in *Food Lipids. Chemistry, Nutrition, and Biotechnology*, Akoh, C.C. and Min, D.B., Eds., Marcel Dekker, New York, NY, 2002.
4. Khan, M.A. and Shahidi, F., Tocopherols and phospholipids enhance the oxidative stability of borage and evening primrose triacylglycerols, *J. Food Lipids*, 7, 143, 2001.
5. Pokorny, J., Phospholipids, in *Chemical and Functional Properties of Lipids*, Sikorski, Z.E. and Kolakowska, A, Eds., CRC Press, Boca Raton, FL, 2002.
6. Lindsay, R.C., Chemical basis of the quality of seafood flavors and aromas, *Mar. Technol. Soc. J.*, 25, 16, 1990.
7. Josephson, D.B., Lindsay, R.C., and Stuiber, D.A., Volatile carotenoid-related oxidation compounds contributing to cooked salmon flavour, *Food Sci. Technol.-LWT*, 24, 424, 1991.
8. Polvi, S.M. et al., Stability of lipids and omega-3-fatty acids during frozen storage of Atlantic salmon, *J. Food Proc. Pres.*, 15, 167, 1991.
9. Milo, C. and Grosch, W., Changes in the odorants of boiled salmon and cod as affected by the storage of the raw material, *J. Agric. Food Chem.*, 44, 2366, 1996.
10. Alasnier, C. and Gandemer, G., Fatty acid and aldehyde composition of individual phospholipid classes of rabbit skeletal muscles is related to the metabolic type of the fibres, *Meat Sci.*, 48, 225, 1998.
11. Andres, A.I. et al., Oxidative stability and fatty acid composition of pig muscles as affected by rearing system, crossbreeding and metabolic type of muscle fibre, *Meat Sci.*, 59, 39, 2002.
12. Hulbert, A.J., Rana, T., and Couture, P., The acyl composition of mammalian phospholipids: an allometric analysis, *Comp. Biochem. Phys. B*, 132, 515, 2002.
13. Muriel, E. et al., Free-range rearing increases (*n*-3) polyunsaturated fatty acids of neutral and polar lipids in swine muscles, *Food Chem.*, 78, 219, 2002.
14. Brosche, T. and Platt, D., The biological significance of plasmalogens in defense against oxidative damage, *Exp. Gerontol.*, 33, 363, 1998.
15. Sindelar, P.J. et al., The protective role of plasmalogens in iron-induced lipid peroxidation, *Free Radic. Biol. Med.*, 26, 318, 1999.
16. Nylander, T., Interactions between proteins and polar lipids, in *Food Emulsions*, Friberg, S.E., Larsson, K., and Sjöblom, J., Eds., Marcel Dekker, New York, NY, 2004.
17. Suzumura, M., Phospholipids in marine environments: a review, *Talanta*, 66, 422, 2005.

18. Wenk, C., Leonhardt, M., and Scheeder, M.R.L., Monogastric nutrition and potential for improving muscle quality, in *Antioxidants in Muscle Foods: Nutritional Strategies to Improve Quality*, Decker, E.A., Faustman, C., and Lopez-Bote, C.J., Eds., Wiley-Interscience, New York, NY, 2000.

19. Ordóñez, J.A. et al., Changes in the components of dry-fermented sausages during ripening, *Crit. Rev. Food Sci. Nutr.*, 39, 329, 1999.

20. Andres, A.I. et al., Lipolysis in dry-cured ham: influence of salt content and processing conditions, *Food Chem.*, 90, 523, 2005.

21. Litchfield, C., Distribution of fatty acids in natural triglyceride mixtures, in *Analysis of Triglycerides*, Academic Press, New York, NY, pp. 233–264, 1972.

22. Breckenridge, W., Stereospecific analysis of triacylglycerols, in *Handbook of Lipid Research, Vol. 1*, Kuksis, A., Ed., Plenum Press, New York, NY, pp. 197–232, 1978.

23. Muriel, M.E. et al., Stereospecific analysis of fresh and dry-cured muscle phospholipids from Iberian pigs, *Food Chem.*, 90, 437, 2005.

24. Perez-Palacios, T. et al., Stereospecific analysis of phospholipid classes in rat muscle, *Eur. J. Lipid Sci. Technol.*, 108, 835, 2006.

25. Christie, W.W., *Lipid Analysis*, The Oily Press, Bridgewater, UK, 2003.

26. Sánchez, V. and Lutz, M., Fatty acid composition of microsomal phospholipids in rats fed different oils and antioxidant vitamins supplement, *J. Nutr. Biochem.*, 9, 155, 1998.

27. Folch, J., Less, M., and Sloane, G.H., A simple method for the isolation and purification of total lipids from animal tissues, *J. Biol. Chem.*, 226, 497, 1957.

28. Bligh, E.G. and Dyer, E.J., A rapid method of total lipid extraction and purification, *Can. J. Biochem. Phys.*, 37, 911, 1959.

29. Bernhardt, T.G. et al., Purification of fatty acid ethyl esters by solid-phase extraction and high-performance liquid chromatography, *J. Chromatogr. B*, 675, 189, 1996.

30. Peterson, B.L. and Cummings, B.S., A review of chromatographic methods for assessment of phospholipids in biological samples, *Biomed. Chromatogr.*, 20, 227, 2006.

31. Brockerhoff, H., Determination of the positional distribution of fatty acid in glycerolipids, *Methods Enzymol.*, 35, 315, 1975.

32. Sherma, J. and Jain, R., Planar chromatography in clinical chemistry, in *Encyclopedia of Analytical Chemistry*, Meyers, R.A., Ed., Wiley, Chichester, UK, pp. 1583–1603, 2000.

33. Lorenz, S. et al., Influence of keeping system on the fatty acid composition in the *longissimus dorsi* muscle of bulls and odorants formed after pressure-cooking, *Eur. Food Res. Technol.*, 214, 112, 2002.

34. Dannenberger, D. et al., Diet alters the fatty acid composition of individual phospholipid classes in beef muscle, *J. Agric. Food Chem.*, 55, 452, 2007.

35. Marmer, W.N. and Maxwell, R.J., Dry column method for the quantitative extraction and simultaneous class separation of lipids from muscle tissue, *Lipids*, 16, 365, 1981.

36. Daza, A. et al., Effects of feeding in free-range conditions or in confinement with different dietary MUFA/PUFA ratios and α-tocopheryl acetate, on antioxidants accumulation and oxidative stability in Iberian pigs, *Meat Sci.*, 69, 151, 2005.

37. Ruiz, J. et al., Improvement of dry-cured Iberian ham quality characteristics through modifications of dietary fat composition and supplementation with vitamin E, *Food Sci. Technol. Int.*, 11, 327, 2005.

38. Ruiz-Gutierrez, V. and Pérez-Camino, M.C., Update on solid-phase extraction for the analysis of lipid classes and related compounds, *J. Chromatogr. A*, 885, 321, 2000.

39. Juaneda, P. and Rocquelin, G., Rapid and convenient separation of phospholipid and non phosphorus lipids from rat heart using silica cartridges, *Lipids*, 20, 40, 1985.

40. Tejeda, J.F., Estudio de la influencia de la raza y la alimentación sobre la fracción lipídica intramuscular del cerdo Ibérico, PhD thesis, University of Extremadura, 1999.

41. Kaluzny, M.A. et al., Rapid separation of lipid classes in high yield and purity using bonded phase columns, *J. Lipid Res.*, 26, 135, 1985.

42. Pinkart, H.C., Devereux, R., and Chapman, P.J., Rapid separation of microbial lipids using solid phase extraction columns, *J. Microbiol. Methods*, 34, 9. 1998.

43. Kim, H.Y. and Salem Jr., N., Separation of lipid classes by solid phase extraction, *J. Lipid Res.*, 31, 2285, 1990.

44. Bateman, H.G. and Jenkins, T.C., Method for extraction and separation by solid phase extraction of neutral lipid, free fatty acids, and polar lipid from mixed microbial cultures, *J. Agric. Food Chem.*, 45, 132, 2001.

45. Ruiz, J. et al., Improvement of a solid phase extraction method for analysis of lipid fractions in muscle foods, *Anal. Chim. Acta*, 520, 201, 2004.

46. Stith, B.J. et al., Quantification of major classes of *Xenopus* phospholipids by high performance liquid chromatography with evaporative light scattering detection, *J. Lipid Res.*, 419, 1448, 2000.

47. Seppanen-Laakso, T. et al., Major human plasma lipid classes determined by quantitative high performance liquid chromatography, their variation and associations with phospholipids fatty acids, *J. Chromatogr. B*, 7542, 437, 2001.

48. Perona, J.S. and Ruiz-Gutierrez, V., Quantitative lipid composition of Iberian pig muscle and adipose tissue by HPLC, *J. Liq. Chromatogr. Related Technol.*, 28, 2445, 2005.

49. Gao, F. et al., Separation of phospholipids by capillary zone electrophoresis with indirect ultraviolet detection, *J. Chromatogr. A*, 1130, 259, 2006.

50. Helmerich, G. and Koehler, P., Comparison of methods for the quantitative determination of phospholipids in lecithins and flour improvers, *J. Agric. Food Chem.*, 51, 6645, 2003.

51. Wang, W. and Gustafson, A., One-dimensional thin-layer chromatographic separation of phospholipids and lysophospholipids from tissue lipid extracts, *J. Chromatogr.*, 581, 139, 1992.

52. Leray, C. et al., Thin-layer chromatography of human platelet phospholipids with fatty acid analysis, *J. Chromatogr.*, 420, 411, 1987.

53. Kaulen, H.D., Separation of phosphatidylserine and -inositol by one-dimensional thin-layer chromatography of lipid extracts, *Anal. Biochem.*, 45, 664, 1972.

54. Dreyfus, H. et al., Successive isolation and separation of the major lipid fractions including gangliosides from single biological samples, *Anal. Biochem.*, 249, 67, 1997.

55. Müthing, J. and Radloff, M., Nanogram detection of phospholipids on thin-layer chromatograms, *Anal. Biochem.*, 257, 67, 1998.

56. Suzuki, E. et al., Improved separation and determination of phospholipids in animal tissues employing solid phase extraction, *Biol. Pharm. Bull.*, 20, 299, 1997.

57. Pietsch, A. and Lorenz, R.L., Rapid separation of the major phospholipid classes on a single aminopropyl cartridge, *Lipids*, 28, 945, 1993.

58. Pérez Palacios, T., Ruiz, J., and Antequera, T., Improvement of a solid phase extraction method for separation of muscle phospholipids classes, *Food Chem.*, 102, 875, 2007.

59. Banni, S. et al., Distribution of conjugated linoleic acid and metabolites in different lipid fraction in the rat liver, *J. Lipid Res.*, 42, 1056, 2001.

60. Breton, L. et al., A new rapid method for phospholipid separation by high-performance liquid chromatography with light-scattering detection, *J. Chromatogr.*, 497, 243, 1989.

61. Gurnnarsson, T. et al., Separation of polyphosphoinositides using normal-phase high-performance liquid chromatography and evaporative light scattering or electrospray mass spectrometry, *Anal. Biochem.*, 254, 293, 1997.

62. Guan, Z. et al., Separation and quantitation of phospholipids and their analogues by high-performance liquid chromatography, *Anal. Biochem.*, 297, 137, 2001.

63. Singh, A.K. and Jiang, Y., Quantitative chromatographic analysis of inositol phospholipids as related compounds, *J. Chromatogr. B*, 671, 255, 1995.

64. Hoving, E.B., Chromatographic methods in the analysis of cholesterol and related lipids, *J. Chromatogr. B*, 671, 341, 1995.

65. Seewald, M. and Eichinger, H., Separation of major phospholipid classes by high-performance liquid chromatography and subsequent analysis of phospholipid-bound fatty acids using gas chromatography, *J. Chromatogr.*, 469, 271, 1989.

66. Yeo, Y.K. and Horrocks, L.A., Analysis of phospholipid classes in various beef tissues by high performance liquid chromatography, *Food Chem.*, 29, 1, 1988.
67. Leseigneur-Meynier, A. and Gandemer, G., Lipid composition of pork muscle in relation to the metabolic type of the fibres, *Meat Sci.*, 29, 229, 1991.
68. Caboni, M.F., Menotta, S., and Lercker, G., High-performance liquid chromatography separation and light-scattering detection of phospholipid from cooked beef, *J. Chromatogr. A*, 683, 59, 1994.
69. Barlett, G.R., Phosphorus assay in column chromatography, *J. Biol. Chem.*, 234, 466, 1959.
70. Marinetti, G.V., Chromatographic separation, *J. Lipid Res.*, 3, 1, 1962.
71. Murphy, J. and Riley, J.P., A modified single solution method for determination of phosphate in natural waters, *Anal. Chem. Acta*, 27, 31, 1962.
72. Gordon, M.H., Polar lipids, in *Encyclopaedia of Analytical Science*, Worsfold, P., Townshend, A., and Poole, C., Eds., Elsevier, Amsterdam, the Netherlands, pp. 88–94, 2005.
73. Olsen, R.E. and Henderson, R.J., The rapid analysis of neutral and polar marine lipids using double-development HPTLC and scanning densitometry, *J. Exp. Mar. Biol. Ecol.*, 129, 189, 1989.
74. Cavalli, R.O. et al., Variations in lipid classes and fatty acid content in tissues of wild *Macrobrachium rosenbergii* (de Man) females during maturation, *Aquaculture*, 193, 311, 2001.
75. Fewster, M., Burns, B.J., and Mead, J.F., Quantitative densitometric TLC of lipids using copper acetate reagent, *J. Chromatogr.*, 43, 120, 1969.
76. White, T. et al., High-resolution separation and quantification of neutral lipid and phospholipid species in mammalian cells and sera by multi-one-dimensional thin-layer chromatography, *Anal. Biochem.*, 258, 109, 1998.
77. Takama, K. et al., Phosphatidylcholine levels and their fatty acid compositions in teleost tissues and squid muscle, *Comp. Biochem. Phys. B*, 124, 109, 1999.
78. Saito, H. et al., High docosahexaenoic acid levels in both neutral and polar lipids of a highly migratory fish: *Thunnus tonggol* (Bleeker), *Lipids*, 40, 941, 2005.
79. Shantha, N.C. and Napolitano, G.E., Lipid analysis using thin-layer chromatography and the Iatroscan, in *Lipid Analysis of Fats and Oils*, Hamilton, J.R., Ed., Blackie Academic and Professional, London, UK, 1998.
80. Muriel, E. et al., Lipolytic and oxidative changes in Iberian dry-cured loin, *Meat Sci.*, 75, 315, 2007.
81. St Angelo, A.J. and James, C., Analysis of lipids from cooked beef by thin-layer chromatography with flame-ionization detection, *J. Am. Oil Chem. Soc.*, 70, 1245, 1993.
82. Baeza, E. et al., Effects of age and sex on the structural, chemical and technological characteristics of mule duck meat, *Br. Poultry Sci.*, 41, 300, 2000.
83. Gunnlaugsdottir, H. and Ackman, R.G., Three extraction methods for determination of lipids in fish meal: evaluation of a hexane:isopropanol method as an alternative to chloroform based methods, *J. Sci. Food Agric.*, 61, 235, 1993.
84. Serot, T., Gandemer, G., and Demaimay, M., Lipid and fatty acid compositions of muscle from farmed and wild adult turbot, *Aquacult. Int.*, 6, 331, 1998.
85. Nordback, J., Lundberg, E., and Christie, W.W., Separation of lipid classes from marine particulate material by HPLC on a polyvinyl alcohol-bonded stationary phase using dual-channel evaporative light-scattering detection, *Mar. Chem.*, 60, 165, 1998.
86. Wang, Y. et al., Derivatization of phospholipids, *J. Chromatogr. B*, 793, 3, 2003.
87. Florin-Christensen, J. et al., A method for distinguishing 1-acyl from 2-acyl lysophosphatidylcholines generated in biological systems, *Anal. Biochem.*, 276, 13, 1999.
88. Aubourg, S.P., Medina, I., and Pérez-Martín, R., Polyunsaturated fatty acids in tuna phospholipids: distribution in the *sn*-2 location and changes during cooking, *J. Agric. Food Chem.*, 44, 585, 1996.
89. Sacchi, R., Medina, I., and Paolillo, L., One- and two-dimensional NMR studies of plasmalogens (alk-1-enyl-phosphalidylethanolamine), *Chem. Phys. Lipids*, 76, 201, 1995.
90. Ng, S., Analysis of positional distribution of fatty acids in palm oil by ^{13}C-NMR spectroscopy, *Lipids*, 20, 778, 1985.

91. Wollenberg, K.F., Quantitative high resolution ^{13}C-nuclear magnetic resonance of the olefinic and carbonyl carbons of edible vegetable oils, *J. Am. Oil Chem. Soc.*, 67, 487, 1990.

92. Bonnet, M., Denoyer, C., and Renou, J., High resolution ^{13}C-NMR spectroscopy of rendered animal fats: degree of unsaturation of fatty acid chains and position on glycerol, *Int. J. Food Sci. Technol.*, 25, 399, 1990.

93. Gunstone, F.D., High resolution NMR studies of fish oils, *Chem. Phys. Lipids*, 59, 83, 1991.

94. Aursand, M. and Grasdalen, H., Interpretation of the ^{13}C-NMR spectra of omega-3 fatty acids and lipid extracted from the white muscle of Atlantic salmon (*Salmo salar*), *Chem. Phys. Lipids*, 62, 239, 1992.

95. Falch, E., Storseth, T.R., and Aursand, A., Multi-component analysis of marine lipids in fish gonads with emphasis on phospholipids using high resolution NMR spectroscopy, *Chem. Phys. Lipids*, 144, 4, 2006.

96. Sacchi, R. et al., Quantitative high-resolution ^{13}C-NMR analysis of lipids extracted from the white muscle of Atlantic tuna (*Thunnus alallunga*), *J. Agric. Food Chem.*, 41, 1247, 1993.

97. Chang, H., *Autooxidation of Unsaturated Lipids*, Academic Press, New York, NY, 1977.

98. Medina, I. and Sacchi, R., Acyl stereospecific analysis of tuna phospholipids via high-resolution ^{13}C-NMR spectroscopy, *Chem. Phys. Lipids*, 70, 53, 1994.

99. Diehl, B.W.K., Multinuclear high-resolution nuclear magnetic resonance spectroscopy, in *Lipid Analysis of Fats and Oils*, Hamilton, J.R., Ed., Blackie Academic and Professional, London, UK, 1998.

100. Culeddu, N., Bosco, M., and Toffanin, R., P^{31}-NMR analysis of phospholipids in crude extracts from different sources: improved efficiency of the solvent system, *Magn. Reson. Chem.*, 36, 907, 1998.

101. Murgia, S., Mele, S., and Monduzzi, M., Quantitative characterization of phospholipids in milk fat via P^{31}-NMR using a monophasic solvent mixture, *Lipids*, 38, 585, 2003.

102. Vecchini, A., Panagia, V., and Binaglia, L., Analysis of phospholipid molecular species, *Mol. Cell. Biochem.*, 172, 129, 1997.

103. Olsson, N.U. and Salem, N., Jr., Molecular species analysis of phospholipids, *J. Chromatogr. B.*, 692, 245, 1997.

104. Wallaert, C. and Babin, P.J., Thermal adaptation affects the fatty acid composition of plasma phospholipids in trout, *Lipids*, 29, 373, 1994.

105. Bell, M.V. and Dick, J.R., The appearance of rods in the eyes of herring and increased di-docosa-hexaenoyl molecular-species of phospholipids, *J. Mar. Biol. Ass. UK*, 73, 697, 1993.

106. Patton, G.M., Fasulo, J.M., and Robins, S.J., Separation of phospholipids and individual molecular species of phospholipids by high-performance liquid chromatography, *J. Lipid Res.*, 23, 190. 1982.

107. Scott, T.W. et al., Effect of fish oil supplementation on the composition of molecular-species of choline and ethanolamine glycerophospholipids in ruminant muscle, *J. Lipid Res.*, 34, 827, 1993.

108. Kuksis, A., Mass spectrometry of complex lipids, in *Lipid Analysis of Fats and Oils*, Hamilton, J.R., Ed., Blackie Academic and Professional, London, UK, 1998.

109. Kerwin, J.L., Tuininga, A.R., and Ericsson, L.H., Identification of molecular species of glycero-phospholipids and sphingomyelin using electrospray mass spectrometry, *J. Lipid Res.*, 35, 1102, 1994.

110. Myher, J.J. et al., Normal phase liquid chromatography/mass spectrometry with electrospray for sensitive detection of oxygenated glycerophospholipids, *INFORM*, 5, 478–479, Abs. 13E, 1994.

111. Kim, H.Y., Wang, T.C., and Ma, Y.C., Liquid chromatography/mass spectrometry of phospholipids using electrospray ionisation, *Anal. Chem.*, 66, 1566, 1994.

Chapter 12

Cholesterol and Cholesterol Oxides in Meat and Meat Products

Neura Bragagnolo

Contents

12.1 Introduction

Cholesterol is one of the most ubiquitous lipids found in all cells, tissues, and nerves of animals and humans. Cholesterol has a number of vital functions: as bile acids, it modulates the absorption of dietary fat in the intestine; it is the precursor for the steroid hormones of the cortex of the suprarenal gland, of male and female sexual hormones, and of vitamin D; it is also essential for the growth and development of young mammals. Although cholesterol is of extreme biological importance, a positive correlation was found between increased serum cholesterol concentration and increased risk of coronary heart diseases, so consumers currently are very conscious of the lipid content of foods in relation to their health. In this context, dietary cholesterol level has become an important issue since in several publications a reduction in cholesterol consumption is recommended as a means of preventing heart disease. Dietary cholesterol and total amount of saturation of fats are thus major areas of interest.

Cholesterol is vulnerable to oxidation since it presents one double bond in its structure, and when subjected to thermal, photocatalyzed, or enzymatic oxidation, several hydroperoxide isomers are produced, which can further be converted into different compounds called cholesterol oxides or oxycholesterols. The temperature and time required to process foods, as well the transport and storage conditions, are some of the factors that can contribute to cholesterol degradation. Other conditions, including pH, light, oxygen, water activity, or a combination of such factors, are also important in oxidative degradation, although heat processing is the most effective factor in the formation of cholesterol oxidation products.[1,2] Several cholesterol oxides have been shown to possess undesirable biological effects, such as sterol metabolism interference, cytoxicity, atherogenicity, mutagenicity, carcinogenicity, and changes in cellular membrane properties.[3,4]

Comparing published data regarding cholesterol and the extension of cholesterol degradation is difficult due to discrepancies in the results found, which can be attributed to natural variation brought about by factors such as age and breed of the animals, diet, and the rearing system. Cholesterol oxides present in meat and meat products can be the result of metabolism of the animal by the enzymatic system or oxidation and especially as a result of processing and storage conditions. In addition, the analytical methods play a special role since advances in analytical chromatography and mass spectrometry now lead to better separation, identification, and quantification of cholesterol and cholesterol oxides to obtain more reliable results.

12.2 Cholesterol

12.2.1 Structure

Sterols are present as the major portion of the unsaponifiable fraction of most edible fats and oils and are minor constituents in the fat of the human diet. Sterols are classified according to their origin as animal, plant, and microorganism sterols. Cholesterol is the principal animal sterol, while sitosterol, campesterol, and stigmasterol are the major plant sterols; ergosterol is the main mycosterol present in the lipid fraction of yeasts and fungi (Figure 12.1). The sterols are compounds consisting of cyclopentanoperhydrofenantreno as the basic structure, with one double bond between carbons 5 and 6, a hydroxyl group at position 3, methyl groups at positions 10 and 13, and a radical at carbon 17. Each ring junction can exist in a cis or trans conformation. The various sterols differ only in their side chain, except for ergosterol, which has another double bond in the ring; this minor differences, however, results in major changes in biological function. Phytosterols contain an extra methyl or ethyl group, or double bond; most phytosterol side chains contain 9 or 10 carbon atoms

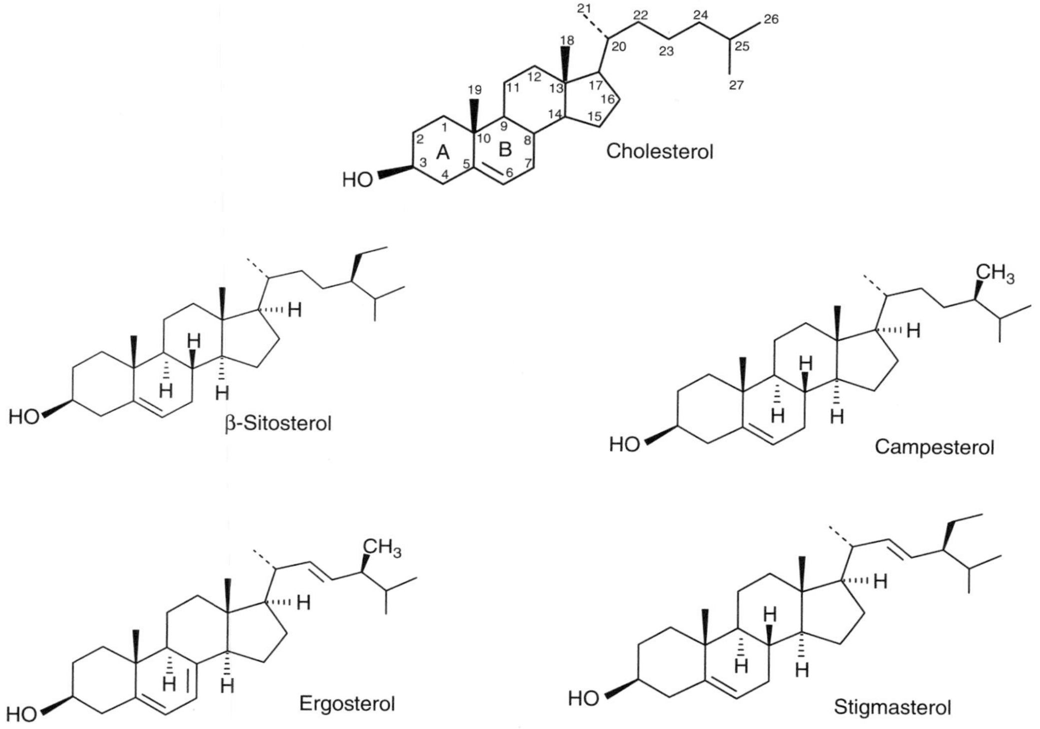

Figure 12.1 Structures and nomenclature of common steroid.

instead of 8, as found in cholesterol. All sterols are widely distributed in nature, and can occur in the free form or combined, most frequently as esters of higher aliphatic acids and glycosides. Cholesterol also occurs in plants, usually in very small quantities, and in marine algae.

12.2.2 Methods of Cholesterol Determination in Food

Various methods have been developed for cholesterol determination in foods due to the necessity of precise and accurate results. The majority of the methods used in food are gravimetric, colorimetric, enzymatic, and chromatographic. The first method used to determine cholesterol in foods was the gravimetric method, in which the cholesterol is separated by precipitation with digitonin or tomatin.[5] This analytical method is time consuming and has low sensitivity. Although colorimetric methods are widely used for measuring the cholesterol extracted from meat,[6-13] they have a tendency to overestimate the cholesterol content of foods, due to the presence of other interfering chromogens present in the samples. Enzymatic methods using cholesterol oxidase have been used to determine cholesterol in meat and meat products.[14-18] As with colorimetric methods, this reaction is also not specific for cholesterol, since all sterols with a 3 β-OH group react with this enzyme, leading to overestimation of the cholesterol content in foods, especially those containing mixtures of animal and vegetable material. The extent of this overestimation, however, would depend on the method used and the sample analyzed. For example, results of the cholesterol obtained from meats by colorimetric methods were comparable to those obtained by gas chromatography

(GC), as pointed by Bohac et al.,[19] or high performance liquid chromatography (HPLC), as demonstrated by Bragagnolo and Rodriguez-Amaya.[20] Using an enzymatic method, Saldanha et al.[14] found no significant difference in bovine meat and milk cholesterol values compared to the results obtained by HPLC. However, Almeida et al.[21] obtained higher cholesterol levels by the enzymatic method than those measured by HPLC in semimembranosus, chicken drumsticks, and chicken thighs, but for biceps femoris no difference was observed between the two methods. However, colorimetric and enzymatic methods need strict control of the analytical conditions to give accurate results.

Currently, chromatographic techniques are preferred for dietary cholesterol analysis, due to their ability to separate and quantify cholesterol specifically. To determine cholesterol in meat and meat products, GC methods have been used[19,22–24] as well as HPLC methods.[25–30] Over the past few years, HPLC has been used as an alternative to GC in cholesterol analysis since the former is carried out at relatively low temperatures, thus preventing the oxidation of cholesterol; cholesterol concentrations obtained by the two methods did not show significant differences.[31] However, in samples containing cholesterol and phytosterols in very small amounts, the GC method is preferred since it is more sensitive.

Normally, the chromatographic cholesterol assay would include the following steps: (1) extraction of total lipids with an organic solvent or solvent mixture, (2) removal of the solvent, (3) alkaline saponification of the total lipids, (4) extraction of the unsaponfiables with an organic solvent, (5) removal of the solvent, (6) derivatization of the unsaponfiable matter if GC is to be used (although Kanada et al.[32] showed that derivatization was not essential), and (7) chromatographic estimation of the analyte.

The most common extraction method is direct saponification of the food, followed by extraction of the unsaponfiable matter with an organic solvent and subsequent determination of cholesterol by GC[15,21,33–35] or HPLC.[21,36] Van Elswyk et al.[37] and Bragagnolo and Rodriguez-Amaya[36] observed that saponification of a lipid extract resulted in a significantly lower value than direct saponification of the sample. The saponification step is very important in the determination of total cholesterol, since in this step the cholesterol esters are converted to free cholesterol; when this step was not carried out, the results must be expressed as free cholesterol or the results of total cholesterol are substituted.

GC detection of cholesterol has also been carried out by flame ionization detector (FID), and separation can be performed in either medium of low or nonpolar columns, but in the majority of studies the slightly polar HP-5, DB-5 fused silica bonded phase capillary column has been used. Identification of the cholesterol was performed by comparison of the retention times of the samples with those of the cholesterol standard, co-chromatography, and confirmed by MS.[15,38] Quantification was done by internal standardization using 5α-cholestane in most studies,[15,31–33,38,39] although the choices depend to a great extent on the type of sample to be analyzed.[40] Thompson and Merola[40] recommended 5α-cholestanol instead of 5α-cholestane because it is an alkane and so does not have the same chemical and physical properties as cholesterol or one of the plant sterols when the samples do not already contain such sterols.

Over the past few years, several HPLC methods for quantitative analysis of cholesterol in meat have been developed using direct saponification[14,21,41] or with extraction of the lipid and subsequent saponification.[27,28,42,43] A reversed-phase (C18, 150 mm, 4.6 mm, and 5 μm) column, with a mixture of acetonitrile and 2-propanol (in the following proportions: 80:20, 70:30, and 60:40) as mobile phases, and ultraviolet (UV) detection set at 210 nm are the HPLC conditions widely used for the analysis of cholesterol in meat. Identification of the cholesterol is performed by comparison of the retention times of the samples with those of the cholesterol standard, co-chromatography, and spectra taken at 190–300 nm with the photodiode array detector. To confirm its identity and purity, cholesterol was accumulated from several HPLC runs of the sample and analyzed by GC-MS.[36] Quantitative determinations are usually made using external standardization or, when possible, internal standardization using 6-ketocholesterol,[14] stigmasterol,[30] or pregnolone.[31]

12.2.3   Cholesterol Content in Meat and Meat Products

Meat of all types contains cholesterol; it is present in lean muscles since cholesterol is the major component of cell membranes and of nerves and is an active metabolite within the cells. Cholesterol occurs in higher concentrations in the organ and glandular meats such as heart, kidney, liver, and sweetbreads than in regular cuts of meat, with or without fat, and is highest in brains.[5] However, this chapter focuses only on the cholesterol content in meat and meat products of different kinds.

12.2.3.1   Cholesterol Content in Bovine Meats

As can be seen in Tables 12.1 and 12.2, the cholesterol content in raw and cooked bovine meats varied from 43 mg/100 g to 84 mg/100 g[6,8,16,19,21,23,27,44–55] and 57 mg/100 g to 101 mg/100 g,[6,16,23,27,33,47,51,54–61] respectively.

Changes in retailing practices and the beef grading system are reflected in the low values of total cholesterol; in fact, Sahasrabudhe and Stewart[23] observed that the values for cholesterol were 18–23% lower than those reported in past years for the same cuts (Table 12.1).

Huerta-Leidenz et al.,[50] using a colorimetric method, did not observe variation of the cholesterol content in response to differences in sex class, age, maturity level, carcass grade, marbling level, or subcutaneous fat thickness. Breed type and dietary energy level effects were nonsignificant in the analysis of cholesterol data for muscle tissue. However, feeding groups were related to a difference in cholesterol content of adipose tissue; cows fed restricted diets were higher than those fed *ad libitum*.[53] Similar results for breed type were found by Bragagnolo and Rodriguez-Amaya,[27] for genetic selection by Morris et al.,[62] for diet by Andrae et al.,[49] for age by Stromer et al.,[63] and for dietary energy level by Abu-Tarbousch and Dawood,[44] who found that the cholesterol content of the adipose tissue obtained from animals fed with a high level of energy was lower than the adipose tissues obtained from animals fed with low levels. Rule et al.[64] suggested that different growing and finishing strategies in beef cattle production systems did not alter cholesterol in meat; but ground carcass values of cholesterol were greater in moderate steers than in high moderate sire growth potential because of the greater fat content in the moderate ground carcass. In addition, Abu-Tarbousch and Dawood[44] showed that adipose tissues of castrated beef carcasses do not contain more cholesterol than those of noncastrated beef carcasses, which is in agreement with the findings of Eichhorn et al.[53]

Several researchers have reported that cholesterol content does not vary among muscle from different anatomical localizations. For example, Bragagnolo and Rodriguez-Amaya[6] analyzed five different beef cuts and found no significant differences; Badiani et al.[47] also found that cholesterol content was similar in three different muscle cuts, which was confirmed by Cifuni et al.[48] in Podolian young bulls and by Bohac and Rhee[52] in steers that were fed diets with varying levels of oleic acid. The values for cholesterol content in bovine muscles with and without external fat obtained by Swize et al.[8] showed no differences in cholesterol level. Similar results were found by Rhee et al.[54] in a study of relationships of eight marbling levels to cholesterol content in beef longissimus muscle; only when practically devoid of marbling did the meat contain significantly

Table 12.1 Cholesterol Content in Raw Bovine Meat

Portions and References	Methods Lipid Extraction	Saponification	Measurement	Cholesterol Level (mg/100 g, Wet Basis)
Adipose tissue[44]	C/M Folch	1 mL 50%, 3 mL Et KOH, 55°C, 90 min	Colorimetric	140–184
Adipose tissue[45]	C/M Folch	8 mL 15% KOH (90% Et), 2 mL 15% PG (90% Et), 80°C, 15 min	Colorimetric	118
Cuts[46]	No	0.5 mL KOH conc., 85°C, 30 min	GC IS: epicholesterol	50–55
Cuts[6]	C/M Folch	10 mL 12% KOH (90% Et), 80°C, 15 min	Colorimetric	50–56
Cuts[23]	C/M	8 mL conc. KOH, 40 mL Et, reflux, 1 h	GC	43–53
Cuts[16]	C/M	0.5 mL 50% KOH, 2 mL 95% Et, 100°C, 1 h	Enzymatic	46–53
Infraspinatu[47] Longissimus lumborum Semitendinosus	No	8 mL 50% KOH, 40 mL Et, reflux, 1 h	GC IS: 5α-cholestane derivatization	48–55
Longissimus dorsi[48] Semimembranosus Semitendinosus	No	8 mL Et with 0.125% BHA, 0.5 mL 40% KOH aqueous, 80°C, 15 min	Enzymatic	50 48 47
Longissimus dorsi[27]	C/M Folch	10 mL 12% KOH (90% Et), 80°C, 15 min	HPLC	40–43
Longissimus muscle[49]	C/M Folch	BT (35%), M (35%), B (30%), 95°C, 45 min	GC IS: stigmasterol	56–57
Longissimus dorsi[50]	C/M Folch	8 mL 15% KOH (90% Et), 2 mL 15% PG (90% Et), 80°C, 15 min	Colorimetric	64–69
Longissimus muscle[8] Semimembranosus muscle	C/M Folch	8 mL 15% KOH (90% Et), 2 mL 15% PG (90% Et), 80°C, 15 min	Colorimetric	63–68 79–84
Longissimus muscle[51]	C/M Folch	8 mL 15% KOH (90% Et), 2 mL 15% PG (90% Et), 80°C, 15 min	Colorimetric	61–64
Longissimus dorsi[52] Semimembranosus Semitendinosus	C/M Folch	10 mL 12% KOH (90% Et) with 3% P, 80°C, 15 min	Colorimetric	56 59 57
Longissimus dorsi[19] Semimembranosus	C/M Folch	10 mL 12% KOH (90% Et) with 3% P, 80°C, 15 min	Colorimetric GC Colorimetric GC	59–63 59–60 57–69 59–69
Longissimus muscle[53] Triceps brachii muscle	C/M	3 mL 15% KOH (90% Et), 90°C, 30–45 min	Colorimetric	54–67 61–63
Longissimus muscle[54]	C/M Folch	5 mL 15% KOH (90% Et), 88°C, 15 min	Colorimetric	52–66
Pectineu[55]	C/M	0.5 mL 50% KOH, 2 mL 95% Et, 100°C, 1 h	GC IS: 5α-cholestane	50
Semimembranosus[21]	No	4 mL 50% KOH, 6 mL Et (95%), 60°C, 10 min	HPLC Enzymatic	52 61

Table 12.1   (Continued)

| Portions and References | Methods | | | Cholesterol Level (mg/100 g, Wet Basis) |
	Lipid Extraction	Saponification	Measurement	
Biceps femoris			HPLC	63
			Enzymatic	63
Tallow[56]	No	5 mL 2 M KOH (Et), dark, room temperature, 18 h	GC-MS	50
Tallow[57]	No	5 mL 2 N KOH (M), room temperature, overnight	GC-MS IS: 5α-cholestane derivatization	71
Tallow[58]	No	100 mL 1 N (Et) KOH, reflux 1 h	GC IS: 5α-cholestane	140

Notes: C/M, extraction with chloroform and methanol, according to Folch et al.[59] or Bligh and Dyer[60]; Et, ethanol; BHA, butylated hydroxy anisole; PG, propyl gallate; P, pyrogallol; BT, boron trifluoride; B, benzene.

Table 12.2   Cholesterol Content in Cooked Bovine Meat

| Portions and References | Methods | | | Cholesterol Level (mg/100 g, Wet Basis) |
	Lipid Extraction	Saponification	Measurement	
Cuts[6]	C/M Folch	10 mL 12% KOH (90% Et), 80°C, 15 min	Colorimetric	66–67
Cuts[23]	C/M	8 mL conc. KOH, 40 mL Et, reflux, 1 h	GC IS: 5α-cholestane derivatization	65–87
Cuts[16]	C/M	0.5 mL 50% KOH, 2 mL 95% Et, 100°C, 1 h	Enzymatic	57–67
Infraspinatus[47] Longissimus lumborum Semitendinosus	No	8 mL 50% KOH, 40 mL Et, reflux, 1 h	GC IS: 5α-cholestane derivatization	67–96
Longissimus dorsi[27]	C/M Folch	10 mL 12% KOH (90% Et), 80°C, 15 min	HPLC	67–70
Longissimus dorsi[61]	C/M Folch	10 mL 15% KOH (90% Et), 80°C, 20 min	Colorimetric	96–101
Longissimus dorsi[51]	C/M Folch	8 mL 15% KOH (90% Et), 2 mL 15% PG (90% Et), 80°C, 15 min	Colorimetric	74–85
Longissimus dorsi[54]	C/M Folch	5 mL 15% KOH (90% Et), 88°C, 15 min	Colorimetric	77–92
Meat cooked[33]	No	1 mL 60% KOH/g, 4 mL Et/g, reflux, 30 min	GC IS: 5α-cholestane	61

Notes: C/M, extraction with chloroform and methanol, according to Folch et al.[59] or Bligh and Dyer[60]; Et, ethanol; PG, propyl gallate.

less cholesterol than did the other seven samples marbling levels. According to Piironen et al.,[46] there was no correlation between the cholesterol content of beef cuts and their fat content. In contrast, Akinwunmi et al.[61] reported that steaks with slight marbling had higher cholesterol than those with modest marbling; Browning et al.[65] found that Supraspinatus and Infraspinatus muscles had higher cholesterol content than other muscles. In addition, the authors showed that lean carcasses had less cholesterol than typical carcasses, although this difference was small. Hood[66] showed a curvilinear relationship between the amount of cholesterol and the percentage of muscle lipids. When the amount of extractable muscle lipids is low, cholesterol concentration is high; therefore, functional membrane lipid contains a higher level of intramuscular adipose tissue lipid.

Adipose tissues contain about 100% as much cholesterol as muscle tissues,[53] or 150% higher than in the muscle.[67] Sweeten et al.[45] found 118 mg/100 g of total cholesterol in intramuscular adipose tissue, with 64 mg/100 g present in the cytoplasm and 54 mg/100 g in the membrane fraction. However, Hoelscher et al.[51] found that the membrane fraction of subcutaneous tissue contained an average of 10 mg/100 g of tissue, while the cytoplasmic fraction contained an average of 90 mg/100 g of tissue. In muscle tissue, Hoelscher et al.[51] noted an opposite distribution, with the membrane containing 76 mg/100 g of tissue in comparison to 23 mg/100 g of tissue in the cytoplasm.

The values found in tallow by Verleyen et al.,[56] 50 mg/100 g, were lower than the results obtained by Park and Addis[57] and Ryan and Gray,[58] which ranged from 70 to 140 mg/100 g. According to Park and Addis,[57] the cholesterol content in tallow is considered to vary depending on the part of beef fat from which commercial products are made.

12.2.3.2 Cholesterol Content in Pork Meat

As shown in Tables 12.3 and 12.4, the levels of cholesterol found in raw and cooked pork meat are in the range of 30–81 mg/100 g[6,8,10,11,17,19,20,28,46,52,68–73] and 56–113 mg/100 g,[6,8–11,17,22,74] respectively.

Compared to bovine meat, there is scant information on the cholesterol content of pork muscle and adipose tissue, and from the results it can be seen that no significant difference was observed in the levels of cholesterol between cuts of pork and beef meat.[6,52] As in bovine meat, results for total cholesterol among pork cuts with different anatomical localization, level of fat, and conventional and new-fashioned cuts were found to be similar.[6,17,28,59,69,75] However, Fernandez et al.[75] reported a significant lower cholesterol content in longissimus lumborum than in semispinalis capitis. Swize et al.[8] showed that longissimus fatless pork chops were lower than the same muscle combined lean and fat; however, semimembranosus muscles did not differ among the fat treatments.

The cholesterol contents in pork meat found by Echarte et al.[70] and Bragagnolo and Rodriguez-Amaya[6,28,68] were lower than the results obtained by Swize et al.[8] and Buege et al.[69] However, Csallany et al.[73] reported lower levels of cholesterol, though these results are for free cholesterol rather than total cholesterol since the saponification step was not carried out; moreover, in pork meat 70% of the cholesterol is in the ester form,[13] which can explain these low results.

Cholesterol content did not differ significantly in the meat of suckling pigs (98 and 95 mg/100 g for 15 and 21 days, respectively), but in adult pork these values were considerably lower (49 mg/100 g for loin and 44 mg/100 g for fresh ham).[28] However, in the skin and backfat the cholesterol content significantly decreases from 15 days (109 mg/100 g) to 21 days

Table 12.3 Cholesterol Content in Raw Pork Meat

| Portions and References | Methods | | | Cholesterol Level (mg/100 g, Wet Basis) |
	Lipid Extraction	Saponification	Measurement	
Cuts[46]	No	0.5 mL KOH conc., 85°C, 30 min	GC IS: epicholesterol	45–54
Cuts[68]	C/M Folch	10 mL 12% KOH (90% Et), 80°C, 15 min	HPLC	42–53
Cuts[28]	C/M Folch	10 mL 12% KOH (90% Et), 80°C, 15 min	HPLC	44–49
Cuts[20]	C/M Folch	10 mL 12% KOH (90% Et), 80°C, 15 min	Colorimetric HPLC	39–43 36–48
Cuts[69]	C/M/W	8 mL conc. KOH, 40 mL Et, reflux, 1 h	GC IS: 5α-cholestane derivatization	59–67
Cuts[6]	C/M Folch	10 mL 12% KOH (90% Et), 80°C, 15 min	Colorimetric	49–54
Cuts[17]	C/M	0.5 mL 50% KOH, 2 mL 95% Et, 100°C, 1 h	Enzymatic	56–54
Fresh loin[70]	No	1 mL 50% KOH, 4 mL 95% Et, 100°C, 1 h	GC IS: 5α-cholestane	56
Longissimus muscle[8] Semimembranosus muscle	C/M Folch	8 mL 15% KOH (90% Et), 2 mL 15% PG (90% Et), 80°C, 15 min	Colorimetric	70–81 74–78
Longissimus dorsi[71]	No	8 mL 50% KOH, 40 mL Et, reflux, 1 h	GC IS: 5α-cholestane TMS	57–59
Longissimus dorsi[72]	No	8 mL 50% KOH, 40 mL Et, reflux, 1 h	GC IS: 5α-cholestane derivatization	51
Longissimus dorsi[73]	C/M Folch	No	HPLC	30
Longissimus dorsi[52] Semimembranosus Semitendinosus	C/M Folch	10 mL 12% KOH (90% Et) with 3% P, 80°C, 15 min	Colorimetric	56 53 60
Longissimus dorsi[19] Semimembranosus	C/M Folch	10 mL 12% KOH (90% Et) with 3% P, 80°C, 15 min	Colorimetric GC Colorimetric GC	64 63 56–68 53–62
Patties[10]	C/M Folch	15 mL 15% KOH (90% Et), 88°C, 10 min	Colorimetric	56–70
Tenderloin steaks[11]	C/M Folch	15 mL 15% KOH (90% Et), 88°C, 10 min	Colorimetric	45

Notes: C/M, extraction with chloroform and methanol, according to Folch et al.[59] or Bligh and Dyer[60]; W, water; Et, ethanol; PG, propyl gallate; P, pyrogallol.

Table 12.4 Cholesterol Content in Cooked Pork Meat

Portions and References	Lipid Extraction	Saponification	Measurement	Cholesterol Level (mg/100 g, Wet Basis)
		Methods		
Cuts[6]	C/M Folch	10 mL 12% KOH (90% Et), 80°C, 15 min	Colorimetric	58–97
Cuts[22]	C/M	8 mL conc. KOH, 40 mL Et, reflux, 1 h	GC IS: 5α-cholestane derivatization	96–104
Cuts[9]	C/M Folch	8 mL 15% KOH (90% Et), 2 mL 15% PG (90% Et), 80°C, 15 min	Colorimetric	80–91
Cuts[17]	C/M	0.5 mL 50% KOH, 2 mL 95% Et, 100°C, 1 h	Enzymatic	83–93
Longissimus thoracis[74] Longissimus lumborum	No	8 mL 50% KOH, 40 mL Et, reflux, 1 h	GC IS: 5α-cholestane derivatization	72–81
Longissimus[8] Semimembranosus	C/M Folch	8 mL 15% KOH (90% Et), 2 mL 15% PG (90% Et), 80°C, 15 min	Colorimetric	86–97 107–113
Patties[10]	C/M Folch	15 mL 15% KOH (90% Et), 88°C, 10 min	Colorimetric	92–101
Tenderloin steaks[11]	C/M Folch	15 mL 15% KOH (90% Et), 88°C, 10 min	Colorimetric	56–78 59–72 66–75

Notes: C/M, extraction with chloroform and methanol, according to Folch et al.[59] or Bligh and Dyer[60]; Et, ethanol; PG, propyl gallate.

(94 mg/100 g). In addition, in the backfat of adult pigs the values of cholesterol were lower than in the skin (102, 79, and 33 mg/100 g, respectively). It is well known that cholesterol has many functions in the animal, being essential for the growth and development of young mammals, which probably is responsible for the lower values of cholesterol in adult pork.

The use of repartitioning agents (compounds that will decrease carcass fat and increase muscle, such as ractopamine hydrochloride) can effectively decrease intermuscular fat. Engeseth et al.[71] reported that the cholesterol content of longissimus muscle from control and ractopamine treated (20 ppm) at various stages of the feeding trial and during the 7 days withdrawal were unchanged. However, Perkins et al.[72] observed a reduction of 9% in cholesterol content of longissimus muscle of pigs treated with 10 and 20 ppm of ractopamine.

12.2.3.3 Cholesterol Content in Poultry Meat

Tables 12.5 and 12.6 show the results found for cholesterol content in raw and cooked poultry meats: 27–90 mg/100 g[7,17,18,21,33,38,42,55,69,76–79] and 59–154 mg/100 g,[7,17,76,79] respectively.

Using the HPLC method, Komprda et al.[42] found that the cholesterol content was lower in chicken breast than in chicken thigh and skin. Similar results were related by Bragagnolo

Table 12.5   Cholesterol Content in Raw Poultry Meat

Portions and References	Methods			Cholesterol Level (mg/100g, Wet Basis)
	Lipid Extraction	Saponification	Measurement	
Chicken				
Breast[76]	No	1 mL 50% KOH, 4 mL 95% Et, 100°C, 1 h	GC IS: 5α-cholestane	79–83
Breast[42]	H/P	10 mL SM, room temperature, 30 min	HPLC	53
Breast[69]	C/M/W	8 mL conc. KOH, 40 mL Et, reflux, 1 h	GC IS: 5α-cholestane derivatization	64
Breast[7]	C/M Folch	10 mL 12% KOH (90% Et), 80°C, 15 min	Colorimetric	58
Breast[18]	C/M	0.5 mL 50% KOH, 2 mL 95% Et, 100°C, 1 h	Enzymatic	50
Thigh[21]	No	4 mL 50% KOH, 6 mL Et (95%), 60°C, 10 min	HPLC/enzymatic	76/99
Thigh[42]	H/P	10 mL SM, room temperature, 30 min	HPLC	83
Thigh[69]	C/M/W	8 mL conc. KOH, 40 mL Et, reflux, 1 h	GC IS: 5α-cholestane derivatization	90
Thigh[7]	C/M Folch	10 mL 12% KOH (90% Et), 80°C, 15 min	Colorimetric	80
Thigh[17]	C/M	0.5 mL 50% KOH, 2 mL 95% Et, 100°C, 1 h	Enzymatic	83
Skin[7]	C/M Folch	10 mL 12% KOH (90% Et), 80°C, 15 min	Colorimetric	104
Wing[69]	C/M/W	8 mL conc. KOH, 40 mL Et, reflux, 1 h	GC IS: 5α-cholestane derivatization	90
Turkey				
Breast[42]	H/P	10 mL SM, room temperature, 30 min	HPLC	53
Breast[77]	C/M Folch	10 mL 1 N KOH (M), 20°C, 18 h	HPLC	27
Breast[78]	No	8 mL conc. KOH, 40 mL Et, reflux, 1 h	GC IS: 5α-cholestane derivatization	54
Breast[79]	C/M Folch	15 mL 15% KOH 90% Et, 88°C, 10 min	Colorimetric	48
Meat[33]	No	1 mL 60% KOH/g, 4 mL Et/g, reflux, 30 min	GC IS: 5α-cholestane	40
Mechanically[38] separated turkey meat	No	5 mL 50% KOH aqueous, 22 mL Et, 23°C, 6 h	CC and MS	144
Thigh[42]	H/P	10 mL SM, room temperature, 30 min	HPLC	62

(Continued)

Table 12.5 (Continued)

Portions and References	Methods			Cholesterol Level (mg/100g, Wet Basis)
	Lipid Extraction	Saponification	Measurement	
Thigh[77]	C/M Folch	10 mL 1 N KOH M, 20°C, 18 h	HPLC	35
Thigh[55]	C/M	0.5 mL 50% KOH, 2 mL 95% Et, 100°C, 1 h	GC IS: 5α-cholestane	37
Thigh[78]	No	8 mL conc. KOH, 40 mL Et, reflux, 1 h	GC IS: 5α-cholestane derivatization	84
Skin[77]	C/M Folch	10 mL 1 N KOH M, 20°C, 18 h	HPLC	81
Skin[78]	No	8 mL conc. KOH, 40 mL Et, reflux, 1 h	GC IS: 5α-cholestane derivatization	87
Visible fat[78]	No	8 mL conc. KOH, 40 mL Et, reflux, 1 h	GC IS: 5α-cholestane derivatization	57
Wing[77]	C/M Folch	10 mL 1 N KOH M, 20°C, 18 h	HPLC	46

Notes: C/M, extraction with chloroform and methanol, according to Folch et al.[59] or Bligh and Dyer[60]; H/P, hexano and 2-propanol; Et, ethanol; SM, sodium methoxide; W, water.

Table 12.6   Cholesterol Content in Cooked Poultry Meat

Portions and References	Methods			Cholesterol Level (mg/100 g, Wet Basis)
	Lipid Extraction	Saponification	Measurement	
Cooked chicken				
Breast[76]	No	1 mL 50% KOH, 4 mL 95% Et, 100°C, 1 h	GC IS: 5α-cholestane	62–75 81–82
Breast[7]	C/M Folch	10 mL 12% KOH (90% Et), 80°C, 15 min	Colorimetric	75
Breast[17]	C/M	0.5 mL 50% KOH, 2 mL 95% Et, 100°C, 1 h	Enzymatic	92
Thigh[7]	C/M Folch	10 mL 12% KOH (90% Et), 80°C, 15 min	Colorimetric	124
Thigh[17]	C/M	0.5 mL 50% KOH, 2 mL 95% Et, 100°C, 1 h	Enzymatic	154
Skin[7]	C/M Folch	10 mL 12% KOH (90% Et), 80°C, 15 min	Colorimetric	139
Cooked turkey				
Breast[79]	C/M Folch	15 mL 15% KOH (90% Et), 88°C, 10 min	Colorimetric	59–66

Notes: C/M, extraction with chloroform and methanol, according to Folch et al.[59] or Bligh and Dyer[60]; Et, ethanol.

and Rodriguez-Amaya[7] using a colorimetric method; by Konjufca et al.[80] and Hutchison et al.[17] using an enzymatic method; and by Ajuyah et al.[81] using GC, although the results obtained in thigh meat by Ajuyah et al.[81] and by Konjufca et al.[80] were higher than the others. It is interesting to note that the results in chicken breast were similar to the results obtained in pork and beef.[6,16–19,23,27,46,52,55,64,68]

As with chicken, highly significant differences exist in the cholesterol levels among the different cuts of turkey meat analyzed. The skin showed the highest levels, whereas the breast meat presented the lowest values.[77] Wong and Sampugna[78] observed that skin and dark meat had similar cholesterol content, and both showed higher cholesterol content than visible fat and breast meat; also, the cholesterol values observed for visible fat and light meat were similar. Thus, on the basis of these results it is clear that white turkey meat is a low-cholesterol alternative among the other meats.

The cholesterol values found in the leg turkey muscle by Baggio et al.[77] were similar to those found by Paleari et al.;[55] higher values were reported by Prusa and Hughes[79] and Wong and Sampugna.[78] However, the cholesterol content in mechanically separated turkey meat[38] was higher than the values obtained in turkey meat.[77–79]

Wong and Sampugna[78] verified considerable variability in cholesterol values among the different ground turkey samples. Ground turkey is commonly a mixture of dark and light meat, including adhering skin and visible fat. Differences in the amounts of these components used by individual processors to prepare the ground turkey would affect the level of cholesterol observed in the final product. Since breast meat is in high demand and commands a premium price, most ground turkey is made from surplus thighs and drumsticks.

Higher cholesterol content in thigh muscle of both poultry species, chicken and turkey, when compared to breast muscle is likely a consequence of substantially higher total lipid content in thigh meat.[42] The study conducted by Smith et al.[82] showed that on pectoralis muscles of ducklings with 16% white fiber have a higher cholesterol content (99 mg/100 g) than in chicken with 100% white fiber (47 mg/100 g). The relationship between fiber type and cholesterol content has been interpreted on physical and metabolic grounds. Duckling pectoralis muscle contains a high proportion of red fibers that are smaller in transverse area than white fibers. Therefore, a duckling pectoralis muscle with weight similar to that of the broiler would be composed, proportionally, of more red fibers. An increase of fiber number within a muscle would increase the ratio of total sarcolemma perimeter to fiber per volume and, therefore, the cholesterol content. In addition, oxidative muscles are known to be richer in phospholipids, and there is a direct relation between the content of phospholipids and cholesterol which seems to be necessary to maintain the membrane fluidity within a narrow range.[83]

The influence of the hens' diet in terms of the cholesterol content of the meat was reported by many researchers. Among these, Ajuyah et al.[84] studied the effect of five diets in the amount of cholesterol in white and dark meat and in whole carcasses, observing that only in the dark meat was the level of cholesterol significantly lower in hens fed with full-fat flax seed than in hens fed with a control diet. Similar results were observed by Ajuyah et al.,[81] testing eight different diets in broiler chicken and finding that only carcasses from birds fed with animal tallow plus canola meal contained less cholesterol than carcasses from birds fed with canola oil plus canola meal. Zanini et al.[85] reported that soybean oil in the diet of the cockerels increases cholesterol content in the thigh meat as compared to diets containing fish and canola oil. In addition, Konjufca et al.[80] found lower cholesterol content in breast and thigh muscles when the chickens were fed with either garlic powder or copper.

Cholesterol content in chicken thigh muscle decreased with increasing live weight until the age of 43 days, but did not change in chicken and turkey breast or turkey thigh muscles.[42]

12.2.3.4 Cholesterol Content in Other Meats

Table 12.7 shows the cholesterol content of other meats: 50–87 mg/100 g in raw[8,29,87–91] and 82–113 mg/100 g in cooked samples.[8]

The cholesterol values obtained in lambs by Perez et al.[29] were similar to the results obtained by Solomon et al.[90] and Lough et al.,[88] but higher values were obtained by Sevi et al.[92] In general, the cholesterol values in lamb and veal were similar to the results obtained in chicken thigh and slightly higher than beef, pork, and breast poultry.

The effect of lamb breeds on cholesterol content was not significant, as shown by Perez et al.[29] and by Sevi et al.,[92] but it was significantly different in the results obtained by Arsenos et al.[43] According to these authors,[43] lamb breed remains the main factor that determines the cholesterol content in carcass fat between nutritional conditions and degree of maturity; in addition, small but significant differences were found between carcasses of male and female lambs. However, there was no difference among sex types in cholesterol content of longissimus dorsi.[90]

Some studies have reported that cholesterol content of muscles increased with increasing age and slaughter weight.[51,83,86] However, results obtained in castrated Boer goats[41] and lamb carcasses[43,29] demonstrated that slaughter weight significantly affected the cholesterol content; it decreased, with a concomitant increase of total lipids, when the slaughter weigh increased. It seems that the cholesterol content is not attributable to differences in the amount of fat in the carcasses. Similarly, Swize et al.[8] observed that longissimus lamb chops with different grades of external fat did not differ in cholesterol content; but semimembranosus muscles with combined lean and fat were higher than the samples without external fat. Significant differences in total cholesterol concentrations among muscles from the castrated Boer goat were found by Werdi Pratiwi et al.;[41] this is in agreement with other work in veal,[91] where the results from different muscles showed significant differences, though this difference is very small. As in other meats, the variation can be explained by differences in fiber composition.[12,82,83]

Arsenos et al.[43] showed that cholesterol contents were lower in carcass fat samples obtained from lambs reared on pasture than those obtained from lambs of the same breed that were reared indoors in individual pens on a diet of restricted amounts of concentrates and *ad libitum* Lucerne hay. Thus, lambs reared on pasture could be a source of low-fat, low-cholesterol carcasses throughout the year. However, Solomon et al.[87] and Lough et al.[88] found that diet had no effect on cholesterol content of lean tissue. Nevertheless, great variability in cholesterol content was reported by Lough et al.[88] in subcutaneous adipose tissue of ram lambs fed with canola seed and soy lecithin in forage diets, and by Solomon et al.[89] in muscles of lambs fed with diets containing rapeseed meal or soybean meal or both.

12.2.3.5 Cholesterol Content in Processed Meat Products

As can be seen in Table 12.8, the lowest value of cholesterol in processed meat products was 23 mg/100 g in cooked chicken frankfurter and highest, 144 mg/100 g, in turkey meat products.[26,33,38,93–97]

Baggio and Bragagnolo[93] analyzed 126 samples, including 45 processed beef products (meatball, hamburger, and jerked beef), 36 pork products (sausage, Spam, and ham), and 45

Table 12.7 Cholesterol Content in Other Meats

Portions and References	Methods			Cholesterol Level (mg/100 g, Wet Basis)
	Lipid Extraction	Saponification	Measurement	
Goat[86] meat	C/M Folch	10 mL 12% KOH (90% Et), 80°C, 15 min	Colorimetric	52–74
Lamb[29]	C/M Folch	10 mL 1 N KOH (M), 20°C, 18 h	C/M Folch	64–75
Lamb[87] Longissimus muscle	C/M Folch	8 mL 3% P (Et), 5 ml sat aqueous KOH, 80°C, 8 min	GC IS: stigmasterol derivatization	66–68
Lamb[88] Longissimus muscle	C/M Folch	8 mL 3% P (Et), 5 mL sat aqueous KOH, 80°C, 8 min	GC IS: stigmasterol derivatization	61–64
Lamb[8] Longissimus Semimembranosus	C/M Folch	8 mL 15% KOH (90% Et), 2 mL 15% PG (90% Et), 80°C, 15 min	Colorimetric	72–77 75–87
Lamb[89] Longissimus muscle Semimembranosus Triceps brachii	C/M Folch	8 mL 3% P (Et), 5 mL sat aqueous KOH, 80°C, 8 min	GC IS: stigmasterol derivatization	70–77 72–78 71–76
Lamb[90] Longissimus muscle	C/M Folch	8 mL 3% P (Et), 5 mL sat aqueous KOH, 80°C, 8 min	GC IS: stigmasterol derivatization	64–70
Veal[91] Longissimus thoracic Longissimus lumborum Semitendinosus Cooked	No	0.2 g L-ascorbic acid, 5.5 mL (11% KOH, 55% Et, 45% H$_2$O), 80°C, 15 min	HPLC	56 52 50
Lamb[8] Longissimus Semimembranosus	C/M Folch	8 mL 15% KOH (90% Et), 2 mL 15% PG (90% Et), 80°C, 15 min	Colorimetric	82–102 97–113

Notes: C/M, extraction with chloroform and methanol, according to Folch et al.[59] or Bligh and Dyer[60]; Et, ethanol; PG, propyl gallate; P, pyrogallol.

mixed meat products (frankfurters, mortadella, and Italian salami). The cholesterol content showed great variability; however, there were no significant differences between meatballs and hamburger, between jerked beef and mortadella, or between Spam and ham. The values obtained by Novelli et al.[98] in mortadella and Milan type salami, by Lercker and Rodriguez-Estrada[99] in salami, by Rodriguez-Estrada et al.[100] in hamburger, by Larkeson et al.[97] in meatballs, by Pereira et al.[101] and Zanardi et al.[95] in sausages, and by Petrón et al.[96] in dry-cured Iberian hams were higher than the values obtained by Baggio and Bragagnolo.[93]

Eight processed turkey meat samples showed no significant difference in cholesterol content,[26] which is in agreement with the values previously reported for turkey sausages by Pereira et al.[101]

Table 12.8 Cholesterol Content in Processed Meat Products

Portions and References	Lipid Extraction	Saponification	Measurement	Cholesterol Level (mg/100 g, Wet Basis)
Bovine products[93]	C/M Folch	10 mL 1 N KOH M, 20°C, 18 h	HPLC	29–43
Chicken products[26]	C/M Folch	10 mL 1 N KOH M, 20°C, 18 h	HPLC	31–64
Cooked chicken frankfurter[94]	C/M Folch	10 mL 12% KOH, 80°C, 15 min	GC IS: 5α-cholestane	23–25
Fermented sausages[95] (pork and beef)	No	8 mL 50% KOH, 40 mL Et, reflux, 1 h	GC IS: 5α-cholestane TMS	66–95
Iberian ham[96]	No	10 mL 1 M KOH, room temperature, 20 h	GC IS:5α-cholestanol derivatization	30–34
Meatballs[97] (pork and beef) Hamburger bovine	H/P	5 mL 2 M KOH in Et, dark, room temperature, 18 h	GC TMS	29–46
Mixed products[93]	C/M Folch	10 mL 1 N KOH M, 20°C, 18 h	HPLC	43–59
Pork products[93]	C/M Folch	10 mL 1 N KOH M, 20°C, 18 h	HPLC	22–32
Turkey meat products[26]	C/M Folch	10 mL 1 N KOH M, 20°C, 18 h	HPLC	32–43
Turkey meat products[38]	No	5 mL 50% KOH aqueous, 22 mL Et, 23°C, 6 h	CC and MS	116–144
Turkey ham[33]	No	1 mL 60% KOH/g, 4 mL Et/g, reflux, 30 min	GC IS: 5α-cholestane	68

Notes: C/M, extraction with chloroform and methanol, according to Folch et al.[59] or Bligh and Dyer[60]; H/P, hexano and 2-propanol; Et, ethanol; TMS, trimethylsilyl ether.

However, higher cholesterol values were found in turkey ham[33] and in some turkey meat processed products by King et al.[38]

In chicken meat products[25] a significant difference was found between two brands of chicken frankfurter but not in other chicken meat products analyzed, with the highest cholesterol content in the Chester* chicken frankfurter and the lowest in the Chester chicken hamburger. Values for cholesterol in chicken and Chester chicken sausages obtained by Pereira et al.[101] were close to those found by Baggio and Bragagnolo.[25]

The greatest variation of the cholesterol content among meat products could be related mainly to the ingredients used in product processing, such as different muscles, red or white, or cuts with low value (from the distal end of the limbs, from the neck, from the head), which tend to be rich in red fibers or ingredients other than muscles, such as offal.[93,102]

The effect of storage time on cholesterol content was investigated in jerked beef, Italian type salami, chicken mortadella, and Chester mortadella; the results obtained showed no significant differences after 90 days at 6°C for mortadella and 120 days at room temperature for the jerked beef or Italian type salami.[103] Similar results were found in restructured beef roast by heat processing (dry and moist) storage at 4°C for 21 days[104] and in raw, broiled, and pressure-cooked mutton, buffalo, and pork after 6 days at 4°C and 90 days at −10°C storage.[105–107] However, the amount of cholesterol decreased during 8 weeks in mechanically separated turkey meat and products,[38] and in raw and cooked ground beef patties it decreased with 30 days of storage at −20°C.[108]

12.2.3.6 Effect of Cooking on Cholesterol Content of Meat and Meat Products

Cholesterol concentrations were higher in cooked than in raw meat when calculated on a wet weight basis.[27,47,70,97,100,105,106,109] As expected, cooking always induced a significant decrease in moisture content, while cholesterol content increased significantly.[27,67] As a consequence of the applied cooking procedures, the cholesterol content became significantly different between cooked muscles cuts,[47] contrary to the results observed by Kritchevsky and Tepper,[13] who found that cooking lowered the cholesterol contents.

Cholesterol retention during cooking varied from 99 to 117% in bovine meat[24,27,47] and from 85 to 105% in processed meat products.[109] Similarly significant loss of 8 to 19% was observed in chicken, beef, and pork meat, which could be due to external levels of fat, type of cooking, and cooking temperature.[6,7] Migration of cholesterol from adipose to muscle tissues as a result of cooking could represent a reliable explanation for cholesterol values higher in cooked cuts than in raw ones, especially when the fat is present in greater amounts in subcutaneous or intramuscular tissues; however, the decrease in the values of cholesterol content in cooking meat can be attributed to migration of the lipids and cholesterol degradation due to the cooking method used.[47] Trimming of fatty tissue before cooking would cause a further reduction in cholesterol content of cooked lean meat,[23] although there was no significant migration of cholesterol from adipose tissue into lean tissue during cooking since no significant difference was observed in cholesterol content among the different fat trim levels.[8,9,27,61,51,109]

In general, the amount of cholesterol in adipose tissue was not affected by cooking. The adipose tissue usually did not differ between raw and cooked tissue, and yet in muscle tissue such a difference was observed. This could be expected due to the low amount of water in adipose tissue, which would make cholesterol more concentrated than it would be in the muscle tissue.[8]

Various investigations[9,11] have demonstrated that the method of cooking did not affect cholesterol levels in meat; however, microwave cooking resulted in slightly higher values than cooking by other methods, which can be attributed to a smaller loss of cholesterol during the short cooking period. Similarly, Badiani et al.[47] found lower cholesterol levels in beef meat in microwaving than in oven roasting. However, Rodriguez-Estrada et al.[100] showed that the total cholesterol present in cooked hamburger varied greatly, being lowest in the boiled sample. Similar results were obtained by Baggio and Bragagnolo[109] in boiled frankfurters; the authors found lower cholesterol values in boiled than in raw materials due to water absorption during cooking, which diluted the medium. In addition, Conchillo et al.[76] found lower concentrations of cholesterol in grilled than in roasted chicken breast and Kesava Rao et al.[105,106] in pressure-cooked than in broiled mutton and pork.

Akinwunmi et al.[61] showed that cholesterol content was not significantly affected by degree of doneness (60, 70, and 77°C). Rhee et al.[54] also showed that longissimus muscle steaks cooked to an internal temperature of 60 or 75°C with eight marbling levels were similar in cholesterol content.

However, Kregel et al.[108] found more cholesterol in patties heated to 77°C than in samples heated to 71°C. Similar results were observed by Prusa and Hughes,[11] with turkey cooked at 82°C having greater amounts of cholesterol than that cooked at 77°C.

12.3 Cholesterol Oxides (COP)

12.3.1 Structure and Mechanism of Formation

Cholesterol oxides, also called oxysterol or oxycholesterol, are compounds with the chemical structure of cholesterol with hydroxyl, ketone, or an epoxide substitution in the ring or with a hydroxyl group in the side chain. Table 12.9 shows the nomenclature and abbreviations of the main cholesterol oxidation products found in foods.

In human and animal tissues, cholesterol oxides may be of exogenous origin, derived from diet, or generated endogenously by enzymatic and nonenzymatic oxidation of cholesterol.[110] The main enzymatic pathway involving cholesterol is its mono-hydroxylation in the synthesis of bile acids and steroid hormones. Two key enzymes of bile acid biosynthesis are 27-hydroxylase and 7α-hydroxylase. In addition, other enzymes appear to be involved in oxysterol production, including 26-hydroxylase, 7-ketone dehydrogenase, and cholesterol-5α,6α-epoxidase.

The nonenzymatic pathway occurs through attack by reactive oxygen species, peroxyl and alkoxyl radicals from lipid peroxidation, and the leukocyte/H_2O_2/HOCl system. Conditions such as light, oxygen, water activity, pH, fatty acids, metal ions, or a combination of such factors are

Table 12.9 Nomenclature and Abbreviations of Cholesterol Oxidation Products

Systematic Name	Trivial Name	Abbreviation
Cholest-5-en-3β-ol	Cholesterol	
3β-Hydroxycholest-5-en-7α-hydroperoxide	7α-Hydroperoxycholesterol	7α-OOH
3β-Hydroxycholest-5-en-7β-hydroperoxide	7β-Hydroperoxycholesterol	7β-OOH
Cholest-5-en-3β,7α-diol	7α-Hydroxycholesterol	7α-HO
Cholest-5-en-3β,7β-diol	7β-Hydroxycholesterol	7β-HO
3β-Hydroxycholest-5-en-7-one	7-Ketocholesterol	7-Keto
5α-Cholestan-3β-5,6β-triol	Cholestanetriol	Triol
5,6α-Epoxy-5α-cholestan-3β-ol	Cholesterol-5α,6α-epoxide	α-Epoxy
5,6β-Epoxy-5β-cholestan-3β-ol	Cholesterol-5β,6β-epoxide	β-Epoxy
Cholest-5-en-3β,19-diol	19-Hydroxycholesterol	19-OH
Cholest-5-en-3β,20α-diol	20α-Hydroxycholesterol	20α-OH
[22R]-Cholest-5-en-3β,22-diol	(22R)-Hydroxycholesterol	22R-OH
[22S]-Cholest-5-en-3β,22-diol	(22S)-Hydroxycholesterol	22S-OH
[24S]-Cholest-5-en-3β,24-diol	(24S)-Hydroxycholesterol	24S-OH
Cholest-5-en-3β,25-diol	25-Hydroxycholesterol	25-OH
[25R]-Cholest-5-en-3β,26-diol	(25R)-26-Hydroxycholesterol or 27-Hydroxycholesterol	25R-26-OH
[25S]-Cholest-5-en-3β,26-diol	(25S)-26-Hydroxycholesterol	25S-26-OH
Cholesta-3,5,7-triene		3,5,7-Triene
Cholesta-5,7-dien-3β-ol		5,7-Diene
Cholesta-3,5-dien-7-one		7-Keto-3,5-diene
Cholesta-3,5-diene		3,5-Diene

important in oxidative degradation of cholesterol.[1,2,111] Initiation phases remain obscure, but their radical character is evidenced by the formation of carbon-centered and peroxyl radicals in crystalline cholesterol;[2] the mechanism is similar to that observed in the oxidation of monounsaturated fatty acids. The hydroperoxides of unsaturated fatty acid formed during lipid oxidation may be necessary to initiate cholesterol oxidation.[2]

As cholesterol has a double bond at C-5, the positions most susceptible to oxidation are C-4 and C-7. However, due to the possible influence of the hydroxyl group at C-3, which can be esterified with fatty acids, and the tertiary C-5, C-4 is rarely attacked by oxygen.[2,112] At low temperature, allylic C-7 radicals formed initially in solid cholesterol will react with triplet oxygen ($^3O_2$) to give 7α- and 7β-hydroperoxicholesterol (7-OOH), which are the first detectable stable products (Figure 12.2). The 7β-OOH is thermodynamically more stable than the 7α-OOH.[113] Thermal decomposition of the 7-OOH results in the formation of the corresponding alcohols, 7α- and 7β-hydroxycholesterol (7α- and 7β-OH), and 7-ketocholesterol (7-keto).[2,113] By heating, dehydration of cholesterol and 7-keto, with the subsequent loss of the OH group from position 3, leads to the formation of a conjugated diene, cholesta-3,5-dien-7-one (3,5-diene), and a conjugated triene with a keto group, cholesta-3,5-dien-7-one (7-keto-3,5-diene); the elimination of the OH at C-3 is favored by the presence of a double bond at C-5. However, elimination of a water molecule from the OH at C-7 of the two 7-OOH gives rise to the conjugated diene, cholesta-5,7-dien-3β-ol (5,7-diene), and subsequently to the conjugated cholesta-3,5,7-triene (3,5,7-triene). These compounds are obtained by means of a monomolecular reaction mechanism.[99]

The epoxidation of cholesterol may occur by attack of 7-hydroperoxides already formed, yielding isomeric 5,6α-epoxycholesterol (α-epoxy) and 5,6β-epoxycholesterol (β-epoxy), since the β-epoxy prevails over the α-epoxy, probably due to steric hindrance of the OH group in position 3. These reactions can also occur by triplet oxygen or other reactive species such as singlet oxygen ($^1O_2$), superoxide ($^•O_2$), peroxide (ROO$^•$), alcoxyl (RO$^•$), and hydroxyl radicals (OH$^•$).[111] Hydratation of the epoxy in acidic environment results in the formation of a triol, the most toxic of the COP tested to date.[2,111]

Oxidations in the side-chain occur mainly at C-25 by the action of the triplet oxygen, yielding 25-OOH, which is decomposed to form 25-OH. From the side-chain, a rich variety of 20-, 22-, 24-, and 26-OOH and their decomposition products such as alcohols, ketones, aldehydes, and carboxylic acids can be formed.[2]

12.3.2 Determination of Cholesterol Oxides

The cholesterol oxide content in foods found in the literature varies widely, due mainly to different analytical methodologies used. The analysis of COP is strongly influenced by their chemical structures, as these compounds have different functional groups that provide different polarity and chemical properties to the molecules. However, some COP are isomers and thus have similar chemical, spectrophotometric, and fragmentation characteristics.[114]

As COP occurs at low levels (micrograms/gram), making its isolation from other interfering substances in foods and quantification of oxycholesterols a difficult analytical problem.[115] The extraction and purification of the COP fraction, which involves saponification and chromatographic fractionation of lipid extracts,[116] is the most critical step for the analytical procedure due to the complexity of the lipid fraction of many foods.[114] Saponification is usually applied to the lipid fraction, although some authors attempted the saponification of the whole food sample directly.[115–118] Dionisi et al.[119] also recommended the use of the direct saponification method, based on recovery, precision, minimal artifact formation, and relative simplicity after comparing

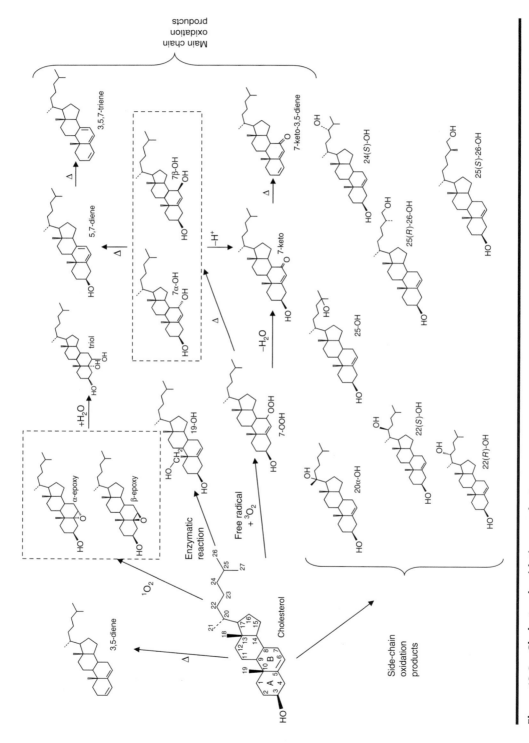

Figure 12.2 Cholesterol oxidation products.

this method with three other methods for extraction and purification of COP from milk powder samples.

Several forms of chromatography are used to enrich and purify COP directly from lipid extract, with some advantages in avoiding the destruction of sensitive COP, as would happen in contact with strong alkali. In contrast, only free COP are recovered. Thus, the analyst needs to select the most suitable procedures, taking into consideration the matrix, the level of COP present in the sample, and the method used for quantification of the COP.[116]

COP are determined by GC or GC coupled to mass spectrometry (GC-MS). Although efficient, this technique can produce artifacts from thermal degradation of cholesterol and its oxides, and it requires silanization of COP to avoid peak tailing and to improve thermal stability of some hydroxycholesterol,[114] making recovery difficult and leading to artifacts.[120] As a consequence of the limitations of GC, HPLC has become the technique of choice, mainly because analysis is carried out at ambient temperature. The COP can be determined by HPLC with several detection systems using the UV detector, which displays high sensitivity and specificity. However, some oxysterols of biological importance, such as α-epoxy, β-epoxy, and triol, do not absorb at UV wavelengths, so other detection methods must be used. The RI detector can be utilized alone or as an online detector with high specificity. Although this detector displays a low sensitivity, the RI, as a universal detector, can be utilized with any type of solvent under isocratic conditions.[114] Reverse-phase columns such as C18 are most widely used in determination of COP; however, the CN column showed better separation of the COP compared to C18 and silica.[117]

Several studies have described the use of HPLC coupled to mass spectrometry to determine cholesterol oxides in foods.[117,118,121,122] However, HPLC-MS has been utilized for identification and quantification of COP to a very limited extension, due partially to the instrumentation costs.[123] Although MS detectors have lower limit detection than UV and RI detectors, comparing the results obtained between HPLC-APCI-MS and HPLC-UV-RI[117,118] showed no significant difference, demonstrating that all these detector systems can be used to quantify cholesterol oxides. However, identification of the COP by MS has become indispensable, since the 20α-OH and 25-OH were identified by HPLC-UV but not confirmed by HPLC-APCI-MS.[118]

12.3.3 Cholesterol Oxides in Meat and Meat Products

Several studies demonstrated that the amount of COP in foods can frequently reach 1% of total cholesterol, but occasionally this value can reach 10% or even more.[124] The cholesterol oxides normally found in meat and meat products are 7α-OH, 7β-OH, 7-keto, α-epoxy, β-epoxy, and triol, as well as oxides formed by oxidation of the cholesterol side chain, such as 20α-OH and 25-OH. Table 12.10 shows the cholesterol oxide content found in some meats and meat products.[26,56,73,77,93,95,96,98,125–136]

Beef tallow was the first sample analyzed for COP in meat and meat products; three COP were tentatively identified, but no quantification values were reported.[137] Verleyen et al.[56] reported only two COP (7β-OH and β-epoxy) in processed and nonprocessed tallow; Park and Addis[138] quantified 7-keto and α-epoxy when heating tallow at 135, 150, 165, and 180°C. Park and Addis[136] did not detect 7-keto in raw beef or beef products, although the authors found COP in freeze-dried pork improperly stored for 3 years, and in broiled beef steak and turkey. COP were not detected in normal fresh pork, in stored frozen pork after 5 months, and in fresh porcine stress syndrome samples;[73] the authors detected COP only in UV-irradiated ground pork tissue, ground pork after 10 days under refrigerated storage, and freeze-fried pork stored for 3 years with periods

Table 12.10 Cholesterol Oxides in Meat and Meat Products

Meat and References	Treatment	Oxysterol (μg/g, Wet Basis)							
		7-Keto	7α-OH	7β-OH	α-Epoxy	β-Epoxy	25-OH	Triol	Others
Beef									
Charqui[125]	Semitendinosus muscle	0.88–5.54	nd	nd–3.16	nd–0.98	nt	nt	nd–3.92	a
Commercial[126]	Freeze-dried	3–27	6–17	9–14	3–21	1–9	nd	2	nt
Ground[127]	Cooked								
	0 d	3.15	2.5	1.8	nd	nd	nd	nd	nd
	2 d	17.2	12.6	6.1	nd	nd	nd	nd	
	4 d	16.0	17.9	9.4					
Ground[128]	Raw	0.4							
	Cooked	0.4–0.6	nt	nt	nt	nt	nt	nt	nt
Hamburger[93]	Commercial	nd–3.2	nd	nd	nd	nd	nd	nt	nt
Hamburger[129]	Retort	9.10	nq	10.92b	5.27	b	nq	3.42	nq
	Raw	9.73		8.06	7.12			2.63	
Minced[130]	Raw	1.12	0.33	0.34	0.42	1.06	0.14		
	Cooked	1.66–2.11	0.51–0.58	0.58–0.67	0.50–0.55	1.29–1.31	0.23–0.34	nd	20-OH
Sliced[131]	Raw 0 d								
	Unirradiated	0.213	nt	nt	0.048	0.130	nt	nt	b
	Irradiated	0.662			0.115	0.322			b
	Raw 2 weeks								
	Unirradiated	1.52			0.182	0.601			b
	Irradiated	4.43			0.577	1.66			b
Tallow[56]	Natural, bleached deodorized	nq	nq	0.17	nq	0.60	nq	nq	nq
				0.49–0.69		0.81–3.37			
				0.60		2.09			
Tenderloin	Raw 0 d	0.83	nt	nt	0.13	0.37	nt	nt	nt
Commercial[132]	Raw 7 d	1.02			0.13	0.44			
	Raw 14 d	1.37			0.37	0.54			
	Raw 21 d	2.55			0.36	0.76			

Product	Treatment								20-OH
Chicken									
Breast[133]									
Cooked	Feed diet								
	With vitamin E 0 d	nd	nd	nd	nd	nd	0.10–0.14	nd	20-OH
	6 d	nd	nt	nt			0.16–0.17	nt	nt
	12 d	0.17–0.39					0.27–1.04	nd	nt
Breast[132]	Commercial	0.13			0.07	0.06	nt	nd	
Commercial[126]	Freeze-dried	4–22	4–8	3–6	2–43	4–27	nd	nt	
Thigh[133]									
Cooked	Feed diet								
	With vitamin E 0 d	nd	nd	nd	nd	nd	0.23–0.56		20-OH
	6 d	nd	nd	nd	nd	nd	0.21–0.87		
	12 d	0.45–0.73					0.74–3.47		
Mixed meat									
Frankfurter[93]	Commercial	nd–2.0					nd		nt
Mortadella		nd–2.6							
Salami Milano[98]	Commercial	nd–1.22	nt	nd–15.82	nd–1.28	nt	nd–2.64	nt	nd
Mortadella		nd–18.69		nd–6.41	nd–6.18		nd		
Salami sausage[129]	Commercial	8.90	nq	3.28c	1.45	c	nq	2.28	nq
Sausage[95]	Fermented	nd	nd	0.10–0.19	0.61–0.64	nd	nd	nd	nd
Pork									
Bacon[129]	Commercial	7.08	nq	1.11b	0.86	b	nq	0.67	nq
Raw ham		10.49		1.56	0.81			0.51	
Roast ham		12.04		2.43	0.83			2.62	
Roast pork		8.14		6.34	3.52			nd	
Sausage		7.29		4.74	3.83			3.68	
Center cut chops[132]	Commercial	0.06	nt	nt	0.02	0.04	nt	nt	nt
Iberian ham[96]	Dry-cured	0.57–0.71	0.36–0.44	0.15–0.27	nt	nt	tr	tr	20-OH
Loin cooked[134]	Feed vitamin E								
	2 d	4.06–7.17	nd	2.86–3.94	nd	3.42–5.05	nd	nd	nt
	4 d	7.85–9.79		4.05–5.07		4.05–5.07			
Meat batter[135]	Unirradiated	nd	17.6	5.3	7.2	11.7	nd	5.0	20-OH
	Irradiated	14.5	72.6	70.5	5.4	8.5	nd	5.4	20-OH
Minced[130]	Raw	0.92	0.19	0.28	0.22	0.35	0.13	0.04	nd
	Cooked	2.25	0.64	0.85	0.39	1.01	0.38	0.06	20-OH
	Refrigerated	1.13	nd	nd	nt	nt	nd	nt	nd
Muscle[73]	Freeze-dried	126.59	28.98	21.05	nt	nt	nd	nt	nt

Table 12.10 (Continued)

Meat and References	Treatment	Oxysterol (µg/g, Wet Basis)							
		7-Keto	7α-OH	7β-OH	α-Epoxy	β-Epoxy	25-OH	Triol	Others
Muscle[136]	Freeze-dried 3 years	259.8	90.9	68.4	12.5	nq	nd	28.4	nd
Sausage[95]	Fermented	nd–0.58	nd	0.09–0.73	0.41–0.75	nd	nd	nd	nd
Sliced[131]	Raw 0 d								
	Unirradiated	0.026	nt	nt	0.019	0.028	nt	nt	b
	Irradiated	0.118			0.045	0.086			b
	Raw 2 weeks								
	Unirradiated	0.051			0.044	0.060			b
	Irradiated	1.08			0.168	0.476			b
Turkey									
Breast[77]	Commercial	0.33–2.52	nd	nd–1.30	nd	nd	nd	nt	nt
Leg		0.69–4.92		nd–1.00					
Skin		1.45–7.65		nd–3.70					
Wing		1.90–3.65		1.10–2.00					
Commercial[126]	Freeze-dried	8–20	nd	nd–18	7–21	6–8	nd	nd	nt
Products[26]	Commercial	nd–1.84	nd	nd	nd	nd–4.50	nd	nt	nt
Veal									
Cuts[132]	Commercial	0.22	nt	nt	0.03	0.09	nt	nt	nt
Minced[130]	Raw	0.71	0.18	0.21	0.17	0.47	0.05	nd	20-OH
	Cooked	1.70	0.61	0.38	0.33	0.84	0.19	0.07	
Sliced[131]	Raw 0 d								
	Unirradiated	0.228	nt	nt	0.066	0.139	nt	nt	b
	Irradiated	0.444			0.114	0.273			b
	Raw 2 weeks								
	Unirradiated	0.316			0.070	0.239			b
	Irradiated	1.83			0.304	0.850			b

a 4β-Hydroxycholesterol.

b 4-Cholesten-3-one, 4,6-cholestadien-3-one, 4-cholestene-3,6-dione.

c 7β-OH + 5β-EP.

Notes: nd, not detected; nt, not tested; nq, no quantitation.

of intermittent exposure to air and light at room temperature. However, several researchers found COP in raw meat[131,132,139,140] or raw frozen meat.[77]

Several studies have evaluated the effect of the heat treatment of meat and processed meat products on the formation of cholesterol oxides, suggesting that time and temperature are determinant factors in this process, directly influencing the rate of oxidation. Larkeson et al.[97] fried minced beef at temperatures ranging from 150 to 160°C, resulting in an increase in the total COP content of about 22%. Pie et al.[130] fried minced beef at 135°C for 3 and 10 min each side and noted an increase in COP of 49–73%, respectively. Pie et al.[130] found a value 150% higher in minced pork fried at 135°C for 10 min compared to fresh samples. Thurner et al.[139] found increases ranging from 6 to 231% in COP content during pan-frying of beef patties, fillets of pork, and braised meat; the lowest increase was found in fried beef patties due to preparation, whereas the highest increase was observed in pork fillets fried without addition oil. Chen et al.[141] observed that the cholesterol oxide content of lard increased with heating time; the 7-keto and cholesta-4,6-dien-3-one levels increased to a maximum after 200 h, while the concentrations of α-epoxy, β-epoxy, and 7β-OH increased during the first 100 h and then decreased; after 20 h of heating of bacon it was already possible to detect the presence of triol; the contents increasing with time, reaching a maximum after 200 h; 25-OH was not found.

Higley et al.[142] found high levels of 7α-OH in cooked frankfurters, triol in cooked frankfurters and raw hamburgers, and 22-OH in cooked frankfurters and raw hamburgers. The exceptionally high triol values obtained by Higley et al.[142] were not confirmed by MS, and Paniangvait et al.[115] suggested that these workers may have overestimated the cholesterol oxide levels, probably due to the analytical methodology used. Osada et al.[129] found the highest level of a mixture of 7β-OH and β-epoxy in retort beef hamburger steak, the α-epoxy in nonheated beef hamburger, and triol in retort beef hamburger steak and pork sausage. However, Rodriguez-Estrada et al.[100] and Lercker and Rodriguez-Estrada[143] showed a decrease in the concentration of 7-keto when hamburgers and beef meat were submitted to heat treatment.

No cholesterol oxides were found in processed meat products (raw or cooked) analyzed by Baggio and Bragagnolo;[109] as all the processed meat products contained the antioxidant sodium erythorbate, spices, and natural condiments in their composition, these products probably protected the cholesterol against oxidation. Spices and natural condiments are natural antioxidants, and their antioxidant properties in foods have long been recognized; the use of antioxidants in the formulation and appropriate packaging materials, presenting physical barriers against air and light, can hinder the formation of cholesterol oxidation products. Torres et al.[125] showed that the concentrations of total cholesterol oxides in jerked beef were 2.5 times lower in samples prepared with refined salt and BHA/BHT than in those prepared with the addition of refined salt without the BHA/BHT. Osada et al.[129] concluded that cholesterol oxidation in sausage was inhibited by the addition of sodium nitrite and apple polyphenols through the stabilization of coexisting polyunsaturated fatty acids and radical scavenging. Britt et al.[127] showed that cherry tissue inhibited COP formation in cooked ground beef storage for 4 days at 4°C. Zanardi et al.[95] observed that the presence of ascorbic acid and nitrite were important for cholesterol protection in fermented sausages.

Regarding different cooking treatments, Rodriguez-Estrada et al.[100] observed that the 7-keto content of the boiled hamburger sample was significantly lower, and the combination of microwave and roasting was higher than in the microwaved, roasted, barbecued, and fried pan samples. Furthermore, no significant differences in the degree of COP were found among roasted, microwaved, barbecued, and boiled beef patties.[128]

Processing procedures could affect cholesterol oxidation in other ways, since heat treatment liberates iron from the heme groups and releases oxygen from oxymyoglobin, creating the conditions for hydrogen peroxide production. Shredding, mincing, and mixing disrupt the muscle structure and, consequently, increase the surface exposed to oxygen and the other oxidation catalysis. Although salt is a necessary ingredient for many meat products, the addition of salt is also known to decrease the oxidative stability of many muscle-based foods,[144] the mechanisms of which are still unclear. In fact, Petrón et al.[96] found 7α-OH, 7β-OH, and 7-keto in dry-cured hams and Sander et al.[126] reported COP in dehydrated beef, chicken, and turkey, with very small differences among the varieties of meat products. Novelli et al.[98] found 7-keto, α-epoxy, 7β-OH, and 25-OH in salami and in mortadella; Schmarr et al.[145] did not find 25-OH in salami but related the presence in small amounts of 7-keto, α-epoxy, β-epoxy, 7α-OH, 7β-OH, and 20α-OH. However, Baggio and Bragagnolo[93] found only 7-keto in one sample of hamburger, two samples of frankfurter, and one sample of mortadella when analyzing 45 samples of processed beef products, 36 pork products, and 45 mixed meat products. In addition, no cholesterol oxides were found in the processed chicken meat products.[26]

Irradiation is another factor that contributes to the formation of COP in meat,[131] the amount of COP increasing with the level of irradiation in raw and cooked chicken meat.[146] Jo et al.[135] verified that the production of COP in raw pork batters irradiated was about fourfold that of the nonirradiated ones. However, Ahn at al.[140] reported that irradiation at 4.5 kGy had no effect on the amounts of COP found in cooked turkey, pork, and beef. The same authors also observed that packaging of cooked meat was more important than irradiation in developing COP in cooking meat during storage; after 7 days of storage, aerobically packaged pork produced 10- to 15-fold higher amounts of total COP than vacuum-packaged pork.

Data reported in several papers showed that storage time increased COP formation. For example, Mahadeo et al.[104] quantified COP in roast beef during heating and storage for 21 days at 4°C, with 7-keto found in higher amounts. Conchillo et al.[147] related that the storage of raw and cooked chicken breast at 4°C resulted in four- and sevenfold increases in COP levels, respectively, and vacuum packaging was very effective in maintaining low levels of COP in the same samples. Boselli et al.[148] showed that raw turkey meat reached maximum COP concentration after 7 days storage, but when meat was exposed to white fluorescent light, the maximum concentration was observed after just 1 day. In addition, the authors demonstrated that a lamp with low emissions in the blue band proved to be useful for lowering COP of turkey meat; this can be a suitable solution for the exhibition of meat products in supermarkets or meat processing factories. Cayuela et al.[149] showed that storage of pork meat for 20 days at 4°C exposed to fluorescent light at high oxygen concentration favors the processes of COP formation, but vacuum packaging was effective in lowering COP formation. However, no cholesterol oxides were formed in Italian type chicken mortadella and Chester mortadella stored for 90 days at 6°C or in jerked beef and salami stored for 120 days at room temperature.[103]

Dietary α-tocopherol supplementation significantly reduced the formation of COP during storage, ranging from 43 to 75% in chicken breast, and 50 to 72% in thigh, at supplementation levels of 200 and 800 mg/kg.[138] In another work, Galvin et al.[150] observed that the vitamin E supplementation did not affect 7-keto formation oxides in longissimus dorsi beef, but significantly reduced concentrations in psoas major beef during refrigerated and frozen storage of cooked beef steaks. Monahan et al.[134] observed that increasing the α-tocopherol content of muscle through diet in pigs can reduce COP in muscle foods. Nevertheless, no difference has been observed by Zanardi et al.[74] on COP formation in cooked chops with increased levels of vitamin E in feed supplementation in pork. Eder et al.[151] also related that the concentrations of COP in fresh pig muscles were not affected by dietary polyunsaturated fatty acids or vitamin E; sausage from pigs which had been fed soybean oil had higher COP contents than sausages from pigs fed palm oil; vitamin E reduced COP levels in sausage from pigs fed soybean oil, but not in sausage from pigs fed palm oil.

12.4 Conclusions

A comparison of data of cholesterol content in meat and meat products showed that the level of fat content present in the meats is not correlated with increased cholesterol content, as is commonly believed. None of the factors investigated, including age, breed, sex, kind of muscle, part of carcasses, degree of marbling, or system of rearing, was shown clearly to have an important association with cholesterol content. However, trimming of subcutaneous fat could be effective in reducing dietary cholesterol and especially fat and calories. Cholesterol content in beef, pork, and breast chicken was similar; thigh chicken and veal are higher and turkey meat lower than beef, pork, and breast chicken.

Among the lipid oxidation products, cholesterol oxides are probably the best known toxic compounds. In foods, normally eight COP are quantified; in meat, the most frequently found were 7-keto, 7α-OH, 7β-OH, β-epoxy, and α-epoxy; the most toxic ones, triol and 25-OH, were found only in some samples. In general, great discrepancies were observed among the COP data obtained from different works, which may be due in part to the samples tested. Differences in extraction procedures and the chromatographic methods used are in urgent need of harmonization. The results in the literature showed that the total concentration of COP in fresh meat and meat products was small; however, some precaution should be taken to prevent or minimize the formation of COP during processing and storage, including the use of vacuum packaging, reduced O_2 atmosphere, addition of natural antioxidants, and storage, the foods in the dark. In addition, it is also important to process foods at low temperature and without light and, of course, to use high-quality raw materials. Oxidation of cholesterol and safety questions for human health should be discussed at greater length, since the relationship between the long-term consumption of cholesterol oxidation products and human health is still not very well established.

Acknowledgments

The author thanks the financial support from the Brazilian Funding Agencies (FAPESP, CNPq, CAPES and FAEPEX-UNICAMP).

References

1. Kim, S.K. and Nawar, W.W., Parameters influencing cholesterol oxidation, *Lipids*, 29, 917, 1993.
2. Smith, L.L., Cholesterol autoxidation, *Chem. Phys. Lipids*, 44, 87, 1987.
3. Guardiola, F. et al., Biological effects of oxysterols: current status, *Food Chem. Toxicol.*, 34, 193, 1996.
4. Bössinguer, S., Luf, W., and Brandl, E., Oxysterols: their occurrence and biological effects, *Int. Dairy J.*, 3, 1, 1993.

5. Sweeney, J.P. and Weihrauch, J.L., Summary of available data for cholesterol in foods and methods for its determination, *Crit. Rev. Food Sci.*, 8, 131, 1976.
6. Bragagnolo, N. and Rodriguez-Amaya, D.B., Teores de colesterol em carne suína e bovina e efeito do cozimento, *Ciênc. Tecnol. Aliment.*, 15, 11, 1995.
7. Bragagnolo, N. and Rodriguez-Amaya, D.B., Teores de colesterol em carne de frango, *Rev. Farm. Bioquím. Univ. S. Paulo*, 28, 122, 1992.
8. Swize, S.S. et al., Cholesterol content of lean and fat from beef, pork, and lamb cuts, *J. Food Comp. Anal.*, 5, 160, 1992.
9. Morgan, J.B., Calkins, C.R., and Mandigo, R.W., Effect of trim level, cooking method, and chop type on lipid retention, caloric content, and cholesterol level in cooked pork, *J. Food Sci.*, 53, 1602, 1988.
10. Reitmeier, C.A. and Prusa, K.J., Cholesterol content and sensory analysis of ground pork as influenced by fat level and heating, *J. Food Sci.*, 52, 916, 1987.
11. Prusa, K.J. and Hughes, K.V., Cholesterol and selected attributes of pork tenderloin steaks heated by conventional, convection and microwave ovens to two internal endpoint temperatures, *J. Food Sci.*, 51, 1139, 1986.
12. Tu, C., Powrie, W.D., and Fennema, O., Free and esterified cholesterol content of animal muscles and meat products, *J. Food Sci.*, 32, 30, 1967.
13. Kritchevsky, D. and Tepper, A.S., The free and ester sterol content of various foodstuffs, *J. Nutr.*, 74, 441, 1961.
14. Saldanha, T., Mazali, M.R., and Bragagnolo, N., Avaliação comparativa entre dois métodos para a determinação de colesterol em carne e leite, *Ciênc. Tecnol. Aliment.*, 24, 109, 2004.
15. Ulberth, F. and Reich, H., Gas chromatographic determination of cholesterol in processed foods, *Food Chem.*, 43, 387, 1992.
16. Hutchison, G.I., Thomas, D.E., and Truswell, A.S., Nutrient composition of Australian beef, *Food Technol. Aust.*, 39, 199, 1987.
17. Hutchison, G.I., Greenfield, H., and Wills, R.B.H., Composition of Australian foods. 35. Pork, *Food Technol. Aust.*, 39, 216, 1987.
18. Hutchison, G.I., Thomas, D.E., and Truswell, A.S., Nutrient composition of Australian chicken, *Food Technol. Aust.*, 39, 196, 1987.
19. Bohac, C.E. et al., Assessment of methodologies for colorimetric cholesterol assay of meats, *J. Food Sci.*, 53, 1642, 1988.
20. Bragagnolo, N. and Rodriguez-Amaya, D.R., Determinação de colesterol em carne: comparação de um método colorimétrico e um método por cromatografia líquida de alta eficiência, *Rev. Inst. Adolfo Lutz*, 60, 53, 2001.
21. Almeida, J.C. et al., Fatty acid composition and cholesterol content of beef and chicken meat in southern Brazil, *Rev. Bras. Ciênc. Farmac.*, 42, 109, 2006.
22. Heymann, H. et al., Sensory and chemical characteristics of fresh pork roasts cooked to different endpoint temperatures, *J. Food Sci.*, 55, 613, 1990.
23. Sahasrabudhe, M.R. and Stewart, L., Total lipid and cholesterol in selected retail cuts of Canadian beef, *Can. Inst. Food Sci. Technol. J.*, 22, 83, 1989.
24. Slover, H.T. et al., Lipids in raw and cooked beef, *J. Food Conp. Anal.*, 1, 26, 1987.
25. Baggio, S.R. and Bragagnolo, N., Fatty acids, cholesterol oxides and cholesterol in Brazilian processed chicken products, *Ital. J. Food Sci.*, 18, 199, 2006.
26. Baggio, S.R., Miguel, A.M.R., and Bragagnolo, N., Simultaneous determination of cholesterol oxides, cholesterol and fatty acids in processed turkey meat products, *Food Chem.*, 89, 475, 2005.
27. Bragagnolo, N. and Rodriguez-Amaya, D.B., New data on the total lipid, cholesterol and fatty acid composition of raw and grilled beef *longissimus dorsi*, *Arch. Latinoamer. Nutr.*, 53, 312, 2003.
28. Bragagnolo, N. and Rodriguez-Amaya, D.B., Simultaneous determination of total lipids, cholesterol and fatty acids in meat and backfat of suckling and adults pigs, *Food Chem.*, 79, 255, 2002.

29. Perez, J.R.O. et al., Efeito do peso ao abate de cordeiros Santa Inês e Bergamácia sobre o perfil de ácidos graxos, colesterol e propriedades químicas, *Ciênc. Tecnol. Aliment.*, 22, 11, 2002.
30. Arneth, W. and Al-Ahmad, H., Cholesterol: its determination in muscle and adipose tissue and in offal using HPLC, *Fleischwirtschaft.*, 75, 1001, 1995.
31. Maraschiello, C., Díaz, I. and García Regueiro, J.A., Determination of cholesterol in fat and muscle of pig by HPLC and capillary gas chromatography with solvent venting injection, *J. High Resolut. Chromatogr.*, 19, 165, 1996.
32. Kanada, T. et al., Quantitative analysis of cholesterol in foods by gas-liquid chromatography, *J. Nutr. Sci. Vitaminol.*, 26, 497, 1980.
33. Al-Hassani, S.M., Hlavac, J., and Carpenter, M.W., Rapid determination of cholesterol in single and multicomponent prepared foods, *J. AOAC Int.*, 76, 902, 1993.
34. Tsui, I.C., Rapid determination of total cholesterol in homogenized milk, *J. Assoc. Off. Anal. Chem.*, 72, 421, 1989.
35. Kovacs, M.I.P., Anderson, W.E., and Ackman, R.G., A simple method for the determination of cholesterol and some plant sterols in fishery-based food products, *J. Food Sci.*, 44, 1299, 1979.
36. Bragagnolo, N. and Rodriguez-Amaya, D.B., Comparison of the cholesterol content of Brazilian chicken and quail eggs, *J. Food Comp. Anal.*, 16, 14, 2003.
37. Van Elswyk, M.E., Schake, L.S., and Hargis, P.S., Research note: evaluation of two extraction methods for the determination of egg yolk cholesterol, *Poult. Sci.*, 70, 1258, 1991.
38. King, A.J. et al., Rapid method for quantification of cholesterol in turkey meat and products, *J. Food Sci.*, 63, 382, 1998.
39. Naeemi, E.D. et al., Rapid and simple method for determination of cholesterol in processed food, *J. AOAC Int.*, 78, 1522, 1995.
40. Thompson, R.H. and Merola, G.V., A simplified alternative to the AOAC official method for cholesterol in multicomponent foods, *J. AOAC Int.*, 76, 1057, 1993.
41. Werdi Pratiwi, N.M., Murray, P.J., and Taylor, D.G., Total cholesterol concentration of the muscles in castrated Boer goats, *Small Rumin. Res.*, 64, 77, 2006.
42. Komprda, T. et al., Cholesterol content in meat of some poultry and fish species as influenced by live weight and total lipid content, *J. Agric. Food Chem.*, 51, 7692, 2003.
43. Arsenos, G. et al., The effect of breed slaughter weight and nutritional management on cholesterol content of lamb carcasses, *Small Rumin. Res.*, 36, 275, 2000.
44. Abu-Tarbousch, H.M. and Dawood, A.A., Cholesterol and fat contents of animal adipose tissues, *Food Chem.*, 46, 89, 1993.
45. Sweeten, M.K. et al., Subcellular distribution and composition of lipids in muscle and adipose tissues, *J. Food Sci.*, 55, 43, 1990.
46. Piironen, V., Toivo, J., and Lampi, A.M., New data for cholesterol contents in meat, fish, milk, eggs and their products consumed in Finland, *J. Food Comp. Anal.*, 15, 705, 2002.
47. Badiani, A. et al., Lipid composition, retention and oxidation in fresh and completely trimmed beef muscles as affected by common culinary practices, *Meat Sci.*, 60, 169, 2002.
48. Cifuni, G.F. et al., Fatty acid profile, cholesterol content and tenderness of meat from Podolian young bulls, *Meat Sci.*, 67, 289, 2004.
49. Andrae, J.G. et al., Effects of feeding high-oil corn to beef steers on carcass characteristics and meat quality, *J. Anim. Sci.*, 79, 582, 2001.
50. Huerta-Leidenz, N. et al., Contenido de colesterol en el músculo longissimus de bovinos venezolanos, *Arch. Latinoamer. Nutr.*, 46, 329, 1996.
51. Hoelscher, L.M. et al., Subcellular distribution of cholesterol within muscle and adipose tissues of beef loin steaks, *J. Food Sci.*, 53, 718, 1988.
52. Bohac, C.E. and Rhee, K.S., Influence of animal diet and muscle location on cholesterol content of beef and pork muscles, *Meat Sci.*, 23, 71, 1988.

53. Eichhorn, J.M. et al., Effects of breed type and restricted versus ad libitum feeding on fatty acid composition and cholesterol content of muscle and adipose tissue from mature females, *J. Anim. Sci.*, 63, 781, 1986.

54. Rhee, K.S. et al., Cholesterol content of raw and cooked beef *longissimus* muscles with different degrees of marbling, *J. Food Sci.*, 47, 716, 1982.

55. Paleari, M.A. et al., Ostrich meat: physico-chemical characteristics and comparison with turkey and bovine meat, *Meat Sci.*, 48, 205, 1998.

56. Verleyen, T. et al., Cholesterol oxidation in tallow during processing, *Food Chem.*, 83, 185, 2003.

57. Park, S.W. and Addis, P.B., Identification and quantitative estimation of oxidized cholesterol derivatives in heated tallow, *J. Agric. Food Chem.*, 34, 653, 1986.

58. Ryan, T.C. and Gray, J.I., Distribution of cholesterol in fractionated beef tallow, *J. Food Sci.*, 49, 1390, 1984.

59. Folch, J., Lees, M., and Sloane Stanley, G.H., A simple method for the isolation and purification of total lipids from animal tissues, *J. Biol. Chem.*, 226, 497, 1957.

60. Bligh, E. and Dyer, W., A rapid method of total lipid extraction and purification, *Can. J. Biochem. Phys.*, 37, 911, 1959.

61. Akinwunmi, I., Thompson, L.D., and Ramsey, C.B., Marbling, fat trim and doneness effects on sensory attributes, cooking loss and composition of cooked beef steaks, *J. Food Sci.*, 58, 242, 1993.

62. Morris, C.A. et al., Meat composition in genetically selected and control cattle from a serial slaughter experiment, *Meat Sci.*, 39, 427, 1995.

63. Stromer, M.H., Goll, D.E., and Roberts, J.H., Cholesterol in subcutaneous and intramuscular lipid depots from bovine carcasses of different maturity and fatness, *J. Anim. Sci.*, 25, 1145, 1966.

64. Rule, D.C., MacNeil, M.D., and Short, R.E., Influence of sire growth potential, time on feed, and growing-finishing strategy on cholesterol and fatty acids of the ground carcass and longissimus muscle of beef steers, *J. Anim. Sci.*, 75, 1525, 1997.

65. Browning, M.A. et al., Physical and compositional characteristics of beef carcasses selected for leanness, *J. Food Sci.*, 55, 9, 1990.

66. Hood, R.L., A note of the cholesterol content of beef rib steaks, *CSIRO Food Res. Q*, 47, 44, 1987.

67. Rhee, K.S., Dutson, T.R., and Smith, G.C., Effect of changes in intermuscular and subcutaneous fat levels on cholesterol content of raw and cooked beef steaks, *J. Food Sci.*, 47, 1638, 1982.

68. Bragagnolo, N. and Rodriguez-Amaya, D.R., Teores de colesterol, lipidios totais e ácidos graxos em cortes de carne suína, *Ciênc. Tecnol. Aliment.*, 22, 98, 2002.

69. Buege, D.R. et al., A nationwide audit of the composition of pork and chicken cuts at retail. *J. Food Composit. Anal.*, 11, 249, 1998.

70. Echarte, M., Ansorena, D., and Astiasaran, I., Fatty acid modifications and cholesterol oxidation in pork loin during frying at different temperatures, *J. Food Prot.*, 64, 1062, 2001.

71. Engeseth, N.J. et al., Fatty acid profiles of lipid depots and cholesterol concentration in muscle tissue of finishing pigs fed ractopamine, *J. Food Sci.*, 57, 1060, 1992.

72. Perkins, E.G. et al., Fatty acid and cholesterol changes in pork longissimus muscle and fat due to ractopamine, *J. Food Sci.*, 57, 1266, 1992.

73. Csallany, A.S. et al., HPLC method for quantitation of cholesterol and four of its major oxidation products in muscle and liver tissues, *Lipids*, 7, 645, 1989.

74. Zanardi, E. et al., Oxidative stability and dietary treatment with vitamin E, oleic acid and copper of fresh and cooked pork chops, *Meat Sci.*, 49, 309, 1998.

75. Fernandez, X. et al., Influence of intramuscular fat content on the quality of pig meat—1. Composition of the lipid fraction and sensory characteristics of *m. longissimus lumborum*, *Meat Sci.*, 53, 59, 1999.

76. Conchillo, A., Ansorena, D., and Astiasarán, I., Intensity of lipid oxidation and formation of cholesterol oxidation products during frozen storage of raw and cooked chicken, *J. Sci. Food Agric.*, 85, 141, 2005.

77. Baggio, S.R., Vicente, E., and Bragagnolo, N., Cholesterol oxides, cholesterol, total lipid and fatty acid composition in turkey meat, *J. Agric. Food Chem.*, 50, 5981, 2002.

78. Wong, M.K. and Sampugna, J., Moisture, total lipids, fatty acids and cholesterol in raw ground turkey, *J. Agric. Food Chem.*, 41, 1229, 1983.

79. Prusa, K.J. and Hughes, K.V., Quality characteristics, cholesterol, and sodium content of turkey as affected by conventional, convection, and microwave heating, *Poult. Sci.*, 65, 940, 1986.

80. Konjufca, V.H., Pest, G.M., and Bakalli, R.I., Modulation of cholesterol levels in broiler meat by dietary garlic and copper, *Poult. Sci.*, 76, 1264, 1997.

81. Ajuyah, A.O. et al., Influence of dietary full-fat seeds and oils on total lipid, cholesterol and fatty acid composition of broiler meats, *Can. J. Anim. Sci.*, 71, 1011, 1991.

82. Smith, D.P. et al., Pekin duckling and broiler chicken pectoralis muscle structure and composition, *Poult. Sci.*, 72, 202, 1993.

83. Alasnier, C., Rémignon, H., and Gandemer, G., Lipid characteristics associated with oxidative and glycolytic fibres in rabbit muscles, *Meat Sci.*, 43, 213, 1996.

84. Ajuyah, A.O. et al., Yield, lipid, cholesterol and fatty acid composition of spent hens fed full-fat oil seeds and fish meal diets, *J. Food Sci.*, 57, 338, 1992.

85. Zanini, S.F. et al., Lipid composition and vitamin E concentration in cockerel meat, *Lebnsm. Wiss. Technol.*, 36, 697, 2003.

86. Madruga, M.S. et al., Castration and slaughter age effects on fat components of "mestiço" goat meat, *Small Rumin. Res.*, 42, 77, 2001.

87. Solomon, M.B., Lynch, G.P., and Lough, D.S., Influence of dietary palm oil supplementation on serum lipid metabolites, carcass characteristics, and lipid composition of carcass tissues of growing ram and ewe lambs, *J. Anim. Sci.*, 70, 2746, 1992.

88. Lough, D.S. et al., Effects of dietary canola seed and soy lecithin in high-forage diets on cholesterol content and fatty acid composition of carcass tissues of growing ram lambs, *J. Anim. Sci.*, 70, 1153, 1992.

89. Solomon, M.B. et al., Influence of rapeseed meal, whole rapeseed and soybean meal on fatty acid composition and cholesterol content of muscle and adipose tissue from ram lambs, *J. Anim. Sci.*, 69, 4055, 1991.

90. Solomon, M.B. et al., Lipid composition of muscle and adipose tissue from crossbred ram, werther and cryptorchid lambs, *J. Anim. Sci.*, 68, 137, 1990.

91. Mestre Prates, J.A. et al., Simultaneous HPLC quantification of total cholesterol, tocopherols and β-carotene in Barrosã-PBO veal, *Food Chem.*, 94, 469, 2006.

92. Sevi, A. et al., Carcass characteristics and cholesterol content of meat and some organs of lambs, *Ital. J. Food Sci.*, 9, 223, 1997.

93. Baggio, S.R. and Bragagnolo, N., Evaluation of the quality of the lipid fraction of meat-based products, *J. Braz. Chem. Soc.*, 2008 (in press).

94. Jeun-Horng, L., Yuan-Hui, L., and Chun-Chin, K., Effect of dietary fish oil on fatty acid composition, lipid oxidation and sensory property frankfurters during storage, *Meat Sci.*, 60, 161, 2002.

95. Zanardi, E. et al., Lipolysis and lipid oxidation in fermented sausages depending on different processing conditions and different antioxidants, *Meat Sci.*, 66, 415, 2004.

96. Petrón, M.J. et al., Identification and quantification of cholesterol and cholesterol oxidation products in different types of Iberian hams, *J. Agric. Food Chem.*, 51, 5786, 2003.

97. Larkeson, B., Dutta, P.C. and Hansson, I., Effects of frying and storage on cholesterol oxidation in minced meat products, *J. Am. Oil Chem. Soc.*, 77, 675, 2000.

98. Novelli, E. et al., Lipid and cholesterol oxidation in frozen stored pork, salame Milano and mortadella, *Meat Sci.*, 48, 29, 1998.

99. Lercker, G. and Rodriguez-Estrada, M.T., Cholesterol oxidation mechanisms; *Cholesterol and phytosterol oxidation products in foods and biological samples: analysis, occurrence and biological effects;* Guardiola, F., Dutta, P., Codony, R., Savage, G.P., Eds.; AOAC Press, USA, 2002, 1–25.

100. Rodriguez-Estrada, M. T. et al., Effect of different cooking methods on some lipid and protein components of hamburgers, *Meat Sci.*, 45, 365, 1997.

101. Pereira, N.R. et al., Proximate composition and fatty acid profile in Brazilian poultry sausages, *J. Food Comp. Anal.*, 13, 915, 2000.

102. Chizzollini, R. et al., Calorific value and cholesterol content of normal and low-fat meat and meat products, *Trends Food Sci. Technol.*, 10, 119, 1999.

103. Baggio, S.R. and Bragagnolo, N., Cholesterol oxide, cholesterol, total lipid and fatty acid contents in processed meat products during storage, *Lebesm. Wiss. Technol.*, 39, 513, 2006.

104. Mahadeo, M. et al., Moist and dry convective heating and chilled storage of restructured beef roasts: effect on cholesterol and cholesterol oxide content, *Lebensm. Wiss. Technol.*, 25, 412, 1992.

105. Kesava Rao, V. et al., Lipid oxidation and development of cholesterol oxidation in pork during cooking and storage, *J. Food Sci. Technol.*, 36, 24, 1999.

106. Kesava Rao, V. et al., Effect of cooking and storage on lipid oxidation and development of cholesterol oxidation products in water buffalo meat, *Meat Sci.*, 43, 179, 1996.

107. Kowale, B.N. et al., Lipid oxidation and cholesterol oxidation in mutton during cooking and storage, *Meat Sci.*, 43, 195, 1996.

108. Kregel, K.K., Prusa, K.J., and Hughes, K.V., Cholesterol content and sensory analysis of ground beef as influenced by fat level, heating, and storage, *J. Food Sci.*, 51, 1162, 1986.

109. Baggio, S.R. and Bragagnolo, N., The effect of heat treatment on the cholesterol oxides, cholesterol, total lipid and fatty acid contents of processed meat products, *Food Chem.*, 95, 611, 2006.

110. Leonarduzzi, G. Sottero, B., and Poli, G., Oxidized products of cholesterol: dietary and metabolic origin, and proatherosclerotic effects (review), *J. Nutr. Biochem.*, 13, 700, 2002.

111. Maeker, G., Cholesterol autoxidation—current status, *J. Am. Oil Chem. Soc.*, 64, 388, 1987.

112. Smith, L.L., Review of progress in sterol oxidations: 1987–1995, *Lipids*, 31, 453, 1996.

113. Kumar, N. and Singhal, O.P., Cholesterol oxides and atherosclerosis: a review, *J. Sci. Food Agric.*, 55, 497, 1991.

114. Guardiola, F. et al., Analysis of sterol oxidation products in foods, *J. AOAC Int.*, 87, 441, 2004.

115. Paniangvait, P. et al., Cholesterol oxides in foods of animal origin, *J. Food Sci.*, 60, 1159, 1995.

116. Ulberth, F. and Buchgraber, M., Extraction and purification of cholesterol oxidation products; *Cholesterol and phytosterol oxidation products in foods and biological samples: analysis, occurrence and biological effects;* Guardiola, F., Dutta, P., Codony, R., Savage, G.P., Eds.; AOAC Press, USA, 2002, 27–49.

117. Saldanha, T. et al., HPLC separation and determination of 12 cholesterol oxidation products in fish: comparative study of RI, UV and APCI-MS detectors, *J. Agric. Food Chem.*, 54, 4107, 2006.

118. Mazalli, M.R. et al., HPLC method for quantification and characterization of cholesterol and its oxidation products in eggs, *Lipids*, 41, 615, 2006.

119. Dionisi, F. et al., Determination of cholesterol oxidation products in milk powders: methods comparison and validation, *J. Agric. Food Chem.*, 46, 2227, 1998.

120. Careri, M. et al., Evaluation of particle beam high performance liquid chromatography–mass spectrometry for analysis of cholesterol oxides, *J. Chromatogr. A*, 794, 253, 1998.

121. Raith, K. et al., A new LC/APCI-MS method for the determination of cholesterol oxidation products in food, *J. Chromatogr. A*, 1067, 207, 2005.

122. Al-Shagir, S. et al., Effects of different cooking procedures on lipid quality and cholesterol oxidation of farmed salmon fish (*Salmo salar*), *J. Agric. Food Chem.*, 49, 5290, 2004.

123. Rodriguez-Estrada, M.T. and Caboni, M.F., Determination of cholesterol oxidation products by high-performance liquid chromatography; *Cholesterol and phytosterol oxidation products in foods and biological samples: analysis, occurrence and biological effects;* Guardiola, F., Dutta, P., Codony, R., Savage, G.P., Eds.; AOAC Press, USA, 2002, 66–100.

124. Kumar, N. and Singhal, O.P., Effect of processing conditions on the oxidation of cholesterol in ghee, *J. Sci. Food Agric.*, 58, 267, 1992.

125. Torres, E. et al., Lipid oxidation in charqui (salted and dried beef), *Food Chem.*, 32, 257, 1989.

126. Sander, B.D. et al., Quantification of cholesterol oxidation products in a variety of foods, *J. Food Prot.*, 52, 109, 1989.

127. Britt, C. et al., Influence of cherry tissue on lipid oxidation and heterocyclic aromatic amine formation in ground beef patties, *J. Agric. Food Chem.*, 46, 4891, 1998.

128. Toschi, T.G. et al., Chromatographic study of cholesterol oxidation in some foods of animal origin, *Ind. Conserve*, 69, 115, 1994.
129. Osada, K. et al., Cholesterol oxidation in meat products and its regulation by supplementation of sodium nitrite and apple polyphenol before processing, *J. Agric. Food Chem.*, 48, 3823, 2000.
130. Pie, J.E., Spahis, K., and Seillan, C., Cholesterol oxidation in meat products during cooking and frozen storage, *J. Agric. Food Chem.*, 39, 250, 1991.
131. Hwang, K.T. and Maerker, G., Quantitation of cholesterol oxidation products in unirradiated and irradiated meats, *J. Am. Oil Chem. Soc.*, 70, 371, 1993.
132. Zubillaga, M.P. and Maerker, G., Quantification of three cholesterol oxidation products in raw meat and chicken, *J. Food Sci.*, 56, 1194, 1991.
133. Galvin, K., Morrissey, P.A., and Buckley, D.J., Cholesterol oxides in processed chicken muscle as influenced by dietary α-tocopherol supplementation, *Meat Sci.*, 48, 1, 1998.
134. Monahan, F.J. et al., Influence of dietary treatment on lipid and cholesterol oxidation in pork, *J. Agric. Food Chem.*, 40, 1310, 1992.
135. Jo, C., Ahn, D.U., and Lee, J.I., Lipid and cholesterol oxidation, color changes, and volatile production in irradiated raw pork batters with different fat content, *Food Qual.*, 22, 641, 1999.
136. Park, S.W. and Addis, P.B., Cholesterol oxidation products in some muscle foods, *J. Food Sci.*, 52, 1500, 1987.
137. Ryan, T.C., Gray, J.I., and Morton, I.D., Oxidation of cholesterol in heated tallow, *J. Sci. Food Agric.*, 32, 305, 1981.
138. Park, S.W. and Addis, P.B., Further investigation of oxidized cholesterol derivatives in heated fats, *J. Food Sci.*, 51, 1380, 1986.
139. Thurner, K. et al., Determination of cholesterol oxidation products in raw and processed beef and pork preparations, *Eur. Food Res. Technol.*, 224, 797, 2007.
140. Ahn, D.U. et al., Effect of irradiation and packaging conditions after cooking on the formation of cholesterol and lipid oxidation products in meats during storage, *Meat Sci.*, 57, 413, 2001.
141. Chen, Y.C., Chiu, C.P., and Chen, B.H., Determination of cholesterol oxides in heated lard by liquid chromatography, *Food Chem.*, 50, 53, 1994.
142. Higley, N.A. et al., Cholesterol oxides in processed meats, *Meat Sci.*, 16, 174, 1986.
143. Lercker, G. and Rodriguez-Estrada, M.T., Cholesterol oxidation: presence of 7-ketocholesterol in different food products, *J. Food Comp. Anal.*, 13, 625, 2000.
144. Bragagnolo, N., Danielsen, B., and Skibsted, L.H., Combined effect of salt addition and high-pressure processing on formation of free radicals in chicken thigh and breast muscle, *Eur. Food Res. Technol.*, 223, 669, 2006.
145. Schmarr, H.G., Gross, H.B., and Shibamoto, T., Analysis of polar cholesterol oxidation products: evaluation of a new method involving transesterification, solid phase extraction, and gas chromatography, *J. Agric. Food Chem.*, 44, 512, 1996.
146. Lee, J.I. et al., Formation of cholesterol oxides in irradiated raw and cooked chicken meat during storage, *Poult. Sci.*, 80, 105, 2001.
147. Conchillo, A., Ansorena, D., and Astiasarán, I., Combined effect of cooking (grilling and roasting) and chilling storage (with and without air) on lipid and cholesterol oxidation in chicken breast, *J. Food Prot.*, 66, 840, 2003.
148. Boselli, E. et al., Photoxidation of cholesterol and lipids of turkey meat during storage under commercial retail conditions, *Food Chem.*, 91, 705, 2005.
149. Cayuela, J.M. et al., Effect of vacuum and modified atmosphere packing on the quality of pork loin, *Eur. Food Res. Technol.*, 419, 316, 2004.
150. Galvin, K. et al., Effect of dietary vitamin E supplementation on cholesterol oxidation in vacuum packaged cooked beef steaks, *Meat Sci.*, 55, 7, 2000.
151. Eder, K. et al., Concentrations of oxysterols in meat and meat products from pigs fed diets differing in the type of fat (palm oil or soybean oil) and vitamin E concentrations, *Meat Sci.*, 70, 15, 2005.

Chapter 13

Determination of Oxidation

Mario Estévez, David Morcuende, and Sonia Ventanas

Contents

13.1 Introduction: Oxidation in Processed Meat and Poultry

Lipid oxidation has been largely recognized as a leading problem in the storage and processing of muscle foods. The oxidative degradation of fatty acids involves several molecular mechanisms that lead to the generation of oxygen-rich precursors of reactive, chain-propagating, free radicals. Though lipid oxidative reactions contribute to certain desirable quality attributes such as the development of pleasant flavors in cooked meats,[1] the overall effect of lipid oxidation is negative, leading to adverse effects on sensory traits, nutritional value, and richness of muscle foods.[2] Recent studies on meat systems have found that muscle proteins can be damaged by reactive-oxygen species (ROS) derived from the decomposition of unsaturated fatty acids.[3] Though the effect of protein oxidation in processed meats is currently poorly understood, great efforts have been made recently to shed light on the mechanisms leading to the oxidative deterioration of muscle proteins.

After slaughter, the intensity of the oxidative deterioration sharply increases during meat processing, because certain operations (size reduction, mixing, etc.) enhance the effective reaction between molecular oxygen and unsaturated lipids, and some others (cooking, pasteurization, etc.) involve the application of high temperatures, increasing the extent of the oxidative reactions.[1,2] The onset of oxidative rancidity in processed poultry and meat products is a major concern nowadays due to the increased demand of precooked meat items.

The relevant role played by oxidative reactions in the loss of quality in muscle foods has challenged scientists to develop a suitable methodology to evaluate the oxidative status of muscle foods. The assessment of oxidative reactions and their unpleasant effects in muscle foods is essential to assure the sensory and nutritional quality of the meat product. Thus, oxidation analyses provide useful information for the development of successful antioxidant strategies and improve recipes and industrial processing of poultry and meat products. However, the complexity of the chemistry involved confronts the possibility of one, general, analytical test for unambiguous evaluation of the oxidative deterioration of muscle foods. However, the deep understanding of lipid oxidation mechanisms and the development of improved techniques for the isolation, identification, and quantification of lipid oxidation products have helped in finding methodology to obtain accurate results.

Great efforts and innovation have also been made to develop suitable methodology to approach the analysis of protein oxidation in poultry and meat products. Since mechanisms and chemical pathways involved in protein oxidation are noticeably different to those previously established for lipid oxidation, the evaluation of these chemical reactions requires a different methodology. In most cases, the analytical methods used in biomedical sciences are being successfully extrapolated to muscle food systems.

13.2 Lipid Oxidation

13.2.1 Mechanisms and Factors

Lipid oxidation is a combination of various chain reactions, consisting of three phases: initiation, propagation, and termination, which take place at the same time excepting the initial step.[4]

During the initial phase, in the presence of initiators (In), unsaturated lipids (RH) lose hydrogen radical (H) to form lipid-free radical ($R^\bullet$):

$$In^\bullet + RH \rightarrow InH + R^\bullet$$

Unsaturated lipids are easily oxidized by the ROS, which include oxygen radicals and nonradical derivatives of oxygen. The susceptibility to be oxidized and rate of oxidation of fatty acids depend on the degree of their unsaturation.[5]

In the propagation stage, the alkyl radical of an unsaturated lipid ($R^\bullet$) containing a labile hydrogen reacts very rapidly with molecular oxygen (O_2) to yield peroxide radicals ($ROO^\bullet$). This reaction is always faster than the following hydrogen transfer reaction with unsaturated lipids to form hydroperoxides (ROOH) that are considered the primary products of lipid oxidation:[4]

$$R^• + O_2 \rightarrow ROO^•$$

$$ROO^• + RH \rightarrow ROOH + R^•$$

The newly formed hydroperoxyl radical can so forth abstract a hydrogen from an adjacent unsaturated fatty acid because the reaction goes through 8–14 propagation cycles before termination. Hydroperoxides are considered to be the most important initial reaction products that are obtained from lipid oxidation; they are labile species, of very transitory nature, which undergo changes and deterioration with the radicals. Their breakage causes secondary products such as pentanal, hexanal, 4-hydroxynonenal, and malonaldehyde (MDA).[6]

In the last stages of oxidation, the radical species react with one another and self-destruct to form nonradical products by different mechanisms. The radical reaction stops when two radicals react and produce nonradical species. At atmospheric pressure, termination occurs first by the combination of peroxyl radicals (ROO•) and an unstable tetroxide intermediate, followed rapidly by their decomposition that yields nonradical products.

$$ROO^• + ROO^• \rightarrow [ROOOOR] \rightarrow \text{nonradical products} + O_2$$

In the case of alkoxyl radicals (RO•), they can react with unsaturated lipids to form stable and innocuous alcohols or undergo transformation into unsaturated aldehydes, such as MDA and other unstable compounds.

$$RO^• + RH \rightarrow ROH + R^•$$

$$RO^• \rightarrow RCOH + R^•$$

13.2.2 Assessment of Lipid Oxidation

From the perspective of the meat product analysis, the detection of lipid oxidation needs to be carried out using specific, sensitive, and rapid methodology. Procedures used for the assessment of lipid oxidation can be classified into two groups based on the analytical approach. The first group measures primary oxidative changes (especially the formation of primary oxidation products) and the second determines the formation of secondary oxidation products, originating from the decomposition of primary oxidation products.

13.2.2.1 Detection of Primary Changes: Hydroperoxides

Hydroperoxides, or simply peroxides, are considered to be the most important initial reaction products that are obtained from lipid oxidation. Measuring the content of primary oxidation products is limited due to the transitory nature of hydroperoxides. After the induction period, the initial rate of hydroperoxide formation is exceeded by its rate of decomposition, but this trend is reversed at later stages, as a consequence of their reaction with other food components[4] or their evolution into secondary oxidation products.[7] When the breakdown of hydroperoxides is as fast as or faster than their formation, lipid hydroperoxides are not good indicators of oxidation. This can occur in meat products, particularly in cooked meats,[4] where the breakdown of hydroperoxides is rapid or in the long-term dry-cured products[8,9] due to the generation of carbonyl and other breakdown products. Thus, a low peroxide value (described later) may represent either early or advanced oxidation. For this reason, when the analysis of hydroperoxides is compared with the measurement of secondary oxidation products (MDA, hexanal), several contradictory reports are found.[9,10] However, it is important to note that there is no ideal chemical method that correlates well with changes in sensory properties of oxidized lipids throughout the entire course of autoxidation. Consequently, both primary and secondary lipid oxidation products, which result from lipid oxidation, should be monitored over time.

The analysis of lipid hydroperoxides in meat products generally requires a previous lipid extraction with solvents, which must be carefully removed to avoid the decomposition of hydroperoxides or loss during solvent evaporation.[4] Various procedures and solvent combinations have been employed to extract lipids in hydroperoxides determination. Usually, the extraction is performed with a mixture of polar and nonpolar solvents, as described by Folch et al.,[11] which employs chloroform/methanol (2:1 v/v), Bligh and Dyer[12] (chloroform/methanol 1:2 v/v), or Radin[13] (hexane/2-propanol 3:2 v/v). Moreover, in the literature other new methods, such as the accelerated solvent extraction (ASE), have been reported.[14] These authors considered that the extraction with ASE is more convenient because it is rapid, solvent saving, and, at the same time, oxidation protecting, given that the extraction occurs in presence of nitrogen.

Numerous analytical procedures for the measurement of lipid hydroperoxides in meats and meat products have been described. Among them, two major groups can be considered: analytical methods for determining the total amount of hydroperoxides and those methods based on chromatographic techniques giving information on the structure and amount of specific hydroperoxides.[15]

13.2.2.1.1 Chemical Methods Based on Redox Reactions

13.2.2.1.1.1 Iodometric Assays

The classical method for quantification of hydroperoxides is the determination of peroxide value (PV or POV). The procedure is described by IUPAC[16] and by AOAC,[17] with those being considered official methods. This iodometric method is based on the titration of iodine released from potassium iodide by lipid hydroperoxides (Equation 13.1). The amount of iodine liberated is proportional to the concentration of hydroperoxides in the sample. Released I_2 is assessed by titration against a standardized solution of sodium thiosulfate using a starch solution as the indicator (Equation 13.2). The peroxide value is generally expressed in terms of milliequivalents (mEq) oxygen/kg of lipid or meat.

$$ROOH + 2H^+ + 2KI \rightarrow I_2 + ROH + H_2O + 2K^+ \tag{13.1}$$

$$I_2 + 2Na_2S_2O_3 \rightarrow Na_2S_4O_6 + 2NaI \tag{13.2}$$

The sensitivity of this method is about 0.5 mEq/kg of lipid. Lower peroxide value cannot be satisfactorily measured by the official methods because of uncertainty with the iodometric titration endpoint, but this drawback can be removed by determining the iodine starch-end potentiometrically at levels as low as 0.06 Eq/kg of lipid.[18] The iodometric method is highly empirical and no change in procedure may cause variation in results.[19,20] The two major sources of error in this method are the liberation of iodine by air oxidation of the potassium iodide and absorption of iodine by fatty acid double bonds. Despite its drawbacks, iodometric determination is one of the most common tests employed to monitor lipid oxidation and has been used for following the stages of oxidation (together with secondary oxidation products detection) in different meat products, such as cooked ground meats,[21] mortadella,[22] dry sausages,[23] and dry-cured ham.[9]

13.2.2.1.1.2 Determination of Hydroperoxides by Measurement of Iron Oxidation

To improve the sensitivity of the iodometric methods, some research has proposed spectrophotometric measurement of hydroperoxides based on the oxidation of Fe^{2+} to Fe^{3+}.[24] This method is based on the ability of peroxides to oxidize ferrous ions to ferric ions. Ammonium thiocyanate reacts with ferric ions, resulting in a colored complex that can be measured spectrophotometrically at 500 nm. Peroxide values as low as 0.1 mEq/kg sample can be determined with this method, providing a distinct advantage over iodometric titration, although the values obtained are higher by a factor of 1.5–2 relative to those of the iodometric methods.[4] These methods have been successfully employed for the determination of lipid stability in cooked minced pork,[10] beef burgers,[25] or fried chicken patties.[26]

Alternatively, the determination of ferric ions can be carried out by ferrous oxidation-xylenol (FOX) orange method.[27,28] This method consists of the peroxide-mediated oxidation of ferrous ions in an acidic medium containing the dye xylenol orange, which binds the resulting ferric ions to produce a blue-purple complex with a maximum absorbance between 550 and 600 nm:

$$Fe^{II}(orange)[+peroxide] \rightarrow Fe^{III}[+xylenol\ orange] \rightarrow Fe^{III} - xylenol\ orange\ (purple) \quad (13.3)$$

The FOX method has been reported to have high sensitivity, being comparable to that of the iodometric assay in meat products.[27,29] The method is simple, easy to apply, fast, and does not depend on the presence of oxygen.[27,28] FOX determination kits for food analysis are also available.[27] Hermes-Lima et al.[30] adapted the FOX method to muscle-based products avoiding previous lipid extraction and thus minimizing the chance of hydroperoxides loss during solvent evaporation. The method is based on the direct reaction of a methanol extract of the tissue with an acidic reaction mixture containing Fe^{II} and xylenol orange. However, the authors emphasized that the volume of the methanol extract and the reaction conditions should be adapted to particular samples for a reliable quantification.

13.2.2.1.2 Analysis of Hydroperoxides by Chromatographic Techniques

In addition to the classical analyses, other procedures with high sensitivity and selectivity, such as gas chromatography-mass spectrometry (GC-MS)[31] and high performance liquid chromatography (HPLC)[32] commonly coupled to chemiluminescence or fluorescence detection, have been developed for the analysis of the structure and amount of specific hydroperoxides. However, these methods are not easily adapted to routine screening of large numbers of muscle-based products, and although some applications in the field of food lipids have been published (for a review, see Ref. 15), these techniques have been more frequently used in model systems and biological studies.

13.2.2.2 Primary Oxidation Products: Conjugated Dienes

During the formation of hydroperoxides from polyunsaturated fatty acids, conjugated dienes are typically formed. The measurement of conjugated dienes is a critical process that follows the early stages of lipid oxidation under conditions in which hydroperoxides undergo little or no decomposition. As with the peroxide value, the conjugated diene determination will reach a maximum during the progress of oxidation and decrease when the rate of decomposition of hydroperoxides exceeds the rate of their formation.[21] The analysis of conjugated dienes is assessed by measuring the change in absorbance at 234 nm for a constant mass of sample. A comprehensible and developed method for conjugated diene analysis can be found in IUPAC Standard Method[33] and in Ref. 19.

In meat products, a previous lipid extraction with hexane/isopropanol (3:2)[34] or chloroform/methanol (2:1)[35] is required. This method has been extensively used to monitor lipid oxidation of fatty acids in meat products, because it is fast and simple, requires no chemical reagents, and can use small samples.[19] However, the usefulness of the detection of conjugated diene in meat products containing elevated unsaturated lipid content (i.e., chicken meat products) is limited, because the value obtained greatly depends on the fatty acid composition of the analyzed sample, as well as on the presence of other conjugated dienes (not hydroperoxides).[7,35] In this sense, Grau et al.[35] found that, in cooked chicken meat, the peroxide value and the thiobarbituric acid (TBA) test techniques showed a higher sensitivity in measuring lipid oxidation than conjugated diene determination as a consequence of the low correlation between diene measurements and the extent of oxidation.

13.2.2.3 Detection of Secondary Changes

The measurement of secondary oxidation products as an index of lipid oxidation is more appropriate than using primary products (hydroperoxides), because secondary products are generally odor-active and stable compounds, whereas primary products are colorless, flavorless, and commonly labile compounds.[20] Therefore, secondary oxidation compounds, such as aldehydes, ketones, hydrocarbons, and alcohols, are more easily detected and quantified and reflect the quality deterioration of a muscle food as a consequence of oxidative reactions. Among these compounds, aldehydes are considered the most important breakdown products because they possess low threshold values and are the major contributors to the development of rancid off-flavors.[36]

13.2.2.3.1 Determination of MDA

MDA (1,3-propanedial) is a three-carbon dialdehyde with carbonyl groups at the C-1 and C-3 positions. During autoxidation of polyunsaturated fatty acids, MDA is produced and this secondary oxidation product is highly reactive and remains bound to other food ingredients. An acid/heat treatment of the food presumably releases the bound MDA.[37] The amount of MDA has been commonly used as an oxidation index in muscle foods and different analytical techniques have been reported in the scientific literature to determine and quantify MDA.

13.2.2.3.1.1 Thiobarbituric Acid Test

The extent of lipid oxidation in muscle foods is commonly assessed by monitoring MDA formation by means of TBA assay.[38]

The TBA test is a colorimetric technique in which the absorbance of a red chromogen formed between TBA and MDA is measured. MDA is the major TBA reactive substance although other

Table 13.1   Processed Meat and Poultry Products and Methodologies Used to Monitor MDA by Means of TBA Test

TBA Test Method	Sample (Meat Product)	Reference
Distillation method	Salame Milano and mortadela	45
	Dry-cured ham	46
	Parma ham	47
	Chicken frankfurters	48
	Bacon	49
	Cooked beef, pork, and chicken	50
Acid extraction method	Liver pâté	51
	Cooked porcine meat patties	52
	Porcine frankfurters	53
	Dry-cured ham	54

oxidation products, such as α,β-unsaturated aldehydes (for instance, 4-hydroxyalkenals) and certain unidentified nonvolatile precursors of these substances may also react with TBA.[39] For this reason, this test is now known as the thiobarbituric acid reactive substances (TBARS) method.

The reaction occurs when the monoenolic form of MDA reacts with the active methylene groups of TBA, leading to the formation of a red-colored complex, the intensity of which is related to the concentration of MDA. Visible and ultraviolet spectrophotometry of the pigment (MDA–TBA adduct) confirms the primary maximum at 532–535 nm and a secondary one at 245–305 nm.[40]

Several methods have been proposed to perform TBA test in meat products.[39] Briefly, the TBA test can be performed (i) by directly heating the sample with TBA, followed by separation of the pink complex produced by centrifugation;[41] (ii) by distillation of the sample, followed by the reaction of the distillate with the TBA;[40] (iii) by extraction of MDA using aqueous thricloroacetic or perchloric acid and subsequent reaction with TBA;[43] and (iv) by extraction of the lipid fraction of the sample with organic solvents and reaction of the extract with the TBA.[44] The most widely used methods for measuring TBARS in processed meat and meat products are the distillation method as well as the acid extraction method (Table 13.1).

The formation of additional MDA and other TBARS during analysis due to heating and acidification can lead to an overestimation of TBARS numbers. Particularly, heating during distillation method enhances the degradation of existing lipid hydroperoxides and thus additional TBARS may be formed even in the presence of metal chelators and phenolic antioxidants.[6]

The TBA test has been largely criticized as TBA also reacts with other lipid oxidation products and nonlipid substances such as sugars and amino acids leading to the generation of interfering colored compounds.[38] These interfering reactions are especially troublesome for samples containing high levels of interfering agents and when the reactions are carried out at high temperatures.[55] Several strategies have been proposed to reduce these interfering reactions. Reaction between MDA and TBA is more specific at room temperature than at boiling temperature but the extent of the reaction should be increased from 30 min (at boiling temperature) to 15 h (at room temperature). Rajarho et al.[56] reported that interfering compounds can be removed by filtering the extract by solid-phase cartridges (C18). Wang et al.[57] described a successfully modified TBA test for measuring the MDA in presence of interfering sugars in cooked beef, by reducing the incubation temperature (40°C), increasing the TBA concentration (80 mM), and adjusting the pH to dissolve the TBA.

Another important interference by residual nitrite has been described in cured meat products because the presence of residual nitrite in cured samples could lead to the nitrosation of MDA, making all or a portion of the MDA unreactive to TBA, which leads to TBA numbers lower than expected. For nitrite-cured products, Zisper and Watts[58] modified the TBA test by adding sulfanilamide (SA) before the distillation step to hinder nitrosation of MDA. Added SA reacts with residual nitrite to yield a diazonium salt and thus permitting MDA to react quantitatively with TBA. However, SA itself may give rise to the formation of condensation products with TBA. When residual nitrite is not present or it is present at a concentration lower than 100 ppm, the SA added may lead to the underestimation of TBA values, since MDA itself takes part in a competition reaction by condensing with one or two molecules of SA.[59]

Considering the aforementioned limitations of the MDA test, Ross and Smith[60] have pointed out that the TBARS procedure may be used to assess the extent of lipid oxidation in general, rather than to quantify MDA.

In addition to TBA, other chromogenic agents have been proposed for measuring MDA by means of spectrophotometric method. Gerard-Monnier et al.[61] found that certain indols, particularly 1-methyl-2-phenyl-indol, react with MDA to form a stable chromophore, which shows maximal absorbance at 580–620 nm, with this reaction being MDA-specific in the presence of hydrochloric acid. Moreover, this reagent can also be used in presence of methanesulfonic acid for quantifying 4-hydroxyalkenal, another secondary oxidation product. Paleari et al.[62] have recently used this method to determine free MDA in a salami product.

13.2.2.3.1.2 Determination of MDA by GC and HPLC

GC methods for the determination of MDA require the formation of a stable derivative of MDA, since free MDA is not suitable for direct GC analysis. Most reported GC methods give a total measure of free MDA and its bound forms because the assay conditions are sufficient to hydrolyze or decompose bound MDA during sample preparation. Hydrazine-based reagents, such as 2,4-dinitrophenylhydrazine and *N*-methylhydrazine, have been preferred for GC-MS analysis because these substances are able to form stable pyrazole derivatives.[63] A developed capillary GC method that allows determination of free MDA has been reviewed by Denis and Shibamotto.[64] The method involves derivatization of MDA to *N*-methylpyrazole under milder reaction conditions than with other GC methods. Recently, Shamberger et al.[65] directly determined MDA levels in cooked meat from different species using a GC.

However, Kakuda et al.[66] originally used HPLC to quantify MDA in distilled water and found a linear correlation between TBARS numbers and HPLC results. Later, Williams et al.[67] used it to evaluate meat oxidation. As described earlier for GC, some authors have reported a derivatization of MDA before HPLC analysis. Sakai et al.[68] described a method to quantify the 3-diethyl-2-thiobarbituric acid (DETBA)–MDA adduct by HPLC in fish meat, which has been lately modified to use it in meat products, particularly in smoked meat products (bacon, ham, and sausage).[69]

13.2.2.3.2 Analysis of Lipid Oxidation–Derived Volatiles

The analysis of individual volatile carbonyl compounds is one of the most widely used techniques to assess lipid oxidation in muscle foods. Lipid oxidation of unsaturated fatty acids results in a wide range of volatile aldehydes, such as hexanal, propanal, or 4-hydroxy-2 nonenal, which have been commonly used as index of rancidity and warmed over flavor (WOF) in meat.

Hexanal is one of the major lipid oxidation products, mainly derived from linoleic and arachidonic fatty acids. Nonaka and Pippe[70] reported the feasibility of using the quantity of hexanal as index of oxidative deterioration in fried chicken. St Angelo et al.[71] reported that hexanal and 2,3-octanedione showed a significant correlation between sensory scores and TBARS numbers in cooked beef. Kerler and Grosch[72] carried out a systematic study to determine the odorants contributing to WOF of cooked beef patties, and indicated that WOF was the result of a combination of a loss of desirable odorants, such as 4-hydroxy-2,5-dimethyl-3(2*H*)-furanone and 3-hydroxy-4,5-dimethyl-2(5*H*)-furanone, along with an increase in lipid peroxidation products, in particular *n*-hexanal and *trans*-4,5-epoxy-(*E*)-2-decenal. Recently, Jayathilakan et al.[73] evaluated the WOF profile, expressed in terms of mg hexanal/100 g fat in cooked meat from three different species (sheep, beef, and pork) following the method described by Benca and Mitchela.[74] In addition to hexanal, other volatile compounds such as propanal, pentanal, octanal, and nonanal have been reported as oxidation markers in meat.[75] Shahidi and Wanasundara[20] suggested that hexanal can be used as an oxidation index when the fats under study are rich in ω-6 fatty acids, whereas propanal would serve as a reliable indicator when fats rich in ω-3 fatty acids are considered. However, several studies have shown that hexanal indicates lipid oxidation of meat more effectively than any other volatile component.[76]

Volatiles arising from lipid oxidation have been extensively used as oxidation markers in a large variety of processed meats, such as cooked meat,[77] dry-cured ham,[54] dry sausages,[78] frankfurters,[79] liver pâtés,[80] and smoked sausages.[81]

The most recognized techniques for the extraction and isolation of lipid oxidation–derived volatiles are solvent extraction, simultaneous distillation extraction (SDE), dynamic headspace or purge and trap methodology, and most recently solid-phase microextration (SPME). Recently, Ross and Smith[60] described the main advantages and disadvantages of these techniques (Table 13.2). The extracted volatiles are separated, identified, and quantified using GC or HPLC usually coupled to MS. Among the techniques for the analysis of lipid-derived volatiles, the use of SPME is growing in popularity due to its sensitivity and ease of use.[60]

13.3 Protein Oxidation

The occurrence, extent, and consequences of the onset of protein oxidation in poultry and meat products are issues of increasing interest among food researchers. After the discovery that myofibril proteins are affected by ROS during meat aging,[87] numerous research studies have dealt with the development of protein oxidation in muscle foods and have tried to shed light on the influence of meat origin and industrial processing on the occurrence and intensity of muscle protein oxidation. Furthermore, it is plausible to consider that the modifications caused by ROS in muscle proteins could be implicated in the loss of their functionality and, therefore, in the loss of the quality in poultry and meat products.[3,53] Therefore, it is essential to develop accurate procedures to assess the oxidative reactions affecting the muscle proteins. To comprehend the methodological approach of the analytical procedures, it is necessary to present a brief overview of the chemical aspects surrounding the oxidation of proteins.

13.3.1 Mechanisms and Factors

The initiation of protein oxidation in poultry and meat products is likely to be affected by the presence of primary and secondary lipid oxidation products, primarily hydroperoxides and aldehydes.[88] Additionally, ROS commonly found in muscle systems such as $^{\bullet}OH$, $O_2^{\bullet}$, $ROO^{\bullet}$ as well

Table 13.2 Summary of Methodologies Used to Analyze Lipid-Derived Volatiles in Processed Meat and Poultry Products

Technique	Advantage(s)	Disadvantage(s)	Sample	Reference
SDE	Isolate and remove in a single step	Solvent evaporation (degradation of volatile compounds)	Cooked chicken	50
	Small volume of solvent		Dry-cured ham	82
	Minimal artifact formation		Cooked ham	83
			Dry-fermented sausages	84
Static headspace	No presence of a solvent peak	Inefficient quantification (establishment of an equilibration state is required)	Dry-cured ham	9
	Low cost per sample		Cooked pork patties	52
	Simple preparation			
	Possibility of automation			
Dynamic headspace	Quite sensitive	Optimization and monitorization of several steps	Cooked beef	85
	There is no need to reach the equilibrium	Require more time per sample comparing static headspace for purging, trapping, and analysis	Dry-cured ham	86
			Smoked sausages	81
SPME	Extraction and concentration without purging the sample	Extraction of variables must be carefully controlled (pH, temperature, time)	Dry-cured ham	54
	Solvent-free sample preparation	High variability of results	Cooked pork	77
			Dry-fermented sausage	78
			Porcine frankfurters	79
			Liver pâté	80

Source: Adapted from Ross, C.F. and Smith, D.M., *Comp. Rev. Food Sci. Food Safety*, 5, 18, 2006.

as metal cations (iron, copper) can catalyze the abstract of a hydrogen from a susceptible amino acid residue, leading to the generation of a protein radical.[89] Heterocyclic amino acids and those containing reactive side chains (sulfhydryl, thioether, aminogroup imidazole, and indole rings) are common targets for ROS due to the presence of OH-, S-, or N-containing groups.[90] According to recent studies, sulfur-containing amino acids, particularly cysteine and methionine, are the most susceptible in muscle proteins.[91]

Oxidation of proteins is manifested as a free radical chain reaction similar to that of lipid oxidation although, in the former, a higher complexity of the pathways and a larger variety of oxidation products have been reported.[90] The abstraction of a hydrogen atom leads to the generation of a protein carbon–centered radical ($P^•$), which is consecutively converted into a peroxyl radical ($POO^•$) in the presence of oxygen and an alkyl peroxide ($POOH$) by abstraction of a hydrogen atom from another molecule. Further reactions with $HO_2^•$ leads to the generation of an alcoxyl radical ($PO^•$) and its hydroxyl derivative (POH). As a direct consequence of the oxidative damage of muscle proteins, the aminoacid residue side chains are modified. These changes include the loss of sulfhydryl groups, the generation of oxidized derivatives (i.e., sulfoxides from methionine), and the conversion of an amino acid residue into a different one.[3] Additionally, the oxidation of side chains of certain amino acids (arginine, lysine, proline, and threonine) leads to the generation of carbonyl residues through deamination reactions. The fragmentation of the peptide backbone, reaction with reducing sugars (via Schiff base formation), and binding nonprotein carbonyl compounds (i.e., MDA) also yield protein carbonyl derivatives.[89] Another relevant consequence of the oxidation of muscle proteins is the formation of aggregates through covalent and noncovalent linkages. The noncovalent forces are promoted by the exposure of nonpolar residues from proteins as a result of an oxidatively induced unfolding, leading to the generation of hydrophobic interactions between protein chains. The formation of noncovalent aggregates is enhanced by the generation of hydrogen bonds and the complexes formed between proteins and oxidized lipids.[89] The ROS-mediated covalent linkages between two amino acid residues lead to the generation of intra- and interprotein cross-linked derivatives. The mechanisms involved in these nondissociable protein aggregates include (i) the direct condensation of two carbon-centered radicals, (ii) the oxidation of cysteine sulfhydryl groups to form disulfide linkages, (iii) the interaction of two oxidized tyrosine residues to yield dityrosines (Figure 13.1), (iv) the reaction between a protein aldehyde and the amino group from a lysine in the same or a different protein, and (v) the reaction of two amino groups (from two lysine residues) with a dialdehyde (i.e., malondialdehyde).[92–95] In recent studies, the development of cross-links between myofibril proteins from chicken beef and porcine muscles subjected to a prooxidant storage is mainly attributed to the generation of disulfide linkages and to a lesser extent, to the presence of dityrosines.[91,96–100]

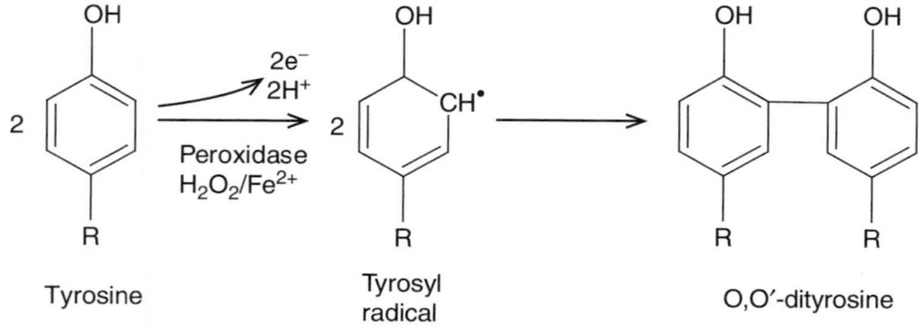

Figure 13.1 Formation of protein cross-linking: dityrosine.

The oxidation of proteins and amino acids is affected by certain environmental factors such as pH, temperature, water activity, and the presence of catalysts or inhibitors. Additionally, the three-dimensional structure of the proteins as well as their amino acid composition influences the susceptibility of proteins to undergo oxidative reactions.[88,91]

13.3.2 Assessment of Protein Oxidation

Innovative procedures and techniques have been used in the past 10 years to evaluate the oxidative damage of muscle proteins in poultry and meat products. These techniques have been mostly adapted from those purposely developed for biomedical research and are focused on (i) proving the modification of oxidized proteins and amino acids and (ii) detecting protein oxidation products.

13.3.2.1 Assessment of the Oxidative Modification of Proteins

13.3.2.1.1 Determination of Protein Thiol Oxidation

Myosin contains approximately 42 thiol (SH) groups,[101] which have been reported to be highly reactive in the presence of ROS. In oxidized systems, cysteine is degraded into cysteine disulfide and sulfenic acid, whereas methionine is readily oxidized to methionine sulfoxide.[90] The loss of free SH groups in muscle proteins is commonly used as an indicator of protein oxidation. Ellman's reagent or 5,5′-dithiobis(2-nitrobenzoate) (DTNB) rapidly forms a disulfide bond with the thiol and releases a thiolate ion (TNB dianion), which is colored and has a maximal absorbance at 412 nm (Figure 13.2). Because the stoichiometry of protein thiol to TNB formed is 1:1, TNB formation can be used to assess the number of thiols present.

Although the original method[102] has been further modified,[103–106] the general procedure is still retained. Muscle proteins are dissolved in a tris-HCl buffer usually containing urea, SDS, and EDTA, and an aliquot of the solution is incubated with a DTNB solution. The reaction is sensitive to alkaline pH (OH competes with R–S), acidic pH (disulfides can be broken), oxygen

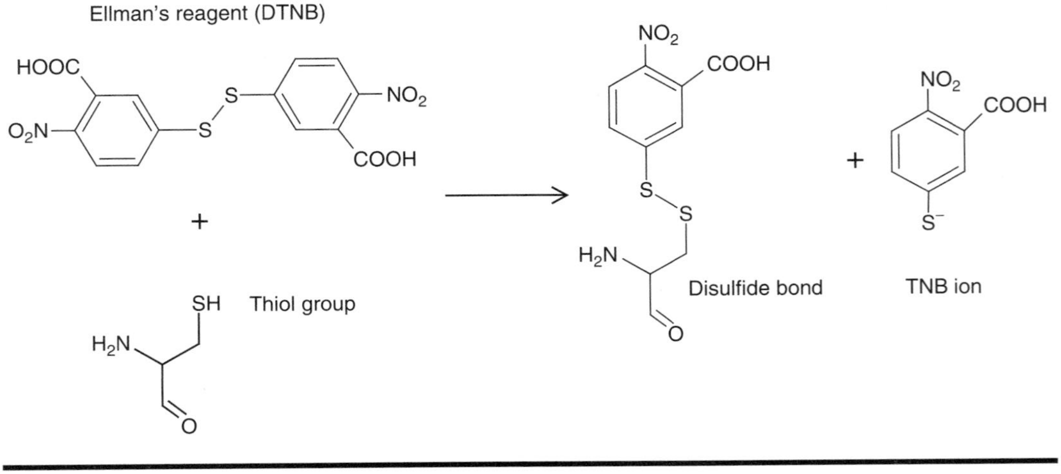

Figure 13.2 Detection of thiol groups using Ellman's reagent.

(reoxidation of R–SH), and temperature (thermochromism). Therefore, the reaction is usually carried out with an excess of DTNB to protein, at neutral pH, fixed temperature, and sometimes under anaerobic conditions. The absorbance at 412 nm can be plotted against a cysteine standard curve to determine the total amount of free thiols in proteins although certain molar extinction coefficients have also been employed for quantitation purposes.[98,105] When selecting the latter option, it is necessary to consider that TNB is sensitive to various buffer ions; thus, the extinction coefficient used to calculate the number of thiols must be properly matched with the reaction conditions. This is also true of reactions carried out in the presence of the denaturants urea or guanidine (GuHCl).

13.3.2.2 Assessment of Protein Oxidation Products

13.3.2.2.1 Detection of Protein Carbonyls

The oxidative modification of amino acids and peptides can yield carbonyl derivatives, and therefore determination of carbonyl content in proteins can be used as a measure of oxidative protein damage.[89] The carbonyl compounds derived from protein oxidation can be detected by measuring fluorescence emitted by protein carbonyls at around 450 nm when they are excited at 350 nm.[88] Recently, Chelh et al.[107] have detected fluorescent protein oxidation products in meat samples using a front-face fluorescence technique. Although the detection of fluorescence using spectrofluorometer is a sensitive assessment of protein oxidation and the procedure is simple and fast, the technique is highly unspecific and no quantitation can be carried out.

The quantitation of carbonyl compounds spectrophotometrically using 2,4 dinitrophenylhydrazine (DNPH) as a marker of the protein oxidation products is one of the widely used methods for evaluating protein oxidation in processed meats. Carbonyl groups react with DNPH to form a 2,4-dinitrophenyl (DNP) hydrazone product, and the amount of hydrazone formed is quantitated spectrophotometrically (Figure 13.3). The original method[108] was developed for using protein carbonyls as markers of the oxidative stress of proteins in age-related diseases. The procedure has been recently implemented by food scientists as a sensitive method for evaluating protein oxidation in a large variety of meat and fat products, such as cooked porcine loin,[109] cooked beef and pork patties,[100,52] cooked sausages,[53] liver pâté,[51] and dry-cured meat products.[54]

The procedure involves a simultaneous determination of carbonyl derivatives and protein content of the meat samples.[108] After the homogenization of samples and the induced precipitation of the muscle proteins with trichloroacetic acid (TCA), meat samples are incubated with a hydrochloric acid solution containing DNPH. After removing the remaining DNPH and muscle lipids

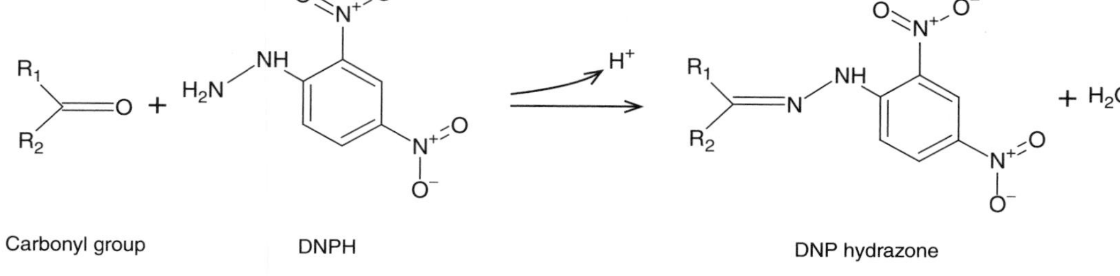

Carbonyl group   DNPH   DNP hydrazone

Figure 13.3   Detection of protein carbonyls using DNPH.

by washing the pellets with ethanol:ethyl acetate (1:1), muscle proteins are finally dissolved in a phosphate buffer containing guanidine hydrochloride. The concentration of DNP hydrazones is calculated by measuring DNPH incorporated on the basis of an absorption of 22,000 $M^{-1}\,cm^{-1}$ at 370 nm. Concentration of protein is determined in a control sample (without added DNPH) at 280 nm using bovine serine albumin as standard. Obtained results are usually displayed as nmols DNP hydrazones per mg of protein.

Modified procedures have been proposed to improve the sensitivity and reliability of the method. Fagan et al.[110] suggested performing additional washes of the pellets with a hydrochloric acid–acetone solution to remove chromophore substances (e.g., hemoglobin, myoglobin, retinoids) that could interfere with the results. Meat researchers have recently employed the modified procedure in porcine muscles with success.[100] The carbonyl assay is a simple, robust, and accurate method for quantifying protein oxidation although some protein oxidative modifications, such as oxidation of histidyl, phenylalanine, and tryptophan residues, do not result in generation of carbonyl compounds.[111] Additionally, the presence of these moieties may also be introduced into proteins by mechanisms that do not involve the oxidation of amino acid residues. For example, certain lipid peroxidation products (i.e., alkenals) may react with sulfhydryl groups of proteins to form stable covalent thioether adducts carrying carbonyl groups.[111] However, the simplicity and convenience of the carbonyl assay has resulted in the widespread measurement of protein carbonyls as a useful and meaningful index of total protein oxidation.

13.3.2.2.2 Detection of Protein Cross-Links

The generation of covalent bonds between amino acids from different proteins is one of the most relevant effects of protein oxidation. Basically there are three methods of interest for the evaluation of protein cross-links in poultry and muscle foods: (1) determination of disulfide bonds, (2) estimation of dityrosine formation, and (3) assessment of the formation of cross-linked myosin heavy chains (CL-MHC).

The procedure for the determination of disulfide bonds was originally described by Thann-hauser et al.[112] and subsequently improved by Damodaran.[113] The method is based on the quantification of 2-nitro-5-thiobenzoate (NTB) formed from the reaction of 2-nitro-5-thiosulfobenzoate (NTSB) with disulfides in the presence of excess sodium sulfite. The NTSB assay solution used must be prepared in advance by treating the Ellman's reagent (DTNB) with excess of sodium sulfite. The amount of the chromophoric derivative is determined through the measurement of absorbance at 412 nm and using a molar extinction coefficient of 13,600 $M^{-1}\,cm^{-1}$. It is recommendable to perform a previous denaturation of the muscle protein using a guanidine salt[112] to enhance the accessibility of the disulfide bonds. It has been highlighted as a sensitive, quantitative, and effortless method although several limitations have also been reported. The NTB undergoes photochemical reaction, which results in the rapid disappearance of absorbance at 412 nm, and therefore the development of the assay in the dark is suggested.[113] However, the method would overestimate the presence of disulfide bonds as NTSB also reacts with free sulfhydryl groups. For accuracy, the presence of sulfhydryl groups could be estimated (i.e., by using the Ellman reagent) and subtracted from the obtained results. Following this method, Liu et al.[96] have detected disulfide bonds in myofibril proteins from chicken muscles.

The detection of dityrosines in processed meat could also be an appropriate method for protein oxidation assessment although it has been scarcely used and the results obtained were at times inconclusive.[91,99] This moiety emits intense 420 nm fluorescence upon excitation within either 315 (alkaline solutions) or 284 nm (acidic solutions) absorption bands.[114] Identification and quantitation of dityrosine in muscle protein hydrolyzates is usually carried out by HPLC.[115] Recently, new attempts for the detection of dityrosines in meat samples have been done by measuring fluorescence through spectrofluorophotometry, with 320 and 420 nm as excitation and emission wavelengths, respectively.[98]

Finally, the formation of protein cross-links can be assessed by the detection of CL-MHC through electrophoresis techniques. SDS-PAGE of oxidized proteins shows the appearance of a band corresponding to molecular weight higher than 500 kDa. This band has been identified as CL-MHC by MS[100] and is usually attributed to the generation of disulfide bonds.[97,116]

References

1. Kanner, J., Hazan, B., and Doll, L., Catalytic 'free' iron ions in muscle foods, *J. Agric. Food Chem.*, 36, 412, 1991.
2. Morrissey, P. A. et al., Lipid stability in meat and meat products, *Meat Sci.*, 49, 73, 1998.
3. Xiong, Y. L., Protein oxidation and implications for muscle foods quality, in *Antioxidants in Muscle Foods*. Decker, E. A., Faustman, C., and Lopez-Bote, C. J., Eds., Wiley, New York, NY, 2000.
4. Frankel, E. N., Free radical oxidation, in *Lipid Oxidation*. Frankel, E. N., Ed., The Oily Press Dundee Ltd., Scotland, 1998.
5. Shahidi, F. *Flavor of Meat, Meat Products, and Seafoods*. Shahihdi, F., Ed., Blackie Academic and Professional, London, UK, 1998.
6. Raharjo, S. and Sofos, J. N., Methodology for measuring malonaldehyde as a product of lipid peroxidation in muscle tissues: a review, *Meat Sci.*, 35, 145, 1993.
7. Moore, K. and Roberts, L. J., Measurement of lipid oxidation, *Free Radic. Res.*, 28, 659–671, 1998.
8. Hernández, P., Navarro, J. L., and Toldrà, F., Lipotytic and oxidative changes in two Spanish pork loin products: dry-cured loin and pickled-cured loin, *Meat Sci.*, 51, 123, 1999.
9. Cava, R. et al., Oxidative and lipolytic changes during ripening of Iberian hams as affected by feeding regime: extensive feeding and alpha-tocopheryl acetate supplementation, *Meat Sci.*, 52, 165, 1999.
10. Nielsen, J. H. et al., Oxidation in pre-cooked minced pork as influenced by chill storage of raw muscle, *Meat Sci.*, 46, 191, 1997.
11. Folch, J., Lees, M., and Stanley, G. H. S., A simple method for the isolation and purification of total lipids from animal tissues, *J. Biol. Chem.*, 226, 497, 1957.
12. Bligh, E. G. and Dyer, W. J., A rapid method of total lipid extraction and purification, *Can. J. Biochem. Phys.*, 37, 911, 1959.
13. Radin, N. S., Extraction of tissue lipids with a solvent of low toxicity, *Methods Enzymol.*, 72, 5, 1981.
14. Saghir, S., Lipid oxidation of beef fillets during braising with different cooking oils, *Meat Sci.*, 71, 440, 2005.
15. Dobarganes, M. C. and Velasco, J., Analysis of lipid hydroperoxides, *Eur. J. Lipid Sci. Technol.*, 104, 420, 2002.
16. IUPAC. Standard Methods for the Analysis of Oils, Fats and Derivatives, IUPAC Standard Method 2.501. International Union of Pure and Applied Chemistry, 7th ed., Pergamon, Oxford, UK, 1992.
17. AOAC. Official Methods of Analysis of AOAC International, Method 41.1.16, AOAC International, 16th ed., Gaithersburg, MD, 1997.
18. Oishi, M. et al., Rapid and simple colorimetric measurements of peroxide value in edible oils and fats, *J. AOAC Int.*, 75, 507, 1992.
19. Pegg, R. B., Measurement of primary lipid oxidation products (Unit D2.1), in *Current Protocols in Food Analytical Chemistry*. Wrolstad, R. E. et al., Eds., Wiley, New York, NY, D2.4.1., 2001.
20. Shahidi, F. and Wanasundara, U. N., Methods for measuring oxidative rancidity in fats and oils, in *Food Lipids-Chemistry, Nutrition, and Biotechnology* (2nd ed.), Akoh, C. C. and Min, D. B., Eds., Marcel Dekker, New York, NY, 465, 2002.

21. Juntachote, T. et al., The antioxidative properties of holy basil and galangal in cooked ground pork, *Meat Sci.*, 72, 446, 2006.
22. Novelli, E. et al., Lipid and cholesterol oxidation in frozen stored pork, salame milano and mortadella, *Meat Sci.*, 48, 29, 1998.
23. Valencia, I. et al., Development of dry fermented sausages rich in docosahexaenoic acid with oil from the microalgae *Schizochytrium* sp.: influence on nutritional properties, sensorial quality and oxidation stability, *Food Chem.*, 104, 1087, 2007.
24. Shantha, N. C. and Decker, E. A., Rapid, sensitive, iron-based spectrophotometric methods for determination of peroxide values of food lipids, *J. AOAC Int.*, 77, 421, 1994.
25. Georgantelis D. et al., Effect of rosemary extract, chitosan and a-tocopherol on lipid oxidation and colour stability during frozen storage of beef burgers, *Meat Sci.*, 75, 256, 2007.
26. Bonoli, M. et al., Effect of feeding fat sources on the quality and composition of lipids of precooked ready-to-eat fried chicken patties, *Food Chem.*, 101, 1327, 2007.
27. Osawa, C. C. et al., Determination of hydroperoxides in oils and fats using kits, *J. Sci. Food Agric.*, 87, 1659, 2007.
28. Grau, A. et al., Lipid hydroperoxide determination in dark chicken meat through a ferrous oxidation–xylenol orange method, *J. Agric. Food Chem.*, 48, 4136, 2000.
29. Eder, K. et al., Concentrations of oxysterols in meat and meat products from pigs fed diets differing in the type of fat (palm oil or soybean oil) and vitamin E concentrations, *Meat Sci.*, 70, 15, 2005.
30. Hermes-Lima, M., Willmore, W. G., and Storey, K. B., Quantification of lipid peroxidation in tissue extracts based on Fe(III) xylenol orange complex formation, *Free Radic. Biol. Med.*, 19, 271, 1995.
31. Hughes, H. et al., Quantification of lipid peroxidation products by gas chromatography-mass spectrometry, *Anal. Biochem.*, 152, 107, 1986.
32. Prior, E. and Loliger, J., Spectrophotometric and Chromatographic Assays in "Rancicity in Foods," Allen, J. C. and Hamilton, R. J., Eds., Blackie Academic & Professional, London, UK, 104, 1994.
33. IUPAC. Standard Methods for the Analysis of Oils, Fats and Derivatives, IUPAC Standard Method 2.505. International Union of Pure and Applied Chemistry, 7th ed., Pergamon, Oxford, UK, 1992.
34. Sirinivasan, S., Xiong, Y. L., and Decker, A., Inhibition of protein and lipid oxidation in beef heart surimi-like material by antioxidants and combinations of pH, NaCl, and buffer type in the washing media, *J. Agric. Food Chem.*, 44, 119, 1996.
35. Grau, A. et al., Evaluation of lipid ultraviolet absorption as a parameter to measure lipid oxidation in dark chicken meat, *J. Agric. Food. Chem.*, 48, 4128, 2002.
36. Ladikos, D. and Lougovois, V., Lipid oxidation in muscle foods: a review, *Food Chem.*, 4, 295, 1990.
37. Ulu, H., Evaluation of three 2-thiobarbituric acid methods for the measurement of lipid oxidation in various meats and meat products, *Meat Sci.*, 67, 683, 2004.
38. Gray, J. L. and Monahan, F. J., Measurement of lipid oxidation in meat and meat products, *Trends Food Sci. Technol.*, 3, 315, 1992.
39. Fernandez, J., Pérez Álvarez, J. A., and Fernández López, J. A., Thiobarbituric acid test for monitoring lipid oxidation in meat, *Food Chem.*, 59, 345, 1997.
40. Sinnhuber, R. O. and Yu, T. C., 2-Thiobarbituric acid method for the measurement of rancidity in fishery products. The quantitative determination of malonaldehyde, *Food Technol.*, 12, 9, 1958.
41. Turner, E. W. et al., Use of the 2-thiobarbuturic acid reagent to measure rancidity in frozen pork, *Food Technol.*, 8, 326, 1954.
42. Tarladgis, B. G. et al., A distillation method for the quantitative determination of malonaldehyde in rancid foods, *J. Am. Oil Chem. Soc.*, 37, 44, 1960.
43. Witte, W. C., Krause, G. F., and Bailey, M. E., A new extraction method for determining 2-thiobarbituric acid values of pork nad beef during storage, *J. Food Sci.*, 35, 582, 1970.
44. Younathan, M. T. and Watts, B. M., Oxidation of tissues lipids in cooked pork, *Food Res.*, 25, 538, 1960.
45. Novelli, E. et al., Lipid and cholesterol oxidation in frozen stored pork, salami Milano and mortadella, *Meat Sci.*, 48, 29, 1998.

46. Vestergaard, C. S., Schivazappa., C., and. Virgili, R., Lipolysis in dry-cured ham maturation, *Meat Sci.*, 55, 1, 2000.
47. Zanardi, E. et al., Oxidative stability of lipids and cholesterol in Salame milano, coppa and parma ham: dietary supplementation with vitamin E and oleic acid, *Meat Sci.*, 55, 169, 2000.
48. Jeun-Horng, L., Yuan-Hui, L., and Chun-Chin, K., Effect of dietary fish oil on fatty acid composition, lipid oxidation and sensory property of chicken frankfurters during storage, *Meat Sci.*, 60, 161, 2002.
49. Coronado, S. A. et al., Effect of dietary vitamin E, fishmeal and wood and liquid smoke on the oxidative stability of bacon during 16 weeks' frozen storage, *Meat Sci.*, 62, 51, 2002.
50. Ajuyah, A. O. et al., Measuring lipid oxidation volatiles in meats, *J. Food Sci.*, 58, 270, 1993.
51. Estévez, M. and Cava, R., Lipid and protein oxidation, release of iron from heme molecule and colour changes during refrigerated storage of liver pâté, *Meat Sci.*, 68, 551, 2004.
52. Salminen, H. et al., Inhibition of protein and lipid oxidation by rapeseed, camelina and soy meal in cooked pork meat patties, *Eur. Food Res. Technol.*, 223, 461, 2006.
53. Estévez, M., Ventanas, S., and Cava, R., Protein oxidation in frankfurters with increasing levels of added rosemary essential oil: effect on colour and texture deterioration, *J. Food Sci.*, 70, 427, 2005.
54. Ventanas, S. et al., Extensive feeding versus oleic acid and tocopherol enriched mixed diets for the production of Iberian dry-cured hams: effect on chemical composition, oxidative status and sensory traits, *Meat Sci.*, 77, 246, 2007.
55. Shlafer, M. and Shepard, B. M., A method to reduce interference by sucrose in the detection of thiobarbituric acid reactives substances, *Anal. Biochem.*, 137, 269, 1984.
56. Rajarho, S., Sofos, J. N., and Schmitd, G. R., Improved speed, specificity and limit of determination of an aqueous acid extraction thiobarbituric acid-C18 method for measuring lipid peroxidation in beef, *J. Agric. Food Chem.*, 40, 2182, 1992.
57. Wang, B. et al., Modified extraction method for determining 2-thiobarbituric acid values in meat with increasing specificity and simplicity, *J. Food Sci.*, 67, 2833, 2002.
58. Zisper, M. W. and Watts, B. M., A modified 2-thiobarbituric acid (TBA) method for the determination of malonaldehyde in cured meats, *Food Tech.*, 16, 102, 1962.
59. Shahidi, F., Pegg, R. B., and Harris R., Effects of nitrite and sulfanilamide on the 2-thiobarbituric (TBA) values in aqueous model and cured meat systems, *J. Muscle Foods*, 2, 1, 1991.
60. Ross, C. F. and Smith, D. M., Use of volatiles as indicators of lipid oxidation in muscle foods, *Comp. Rev. Food Sci. Food Safety*, 5, 18, 2006.
61. Gerard-Monnier, D. et al., Reactions of 1-methyl-2-phenylindole with malondialdehyde and 4-hydroxyalkenals, analytical applications to a colorimetric assay of lipid peroxidation, *Chem. Res. Toxicol.*, 11, 1176, 1998.
62. Paleari, M. A. et al., Characterisation of a lard cured with spices and aromatic herbs, *Meat Sci.*, 67, 549, 2004.
63. Miyake, T. and Sibamoto, T., Simultaneous determination of acrolein, malonaldehyde and 4-hydroxy-2-nonenal produced from lipids oxidized with Featon's reagent, *Food Chem. Toxicol.*, 34, 109, 1996.
64. Denis, K. J. and Shibamotto, T., Gas chromatographic determination of malonaldehyde formed by lipid peroxidation, *Free Radic. Biol. Med.*, 7, 187, 1989.
65. Shamberger, R. J., Shamberger, B. A. and Willis, E. C., Malonaldehyde content in food, *J. Nutr.*, 22, 1404, 2007.
66. Kakuda, Y., Stanley, D. W., and Van de Voort, F. R., Determination of TBA number by high performance liquid chromatography, *J. Am. Oil Chem. Soc.*, 58, 773, 1981.
67. Williams, J. C. et al., Evaluation of TBA methods for determination of lipid oxidation in red meat from four species, *J. Food Sci.*, 48, 1776, 1983.
68. Sakai, T. et al., Effect of NaCl on lipid peroxidation-derived aldehyde, 4-hydroxy-2-nonenal formation in minced pork and beef, *Meat Sci.*, 66, 789, 2004.
69. Munasinghe, D. M. et al., Lipid peroxidation-derived cytotoxic aldehyde, 4-hydroxy-2-nonenal in smoked pork, *Meat Sci.*, 63, 377, 2004.

70. Nonaka, M. and Pippe, E. L., Volatiles and oxidative flavor deterioration in fried chicken, *J. Agric. Food Chem*, 14, 2, 1965.

71. St Angelo, A. J. et al., Chemical and instrumental analysis of warmed over flavor in beef, *J. Food Sci.*, 55, 1501, 1987.

72. Kerler, J. and Grosch, W., Odorants contributing to warmed-over flavor (WOF) of refrigerated cooked beef, *J. Food Sci.*, 61, 1271, 1996.

73. Jayathilakan, K. et al., Antioxidant potential of synthetic and natural antioxidants and its effect on warmed over flavor in different species of meat, *Food Chem.*, 105, 908, 2007.

74. Benca, H. M. F. and Mitchela, J. H., Estimation of carbonyls in fats and oils, *J. Am. Oils Chem. Soc.*, 31, 88, 1954.

75. Lamikanra, V. T. and Dupuy, H. T., Analysis of volatiles related to warmed over flavor of cooked chevon, *J. Food Sci.*, 55, 861, 1990.

76. Dupuy, H. P. et al., Instrumental analysis of volatiles related to warmed-over flavor of cooked meats, in *Warmed-Over Flavor of Meat*. St. Angelo, A. J. and Bailey, M. E., Eds., Academic Press, Orlando, FL, 165, 1987.

77. Estévez, M. et al., Analysis of volatiles in meat from Iberian and lean pigs after refrigeration and cooking by using SPME-GC-MS, *J. Agric. Food Chem.*, 51, 3429, 2003.

78. Marco, J., Navarro, L., and Flores, M., Volatile compounds of dry-fermented sausages as affected by solid-phase microextraction (SPME), *Food Chem.*, 84, 633, 2004.

79. Estévez, M. et al., Influence of the addition of rosemary essential oil on the volatiles pattern of porcine frankfurters, *J. Agric. Food Chem.*, 53, 8317, 2005.

80. Estévez, M. et al., Analysis of volatiles in porcine liver pâtés with added sage and rosemary essential oils by using SPME-GC-MS., *J. Agric. Food Chem.*, 52, 5168, 2004.

81. Olsen, E. et al., Analysis of the early stages of lipid oxidation in freeze-stored pork back fat and mechanically recovered poultry meat, *J. Agric. Food Chem.*, 53, 338, 2005.

82. Dirinck, P., Van Opstaele, F., and Vandendriessche, F., Flavour differences between northern and southern European cured hams, *Food Chem.*, 59, 511, 1997.

83. De Winne, A. and Dirinck, P., Studies on vitamin e and meat quality. 3. Effect of feeding high vitamin e levels to pigs on the sensory and keeping quality of cooked ham, *J. Agric. Food Chem.*, 45, 4309, 1997.

84. Ansorena, D. et al., Analysis of volatile compounds by GC–MS of a dry fermented sausage: chorizo de Pamplona, *Food Res. Int.*, 34, 67, 2001.

85. Elmore, J. S. et al., Effect of the polyunsaturated fatty acid composition of beef muscle on the profile of aroma volatiles, *J. Agric. Food Chem.*, 47, 1619, 1999.

86. Carrapiso, A. I., Ventanas, J., and Garcia, C., Characterization of the most odor-active compounds of iberian ham headspace, *J. Agric. Food Chem.*, 50, 1996, 2002.

87. Martinaud, A. et al., Comparison of oxidative processes on myofibrillar proteins from beef during maturation and by different model oxidation systems, *J. Agric. Food Chem.*, 45, 2481, 1997.

88. Viljanen, K., Kivikari, R., and Heinonen, M., Protein-lipid interactions during liposome oxidation with added anthocyanin and other phenolic compounds, *J. Agric. Food Chem.*, 52, 1104, 2004.

89. Butterfield, D. A. and Stadtman, E. R., Protein oxidation processes in aging brain, *Adv. Cell Aging Gerontol.*, 2, 161, 1997.

90. Stadtman, E. R. and Levine, R. L., Free radical-mediated oxidation of free amino acids and amino acid residues in proteins, *Amino Acids*, 25, 207, 2003.

91. Park, D., and Xiong, Y. L., Oxidative modification of amino acids in porcine myofibrillar protein isolates exposed to three oxidizing systems, *Food Chem.*, 103, 607, 2007.

92. Garrison, W. M., Reaction mechanism in the radiolysis of peptides, polypeptides, and proteins, *Chem. Rev.* 87, 381, 1987.

93. Bhoite-solomon, V., Kessler-icekson, G., and Shaklai, N., Peroxidative crosslinking of myosins, *Biochem. Int.*, 26, 181, 1992.

94. Decker, E. A. et al., Chemical, physical and functional properties of oxidized turkey white muscle myofibrillar proteins, *J. Agric. Food Chem.*, 41, 186, 1993.

95. Srinivasan, S. and Xiong, Y. L., Gelation of beef heart surimi as affected by antioxidants, *J. Food Sci.*, 61, 707, 1996.

96. Liu, G., Xiong, Y. L., and Butterfield, D. A., Chemical, physical, and gel-forming properties of oxidized myofibrils whey- and soy-protein isolates, *J. Food Sci.*, 65, 811, 2000.

97. Ooizumi, T. and Xiong, Y. L., Biochemical susceptibility of myosin in chicken myofibrils subjected to hydroxyl radical oxidizing systems, *J. Agric. Food Chem.*, 52, 4303, 2004.

98. Morzel, M. et al., Chemical oxidation decreases proteolytic susceptibility of skeletal muscle myofibrillar proteins, *Meat Sci.*, 73, 536, 2006.

99. Park, D. et al., Biochemical changes in myofibrillar protein isolates exposed to three oxidizing systems, *J. Agric. Food Chem.*, 54, 4445, 2006.

100. Lund, M. N., High-oxygen packaging atmosphere influences protein oxidation and tenderness of porcine longissimus dorsi during chill storage, *Meat Sci.*, 77, 295, 2007.

101. Lowey, S. et al., Substructure of the myosin molecule. I. Subfragments of myosin by enzymic degradation, *J. Mol. Biol.*, 42, 1, 1969.

102. Ellman, G. L., Tissue sulfhydryl groups, *Arch. Biochem. Biophys.*, 82, 70, 1959.

103. Simplicio, P. D., Chesseman, K. H., and Slater, T. F., The reactivity of the SH group of bovine serum albumin with free radicals, *Free Radic. Res. Commun.*, 14, 253, 1991.

104. Benjakul, S. et al., Physicochemical changes in Pacific whiting muscle proteins during iced storage, *J. Food Sci.*, 62, 729, 1997.

105. Srinivasan, S. and Hultin, H. O., Chemical, physical, and functional properties of cod proteins modified by a nonenzymic free-radical-generating system, *J. Agric. Food Chem.*, 45, 310, 1997.

106. Mercier, Y. et al., Lipid and protein oxidation in microsomal fraction from turkeys: influence of dietary fat and vitamin E supplementation, *Meat Sci.*, 58, 125, 2001.

107. Chelh, I., Gatellier, P., and Santé-Lhoutellier, V., Characterisation of fluorescent Schiff bases formed during oxidation of pig myofibrils, *Meat Sci.*, 76, 210, 2007.

108. Oliver, C. N. et al., Aged-related changes in oxidized proteins, *J. Biol. Chem.*, 262, 5488, 1987.

109. Haak et al., Effect of dietary antioxidant and fatty acid supply on the oxidative stability of fresh and cooked pork, *Meat Sci.*, 74, 476, 2006.

110. Fagan, J. M., Sleczka, B. G. and Sohar, I., Quantitation of oxidative damage to tissue proteins, *Int. J. Biochem, Cell Biol.*, 31, 751, 1999.

111. Requena, J. R., Levine, R. L., and Stadtman, E. R., Recent advances in the analysis of oxidized proteins, *Amino Acids*, 25, 221, 2003.

112. Thannhauser, T. W., Konishi, Y., and Scheraga, H. A., Sensitive quantitative analysis of disulfide bonds in polypeptides and proteins, *Anal. Biochem.*, 138, 181, 1984.

113. Damodaran, S., Estimation of disulfide bonds using 2-nitro-5-thiosulfobenzoic acid: limitations, *Anal Biochem.*, 145, 200, 1985.

114. Giulivi, C. and Davies, K. J. A., Dityrosine: a marker for oxidatively modified proteins and selective proteolysis, *Methods Enzymol.*, 233, 363, 1994.

115. Bertram, H. C., Andersen, R. H., and Andersen, H. J., Development in myofibrillar water distribution in two pork qualities during 10-month freezer storage, *Meat Sci.*, 75, 128, 2007.

116. Ooizumi, T. and Xiong, Y. L., Identification of crosslinking site(s) of myosin heavy chains in oxidatively stressed chicken myofibrils, *J. Food Sci.*, 71, 196, 2006.

Chapter 14

Aroma

J. Stephen Elmore

Contents

14.1 Introduction

More than 1000 compounds have been identified in cooked meat aroma [1], with beef being by far the most widely studied meat. These compounds are of low molecular weight (<300 Da), volatile, and present at very low concentrations (parts per million or less). Most are of limited solubility in water, with the exception of volatile organic acids, for example, butyric acid; aromatic phenols, for example, eugenol; and sugar-derived compounds, such as diacetyl, maltol, and furaneol.

It is often necessary to characterize meat aroma for reasons that will be described later in the chapter. In such a case, one would extract the volatile material from the meat matrix, concentrate it, and attempt to separate and identify the individual components. By far, the most widely used technique for the separation and identification of aroma compounds in extracts is gas chromatography–mass spectrometry (GC-MS), and unless stated otherwise, it can be assumed that, in all examples discussed in this chapter, GC-MS was used. As far as possible, the extract should contain only the volatile components, in the same relative proportions as in the meat itself, without the introduction of artifacts. However, to obtain such an extract is an extremely difficult task; therefore, numerous complementary extraction techniques exist, which allow the flavor chemist to obtain a complete knowledge of compounds that are present in meat aroma. The components of meat, which are responsible for its aroma, are present in extremely small quantities, compared with the major constituents, which include primarily water. A number of isolation techniques exist, all based on utilizing the physical properties of the aroma compounds to separate them from the food matrix and from water. This chapter describes the most widely used isolation techniques and their application in the analysis of meat aroma.

Often, a detailed analysis of individual compounds may not be necessary. An analytical technique that provides a "fingerprint" of the sample under study may be all that is required, for example, to distinguish meat from different breeds or diets, or to determine whether a piece of meat is of sufficient freshness. Electronic noses are detectors that use gas sensor arrays or mass spectrometers, the latter sometimes known as MS-noses, to discriminate between samples, without the need for an extraction or separation. This chapter also discusses the use of electronic noses to discriminate between meat samples.

14.2 Reasons for Studying Meat Aroma

The meat industry, like other food producers, is interested in reducing production costs, increasing throughput, and maximizing profit. As a result of increased consumer awareness, producers have been obliged to address issues of sensory and nutritional quality; legislation regarding, for example, animal welfare, also has to be observed. These issues have provided the background for much of the published work on meat aroma analysis.

Studies on meat aroma can be broadly divided into four areas:

1. Identification of those compounds that are important in desirable cooked meat flavor
2. Identification of compounds that give undesirable aroma and flavor to cooked meat
3. Examination of how different pre- and postslaughter treatments may affect aroma volatiles
4. Examination of how different cooking treatments may affect aroma volatiles

Furthermore, much work has been carried out using model systems. These are simple combinations of chemicals, which are regarded as important in meat aroma formation; of particular interest is the reaction between cysteine and ribose.

14.2.1 Identification of Those Compounds That Are Important in Desirable Cooked Meat Flavor

Much of the work in this area has used the technique of aroma extract dilution analysis (AEDA), which is described later in the chapter. 2-Methyl-3-furanthiol and its corresponding disulfide, (*E*)-2-nonenal and (*E,E*)-2,4-decadienal, have been shown to be important in cooked beef flavor [2]; 2-methyl-3-furanthiol and (*E,E*)-2,4-decadienal were also found to be the major character impact components of cooked chicken aroma [3]; 4-hydroxy-2,5-dimethyl-3(2*H*)-furanone (furaneol) and 12-methyltridecanal are key aroma components of stewed beef juice [4], whereas furaneol, along with 2-acetyl-2-thiazoline, is among the character impact compounds of roast beef [5]; alkylpyrazines play an important part in fried beef flavor [6]. Methanethiol has been shown to be important in cooked beef, pork, and chicken, whereas 2-methyl-3-furanthiol and 2-furanmethanethiol were both important contributors to cooked beef and pork aroma [7]. Rota and Schieberle [8] showed that furaneol was a key contributor to cooked mutton aroma, but did not contribute to raw mutton aroma, whereas 4-ethyloctanoic acid and *trans*-4,5-epoxy-(*E*)-2-decenal were important in both raw and cooked mutton aroma.

14.2.2 Identification of Compounds That Give Undesirable Aroma and Flavor to Cooked Meat

Three compounds that have been widely studied, with regard to their contribution to the phenomenon known as boar taint, an unpleasant odor of boar found in pork from a proportion of uncastrated male pigs, are 5α-androst-16-en-3-one, indole, and skatole (3-methylindole). These compounds are primarily located in adipose tissue and, as they have a relatively high boiling point, have been analyzed by both GC and high-performance liquid chromatography (HPLC), a technique more usually applied to nonvolatile compounds [9]. Skatole and indole have also been implicated in the unpleasant "pastoral" aromas of grass-fed animals, along with (*Z*)-4-heptenal and 4-methylphenol [10].

Branched-chain fatty acids, particularly 4-methyloctanoic and 4-ethyloctanoic, have been shown to contribute a sweaty aroma in both sheep and goat meat [11], which may be responsible for the limited appeal of these meats, compared to beef, where these compounds are absent. These components are usually isolated by techniques appropriate for fatty acids, for example, solvent extraction followed by preparation of methyl esters.

Warmed-over flavor is a phenomenon observed in reheated meat, the aroma of which is characterized as "rancid" and "stale." Konopka and Grosch [12] reported that warmed-over flavor in boiled beef, stored for 48 h at 4°C, was due primarily to *trans*-4,5-epoxy-(*E*)-2-decenal and (*E,E*)-2,4-decadienal; Kerler and Grosch [13] identified (*E,E*)-2,4-decadienal, butyric acid, and furaneol as key compounds in fresh boiled chicken aroma, whereas hexanal, butyric acid, and 1-octen-3-one were the key compounds in chicken reheated after 48 h at 4°C.

14.2.3 Measuring the Effect of Pre- and Postslaughter Treatments

Many researchers have examined the effect of different diets on cooked meat composition, particularly comparing the effects of forage and cereal diets on ruminants, and several reviews have been published in this area [14–17]. Forage-based diets (grass, silage) are relatively high in n-3 polyunsaturated fatty acids (PUFA), whereas cereal-based diets are relatively high in n-6 PUFA, and this is reflected in the aroma composition of the cooked meat. The breakdown products of n-6 PUFA, such as 1-octen-3-ol and 2-pentylfuran, are at elevated levels in grilled meat from cereal-fed animals, whereas breakdown products of n-3 PUFA, such as 1-penten-3-ol and 2-ethylfuran, are at elevated levels in grilled meat from cereal-fed animals [18]. The production of meat high in n-3 PUFA has been achieved by the addition of marine oils to animal feed, and although significant improvements in the fatty acid profile of the meat have been achieved, these are accompanied by undesirable flavor notes, caused by breakdown of PUFA on cooking [19].

The effect of breed on cooked meat aroma composition is generally small, compared to the effect of diet. However, relatively large effects in aroma compounds were observed when the pressure-cooked meat from conventional Suffolk crossbred lambs was compared with that from Soay lambs, which is a semiferal breed, prized for its flavor [20].

Other preslaughter parameters, which have been studied with regard to their effect on cooked meat aroma, are castration and age at slaughter. Animals may become stressed at slaughter, which affects the pH of their meat. This has been shown to have an effect on aroma composition [21]. Post slaughter, meat is aged for up to 21 days, at 4°C, which results in changes in flavor precursors, such as sugars and free amino acids. Preservation techniques, such as curing and smoking, will also have a profound effect on meat aroma.

14.2.4 Effect of Different Cooking Treatments on Aroma Volatiles

The group of MacLeod, at King's College, London, published widely on the effect of different cooking treatments on beef aroma, examining boiled, microwaved, and dry-heated beef [22,23]. Mottram [24] compared the headspace volatiles from grilled, boiled, and roast pork, and found that lipid-derived volatiles predominated in meat cooked under mild conditions, whereas compounds derived from the Maillard reaction were at high concentrations in well-cooked meat.

14.2.5 Reactions Modeling Meat Aroma Formation

Many of the papers published, examining how meat flavor is formed during cooking, have studied the reaction between cysteine and ribose and measured the aroma compounds formed from this reaction. The compounds formed from this reaction appear to be particularly important for the characteristic aroma of meat [1]. Other workers have used this reaction as a starting point, adding compounds, such as PUFA or phospholipids (intramuscular structural lipids of key importance in aroma formation) [25], to study how the effects on meat aroma volatiles caused by dietary changes may be explained [26,27].

14.3 Sample Preparation

Cooked meat flavor has been studied far more than raw meat flavor, although a work by Rota and Schieberle [8] showed that many of the aroma compounds identified as being important in cooked mutton flavor could be isolated from the uncooked meat. Work in the early 1960s suggested that the components contributing to the development of the characteristic species-specific aromas of cooked meat are located in the lipid fraction, whereas the water-soluble fraction contains components contributing to the development of "meaty" aroma and flavor, such as amino acids, peptides, and carbohydrates, particularly ribose [1]. When heated, these compounds participate in the Maillard reaction, generating the color of cooked meat, as well as its aroma [28].

A large number of factors need to be considered when analyzing meat aroma. Meat has a heterogeneous structure. Adipose tissue surrounds muscle, which contains muscle fibers, marbling fat, and connective tissue. When meat is cooked with adipose tissue attached, large amounts of lipid-derived volatiles are generated, which may obscure volatiles responsible for characteristic meatiness. If volatiles derived from the Maillard reaction are being studied, then the meat could be cooked with the adipose tissue removed. More reproducible results might be obtained if a muscle is minced and turned into a burger. It is important to consider sample size before cooking. Although the major muscle of a rib steak from a steer may weigh approximately 120 g after grilling and removal of adipose tissue, the equivalent from that of a lamb may only weigh approximately 25 g. This may not be a large enough sample for the extraction technique that is being used.

Grilling, boiling, pressure-cooking, roasting, and frying are some of the cooking processes, which could be used. The sample could be cooked for a constant time or to a constant internal temperature, or some visual aspect of the cooked meat could be used as a marker; for example, the surface color could be compared with a color chart. A well-done steak will have an aroma profile different from that of a rare steak [24], and the degree of cooking should be considered before the analysis commences. It is important to employ reproducible methods when cooking the sample. This may be difficult when grilling or frying. Pressure-cooking, although not as commonly used in the kitchen as some of the other methods, has an advantage in that there is no sample loss and temperature control is straightforward.

When a cooked sample is analyzed, it is important to extract the sample as soon after cooking as possible. Otherwise, warmed-over flavor may be generated when the meat is reheated. There is no reason why the extraction cannot be carried out at room temperature, although at higher temperatures the amount of volatile material extracted will increase. Meat is normally eaten at around 60°C. Variation within the sample should be minimized by chopping, or mincing or homogenization and appropriate replication should be performed, so that effective comparison between treatments may be achieved. Large variations may exist between the aroma compositions of meat samples from animals fed the same controlled diet, and at least four analyses per treatment would always be appropriate when studying cooked meat.

14.4 Aroma Extraction Methods

Several methods exist to isolate the aroma volatile from a food. Of course, some will be more appropriate for the analysis of meat than others. Reineccius [29] provides a comprehensive discussion of all the commonly used aroma extraction techniques. Some of these techniques are not suitable for meat aroma analysis. For example, stir bar sorptive extraction and immersion solid-phase microextraction (SPME) are only suitable for low-fat liquids or slurries, and are not discussed in this chapter.

14.4.1 Choosing an Extraction Method

When choosing an appropriate extraction technique for a particular analysis, there are numerous considerations. First and foremost, what is the purpose of the analysis? If accurate quantification of one particular aroma compound is required, whether it is a desirable aroma attribute, or a taint or off-flavor, the extraction method should be selected to maximize the extraction of that particular compound, without generating that compound *in situ*. However, if the full aroma profile of the cooked meat is required, another extraction method may be more suitable. Are polar volatiles of interest? Perhaps, very low-boiling compounds of crucial importance to meat aroma, such as ammonia and hydrogen sulfide, may be those that need to be studied.

What do you want to do with your aroma extract once you have it? Some extraction techniques only provide enough extract for one analysis, whereas others provide a liquid sample that can be used for several experiments. Maybe the laboratory budget can stretch to sophisticated automated extraction systems, or precision glassware and high vacuum pumps, GC-MS systems with autosamplers, etc. Are you looking for particular compounds present in cooked meat at very low levels or perhaps a simple manual extraction may provide all you need to know? Furthermore, what experience does the scientist performing the extraction have? Some extraction techniques are far simpler to use than others.

14.4.2 Solvent Extraction

Direct extraction of meat with an organic solvent is of limited use because the extract will contain much nonvolatile matter, particularly lipid. However, some workers have used supercritical carbon dioxide as a solvent for the extraction of aromas. Its solvating qualities can be altered by changing the pressure or temperature at which the extraction takes place, and under ideal conditions, supercritical carbon dioxide exhibits a strong affinity for most aroma compounds, while most nonvolatile constituents are insoluble. The ease of removal of the solvent, after extraction, to give a concentrated aroma extract is another attractive feature of supercritical carbon dioxide extraction. Um et al. [30] quantified 66 compounds in heated beef fat, including many relatively high-boiling compounds, such as long-chain lactones, methyl esters, alkanes, aldehydes, and ketones. King et al. [31] used supercritical carbon dioxide extraction to obtain a fraction from raw beef, which was assumed to be lipid. The volatile compounds from this fraction were immediately swept onto a Tenax trap for analysis by GC and GC-MS. Eighty-six compounds were identified, many similar to those reported by Um et al. [30].

14.4.3 Simultaneous Distillation Extraction

Steam distillation finds application in the analyses of volatiles from beverages and high-water-content foods, although it is less applicable to fats and oils. It has the disadvantage that large quantities of aqueous distillate require further extraction with a solvent, to separate the volatiles from the water. Concentration of the extract is then necessary. The formation of artifacts may also be a problem.

One of the most widely used techniques in aroma analysis combines steam distillation with solvent extraction in a Likens–Nickerson apparatus (Figure 14.1), which was first reported in 1964 for the extraction of hop oil [32]. The extracting solvent is immiscible with and less dense than water. Upon heating, volatile compounds in the steam are transferred to the solvent and both liquids condense. The glassware is constructed so that both solvent and water are returned to their starting vessels. After an extraction time of 1–12 h, the extract is collected and dried, either using anhydrous sodium sulfate, or by freezing and decanting the solvent from the ice. The extract is then concentrated before analysis to a volume of approximately 0.1 mL; a low-boiling extracting solvent is therefore desirable, so that it can be removed without substantial losses of compounds of interest. In addition, the solvent should be of high purity, so that impurities do not become major chromatographic peaks, when the extract is concentrated. Appropriate solvents, which have been widely used, are pentane, diethyl ether, or a combination of the two. Solvents denser than water, for example, dichloromethane, could be used in a modified apparatus.

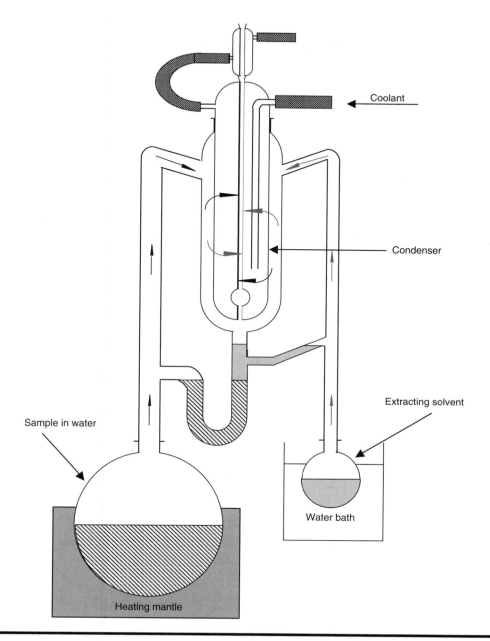

Coolant

Condenser

Extracting solvent

Sample in water

Water bath

Heating mantle

Figure 14.1   Likens–Nickerson apparatus for simultaneous distillation-extraction.

Simultaneous distillation-extraction (SDE) has been widely used for the analysis of cooked meat volatiles. In SDE, the sample is boiled for 1–2 h and hence precooking may not be necessary; the meat is usually minced, to maximize the surface area for extraction. MacLeod and Coppock used SDE, with 2-methylbutane and a range of extraction times, to examine the aroma of cooked beef [22,33]. Mussinan et al. [34] used SDE with diethyl ether to isolate 33 pyrazines from pressure-cooked beef, whereas Ohnishi and Shibamoto [35] used SDE with dichloromethane to identify 112 compounds in heated beef fat. Raes et al. [36] also used SDE with dichloromethane and showed that they could distinguish four breeds of cattle by the volatile compounds from their grilled steaks. Ramarathnam et al. [37] using pentane as extracting solvent, showed that the aroma extracts of cured beef, pork, and chicken contained far lower levels of lipid oxidation products than their uncured equivalents. Madruga et al. [38] studied the effects of castration and age of slaughter on goat meat aroma, extracting with pentane:ether (9:1) for 2 h and identifying 108 volatile compounds.

Although SDE has been used for many years, it has several advantages compared to other commonly used extraction techniques, such as those involving headspace collection on a polymer, followed by heat desorption into a GC injection port. Efficient stripping of volatiles from foods allows quantitative recoveries to be achieved for many compounds [39]. The aroma extract is obtained in a solvent; therefore, many injections can be performed from one extraction. Hence, one sample could provide material for GC, GC-MS, and quantitative GC-olfactometry (GC-O) techniques, such as CharmAnalysis™ (Datu Inc., Geneva, New York) and AEDA [40]. Fractionation of the extract can be carried out, resulting in increased separation of the components in the extract, facilitating the identification of minor components of the extract. Werkhoff et al. [41] used medium-pressure liquid chromatography to separate and identify numerous sulfur compounds in cooked beef, pork, and chicken. Gasser and Grosch [2] washed an SDE extract of cooked beef with acidic and basic buffer solutions to obtain three fractions. The neutral fraction was fractionated further, using silica gel chromatography. The same authors used similar methodology [3] to identify the key odorants of cooked chicken.

As with all aroma extraction techniques, SDE has drawbacks. When the extract is concentrated, by distilling off the solvent, low-boiling volatile compounds can be lost. These compounds include some present at high levels in headspace extracts of cooked meat, such as 2-butanone, 2-pentanone, 2- and 3-methylbutanal, diacetyl, 1-propanol, and 1-penten-3-ol. Artifacts can be formed as a result of the high temperatures used. Mottram and Puckey [42] found 2-methyl-3-nitro-2-butanol and 2-methyl-3-nitro-2-butyl nitrate in an SDE extract of bacon. These compounds were formed from the reaction between 2-methyl-2-butene, a solvent impurity, and breakdown products from nitrite, the curing agent. In addition, volatiles can be generated when samples are overcooked during extraction, for example, enhanced lipid oxidation [43].

If SDE is carried out under reduced pressure, thermal degradation of labile components can be diminished. By maintaining the system under a static vacuum, sample loss is reduced and higher boiling solvents, such as heptane and octane, can be used. In fact, extraction at room temperature is possible. An excellent discussion of the SDE technique is available in Ref. 39.

14.4.4 High Vacuum Distillation/Solvent-Assisted Flavor Evaporation

Because SDE may lead to artifact formation and overcooking, a high vacuum transfer technique was developed in the early 1990s as a way of removing the volatile aroma compounds from solvent extracts of food materials. A series of cold traps was used to collect the volatile material after it had sublimed. This technique was shown to be more effective than SDE for the collection of high-boiling polar compounds, such as furaneol. It was used to determine the potent odorants of roast beef [5] and also to examine the volatile compounds responsible for warmed-over flavor in beef, pork, and chicken [12,44]. However, there were numerous drawbacks with the technique [45] and a robust alternative known as solvent-assisted flavor evaporation (SAFE) was developed to supersede it (Figure 14.2). Although high vacuum transfer and SAFE are similar techniques in principle, greater thermal control and a more compact arrangement of the glassware mean that

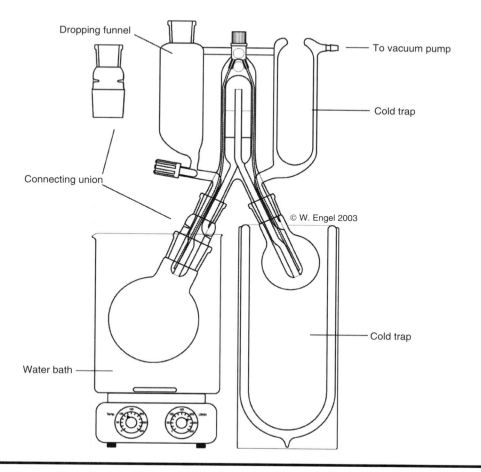

Dropping funnel

To vacuum pump

Cold trap

Connecting union

© W. Engel 2003

Water bath

Cold trap

Figure 14.2   Solvent-assisted flavor evaporation (SAFE) apparatus.

SAFE is more efficient than high vacuum transfer, resulting in higher yields of high-boiling and polar compounds. It can also be used directly on the food, with no need of a solvent extraction step, producing an extract with typical aroma. Hence, the time of extract preparation can be substantially reduced.

Since SAFE has only been recently developed, it has not been used regularly for the analysis of meat aroma [8]. However, it is proving to be a popular technique for aroma analysis and may well be the most effective aroma isolation technique available. The main drawback to SAFE is the high cost, primarily due to the need for a high vacuum (10^{-3} Pa) pump.

14.4.5   Headspace Analysis

The concentration of volatiles in the headspace vapor above a food or beverage is very low. Hence, injecting the vapors above a food sample onto a GC column is rarely used. Furthermore, water from the sample will often cause interference. Commercial automated headspace analyzers are available, usually for quality control purposes, allowing unattended operation of up to 50 samples. These systems usually involve a pressurization step that increases the amount of volatile material in the headspace of the sample vial. Such a system was used by Guth and Grosch [4] to identify the character impact odorants of stewed beef juice [4]. An automated headspace sampler was used to measure quantitative differences between cooked beef samples from animals fed grain or forage [46], and also to examine the effect of dietary linoleic acid on the composition of cooked pork aroma [47].

A concentrated headspace extract can be obtained by passing a stream of inert gas (nitrogen or helium) over the sample and condensing the volatiles in a series of traps cooled by ice, solid carbon dioxide, or liquid nitrogen. Extraction of the condensate with a small amount of a suitable solvent provides an aroma extract suitable for chromatographic analysis. This technique, with pentane as solvent, was used to examine cured and uncured pork, beef, and chicken [48,49]. The solvent extract obtained using this technique can be fractionated; 130 volatile compounds were thus identified in a fried chicken extract [50]. Brennand and Lindsay [51] used a vacuum pump rather than an inert gas flow to examine volatile free fatty acids in cooked mutton. Maruri and Larick [52] examined the effects of diet on beef aroma by sweeping the volatiles from heated beef fat directly onto the front of a GC column, which was cooled to −30°C.

14.4.6   Adsorption

Headspace aroma volatiles can also be collected on suitable adsorbent materials, the most widely used of which is Tenax TA, a porous polymer resin based on 2,6-diphenylene oxide. These materials readily adsorb volatiles, although having little affinity for water, making them useful particularly in the analysis of samples with high water content.

In a typical collection, purified inert gas sweeps the volatiles from the sample flask into a small tube containing 10–200 mg of the adsorbent (Figure 14.3), which is usually called a "trap." Typically, at least five volumes of headspace should be collected from the sample vessel. Adsorbed volatiles can be heat desorbed directly onto a gas chromatographic column by placing the trap

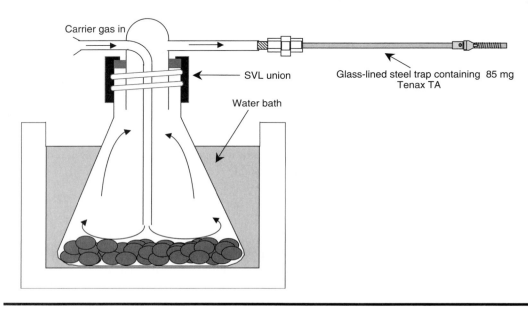

Figure 14.3   Headspace adsorption on Tenax TA.

in a specially modified injection port, thus avoiding loss of components or unnecessary dilution. Cooling the front of the column with solid carbon dioxide or liquid nitrogen during this desorption will avoid any loss in chromatographic resolution. Alternatively, volatiles may be removed from the trap by passing a small amount of a low-boiling solvent through the adsorbent. Concentration of the collected solution by evaporation of the solvent provides an aroma extract for GC.

Variations of the headspace adsorption technique exist. For liquids or porous samples, the collecting gas could be passed through the sample. Closed-loop stripping is another adsorbent-based technique, featuring a closed system where a pump draws aroma compounds from above the sample and onto the adsorbent trap. A battery-powered pump can be used to analyze samples outdoors, such as vapors from an outflow or fruit on the tree.

Headspace adsorption has been used to study cooked meat aroma since the 1980s [53]. It has probably been used more than any other aroma extraction technique for the analysis of meat aroma and continues to be widely used. Headspace adsorption on Tenax TA was used to identify 2-methyl-3-(methylthio)furan, a meaty character impact compound, for the first time in cooked beef [54]. Tenax TA was also used to examine the effects of diet, breed, and slaughter age on the aroma volatiles of grilled and pressure-cooked beef [18,19,55], and the effects of diet and breed on the aroma volatiles of grilled and pressure-cooked lamb [20,56]. Braggins [21] showed that 57 volatiles decreased in sheep meat with increasing pH at slaughter [21]; high pH can be caused by preslaughter stress and was associated with undesirable aroma. Cooked beef was "chewed" in a model mouth and the volatiles swept onto Tenax TA. The method was used to compare two diets and three breeds [57]; it was also used to compare beef from conventionally reared animals with beef from animals designated "organic." [58]. Numerous studies on cured meat have been carried out using Tenax TA. Pork sausages cured with nitrite were shown to contain higher amounts of important branched-chain aroma compounds than those cured in nitrite. Also, the type and concentration of starter culture used had a significant effect on the formation of some of these compounds [59,60].

Tenax GC, the forerunner of Tenax TA, was used to examine the effect of diet on the volatile compounds in heated lamb fat [61,62] and was also used to determine the effect of castration on cooked lamb meat and adipose tissue [63]. Headspace adsorption onto Tenax GC, using a vacuum pump, was used to identify numerous heteroatomic compounds in cooked beef and a commercial beef flavor concentrate [64]. In most cases, heat desorption is used to remove the aroma compounds from the trap. However, adsorption on Tenax GC, followed by solvent desorption with hexane, was used to compare the aroma compounds in cooked steaks from two herds of the premium beef breed Wagyu with those of the cooked steaks from three regular U.S. breeds. The effects of refrigerated storage and cooking method were examined for each sample set [65].

Automated devices exist for headspace adsorption, followed by thermal desorption in a dedicated injection port. Where samples have been extracted manually, multitrap autosamplers can be used to heat desorb the volatile compounds, allowing round-the-clock GC-MS analysis. An automated device was used to detect more than 250 volatiles in rendered lamb fat [66] from animals fed pasture and grain diets. Trapping on Tenax TA, followed by desorption on a multi-trap automatic thermal desorption system, was used to compare the meat from two pig genotypes, aged for 2, 15, and 22 days, and fried at two temperatures [67].

Headspace adsorption on Tenax is a desirable technique because it is sensitive and extracts a wide boiling point range of volatile compounds; artifact formation is minimal, as extraction is carried out under inert gas flow. However, a dedicated injection system for traps may be expensive, especially if automated extraction is desired. Normally, only one GC analysis is obtained from each extraction, unless solvents are used to desorb the contents of the traps.

14.4.7 Solid-Phase Microextraction

A very popular and simple-to-use technique, SPME, uses a small fused silica fiber, coated in an adsorbent material, mounted inside a syringe-like device (Figure 14.4). The needle is pushed through a septum, and the fiber is exposed to the headspace above the food or beverage sample, which is sealed in a suitable container. Volatile compounds are adsorbed onto the fiber and at the end of the extraction, the fiber can be removed from the sample vessel and directly desorbed into

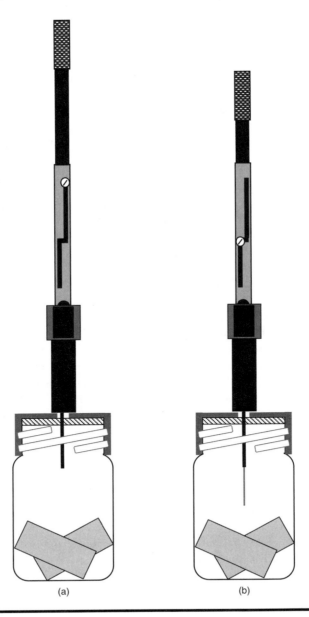

(a) (b)

Figure 14.4   Solid phase microextraction (SPME): (a) fiber inside syringe barrel; (b) fiber exposed during extraction.

the split/splitless injector of a gas chromatograph. The injector of the GC contains a very narrow quartz liner, which helps to focus the volatile compounds at the front of the GC column. Alternatively, cryofocusing of the aroma volatiles on the front of the column can be performed to prevent peak broadening. SPME has evolved rapidly since its introduction in the early 1990s [68]. It can be automated, and its ease of use, relatively low cost, and affinity for a large range of compounds mean that it has become a widely used technique for the isolation of aroma volatiles. The technique is evolving and, in recent years, related techniques, such as stir-bar sorptive extraction and solid-phase dynamic extraction, have been developed [69].

Numerous stationary phases have been used as coatings for SPME fibers, either on their own or in combination. Coatings may be absorptive—volatiles are bound to the surface of the fiber—or adsorptive—volatiles are trapped within pores in the stationary phase [70]. Popular absorptive stationary phases include polydimethylsiloxane (PDMS), whereas Carboxen and divinylbenzene are adsorptive phases.

Fibers coated with a combination of Carboxen and PDMS were shown to be effective for the examination of pressure-cooked beef flavor. Although less sensitive than headspace adsorption on Tenax TA, SPME with this fiber coating extracted polar volatiles, such as furaneol, which could not be extracted by headspace adsorption [71]. SPME with a Carboxen/PDMS fiber was compared with SDE, using diethyl ether as extracting solvent, and headspace adsorption on Tenax TA, for the analysis of traditional Chinese Nanjing-marinated duck [72]. In this case, SPME was more sensitive than headspace adsorption; the SDE extract contained a higher proportion of high-boiling volatile compounds than SPME and it was suggested that SDE and SPME would complement each other as meat aroma analysis techniques.

The aroma of pressure-cooked pork was analyzed using two SPME fibers simultaneously: a Carboxen/PDMS fiber, which absorbed a relatively high proportion of low-boiling volatile compounds, and a divinylbenzene/Carboxen/PDMS fiber, which absorbed a relatively high proportion of higher-boiling volatile compounds. The fibers were desorbed sequentially onto the front of a GC column; the aroma compounds were held at the start of the column by cryofocusing, producing a GC trace that contained a wider range of volatile compounds than when each fiber was used on its own [73]. The same two fibers were compared for the analysis of cooked beef aroma, and again a relatively high proportion of higher-boiling volatile compounds were absorbed by divinylbenzene/Carboxen/PDMS fiber [74].

SPME with a PDMS fiber was used to analyze sulfur compounds in the condensate collected from a cooking beef roast. Using two-dimensional GC and time-of-flight (TOF) MS, more than 50 sulfur compounds were identified, seven for the first time in cooked beef [75]. The important carbonyl compounds (aldehydes and ketones) in cooking beef were analyzed in a similar way, using a divinylbenzene/PDMS fiber. In addition, the aldehydes were also studied as their (2,3,4,5, 6-pentafluorophenyl)hydrazone derivatives [76]. This was achieved by suspending the SPME fiber, before extraction, above a solution of the derivatizing agent (2,3,4,5,6-pentafluorophenyl) hydrazine. The derivatizing agent was adsorbed onto the fiber and then reacted with the aldehydes in the headspace of the cooked beef. The benefits of this procedure were that the carbonyl compounds were well separated from the other compounds in the GC-MS chromatogram. Furthermore, the derivatized compounds possessed mass spectra with stronger molecular ions than the underivatized carbonyls, where a molecular ion is often absent. This meant that unknown aldehydes were easier to identify, as their molecular weights could be ascertained. As the derivatized compounds possessed fragment ions with relatively high masses, they were easier to distinguish from coeluting compounds and background noise, which meant that derivatization was more sensitive for carbonyls than SPME without derivatization.

SPME has been widely used for the analysis of cured meat, particularly dry-cured ham. The effects of pig rearing system and ham ripening time on amino acid-derived volatiles were examined in Iberian ham. All studied volatiles increased with ripening time, although the effect of rearing system was comparatively small [77]. Sánchez-Pena et al. [78] showed that the levels of methylbenzene and 1-octanol in subcutaneous fat, and the levels of 2-octanone and 2-butanone in muscle, could be used to provide unequivocal discrimination of French and Spanish dry-cured hams. Furthermore, levels of 1-octanol in subcutaneous fat and 3-methyl-1-butanol in muscle could be used to distinguish hams of Iberian pigs from those of Large White pigs. N-Nitrosamines are potentially carcinogenic compounds that are formed in meat cured using nitrites. An SPME method for the analysis of nitrosamines in pork sausages, using GC with a thermal energy analyzer, rather than a mass spectrometer, as a detector [79], has been developed.

SPME can be used to analyze off-odors in meat. It was used to show that pigs fed with a diet high in Brussels sprouts produced pork that contained very high levels of dimethyl sulfide, dimethyl disulfide, and dimethyl trisulfide, when compared with the meat of pigs fed a control diet [80]. SPME has also been used to monitor warmed-over flavor formation in turkey breast [81] and was used to measure amines formed by bacterial spoilage in Iberian hams [82].

Like headspace adsorption on Tenax, SPME is a desirable technique because it is sensitive, and extracts a wide boiling point range of volatile compounds. Again artifact formation is minimal. Its major advantage over headspace adsorption is that it can be used with any GC. It is also the easiest aroma extraction technique to use, requiring little training. Method development is straightforward and a choice of stationary phases means that the extraction can be tailored to maximize extraction of desired compounds. Like headspace adsorption with thermal desorption, only one GC analysis is obtained from each extraction.

14.5 Separation and Identification of Aroma Components

To determine the important compounds in an aroma extract, the complex mixture needs to be separated into its components. The amount of isolate is usually small containing many compounds of diverse chemical structures, varying greatly in concentration, and important components are often present in extremely low amounts. The success of any aroma analysis depends primarily on the efficiency of separation and the sensitivity of detection. GC, using bonded phase fused silica capillary columns, is universally used as the separation method in aroma analysis. Such columns can separate complex mixtures and the most commonly used stationary phases are Carbowax 20M, a polar phase, and the two nonpolar phases, 100% poly(dimethylsiloxane) and poly(5% diphenylsiloxane/95% dimethylsiloxane). The retention times of an aroma compound on two columns with different stationary phases, relative to the retention times of a series of straight-chain alkanes, can be helpful in its identification; databases containing retention data for volatile compounds are available [83]. GC is a widely used technique and is not discussed in this chapter.

Structure elucidation of the chromatographically separated components is the next step in the analysis of an aroma isolate. GC-MS allows direct analysis of the separated components and provides the most efficient means of volatile identification. Compounds eluting from the GC column enter the ion source of the mass spectrometer, where they are bombarded with electrons. A compound will fragment and the fragments are separated by their mass-to-charge ratio, resulting in a characteristic spectrum, which will provide structural information.

Several types of mass spectrometer are suitable for the identification of meat aroma compounds, although most of the work discussed in this chapter was performed using quadrupole mass spectrometers, which are the most widely used type for aroma analysis. Quadrupole mass spectrometers, obtaining at least one spectrum per second, are ideal for low-resolution GC-MS of aroma extracts.

Although quadrupoles are by far the most common type of mass spectrometer for GC-MS, some other types are also used [84]. Ion traps offer all of the facilities of the quadrupole, plus MS^n, that is, the trapping of fragments from the first ionization, for further fragmentation, to yield more structural information about unknown compounds. Double-focusing magnetic sector mass spectrometers can acquire accurate mass data, allowing the calculation of empirical formulae for ions in a spectrum, which can be a great asset in the identification of unknown compounds. These machines are relatively expensive compared to quadrupoles and less robust; therefore, they are used far less often for routine flavor analysis. Recently, TOF machines have become increasingly popular as mass spectrometric detectors, with newer models offering rapid scan speeds (up to 500 spectra per second) and high resolution capabilities. High scan speeds are necessary when using fast GC techniques, such as two-dimensional GC (GC × GC). These machines are robust but relatively expensive, although, as they have been introduced relatively recently, they may become cheaper as they become more popular [85].

The characterization of unknown compounds is greatly facilitated by comparing their mass spectra with those of known compounds in compiled libraries, which are supplied with the GC-MS data system. Confirmation of the identity of compounds should always be carried out, preferably by comparing their mass spectra and GC retention times with those of authentic samples.

14.6 Quantification of Aroma Components

Often quantitative information on aroma compounds in a food is needed, for example, when using AEDA to determine the key compounds contributing to the aroma of a food. Quantification is rarely simple because most extraction techniques only remove a proportion of the aroma from the food. Sometimes, the extraction technique may generate aroma compounds, and difficulties in quantification may arise when the compounds are not resolved by GC.

The most effective means of quantification is isotope dilution assay using GC-MS. A known amount of a ^{13}C- or ^{2}H-labeled internal standard is added to a slurry of the food under study to quantify its nonlabeled equivalent. As the labeled and unlabeled aroma compounds possess similar physical properties, the proportion of each extracted from the food will be the same. The relationship between the labeled standard and the compound of interest can be used to calculate accurately the amount of the compound of interest in the food. If the labeled standard is efficiently mixed into the food, so that it is homogeneously distributed, then quantitative extraction of the compound under study is not necessary [86].

Four sulfur-containing compounds with very low odor thresholds were quantified in pressure-cooked beef, pork, lamb, and chicken, using an isotope dilution assay [87]. Isotope dilution assay was also used to quantify eight volatile compounds known to be important in cooked beef aroma in the pressure-cooked meat of pasture-fed and concentrate-fed cattle [88]. 1-Octen-3-one and (E,E)-2,4-decadienal were key odorants in both types of meat, although both were present at higher levels in beef from animals fed on concentrates.

Other quantification methods include the addition of an internal standard, not present in the food, of a chemical composition similar to that of the compound of interest; for example, 2-methylpentanal could be used to quantify hexanal. If the extraction of the compound of interest is quantitative, then solutions of the compound of interest can be used to plot a calibration curve (an external standard), which can then be used to quantify the compound of interest.

Conversely, standards could be added to the extract. The peak area of a known concentration of a standard added to an extract can be compared with the peak areas of all the compounds in the extract, to give an approximate concentration for all of the compounds in the extract. Standards can be injected into traps containing adsorbent and, in the case of SPME, injected onto the GC column immediately before desorption of the fiber. Methanol is a useful solvent for such standards; it can be easily purged from the trap as its affinity for Tenax is very low and as its molecular weight is 32. Data acquisition down to *m/z* 33 will provide enough mass spectral data for successful library searching, without peaks of interest being hidden by a solvent peak.

14.7 Detection of Components of Sensory Significance

GC-O is a widely-used technique for determining components that contribute to aroma. The column effluent is split between a conventional GC detector and a vent to the outside of the oven, where the emerging odors can be smelled and described.

AEDA is a quantitative GC-O technique that has been used many times to estimate the relative contributions of volatile components toward the total aroma quality of cooked meat. The aroma extract under study is diluted twofold and analyzed by GC-O, and then diluted twofold again and again. After a certain number of dilutions of the extract, no aromas will be perceived. The flavor dilution factor for a particular compound is defined as the highest dilution at which that compound can be perceived by GC-O. For example, if the concentration of the extract was halved at each dilution and the seventh dilution was the last at which the compound could be detected, its flavor dilution factor would be 2^7 (128). Hence, if the aroma extract is representative of the food from which it is derived, the most important contributors to the aroma of the food are those with the highest flavor dilution factors. It should be noted that components with high flavor dilution factors might not give GC peaks of any significant size. These flavor dilution factors can be plotted against retention time to give an aromagram for a particular extract [2].

Another way of estimating the contribution of a volatile compound to the aroma of a food is by calculating its odor unit number. The odor unit number of an aroma compound in a food is defined as the *ratio between its concentration in the food and its odor threshold, as measured by sensory evaluation*. Hence, the higher the odor unit number of the compound, the more likely it is to contribute to the overall aroma of the food [89]. Odor unit values were calculated for the potent odorants of stewed beef juice. Twelve of these compounds were combined together in an oil-in-water emulsion to give an aroma similar to that of stewed beef juice [4].

14.8 The Electronic Nose

Originally, the term "electronic nose" was used to describe an array of chemical sensors, connected to a pattern recognition system, which responded to odors passing over it. Different odors cause different responses in the sensors and these responses provide a signal pattern, characteristic of a particular aroma. The computer evaluates the signal pattern and can compare the aromas of different samples, using pattern recognition. Sensors are usually made of metal oxides or organic polymers, although more recently surface acoustic waves and piezoelectric crystals have been used. Difficulties may arise when samples with a high water content are analyzed, as many of the sensors respond strongly toward water, preventing any sample differences from being observed.

More recently, electronic noses based on MS have been developed, which are also known as mass sensors or MS-noses. Volatile compounds are introduced directly into the mass spectrometer, without any preseparation. With these instruments, each mass scanned by the mass spectrometer can be described as a sensor, which detects any ion fragment with that mass. In fact, the mass sensor takes all of the scans that make up a GC-MS run and, then combining these scans, provides a fingerprint of the food under study. The advantage of these machines over conventional electronic noses is that they are less prone to sensor poisoning (due to excess sample), moisture effects, and nonlinearity of signals [90].

Metal oxide sensors successfully discriminated cured pork products containing different microorganisms, either as components of the curing agent or as contaminants [91]. Metal oxide sensors were also able to separate Iberian hams from animals fed forage or acorns, and hams matured for 10 months from those matured for 13 months [92]. Sensor response correlated positively with microbial spoilage in uncooked chicken [93] and vacuum-packed beef [94].

An electronic nose with conducting polymer sensors could be used to measure the degree of boar taint in cooked pork samples [95] and was also used to measure warmed-over flavor in reheated cooked beef [96]. Conducting polymers were also used to distinguish cooked meats from alpaca and llama [97]. Siegmund and Pfannhauser [97], by performing AEDA on an SDE extract, found that hexanal and (*E,E*)-2,4-decadienal were key aroma compounds in warmed-over chicken. They found that GC-MS results obtained for chicken stored over four different time periods correlated well with the results obtained for the same samples using a conducting polymer electronic nose [98].

An ion mobility spectrometer was used as an electronic nose to measure warmed-over flavor in pork. The pork was also analyzed by GC-MS and sensory analysis. Electronic nose data and sensory data correlated well. Linoleic acid breakdown products were shown to be good indices of lipid oxidation [99]. The same electronic nose was also used to measure boar taint in entire male pigs. There were direct relationships between electronic nose data, sensory data, and the concentrations of androstenone and skatole in the samples [100].

Vasta et al. [101] compared the volatiles of raw lamb from animals fed pasture or concentrates using headspace adsorption on Tenax TA and observed more than 30 compounds that were affected by diet. They also combined all the mass spectral data from each GC-MS trace to give what was described as a dynamic headspace-MS fingerprint, which would be similar to the fingerprint obtained by a mass sensor. The fragment ions were then examined in each fingerprint; 60 mass fragments were significantly affected by diet.

14.9 Future Developments

Since the mid-1990s, it has been possible to measure the release of aroma volatiles from chewed food in real time, using MS. As many volatile compounds enter the mass spectrometer at the same time, a soft ionization technique is used, that is, one that favors the formation of a protonated molecular ion, with little additional fragmentation. The two processes most commonly used to achieve this are atmospheric pressure chemical ionization mass spectrometry (APCI-MS) [102] and proton-transfer reaction mass spectrometry (PTR-MS) [103]. A plastic tube is inserted into one nostril and exhaled air passes directly into the mass spectrometer, giving a characteristic sigmoid trace, with troughs during inhalation and peaks during exhalation.

Specific ions are assumed to be associated with particular compounds. Using PTR-MS, 12 ions were monitored in the exhaled air from both the oral and the nasal cavities of a panelist who was eating cooked Wagyu beef. It was found that polar volatiles, such as acetic acid and ethyl acetate, were relatively high in the breath, compared to their amounts in a solvent extract of

Wagyu beef [104]. PTR-MS has also been used to measure volatile indicators of spoilage in beef and pork online, monitoring ions for four-carbon esters (*m/z* 89) and ethanol (*m/z* 47) [105].

The recent paper by Rochat et al. [75] showed the potential of two-dimensional GC, hyphenated to a TOF MS, as an unrivaled technique for the separation of complex mixtures. The technique is extremely sensitive, as a result of low background and exceptionally high peak resolution, allowing thousands of peaks to be separated in one GC-MS trace. At present, the cost of such equipment may place it beyond the reach of most analytical laboratories, but its potential is quite evident.

SAFE appears to be the most effective aroma extraction technique currently available; yields are higher than any other technique, and sample degradation does not occur readily. Its potential has been shown in the analysis of numerous foods, and it is inevitable that it will be used for the study of meat aroma in the near future.

It is likely that the three techniques described in this section will become increasingly popular for the analysis of meat aroma, although many of the other techniques described in this chapter will continue to provide useful information on meat aroma for many years to come.

References

1. Mottram, D.S., Flavour formation in meat and meat products: a review, *Food Chem.*, 62, 415, 1998.
2. Gasser, U. and Grosch, W., Identification of volatile flavour compounds with high aroma values from cooked beef, *Z. Lebensm. -Unters. -Forsch.*, 186, 489, 1988.
3. Gasser, U. and Grosch, W., Primary odorants of chicken broth—a comparative study with meat broths from cow and ox, *Z. Lebensm. -Unters. -Forsch.*, 190, 3, 1990.
4. Guth, H. and Grosch, W., Identification of the character impact odorants of stewed beef juice by instrumental analyses and sensory studies, *J. Agric. Food Chem.*, 42, 2862, 1994.
5. Cerny, C. and Grosch, W., Evaluation of potent odorants in roasted beef by aroma extract dilution analysis, *Z. Lebensm. -Unters. -Forsch.*, 194, 322, 1992.
6. Specht, K. and Baltes, W., Identification of volatile flavor compounds with high aroma values from shallow-fried beef, *J. Agric. Food Chem.*, 42, 2246, 1994.
7. Kerscher, R. and Grosch, W., Comparison of the aromas of cooked beef, pork and chicken, in *Frontiers of Flavour Science*, Schieberle, P. and Engel, K.-H., Eds., Deutsche Forschunganstalt für Lebensmittelchemie, Garching, 2000, 17.
8. Rota, V. and Schieberle, P., Changes in key odorants of sheep meat induced by cooking, in *Food Lipids: Chemistry, Flavor, and Texture (ACS Symposium Series 920)*, Shahidi, F. and Weenen, H., Eds., American Chemical Society, Washington, DC, 2006, 73.
9. Rius, M.A. and García-Regueiro, J.A., Skatole and indole concentrations in longissimus dorsi and fat samples of pigs, *Meat Sci.*, 59, 285, 2001.
10. Young, O.A. et al., Animal production origins of some meat color and flavor attributes, in *Quality Attributes of Muscle Foods*, Xiong, Y.L., Ho, C.-T., and Shahidi, F., Eds., Kluwer Academic, New York, 1999, 11.
11. Young, O.A. and Braggins, T.J., in *Flavor of Meat, Meat Products and Seafood*, Shahidi, F., Ed., Blackie Academic and Professional, London, 1998, 101.
12. Konopka, U.C. and Grosch, W., Potent odorants causing the warmed-over flavor in boiled beef, *Z. Lebensm. -Unters. -Forsch.*, 193, 123, 1991.
13. Kerler, J. and Grosch, W., Character impact odorants of boiled chicken: changes during refrigerated storage and heating, *Z. Lebensm. -Unters. -Forsch.*, 205, 232, 1997.
14. Melton, S.L., Effects of feeds on flavor of red meat: a review, *J. Anim. Sci.*, 68, 4421, 1990.
15. Muir, P.D., Deaker, J.M., and Bown, M.D., Effects of forage- and grain-based feeding systems on beef quality: a review, *N. Z. J. Agric. Res.*, 41, 623, 1998.

16. Moloney, A.P. et al., Producing tender and flavoursome beef with enhanced nutritional characteristics, *Proc. Nutr. Soc.*, 60, 221, 2001.
17. Vasta, V. and Priolo, A., Ruminant fat volatiles as affected by diet. a review, *Meat Sci.*, 73, 218, 2006.
18. Elmore, J.S. et al., A comparison of the aroma volatiles and fatty acid compositions of grilled beef muscle from Aberdeen Angus and Holstein-Friesian steers fed diets based on silage or concentrates, *Meat Sci.*, 68, 27, 2004.
19. Elmore, J.S. et al., Effect of the polyunsaturated fatty acid composition of beef muscle on the profile of aroma volatiles, *J. Agric. Food Chem.*, 47, 1619, 1999.
20. Elmore, J.S. et al., The effects of diet and breed on the major volatiles present in lamb aroma, *Meat Sci.*, 55, 149, 2000.
21. Braggins, T.J., Effect of stress-related changes in sheepmeat ultimate pH on cooked odor and flavor, *J. Agric. Food Chem.*, 44, 2352, 1996.
22. MacLeod, G. and Coppock, B.M., A comparison of the chemical composition of boiled and roast aromas of heated beef, *J. Agric. Food Chem.*, 25, 113, 1977.
23. MacLeod, G. and Ames, J.M., Capillary gas chromatography-mass spectrometric analysis of cooked ground beef aroma, *J. Food Sci.*, 51, 1427, 1986.
24. Mottram, D.S., The effect of cooking conditions on the formation of volatile heterocyclic compounds in pork, *J. Sci. Food Agric.*, 36, 377, 1985.
25. Mottram, D.S. and Edwards, R.A., The role of triglycerides and phospholipids in the aroma of cooked beef, *J. Sci. Food Agric.*, 34, 517, 1983.
26. Elmore, J.S. et al., Effect of lipid composition on meat-like model systems containing cysteine, ribose, and polyunsaturated fatty acids, *J. Agric. Food Chem.*, 50, 1126, 2002.
27. Farmer, L.J., Mottram, D.S., and Whitfield, F.B., Volatile compounds produced in Maillard reactions involving cysteine, ribose and phospholipid, *J. Sci. Food Agric.*, 49, 347, 1989.
28. Nursten, H.E., *The Maillard Reaction: Chemistry, Biochemistry and Implications*, Royal Society of Chemistry, Cambridge, 2005.
29. Reineccius, G., *Flavor Chemistry and Technology*, 2nd edn., Taylor & Francis, London, 2005.
30. Um, K.W. et al., Concentration and identification of volatile compounds from heated beef fat using supercritical CO_2 extraction-gas liquid chromatography/mass spectrometry, *J. Agric. Food Chem.*, 40, 1641, 1992.
31. King, M.-F. et al., Isolation and identification of volatiles and condensable material in raw beef with supercritical carbon dioxide extraction, *J. Agric. Food Chem.*, 41, 1974, 1993.
32. Likens, S.T. and Nickerson, G.B., Detection of certain hop oil constituents in brewing products, *Proc. Am. Soc. Brew. Chem.*, 1964, 5.
33. MacLeod, G. and Coppock, B.M., Volatile flavour components of beef boiled conventionally and by microwave radiation, *J. Agric. Food Chem.*, 24, 835, 1976.
34. Mussinan, C.J., Wilson, R.A., and Katz, I., Isolation and identification of pyrazines present in pressure-cooked beef, *J. Agric. Food Chem.*, 21, 871, 1973.
35. Ohnishi, S. and Shibamoto, T., Volatile compounds from heated beef fat and beef fat with glycine, *J. Agric. Food Chem.*, 32, 987, 1984.
36. Raes, K. et al., Meat quality, fatty acid composition and flavour analysis in Belgian retail beef, *Meat Sci.*, 65, 1237, 2003.
37. Ramarathnam, N., Rubin, L.J., and Diosady, L.L., Studies on meat flavor. 2. A quantitative investigation of the volatile carbonyls and hydrocarbons in uncured and cured beef and chicken, *J. Agric. Food Chem.*, 39, 1839, 1991.
38. Madruga, M.S. et al., Castration and slaughter age effects on panel assessment and aroma compounds of the "Mestico" goat meat, *Meat Sci.*, 56, 117, 2000.
39. Chaintreau, A., Simultaneous distillation-extraction: from birth to maturity—review, *Flavour Fragr. J.*, 2001, 16, 136.

40. Acree, T.E., Bioassays for flavour, in *Flavor Science: Sensible Principles and Techniques*, Acree, T.E. and Teranishi, R., Eds., American Chemical Society, Washington, DC, 1993, 1.

41. Werkhoff, P.B. et al., Flavor chemistry of meat volatiles: new results on flavor components from beef, pork and chicken, in *Recent Developments in Flavor and Fragrance Chemistry*, Hopp, R. and Mori, K., Eds., VCH, Weinheim, 1993, 183.

42. Mottram, D.S. and Puckey, D.J., Artefact formation during the extraction of bacon volatiles in a Likens-Nickerson apparatus, *Chem. Ind.*, 1978, 385, 1978.

43. Elmore, J.S., Mottram, D.S., and Dodson, A.T., Meat aroma analysis: problems and solutions, in *Handbook of Flavor Characterization: Sensory Analysis, Chemistry, and Physiology*, Deibler, K.D. and Delwiche, J., Eds., Marcel Dekker, New York, 2004, 295.

44. Konopka, U.C., Guth, H., and Grosch, W., Potent odorants formed by lipid peroxidation as indicators of the warmed-over flavour (WOF) of cooked meat, *Z. Lebensm. -Unters. -Forsch.*, 201, 339, 1995.

45. Engel, W., Bahr, W., and Schieberle, P., Solvent assisted flavour evaporation—a new and versatile technique for the careful and direct isolation of aroma compounds from complex food matrices, *Eur. Food Res. Technol.*, 209, 237, 1999.

46. Larick, D.K. and Turner, B.E., Headspace volatiles and sensory characteristics of ground beef from forage-fed and grain-fed heifers, *J. Food Sci.*, 55, 649, 1990.

47. Larick, D.K. et al., Volatile compound content and fatty acid composition of pork as influenced by linoleic acid content of the diet, *J. Animal Sci.*, 70, 1397, 1992.

48. Ramarathnam, N., Rubin, L.J., and Diosady, L.L., Studies on meat flavor. 3. A novel method for trapping volatile components from uncured and cured pork, *J. Agric. Food Chem.*, 41, 933, 1993.

49. Ramarathnam, N., Rubin, L.J., and Diosady, L.L., Studies on meat flavor. 4. Fractionation, characterization and quantitation of volatiles from uncured and cured beef and chicken, *J. Agric. Food Chem.*, 41, 939, 1993.

50. Tang, J. et al., Isolation and identification of volatile compounds from fried chicken, *J. Agric. Food Chem.*, 31, 1287, 1983.

51. Brennand, C.P. and Lindsay, R.C., Influence of cooking on concentrations of species-related flavor compounds in mutton, *Lebensm.-Wiss. –Technol.*, 25, 357, 1992.

52. Maruri, J.L. and Larick, D.K., Volatile concentration and flavor of beef as influenced by diet, *J. Food Sci.*, 57, 1275, 1992.

53. Galt, A.M. and MacLeod, G., Headspace sampling of cooked beef aroma using Tenax GC, *J. Agric. Food Chem.*, 32, 59, 1984.

54. MacLeod, G. and Ames, J.M., 2-Methyl-3-(methylthio)furan: a meaty character impact aroma compound identified from cooked beef, *Chem. Ind.*, 175, 1986.

55. Elmore, J.S. et al., The effect of diet, breed and age of animal at slaughter on the volatile compounds of grilled beef, in *Food Lipids: Chemistry, Flavor, and Texture (ACS Symposium Series 920)*, Shahidi, F. and Weenen, H., Eds., American Chemical Society, Washington, DC, 2006, 35.

56. Elmore, J.S. et al., Dietary manipulation of fatty acid composition in lamb meat and its effect on the volatile aroma compounds of grilled lamb, *Meat Sci.*, 69, 233, 2005.

57. Machiels, D. et al., Gas chromatography-olfactometry analysis of the volatile compounds of two commercial Irish beef meats, *Talanta*, 60, 755, 2003.

58. Machiels, D., Istasse, L., and van Ruth, S.M., Gas chromatography-olfactometry analysis of beef meat originating from differently fed Belgian Blue, Limousin and Aberdeen Angus bulls, *Food Chem.*, 86, 377, 2004.

59. Tjener, K. et al., Growth and production of volatiles by *Staphylococcus carnosus* in dry sausages: influence of inoculation level and ripening time, *Meat Sci.*, 67, 447, 2004.

60. Olesen, P.T., Meyer, A.S., and Stahnke, L.H., Generation of flavour compounds in fermented sausage—the influence of curing ingredients, *Staphylococcus* starter culture and ripening time: influence of inoculation level and ripening time, *Meat Sci.*, 67, 675, 2004.

61. Suzuki, J. and Bailey, M.E., Direct sampling capillary GLC analysis of flavor volatiles from ovine fat, *J. Agric. Food Chem.*, 33, 343, 1985.

62. Bailey, M.E. et al., Influence of finishing diets on lamb flavor, in *Lipids in Food Flavors (ACS Symposium Series 920)*, Ho, C.-T. and Hartman, T.G., Eds., American Chemical Society, Washington, D.C., 1994, 2.

63. Sutherland, M.M. and Ames, J.M., The effect of castration on the headspace aroma components of cooked lamb, *J. Sci. Food Agric.*, 69, 403, 1995.

64. Vercellotti, J.R. et al., Analysis of volatile heteroatomic meat flavor principles by purge-and-trap/gas chromatography-mass spectrometry, *J. Agric. Food Chem.*, 35, 1030, 1987.

65. Boylston, T.D. et al., Volatile lipid oxidation products of Wagyu and domestic breeds of beef, *J. Agric. Food Chem.*, 44, 1091, 1996.

66. Young, O.A. et al., Fat-borne volatiles and sheepmeat odour, *Meat Sci.*, 45, 183, 1997.

67. Meinert, L. et al., Chemical and sensory characterisation of pan-fried pork flavour: interactions between raw meat quality, ageing and frying temperature, *Meat Sci.*, 75, 229, 2007.

68. Zhang, Z.Y. and Pawliszyn, J., Headspace solid-phase microextraction, *Anal. Chem.*, 41, 809, 1993.

69. Kataoka, H., Recent advances in solid-phase microextraction and related techniques for pharmaceutical and biomedical analysis, *Curr. Pharm. Anal.*, 1, 65, 2005.

70. Shirey, R.E., Optimization of extraction conditions for low-molecular-weight analytes using solid-phase microextraction, *J. Chromatogr. Sci.*, 38, 109, 2000.

71. Elmore, J.S., Papantoniou, E., and Mottram, D.S., A comparison of headspace entrainment on Tenax with solid-phase microextraction for the analysis of the aroma volatiles of cooked beef, in *Headspace Analysis of Foods and Flavors: Theory and Practice*, Rouseff, R.L. and Cadwallader, K.R., Eds., Kluwer Academic/Plenum Publishers, New York, 2001, 125.

72. Liu, Y., Xu, X.-L., and Zhou, G.-H., Comparative study of volatile compounds in traditional Chinese Nanjing marinated duck by different extraction techniques, *Int. J. Food Sci. Technol.*, 42, 543, 2007.

73. Elmore, J.S., Mottram, D.S., and Hierro, E., Two-fibre solid-phase microextraction combined with gas chromatography-mass spectrometry for the analysis of volatile aroma compounds in cooked pork, *J. Chromatogr. A*, 905, 233, 2001.

74. Machiels, D. and Istasse, L., Evaluation of two commercial solid-phase microextraction fibres for the analysis of target aroma compounds in cooked beef meat, *Talanta*, 61, 529, 2003.

75. Rochat, S., de Saint Laumer, J.-Y., and Chaintreau, A., Analysis of sulfur compounds from the in-oven roast beef aroma by comprehensive two-dimensional gas chromatography, *J. Chromatogr. A*, 1147, 85, 2007.

76. Rochat, S. and Chaintreau, A., Carbonyl odorants contributing to the in-oven roast beef top note, *J. Agric. Food Chem.*, 53, 9578, 2005.

77. Jurado, A. et al., Effect of ripening time and rearing system on amino acid-related flavour compounds of Iberian ham, *Meat Sci.*, 75, 585, 2007.

78. Sánchez-Pena et al., Characterization of French and Spanish dry-cured hams: influence of the volatiles from the muscles and the subcutaneous fat quantified by SPME-GC, *Meat Sci.*, 69, 635, 2005.

79. Andrade, R., Reyes, F.G.R., and Rath, S., A method for the determination of volatile N-nitrosamines in food by HS-SPME-GC-TEA, *Food Chem.*, 91, 173, 2005.

80. Jensen, M.T., Hansen, L.L., and Andersen, H.J., Transfer of the meat aroma precursors (dimethyl sulfide, dimethyl disulfide and dimethyl trisulfide) from feed to cooked pork, *Lebensm. –Wissen. –Technol.*, 35, 485, 2002.

81. Brunton, N.P. et al., A comparison of solid-phase microextraction (SPME) fibres for measurement of hexanal and pentanal in cooked turkey, *Food Chem.*, 68, 339, 2000.

82. Jones, P.R.H., Ewen, R.J., and Ratcliffe, N.M., Simple methods for the extraction and identification of amine malodours from spoiled foodstuffs, *J. Food Compd. Anal.*, 11, 274, 1998.

83. Kondjoyan, N. and Berdagué, J.-L., *A Compilation of Relative Retention Indices for the Analysis of Aromatic Compounds*, INRA de Theix, Saint Genes Champanelle, France, 1996.

84. Mukhopadhyay, R., Old reliable benchtop GC/MS, *Anal. Chem.*, 76, 213A, 2004.

85. Čajka, T. and Hajslová, J., Gas chromatography–time-of-flight mass spectrometry in food analysis, *LC-GC Europe*, 2007, 25, 2007.

86. Milo, C. and Blank, I., Quantification of impact odorants in food by isotope dilution assay: strengths and limitations, in *Flavor Analysis: Developments in Isolation and Characterization*, Mussinan, C.J. and Morello, M.J., Eds., American Chemical Society, Washington, DC, 1998, 69–77.

87. Kerscher, R. and Grosch, W., Quantification of 2-methyl-3-furanthiol, 2-furfurylthiol, 3-mercapto-2-pentanone, and 2-mercapto-3-pentanone in heated meat, *J. Agric. Food Chem.*, 46, 1954, 1998.

88. Lorenz, S. et al., Influence of keeping system on the fatty acid composition in the *longissimus* muscle of bulls and odorants formed after cooking, *Eur. Food Res. Technol.*, 214, 112, 2002.

89. Teranishi, R. et al., Use of odor thresholds in aroma research, *Lebensm. –Wiss. u. –Technol.*, 24, 1, 1991.

90. Pavón, J.L.P. et al., Strategies for qualitative and quantitative analyses with mass-spectrometry-based electronic noses, *Trends Anal. Chem.*, 25, 257, 2006.

91. Vernat-Rossi, V. et al., Rapid discrimination of meat products and bacterial strains using semiconductor gas sensors, *Sensors Actuat. B*, 37, 43, 1996.

92. Santos, J.P. et al., Electronic nose for the identification of pig feeding and ripening time in Iberian hams, *Meat Sci.*, 66, 727, 2004.

93. Rajamaki, T. et al., Application of an electronic nose for quality assessment of modified atmosphere packaged poultry meat, *Food Control*, 17, 5–13, 2006.

94. Blixt, Y. and Borch, E., Using an electronic nose for determining the spoilage of vacuum-packaged beef, *Int. J. Food Microbiol.*, 46, 123, 1999.

95. Annor-Frempong, I.E. et al., The measurement of the responses to different odour intensities of 'boar taint' using a sensory panel and an electronic nose, *Meat Sci.*, 50, 139, 1998.

96. Grigioni, G.M. et al., Warmed-over flavour analysis in low temperature-long time processed meat by an "electronic nose", *Meat Sci.*, 56, 221, 2000.

97. Neely, K. et al., Assessment of cooked alpaca and llama meats from the statistical analysis of data collected using an 'electronic nose', *Meat Sci.*, 58, 53, 2001.

98. Siegmund, B. and Pfannhauser, W., Changes of the volatile fraction of cooked chicken meat during chill storing: results obtained by the electronic nose in comparison to GC-MS and GC olfactometry, *Z. Lebensm.-Unters. –Forsch.*, 208, 336, 1999.

99. Vestergaard, J.S., Haugen, J.E., and Byrne, D.V., Application of an electronic nose for measurements of boar taint in entire male pigs, *Meat Sci.*, 74, 564, 2006.

100. O'Sullivan, M.G. et al., A comparison of warmed-over flavour in pork by sensory analysis, GC/MS and the electronic nose, *Meat Sci.*, 65, 1125, 2003.

101. Vasta, V., Ratel, J., and Engel, E., Mass spectrometry analysis of volatile compounds in raw meat for the authentication of the feeding background of farm animals, *J. Agric. Food Chem.*, 55, 4630, 2007.

102. Taylor, A.J. and Linforth, R.S.T., Atmospheric pressure ionisation mass spectrometry for in vivo analysis of volatile flavour release, *Food Chem.*, 71, 327, 2000.

103. Blake, R.S. et al., Demonstration of proton-transfer reaction time-of-flight mass spectrometry for real-time analysis of trace volatile organic compounds, *Anal. Chem.*, 76, 3841, 2004.

104. Odake, S. et al., Volatile compounds of Wagyu (Japanese black cattle) beef analysed by PTR-MS, in *Flavour Science: Recent Advances and Trends*, Bredie, W.L.P. and Petersen, M.A., Eds., Elsevier, Amsterdam, 2006, 29.

105. Mayr, D. et al., Determination of the spoilage status of meat by aroma detection using Proton-Transfer-Reaction Mass Spectrometry, in *Flavour Research at the Dawn of the Twenty-First Century*, Le Quere, J.L. and Etievant, P.X., Eds., Editions Tec & Doc, Paris, 2003, 757.

Chapter 15

Carbohydrates

Jimmy T. Keeton, Hakan Benli, and Amy E. Claflin

Contents

15.1 Introduction

Skeletal muscles in animals serve as a means of locomotion, and in the case of many domesticated species, muscles and their associated tissues are an important food source. With the cessation of blood flow at death, loss of oxygen, depletion of glycogen, and the accumulation of metabolic by-products, muscle function ceases, allowing its conversion to meat.[1] The conversion of muscle to meat is a complex metabolic event involving numerous transitions in cellular biochemical processes, changes in muscle ultrastructure, and alterations in physical characteristics that contribute to attributes such as desirable color, water-holding capacity, protein functionality, and palatability. Glycogen is the predominant carbohydrate in muscle, constituting 1–2% of the total muscle mass antemortem. However, postmortem catalytic events such as the hydrolysis of nonmitochondrial adenosine triphosphate (ATP) with subsequent release of H^+ protons and the accumulation of lactate cause significant changes in meat tissues by decreasing pH, changing the ionic environment, altering enzymatic activity, and increasing protein denaturation. This chapter will provide an overview of the principal muscle tissue carbohydrates and the analytical methods used to quantitate their presence in muscles destined for meat and meat products.

15.1.1 Carbohydrates in Mammalian Tissues

Simple carbohydrates serve as the primary source of readily available energy for muscle cell metabolism, while more complex forms serve as structural components for a variety of tissues, organs, and organelles. Carbohydrates provide a framework for connective tissue, act as lubricants, and are essential components of nucleic acids, lipids, and proteins. In mammalian tissues, carbohydrates occur in the form of D-polyhydroxyaldehydes or D-polyhydroxyketones, and are classified as monosaccharides, oligosaccharides, and polysaccharides. Monosaccharides (glyceraldehyde, threose, ribose, xylose, glucose, galactose, fructose, mannose) are single polyhydroxy aldehyde or ketone units three to seven carbons in length, whereas oligosaccharides (sucrose, lactose, maltose) are composed of 2–20 monosaccharide units joined by an α or β glycosidic linkage. Polysaccharides or glycans are covalently linked polymeric monosaccharide units that can be hundreds of units in length, and are further subdivided into homopolysaccharides (glycogen, starch, cellulose) and heteropolysaccharides (glycoproteins, proteoglycans, glycosaminoglycans, glycolipids). Other structural carbohydrates include such polysaccharides as cellulose in plant cell walls, which may have up to 15,000 D-glucose units linked by a β(1 → 4) glycosidic bond, and chitin in the exoskeleton of invertebrates, which consists of β(1 → 4) linked *N*-acetyl-D-glucosamine residues.

Glycosaminoglycans or mucopolysaccharides are an important class of polymerized disaccharide repeating units covalently attached to a core protein and are major structural components of connective tissues (collagen, elastin, ground substance), viscous fluids (synovial fluid, vitreous humor), and extracellular matrices (cartilage, bone, skin, blood vessels, tendon, intestinal mucosa, cornea, intervertebral disks). Historically, the term "mucopolysaccharide" was given to these substances because of their ability to yield a viscous, slimy, mucin-like solution; they were later renamed "glycosaminoglycans" to indicate that they are polysaccharides (glycans) that contain glycosamines.[2] In solution, glycosaminoglycans are extremely hydrophilic, viscous, mucus-like, exhibit shear thinning, and are elastic in nature. Glycosaminoglycans can be divided into subgroups, called galactosaminoglycans (chondroitins and dermatan sulfate) and glucosaminoglycans (hyaluronic acid, keratan sulfate, heparin sulfate, and heparin). Compositional properties of the major glycosaminoglycans are shown in Table 15.1.[3,4]

15.1.1.1 Carbohydrates in Muscle Tissues

Carbohydrates, fats, and proteins are the principal dietary sources of energy for cellular metabolism in mammals. Glucose is the primary carbohydrate energy source; it can be synthesized from

Table 15.1   Composition and Properties of Common Glycosaminoglycans

Glycosaminoglycan	Disaccharide Units		Tissue Component
	Amino Sugar + Uronic Acid		
Hyaluronate (highly hydrated shock absorber, lubricant)	N-Acetyl-D-glucosamine	D-Glucuronate	Synovial fluid, vitreous humor, ground substance
Chrondroitin-4-sulfate	N-Acetyl-D-glucosamine-4-sulfate	D-Glucuronate	Cartilage, bone, skin, blood vessels
Chrondroitin-6-sulfate	N-Acetyl-D-glucosamine-6-sulfate	D-Glucuronate	Cartilage, bone, skin, blood vessels
Dermatan sulfate	N-Acetyl-D-glucosamine-4-sulfate	L-Iduronate, D-glucuronate	Skin, heart valve, tendon
Keratan sulfate	N-Acetyl-D-glucosamine-6-sulfate	D-Galactose	Cartilage, cornea, disks
Heparin (most negatively charged polyelectrolyte in mammalian tissues)	N-Sulfo-D-glucosamine-6-sulfate	L-Iduronate (most), D-glucuronate	Mast cells, arterial walls, liver, lungs, skin, inhibits clotting

Source: Bhagavan, N.V., in *Medical Biochemistry*, Harcourt/Academic Press, New York, NY, 2001; Voet, D. and Voet, J.G., in *Biochemistry*, Wiley, New York, NY, 1995.

amino acids and glycerol, but with greater expenditure of energy than that of dietary glucose. In antemortem muscle tissue and liver, the predominant storage form of glucose is the branched polysaccharide, glycogen that provides an easily metabolized energy source for muscle contraction and relaxation. Contraction is fueled by large quantities of ATP generated by glycogen degradation to pyruvate through aerobic glycolysis and subsequent metabolism in the citric acid cycle and electron transport chain. Much smaller amounts of ATP are generated through anaerobic glycolysis.

Glycogen consists of a core protein, glycogenin (37,300 Da), which serves as a catalyst for glycogen synthase activity, and branched glucosyl units containing approximately 55,000 glucose molecules having a molecular weight of 9–10 million Da. These cytoplasmic granules also contain 40–50 phosphorylase dimers that regulate glycogen degradation. Glycogen granules exist in two forms, one consisting of a small molecular weight fraction called proglycogen (PG; 400,000 Da), and a larger molecule recognized as "classic" glycogen or macroglycogen (MG; 10,000,000 Da).[5] When needed for muscle contraction, stored glycogen is readily converted to glucose-6-phosphate for further hydrolysis to glucose and entry into the glycolysis pathway. Subsequent metabolism follows either through an oxidative or anaerobic pathway. At death, glycogen undergoes anaerobic glycolysis and terminates with the formation of lactate + oxidized nicotinamide adenine dinucleotide (NAD[+]) from pyruvate and NADH (reduced form) via lactate dehydrogenase (LDH).

The predominant structural carbohydrates in muscle tissue are the glycoproteins, proteoglycans, glycosaminoglycans, and glycolipids. Their location, structure, and function are discussed in Section 15.2.

15.1.1.2   Carbohydrate Concentrations in Muscle Tissues

The concentration of glycogen and its metabolites varies in bovine, porcine, ovine, and avian muscle tissues as a consequence of differences in physiological fiber type, species, physical activity,

Table 15.2 Selected Glycolytic Metabolite Concentration (µM/g) and pH Values of Muscle Tissue from Bovine, Porcine, and Avian Species

	Bovine			Porcine				Avian		
		Post-mortem		Ante-mortem		Post-mortem		Ante-mortem	Post-mortem	
	Ante-mortem	Normal	DFD	Normal	PSE	Normal	PSE	Normal	Normal	PSE
Glycogen	60–100	16–37	1–10	52–85	37	1–10	2–3	37–56	0–7	0–1
Lactate	10–16	72–100	5–10	16–28	41–42	79–97	90–111	10–40	89–120	<87
pH	7.0–7.1	5.6–5.8	6.3–6.7	6.7–7.1	6.3–7.0	5.5–5.7	<5.4	7.0–7.3	5.7–6.0	<5.6

Note: PSE: pale, soft, and exudative; DFD: dark, firm, dry.

Source: Pösö, A.R, Puolanne, E., and Viikki, E.E., *Meat Sci.*, 70, 423, 2005; Bee, G.C. et al., *J. Anim. Sci.*, 84, 191, 2006; Klont, R.E. and Lambooy, E., *J. Anim. Sci.*, 73, 108, 1995; Nuss, J.I. and Wolfe, F.H., *Meat Sci.*, 5, 201, 1981; Romans, J.R. et al., in *The Meat We Eat*, Interstate Publishers, Danville, IL, 2001, 903; Barbut, S., *J. Muscle Foods*, 9, 35, 1998; El Rammouz, R., Babilé, R., and Fernandez, X., *Poult. Sci.*, 83, 1750, 2004; Fernandez, X. et al., *Br. Poult. Sci.*, 42, 462, 2001; Hambrecht, E. et al., *J. Anim. Sci.*, 82, 1401, 2004; Leheska, J.M., Wulf, D.M., and Maddock, R.J., *J. Anim. Sci.*, 80, 3194, 2002; Lowe, T.E. et al., *Meat Sci.*, 67, 251, 2004; Sosnicki, A.A. et al., *J. Muscle Foods*, 9, 13, 1998; Van Laack, R.L.J.M. et al., *Poult. Sci.*, 79, 1057, 2001.

stress, exercise, fasting state, cellular metabolic state (antemortem versus postmortem), environmental conditions prior to and after death (hot or cold temperatures, electrical stimulation), and genetic mediated control of glycolysis. Glycogen concentrations in the longissimus muscle of live cattle can range from 60 to 100 µM/g, while in pigs the value may be 85 µM/g of tissue.[6] Glycolytic metabolites that are often measured to evaluate treatment effects on postmortem muscle tissue are glycogen concentration, residual ATP, lactate concentration, ultimate pH, glycolytic potential (GP = 2 [glycogen + glucose + glucose-6-phosphate] + lactic acid), glycogen intermediates (GI = glycogen + glucose + glucose-6-phosphate), and glycolytic enzyme (phosphorylase) activity. Examples of some metabolic values for different meat species are shown in Table 15.2.[6–18]

15.1.1.2.1 Influence of Muscle Type and Fiber Type

Individual muscles and muscle fibers have been classified according to their physiological activity (e.g., movement, posture) in the animal, and enzyme activity (ATPase) as it relates to fiber twitch speed (slow-twitch versus fast-twitch) and type of energy metabolism (oxidative versus glycolytic). Individual muscles are generally classified as either fast-twitch or slow-twitch, whereas individual fibers are generally recognized as Type I (slow-twitch oxidative), Type IIA (fast-twitch oxidative), and Type IIB (fast-twitch glycolytic) fibers. Alternative names for these muscle fibers are red (βR), intermediate (βW), and white (αW), respectively.[19]

Fast-twitch and slow-twitch muscles differ metabolically, but both types are important to the overall function of the animal and from a meat quality perspective. Fast-twitch muscles, which are white or light in appearance, assist the animal with locomotion and provide other forceful movements when necessary. In contrast, slow-twitch muscles have a darker red color due to higher myoglobin concentration and are used more for slow, sustained movements and posture. Glycogen and glycolytic enzymes are more abundant in white fibers than in red fibers, enabling a rapid source of energy for contraction. Fast-twitch muscles have a lower ultimate pH than slow-twitch muscles. The fast-twitch muscles in beef, chicken, and turkey have ultimate pH values of 5.5, 5.5, and 5.7, respectively, whereas the slow-twitch muscles in these three species have ultimate pH values of 6.3, 6.1, and 6.4, respectively. Muscles in which white fibers predominate appear to be less susceptible to cold shortening (the increased contraction of prerigor muscle when exposed to temperatures of 0–10°C) due to rapid chilling as compared to muscles comprised mostly of red fibers.[19]

15.1.1.2.2 Influence of PSE and DFD Conditions

Two physiological abnormalities that negatively affect muscle quality are the pale-soft-exudative (PSE) condition noted primarily in pork and turkey, and the dark-firm-dry (DFD) condition occurring mostly in beef. Conditions affecting the glycogen content of muscle tissue within the hours prior to harvest of the animal, such as fasting conditions, stress, and genetic predisposition of the animal (e.g., the RN or Napole gene of pork), contribute to these two conditions. Animals susceptible to the PSE condition incur an accelerated glycolytic rate immediately postmortem, resulting in a rapid pH decline and a lower than normal ultimate pH (24 h). As a consequence, protein denaturation occurs, causing myoglobin to exhibit an atypical pale pink muscle color and reducing myofibrillar protein binding ability, thus weakening the structure (soft). More importantly, the muscle tissue's water-holding capacity is diminished due to the pH being near the myofibrillar protein's isoelectric point (pH 5.1). Tenderness may also decrease due to moisture loss and protein denaturation. The pH of normal pork 45 min postmortem ranges from 6.4 to 6.5, but PSE pork will usually be less than 6.0 with an ultimate 24 h pH of 5.2–5.4 in comparison to a normal range of 5.5–5.7.[20]

The DFD, or dark-cutting condition, is characterized by a dark red muscle color, a dry surface, a higher ultimate pH, and a higher than normal water-holding capacity. Harvesting an animal shortly after it has undergone stress (vigorous exercise, rough handling) is the primary cause of DFD meat. Antemortem stress depletes muscle glycogen reserves and limits postmortem glycolysis, which, in turn, limits postmortem lactic acid production, resulting in a higher ultimate pH (6.3–6.7 for beef, 6.0–6.5 for pork),[10] more favorable conditions for bacterial growth, and a reduction in shelf life.

15.2 Role of Carbohydrates

15.2.1 *Cellular Metabolism*

15.2.1.1 *Glycogen Metabolism*

Glycogen is an $\alpha(1 \rightarrow 4)$, $\alpha(1 \rightarrow 6)$ linked D-glucose polymer that serves as a nutrient source for cellular metabolism in animals. Starch, likewise, serves as a nutrient reserve in plants and is stored as insoluble granules of α-amylose and amylopectin polymers. Glycogen is the predominant storage polysaccharide most prevalent in antemortem skeletal muscle tissue (1–2%) and the liver (up to 10%). It resembles amylopectin, but is more highly branched, having branch points at every 8–12 glucose residues.[4] A glycogen molecule weighing 9–10 million Da contains approximately 55,000 glycogen residues having an estimated 2,100 nonreducing ends.[6] Glycogen exists in two forms, PG and MG. In rested muscle, acid-insoluble PG is approximately 60% of the total glycogen, but in stressed muscle containing less glycogen, it constitutes up to 90% of the total glycogen. Glycogen degradation, and the inhibition of glycogen synthesis, is initiated by catecholamine release from sympathetic neurons and the adrenal medulla to activate gylcogen phosphorylase, which is under allosteric AMP and Ca^{2+} activated control. Glycogen phosphorylase cleaves the $\alpha(1 \rightarrow 4)$ linkage inward from the nonreducing outer polymer ends to yield glucose-1-phosphate, while the $\alpha(1 \rightarrow 6)$ linkage is cleaved by the glycogen debranching enzyme. It is estimated that 10–20 μM/g residual glycogen is left in muscles postmortem and that residues of up to 80 μM/g may remain in unstressed, well-fed animals with an ultimate pH of 5.5.[6]

Under aerobic oxidative conditions, glycogen is converted to two pyruvate molecules by LDH at the endpoint of aerobic glycolysis. These molecules then enter the citric acid cycle and electron transport pathway to ultimately generate 38 ATPs. However, under vigorous activity and anaerobic conditions that deplete oxygen, lactate is the end product of glycolysis, producing only two molecules of ATP. The lactate produced in antemortem muscle is recirculated in the blood to the liver for conversion back to glucose and storage as glycogen, thus maintaining cellular homeostasis. At death, anaerobic glycolysis continues as long as glycogen remains available or until catalysis is inhibited. Anaerobic glycolysis depletes muscle glycogen, as well as ATP and creatine phosphate, leading to an accumulation of lactate (pK_a 3.86). An increase in H^+ protons occurs due to lactate dissociation with declining pH or, as has been theorized, nonmitochondrial hydrolysis of ATP ($ATP \rightarrow ADP + Pi + H^+$).[21] The inability of the muscle to buffer and remove protons allows for the accumulation of lactate, loss of buffering capacity, inhibition of glycogen phosphorylase and phosphofructokinase, and the rephosphorylation of ADP to ATP. All of these events contribute to halting glycolysis and ultimately lead to irreversible muscle contraction and rigor mortis. The reduction in muscle pH corresponds with a reduction in glycogen, ATP, and creatine phosphate and the subsequent build up of lactate in the tissue. Typically, the pH declines in postmortem muscle from 7.1–7.3 to 5.4–5.6 due to H^+ proton accumulation. The ultimate pH varies due to factors previously mentioned, such as physiological fiber type, species, physical activity, stress, exercise, fasting state, cellular metabolic state (antemortem versus postmortem), environmental conditions prior to and after death (hot or cold temperatures, electrical stimulation), and genetic mediated control of glycolysis.

15.2.2 Structural Carbohydrates

15.2.2.1 Glycoproteins

Glycoproteins are a diverse category of molecules that provide structural support in tissues, cells, and organelles (collagen, erythrocytes, mitochondria, etc.), are an essential constituent of cell membranes (membrane receptors, cell recognition molecules) and intercellular matrices (epithelial mucins, synovial fluid), and moderate regulatory functions in cellular fluids (endocrine regulators, catalysis enzymes) or plasma (antigens, immunoglobulins, clotting factors). Glycoproteins are made up of a protein covalently linked to a carbohydrate residue by either an *N*- or *O*-glycosidic bond. An *N*-glycosidic linkage occurs when the amide nitrogen of asparagine (Asp) is β-linked to an oligosaccharide moiety through the *N*-acetylglucosamino-Asp bond. The *O*-glycosidic bond combines a disaccharide, β-glactosyl-$(1 \rightarrow 3)$-α-acetylgalactosamine, α-linked to the OH group of either serine (Ser) or threonine (Thr), and in the special case of collagen, hydroxylysine or hydroxyproline. The synthesis of a glycoprotein is initiated with the formation of the polypeptide component by membrane-bound ribosomes in the rough endoplasmic reticulum. The carbohydrate chains are added by glycosyltransferases in the endoplasmic reticulum and subsequent passage through the Golgi apparatus. Glycoproteins may also be formed by nonenzymatic glycation.

In muscle tissue, structural glycoproteins are dominant in connective tissue, especially collagen, and to some extent elastin, which contains fibrillin.[3] The hydroxylysyl residues of the collagen polypeptide provides sites for attachment of glucose, galactose, or an $\alpha(1 \rightarrow 2)$ glucosylgalactose carbohydrate by a β-*O*-glycosidic linkage. Collagen synthesis is considered to occur intracellularly and extracellularly. The first stage involves hydroxylation, glycosylation, formation of a triple helix, and secretion of a polypeptide chain. The second stage converts procollagen to tropocollagen fibrils by limited proteolysis and cross-linking of the chains. The unique properties of collagen such as flexibility, rigidity, elasticity, and strength are dependent upon its fibrous nature and the composition of other connective tissue glycoprotein components.

Proteoglycans or mucoproteins are specialized gel-like, high molecular weight polyanionic glycoproteins that contain multiple carbohydrate units, namely glycosaminoglycans (Table 15.1), covalently attached to a core protein. These substances are found embedded in the ground substance of connective tissue, tendons, cartilage, cell surface receptors, and mucus. Proteoglycans have large clusters of carbohydrates, the most abundant being *N*-acetylglucosamine (NAG), attached to a protein backbone, giving a filamentous, bottle-brush appearance. The negatively charged carboxylate and sulfate groups attached to the glycosaminoglycans make them extremely hydrophilic and, as a consequence, they hold most of the water in the intracellular matrix of connective tissues. Their function is best illustrated when cartilage undergoes compression, resulting in water being expressed from the carbohydrate moiety; when the pressure is released, the water returns and reassociates with the carbohydrate molecules. Likewise, tendons that connect skeletal muscle and bone have a high collagen content and contain chondroitin and dermatan sulfates, two glycosaminoglycans that allow flexibility along the long axis of the tendon.

15.2.2.2 Glycolipids

Eukaryotic cell and organelle membranes are composed of phospholipids arranged in a bilayer configuration in which the fatty acyl tails form the hydrophobic interior of the bilayer while their polar, hydrophilic head groups line the internal and external surface of the cell or organelle. Other essential membrane components include sphingoglycolipids that provide added structure to the bilayer; integral proteins that may be embedded into either side of the cell surface or transverse the bilayer to allow exchange of metabolites; peripheral proteins that interact with the inner membrane surface or are free in the cytoplasm; cholesterol that is interspersed within the bilayer to assist in the regulation of fluidity; and cell surface glycolipids and glycoproteins that participate in antigen–antibody reactions, hormone function, enzyme catalysis, and cell recognition reactions. Carbohydrates that are bound to glycolipids and glycoproteins increase the hydrophilic character of the lipids and proteins, and stabilize the conformation of the membrane proteins. Owing to the anionic character of the polar heads of the phospholipids, the surface of a cell membrane has a net negative charge.

Glucosylcerebroside and galactocerebroside, the simplest sphingoglycolipids, consist of ceramide formed from sphingosine and oleic acid linked to a single glucose or galactose residue. These glycolipids are abundant in the myelin sheath and neuronal cell membranes of the brain, respectively. Gangliosides are ceramide oligosaccharides that have up to four sugar residues that extend beyond the cell membrane surface, and serve as recognition receptors for bacterial protein toxins.

15.3 Carbohydrate Analyses

15.3.1 Methods of Analysis

Muscle glycogen concentration in antemortem muscle is regulated by cellular glycogenesis and glycolysis in response to contractile requirements of the tissue and the physiological state of the muscle. Postmortem glycogenolysis and glycolysis regulate lactate and H^+ proton accumulation, which in turn affects qualitative muscle characteristics such as color, appearance, water-holding capacity, textural properties, myofibrillar protein functionality, and tissue interactions with non-meat ingredients. The initial concentration of intercellular glycogen has a direct effect on the ultimate concentration of lactate in muscle, which in turn affects final tissue pH, degree of protein denaturation, and myofibrillar degradation due to calpain-mediated proteolysis. Other factors such as antemortem stress, carcass chilling rate, ultimate carcass temperature, electrical stimulation postslaughter, and other processing conditions also have an effect on glycogen metabolism. Thus, the dominant carbohydrate targeted for analyses in muscle tissue, whether ante- or post mortem, is glycogen and its associated metabolites. Various methods of sample preparation and subsequent techniques for glycogen analysis are given in the following sections.

Glycolytic enzymes that enable glycolysis constitute the major portion of the sarcoplasmic proteins in muscle tissue and are often monitored to determine the glycolytic state of the tissue. Portions of some of these enzymes are bound to myofibrillar proteins and, upon electrochemical stimulation by an action potential from a nerve signal, increase enzymatic binding to myofibrillar proteins to stimulate glycolysis. When stimulation ceases, enzymatic binding is reduced. Glycolytic enzymes also bind to other sites in the cell, such as the sarcolemma, sarcoplasmic reticulum, nuclear membranes, and mitochondria.

15.3.1.1 Metabolic Carbohydrates (Glycogen)

Glycogen is generally analyzed by acid or enzymatic hydrolysis followed by glucose determination. Slight modifications in both sample preparation (or pretreatment) and hydrolysis to glucose have been reported by numerous scientists. Following is a summary of the most used glycogen analyses.

15.3.1.1.1 Preparation of Muscle Tissue

Specific preparation procedures for glycogen analysis of muscle tissues may vary slightly, but typically, excised samples are frozen immediately in liquid nitrogen and stored at −80°C until analyzed. Samples destined for acid hydrolysis are usually freeze-dried (−35 to 22°C for 24–168 h), and dissected free of visible blood, fat, and connective tissue prior to hydrolysis. Frozen samples intended for enzyme hydrolysis are degraded proteolytically without removal of any tissue constituents. One to 3 mg of freeze-dried tissue is sufficient to determine the glycogen content of a sample, and a pretreatment step is generally performed just before acid or enzymatic hydrolysis. For acid hydrolysis, extraction of total glycogen is achieved by placing the sample in 200 µL of ice-cold 1.5 M perchloric acid, crushing the sample with a glass rod, extracting on ice for 20 min, and centrifuging the extract at 2000–3200 × *g* for 10–15 min at 4°C. The supernatant fraction is used for MG or free glycogen analysis, while the pellet is used for PG analysis.

Total glycogen in the tissue is composed of two fractions, PG and MG, which are characterized by their solubility in acid. PG has a smaller molecular weight (400,000 Da), is acid-insoluble,

Table 15.3   Muscle Tissue Preparation for Glycogen Analysis by Acid or Enzymatic Hydrolysis

Sample Preparation Protocol	Analytical Method	Reference
Ice-cold 1.5 M perchloric acid extraction for 20 min	Acid hydrolysis	22, 24–27
Ice-cold 1.5 M perchloric acid extraction for 20 min	Enzymatic hydrolysis	5, 24, 28
Homogenization in 0.03 N HCl and heating 10 min at 100°C	Acid hydrolysis	29
Homogenization in 0.03 N HCl and heating 10 min at 100°C	Enzymatic hydrolysis	23, 29, 30
Homogenization in ice-cold 6% (w/v) perchloric acid	Enzymatic hydrolysis	31
Homogenization in a 50% glycerol, 20 mM Na_2HPO_4 buffer (50:1 w/v, pH 7.4) that contains 0.5 mM EDTA, 0.02% bovine serum albumin, and 5 mM β-mercaptoethanol	Acid hydrolysis	32
Homogenization in hot 1 M LiCl and heating 5 min at 100°C	Acid hydrolysis, enzymatic hydrolysis	33
Homogenization in hot water and heating 5 min at 100°C	Enzymatic hydrolysis	34
Incubation with 0.1 M NaOH at 80°C for 10 min and neutralized by a combination of 0.1 M HCl, 0.2 M citric acid, and 0.2 M Na_2HPO_4	Enzymatic hydrolysis	5
Digestion with 1 M KOH	Enzymatic hydrolysis	24

and is relatively rich in protein or has a low ratio of carbohydrate to protein (approximately 10% protein). MG has a larger molecular weight (10,000,000 Da), is acid-soluble, and has a high ratio of carbohydrate to protein (approximately 0.4% protein) in its structure.[5,22]

Sample pretreatment procedures for enzymatic hydrolysis of muscle tissue require several steps, including (1) homogenization of the sample in 0.03 N HCl and heating 10 min at 100°C, (2) addition of 0.1 N NaOH to 2–3 mg dry muscle tissue and incubation for 10 min at 80°C (to destroy background glucose and hexose monophosphates, followed by a neutralization with 0.1 M HCl, 0.2 M citric acid, and 0.2 M Na_2HPO_4), or (3) digestion of a whole sample in 1 M KOH prior to enzymatic hydrolysis, thereby making it unnecessary to remove the insoluble residue.[5,23,24] Examples of pretreatment procedures are summarized in Table 15.3.

15.3.1.1.2   Acid Hydrolysis

Acid hydrolysis is the most widely used technique for assessing muscle glycogen content. Following pretreatment of 1–3 mg of tissue, the extracted glycogen is hydrolyzed by the addition of 1–2 M HCl acid and heating at 85–100°C for 2–3 h. If perchloric acid extraction has been applied to samples prior to acid hydrolysis, and the supernatant and pellet have been separated by centrifugation, each fraction is hydrolyzed separately to determine MG (free glycogen), PG, and total glycogen (MG + PG) contents.[5,7,22,24,26,35] After acid hydrolysis, additional treatments such as neutralization with 2 M NaOH and centrifugation at 1500 × *g* for 10 min at 4°C can also be performed before glucose determination by spectrophotometric or fluorometric techniques, as described by Lowry and Passonneau,[36] Bergmeyer et al.,[37] and Lachenicht and Bernt,[38] or by the use of commercially available assay kits. [39]

15.3.1.1.3 Enzymatic Hydrolysis

Enzymatic hydrolysis of glycogen has been presented as an alternative method of glucose analysis; it has been reported not to differ from acid hydrolysis in determining glycogen content.[5,24] Enzymatic hydrolysis can be performed by two different methods. The first approach hydrolyzes glycogen with amyloglucosidase, while the second employs phosphorylase and debranching enzymes.[5,23,24,29,36] As with acid hydrolysis, the glucose concentration can be measured spectrophotometrically or fluorometrically.[5,24,29,37,38]

Amyloglucosidase, produced by *Aspergillus niger* (1,4-α-D-glucan glucohydrolase), is an exoglucosidase that can hydrolyze glycogen from the α-D-(1 → 4) and α-D-(1 → 6) linkage. Glycogen isolation steps can be omitted from this procedure because of the specificity of amyloglucosidase. The optimum conditions for maximizing enzyme activity are achieved at pH 4.8 with 2 h of continuous shaking at 40°C. Following enzymatic hydrolysis, the glucose is liberated with hexokinase (D-hexose-6-phosphotransferase) and glucose-6-phosphate dehydrogenase (D-glucose-6-phophate) to form NADPH.

Enzymatic reactions that demonstrate this sequence are

$$\text{Glycogen} + (H_2O)_{n-1} \xrightarrow[\text{pH 4.8}]{\text{amyloglucosidase}} (\text{glucose})_n$$

$$\text{Glucose} + \text{ATP} \xrightarrow[Mg^{2+}, \text{ pH 7.5}]{\text{hexokinase}} \text{ADP} + \text{glucose-6-phosphate}$$

$$\text{Glucose-6-phosphate} + \text{NADP}^+ \xrightarrow[\text{dehydrogenase}]{\text{glucose-6-phosphate}} \text{6-phosphogluconolactone} + \text{NADPH} + \text{H}^+$$

The glucose concentration after hydrolysis is proportional to the concentration of NADPH formed, which is measured spectrophotometrically at 340 nm (334–365 nm) or fluorimetrically at 485 nm.[37,38]

A second example of the enzymatic hydrolysis and analysis of glycogen is described by Keppler and Decker.[40] In this procedure, the deep-frozen tissue sample is thoroughly homogenized at high speed in 5 parts (by weight) ice-cold perchloric acid (0.6 N). After homogenization, 0.2 mL of the homogenate is transferred to a glass centrifuge tube and kept in an ice bath. The homogenate is then incubated in a 0.1 mL potassium hydrogen carbonate solution (1 M) with 2.0 mL of amyloglucosidase solution (10 mg/mL) in a stoppered centrifuge tube. The tubes are then incubated for 2 h at 40°C with constant shaking. Enzymatic hydrolysis is stopped by addition of 1.0 mL perchloric acid and the tubes are centrifuged briefly to separate the supernatant and pellet. Afterward, the supernatant is measured spectrophotometrically at 340 nm to determine glucose concentration.

Glycogen can also be determined in a single analytical step by hydrolyzing with phosphorylase *a* (rabbit muscle) and debranching enzymes such as amylo-1,6-glucosidase, transglucosylase (oligo-1,4 → 1,4-glucantransferase), P-glucomutase (rabbit muscle), and glucose-6-P dehydrogenase (yeast). The following reactions illustrate the products formed from the multienzyme hydrolysis:

$$\text{Glycogen} + \text{Pi} \xrightarrow[\text{dehydrogenase}]{\text{glucose-6-P}} \text{glucose-1-P (92\%)} + \text{glucose (8\%)}$$

$$\text{Glucose-1-P} \xrightarrow{\text{P-glucomutase}} \text{glucose-6-P}$$

$$\text{Glucose-6-P} + \text{TPN}^+ \xrightarrow[\text{dehydrogenase}]{\text{glucose-6-P}} \text{6-gluconolactone} + \text{TPNH}$$

The glucose concentration after the hydrolysis is proportional to the concentration of triphosphopyridine nucleotide (TPNH) formed, which is measured spectrophotometrically at 340 nm (334–365 nm) with a sensitivity of 20–200 μM of glucose-6-P or fluorimetrically with a sensitivity of 0.2–10 μM of glucose-6-P.[36,37,38,41]

15.3.1.2 Metabolic Carbohydrates (L-(+)-Lactate)

Muscle samples are prepared for L-(+)-lactate analysis by homogenizing in a mixer, mincer, or mortar; extracting or dissolving with water; and filtering to remove solid material. If the sample contains fat, warm water is used during homogenization, the homogenate is allowed to cool to solidify the fat, and the solid material is trapped by filtering. The supernatant is deproteinized by adding ice-cold 1 N perchloric acid in a 1:1 ratio, centrifuging to remove the protein precipitate, and neutralizing the supernatant to a salmon-pink endpoint (pH ~3.5) with a 5 M potassium carbonate solution containing methyl orange indicator. The sample solution should be clear or at most slightly opalescent and must not contain more than 0.1 g L-lactate per liter; otherwise, a serial dilution is necessary. The L-(+)-lactate extract is enzymatically oxidized to pyruvate by NAD^+ and catalyzed by LDH.

$$\text{L-(+)-Lactate} + \text{NAD}^+ \underset{}{\overset{\text{LDH}}{\rightleftharpoons}} \text{pyruvate} + \text{NADH} + \text{H}^+$$

The NADH formed is proportional to the amount of lactate present and is measured spectrophotometrically at 340 nm.[42] The increase in NADH concentration can also be measured fluirimetrically at a primary wavelength of 360 nm and a secondary wavelength of 460 nm.[41]

Lactate determination is also possible using other enzymatic reactions, such as with 3-acetyl analogs of NAD (APAD) and LDH, or coupling the oxidative reaction of LDH with the transamination reaction of glutamate–pyruvate transaminase.[43,44]

15.3.1.3 Structural Carbohydrates (Glycoproteins, Proteoglycans)

Proteoglycans are ubiquitous in body tissues, but are most prevalent in connective tissues and cartilage. The proteoglycans in cartilage have been researched extensively because of their presence in relatively high amounts (up to 10% of cartilage dry weight), but researchers have identified and investigated proteoglycans in other human and animal tissues, including but not limited to the cornea of the eye, mast cells, skin, organs (liver, brain, kidneys), basement membranes, human platelets, mammary epithelial cells, and proteoglycans in the intracellular and extracellular matrices. Several techniques and procedures employed for the analysis of proteoglycans will be summarized in the following sections, with emphasis given to proteoglycans associated with muscle and connective tissues.[45,46]

15.3.1.3.1 Extraction of Proteoglycans

Extraction of glycosaminoglycans and proteoglycans from connective tissue has been conducted using methods which utilize proteases that cleave peptide chains or alkali reagents that cleave Xyl-Ser protein–carbohydrate linkages. Extraction of these compounds from dense connective tissue, such as tendon, requires harsher conditions to release the proteins and carbohydrates, which are tightly embedded within the tissue. Proteoglycans can be extracted from dense connective tissue using a solution of 4 M guanidine that will denature the collagen and disrupt the noncovalent interactions, allowing the soluble proteins such as proteoglycans to be separated from the bulk of the tissue upon centrifugation. The anionic nature of the glycosaminoglycan chain promotes the separation of proteoglycans from other extracted proteins when ion-exchange chromatography is performed.[46]

Glycosaminoglycans in dense connective tissue can be quantified using a method in which the tendon tissue is lyophilized, digested with a papain solution (50 mg dry tissue/mL enzyme), incubated until the tissue disappears (at 65°C), and then centrifuged. The supernatant is then removed, diluted with equal parts distilled water, and applied to a diethylaminoethyl (DEAE) cellulose column. The glycosaminoglycans are then eluted with 1 M HCl, a portion mixed with orcinol reagent, heated to 100°C, and then read spectrophotometrically at 670 nm within 30 min. The carbohydrate content is then determined by calculating the micrograms of uronic acid per milligram of tissue (dry weight basis) and tendon uronic acid content.[46]

15.3.1.3.2 Isolation of Proteoglycans

Extraction of glycoproteins and proteoglycans from tissues and cells in a soluble and undegraded form is important when isolating these molecules for purification and other analysis. Removal of contaminants using fractionation procedures aids in reducing or eliminating unnecessary substances that could interfere with the analysis. Traditional protein and carbohydrate techniques of selective precipitation, ion-exchange chromatography, zone electrophoresis, and gel filtration have been applied to fractionation methods for glycoproteins and proteolgycans. Buoyant density ultracentrifugation may be employed as a preparative method for isolation in circumstances involving molecules containing a high carbohydrate content.

15.3.1.3.3 Analysis of Carbohydrate Isolates

Hydrolysis can be used for the initial analysis of neutral sugars that are a part of proteoglycans. Hydrolysis involves adding 1 mL of HCl to a sample in a screw-capped lid, flushing the tube with oxygen-free nitrogen before sealing the tube, and allowing hydrolysis to occur for 4–6 h at 100°C. The sample is then cooled, 1.5 mL of water added, the contents of the tube passed through a Dowex resin column (200–400 mesh, CO_3 form) to neutralize the sample, the eluate collected, and the sugars present identified and quantitated using gas–liquid chromatography (GLC), ion-exchange chromatography, or paper chromatography.

Analysis of the carbohydrate constituents of proteoglycans or glycoproteins can be undertaken once a homogeneous preparation has been obtained. GLC following methanolysis as well as automated ion-exchange methods can be used to separate and quantify the neutral sugars in a sample. These two methods are best used if the carbohydrate content in the sample is relatively large (>10 μM), though other methods of analysis would provide better results when analyzing samples containing smaller amounts of carbohydrates. If equipment necessary for performing these methods of analysis is not available, then paper chromatography and thin-layer chromatography can serve as auxiliary methods.

Colorimetric assays have been developed that are suitable for measuring the neutral sugars (D-mannose, D-galactose, D-glucose, D-xylose) found in proteoglycans. Such assays are generally carried out in strongly acidic conditions and are applicable for the determination of free monosaccharides and unhydrolyzed samples of proteoglycans. Colorimetric methods offer simplicity, quantitative analysis of sugar residues, and a means of monitoring the purification of proteoglycans, but more sensitive and accurate methods are becoming available. One colorimetric method that has been used to quantitate the neutral sugars in proteoglycans is the phenol–H_2SO_4 assay for neutral sugars. This involves the addition of a phenol reagent (5% w/v aqueous solution of colorless crystals of phenol) and concentrated H_2SO_4 to a sample while mixing. Heat generated by mixing the acid with water is essential for color development of the sample, which can be read at an absorbance of 484 nm after letting it stand at room temperature for 30 min.

Enzyme assays have been developed for the specific assay of D-mannose, D-glucose, D-galactose, D-glucosamine, and sialic acid in sugar mixtures and are applicable to the analysis of carbohydrates in proteoglycans. These methods are highly specific, and the fluorimetric versions of the assay is extremely sensitive. Commercial kits are available for the spectrophotometric assay of galactose.

15.4   Summary

Simple carbohydrates serve as the primary source of readily available energy for muscle cell metabolism, while more complex forms serve as structural components for a variety of tissues, organs, and organelles. Polysaccharides or glycans are covalently linked polymeric monosaccharide units that can be hundreds of units in length, and are further subdivided into homopolysaccharides (glycogen, starch, cellulose) and heteropolysaccharides (glycoproteins, proteoglycans, glycosaminoglycans, glycolipids). In antemortem muscle tissue and liver, the predominant storage form of glucose is the branched polysaccharide glycogen, which provides an easily metabolized energy source for muscle contraction and relaxation. When needed for muscle contraction, stored glycogen is readily converted to glucose-6-phosphate for further hydrolysis to glucose and entry into the glycolysis pathway. Subsequent metabolism follows through either an oxidative or an anaerobic pathway. At death, glycogen undergoes anaerobic glycolysis, terminating with the formation of lactate + oxidized nicotinamide adenine dinucleotide (NAD^+) from pyruvate and NADH (reduced form) via LDH. A corresponding pH decline from approximately 7.0 to 5.6–5.8 occurs as a consequence of the buildup of lactate and hydrogen ions. Glycogen is generally analyzed by acid or enzymatic hydrolysis followed by glucose determination. L-(+)-Lactate is typically enzymatically oxidized to pyruvate by NAD^+ and catalyzed by LDH. The NADH formed is proportional to the amount of lactate present and is measured spectrophotometrically or fluorimetrically. Traditional protein and carbohydrate techniques of selective precipitation, ion-exchange chromatography, zone electrophoresis, and gel filtration have been applied to fractionation methods for glycoproteins and proteolgycans.

References

1. Greaser, M.L., Conversion of muscle to meat, in *Muscle as a Food*, Bechtel, P.J., Ed., Academic Press, New York, NY, 1986, 37.
2. Bhavanandan, V.P. and Davidson, E.A., Proteoglycans, structure, synthesis, function, in *Glyconconjugates: Composition, Structure and Function*, Allen, H.J. and Kisailus, E.C., Eds., Marcel Dekker, New York, NY, 1992.
3. Bhagavan, N.V., *Medical Biochemistry*, 4th ed., Harcourt/Academic Press, New York, NY, 2001.
4. Voet, D. and Voet, J.G., *Biochemistry*, Wiley, New York, NY, 1995.
5. Adamo, K.B. and Graham, T.E., Comparison of traditional measurements with macroglycogen and proglycogen analysis of muscle glycogen, *J. Appl. Physiol.*, 84(3), 908, 1998.
6. Pösö, A.R. Puolanne, E., and Viikki, E.E., Carbohydrate metabolism in meat animals—a review, *Meat Sci.*, 70(3), 423, 2005.
7. Bee, G.C. et al., Effects of available dietary carbohydrate and preslaughter treatment on glycolytic potential, protein degradation, and quality traits of pig muscles, *J. Anim. Sci.*, 84, 191, 2006.
8. Klont, R.E. and Lambooy, E., Effects of preslaughter muscle exercise on muscle metabolism and meat quality studied in anesthetized pigs of different halothane genotypes, *J. Anim. Sci.*, 73, 108, 1995.
9. Nuss, J.I. and Wolfe, F.H., Effect of post-mortem storage temperatures on isometric tension, pH, ATP, glycogen and glucose-6-phosphate for selected bovine muscles, *Meat Sci.*, 5, 201, 1981.
10. Romans, J.R. et al., *The Meat We Eat*, 14th ed., Interstate Publishers, Danville, IL, 2001, 903.
11. Barbut, S., Estimating the magnitude of the PSE problem in poultry, *J. Muscle Foods*, 9, 35, 1998.
12. El Rammouz, R., Babilé, R., and Fernandez, X., Effect of ultimate pH on the physicochemical and biochemical characteristics of turkey breast muscle showing normal rate of postmortem pH fall, *Poult. Sci.*, 83, 1750, 2004.
13. Fernandez, X. et al., Post mortem muscle metabolism and meat quality in three genetic types of turkey, *Br. Poult. Sci.*, 42, 462, 2001.
14. Hambrecht, E. et al., Preslaughter stress and muscle energy largely determine pork quality at two commercial processing plants, *J. Anim. Sci.*, 82, 1401, 2004.
15. Leheska, J.M., Wulf, D.M., and Maddock, R.J., Effects of fasting and transportation on pork quality development and extent of postmortem metabolism, *J. Anim. Sci.*, 80, 3194, 2002.
16. Lowe, T.E. et al., The relationship between postmortem urinary catecholamines, meat ultimate pH, and shear force in bulls and cows, *Meat Sci.*, 67, 251, 2004.
17. Sosnicki, A.A. et al., PSE-like syndrome in breast muscle of domestic turkeys: a review, *J. Muscle Foods*, 9, 13, 1998.
18. Van Laack, R.L.J.M. et al., Characteristics of pale, soft, exudative broiler breast meat, *Poult. Sci.*, 79, 1057, 2001.
19. Foegeding, E.A., Lanier, T.C., and Hultin, H.O., Characteristics of edible muscle tissues, in *Food Chemistry*, 3rd ed., Fennema, O.R., Ed., Marcel Dekker, New York, NY, 1996, 879.
20. Faustman, C., Postmortem changes in muscle foods, in *Muscle Foods*, Kinsman, D.M., Kotula, A.W., and Breidenstein, B.C., Eds., Chapman and Hall, New York, NY, 1994, 68.
21. Robergs, R.A., Ghiasvand, F., and Parker, D., Biochemistry of exercise-induced metabolic acidosis, *Am. J. Physiol. Regul. Integr. Comp. Physiol.*, 287, R502, 2004.
22. Rosenvold, K., Essén-Gustavsson, B., and Andersen, H.J., Dietary manipulation of pro- and macroglycogen in porcine skeletal muscle, *J. Anim. Sci.*, 81, 130, 2003.
23. Passonneau, J.V. et al., An enzymic method for measurement of glycogen, *Anal. Biochem.*, 19, 315, 1967.
24. Jansson, E., Acid soluble and insoluble glycogen in human skeletal muscle, *Acta Physiol. Scand.*, 113, 337, 1981.
25. Essén-Gustavsson, B.M. et al., Effect of exercise on proglycogen and macroglycogen content in skeletal muscles of pigs with the Rendement Napole mutation, *Am. J. Vet. Res.*, 66, 1197, 2005.

26. Derave, W., Gao, S., and Richter, E.A., Pro- and macroglycogenolysis in contracting rat skeletal muscle, *Acta Physiol. Scand.*, 169, 291, 2000.
27. Robergs, R.A. et al., Muscle glycogenolysis during differing intensities of weight-resistance exercise, *J. Appl. Physiol.*, 70, 1700, 1991.
28. Marchand, I.K. et al., Quantification of subcellular glycogen in resting human muscle: granule size, number, and location, *J. Appl. Physiol.*, 93, 1598, 2002.
29. Passonneau, J.V. and Lauderdale, V.R., A comparison of three methods of glycogen measurement in tissues, *Anal. Biochem.*, 60, 405, 1974.
30. Ivy, J.L. et al., Muscle glycogen synthesis after exercise: effect of time of carbohydrate ingestion, *J. Appl. Physiol.*, 64, 1480, 1988.
31. Ferreira, L.D.M. et al., Effect of streptozotocin-induced diabetes on glycogen resynthesis in fasted rats post-high-intensity exercise, *Am. J. Physiol. Endocrinol. Metab.*, 280, E83, 2001.
32. Zawadzki, K.M., Yaspelkis III, B.B., and Ivy, J.L., Carbohydrate–protein complex increase the rate of muscle glycogen storage after exercise, *J. Appl. Physiol.*, 72, 1854, 1992.
33. Johson, J.A. and Fusaro, R.M., Enzymic analysis of rabbit skeletal muscle carbohydrate, *Anal. Biochem*, 37, 298, 1970.
34. Johnson, J.A. and Fusaro, R.M., Use of purified amyloglucosidase for the specific determination of total carbohydrate content of rat liver homogenate in a single step, *Anal. Biochem.*, 98, 47, 1979.
35. Shearer, J.I. et al., Pro- and macroglycogenolysis during repeated exercise: roles of glycogen content and phosphorylase activation, *J. Appl. Physiol.*, 90, 880, 2001.
36. Lowry, O.H. and Passonneau, J.V., *A Flexible System of Enzymatic Analysis*, Academic Press, New York, NY, 1972.
37. Bergmeyer, H.U. et al., D-Glucose determination with hexokinase and glucose-6-phosphate dehydrogenase, in *Methods of Enzymatic Analysis*, 2nd ed., Bergmeyer, H.U., Ed., Academic Press, New York, NY, 1974. 1196.
38. Lachenicht, R. and Bernt, E., Fluorimetric determination in blood with automatic analysers, in *Methods of Enzymatic Analysis*, 2nd ed., Bergmeyer, H.U., Ed., Academic Press, New York, NY, 1974, 1201.
39. Immonen, K. and Poulanne, E., Variation of residual glycogen-glucose concentration at ultimate pH values below 5.75, *Meat Sci.*, 55, 279, 2000.
40. Keppler, D. and Decker, K., Glycogen determination with amyloglucosidase, in *Methods of Enzymatic Analysis*, 2nd ed., Bergmeyer, H.U., Ed., Academic Press, New York, NY, 1974, 1127.
41. Passonneau, J.V., Fluorimetric method, in *Methods of Enzymatic Analysis*, 2nd ed., Bergmeyer, H.U., Ed., Academic Press, New York, NY, 1974, 1468.
42. Gutmann I. and Wahlefeld, A.W., L-(+)-Lactate determination with lactate dehydrogenase and NAD, in *Methods of Enzymatic Analysis*, 2nd ed., Bergmeyer, H.U., Ed., Academic Press, New York, NY, 1974, 1465.
43. Maurer, C. and Poppendiek, B., Determination with lactate dehydrogenase and APAD, in *Methods of Enzymatic Analysis*, 2nd ed., Bergmeyer, H.U., Ed., Academic Press, New York, NY, 1974, 1472.
44. Noll, F., Determination with LDH, GPT and NAD, in *Methods of Enzymatic Analysis*, 2nd ed., Bergmeyer, H.U., Ed., Academic Press, New York, NY, 1974, 1475.
45. Beeley, J.G., Glycoprotein and proteoglycan techniques, in *Laboratory Techniques in Biochemistry and Molecular Biology*, vol. 16, Burdon, R.H. and van Knippenberg, P.H., Eds., Elsevier Science Publishers (Biomedical Division), Amsterdam, The Netherlands, 1985.
46. Vogel, K.G. and Peters, J.A., Isolation of proteoglycans from tendon, in *Proteoglycan Protocols*, Iozzo, R.V., Ed., Humana Press, Totowa, NJ, 2001, 9.

Chapter 16

Nucleotides and Its Derived Compounds

María-Concepción Aristoy and Fidel Toldrá

Contents

16.1 Introduction

Adenosine triphosphate (ATP) depletion is the actual cause of the onset of rigor mortis.[1] Once the animal is slaughtered, ATP is rapidly split into adenosine diphosphate (ADP) and subsequently into adenosine monophosphate (AMP) and other derived compounds. The rate of ATP disappearance depends on several factors during postmortem, the muscle metabolism status being the most important. In fact, ATP turnover in exudative meats is two- to threefold faster than in normal meats. This means that normal muscle needs about 4–6 h to reach the same concentration of nucleotides and nucleosides that are already formed in the exudative samples at

279

2 h postmortem.[2,3] The meat industry demands high, predictable, and consistent meat quality. This chapter presents typical procedures for the extraction and analysis of nucleotides and nucleosides in meat. The analysis of these compounds may help for a better classification of meat, which is also described in this chapter.

16.2 Nucleotide Breakdown in Postmortem Muscle

ATP is the main source of energy in the muscle for the biochemical reactions. ATP is necessary in the muscle to provide the required energy to drive the Na/K pump of the membranes, to drive the calcium pump in the sarcoplasmic reticulum, and to provide energy for muscle contraction and relaxation.[4] The ATP concentration is usually approximately 5–8 µmol/g in resting muscle but this amount drops to final negligible values in a few hours postmortem, when the rigor state is reached. This is related to the formation of actomyosin and the loss of extensibility.[5] At very early postmortem, there is an apparent stability of ATP due to its formation from creatine phosphate through creatine kinase and also through anaerobic glycolysis.[1,3,6] But the efficiency is not so good as only 2 moles of ATP are produced per mole of glucose through anaerobic glycolysis versus 12 moles that are produced under aerobic conditions (living state). Once both creatine phosphate and glycogen are exhausted, or the involved enzymes inactivated, ATP rapidly drops within a few hours postmortem by converting into ADP, AMP, and other derived compounds (see Figure 16.1).[4] ADP and AMP act as intermediate compounds and both decrease to negligible values after 24–48 h postmortem. ADP is transformed into inosine diphosphate (IDP) that is depleted in a few days, and AMP that is deaminated by the action of AMP deaminase into 5′-inosine monophosphate (IMP). Then, IMP can be further degraded into inosine and hypoxanthine (Hx) that experience a substantial increase as muscle is aged.[7] In a similar way, although at lower concentrations, 5′-guanosin monophosphate (GMP) can be degraded into guanosine and guanine. Common end product from both groups of reactions is xanthine that can be further hydrolyzed to uric acid,[8] more frequently associated to the presence of microbial flora.

The reaction rates depend on the metabolic status of the animal before slaughter and the postmortem conditions, especially the pH and the temperature of the meat. For instance, ATP can be almost completely depleted very rapidly in pale, soft, and exudative (PSE) muscles. In general, the rate in the muscle postmortem proceeds much faster in PSE meats than in normal ones.[3,9–12] The rate is also affected by the pH and the temperature of the meat.[12,13] For instance, the ATP content in beef *sternomandibularis* drops very fast when kept at 38°C (0.5 µmol/g at 6–7 h postmortem) but lasts longer when kept at lower temperatures, for example 10–15°C (approximately 3.5 µmol/g at 8–9 h postmortem).[14]

16.3 Sample Preparation

Early postmortem muscle samples are very sensitive to the temperature. ATP is rapidly degraded during the first 8 h postmortem and this means that samples must be removed as soon as possible from the carcass and immediately frozen with liquid nitrogen to stop all enzymatic reactions. To achieve this rapid freezing, it is advisable to collect small tissue samples and immerse them into the liquid nitrogen. This facilitates its immediate freezing. Tissue samples may be obtained from the fifth vertebra of the muscle *longissimus* dorsi or any other location of interest (i.e., *semimembranosus* muscle for ham producers).

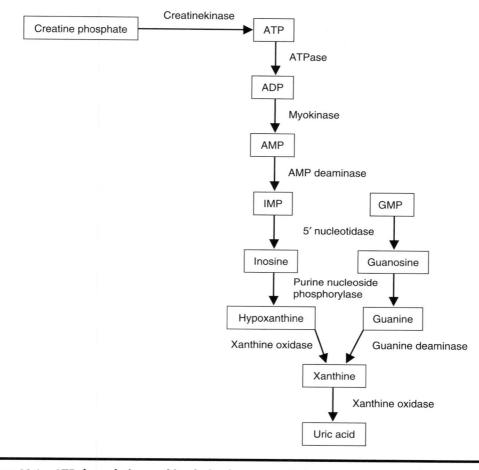

Figure 16.1 ATP degradation and its derived compounds in postmortem muscle.

An appropriate sampling time to discriminate between high and low qualities may vary depending on the slaughterhouse practice performed in each country. For instance, pork slaughter and evisceration in Spanish slaughterhouses usually take about 30 min and afterward carcasses spend near 1.5 h in the freezing tunnel. This makes a total of 2 h postmortem until they leave the freezing tunnel.

16.4 Extraction of Nucleotides and Nucleosides

A typical extraction procedure is the following: Five grams or less of muscle tissue are excised and rapidly frozen with liquid nitrogen. The frozen tissue is minced, avoiding any thawing, added 3 vol. cold 0.6 M perchloric acid and homogenized with a stomacher-type homogenizer for 4 min at 4°C. Then, the extract is centrifuged at 15,000 × g for 20 min at 4°C and the supernatant is filtered through glass wool and neutralized by adding solid potassium carbonate.[15] This neutralized extract is kept in ice bath for 5 min and centrifuged again at 15,000 × g for 10 min. The supernatant is filtered through a 0.2 μm membrane filter and stored in frozen storage (at temperatures below −25°C; if possible −80°C) until analysis.

16.5 Chromatography Analysis

Several methodologies are available for the analysis of nucleotides and nucleosides in muscle. Initially, these compounds were determined by classical ion exchange chromatography,[9] but these methodologies were tedious and gave poor resolution of the analyzed compounds. With the development of HPLC, there were new possibilities and anion exchange HPLC was used for the analysis of nucleotides.[16,17] More recently, other methods based on reverse-phase HPLC with the optional addition of ionic pairs have given good analytical separations and recoveries. These methods are described in the following sections.

16.5.1 Reversed-Phase HPLC

The chromatographic analysis is performed in a liquid chromatograph equipped with a UV detector (254 nm). The column used is a reverse-phase RP-18 (5 µm), 250, or 125 mm × 4 mm while temperature does not need to be controlled.[18] There are many approaches to analyze nucleotides and nucleosides by this technique that vary primarily on the pH of the mobile phase. The best option because of its resolution and simplicity is employing a mobile phase gradient between two solvents: phosphate buffer at pH 7 and methanol.[19] The identification of the chromatographic peaks can be performed by comparing the peak retention times and spectral characteristics with those of standards. Quantitative analysis can be performed using an external standard method. A chromatogram obtained with pure standard compounds is shown in Figure 16.2, whereas Figure 16.3 shows an example of this method applied to an early postmortem (2 h postmortem) and to an aged (5 days) pork meat. As can be observed, the peak

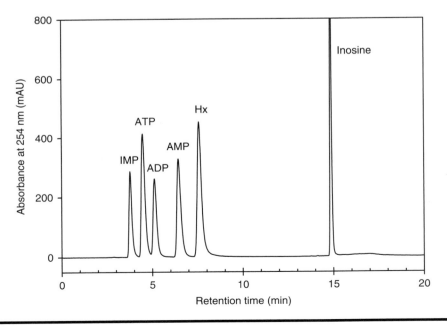

Figure 16.2   Reverse-phase chromatography of pure standards: ATP, ADP, AMP, IMP, Inosine, and Hx.

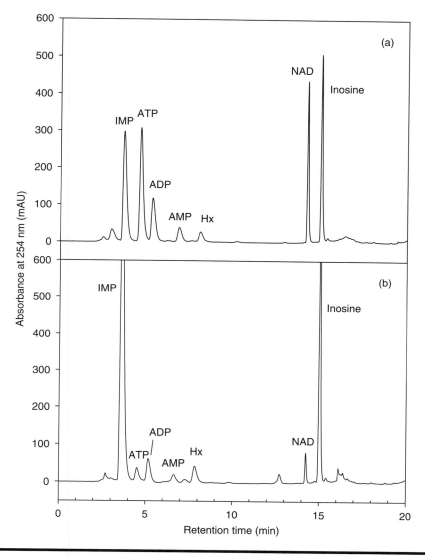

Figure 16.3 **Reverse-phase chromatography of ATP and its degradation compounds present in early postmortem pork meat sampled at 2 h (a) and 8 h (b) postmortem.**

corresponding to ATP decreases dramatically during maturation in cold storage, whereas peaks corresponding to inosine and IMP increase considerably.

16.5.2 *Ion-Pair Reversed-Phase HPLC*

The use of ionic suppressor is reported to increase the retention of nucleotides (in special tri- and di-nucleotides) in reverse-phase columns, improving the resolution[20] and making their analysis easier. Ionic suppression is performed by the addition to the mobile phase of a counterion opposite to that of the molecule to be analyzed. In this way, two kinds of ionic suppressors are used depending on the aim of the analysis. To analyze nucleotides (mono-, di-, and tri-nucleotides), quaternary ammonium salts such as tetrabutylammonium hydrogen phosphate or sulfate are used. The separation, identification, and quantification of compounds are performed as described in Section 16.5.1. Some examples of this method have been applied to the determination of nucleotides in meat,[13] fish,[19] or human tissues.[21] However, to analyze purine and pyrimidine bases, together with nucleosides, anionic counterion showed to be useful. It is the case of 1-octane sulfonic used as ion pair to separate free purines and pyrimidine bases and metabolites in chicken and turkey meat extracts.[22]

Figure 16.4 shows the liquid chromatograms of a standard solution (a) and 2 h postmortem pork meat (b) obtained by using tetrabutylammonium sulfate as paired ion. Chromatographic conditions are those described in Ref. 12.

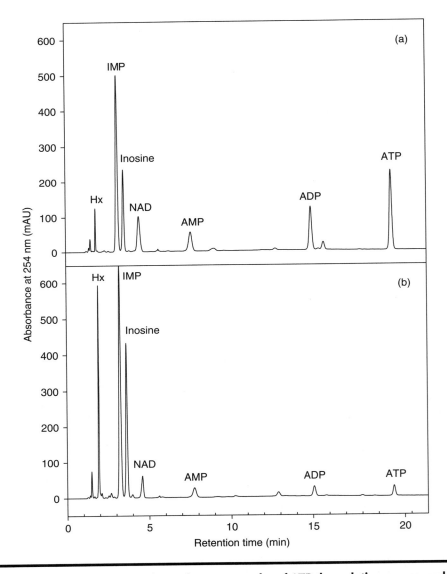

Figure 16.4 **Ion-paired reverse-phase chromatography of ATP degradation compounds present in pork meat at 2 h postmortem (a) and after 6 days of aging (b).**

16.6 Applications to Meat Quality

16.6.1 Prediction of Meat Quality

The high incidence of PSE meats results in poor quality, high dripping losses, and shrinkage during storage, making these types of meats unattractive to consumers.[23] Initially, the measurement of pH at 45 min postmortem was considered as a valid detection,[24] but this measurement alone is subjected to gross classification errors. The availability of rapid methods for early postmortem detection of these meats is very important for the meat industry because of the substantial loss in sensory quality and its strong economical significance. This has prompted the development of several methods based on the analysis of nucleotides and nucleosides that are briefly described.

The breakdown of ATP and the formation of IMP and inosine occur more rapidly in PSE muscles but slower in dark, firm, and dry (DFD) muscles than in normal ones. This means that PSE meats contain very low levels of ATP and high levels of IMP at few hours postmortem, whereas normal meats contain relatively high concentrations of ATP and low amounts of IMP.[9–11] Tables 16.1 and 16.2 show examples of the evolution of ATP-related compounds in normal and exudative postmortem pork meats, respectively. The different ATP degradation rates are evident while comparing both tables. This observation has constituted the basis for different methods of early postmortem detection of PSE and DFD meats. One of these methods

Table 16.1 Example of Typical Progress of ATP and Its Major Degradation Compounds in Postmortem Pork Meat of Normal Quality

Compound	2 h Postmortem	4 h Postmortem	8 h Postmortem
ATP	4.39	3.10	0.29
ADP	1.08	0.96	0.45
AMP	0.14	0.42	0.67
IMP	0.62	2.80	4.91
Inosine	0.15	0.40	0.69
Hypoxanthine	0.05	0.12	0.19

Note: Concentrations expressed as µmol/g muscle.
Source: Adapted from Batlle, N., Aristoy, M-C., and Toldrá, F., *J. Food Sci.*, 66, 68–71, 2001.

Table 16.2 Example of Typical Progress of ATP and Its Major Degradation Compounds in Postmortem PSE Pork Meat

Compound	2 h Postmortem	4 h Postmortem	8 h Postmortem
ATP	1.39	0.39	0.21
ADP	0.91	0.38	0.23
AMP	0.38	0.36	0.22
IMP	3.31	5.28	5.90
Inosine	0.61	0.92	1.31
Hypoxanthine	0.13	0.27	0.36

Note: Concentrations expressed as µmol/g muscle.
Source: Adapted from Batlle, N., Aristoy, M-C., and Toldrá, F., *J. Food Sci.*, 66, 68–71, 2001.

consists in the measurement of the R value. This value is the absorption ratio at 250 and 260 nm and only takes 3–4 min, making its use attractive for quality control at the slaughter line.[24]

The R' value is a variation of the R value and consists of the rate between the inosine-related compounds and the adenosine-related compounds. It was used by Honikel and Fischer,[25] based on the ratio of absorbance at 250 and 260 nm, to distinguish among PSE, normal, and DFD meats at 45 min postmortem. This value has shown utility to discriminate between qualities also at 2 h and up to 8 h but fails the discrimination when reaching 24 h.[12] Another simple ratio is the IMP/ATP ratio that has also been proposed as a simpler value to detect exudative pork meats. This index is very useful for discrimination between qualities at 2 h postmortem, but is not useful for aged meat because of its high standard deviation.

The K_0 value, which is the ratio of the inosine plus Hx contents to the total ATP metabolites content $\times$ 100, was introduced as an indicator of fish freshness[26] and afterward as an index of freshness in beef and rabbit meat.[27] This index was applied to pig and poultry muscles.[28] However, the procedure does not allow a rapid detection of exudative meats.

In general, all these ratios drastically increase when pH drops, especially at pH lower than 6. To summarize, the analysis of ATP-related compounds at 2 h postmortem is considered the best time to discriminate between exudative and nonexudative meats.[13] The other analysis can be achieved by analyzing ATP within 6 h postmortem, ADP between 4 and 6 h, IMP within 4 h, inosine or R' value within 8 h, Hx within the first day postmortem, and the K_0 value during the entire seven days of aging.

The analysis of IMP, inosine, and Hx may also be useful to evaluate the time postmortem of a refrigerated meat. In fact, these values may give an indication of the freshness of meat by determination of the approximate number of days postmortem.

16.6.2 Flavor Contribution to Meat

Umami taste is defined as the taste resulting from some compounds like 5′-inosine monophosphate and 5′-guanosin monophosphate. These compounds also constitute important flavor enhancers or potentiators.[29] Nucleotides are present in larger amounts in meat and have an important contribution to meat flavor. So, the contents of flavor nucleotides have been studied in different meats to evaluate their contribution to the final flavor. Some examples of analysis performed in a variety of meats are electrical stimulated beef muscle,[30] ingredients for meats,[31] chicken muscle,[32] and cooked duck.[33]

References

1. Greaser, M.L. Conversion of muscle to meat. In: *Muscle as Food*, P.J. Bechtel, Ed. Orlando, FL.: Academic Press. 1986, 37–102.
2. Jensen, P. and Barton-Gade, P. In: *Stress Susceptibility and Meat Quality in Pigs*. J.B. Ludvigsen, Ed. *EAAP Publication No. 33*, Jelling Bogtrykkeri A/S, Roskilde, Denmark, 1985, 80.
3. Moesgaard, B. et al. Differences of *post-mortem* ATP turnover in skeletal muscle of normal and heterozygote malignant-hyperthermia pigs: Comparison of PNMR and analytical biochemical measurements. *Meat Sci*. 39, 43–57, 1995.
4. Toldrá, F. Meat: Chemistry and biochemistry. In: *Handbook of Food Science, Technology and Engineering*. Y.H. Hui, J.D. Culbertson, S. Duncan, I. Guerrero-Legarreta, E.C.Y. Li-Chan, C.Y. Ma, C.H. Manley, T.A. McMeekin, W.K. Nip, L.M.L. Nollet, M.S. Rahman, F. Toldrá, and Y.L. Xiong, Eds. Volume 1. CRC Press, Boca Raton, FL, 2006, 28-1–28-18.

5. Eskin, N.A.M. Biochemical changes in raw foods: Meat and fish. In: *Biochemistry of Foods*, 2nd ed. Academic Press, San Diego, CA, 1990, 3–68.

6. Kocwin-Podsiadla, M. et al. Muscle glycogen level and meat quality in pigs of different halothane genotypes. *Meat Sci.* 40, 121–125, 1995.

7. Burns, B.G. and Ke, P.J. Liquid chromatographic determination of hypoxanthine content in fish tissue. *J. Assoc.Off. Anal. Chem.* 68, 444–448, 1985.

8. Urich, K. Small nitrogen compounds. In: *Comparative Animal Biochemistry*. Springer-Verlag, Berlin, 1990, 403.

9. Tsai, R. et al. Studies on nucleotide metabolism in porcine longissimus muscle postmortem. *J. Food Sci.* 37, 612–616, 1972.

10. Lundstöm, K. et al. Effect of halothane genotype on muscle metabolism at slaughter and its relationship with meat quality: A within-litter comparison. *Meat Sci.* 25, 251–263, 1989.

11. Essén-Gustavsson, B. et al. Adenine nucleotide breakdown products in muscle at slaughter and their relation to meat quality in pigs with different halothane genotypes. Proc. Int. Congr. Meat Sci. Technol., 37th , Vol 1, Kulmback, Germany, 1991.

12. Batlle, N., Aristoy, M-C., and Toldrá, F. Early postmortem detection of exudative pork meat based on nucleotide content. *J. Food Sci.* 65, 413–416, 2000.

13. Batlle, N., Aristoy, M-C., and Toldrá, F. ATP metabolites during aging of exudative and non-exudative pork meats. *J. Food Sci.* 66, 68–71, 2001.

14. Toldrá, F. Biochemistry of processing meat and poultry. In: *Food Biochemistry and Food Processing.* Y.H. Hui, W.K. Nip, M.L. Nollet, G. Paliyath, and B.K. Simpson, Eds. Blackwell Publishing, Ames, IA, 2006, 315–335.

15. Lehninger, A.L., Nelson, D.L., and Cox, M.M. *Principles of Biochemistry*, 2nd ed., Worth Publishers, New York, 1993, 727–729.

16. Iwamoto, M. et al. Effect of storage temperature on rigor mortis and ATP degradation in plaice *Paralichtithys olivaceus* muscle. *J. Food Sci.* 52, 1514–1517, 1987.

17. Zhang, Y.W., Lu, H.J., and Wang, J.S. Determination of nucleotides in foods by HPLC. *Food Sci. China* 6, 59–62, 1994.

18. Meynial, I., Paquet, V., and Combes, D. Simultaneous separation of nucleotides and nucleotide sugars usind ion-pair reversed-phase HPLC: application for assaying glycosyltransferase activity. *Anal. Chem.* 67, 1627–2631, 1995.

19. Veciana-Nogués, M.T., Izquierdo-Pulido, M., and Vidal-Carou, M.C. Determination of ATP related compounds in fresh and canned tuna fish by HPLC. *Food Chem.* 59, 467–472, 1997.

20. Murray, J. and Thomson, A.B. Reverse phase ion pair separation of nucleotides and related products in fish muscle. *J. High Res. Chromatogr. Commun.* 6, 209–210, 1983.

21. Bernochi, P. et al. Extraction and assay of creatine phosphate, purine, and pyridine nucleotides in cardiac tissue by reversed-phase high-performance liquid chromatography. *Anal. Biochem.* 222, 374–379, 1994.

22. Scarborough, A. et al. Investigation of the levels of free purine and pyrimidine bases and metabolites in mechanically recovered meats. *Meat Sci.* 33, 25–40, 1993.

23. Flores, M. et al. Sensory characteristics of cooked pork loin as affected by nucleotide content and post-mortem meat quality. *Meat Sci.* 51, 53–59, 1999.

24. Warris, P.D. The relationship between pH45 and drip loss in pig muscle. *J. Food Technol.* 17, 573–578, 1982.

25. Honikel, K.O. and Fischer, C. A rapid method for the detection of PSE and DFD porcine muscle. *J. Food Sci.* 42, 1633–1636, 1977.

26. Saito, T., Arai, K., and Matsuyoshi, M. A new method for estimating the freshness of fish. *Bull. Jpn. Soc. Sci.* 24, 749–752, 1959.

27. Nakatani, Y. et al. Changes in ATP related compounds of beef and rabbit muscles and a new index of freshness of muscle. *Agric. Biol. Chem.* 50, 1751–1756, 1986.

28. Fujita, T. et al. Applicability of the K_0 value as an index of freshness for porcine and chicken muscles. *Agric. Biol. Chem.* 52, 107–112, 1988.

29. Maga, J.A. Umami flavor of meat. In: Flavor of Meat and Meat Products. F. Shahidi, Ed. Chapman and Hall, NY, 1994, 98–115.

30. Calkins, C.R. et al. Concentration of creatine phosphate, adenine nucleotides and their derivatives in electrically stimulated and nonstimulated beef muscle. *J. Food Sci.* 47, 1350–1353, 1982.

31. Fish, W.W. A method for the quantitation of 5′ mononuleotides in foods and food ingredients. *J. Agric. Food Chem.* 39, 1098–1101, 1991.

32. Aliani, M. and Farmer, L.J. Precursors of chicken flavor. I. Determination of some flavor precursors in chicken muscle. *J. Agric. Food Chem.* 53, 6067–6072, 2005.

33. Liu, Y., Xu, X., and Zhou, G. Changes in taste compounds of duck during processing. *Food Chem.* 102, 22–26, 2007.

PROCESSING CONTROL

Chapter 17

Basic Composition: Rapid Methodologies

Alfredo Teixeira

Contents

17.1 Introduction

Meat is a protein source with an important role in the human diet. Despite the great number of possible sources of meat, cattle, sheep, goats, pigs, and poultry are the most important meat producers' species. Differences between species exist in terms of the principal tissues (muscle, fat, and bone), and within the same species the content of these components differs according mainly to breed, age, sex, commercial category, and production or feed system.

Normally the word meat means the flesh of animals used as food for consumers. Some consumers associate meat with the negative image of fat consumption, high cholesterol levels, and heart disease, but adipose tissue in meat is not altogether undesirable or wasteful. Subcutaneous fat in appropriate quantity is desirable in terms of carcass conformation; also, for example, the fat deposited inside the muscles, the intramuscular fat commonly known as marbling, confers juiciness, improves the flavor, and makes meat tender and more succulent when cooked. Although a minimum level of fat is required to assure juiciness and flavor, from the point of view of human health, consumers have an increased concern for their diet and in general show preferences for leaner meat, with less or no fat. Consequently, in terms of carcass value, a knowledge of tissue composition, distribution, and partitioning of fat and muscle units has become more and more important for consumers, packers, processors, retailers, and producers.

"Meat quality" animals should be evaluated with reference to two important factors:

1. Quality parameters such as tenderness, muscle and fat color, flavor, and marbling
2. Composition such as saleable meat or tissue proportions (muscle, fat, and bone)

The attempt to assess body or carcass composition has a long history. Several methodologies have been tried and tested, mainly with objectives related to genetic improvement and commercial carcass classification.

17.2 General Methodologies

Body and carcass composition, particularly fat deposits, can be assessed using several methodologies, including subjective measurements, live or carcass weight, linear measurements, the use of carcass joint compositions as predictors, the number and size of adipocytes, dilution techniques, underwater weighing, optical probes, video image analysis (VIA), total body electrical conductivity, bioelectrical impedance analysis (BIA), ultrasounds, computer tomography, and magnetic resonance imaging. Although there are different levels of accuracy in tissue prediction and different relations between weak and strong points, all methodologies are valuable; however, only some of them can be considered rapid methodologies.

The dilution techniques for estimating body water using radionucleotides [1] or urea [2] as well as the method of underwater weighing [3], deuterium oxide [4], or the number and size of adipocytes [5] require an amount of time only suitable under specific experimental conditions and research studies.

The dissection of small carcass joints as predictors of carcass or body composition [6–8], while precise, requires time, and the regression equations to predict overall composition should be determined for each breed and should be used only under the same environmental and experimental conditions.

Of all the cited methods, computer tomography and magnetic resonance imaging are the most accurate in predicting composition in live animals, but the high cost of the equipment relative to other methods, and the exposure of the subject to radiation, limit the use of these technologies in animal science. Furthermore, the use online of computer tomography or magnetic resonance imaging takes time not compatible with a slaughter line in an abattoir. Nevertheless, in breeding programs, computer tomography scanning of elite animals presents a clear and convincing case for supplementation of ultrasound scanning, improving the genetic process [9,10].

To respond to consumer demand, it is becoming increasingly important to provide adequate information about carcass composition such as saleable meat, fat, muscle, and bone composition. There is an increasing need for more accurate, rapid, and inexpensive methods to assess body and carcass composition for all species. It is the purpose of this chapter to review the methodologies used to assess the basic body or carcass composition in such a way as to find the most appropriate, efficient, rapid, and inexpensive procedure.

17.3   Rapid Methodologies

Efforts to predict body or carcass composition of live animals have been made over a long period of time. The first attempts to determine the general condition of animals amounted to the use of methodologies such as subjective measurements of live weight.

The preliminary use of new electronic methods to predict body or carcass composition of live animals has had limited success because of the dynamic of body composition as the result of growth, feeding systems, and variability between and within species.

Our understanding and quantification of the rapid advances in electronic and computer sciences have advanced and significantly stimulated some methods as tools to assess carcass or body composition in live animals. The relation between cost and ease of use, and the suitability to the meat industry or slaughter line of the abattoir, are probably the most important factors for the success of a methodology to quickly assess basic composition.

17.3.1   Subjective Measurements

Visual assessment, body dimensions, and handling of the live animals in association with a body condition score are the most rapid and cheapest methods for predicting the body composition *in vivo* [11,12]. These do not require transportation of the animals or the use of special equipment. They are useful for all people interested in meat production and commercialization, and are practised under several environmental conditions, including in the field under farm conditions [6,8].

The major problems with these techniques are related to the difficulty of distinguishing between lean and fat, and of assessing fat deposition in the different physiological conditions of the animal, especially in those that are associated with mountain grazing, which are involved in fat mobilization or deposition periods, and lactation or pregnancy periods. The precision obtained by these techniques is very variable and is particularly influenced by the degree of fatness or the conformation of the animals. As several authors have demonstrated, conformation is related to fatness, and normally animals with good conformation are fatter than others. In spite of the number of standardized methods developed, these methodologies still have great variations due to the subjectivity of the operators, low incidence of repeat measurements between different operators, and the accuracy with which the measurements or the evaluations are taken.

The visual assessment of fatness using different photographic reference scales is a cheap method to predict the fat content of beef and sheep carcasses, and it is used in commercial abattoirs in several countries as a method of commercial carcass classification. Despite all attempts at standardization, the accuracy of the method as a predictor of carcass composition is still largely subjective, mainly due to the experience of the judges and the environmental conditions under which evaluations are made. As a result, visual scoring assessments are only appropriate under commercial conditions to classify carcasses for payment.

17.3.2 Live and Carcass Weight

The importance of live weight stems from the general knowledge that as an animal grows, its body or carcass composition changes; for example, fat normally increases in relation to muscle and bone. In several experiments across all species, and according to different feed or rearing systems, live weight shows a positive correlation with fatness independent of the variation in live weight or body condition [6].

Live weight is a predictor with negligible cost, is available in many circumstances in which it is easy to use, and is included in prediction equations that use other variables as predictors. In most experimental situations, live weight is the first variable in association with other variables in multiple regression equations. One of the most important problems affecting the use of live weight is its dependence on the way the animals are fed, the environmental conditions in which they are reared, the existence of any disease that alters the growth rate, the gut content (there is a close relationship between level of feed intake and the live weight gain/loss and the weight of some noncarcass organs according to Aziz et al. [13]) and the stage of growth of the animal.

In addition to live weight, carcass weight can be used as a predictor of carcass composition. Easy and cheap to use in abattoir conditions, carcass weight is normally included as a first independent variable in association with other kinds of predictors in multiple regression equations.

17.3.3 Electronic Technologies

Optical reflectance probes [14], BIA [15–19], electromagnetic scanning [20,21], fiber-optic spectroscopy [22–24], and real-time ultrasonography (RTU) [25–29] have been tested in several species for assessing body or carcass composition. Most of these technologies have been used mainly to assess carcass composition; Table 17.1 shows the accuracy obtained with different methods in several species. Cross and Belk [30] reviewed several technologies and objective measurements of carcass and meat quality and stated that some may be ready for commercial use, but that others required further evaluation. To evaluate fat and muscle quality for market, a rapid and cheap method is required. Further tests of all of these technologies should be made in an industrial setting before the adoption of any particular one, testing to their usefulness and adaptability to the slaughter line in an abattoir, cost of the equipment, scanning time required, speed, repeatability, and accuracy of estimation.

Recently, whether to assess body composition in live animals or carcass composition, new methods (sometimes based on old ones), easily performed and nondestructive to live animals, have been developed for rapid evaluation. In addition to the use of the correct technology, the use of advanced statistical analyses such as multiple regression, neural networks application [31,32], or support vector machines [33] may also improve the precision of predictions.

Backfat probes have been used, first in live pigs, to measure fat thickness. The oldest techniques used on live hogs are scalpel and ruler; a small incision on the back of the animal is required to insert a metal ruler. For animal welfare reasons, this procedure is no longer permitted in several countries.

The increasing need for an objective carcass grading system for market evaluation, to estimate carcass composition, and to assess fat and meat quality has led in recent years to the development of equipment that is objective, rapid, compatible with slaughter time, and able to measure fat and muscle thickness in intact carcasses. Several technologies have been evaluated to determine their accuracy and precision in predicting carcass components; some of them are utilized as probes in commercial carcass grading and classification systems such as (see Figures 17.1 and 17.2): hand-held

Table 17.1   Different Rapid Methodologies to Assess Body and Carcass Composition in Several Species

References	Animals	Methodology	Prediction	R^2	Error
Gresham et al. [64]	Swine	Ultrasound + BW	Body lean (live)	.56	RSD = 3.84
			Body lean (carcass)	.61	RSD = 3.61
			TOTLLEAN (live)	.60	RSD = 3.70
			TOTLLEAM (carcass)	.64	RSD = 3.52
			KGEEFAT (live)	.57	RSD = 2.75
			KGEEFAT (carcass)	.60	RSD = 2.65
Berg et al. [20]	Lambs	BIA + BW	FFM (live)	.78	RMSE = 1.04
			FFM (hot)	.78	RMSE = 1.04
Berg and Marchello [15]	Lambs	BIA + BW	FFST (live)	.78	RMSE = 1.77
			FFST (hot)	.79	RMSE = 0.92
Berg et al. [20]	Lambs	TOBEC + HCW	Lean weight	.98	RMSE = 0.35
			% Dissected lean	.79	RMSE = 1.39
Kirton et al. [35]	Lambs	HGP + HCW	% Fat	.47	RSD = 3.07
			% Water	.44	RSD = 2.39
		Aus-Meat + HCW	% Fat	.38	RSD = 3.25
			% Water	.34	RSD = 2.62
		Ruakura probe + HCW	% Fat	.31	RSD = 3.46
			% Water	.28	RSD = 2.77
Berg et al. [26]	Lambs	Ultrasound + HCW	Lean weight	.48	RMSE = 1.25
			Fat-free lean weight	.41	RMSE = 1.29
		Optical grading probe + HCW	Total dissected lean	.55	RMSE = 1.02
			Fat-free lean	.53	RMSE = 1.06
		BIA + HCW	Total dissected lean	.81	RMSE = 0.89
			Fat-free lean	.78	RMSE = 0.94
		TOBEC + HCW	Total dissected lean	.88	RMSE = 0.73
			Fat-free lean	.88	RMSE = 0.71
Swantek et al. [18]	Pigs	BIA + BW	Fat-free mass (live)	.98	RMSE = 2.83
Velazco et al. [19]	Steers	BIA + BW	Fat-free mass (live)	.98	SEE = 9.03
Irie [22]	Porcine	Fiber-optic spectroscopy	SFAC	.73	
			MFAC	.68	
			PFAC	−.76	
Marchello et al. [48]	Beef	BIA + BW	% Fat of beef trim	.80	RMSE = 6.64
Delfa et al. [102]	Goats	Ultrasound + BW	Carcass fat weight	.92	RSD = 0.22
			Body fat weight	.94	RSD = 2.01
Schinckel et al. [43]	Pork	Ultrasound + BW	Fat-free mass weight	.83	RSD = 2.58
			Lipid-free soft mass	.91	RSD = 2.13
			Dissected lean mass	.90	RSD = 1.76
			Total fat mass	.88	RSD = 3.20
Steiner et al. [117]	Beef	VIA	M. longissimus area	.93	RSD = 3.48
Steiner et al. [118]	Beef	CVS	Fat PSPC	.51	RSD = 0.022
		VIASCAN	Fat PSPC	.46	RSD = 0.023
Johnson et al. [40]	Pork	FOM + HCW	Fat-free lean weight		RSD = 3.57
		UFOM			RSD = 3.62
		Ultrasound scan			RSD = 3.06
		AUS			RSD = 3.46

Table 17.1 (Continued)

References	Animals	Methodology	Prediction	R^2	Error
Silva et al. [28]	Lambs	Ultrasound + BW	Fat weight	.90	RSD = 0.76
			Protein	.83	RSD = 0.40
Bergen et al. [82]	Beef bulls	Ultrasound	Lean meat yield	.63	RSD = 18.9
Teixeira et al. [29]	Lambs	Ultrasound + BW	Muscle weight	.96	RSD = 0.21
			Carcass fat weight	.88	RSD = 0.07
Aass et al. [85]	Cattle	Ultrasound	CHIMF %	.48	RMSE = 0.46

Note: BW, body weight; TOTLLEAN, fat-free soft tissue; KGEEFAT, total ether-extractable fat; BIA, bioelectrical impedance analysis; FFM, fat-free mass; FFST, fat-free mass soft tissue; TOBEC, total body electrical conductivity; HCW, hot carcass weight; HGP, Hennessy grading probe; SFAC, saturated fatty acid content; MFAC, monounsaturated fatty acid content; PFAC, polyunsaturated fatty acid content; VIA, video image analysis; CVS, computer vision system; Fat PSPC, fat from the subprimal cuts; FOM, Fat-O-Meter; UFOM, UltraFOM; AUS, automated ultrasonic system; CHIMF, chemical analysis of intramuscular content.

Figure 17.1 Different electronic probes: Hennessy Grading Probe (by Hennessy Grading System Ltd., with permission), CGM Probe (by Sydel, with permission), and Fat-O-Meter (by SFK Technologies, with permission).

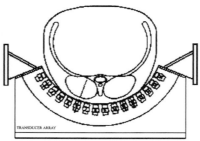

Figure 17.2 Ultrasound probes: UltraFOM 300 and the AutoFOM scheme by SFK Technologies, with permission.

optical probes (Fat-O-Meter [FOM]; SFK Technologies, Herlev, Denmark); the reflective spectroscopy probes, such as the Hennessy Probe (Hennessy Grading Systems Ltd., Auckland, NZ); the PG-100 electronic Pork Grader and the CGM version 01-A by Sydel; the AUS-Meat Sheep Probe (SASTEK, Hamilton, Queensland, Australia); the automated ultrasound scanning devices (AutoFOM; SFK Technologies); and bioelectrical impedance, total body electromagnetic conductivity (TOBEC), and ultrasonic scanning. Some of these have been tested at slaughter chains as noninvasive systems.

TOBEC (Meat Quality Inc., Springfield, IL) is a large and relatively expensive piece of equipment that requires a large space on the line where carcasses are scanned; it may require operators to detach and attach the gambrels when the animals enter and leave the equipment [34]. These considerations limit the potential of the system to assess basic composition as a rapid, automated, and inexpensive method.

17.3.3.1 Electronic Probes

Optical probes objectively measure fat and muscle depth and improve quality control and the grading of fresh meat because they are better than electrical probes [21]. They are mainly used to assess meat quality in swine carcasses for backfat and longissimus dorsi muscle measurements (see Table 17.1). A great variety of instruments have been used in various circumstances, as can be seen on Figures 17.1 and 17.2. Fiber-optic methodologies are useful for the evaluation of porcine fat quality [22] and for estimation of bovine fat quality [24] using the optical method (HRS-6500, Optoelectronics, Tokyo, Japan); this was the same system as used previously in the study of porcine fat.

Some probes, such as the UltraFOM and CSB Ultra-Meter, ultrasonically measure backfat and longissimus dorsi depth (eye muscle) in pig carcasses. The AutoFOM ultrasound (SFK Technology) (see Figure 17.2) is a completely automatic scanner system based on ultrasound technology, providing a sophisticated system of pig carcass classification with great accuracy in assessing lean meat distribution and the different fat depots throughout the carcass. In pig carcasses with a clear separation between fat and lean compared to other species, this machine is very efficient.

Some of these probes have been used for sheep carcass evaluation. Three commercial probes—the Hennessy Grading Probe, the AUS-Meat Sheep Probe, and the FTC Lamb Probe (FTC Sweden, Upplands Väsby, Sweden)—were used by Kirton et al. [35] for classifying lamb carcasses and to compare accuracies in predicting GR fat measurement and carcass composition. Small differences were found between probes overseas, where probes are in use for objectively grading the carcasses of meat animals [35]. Jones et al. [36] used the Hennessy lamb grading probe for online grading in Canada. The AUS-Meat Sheep Probe (SASTEK) was tested for its ability to test 9–10 carcasses per minute [37]. Another probe tested in sheep carcass evaluation [35] was the Ruakura GR Lamb Probe (Hamilton, New Zealand). Hopkins [38] tested the Hennessy Grading Probe for measuring fat depth in beef carcasses. The fat probes created and used for pork meat evaluation have been of little use for beef or for sheep. In fact, the irregularity of the fat layer in these species is very high in comparison to pigs, and Chadwick and Kempster [39] have applied different probing instruments in subcutaneous fat measurements taken on the intact carcass, using them to estimate beef carcass composition. These authors concluded that the best fat thickness measurements taken by probe can provide a prediction of carcass lean percentage as precise as visual fat scores given by experienced operators. Later on, new techniques were found to assess rapidly beef carcass quality or composition, including BIA, RTU, and, more recently, VIA.

According to Swatland [21], one problem in predicting meat yield from fatness is to find a simple yet reliable measure of carcass fatness. Most of the backfat or longissimus dorsi measurements are principally used in prediction equations to assess the quality of meat; recent studies have shown that there could be more potential if they were taken after carcass chilling. Fat depth is a simple linear measurement, while total carcass fat is a complex anatomical volume, and it would be surprising if the former were a perfect indicator of the latter [21]. Johnson et al. [40] have developed different equations for predicting fat-free lean in swine carcasses and have estimated the prediction bias due to genetic group, sex, and dietary lysine level. They concluded that research is needed to develop new procedures and additional variables that can be measured at normal line speeds of packing plants to decrease the bias in prediction. In this regard, Schinckel [41], in a critique of evaluation procedures to predict fat-free lean in swine carcasses, said that the magnitude of the biases must be compared with the actual genetic population, gender, or treatment differences. Furthermore, as Gu et al. [42] and Schinckel et al. [43] reported, if the equations are to be used in the future, it is important to know what percentage of the total variation among the genetic populations is expected to be predicted, and also whether the prediction equations developed over larger body weight ranges can have a greater magnitude of body weight range and interaction between live weight and sex or genetic population range biases.

17.3.3.2 Bioelectrical Impedance Analysis

BIA is a simple, objective, and inexpensive method of analyzing body composition and measuring body fluid volumes. Essentially, the method uses the resistance of electrical flow through the body to estimate body fat and measure lean content, depending on the different electrical properties of lean and fat tissues. In the beginning, BIA was not generally accepted as an accurate methodology to assess body composition. The technological improvements made at the end of 1980s and the beginning of the 1990s made BIA a more reliable and acceptable methodology to predict body composition (see Table 17.1).

The four-terminal plethysmograph (BIA; Model BIA-101, RJL Systems, Detroit, MI) introduces a constant alternating current that provides a deep homogeneous electrical field in the body. A constant alternating current of 800 µA at 50 kHz is introduced into the body via transmitter terminals and received by detector terminals [44]. The alternating current is transmitted from each of the two outer electrodes to its respective opposite inner detector electrode. The voltage drop is measured with a high-input impedance amplifier [45]. The electrodes (21-gauge needles) may be located in an anterior to posterior sequence along the full length of the animal's back with 10 cm separation between transmitter and detector electrodes at each end [16].

Swantek et al. [46] evaluated the bioelectrical impedance in swine to predict body and carcass composition using the four-terminal plethysmograph (BIA; Model BIA-101, RJL Systems) with results indicating the excellent potential of BIA as a rapid and nondestructive procedure for the prediction of fat-free mass in live pigs and chilled pork carcasses. Berg and Marchello [15] reported that initial findings indicate that BIA had excellent potential as a way of predicting lean body mass in commercial situations, given its precision, simplicity, and portability. Marchello and Slanger [47] researched the potential of BIA to predict muscle and fat-free weight of beef cows, concluding that the system could be used as a value-based marketing tool and had potential in the genetic selection of superior animals.

The adequacy of the method in evaluating body composition in abattoir conditions has been debated by several authors. Working with lambs, Berg and Marchello [15], using BIA, concluded that the system could be incorporated easily into existing industrial packing plants; if adapted for online usage, the system could be more rapid for the analysis of carcasses than probing for fat depth. The prediction equations, as well as other methodologies, use live weight and carcass weight as independent variables in the regression models to estimate fat-free mass or fat-free soft tissue, and although higher coefficients of determination are reported and lower residual mean square errors are much larger, their total sample variation was explained by live weight or carcass weight. Swatland [21] pointed out that some type of penetration electrode was required in online measurements of carcasses. According to this author, two parallel needle electrodes (two pairs of detector and transmitter electrodes) are inserted in such a way that, in a system for online use, it is important to find the electrode orientation that responds most readily to the subject of measurement and then to standardize it rigorously. Berg et al. [17,26] said that one of the advantages of BIA is that measurements can be made in live animals and on carcasses, but the invasiveness of the procedure, as well as its low precision, would not favorably compare with other relatively inexpensive *in vivo* methods such as ultrasound. The usefulness of BIA methodology was examined by Velazco et al. [19] in determining the soft tissue composition of Holstein steers; they concluded that more research was needed to determine the effects of electrode placing as well as the magnitude and type of the electric impulse. Studying whether bioelectrical impedance could predict fat content of ground beef and pork, Marchello et al. [48] verified that it provides a system whereby a company can use marketing strategies to provide consumers with the type of product they want to purchase.

17.3.3.3 Ultrasound

Like other technologies such as bioelectrical impedance, ultrasound was first developed for human medicine. The first application of ultrasound technology in animals was probably for medical reasons, before it was recognized for its potential to predict carcass composition in animals and used extensively in pig breeding schemes and research. Stouffer [49] reported pioneering work in measuring the backfat thickness on beef cattle with a somascope unit by Temple et al. [50] and in pigs using an A-mode ultrasonic metal flaw detection device by Dumont [51] and Claus [52]. In 1959 Stouffer [53] recognized the necessity of measuring muscle mass (rib eye depth or area) in addition to backfat thickness for improving accuracy in predicting body composition in a series of A-mode measurements with the Sperry Reflectoscope (Sperry Company, Danbury, CT) in beef cattle.

With constant advances in technology, a range of ultrasonic equipment, from simple machines to more complex ones, have been developed and tested by researchers after being commercialized by the industry. Essentially, an ultrasonic machine consists of a pulse generator, a transducer probe, receiving circuits, and a display [54]. The transducer converts electrical pulses from ultrasound pulses that are traveling through the body with a velocity characteristic of the tissues they pass through, and sends a reflected pulse to the transducer when an interface between two different tissues (fat and muscle) is reached. The transducer reconverts this reflected pulse into an electrical pulse. The reflected waves picked up by the probe are relayed to the machine, which display the distances and the intensities of the echoes on the screen, forming a two-dimensional image. Generally two types of ultrasound equipments are used:

1. A-mode machines, which measure the amplitude of echo in function of time. These are now obsolete in medical imaging.
2. B-mode or real-time machines, which measure echo intensity in a two-dimensional scan, showing all the tissues screened by the ultrasound beam. Some machines have incorporated another imaging mode, the M-mode, which is another method to visualize the movement,

in which the result is a line with an abdominal standard sound with a high sampling frequency useful in cardiac and fetal imaging.

In fact, RTU is a version of B-mode, creating images which are seen instantaneously and changing as the position of the transducer changes. Several electronic companies have produced different scanners for human medicine that can be adapted and used for scanning farm livestock, machines that are cost-effective, reasonably robust, have great versatility, portability, are capable of automatic calibration, wide probe ranging, and provided with functions of measurement and calculation. Modern machines are equipped with some practical functions such as the split screen (allowing the observation of two images simultaneously or, for example, the observation of the longissimus muscle in beef cattle, which is greater than the width of the transducer), freeze frame (the possibility of freezing one image for a detailed study later), and a disk storage device. Most studies have been using another computer with an interpreting and analyzing image system, although modern machines have this facility built in.

RTU can be used in live animal evaluation because it uses a noninvasive technology, and carcass quality and composition can be assessed without damage to the carcass and the corresponding reduction in its market value.

To operate ultrasound equipment there are some factors that should be taken into account, including:

1. Animal conditioning is one of the most important factors in obtaining a good-quality ultrasound image. If the objective is to scan a live animal, it is essential to provide a system to keep the animal relaxed and not to modify the normal anatomical structure of the animal tissues. The animal surface must be completely clean of dirt or foreign material as well as air bubbles to avoid interference with the acoustics of the sound waves.

2. The operators should have a good knowledge of the anatomical points where the transducer will be placed. Live animal or carcass evaluations involve several fat thickness measurements in different parts of the body (the lumbar, rump, or brisket regions) and muscle area determinations in distinct anatomical points, and the accuracy and repeatability of these measurements depend upon a thorough knowledge of animal morphology and anatomy on the part of the operators. The placement of the transducer is an important factor to be taken into account, and operators should be able to find the rib and the corresponding vertebrae to place the transducer parallel to the ribs and close to the backbone, rump, or parallel to the brisket in each point they want to measure fat thickness, the area of the longissimus dorsi muscle (rib eye area), or to assess intramuscular fat deposition.

3. The transducer is one of the most important components of the ultrasound equipment. Modern transducers using piezoelectric material convert electrical energy to ultrasound. Then they transmit ultrasound and receive the reflected waves. Depending on the species (cattle, swine, sheep, or goats) the market offers a wide range of sizes and frequencies. A 7.5 MHz device has a short wavelength, low tissue penetration, and high resolution; it is used mainly to measure slight body tissues such as the subcutaneous fat thickness in sheep and goats. Otherwise, a 3.5 MHz transducer has a long wave with deep penetration and is normally used for live beef cattle carcass imaging and to collect images for estimation of carcass composition.

4. A couplant agent should be applied between the transducer and the tissue or body surface to be scanned to provide an efficient medium of transmission of sound waves and obtain a good quality image. Many couplant agents can be used, including a simple water bag, vegetable oil, paraffin oil, or a specific ultrasonic gel. The couplant should be at the same temperature as the external body temperature of the animal.

5. The velocity or speed of ultrasound waves increases with tissue density. The denser tissues such as bone reflect more of the sound waves than soft tissues such as fat and muscle. Most RTU machines are calibrated for average velocity in soft tissues or water.

6. Overall gain adjusts for the brightness of the image. Machines have two gain settings, proche/near and distal/far, and are normally automatic. Nevertheless, if the objective is to assess marbling in beef or intramuscular fat, gain settings should be standardized.

Basically, ultrasounds are used to assess carcass composition and quality, assessing the following factors:

1. Backfat determination on subcutaneous fat thickness
2. Loin eye muscle area or longissimus dorsi area
3. Percentage of lean
4. Percentage of carcass fat
5. Intramuscular fat estimation or muscle quality (marbling)

Real-time scanning has been one of the most frequently used technologies in recent years, as is shown in Table 17.1. Developed for use in human medicine to enable the possibility of rapid observations of internal physiological movements of organs, tissues, or fluids, such as the growth of a fetus inside a uterus, blood flow, or the beating of the heart, it was used for the first time to scan livestock by Horst [55], according to Kempster et al. [54], despite the fact that Stouffer et al. in 1961 [56] had shown the superior performance of the mechanical B-scan over A-mode technology for measuring fat thickness and the rib eye area in cattle and hogs.

Over the history of the application of ultrasonic technology to livestock, it has become evident that it has not always been either useful or applicable to assessing body or carcass composition. In the beginning some difficulties were found, related to the performance of the machines, particular circumstances of experiments, the scanning technique used, and the experience of the operator, but mainly with the animal species being studied. The effectiveness of ultrasound was not the same for all animal species. Houghton and Turlington [57] have published a review article about the application of ultrasound for feeding and finishing animals; they indicate that ultrasound is of potential use in educational and research efforts for swine, sheep, and beef cattle.

The results obtained in sheep have not been as successful as with other species as swine and beef. Particularly in goats, we had to wait until 1995 to see the first papers published with reference to the use of ultrasound to predict carcass composition [58–62]. Most of the accuracy of the use of ultrasonics in live animals depends on the relationship between small sections of the animal body and overall carcass composition.

In summary, advances in technology have meant that RTU is now the most common method used in livestock.

17.3.3.3.1 RTU in Swine

Houghton and Turlington [57] have suggested that ultrasound is useful in swine under field conditions. Producers, abattoirs, and retailers have been interested for many years in the ability to produce carcasses with a composition and quality consistent with consumer interests. Studying the commercial adaptability of ultrasonography to predict pork carcass composition from live animal and carcass measurements, Gresham et al. [63] have calculated regression equations for predicting carcass composition from ultrasonic carcass measurements and also from live animal ultrasonic measurements. The authors concluded that ultrasonography can have a place in a value-based marketing system, and the results have shown the potential of the methodology to estimate composition of both the live animal and the carcass. Using a single longitudinal ultrasonic scan, Gresham et al. [64] confirmed the accuracy (see Table 17.1) of the technique to be automated in providing information to predict carcass composition on live pigs as well as in the carcass. Ultrasound measurements on live pigs are taken anywhere from the first rib to the last lumbar vertebrae, with carcass measurements at the 10th or last rib being the most often used [64–66]. But Johnson et al. [40], in a recent study of predicting fat-free lean in pork carcasses, verified that the use of ultrasound and optical probe instruments to measure fat and muscle depth off the midline were more reliable than a single measurement of fat depth at the last rib.

17.3.3.3.2 Use of RTU in Cattle

During the last 20 years, a considerable amount of research and practical work with RTU has been made in cattle to assess composition and quality of live animals and carcasses. As in other species, the technology has used an offline computer image interpreting system for determining composition [67,68] and muscle quality [31,69]. Whittaker et al. [69] have also found that ultrasound was a promising technology for the development of an automated, quantitative grading system for beef animals. The effectiveness of ultrasound for measuring fat thickness, and the fact that this measurement can be combined with other live measurements to estimate percentage of fat, weight of carcass fat and lean, and percentage of carcass bone in beef finishing programs, was pointed out by Houghton and Turlington [57]. During the past 10 years, RTU has been used in beef cattle research and industry, in the evaluation of fat thickness and rib eye area, to predict intramuscular fat percentage in live animals, to predict carcass retail products, or to predict tenderness in carcasses [70–76]. Herring et al. [77] compared different RTU systems for predicting intramuscular fat in beef cattle; the most precise were the CPEC (Oakley, Kansas, developed by Kansas State University) and the CVIS (Critical Vision, Inc., Atlanta, GA). The potential for the use of ultrasound as a predictor of carcass quality or composition, and the application of ultrasound in high-speed slaughter operations, depends on the automated ultrasonic measuring equipment, as was concluded by Griffin et al. [27] in their study with beef cattle. The RTU offers the ability to assess accurately subcutaneous fat, which is the prime contributor to variations in lean composition of animals of similar weights [78]. Therefore, Tait et al. [79] have found that the inclusion of a linear measurement, gluteus medius muscle depth, could help the prediction of retail product with the aim of looking at the prediction of lean in the carcass using live ultrasound measurements. One of the most important factors affecting meat quality is related to the quantity of intramuscular fat. Two types of RTU equipment were evaluated to predict the percentage of intramuscular fat in live cattle [80]; the authors have verified that the technology tested did not differ in the accuracy of the prediction, which was accurately done. Discussion around the accuracy of predicting carcass composition using ultrasound and live animal measurements was conducted by Greiner et al. [81], who found that ultrasonic measurement of rump fat and body wall thickness are easy to obtain on the live animal, increasing the capability of the other traditional ultrasound measurements, such as the fat thickness on 12th rib or longissimus muscle area. In the same vein, Bergen et al. [82] have concluded that equations based on those live measurements provide more accuracy to predict the lean meat yield of young beef bulls. Ribeiro et al. [83] concluded from the results obtained in their experiment that RTU is an accurate tool to measure body composition in beef feedlot heifers, as did Wall et al. [84] in their study using ultrasound in the feedlot to predict body composition changes in steers at extended periods before slaughter. Knowledge of intramuscular fat is one of the most important factors related to meat quality, because it is the best indicator of marbling. Aass et al. [85]

conducted a study of the accuracy and precision of intramuscular fat prediction by ultrasound in live cattle with low fat levels; the results were promising and strongly indicate the potential of the technology for this purpose.

In conclusion, and as a result of rapid advances in the last two decades, ultrasonic technology has become more and more important in all sectors of the beef industry:

- As a noninvasive method for live estimating of fat and muscle deposition and body composition
- To evaluate and estimate carcass quality and composition
- To predict breeding values for carcass traits
- To collect data on live animals through progeny testing
- As a tool to evaluate growth and predict the optimal time for slaughter

According to Williams [86], the most common and accurate measurements include

- Backfat thickness: subcutaneous fat thickness between 12th and 13th rib over the longissimus muscle
- Longissimus muscle area: a cross-sectional area of longissimus muscle at a point between the 12th and 13th rib
- Intramuscular fat: percentage of intramuscular fat measured in the longitudinal image of the longissimus muscle over the 11th, 12th, and 13th rib, providing an estimation of the degree of marbling

Greiner et al. [87] have suggested the following measurements:

- Rump fat thickness: rump fat thickness taken parallel to the vertebral column in the junction of the biceps femoris and gluteus medius between the ischium and ilium
- Gluteus medius muscle depth: measurement taken from the same image of rump fat thickness
- Body wall thickness: fat thickness between the 12th and 13th ribs 4 cm ventral to the longissimus muscle, perpendicular to the external body surface

17.3.3.3.3 Use of RTU in Sheep and Goats

In the beginning, the use of ultrasound in sheep was not as successful as in other species. The early studies for predicting body or carcass composition [88–91] were not promising. The small amount of subcutaneous fat, the great mobility of the skin, the lack of sufficient variation in the fat thickness, and the presence of wool were the main limitations to the use of ultrasound technology. Initially, the measurements taken, as in other species, were of the fat thickness over the m. longissimus between the 12th and 13th ribs, and longissimus muscle depth and area at the same point. The prediction models included the live or carcass weight as independent variables in multiple regression equations and, although reflecting acceptable accuracy, they indicated the necessity of finding and investigating other areas of the animal. In spite of this, subcutaneous fat thickness measured between the 3rd and 4th lumbar vertebrae was used as a predictor of carcass composition in a study of the lamb and mutton carcass grading system in South Africa by Bruwer et al. [92]. In the same vein, Stanford et al. [62] found a good predictor of saleable meat yield using an ultrasound measurement of subcutaneous depth taken at the first lumbar vertebrae. Also, Young et al. [93], studying the factors affecting the repeatability of measuring tissue depth by real-time ultrasound, concluded that the measurements could be accurately assessed from one ultrasound measurement. Previous studies in some Mediterranean sheep breeds [29,94–100] have shown the usefulness of ultrasound measurements taken between the 3rd and 4th lumbar vertebrae to predict carcass composition and suggested that ultrasound fat thickness measurements with live weight could be good predictors of carcass and body composition.

In the several studies, measurements were taken in a position perpendicular to the dorsal medium line between 12th and 13th ribs and in the 1st or between the 3rd and 4th lumbar vertebrae at the level of the largest depth of the muscle or at 3/4 position of the carcass midline. Frequencies of 5 and 7.5 MHz were the probes commonly used.

One of the principal factors influencing the accuracy of the different predictions obtained was the different ultrasonic measurement procedures related to the wool. Undoubtedly, to obtain better images, it is necessary to shear and clip the animals' wool, as per the procedure proposed by Silva et al. [100]. Nevertheless, Teixeira et al. [29], working with unshaven animals and using portable RTU equipment, obtained, *in vivo*, good predictions of fat thickness of the carcass, reporting that it is possible to do the work at the abattoir as well as under field conditions (Figure 17.3). The procedure of shearing the wool, clipping, or shaving the skin is incompatible with evaluation in a commercial slaughterhouse.

In goats, the first studies of the use of real-time ultrasound to predict carcass or body composition were published in 1995 by Delfa et al. [58] and Stanford et al. [62]. In addition to the normal anatomical points for taking the subcutaneous fat measurements (between the 12th and 13th ribs or lumbar vertebras) [101,102], the sternum region has been suggested as the most useful part of the body to assess subcutaneous fat in goats. Indeed, goats have a lower fat deposition on the back in contrast to the breastbone, where the amount of subcutaneous fat is considerably deeper and variable [7]. Furthermore, the breast of goats is practically hairless and perfectly suitable for placing the probe, allowing complete contact with the body surface, so that a fat measurement can be taken rapidly and efficiently.

Measurements taken between the 3rd and the 4th sternebrae were the most suitable for assessing fat thickness [59,60]. To predict muscle weight, Delfa et al. [58,61] suggested the longissimus muscle depth measurement be taken between 3rd and 4th lumbar vertebrae in multiple regression with live weight. Nevertheless, for young animals with less fat deposition, the muscle depth assessed in the sternum region is a better predictor than the other muscle measurements [102]. Globally, authors consider that RTU fat thickness measurements taken in the breastbone could be useful in the prediction of carcass and body composition of goats, improving the accuracy of estimates with live or carcass weight as independent variables in multiple regression equations.

17.3.3.4 Video Image Analysis

The use of VIA to predict carcass composition has recently been researched. Recent advances in computer science and video processing have generated new ways to analyze and monitor quality in the meat industry. In the beginning, VIA was a system described as capable of objectively assessing carcass conformation and evaluating beef quality, particularly to determine marbling and the color of meat.

Essentially, VIA involves taking an image of the whole carcass with a video camera and then analyzing the image for dimensional measurements related to carcass quality or composition. The images are transmitted to a computer to be digitized, and fat and muscle measurements are taken electronically.

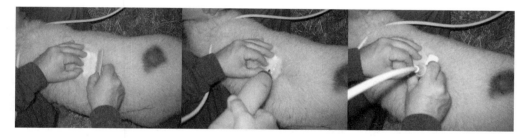

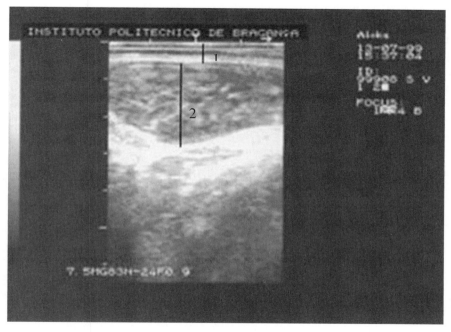

Figure 17.3 **Example of sheep RTU measurements of fat thickness (1) and longissimus dorsi depth (2) between 12th and 13th rib with 7.5 MHz probe with an ALOKA SSD-500V scanner. (Adapted from Teixeira, A., Matos, S., Rodrigues, S., Delfa, R., and Cadavez, V., *Meat. Sci.* 74, 289, 2006.)**

Images taken in the slaughter line at an abattoir could also be used for a carcass grading system classification. An experiment conducted by Gerrard et al. [103] has reported that image analysis could be used to quantify the marbling and determine the color of the beef longissimus muscle. Basset et al. [104] have used photographic image analyses for the classification of bovine meat according to such factors as muscle type, age, and breed. A study by Shackelford et al. [105] to determine whether image analysis of the 12th rib cross-section, used for tenderness classification of beef, could accurately evaluate carcass cutability, longissimus area, and subprimal cut weights; they concluded that the technology could be used in the beef industry. Later, the same authors [106] evaluated under commercial beef processing conditions, with some success, the ability of the U.S. Meat Animal Research Center's beef carcass image analysis system to predict calculated yield grade, longissimus muscle area, preliminary yield grade, and adjusted preliminary yield grade, but with low expectation of accuracy in the prediction of marbling score.

The VIAscan® was an objective grading tool utilizing the VIA technology to rapidly and accurately assess lamb and beef carcass characteristics. VIAscan was developed by Systems Intellect Pty. Ltd. and VQA Australasia as part of the Australian Meat Research Corporation's Objective Carcase Measurement Program [107,108]. The system analyzed a video image of a whole carcass and of a cross-section at the ribbing point to calculate fat content and meat color, from which yield and quality were predicted. The system has been developed to assess carcass quality attributes as well as saleable meat yield, based on analysis by computer video images. To evaluate beef carcasses there are two versions: a video camera (chiller assessment system [CAS]) to collect images of the rib area of the carcass such as rib eye area, fat thickness, marbling, and fat or muscle color; and another video camera (hot assessment system [HAS]) to take images online at slaughter of the surface fatness of the carcass, carcass measurements, and color of fat and muscle [109]. Hopkins [110] used the VIAscan system to predict lamb carcass muscularity and confirmed that the system had potential for online classification of lamb carcasses and could be used to predict the proportion of leg and shoulder primal cuts [111]. Later, Hopkins et al. [112], working at an Australian commercial abattoir, demonstrated that the system offers potential for predicting meat yield, using computing facilities allowing predictions in real time and a link to carcass ticketing technology.

At Colorado State University and Hunter Associates Laboratory (Reston, VA), they have developed a prototype video imaging system (prototype BeefCam) as a noninvasive technology to analyze beef carcasses. Wyle et al. [113] published the results of a study conducted to determine the effectiveness of this equipment in classifying beef carcasses by palatability and concluded that further development of a commercial BeefCam system was warranted. Research Management Systems USA (RMS Inc., Fort Collins, CO) has incorporated into the system a Computer Vision System (CVS); Cannell et al. [114,115] have tested its usefulness in predicting the composition of beef carcasses under commercial conditions, using the Dual-Component CVS, which consists of one video camera (Hot camera) to obtain images of the outside surface and for computer analysis of carcass shape, dimensions, and fat distribution; and a second video camera (Cold camera) that records images of the exposed 12th–13th rib interface, allowing computer analysis of different fat thickness measurements and rib eye area. The two video image analyses together correspond to CVS Dual-Component VIA System developed by RMS Inc. As a predictor of beef carcass meat yield and for an "augmentation application" of USDA yield grades, the cited authors [114], tested the VIAscan in its two versions (Dual-Component VIA System). With the same objectives, Cannell et al. [115] tested the system of image analysis in the CVS, studying the ability of the equipment to predict beef carcass red meat yields, and the "augmentation application" of USDA yield grades to beef carcasses, not only at chain speeds, but also as a fully online, installed commercial system in a commercial packing plant. The authors found in both studies that the two systems (VIAscan and CVS) allowed a more accurate prediction of yields than those achieved by online whole number yield grades alone. In commercial packing plants, Vote et al. [116] have evaluated online the effectiveness of the CVS BeefCam in predicting the tenderness of beef steaks, verifying that the system was useful. In subsequent studies [117,118], Steiner and collaborators used the two systems (VIAscan and CVS) to ascertain whether the accuracy of USDA yield grade at chain speeds could be improved, reporting that the VIA systems could operate with accuracy and could be used to assess longissimus muscle area with high levels of accuracy and repeatability.

The VIA has not been used to predict pork carcass composition as other methodologies discussed before. To predict pork carcass cutability, McClure et al. [119] have tested the video image system VCS2001 (E+V, Oranienburg, Germany). Authors found that the VCS2001 is similar to Fat-O-Meter in predicting the weight of total saleable meat but did not provide an estimate of percentage of carcass lean as the Fat-O-Meter did; further development is needed to make the VCS2001 a viable commercial tool.

Colorado State University and Mountain States Lamb Cooperative developed the lamb carcass scanning hardware and software for a Lamb Vision System (LVS; RMS Inc.); Brady et al. [120] investigated whether the LVS could be used for accurate prediction of lamb carcass cutability, and subsequently carcass value, in a commercial setting. The authors found that the online LVS combined with hot carcass weight could be used to accurately sort carcasses into cutability classes—and do better in the prediction of saleable meat yield than expert USDA graders. In another study, also by Brady et al. [121], to determine if the LVS saleable meat yield prediction could be used to predict carcass value with accuracy and precision, they concluded that LVS provides lamb producers and packers with several opportunities to objectively assess lamb carcass value. Later, Cunha et al. [122] validated the regression equations developed by Brady et al. [120] to predict lamb carcass fabrication yields and identified possible improvements to the accuracy of the equations using the two components of LVS: the hot carcass (LVS-HCC) and the chilled imaging system (LVS-CCC). The authors referred to also assessed the repeatability of longissimus muscle area using the LVS system. The results suggested that the LVS system could be a valuable tool for assisting in a value-based pricing system for sheep in the United States.

The carcass assessment unit of the LVS consists of a stationary camera with a lighting processor, a computer processor, and a monitor housed in a stainless steel cabinet. LVS software operates by (1) recording an image of a background, (2) recording an image of the carcass, and (3) subtracting the carcass image from the background image to provide a defined image of the carcass [120]. The software recognizes all anatomical points that are needed to make carcass measurements, and online images are obtained with a speed of approximately 450 carcasses/h [122]. Carcass measurements were used to describe shape and size of the carcass as well as muscularity and fat or lean proportions and are used as independent variables in different regression models.

The different studies on the use of VIA suggest that it is a valuable and accurate tool to assess carcass quality, and further development and improvement of hardware and software used in taking and interpreting the images are needed.

17.4 General Conclusion

All methodologies for assessing carcass quality or composition have strong and weak points, and, depending on the species concerned and different work conditions, the selection of the appropriate one should be made with reference to the following factors: (1) accuracy of the method in making different predictions; (2) precision of prediction models; (3) appropriateness to working conditions such as working in the field, in the abattoir, on slaughter or cutting lines, or in an industrial plant; (4) rapidity of the methodology and speed of measurement; (5) cost of the equipment; (6) ease and ability to operate with good repeatability; (7) possibility of standardization; and (8) robustness and portability of the equipment.

TOBEC, ultrasound, and VIA would seem to be the most suitable technologies for commercial classification. VIA is a recent technology that has been seen as the most promising by many scientists and meat industry policy makers.

References

1. Robelin, J. Prediction of body composition in vivo by dilution technique. In *In Vivo Measurement of Body Composition in Meat Animals*, Lister, D., Ed., 106–112. Elsevier, New York, NY. 1984.
2. Jones, S. D. M., J. S. Walton, J. W. Wilton, and J. E. Szkotnicki. The use of urea dilution and ultrasonic backfat thickness to predict the carcass composition of live lambs and cattle. *Can. J. Anim. Sci.* 69: 641–648. 1982.
3. Wang, J., S. B. Heymsfield, M. Aulet, J. C. Thorton, and R. N. Pierson. Body fat from body density: underwater weighing vs. dual-photon absorptiometry. *Am. J. Phys.* 256: 829–834. 1989.
4. Rozeboom, D. W., J. E. Pettigrew, R. L. Moser, S. G. Cornelius, and S. M. el Kandelgy. In vivo estimation of body composition of mature gilts using live weight, backfat thickness, and deuterium oxide. *J. Anim. Sci.* 72: 355–366. 1994.
5. Robelin, J. and J. Agabriel. Estimation de l'état d'engraissement des bovins vivants à partir de la taille des cellules adipeuses. *Bull. Tech. CRZV Theix. INRA.* 66: 37–41. 1986.
6. Teixeira, A., R. Delfa, and F. Colomer-Rocher. Relationships between fat depots and body condition score or tail fatness in the Rasa Aragonesa breed. *Anim. Prod.* 49(Part 2): 275–280. 1989.
7. Teixeira, A., R. Delfa, C. Gonzalez, L. Gosalvez, and M. Tor. The use of three joints as predictors of carcass and body fat depots in adult Blanca Celtibérica goats. *Opt. Médit. Sér. Sémin.* 27: 121–132. 1995.
8. Delfa, R., A. Teixeira, and F. Colomer-Rocher. A note on the use of a lumbar joint as a predictor of body fat depots in Aragonesa ewes with different body condition scores. *Anim. Prod.* 49(Part 2): 327–329. 1989.
9. Simm, G. and W. S. Dingwall. Selection indices for lean meat production in sheep. *Livest. Prod. Sci.* 21: 223–233. 1989.
10. Jopson, N. B., J. C. McEwan, K. G. Dodds, and M. J. Young. Economic benefits of including computed tomography measurements in sheep breeding programmes. *Proc. Aust. Assoc. Anim. Breed. Genet.* 11: 194–197. 1995.
11. Kempster, A. J., D. Arnall, J. C. Alliston, and J. D. Barker. An evaluation of two ultrasonic machines (scanogram and danscanner) for predicting the body composition of live sheep. *Anim. Prod.* 34: 249–255. 1982.
12. Kempster, A. J. Cost-benefit analysis of in vivo estimates of body composition in meat animals. In *In Vivo Measurement of Body Composition in Meat Animals*, Lister, D., Ed., 191–203. Elsevier, New York, NY. 1984.
13. Aziz, N. N., D. M. Murray, and R. O. Ball. The effect of live weight gain and live weight loss on body composition of merino wethers: noncarcass organs. *J. Anim. Sci.* 71: 400–407. 1993.
14. Swatland, H. J., S. P. Ananthanarayanan, and A. A. Goldenberg. A review of probes and robots: implementing new technologies in meat evaluation. *J. Anim. Sci.* 72: 1475–1486. 1994.
15. Berg, E. P. and M. J. Marchello. Bioelectrical impedance analysis for the prediction of fat-free mass in lambs and lamb carcasses. *J. Anim. Sci.* 72: 322–329. 1994.
16. Slanger, W. D., M. J. Marchello, and J. R. Busboom. Predicting total weight of retail-ready lamb cuts from bioelectrical impedance measurements taken at the processing plant. *J. Anim. Sci.* 72: 1467–1474. 1994.
17. Berg, E. P., M. K. Neary, J. C. Forrest, D. L. Thomas, and R. G. Kauffman. Assessment of lamb carcass composition from live animal measurement of bioelectrical impedance ultrasonic tissue depths. *J. Anim. Sci.* 74: 2672–2978. 1996.
18. Swantek, P. M., M. J. Marchello, J. E. Tilton, and J. D. Crenshaw. Prediction of fat-free mass of pigs from 50 to 130 kilograms live weight. *J. Anim. Sci.* 77: 893–897. 1999.
19. Velazco, J., J. L. Morrill, and K. K. Grunewald. Utilization of bioelectrical impedance to predict carcass composition of Holstein steers at 3, 6, 9, and 12 months of age. *J. Anim. Sci.* 77: 131–136. 1999.
20. Berg, E. P., J. C. Forrest, D. L. Thomas, N. Nusbaum, and R. G. Kauffman. Electromagnetic scanning to predict lamb carcass composition. *J. Anim. Sci.* 72: 1728–1736. 1994.

21. Swatland, H. J. *On-Line Evaluation of Meat.* Technomic, Lancaster, Basel. 1995.
22. Irie, M. Evaluation of porcine fat with fiber-optic spectroscopy. *J. Anim. Sci.* 77: 2680–2684. 1999.
23. Irie, M. Optical evaluation of factors affecting appearance of bovine fat. *Meat Sci.* 57: 19–22. 2001.
24. Irie, M., A. Oka, and F. Iwaki. Fiber-optic method for estimation of bovine fat quality. *J. Sci. Food Agric.* 83: 483–486. 2003.
25. Greiner, S. P. The use of real-time ultrasound and live animal measurements to predict carcass composition in beef cattle. PhD thesis. Iowa State University, Ames. 1997.
26. Berg, E. P., M. K. Neary, J. C. Forrest, D. L. Thomas, and R. G. Kauffman. Evaluation of electronic technology to assess lamb carcass composition. *J. Anim. Sci.* 75: 2433–2444. 1997.
27. Griffin, D. B., J. W. Savell, H. A. Recio, R. P. Garrett, and H. R. Cross. Predicting carcass composition of beef cattle using ultrasound technology. *J. Anim. Sci.* 77: 889–892. 1999.
28. Silva, S. R., M. J. Gomes, A. Dias-da-Silva, L. F. Gil, and J. M. T. Azevedo. Estimation in vivo of the body and carcass chemical composition of growing lambs by real-time ultrasonography. *J. Anim. Sci.* 83: 350–357. 2005.
29. Teixeira, A., S. Matos, S. Rodrigues, R. Delfa, and V. Cadavez. In vivo estimation of lamb carcass composition by real-time ultrasonography. *Meat. Sci.* 74: 289–295. 2006.
30. Cross, H. R. and K. E. Belk. Objective measurements of carcass and meat quality. *Meat Sci.* 36: 191–202. 1994.
31. Brethour, J. R. Estimating marbling score in live cattle from ultrasound images using pattern recognition and neural network procedures. *J. Anim. Sci.* 72: 1425–1482. 1994.
32. Li, J., J. Tan, F. A. Martz, and H. Heymann. Image texture features as indicators of beef tenderness. *Meat Sci.* 53: 17–22. 1999.
33. Cortez, P., M. Portelinha, S. Rodrigues, V. Cadavez, and A. Teixeira. Lamb meat quality assessment by support vector machines. *Neural Process. Lett.* 24: 41–51. 2006.
34. Allen, P. and B. McGeehin. Measuring the lean content of carcasses using TOBEC. Irish Agriculture and Food Development Authority. The National Food Centre. Teagasc Research Report. 7 pp. 2001.
35. Kirton, A. H., G. J. K. Mercer, D. M. Duganzich, and A. E. Uljee. Use of electronic probes for classifying lamb carcasses. *Meat Sci.* 39: 167–176. 1995.
36. Jones, S. D. M., L. E. Jeremiah, A. K. W. Tong, W. M. Robertson, and L. L. Gibson. Estimation of lamb carcass composition using an electronic probe, a visual scoring system and carcass measurements. *Can. J. Anim. Sci.* 72: 237–244. 1992.
37. Hopkins, D. L., M. A. Anderson, J. E. Morgan, and D. G. Hall. A probe measure GR in lamb carcasses at chain speed. *Meat Sci.* 39: 159–165. 1995.
38. Hopkins, D. L. An evaluation of the Hennessy Grading Probe for measuring fat depth in beef carcasses. *Aust. J. Exp. Agric.* 29(6): 781–784. 1989.
39. Chadwick, J. P. and A. J. Kempster. The estimation of beef carcass composition from subcutaneous fat measurements taken on the intact side using different probing instruments. *J. Agric. Sci.* 101: 241–248. 1983.
40. Johnson, R. K., E. P. Berg, R. Goodwin, J. W. Mabry, R. K. Miller, O. W. Robison, H. Sellers, and M. D. Tokach. Evaluation of procedures to predict fat-free lean in swine carcasses. *J. Anim. Sci.* 82: 2428–2441. 2004.
41. Schinckel, A. P. Critique of Evaluation of procedures to predict fat-free lean in swine carcasses. *J. Anim. Sci.* 83: 2719–2720. 2005.
42. Gu, Y., A. P. Schinckel, T. G. Martin, J. C. Forrest, C. H. Kuei, and L. E. Watkins. Genotype and treatment biases in estimation of carcass lean of swine. *J. Anim. Sci.* 70: 1708–1718. 1992.
43. Schinckel, A. P., J. R. Wagner, J. C. Forrest, and M. E. Einstein. Evaluation of alternative measures of pork carcass composition. *J. Anim. Sci.* 79: 1093–1119. 2001.
44. Lubaski, H. C., P. E. Johnson, W. W. Bolonchuk, and G. I. Lykken. Assessment of fat-free mass using bioelectrical impedance measurements of the human body. *Am. J. Clin. Nutr.* 41: 810. 1985.
45. Twyman, D. L. and R. J. Liedtke. *Bioelectrical Impedance Analysis of Body Composition.* RJL Systems Inc., Detroit, MI. 1987.

46. Swantek, P. M., J. D. Crenshaw, M. J. Marchello, and H. C. Lubaski. Bioelectrical impedance: a nondestructive method to determine fat-free mass of live market swine and pork carcasses. *J. Anim. Sci.* 70: 169–177. 1992.

47. Marchello, M. J. and W. D. Slanger. Bioelectrical impedance can predict skeletal muscle and fat-free skeletal muscle of beef cows and their carcasses. *J. Anim. Sci.* 72: 3118–3123. 1994.

48. Marchello, M. J., W. D. Slanger, and J. K. Carlson. Bioelectrical impedance: fat content of beef and pork from different size grinds. *J. Anim. Sci.* 77: 2464–2468. 1999.

49. Stouffer, J. R. History of ultrasound in animal science. *J. Ultrasound Med.* 23: 577–584. 2004.

50. Temple, R. S., H. H. Stonaker, D. Howry, G. Posankony, and M. H. Hazeleus. Ultrasonic and conductivity methods for estimating fat thickness in live cattle. *Proc. West. Am. Soc. Anim. Prod.* 7: 477–481. 1956.

51. Dumont, B. L. Nouvelles méthodes pour l'estimation de la qualité des carcasses sur les porcs vivants. Paper presented at Joint Food and Agriculture Organization of the United Nations/EAAP meeting. Copenhagen, Denmark. 1957.

52. Claus, A. Die messung natürlicher granzflachen im schweinkorper mit ultraschall. *Dtsch. Fleischwirtschaft.* 9: 552–553. 1957.

53. Stouffer, J. R. Status of the application of ultrasonics in meat animal evaluation. *Proc. Recip. Meat Conf.* 12: 161–169. 1959.

54. Kempster, A. J., A. Cuthbertson, and G. Harrington. *Carcass Evaluation in Livestock Breeding, Production and Marketing.* Granada, London, UK. 306 pp. 1982.

55. Horst, P. Erste Untersuchungsergebnisse über den Einsatz des "Vidoson" Schnittbild-gerätes beim Schwein. *Zuchtungeskunde.* 43: 208–218. 1971.

56. Stouffer, J. R., M. V. Wallentine, G. H. Wellington, and A. Diekmann. Development and application of ultrasonic method for measuring fat thickness and rib-eye area in cattle and hogs. *J. Anim. Sci.* 20: 759–767. 1961.

57. Houghton, P. L. and L. M. Turlington. Application of ultrasound for feeding animals: a review. *J. Anim. Sci.* 70: 930–941. 1992.

58. Delfa, R., A. Teixeira, and C. Gonzalez. Ultrasonic measurements of fat thickness and longissimus dorsi depth for predicting carcass composition and body fat depots of live goats. In *46th Annual Meeting of EAAP.* Abstracts, S.5.15, Vol. 1, 276. 1995.

59. Delfa, R., A. Teixeira, C. Gonzalez, and E. Vijil. Ultrasonic measurements for predicting carcass quality in live goats. In *47th Annual Meeting of EAAP.* Abstracts, S.5.12, Vol. 2, 272. 1996.

60. Delfa, R., C. Gonzalez, A. Teixeira, and E. Vijil. Ultrasonic measurements in live goats. Prediction of weight of carcass Joints. In *47th Annual Meeting of EAAP.* Abstracts, S.5.13, Vol. 2, 273. 1996.

61. Delfa, R., A. Teixeira, C. Gonzalez, and E. Vijil. Ultrasonic measurements for predicting carcass quality in live goats. In *47th Annual Meeting of EAAP.* Abstracts, S.5.12, Vol. 2, 272. 1996.

62. Stanford, K., T. A. McAllister, M. MsDougall, and D. R. C. Bailey. Use of ultrasound for prediction of carcass characteristics in Alpine goats. *Small Rumin. Res.* 15: 195–201. 1995.

63. Gresham, J. D., S. R. McPeake, J. K. Bernard, and H. H. Henderson. Commercial adaptation of ultrasonography to predict pork carcass composition from live animal and carcass measurements. *J. Anim. Sci.* 70: 631–639. 1992.

64. Gresham, J. D., S. R. McPeake, J. K. Bernard, M. J. Rieman, R. W. Wyatt, and H. H. Henderson. Prediction of live and carcass characteristics of market hogs by use of a single longitudinal scan. *J. Anim. Sci.* 72: 1409–1416. 1994.

65. Moeller, S. J., L. L. Christian, and R. N. Goodwin. Development of adjustment factors for backfat and loin muscle area from serial real-time ultrasonic measurements on purebred lines of swine. *J. Anim. Sci.* 76: 2008–2016. 1998.

66. Youssao, I., V. Verleyen, C. Michaux, and P. L. Leroy. Choice of probing site for estimation of carcass lean percentage in Piétrain pig using the real-time ultrasound. *Biotechnol. Agron. Soc. Environ.* 6(4): 195–200. 2002.

67. Hamlin, K. E., R. D. Green, T. L. Perkins, L. V. Cundiff, and M. F. Miller. Real-time ultrasonic measurement of fat thickness and longissimus area: I. Description of age and weight effects. *J. Anim. Sci.* 73: 1713–1724. 1995.

68. Hamlin, K. E., R. D. Green, L. V. Cundiff, T. L. Wheeler, and M. E. Dikeman. Real-time ultrasonic measurement of fat thickness and longissimus area: II. Relationship between real-time ultrasound measures and carcass retail yield. *J. Anim. Sci.* 73: 1725–1734. 1995.

69. Whittaker, A. D., B. Park, B. R. Thane, R. K. Miller, and J. W. Savell. Principles of ultrasound and measurement of intramuscular fat. *J. Anim. Sci.* 70: 942–952. 1992.

70. Greiner, S. P., G. Rouse, D. E. Wilson, and L. Cundiff. Predicting beef carcass retail product using real-time ultrasound and live animal measures: progress report. Beef and Sheep Research Report, Iowa State University, Ames, IA, A. S. Leaflet R1327. 1996.

71. Izquierdo, M., V. Amin, D. E. Wilson, and G. H. Rouse. Models to predict intramuscular fat percentage in live beef animals using real time ultrasound and image parameters: report on data from 1991–1994. Beef Research Report, Iowa State University, Ames, IA, A. S. Leaflet R1324, pp. 3–6. 1996.

72. Hassen, A., D. E. Wilson, G. H. Rouse, A. Trenkle, R. L. Willham, D. Beliele, C. Crawley, and J. C. Iiams. Evaluation of ultrasound measurements of fat thickness and ribeye area, I. Assessment of technician effect on accuracy. Beef and Sheep Research Report, Iowa State University, Ames, IA, A. S. Leaflet R1329. 1996.

73. Rouse, G. H., D. Wilson, M. Bilgen, V. Amin, and R. Roberts. Prediction of tenderness in beef carcasses by combining ultrasound and mechanical techniques. Beef and Sheep Research Report, Iowa State University, Ames, IA, A. S. Leaflet R1333. 1996.

74. Iiams, J. C. and A. Trenkle. Estimating ribeye area using a longitudinal ultrasound image of the 12th and 13th rib section. Beef and Sheep Research Report, Iowa State University, Ames, IA, A. S. Leaflet R1334. 1996.

75. Wilson, D. E., G. H. Rouse, and S. Greiner. Relationship between chemical percentage intramuscular fat and USDA marbling score. Beef and Sheep Research Report, Iowa State University, Ames, IA, A. S. Leaflet R1529. 1998.

76. Wilson, D. E., G. H. Rouse, G. H. Graser, and V. Amin. Prediction of carcass traits using live animal ultrasound. Beef and Sheep Research Report, Iowa State University, Ames, IA, A. S. Leaflet R1530. 1998.

75. Wilson, D. E., G. H. Rouse, and S. Greiner. Relationship between chemical percentage intramuscular fat and USDA marbling score. Beef and Sheep Research Report, Iowa State University, Ames, IA, A. S. Leaflet R1529. 1998.

76. Wilson, D. E., G. H. Rouse, G. H. Graser, and V. Amin. Prediction of carcass traits using live animal ultrasound. Beef and Sheep Research Report, Iowa State University, Ames, IA, A. S. Leaflet R1530. 1998.

77. Herring, W. G., L. A. Kriese, J. K. Bertrand, and J. Crouch. Comparison of four real-time ultrasound systems that predict intramuscular fat in beef cattle. *J. Anim. Sci.* 76: 364–370. 1998.

78. Faulkner, D. B., D. F. Parrett, F. K. McKeith, and L. L. Berger. Prediction of fat cover and carcass composition from live and carcass measurements. *J. Anim. Sci.* 68: 604–610. 1990.

79. Tait, J. R., G. H. Rouse, D. E. Wilson, and C. L. Hays. Prediction of lean in the round using ultrasound measurements. Beef and Sheep Research Report, Iowa State University, Ames, IA, A. S. Leaflet R1733. 2000.

80. Hassen, A., D. E. Wilson, V. R. Amin, G. H. Rouse, and C. L. Hays. Predicting percentage on intramuscular fat using two types of real-time ultrasound equipment. *J. Anim. Sci.* 79: 11–18. 2001.

81. Greiner, S. P., G. H. Rouse, D. E. Wilson, L. V. Cundiff, and T. L. Wheeler. Accuracy of predicting weight and percentage of beef carcass retail product using ultrasound and live animal measures. *J. Anim. Sci.* 81: 466–473. 2003.

82. Bergen, R., S. P. Miller, I. B. Mandell, and W. M. Robertson. Use of live ultrasound, weight and linear measurements to predict carcass composition of young beef bulls. *Can. J. Anim. Sci.* 85: 23–35. 2005.

83. Ribeiro, F., J. R. Tait, G. Rouse, D. Wilson, and D. Busby. The accuracy of real-time ultrasound measurements for body composition traits with carcass traits in feedlot heifers. Beef and Sheep Research Report, Iowa State University, Ames, IA, A. S. Leaflet R2072. 2006.

84. Wall, P. B., G. H. Rouse, D. E. Wilson, R. G. Tait, Jr., and W. D. Busby. Use of ultrasound to predict body composition changes in steers at 100 and 65 days before slaughter. *J. Anim. Sci.* 82: 1621–1629. 2004.

85. Aass, L., J. D. Gresham, and G. Klemetsdal. Prediction of intramuscular fat by ultrasound in lean cattle. *Livest. Sci.* 101: 228–241. 2006.

86. Williams, A. R. Ultrasound applications in beef cattle carcass research and management. *J. Anim. Sci.* 80(E. Suppl. 2): E183–E188. 2002.

87. Greiner, S. P., G. H. Rouse, D. E. Wilson, L. V. Cundiff, and T. L. Wheeler. The relationship between ultrasound measurements and carcass fat thickness and longissimus muscle area in beef cattle. *J. Anim. Sci.* 81: 676–682. 2003.

88. Edwards, J. W., R. C. Cannell, R. P. Garrett, J. W. Savell, H. R. Cross, and M. T. Longnecker. Using ultrasound, linear measurements and live fat thickness estimates to determine the carcass composition of market lambs. *J. Anim. Sci.* 67: 3322–3330. 1989.

89. Fortin, A. and J. N. B. Shrestha. In vivo estimation of carcass meat bu ultrasound in ram lambs slaughtered at average live weight of 37 kg. *Anim. Prod.* 43: 469–475. 1986.

90. Hamby, P. L., J. R. Stouffer, and S. B. Smith. Muscle metabolism and real-time ultrasound measurement of muscle and subcutaneous adipose tissue growth in lambs fed diets containing a beta-agonist. *J. Anim. Sci.* 63: 1410–1417. 1986.

91. McLaren, D. G., J. Novakofski, D. F. Parrett, L. L. Lo, S. D. Singh, K. R. Neumann, and F. K. Keith. A study of operator effects on ultrasonic measures of fat depth and longissimus muscle area in cattle, sheep and pigs. *J. Anim. Sci.* 69: 54–56. 1991.

92. Bruwer, G. G., R. T. Naude, M. M. Du Toit, A. Colete, and W. A. Vosloo. An evaluation of lamb and mutton carcass grading system in the Republic of South Africa. 2. The use of fat measurements as predictors of carcass composition. *S. Afr. J. Anim. Sci.* 17: 85–89. 1987.

93. Young, M. J., S. J. Nsoso, C. M. Logan, and P. R. Beatson. Prediction of carcass tissue weight in vivo using live weight, ultrasound or X-ray CT measurements. *Proc. N. Z. Soc. Anim. Prod.* 56: 205–211. 1996.

94. Delfa, R., A. Teixeira, F. Colomer-Rocher, and I. Blasco. Ultrasonic estimates of fat thickness, *C* measurement and longissimus dorsi depth in Rasa Aragonesa ewes with same body condition score. *Opt. Médit. Sér. Sémin.* 13: 25–30. 1991.

95. Delfa, R., A. Teixeira, C. Gonzalez, and I. Blasco. Ultrasound estimates of carcass composition of live Aragon lambs. In *43rd Annual Meeting of EAAP*. Abstracts, SV.6, Vol. 2, 364. 1992.

96. Delfa, R., A. Teixeira, C. Gonzalez, and I. Blasco. Ultrasonic estimates of fat thickness and longissimus dorsi muscle depth for predicting carcass composition of live Aragon lambs. *Small Rumin. Res.* 16: 159–164. 1995.

97. Delfa, R., C. Gonzalez, E. Vijil, A. Teixeira, M. Tor, and L. Gosalvez. Ultrasonic measurements for predicting carcass quality and body fat depots in ternasco of Aragon—Spain. In *47th Annual Meeting of EAAP*. Abstracts, S.5.11, Vol. 2, 272. 1996.

98. Delfa, R., M. Joy, A. Sanz, B. Panea, and A. Teixeira. In vivo ultrasonic measurements and live weight for predicting carcass quality in Churra Tensina Mountain breed lambs. In *New Developments in Evaluation of Carcass and Meat Quality in Cattle and Sheep*, C. Lazzaroni, Ed., EAAP Scientific Series. Wageningen Academic Publishers, Wageningen, The Netherlands. 2007.

99. Teixeira, A. and R. Delfa. The use of ultrasonic measurements assessed with two probes in live lambs for prediction the carcass composition. In *48th Annual Meeting of EAAP*. Abstracts, S3.18, Vol. 3, 295. 1997.

100. Silva, S. R., J. J. Afonso, V. A. Santos, A. Monteiro, C. M. Guedes, J. M. T. Azevedo, and A. Dias-da-Silva. In vivo estimation of sheep carcass composition using real-time ultrasound with two probes of 5 and 7.5 MHz and image analysis. *J. Anim. Sci.* 84: 3433–3439. 2006.

101. Delfa, R., A. Teixeira, V. Cadavez, C. Gonzalez, and I. Sierra. Relationships between ultrasonic measurements in live goats and the same measurements taken on carcass. In *Proceedings of the 7th International Conference on Goats*, Vol. II, 833–834. 2000.

102. Delfa, R., A. Teixeira, C. Gonzalez, V. Cadavez, and I. Sierra. Use of ultrasound measurements for predicting body fat depots in live goats. In *Proceedings of the 7th International Conference on Goats*, Vol. II, 835–836. 2000.

103. Gerrard, D. E., X. Gao, and J. Tan. Beef marbling and color score determination by image processing. *J. Food. Sci.* 61: 145–147. 1996.

104. Basset, O., B. Buquet, S. Abouelkaram, G. Gimenez, and J. Culioli. Photographic image analysis for the classification of bovine meat. In *New Developments in Guaranteeing the Optimal Sensory Quality of Meat*. Toldrá and Troy, Madrid. 1999.

105. Shackelford, S. D., T. L. Wheeler, and M. Koohmaraie. Coupling of image analysis and tenderness classification to simultaneously evaluate carcass cutability, longissimus area, subprimal cut weights and tenderness of beef. *J. Anim. Sci.* 76: 2631–2640. 1998.

106. Shackelford, S. D., T. L. Wheeler, and M. Koohmaraie. On-line prediction of yield grade, longissimus muscle area, preliminary yield grade, adjusted preliminary yield grade, and marbling score using the MARC beef carcass image analysis system. *J. Anim. Sci.* 81: 150–155. 2003.

107. Ferguson, D. M., J. M. Thompson, and P. Cabassi. Video Image Analysis. Meat '95. Paper presented at the Australian Meat Industry Research Conference, September 1995, 2 pp. 1995.

108. Ferguson, D. M., J. M. Thompson, D. Barrett-Lennard, and B. Sorensen. Prediction of beef carcass yield using whole carcass VIASCAN. Paper presented at International Congress of Meat Science and Technology, San Antonio, TX, August 1995, 183–184. 1995.

109. McIntyre, B. Yield of saleable meat in beef cattle. Government of Western Australia. Department of Agriculture. *Farmnote.* 26: 3. 2004.

110. Hopkins, D. L. The relationship between muscularity, muscle:bone ratio and cut dimensions in male and female lamb carcasses and measurement using image analysis. *Meat Sci.* 43: 307–317. 1996.

111. Stanford, K., S. D. M. Jones, and M. A. Price. Methods of predicting lamb carcass composition: a review. *Small Rumin. Res.* 29: 241–254. 1998.

112. Hopkins, D. L., E. Safari, J. M. Thompson, and C. R. Smith. Video image analysis in the Australian meat industry—precision and accuracy of predicting lean meat yield in lamb carcasses. *Meat Sci.* 67: 269–274. 2004.

113. Wyle, A. M., D. J. Vote, D. L. Roeber, R. C. Cannell, K. E. Belk, J. A. Scanga, M. Goldberg, J. D. Tatum, and G. C. Smith. Effectiveness of the SmartMV prototype BeefCam System to sort beef carcasses into expected palatability groups. *J. Anim. Sci.* 81: 441–448. 2003.

114. Cannell, R. C., J. D. Tatum, K. E. Belk, J. W. Wise, R. P. Clayton, and G. C. Smith. Dual-component video image analysis system (VIASCAN) as a predictor of beef carcass red meat yield percentage and for augmenting application of USDA yield grades. *J. Anim. Sci.* 77: 2942–2950. 1999.

115. Cannell, R. C., K. E. Belk, J. D. Tatum, J. W. Wise, P. L. Chapman, J. A. Scanga, and G. C. Smith. Online evaluation of a commercial video image analysis system (Computer Vision System) to predict beef carcass red meat yield and for augmenting the assignment of USDA yield grades. *J. Anim. Sci.* 80: 1195–1201. 2002.

116. Vote, D. J., K. E. Belk, J. D. Tatum, J. A. Scanga, and G. C. Smith. Online prediction of beef tenderness using a computer vision system equipped with a BeefCam module. *J. Anim. Sci.* 81: 457–465. 2003.

117. Steiner, R., D. J. Vote, K. E. Belk, J. A. Scanga, J. W. Wise, J. D. Tatum, and G. C. Smith. Accuracy and repeatability of beef carcass longissimus muscle area measurements. *J. Anim. Sci.* 81: 1980–1988. 2003.

118. Steiner, R., A. M. Wyle, D. J. Vote, K. E. Belk, J. A. Scanga, J. W. Wise, J. D. Tatum, and G. C. Smith. Real-time augmentation of USDA yield grade application to beef carcasses using video image analysis. *J. Anim. Sci.* 81: 2239–2246. 2003.

119. McClure, E. K., J. A. Scanga, K. E. Belk, and G. C. Smith. Evaluation of the E+V video image analysis system as a predictor of pork carcass meat yield. *J. Anim. Sci.* 81: 1193–1201. 2003.

120. Brady, A. S., K. E. Belk, S. B. LeValley, N. L. Dalsted, J. A. Scanga, J. D. Tatum, and G. C. Smith. An evaluation of the lamb vision system as a predictor of lamb carcass red meat yield percentage. *J. Anim. Sci.* 81: 1488–1498. 2003.
121. Brady, A. S., B. C. N. Cunha, E. Belk, S. B. LeValley, N. L. Dalsted, J. D. Tatum, and G. C. Smith. Use of the lamb vision system to predict carcass value. *Sheep Goat Res. J.* 18: 4. 2003.
122. Cunha, B. C. N., K. E. Belk, J. A. Scanga, S. B. LeValley, J. D. Tatum, and G. C. Smith. Development and validation of equations utilizing lamb vision system output to predict lamb carcass fabrication yields. *J. Anim. Sci.* 82: 2069–2076. 2004.

Chapter 18

Moisture and Water-Holding Capacity

Karl O. Honikel

Contents

18.1   Introduction

Lean muscular tissue contains approximately 75% water. The water is bound to approximately 95% within the muscle cells. "Binding" means that the water is restricted in its molecular movements; it is immobilized by charged or hydrophilic side chains of amino acids and capillary forces. A very small amount of water is bound very tightly to proteins like crystal water in salt (protein-bound water; Table 18.1). Approximately 80% water is immobilized by the myofibrillar and cytoskeletal proteins (intrafilamental; Table 18.1). In the sarcoplasm with its soluble (sarcoplasmic) proteins between the fibres (interfibrillar; Table 18.1) approximately 15% of the water is partially immobilized by the protein surfaces, water–solute, and water–water interactions. A part of this water is "free"—it means unbound by protein side chains, ions, or capillary forces. Nevertheless, this water is inhibited from free flowing out of the cell by the cellular and subcellular lipid bilayer membranes.[1–4]

After death, with the breakdown of glycogen and the formation of lactic acid (lactate), the pH within the cell falls from approximately pH 7 in the muscle of a rested animal to approximately pH 5.3–5.8. This ultimate pH is near the isoelectric point of meat proteins. These chemical changes postmortem cause physical changes because the swelling of the myofibers is reduced and is at a

Table 18.1   Water Distribution in Muscles of Live Animals (pH 7) and Meat (pH 5.3–5.8)

	% Water	
	Muscle	*Meat*
Protein-bound water	~1	~1
Intrafilamental	~80	~75
Interfibrillar	~15	~10
Extracellular water	~5	~15

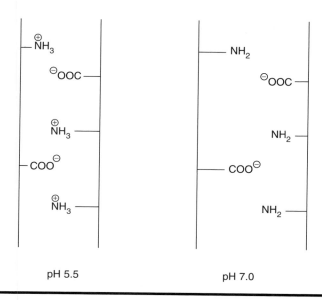

pH 5.5   pH 7.0

Figure 18.1 **Scheme of protein shrinkage by changes in pH from muscle (pH 7) to meat (pH 5.5). Owing to the increase of positive charges of side chains ($-NH_2$ → $-NH_3^+$) the meat proteins are approaching the isoelectric point at 5.2/5.3 (see Figure 18.3). The structure shrinks due to the attraction of the opposite electric charges of the side chains and it leaves less space for water molecules in between.**

minimum at the isoelectric point (Figure 18.1). Also, the rigor mortis occurs after the exhaust of convertible energy in muscle cells postmortem with its permanent interaction of myosin and actin in the filaments. This shrinkage, due to both events, means that less water is immobilized within the myofibers, and by the shrinkage water appears slowly outside the cell. Approximately 15% of the water is located finally outside the muscle cells and kept within the meat by capillary force (extracellular water; Table 18.1). But slowly this water evaporates into the environment or is lost as drip loss.

The attractive forces, however, under which the water is bound and immobilized by muscle proteins, not only change during the postmortem processes but also during the processing of meat, manufacturing of meat products, and preparation of meat for eating. Processing involves heating, chilling, cold storage, freezing, thawing, drying, and mincing; furthermore, manufacturing includes salting (curing), fermenting, and comminuting. All these procedures change more or less the water retention of meat—the water-holding capacity (WHC). Thus, the knowledge of factors influencing the WHC of meat handled under various conditions is important for the quality of meat and is also of considerable economic interest. Changes in WHC are initiated by postmortem changes and are very sensitive indicators of changes in the structure of myofibrillar proteins.[5,6]

This chapter discusses only selected aspects of water in meat, WHC, and its importance in quality aspects. Several important factors related to WHC that will not be considered are the influence of animal-specific factors such as species, sex, age, muscle type, and treatment of animals before slaughter.

18.1.1 General Principles for All Methods Regarding Meat Characteristics

The origin and husbandry of the live animal, slaughtering procedures, and the postmortem handling of the carcass are described as precisely as possible. The description includes species, breed, sex, age, feeding regime, transport and preslaughter handling, slaughter conditions, chilling, and aging regime. The rate of pH and temperature decline postmortem, together with the final pH of the muscle, are reported. The history of the animal should preferably be known, although it is not always important. If it is known, it should be reported.

18.2 Moisture of Meat

18.2.1 Principle of Measurement

Moisture is, in general terms, the total water content of meat. If a piece of lean meat contains, analytically, 75% water then this amount is determined as moisture. But one must keep in mind that a small portion of water, approximately 1%, is tightly bound to protein (Table 18.1) and salt structures and behaves like crystal water, which is released from salt only under extreme conditions, for example, high degree of heating.

In meat, the degree of heating for moisture determination is limited as other compounds such as fat or proteins may disintegrate on higher temperatures and release decomposition compounds in gaseous form, which would falsify the results. Thus the determination of moisture in meat is usually not carried out above 105°C. This drying process, however, is slow and must be repeated until a constant weight is reached and so requires long hours to complete.

This may be too time-consuming especially within or before a manufacturing process. For these reasons, rapid methods for moisture determination have been developed. Faster heating, for example, by microwave, however, has the disadvantage of a higher variability that may be acceptable in a processing line. A faster method from AOAC International[7] (method 950.46) recommends the application of 125°C in a convection oven.

Near-infrared methods (NIR, NIT), nuclear magnetic resonance (NMR), and guided microwave spectroscopy (GMS) are few other methods that, however, require a rather laborious calibration for each instrument, often also for various animal species, and also for fatter or leaner cuts and processed (raw, salted, heated) products.

Meat consists primarily of water, protein, fat, and approximately 1–2% other substances, such as salts (ca. 1%), carbohydrates (<0.5%), vitamins, adenine nucleotides, and DNA in even smaller amounts. The four major components—water, fat, protein, and salt (the latter is usually measured and expressed as ash)—are present in fresh meat of all common meat species before storage, that is, without water loss, a close relationship.

The correlation coefficient of water and fat in freshly slaughtered meat is usually >0.99, the coefficients between water and protein are lower with 0.97, and that of water with ash is 0.98.[8–12] Therefore, a water determination allows the calculation of fat content or a fat determination the calculation of moisture. But this is valid for fresh meat only that has not lost moisture (drip, purge, evaporation) during storage.

18.2.2 Determination of Moisture

One of the standard reference methods for the determination of moisture is the oven drying method by AOAC International[7] (method 950.46). Similar ones exist in the collection of several analytical methods, such as ISO, EN, or DIN. In AOAC method 950.46, it is described how meat samples are dried after preparation according to AOAC method 983.18.

18.2.2.1 Equipment

Food chopper with 3 mm plate or a bowl chopper, glass or alumina dishes 50–100 m diameter and 20–40 mm deep, desiccator, balance, air oven or mechanical convection oven, accurate ±1°C at 100°C.

18.2.2.2 Sample Preparation for Drying

a. Separate meat as completely as possible from any bone; pass rapidly three times through grinder with plate openings 3 mm, mix thoroughly after each grinding; and begin all determinations promptly. If any delay occurs, chill the sample to inhibit decomposition. Alternatively, use a bowl cutter for sample preparation. Chill all cutter parts before preparation of each sample.

b. Precut sample, up to 1 kg, to a maximum dimension <5 cm, and transfer to bowl for processing. Include any separated liquid. Process for 30 s, then wipe down inner side wall and bottom of bowl with spatula (use household plastic or rubber spatula with ca. 5–10 cm straight-edge blade) and transfer gathered material to the body of sample. Continue processing for another 30 s and wipe down as before. Repeat sequence to give total of 2 min processing and 3 wipe-downs.

Take particular care with certain meat types such as ground beef to assure uniform distribution of fat and connective tissue. At each wipe-down interval, reincorporate these into sample by using spatula to remove fat from inside surfaces of bowl and connective tissue from around blades. If sample consolidates as ball above blades, interrupt processing and press sample to bottom of bowl with spatula before continuing.

18.2.2.3 Drying

a. With lids removed, dry sample containing ca. 2 g dry material (equivalent to approximately 8–10 g of fresh meat) 16–18 h at 100–102°C in air oven (mechanical convection preferred). Use covered Al dish with diameter ≥50 mm and depth ≤40 mm. Cool in desiccator and weigh. Report loss in weight as moisture.

b. With lids removed, dry sample containing ca. 2 g dry material to constant weight (2–4 h depending on product) in mechanical convection oven or in gravity oven with single shelf at ca. 125°C. Use covered Al dish with diameter ≥50 mm and depth ≤40 mm. Avoid excessive drying. Cover, cool in desiccator, and weigh. Report loss in weight as moisture. (Dried sample is not satisfactory for subsequent fat determination.)

c. AOAC Official Method 985.14, Moisture in Meat and Poultry Products, Rapid Microwave Drying Method.

Moisture is removed (evaporated) from sample by using microwave energy. Weight loss is determined by electronic balance readings before and after drying and is converted into moisture content by microprocessor with digital percent readout.

A ground or comminuted representative sample is dried until all water has been removed. In the discussed method, it is recommended to wait for 16–18 h. In other methods such as the German method[13] L 06.00-3, two variations are recommended for the determination of moisture:

i. The ground meat is mixed with about three to five times the amount of clean seasand.
ii. The sample is weighed after 4 h. Experienced analysts reheat afterward to control weight stability.

18.2.2.4   Method L 06.00-3, Drying of Meat

A sample of 5–10 g of ground meat is weighed accurately and mixed in a steel or glass vessel with 35 g of dried seasand. The material is put into an air-drying oven at 103 ± 2°C for 4 h. After this, it is put in a desiccator and weighed after cooling. The procedure of drying, cooling, and weighing is repeated (usually once) until constant weight (difference lower than 0.1% of meat) is reached. The whole procedure may take 5–6 h.

Higher temperatures are not recommended because volatile substances present or which may be formed by disintegration of other ingredients, such as organic acid, or fat degradation products, may leave the sample and are counted as water loss.

Vacuum-drying (~100 mmHg) at 95–100°C may be used. This will speed up the time to constant weight. But it is not recommended for samples with increased fat content (AOAC International,[7] method 950.46).

Microwave drying is also used in practice. CEM designed a microwave oven specifically for moisture in foods. Samples are placed between glass fiber pads, and dried for 3–5 min, after which they are reweighed. This is published as official AOAC method (AOAC International,[7] AOAC 985.14).

18.2.2.5   Near-Infrared Method

NIR is also applied for moisture determination. Quantitative measurement of meat components requires measurement of several known samples for calibration. Unknown samples of similar type can then be scanned and components can be determined by comparing the response to the calibration data.[14,15] Once calibration is complete, the method provides a simultaneous measure of fat, moisture, and protein that is extremely fast and nondestructive.

The availability of economical microprocessors that provide for easy calibration using artificial neural networks (ANN) has made NIR instruments commonly available for meat analysis. Examples of available instruments include those from Infratec and Foss Electric (FoodScan). It is important to note that NIR analysis is highly dependent on proper calibration of instruments with samples similar to the unknowns to be measured. Recalibration is necessary for any change in sample material that is outside the range of properties of the samples used for calibration. The need for careful and proper calibration is viewed by some analysts as a disadvantage for this method. Recent developments in technology have resulted in changes of most instruments from NIR reflectance measurement to NIR transmission (NIT). The transmission measurements utilize greater sample volume, which improves results. Correlations between NIT measurements and AOAC methods have been reported as 0.984–0.995, 0.987–0.992, and 0.949–0.957 for fat, moisture, and protein, respectively, in meat. Repeatability ranges from 0.42 to 0.50%, 0.32 to 0.36%, and 0.53 to 0.54% for fat, moisture, and protein, respectively, have also been reported.[16] Furthermore, AOAC reports that the FOSS Food Scan™ with ANN Method has been granted AOAC

Official Method[SM] status.[16] A collaborative study was conducted to evaluate the repeatability and reproducibility of the FOSS FoodScan NIR spectrophotometer with ANN calibration model and database for the determination of fat, moisture, and protein in meat and meat products. Representative samples were homogenized by grinding according to AOAC[16] Method 983.18.

18.2.2.5.1 Description of Method

Approximately 180 g ground sample was placed in a 140 mm round sample dish, and the dish was placed in the FoodScan. Results were displayed for percent (g/100 g) fat, moisture, and protein. Ten blind duplicate samples were sent to 15 collaborators in the United States. The within-laboratory (repeatability) relative standard deviation (RSD$_r$) ranged from 0.23 to 0.92% for moisture. The between-laboratories (reproducibility) relative standard deviation (RSD$_r$) ranged from 0.39 to 1.55% for moisture. The hardware is described in the following.

From a tungsten-halogen lamp housed at the back of the instrument, light is guided through an optical fiber into the internal moving-grating monochromator, which provides monochromatic light in the spectral region between 850 and 1050 mm. Through a second optical fiber, light is then guided through a collimator lens positioned over the sample cup in the sample chamber. The light is transmitted through the sample, and the unabsorbed light strikes a detector. The detector measures the amount of light and sends the result to the digital signal processor, which communicates with the personal computer (PC) where the final results are calculated.

The sample is placed in a cup and positioned inside the FoodScan sample chamber. The sample cup is rotated during the analysis process to subscan various zones of the test sample that are then used to calculate the final result. This procedure provides a more representative result from potentially nonhomogeneous samples. ANN calibration is a technique designed to emulate the basic function of the human brain to solve complex problems. The ANN model has the ability to describe both linear and nonlinear relationships between spectral characteristics and compositional analysis.

The study samples were chosen to represent the majority of products from the commercial meat industry (beef, pork, and poultry) and included raw meats, emulsions, and finished products. All samples were natural and real-world type, and none were adulterated. The collaborative study samples consisted of 10 meat study samples prepared as blind duplicate pairs, resulting in 20 test samples. The method is applicable to the simultaneous determination of fat, moisture, and protein in meat and meat products (fresh meat, beef, pork, and poultry, emulsions, and finished products) in the constituent ranges of 1–43% fat, 27–74% moisture, and 14–25% protein.

There is an NIT system called the Continuous Fat Analyzer (CFA) for use on mixers and grinders. The CFA utilizes a 850–1050 nm wavelength range to continuously monitor fat, moisture, and protein content during mixing, and composition can be checked and adjusted on the spot. Standard deviations of 0.3% for the measurements have been reported.[17]

18.2.2.6 *Guided Microwave Spectrometry*

Guided microwave spectrometry has not been studied as extensively as NIR systems, but this approach has been developed to the point of being offered as part of meat processing equipment, similar to NIR. The GMS measurement is based on microwave energy absorption, which is used to measure differences in conductivity and dielectric constant of water. The conductivity and dielectric constant are then used for determination of sample fat, moisture, and protein content. Protein and fat are indirect measures with this method. Calibration with known samples is necessary for GMS measurements. These systems have been reported to result in measurements with standard deviations of 0.3%.

While there is far less information available on the GMS systems than for most other analytical methods, one of the limitations appears to be matrix sensitivity. For example, presence of air or ice crystals has been reported to have a significant effect on results.[18]

18.2.2.7 Nuclear Magnetic Resonance

NMR is the most recent development in commercial instruments for fat analysis in meat. NMR data distinguish between protons from different molecular sources and can provide sharp contrasts between meat components, such as fat and lean. Correlations between NMR measurements and known fat content in meat have been reported to be 0.967. Refs 19–21 describe the method in detail.

A peer-verified method is presented for the determination of percent moisture and fat in meat products by microwave drying and NMR analysis. The method involves determining the moisture content of meat samples by microwave drying and using the dried sample to determine the fat content by NMR analysis. Both the submitting and peer laboratories analyzed five meat products by using the CEM SMART system (moisture) and the SMART Trac (fat). The samples, which represented a range of products that meat processors deal with daily in plant operations, included the following: (1) fresh ground beef, high-fat; (2) deboned chicken with skin; (3) fresh pork, low-fat; and (4) all-beef hot dogs. The results were compared with moisture and fat values derived from AOAC International, methods 950.46[7] (Forced Air Oven Drying) and 960.39 (Soxhlet Ether Extraction).

As shown in Figure 18.2, the relationship between fat and water in meat is very close ($r = 0.99$); the moisture content can be measured.

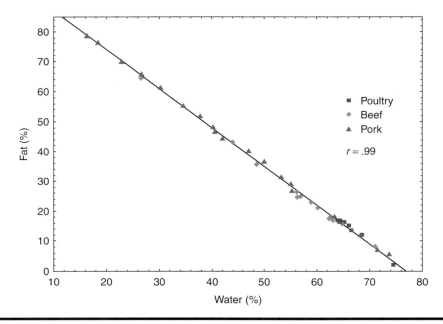

Figure 18.2 Relationship of water and fat content of fresh beef, pork, and chicken meat (total N = 45); equation: % fat = 100.02 × % water. (From Arneth, W., *Kulmbacher Reihe Band*, 16, 65, 1999.)

18.2.2.8 Summary for Moisture Determination

A wide range of methods are available for measuring the composition of meat raw materials—in this case, the water content. These methods range from traditional slow wet chemistry methods that have been in use for decades to extremely rapid, in-line multicomponent analyses that have been recently developed. But repeatability, reproducibility, and bias must be determined to permit selection of a method that will meet the expectations of the analytical laboratory applying them.

18.3 Water-Holding Capacity

18.3.1 General Remarks

WHC is the ability of meat to hold all or part of its own water. This ability depends on the way of handling and the state of the system. As the state of meat and the treatments vary considerably, the meaning of WHC varies to a large extent. Therefore, the methods applied and the state of meat at the time of measurement must be exactly defined to obtain comparable results. In spite of all the variations of methods used and reported in literature there are three major ways of treatment, which can be divided into three different basic methods of measuring WHC.

18.3.2 Methods of Measurement

18.3.2.1 Applying No External Force

To this group belongs the measurement of evaporation and weight loss, free drip, bag drip,[22-25] cube drip, and related methods,[26] whereby the meat is left to itself under different environmental conditions. These methods are very sensitive but time-consuming (one to several days) and are often sped up by the following methods.

18.3.2.2 Applying External Mechanical Force

WHC of meat can be detected within a few minutes or an hour by using positive or negative pressure. A few similar methods include centrifugation methods,[24,27] filter paper press method (FPPM),[28,29] and suction loss methods.[30-32] The amount of water released with these methods is far higher than with those without external force because the pressure applied enforces the release of water from the intra- and extracellular space of the muscle structure (Table 18.2). In drip loss measurements (usually carried out for 1–2 days) only extracellular water exudates from the meat.[33-35] Therefore, a factor must be known to evaluate the actual drip loss of the meat, for instance, within seven days. The matter becomes even more complicated as the state of meat changes during the time of conditioning and aging, which also influences the WHC. Therefore, methods applying mechanical force reveal only the tendency of how the meat may behave in the following days but the absolute values are not directly comparable with the drip loss measurements.

18.3.2.3 Applying Thermal Force

As meat is consumed usually after heating, the WHC of meat on cooking is of interest. The cooking loss is measured by a wide range of methods.[25,36] During heating, the meat proteins denature and the cellular structures are disrupted, which have a strong influence on the WHC of meat. Extra- and intracellular water are released by the meat sample on cooking. The influence of

Table 18.2 Water Loss by Different Treatments of Meat (Lean Meat 72–76% Water)

	% Water Loss
Evaporation loss during storage (up to 7 days)	0.5–3
Drip loss of normal meat (1–2 days measurement)[a]	2–5
Drip loss of PSE pork (1–2 days measurement)[b]	10–15
Centrifugation loss (~500 × g) for 20 min	20–30
Heating (cooking) loss (maximum 75°C)	20–30
Cooking (100°C)	35–45
Filter paper press method	60

[a] Slow pH fall (beef, pork 10 resp. 6 h before ultimate pH is reached).
[b] PSE-pork; ultimate pH is reached within 1 h postmortem.

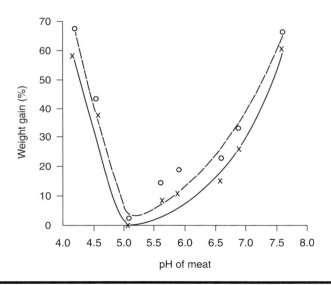

Figure 18.3 WHC of cubes of beef by inserting into water solutions at different pH values. (For details see text.) (From Grau, R., Hamm, R., and Baumann, A., *Biochemische Zeitschrift*, 325, 1, 1953.)

the way of the heating and the final temperature are great on the WHC (Table 18.2 shows two examples of heating at 75 and 100°C).

These methods of measurement of WHC in research and practice are due to the different interests of people who handle meat. It becomes evident that WHC also means different things to different people.

18.3.3 Factors That Influence the Measurement of WHC Applying the Different Methods

The WHC of all methods used depends strongly on the pH of the meat, which changes after death by the formation of lactic acid. Figure 18.3 shows one of the first examples of measuring the WHC of meat at different pH values.[28] Cubed beef was immersed in buffer solutions at various pH values, and the weight and volume gain were measured. The minimum % weight gain is approximately pH 5.2 at the isoelectric point of meat proteins where they form the closest structure by attraction of the equal amount of positively and negatively charged side chains of the amino acids in proteins. The changes in charges of side chains of amino acids are shown in Figure 18.1. Furthermore, the WHC depends on the muscle type and species of animal due to their varying composition and structure.

Above this, *the evaporation loss depends on* the surface cover of the muscle tissue such as adipose tissue or wrapping material, the size of the sample (big or small), the length of the measuring period, the temperature of meat and chilling room, and the air speed and air humidity within the chiller.

The drip loss depends on the size of sample, the shape of sample, the treatment during the conditioning period (Figure 18.4), the chilling temperature, and the duration of chilling period (Figure 18.5). Drip loss in meat samples can be measured under defined conditions; a procedure has been recommended.[25] The sum of evaporation and drip loss is the weight loss, which the individual is interested in.

Centrifugation method of uncooked meat: besides the factors influencing weight loss, centrifugation loss depends additionally on the speed of centrifugation ($\times g$), the time of centrifugation, and the plasticity of the meat influenced by state of meat and additives.

FPPM:[29] this method is very easy to handle and fast and thus is widely used. The results depend on the pressure applied, the time of pressure applied, and the plasticity of the meat, that is, prerigor state (pH) (Figure 18.6). But there is a poor relationship with drip loss (Figure 18.7).

In Figure 18.6 a further development of the FPPM is described. Hofmann[32] and Hofmann et al.[37] describe an advanced evaluation by determining the quotient of the meat (M) to the total area (meat + surrounding fluid: T). The influence of pH is clearly shown. Reuter[38] proposed a template method (rings of different diameters), speeding up the method of Hofmann et al.[37] who still used a planimeter for determining the size of the area.

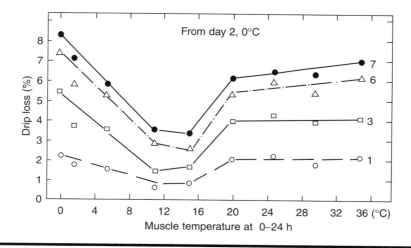

Figure 18.4   Relationship between temperature of incubation during the first 24 h postmortem of m. cleidomastoideus of beef and drip loss of muscle cubes (approximately 30 g) after storage for 1, 3, 6, and 7 days postmortem. From day 2 the samples were stored at 0°C.

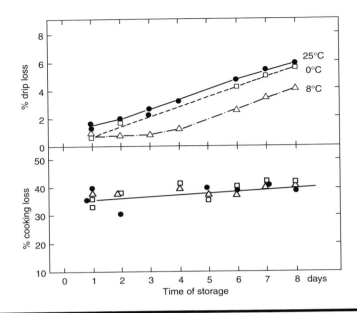

Figure 18.5 Drip and cooking loss during storage of beef m. sternomandibularis. The muscles were excised 45 min postmortem and stored immediately in a water bath at 0, 8, and 25°C. After 24 h the storage temperature was 5°C. The packing was in plastic pouches without vacuum. Drip loss was measured as weight loss of the whole muscle (300–400 g) related to original weight on day 0. Cooking loss was the loss of a cube sample of 30 g, heated in a plastic pouch for 30 min in a water bath of 75°C. Each cooking sample was cut off the meat piece freshly every day just before measurements.

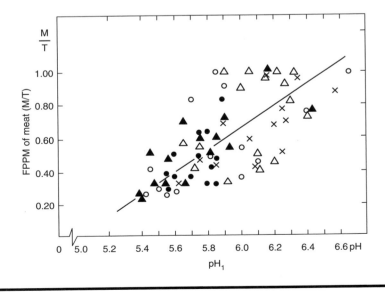

Figure 18.6 Results of filter paper press method (FPPM) expressed as M/T (meat area/total fluid area, see text) in dependence of various pork muscles (different symbols) and the early postmortem pH fall at 45 min (pH$_1$). (From Hofmann, K., *Fleischwirts*, 62, 1604, 1982.)

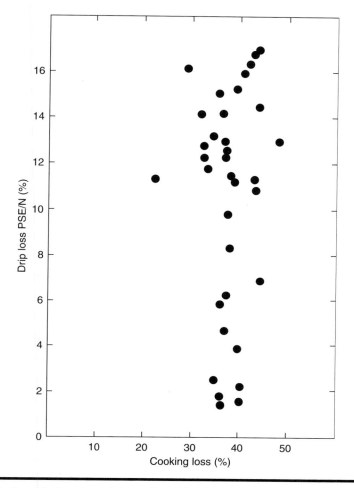

Figure 18.7   **Relationship between drip and cooking loss of pork muscles at ultimate pH value
of 5.4–6.2, including normal and fast-glycolyzing PSE muscles (pH$_1$ was 5.5–6.4). The muscles
(ca. 500 g) were cut off the carcasses at 24 h and stored at 5°C in vacuum packs. Drip loss
was measured between the fifth and sixth day postmortem for 24 h at 80 g slices, cooking loss
after drip loss measurement in the same samples heated to 75°C similar to the description in
Figure 18.8.**

Inspired by the FPPM, Monin et al.[39] proposed an imbibition method, measuring the fluid
absorption within a 3 min period on a muscle surface, immediately after a fresh cut.

Kauffman et al.[40] proposed similarly the absorption of excess fluid of a muscle surface on filter
paper for 2 s by either weighing the filter paper or visual score. Both evaluation systems resulted in
very high correlations to drip loss in the 4 days after the measurement of $r > 0.90$.

A similar method was the capillary volumeter by Hofmann[30,32] and Fischer et al.[31] where the
uptake of fluid by a gypsum body put on a freshly cut meat surface was measured. The instrument
does no longer exist on the market.

Cooking loss depends on the shape and size of sample, the temperature profile during cooking, the final temperature of cooking (Figure 18.8), and the environment during cooking (in water, salt solution, air, wrapping). Whereas the drip loss changes with time postmortem, the cooking loss remains constant (Figures 18.5 and 18.7) due to the fact that the ultimate pH does not change any longer and the disintegration of meat during aging is a very slow process.

Figure 18.7 shows clearly that there is no relationship between drip loss and cooking loss at 6 days postmortem. Table 18.3 shows the correlation coefficients among different methods for determining WHC. High correlation coefficients show drip losses measured on different days. Drip loss does not have close relationships with other methods.

A standardized procedure for cooking concerning shape, size, and environment is possible and has been published.[25]

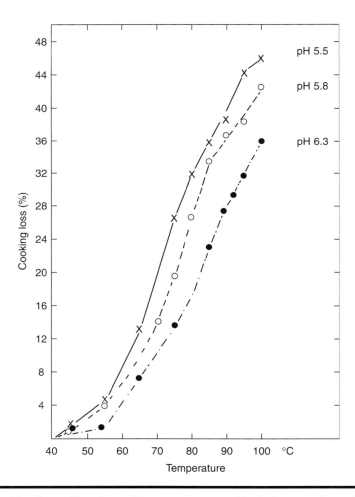

Figure 18.8 Cooking loss of beef muscles (m. sternomandibularis) at various ultimate pH values at different end points of heating. Muscle cubes of 30 g in plastic pouches were inserted in a water bath at ambient temperature and heated at approximately 2.5°C/min to the various final temperatures where the meat was kept for 5 min. Then the samples were cooled in water at ambient temperature for 30 min and the cooking loss determined.

Table 18.3 Correlation Coefficient between Different Methods for Determining WHC on Beef Muscles (N = 415)

Method	Storage Time (Days)	Correlation Coefficient to Drip Loss at	
		3 Days	*6 Days*
Drip loss	6	0.91	—
Drip loss	8	0.86	0.96
Drip loss	14	0.96	0.91
Centrifugation loss	2	0.38	0.33
Suction method	2	0.31	0.27
FPPM	2	−0.15	−0.21
Cooking loss (75°C)	5	0.04	0.19

Notes: For suction method, see Hofmann, K., *Fleischwirts*, 62, 1604, 1982; for FPPM, see Grau, R. and Hamm, R., *Z. Lebensm. Unters. Forsch.*, 105, 446, 1957.

Table 18.4 Cooking Loss (%) of Beef Muscle at Different pH and Temperatures Measured at the Time Postmortem When the pH Was Obtained

Temperature (°C)	pH 6.8	pH 6.1	pH 5.9	pH 5.5
0.5	34	42	44	—
4	34	40	41	44
5	36	41	42	45
7.5	37	40	42	45
10	27	38	40	45
14	33	39	41	44
17	32	40	43	44
20	37	42	43	44
20	35	39	41	43
23	33	40	42	44
24	37	44	45	43
27	37	39	40	42
30	34	41	42	45
$\bar{x}$	34.4	40.4	42.1	44.8
S.D.	±2.8	±1.6	±1.7	±2.7

18.3.4 Summary of Principles

It must be kept in mind that of all chemical and physical changes in muscles forming meat postmortem the changes of pH have the biggest influence as Figures 18.1, 18.3, 18.6, 18.8, and Table 18.4 show. Additionally, drip loss temperature at the time early postmortem and the time of measurement influence the results (Figure 18.4). This influence of temperature and time does not exist with cooking loss (Figure 18.5 and Table 18.4)—there is a strong influence of pH.

Many proposals have been made to measure WHC of meat. Kauffman et al.[41] described the methods that are being used now. Still, a consensus is lacking in the scientific community regarding methods of measuring the WHC of meat. Many methods have been published but only one procedure, to our knowledge, has been agreed upon internationally, and that too for beef only.[42]

However, standardization of methods is essential if investigations carried out by different groups are to be compared directly. Thus, some agreement should be made regarding methods measuring physical quality characteristics in meat and meat products. The lack of standard measurement is in contrast to accepted methods of measuring the chemical components of meat and meat products, such as the determination of moisture.

In considering reference methodology, it was recognized that the techniques used to evaluate physical characteristics such as WHC could be applied for at least three different reasons:

1. As a quality assurance (QA) toll, within a processing operation
2. As an assessment of the effectiveness of production and processing treatments where there may be an interest in being able to compare results among laboratories or countries
3. As a research tool, in fundamental structural studies of muscle and meat

In the first case, a common methodology should be appropriate for the plant or group of plants being controlled by specific QA programs. The methods used should measure the desired characteristics necessary to monitor the process, but need not be comparable with other laboratories, where different criteria may be important. Where international comparison is important, it is essential that methodologies be standardized. This would include all aspects of the testing procedure and this is the area at which the reference methods are primarily directed.

In contrast, where direct assessments are being made of the physical properties of meat as a function of structural (chemical or physical) changes, the experimental methodologies should not be constrained by reference methods. Instead, researchers are encouraged to develop and use methodologies that enhance the precision and accuracy of testing methods, leading to an understanding of the basic mechanisms. It is likely that new understanding will lead, eventually, to methods that more accurately predict consumer assessments of meat characteristics.

As there is a multitude of procedures for measuring the WHC of meat, we restrict the description to the following points:

1. Drip loss in raw, whole meat
2. Water loss in cooked, whole meat

18.3.5 Drip Loss in Raw, Whole Meat

18.3.5.1 Principle

The mechanism of drip formation in raw, whole meat has been reviewed by Offer and Knight.[34,35] Water losses originate from volume changes of myofibrils induced by prerigor pH fall and the attachment of myosin heads to actin filaments at rigor where myofibrils shrink owing to pH fall. Denaturation of proteins may also contribute to a reduction in WHC, particularly in conditions of rapid prerigor pH fall. The fluid thus expelled accumulates between fiber bundles. When a muscle is cut, this fluid is low enough and capillary forces do not retain it.

This means that the methods chosen for measuring drip loss must conserve the integrity of the muscle before the sampling takes place to avoid external forces other than gravity. Orientation of the fibers with respect to cut is also important and should be taken into consideration. Surface evaporation has to be prevented and the method of supporting the meat piece should minimize tension (suspended from above) or compression (supported from below). For standardized meat samples, the following should be described: type of muscle, the muscle where the sample is taken from, muscle fiber orientation, surface area to weight ratio, time postmortem, temperature, and pH.

18.3.5.2   Equipment

The equipment required is a balance of sufficient accuracy (±0.05 g), a sealable, water-impermeable container (or plastic bag), sample support that allows the escape of fluid (plastic net bag or perforated support), and a temperature-controlled environment.

18.3.5.3   Procedure

Meat samples are cut from the carcass and immediately weighed. A sample weight of approximately 80–100 g is recommended but other sample sizes may also be used. The samples are either placed in a netting and then suspended in an inflated bag, ensuring that the sample does not make contact with the bag, or placed within the container on the supporting mesh and sealed. After a storage period (usually 24–48 h) at chill temperatures (1–5°C), samples are again weighed. The same samples can be used for further drip loss measurements, for instance, after 2 or 7 days, but in every case the initial weight is used as the reference point. At the time of measurement, samples should be taken immediately from the containers, gently blotted dry and weighed.

18.3.5.4   Calculation

Drip loss is expressed as a percentage of the initial weight.

18.3.5.5   Evaluation

At least two adjacent samples from the same muscle of similar weight and shape should be used. Triplicates are recommended.

18.3.6   *Cooking Loss in Whole Meat*

18.3.6.1   Principle

During heating, the different meat proteins denature at varying temperatures (37–75°C). Denaturation causes structural changes such as the destruction of cell membranes, transverse and longitudinal shrinkage of muscle fibers, the aggregation of sarcoplasmic proteins, and shrinkage of the connective tissue. All these events, particularly the connective tissue changes, result in cooking losses in meat. Hamm[2] and Offer[33] provide relevant reviews on the effect of heat on muscle proteins and structure.

Precautions taken regarding the geometry of the sample for the measurement of drip loss apply also to the cooking loss. Cooking conditions must be defined and controlled (heating rate and the end-point temperature at the thermal center).

18.3.6.2   Equipment

The equipment required is a balance of sufficient accuracy (±0.05 g), a temperature-controlled water bath, thin-walled polyethylene bags, and thermocouples to allow temperature recording in the center of each sample.

18.3.6.3 Procedure

Samples should be freshly cut and weighed (initial weight). Individual standardized slices of 50 mm thick (maximum) and of a standard weight in thin-walled plastic bags are placed in a continuously boiling water bath (>95°C), with the bag opening extending above the water surface. Samples should be cooked to a defined internal temperature; 75°C is recommended. If other temperatures are used, these must be defined in the methodology. When the end-point temperature has been attained, samples should be removed from the water bath, cooled in an ice slurry, and held in chill conditions (1–5°C) until equilibrated. The meat is then taken from the bag, blotted dry, and weighed.

18.3.6.4 Calculation

The cooking loss is expressed as a percentage of the initial sample weight.

18.3.6.5 Evaluation

At least two samples of adjacent positions and similar weight and shape should be used. Triplicates are recommended.

18.4 Conclusion

It is essential that relevant factors that can affect the WHC values are defined as far as possible. These include the type of muscle and sample location within the muscle, meat quality parameters, such as the rate of pH decline, ultimate pH, and details of the temperatures at which the samples or carcasses were maintained. The carcass chilling process (which affects chilling losses) is particularly important.

References

1. Hamm, R. Water-holding capacity of meat. In: *Meat*, D.D.A. Cole and R.A. Lawrie (Eds), Butterworth, London, pp. 321–338, 1975.
2. Hamm, R. Functional properties of the myofibrillar system and their measurement. In: *Muscle as Food*, P.J. Bechtel (Ed), Academic Press, New York, pp. 135–199, 1986.
3. Fennema, O.R. Water and protein hydration. In: *Food Proteins*, J.R. Whitaker and S.R. Tannenbaum (Eds), AVI Publ. Co., Westport, CT, pp. 50–90, 1977.
4. Honikel, K.O. The meat aspects of water and food quality. In: *Water and Food Quality*, T.M. Hardman (Ed), Elsevier Appl. Sci., London, NY, pp. 277–303, 1989.
5. Hamm, R. Biochemistry of meat hydration. *Adv. Food Res.*, 10: 355–463, 1960.
6. Honikel, K.O. et al. Sarcomere shortening and their influence on drip loss. *Meat Sci.*, 16: 267–282, 1986.
7. AOAC International Official methods of analysis of AOAC International. W. Horwitz (Ed), 18th ed. Gaithersburg, ISBN: 0-935584-75-7, 2005.
8. Keeton, J.T. and Eddy, S. Chemical and physical characteristics of meat. In: *Encyclopedia of Meat Sciences*, W.K. Jensen, C. Devine, and M. Dikeman (Eds), Elsevier, Academic Press, Amsterdam, Vol. I, pp. 210–218, 2004.

9. Arneth, W. Chemische Untersuchungsmethoden für Fleisch und Fleischerzeugnisse. *Fleischwirts.* 76: 120–123, 1996.

10. Arneth, W. Beispiele chemischer Schnellmethoden. In: Analytik bei Fleisch, Schnell-, Schätz- und Messmethoden. *Kulmbacher Reihe, Band* 16: 65–82, 1999.

11. Arneth, W. Beispiele physikalisch-chemischer Schnellmethoden zur Fett- und Wasseranalyse. *Fleischwirts.* 81: 75–77, 2001.

12. Berg, H. and Kolar, K. Schnellmethode: Überprüfung des Infratec Food and Feed Analyzers zur Schnellbestimmung von Wasser, Fett, Roheiweiß und Hydroxyprolin in Rind- und Schweinefleisch. *Fleischwirts.* 71: 765–769, 1991.

13. Amtliche Sammlung von Untersuchungsverfahren nach § 64 Lebensmittel- und Bedarfsgegenstände- und Futtermittelgesetzbuch (LFGB), § 35 vorläufiges Tabakgesetz, § 28b GenTG, Deutschland (2007), Band I, (L) 0.600-3: Hrsg.: Bundesamt für Verbraucherschutz und Lebensmittelsicherheit (BVL), Beuth Verlag, Berlin, Wien, Zürich, 2007.

14. Freudenreich, P. Rapid simultaneous determination of fat, moisture, protein and colour in beef by NIT-analysis. Proceedings of 38th ICoMST, Clermont-Ferrand, France, Vol. 5, S895–898, 1992.

15. Freudenreich, P. and Wagner, E. Analysis of Meat Products (Frankfurter Sausages) by NIT-Spectrometry. Poster Proceedings "Meat for the Consumer". 42nd ICoMST, Lillehammer 1996, Norway, Published by MATFORSK, Norwegian Food Research Institute, S258–259, 1996.

16. http://www.aoac.org/ILM/jul_aug_07/foodscan.htm

17. King-Brink, M., DeFreitas, Z., and Sebranek, J.G. Use of near infrared transmission for rapid analysis of meat composition. In: *Near Infrared Spectroscopy: The Future Waves*, A.M.C. Davies and P. Wiliams (Eds), pp. 142—148, NIR Publications, Chichester, West Sussex, 1996.

18. Sebranek, J.G. Raw material composition analysis. In: *Encyclopedia of Meat Sciences*, W.K. Jensen, C. Devine, and M. Dikeman (Eds), Elsevier Appl. Sci., Amsterdam, Vol. I, pp. 173–179, 2004.

19. Pedersen, J.T., Berg, H., Lundby, F., and Balling-Engelsen, S. The multivariate advantage in fat determination in meat by bench-top NMR. *Innov. Food Sci. Emerg. Technol.* 2: 87–94, 2001.

20. Keeton, J.T. et al. Rapid determination of moisture and fat in meats by microwave and nuclear magnetic resonance analysis. *J. AOAC Int.* 86: 1193–1202, 2003.

21. Sørland, G.H. et al. Determination of total fat and moisture content in meat using low field NMR. *Meat Sci.* 66: 543–550, 2004.

22. Penny, J.F. The effect of temperature on drip, denaturation, and extracellular space of pork longissimus dorsi muscle. *J. Sci. Food Agric.*, 28: 329–338, 1977.

23. Honikel, K.O. Characteristics of meat important to product quality. Basic Concepts of water-holding capacity in meat and meat products. In: *The Functionality of Meat Compounds*, J.D. Buckley, K.O. Honikel, and F.J.M. Smulders (Eds). ECCEAMST Foundation, Utrecht, The Netherlands, pp. 23–48, 1998.

24. Honikel, K.O. and Hamm, R. Critical evaluation of methods detecting effects of processing on meat protein characteristics. In: *Chemical Changes During Food Processing*, Vol. II, S. Bermell (Ed), Proc. IUFoST Symp., Consejo Superior de Investigaciones Científicas, Valencia, Spain, pp. 64–82, 1986.

25. Honikel, K.O. Reference methods for the assessment of physical characteristics of meat, *Meat Sci.* 49: 447–457, 1998.

26. Howard, A. and Lawrie, R.A. *Studies on Beef Quality*, Part I-III, Dept. of Scient. and Industr. Research, Food Investigation, Her Majesty's Stationery Office, London, 1956.

27. Wierbicki, E. and Deatherage, F.F. Determination of water holding capacity in fresh meats. *J. Agric. Food Chem.* 6: 387–392, 1958.

28. Grau, R., Hamm, R., and Baumann, A. Über das Wasserbindungsvermögen des toten Säugetiermuskels. I. Mitteilung. *Biochemische Zeitschrift.* 325: 1–11, I, 1953.

29. Grau, R. and Hamm, R. Über das Wasserbindungsvermögen des Säugetiermuskels. II. Mitteilung. *Z. Lebensm. Unters. Forsch.* 105: 446–460, 1957.

30. Hofmann, K. Ein neues Gerät zur Bestimmung der Wasserbindung des Fleisches: Das Kapillar-Volumeter. *Fleischwirts.* 55: 25–30, 1975.

31. Fischer, C., Hofmann, K., and Hamm, R. Erfahrungen mit der Kapillarvolumeter-Methode nach Hofmann zur Bestimmung des Wasserbindungsvermögens von Fleisch. *Fleischwirts.* 56: 91–95, 1976.

32. Hofmann, K. Optimierte Anwendung des Kapillarvolumeters zur Bestimmung der Wasserbindung des Fleisches. Eliminierung von Fehlerquellen und Erneuerung der Messkörper. *Fleischwirts.* 62: 1604–1608, 1982.

33. Offer, G. Progress in the biochemistry, physiology and structure of meat. Proceedings of 30th European Meeting of Meat Research Workers, Bristol, pp. 87–94, 1984.

34. Offer, G. and Knight, P. The structural basis of water-holding in meat. I. General principles and water uptake in meat processing. In: *Developments in Meat Science* 4, R. Lawrie (Ed), Elsevier Applied Science, London, NY, pp. 63–171, 1988.

35. Offer, G. and Knight, P. The structural basis of water-holding in meat. II. Drip loss. In: *Developments in Meat Science* 4, R. Lawrie (Ed), Elsevier Applied Science, London, NY, pp. 173–243, 1988.

36. Bendall, J.R. and Restall, D.J. The cooking of single myofibres, small myofibre bundles, and muscle strips of beef M. psoas and M. sternomandibularis of varying heating rates and temperatures. *Meat Sci.* 8: 93–117, 1983.

37. Hofmann, K., Hamm, R., and Blüchel, E. Neues über die Bestimmung der Wasserbindung des Fleisches mit Hilfe der Filterpapierpreßmethode. *Fleischwirts.* 62: 87–94, 1982.

38. Reuter, G. Verfahren zur Erkennung von Fleischqualitätsabweichungen bei Schlachttierkörpern. *Fleischwirts.* 62: 1153–1160, 1982.

39. Monin, G. et al. Carcass characteristics and meat quality of halothane negative and halothane positive pietrain pigs. *Meat Sci.* 5: 413–423, 1981.

40. Kauffman, R.G. et al. The Use of filter-paper to estimate drip loss of porcine musculature. *Meat Sci.* 18: 191–200, 1986.

41. Kauffman, R.G. et al. A comparison of methods to estimate water-holding capacity in post-rigor porcine muscle. *Meat Sci.* 18: 307–322, 1986.

42. Boccard, R. et al. Procedures for measuring meat quality characteristics in beef production experiments: report of a working group in the Commission of the European Communities (CEC) beef production research programme. *Livestock Prod. Sci.* 8: 385–397, 1981.

Chapter 19

Microstructure of Muscle Foods

I. Pérez-Munuera, V. Larrea,
A. Quiles, and M.A. Lluch

Contents

19.1 Introduction on Food Microstructure

The chemical components of a food are organized into a characteristic microscopic pattern that constitutes its microstructure. The food's microstructure is currently being presented as a new challenge in multidisciplinary studies of foods and a way of contributing to our understanding of scientific and technological progress.

In foods from animal and plant tissues, macromolecules such as proteins, lipids, and polysaccharides and their interactions with other chemical components, including water, perform important functional roles. These chemical components constitute the fundamental structural components from which the characteristic microstructures of each food are built. For instance, within muscle foods such as Spanish "serrano" dry-cured ham, the organization of the muscle fibers, with their myofibrils and free fat infiltration among the fibers, constitutes the microstructural basis for the characteristic texture and sensory quality that consumers appreciate.

The different stages of the processing and storage of foods cause modifications that are manifested by macroscopic changes that impart the characteristic texture, color, flavor, etc., to the finished food. The microstructure is directly affected by the chemical, biochemical, and physical changes undergone by the foods during processing, and the final properties of the elaborated foods, desirable or not, depend on their microstructure. For example, there are foods where a total microstructural change is desired and all the original plant or animal tissue organization disappears (products such as patés or vegetable purees). In other cases, a predefined microstructure is imitated to obtain processed foods such as surimi, artificially reproducing a particular tissue organization. However, the original submicroscopic structure in processed foods, where the appearance of a "fresh food" has to be retained, needs to be preserved as much as possible during the various treatments (fresh-cut fruits and vegetables, refrigerated fresh meat packed in inert atmospheres, marinated fish, and so on). It is also possible to seek an optimum microstructure, which is defined both by the arrangement of the chemical components and by the size and distribution of the gaps or air bubbles, in bubble foods such as certain emulsions and foams or in bakery products, for example. Additionally, and always depending on the temperature, the physical state of the chemical compounds also affects the final structure. For instance, the solidification of the ingredients and the even distribution of the air bubbles in ice creams are key aspects of their microstructure. The presence of microscopic foreign bodies can occasionally lead to the detection of unsuitable practices or even fraud.

However, the microstructure of the food plays an important role in many processes where transport takes place within the food: (i) saline agents and other preservatives are carried through the intercellular spaces and through the cell walls and membranes when pickling vegetables, curing hams, and salting and smoking fish, (ii) the heat transmission in thermal processes such as sterilization, blanching, freezing, etc., (iii) water/solutes move intra- and intercellularly during dehydration by hot air or by osmosis, and (iv) matter is transferred intra- and intercellularly in solid/liquid extraction of oils and fats from food tissues, etc. The establishment of mathematical transport models, which provides sufficiently accurate predictions, requires a real understanding of the mechanisms that govern them. These mechanisms are influenced by the chemical and biochemical components and their structural distribution within the food. Consequently, a better understanding of the microstructure of foods should, in turn, allow for better modeling of the processes.

Understanding the organization of chemical components is key to describing, predicting, or controlling the changes undergone by foodstuffs during processing. Conceptually, this adds a new viewpoint to the classic chemical or physical aspect of foods; microstructure is thus seen as a fundamental aspect of the combined, multidisciplinary study of food.

Nowadays, different microscopy techniques, particularly electron microscopy (scanning or SEM, transmission or TEM), offer powerful tools for clarifying the microstructure of foods and establishing its interrelationships with their physical, chemical, and biochemical behaviors. Coupled with image analysis, it also allows the quantification of the morphological changes that take place during food processing. Additionally, current X-ray detection systems (EDX, WDX) make it possible to microanalyze foods at submicroscopic levels. Above all, broad horizons have opened up in the field of food microstructure and its multidisciplinary study. It could be interesting to take research carried out in this field as a complement for solving problems and developing new products in the food industry. Furthermore, this could also be a very useful approach defining the properties of functional foods and the importance of their components for the health of the consumers.

As the aim of this chapter is to give a brief overview of these issues, certain interesting aspects concerning the microstructure of different foods of animal origin with, for example, native, modified or altered, destroyed, restructured, or imitated tissue structures will be discussed.

19.2 Main Microscopy Strategies for Studying the Structure of Muscle Foods

The invention of the light microscope (LM) at the end of the sixteenth century provided food scientists with the first tool that allowed them to observe samples magnified up to 20 or 30 times their original size. LM uses visible light as its illumination source and was used to study food microstructure despite its low resolution, which is limited due to the wavelength of the light (Figure 19.1a). Nowadays, modern LMs have a resolution (200–500 nm) that is about 10^3 times that of the human eye and they produce an image that is magnified 4–1500 times (4–1500×) (Figure 19.2). The LM is a versatile and useful tool that works in different applications such as *bright field*, which is the most common application for muscle foods; *phase contrast or differential interference contrast (Nomarski)*, where the contrast of unstained tissues is enhanced; *polarizing microscopy*, where a plane-polarized light is used to illuminate the sample (useful with birefringent structures); or *fluorescence microscopy*, which is the basis of the modern technique of confocal laser scanning microscopy (CLSM). There are different ways of preparing samples for LM observation (Figure 19.3); the only essential requirement is that the sample be translucent, in other words, sufficiently thin to allow light to pass through it when mounted in a suitable medium. Consequently, depending on the physical characteristics of the sample, different methods are available [1]. Powdered foods (normally under 200 μm) are fine enough to be observed directly, mounted in an inert oil or an appropriate dye solution. Fluid foods such as emulsions and sauces are spread

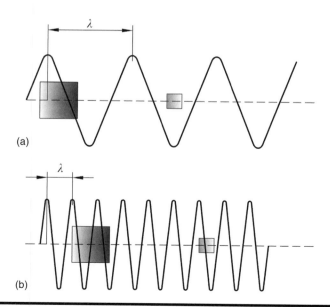

Figure 19.1 **(a) Visible light, high wavelength, only interferes with big objects. (b) Accelerated electrons, short wavelength, only interfere with very small objects.**

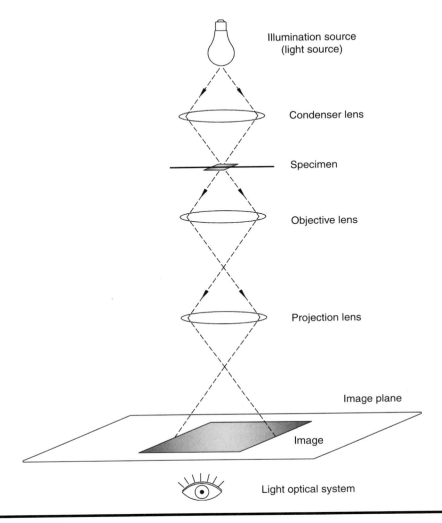

Illumination source
(light source)

Condenser lens

Specimen

Objective lens

Projection lens

Image plane

Image

Light optical system

Figure 19.2   Diagram of a light microscope (LM).

on the slide as translucent films. Lastly, solid foods are prepared in thin sections. The sections can be obtained after embedding the food in paraffin or resins, but the simplest and quickest way is to use a cryotome, placed in a freezer, to obtain cryosections of samples after having frozen their constituent water along with their other chemical constituents. In this method, the solid water acts as the embedding medium. The use of a hot stage coupled to the microscope helps to reproduce food processes that include a heating step.

In recent years, an important advance in the field of food structure has been the introduction of CLSM. In this technique, as the muscle food materials of structural interest are not auto-fluorescent, the sample should be stained with fluorescent dyes before examination under the microscope. Optical sectioning is one of the major advantages of this technique, since physical slicing is not necessary. This leads to the possibility of generating three-dimensional (3D) images of biological cells and tissues [2]. In CLSM, the illumination of the sample is restricted to a

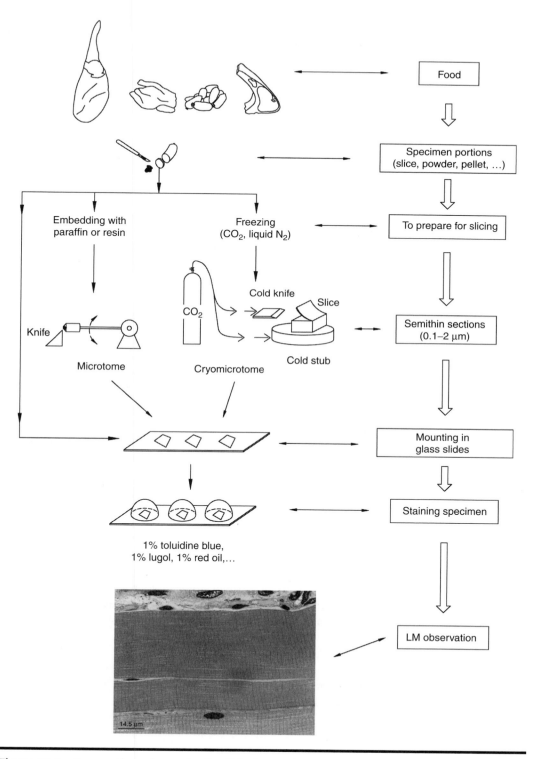

Figure 19.3   Preparation of samples for LM observation.

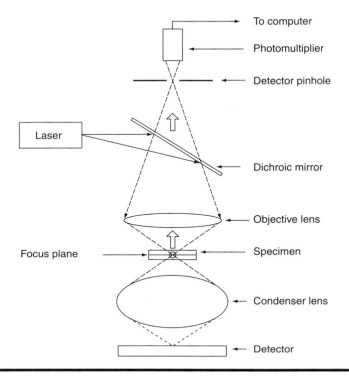

Figure 19.4   Diagram of a confocal laser scanning microscope (CLSM).

single point (or an array of points), which is scanned to produce a complete image (Figure 19.4). Another major advantage of CLSM is the exclusion of out-of-focus blur, since the fluorescence emissions from the illuminated regions of the sample above and below the focal plane are not allowed to reach the photomultiplier and form an image.

The development of electron microscopy (EM) has allowed food structure to be studied in greater detail. The electron microscope takes advantage of the much shorter wavelength of the accelerated electron compared to light (Figure 19.1b). This makes a 1000-fold increase in magnification possible, together with a parallel increase in resolution. There are two microscopes that use electron beams as source of illumination: transmission electron microscope (TEM) and scanning electron microscope (SEM) (Figure 19.5). In both methods, the samples first need to be prepared. Both the working conditions (the electron microscope works under high vacuum) and the nature of the sample itself (in the case of muscle foods, a biological tissue) make long and occasionally tedious sample preparation protocols essential.

The TEM projects accelerated electrons through a very thin slice of specimen [3]; the transmitted electrons produce a 2D image on a phosphorescent screen (Figure 19.6). TEM gives an image with better resolutions (0.2–1 nm) and higher magnifications (200–300,000×) than LM. The steps in the preparation of the sample (Figure 19.7) are primary fixation, washing, secondary fixation, dehydration, infiltration with resin, embedding, cutting ultrathin sections (100–500 Å) in an ultramicrotome, and staining with heavy metals. The resin permeates the sample, replacing all the water within the structural components and making the sample firm enough for sections

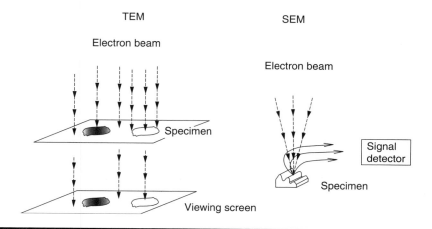

Figure 19.5   Two basic electron microscopes. TEM: image is formed from the electrons transmitted by the electron beam. SEM: image is formed from the signal (secondary electrons, etc.) emitted by the sample scanned by the electron beam.

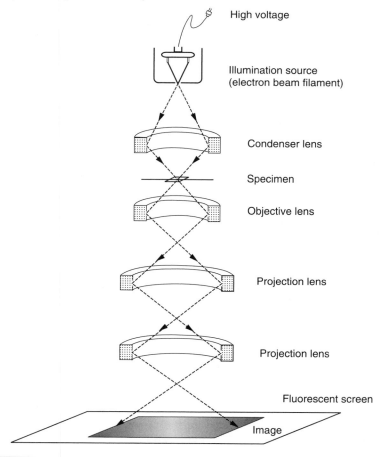

Figure 19.6   Diagram of a transmission electron microscope (TEM).

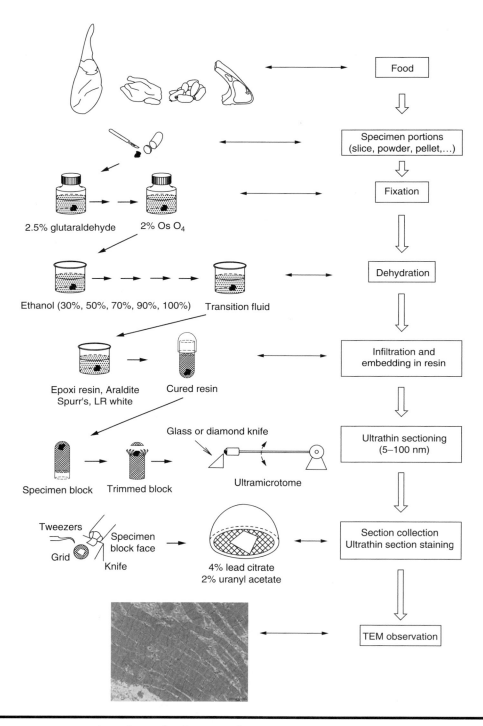

Figure 19.7 Preparation of samples for TEM observation.

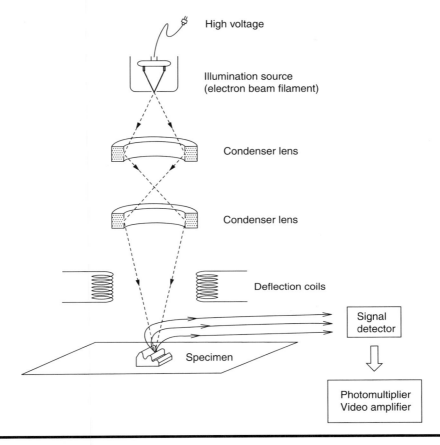

High voltage

Illumination source
(electron beam filament)

Condenser lens

Condenser lens

Deflection coils

Signal
detector

Specimen

Photomultiplier
Video amplifier

Figure 19.8   Diagram of a scanning electron microscope (SEM).

to be cut. This technique is very useful for observing the characteristic banding of animal muscle fibers and the ultrastructure of the cells, differentiating the cell organelles.

The SEM uses a spot of electrons that scans the surface of the specimen to generate secondary electrons, which are the primary signal emitted by the specimen and form a 3D image (Figure 19.8). SEM gives resolutions (3–4 nm) and magnifications (20–100,000×) that are intermediate between those of LM and TEM. The SEM method observes the surface of the sample, so there is no need to section it. Surface and internal structures can be observed, depending on the preparation techniques used, and these are much easier than the TEM. In essence, there are two ways of preparing samples for SEM: chemical fixing and physical fixing (Figure 19.9). In the former, the sample preparation steps are chemical fixation, dehydration in a series of ethanol dilutions of increasing concentration, change to a transition fluid (acetone), critical point drying, mounting, and coating with a conducting metal or carbon. When physical fixation is used, the sample is frozen in liquid nitrogen and then freeze-dried before being mounted, coated, and observed. In recent years, considerable progress has been made in the field of SEM through vitrification techniques. In cryo-SEM, the sample is frozen in slush nitrogen and quickly transferred to a cold stage fit on a microscope where the frozen sample is coated and observed (Figure 19.10); in this way, the sample can be observed with all its constituent water.

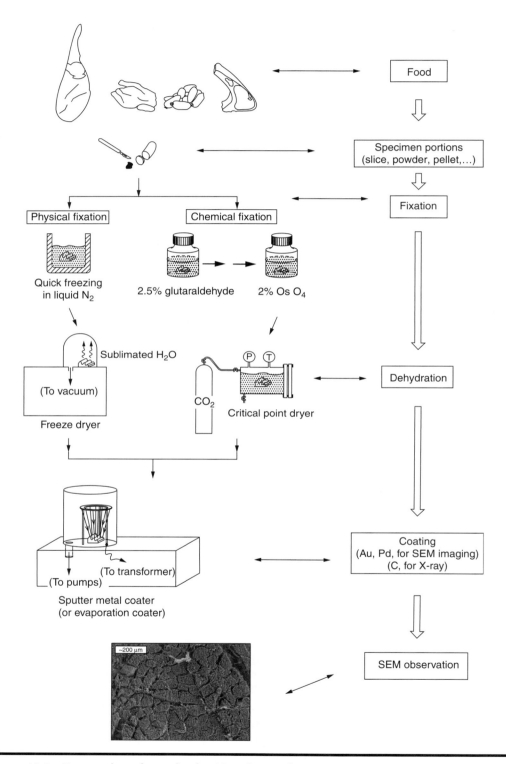

Figure 19.9   Preparation of samples for SEM observation.

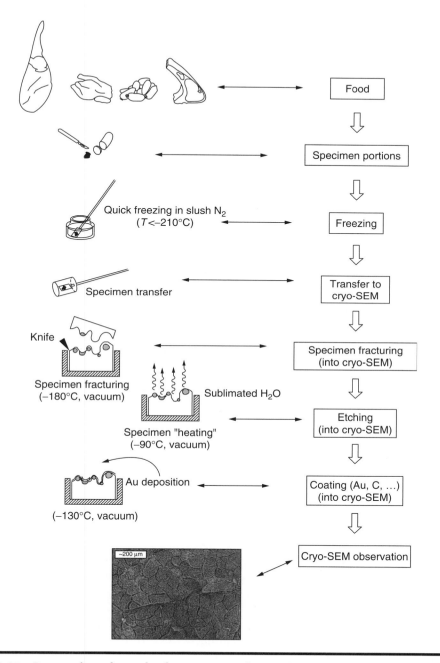

Figure 19.10   Preparation of samples for cryo-SEM observation.

Besides the secondary electrons, other emanations or signals such as X-rays and backscattered electrons may be generated as a result of the electron beam striking the specimen [3]; the different signals can be captured by the appropriate detector in each case. In this way, ions or molecules can be identified and quantified *in situ* using specific detectors coupled to the electron microscope

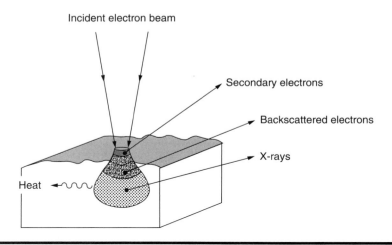

Figure 19.11   Different signals emitted by the sample when scanned by the electron beam.

(X-ray microanalysis) (Figure 19.11); this can be useful, for example, in meat products where the curing process involves the use of salts, such as Spanish "serrano" ham.

Finally, image analysis relies heavily on computer technology to obtain quantitative results from microscopy observation [4], for example, shortening or lengthening of muscle fibers during meat processing.

19.3   The Muscle Structure of Meat

The muscle structure of meat is similar in all species. Although meat includes a number of different tissues, the majority is skeletal muscle tissue. The muscle mass is elongated and covered by a connective tissue termed the epimysium which binds both individual fiber bundles and groups of muscle bundles together [5]. It tapers into a tendon that connects the muscle to the skeletal structure. The perimysium forms partitions within the muscle at irregular intervals to form both primary and secondary muscle bundles. Within the perimysium, the sarcolemma or cell membrane retains the sarcoplasmic fluids, which bathe each muscle fiber. The muscle fiber is surrounded by a connective tissue called the endomysium, and inside the cells there are numerous myofibrils (Figure 19.12).

The contractile unit of the myofibril is known as sarcomere. The dark band is called the A-band, the light band is called the I-band, and the dark line bisecting the light band is called the Z-disk. The sarcomere is the distance from one Z-disk to the next (1.5–2.5 μm in length, depending on the state of contraction) and is made up of two sets of interdigitating filaments: thin and thick. The thin filaments are predominantly composed of actin molecules, wrapped around one another, and of tropomyosin and troponin, connected to the Z-disk; the thick filaments are composed of densely packed myosin proteins, running from the center of the sarcomere toward the Z-disk (Figure 19.12). The myofibrils are linked to the sarcolemma by filamentous structures called costameres; the protein constituents of the costameres (desmin, actin, vinculin, talin, etc.) extend into the muscle cell where they encircle the myofibrils at the Z-disk and run from myofibril to myofibril and from myofibril to sarcolemma [6].

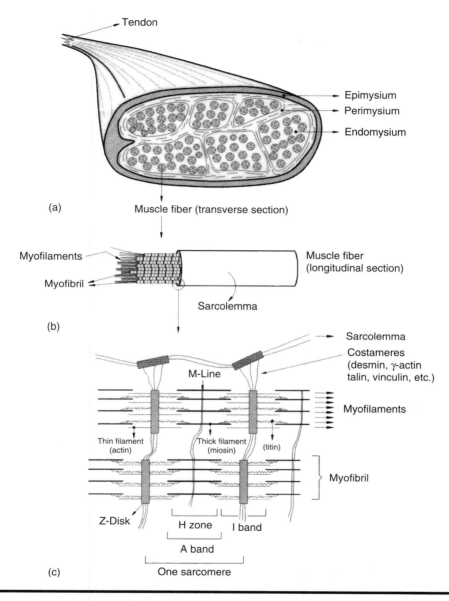

Figure 19.12 Schematic representation of the skeletal muscle. (a) A transverse section of the muscle bundles and the connective tissues: epimysium, perimysium, and endomysium. (b) Longitudinal muscle fiber composed of myofibrils, which are composed of groups of myofilaments and surrounded by the sarcolemma. (c) Muscle proteins organized into ultrastructural features or sarcomeres. Costameres are protein filaments anchored in the Z-disk and run from myofibril to myofibril and from myofibril to sarcolemma. (From Lluch et al., *Chemical and Functional Properties of Food Proteins*, Technomic, Lancaster, PA, 2001. With permission.)

19.4 Influence of Postmortem Processing on Meat Microstructure

Different microscopy techniques have been used to characterize meats and to analyze the changes in muscle tissue that take place under different postmortem storage or cooking conditions. Some authors [7,8] have observed the microstructure of rabbit semimembranosus muscles by LM. They found that the muscle tissue of the dead animal presented a type of structure in which the myofibrils were perfectly packed and the intercellular connections of the connective tissue remained unchanged; as the postmortem progressed, the structure of the muscle fibers and endomysial connective tissue gradually broke down. Transverse sections of rabbit semimembranosus muscle showed the typical muscle tissue structure (Figure 19.13a); at 24 h postmortem, gaps caused by the destruction of the perimysium were observed in the muscle bundles in some zones and the fibers inside each cell bundle appeared fairly separate due to degradation of the endomysial connective tissue (Figure 19.13b).

SEM was employed by Palka [9] to study the influence of postmortem aging of bovine semitendinosus muscle. The structure of the intramuscular connective tissue and myofibrillar structure of the meat after five days of aging was regular. In 12-day-old samples, the fibrous and myofibrillar structures were less distinct, damage appeared in the endomysium tubes, and the fibers of the perimysium were swollen.

CLSM was used by Straadt et al. [10] to study aging-induced changes in microstructure and water distribution in fresh pork. At day 1 (24 h postmortem) a few muscle fibers appeared to be swollen.

Larrea et al. [11,12] have used several techniques to observe muscle in pork meat. No structural differences were observed between the biceps femoris and semimembranosus muscle tissues. A cross section of the semimembranosus muscle from a sample of pork meat observed by SEM shows the perimysial connective tissue that separates the bundles of fibers (Figure 19.14a). The cells are surrounded by endomysial connective tissue, a reticular structure primarily composed of collagen, which keeps the muscle fibers firmly attached to one another. This technique provides a clearer picture of the myofibrils inside the cells, where they are strongly attached to one another and to the sarcolemma (Figure 19.14b).

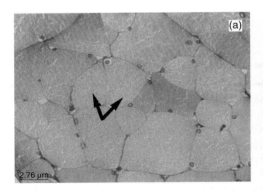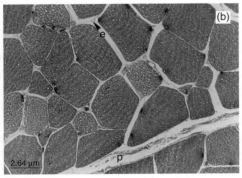

Figure 19.13 Transverse sections of rabbit meat at 1 h postmortem (a) and 24 h postmortem (b) observed by light microscopy (LM); e: endomisyal conjuctive tissue; p: perimysial conjuctive tissue. (From Sotelo, I., Pérez-Munuera, I., Quiles, A., Hernando, I., Larrea, V., and Lluch, M.A., *Meat Sci.*, 66, 823, 2004. With permission.)

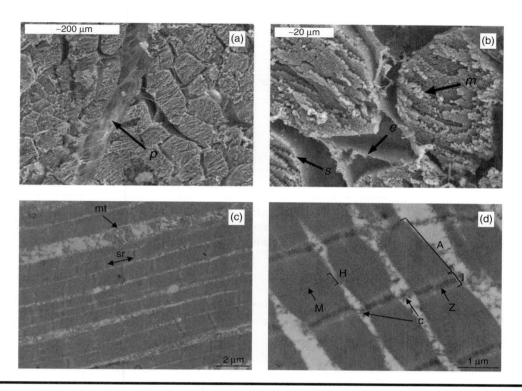

Figure 19.14   Cross section of semimembranosus muscle of raw ham observed by SEM (a: 250× and b: 1500×) and longitudinal section of biceps femoris muscle of raw ham observed by TEM (c: 10,000× and d: 25,000×). A: A band; c: costameres; e: endomysial connective tissue; H: H-zone; I: I band; M: M-line; m: myofibrils; mt: mitochondria; p: perimysial connective tissue; S: sarcolemma; sr: sarcomere; Z: Z-disk. (From Larrea, V., Pérez-Munuera, I., Hernando, I., Quiles, A., Llorca, E., and Lluch, M.A., *Meat Sci.*, 76, 574, 2007. With permission.)

When ultra-thin sections of pork muscle tissue are studied by TEM, it is possible to observe ultrastructural details that pass unnoticed with other methods. Figure 19.14c shows the inside of a muscle fiber with perfectly bundled myofibrils. In this figure, several mitochondria can be observed between the myofibrils. Figure 19.14d shows a more detailed view of a number of myofibrils; it is even possible to distinguish their component sarcomere filaments.

The microstructure of broiler chicken muscles was studied by Wattanachant et al. [13] using SEM. The thickness of the perimysium directly contributes to the texture of raw chicken muscles, whereas the cross-linked collagen content does not.

The amount of fat in different meats and muscles depends on a wide range of factors. Fat is located in the adipocytes, in the adipose tissue. Figures 19.15a and 19.15b show the intramuscular fat of pork meat observed by SEM and cryo-SEM [11]. As observed by SEM, it is made up of spherical adipocytes closely arranged and surrounded by a membrane, with perimysial connective tissue fibers among them.

Nishimura et al. [14] used SEM to investigate changes in the structures and mechanical properties of intramuscular connective tissue during the fattening of steers. They observed that the

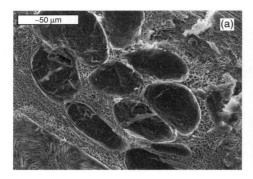
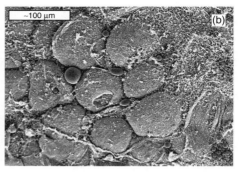

Figure 19.15 **Transversal section of intramuscular fat in biceps femoris of raw ham observed by cryo-SEM (a) and SEM (b) a: adipocytes; ct: connective tissue; m: muscular tissue. (From Larrea, V., Pérez-Munuera, I., Hernando, I., Quiles, A., and Lluch, M.A., *Food Chem.*, 101, 1327, 2007. With permission.)**

development of adipose tissue in longissimus muscle appears to disorganize the structure of the intramuscular connective tissue and contributes to tenderization of highly marbled beef during the late fattening period.

19.5 Influence of Processing on Meat Microstructure

The effect of cooking on different types of meat has been studied by several authors [9]. Roasting to an internal temperature of 50°C slightly affected the structure of bovine meat. During roasting to an internal temperature of 60–90°C, significant changes occurred both in the myofibrils and in the intramuscular connective tissue. The degree of postmortem aging of meat had a significant effect on the thermal stability of tissue structures. The changes in myofibril structure during roasting were considerably smaller in meat aged for 5 days than in that aged for 12 days postmortem.

Using SEM, Wattanachant et al. [13] have studied the effect on broiler muscle fibers after cooking (80°C/10 min). The broiler muscle fibers shrank more in a parallel than in a transverse direction to the fiber axis and expanded transversally after cooking, resulting in increasing muscle tenderness.

Sorheim et al. [15] have studied the effects of carbon dioxide on the microstructure of cooked beef by LM. Beef meat, previously stored in an environment containing 20–100% CO_2, had higher cooking losses compared to meat stored in 100% N_2 and under vacuum. Decreases in the pH of raw meat and the formation of small pores and microstructural changes upon heating due to CO_2 exposure were likely to cause the increased cooking losses. During cooking, structural changes in the meat protein matrix showed an increase in water loss due to pores and gaps in the matrix created by heat denaturation of the myofibrillar proteins and collagen.

Ueno et al. [16] have studied the effect of high-pressure treatments on intramuscular connective tissue from beef using SEM; it seems that high pressure may have a different effect from aging on the intramuscular connective tissue membrane structure. Structural changes in the endomysium and perimysium occurred and disruption of the honeycomb-like structure of the endomysium was observed.

19.6 Microstructure of Model Systems Based in Meat Components

In recent years, a huge number of works have focused on the effects of using different ingredients in meat product processing, employing model systems. These lines of research help to elucidate the interactions between the ingredients and the main meat components.

Iwasaki et al. [17] have studied the effect of hydrostatic pressure pretreatment on thermal gelation of chicken myofibrils using SEM and TEM. The structure of the myofibrils observed by SEM was disrupted at 200 MPa and the myofilaments were dispersed. The myofibril debris was observed at 300 MPa and the aggregated structures were observed in the debris periodically. The dispersed myofilaments became short at 300 MPa. The structures and characteristics of pressure/heat-induced gels of chicken myofibrils were investigated by TEM. The M-line and the Z-line in the chicken myofibril in the 0.2 M NaCl treatment were seen to be disrupted, and both the thin and the thick filaments were dissociated by the pressure treatment. The microstructure of pressure/heat-induced chicken myofibrillar gel was composed of a 3D matrix of fine strands.

Chen et al. [18] investigated the microstructure of salt-soluble meat protein (SSMP) and flaxseed gum (FG) mixtures by using SEM. The addition of FG changes the microstructure of SSMP gels. SEM observations showed that an interaction between FG and SSMP might occur. The results of adding a destabilizer to SSMP gels indicated that electrostatic forces seemed to be the primary force involved in the formation and stability of a protein-polysaccharide gel. The structure of SSMP gels showed a granular aggregated structure of large open spaces within the matrix, which seemed to show less linkage and be somewhat discontinuous between protein strands. The structure of SSMP–FG gels showed a well-structured matrix with a highly interconnected network of strands that may cause more resistance to applied stress and impart great water-holding capacity. These microstructural changes helped to explain functionality differences among the gels. A fine, uniform structure with numerous small pores would probably result in a higher absorption capacity and better retention of fat and water compared to a coarse structure with large pores.

References

1. Flint, O., in *Food Microscopy: A Manual of Practical Methods, Using Optical Microscopy*, Acribia S.A., Ed., Bio Scientific Publishers Limited, Zaragoza, 1994, chap. 4.
2. Sheppard, C.J.R. and Shotton, D.M., *Confocal Laser Scanning Microscopy*, Bio Scientific Publishers Limited, Springer-Verlag, New York, 1997, chap. 1.
3. Bozzola, J.J. and Russell, L.D., *Electron Microscopy*, Jones and Barlett, Boston, MA, 1992, chap. 1.
4. Aguilera, J.M. and Stanley, D.W., *Microstructural Principles of Food Processing and Engineering*, 2nd ed., Aspen Publishers, Inc., Gaithersburg, MD, 1999, chap. 1.
5. Hulting, H.O., Characteristics of muscle tissue, in *Principles of Food Science I. Food Chemistry*, Fennema, O.R., Eds., Marcel Dekker, New York, 1976.
6. Lluch, M.A., Pérez-Munuera, I., and Hernando, I., Proteins in food structures, in *Chemical and Functional Properties of Food Proteins*, Sikorski Z.E., Eds., Technomic, Lancaster, PA, 2001, chap. 2.
7. Pérez-Munuera, I., Sotelo, I., Hernando, I., and Lluch, M.A., Degradación postmorten en músculo de conejo: Evolución de la microestructura, *Alimentaria*, 99, 19, 1999.
8. Sotelo, I., Pérez-Munuera, I., Quiles, A., Hernando, I., Larrea, V., and Lluch, M.A., Microstructural changes in rabbit meat wrapped with *Pteridium aquilinum* fern during postmortem storage, *Meat Sci.*, 66, 823, 2004.

9. Palka, K., The influence of post-mortem ageing and roasting on the microstructure, texture and collagen solubility of bovine *semitendinosus* muscle, *Meat Sci.*, 64, 191, 2003.

10. Straadt, I.D., Rasmussen, M., Andersen, H.J., and Bertram, H.C., Aging-induced changes in microstructure and water distribution in fresh and cooked pork in relation to water-holding capacity and cooking loss—a combined confocal laser scanning micrsocopy (CLSM) and low-field nuclear magnetic resonance relaxation study, *Meat Sci.*, 75, 687, 2007.

11. Larrea, V., Pérez-Munuera, I., Hernando, I., Quiles, A., and Lluch, M.A., Chemical and structural changes in lipids during the ripening of Teruel dry-cured ham, *Food Chem.*, 101, 1327, 2007.

12. Larrea, V., Pérez-Munuera, I., Hernando, I., Quiles, A., Llorca, E., and Lluch, M.A., Microstructural changes in Teruel dry-cured ham during processing, *Meat Sci.*, 76, 574, 2007.

13. Wattanachant, S., Benjakul, S., and Ledwardt, D.A., Microstructure and thermal characteristics of Tai Indigenous and broiler chicken muscles, *Poultry Sci.*, 84, 328, 2005.

14. Nishimura, T., Hattori, A., and Takahashi, K., Structural changes in intramuscular connective tissue during the fattening of Japanese black cattle: Effect of marbling on beef tenderization, *J. Anim. Sci.*, 77, 93, 1999.

15. Sorheim, O., Ofstad, R., and Lea, P., Effects of carbon dioxide on yield, texture and microstructure of cooked ground beef, *Meat Sci.*, 67(2), 231, 2004.

16. Ueno, Y., Ikeuchi, Y., and Suzuki, A., Effects of high pressure treatments on intramuscular connective tissue, *Meat Sci.*, 52, 143, 1999.

17. Iwasaki, T., Noshiroya, K., Saitoh, N., Okano, K., and Yamamoto, K., Studies of the effect of hydrostatic pressure pretreatment on thermal gelation of chicken myofibrils and pork meat patty, *Food Chem.*, 95, 474, 2006.

18. Chen, H.H., Xu, S.Y., and Wang, Z., Interaction between flaxseed gum and meat protein, *J. Food Eng.*, 80, 1051, 2007.

Chapter 20

Physical Sensors and Techniques

Jean-Pierre Renou

Contents

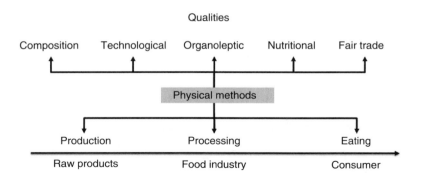

Figure 20.1 Physical methods and meat qualities.

20.1   Introduction

Although criteria of meat quality depend on levels of socio-economic development, meat purchasers everywhere seek an optimal quality/price ratio. Product quality depends on technological characteristics and sensory and nutritional values. Geographic origin and animal welfare are important in fair trading.

Producers today are interested in fat and lean composition of carcasses because this criterion underlies the predominant pricing system. Technological characteristics such as texture and water-holding capacity (WHC) are very important factors that affect acceptance by both the meat retailer and the consumer. Although sensory qualities vary according to the consumer, there are physical measurements that can objectively gauge aroma, toughness, and color. Recent health scares in the food industry (e.g., dioxin-contaminated poultry, BSE, and illegal pesticide residues) have heightened consumer awareness of all aspects of health quality and microbiological risks.

This chapter deals with biophysical sensors and their application in the study of meat quality characteristics (Figure 20.1).

20.2   Composition of Meat

20.2.1   Carcass

The accurate, objective prediction of carcass composition is an issue of great interest to livestock producers and retailers (Table 20.1). Because of the pricing system for marketing swine, fat and lean composition is determined on the live animal. Ultrasonic techniques have been applied to both live pigs and carcasses and provide the basis for a uniform grading system for swine before and after slaughter. Ultrasound instruments are commercially available. However, predicted lean meat yield can be underestimated by ultrasonic methods compared with dissection [1]. Ultrasonic scanners provide video images that make sensor positioning easier for the estimation of abdominal fat in poultry carcasses, but the estimation of breast muscle is poor [2]. The CGM optical probe (fat/lean meter) is also routinely used to estimate carcass lean content. However, some differences have been found between CGM and ultrasonic methods [3]. Noninvasive methods such as X-ray computerized tomography (CT), magnetic resonance imaging (MRI), and dual-energy X-ray absorptiometry (DEXA) can significantly improve carcass composition. This technology is very costly to implement but it can give a better estimation by improving the positioning of optical or ultrasonic sensors [4].

Table 20.1 Review of the References for the Determination of Carcass and Meat Composition

Composition	Visible Fluorescence	IR NIR	X-Ray	NMR	Electro-magnetic	Ultrasound
Carcass	Youssao et al. [3]	—	Mitchell et al. [11] Marcoux et al. [10] Bartle et al. [12]	Mitchell et al. [8] Fowler et al. [7] Davenel et al. [4] Collewet et al. [6] Monziols et al. [9]	Lin et al. [14] Berg et al. [13] Lin et al. [15]	Hoereth and Branscheid [1] Gillis et al. [2]
Meat	Skjervold et al. [19]	Isaksson et al. [16] Togersen et al. [18] Togersen et al. [17] Gonzalez-Martin et al. [26] Gonzalez-Martin et al. [25] Olsen et al. [28] Gonzalez-Martin et al. [29] Flatten et al. [27]	Kroger et al. [20]	Renou et al. [21,34] Renou et al. [22] Foucat et al. [23] Antequera et al. [24]	—	—

The use of CT is not easy in practice, and the cost of the complete chain of analysis severely restricts the application of this tool in the meat industry [5]. There is a large body of work on MRI for the characterization of carcass composition [4,6–9]. All these studies underline the fact that the MRI measurements give accurate quantification for different tissues that closely match the values produced by dissection and chemical analysis.

DEXA is cheaper, faster, and easier to use than MRI or CT. It has been successfully used to predict the chemical composition of pork carcasses [10,11]. This technology seemed, however, to be too slow for routine in-line use in the meat industry. Recent work shows that the new dual beam scanner system may improve the scanning rate [12].

Electromagnetic scanning has also been reported to be an effective and reliable means to estimate carcass composition of pork [13] and beef [14] and carcass lean of broilers [15]. The measured electromagnetic parameters were closely correlated with lean content determined by dissection.

20.2.2 Meat

Progress in spectroscopy allows the rapid, nondestructive measurement of meat composition through the development of new sensor technology, including cameras and spectrophotometers. Near-infrared (NIR) spectroscopy has been used successfully to quantify composition [16–18]. However, the results depend on the calibration model. Rapid, noninvasive quantification of meat composition can also be performed by fluorescence. In meat, specific wavelengths of 290/332 nm, 322/440 nm (or 322/405 nm), and 380/440 nm can be used respectively for autofluorescence of fat, myofibers, and connective tissue [19]. These wavelengths provide segmented images of these tissue components that yield quantitative data useful for meat composition determination.

Unlike NIR spectroscopy, X-ray scanning is insignificantly influenced by ambient conditions. The instrumentation based on DEXA was developed to determine the chemical lean in boneless meat that is packed in standard size meat boxes. This method was successfully installed in industry, scanning nearly continuously several thousand meat boxes each day [20].

NMR relaxometry [21] or spectroscopy [22] at low field determines fat content with great accuracy. The NMR results have always been closely related with the reference chemical methods. The main characteristics of the NMR equipment dedicated to the determination of fat content in ground beef are defined [23].

Three classes of dry-cured loin were assessed for each sensory and chemical trait by a trained panel. Using the texture features from texture analysis of magnetic resonance images, a classification of loins was obtained. These results point to the potential utility of MRI for *in situ* automatic, objective, nondestructive measurements [24].

The composition of dietary fat has received increased attention in recent years because it influences human health. The more saturated fat we consume, the higher our blood cholesterol levels are likely to be. NIR spectroscopy, with a remote reflectance fiber-optic probe, has been used to determine fatty acid composition in subcutaneous and intramuscular fat of Iberian pork loin [25,26]. Recent results indicate the feasibility of using Raman spectroscopy as a rapid nondestructive method to assay iodine, polyunsaturated fatty acids, monounsaturated fatty acids, and saturated fatty acids measured directly on pork adipose tissue and in melted fat from the same tissue [27]. Fatty acid composition of pork adipose tissue and characteristics of pork meat also change according to the dietary alterations of these fatty acids. Fourier transform-infrared spectroscopy (FT-IR) has been used to examine different components in fats and oils. FT-IR spectroscopy has been used to determine the degree of unsaturation and solid fat index in pork adipose tissue [28]. The contents of mineral elements Fe, Ca, Zn, Na, and K have been predicted by near-infrared spectrometry (NIRS) in ground and intact pork loin. After calibration, the best results were obtained for the elements K, Na, and Zn [29].

20.3 Technological Qualities

20.3.1 WHC

Pork quality is usually classified into three categories based on color, texture, and WHC (Table 20.2). Pale, soft, and exudative (PSE) meat has an undesirable color and lacks firmness owing to excessive drip loss, while dark, very firm, and dry (DFD) meat has a firm, sticky texture with high WHC. Pork with a desirable color (reddish-pink), normal texture, and normal WHC is sought.

Table 20.2   Review of the References for the Determination of Meat Technological Qualities

Technological Qualities	Visible Fluorescence	NMR	Electromagnetic
WHC	Qiao et al. [30,31]	Renou et al. [21,34]	—
	Qiao et al. [30,31]	Brøndum et al. [33]	
		Bertram et al. [32]	
Process	O'Farrell et al. [38]	Antequera et al. [36]	Clerjon and Damez [35]
	Allais et al. [37]		

Hyperspectral imaging can collect image data simultaneously in narrow adjacent spectral bands. These measurements make it possible to record a continuous spectrum for each image cell. Recent studies claim its ability to predict technological qualities of pork such as drip-loss (WHC) and pH [30].

WHC has been widely studied by low-field NMR [21,32–34]. NMR relaxation data of water were related to water interactions in muscle. There were highly significant relationships between NMR relaxation parameters and some meat characteristics such as pH, cooking yield, and protein denaturing.

20.3.2   Process

Muscles are very highly anisotropic dielectric materials. Changes occurring during the transformation from muscle to meat (rigor and maturation) lower dielectric anisotropy. The microwave polarimetric method allows this decrease to be followed during muscle aging [35]. A combination of frequencies could improve sensitivity, and some developments are required for the absolute rather than relative measurement of anisotropy during aging. This method could predict the best aging time for tenderness and also help to cut storage costs.

MRI combined with a fully automated image analysis method was used to identify some muscles in ham and follow their evolution (moisture and weight) during maturation. Loss of weight was closely correlated with decrease in size, as measured by computer visioning. It can be used as a noninvasive addition to the traditional processes of ham weighing and moisture estimation [36]. Front-face fluorescence spectroscopy provides information on the presence of fluorescent molecules and on their environment in biological and food samples. It was used to determine the optimal state of the batter and to decide when to stop the chopping process. Chopping is a key stage in the processing of meat emulsions such as frankfurters as it influences the properties of the final product. The results suggest that it may be possible to predict the texture of a frankfurter from fluorescence spectra recorded on the batter [37].

An optical fiber-based sensor system (visible light) has also been developed for accurate classification (100%) of food products (roast whole and pieces of chicken) in terms of their cooking state [38].

20.4   Organoleptic Qualities

20.4.1   Tenderness

Tenderness is a major quality factor in meat and is therefore limiting for consumer acceptance (Table 20.3). The tenderness of cooked meat is determined by the relative strength of the connective tissue network and the myofibrillar structure and their interactions. These relative contributions

Table 20.3 Review of the References for the Determination of Meat Organoleptic Qualities

Organoleptic Qualities	Visible Fluorescence	IR NIR	X-Ray	NMR	Rheology
Tenderness	—	Park et al. [42] Rodbotten et al. [43] Prevolnik et al. [44]	Kroger et al. [45]	Mahmoud-Ghoneim et al. [41]	Lepetit and Culioli [39] Shackelford et al. [40]
Color	Gerrard et al. [48] Lu et al. [47]	—	—	—	—

to meat tenderness can be evaluated differently depending on the mechanical test used [39]. The two measurements used most often are shear force, which measures the toughness due to both muscle fibers and connective tissue, and compression, which estimates the contribution of the connective tissue to meat tenderness.

Approaches for the determination of mechanical properties of food products require contact with the food and are mostly destructive. Shear force can be measured using the Warner-Bratzler or Mirinz devices. These standardized procedures determine shear forces accurately without taking into account intramuscular variability [40]. Imaging is an effective way to obtain texture features. Texture analysis combined with dedicated gradient echo MRI at high field provided specific parametric maps of connective tissue and allowed statistical analysis of the resulting texture [41]. NIR spectroscopy has been also used to predict beef tenderness. The best performance was obtained with beef longissimus thoracis steaks using the wavelengths of the NIR spectrum [42]. The correlations between NIR data and shear force varied according to time postmortem [43]. The results of various studies [44] show that prediction is better when a classification into two or three groups (tough, medium, and tender) is required.

DEXA is another tool used to estimate tenderness [45,46]. The study focused on the myofibrillar contribution to tenderness. The results are better for large-sized samples and for raw steaks. For cooked and frozen meat samples, low correlation ($R^2 = .12$) was found with the peak shear force.

20.4.2 Color

Among the pork quality attributes, color is of special significance because it is critically appraised by consumers. Fresh pork loin color has been evaluated from red, green, and blue bands of muscle area obtained from the segmented pork loin images. Results of this study showed that an image processing system in conjunction with a neural network is an effective tool for evaluating fresh pork color [47]. Image processing can also be an effective tool for determining marbling and color of fresh beef [48].

20.5 Nutritional Qualities

20.5.1 Freshness

A calibration model using absorption NIRS spectra can predict the freshness of beef through the indirect measurement of K-value related to ATP degradation [49] (Table 20.4).

Table 20.4   Review of the References for the Determination of Meat Nutritional Qualities

Nutritional Qualities	Visible Fluorescence	IR NIR	X-Ray	NMR	SM	ESR
Freshness	—	Bae et al. [49]	—	Yano et al. [52] Mortensen et al. [51] Bertram et al. [50]	—	—
Irradiation	—	—	—	—	—	Chawla and Thomas [53] Marchioni et al. [54]
Contaminant		Ellis et al. [61] Ellis et al. [60]	Schatzki et al. [56] Gupta [55] Davies and Board [58] McFarlane et al. [57]	—	Harper [59]	—
Adulteration	Al-Jowder et al. [63] Al-Jowder et al. [62] Downey et al. [64] Al-Jowder et al. [65]	—	—	—	—	—
Xenobiotic	Zhao et al. [71]	—	—	—	Ninh et al. [68] Weiss et al. [70] Rubies et al. [69]	—

Other methods have been used to detect thawed meat. The water fractions, as determined by the NMR relaxation times, were found to vary between fresh and frozen-thawed meat [50–52]. This measurement is relative. At present it is not a specific marker, because this water fraction also varies with other meat traits [34].

Food irradiation is useful to prevent food-borne illness because it efficiently reduces the population of pathogens and parasites. In the meat industry, this use is limited to ground beef meat and poultry. ESR is a spectroscopic method that allows the detection of unpaired electrons; it can be used as a detection test if the radicals induced by irradiation are stable during the commercial storage life of the food, and the corresponding signals are clearly distinguishable from those of the nonirradiated material (EN 1786: Anonymous. *Foodstuffs. Detection of irradiated food containing bones. Method by ESR spectroscopy.* In EN 1786; European Committee for Standardization: Brussels, Belgium, 1996).

Bone tissue is composed of approximately 60% hydroxyapatite with most of the remaining fraction as collagen. Hydroxyapatite is known to induce stable and specific ESR signals upon irradiation. The effect of storage and cooking was also studied for lamb meat. The decrease in the intensity of the ESR signal varied with time and cooking conditions. However, it allowed the detection of irradiated meat on bone [53].

The detection of ingredients containing irradiated bone in low concentrations in nonirradiated food requires an enzymatic hydrolysis method for the extraction of the bone fraction and purification of the extracts. This method made possible the detection of irradiated, mechanically recovered poultry meat at very low inclusions (0.5% w/w by ESR) in various meat products, the ESR signal amplitude being 11 times higher than the background noise [54].

20.5.2 Contaminants

20.5.2.1 Bone

X-ray technology is used in the food industry [55] to detect broken pits in fruits and bones or other foreign bodies in meat. In meat packages less than 6 cm thick, 2 mm bone chips, and 1 mm glass splinters are easily recognized [56]. The main problem is to achieve enough image contrast for the small foreign body to be detectable in a heterogeneous sample. X-ray Compton backscatter allows detection of a fragment of bone 0.1 cm below the surface of chicken breast [57].

20.5.2.1.1 Microbial Spoilage

Meat is a highly perishable food product that, unless correctly stored, processed, packaged, and distributed, spoils quickly and becomes a health hazard owing to microbial growth. The presence of pathogens in the food supply, even in low numbers, is undesirable and is considered a major cause of gastrointestinal disease. Numerous methods have been proposed to measure and detect bacterial spoilage in meats [58]. However, they are time consuming. A major problem in the food industry is that monitoring procedures, such as in the hazard analysis critical control point (HACCP) system, need to give results in real time to enable corrective action to be taken as soon as possible in busy and highly automated processing environments. The ideal method for online microbiological analysis of meat should be rapid, nondestructive, reagentless, quantitative, and relatively inexpensive, but no such method exists in the meat industry at present.

Electronic noses, made from an array of electronic chemical sensors with partial specificity and an appropriate pattern recognition system, are capable of recognizing simple or complex odors [58]. Rapid quantitative detection of microbial volatiles associated with muscle food spoilage has already been performed. However, this method shows some weaknesses, such as loss of sensitivity in humid conditions, very significant instrumental drift, and inability to provide quantitative data for aroma differences [59].

FT-IR was used directly on the surface of meat to produce fingerprints. Using these fingerprints and partial least squares regression, quantitative data on food sample loads can be obtained accurately, noninvasively, and rapidly in 60 s. The prediction of bacterial spoilage in poultry is more accurate than in beef [60,61], which may reflect differences in the respective spoilage processes. However, the bacterial contamination load in beef was significantly lower than that observed in chicken. This FT-IR process could be incorporated online in the production process.

20.5.2.2 Adulteration

A number of studies have applied the FT-IR technique to the discrimination and adulteration of meats [62–64]. This methodology seems robust and reproducible and an IR absorbance spectrum is collected in just a few seconds.

Using mid-infrared spectroscopy, discrimination between pure beef and beef containing 20% w/w of a range of potential adulterants (heart, tripe, kidney, and liver) is possible for raw and cooked samples [65]. At the highest cooking level, however, the spectra become more variable.

20.5.2.3 Xenobiotics

The detection of residues of pesticides or veterinary drugs as well as heavy metal pollutants in meat products is a major challenge for consumers. The transfer of contaminants from animal feed to animal products has been calculated [66]. Most studies designed to detect xenobiotics quantitatively are performed with analytical tools such as liquid chromatography coupled with NMR and mass spectrometry [67–70] or HPLC/fluorescence detection [71].

20.5.2.4 Fair Trade

Isotope ratio mass spectrometry (IRMS) and high-resolution and ^{13}C-NMR spectroscopy are accurate and robust techniques (Table 20.5). Stable isotope ratios δ^{18}O, δ^2H, ..., measured on the whole sample by IRMS, and the site-specific ratio determined by ^{2}H- and ^{13}C-NMR are the main isotopic techniques used to characterize the geographical origin of beef [72], chicken, [73] and lamb [74]. Geographical origin can be authenticated, as can type of diet, that is, organic as against conventional farming. Extracted volatile components and fatty acid composition shared a relationship with diet, while water ^{18}O and ^{2}H enrichment were more closely related to geographical origin. The differentiation as a function of animal diet was performed from isotope analysis of δ^{13}C in adipose tissue from pork meat [75]. Joint analysis of δ^{13}C and δ^{34}S in liver tissue have permitted the differentiation of swine of different breeds receiving different diets [76].

More recently, by extraction of the volatile compounds from ruminant raw muscles trimmed of fat, the authentication of the type of feeding has been assessed by mass spectrometry-based techniques [77].

Table 20.5 Review of the References for the Determination of Fair Trade

Fair Trade	NMR	SM
Geographic origins	Piasentier et al. [74]	Piasentier et al. [74]
Feeding	Renou et al. [72]	Gonzalez-Martin et al. [75]
	Franke et al. [73]	Gonzalez-Martin et al. [76]
		Vasta et al. [77]
		Franke et al. [73]

20.6   Conclusion

There is growing interest in applying techniques in real time on/in-line for quality control. Physical sensors can determine some traits of meat quality. The more accurate they are, the more they cost, and the less easy they are to install in an industrial plant. The added value provided by the determination of quality traits may not be great enough to offset the cost of the sensors. Used as research tools, they help to improve our understanding of the different mechanisms underlying meat quality. They are also a reference tool useful for the development of other, less expensive but lower-performing sensors that may respond better to industrial constraints (e.g., rapid measurement without sample preparation, and simultaneous determination of different attributes). Their weaknesses are their specificity, which requires previous calibration combining statistical analysis with artificial neural networks, weak sensitivity to minor constituents, and spectral data interpretation.

The implantation of physical sensors in the meat industry remains too limited and requires developments toward continuous monitoring. The results have to be validated by reference, and calibration has to be performed where all known sources of error are taken into account. These improvements could promote the use of sensors in the meat industry. Needs vary along the meat supply chain, from slaughtering to meat cutting to distribution. At these various stages, different meat qualities (technological, organoleptic, nutritional, etc.) are expected for different sizes: Carcass meat cuts from raw material vary widely.

References

1. Hoereth R, and Branscheid W. 2006. Grading methods. Possible modifications. Are new prediction formulae for grading of slaughter swine needed? *Fleischwirtschaft* 86:14–7.
2. Gillis WA, Orr HL, and Usborne WR. 1973. Ultrasonic estimation of carcass yield in turkey broilers. *Poultry Science* 52:1439–45.
3. Youssao IAK, Verleyen V, Michaux C, and Leroy PL. 2002. A comparison of the fat lean meter (CGM), the ultrasonic device Pie Medical 200 and the Piglog 105 for estimation of the lean meat proportion in Pietrain carcasses. *Livestock Production Science* 78:107–14.
4. Davenel A, Seigneurin F, Collewet G, and Remignon H. 2000. Estimation of poultry breastmeat yield: Magnetic resonance imaging as a tool to improve the positioning of ultrasonic scanners. *Meat Science* 56:153–8.
5. Szabo C, Babinszky L, Verstegen MWA et al. 1999. The application of digital imaging techniques in the in vivo estimation of the body composition of pigs: A review. *Livestock Production Science* 60:1–11.
6. Collewet G, Bogner P, Allen P et al. 2005. Determination of the lean meat percentage of pig carcasses using magnetic resonance imaging. *Meat Science* 70:563–72.
7. Fowler PA, Fuller MF, Glasbey CA, Cameron GG, and Foster MA. 1992. Validation of the in vivo measurement of adipose tissue by magnetic resonance imaging of lean and obese pigs. *American Journal of Clinical Nutrition* 56:7–13.
8. Mitchell AD, Wang PC, Rosebrough RW, Elsasser TH, and Schmidt WF. 1991. Assessment of body composition of poultry by nuclear magnetic resonance imaging and spectroscopy. *Poultry Science* 70:2494–500.
9. Monziols M, Collewet G, Bonneau M et al. 2006. Quantification of muscle, subcutaneous fat and intermuscular fat in pig carcasses and cuts by magnetic resonance imaging. *Meat Science* 72:146–54.
10. Marcoux M, Bernier JF, and Pomar C. 2003. Estimation of Canadian and European lean yields and composition of pig carcasses by dual-energy X-ray absorptiometry. *Meat Science* 63:359–65.
11. Mitchell AD, Scholz AM, Pursel VG, and Evock-Clover CM. 1998. Composition analysis of pork carcasses by dual-energy X-ray absorptiometry. *Journal of Animal Science* 76:2104–14.

12. Bartle CM, Kroger C, and West JG. 2004. New uses of X-ray transmission techniques in the animal-based industries. *Radiation Physics and Chemistry* 71:843–51.
13. Berg EP, Asfaw A, and Ellersieck MR. 2002. Predicting pork carcass and primal lean content from electromagnetic scans. *Meat Science* 60:133–9.
14. Lin RS, Forrest J, Judge M, Lemenager R, and Grant A. 2001. Electromagnetic scanning of beef hindquarters for estimating beef carcass composition. *Journal of the Chinese Society of Animal Science* 30:115–27.
15. Lin RS, Chen LR, Huang SC, and Liu CY. 2002. Electromagnetic scanning to estimate carcass lean content of Taiwan native broilers. *Meat Science* 61:295–300.
16. Isaksson T, Nilsen BN, Togersen G, Hammond RP, and Hildrum KI. 1996. On-line, proximate analysis of ground beef directly at a meat grinder outlet. *Meat Science* 43:245–53.
17. Togersen G, Arnesen JF, Nilsen BN, and Hildrum KI. 2003. On-line prediction of chemical composition of semi-frozen ground beef by non-invasive NIR spectroscopy. *Meat Science* 63:515–23.
18. Togersen G, Isaksson T, Nilsen BN, Bakker EA, and Hildrum KI. 1999. On-line NIR analysis of fat, water and protein in industrial scale ground meat batches. *Meat Science* 51:97–102.
19. Skjervold PO, Taylor RG, Wold JP et al. 2003. Development of intrinsic fluorescent multispectral imagery specific for fat, connective tissue, and myofibers in meat. *Journal of Food Science* 68:1161–8.
20. Kroger C, Bartle CM, and West JG. 2005. Non-invasive measurements of wool and meat properties. *Insight* 47:25–8.
21. Renou J-P, Kopp J, and Valin C. 1985. Use of low resolution NMR for determining fat content in meat products. *Journal of Food Technology* 20:23–9.
22. Renou J-P, Briguet A, Gatellier P, and Kopp J. 1987. Technical note: Determination of fat and water ratios in meat products by high resolution NMR at 19.6 MHz. *International Journal of Food Science and Technology* 22:169–72.
23. Foucat L, Donnat JP, Humbert F, Martin G, and Renou J-P. 1997. On-line determination of fat content in ground beef. *Journal of Magnetic Resonance Analysis* 8:108–12.
24. Antequera T, Muriel E, Rodriguez PG, Cernadas E, and Ruiz J. 2003. Magnetic resonance imaging as a predictive tool for sensory characteristics and intramuscular fat content of dry-cured loin. *Journal of the Science of Food and Agriculture* 83:268–74.
25. Gonzalez-Martin I, Gonzalez-Perez C, Alvarez-Garcia N, and Gonzalez-Cabrera JM. 2005. On-line determination of fatty acid composition in intramuscular fat of Iberian pork loin by NIRs with a remote reflectance fibre optic probe. *Meat Science* 69:243–8.
26. Gonzalez-Martin I, Gonzalez-Perez C, Hernandez-Mendez J, and Alvarez-Garcia N. 2003. Determination of fatty acids in the subcutaneous fat of Iberian breed swine by near infrared spectroscopy (NIRS) with a fibre-optic probe. *Meat Science* 65:713–9.
27. Flatten A, Bryhni EA, Kohler A, Egelandsdal B, and Isaksson T. 2005. Determination of C22:5 and C22:6 marine fatty acids in pork fat with Fourier transform mid-infrared spectroscopy. *Meat Science* 69:433–40.
28. Olsen EF, Rukke EO, Flatten A, and Isaksson T. 2007. Quantitative determination of saturated-, monounsaturated- and polyunsaturated fatty acids in pork adipose tissue with non-destructive Raman spectroscopy. *Meat Science* 76:628–34.
29. Gonzalez-Martin I, Gonzalez-Perez C, Hernandez-Mendez J, and Alvarez-Garcia N. 2002. Mineral analysis (Fe, Zn, Ca, Na, K) of fresh Iberian pork loin by near infrared reflectance spectrometry: Determination of Fe, Na and K with a remote fibre-optic reflectance probe. *Analytica Chimica Acta* 468:293–301.
30. Qiao J, Wang N, Ngadi MO et al. 2007. Prediction of drip-loss, pH, and color for pork using a hyperspectral imaging technique. *Meat Science* 76:1–8.
31. Qiao J, Ngadi MO, Wang N, Gariepy C, and Prasher S. 2007. Pork quality and marbling level assessment using a hyperspectral imaging system. *Journal of Food Engineering* 83:10–6.
32. Bertram HC, Andersen HJ, and Karlsson AH. 2001. Comparative study of low-field NMR relaxation measurements and two traditional methods in the determination of water holding capacity of pork. *Meat Science* 57:125–32.
33. Brøndum J, Munck L, Henckel P et al. 2000. Prediction of water-holding capacity and composition of porcine meat by comparative spectroscopy. *Meat Science* 55:177–85.

34. Renou J-P, Monin G, and Sellier P. 1985. Nuclear magnetic resonance measurements on pork of various qualities. *Meat Science.* 15:225–33.

35. Clerjon S, and Damez JL. 2007. Microwave sensing for meat and fish structure evaluation. *Measurement Science and Technology* 18:1038–45.

36. Antequera T, Caro A, Rodriguez PG, and Perez T. 2007. Monitoring the ripening process of Iberian ham by computer vision on magnetic resonance imaging. *Meat Science* 76:561–7.

37. Allais I, Dufour E, Pierre A et al. 2003. Monitoring the texture of meat emulsions by front-face fluorescence spectroscopy. *Sciences Des Aliments* 23:128–31.

38. O'Farrell M, Lewis E, Flanagan C, Lyons WB, and Jackman N. 2005. Combining principal component analysis with an artificial neural network to perform online quality assessment of food as it cooks in a large-scale industrial oven. *Sensors and Actuators B: Chemical* 107:104–12.

39. Lepetit J, and Culioli J. 1994. Mechanical properties of meat. *Meat Science* 36:203–37.

40. Shackelford SD, Wheeler TL, and Koohmaraie M. 1997. Repeatability of tenderness measurements in beef round muscles. *Journal of Animal Science* 75:2411–6.

41. Mahmoud-Ghoneim D, Bonny JM, Renou J-P, and de Certaines JD. 2005. Ex-vivo magnetic resonance imaging texture analysis can discriminate genotypic origin in bovine meat. *Journal of the Science of Food and Agriculture* 85:629–32.

42. Park B, Chen YR, Hruschka WR, Shackelford SD, and Koohmaraie M. 1998. Near-infrared reflectance analysis for predicting beef longissimus tenderness. *Journal of Animal Science* 76:2115–20.

43. Rodbotten R, Mevik BH, and Hildrum KI. 2001. Prediction and classification of tenderness in beef from non-invasive diode array detected NIR spectra. *Journal of Near Infrared Spectroscopy* 9:199–210.

44. Prevolnik M, Candek-Potokar M, and Skorjanc D. 2004. Ability of NIR spectroscopy to predict meat chemical composition and quality—a review. *Czech Journal of Animal Science* 49:500–10.

45. Kroger C, Bartle CM, West JG, Purchas RW, and Devine CE. 2006. Meat tenderness evaluation using dual energy X-ray absorptiometry (DEXA). *Computers and Electronics in Agriculture* 54:93–100.

46. PCT. 2001. A method for the non-invasive measurement of properties of meat PCT, International Patent Application. PCT/NZ01/00108, and issued 2001.

47. Lu J, Tan J, Shatadal P, and Gerrard DE. 2000. Evaluation of pork color by using computer vision. *Meat Science* 56:57–60.

48. Gerrard DE, Gao X, and Tan J. 1996. Beef marbling and color score determination by image processing. *Journal of Food Science* 61:145–8.

49. Bae YM, Cho SI, Kim YY, Park IS, and Hwang KY. 2006. Estimation of freshness of beef using near-infrared spectroscopy. *Transactions of the Asabe* 49:557–61.

50. Bertram HC, Andersen RH, and Andersen HJ. 2007. Development in myofibrillar water distribution of two pork qualities during 10-month freezer storage. *Meat Science* 75:128–33.

51. Mortensen M, Andersen HJ, Engelsen SB, and Bertram HC. 2006. Effect of freezing temperature, thawing and cooking rate on water distribution in two pork qualities. *Meat Science* 72:34–42.

52. Yano S, Tanaka M, Suzuki N, and Kanzaki Y. 2002. Texture change of beef and salmon meats caused by refrigeration and use of pulse NMR as an index of taste. *Food Science and Technology Research* 8:137–43.

53. Chawla SP, and Thomas P. 2004. Identification of irradiated chicken meat using electron spin resonance spectroscopy. *Journal of Food Science and Technology, Mysore* 41:455–458.

54. Marchioni E, Horvatovich N, Charon H, and Kuntz F. 2005. Detection of irradiated ingredients included in low quantity in non-irradiated food matrix. 1. Extraction and ESR analysis of bones from mechanically recovered poultry meat. *Journal of Agricultural and Food Chemistry* 53:3769–73.

55. Gupta NK. 1995. X-ray system for on-line detection of foreign objects in food. In proceedings of the Food Processing Automation Conference FPAC IV, Paper presented at FPAC IV, 3–5 November 1995, USA, Chicago, IL, pp. 7–13.

56. Schatzki TE, Young R, Haff RP, Eye J, and Wright G. 1995. Visual detection of particulates in processed meat products by x ray. Paper presented at Optics in Agriculture, Forestry, and Biological Processing.

57. McFarlane NJB, Speller RD, Bull CR, and Tillett RD. 2003. Detection of bone fragments in chicken meat using X-ray backscatter. *Biosystems Engineering* 85:185–99.

58. Davies A, and Board R. 1998. *Microbiology of Meat and Poultry*. London: Springer-Verlag.
59. Harper WJ. 2001. *The Strengths and Weaknesses of the Electronic Nose*. In Headspace analysis of foods and flavors: Theory and Practice Series: Advances in Experimental Medicine and Biology, Vol. 488, Eds: Rouseff Russall, L., Cadwalleder keith R., 2001, pp. 59–71.
60. Ellis DI, Broadhurst D, and Goodacre R. 2004. Rapid and quantitative detection of the microbial spoilage of beef by Fourier transform infrared spectroscopy and machine learning. *Analytica Chimica Acta* 514:193–201.
61. Ellis DI, Broadhurst D, Kell DB, Rowland JJ, and Goodacre R. 2002. Rapid and quantitative detection of the microbial spoilage of meat by Fourier transform infrared spectroscopy and machine learning. *Applied and Environmental Microbiology* 68:2822–8.
62. Al-Jowder O, Defernez M, Kemsley EK, and Wilson RH. 1999. Mid-infrared spectroscopy and chemometrics for the authentication of meat products. *Journal of Agricultural and Food Chemistry* 47:3210–8.
63. Al-Jowder O, Kemsley EK, and Wilson RH. 1997. Mid-infrared spectroscopy and authenticity problems in selected meats: A feasibility study. *Food Chemistry* 59:195–201.
64. Downey G, McElhinney J, and Fearn T. 2000. Species identification in selected raw homogenized meats by reflectance spectroscopy in the mid-infrared, near-infrared, and visible ranges. *Applied Spectroscopy* 54:894–9.
65. Al-Jowder O, Kemsley EK, and Wilson RH. 2002. Detection of adulteration in cooked meat products by mid-infrared spectroscopy. *Journal of Agricultural and Food Chemistry* 50:1325–9.
66. Leeman WR, Van den Berg KJ, and Houben GF. 2007. Transfer of chemicals from feed to animal products: The use of transfer factors in risk assessment. *Food Additives and Contaminants* 24:1–13.
67. Jestoi M, Rokka M, and Peltonen K. 2007. An integrated sample preparation to determine coccidiostats and emerging *Fusarium*-mycotoxins in various poultry tissues with LC-MS/MS. *Molecular Nutrition and Food Research* 51:625–37.
68. Ninh TD, Nagashima Y, and Shiomi K. 2006. Quantification of seven arsenic compounds in seafood products by liquid chromatography/electrospray ionization-single quadrupole mass spectrometry (LC/ESI-MS). *Food Additives and Contaminants* 23:1299–307.
69. Rubies A, Vaquerizo R, Centrich F et al. 2007. Validation of a method for the analysis of quinolones residues in bovine muscle by liquid chromatography with electrospray ionisation tandem mass spectrometry detection. *Talanta* 72:269–76.
70. Weiss C, Conte A, Milandri C et al. 2007. Veterinary drugs residue monitoring in Italian poultry: Current strategies and possible developments. *Food Control* 18:1068–76.
71. Zhao SJ, Jiang HY, Li XL et al. 2007. Simultaneous determination of trace levels of 10 quinolones in swine, chicken, and shrimp muscle tissues using HPLC with programmable fluorescence detection. *Journal of Agricultural and Food Chemistry* 55:3829–34.
72. Renou J-P, Bielicki G, Deponge C et al. 2004. Characterization of animal products according to its geographic origin and feeding diet using nuclear magnetic resonance and isotope ratio mass spectrometry Part II: Beef meat. *Food Chemistry* 86:251–6.
73. Franke BM, Koslitz S, Micaux F et al. 2007. Tracing the geographic origin of poultry meat and dried beef with oxygen and strontium isotope ratios. *Eur. Food Res. Technol.* 266:761–769.
74. Piasentier E, Valusso R, Camin F, and Versini G. 2003. Stable isotope ratio analysis for authentication of lamb meat. *Meat Science* 64:239–47.
75. Gonzalez-Martin I, Gonzalez-Perez C, Hernandez Mendez J, Marques-Macias E, and Sanz Poveda F. 1999. Use of isotope analysis to characterize meat from Iberian-breed swine. *Meat Science* 52:437–41.
76. Gonzalez-Martin I, Gonzalez Perez C, Hernandez Mendez J, and Sanchez Gonzalez C. 2001. Differentiation of dietary regimen of Iberian swine by means of isotopic analysis of carbon and sulphur in hepatic tissue. *Meat Science* 58:25–30.
77. Vasta V, Ratel J, and Engel E. 2007. Mass spectrometry analysis of volatile compounds in raw meat for the authentication of the feeding background of farm animals. *Journal of Agricultural and Food Chemistry* 55:4630–9.

NUTRITIONAL QUALITY

Chapter 21

Composition and Calories

Karl O. Honikel

Contents

21.1 Introduction

The food we eat provides us with nutritive substances to build the body structures and to let them function properly. Food also provides the body with energy from reduced organic compounds, which undergo oxidation releasing chemical energy and heat. The main energy-delivering nutrients are carbohydrates, fat, and to a smaller extent, protein. In a live animal, these compounds can be oxidized to H_2O and CO_2; protein releases additionally NH_3, which is excreted as urea or uric acid. A part of these reduced organic compounds are stored like fat in adipose tissue and carbohydrates in glycogen globules.

The outstanding constituent of muscle foods is the protein, which in its content varies among the meat cuts: the meat animal species, fish species, and other seafood. Its content ranges between 13 and 23% of the fresh weight.[1,2] The main constituent of all muscle food species is still water, which varies from 61 to 80% in sea(water) food and 68 to 76% in meat except pure adipose (fatty) tissue such as backfat of pork. The third macronutrient is fat, which may be as low as <1%. It may reach around 25% in fat fish and around 30% in pork belly (Table 21.1).[3] It may even reach 90% in pure backfat of pork. The overall mean fat content is around 10% in meat cuts of the commonly eaten meat animals sold as fresh meat, whereas it is <10% in fish due to the majority being of very lean type.

All muscle foods contain cholesterol, again varying in a wide range from 30 mg/100 g in some fish to as high as >100 mg/100 g in many crustaceans and mollusks, 40–90 mg/100 g in meat, and several hundred mg/100 g in offals such as liver, kidney, or brain.

All muscle foods also contain between 1 and 2% minerals including trace elements and different concentrations of many vitamins (shown for various meat species in Table 21.2).[4]

Table 21.1 Composition of German Pork Cuts—g/100 g Edible Raw Meat ($N = 9$–18)

Cut	Protein (Mean)	Fat (Mean)	Fat Min	Fat Max	Energy (Mean) (kJ/kcal per 100 g)
Thick flank, knuckle	21.75	1.3	1	2	415/99
Topside	22.2	1.9	1.5	2.5	445/106
Tenderloin, fillet	22.0	2.0	1.5	2.5	445/106
Loin boneless	22.45	2.1	1	3	455/108
Chump	21.75	2.34	1.5	3	455/108
Loin bone-in, caudal (chops, cutlet)	21.6	7.5	3.5	14	645/154
Neck	19.7	9.6	7	11.5	660/165
Shoulder with rind	20.2	9.65	7	12	670/168
Loin bone-in, cranial (chops)	21.25	9.85	8	15	720/174
Forequarter hock, shank	20.35	10.8	9	11.5	752/179
Hindquarter hock, shank	18.95	12.2	9	14.5	780/186
Silverside with rind	19.2	15.35	11.5	17.5	903/215
Riblets	18.3	15.6	12	18	897/214
Shoulder blade	17.55	16.8	13.5	19.5	928/221
Belly rindless	17.8	21.1	14	22	1097/261
Belly rind-on	15.75	29.0	23	32	1361/324
Overall mean ± standard deviation of meat cuts	19.95 ± 1.96	10.5 ± 7.73			175/733

Source: Honikel, K.O. and Wellhäuser, R., *Fleischwirts*, 73, 863, 1993. With permission.

Table 21.2   Content of Protein, Fat, Vitamins, and Minerals in Different Species of Meat

	Pork (Lean Meat) (%)	Beef (Lean Meat) (%)	Poultry (Carcass with Skin) (%)[a]	Liver (Calf) (%)
Protein	22.0	22.0	21.2	19.2
Fat	1.9	1.9	8.2	4.1

	Recommended Daily Allowance 90/496 (EEC[5]) or D-A-CH Reference Values[6]	Pork (mg/100 g)	Beef (mg/100 g)	Poultry (mg/100 g)[a]	Liver (mg/100 g)
Thiamin (B1)	1.4	0.9	0.06	0.08	0.28
Riboflavin (B2)	1.6	0.23	0.26	0.15	2.6
Niacin	18	6.9	5.0	7.3	12.5
Pyridoxin (B6)	2	0.6	0.24	0.53	0.17
Panthotenic acid	6	0.7	0.31	0.89	7.9
Biotin	0.15	0.005	0.003	0.002	0.08
Cobalamin (B12)	0.001	0.002	0.005	0.0005	0.06
Vitamin D	0.005	0.0005	0.0005	0.0005	0.0003
Sodium (Na)	<2400	69	66	76	87
Potassium (K)	2000[b]	397	358	289	316
Phosphorus (P)	800	192	190	202	306
Magnesium (Mg)	300	26	23	24	19
Iron (Fe)	14	1.4	2.4	1.4	7.9
Zinc (Zn)	15	2.4	4.3	1.6	8.4
Selenium (Se)	0.05[b]	0.012	0.006	0.01	0.024
Chrome (Cr)	0.065[b]	0.003	0.005	0.002	K.A.

[a] Mean values of chicken and turkey.
[b] Values are in milligrams.
Source: Bauer, F., Personal communication, 2007.

Some constituents of meat and seafood metabolize partially due to slaughter and storage conditions.[7,8] During preparation they degrade further. On the one hand, this may enhance the flavor and, on the other hand, it may deteriorate the food to rancid or fishy flavor.[9] Some of the degradation products such as biogenic amines may cause health problems.[10] Owing to the differences in composition and changes postmortem, the composition of meat from land animals and seafood has to be addressed separately.

21.2   Composition

21.2.1   Composition of Meat

According to a definition accepted worldwide, the term "meat" signifies all edible parts of the carcass of a slaughtered animal. This includes the lean muscular tissue, adjacent (intra- and intermuscular) fat, and adipose tissue usually called fatty tissue. This part of "meat," including skin in some species, will be the subject of this chapter. Pork cuts are presented in Table 21.1 as an example for meat composition in the two most important ingredients.

Table 21.3 Composition of Lean Muscle in Live Animal and Meat 2 Days after Slaughter

Compound	Muscle in Live Animal (%)	Meat 2 Days Postmortem (%)
Protein	22	22
Fat	1	1
Carbohydrate	~1 (glycogen)	<0.5 (glycogen + glucose)
Organic acids	0.2 (lactate)	>0.8 (lactate + lactic acid)
pH	7.0	5.2–5.8

The other "meat parts" such as blood, organs (offal), and intestines cannot be addressed here. Additionally, consumers understand the intended restrictions to meat and fat.

21.2.1.1 Changes in Meat Postmortem

After slaughter, which according to the rules of most countries must be carried out after stunning (unconsciousness) followed by debleeding, the composition of the muscular tissues changes and the muscle is converted into meat (Table 21.3). For a more detailed description of these changes see, for example, *Lawrie's Meat Science*.[11]

At the time of death the oxygen and nutrient supply via bloodstream to the muscle cells, as well as the transport of degradation products from the cells, stops. Owing to the lack of oxygen the reduced organic matter in fatty acids cannot be used at all, as the oxidative phosphorylation and the citric acid cycle do not function any longer. However, carbohydrates 95% of which are stored as glycogen in the muscle cells of the living organism, which is a polymer of glucose, can be used anaerobically. The C-6 unit of glucose is broken down into two units of lactic acid, delivering energy for the phosphorylation of three ADP to three ATP molecules per glucose unit in glycogen. Free glucose itself would provide energy for two ATP molecules only. With oxygen in a live animal about 36 molecules of ATP could be phosphorylated from ADP by one glucose unit and the carbohydrates would be oxidized to CO_2 and H_2O. Owing to this considerable increase in lactic acid (0.1 M increase), the pH falls from about 7.0 to 5.3–5.8 under normal conditions. In exhausted animals, for example, after long transport and immediate slaughter or after fights in newly mixed groups in the lairage, the glycogen may be partially or nearly totally depleted before slaughter and the lactic acid formation may be limited. The pH fall may end at any value between >5.8 and 7.0. This meat is of minor quality as it lacks flavor and gets spoiled easily by microbial action. It is called DFD (dark, firm, dry) meat. In any case, the meat has less than 0.5% carbohydrates (Table 21.3). The exhaustion of glycogen and the fall in pH may last between 1 and 30 h depending on muscle meat species and slaughter conditions. In stressed porcine muscles, the ultimate pH is reached in 60–90 min, in chicken breast in 2–3 h, in unstressed porcine muscles in 5–8 h, and in bovine muscles in 15–30 h.

The fall in pH and the exhaustion of glycogen (no further formation of ATP possible) leads to the onset of rigor mortis where the meat becomes stiff. On preparation in this state the meat remains tough. It needs an aging (tenderizing) process that lasts at the freezing temperatures (<7°C) in poultry for >1 day; in pork >2 days; and in all other species such as veal, lamb, sheep, beef, and venison >7 days.[8] During this period, the highly ordered structures of myofibers are partially broken down by endo- and exoproteases. There is, however, only a small increase in the originally low concentration of free amino acids. Therefore, in fresh meat before spoilage biogenic amines, which are decarboxylation products of free amino acids, are not formed in significant amounts that are.

For sensorial reasons, it is worthwhile reporting that with the exhaustion of the glycogen → lactic acid turnover, ATP is degraded to ADP, and then to AMP. AMP is deaminated (AMP → IMP + NH_3) to IMP, which like lactic acid is a flavor enhancer of meat (see Chapter 8).

Nearly all of these postmortem changes occur similar to the muscle cell with the help of proteases, which are activated postmortem due to pH fall, changes in cation concentrations, or degradation of subcellular membranes.[12] The connective tissue, which is important for meat tenderness, is located outside the cellular structures.[13] As the proteases appear slowly outside the cellular membranes, the tenderizing process may take 2 weeks, for example, in cuts of adult bovine animals.

As mentioned in the foregoing text, in contrast to carbohydrates and protein, the fat in meat during the postmortem changes and aging does not metabolize much. The main changes are the oxidation processes of unsaturated fatty acid chains. The hydrolysis of triacylglycerols and phospholipids to mono- and diacylglycerols and free fatty acids during this period of aging is very slow.

Lean meat of all common meat species in Europe contains more than 50% unsaturated fatty acids. The main lipid constituent of lean meat is the phospholipids that contain a high proportion of unsaturated fatty acids. But even in meat with 10–20% fat the unsaturation of fatty acids is above 50%. The main single fatty acid with few exceptions is oleic acid (up to 47%).[1] The unsaturation increases from about 50% in lamb to >65% in poultry, in the following order: lamb > cattle > pigs > poultry.[14]

21.2.1.2   Summary

In meat of the commonly eaten warm-blooded animals (mammals and birds) the breakdown of carbohydrates (glycogen) leads to the formation of lactic acid and a fall in pH. Together with the ATP depletion and IMP formation these are the most pronounced chemical changes in fresh meat. The most important physical process is the onset of rigor and the following aging (tenderizing) process. This does not lead to far-reaching chemical changes in ingredients. The smallest chemical changes in fat occur by hydrolysis and oxidation.

21.2.2   Composition of Seafood

The proximate composition of some fish is shown in Table 21.4.[15] In contrast to the various cuts of meat animals, the composition of fish is usually described as that of a whole edible part of fish. The three main ingredients and the total mineral content vary in a similar range as in meat cuts.

The composition of seafood other than fish is shown in Table 21.5.[1] There are differences compared to that of fish and meat. The water content is higher than in any meat, whereas the protein

Table 21.4   Composition of Some Fish Species

Species	Water (%)	Protein (%)	Fat (%)	Minerals (%)
Perch	80	18	0.8	1.3
Halibut	75	19	1.7	1.3
Herring (Atlantic)	63	17	18	1.3
Tuna	62	22	16	1.1
Eel	61	15	26	1.0

Source: Adapted from Belitz, H.D., Grosch, W., and Schieberle, P. in *Food Chemistry*, Springer Verlag, Berlin, 2004, 627.

Table 21.5 Composition of Some Crustaceans and Mollusks

Species	Water (%)	Protein (%)	Fat (%)	Carbohydrate (%)	Minerals (%)	Cholesterol (mg/100 g)
Lobster	80	16	2	—	2.1	90
Shrimp	78	18.5	1.4	—	1.4	138
Crab	83	15	0.5	—	1.3	100
Oyster	83	9.	1.2	4.8	2.0	125
Mussels	83	10.2	2.0	2.4	1.5	125
Octopus	81	16.1	0.9	—	1.0	275

Source: Adapted from Souci, Fachmann, and Kraut. in *Der kleine SFK, Lebensmitteltabelle für die Praxis*, Verlagsgesellschaft, Stuttgart, Germany, 2004.

content is lower and the fat content is very low. The mineral content varies widely than in meat and fish. The cholesterol content is higher than in the muscular or fatty tissue of meat and fish.

This low fat but high cholesterol concentration shows that, in general, in cellular tissues both constituents are independent of each other. Cholesterol serves the purpose of membrane functionality, whereas fat (triacylglycerols) is an energy store.

The connective tissue protein in fish is around 1.3%.[15] In meat the connective tissue concentration varies[11] from as low as 0.5% in cuts of the upper part of hind leg of pork and pork fillet[3] to 2–4% in the lower cuts of legs near the feet. The mean value is around 1.4%. The fish connective tissue denatures (shrinks) at around 45°C and dissolves more easily (gelatinizes)[16] than the connective tissue of meat, which shrinks at 55–65°C depending on the species and the age.[13]

A number of fish, for example, shark and rays, contain urea in their meat (1.3–2.1 g/kg),[16] which is broken down during storage by the enzyme urease forming NH_3. Besides ammonia, biogenic amines also occur as amino acid degradation products in fish meat, as fish contains a higher concentration of free amino acids than meat. The total volatile base nitrogen (TVB-N) value (mg N/100 g fish) is an indicator of freshness.[17]

The mineral content also varies between fish species like in meat but fish usually contains higher iodine concentrations than meat (0.3–8 mg/kg).[18] Vitamin concentrations are comparable to meat.

Fish meat is also low in carbohydrates. Fish contains <0.3% carbohydrates, whereas mollusks may contain some (Table 21.5).

Fish, like meat, contain creatine and also purines, about 300 mg/kg.[15] In meat the purine content is about 1500 mg/kg.[19]

21.2.2.1 Changes in Fish Postmortem

The pH fall postmortem in fish is usually small and ends at 6.3–6.5, a full pH unit higher than in meat.[16] Owing to this higher ultimate pH value, spoilage starts early and a number of degradation products, especially of proteins and fat, appear in higher concentrations than in meat.

Another compound in fish involved in the regulation of osmotic pressure is trimethylamine oxide (TMAO, 40–120 mg/kg sea fish[16]). Postmortem it is decomposed by bacteria to trimethylamine, which is the typical fishy smell of sea fish. Freshwater fish exhibit much lower concentrations of trimethylamine.[16]

The fat of fish shows a higher concentration of polyunsaturated fatty acid (PUFA) with four to six double bonds. Many of them are ω-3 fatty acids such as C20:5 ω3 and C22:6 ω3. Most fatty acids have chain length of 16, 18, and 20 C atoms. The saturated fatty acids are low and in the range of 20–40% of the fatty acids.

These PUFA are easily oxidized postmortem and cause, on one hand, specific flavors, and on the other, off-flavors at higher concentrations.[15]

21.2.2.2 Summary

Fish meat and other seafood are comparable to meat in the amount of the main nutrient groups. There exist differences in fatty acid composition, mineral content, and carbohydrates in nonfish seafood. But most pronounced are the faster chemical changes postmortem due to the higher ultimate pH of about 6.3–6.5 and other enzymatic activities in comparison to meat.

21.3 Calories = Physiological Energy

Living animals and human beings consist of chemotrophic or heterotrophic cells. It main that they have to take in reduced organic matter as feed/food that is oxidized by the oxygen in air or water to CO_2 and H_2O.[20] The latter compounds are the most oxidized compounds of carbon (C) or hydrogen (H). Between the most reduced compound CH_4 (methane) and the oxidation end-products CO_2 and H_2O of the food constituents, the lipids, the carbohydrates, the organic acids such as lactic acid, alcohol, and amino acids (protein), and nucleotides are situated. Amino acids and nucleotides additionally contain nitrogen. Besides the organic matter, water and minerals must also be part of the food/feed. Minerals and water, however, do not deliver energy to the body.

All these compounds serve more than one purpose in the body. On the one hand, they provide energy to keep the physiological equilibrium of life in a steady, well-organized but energy-consuming state. A varying part of these energy-delivering compounds can be stored as fat in adipose tissue or as carbohydrates in glycogen globules. On the other hand, they are construction material for the body, like proteins for muscles, fat for bilayers of cellular membranes, and minerals for bones and teeth. They may also function as catalysts for the physiological equilibrium.

All this happens in live animals in an aqueous environment. Water is the main constituent of animal tissues and, in this sense, also a nutrient. This also applies to oxygen. But both themselves do not deliver energy for the body. They are either a reaction partner (oxygen) or, like water, solvent, transport medium, and end product. Owing to the different use of the organic nutrients, the chemical energy value on oxidation can differ from the physiological energy value, which is the energy of the food components the body may use.

Traditionally the energy is expressed in calories (cal), but another energy unit, Joule (J), is recommended nowadays (Table 21.6).[21] Despite their different purposes and use in the body, the physiological energy of fat and carbohydrates is equal to the chemical energy value, with the argument that at the end all fats and carbohydrates are oxidized to CO_2 and H_2O. With proteins it is different. There is a considerable part used in the construction and function of cells. This has been taken into account while setting the physiological energy value of proteins. The physiological values have been worldwide accepted and are laid down in the EU directive,[5] see Table 21.7.

Owing to the composition of muscle foods the energy of only fat and protein has to be taken into account. The amount of carbohydrates and organic acids (~1%) would account maximally for about 3–4 kcal (13–17 kJ) in 100 g of muscle foods and can remain unconsidered. The vitamins may be oxidized, but due to their concentrations, the energy input can also be overlooked.

Table 21.6   Energy Units

Energy is equal to work
1 Joule (J) = 1 Newton (N) × m = 1 Watt (W) × s = 10^7 erg = 0.239 cal
1 Calorie (cal) = 4.1868 J

Source: IUPAC Compendium of Chemical Terminology, 1997, and www.IUPAC.
org/Goldbook/C00784.pdf and www.IUPAC.org/goldbook/J03363.pdf.

Table 21.7   Energy Units of Food Compounds

Compound	kJ/g	kcal/g
Carbohydrate	17	4
Multiple alcohols	10	2.4
Ethyl alcohol	29	7
Organic acids	13	3
Fat	37	9
Protein	17	4

Source: Adapted from EU Council Directive 90/496/EEC on Nutrition
Labelling for Foodstuffs, O.J. L276, of 06/10/1990, 1990, 40–44.

21.4   Analytical Methods

21.4.1   Analysis of Water

In Chapter 18 the determination of water is described in detail. There exist two principal methodologies:

1. The specific absorbance wavelengths of water molecules are measured with infrared, NMR, and other spectroscopic methods, calculated and quantified by the existing calibration modules.
2. The food sample is heated until all water has been evaporated and the weight difference is determined. But there are limits in heating as other components in the sample, like fat, may decompose forming volatile compounds that lower the remaining dry weight. Thus, the temperature should not exceed 105°C in a slow drying and 125°C in a fast drying process. For details see Chapter 18. The water content is obtained by weight difference; see, for instance, Refs 22 and 23.

21.4.2   Analysis of Fat

Fat, or to be more correct lipids, of muscle foods can be divided into four main groups:[20]

a. *Triacylglycerols*: These compounds consist of the tri-alcohol glycerol to which three fatty acids are bound with ester bonds. These components are the main constituents of storage fat in intra- and intermuscular fat (marbling) or in adipose (fatty) tissue. They are rather lipophilic.
b. *Phospholipids (phosphoglycerols)*: Glycerol is bound to two fatty acids. The third OH-group of glycerol is bound to phosphate, which again binds various alcoholic compounds; some of

these even contain –NH$_2$ groups. The best known phospholipids are the lecithins. Contrary to triacylglycerols, which are very lipophilic, the phosphate–alcohol part enhances the hydrophilicity of phospholipids. They may act as emulsifiers in water–fat mixtures. These compounds mainly occur in cellular membranes of animals where they make up 25–35% of the lipid fraction.

c. *Sphingo- and glycolipids*: Sphingolipids do not contain glycerol. They consist of fatty acids, amino, and sometimes phosphate groups. Some exchange amino or phosphate groups for carbohydrates. They are as hydrophilic as phospholipids and occur in the range of phospholipids in cellular membranes.

d. *Cholesterol (ester)*: Cholesterol is the main constituent of animal cellular membranes. About 40–55% of the membrane lipids are cholesterol. Cholesterol is a sterol compound and cholesterol itself and its esters, which occur in smaller amounts, are strongly lipophilic.

All groups (a)–(d) occur in meat, compounds (b)–(d) mainly in membranes. Membranes amount to 0.6–1.0% of the lean muscle weight. Therefore, lean meat primarily contains compounds (b)–(d). For calculation of the energy value of the various lipids, only one value from Table 21.7 is used for fat, which is the value for triacylglycerols that constitute, except in very lean meat, the main part of fat. Compounds (b) and (c) are more hydrophilic and can be extracted from food separately by less lipophilic solvents.

With *n*-hexane or petroleum ether groups (a) and (d) are primarily extracted. The use of diethylether or a mixture of methanol–methylene chloride in lean meat leads to a higher fat content than using *n*-hexane or petroleum ether for extraction, as now all four groups (a)–(d) are extracted. All the lipids extracted by the different solvents are commonly called "crude" fat.

21.4.2.1 Analysis by Extraction

A prerequisite for the extraction with lipophilic substances is the drying of the product, as described under Section 18.2.2.

In the classical Soxhlet extraction, the sample is extracted many times by reflux with a solvent such as diethylether, *n*-hexane, or petroleum ether; the solvent is evaporated; and the extracted fat is weighed (AOAC 960.39[22] and AAC 991.36[22]) or the specific gravity of the extract is measured (AOAC 976.21[22]), for example, with the Foss-Let fat analyzer.

Another classical method is the Folch extraction, where the fat is extracted with a 3:1 mixture of chloroform:methanol. This more polar mixture of solvents extracts all the lipids, including structural lipids, as well as other components. Therefore, with this method higher results are expected for total fat. However, due to the use of hazardous organic chlorine solvents this method of extraction is not preferred nowadays.

With the Schmid–Bondzynski–Ratzlaff (SBR) method the sample is boiled with hydrochloric acid to hydrolyze triacylglycerols and phospholipids, and break down lipoproteins, glycoproteins, and protein; it is then extracted with a mixture of diethylether–petroleum ether. The solvents are evaporated and the extracted fat is weighed. With this method practically all the lipid material in the sample is extracted, but sometimes nonlipid material is extracted as well. This method therefore tends to give higher results for total fat.

The similar Weibull–Stoldt method also involves hydrolysis with hydrochloric acid, but the fat is filtered off on a fat-tight filter paper, washed, and extracted. This method gives results that are comparable to or slightly lower than those obtained using the above-mentioned method, but the filtering step makes it more laborious.

The newest extraction technique is accelerated solvent extraction (ASE), which reduces extraction times considerably and uses very small amounts of solvent under pressure. Automated methods deliver the same results as the classical methods like Soxhlet.

Fat can be extracted using supercritical fluid extraction (SFE), with carbon dioxide as solvent. Because this solvent is nonpolar like petroleum ether, mainly the triacylglycerols and cholesterol are extracted.

21.4.2.2 Analysis with Spectrophotometric Methods

Near-infrared transmission (NIT) and near-infrared reflectance (NIR) are indirect methods based on absorption of light by the sample in the near-infrared range from 800 to 2500 nm. A calibration by a number of similar samples, a comparison with a reference method, and calculations with multivariate statistics are necessary. Equipment can be bought with built-in broad calibrations. However, it is necessary to make sure that the calibration data set covers all types of samples that are to be analyzed with NIR/NIT, as changing the matrix, for instance from pork to beef or poultry, will also change the calibration.

Nuclear magnetic resonance (NMR) is another indirect method based on the measurement of a spin echo of protons in a magnetic field. The samples must be heated to ensure that the entire fat phase is liquid. It is also necessary to dry the samples, as protons from water will give a signal. A linear regression with results from a reference method must be used for calibration; hence, the NMR technique has the same limitations as those mentioned in the foregoing for NIR/NIT.

All these methods are also valid for fish. For more detailed information on meat refer to Ref. 24. In Figure 18.2, a very close relationship between water and fat in fresh meat ($r > .99$) is shown.[14,25] It permits to measure moisture (water) content and calculate the fat content without further analysis. As mentioned in Chapter 18, the close relationship is only valid in meat that has not yet experienced drip or evaporation loss.

21.4.2.3 Free Fatty Acid Content

Meat contains a minor amount of free fatty acids. But it can be higher in fish. AOAC[22] has published a method for measurement of free fatty acids in food (AOAC official method Ca5a-40). In this context, the American definition of fat for nutrition labeling purposes should be mentioned. According to this definition, total fat is the total lipid fatty acids, that is, the sum of fatty acids from mono-, di-, and triacylglycerols; free fatty acids; phospholipids fatty acids; and sterol fatty acids calculated as the amount of triacylglycerols, which will give the sum of fatty acids. Total fat could be calculated from a fatty acid determination. The result, however, will be lower than that obtained with the acid hydrolysis methods, but higher or comparable to that obtained through Soxhlet extraction.

The usual methods for the detection of fatty acid are gas or liquid chromatography. The fatty acids are isolated either as total fat that must be hydrolyzed, or as free fatty acids. The free fatty acids are usually methylated before analysis. For details see Ref. 26.

21.4.2.4 Cholesterol (Ester)

The lipophilic cholesterol can be determined after separation from other lipids by enzymatic analysis, gas chromatography, or HPLC measurements.[27] All these methods can be applied to meat, fish, and seafood.

Table 21.8 Amino Acid Composition of Beef Muscle and Cod Muscle

Compound	Beef Muscle (%)[a]	Cod Muscle (%)[a]
Aspartic acid/asparagine	4.0	6.8
Threonine	3.7	3.4
Serine	4.6	3.6
Glutamic acid/glutamine	**9.3**	**8.8**
Proline	4.3	3.4
Glycine	6.0	5.8
Alanine	4.9	5.9
Cystine	*0.8*	*2.5*
Valine	3.7	*2.5*
Methionine	*2.2*	*2.0*
Isoleucine	4.2	*2.7*
Leucine	5.1	5.1
Tyrosine	*2.1*	*1.7*
Phenylalanine	*2.7*	*2.1*
Tryptophan	*1.2*	*1.1*
Lysine	**9.8**	**11.7**
Histidine	4.9	3.5
Arginine	**14.5**	**13.2**

[a] All amino acids >7% are shown in bold, all <3% in italics.

Source: Adapted from Belitz, H.D., Grosch, W., and Schieberle, P. in *Food Chemistry*, Springer Verlag, Berlin, 2004, 627.

21.4.3 Analysis of Protein

Proteins are composed in general of 20 L-amino acids. The percentage of these 20 amino acids in proteins varies. The amino acid sequence is laid down in the genetic information and is specific for the individual being. Table 21.8 shows that nearly none of the amino acids has the (100:20 = 5) 5% average value in beef or in cod muscle (only alanine in beef and leucine in beef and cod come near this value).[15] Many of those amino acids that carry a negative or positive charge on a side chain occur in percentages higher than 5%. The sulfur-containing amino acids show the lowest percentage in bovine and in cod muscle.

The protein concentration in a food is determined by its N-content. As the molecular weights (MW) of the amino acids vary widely (MW of glycine is 75 Da and of tryptophan 204 Da), the percentage of N on the total weight also varies. It ranges from about 8% in tyrosine to 19% in glycine and lysine. Hence, the conversion factor from N-content to amino acid content ranges between 5.2 and 12.5.

The total protein of meat and fish determined through its N-content is composed of myofibrillar protein + sarcoplasmic protein + other nonprotein nitrogen compounds + connective tissue protein. In analytical food chemistry it is called "crude" protein. The proportion of the various N-containing groups in meat is shown in Table 21.9.[11] In fish the percentage of nonprotein nitrogen compounds is often somewhat higher but the ratios are in principle similar.

The variation in amino acid composition of different foods leads to different conversion factors of calculation from the N-determination to protein content,[28] as shown in Table 21.10. The factor for meat and seafood is 6.25, which is now according to EU[5] valid for all foods.

Table 21.9 Proportion of N-Containing Compound Groups in Meat

N-Compound Group	Percentage of Total
Total protein	100
Myofibrillar protein	60–65
Sarcoplasmic protein	30–32
Nonprotein nitrogen compounds	~1.5
Connective tissue protein	2.5–12

Source: Compiled from Lawrie, R.A. in *Lawrie's Meat Science*, Woodhead Publishing Ltd, Cambridge, UK, 1998, 59.

Table 21.10 Conversion Factors from N-Determination to Protein for Different Foods

Food	Conversion Factor
Grain	5.80
Oil seeds	5.30
Milk	6.38
Mushrooms	4.17
Meat/fish/seafood	6.25
Vegetables/fruits	6.25

Source: Adapted from N.N., *Lebensmittelchem. Ger. Chemie*, 39, 59, 1985.

The connective tissue of meat contains a special amino acid, the 4-hydroxyproline. Its content in connective tissue is 12.4%. After measuring the hydroxyproline content in a reaction as a colored pigment, it is multiplied by 8 to calculate the connective tissue content.

The connective tissue content is decisive for the tenderness of meat. As mentioned in the foregoing text, the aging tenderizing effect takes place at first within the cell. The connective tissue, however, is located extracellularly. Additionally, the most esteemed (expensive) cuts of meat in a carcass are those that are low in connective tissue. That is why in many countries, besides the total protein content a value for connective tissue is determined, at least in meat, and used as a part of quality characteristics of the cut.

To measure not the "crude" but the "real" protein content, that is, the myofibrillar, sarcoplasmic, and connective tissue protein of meat or fish, a homogenate of tissue must be acidified; usually perchloric or trichloroacetic acid is used. Under these conditions the high-molecular proteins precipitate. Small peptides such as carnosine or anserine and glutathione, amino acids, and nucleotides remain in solution.

But by a century-old tradition the "crude" protein is determined and the small error in meat/fish is neglected. In products, however, the error may be considerable when other N-containing ingredients are added, and hence these nonmeat/fish compounds must be determined otherwise. High-molecular proteins of, for example, plant origin may be added, which cannot be differentiated by precipitation. But immunological or chromatographic methods can be used for determination.

21.4.3.1 Analytical Methods

21.4.3.1.1 Crude Protein Content

The method used for over a century is the Kjeldahl method where all nitrogen in the sample is reduced to NH_3 by heating the sample in acid with a catalyst. After alkalization the NH_3 is distilled with water vapor and titrated with acid. The nitrogen content (not NH_3) is multiplied by 6.25 to obtain "crude" protein (AOAC 981.10[22]). As the heavy metal catalyst causes concern and for reasons of automated and miniaturized methods, equipment, for example, from Kjeltec, Labconco, and Tecator, is offered on the market. Some are part of official methods, for example, in AOAC 928.08, AOAC 960.52, AOAC 970.42, AOAC 977.14, and AOAC 981.10.[22]

Another Kjeldahl method is a combustion method reported for nitrogen in fish.[29] A sample is heated to about 430°C. The advantage seems to be the short digestion time.

A different approach to the protein content of food but also via the nitrogen content is the so-called Dumas method. At temperatures above 850°C a small sample (200–300 mg) is incinerated in an oxygen stream. The H_2O and CO_2 formed are absorbed and the nitrogen is determined by its thermal conductivity (AOAC 992.15[22]). For this a calibration is required. AOAC recommends EDTA with a nitrogen content of 9.59%.

The recent developments by NIR, NIT, and NMR permit the determination of moisture, protein, and fat by spectroscopic method. The principle for moisture determination is described in Chapter 18. In all cases moisture, protein, and fat contents are determined in one analysis.

In Figure 18.2 it is reported that a close relationship exists between water and fat in fresh meat with $r > .99$. There also exists a close relationship between water and protein with $r = .99$. Keeton and Eddy[14] report a linear relationship between fat and protein as follows: % protein $= -0.2137\%$ fat $+21.296$.

21.4.3.1.2 Connective Tissue Protein

Connective tissue is an extracellular network of proteins (mainly collagen), which in meat is also decisive for its tenderness. The content varies (Table 21.9) and the cross-linking between amino acids in the triple helix of collagen or between fibers enhances the toughness.[30] With cross-linking the solubility in hot water or acidic solution is reduced; this property is sometimes used for determining the soluble collagen by various nonofficial methods.[31] The total collagen, however, is officially determined via the determination of the specific amino acid 4-hydroxyproline, which is exclusively present in collagen.

Meat/fish samples (ca. 4 g weighed in accurate to the mg) are hydrolyzed in acid (3.5 M H_2SO_4 or HCl) at ~105°C. The final solution is filtered and the 4-hydroxyproline is oxidized with chloramine-T to a pyrrole. With 4-dimethylaminobenzaldehyde a red color develops, which is measured spectrophotometrically at 560 nm. A calibration curve is required. With the usual $N \times 6.25$ protein factor it is assumed that collagen contains 12.4% 4-hydroxyproline. Collagen is therefore calculated by $8 \times$ the concentration of 4-hydroxyproline (AOAC 990.26[22] or LFGB[23]).

21.4.4 Free Amino Acids

In meat and fish like in all beings, the amino acids sequence of the proteins and hence its percentage in the protein is genetically determined. These characteristics can be used to identify or determine a protein, for example, by its size or its electric charges of the side chains, and its hydrophilicity or hydrophobicity at the surface in the chosen solution. Absorbance or fluorescence in the UV or visible range of wavelengths, ion-exchange or exclusion chromatography, and electrophoretic methods or antibody/antigen reactions are common methods, which cannot be discussed here.

As mentioned, fresh meat contains a low amount of free amino acids, which is somewhat higher in fish. The determination of the free ones is therefore not very sensuous. The total amino acid composition after hydrolysis, however, is quite common especially with regard to the nutritional value of proteins, which is expressed in its biological value for human beings related to the essential amino acid content in a protein.[32] The reference value is that of whole chicken egg, equal to 100.

The protein, for this purpose, is hydrolyzed with acid for a long time and chromatographically (ion-exchange or HPLC) separated after a color reaction with ninhydrin reagent or fluorescamine spectrophotometrically determined.[33] Some amino acids such as tryptophan are destroyed by the acidic hydrolysis. Some other amino acids such as serine, threonine, valine, leucine, and isoleucine are partially destroyed. An alkaline hydrolysis is only partially a solution.[24] Thus, the measured free amino acids do not add up to what is measured by the Kjeldahl or Dumas methods.

21.4.5 Mineral Content

The whole mineral content is usually determined as ash. Ash includes the oxides of all nonvolatile oxides of constituents of meat. It means that the oxides of H (i.e., H_2O), C (i.e., CO_2), and N (i.e., NO_2) evaporate at the temperature of incineration, and only inorganic matter, mainly metals, are present as oxides.

21.4.5.1 Analysis of Ash

For determination meat or fish has to be ground to homogeneity, distributed in a thin layer on a metal (stainless steal, nickel) dish, and dried at 100–105°C, which is the same method as used for moisture determination (see Chapter 18). To check the homogeneity of the material during grinding charcoal powder may be added. The charcoal is oxidized to CO_2 and evaporates; hence, it is not to be taken into account. If the sample is to be used for moisture determination beforehand, the charcoal powder has to be deducted.[34] Some methods (AOAC 920.153[22]) add a defined amount of magnesium acetate to the ground mixture of meat or fish. After incineration the molar equivalent of MgO has to be deducted.

The dry material is slowly heated to about 550°C for 5–6 h. After cooling down in a desiccator the sample weight and the ash content are calculated. In meat and fish it is around 0.8–1.3% of the fresh weight depending on the fat content and the species.

This ash content is by no means comparable to the salt content of a meat product, where salt refers to NaCl. The ash content of a meat/fish is equal to the physiologically present oxidized inorganic matter, which is equal to about 0.17 M of monovalent ions such as NaCl: 1% NaCl = 10 g/kg is equivalent to 10/58.5 = 0.17 M NaCl. This means that in fresh meat the mineral content (only 0.7 g/kg is sodium) is about as high as it would be on the addition of 1% NaCl.

21.4.5.2 Iodine in Fish

This chapter focuses on composition and calories and relates to nutritional questions, which necessitates the discussion on iodine in fish and its detection. Contrary to meat if there is no considerable addition of iodine to the feed, seafood has a much higher iodine content, which is important for human nutrition as in many countries the iodine intake with food is too low.

The concentration varies. But as a rule of thumb a concentration of 100 μg I/100 g fresh tissue can be assumed. It may go up to 1000 μg I/100 g fresh weight in some locations or season and species.[35] Iodine is determined usually by oxidation followed by ICP measurements or it is bound to 2-iodo-pentane and analyzed by gas chromatography. Also, iodine-sensitive electrodes exist. Karl et al.[35] describe a method for fish tissues.

21.5 Conclusion

Meat, fish, and seafood are very similar in composition, with the main ingredients, in descending order, of water, protein, fat, minerals, and vitamins. With few exceptions in seafood other than fish the carbohydrate content is very low. Thus, energy for the body is provided in these foods by fat and protein only. From a nutritional point of view, however, protein, fatty acids, minerals, and vitamins are important constituents of these foods.

In central Europe, the average consumer eats about 150 g meat and 20 g fish per day. This amount of intake contains roughly about 30 g fat and about 30 g of protein. The energy from these two energy-delivering ingredients adds up to 390 kcal/1630 kJ (equivalent to 230 kcal/100 g), which is with recommended 2400 kcal/day for a male person about 17% of the proposed daily energy intake. Owing to rather low energy value and with the high mineral and some vitamin contents meat and fish are nutrient-dense foods.

References

1. Souci, Fachmann, and Kraut. *Der kleine SFK, Lebensmitteltabelle für die Praxis*, 3rd edition, Wiss. Verlagsgesellschaft, Stuttgart, Germany, 2004.
2. McCance, R.A. and E.M. Widdowsen. *The Composition of Foods*, 5th revised edition. Royal Society of Chemistry, Cambridge, UK.
3. Honikel, K.O. and R. Wellhäuser. Zusammensetzung verbrauchergerecht zugeschnittener Schweinefleischteilstücke. *Fleischwirts* 73: 863–866, 1993.
4. Bauer, F. Personal communication, 2007.
5. EU Council Directive 90/496/EEC on Nutrition Labelling for Foodstuffs, O.J. L276, of 06/10/1990, pp. 40–44, 1990.
6. D-A-CH. *Reference Values for Nutrient Intake*, 1st edition. Umschau, Braus Publisher, Frankfurt/Main, Germany, 2002.
7. Honikel, K.O. Glycolysis. In: W.K. Jensen, C. Devine, and M. Dikeman (eds), *Encyclopedia of Meat Science*. Elsevier, Academic Press, Oxford, UK, pp. 314–323, 2004.
8. Devine, C. Ageing. In: W.K. Jensen, C. Devine, and M. Dikeman (eds), *Encyclopedia of Meat Science*. Elsevier, Academic Press, Oxford, UK, pp. 330–338, 2004.
9. Aalhus, J.L. and M.E.R. Dugan. Oxidative and enzymatic. In: W.K. Jensen, C. Devine, and M. Dikeman (eds), *Encyclopedia of Meat Science*. Elsevier, Academic Press, Oxford, UK, pp. 1330–1336, 2004.
10. Bauer, F. Residues associated with meat production. In: W.K. Jensen, C. Devine, and M. Dikeman (eds), *Encyclopedia of Meat Science*. Elsevier, Academic Press, Oxford, UK, pp. 1187–1192, 2004.
11. Lawrie, R.A. *Lawrie's Meat Science*, 6th edition. Woodhead Publishing Ltd, Cambridge, UK, p. 59, 1998.

12. Ouali, A. Meat tenderization: possible causes and mechanisms. A review. *J. Muscle Foods* 1: 129–165, 1990.

13. Taylor, R.G. Connective tissue structure, function and influence on meat quality. In: W.K. Jensen, C. Devine, and M. Dikeman (eds), *Encyclopedia of Meat Science*. Elsevier, Academic Press, Oxford, UK, pp. 306–313, 2004.

14. Keeton, J.T. and S. Eddy. Chemical and physical characteristics of meat. In: W.K. Jensen, C. Devine and M. Dikeman (eds), *Encyclopedia of Meat Science*. Elsevier, Academic Press, Oxford, UK, pp. 210–218, 2004.

15. Belitz, H.D., W. Grosch, and P. Schieberle. *Food Chemistry*. Springer Verlag, Berlin, p. 627, 2004.

16. Römpp Lexikon. In: G. Eisenbrand and P. Schreier (eds), *Lebensmittelchemie*, 2nd edition. Georg-Thieme-Verlag, Stuttgart, Germany, pp. 355–364, 2006.

17. Oetjen, K. and H. Karl. Improvement of gas chromatographic determination methods of volatile amine in fish and fish products. *Dt. Lebensmittel-Rundschau* 95: 403–406, 1999.

18. Karl, H., W. Münkner, S. Krause, and I. Bagge. Determination, spatial variation and distribution of iodine in fish. *Dt. Lebensmittel-Rundschau* 97: 89–96, 2001.

19. Seuß-Baum, I. Ernährungsphysiologische Bedeutungen von Fleisch und Fleischerzeugnissen. In: W. Branscheid, K.O. Honikel, G. von Lengerken, and K. Troeger (eds), *Qualität von Fleisch und Fleischwaren*, 2nd edition. Dt. Fachverlag, Frankfurt/Main, pp. 755–778, 2007.

20. Lehninger, A.L., D.L. Nelson, and M.M. Cox. *Principles of Biochemistry*, 4th edition. Freeman, New York, NY, 2005.

21. IUPAC Compendium of Chemical Terminology, 1997, and www.IUPAC.org/Goldbook/C00784. pdf and www.IUPAC.org/goldbook/J03363.pdf

22. AOAC International. In: W. Horwitz (ed.), *Official Methods of Analysis of AOAC International*, 18th edition. AOAC International, Gaithersburg, MD, Methods 950.46 and 985.14, 2005.

23. Amtliche Sammlung von Untersuchungsverfahren nach § 64 Lebensmittel- und Bedarfsgegenstände- und Futtermittelgesetzbuch (LFGB), § 35 vorläufiges Tabakgesetz, § 28b GenTG, Deutschland (2007), Band I, (L) 0.600-3: Hrsg.: Bundesamt für Verbraucherschutz und Lebensmittelsicherheit (BVL), Beuth Verlag, Berlin, Germany.

24. Leth, T. Major meat components. In: W.K. Jensen, C. Devine, and M. Dikeman (eds), *Encyclopedia of Meat Science*. Elsevier, Academic Press, Oxford, UK, pp. 185–190, 2004.

25. Arneth, W. Chemische Untersuchungsmethoden für Fleisch und Fleischerzeugnisse, *Fleischwirts* 76: 120–123, 1996.

26. Christie, W.W. Gas-chromatography—mass spectrometric method for structural analysis of fatty acids. *Lipids* 33: 343–353, 1998.

27. Arneth, W. and Hussein, A. Cholesterol—its determination in muscle and adipose tissue and in offal using HPLC. *Fleischwirts* 75: 1001–1003, 1995.

28. N.N. *Lebensmittelchem. Ger. Chemie* 39: 59–61, 1985.

29. Oehlenschläger, J. WEFTA interlaboratory comparison on nitrogen determination by Kjeldahl digestion in fishery products and standard substances. *Fleischwirts* 44: 31–37, 1997.

30. Miller, R.K. Palatability. In: W.K. Jensen, C. Devine, and M. Dikeman (eds), *Encyclopedia of Meat Science*. Elsevier, Academic Press, Oxford, UK, pp. 256–266, 2004.

31. Bailey, A.J. and N.D. Light. *Connective Tissue in Meat and Meat Products*. Elsevier Applied Science, London, UK, 1989.

32. Kofranyi, E. and W. Wirths. *Einführung in die Ernährungslehre*. Umschau Verlag, Frankfurt/Main, Germany, 1994.

33. Kivi, J.T. Amino acids. In: L.M.L. Nollet (ed.), *Food Analysis by HPLC*. Marcel Dekker, New York, NY, pp. 55–97, 2000.

34. Kolar, K. Gravimetric determination of moisture and ash in meat and meat products: NMKL interlaboratory study. *J. AOAC Int.* 75: 1016–1022, 1992.

35. Karl, H., W. Münkner, S. Krause, and I. Bagge. Determination, spatial variation and distribution of iodine in fish. *Dt. Lebensmittel-Rundschau* 97: 89–96, 2001.

Chapter 22

Essential Amino Acids

María-Concepción Aristoy and Fidel Toldrá

Contents

22.1   Introduction

An essential or indispensable amino acid cannot be synthesized *de novo* by an organism and therefore must be supplied in the diet. This is the case with the amino acids valine, leucine, isoleucine, phenylalanine, tryptophan, threonine, methionine, lysine, and histidine, which are generally regarded as essential for humans (Table 22.1). The importance of essential amino acids in nutrition and health makes their analysis of high relevance. The quality of a given protein depends on the balance among individual amino acids and is limited by the content of some of them, which are essential amino acids, also known as "limiting amino acids." A limiting amino acid is the

Table 22.1 Molecular Mass and Chemical Structure of Essential Amino Acids

Name	Abbreviations	Molecular Mass	Structure
Valine	Val, V	117.15	H_3C $\diagdown$ $CH-CH-COOH$ / H_3C ; NH_2
Leucine	Leu, L	131.17	CH_3-CH_2 $\diagdown$ $CH-CH-COOH$ / H_3C ; NH_2
Isoleucine	Ile, I	131.17	H_3C $\diagdown$ $CH-CH_2-CH-COOH$ / H_3C ; NH_2
Phenylalanine	Phe, F	165.19	$C_6H_5-CH_2-CH-COOH$; NH_2
Tryptophan	Trp, W	204.22	indole$-CH_2-CH-COOH$; NH_2
Threonine	Thr, T	119.12	$H_3C-CH-CH-COOH$; OH ; NH_2
Methionine	Met, M	149.21	$H_3C-S-CH_2-CH_2-CH-COOH$; NH_2
Lysine	Lys, K	146.19	$H_2N-CH_2-CH_2-CH_2-CH_2-CH-COOH$; NH_2
Histidine	His, H	155.16	imidazole$-CH_2-CH-COOH$; NH_2

Source: Aristoy, M.C. and Toldrá, F. in *Handbook of Food Analysis*, 2nd ed. Marcel-Dekker Inc., New York, 2004, 83–123.

essential amino acid found in the smallest amount in a foodstuff. This is the case with lysine in cereals, and methionine or cysteine in legumes and milk or whey from bovines. The quality of a protein may be evaluated through different methods, including the amino acid score, the biological value, the net protein utilization, the protein digestibility corrected amino acid score, or the protein efficiency rate (PER).[1] For instance, PER values below 1.5 indicate a low quality protein, between 1.5 and 2.0 an intermediate quality, and above 2.0 a good quality. Meat proteins have calculated PER values higher than 2.7, indicating a high quality, and are considered as high biological value because they contain large amounts of all essential amino acids.[2,3] Nevertheless, not all meat proteins have the same nutritive value. This is the case with the connective tissue proteins collagen and elastine, which are rich in proline, glycine, and hydroxyproline but poor in tryptophan and

Table 22.2 Molecular Mass and Chemical Structure of Other Nutritionally Important Amino Acids

Name	Abbreviations	Molecular Mass	Structure
Arginine	Arg, R	174.20	$\underset{\overset{\shortparallel}{\text{NH}}}{}$ $\qquad$ $\underset{\overset{\shortmid}{\text{NH}_2}}{}$ $\text{H}_2\text{N—C—NH—CH}_2\text{—CH}_2\text{—CH}_2\text{—CH—COOH}$
Glutamine	Gln, Q	146.15	$\text{H}_2\text{N—C—CH}_2\text{—CH}_2\text{—CH—COOH}$ (with O and NH$_2$)
Cysteine	Cys, C	121.16	$\text{HS—CH}_2\text{—CH—COOH}$ (with NH$_2$)
Tyrosine	Tyr, Y	181.19	$\text{HO—}\langle\text{ring}\rangle\text{—CH}_2\text{—CH—COOH}$ (with NH$_2$)
Taurine	Tau	125.14	$\text{H}_2\text{N—CH}_2\text{—CH}_2\text{—SO}_3\text{H}$

sulfur-containing amino acids, and are thus of low biological value. Their presence in meat or meat products decreases those products' value.

Other amino acids, such as arginine, glutamine, and tyrosine, are considered conditionally essential (Table 22.2). These amino acids are not normally required in the diet intended for general population, but must be supplemented in diets targeted for specific groups that are unable to synthesize them in adequate amounts. The distinction between essential and nonessential amino acids is somewhat unclear, as some amino acids, such as tyrosine and cysteine (Table 22.2), may be produced from others that are considered essentials. Arginine is classified as a "semiessential" or "conditionally essential amino acid," depending on the developmental stage and health status of the individual. Infants are unable to effectively synthesize arginine, making it nutritionally essential for infants. Adults, however, are able to synthesize arginine from ornithine and citrulline, through the urea cycle. These three amino acids, which are interconvertible, may be considered a single group. Glutamine has also been proposed as a "conditionally essential amino acid" due to its importance under exceptionally severe stress conditions, such as very intense exercise, infectious disease, surgery, burn injury, or any other acute trauma that might lead to glutamine depletion and consequent immune dysfunction, intestinal problems, and muscle wasting. In all these cases, supplementation with glutamine can be a matter of life or death. Tyrosine is a precursor of the neurotransmitters epinephrine, norepinephrine, and dopamine, all of which are extremely important for brain functions such as transmission of nerve impulses and prevention of depression. The essential amino acid phenylalanine is the precursor for tyrosine synthesis, and thus for those patients suffering from phenylketonuria, who must keep their intake of phenylalanine extremely low to prevent mental retardation and other metabolic complications, tyrosine becomes essential in their diet.

Cysteine is not classified as an essential amino acid because it can usually be synthesized by the human body under normal physiological conditions if sufficient amounts of other sulfur-amino acids (i.e., methionine and homocysteine) are available. So, for convenience, sulfur-containing amino acids are sometimes considered a single pool of nutritionally equivalent amino acids.

Essential amino acids vary from species to species because different metabolisms are able to synthesize different substances. For instance, taurine (which is not, by strict definition, an amino acid) is essential for cats, but not for dogs. Thus, dog food is not nutritionally sufficient for cats, and taurine is supplemented in commercial cat foods, but not in dog foods.

Essential amino acids in meat can be analyzed following the procedures described in Chapter 2. Nevertheless, some extra careful attention must be given to some of them, especially when hydrolysis is required. Furthermore, cysteine requires specific methodologies for its analysis. All these methods are described in detail in this chapter.

22.2 Sample Preparation

22.2.1 Free Essential Amino Acids

No special care is required in the extraction of free amino acids and sample deproteinization. Procedures for both extraction and deproteinization are fully described in Chapter 2.

22.2.2 Total Essential Amino Acids

Sample preparation for the analysis of total amino acids includes the hydrolysis of proteins and peptides as a first step. A quantitative hydrolysis may be difficult to achieve for some essential amino acids. The main hydrolysis methods are described in Chapter 2, but special caution should be taken depending on the target amino acid, as described below.

22.2.2.1 Acid Hydrolysis

Acid hydrolysis is the most common method for hydrolyzing proteins. This consists of an acid digestion with constant boiling 6 N hydrochloric acid in an oven at around 110°C for 20–96 h. The hydrolysis must be carried out in sealed vials under nitrogen atmosphere and the presence of antioxidants/scavengers to minimize the degradation suffered by some especially labile amino acids (tyrosine, threonine, serine, methionine, and tryptophan) in such acidic and oxidative medium. Phenol (up to 1%) or sodium sulfite (0.1%) are typical protective agents currently used to improve the recovery of essential amino acids and are effective for nearly all of them except tryptophan and cysteine. Hydrolysis may also be improved by optimizing the temperature and time of incubation.[5] As mentioned above, tryptophan is often completely destroyed by hydrochloric acid hydrolysis, although considerable recoveries have been reported in the presence of phenol when using liquid-phase hydrolysis[6] or tryptamine when using gas-phase hydrolysis,[5] and in the absence of oxygen (see Chapter 2). Nevertheless, hydrolysis with 4 M methanesulfonic acid (115°C for 22–72 h or 160°C for 45 min, under vacuum) has been preferred to improve tryptophan recovery.[7,8] In this case, the hydrolysis is possible only in the liquid phase due to the high boiling point of the reagent and the use of protective reagents against oxidation such as tryptamine[9–11] or thioglycolic acid,[12,13] is advised. Furthermore, nitrogen flush, used in preparation for methanesulfonic acid hydrolysis, significantly increases the recovery of all amino acids in general, and especially cysteine, methionine, and tyrosine.

An important fact to consider is the impossibility of evaporating methanesulfonic acid after the hydrolysis. This means that the hydrolyzate can be used for chromatographic analysis only by pH adjustment and dilution. This drawback makes the fluorescence (more sensitive than ultraviolet) the detection of choice. This procedure, which is generally applied solely to the determination of

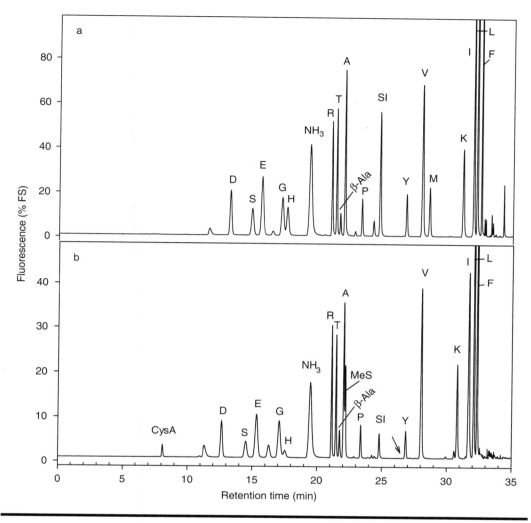

Figure 22.1 **AQC-amino acids from hydrolyzed pork meat without (a) and after performic acid oxidation (b). Arrow points at cysteine elution position. For amino acids identification see Tables 22.1 and 22.2. Cys A: Cysteic acid, MeS: Methionine sulfone.**

tryptophan, is used in conjunction with the derivatization with DABS-Cl[10] (dabsyl chloride) or FMOC[11] (9-fluorenylmethylchloroformate), resulting in very good recoveries for all amino acids, including tryptophan. The hydrolysis with 3 M mercaptoethanesulfonic acid at high temperature for a short time (160–170°C for 15–30 min) also improves tryptophan and methionine recoveries.[14]

Cyst(e)ine is partially oxidized during acid hydrolysis, yielding several adducts (cystine, cysteine, cystein sulfinic acid, and cysteic acid), making its analysis rather difficult. Several procedures have been proposed to analyze cyst(e)ine after acid hydrolysis. The previous performic acid oxidation of cysteine to cysteic acid, in which methionine is also oxidized to methionine sulfone,[15–22] improves cysteine (and methionine) recoveries, making its analysis easier. An example of the analysis of pork meat amino acids with and without performic acid oxidation previous to acid hydrolysis is shown in Figure 22.1.

The use of alkylating agents to stabilize cysteine before or after hydrolysis has been used as a valid alternative. Good recoveries have been reported using 3-bromopropionic acid,[23] 3-bromopropylamine,[24] iodoacetic acid,[25] and 3,3′-dithiodipropionic acid.[26–29]

Finally, glutamine is deamidated during hydrolysis and co-determined with glutamic acid as glx. Some methods have been proposed to analyze it based on the blockage of the amide residue with (*bis*[trifluoroacetoxy]iodo) benzene (BTI) to acid stable L-2,4-diaminobutyric acid (DABA), before hydrolysis. Owing to a lack of reproducibility,[30] this methodology has to be optimized for the analysis of glutamine in peptides and proteins by careful control of temperature, time, and BTI/sample ratio on the conversion of bound glutamine to the stable DABA.[31]

Another problem to take into account for a reliable essential amino acid determination in meat proteins consists in the remaining intact peptidic bonds left after 24 h of hydrolysis. These peptidic bonds are mainly formed by the essential hydrophobic amino acids such as valine, leucine, isoleucine, and phenylalanine. They are more resistant to hydrolysis, requiring longer hydrolysis times (up to 96 h), but such a long hydrolysis time causes some other amino acids to degrade. Many authors[32,33] overcome this problem by calculating the averages of data obtained at 24, 48, 72, and 96 h of hydrolysis for valine, leucine, isoleucine, and phenylalanine and obtaining the data for the most labile amino acids (methionine, threonine, and tyrosine) from the average of values extrapolated to zero time of hydrolysis. These complex hydrolysis procedures are really not practical for the meat industry.

As can be observed in this section, no single set of conditions will yield an accurate determination of all amino acids. In fact, it is a compromise of conditions that offers the best overall estimation for the largest number of amino acids. In general, the 22–24 h acid hydrolysis at 110°C (vapor-phase hydrolysis, preferably), with the addition of a protective agent such as phenol, yields acceptable results for the majority of essential amino acids, enough in any case for the requirements of any food control laboratory. However, when analysis of tryptophan and cyst(e)ine is necessary, adequate special hydrolysis procedures such as those described above should be performed. When high sensitivity is required, pyrolysis, from 500°C for 3 h[34] to 600°C overnight,[10] of all glass material in contact with the sample is advisable, as well as analysis of some blank samples to control the level of background present. The optimization of conditions for hydrolysis based on the study of hydrolysis time and temperature, acid-to-protein ratio, presence and concentration of oxidation protective agents, importance of a correct deairation, etc. has been extensively reported in manuscripts[35–39] and books.[40,41]

22.2.2.2 Alkaline Hydrolysis

Alkaline hydrolysis with 4.2 M of either NaOH, KOH, LiOH, or BaOH, with or without the addition of 1% (w/v) thiodiglycol for 18 h at 110°C is recommended by some authors[10,20,42–45] for a better tryptophan determination. This would be the method of choice in food samples containing high sugar concentration but this is not the case of meat.

22.2.2.3 Enzymatic Hydrolysis

Enzymatic hydrolysis with proteolytic enzymes such as trypsin, chymotrypsin, carboxypeptidase, papain, thermolysin, and pronase has been used to analyze specific amino acid sequences or single amino acids because of their specific and well-defined activity. Using this method, tryptophan content was analyzed in soy- and milk-based nutritional products by enzymatic (pronase) digestion of the protein to release tryptophan, which was further analyzed by isocratic reversed-phase liquid chromatography with fluorescence detection. Enzymatic digestion was completed in less than 6 h and was accomplished under chemically mild conditions (pH 8.5, 50°C), which did not significantly degrade tryptophan.[46] Although promising, this method has not been applied to meat samples.

22.3   Analysis

After sample preparation, target essential amino acids may be analyzed either by direct spectro-photometric or by chromatographic methods.

22.3.1   Direct Spectrophotometric Methods

Direct determinations of tryptophan without separation or even without hydrolysis of the sample are based on the acid ninhydrin method[47] or on the direct measurement of tryptophan fourth derivative ultraviolet absorption spectrum.[48]

During the acid hydrolysis used in amino acid analysis, some of the essential amino acids that are found blocked in their native proteins revert back to the parent amino acid, leading to errors in estimates of both the amino acid content of foods and amino acid digestibility. This is a particular concern for the amino acid lysine in damaged food proteins. To overcome this problem, methods analyzing free NH_2-lys residues that do not require the previous sample hydrolysis have been developed. These methods use trinitrobenzenesulfonic acid (TNBS)[49–51] or *o*-phthalaldehyde (OPA) as derivatizing reagents,[52] with significant advantages compared to the longer and more tedious method consisting in the hydrolysis of proteins and analysis of the free lysine. This analysis, which is very often used, in cereals, is not usual in meat, where lysine is not a limiting amino acid.[51,53]

22.3.2   Chromatographic Methods

Chromatographic methods for the analysis of amino acids are widely described elsewhere[4] and in Chapter 2; these also apply to the analysis of essential amino acids.

22.3.2.1   Derivatization

Pre- or postcolumn derivatization with the reagents described in Chapter 2 are also useful for essential amino acids, with some exceptions. Essential amino acids, including histidine, lysine, tryptophan, cysteine, and sometimes methionine, present some difficulties. Some derivatives may be used to analyze some essential amino acids without separation; these are described in the following text.

The reproducibility of the phenylisothiocyanate (PITC) derivatization method is very good (2.6–5.5%) for all amino acids except histidine (6.3%) and cystine (10%), which are very poor. PITC–cystine shows poor linearity, making the quantitation of free cystine by this method infeasible.[54]

The reaction conditions (pH, temperature, and excess of reagent) to obtain dansyl derivatives must be carefully fixed to optimize the product yield and to minimize secondary reactions.[55,56] Even so, this will commonly form multiple derivatives with histidine, lysine, and tyrosine. Histidine gives a very poor fluorescence response (10% of the other amino acids), reinforcing the poor reproducibility of its results.[54] However, this methodology reveals excellent linearity for cystine and also cystine-containing short chain peptides.[54,57]

FMOC derivatives of tryptophan, histidine, and cyst(e)ine adducts fluoresce weakly and with high C.V. (by 10%).

The OPA yield derivatization with lysine and cysteine is low and variable. The addition of detergents, such as Brij 35, to the derivatization reagent seems to increase the fluorescence response of lysine.[58–60] In the case of cysteine, several methods have been proposed before derivatization, including the conversion of cysteine and cystine to cysteic acid by oxidation with performic acid; carboxymethylation[61] of the sulfhydryl residues with iodoacetic acid;[22] and the formation of mixed disulfide *S*-2-carboxyethylthiocysteine (Cys-MPA) from cysteine and cystine using 3,3'-dithiodipropionic acid,[62] as incorporated by Godel et al.[63] into the automatic sample preparation protocol described by Schuster.[64] In these methods, cysteine and cystine are quantified together. Another proposal consists of a slight modification in the OPA derivatization method using 2-aminoethanol as a nucleophilic agent[65] and altering the order of addition of reagents in the automated derivatization procedure.[64]

The 6-aminoquinolyl-*N*-hydroxysuccinimidyl carbamate (AQC) amino acid derivatization reaction at 55°C for 10 min ensures the tyrosine mono-derivative formation, avoiding the mixture of both mono- and diderivatives, which are the initial adducts from tyrosine. Fluorescence of tryptophan derivative is very poor, but UV detection at 254 nm may be used for its analysis. In this case, the 6-aminoquinoline (AMQ) peak is very large at the beginning of the chromatogram and may interfere with the first eluting amino acid peak.[66]

Some special derivatives are also proposed for the determination of cyst(e)ine. 7-Halogenated-4-nitrobenzo-2-oxa-1,3-diazoles can be used in the quantitative estimation of thiols and amines. For instance, 7-chloro-4-nitrobenzo-2-oxa-1,3-diazole (NBD-Cl) has been used for the analysis of cysteine and cystine in foods by Akinyele et al.[22] This reagent reacts with cyst(e)ine in acidic medium (0.2 M sodium acetate/HCl buffer, pH 2.0), yielding a greenish product showing a maximum of absorbance at 410 nm. This method is highly specific for cysteine and does not require a posterior chromatographic separation. 5,5'-Dithio-*bis*-nitrobenzoic acid (DTNB) is used for the precolumn derivatization of sulphydryl and disulfide amino acids.[67]

Fluorescamine, which yields fluorescent derivatives with amino acids, has been used in precolumn derivatization of taurine. The eluent was monitored at 480 nm (emission) after excitation at 400 nm.[68]

22.3.2.2 Separation and Detection

The separation techniques more often used for essential amino acid analysis are cation-exchange and reversed-phase HPLC (RP-HPLC). Cation-exchange, in which nonderivatized amino acids are separated, is used in postcolumn derivatization (see Chapter 2), while RP-HPLC is mainly used to separate precolumn derivatized amino acids. Nevertheless, reversed phase has also been used to separate some underivatized amino acids such as methionine, which is further detected at 214 nm,[69] or the aromatic amino acids Tyr, Phe, and Trp, which can be detected at 214 nm but also at 260 or 280 nm. Indeed, Phe presents a maximum absorption at 260 nm, Tyr at 274.6 nm, and Trp at 280 nm. A chromatogram of Tyr, Phe, and Trp from a meat extract is shown in Figure 22.2. The separation was achieved in an octadecyl silane column by using a gradient between 0.1% trifluoro-acetic acid (TFA) in water and 0.08% TFA in acetonitrile:water (60:40). Absorption spectra from these amino acids are also shown in the same figure. For the rest of the amino acids, the detector

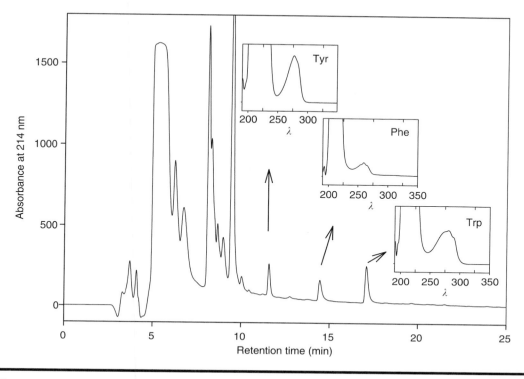

Figure 22.2 **RP-HPLC chromatogram of meat extract showing the separation of Tyr, Phe, and Trp, and their respective UV spectra.**

used depends on the chosen derivative, but it is worthwhile to take into account the previous section concerning derivatization, because some derivatives from some specific amino acids have a poor response.

There are many different techniques for the analysis of any amino acid. An example is tryptophan, which has been analyzed by cation-exchange chromatography with postcolumn derivatization with OPA and fluorescence detection[70] and by RP-HPLC without derivatization and UV or fluorescence detection.[71]

GLC techniques are, in general, not recommended for some of the essential amino acids, such as cysteine, tryptophan, and methionine. Nevertheless, a method of analysis for tryptophan in proteins was developed by Danielson and Rogers based upon the GLC separation of skatole produced by pyrolysis of tryptophan at 850°C.[72] Sample pretreatment for this method is limited to sample lyophilization to form a dry solid, and hydrolysis is not required. The most interesting alternative is the method EZ:fast, which is a patent pending method (Phenomenex, Torrance, CA, USA) to analyze protein hydrolizates and physiological amino acids from serum, urine, beer, wine, feeds, fermentation broths, and foodstuffs. Applications in meats have not been described yet. This method includes a derivatization reaction (proprietary) in which both the amine and carboxyl groups of amino acids are derivatized. Derivatives are stable for up to 1 day at room temperature and for several days if refrigerated; they are further analyzed by GC with a flame ionization detector (FID), GC with a nitrogen phosphorous detector (NPD), GC/MS, and LC/MS. Results are obtained in about 15 min when using the GC method or 24 min using the LC method.

References

1. Reig, M., Toldrá, F. Protein nutritional quality of muscle foods. *Recent Res. Dev. Agric. Food Chem.*, 2: 71–78, 1998.
2. Aristoy, M.C., Toldrá, F. Deproteinization techniques for HPLC amino acids analysis in fresh pork muscle and dry-cured ham. *J. Agric. Food Chem.*, 39: 1792–1795, 1991.
3. Zarkadas, C.G. Assessment of the protein quality of selected meat products based on their amino-acid profiles and their myofibrillar and connective-tissue protein contents. *J. Agric. Food Chem.*, 40: 790, 1992.
4. Aristoy, M.C., Toldrá, F. Amino acids. In L.M.L. Nollet, ed., *Handbook of Food Analysis*, 2nd ed., New York: Marcel-Dekker Inc., 2004, pp. 83–123.
5. Molnár-Perl, I., Pinter-Szakács, M., Khalifa, M. High-performance liquid-chromatography of tryptophan and other amino-acids in hydrochloric-acid hydrolysates. *J. Chromatogr.*, 632: 57–61, 1993.
6. Muramoto, K., Kamiya, H. The amino-acid sequence of multiple lectins of the acorn barnacle megabalanus-rosa and its homology with animal lectins. *Biochim. Biophys. Acta*, 1039: 42–51, 1990.
7. Simpson, R.J., Neuberger, M.R., Liu, T.Y. Complete amino-acid analysis of proteins from a single hydrolysate. *J. Biol. Chem.*, 251: 1936–1940, 1976.
8. Chiou, S.H., Wang, K.T. Peptide and protein hydrolysis by microwave irradiation. *J. Chromatogr. Biomed. Appl.*, 491: 424–431, 1989.
9. Umagat, H., Kucera, P., Wen, L.F. Total amino-acid analysis using precolumn fluorescence derivatization. *J. Chromatogr.*, 239: 463–474, 1982.
10. Stocchi, V. et al. Reversed-phase high-performance liquid-chromatography separation of dimethyl-aminoazobenzene sulfonylaminoazobenzene and dimethylaminoazobenzene thiohydantoin amino-acid derivatives for amino-acid analysis and microsequencing studies at the picomole level. *Anal. Biochem.*, 178: 107–117, 1989.
11. Malmer, M.F., Schroeder, L.A. Amino acid analysis by high-performance liquid chromatography with methanesulfonic acid hydrolysis and 9-fluorenylmethylchloroformate derivatization. *J. Cromatogr.*, 514: 227–239, 1990.
12. Ashworth, R.B. Amino-acid analysis for meat protein evaluation. *J. Assoc. Off. Anal. Chem.*, 70: 80–85, 1987.
13. Ashworth, R.B. Ion-exchange separation of amino acids with postcolumn orthophthalaldehyde detection. *J. Assoc. Off. Anal. Chem.*, 70: 248–252, 1987.
14. Csapo, J. Determination of tryptophan and methionine by mercaptoethanesulfonic acid hydrolysis at elevated temperature. *Acta Alimen.*, 23: 257–266, 1994.
15. Moore, S. On the determination of cystine and cysteic acid. *J. Biol. Chem.*, 243: 235–237, 1963.
16. Hirs, C.H.W. Performic acid oxidation. *Methods Enzymol.*, 11: 197–199, 1967.
17. MacDonald, J.L., Krueger, M.W., Keller, J.H. Oxidation and hydrolysis determination of sulfur amino acids in food and feed ingredients: collaborative study. *J. Assoc. Off. Anal. Chem.*, 68: 826–829, 1985.
18. Elkin, R.G., Griffith, J.E. Hydrolysate preparation for analysis of amino acids in sorghum grains: effect of oxidative pre-treatment. *J. Assoc. Off. Anal. Chem.*, 36, 1117–1121, 1985.
19. Gehrke, C.W. et al. Quantitative analysis of cystine, methionine, lysine, and nine other amino acids by a single oxidation-4 hours hydrolysis method. *J. Assoc. Off. Anal. Chem.*, 70: 171–174, 1987.
20. Meredith, F.I., McCarthy, M.A., Leffler, R. Amino acid concentrations and comparison of different hydrolysis procedures for American and foreign chestnuts. *J. Agric. Food Chem.*, 36: 1172–1175, 1988.
21. Alegría, A. et al. HPLC method for cyst(e)ine and methionine in infant formulas. *J. Food Sci.*, 61: 1132–1135, 1170, 1996.
22. Akinyele, A.F., Okogun, J.I., Faboya, O.P. Use of 7-chloro-4-nitrobenzo-2-oxa-1,3-diazole for determining cysteine and cystine in cereal and legume seeds. *J. Agric. Food Chem.*, 47, 2303–2307, 1999.

23. Bradbury, A.F., Smith, D.G. The use of 3-bromopropionic acid for the determination of protein thiol groups. *J. Biochem.*, 131: 637–642, 1973.

24. Hale, J.E., Beidler, D.E., Jue, R.A. Quantitation of cysteine residues alkylated with 3-bromopropylamine by amino acid analysis. *Anal. Biochem.*, 216: 61–66, 1994.

25. Pripis-Nicolau, L. et al. Automated HPLC method for the measurement of free amino acids including cysteine in musts and wines; first applications. *J. Sci. Food Agric.*, 81: 731–738, 2001.

26. Strydom, D.J., Cohen, S.A. Comparison of amino acid analyses by phenylisothiocyanate and 6-aminoquinolyl-*N*-hydroxysuccinimidyl carbamate precolumn derivatization. *Anal. Biochem.*, 222: 19–28, 1994.

27. Krause, I. et al. Simultaneous determination of amino acids and biogenic-amines by reversed-phase high-performance liquid-chromatography of the dabsyl derivatives. *J. Chromatogr. A*, 715: 67–79, 1995

28. Tuan, Y.H., Phillips, R.D. Optimized determination of cystine/cysteine and acid-stable amino acids from a single hydrolysate of casein- and sorghum-based diet and digesta samples. *J. Agric. Food Chem.*, 45: 3535–3540, 1997.

29. Ballin, N.Z. Estimation of whey protein in casein coprecipitate and milk powder by high-performance liquid chromatography. Quantification of cysteine. *J. Agric. Food Chem.*, 54: 4131–4135, 2006.

30. Fouques, D., Landry, J. Study of the conversion of asparagines and glutamine of proteins into diamino-propionic and diaminobutyric acid using with [*bis*(trifluoroacetoxy)iodo] benzene prior to amino acid determination. *Analyst*, 116: 529–531, 1991.

31. Khun, K.S., Stehle, P., Fürst, P. Quantitative analyses of glutamine in peptides and proteins. *J. Agric. Food Chem.*, 44: 1808–1811, 1996.

32. Nguyen, Q. et al. Comparison of the amino acid composition of two commercial porcine skins (Rind). *J. Agric. Food Chem.*, 34: 565–572, 1986.

33. Nguyen, Q., Zarkadas, C.G. Comparison of the amino acid composition and connective tissue protein contents of selected bovine skeletal muscles. *J. Agric. Food Chem.*, 37: 1279–1286, 1989.

34. Knecht, R., Chang, J.Y. Liquid chromatographic determination of amino acids after gas-phase hydrolysis and derivatization with (dimethylamino)azobenzenesulfonyl chloride. *Anal. Chem.*, 58: 2375–2379, 1986.

35. Lucas, B., Sotelo, A. Amino acid determination in pure proteins, foods, and feeds using two different acid hydrolysis methods. *Anal. Biochem.*, 123: 349–356, 1982.

36. Gehrke, C.W. et al. Sample preparation for chromatography of amino acids: acid hydrolysis of proteins. *J. Assoc. Off. Anal. Chem.*, 68: 811–821, 1985.

37. Zumwalt, R.W. et al. Acid hydrolysis of proteins for chromatographic analysis of amino acids. *J. Assoc. Off. Anal. Chem.*, 70: 147–151, 1987.

38. Molnár-Perl, I., Khalifa, M. Tryptophan analysis simultaneously with other amino acids in gas phase hydrochloric acid hydrolyzates using the Pico-Tag™ Work Station. *Chromatographia*, 36: 43–46, 1993.

39. Albin, D.M., Wubben, J.E., Gabert, V.M. Effect of hydrolysis time on the determination of amino acids in samples of soybean products with ion-exchange chromatography or precolumn derivatization with phenyl isothiocyanate. *J. Agric. Food Chem.*, 48: 1684–1691, 2000.

40. Ambler, R.P., Standards and accuracy in amino acid analysis. In J.M. Rattenbury, ed. *Amino Acid Analysis*, Chichester, UK: Ellis Horwood, 1981, pp. 119–137.

41. Williams, A.P. Determination of amino acids and peptides. In R. McCrae, ed. *HPLC in Food Analysis*, New York: Academic Press, 1982, pp. 285–311.

42. Hugli, T.E., Moore, S. Determination of the tryptophan content of proteins by ion-exchange chromatography of alkaline hydrolysates. *J. Biol. Chem.*, 247: 2828–2834, 1972.

43. Slump, P., Flissebaalje, T.D., Haaksman, I.K. Tryptophan in food proteins: a comparison of two hydrolytic procedures. *J. Sci. Food Agric.*, 55: 493–496, 1991.

44. Zarkadas, C.G. et al. Rapid method for determining desmosine, isodesmosine, 5-hydroxylysine, tryptophan, lysinoalanine and the amino-sugars in proteins and tissues. *J. Chromatogr.*, 378: 67–76, 1986.

45. Viadel, B. et al. Amino acid profile of milk-based infant formulas. *Int. J. Food Sci. Nut.*, 51: 367–372, 2000.
46. García, S.E., Baxter, J.H. Determination of tryptophan content in infant formulas and medical nutritionals. *J. Assoc. Off. Anal. Chem. Int.*, 75: 1112–1119, 1992.
47. Pinter-Szakacs, M., Molnar-Perl, I. Determination of tryptophan in unhydrolyzed food and feedstuffs by the acid ninhydrin method. *J. Agric. Food Chem.*, 38: 720–726, 1990.
48. Fletouris, D.J. et al. Rapid determination of tryptophan in intact proteins by derivative spectrophotometry. *J. Assoc. Off. Anal. Chem.*, 76(6): 1168–1173, 1993.
49. Holguin, M., Nakai, S. Accuracy and specificity of the dinitrobenzenesulfonate methods for available lysine in proteins. *J. Food Sci.*, 45: 1218–1222, 1980.
50. Obi, I.U. Application of the 2,4,6-trinitrobenzene-1-sulfonic acid (TNBS) method for determination of available lysine in maize seed. *Agric. Biol. Chem.*, 46: 15–20, 1982.
51. van Staden J.F., McCormack, T. Sequential-injection spectrophotometric determination of amino acids using 2,4,6-trinitrobenzenesulphonic acid. *Anal. Chim. Acta*, 369: 163–170, 1998.
52. Bertrand-Harb, C. et al. Determination of alkylation degree by three colorimetric methods and amino acid analysis: a comparative study. *Sci. Aliments*, 13: 577–584, 1993.
53. Moughan, P.J. Absorption of chemically unmodified lysine from proteins in foods that have sustained damage during processing or storage. *J. Assoc. Off. Anal. Chem. Int.*, 88: 949–954, 2005.
54. Fürst, P. et al. HPLC analysis of free amino acids in biological material—an appraisal of four pre-column derivatization methods. *J. Liq. Chromatogr.*, 12: 2733–2760, 1989.
55. Martín, P. et al. HPLC analysis of 19 amino acids. *Anal. Bromatol.*, 32: 289–293, 1980.
56. Prieto, J.A., Collar, C., Benedito de Barber, C. Reversed-phase high-performance liquid chromatographic determination of biochemical changes in free amino acids during wheat flour mixing and bread baking. *J. Chromatogr. Sci.*, 28, 572–577, 1990.
57. Stehle, P. et al. In vivo utilization of cystine-containing synthetic short chain peptides after intravenous bolus injection in the rat. *J. Nutr.*, 118, 1470–1474, 1988.
58. Gardner, W.S., Miller, W.H. Reverse-phase liquid-chromatographic analysis of amino acids after reaction with ortho-phthalaldehyde. *Anal. Biochem.*, 101: 61–65, 1980.
59. Jones, B.N., Gilligan, J.P. Ortho-phthaldialdehyde precolumn derivatization and reversed-phase high-performance liquid-chromatography of polypeptide hydrolysates and physiological fluids. *J. Chromatogr.*, 266: 471–482, 1983.
60. Jarrett, H.W. et al. The separation of ortho-phthalaldehyde derivatives of amino acids by reversed-phase chromatography on octylsilica columns. *Anal. Biochem.*, 153: 189–198, 1986.
61. Gurd, F.R.N. Carboxymethylation. *Meth. Enzymol.*, 25: 424–438, 1972.
62. Barkholt, V., Jensen, A.L. Amino acid analysis: determination of cysteine plus half-cystine in proteins after hydrochloric acid hydrolysis with a disulfide compound as additive. *Anal. Biochem.*, 177: 318–322, 1989.
63. Godel, H., Seitz, P., Verhoef, M. Automated amino acid analysis using combined OPA and FMOC-Cl precolumn derivatization. *LC-GC Intl.*, 5: 44–49, 1992.
64. Schuster, R. Determination of amino acids in biological, pharmaceutical, plant and food samples by automated precolumn derivatization and high-performance liquid chromatography. *J. Chromatogr.*, 431: 271–284, 1988.
65. Park, S.K., Boulton, R.B., Noble, A.C. Automated HPLC analysis of glutathion and thiol-containing compounds in grape juice and wine using pre-column derivatization with fluorescence detection. *Food Chem.*, 68: 475–480, 2000.
66. Bosch, L., Alegría, A., Farré, R. Application of the 6-aminoquinolyl-*N*-hydroxysuccinimidyl carbamate (AQC) reagent to the RP-HPLC determination of amino acids in infant formula. *J. Chromatogr. B*, 831: 176–183, 2006.
67. Hofmann, K. The function of the amino acids cysteine and cystine in meat and methods for determining their sulphydryl (SH) and disulphide (SS) groups. *Fleischwirtschaft*, 57: 2225–2228; 2231–2237; 2169; 2172–2173, 1977.

68. Sakai, T., Nagasawa, T. Simple, rapid and sensitive determination of plasma taurine by high-performance liquid-chromatography using precolumn derivative formation with fluorescamine. *J. Chromatogr. Biomed. Appl.*, 576: 155–157, 1992.
69. Johns, P., Phillips, R., Dowlati, L. Direct determination of free methionine in soy-based infant formula. *J. Assoc. Off. Anal. Chem. Int.*, 87: 123–128, 2004.
70. Ravindran, G., Bryden, W.L. Tryptophan determination in proteins and feedstuffs by ion exchange chromatography. *Food Chem.*, 89: 309–314, 2005.
71. Delgado-Andrade, C. et al. Tryptophan determination in milk based ingredients and dried sports supplements by liquid chromatography with fluorescence detection. *Food Chem.*, 98: 580–585, 2006.
72. Danielson, N.D., Rogers, L.B. Determination of tryptophan in proteins by pyrolysis gas chromatography. *Anal. Chem.*, 50: 1680–1683, 1978.

Chapter 23

Trans and n-3 Fatty Acids

José A. Mestre Prates and Rui J. Branquinho Bessa

Contents

23.1 Introduction

Specific guidelines for fat in the human diet have been developed by various health institutions. Among them, reducing the intake of saturated fatty acids (SFA; <10% of caloric intake) and trans fatty acids (TFA; <1%), as well as increasing the intake of n-3 polyunsaturated fatty acids (PUFA; 1–2%), is particularly encouraged [1]. Fatty acid composition of meat has been extensively studied due to its implications for human health. In fact, dietary TFA at sufficiently high levels have been found to increase the ratio of LDL-cholesterol/HDL-cholesterol, which is unfavorable for human health (see Section 23.2.1). In contrast, some conjugated linoleic acid (CLA) isomers may have beneficial effects on the prevention and treatment of several pathologies (see Section 23.3.1). However, the major TFA in ruminant meat, vaccenic acid (VA, 18:1t11), is the precursor of c9,t11 CLA isomer in animal tissues. In addition, it is well-known that the lower PUFA/SFA and higher n-6/n-3 ratios of some meats contribute to the imbalance in the fatty acid intake of today's consumers (see Section 23.4.1).

This chapter provides an overview of the analytical techniques and methodologies available for the analysis of TFA (see Section 23.2.2), CLA (see Section 23.3.2), and n-3 PUFA (see Section 23.4.2) in meat. The most common methodologies used for the analysis of TFA—gas chromatography (GC), silver ion chromatography, and spectroscopy—are described. A summary of complementary GC and silver ion high-performance liquid chromatography (Ag⁺-HPLC) required for the analysis of individual CLA isomers, as well as some additional methodologies, is presented. Although most of CLA isomers are polyunsaturated TFA (excluding cis,cis isomers), with a conjugated structure, they are separately studied in this chapter due to their specific biological effects and analytical methodologies. Finally, some particularities of GC analysis of n-3 PUFA in meat are considered.

Moreover, examples of the application of these methodologies for the evaluation of nutritional quality of meat fat are provided (see Sections 23.2.3, 23.3.3, and 23.4.3). Owing to the above-mentioned imbalance in the fatty acids of human diet, considerable attention has been given to improving the nutritional and health value of meat fat. Some of the strategies used for this purpose are illustrated in this chapter. In ruminant animals, the high saturation of edible fats has raised concerns about its contribution to the increase in the risk of cardiovascular diseases (CVD) and the metabolic syndrome. Thus, manipulation of fatty acid composition of ruminant meat, to reduce the SFA level and n-6/n-3 PUFA ratio and increase the PUFA and CLA contents, is a major target in ruminant meat research. In monogastric animals and poultry, the drive has been to increase n-3 PUFA in meat, which can be achieved by including sources such as linseed in the diet. Finally, it has been shown that in nonruminant animals the supplementation of CLA itself or its precursor VA is effective in elevating CLA content in meat fat.

23.2 Trans Fatty Acids

TFA are unsaturated fatty acids that have at least one double bond in the trans configuration. However, the chemical definition of Food and Drug Administration (FDA) is more limited, considering TFA as unsaturated fatty acids that contain one or more isolated (i.e., nonconjugated) double bond in the trans configuration [2]. Chapter 9 discusses the chemistry and biochemistry of fatty acids in greater detail. In ruminant meat fat, the main TFA are trans octadecenoates (trans 18:1), with VA predominating. In addition, trans 16:1 and trans PUFA (18:2 and 18:3) isomers are also usually present. Apart from cis,cis geometric isomers, CLA isomers are polyunsaturated TFA with a conjugated structure. This mixture of TFA in ruminant animals arises as intermediates of the ruminal biohydrogenation of dietary unsaturated fatty acids. TFA in foods are of rising importance and knowledge of the distinct effects of the individual trans isomers is increasing.

23.2.1 Nutritional Value of Trans Fatty Acids

The association between the intake of TFA (4–6% of diet energy) and CVD risk has been indicated in five prospective cohort studies [3,4]. There is increasing evidence that different trans 18:1 isomers have differential effects on plasma ratio of LDL-cholesterol/HDL-cholesterol, which is an area of active investigation [5]. For instance, there is evidence that trans9 and trans10 18:1 isomers are more powerful in increasing plasma cholesterol ratio than VA [5]. In addition, epidemiological evidence for a possible relationship between TFA intake and cancer, type 2 diabetes, or allergy is weak or inconsistent [3]. VA, the major TFA in most ruminant meats and the precursor for the tissue c9,t11 CLA isomer in both animals and man, should be considered as a neutral or beneficial trans isomer [6]. Rats given supplements of c9,t11 CLA isomer indicate that VA could also add to the beneficial effects of c9,t11 CLA on cancer [7] and atherogenesis [8]. In addition, it was very recently reported that VA, in contrast to its 18:1t10 isomer, is neutral or beneficial to aortic lipid deposition in rabbits [9]. However, in most of the human intervention studies, monounsaturated TFA from hydrogenated vegetable oils were evaluated. Thus, very little information is available on the effect of natural sources of TFA coming from ruminant fats. Therefore, differences between the effects of TFA from ruminant fats and those from hydrogenated vegetable oils on metabolic risk parameters are yet to be determined. Nevertheless, an ongoing research aims to assess and compare the effect of these two dietary sources of TFA on CVD risk factors in humans [10].

It was recommended, by a joint WHO/FAO expert consultation [1], that TFA should contribute less than 1% of total energy intake of the human diet. Most health professional organizations and some governments now recommend to reduce the consumption of foods containing TFA [4]. In Canada and the United States, it is even mandatory to declare the TFA content of food since January 2006. The opinion of European Food Safety Authority (EFSA) about the presence of TFA in foods and their effect on human health was reported in 2004 [3]. The intake estimations of TFA vary from 1.2 g/day (0.5% of diet energy) in Spain to 5.3 g/day (2.6%) in the United States (reviewed by Hunter [4] and Craig-Schmidt [11]). The estimated values indicate that the TFA intake in the United States is higher compared with that of European countries, probably due to the higher total fat intake. Moreover, individuals from Mediterranean countries seem to have a lower TFA intake relative to those from Northern European countries (ranging from 0.5 to 2.1%). Trans 18:1 contribute 54–82% of the total TFA intake. Nowadays, the TFA intakes have a tendency to decrease primarily due to industry efforts to replace or reduce TFA contents in foods [11].

The TFA contents in meat and meat products were reviewed by Fritsche and Steinhart [12] and Valsta et al. [13]. Meat from nonruminants, such as pork or poultry, presents distinct lower TFA contents than that from ruminants. The values described range from 0.20% in horse meat up to 10.6% in ovine meat. However, TFA in ruminant meats usually represent 2–4% of total fatty acids. According to the same authors, the TFA content in meat products varies from 0.2 to 3.4% of total fat, which are intermediate values between meat from ruminants and nonruminants. The trans 18:1 isomers dominate in meat and meat products (~80% of total TFA), in which VA predominates (~60% of total trans 18:1 isomers). The major sources of TFA in human diet are industrial and ruminant fats. The contribution of TFA from ruminant fat ranges from 30 to 80% of total TFA, corresponding to 0.3–0.8% of diet energy [11]. It was estimated that meat and meat products may contribute from 10 to 30% of TFA (higher in Mediterranean countries than in other countries) for human diet [14].

23.2.2 Analytical Methodology

The most common derivatives used to analyze fatty acids in meat, either by GC or by HPLC, are methyl esters (FAME) prepared by reaction with excess of methanol in the presence of catalytic amounts of acid or base [15,16]. The usual analytical techniques for the determination of total and individual TFA include GC, thin-layer chromatography (TLC), infrared spectroscopy (IR), attenuated total reflectance (ATR), mass spectrometry (MS), and a combination of hyphenated techniques.

23.2.2.1 Gas Chromatography

GC is the most common method for the analysis of fatty acids in meat fat. Most of the detailed reports on chromatographic resolution of TFA in animal products use milk fat matrices, but they are generally applicable to intramuscular fat. The fatty acids are usually analyzed as their FAME using long, highly polar fused silica capillary columns coated with cyanoalkyl polysiloxane stationary phases. The 100 m CP-Sil 88™ capillary column is the most commonly used for the trans/cis FAME separation, although SP 2380™, SP 2560™, and BPX-70™ columns have been frequently used [15,17]. However, the complete resolution of trans 18:1 isomers, as well as of trans 16:1, trans 18:2, and trans 18:3 isomers, is rather complex. Depending on the analytical conditions, numerous overlaps may occur, leading to either underestimations of trans 18:1 and 18:2 contents [18] or overestimations of trans 16:1 levels [19].

The elution times for both trans and cis 18:1 isomers increase with the position of the double bond along the carbon chain, eluting the trans isomers before the cis isomers on cyanoalkyl polysiloxane stationary phases. The major overlaps take place on 18:1 region, where trans12 to trans16 isomers coelute with those 18:1 of cis configuration. Overloaded column or large amounts of cis isomers affect the resolution and quantification of trans isomers. Moreover, column temperature affects greatly the cis/trans isomer resolution [20].

To maximize the separation of the TFA, low-temperature isothermical programs have been applied successfully by several authors [21,22]. Kramer et al. [22] used isothermal conditions at 120°C to separate the 16:1 isomers, 150°C for 18:1 isomers, 175°C for 20:1 isomers, and 220°C for 22:1 and 24:1 isomers, using a CP-Sil 88 capillary column. However, even when using stepwise isothermal conditions, resolution of all 18:1 isomers could not be achieved accurately without prior separation of cis and trans isomers by silver ion chromatography. Recently, a GC method from American Oil Chemist Society (AOCS) for measuring TFA in animal and vegetable fats has been reported (AOCS method Ce1h-05), which meets the current FDA regulations.

23.2.2.2 Silver Ion Chromatography

The complete analysis of all TFA can only be accomplished by previous separation of cis and trans isomers. Silver ion chromatography has been applied in the separation of fatty acids according to both the number and the configuration of their double bond [15]. Ag⁺-TLC, Ag⁺-HPLC, or Ag⁺-SPE fractionation, followed by analysis of the fraction by GC, allows complete and accurate analysis of TFA, especially the mono- and diethylenic isomers [23]. Common TLC solvent systems are hexane:diethyl ether (90:10), benzene for cis/trans methyl linolenate, and chloroform:acetone: acetic acid (96:5:0.5) for cis/trans methyl arachidonate fractionation [12,15]. The amount of TFA can be determined when samples are analyzed by GC before and after silver ion separation. The quantification of trans 18:1 isomers may be easily achieved by using the area eluting before trans12 in the nonfractionated chromatogram and the total trans 18:1 area in silver ion fractionated chromatogram, as explained by Kramer et al. [24]. Otherwise, the incorporation of a known amount of methyl penta- or heptadecanoate as internal standard into trans and cis 18:1 fractions allows quantification by GC [23]. This procedure was adopted by the International Union of Pure and Applied Chemistry (IUPAC) for the quantification of TFA in natural and hydrogenated animal and vegetable oils and fats (IUPAC method 2.302). Christie [15] proposed a method where fractions containing both trans monoenes and saturated FAME are collected together, whereas the proportion of the trans monoenes is calculated by reference to one or all of the saturated compounds.

Cruz-Hernandez et al. [20] reported a complete methodology by using argentation TLC and GC in the analysis of trans 18:1 isomers in dairy fats. The quantification of the different trans isomers of 18:1, 18:2, and 18:3 and their overestimation have been widely studied by Precht and Molkentin [18] in milk fat. These authors proposed GC under isothermal conditions at a low temperature (125°C) to achieve the separation of trans13 and trans14 18:1 isomers, as well as trans11 and trans12 16:1 isomers [25].

23.2.2.3 Spectroscopy

The measurement of the intensity of a characteristic absorption band, at 966 cm^{-1}, under a defined set of analytical conditions constitutes the basis for the various IR methods used for the determination of total trans unsaturation in fats [23]. The older standard official methods have been subjected to modifications to improve their detected limitations. Improvements were made by correcting the various background interferences, calibrations, and reference materials in AOCS and AOAC International methods. However, results from IR are biased, in particular those concerning fat products with little TFA [26].

Advances in equipment technology introduced the Fourier-transform infrared (FT-IR) and near infrared (FT-NIR), and ATR spectroscopic instruments. TFA formed during biohydrogenation of ruminant animals were quantified by Ulberth and Henninger [27] in milk fat using the FT-IR methodology and are only slightly higher than those obtained by the same authors using the Ag^{+}-TLC-GC technique. The ATR-IR method has been widely studied for the determination of total TFA content in food products [28,29], which has been adopted by AOCS (official method AOCS Cd 14d-99) and AOAC International (official method AOAC 2000.10). Recently, Azizian and Kramer [30] described a rapid FT-NIR method coupled to chemometric calibration techniques, which allows qualitative and quantitative information of most fatty acids present in oils and fats, including TFA. None of these spectroscopic methods are directly applied to meat matrices but only to oils, extracted fats, or FAME mixtures.

23.2.2.4 Other Methods

Few hyphenated techniques have been developed for the analysis of TFA. Coupling capillary GC to FT-IR spectrometers led to some advantages in the determination of trans monounsaturated FAME, particularly the measurement of subnanogram quantities of analyte, in contrast to those required for the standard IR equipments [31]. The online measurement of FT-IR spectra of compounds eluting from the capillary column allows the lack of interferences, particularly due to partial overlap of adjacent trans and cis isomers [12]. Although the GC-FT-IR technique can provide information about double bond geometry of fatty acids, it is relatively expensive for routine analysis [31].

23.2.3 Applications for Nutritional Quality of Meat Fat

Although the recommendations of international health institutions are to reduce the content of total TFA in foods, there is increasing evidence that different trans 18:1 isomers have distinct biological effects. The methodologies described in this context are focused essentially on assessing the content and profile of TFA in meat fats, especially those from ruminant animals fed diets supplemented with polyunsaturated oils. Moreover, these methods are further required to differentiate natural TFA in ruminant meats from the synthetic ones produced by industrial hydrogenation of polyunsaturated vegetable or fish oils.

Dietary α-linolenic acid (ALA, 18:3n-3) increases the diversity of rumen biohydrogenation TFA relative to dietary linoleic acid (LA, 18:2n-6) [32]. In addition, the effect of PUFA-rich oil supplementation on the fatty acid pattern of muscle lipids strongly depends on the basal diet, and oil exacerbates these differences [33]. The VA is the major trans 18:1 isomer of ruminant products in most situations, with the exception of low-fiber diets supplemented with polyunsaturated oils, in which 18:1t10 predominates. However, it was reported that 18:1t10 is the predominant isomer in lamb meat from animals fed with concentrate even without the inclusion of oil [33]. An increase in the percentages of forage in a cattle diet with polyunsaturated oil increases the levels of VA and decreases those of 18:1t10 in beef [34].

The trans 18:1 occurring in foods consist of mixtures of different positional isomers, which can be completely separated by using the methodologies described earlier. More than ten trans 18:1 isomers have been identified, both in partially hydrogenated fats and in ruminant fats [21]. The percentage distribution of the various isomers differs between ruminant and hydrogenated fats and also within those fats. In partially hydrogenated fats, relative to ruminant fats, the position of the double bond is more evenly distributed along the carbon chain [21]. In addition, the 18:1t9 constitutes typically 20–30% of the trans 18:1 isomers and VA 10–20%. In partially hydrogenated fish oils, a range of different trans isomers of long-chain fatty acids also occurs. Although some strategies for the estimation of the relative contributions of naturally occurring TFA and of industrially produced TFA in a food matrix have been proposed, these approaches cannot be applied with confidence to a wide range of foods [4].

23.3 Conjugated Isomers of Linoleic Acid

Conjugated isomers of LA, commonly referred to as CLA, are a group of fatty acids, composed of positional (from carbons 6,8 to 12,14) and geometric (trans,trans, trans,cis, cis,trans, and cis,cis) isomers of LA, that contain a conjugated double-bond system [35]. Twenty-four different CLA isomers have been reported as occurring naturally in food, especially in ruminant fat [20]. The major CLA isomer in ruminant meat (rumenic acid, 18:2c9,t11), like the usual second-most prevalent t7,c9 isomer, is primarily produced in the tissues through delta9 desaturation of VA [36]. The origin of all other CLA isomers is ruminal biohydrogenation of dietary C18 PUFA [37]. Interest in these compounds has expanded since CLA was found to be a naturally occurring compound, present in animal-derived foods, associated with a multitude of health benefits.

23.3.1 Biological Effects of Conjugated Isomers of Linoleic Acid

In animal trials, some CLA isomers (0.1–1% in the diet) exert positive effects on cancer, CVD, diabetes, body composition, immune system, and bone health (reviewed by Wahle et al. [35] and Bhattacharya et al. [38]). For instance, the t10,c12 isomer has anticarcinogenic, antiobesity, and antidiabetogenic effects, whereas the c9,t11 isomer exerts an anticancer effect [39]. The National Academy of Sciences of USA recognized CLA as the only fatty acid that has been shown unequivocally to inhibit carcinogenesis in experimental animals [40]. The mechanism of modulation of carcinogenesis by CLA is not completely understood, although it may be related to its antioxidative properties or with the induction of apoptotic cell death and cell cycle regulation [41]. However, the biological effects of these CLA isomers in humans remain to be established [41]. Few studies have investigated the health effects in humans of naturally occurring CLA from foods, and evidence is weak and conflicting with respect to any health effects at current levels of intake. CLA research in humans has focused on the effects of t10,c12 isomer, which is a minor isomer in foodstuffs, rather than on the major isomer c9,t11. This is because commercial CLA preparations consist of an equal mixture of t10,c12 and c9,t11 isomers. Thus, research to assess the benefits of CLA isomers for humans will constitute a unique challenge in the coming years.

Although optimal dietary intake in humans is not known, on the basis of the anticancer effects of CLA in rats, a daily consumption of 0.8–3.0 g of CLA might provide a significant health benefit to humans [42]. Depending on the country, the estimation so far of average total CLA or c9,t11 isomer consumption ranges between 95 and 440 mg (reviewed by Collomb et al. [43]). However, in some countries (e.g., Australia) consumption of CLA may reach 1.5 g/day [44].

The CLA contents in meat and meat products were reviewed by Parodi [45] and Schmid et al. [42]. The highest CLA concentrations were found in lamb (4.3–19.0 mg/g lipid), with slightly lower concentrations in beef (1.2–10.0 mg/g lipid), representing 0.5–2% of fatty acids. The CLA content of pork, chicken, and meat from horses is usually lower than 1 mg/g lipid. Furthermore, CLA contents (mg/g fat) vary substantially not only among species but also from animal to animal, and within an animal too in different tissues. The main CLA isomer in meat, c9,t11 (~80% of CLA isomers), is primarily associated with the triacylglycerol lipid fraction and therefore is positively correlated with the level of fatness. Meat and meat products contribute approximately 25–30% of the total human CLA intake in Western populations. Concerning the estimation of daily average intake of CLA, the proposed range should only be achieved through supplementation. It is noteworthy that dietary supplements have different CLA isomeric profile compared to foodstuffs. The main difference lies in the high percentage of t10,c12 in supplements, which may have adverse effects on human health, as it has been recently suggested [46].

23.3.2 Analytical Methodology

In animal tissues, natural CLA is in esterified form and is present at relatively low levels. Thus, a preconcentration step is required, which can be accomplished either by reversed-phase HPLC or by silver ion chromatography or, improving the efficacy of the process, by using the two techniques sequentially [15]. In addition, there is no single methylation procedure adequate for CLA analysis in meat lipids: base-catalyzed methylation transesterifies acyl lipids but not *N*-acyl or alk-1-enyl acyl lipids; and acid-catalyzed methylation isomerizes cis/trans to trans/trans CLA and produces methoxy artifacts [20]. Therefore, base- and acid-catalyzed methylations must be used, in sequence or separated, for the derivatization of meat lipids [47]. Finally, recent reviews on CLA isomer analysis have been reported elsewhere [17,20,47].

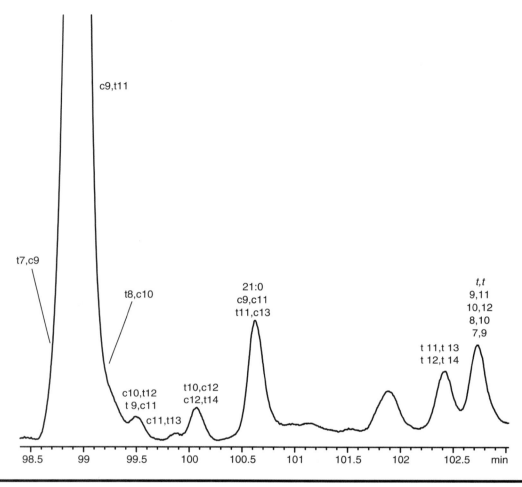

Figure 23.1 Partial GC-FID chromatogram (CLA region) of a total FAME mixture from lamb meat, obtained using a CP-Sil 88™ capillary column (100 m × 0.25 mm i.d. × 0.25 μm film thickness) and He as the carrier gas. The column temperature ranged from 100 to 200°C (held 15 min at 100°C, increased to 150°C at a rate of 10°C/min, held for 5 min, increased to 158°C at 1°C/min, held for 30 min, increased to 200°C at a rate of 1°C/min, and maintained for 65 min). The injector and detector temperatures were set at 250 and 280°C, respectively.

23.3.2.1 Gas Chromatography

GC, the standard technique in most laboratories for routine analysis of fatty acids, is of limited value for the analysis of CLA. The various types of geometrical isomers give distinct peaks, but within these groups, positional isomers are not completely resolved. GC with 100 m highly polar capillary columns, for example, CP-Sil 88, SP 2560, HP88™ or BPX-70 must be used to achieve improved separations of CLA methyl esters [22]. However, some FAME (e.g., 21:0 and 20:2 isomers) elute in the same region as the CLA isomers even when these highly polar GC columns are used [20]. The extent of the overlap is shown in a typical separation of meat fat total FAME in the 18:0–18:2n-6 region (Figure 23.1). Moreover, the relative elution of these FAME differs slightly among columns, even from the same supplier, and depends on the age of the column [17].

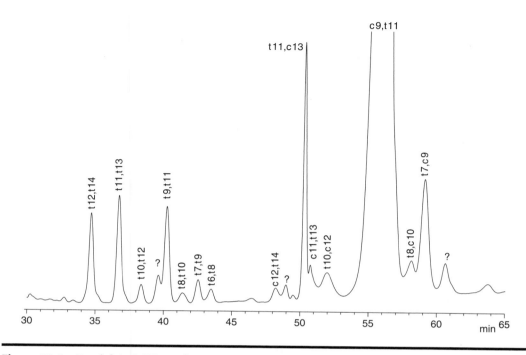

Figure 23.2   Partial Ag⁺-HPLC chromatogram (trans,trans and cis/trans regions) of a total CLA methyl ester mixture from a bovine meat fat, obtained using three ChromSpher 5 Lipids™ columns (25 cm × 4.6 mm i.d. × 5 μm particle size) in series, maintained at 25°C. The mobile phase was *n*-hexane containing 0.1% acetonitrile and 0.5% diethyl ether, at a flow rate of 1 mL/min, and the diode array detector was adjusted to 233 nm.

Some individual (c9,t11, t9,c11, c11,t13, t10,c12, and t9,t11) and mixtures (from positions 8,10 to 11,13) of CLA isomers, as free acids or methyl esters, are commercially available from Matreya Inc. (Pleasant Gap, Pennsylvania), Nu-Check Prep (Elysian, Minnesota) or Sigma Inc. (St. Louis, Missouri). Additional standards of mixtures (cis,trans, trans,cis, and trans,trans from 7,9 to 12,14) of CLA isomers as methyl esters can be prepared according to the procedure described by Destaillats and Angers [48].

A definitive identification of the interfering FAME requires, instead of GC-flame ionization detection (FID), GC-MS using the 4,4-dimethyloxazoline (DMOX) derivatives or 4-methyl-1,2,4-triazoline-3,5-dione (MTAD) adducts as invaluable adjuncts [49].

23.3.2.2   Silver Ion High-Performance Liquid Chromatography

Ag⁺-HPLC, which uses columns packed with ion-exchange media loaded with silver ions (ChromSpher 5 Lipids™), has proved to be very useful for the separation of geometrical and positional isomers of CLA [50]. The chromatographic separation is based on the interaction between the π-electrons of double bonds and the empty d orbitals on the silver. Improved separations of CLA isomers have been obtained using three of these argentated columns in series [51]. The mobile phase is *n*-hexane containing 0.1% acetonitrile and 0.5% diethyl ether, with specific detection of the conjugated double bonds of CLA methyl esters at 233 nm. Trans,trans isomers elute first, followed by cis/trans, and then cis,cis, and within each group most positional isomers are clearly resolved. As an example, Figure 23.2 illustrates a separation of CLA isomers in a bovine meat sample. However, the separation of CLA isomers using Ag^+-HPLC still has two major drawbacks: the reproducibility of the retention times among chromatographic runs and the lack of an appropriate internal standard for the quantification of CLA isomers.

Therefore, the best strategy for the analysis of the individual CLA isomers in ruminant fats is the combination of GC, using 100 m highly polar capillary columns, and Ag^+-HPLC, using three 25 cm columns in series [52]. The CLA peaks are quantified by GC analysis of total FAME, and the relative concentrations obtained by Ag^+-HPLC are used to calculate the unresolved peaks in the GC chromatogram. Kraft et al. [53] performed the calculation by comparing the HPLC areas of c9,t11, t7,c9, and t8,c10 isomer peaks with the peak area of the three coeluted isomers from GC chromatogram. Quantification of the other CLA isomers was assessed from their Ag^+-HPLC areas relative to the area of the main isomer c9,t11. A detailed description of the quantification process of individual CLA isomers using these two complementary methods was described by Cruz-Hernandez et al. [47].

23.3.2.3 Other Methods

^{13}C NMR spectroscopy, the most complete single analysis for commercial CLA preparations, requires a substantial amount of sample and is not likely to be applicable to tissue extracts at natural levels [54]. Reversed-phase HPLC can also be used for CLA analysis, although it is mainly useful for the analysis of CLA metabolites [55]. However, when combined with the second derivative of UV absorbance it is possible to analyze CLA and its metabolites [56]. FT-NIR requires the extraction of fat from meat and does not distinguish between cis,trans and trans,cis isomers [30]. Finally, chemical ionization tandem mass spectrometry (CI-MS and CI-MS-MS) with acetonitrile as reagent of chemical ionization provides a rapid alternative to conventional CLA analysis. This technique, in combination with GC, enables rapid and positive identification of double bond position and geometry in most CLA methyl esters [57].

23.3.3 Applications for Nutritional Quality of Meat Fat

It is increasingly evident that different CLA isomers have distinct physiologic properties and that diet intake should increase to achieve the potential beneficial values. Thus, these methodologies are essential to evaluate meat fat for its CLA content, to design experimental diets to increase the amount of CLA in meat fats, and to determine the CLA profile in these CLA-enriched meat fats.

The diet is the major factor affecting the content and profile of CLA in ruminant meats [42,58]. Furthermore, it is also well-known that many of the differences in CLA profile appear to be related to pasture *versus* concentrate feeding. French et al. [59] reported that meat fat from grazing steers has greater CLA contents (10.8 mg/g FAME) than that from animals fed concentrate (3.7 mg/g FAME). Regarding CLA profile, pasture feeding as compared with concentrate feeding primarily increases the proportion of the t11,c13 isomer in beef lipids (up to 18.5% of total CLA), with a decrease of t7,c9 isomer (down to 4.1% of total CLA), while increasing the percentages of t11,t13 and t12,t14 isomers [52]. These results suggested that the t11,c13, t12,t14, and t11,t13 CLA isomers are sensitive grass intake indicators. Finally, the effects of the slaughter season and muscle type on CLA contents and profile in beef [60,61] and veal [62,63] have also been assessed by using these methodologies.

Polyunsaturated oil supplementation of ruminant diets is the most straightforward method to modify CLA content and proportion in ruminant meats [42,58]. However, rumen biohydrogenation of dietary PUFA is modulated by several factors, such as the amount and type of lipid supplement and the basal diet, resulting in differences in the amount of PUFA that escapes from rumen biohydrogenation and in the type and distribution of biohydrogenation intermediates. Supplementation of dehydrated lucerne up to 10% of LA-rich oils has been effective in achieving high concentrations of rumenic acid and its precursor VA in lamb meat [33], although feeding linseed oil (rich in ALA) or processed linseeds is an effective approach to increase both CLA and n-3 PUFA in ruminant meat [42]. In addition, linseed oil supplementation of forage-based diets results in a lower response in the levels of c9,t11, when compared with sunflower oil (rich in 18:2n-6), although a greater CLA isomer diversification is observed, resulting in an important increase in t11,c13 and t11,t13 [32].

These analytical techniques are further required to evaluate the CLA profile in monogastric animals fed commercial CLA preparations for enrichment of animal products in these fatty acids. In contrast to the enrichment of CLA in ruminant-derived foods, which is achieved primarily by supplementing animals with CLA precursors, its increase in monogastric-derived foods is usually obtained by direct feeding. This is particularly important because absorption and metabolism will alter the ingested-CLA profile in the animal fed [20]. For instance, Kramer et al. [17] used these techniques to evaluate the CLA profile in pigs fed different commercial CLA mixtures.

23.4 n-3 Fatty Acids

The balance between n-3 and n-6 PUFA in the diet is required for human health due to the competition for the same elongation and desaturation enzymes [64]. ALA, an essential fatty acid for humans, is the precursor for long-chain (C20-22) n-3 PUFA, including eicosapentaenoic acid (EPA, 20:5n-3) and docosahexaenoic acid (DHA, 22:6n-3). EPA and DHA are important constituents of the phospholipids in animal tissues, especially in brain, and the precursors of docosanoids [15]. The 18:4n-3, 22:3n-3, and 22:5n-3 PUFA were also found in animal tissues. Interest in n-3 PUFA has increased since it was found that their consumption in most Western populations, particularly those of EPA and DHA, is suboptimal for protection against the most prevalent chronic diseases.

23.4.1 Nutritional Value of n-3 Fatty Acids

Dietary n-3 PUFA have effects on diverse physiological processes impacting normal health and chronic diseases, such as the regulation of plasma lipid levels, cardiovascular and immune function, insulin action, and neuronal development and visual function [64]. In addition, the very high n-6/n-3 ratio of typical Western diets (15–17/1) favors the development of CVD, cancer, and inflammatory and autoimmune diseases [65]. In fact, the balance of n-6 and n-3 PUFA in the human diets is crucial because the eicosanoid metabolic products from n-6 and n-3 PUFA have opposing biological properties. Current nutritional recommendations are that the n-6/n-3 ratio in human diets should not exceed 4.0 [66]. In recent years, there has been much interest in the beneficial effects of the long-chain n-3 PUFA, in particular EPA and DHA. Indeed, it has been shown that consumption of EPA and DHA may reduce the risk of CVD, as well as some inflammatory and neurological diseases [67]. Moreover, EPA and DHA may also have roles in reducing the incidences of cancer, obesity, and type 2 diabetes (WHO, 2003). Thus, SACN/COT [67] recommended that the intake of long-chain n-3 PUFA in the United Kingdom should be increased from the current value of approximately 200 to 450 mg/day. The principal biological role of ALA seems to be as a precursor for EPA synthesis [68]. Crucially, it is now evident that *in vivo* synthesis of EPA and DHA from dietary ALA is very limited in adult humans. However, several studies suggest that high intakes of ALA can beneficially affect a number of CVD risk factors, including LDL-cholesterol [69]. Thus, more studies are needed to determine the potential health benefits of ALA intake.

The most recent nutrient intake guidelines published by a joint WHO/FAO expert consultation [1] recommended that n-3 PUFA should represent 1–2% of diet energy. In addition, further recommendations for PUFA, reviewed by Simopoulos [70], are of 1% of diet energy for ALA and 0.3% for EPA + DHA, with at least 0.1% of each long-chain n-3 PUFA. A large variation was found in the intake estimations of ALA and long-chain n-3 PUFA across several studies and countries [71]. In France, for instance, the intake of ALA is 0.74–0.94 g/day, representing 0.37% of energy intake. The mean intakes of EPA and DHA are 118–150 and 226–273 mg/day, respectively.

Ruminant meats usually have a more favorable n-6/n-3 ratio, when compared with monogastric animals, due to the relatively high levels of ALA in green pastures and forages. In fact, it is well established that the values of n-6/n-3 ratio for meat from grass-fed cattle (2.0–2.3) are much lower than those from concentrate-fed cattle (15.6–20.1) [72]. The contents of n-3 PUFA in meat and meat products were reviewed by Givens et al. [73]. In animal tissue lipids, ALA tends to be a minor component (<1.5% of total fatty acids), the exception being grazing nonruminants where it can amount to 10% of the adipose tissue lipids. The highest concentrations of EPA + DHA were found in poultry meat (~2%), followed by sheep meat (~1%), pork (~0.6%), and beef and veal (~0.5%). The 22:5n-3 is also a long-chain n-3 PUFA present in ruminant and poultry meats at significant levels (~0.5%). A survey of European diets revealed that 21% of total fat intake comes from meat and meat products [74]. It is believed that the lower PUFA/SFA and higher n-6/n-3 ratios of some meats contribute to the imbalance in the fatty acid intake of today's consumers [73,75]. However, meat is an important source of n-3 PUFA in human diets [6]. Moreover, although the main source of EPA and DHA (and of total long-chain n-3 PUFA) is fish and seafood, the major sources of 22:5n-3 are ruminant and poultry meats. It was estimated that in the United Kingdom, meat and meat products contribute 36 mg/day of long-chain n-3 fatty acids in adults, with poultry meat contributing 26 mg/day [73]. Howe et al. [76] estimated that almost half of the average Australian adult intake of long-chain n-3 PUFA (55 mg/day) originates from meat sources. Finally, for people who do not consume fish, meat is the only source of EPA and DHA in the diet.

23.4.2 Analytical Methodology

GC-FID is the usual analytical technique for the determination of n-3 PUFA in fats, including meat fats.

23.4.2.1 Gas Chromatography

GC-FID has been the method of choice for the analysis of n-3 PUFA in fats. There is an official method that uses this technique as a method for the determination of long-chain n-3 fatty acids in marine oils (AOCS Ce 1b-89). Although they are generally analyzed as their methyl ester derivatives, occasionally they can also be analyzed as DMOX or picolinyl derivatives by GC in combination with MS. Several capillary columns have been developed especially for the separation and analysis of long unsaturated fatty acids, such as DB-Wax™, SupelcoWax™, and OmegaWax™.

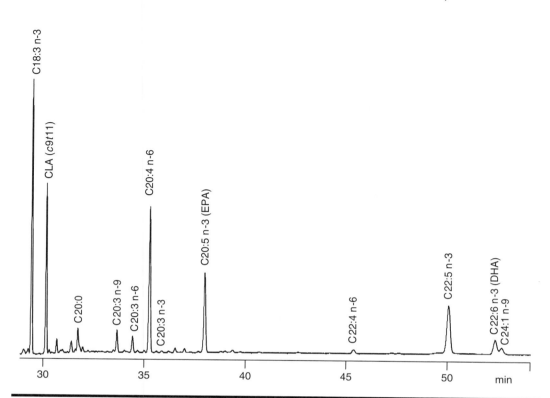

Figure 23.3 **Partial GC-FID chromatogram (PUFA region) of a total FAME mixture from lamb meat fat, obtained using an OmegaWax™ 250 capillary column (30 m × 0.25 mm i.d. × 0.25 µm film thickness) and He as the carrier gas. The column temperature was held at 150°C for 11 min, increased to 220°C at 3°C/min, and held at 220°C for 20 min.**

In meat, good resolution of the major n-3 PUFA can be achieved with 30 m columns (Figure 23.3). These polar polyethylene glycol stationary phases combine the advantage of a relatively high resolution capability with that of a relatively high thermal stability [77], although highly polar cyanoalkyl polysiloxane stationary phase has also been frequently used. The 23:0 methyl ester has been extensively used as an internal standard for the analysis of the highly unsaturated omega-3 fatty acids EPA, DHA, and 22:5n-3 [77].

The chromatographic conditions for long-chain PUFA analysis usually require high oven temperatures (>200°C). However, the damage of DHA by the use of GC high temperatures has been reported recently [78]. It was recommended by the same author that the FID response correction factors for these analytical conditions should be used for the accurate determination of long-chain n-3 PUFA.

23.4.2.2 Other Methods

Analytical approaches common to other fatty acid analysis could also be applied for the determination of n-3 PUFA, particularly the FT-NIR coupled with chemometric calibration procedures, as described by Azizian and Kramer [30].

23.4.3 Applications for Nutritional Quality of Meat Fat

It is widely acknowledged that there is an urgent need to return to a balanced fatty acid diet by improving the intake of n-3 PUFA, especially EPA and DHA. The methodologies reviewed here are required to assess the composition of n-3 fatty acids in meat fat, to design experimental diets for increasing the levels of these fatty acids, and to determine the incorporation of n-3 PUFA in meat fats of animals fed such diets. In addition, attention has also been focused on the extent to which consumption of the precursor ALA can provide sufficient amounts of tissue EPA and DHA through the n-3 PUFA elongation–desaturation pathway.

The aim in most feeding studies is to enrich meat fats in n-3 PUFA and improve its n-6/n-3 ratio. The most effective means of manipulating the fatty acid composition of ruminant meat is through nutrition with strategic use of forages and dietary lipids, often with the aim of bypassing the rumen biohydrogenation process (reviewed by Scollan et al. [6]). Specifically, feeding grass and concentrate-containing linseed or fish oils results in important beneficial responses in the content of n-3 PUFA in beef, although too-high levels cause adverse flavor and color changes. However, the enrichment of ruminant meats with most PUFA will also give rise to increased contents of TFA and CLA in meat. The benefits or drawbacks arising from this are not yet clear and further work is needed in this field.

Howe et al. [76] concluded that red meats are a valuable source of PUFA, particularly of long-chain n-3 fatty acids, for the human diet. By using these methodologies, Nuernberg et al. [79] reported that bovines grazing on pasture (rich in ALA) have a threefold higher concentration of n-3 fatty acids (ALA, EPA, and DHA) in their meat, with an important decrease of the n-6/n-3 fatty acid ratio, compared to concentrate-fed (rich in n-6 PUFA) cattle. However, the highest increases in n-3 fatty acids in meat from cattle [80] and lambs [81] are obtained using supplements of linseed oil (the plant seed oil richest in ALA).

Moreover, the use of diets supplemented with fish or algae oils approximately double EPA + DHA contents in meat [73]. Studies using ruminally protected lipids revealed that the muscle does have a high capacity to deposit n-3 PUFA, suggesting that strategies to address the high degree of biohydrogenation of dietary PUFA in the rumen are required [6].

Finally, these analytical techniques are also of great value for the determination of n-3 PUFA profile in meat fats from poultry and pigs fed different diets. In fact, due to the low fish consumption in Western societies, poultry meat (41% of the meat intake in the United Kingdom) has been considered as one of the main sources of PUFA, in particular n-3 PUFA, for human diets [82]. Interestingly, the content of poultry meat in n-3 fatty acids, particularly in ALA, can be readily improved by increasing the levels of n-3 PUFA in poultry diets through the incorporation of vegetable oils [83] and oily fish by-products [84]. In pigs, the drive has also been to increase n-3 PUFA in meat by including sources such as linseed in the diet [75].

23.5 Conclusions

This chapter presents an overview of the analytical techniques and methodologies available for the analysis of TFA, CLA, and n-3 PUFA in meat from nonruminant and ruminant animals. The most used methodologies are illustrated with applications to assess the nutritional quality of fat in those meats. For the detailed TFA analysis in meat a previous silver ion chromatographic cis/trans fractionation, followed by GC analysis with long (100 m) highly polar capillary columns, is required. Regarding the analysis of the individual CLA isomers in meat, a combination of GC and Ag$^+$-HPLC should be used. The CLA peaks are quantified by GC analysis of total FAME, and the relative concentrations obtained by Ag$^+$-HPLC (three 25 cm columns in tandem) are used to calculate the unresolved peaks in the GC chromatogram. Finally, the analysis of n-3 PUFA in meat can be performed by the straight GC separation of FAME with short (30 m) polyethylene glycol capillary columns.

Acknowledgments

Some of the experimental work mentioned in this chapter was supported by Ministério da Agricultura grant AGRO/2003/512 and FCT grant POCTI/CVT/2002/44750.

References

1. WHO diet, nutrition and the prevention of chronic diseases. Report of a joint WHO/FAO expert consultation. Geneva: World Health Organization, 2003.
2. DHHS/FDA Food Labeling: Trans fatty acids in nutrition labeling; nutrient content claims, and health claims: final rule, July 11. *Fed. Regist.* 68:41433–41506, 2003.
3. EFSA Request N° EFSA-Q-2003-022 Opinion of the Scientific Panel on Dietetic Products, Nutrition and allergies on a request from the commission related to the presence of trans fatty acids in foods and the effect on human health of the consumption of trans fatty acids. *EFSA J.* 81:1–49, 2004.
4. Hunter, J. E. Dietary trans fatty acids: Review of recent human studies and food industry responses. *Lipids* 41:967–992, 2006.
5. Willett, W. C. The scientific basis for TFA regulations—Is it sufficient? Comments from the USA. *Atheroscler. Suppl.* 7:69–71, 2006.
6. Scollan, N., Hocquette, J. F., Nuernberg, K., Dannenberger, D., Richardson, I., and Moloney, A. Innovations in beef production systems that enhance the nutritional and health value of beef lipids and their relationship with meat quality. *Meat Sci.* 74:17–33, 2006.
7. Corl, B. A., Barbano, D. M., Bauman, D. E., and Ip, C. Cis-9, trans-11 CLA derived endogenously from trans-11 18 : 1 reduces cancer risk in rats. *J. Nutr.* 133:2893–2900, 2003.
8. Valeille, K., Ferezou, J., Amsler, G., Quignard-Boulange, A., Parquet, M., Gripois, D., Dorovska-Taran, V., and Martin, J. C. A cis-9,trans-11-conjugated linoleic acid-rich oil reduces the outcome of atherogenic process in hyperlipidemic hamster. *Am. J. Physiol. Heart C.* 289:H652–H659, 2005.
9. Roy, A., Chardigny, J. M., Bauchart, D., Ferlay, A., Lorenz, S., Durand, D., Gruffat, D., Faulconnier, Y., Sebedio, J. L., and Chilliard, Y. Butters rich either in trans-10-C18:1 or in trans-11-C18:1 plus cis-9,trans-11 CLA differentially affect plasma lipids and aortic fatty streak in experimental atherosclerosis in rabbits. *Animal* 1:467–476, 2007.
10. Chardigny, J. M., Malpuech-Brugere, C., Dionisi, F., Bauman, D. E., German, B., Mensink, R. P., Combe, N., Chaumont, P., Barbano, D. M., Enjalbert, F., Bezelgues, J. B., Cristiani, I., Moulin, J., Boirie, Y., Golay, P. A., Giuffrida, F., Sebedio, J. L., and Destaillats, F. Rationale and design of the TRANSFACT project phase I: A study to assess the effect of the two different dietary sources of trans fatty acids on cardiovascular risk factors in humans. *Contemp. Clin. Trials* 27:364–373, 2006.
11. Craig-Schmidt, M. C. World-wide consumption of trans fatty acids. *Atheroscler. Suppl.* 7:1–4, 2006.
12. Fritsche, J. and Steinhart, H. Analysis, occurrence, and physiological properties of trans fatty acid (TFA) with particular emphasis on conjugated linoleic acid isomers (CLA)—A review. *Fett-Lipid* 100:190–210, 1998.
13. Valsta, L. M., Tapanainen, H., and Mannisto, S. Meat fats in nutrition. *Meat Sci.* 70:525–530, 2005.

14. Wolff, R. L., Precht, D., and Molkentin, J. Occurrence and distribution profiles of trans-18:1 acids in edible fats of natural origin. In: *Trans Fatty Acids in Human Nutrition*, ed. J. L. Sebedio and W. W. Christie, 1–33. Dundee, Scotland, UK: The Oily Press, 1998.

15. Christie, W. W. *Lipid Analysis—Isolation, Separation, Identification and Structural Analysis of Lipids*. Dundee, Scotland: The oily press, 2003.

16. Aldai, N., Osoro, K., Barron, L. J. R., and Najera, A. I. Gas-liquid chromatographic method for analysing complex mixtures of fatty acids including conjugated linoleic acids (cis9trans11 and trans10cis12 isomers) and long-chain (n-3 or n-6) polyunsaturated fatty acids - Application to the intramuscular fat of beef meat. *J. Chromatogr. A* 1110:133–139, 2006.

17. Kramer, J. K. G., Cruz-Hernandez, C., Deng, Z. Y., Zhou, J. Q., Jahreis, G., and Dugan, M. E. R. Analysis of conjugated linoleic acid and trans 18 : 1 isomers in synthetic and animal products. *Am. J. Clin. Nutr.* 79:1137S–1145S, 2004.

18. Precht, D. and Molkentin, J. Overestimation of linoleic acid and trans-C18: 2 isomers in milk fats with emphasis on trans Delta 9,trans Delta 12-octadecadienoic acid. *Milchwiss* 58:30–34, 2003.

19. Molkentin, J. and Precht, D. Occurrence of trans-C16 : 1 acids in bovine milkfats and partially hydrogenated edible fats. *Milchwiss* 52:380–385, 1997.

20. Cruz-Hernandez, C., Deng, Z. Y., Zhou, J. Q., Hill, A. R., Yurawecz, M. P., Delmonte, P., Mossoba, M. M., Dugan, M. E. R., and Kramer, J. K. G. Methods for analysis of conjugated linoleic acids and trans-18 : 1 isomers in dairy fats by using a combination of gas chromatography, silver-ion thin-layer chromatography/gas chromatography, and silver-ion liquid chromatography. *J. AOAC Int.* 87:545–562, 2004.

21. Precht, D., Molkentin, J., Destaillats, F., and Wolff, R. L. Comparative studies on individual isomeric 18 : 1 acids in cow, goat, and ewe milk fats by low-temperature high-resolution capillary gas-liquid chromatography. *Lipids* 36:827–832, 2001.

22. Kramer, J. K., Blackadar, C. B., and Zhou, J. Evaluation of two GC columns (60-m SUPELCOWAX 10 and 100-m CP Sil 88) for analysis of milkfat with emphasis on CLA, 18:1, 18:2 and 18:3 isomers, and short- and long-chain FA. *Lipids* 37:823–835, 2002.

23. Ratnayake, W. M. Analysis of trans fatty acids. In: *Trans Fatty Acids in Human Nutrition*, ed. J. L. Sebedio and W. W. Christie, 115–161. Dundee, Scotland, UK: The Oily Press, 1998.

24. Kramer, J. K. G., Cruz-Hernandez, C., and Zhou, J. Q. Conjugated linoleic acids and octadecenoic acids: Analysis by GC. *Eur. J. Lipid Sci. Technol.* 103:600–609, 2001.

25. Precht, D. and Molkentin, J. C18:1, C18:2 and C18:3 trans and cis fatty acid isomers including conjugated cis delta 9, trans delta 11 linoleic acid (CLA) as well as total fat composition of German human milk lipids. *Nahrung* 43:233–244, 1999.

26. Precht, D. and Molkentin, J. Rapid analysis of the isomers of trans-octadecenoic acid in milk fat. *Int. Dairy J.* 6:791–809, 1996.

27. Ulberth, F. and Henninger, M. Quantitation of trans fatty acids in milk fat using spectroscopic and chromatographic methods. *J. Dairy Res.* 61:517–527, 1994.

28. Adam, M., Mossoba, M. M., and Lee, T. Rapid determination of total trans fat content by attenuated total reflection infrared spectroscopy: An international collaborative study. *J. Am. Oil Chem. Soc.* 77:457–462, 2000.

29. Mossoba, M. M., Yurawecz, M. P., Delmonte, P., and Kramer, J. K. G. Overview of infrared methodologies for trans fat determination. *J.AOAC Int.* 87:540–544, 2004.

30. Azizian, H. and Kramer, J. K. G. A rapid method for the quantification of fatty acids in fats and oils with emphasis on trans fatty acids using Fourier transform near infrared spectroscopy (FT-NIR). *Lipids* 40:855–867, 2005.

31. Mossoba, M. M., McDonald, R. E., Yurawecz, M. P., and Kramer, J. K. G. Application of on-line capillary GC-FTIR spectroscopy to lipid analysis. *Eur. J. Lipid Sci. Technol.* 103:826–830, 2001.

32. Bessa, R. J. B., Alves, S. P., Jerónimo, E., Alfaia, C. M. M., Prates, J. A. M., and Santos-Silva, J. Effect of lipid supplements on ruminal biohydrogenation intermediates and muscle fatty acids in lambs. *Eur. J. Lipid Sci. Technol.* 109:868–878, 2007.

33. Bessa, R. J. B., Portugal, P. V., Mendes, I. A., and Santos-Silva, J. Effect of lipid supplementation on growth performance, carcass and meat quality and fatty acid composition of intramuscular lipids of lambs fed dehydrated lucerne or concentrate. *Livestock Prod. Sci.* 96:185–194, 2005.

34. Sackmann, J. R., Duckett, S. K., Gillis, M. H., Realini, C. E., Parks, A. H., and Eggelston, R. B. Effects of forage and sunflower oil levels on ruminal biohydrogenation of fatty acids and conjugated linoleic acid formation in beef steers fed finishing diets. *J. Anim. Sci.* 81:3174–3181, 2003.

35. Wahle, K. W., Heys, S. D., and Rotondo, D. Conjugated linoleic acids: are they beneficial or detrimental to health? *Prog. Lipid Res.* 43:553–587, 2004.

36. Palmquist, D. L., St-Pierre, S., and McClure, K. E. Tissue fatty acid profiles can be used to quantify endogenous rumenic acid synthesis in lambs. *J. Nutr.* 134:2407–2414, 2004.

37. Collomb, M., Sieber, R., and Butikofer, U. CLA isomers in milk fat from cows fed diets with high levels of unsaturated fatty acids. *Lipids* 39:355–364, 2004.

38. Bhattacharya, A., Banu, J., Rahman, M., Causey, J., and Fernandes, G. Biological effects of conjugated linoleic acids in health and disease. *J. Nutr. Biochem.* 17:789–810, 2006.

39. Nagao, K. and Yanagita, T. Conjugated fatty acids in food and their health benefits. *J. Biosci. Bioeng.* 100:152–157, 2005.

40. National Research Council. *Carcinogens and Anticarcinogens in the Human Diet.* Washington, DC: National Academy Press, 1996.

41. Yamasaki, M., Miyazaki, H. C., and Yamada, K. Growth inhibition and apoptotic cell death of cancer cells induced by conjugated linoleic acid. In: *Advances in Conjugated Linoleic Acid Research* (Vol. 3), ed. M. P. Yurawecz, J. K. G. Kramer, O. Gudmundsen, M. W. Pariza, and S. Banni, 141–152. Champaign, IL: AOCS Press, 2006.

42. Schmid, A., Collomb, M., Sieber, R., and Bee, G. Conjugated linoleic acid in meat and meat products: A review. *Meat Sci.* 73:29–41, 2006.

43. Collomb, M., Schmid, A., Sieber, R., Wechsler, D., and Ryhanen, E. L. Conjugated linoleic acids in milk fat: Variation and physiological effects. *Int. Dairy J.* 16:1347–1361, 2006.

44. Parodi, P. W. Conjugated linoleic acids—An anticarcinogenic fatty acid present in milk fat. *Aust. J. Dairy Technol.* 49:93–97, 1994.

45. Parodi, P. W. Conjugated linoleic acid in food. In: *Advances in Conjugated Linoleic Acid Research* (vol. 2), ed. J. L.Sebedio, W. W. Christie, and R. Adlof, 101–122. Champaign, IL: AOCS Press, 2003.

46. Zee, M., O'Hagan, S., and Menzel, A. Safety data on conjugated linoleic acid from animal studies: an overview. In: *Advances in Conjugated Linoleic Acid Research* (Vol. 3), ed. M. P. Yurawecz, J. K. G. Kramer, O. Gudmundsen, M. W. Pariza, and S. Banni, 239–256. Champaign, IL: AOCS Press, 2006.

47. Cruz-Hernandez, C., Kramer, J. K. G., Kraft, J., Santercole, V., Rashid, M., Deng, Z. Y., Dugan, M. E. R., Delmonte, P., and Yurawecz, M. P. Systematic analysis of trans and conjugated linoleic acids in the milk and meat of ruminants. In: *Advances in Conjugated Linoleic Acid Research* (Vol. 3), ed. M. P. Yurawecz, J. K. G. Kramer, O. Gudmundsen, M. W. Pariza, and S. Banni, 45–93. Champaign, IL: AOCS Press, 2006.

48. Destaillats, F. and Angers, P. Directed sequential synthesis of conjugated linoleic acid isomers from d7,9 to d12,14. *Eur. J. Lipid Sci. Technol.* 105:3–8, 2003.

49. Dobson, G. Gas chromatography-mass spectrometry of conjugated linoleic acids and metabolites. In: *Advances in Conjugated Linoleic Acid Research* (vol. 2), ed. J. L. Sebedio, W. W. Christie, and R. Adlof, 13–36. Champaign, IL: AOCS Press, 2003.

50. Sehat, N., Kramer, J. K., Mossoba, M. M., Yurawecz, M. P., Roach, J. A., Eulitz, K., Morehouse, K. M., Ku, Y., Wander, R. C., Du, S. H., and Thomas, D. R. Identification of conjugated linoleic acid isomers in cheese by gas chromatography, silver ion high performance liquid chromatography and mass spectral reconstructed ion profiles. Comparison of chromatographic elution sequences. Influence of long-chain polyunsaturated fatty acids on oxidation of low density lipoprotein. *Prostag. Leuckotr. Ess.* 59:143–151, 1998.

51. Sehat, N., Rickert, R., Mossoba, M. M., Kramer, J. K. G., Yurawecz, M. P., Roach, J. A. G., Adlof, R. O., Morehouse, K. M., Fritsche, J., Eulitz, K. D., Steinhart, H., and Yuoh, K. Improved separation of conjugated fatty acid methyl esters by silver ion–high-performance liquid chromatography. *Lipids* 34:407–413, 1999.

52. Dannenberger, D., Nuernberg, K., Nuernberg, G., Scollan, N., Steinhart, H., and Ender, K. Effect of pasture vs. concentrate diets on CLA isomer distribution in different tissue lipids of beef cattle. *Lipids* 40:589–598, 2005.

53. Kraft, J., Collomb, M., Mockel, P., Sieber, R., and Jahreis, G. Differences in CLA isomer distribution of cow's milk lipids. *Lipids* 38:657–664, 2003.

54. Davis, A. L., McNeill, G. P., and Caswell, D. C. Identification and quantification of conjugated linoleic acid isomers in fatty acid mixtures by 13C NMR spectroscopy. In: *Advances in Conjugated Linoleic Acid Research* (Vol. 1), ed. M. P. Yurawecz, M. Mossoba, J. K. G. Kramer, G. Nelson, and M. W. Pariza, 164–179. Champaign, IL: AOCS Press, 1999.

55. Delmonte, P., Kramer, J. K. G., Banni, S., and Yurawecz, M. P. New developments in silver ion and reverse phase HPLC of conjugated linoleic acid. In: *Advances in Conjugated Linoleic Acid Research* (Vol. 3), ed. M. P. Yurawecz, J. K. G. Kramer, O. Gudmundsen, M. W. Pariza, and S. Banni, 95–118. Champaign, IL: AOCS Press, 2006.

56. Banni, S. and Martin, J. C. Conjugated linoleic acid and metabolites. In: *Trans Fatty Acids in Human Nutrition*, ed. J. L. Sebedio and W. W. Christie, 262–302. Dundee, Scotland, UK: The Oily Press, 1998.

57. Michaud, A. L. and Brenna, J. T. Structural characterization of conjugated linoleic acid methyl esters with acetonitrile chemical ionization tandem mass spectrometry. In: *Advances in Conjugated Linoleic Acid Research* (Vol. 3), ed. M. P. Yurawecz, J. K. G. Kramer, O. Gudmundsen, M. W. Pariza, and S. Banni, 119–138. Champaign, IL: AOCS Press, 2006.

58. Raes, K., de Smet, S., and Demeyer, D. Effect of dietary fatty acids on incorporation of long chain polyunsaturated fatty acids and conjugated linoleic acid in lamb, beef and pork meat: A review. *Anim. Feed Sci. Technol.* 113:199–221, 2004.

59. French, P., Stanton, C., Lawless, F., O'Riordan, E. G., Monahan, F. J., Caffrey, P. J., and Moloney, A. P. Fatty acid composition, including conjugated linoleic acid, of intramuscular fat from steers offered grazed grass, grass silage, or concentrate-based diets. *J. Anim. Sci.* 78:2849–2855, 2000.

60. Alfaia, C. M. M., Quaresma, M. A. G., Castro, M. L. F., Martins, S. I. V., Portugal, A. P. V., Fontes, C. M. G. A., Bessa, R. J. B., and Prates, J. A. M. Fatty acid composition, including isomeric profile of conjugated linoleic acid, and cholesterol in Mertolenga-PDO beef. *J. Sci. Food Agric.* 86:2196–2205, 2006.

61. Alfaia, C. M. M., Ribeiro, V. S. S., Lourenco, M. R. A., Quaresma, M. A. G., Martins, S. I. V., Portugal, A., Fontes, C. M. G. A., Bessa, R. J. B., Castro, M. L. F., and Prates, J. A. M. Fatty acid composition, conjugated linoleic acid isomers and cholesterol in beef from crossbred bullocks intensively produced and from Alentejana purebred bullocks reared according to Carnalentejana-PDO specifications. *Meat Sci.* 72:425–436, 2006.

62. Alfaia, C. M. M., Castro, M. L. F., Martins, S. I. V., Portugal, A. P. V., Alves, S. P. A., Fontes, C. M. G. A., Bessa, R. J. B., and Prates, J. A. M. Effect of slaughter season on fatty acid composition, conjugated linoleic acid isomers and nutritional value of intramuscular fat in Barrosa-PDO veal. *Meat Sci.* 75:44–52, 2007.

63. Alfaia, C. P. M., Castro, M. L. F., Martins, S. I. V., Portugal, A. P. V., Alves, S. P. A., Fontes, C. M. G. A., Bessa, R. J. B., and Prates, J. A. M. Influence of slaughter season and muscle type on fatty acid composition, conjugated linoleic acid isomeric distribution and nutritional quality of intramuscular fat in Arouquesa-PDO veal. *Meat Sci.* 76:787–795, 2007.

64. Jump, D. B. The biochemistry of n-3 polyunsaturated fatty acids. *J. Biol. Chem.* 277:8755–8758, 2002.

65. Simopoulos, A. P. Omega-6/omega-3 essential fatty acid ratio and chronic diseases. *Food Rev. Int.* 20:77–90, 2004.

66. British Department of Health. Nutritional aspects of cardiovascular disease. London: HMSO, 1994.

67. SACN/COT Scientific Advisory Committee on Nutrition (SACN) and Committee on Toxicity (COT), Advice on fish consumption: benefits and risks. Norwich: TSO, 2004.

68. Burdge, G. C. and Calder, P. C. α-Linolenic acid metabolism in adult humans: the effects of gender and age on conversion to longer-chain polyunsaturated fatty acids. *Eur. J. Lipid Sci. Technol.* 107:426–439, 2005.

69. Zhao, G. X., Etherton, T. D., Martin, K. R., West, S. G., Gillies, P. J., and Kris-Etherton, P. M. Dietary alpha-linolenic acid reduces inflammatory and lipid cardiovascular risk factors in hypercholesterolemic men and women. *J. Nutr.* 134:2991–2997, 2004.

70. Simopoulos, A. P. Human requirement for n-3 polyunsaturated fatty acids. *Poultry Sci.* 79:961–970, 2000.

71. Sioen, I. A., Pynaert, I., Matthys, C., De Backer, G., Van Camp, J., and De Henauw, S. Dietary intakes and food sources of fatty acids for Belgian women, focused on n-6 and n-3 polyunsaturated fatty acids. *Lipids* 41:415–422, 2006.

72. Enser, M., Hallett, K. G., Hewett, B., Fursey, G. A. J., Wood, J. D., and Harrington, G. Fatty acid content and composition of UK beef and lamb muscle in relation to production system and implications for human nutrition. *Meat Sci.* 49:329–341, 1998.

73. Givens, D. I., Khem, K. E., and Gibbs, R. A. The role of meat as a source of n-3 polyunsaturated fatty acids in the human diet. *Meat Sci.* 74:209–218, 2006.

74. Hulshof, K. F., van Erp-Baart, M. A., Anttolainen, M., Becker, W., Church, S. M., Couet, C., Hermann-Kunz, E., Kesteloot, H., Leth, T., Martins, I., Moreiras, O., Moschandreas, J., Pizzoferrato, L., Rimestad, A. H., Thorgeirsdottir, H., van Amelsvoort, J. M., Aro, A., Kafatos, A. G., Lanzmann-Petithory, D., and van Poppel, G. Intake of fatty acids in western Europe with emphasis on trans fatty acids: the TRANSFAIR Study. *Eur. J. Clin. Nutr.* 53:143–157, 1999.

75. Wood, J. D., Richardson, R. I., Nute, G. R., Fisher, A. V., Campo, M. M., Kasapidou, E., Sheard, P. R., and Enser, M. Effects of fatty acids on meat quality: A review. *Meat Sci.* 66:21–32, 2004.

76. Howe, P., Meyer, B., Record, S., and Baghurst, K. Dietary intake of long-chain omega-3 polyunsaturated fatty acids: contribution of meat sources. *Nutrition* 22:47–53, 2006.

77. Ackman, R. G. The gas chromatograph in practical analyses of common and uncommon fatty acids for the 21st century. *Anal. Chim. Acta* 465:175–192, 2002.

78. Ackman, R. G. Losses of DHA from high temperatures of columns during GLC of methyl esters of long-chain omega-3 fatty acids. *J. Am. Oil Chem. Soc.* 83:1069–1070, 2006.

79. Nuernberg, K., Nuernberg, G., Ender, K., Lorenz, S., Winkler, K., Rickert, R., and Steinhart, H. n-3 fatty acids and conjugated linoleic acids of longissimus muscle in beef cattle. *Eur. J. Lipid Sci. Technol.* 104:463–471, 2002.

80. Scollan, N. D., Dhanoa, M. S., Choi, N. J., Maeng, W. J., Enser, M., and Wood, J. D. Biohydrogenation and digestion of long chain fatty acids in steers fed on different sources of lipid. *J. Agric. Sci.(Camb.)* 136:345–355, 2001.

81. Cooper, S. L., Sinclair, L. A., Wilkinson, R. G., Hallett, K. G., Enser, M., and Wood, J. D. Manipulation of the n-3 polyunsaturated fatty acid content of muscle and adipose tissue in lambs. *J. Anim. Sci.* 82:1461–1470, 2004.

82. Rymer, C. and Givens, D. I. n-3 fatty acid enrichment of edible tissue of poultry: A review. *Lipids* 40:121–130, 2005.

83. Lopez-Ferrer, S., Baucells, M. D., Barroeta, A. C., Galobart, J., and Grashorn, M. A. n-3 enrichment of chicken meat. 2. Use of precursors of long-chain polyunsaturated fatty acids: Linseed oil. *Poultry Sci.* 80:753–761, 2001.

84. Lopez-Ferrer, S., Baucells, M. D., Barroeta, A. C., and Grashorn, M. A. n-3 enrichment of chicken meat. 1. Use of very long-chain fatty acids in chicken diets and their influence on meat quality: Fish oil. *Poultry Sci.* 80:741–752, 2001.

Chapter 24

Vitamins

Young-Nam Kim, David W. Giraud, and Judy A. Driskell

Contents

Vitamins are a group of complex organic compounds that are essential to normal functioning and metabolic reactions in the body. Vitamins are not utilized as a source of energy or structural tissue components, but rather as cofactors or coenzymes in biochemical reactions. Vitamins are divided into two categories based on their solubility—those soluble in fat solvents are known as fat-soluble vitamins and those soluble in water are known as water-soluble vitamins. Table 24.1 lists the

Table 24.1 Composition of Selected Vitamins in Muscle Foods

Food Product	Vitamin A μg RAE[b]/100 g	Vitamin A % DV[c]	Vitamin E[a] mg/100 g	Vitamin E % DV	Thiamin mg/100 g	Thiamin % DV	Riboflavin mg/100 g	Riboflavin % DV	Niacin mg/100 g	Niacin % DV	Vitamin B_6 mg/100 g	Vitamin B_6 % DV	Folate μg/100 g	Folate % DV	Vitamin B_{12} μg/100 g	Vitamin B_{12} % DV	Pantothenic Acid mg/100 g	Pantothenic Acid % DV
Beef, top round, trimmed to 0 in. fat, braised	0	0	0.15	1	0.070	5	0.250	15	3.760	19	0.280	14	9	2	2.68	45	0.370	4
Bison, top round, lean only, broiled	0	0	0.21	1	0.165	11	0.365	22	6.475	32	0.657	33	19	5	1.81	30	1.448	14
Pork, top loin, boneless, braised	2	<1	0.24	1	0.552	37	0.256	15	4.540	23	0.329	16	4	1	0.46	8	0.638	6
Lamb, loin, trimmed to 1/8 in. fat, roasted	0	0	—[d]	—	0.100	7	0.240	14	7.050	35	0.130	7	20	5	2.20	37	0.660	7
Veal, top round, braised	0	0	0.49	2	0.060	4	0.350	21	10.560	53	0.360	18	18	5	1.17	20	1.020	10
Chicken, breast, roasted	28	2	0.27	1	0.066	4	0.119	7	12.710	64	0.560	28	4	1	0.32	5	0.936	9
Turkey, light meat, roasted	0	0	0.09	<1	0.061	4	0.129	8	6.838	34	0.540	27	6	2	0.37	6	0.677	7

[a] Milligrams of α-tocopherol.
[b] Retinol activity equivalent.
[c] Percent Daily Value, established by the U.S. Food and Drug Administration to help consumers understand how the nutrient reference value for that nutrient fits into their overall diets [2].
[d] No composition data provided by USDA.

Note: No significant amounts of vitamin D, vitamin K, biotin, and vitamin C are found in muscle foods.

Source: Data obtained from U.S. Department of Agriculture, Agricultural Research Service, USDA National Nutrient Database for Standard Reference, Release 19, Nutrient Data Laboratory, 2006.

composition of selected vitamins in muscle foods according to the U.S. Department of Agriculture National Nutrient Database [1]. Table 24.1 also includes percentages of the Daily Value (DV) of the vitamin estimated to be in 100 g, a typical serving size, of the meats. The DV, established by the U.S. Food and Drug Administration, is a nutrient reference value intended to help consumers understand how foods fit into their overall diets [2]. Foods containing 20% or more of the DV of nutrients per reference amount are indicated to be "high," "rich," or "excellent" sources of the nutrients. Foods containing 10–19% of the DV are categorized as "good" sources. Muscle foods are good to excellent sources of most of the B-vitamins as defined by the U.S. Food and Drug Administration [3]. Pork contains 3–10 times as much thiamin per 100 g as other muscle foods such as beef, lamb, and chicken. However, small amount of the fat-soluble vitamins are found in muscle foods.

Generally, the methodologies used for determining the composition of the various vitamins in meats are the same as those used for other foods. Internal standards are frequently used in the analytical methods for determination of the vitamins. Several vitamins are lost or interconverted into their isomers in extraction and purification procedures during analyses.

24.1 Fat-Soluble Vitamins

24.1.1 Vitamin A and Carotenoids

Vitamin A is a fat-soluble vitamin that is essential for humans and other vertebrates. Vitamin A as retinoids, primarily retinyl esters, is abundant in some animal-derived foods, whereas carotenoids are abundant in plant foods as pigments. Animals cannot synthesize carotenoids. They obtain carotenoids from plant sources in their diets such as corn, green feedstuffs, vegetables, and fruits. Provitamin A carotenoids (α-carotene, β-carotene, and β-cryptoxanthin) are converted into vitamin A in the animal body. β-Carotene is the most prevalent carotenoid in meats, but frequently it is poorly converted into vitamin A, especially in ruminants [4]. The liver of animals is an excellent source of vitamin A. In foods, retinyl esters and carotenoids are vulnerable to oxidation. Exposure to air, heat, and storage time also affect the destruction of vitamin A compounds. Thus, overcooking can cause loss of retinyl esters and provitamin A carotenoids in foods.

Reversed-phase high-performance liquid chromatography (HPLC) followed by UV detection for retinoids and carotenoids is the most common method of analysis. HPLC methodologies are given in AOAC Official Methods 2001.13 and 2005.07 [5]. HPLC procedures for determination of retinoids and carotenoids in muscle foods [6–8] also have been published. During sample preparation and analysis, samples should be protected from heat, light, and oxidizing substances to avoid destructions and isomerizations of the retinoids and carotenoids. Antioxidants such as butylated hydroxytoluene (BHT), pyrogallol, or ascorbyl palmitate are used to prevent oxidation of retinoids and carotenoids.

Alkali hydrolysis (saponification) is routinely used to extract retinoids and carotenoids from muscle foods. Saponification removes chlorophylls, unwanted lipids, and other materials, which may interfere with the chromatographic separation. Retinyl esters and carotenoid esters in foods are converted into retinol and carotenoids during saponification. However, the degradation and isomerization of retinol and carotenoids may occur during saponification. This is greater with higher concentrations of alkali and higher temperatures [9]. Hexane, petroleum ether, diethyl ether, dichloromethane, or mixtures of these solvents are common extracting solvents. The reverse-phase C_{18} column is commonly used to resolve retinoids and carotenoids in foods. The polymeric

C_{30} column designed at the U.S. National Institute of Standards and Technology [10] provides high absolute retention and resolution of *cis* and *trans* isomers of carotenoids [11]. Maximum absorbance is 320–380 nm for retinoids and 400–500 nm for carotenoids. Thus, to perform simultaneous HPLC analysis of retinoids and carotenoids, a photodiode array detector is essential to establish the identity of the compound in each peak and validate homogeneity.

Several units are used for expressing vitamin A contents in foods. International units (IU) are defined by the relationship of 1 IU = 0.3 µg of all-*trans*-retinol or 0.6 µg of β-carotene. The biological activity of vitamin A is quantified by conversion of retinol and provitamin A carotenoids into retinol equivalents (RE). One RE is defined as 1 µg of retinol, 6 µg of β-carotene, or 12 µg of other provitamin A carotenoids. Therefore, 1 RE is equal to 3.33 IU based on retinol. In 2001, the U.S. Institute of Medicine [12] proposed the new vitamin A unit, retinol activity equivalents (RAE). One RAE is equivalent to 1 µg of retinol, which is nutritionally equivalent to 12 µg of β-carotene or 24 µg of other provitamin A carotenoids.

24.1.2 Vitamin D

Vitamin D is a fat-soluble vitamin found in food and also synthesized in the body after exposure to UV rays from the sun. Several forms of vitamin D have been described, but the two major physiologically relevant ones are vitamin D_2 and vitamin D_3 [13]. Vitamin D_2 (ergocalciferol) is a synthetic form of vitamin D that is produced by irradiation of plant and yeast steroid ergosterol. Vitamin D_3 (cholecalciferol) is the naturally occurring form of vitamin D produced from 7-dehydrocholesterol when the skin of animals and humans is exposed to sunlight, specifically UV B radiation. Vitamin D is biologically inactive and is metabolized to 25-hydroxyvitamin D [25(OH)D] in the liver, which is the most abundant form of vitamin D in the circulatory system. This circulating metabolite is hydroxylated again to form its biologically active hormone, 1,25-dihydroxyvitamin D [1,25(OH)$_2$D], which acts as a hormone in controlling calcium homeostasis and regulating the growth of various cell types [14]. Beef and pork livers and other organs contain considerable quantities of vitamin D, whereas concentrations of vitamin D of other meats and meat products are low and therefore are often not included in food composition tables. Most of vitamin D in animal foods is vitamin D_3 and 25(OH)D_3. Therefore, to estimate vitamin D values in muscle foods, analyses of 25(OH)D_3 should be included. Food processing, cooking, and storage of foods do not generally affect the concentration of vitamin D [14].

HPLC methods using a UV absorbance detector are available for the quantitation of vitamin D from most food matrices. Several HPLC methodologies for vitamin D analyses are provided in published articles [15–17] including AOAC Official Method 995.05, which is the method for determination of vitamin D in infant formulas and enteral products [5]. Vitamin D oxidation can occur during the oxidation of fats. Thus, an antioxidant such as pyrogallol or ascorbic acid is added when analyzing food samples.

Estimation of the low concentrations of vitamin D in muscle foods is often difficult due to interfering substances such as fats, cholesterol, vitamin A, and vitamin E. To remove fats, saponification and cleanup procedures should be applied. Hot saponification promotes thermal isomerization of vitamin D with the formation of previtamin D. Hence, several methods [15,18] have been used for saponification at ambient temperature overnight. After saponification, unsaponified lipids including vitamin D are extracted with diethyl ether:petroleum ether used in ratio of 1:1 [15–17]. Vitamin A, vitamin E, sterols, and other interfering components in the unsaponified fraction are removed using silica solid-phase extraction [15–17,19,20]. Both reversed- and normal-phase systems offer efficient resolution of vitamin D, 7-dehydrocholesterol, and hydroxylated metabolites. However, reversed-phase chromatography can separate vitamin D_2 from vitamin D_3 [16,21]. Vitamin D shows identical UV absorption spectra with λ_{max} at 265 nm, which is sensitive enough for the detection of vitamin D_2, vitamin D_3, and their metabolites.

Vitamin D content in foods is expressed in either IU or micrograms of vitamin D. One IU of vitamin D is the activity obtained from 0.025 µg of cholecalciferol in bioassays. The activity of 25(OH)D is five times more potent than cholecalciferol. Therefore, the biological activity of 1 µg of vitamin D is 40 IU, and 1 IU is 0.005 µg of 25(OH)D [13].

24.1.3 Vitamin E

Vitamin E is the most effective fat-soluble antioxidant known to occur in the human body. Natural vitamin E exists in eight different forms: four tocopherols (α-, β-, γ-, and δ-tocopherols) and four tocotrienols (α-, β-, γ-, and δ-tocotrienols). Vitamin E is found in plant and animal foods. Meats and meat products provide minute amounts of vitamin E. The concentration of vitamin E in meats depends on the amount of the vitamin in the animal's diet [22–24]. In raw meat, most of the vitamin E is in the form of α-tocopherol. During processing, vitamin E is lost quite rapidly. Loss is accelerated by oxygen, light, heat, and various metals, primarily iron and copper, and by the presence of free radicals in the fat that can initiate autoxidation.

For vitamin E assay in foods, HPLC methods using fluorescence or UV detection have largely replaced the colorimetric and polarimetric procedures of AOAC Official Methods 948.26, 971.30, and 975.45 [5]. Although gas chromatography (GC) methodologies (AOAC Official Methods 988.14 and 989.09) were developed to increase the precision of vitamin E quantification before the advent of HPLC procedures, HPLC is considered a suitable method for measurement of the individual tocopherols and tocotrienols because GC methodologies can be time-consuming. HPLC methodologies are provided in AOAC Official Method 992.03 [5] and other published articles [6,7,25,26].

Alkaline hydrolysis of muscle foods is obligatory to release the α-tocopherol. Hydrolysis results in cleavage of the ester linkages of lipids in food samples, destroys pigments, and disrupts the sample matrix that facilitates vitamin E extraction. Following saponification, the digest is extracted with ether, petroleum ether, hexane, ethyl acetate in hexane, or other organic solvent mixtures. During saponification, losses of vitamin E can be prevented by addition of antioxidants, pyrogallol, ascorbic acid, or BHT into extraction solvents, and by protection from light. Both reversed- and normal-phase systems are useful for the resolution of vitamin E. The advantage of normal-phase HPLC systems is the ability to separate the eight tocopherols and tocotrienols that occur in nature. Reversed-phase systems cannot resolve the β- and γ-isomers [27]. However, reversed-phase HPLC is the preferred method for the determination of α-tocopherol with retinol and carotenoids in muscle foods. Fluorescence detection compared to UV detection provides better sensitivity and specificity and cleaner chromatograms [28]. Although the UV absorbance of tocopherols and tocotrienols is relatively weak, detection at the absorption maximum (292–298 nm) using a variable wavelength detector affords sufficient sensitivity for most applications [29]. For simultaneous detection of vitamin E with other fat-soluble vitamins and carotenoids, a multichannel UV detector or a photodiode array detector may be useful in a single sample assay.

IUs of vitamin E activity (1 mg of α-tocopherol = 1.49 IU) may be used in food composition tables, but the IU is not commonly used. The vitamin E composition in foods is often expressed as milligrams of α-tocopherol equivalents based on the biological activity of the various forms of vitamin E [30]. However, the U.S. Institute of Medicine [31] has indicated that α-tocopherol is the only form of the tocopherols and tocotrienols that has vitamin E activity in humans.

24.1.4 Vitamin K

Vitamin K is a fat-soluble vitamin. Two forms of vitamin K exist in nature: phylloquinone and mena-quinones. Phylloquinone, known as vitamin K_1, is synthesized by plants. Menaquinones, known as vitamin K_2, are produced by bacteria and contain a polyisoprenyl side chain at the 3-position [32]. However, one of menaquinones, menaquinone (MK)-4, is not a major bacterial product, but is synthesized by animals from phylloquinone [33,34]. Menadione, vitamin K_3, is a synthetic form of vitamin K, which has been used in commercial animal feeds. Menadione can be also converted into MK-4 in animal tissues [33,35]. Several studies [36–39] have reported that relatively low quantities of phylloquinone as well as small amounts of menaquinones, especially MK-4, are found in meats and meat products. Vitamin K is quite stable to oxidation and most food processing and food prepa-ration procedures, whereas it is unstable to light, alkali, strong acid, and reducing agents [40].

Current methods to determine vitamin K in foods are HPLC procedures that use fluorescence or electrochemical detection systems. AOAC Official Methods 992.27 and 999.15 [5] for phyllo-quinone determination in infant formulas and several HPLC methodologies [36–39] determining simultaneously phylloquinone and menaquinones have been developed. GC procedures have been described, but are not routinely used due to long retention times and the potential for on-column degradation from high column temperatures.

When analyzing vitamin K, saponification cannot be applied to remove fats and other com-ponents because of the instability of vitamin K under alkaline conditions. Vitamin K is extracted from meats with organic solvents and then purified by solid-phase extraction before the resolution of vitamin K by reversed-phase HPLC. Phylloquinone and menaquinones are detected by UV detection, but lipids and other interfering compounds that remain in extract solutions make UV detection unworkable for muscle foods. Although vitamin K does not fluoresce, the quinones are reduced to hydroquinones by the addition of zinc chloride using a postcolumn zinc metal reduction column, so that fluorescence detection provides a highly specific detection system for vitamin K determination in foods. Electrochemical detection is also used by adding an elec-trolyte such as sodium acetate or perchlorate in the mobile phase to support the conductivity. For the determination of phylloquinone and menaquinones in animal products, Koivu-Tikkanen et al. [36] extracted the samples with 2-propanol/hexane after adding internal standards. Sample extracts were purified by normal-phase HPLC and the fraction containing vitamin K was ana-lyzed by reversed-phase HPLC using postcolumn reduction and fluorescence detection. For the determination of phylloquinone, dihydrophylloquinone, and MK-4 in meats, Elder et al. [37] used a reversed-phase HPLC procedure with silica solid-phase extraction and postcolumn reduc-tion. The vitamin K composition of foods is commonly expressed as micrograms of vitamin K.

24.2 Water-Soluble Vitamins

24.2.1 Thiamin (Vitamin B_1)

Thiamin, known as vitamin B_1, is one of the B-vitamins. Thiamin exists in interconvertible phos-phorylated forms in nature: thiamin monophosphate, thiamin pyrophosphate, and thiamin tri-phosphate. The major form of thiamin in muscle foods is protein-bound thiamin pyrophosphate.

Because thiamin is a water-soluble vitamin, it is found in the lean portions of muscle foods. Pork is an excellent source of thiamin, and contains 3–10 times as much thiamin per 100 g as other muscle foods. Thiamin is most stable between pH 2 and 4, and unstable at alkaline pH [41]. In alkaline solution, thiamin is readily oxidized, even at room temperature. It is the most heat-labile of the B-vitamins, with its decomposition dependent on pH and exposure time to heat. Hence, an alkaline environment during processing or cooking will promote loss of thiamin.

Several different analytical methodologies have been used to determine the thiamin content of foods. These methodologies include the fluorometric methods of AOAC Official Methods 942.23, 953.17, 957.17, and 986.27 [5], microbiological analyses [42–44], HPLC methods [45–48], and GC procedures [49,50].

Extraction procedures for the microbiological, HPLC, and GC analyses generally follow the thiochrome analysis procedures of AOAC Official Method 942.23 [5]. Because thiamin is stable under acidic conditions, hot acid hydrolysis with HCl is used to release the thiamin and thiamin phosphate esters from their associations with proteins, followed by enzyme hydrolysis of the phosphorylated thiamin to free thiamin using takadiastase, Mylase 100® (U.S. Biochemical Corp.), or α-amylase. The use of the same extraction procedure allows both thiamin and riboflavin in foods to be separated and quantitated by HPLC simultaneously.

Microbiological analyses depend on the extent of growth of a thiamin-dependent organism such as *Lactobacillus fermentum* and *L. viridescens*. *L. fermentum* is susceptible to matrix effects such as carbohydrates, fats, and some minerals in a growth medium [51], but *L. viridescens* is more specific for thiamin and not susceptible to matrix effects. Microbiological assay is still used in food analysis, but AOAC International [5] does not provide an official method for thiamin determination using microorganisms.

HPLC methods have been recently developed to allow the rapid, sensitive, and specific analysis of thiamin and its phosphorylated forms in drugs and biological materials. Thiamin is measured itself with absorbance detection, usually at 245–254 nm, or with fluorescence detection systems after conversion into thiochrome. Although absorbance detection has sensitivity for high-thiamin-containing foods, such as pork, it is not appropriate for other muscle foods. Fluorescence detection is much more sensitive than absorbance detection. Therefore, HPLC with fluorescence detection has been widely used for the determination of thiamin in foods including muscle foods. Thiamin itself does not fluoresce, so thiamin should be converted into thiochrome using reagents for alkaline oxidation by either post- or precolumn derivatization. The maximum fluorescence of thiochrome is excitation λ 365–375 nm and emission λ 425–435 nm. The mobile phase pH should be kept above 8, because the fluorescence intensity of thiochrome is pH dependent and reaches a steady state at a pH above 8 [52]. Milligrams of thiamin are frequently used for expressing the content of thiamin in foods.

24.2.2 Riboflavin (Vitamin B₂)

Riboflavin, known as vitamin B_2, is a water-soluble vitamin naturally found in foods. Riboflavin acts as an integral component of two coenzymes: flavin adenine dinucleotide (FAD) and flavin mononucleotide (FMN). Riboflavin occurs naturally in foods as free riboflavin and as the protein-bound coenzymes, FAD and FMN. Organ meats (liver, kidney, and heart) are considered rich sources of riboflavin. Riboflavin is stable to heat, acidic conditions, and oxidation if light is excluded. Riboflavin is destroyed by exposure to UV and visible light within the range of 420–560 nm. The rate of destruction is accelerated by increasing temperature and pH. Thus, riboflavin is generally stable during heat processing and normal cooking of foods if protected from light.

Several methods have been proposed for the determination of riboflavin in foods, usually involving the conversion of FAD and FMN into free riboflavin. The fluorometric methods of AOAC Official Methods 970.65 and 981.15 [5], microbiological assays [53], and HPLC methodologies using fluorescence detection [48,54–56] are used for measuring total riboflavin in foods. HPLC can separate individual free riboflavin, FAD, and FMN in foods.

To measure total riboflavin contents such as free riboflavin in foods using these methods, acid hydrolysis with autoclaving is used to release the riboflavin from association with proteins and to convert FAD and FMN into free riboflavin. However, to complete the conversion of FMN into free riboflavin, enzyme hydrolysis is required with diastatic enzymes after acid hydrolysis [57]. Combined extractions for thiamin and riboflavin assays have been usually used for food analysis [44,47,48].

AOAC Official Method 940.33 [5] is an approved microbiological method only for riboflavin in vitamin preparations. However, microbiological assay applying *L. rhamnosis* (formerly *L. casei*) has been used to determine total riboflavin content in foods. Because lactic acid bacteria utilize riboflavin and FMN, but not FAD, an acid hydrolysis step is necessary to convert FAD and FMN into free riboflavin, but enzyme hydrolysis for completing the conversion of FMN into riboflavin is not required. *L. rhamnosis* is affected by starch and fatty acids. The matrix effect by starch in the media is eliminated by acid hydrolysis. Fatty acids stimulate or inhibit the growth of *L. rhamnosis*. Hence, for fat-containing foods such as muscle foods, the fat extraction step with petroleum ether or hexane should be conducted before acid hydrolysis.

Reversed-phase HPLC with fluorescence detection is frequently utilized for chromatography of riboflavin assays in foods. Extraction procedures for total riboflavin analyses include acid and enzyme hydrolysis. Chromatography is capable of separating FMN and free riboflavin, and the concentrations of FMN and free riboflavin determined by HPLC are summed to obtain total riboflavin concentration; therefore, in this case, enzyme hydrolysis can be skipped. To quantify individual free riboflavin, FAD, and FMN in foods, Viñas et al. [56] used an extraction method utilizing acetonitrile without acid hydrolysis. Solid-phase cleanup procedures are often used before injection to remove some of the interfering materials. UV detection with reversed-phase HPLC has been used for riboflavin analyses in meats including liver at 254 nm [46]; however, fluorescence detection (excitation λ: 440–500 nm, emission λ: 520–530 nm) is more sensitive and specific for riboflavin quantitation than UV detection [46–48]. Riboflavin content in foods is commonly expressed in milligrams of riboflavin.

24.2.3 Niacin (Vitamin B_3)

Niacin is one of the water-soluble B-vitamins known as vitamin B_3. The term niacin is the generic descriptor for nicotinic acid and nicotinamide, which are essential for formation of the coenzymes, nicotinamide adenine dinucleotide (NAD), and nicotinamide adenine dinucleotide phosphate (NADP), in the body. Niacin can be biosynthesized from the amino acid tryptophan. Meats including organ meats contain large amounts of niacin and tryptophan. Nicotinic acid is found mainly in plant foods, but animal foods contain nicotinamide, which is bioavailable. In uncooked foods, niacin is present as NAD and NADP, but these nucleotides may be hydrolyzed to nicotinamide by cooking [58]. Niacin is not affected by thermal processing, light, oxygen, and pH. It is stable during processing, storage, and cooking of foods. Thus, acid or alkali hydrolysis can be used for extraction of niacin from food samples.

Niacin in foods can be determined by the colorimetric methods of AOAC Official Methods 961.14 and 975.41 [5] using the König reaction in which nicotinic acid and nicotinamide react with cyanogen bromide and the aromatic amine, sulfanilic acid. Microbiological assays [59], including AOAC Official Method 985.34 [5], and HPLC methodologies [60–63] are also currently used for determining niacin in meats.

Because niacin in muscle foods is present in free forms (nicotinic acid and nicotinamide) and bound forms (NAD and NADP), hydrolysis procedures are required. Acid hydrolysis with HCl or H_2SO_4 or alkaline hydrolysis with NaOH or $Ca(OH)_2$ is used as the initial step in niacin extraction procedures. Acid hydrolysis liberates nicotinamide from bound forms and hydrolyzes it into nicotinic acid; however, nicotinic acid, mainly distributed in cereal products and biologically unavailable niacin, is not completely liberated from bound forms by acid hydrolysis. Alkaline hydrolysis liberates most bound forms and provides a measure of total niacin in cereal products [64].

Microbiological assay is used for the determination of total niacin using *L. plantarum*, which responds to nicotinic acid, nicotinamide, nicotinuric acid, and NAD. AOAC Official Methods 944.13 and 985.34 [5] are used for the determination of total niacin concentration using *L. plantarum* in vitamin preparations and ready-to-feed milk-based infant formula, respectively. Solve et al. [65] developed an automated microplate method with *L. plantarum*, which reduced time expenditure and materials compared to conventional microbiological procedures.

HPLC determination of niacin has generally been carried out with ion-pairing or reversed-phase chromatography with UV detection. To measure total niacin in muscle foods, either acid or alkali hydrolysis is used, which liberates free niacin from bound forms. Water or methanol is used to measure only free forms of nicotinic acid and nicotinamide in muscle foods by HPLC [63,66]. Cleanup procedures such as cartridge extractions and column switching are usually performed to improve selectivity and sensitivity by eliminating interfering materials before HPLC. Most studies have used UV absorbance detection of nicotinic acid and nicotinamide at 254 or 264 nm. Fluorescence detection (excitation λ: 322 nm, emission λ: 380 nm) may be used to increase specificity and sensitivity of HPLC. Niacin is not naturally fluorescent, but fluorescent derivatives can be formed using cyanogen bromide and p-aminophenol [67]. A postcolumn UV irradiation in the presence of hydrogen peroxide and copper (II) ions also induces fluorescence [59,68]. Lombardi-Boccia et al. [47] determined niacin, thiamin, and riboflavin together in meats by using reversed-phase HPLC with a photodiode array detector after acid and enzyme hydrolysis.

Niacin content in foods is commonly expressed in milligrams of niacin equivalents (NE). Sixty milligrams of dietary tryptophan is considered equivalent to 1 mg of niacin [69]. Thus, 1 mg NE is equal to 1 mg of niacin or 60 mg of dietary tryptophan.

24.2.4 Vitamin B₆

Vitamin B_6 is a water-soluble vitamin. Vitamin B_6 consists of derivatives of 3-hydroxy-2-methylpyridine, that is, pyridoxal (PL), pyridoxine (PN), pyridoxamine (PM), and their respective 5′-phosphates (PLP, PNP, and PMP). PLP is a metabolically active B_6 vitamer. Vitamin B_6 in free and bound forms is found in a wide variety of foods including meats. PLP, bound to the apoenzyme by a Schiff base in animal tissues, is the predominant B_6 vitamer in muscle foods, which is bioavailable. PMP is also found in muscle foods. Hence, PL, PLP, PM, and PMP are determined in muscle foods as a result of interconversion of aldehyde and amine forms during processing and storage. However, PN and PNP found in plants are not detected in animal foods. Vitamin B_6 is unstable to light, both visible and UV. The vitamin is stable in acidic conditions if protected from light. The stability of vitamin B_6 during heat treatment, processing, and storage depends on the pH of the media. Losses of the vitamin increase as pH increases. PN is more stable to heat than PL and PM.

Several different methodologies have been developed to analyze vitamin B_6 in foods. These include animal growth, microbiological, enzymatic, fluorometric, GC, and HPLC assays [70].

Currently, microbiological assays and HPLC methods are often used for the determination of vitamin B_6 in foods. AOAC International provides both microbiological (Official Methods 961.15 and 985.32) and HPLC (Official Method 2004.07) methods to measure total vitamin B_6 in ready-to-feed milk-based infant formula and reconstituted infant formula, respectively [5]. Also, several HPLC methodologies have been reported for determining total vitamin B_6 as well as individual B_6 derivatives in meats and meat products [71–73].

Before being analyzed by microbiological, chromatographic, and other methods, muscle foods are usually extracted by acid hydrolysis with autoclaving in HCl or H_2SO_4 to dissociate vitamin B_6 from proteins. The phosphate esters of PNP, PLP, and PMP are also hydrolyzed by this procedure. In the AOAC microbiological methods [5] for measuring total vitamin B_6, sample foods are autoclaved with 0.055 N HCl for 5 h at 121°C. However, the AOAC liquid chromatographic method for total vitamin B_6 uses enzymatic hydrolysis using acid phosphatase followed by a reaction with glyoxylic acid in the presence of a Fe^{2+} catalyst to transform PM into PL [5]. For separation of individual B_6 vitamers, metaphosphoric acid, perchloric acid, trichloroacetic acid, and sulfosalicylic acid are used as deproteinating agents. These agents are used to preserve phosphorylated forms in foods.

The total vitamin B_6 composition in foods is usually estimated microbiologically using a turbidimetric assay. *Saccharomyces uvarum* (formerly *S. carlsbergenesis*) is the commonly used microorganism, which is also used in the AOAC method. Acid hydrolysis is necessary for determining total vitamin B_6 because the microorganism utilizes only nonphosphorylated B_6 vitamers. *S. uvarum* responds unequally to PL, PM, and PN. The growth response of *S. uvarum* to PL relative to that to PN is practically equal, but the response of the microorganism to PM is frequently 60–80% of that to PL and PN [74]. The problem of differential response of *S. uvarum* can be overcome by separating PL, PM, and PN chromatographically and analyzing each form of the vitamin. Cation-exchange chromatography on Dowex AG 50 W-X8 resin resolves the vitamers in the acid extractant and allows their individual quantitation [75,76]. *Kloechera brevis* (formerly *K. apiculata*) may be used for vitamin B_6 assay. However, the results of responses of the organism to PL, PM, and PN are not consistent [70], and *K. brevis* has not seen wide usage in food analysis.

Ion-exchange and reversed-phase HPLC with fluorescence detection has been used for quantitative determination of vitamin B_6 in foods. HPLC has the ability to separate and quantitate PL, PM, PN, and their 5'-phosphate esters, and also vitamin B_6 metabolites such as 4-pyridoxic acid. HPLC provides high resolution and high sensitivity to the B_6 vitamers. To determine total vitamin B_6 by quantitating PL, PM, and PN, hydrolysis of the phosphate esters is usually completed with a commercial phosphatase or treatment with H_2SO_4. For preservation of the phosphorylated vitamers and metabolites, deproteinizing agents are used. Because of the native fluorescence of PL, PM, PN, and their 5'-phosphorylated derivatives and a relatively low UV detection sensitivity, most of the HPLC methods have used fluorescence detection. The intensity of fluorescence among the B_6 vitamers is pH dependent. The B_6 vitamers are suited to ion exchange because of their pH-dependent ionic nature. In reversed-phase HPLC, sample components are separated according to their relative affinity for a nonpolar-bounded stationary phase and a polar mobile phase [77]. HPLC with fluorescence detection has been recommended for quantitative determination of vitamin B_6 in foods because individual vitamers can be determined [78,79], but microbiological assays are still used in total vitamin B_6 measurements in foods. The content of vitamin B_6 in foods is generally expressed in milligrams of vitamin B_6.

24.2.5 Folate

Folate is a generic term for a water-soluble vitamin and includes naturally occurring food folates and folic acid found in dietary supplements and used in food fortification. Folate can vary in structure by reduction of the pteridine moiety to dihydrofolic acid and tetrahydrofolic acid (THF). Folate exists predominantly as polyglutamyl forms of THF, which are biologically active folate coenzymes in the body. Liver is considered to be a good source of folate, but meats and meat products are poor sources compared to plant-based foods. Major folate vitamers in animal tissues are polyglutamyl forms of THF, 5-methyl-THF, and 10-formyl-THF [80,81]. The composition of 5-formyl-THF in animal tissues is low, but heating can increase it by isomerization of 10-formyl-THF [82]. Folate is sensitive to heat, acids, oxidation, and light. Folate losses during food processing and storage vary according to food matrices, oxygen availability, heating times, and forms of folate in foods. Folate is quite stable in dry products if protected from light and oxygen, but folate losses are large in water. Reducing agents such as ascorbic acid increase folate retention, whereas metals such as Fe^{2+} increase folate losses. Folic acid is generally more stable than naturally occurring folates.

Methodologies for the determination of folates in foods are microbiological assay [83] including AOAC Official Method 2004.05 for total folates in cereals and cereal foods [5]; HPLC with UV, fluorescence, or electrochemical detection [80,81,84,85]; and HPLC or GC with mass spectrometric (MS) methods [86–88] using stable isotopes. Microbiological assay for total folate determination is still the most widely used procedure. HPLC methods can measure each form of folates.

The traditional food folate extraction method includes heat treatment to release folate from its binding proteins and folate conjugase treatment to hydrolyze polyglutamyl folate to di- or monoglutamyl folate. Insufficient enzymatic deconjugation may result in underestimation of folate when measured by either microbiological or HPLC methods using single-enzyme digestion. Several studies have reported that treatment of food homogenates with α-amylase, protease, and folate conjugase (trienzyme extraction) enhances the yield of measurable folate in folate assays [89–91]. The use of α-amylase and protease allows for a more complete extraction of folate trapped in carbohydrates or proteins in foods. This trienzyme extraction method has become widely used in the extraction of folate from food samples. To prevent the destruction of labile folates during heat treatment, antioxidants, both ascorbic acid and mercaptoethanol, should be added.

For routine food analysis purposes, microbiological assay with *L. rhamnosus* (formerly *L. casei*) after extraction with folate conjugase is used for determination of folate. *L. rhamnosus* has greater capacity for response to the γ-glutamyl folate polymers compared to the other assay organisms. Although *L. rhamnosus* is the commonly used and accepted organism for folate analysis in foods, its ability to respond on an equimolar basis to metabolically active folates has been controversial [92]. Chicken pancreas conjugase is used to hydrolyze polyglutamates to di- and monoglutamyl folates, which are used by *L. rhamnosus*. AOAC Official Method 2004.05 [5] is the microbiological method that uses *L. rhamnosus* after extracting samples by the trienzyme procedure. This method can determine turbidity semiautomatically by using 96-well microtiter plates [5].

The major advantage of HPLC analysis is the ability to quantify the specific folate forms. Current HPLC systems for separating folates use either ion-pair or reversed-phase chromatography with UV, fluorometric, or electrochemical detection. Trienzyme extraction procedures are commonly used for HPLC analysis of food folates. Because HPLC systems for determining folate are able to detect only monoglutamates, human or rat plasma conjugase, not chicken pancreas conjugase, is used to deconjugate polyglutamyl folates to monoglutamyl folates. To remove interfering substances in extracts, purification is recommended with affinity chromatography using immobilized folate-binding protein or solid-phase extraction using silica-based strong anion-exchange cartridges. Owing to its sensitivity and selectivity, fluorescence detection is most commonly used, particularly for reduced folate forms. UV detection is useful in detecting the folic acid found in fortified foods, but not for naturally occurring folates due to a lack of sensitivity. UV spectra, fluorescence excitation, and emission spectra for the different forms of folates have been published by Ball [93]. Electrochemical detection has sensitivity for 5-methyl-THF; however, it has not been widely used in food analysis. Mass spectrometer with an HPLC system using stable isotope-labeled analyte standards has been used for folate detection to improve sensitivity and selectivity.

Folate content of foods is expressed in either milligrams or micrograms of naturally occurring folate and fortified folic acid in the foods or dietary folate equivalents (DFE). Micrograms of DFE are calculated based on micrograms of food folate plus fortified folic acid multiplied by the factor 1.7 [94].

24.2.6 Vitamin B_{12}

Vitamin B_{12} is a water-soluble vitamin and a family of compounds called cobalamins which includes cyanocobalamin, hydroxocobalamin, and the two coenzyme forms 5′-deoxyadenosylcobalamin (adenosylcobalamin) and methylcobalamin. Cyanocobalamin and hydroxocobalamin are used in most dietary supplements, and are converted into adenosylcobalamin and methylcobalamin in the body. Vitamin B_{12} found in nature appears to be from microbial synthesis. Meats and meat products, especially liver and kidney, are rich sources of vitamin B_{12}. In animal foods, vitamin B_{12} is derived from the animals' diets or from synthesis by the intestinal microflora. Thus, livers and kidneys of ruminants contain more vitamin B_{12} than those of nonruminants. The most prevalent forms of vitamin B_{12} in animal foods are adenosylcobalamin, methylcobalamin, and hydroxocobalamin. Vitamin B_{12} is generally stable if protected from light. Cobalamins are considered to be stable to thermal processing, but large losses of the vitamin occur by leaching into the cooking water. Cyanocobalamin is the most stable form of vitamin B_{12}. Strong alkaline and acid conditions, intense visible light, and oxidizing agents inactivate the vitamin.

The determination of total vitamin B_{12} may be performed by microbiological assays, including AOAC Official Methods 952.30 and 986.23 [5], radioisotope dilution methods [95,96], and HPLC methods [97,98]. Microbiological assay is most widely used for the determination of total vitamin B_{12} in foods. Radioassay kits for clinical samples are not useful for analysis of food samples. Radioisotope dilution methods lack selectivity because the intrinsic factor used for the assay could also bind other cobalamins or analogs [99]. Currently, these methods are not routinely used for analysis of vitamin B_{12} in foods. The HPLC method lacks the sensitivity to measure vitamin B_{12} in nonfortified food products.

The extraction procedures of the AOAC microbiological methods [5] are usually utilized for determining total vitamin B_{12} content in foods. Extraction procedures liberate cobalamins from protein and convert the naturally occurring labile forms into a single, stable form, which is cyanocobalamin or sulfitocobalamin [100]. To protect cobalamins in samples, metabisulfite or ascorbic acid is added to the extracting solutions.

L. delbrueckii subsp. *lactis* (*L. leichmannii*) is frequently used for determination of vitamin B_{12} in foods. *L. delbrueckii* has a similar response to nitritocobalamin, hydroxocobalamin, dicyanocobalamin, and sulfitocobalamin. However, adenosylcobalamin produces a greater response and methylcobalamin, a lesser growth response. If the sample extracts are exposed to light before analysis, adenosylcobalamin and methylcobalamin are completely converted into hydroxocobalamin, so that vitamin B_{12} activity can be measured accurately [101]. *L. delbrueckii* responds to deoxyribonucleosides. Treatment of the sample with alkali and heat destroys the vitamin cobalamins, leaving the deoxyribonucleosides intact; thus, the activity attributable to deoxyribonucleosides can be determined. Dilution of deoxyriboside concentrations to less than 1 μg/mL of assay solution can also eliminate the effect [102]. The vitamin B_{12} concentrations in foods are expressed in milligrams or micrograms of vitamin B_{12}.

24.2.7 *Pantothenic Acid (Vitamin B₅)*

Pantothenic acid, also known as vitamin B_5, occurs primarily bound as part of coenzyme A and acyl-carrier proteins. Pantothenic acid is found in plant- and animal-derived foods, but its contents in foods are relatively low. Liver and kidney are major sources of pantothenic acid among the muscle foods. Pantothenic acid is also a component of coenzyme A and acyl-carrier proteins, but the free form of pantothenic acid is rare in muscle foods. Pantothenic acid is stable to atmospheric oxygen and light, whereas large losses of the vitamin can occur in the blanching and boiling of foods. The stability of pantothenic acid is highly pH dependent. The vitamin is stable in slightly acidic solutions at pH 5–7.

Microbiological assay [103] is most commonly used for determining pantothenic acid in foods. Microbiological assay has been accepted by AOAC (Official Methods 945.74 and 992.07) [5] for quantification of pantothenic acid in vitamin preparations and milk-based infant formula. Other methodologies for determination of the vitamin in foods include radiometric microbiological assay [104], radioimmunoassay [105], enzyme-linked immunosorbent assay [106], optical biosensor inhibition immunoassay [107], capillary electrophoresis [108], GC-MS using stable isotope dilution [109], and HPLC methods [110–112]. Methods for the pantothenic acid determination in foods vary widely in approach, because the methodologies for pantothenic acid determination in food products remain limited by their low sensitivity and poor selectivity. And, the applications of the methods have several drawbacks, such as time-consuming, use of radioisotopes and scintillation counting, and sample derivatization procedures.

Although the microbiological assay is time-consuming and lacks specificity [113], determination of pantothenic acid in foods has most frequently been accomplished by this type of assay. The commonly used microorganism is *L. plantarum*. This microorganism does not respond to phosphopantetheine or bound forms of pantothenic acid. Owing to instability of the vitamin in acid and alkaline conditions, enzyme hydrolysis should be utilized to obtain free pantothenic acid and pantetheine. Intestinal phosphatase, which cleaves the phosphate linkage, and avian liver peptidase, which breaks the linkage between mercaptoethylamine and pantothenic acid, are used in the enzyme hydrolysis. This is also utilized in the AOAC microbiological method for milk-based infant formula [5]. Fatty acids in foods stimulate the growth of *L. plantarum*; therefore, a fat extraction step may be necessary before an enzyme hydrolysis in fat-containing foods such as muscle foods [103]. The pantothenic acid content of foods is commonly expressed in milligrams of pantothenic acid.

24.2.8 *Biotin*

Biotin is a water-soluble B-vitamin, which contains sulfur. The biotin molecule contains three asymmetric carbon atoms, and therefore eight different isomers are possible. Of these isomers, only the dextrorotatory (+) *d*-biotin possesses biotin activity as a coenzyme. Biotin is widely distributed in many foods, but its concentrations are very low compared to that of other water-soluble vitamins. Liver contains considerable amounts of biotin. Most of biotin in animal products is in a protein-bound form. Biotin is generally stable to heat, but is gradually destroyed by UV light. In strong acidic and alkaline solutions, biotin is unstable to heating. The sulfur atom in biotin is susceptible to oxidation with the formation of biotin sulfoxide and biotin sulfone during food processing, which leads to loss of biotin activity.

Microbiological assays [114], as well as protein-binding assays [115,116], and a biosensor-based immunoassay [117] have been developed for the determination of biotin content of foods.

HPLC methods [118–120] have been also used for quantification of biotin and biocytin in foods. Currently an AOAC Official Method for biotin determination does not exist [5].

The most widely used method for determination of biotin in foods is a microbiological assay using *L. plantarum*. This microorganism, which requires biotin for growth and reproduction, is incubated with diluted sample extracts. The resulting increased turbidity of the extract is measured and correlated with the biotin content of the sample. The microorganism cannot utilize bound forms of biotin including biocytin; thus, acid hydrolysis with autoclaving is required to liberate biotin completely from food samples. Although *L. plantarum* is more specific for biotin active forms than other biotin requiring organisms, *L. plantarum* responds to dethiobiotin that spares biotin and thus overestimates biotin content [121].

HPLC can separate free biotin, biotin sulfoxides, biotin sulfones, and other biotin analogs. The biotin molecule does not have enough UV absorbance and native fluorescence; thus UV and fluorescence detection are not useful. However, avidin can be labeled with a fluorescent marker and the complex used as a postcolumn derivatizing agent. The fluorescence of the labeled protein is enhanced on binding of its specific ligands, biotin and biocytin. HPLC methods with avidin-binding detection, formation of fluorescent derivatives, or usage of other detection systems, such as MS and electrochemical detection, increase the sensitivity of the detection after HPLC separation. Analytical results are expressed in micrograms of biotin.

24.2.9 Vitamin C

Vitamin C is a water-soluble vitamin and occurs in two forms, the reduced ascorbic acid and the oxidized dehydroascorbic acid. Ascorbic acid is reversibly oxidized to dehydroascorbic acid. Further oxidations convert dehydroascorbic acid into the inactive and irreversible compound diketoglutamic acid. There are two enantiomeric pairs, L- and D-ascorbic acid and L- and D-isoascorbic acid. L-Ascorbic acid and D-isoascorbic acid (known as D-araboascorbic acid and erythobic acid) have the biological activity of vitamin C. D-Isoascorbic acid has 1/20 the activity of L-ascorbic acid [122]. L-Ascorbic acid and dehydroascorbic acid are naturally occurring forms of vitamin C, but D-isoascorbic acid is not found in foods. D-Isoascorbic acid is synthesized commercially and used as an antioxidant in foods, usually processed and canned meats. Muscle meats except organ meats do not contain significant amounts of vitamin C. Liver, kidney, and heart are fair to good sources of the vitamin. Vitamin C is very susceptible to oxidation during the processing, storage, and cooking of foods, especially under alkaline conditions. Vitamin C losses by cooking are dependent on the degree of heating, surface area exposed to water and oxygen, pH, and leaching into the cooking medium.

Methodologies for determination of total vitamin C in foods are the titrimetric method using oxidation–reduction indicators [5], the fluorometric method including derivatization procedures [5], enzymatic methods [123], and electrochemical procedures [124]. Total vitamin C content is also usually determined by HPLC using UV, fluorescence, or electrochemical detection [125,126]. Only HPLC procedures simultaneously separate L-ascorbic acid, dehydroascorbic acid, and D-isoascorbic acid in foods [127–130]. The titrimetric method employed in AOAC Official Methods 967.21 and 985.33 [5] uses 2,6-dichloroindophenol in measuring ascorbic acid, not dehydroascorbic acid. This method cannot distinguish between L-ascorbic acid and D-isoascorbic acid; thus, titrimetric method cannot be used for processed meat products containing D-isoascorbic acid. The fluorometric method of AOAC Official Methods 967.22 and 984.26 [5] determines total vitamin C content in vitamin preparations and foods, respectively, by derivatization of dehydroascorbic acid using the *o*-phenylenediamine condensation reaction.

Because vitamin C is destroyed easily, extraction procedures should be conducted to stabilize the vitamin. Although the choice of extracting solution is dependent on the sample matrix and determination method, the solutions should maintain an acidic environment, chelate metals, inactivate ascorbic acid oxidase, and precipitate proteins and starches [131]. The extracting solution usually is 3–6% metaphosphoric acid dissolved in 8% glacial acetic acid or ethylenediaminetetraacetic acid (EDTA). Metaphosphoric acid prevents metal catalysis and activation of ascorbic acid oxidase, and precipitates proteins. EDTA chelates metals, and acetic acid precipitates starches in extractants.

HPLC methodologies are widely used for determining ascorbic acid and its degradation products in foods. Reversed-phase chromatography with and without ion suppression, and ion-pair reversed-phase chromatography with C_{18} columns, and ion exchange chromatography are currently employed for analysis of vitamin C. UV, electrochemical, and fluorescence detection are commonly used for quantitation of L-ascorbic acid, dehydroascorbic acid, and D-isoascorbic acid in foods. UV detection is not suitable for low vitamin C content foods due to poor sensitivity of dehydroascorbic acid. Dehydroascorbic acid is electrochemically inactive; therefore, to measure total vitamin C contents, it is reduced to L-ascorbic acid before electrochemical detection. Fluorescence detection is accomplished after *o*-phenylenediamine derivatization using pre- or postcolumns. Both electrochemical and fluorescence detection have excellent selectivity and sensitivity to vitamin C analysis by HPLC. The vitamin C contents in foods are commonly expressed in milligrams of vitamin C.

24.3 Summary

Vitamins are classified according to their solubility in fat solvents (fat-soluble vitamins) or water (water-soluble vitamins). The solubility properties are related to the distribution of vitamins in foods as well as the analytical methods employed. Although muscle foods contain limited amounts of fat-soluble vitamins, they are good to excellent dietary sources of most of the B-vitamins. Owing to their solubility in water, B-vitamins are found principally in the lean portions of the meat. Vitamins are generally susceptible to oxidation, heat, pH, moisture, light, degradative enzymes, and metal trace elements.

To liberate vitamins bound in lipid or protein fractions, food samples may need to be hydrolyzed using acids, alkalines, and enzymes, or extracted directly with solvents without hydrolysis. The extract solutions may require some forms of cleanup before the vitamins are measured, removing interfering substances and improving the sensitivity and selectivity of the analytical methods. Most of the vitamins are very liable to light, and therefore food samples must be protected from light throughout the analysis. Antioxidants such as BHT, pyrogallol, or ascorbic acid are frequently added into extraction solvents for preventing oxidation and conversion of vitamins.

Various methodologies, including colorimetric, fluorometric, titrimetric, and spectrophotometric methods, have been developed and used for determining the vitamins in foods. However, microbiological and HPLC methods are most frequently used in estimating the naturally occurring vitamins in foods, because these methods have sufficient sensitivity and selectivity to quantitate low concentrations of naturally occurring vitamins. Microbiological assay can be applied to all the B-vitamins. Lactic acid bacteria are suitable for determining B-vitamins turbidimetrically, except for vitamin B_6, which may be determined using yeast. Fat-soluble vitamins and vitamin C are most commonly determined by HPLC. HPLC methods can also be used for estimating the B-vitamins. HPLC distinguishes between naturally occurring and added (fortified or enriched) vitamins, and also separates the individual forms of vitamins. Some vitamins having low UV absorbance or fluorescence responses can be determined after conversion into fluorescent derivatives using pre- or postcolumn derivatization. Biospecific methods for determining some of the water-soluble vitamins include immunoassays and protein-binding assays. Newer techniques continue to be developed for quantitating the concentrations of the various vitamins in foods of all types, including muscle foods.

References

1. U.S. Department of Agriculture, Agricultural Research Service, USDA National Nutrient Database for Standard Reference, Release 19, Nutrient Data Laboratory, 2006.
2. Kurtzweil, P., 'Daily Values' Encourage Healthy Diet, in *Focus on Food Labeling*, U.S. Food and Drug Administration, Rockville, MD, 1993, updated 2003.
3. FDA/CFSAN, *A Food Labeling Guide: Appendix B Relative or Comparative Claims*, U.S. Department of Health and Human Services, Washington, DC, 1994.
4. Carlson, C.W., Greaser, M.L., and Jones, K.W., *The Meat We Eat*, Interstate Publishers, Inc., Danville, IL, 2001, chap. 23.
5. *Official Methods of Analysis of AOAC International*, 18th ed., revision 1, AOAC International, Gaithersburg, MD, 2006.
6. Driskell, J.A., Marchello, M.J., Giraud, D.W., and Sulaeman, A., Vitamin and selenium content of ribeye cuts from grass- and grain-finished bison of the same herd, *J. Food Qual.*, 27, 3, 2004.
7. Epler, K.S., Ziegler, R.G., and Craft, N.E., Liquid chromatographic method for the determination of carotenoids, retinoids and tocopherols in human serum and in food, *J. Chromatogr.*, 619, 37, 1993.
8. Kang, K.R., Cherian, G., and Sim, J.S., Tocopherols, retinol and carotenes in chicken egg and tissues as influenced by dietary palm oil, *J. Food Sci.*, 63, 592, 1998.
9. Kimura, M., Rodriguez-Amaya, D.B., and Godoy, H.T., Assessment of the saponification step in the quantitative determination of carotenoids and provitamin A, *Food Chem.*, 35, 187, 1990.
10. Sander, L.C., Sharpless, K.E., Craft, N.E., and Wise, S.A., Development of engineered stationary phases for the separation of carotenoid isomers, *Anal. Chem.*, 66, 1667, 1994.
11. Emenhiser, C., Simunovic, N., Sander, L.C., and Schwartz, S.J., Separation of geometrical carotenoid isomers in biological extracts using a polymeric C_{30} column in reversed-phase liquid chromatography, *J. Agric. Food Chem.*, 44, 3887, 1996.
12. Institute of Medicine, *Dietary Reference Intakes for Vitamin A, Vitamin K, Arsenic, Boron, Chromium, Copper, Iodine, Iron, Manganese, Molybdenum, Nickel, Silicon, Vanadium, and Zinc*, National Academy Press, Washington, DC, 2001, chap. 4.
13. Institute of Medicine, *Dietary Reference Intakes for Calcium, Phosphorous, Magnesium, Vitamin D, and Fluoride*, National Academy Press, Washington, DC, 1997, chap. 7.
14. Ball, G.F.M., *Vitamins in Foods: Analysis, Bioavailability, and Stability*, CRC Press, Boca Raton, FL, 2006, chap. 4.
15. Purchas, R., Zou, M., Pearce, P., and Jackson, F., Concentrations of vitamin D_3 and 25-hydroxyvitamin D_3 in raw and cooked New Zealand beef and lamb, *J. Food Comp. Anal.*, 20, 90, 2007.
16. Jakobsen, J., Clausen, I., Leth, T., and Ovensen, L., A new method for the determination of vitamin D_3 and 25-hydroxyvitamin D_3 in meat, *J. Food Comp. Anal.*, 17, 777, 2004.
17. Clausen, I., Jakobsen, J., Leth, T., and Ovesen, L., Vitamin D_3 and 25-hydroxyvitamin D_3 in raw and cooked pork cuts, *J. Food Comp. Anal.*, 16, 575, 2003.
18. Mattila, P., Piironen, V., Bäckman, C., Asunmaa, A., Uusi-Rauva, E., and Koivistoinen, P., Determination of vitamin D_3 in egg yolk by high-performance liquid chromatography with diode array detection, *J. Food Comp. Anal.*, 5, 281, 1992.
19. Bui, M.H., Sample preparation and liquid chromatographic determination of vitamin D in food products, *J. Assoc. Off. Anal. Chem.*, 70, 802, 1987.

20. Thompson, J.N., and Plouffe, L., Determination of cholecalciferol in meat and fat from livestock fed normal and excessive quantities of vitamin D, *Food Chem.*, 46, 313, 1993.

21. Sliva, M.G., Green, A.E., Sanders, J.K., Euber, J.R., and Saucerman, J.R., Reversed-phase liquid chromatographic determination of vitamin D in infant formulas and enteral nutritionals, *J. AOAC Int.*, 75, 566, 1992.

22. Wulf, D.M., Morgan, J.B., Sanders, S.K., Tatum, J.B., Smith, G.C., and Williams, S., Effect of dietary supplementation of vitamin E on storage and caselife properties of lamb retail cuts, *J. Anim. Sci.*, 73, 399, 1995.

23. Engeseth, N.J., Gray, J.I., Booren, A.M., and Asghar, A., Improved oxidative stability of veal lipids and cholesterol through dietary vitamin E supplementation, *Meat Sci.*, 35, 1, 1993.

24. Monahan, F.J., Buckley, D.J., Gray, J.I., and Morrissey, P.A., Effect of dietary vitamin E on the stability of raw and cooked pork, *Meat Sci.*, 27, 99, 1990.

25. Bosco, A.D., Castellini, C., and Bernardini, M., Nutritional quality of rabbit meat as affected by cooking procedure and dietary vitamin E, *J. Food Sci.*, 66, 1047, 2001.

26. Chun, J., Lee, J., Ye, L., Exler, J., and Eitenmiller, R.R., Tocopherol and tocotrienol contents of raw and processed fruits and vegetables in the United States diet, *J. Food Comp. Anal.*, 19, 196, 2006.

27. Bonvehi, J.S., Coll, F.V., and Rius, I.A., Liquid chromatographic determinations of tocopherols and tocotrienols in vegetable oils, formulated preparations, and biscuits, *J. AOAC Int.*, 83, 627, 2000.

28. Thompson, J.N., and Hatina, G., Determination of tocopherols and tocotrienols in foods and tissues by high-performance liquid chromatography, *J. Liq. Chromatogr.*, 2, 327, 1979.

29. Nelis, H.J., D'Haese, E., and Vermis, K., Vitamin E, in *Modern Chromatographic Analysis of Vitamins*, 3rd ed., De Leenheer, A.P., Lambert, W.E., and van Bocxlaer, J.F., Eds., Marcel Dekker, New York, NY, 2000, chap. 2.

30. National Research Council, *Recommended Dietary Allowance*, 10th ed., National Academy of Sciences, Washington, DC, 1989, p. 78.

31. Institute of Medicine, *Dietary Reference Intakes for Vitamin C, Vitamin E, Selenium, and Carotenoids*, National Academy Press, Washington, DC, 2000, chap. 6.

32. Ferland, G., Vitamin K, in *Present Knowledge in Nutrition*, 9th ed., Bowman, B.A., and Russell, R.M., Eds., ILSI Press, Washington, DC, 2006, chap. 16.

33. Thijssen, H., and Drittij-Reijnders, J., Vitamin K distribution in rat tissues: Dietary phylloquinone is a source of tissue menaquinone-4, *Br. J. Nutr.*, 72, 415, 1994.

34. Davidson, R.T., Foley, A.L., Engelke, J.A., and Suttie, J.W., Conversion of dietary phylloquinone to tissue menaquinone-4 in rats is not dependent on gut bacteria, *J. Nutr.*, 12, 220, 1998.

35. Suttie, J., The importance of menaquinones in human nutrition, *Annu. Rev. Nutr.*, 15, 399, 1994.

36. Koivu-Tikkanen, T.J., Ollilainen, V., and Piironen, V., Determination of phylloquinone and menaquinones in animal products with fluorescence detection after postcolumn reduction with metallic zinc, *J. Agric. Food Chem.*, 48, 6325, 2000.

37. Elder, S.J., Haytowitz, D.B., Howe, J., Peterson, J.W., and Booth, S.L., Vitamin K contents of meat, dairy, and fast food in the U.S. diet, *J. Agric. Food Chem.*, 54, 463, 2006.

38. Weizmann, N., Peterson, J.W., Haytowitz, D., Pehrsson, P.R., de Jesus, V.P., and Booth, S.L., Vitamin K content of fast foods and snack foods in the US diet, *J. Food Comp. Anal.*, 17, 379, 2004.

39. Booth, S.L., Sadowski, J.A., and Pennington, A.T., Phylloquinone (vitamin K_1) content of foods in the U.S. Food and Drug Administration's Total Diet Study, *J. Agric. Food Chem.*, 43, 1574, 1995.

40. Eitenmiller, R.R., and Landen, W.O. Jr., *Vitamin Analysis for the Health and Food Sciences*, CRC Press, Boca Raton, FL, 1999, chap. 4.

41. Bates, C.J., Thiamin, in *Present Knowledge in Nutrition*, 9th ed., Bowman, B.A., and Russell, R.M., Eds., ILSI Press, Washington, DC, 2006, chap. 18.

42. Defibaugh, P.W., Smith, J.S., and Weeks, C.E., Assay of thiamin in foods using manual and semiautomated fluorometric and microbiological methods, *J. Assoc. Off. Anal. Chem.*, 60, 552, 1977.

43. Bui, M.H., A microbiological assay on microtitre plates of thiamine in biological fluids and foods, *Int. J. Vitam. Nutr. Res.*, 69, 362, 1999.

44. Ollilainen, V., Finglas, P.M., van den Berg, H., and de Froidmont-Görtz, I., Certification of B-group vitamins (B$_1$, B$_2$, B$_6$, and B$_{12}$) in four food reference materials, *J. Agric. Food Chem.*, 49, 315, 2001.

45. Fellman, J.K., Artz, W.E., Tassinari, P.D., Cole, C.L., and Augustin, J., Simultaneous determination of thiamin and riboflavin in selected foods by high-performance liquid chromatography, *J. Food Sci.*, 47, 2048, 1982.

46. Barna, É., and Dworschák, E., Determination of thiamine (vitamin B$_1$) and riboflavin (vitamin B$_2$) in meat and liver by high-performance liquid chromatography, *J. Chromatogr.*, 668, 359, 1994.

47. Lombardi-Boccia, G., Lanzi, S., and Aguzzi, A., Aspects of meat quality: Trace elements and B vitamins in raw and cooked meats, *J. Food Comp. Anal.*, 18, 39, 2005.

48. Tang, X., Cronin, D.A., and Brunton, N.P., A simplified approach to the determination of thiamine and riboflavin in meats using reverse phase HPLC, *J. Food Comp. Anal.*, 19, 831, 2006.

49. Echols, R.E., Miller, R.H., Winzer, W., Carmen, D.J., and Ireland, Y.R., Gas chromatographic determination of thiamine in meats, vegetables and cereals with a nitrogen-phosphorus detector, *J. Chromatogr.*, 262, 257, 1983.

50. Echols, R.E., Miller, R.H., and Foster, W., Analysis of thiamine in milk by gas chromatography and nitrogen-phosphorus detector, *J. Dairy Sci.*, 69, 1246, 1986.

51. Voigt, M.N., and Eitenmiller, R.R., Comparative review of the thiochrome, microbial and protozoan analysis of B-vitamins, *J. Food Prot.*, 41, 730, 1978.

52. Kawasaki, T., and Egi, Y., Thiamine, in *Modern Chromatographic Analysis of Vitamins*, 3rd ed., De Leenheer, A.P., Lambert, W.E., and van Bocxlaer, J.F., Eds., Marcel Dekker, New York, NY, 2000, chap. 8.

53. Kornberg, H.A., Langdon, R.S., and Cheldelin, V.H., Microbiological assay for riboflavin, *Anal. Chem.*, 20, 81, 1948.

54. Andrés-Lacueva, C., Mattivi, F., and Tonon, D., Determination of riboflavin, flavin mononucleotide and flavin-adenine dinucleotide in wine and other beverages by high-performance liquid chromatography with fluorescence detection, *J. Chromatogr. A*, 823, 355, 1998.

55. Russell, L.F., and Vanderslice, J.T., Non-degradative extraction and simultaneous quantitation of riboflavin, flavin mononucleotide, and flavin adenine dinucleotide in foods by HPLC, *Food Chem.*, 43, 151, 1992.

56. Viñas, P., Balsalobre, N., López-Erroz, C., and Hernández-Córdoba, M., Liquid chromatographic analysis of riboflavin vitamers in food using fluorescence detection, *J. Agric. Food Chem.*, 52, 1789, 2004.

57. Nielsen, P., Rauschenbach, P., and Bacher, A., Preparation, properties, and separation by high-performance liquid chromatography of riboflavin phosphates, *Methods Enzymol.*, 122, 209, 1986.

58. Ball, G.F.M., *Vitamins in Foods: Analysis, Bioavailability, and Stability*, CRC Press, Boca Raton, FL, 2006, chap. 9.

59. Rose-Sallin, C., Blake, C.J., Genoud, D., and Tagliaferri, E.G., Comparison of microbiological and HPLC-fluorescence detection methods for determination of niacin in fortified food products, *Food Chem.*, 73, 473, 2001.

60. Tyler, T.A., and Genzale, J.A., Liquid chromatographic determination of total niacin in beef, semolina and cottage cheese, *J. Assoc. Off. Anal. Chem.*, 73, 467, 1990.

61. Ward., C.M., and Trenerry, V.C., The determination of niacin in cereals, meat and selected foods by capillary electrophoresis and high performance liquid chromatography, *Food Chem.*, 60, 667, 1997.

62. Vidal-Valverde, C., and Reche, A., Determination of available niacin in legumes and meat by high performance liquid chromatography, *J. Agric. Food Chem.*, 39, 116, 1991.

63. Hamano, T., Mitsuhashi, Y., Aoki, N., Yamamoto, S., and Oji, Y., Simultaneous determination of niacin and niacinamide in meats by high performance liquid chromatography, *J. Chromatogr.*, 457, 403, 1988.

64. Ball, G.F.M., *Vitamins in Foods: Analysis, Bioavailability, and Stability*, CRC Press, Boca Raton, FL, 2006, chap. 16.

65. Solve, M., Eriksen, H., and Brogren, C.H., Automated microbiological assay for quantitation of niacin performed in culture microplates read by digital image processing, *Food Chem.*, 49, 419, 1994.

66. Takatsuki, K., Suzuki, S., Sato, M., Sakai, K., and Ushizawa, I., Liquid chromatographic determination of free and added niacin and niacinamide in beef and pork, *J. Assoc. Anal. Chem.*, 70, 69, 1987.
67. Krishnan, P.G., Mahmud, I., and Matthees, D.P., Postcolumn fluorimetric HPLC procedure for determination of niacin content of cereal, *Cereal Chem.*, 76, 512, 1999.
68. Lahél, S., Bergaentzlé, M., and Hasselmann, C., Fluorimetric determination of niacin in foods by high-performance liquid chromatography with post-column derivatization, *Food Chem.*, 65, 129, 1999.
69. Horwitt, M.K., Harper, A.E., and Henderson, L.M., Niacin-tryptophan relationships for evaluating niacin equivalents, *Am. J. Clin. Nutr.*, 34, 423, 1981.
70. Driskell, J.A., Vitamin B$_6$, in *Encyclopedia of Food Sciences and Nutrition*, 2nd ed., Caballero, B.L., Trugo, L., and Ginglas, P.M., Eds., Elsevier, Kent, UK, 2003, p. 6012.
71. Valls, F., Sancho, M.T., Fernández-Muiño, M.A., and Checa, M.A., Determination of vitamin B$_6$ in cooked sausages, *J. Agric. Food Chem.*, 49, 38, 2001.
72. Esteve, M.J., Farré, R., Frígola, A., and Carcía-Cantabella, J.M., Determination of vitamin B$_6$ (pyridoxamine, pyridoxal and pyridoxine) in pork meat and pork meat products by liquid chromatography, *J. Chromatogr. A*, 795, 383, 1998.
73. van Schoonhoven, J., Schrijver, J., van den Berg, H., and Haenen, G.R.M.M., Reliable and sensitive high-performance liquid chromatographic method with fluorometric detection for the analysis of vitamin B$_6$ in foods and feeds, *J. Agric. Food Chem.*, 42, 1475, 1994.
74. Gregory, J.F. III, Relative activity of the nonphosphorylated B-6 vitamers for *Saccharomyces uvarum* and *Kloeckera brevis* in vitamin B-6 microbiological assay, *J. Nutr.*, 112, 1643, 1982.
75. Toepfer, E.W., and Polansky, M.M., Microbiological assay of vitamin B-6 and its components, *J. Assoc. Off. Anal. Chem.*, 53, 546, 1970.
76. Toepfer, E.W., and Lehmann, J., Procedure for chromatographic separation and microbiological assay of pyridoxine, pyridoxal, and pyridoxamine in food extracts, *J. Assoc. Off. Anal. Chem.*, 44, 426, 1961.
77. Gregory, J.F. III, Methods for determination of vitamin B$_6$ in foods and other biological materials: A critical review, *J. Food Comp. Anal.*, 1, 105, 1988.
78. Gregory, J.F., and Feldstein, D., Determination of vitamin B$_6$ in foods and other biological materials by paired ion high performance liquid chromatography, *J. Agric. Food Chem.*, 33, 359, 1985.
79. Morrison, L.A., and Driskell, J.A., Quantities of B$_6$ vitamers in human milk by high-performance liquid chromatography: Influence of maternal vitamin B$_6$ status, *J. Chromatogr. Biomed. Appl.*, 337, 249, 1985.
80. Vahteristo, L.T., Ollilainen, V., and Varo, P., Liquid chromatographic determination of folate monoglutamates in fish, meat, egg, and dairy products, *J. AOAC Int.*, 80, 373, 1997.
81. Vahteristo, L.T., Ollilainen, V., and Varo, P., HPLC determination of folate in liver and liver products, *J. Food Sci.*, 61, 524, 1996.
82. Gregory, J.F. III, Chemical and nutritional aspects of folate research: Analytical procedures, methods of folate synthesis, stability, and bioavailability of dietary folate, *Adv. Food Nutr. Res.*, 33, 1, 1989.
83. Hyun, T.H., and Tamura, T., Trienzyme extraction in combination with microbiologic assay in food folate analysis: An updated review, *Exp. Biol. Med.*, 230, 444, 2005.
84. Póo-Prieto, R., Haytowitz, D.B., Holden, J.M., Rogers, G., Choumenkovitch, S.F., Jacques, P.F., and Selhub, J., Use of the affinity/HPLC method for quantitative estimation of folic acid in enriched cereal-grain products, *J. Nutr.*, 136, 3079, 2006.
85. Konings, E.J.M., Roomans, H.H.S., Dorant, E., Goldbohm, R.A., Saris, W.H.M., and van den Brandt, P.A., Folate intake of the Dutch population according to newly established liquid chromatography data for foods, *Am. J. Clin. Nutr.*, 73, 765, 2001.
86. Rychlik, M., and Freisleben, A., Quantification of pantothenic acid and folates by stable isotope dilution assays, *J. Food Comp. Anal.*, 15, 399, 2002.
87. Stokes, P., and Webb, K., Analysis of some folate monoglutamates by high-performance liquid chromatography-mass spectrometry, *J. Chromatogr.*, 864, 59, 1999.

88. Santhosh-Kumar, C.R., and Kolhouse, N.M., Molar quantitation of folates by gas chromatography-mass spectrometry, *Methods Enzymol.*, 281, 26, 1997.

89. Tamura, T., Mizuno, Y., Johnston, K.E., and Jacob, R.A., Food folate assay with protease, α-amylase, and folate conjugase treatments, *J. Agric. Food Chem.*, 45, 135, 1997.

90. Johnston, K.E., DiRienzo, D.B., and Tamura, T., Folate concentrations of dairy products measured by microbiological assay with trienzyme treatment, *J. Food Sci.*, 67, 17, 2001.

91. Yon, M., and Hyun, T.H., Folate content of foods commonly consumed in Korea measured after trienzyme extraction, *Nutr. Res.*, 23, 735, 2003.

92. Ball, G.F.M., *Vitamins in Foods: Analysis, Bioavailability, and Stability*, CRC Press, Boca Raton, FL, 2006, chap. 18.

93. Ball, G.F.M., *Vitamins in Foods: Analysis, Bioavailability, and Stability*, CRC Press, Boca Raton, FL, 2006, chap. 21.

94. Institute of Medicine, *Dietary Reference Intakes for Thiamin, Riboflavin, Niacin, Vitamin B_6, Folate, Vitamin B_{12}, Pantothenic acid, Biotin, and Choline*, National Academy Press, Washington, DC, 2000, chap. 8.

95. Casey, P.J., Speckman, K.R., Ebert, F.J., and Hobbs, W.E., Radioisotope dilution technique for determination of vitamin B_{12} in foods, *J. Assoc. Off. Anal. Chem.*, 65, 85, 1982.

96. Osterdahl, B.G., and Johanosson, E., Radioisotope dilution determination of vitamin B_{12} in dietary supplements, *Int. J. Vitam. Nutr. Res.*, 58, 300, 1988.

97. Heudi, O., Kilinç, T., Fontannaz, P., and Marley, E., Determination of vitamin B_{12} in food products and in premixes by reversed-phase high performance liquid chromatography and immunoaffinity extraction, *J. Chromatogr. A*, 1101, 63, 2006.

98. Choi, Y.J., Jang, J.H., Park, H.K., Koo, Y.E., Hwang, I.K., and Kim, D.B., Determination of vitamin B_{12} (cyanocobalamin) in fortified foods by HPLC, *J. Food Sci. Nutr.*, 8, 301, 2003.

99. Indyk, H.E., Persson, B.S., Caselunghe, M.C., Moberg, A., Filonzi, E.L., and Woollard, D.C., Determination of vitamin B_{12} in milk products and selected foods by optical biosensor protein-binding assay: Method comparison, *J. AOAC Int.*, 85, 72, 2002.

100. Ball, G.F.M., *Vitamins in Foods: Analysis, Bioavailability, and Stability*, CRC Press, Boca Raton, FL, 2006, chap. 17.

101. Muhammad, K., Briggs, D., and Jones, G., The appropriateness of using cyanocobalamin as calibration standard in *Lactobacillus leichmannii* ATCC 7830 assay of vitamin B_{12}, *Food Chem.*, 4, 427, 1993.

102. Skeggs, H.R., *Lactobacillus leichmannii* assay for vitamin B_{12}, in *Analytical Microbiology*, Kavanagh, F., Ed., Academic Press, New York, 1963, chap. 7.

103. Skeggs, H.R., and Wright, L.D., The use of *Lactobacillus arabinosus* in the microbiological determination of pantothenic acid, *J. Biol. Chem.*, 156, 21, 1944.

104. Guilarte, T.R., Radiometric microbiological assay of B vitamins. II, Extraction methods, *J. Nutr. Biochem.*, 2, 399, 1991.

105. Walsh, J.H., Wyse, B.W., and Hansen, R.G., A comparison of microbiological and radioimmunoassay methods for the determination of pantothenic acid in foods, *J. Food Biochem.*, 3, 175, 1979.

106. Gonthier, A., Boullanger, P., Fayol, V., and Hartmann, D.J., Development of an ELISA for pantothenic acid (vitamin B_5) for application in the nutrition and biological fields, *J. Immunoassay*, 19, 167, 1998.

107. Haughey, S.A., O'Kane, A.A., Baxter, G.A., Kalman, A., Trisconi, M.J., Indyk, H.E., and Watene, G.A., Determination of pantothenic acid in foods by optical biosensor immunoassay, *J. AOAC Int.*, 88, 1008, 2005.

108. Kodama, S., Yamamoto, A., and Matsunaga, A., Direct chiral resolution of pantothenic acid using 2-hydroxypropyl-β-cyclodextrin in capillary electrophoresis, *J. Chromatogr. A*, 811, 269, 1998.

109. Rychlik, M., Quantification of free and bound pantothenic acid in foods and blood plasma by a stable isotope dilution assay, *J. Agric. Food Chem.*, 48, 1175, 2000.

110. Pakin, C., Bergaentzlé, M., Hubscher, V., Aoudé-Werner, D., and Hasselmann, C., Fluorimetric determination of pantothenic acid in foods by liquid chromatography with post-column derivatization, *J. Chromatogr. A*, 1035, 7, 2004.

111. Mittermayr, R., Kalman, A., Trisconi, M.-J., and Heudi, O., Determination of vitamin B_5 in a range of fortified food products by reversed-phase liquid chromatography-mass spectrometry with electrospray ionization, *J. Chromatogr. A*, 1032, 1, 2004.

112. Woollard, D.C., Indyk, H.E., and Christiansen, S.K., The analysis of pantothenic acid in milk and infant formulas by HPLC, *Food Chem.*, 69, 201, 2000.

113. Tanner, J.T., Barnett, S.A., and Mountford, M.K., Analysis of milk-based infant formula. Phase V. Vitamins A and E, folic acid, and pantothenic acid: Food and Drug Administration-Infant Formula Council: Collaborative study, *J. AOAC Int.*, 76, 399, 1993.

114. Livaniou, E., Costopoulou, D., Vassiliadou, I., Leondiadis, L., Nyalala, J.O., Ithakissios, D.S., and Evangelatos, G.P., Analytical techniques for determining biotin, *J. Chromatogr. A*, 881, 331, 2000.

115. Bitsch, R., Salz, I., and Hötzel, D., Biotin assessment in foods and body fluids by a protein binding assay (PBA), *Int. J. Vitam. Nutr. Res.*, 59, 59, 1989.

116. Reyes, F.D., Romero, J.M.F., and de Castro, M.D.L., Determination of biotin in foodstuffs and pharmaceutical preparations using a biosensing system based on the streptavidin-biotin interaction, *Anal. Chim. Acta*, 436, 109, 2001.

117. Indyk, H.E., Evans, E.A., Caselunghe, M.C.B., Persson, B.S., Finglas, P.M., Woollard, D.C., and Filonzi, E.L., Determination of biotin and folate in infant formula and milk by optical biosensor-based immunoassay, *J. AOAC Int.*, 83, 1141, 2000.

118. Staggs, C.G., Sealey, W.M., McCabe, B.J., Teague, A.M., and Mock, D.M., Determination of the biotin content of select foods using accurate and sensitive HPLC/avidin binding, *J. Food Comp. Anal.*, 17, 767, 2004.

119. Nelson, B.C., Sharpless, K.E., and Sander, L.C., Improved liquid chromatography methods for the separation and quantification of biotin in NIST Standard Reference Material 3280: Multivitamin/multielement tablets, *J. Agric. Food Chem.*, 54, 8710, 2006.

120. Höller, U., Wachter, F., Wehrli, C., and Fizet, C., Quantification of biotin in feed, food, tablets, and premises using HPLC-MS/MS, *J. Chromatogr. B*, 31, 8, 2006.

121. Eitenmiller, R.R., and Landen, W.O. Jr., *Vitamin Analysis for the Health and Food Sciences*, CRC Press, Boca Raton, FL, 1999, chap. 12.

122. McDowell, L.R., *Vitamins in Animal and Human Nutrition*, 2nd ed., Iowa State University Press, Ames, IA, 2000, chap. 15.

123. Tsumura, F., Ohsako, Y., Haraguchi, Y., Kumagai, H., Sakurai, H., and Ishii, K., Rapid enzymatic assay for ascorbic acid in various foods using peroxidase, *J. Food Sci.*, 58, 619, 1993.

124. Davey, M.W., Bauw, G., and van Montagu, M., Analysis of ascorbate in plant tissue by high performance capillary zone electrophoresis, *Anal. Biochem.*, 239, 8, 1996.

125. Dodson, K.Y., Young, E.R., and Soliman, A.G.M., Determination of total vitamin C in various food matrixes by liquid chromatography and fluorescence detection, *J. AOAC Int.*, 75, 87, 1992.

126. Brause, A.R., Woollard, D.C., and Indyk, H.E., Determination of total vitamin C in fruit juices and related products by liquid chromatography: Interlaboratory study, *J. AOAC Int.*, 86, 367, 2003.

127. Hidiroglou, N., Madere, R., and Behrens, W., Electrochemical determination of ascorbic acid and isoascorbic acid in ground meat and in processed foods by high pressure liquid chromatography, *J. Food Comp. Anal.*, 11, 89, 1998.

128. Nisperos-Carriedo, M.O., Buslig, B.S., and Shaw, P.E., Simultaneous detection of dehydroascorbic, ascorbic, and some organic acid in fruits and vegetables by HPLC, *J. Agric. Food Chem.*, 40, 1127, 1992.

129. Kutnink, M.A., and Omaye, S.T., Determination of ascorbic acid, erythorbic acid, and uric acid in cured meats by high-performance liquid chromatography, *J. Food Sci.*, 52, 53, 1987.

130. Doner, L.W., and Hicks, K.B., High-performance liquid chromatographic separation of ascorbic acid, erythobic acid, dehydroascorbic acid, dehydroerythobic acid, diketogulonic acid, and diketogluconic acid, *Anal. Biochem.*, 115, 225, 1981.

131. Eitenmiller, R.R., and Landen, W.O. Jr., *Vitamin Analysis for the Health and Food Sciences*, CRC Press, Boca Raton, FL, 1999, chap. 6.

Chapter 25

Minerals and Trace Elements

A. Alegría, R. Barberá, M.J. Lagarda, and R. Farré

Contents

25.1 Introduction

Muscle foods (meat, poultry, and seafood) constitute an important part of the human diet, and their mineral content influences the quality of the final product and hence the acceptance by the consumers. Meat contains all mineral substances required by the human body, and is a major dietary source of iron, zinc, phosphorus, and magnesium. It is relatively poor in sodium, potassium, and calcium. Although there are no significant differences among animal species, veal and game are known to be rich in phosphorus. Nevertheless, the nutritional importance of foods of animal origin lies in their essential trace element contents, especially iron that is present in the form of heme (hemoglobin and muscle myoglobin) and hence accounts for its high bioavailability in red meats. No other food contributes such high iron bioavailability. Moreover, owing to its color heme iron helps to improve the organoleptic properties of meat [1,2]. Heme iron accounts for approximately 50–60% of the iron in beef, lamb, and chicken, and 30–40% of that in pork, liver, and fish. Humans absorb heme iron at a rate that is about 5–10 times greater than that for nonheme iron. In addition, the mixing of heme iron with sources of nonheme iron (e.g., small amounts of meat, poultry, and fish in a vegetable mixture) improves the absorption of nonheme iron [3]. Beef is richer in iron than veal or pork, whereas in avian species iron is more abundant in turkey than in chicken. The internal organs such as the liver (particularly pork liver), spleen, and blood products contain more iron than lean meat [1,2].

Meat contributes 50–70% of ingested zinc, which (particularly in beef) is more bioavailable than the zinc from vegetables. Absorption of zinc may be as high as 40% from meat, poultry, and fish. Nearly half of the amount of zinc recommended for adult women can be met by only three ounces of lean beef [3]. Despite the low selenium content of meat, its bioavailability is high because the element is present in the form of selenomethionine and selenocysteine [1,2].

Fish is also an important source of phosphorus, iron, and copper. The edible portion of fish has high iodine content due to the seaweeds consumed by fish. The highest iodine content corresponds to marine species (>100 µg/100 g). Despite the marine origin the sodium content of fish is in the same range as that found in meat. However, fish is a good source of potassium, with contents that exceed the amounts in meat by two to threefold [2,4]. Mollusks and crustaceans are a good source of elements such as zinc, iron, copper, magnesium, iodine, and selenium.

The percentage ash of meat, an indicator of mineral content, ranges from 1 to 1.5 g/100 g for fresh meat [5] whereas in seafood (fish and shellfish) it ranges from 0.9 to 2.1 g/100 g [6].

In most foods of animal origin, including meat and fish, sodium and magnesium are found at levels between 50–100 and 20–30 mg/100 g, respectively. Meat and fish generally have low calcium contents, around 10 mg/100 g. Lean meat (muscle) has around 180 mg of phosphorus per 100 g, whereas the corresponding value for liver is 370 mg/100 g. Lean meat contains between 2 and 4 mg iron/100 g, mostly as myoglobin. Chicken, lamb, and ox liver have rather more, around 9 mg iron/100 g, and pig liver around twice as much. Mammalian liver is exceptionally high in copper (around 8 mg/100 g). Animal tissues such as lean meat have high zinc levels of around 4 mg/100 g. Seafood is the only consistently useful source of minerals, with some popular fish species containing several thousand µg/kg [7,8].

25.2   Sample Treatment

25.2.1   *Organic Matter Destruction—Mineralization*

When measuring trace elements in solid samples such as muscle foods, a prior sample pretreatment stage is mandatory. Owing to the absence of insoluble substances in these samples compared with other environmental samples, wet or dry decomposition procedures can be successfully applied. However, the time for sample pretreatment exceeds the time needed for measurement, and therefore this preliminary step of the analytical method should be carefully studied.

The choice of procedure depends on the element to be determined and the analytical technique applied, as well as on the involvement of multiple-element analysis if any. It should be mentioned that sample preparation is time limiting, requiring about 61% of the total time needed to perform the complete analysis, and is responsible for 30% of the total analytical error.

25.2.1.1   *Wet Ashing*

Wet digestion with concentrated acids is the most common sample pretreatment for element determination in muscle foods. Microwave energy reduces digestion time. Different acid digestion procedures have been developed using concentrated acid, which completely destroy the solid sample matrix; therefore, the digest can be easily analyzed using conventional sample introduction systems in different techniques for mineral determination. The main limitations of these procedures are matrix interferences due to the use of concentrated acid in atomic absorption spectrometry (AAS), the formation of highly carcinogenic nitrous vapors, high blank values, the possibility of sample contamination, long cooling times required before opening the low- or high-pressure pumps, etc. [9–11]. Treatments based on wet digestions have replaced dry ashing methods, and microwave ovens have replaced conventional wet digestion procedures. The methods described for marine organisms cover different types of sample decomposition using microwave energy, vapor-phase acid, open vessels, semiclosed vessels, and focused microwave technology [12].

New sample pretreatment methods such as acid leaching assisted by microwave or ultrasonic bath have been developed. The procedure involves the solubilization of minerals in the leaching solvent (an acid or oxidant agent) without sample matrix decomposition. This type of procedure minimizes the main limitations of wet digestion methods.

The parameters associated with ultrasonic solid–liquid extraction (ultrasonic bath or ultrasonic probe, particle size, acid concentration, sonication time and sonication amplitude, and analyte–matrix binding) for elemental analysis have been recently reviewed [13].

Although wet digestion is the most widely used organic material destruction technique for muscle foods, only few studies involving muscle foods compare and validate different procedures for sample preparation.

Five previously optimized sample pretreatment procedures (microwave-assisted acid digestion, microwave-assisted-acid leaching, slurry sampling, ultrasound assisted-acid leaching, and Pronase E-enzymatic hydrolysis) are compared after being applied to DORM-1 and DOLT-1 for the determination of Cr, Cu, Fe, Mn, Mg, Se, and Zn. Microwave-assisted acid digestion, microwave-assisted acid leaching, and slurry sampling procedures have been found to be adequate as sample pretreatment methods for multielement determination in seafood products. Acid leaching induced by a low-energy source such as ultrasound has not led to accurate results for some elements such as Fe and Se, though it is suitable for leaching most of them. Bermejo-Barrera et al. [11] have shown that pronase E enzymatic hydrolysis is unsuitable for extracting most of the elements studied. In earlier studies, they have compared microwave acid digestion with ultrasound bath–induced acid leaching or microwave-assisted acid leaching validated in DOLT-1, DORM-1, and TORT-1 for the determination of Ca, Co, Cr, Cu, Fe, Mg, Mn, Se, and Zn. This new acid-leaching technique, involving microwaves or ultrasound, yields good results once optimized (type and concentration of acid, temperature, time, power rating, etc.). However, the application of ultrasound bath–induced acid leaching to selenium in mussel tissue requires high sonication temperatures and longer sonication exposure times. A more oxidative acid solution is needed to extract Fe, and hydrogen peroxide is found to be necessary to secure quantitative Fe recoveries [9,10].

Five extraction procedures—acid digestion (EPA Method 3050 B2 as reference method), Soxhlet extraction, room-temperature mixing and sonication extraction, microwave-assisted extraction, and supercritical fluid (CO_2)—were applied to TORT-2 lobster hepatopancreas and fresh, cooked lobster tissue for Co, Mo, and Se determination. The fourth process, microwave-assisted extraction, was selected for lobster tissue extraction, because it uses less solvent than Soxhlet, yields improved recoveries for all analytes (similar to or higher than the mixture at room temperature with or without sonication), and is more reproducible with greater recoveries than critical fluid extraction. Furthermore, the procedure is less expensive, faster, and more reproducible than the second and the third techniques [14].

For the determination of selenium in fish species (NIST-RM-50 albacore tuna), different digestion procedures, such as wet and dry ashing methods, oxygen pump digestion, ultraviolet digestion, and methods involving elevated pressures, have been compared in Ref. 15. AOAC 986.15 is the only method (AOAC—Association of Official Analytical Chemists) that has reliably produced results. This method is the combination of wet and dry ashing methods incorporating elevated pressure, and is recommended by the Association of Official Analytical Chemists [16].

Ultrasound-assisted acid leaching extraction applied to fish and mussel for Cu and Zn determination was optimized and compared with microwave digestion. The precision and accuracy (DORM-2 dogfish muscle tissue) for both techniques were not significantly different. It has been demonstrated that acid-leaching extraction is a more rapid sample pretreatment procedure than microwave digestion, and compared with similar procedures by ultrasound-induced acid leaching reported earlier reported earlier [10,17], it reduces both the use of concentrated reagent and sonication time (120 to 30 min). Furthermore, high ultrasonic energy is not required [18].

In raw meat, focused microwave-assisted digestion and ultrasound leaching have been applied for the extraction of Pb, Cd, Cr, Cu, Fe, Zn, Ca, and Mg [19]. Microwave-assisted digestion provides more accurate and precise results than ultrasound leaching, though for iron the recovery rate is only 80%. In the case of ultrasound leaching, only the extraction of Pb, Cu, and Ca has proved quantitative.

Bou et al. [20] have used open-vessel wet mineralization/digestion block and closed-vessel wet mineralization/quartz vessels and microwave ovens for Fe, Cu, Zn, and Se determination in chicken meat. The first method is suitable for Zn and Fe determinations, but it is not appropriate for Se quantification because of the volatilization losses that occur. Hence, the second method, which is suitable for Zn, Fe, and Se determinations, is proposed. Although the Cu contents determined with this method presented a relatively high RSD, this variability does not exceed the AOAC recommendation. In addition, the results obtained for this analyte in the certified reference material (CRM 184 bovine muscle) are in close agreement with the certified values and have shown good recovery for this element [20].

A continuous ultrasound-assisted extraction system connected to a flow injection manifold has been used for the on-line determination of zinc in meat samples by flame AAS. Flow injection ultrasound-assisted extraction methodology presents several advantages over its off-line counterpart: sonication time is shorter (reduced by a factor of 6–12); and the centrifugation step to separate the liquid phase (10–20 min) is avoided. Hence, the analytical process is considerably simplified [21].

The official AOAC methods [16] included different wet ashing procedures for the determination of Zn, Cu, and Fe (999.10), Zn (969.32), Se and Zn (986.15), and Se (974.15) in food; Na and K in seafood (969.23); P in meat (969.31); and Ca in poultry and beef (983.19).

25.2.1.2 Dry Ashing

This organic material destruction procedure is rarely used for determining mineral elements in muscle foods. The official AOAC methods [16] include some ashing procedures for the determination of Zn, Cu, and Fe in foods (liver paste and minced fish) (AOAC 999.11); Zn in food (AOAC 969.32); Na and K in seafood (AOAC 969.23); and total P in meat (969.31).

25.2.2 Other Sample Pretreatments

Enzymatic hydrolysis is the process of hydrolyzing certain proteins or lipids of the sample by using certain enzymes such as protease, trypsin, pronase, and lipase. After the breakdown of these large molecules, a variable fraction of minerals is freed in solution and measured after a centrifugation step. This procedure has few advantages such as preventing losses of certain volatile minerals, and due to the selectivity of the enzymes used, fractions of elements bonded to the different components of the sample can be distinguished [10,11].

The use of slurries is an alternative to sample digestion before mineral determination. Slurry sampling offers the advantages of the direct solid and liquid sampling methods. The conventional atomizers and injection systems used for liquid samples can also be used in slurry samples, thus allowing the use of samples of higher weight than those used in solid sample analysis. It is even possible to carry out dilutions. The major factor is the need to maintain the stability of the slurry until sample injection—a factor that affects the variability of the procedure. Different solvents and stabilizers such as HNO_3, H_2O_2, Triton X-100, ethanol, and so on are used in the preparation of the slurries. The use of slurry sampling and electrothermal atomic absorption spectrometry (ETAAS) determination, involving ultrasonic homogenization devices, overcomes some of the inherent drawbacks of the traditional sample decomposition methods [22]. Cal-Prieto et al. [23] reviewed the use of slurry sampling and ETAAS, including animal tissues.

Pereira-Filho and Arruda [24,25] developed a mechanized procedure for on-line slurry food sample digestion that allows the digestion of simultaneous complex matrixes such as mussel, fish, and bovine liver by using three poly(tetrafluoroethylene) (PTFE) coils inserted into the microwave oven cavity. The off-line spectrophotometric determination of cobalt and iron was performed by using a spectrophotometric monosegmented system. The method was validated by using certified reference material and ETAAS.

A solid sampling ETAAS has been applied to Cu and Zn determination in bovine liver. The main parameters investigated were sample drying, grinding process, particle size, sample size, microsample homogeneity, and their relationship to precision and accuracy of the method. The advantages of this sample preparation are high sample introduction efficiency, low sample consumption, *in situ* analyte–matrix separation during the pyrolysis stage, a rapid and low-cost analytical procedure, high selectivity and sensitivity and long residence time, and improvement of atomization efficiency [26].

25.2.2.1 Determination Methods

25.2.2.1.1 Gravimetric Methods

At present, the determination of phosphorus in meat is the only gravimetric method proposed by the AOAC [16] (AOAC 969.31) that has found application in bovine muscle (longissimus dorsi) [27]. Phosphorus can be precipitated as $Mg_2P_2O_7$ (reported as P_2O_5) or $(C_9H_7N_3)H_3(PO_4 12MoO_3)$ in the AOAC method or the quinolimium molybdophosphate method, respectively. The precipitate is first dried in a forced-draft oven at 125°C for 30 min and then ashed in a furnace at 550°C.

25.2.2.1.2 Titrimetric Methods

Complexometric methods with ethylenediamine tetraacetic acid disodium salt (Na_2EDTA) have been used in the determination of calcium (collaborative study) in poultry beef [28] and magnesium in fish [29]. The endpoint is detected by a color change. The method is recommended for the determination of bone in poultry (in the absence of phosphates) and calcium in red meat [28].

Precipitation titration (Mohr and Volhard methods) is commonly applied to chloride determination in foods [22]. However, in muscle foods the applications of this methodology are limited. It has been used to measure NaCl by titrimetry with $AgNO_3$ in processed seafood and meat.

Table 25.1   Minerals Determination in Muscle Foods by Titrimetric Methods

Element	Food	Sample Treatment/Reagent	Reference
Complexometric			
Ca	Poultry and beef (collaborative study)	HCl/hot plate 20 min cover with watch glass/pH 12.5. EDTA-Na$_2$ addition. Back-titration with CaCO$_3$. Indicator: hydroxy naphtol blue	28
		AOAC Method (983.19)	16
Mg	Fish	Dry ashing: 520°C, 2 h/EDTA titration	29
Precipitation			
Cl$^-$	Meat: fresh ham, beef	Precipitated with AgNO$_3$	30
Cl$^-$	Meat, seafood	Precipitated with AgNO$_3$. Titration with NH$_4$SCN (AOAC 935.47, 937.09)	16
Alkalimetric			
P	Meat	HNO$_3$–HClO$_4$/precipitate with acidified molybdate/dissolution with small standard alkali (NaOH)/titrate with HCl or HNO$_3$ and phenolphthalein (969.31)	16

In meat, this procedure has been used as the reference method to validate NaCl determination by the near-infrared (NIR) method [30]. An alkalimetric ammonium molybdophosphate method is proposed by the AOAC [16] to measure the chloride content in foods.

The application of this technique is shown in Table 25.1.

25.2.2.1.3   Visible Spectrophotometric Methods

This technique is the most widely used option for measuring the phosphorus content in foods, including muscle foods. Some of the visible spectrophotometric methods applied to muscle foods are reported in Table 25.2.

A semiautomated photocolorimetric method such as the Spectroquant® Photometer SQ 118 has been applied to Ca, P, Mn, Fe, and Mg determination in fish species [45,46]. This system has been developed to analyze surface, ground, and drinking water (http://www.merck.de/servlet/PB/menu/1282690/index.html). However, with appropriate sample preparation, it can be applied to the analysis of more complex matrixes. A dry ashing treatment at 450°C for 12 h could be appropriate.

Phosphates are important components of meat. They are also an integral part of some proteins and cell walls, and of the energy generation process (e.g., ATP, ADP, and phosphocreatine). In the meat industry, several polyphosphates are used, although the amount of added phosphates is limited because they may chelate important metal ions (calcium and magnesium). The amount of added phosphate is determined by the difference between the amount of total phosphate and protein-bound phosphate. Sample treatment by dry ashing followed by spectrophotometric measurement is the technique that is most frequently applied for the determination of phosphate in foods. The amount of protein-bound phosphate is calculated by taking into account the nitrogen

Table 25.2 Minerals and Trace Elements Determination in Muscle Foods by Visible Spectrophotometric Methods

Element	Food	Method	Reference
Fe	Seafoods, oysters, beef, and bovine liver[a]	(a) Dry ashing 550°C; (b) wet digestion: HNO_3 + $HClO_4$ ferrozine/NH_4OH/mercaptoacetic/NaOAc (pH = 4.5)/562 nm	31
	Fish and meats	Dry ashing: 550°C/1,10-phenanthroline hydrochloride/510 nm	32
	Fish, oyster tissue,[a] fish homogenate,[a] and mussel[a]	On-line microwave slurry wet digestion: HNO_3/H_2O_2. Off-line monosegmented system: ascorbic acid and 1,10-phenanthroline solution/acetate buffer pH 3.7, 512 nm	24
Nonheme iron	Beef muscle and raw meats	$NaNO_2$/HCl–TCA/65°C 20 h/bathophenanthroline disulfonate, 540 nm	33,34
	Raw and cooked turkey breast, leg, and mechanically deboned meat	(a) $NaNO_2$/HCl–TCA/65°C 20 h/bathophenanthroline disulfonate, 540 nm; (b) HCl–TCA/65°C 20 h/bathophenanthroline disulfonate, 540 nm; (c) citrate phosphate buffer pH 5.5/ascorbic acid in 0.2 N HCl 22°C 15 min/TCA/ammonium acetate and Ferrozine, 562 nm	35
	Beef, pork, and lamb	Nonheme iron: HCl–TCA/65°C 20 h/bathophenanthroline disulfonate, 540 nm	36
Heme, nonheme, and total iron	Ready-to-serve beef products, bovine liver[a]	*Total iron:* HNO_3 + H_2O_2/Ferrozine/NH_4OH/hydroxylamine hydrochloride/ammonium acetate (pH = 8)/562 nm *Heme iron:* HCl: water:acetone 40:4:1/heme iron convert acid hematin 640 nm	37
	Meat, fish, and shellfish	Heme: HCl: water: acetone 40:4:1/heme iron convert acid hematin 640 nm Nonheme: citrate phosphate buffer pH 3.6/ammonium acetate/FerroZine, 562 nm	38,39
P	Chicken meat and tissues	Oven-drying of samples	40
	Broiler chicken carcasses	Ashing 450°C/12 h. Molybdovanadphosphate	41
	Meat and meat products	H_2SO_4/cyclohexanediamine/tetraacetic acid/ammonium molybdate tetrahydrate/ascorbic acid/dimethyl sulfoxide/890 nm (AOAC, 991.27)	16
	Meat	Automated method (972.22)	16
	Meat	Molybdate/H^+/ascorbic acid/890 nm	42
	Fish tissue and oyster tissue[a]	Microwave digestion (HNO_3 or HNO_3/H_2O_2) versus ashing digestion vanadomolybdophosphoric acid method	43
	Crab	Wet digestion: HNO_3/$NClO_4$. Barton solution	44
Zn	Food	Dithizone/CCl_4/540 nm (AOAC 991.27)	16
Co	Fish, fish homogenate,[a] mussel,[a] and bovine liver[a]	On-line microwave slurry wet digestion: HNO_3/H_2O_2. Off-line monosegmented system: Tiron or 4,5-dihydroxy-1,3-benzenedisulfonic acid, disodium salt monohydrate and hydrogen peroxide, 426 nm	25

[a] Standard reference material.

content (Kjeldahl method) and the known ratio of phosphate to protein (0.0106 g P/g protein). However, a new ratio of free (nonprotein) phosphate/protein (16 ± 2 mg/g) has been calculated in meat and meat products. The amount of added phosphates can be calculated from the level of soluble phosphate determined by capillary isotachophoresis and the protein content determined by the Kjeldahl method. The method was applied to samples of meat products and demonstrated good agreement with the spectrophotometric dry-ashing method [47].

Visible spectrophotometry has also been used to determine minor elements such as iron in seafood [31], fish, and meats [48]; in both cases the results obtained have been compared with those recorded with AAS. In general, good agreement is found between the results obtained by both methods in meat, fish, and fish products, despite statistically significant differences observed in the meat group [48]. Both methods were found to yield satisfactory recovery values, though variance was slightly higher for the spectrophotometric method than for the AAS method. However, the colorimetric method is less accurate when iron levels in all seafood, 6.1–44 μg/g, are pooled [31]. In this low iron content/sample range, the colorimetric method yields results that are significantly lower (8%) than those obtained by AAS.

In the speciation of iron from raw meats, the nonheme iron content has also been determined by colorimetry (bathophenantroline). This method has been used to validate a near-infrared spectrometry (NIRS) method [33]. Colorimetric methods can be applied to iron speciation such as heme and nonheme iron fractions in meat and fish (see Table 25.2).

Iodine is an essential nutrient for which no AOAC [16] or other generally accepted standard assay exists. A comparison was made of two colorimetric iodine assays based on different chemistries, in samples that had been previously subjected to alkaline dry ashing (first alkaline dry ashing of the sample, and then application of the colorimetric method to the ashes). One assay kinetically measured the initial iodine catalysis of the redox reaction between Ce^{4+} and As^{3+} at 380 nm. The other assay monitored iodine catalysis of the reaction between thiocyanate and nitrite by measuring absorbance after 20 min at 450 nm. Both methods were accurate when tested with reference sample of oyster tissue. The thiocyanate–nitrite assay consistently showed greater sensitivity and precision, and less variability [49].

A method for the spectrophotometric determination of iron [24] and cobalt [25] in seafood and reference materials for online slurry food sample digestion has been proposed. After microwave sample preparation and collection of the digested samples, the latter were introduced into an off-line monosegmented system for spectrophotometric determination. The proposed method covered the 10–200 ng cobalt/L or 50–200 μg iron/L concentration ranges, with quantification limits of 5.5 ng cobalt/L and 24.4 μg iron/L, respectively.

25.2.2.1.4 Spectrofluorometry

Fluorimetric methods have been proposed for determining the selenium (Se) content in meat, chicken and fish [50], pork [51], and food [16].

Generally, following wet digestion, Se is converted into Se(IV) by boiling with hydrochloric acid, and determined by measuring the fluorescence of the piazselenol formed on reaction with 2,3-diaminonaphthalene (DAN) or 3,3′-diaminobenzidine (DAB). DAN is the reagent of choice because the complex is extractable into an organic solvent and the sensitivity of fluorescence is higher than that of the DAB complex [22].

When compared to ETAAS that is used for the determination of selenium in foods, spectrofluorometry offers minimal matrix interferences and relatively inexpensive equipment requirements. The detection limit in selenium spectrofluorometry determination (0.001 μg/g) is lower than that in hydride generation AAS (HGAAS) (0.033 µg/g). It is clear that spectrofluorometry is more sensitive than HGAAS [50].

Ragos et al. [52] describes a high-sensitivity spectrofluorometric method for the determination of micromolar concentrations of iron (III) in bovine liver with 4-hydroxyquinoline as fluorescent agent. Following ultrasonic wet digestion with hydrochloric acid, cobalt was determined in fish liver and muscle by a method based on its significant catalytic effect on the luminol-dissolved oxygen chemiluminescence reaction in the flow injection analysis system (FIAS) with a fentogram-detection limit (4 fg/mL) [53].

25.2.2.1.5 Neutron Activation Analysis

Neutron activation analysis (NAA) techniques have been described since 1975 in the multielement analysis of over 20 elements in foodstuffs. NAA requires a source of neutrons; therefore, a nuclear reactor is needed nearby. NAA methods offer specificity, freedom from blank errors, and easy sampling (direct analysis without any mineralization or dissolution). This simplifies the analysis and minimizes the risk of contamination; moreover, there are relatively few interferences and matrix effects. Little sample preparation is needed; hence, it is more economical and faster than atomic methods. NAA, together with atomic techniques, is the method of choice for trace element measurements [54–56].

NAA is a powerful analytical method for the determination of minerals in food. However, if the food samples contain high concentrations of elements generating high activities when irradiated with thermal neutrons, the technique proves less successful. In such situations, it may be worthwhile to use epithermal rather than thermal neutron activity [57].

Instrumental thermal neutron capture prompt gamma-ray activation analysis (PGAA) requires no matrix alteration (e.g., drying or digestion—either of which may lead to analyte loss) before analysis, and is nondestructive because test portions of foods are subjected to neutron doses four or five orders of magnitude lower than those required for conventional or delayed gamma-ray neutron activation analysis [58]. This technique has been redesigned and improved to obtain lower background radiation levels and improved detection limits (by a factor of two or more) for Na, P, Cl, K, Ca, and Fe [59].

Selenium is present in foods at low concentrations; therefore, sensitive techniques are required. NAA with its different modes (short- and long-term) proves quite suitable. For Se, better counting statistics and more than one order of magnitude lower detection limits (down to 2 µg/kg) have been achieved using long irradiation with radiochemical separation [55].

The application of these techniques is shown in Table 25.3.

25.2.2.1.6 X-ray Fluorescence

X-ray fluorescence (XRF) spectroscopy, with its capacity to detect several elements simultaneously at very low concentrations, offers an excellent tool for trace element analysis. In addition, the ease of sample preparation in dried and pelletized food samples without dissolution, ashing, or other destructive preparation techniques—this in turn reducing dilution and contamination problems—and the small amount of sample needed make this technique particularly suitable for samples of biological origin. The sensitivity of the method is approximately 1 µg/g for Mn, Fe, Co, Ni, Cu, and Zn, and 20 µg/g for K and Ca [72].

These methods use a prior system of calibration of element sensitivities and fundamental parameters of x-ray physics for mineral determination. This approach eliminates the need for calibration standards of composition similar to the samples, and allows the use of standard reference

Table 25.3 Minerals and Trace Elements Determination in Muscle Foods by Neutron Activation Analysis

Element	Food	Method	Reference
Na, K, Cl	Unsalted foods (fish), bovine liver[a]	INAA: irradiation time 10 min. Thermal neutron flux 9×10^{12} n/(cm^2 s) Counted time 5 min. ^{38}Cl, ^{24}Na, ^{42}K, Detector Ge(Li)	60
Na	Bovine liver[a], oyster tissue[a]	Thermal neutron flux 10^{16}–10^{18} n/(m^2 s). ^{24}Na. Detector Ge or Ge(Li)	61
Se	Bovine liver[a], roast beef	CINAA: Se (irradiation time 20 s, decay time 10 s, counting time 20 s) number cycles 4. Thermal flux 5×10^{11} n/(cm^2 s)	62
I	Bovine liver[a], roast beef	EINAA: I (irradiation time 30 min, decay time 1–6 min, counting time 30 min)	
Co, Cu, Fe, Se, Zn	Oyster tissue[a]	RNAA: irradiation 24 h and posterior coprecipitation of Cu, Se at pH 2–3 in CuS and Co, Fe, Zn at pH 8–9 in ZnS. Detectors: Ge and Ge(Li)	56
Br, Ca, Cl, Co, K	Beef, chicken, pork, lamb	Neutron flux 5×10^{11} n/(cm^2 s). Irradiation (60–200 s). Detector: Ge	59
Na, Al, P, S, Cl, K, Ca, Fe	Bovine liver[a]	PGAA: thermal neutron fluence rate 3.0×10^{8} n/(cm^2 s). Detector: Ge	
Mg, Cl, Mn, K, Zn, Co, Fe, Cu	Marine bivalves, bovine liver[a]	INAA: thermal flux 4×10^{12} n/(cm^2 s). Irradiation 5 min (Mg, Cl, Mn, K); 10 h (Zn, Co, Fe, Cu). Detector: Ge	63
Se	Bovine liver[a], beef	INAA (short-term irradiation), INAA and RNAA (long-term radiation). Fluence rates 8×10^{13} n/(cm^2 s), Detector HPGe coaxial	55
K	Oyster tissue[a], beef	ENAA: epithermal flux 1.3×10^{11} n/(cm^2 s). Detector: Ge	64
Se	Oyster tissue[a], bovine liver[a], fish homogenate, animal muscle	CINAA thermal neutron flux 5×10^{11} n/(cm^2 s). Irradiation time 20 s, decay time 10 s, counting time 20 s. Detectors: Ge(Li)	65
Se, Zn	Fish muscle	NAA thermal neutron flux 3×10^{12} n/(cm^2 s); irradiation time 8–36 h	66

[a] Standard reference materials.

Table 25.4   Minerals and Trace Elements Determination in Muscle Foods by X-ray Fluorescence Methods

Elements	Food	Technique	Reference
Ca, Cu, Fe, K, Mo, Mn, Ni, Se, Zn	Fish	PIXE and XRF	67
Ca, Cl, Fe, K, Mn, P, Zn	Bovine liver,[a] oyster tissue[a]	XRF	68
P, Cl, K, Ca, Cr, Mn, Fe, Ni, Cu, Zn	Fish	TXRF	69
K, Ca, Fe, Se	Fish muscle	EDXRF	70
Ca, Mn, Fe, Ni, Cu, Zn, Se	Beef, mutton, chicken	PIXE and XRF	71
K, Ca, Mn, Fe, Co, Ni, Cu, Zn, Se	Fish (muscle, liver, fat tissue, and skin)	XRF	72
Mn, Fe, Cu, Zn	Fish	EDXRF	73

[a] Standard reference materials.

materials only to monitor accuracy and precision. Since the analyses are nondestructive, samples can be repeatedly reanalyzed, adding flexibility to the experimental designs [68].

The main limitations of the XRF methods are the reduced number of elements to which the method is applicable—only elements having an atomic number between 22 (titanium) and 55 (cesium)—and the need to work under special conditions or with low sensitivity when light elements such as magnesium are to be measured. Other drawbacks of the XRF methods are high cost, important matrix effects when the salt contents are high, and the fact that the radiation intensity is not a lineal function of the element or element contents [22].

Comparisons of XRF results with AAS analyses in fast-food samples, including beef, chicken, and fish, indicated generally good agreement, with correlation coefficients from XRF versus AAS plots for manganese, iron, and zinc of .94, .97, and .97, respectively [68].

Energy-dispersive x-ray fluorescence (EDXRF) and flame atomic absorption spectroscopy (FAAS) have been applied to K, Ca, Fe, Zn, and Se determination in edible fish muscle. The advantages of EDXRF versus FAAS are simultaneous determination of elements and also the fact that there is no need for chemical sample preparation. However, the technique is restricted by its detection limits (>1 µg/g) [70].

The application of these techniques is shown in Table 25.4.

25.2.2.1.7   Near-Infrared Spectrometry

This technique is based on the use of selective absorption of electromagnetic radiation from 800/1000 to 2500 nm in accordance with the characteristic vibration frequencies of functional groups (CH, NH, OH bonds) [41]. There are no absorption bands for minerals in the near infrared region, but organic complexes and chelates may be detected, so NIRS can be used for determining mineral concentration. In the case of meat, the mineral components are associated with the water and protein fractions so that the differences seen in the NIR spectra of meat are mainly due to those found in proteins and to humidity [74]. Thus, the spectral information defined absorption bands at 1114, 1230, and 1474 nm for total iron, absorption for total iron to the first overtone of NH stretching vibration in peptide, 1964 nm for nonheme iron to the combination NH stretching vibration and amide II in amide, and 1574 nm for heme iron to the first overtone of NH stretching vibration in peptide [33]. The absorption bands at 1510, 2060, and 2172–2186 nm, related to the protein content, would be associated with elements such as Ca and Zn. In the case of K, there is a direct relation to the amount of water retained by the meat [74].

The use of NIRS for mineral determination offers many benefits: there is no production of chemical waste associated with more traditional digestion-type assays in use today; no extensive sample preparation (only drying and grinding to uniform size) is required; and the technique can simultaneously and rapidly determine several analytes. Furthermore, it is cost-effective, reproducible, and much more suitable analytical procedure for real-time analysis. NIRS can be an effective alternative to mineral determination by standard methods because it seems possible to develop a robust calibration with the inclusion of more samples in the calibration set, with some refinement in the sampling technique [41].

NIRS has not been applied to mineral determination in seafood. In meats, Begley et al. [30] describe the use of NIR reflectance to measure the amount of salt/NaCl added to meat products on the basis of the change in water component of the meat spectrum, using canned cured ham to develop a calibration of NIRS data to chemically determine salt values. The concentrations of sodium chloride in meats are measured from the absorption intensity at wavelengths where the hydroxyl group of the water molecule produces absorption.

NIR scanning from 1100 to 2500 nm was applied for the determination of heme and non-heme iron in raw muscle meats by the absorption attributable to the iron-binding ligand in different proteins (heme iron–myoglobin–hemoglobin and cytochromes, and nonheme compounds comprising ferritin and hemosiderin). The best multiple regression equations were obtained with the NIRS data recorded at wavelengths where NH groups in protein molecules, C–C and CH in the porphyrin ring of heme (1114 and 1118 nm to total Fe and heme Fe), and iron-sulfur clusters in iron-storage proteins (1738 nm to nonheme-Fe) yield absorption. The correlation coefficients between the predicted and the reference data were 0.991, 0.906, and 0.991 for the contents of heme, nonheme, and total iron, respectively [33].

An analytical procedure was set up using NIR techniques for the determination of Fe, Zn, Ca, Na, and K in fresh Iberian breed pork loin (longissimus dorsi muscle) using ground samples and the determination of Fe, Na, and K by direct application to the loin samples of a remote fiberoptic reflectance probe. Using intact samples measurement with the fiberoptic probe, it was only possible to obtain calibration equations for Fe, Na, and K. The results obtained in the determination of the mineral composition are quite acceptable for Fe, Zn, Ca, and Na, but not so good for K. It is concluded that the NIR technique may be an excellent alternative for the determination of elements such as Fe, Zn, Ca, Na, and K in samples of Iberian pork loin [74].

Data on NIRS calibration to predict calcium and phosphorus in poultry meat are scarce, though it has been reported that NIRS is a good predictor for calcium ($R^2 = .90$) and phosphorus ($R^2 = .91$) in accurately, rapidly, and cost-effectively predicting their contents in whole broiler chicken carcass. The calibration set should be valid for samples within the ranges of approximately 1–4% Ca and 1–2% P. Compared with the AOAC reference method (dry ashing and Ca and P determination by AAS or the spectrophotometric molybdovanadphosphate method), NIRS requires less time and cost of chemicals for the determination of Ca and total P [41].

25.2.2.1.8   Electroanalysis

Voltammetry, together with appropriate pretreatment of the sample (digestion methods that are suitable for analysis by atomic spectrometry in which the digest is atomized but may not be suitable for voltammetric analysis), has been shown to be a valid analytical procedure that is certainly applicable for simultaneous metal determinations in multicomponent complex matrixes.

Electroanalytical methods have been applied to Cu, Zn, and Se determination in muscle foods. In the case of Cu and Zn, differential pulse anodic stripping voltammetry (DPASV) is the method usually applied, but Se has been determined by different procedures—cathodic stripping voltammetry (CSV), differential pulse cathodic stripping voltammetry (DPCSV), and differential pulse polarography (DPP). Selenium occurs in foodstuffs as organic and inorganic compounds in different oxidation states (-2, 0, $+4$, and $+6$), and must be converted to electrochemically active selenium (IV) with HCl.

The detection limits ($\mu g/g$) in different reference materials such as mussel tissue, cod muscle, and oyster tissue obtained by differential pulse stripping voltammetry are the following: Se (0.024–0.041), Cu (0.024–0.032), Zn (0.069–0.072) [75].

When potentiometric stripping analysis (PSA) is used as electrochemical method for analysis, total organic matter destruction is unnecessary. This technique is an interesting alternative when other electrochemical methods (such as anodic stripping or DPP) fail. In PSA, the potential change of the working electrode is measured as a function of time, and the determination is not based on current measurements; this eliminates measurement errors caused by the liquid resistance between the electrodes and the potential drop over the electrical double layer around the working electrode. The no-current measurement reduces the background signal and provides the possibility of analyzing samples with a high content of organic redox compounds without these interfering with the metal determination. The process can be conducted, from the lyophilized sample, in approximately 3–4 h of operation [76].

The application of these techniques is shown in Table 25.5.

Table 25.5 Minerals and Trace Elements Determination in Muscle Foods by Electrochemical Techniques

Element	Food	Sample Treatment	Technique	Reference
Cu	Finfish	Wet digestion (HNO_3 + HCl)	DPASV	77
Cu, Zn	Mussels, clams, fish	H_2SO_4–HNO_3	DPASV	78
Cu, Zn/Se	Oyster tissue,[a] mussel tissue,[a] cod muscle[a]	H_2SO_4–HNO_3	DPASV/DPCSV	75
Cu, Zn	Liver, fish, mussel	HNO_3–$HClO_4$ (microwave radiation)	DPASV	79
Se	Pork meat	$HNO_3/Mg(NO_3)_2/H_2O_2$ Wet/dry ashing (450°C)	CSV	80
Se	Albacore tuna[a]	HNO_3 (closed digestion)/$Mg(NO_3)_2$ Wet/dry ashing (500°C)	DPCSV	15
Se	Fish	$HNO_3/H_2SO_4/H_2O_2$ Closed PTFE/microwave	DPP	81
Cu, Zn	Bovine liver[a]	Dry ashing	DPASV	82
Cu	Heifer and fish liver tissues	(a) Sodium dodecylsulfate and sodium dexoycholic acid + sonication; (b) low-temperature ashing	PSA	76

[a] Standard reference materials.

25.2.2.1.9 Atomic Spectrometry

Atomic spectroscopic techniques are the best choice for elemental analysis because of their wide-spread availability and ease of use. The extensive annual literature reviews in the section entitled "Atomic Spectrometry Update—Clinical and Biological Materials, Food, and Beverages" of the *Journal of Analytical Atomic Spectroscopy* reflects the major role that atomic spectroscopy has played in the development of the current databases on minerals in foods. Comprehensive reviews of literature reports on methods focus on the progress of single and multielement analysis, sampling and sample preparation, references materials, developments in analytical methodology and instrumentation, and applications in foods, including muscle foods.

On-line sample dissolution and continuous separation have evolved into two of the most prominent contributions of the flow injection technique to atomic spectroscopy. An excellent review illustrating the advantages involved in the use of continuous flow systems for the determination of trace metals in seafood samples has been published by Yebra-Biurrun and García-Garrido [83].

In general, flame AAS and inductively coupled plasma atomic emission spectrometry (ICP-AES) provide sufficient detection capability for the quantitation of elements of interest in foods. Although both techniques provide good precision over their entire calibration range, the dynamic range for ICP-AES is clearly larger, making the need for sample dilution less likely. ICP-AES is less prone to chemical matrix interference due to molecule formation than flame AAS, but there can be problems with spectral line overlaps. Flame AAS is simple to perform, and if only one or two elements are needed, flame AAS is usually faster. ICP-AES covers a wider concentration range, offers the benefit of not requiring routine dilution of samples, and allows simultaneous multielement determination; however, it is more expensive.

It has been reported that for Ca and Mg, the detection limits are much poorer for ICP-AES than for flame AAS. However, in application to Cu, Cr, Fe, Mn, and Zn, the limits are slightly lower for ICP-AES. When increased sensitivity is required, ETAAS and inductively coupled plasma mass spectrometry (ICP-MS) are used [84].

ICP-MS is a fast and multielement technique with a wide dynamic range and excellent detection limits for trace element analysis in muscle foods. In chicken meat and bovine muscle (CRM 184), Bou et al. [20] report that it is much more sensitive than ICP-AES, and the detection limits for Cu and Se via ICP-MS or hydride generation inductively coupled plasma mass spectrometry (HG-ICP-MS) are 67.9 and 27.30 μg/kg, respectively. The major drawback is that it suffers from spectral and nonspectral interferences. The former problem may lead to severe systematic errors; Cubadda et al. [85] report that C, Cl, and Ca are the main sources of spectral interference, especially when multielement analyses of seafood are made.

Some of the recent research into the application of multielement and multiisotopic measurements to determine geographical origin has focused on beef and lamb [86].

The application of these techniques is shown in Table 25.6.

25.2.2.1.10 Speciation

In meat, fish, and poultry, 40% of the iron is heme and 23% of dietary heme versus 2–8% of dietary nonheme iron is absorbable [34]. Accurate estimates of nonheme iron content are important not only because of their different bioavailability but also because nonheme iron is a major catalyst of lipid oxidation in cooked meat [101].

Table 25.6   Minerals and Trace Elements Determination in Muscle Foods by Atomic Spectrometric Methods

Element/Food	Sample Treatment/Comments	Technique	Reference
Ca, Cu, Fe, Mg, Mn, Zn/Mussel	(a) Acid leaching: $HNO_3 + HCl + H_2O_2$ (b) Microwave digestion: $HNO_3 + H_2O_2$	FAAS, ETAAS ICP-MS	17
Fe, Zn, Cu/Cu/Se/chicken meat, and bovine muscle[a]	Comparison wet digestion methods	ICP-AES/ICP-MS/HG-ICP-MS	20
Cu, Zn, Cr, Ni/Mussel, and oyster tissue[a]	(a) Microwave digestion: HNO_3 (b) Wet digestion with HNO_3	FAAS ETAAS	12
Ca, Cu, Cr, Fe, Mg, Mn, Zn/Ca, Cu, Co, Cr, Fe, Mg, Mn, Ni, P, Zn/bovine liver[a]	Wet ashing: $HNO_3 + H_2O_2$ Dry ashing: 480°C	FAAS or ETAAS/ICP-AES	84
Co, Cr, Cu, Fe, Mg, Mn, Mo, Ni, Se, Zn/meat (beef), Fish (trout), lobster hepatopancreas[a], mussel tissue[a], cod muscle[a]	Microwave-assisted drying: $HNO_3 + HCl$	ICP-AES/ETAAS	87
Zn, Fe, Mg, Ca/Cr, Cu/raw pork meat	(a) Focused microwave-assisted digestion with HNO_3 (b) Ultrasound leaching HNO_3	FAAS/ETAAS	19
Zn, Cu, Mn/Se/Cr, Ni, Co/meat, liver, and kidney of pigs and cattle	Dry ashing: 450°C Se: Dry ashing: 450°C ($MgNO_3 + OMg$)	FAAS/HGAAS/ETAAS	88
Zn, Cu, Ni, Cr, Mo/fish, baikal perch[a], baikal roach[a]	Dry ashing (450°C)	AAS, AES	89
Cu, Zn/fish, mussel, and dogfish muscle tissue[a]	(a) Microwave digestion ($HNO_3 + HCl$) (b) Acid leaching extraction ($HNO_3 + HCl + H_2O_2$)	ETAAS FAAS	18
Cu, Fe, Mg, Zn/As, Cr, Ni, Mn, Se/dogfish muscle[a] and dogfish liver[a]	Comparison wet digestion method and enzymatic hydrolysis	FASS/ETAAS	11

(Continued)

Table 25.6 (Continued)

Element/Food	Sample Treatment/Comments	Technique	Reference
Ca, Cu, Fe, Mg, Zn/Co, Cr, Mn, Se/mussel, dogfish muscle[a], lobster hepatopancreas[a]	(a) Microwave acid digestion (HNO_3 + H_2O_2) (b) Microwave pseudodigestion	FASS/ETAAS	9
Mn, Cr/meats and bovine liver[a]	Wet digestion (HNO_3 + H_2SO_4 + $HClO_4$) versus (HNO_3 + H_2SO_4). HF was used for Cr	FASS/ETAAS	90
Zn, Mn, Fe, Cu, Na, K, Co/fish	Dry ashing (540°C)	AAS	91
Na, K, Ca, Mg, Fe, Zn, Mn, Cu/crab	Wet digestion (HNO_3 + HCl)	AAS	44
Fe/seafoods and bovine liver[a]	(a) Wet digestion (HNO_3 + HCl) (b) Dry ashing (550°C)	AAS	31
Se/fish tissue[a], mussel tissue[a], bovine liver[a]	Microwave digestion/ diethyldithiophosphate preconcentration	ETAAS	92
Cu, Zn/lobster	Microwave digestion (HNO_3)	FAAS	93
Fe/seafoods	Dry ashing (550°C), wet ashing (HNO_3 + $HClO_4$)	FAAS	31
Cu/mussel and Mn/mussel and clam	Ultrasound acid leaching (HCl + HNO_3)	FIA-FAAS	94,95
Se/meat organs and tissue	Wet digestion: HNO_3	HGAAS	96
Se/fish, mussel tissue[a], cod muscle[a]	Wet digestion: HNO_3	HGAAS	97
Se, Cu, Ni, Zn, Mn, Fe/fish and meat	Microwave acid digestion: HCl	ICP-AES	98
I/bovine muscle[a]	Wet ashing: HNO_3	ICP-MS	99
Zn/fish	Wet ashing (HNO_3 + H_2O_2)	ICP-MS	100
Co, Mo, Se/lobster hepatopancreas[a], fresh, cooked lobster tissue	Comparison five extraction procedure	ICP-MS	14

[a] Standard reference materials.

The available heme iron content data are largely centered on raw meat, and the determined values vary widely. However, there is limited information regarding meat products (i.e., generally cooked). Cooking time and the source of meat affect heme iron content [36,102].

Schricker et al. [103] have published a colorimetric method for measurement of nonheme and total iron content in muscle. The major deficiency of this method is the overestimation of nonheme iron because of the iron released from heme pigments during long incubation in acid. Rhee and Ziprin [101] have modified the Schricker method to minimize pigment effects in determining nonheme iron in meat; the use of $NaNO_2$ to stabilize heme pigments reduces the pigment effect in red meat. The ferrozine method [35] uses milder assay conditions than the other two methods, and an incubation period is not required. Ahn et al. [37] have compared the amount of nonheme iron in turkey meat using the three analytical methods mentioned. The apparent content of nonheme iron was lower in the case of ferrozine method, and the results obtained by the Schricker method were no different.

Kalpalathika et al. [38] have determined total iron, heme iron, nonheme iron, and heme iron percent contents of beef steak, beef burger, roast beef, and frankfurter by colorimetric methods. Percentage heme iron ranged from 50.2 to 63.8% for these meat foods.

The determination of the heme and nonheme iron fractions in raw and cooked beef steak, by using spectrophotometric methods and high-performance liquid chromatography coupled to a double-focusing sector field inductively coupled plasma mass spectrometer, has been reported by Harrrington et al. [39,104]. Accordingly, NIR has been used in muscle meats [33].

Brisbin and Caruso [14] have evaluated the chemical forms of Co, Mo, and Se in TORT-2 and lobster organs and tissue through extraction with solvents of different polarity and spectrometric techniques. In TORT-2 the following two forms are found: nonpolar organic and polar inorganic. However, in lobster organs and tissue, Co is in polar inorganic form, Mo is in less polar organic form, and Se is found in both forms.

The hyphenated techniques such as liquid chromatography (reversed phase-RP, ion exchange-IE, size exclusion chromatography-SEC), and the detectors of the three major atomic spectrometric techniques—flame or ETAAS, NAA, ICP-AES, and ICP-MS—for trace element speciation have been applied to muscle foods. A review of speciation analysis by hyphenated technique has been published by Szpunar [105].

Some recent studies on the speciation of elements of nutritional interest in muscle foods by using hyphenated techniques comprise the following: SEC with off-line ETAAS has been used to study specific seleno-protein in fish species [106]; SEC-RP on-line ICP-MS or off-line ESI-MS has been used for Fe, Cu, Zn, and Mn in species from porcine liver [107]; SEC-ICP-MS has been used with Zn, Cu, and Se; and IC-ICP-MS for Se in fish [108] or oyster [109]. In turn, SEC or RP with on-line ICP-MS has been applied to Se in cold muscle [110], and SEC-IC-ICP-MS has been used for Cu, Zn, Cr, Co, Mn, and Mo in mussel.

25.3 Conclusions

The key step in mineral element analysis in most of the techniques is sample preparation. Therefore, the principal advances aim to simplify these procedures, reducing the analytical time, and the risk of losses or contamination—with the resulting benefits for environmental preservation.

The most widely used techniques for the determination of mineral contents in muscle foods are the various spectroscopic methods. The spectrophotometric techniques remain valid for the determination of certain elements such as P and Fe, fundamentally in meats. Other available techniques such as NIR, NAA, and XRF are sufficiently sensitive and only very simple sample treatment is needed.

At present, there is growing interest in speciation because the physiological role of a given element depends on the form in which it occurs, the hyphenated techniques being the most commonly used options for this purpose.

25.4 Abbreviations

AAS	atomic absorption spectroscopy
AOAC	Association Official of Analytical Chemist
ASV	anodic stripping voltammetry
CINAA	cyclic instrumental neutron activation analysis
CSV	cathodic stripping voltammetry
DPASV	differential pulse anodic stripping voltammetry
DPCSV	differential pulse cathodic stripping voltammetry
DPP	differential pulse polarography
EDTA	ethylenediamine tetraacetic acid
EDXRF	energy-dispersive x-ray fluorescence
EINAA	epithermal instrumental neutron activation analysis
ETAAS	electrothermal atomic absorption spectroscopy
FAAS	flame atomic absorption spectroscopy
FIAS	flow injection analysis system
HGAAS	hydride generation atomic absorption spectroscopy
HG-ICP-MS	hydride generation inductively coupled plasma mass spectrometry
ICP-AES	inductively coupled plasma atomic emission spectrometry
ICP-MS	inductively coupled plasma mass spectrometry
INAA	instrumental neutron activation analysis
NAA	neutron activation analysis
NIRS	near-infrared spectrometry
PGAA	prompt gamma-ray activation analysis
PIXE	proton-induced x-ray emission
PTFE	poly(tetrafluoroethylene) bombs
PSA	potentiometric stripping analysis
RNAA	radiochemical neutron activation analysis
TCA	trichloroacetic acid
TXRF	total reflection x-ray fluorescence
XRF	X-ray fluorescence

References

1. Bello, J. 2000. Carnes y derivados. In I. Astiasarán and J.A. Martínez (eds.) *Alimentos, Composición y propiedades*, McGraw-Hill Interamericana, Madrid.
2. Ros, G. and C. Martínez. 2005. Calidad y composición nutritiva de la carne, el pescado y el marisco. In A. Giland and M.D. Ruiz (eds.) *Tratado de Nutrición. Tomo II. Composición y Calidad Nutritiva de los Alimentos*, Acción Médica, Madrid.
3. Lupton, J.R. and H.R. Cross. 1994. The contributions of meat, poultry, and fish to the health and well being of man. In A.M. Pearson and T.R. Dutson (eds.) *Quality attributes and their measurement in meat, poultry, and fish products. Advances in Meat Research Series*, Vol. 9, Blackie Academic and Professional, Glasgow.

4. Primo, E. 1998. *Química de los Alimentos*, Vol. 1, Síntesis, Madrid.

5. Schweigert, B.S. 1994. Contenido en nutrientes y valor nutritivo de la carney los productos cárnicos. In J.F. Price and B.S. Schweigert (eds.) *Ciencia de la Carne y de los Productos Cárnicos,* 2nd ed., Acribia, Zaragoza.

6. Madrid, A., J.M. Vicente, and R. Madrid. 1999. Composición y valor nutritivo del pescado y productos derivados. In *El Pescado Y Sus Productos Derivados*, 2nd ed., Mundi Prensa, Madrid.

7. Coultate, T.P. 2002. *Food: The Chemistry of its components*, 4th ed., The Royal Society of Chemistry, Cambridge.

8. Roe, M.A., P.M. Finglas, and A.M. Church. 2002. *McCance and Widdowson's The Composition of Foods*, 6th Summary Ed., Royal Society Chemistry and Food Standards Agency.

9. Bermejo-Barrera, P. et al. 2000. Optimization of microwave-pseudo digestion procedure by experimental designs for the determination of trace elements in seafood products by atomic absorption spectrometry. *Spectrochim. Acta B*, 55: 1351.

10. Bermejo-Barrera, P. et al. 2001. The multivariate optimisation of ultrsonic bath-induced acid leaching for the determination of trace elements in seafood products by atomic absorption spectrometry. *Anal. Chim. Acta*, 439: 211.

11. Bermejo-Barrera, P. et al. 2002. Sample pretreatment methods for the trace elements determination in seafood products by atomic absorption spectrometry. *Talanta*, 57: 969.

12. Saavedra, Y. et al. 2004. A simple optimized microwave digestion method for multielement monitoring in mussel simples. *Spectrochim. Acta B*, 59: 533.

13. Capelo, J.L., C. Maduro, and C. Vilhena. 2005. Discussion of parameters associated with the ultrasonic solid-liquid extraction for elemental analysis (total content) by electrothermal atomic absorption spectrometry. An overview. *Ultrason. Sonochem.*, 12: 225.

14. Brisbin, J.A. and J.A. Caruso. 2002. Comparison of extraction procedures for the determination of arsenic and other elements in lobster tissue by inductively coupled plasma mass spectrometry. *Analyst*, 127: 921.

15. Lambert, D.F. and N.J. Turoczy. 2000. Comparison of digestion methods for the determination of selenium in fish tissue by cathodic stripping voltammetry. *Anal. Chim. Acta*, 408: 97.

16. AOAC. 2002. In W. Horwitz (ed.) *Oficial Methods of Analysis of AOAC International*, 17th ed., Vols. 1 and 2, AOAC, Gaithersberg, MD.

17. El Azouzi, H., M.L. Cervera, and M. De la Guardia. 1998. Multielemental analysis of mussel samples by atomic absorption spectrometry after room temperature sonication. *J. Anal. Atom. Spectrom.*, 13: 533.

18. Manutsewee, N. et al. 2007. Determination of Cd, Cu, and Zn in fish and mussel by AAS after ultrasound-assisted acid leaching extraction. *Food Chem.*, 101: 817.

19. Garcia-Rey, R.M., R. Quiles-Zafra, and M.D. Luque de Castro. 2003. New methods for acceleration of meat sample preparation prior to determination of the metal content by atomic absorption spectrometry. *Anal. Bioanal. Chem.*, 377: 316.

20. Bou, R. et al. 2004. Validation of mineralization procedures for the determination of selenium, zinc, iron, and copper in chicken meat and feed samples by ICP-AES and ICP-MS. *J. Anal. Atom. Spectrom.*, 19: 1361.

21. Yebra-Biurrun, M.C., A. Moreno-Cid, and S. Cancela-Pérez. 2005. Faster on-line ultrasound-assisted extraction coupled to a flow injection-atomic absorption spectrometric system for zinc determination in meat samples. *Talanta*, 66: 691.

22. Alegría, A. et al. 2004. Inorganic nutrients. In L.M.L. Nollet (ed.) *Handbook of Food Analysis*, 2nd ed., Vol. 1, Marcel Dekker, New York.

23. Cal-Prieto, M.J. et al. 2002. Slurry sampling for direct analysis of solid materials by electrothermal atomic absorption spectrometry (ETAAS). A literature review from 1990 to 2000. *Talanta*, 56:1.

24. Pereira, E.R., J.J.R. Rohwedder, and M.A.Z. Arruda. 1998. On-line microwave slurry sample digestion using flow systems for the spectrophotometric determination of iron in seafood. *Analyst*, 123: 1023.

25. Pereira-Filho, E.R. and M.A.Z. Arruda. 1999. Mechanised flow system for on-line microwave digestion of food samples with off-line catalytic spectrophotometric determination of cobalt at ng l^{-1} levels. *Analyst*, 124: 1873.

26. Nomura, C.S. 2005. Bovine liver sample preparation and micro-homogeneity study for Cu and Zn determination by solid sampling electrothermal atomic absorption spectrometry. *Spectrochim. Acta B*, 60: 673.

27. Huerta-Leidenz, N. et al. 2003. Composición mineral del músculo longissimus crudo derivado de canales bovinas producidas y clasificadas en Venezuela. *Arch. Latinoam. Nutr.*, 53: 96.

28. Corrao, P.A. et al. 1983. Titrimetric determination of calcium in mechanically separated poultry and beef: collaborative study. *J. AOAC Int.*, 66: 989.

29. Ipinmoroti, K.O. 1993. Determination of trace-metals in fish, associated waters, and soil sediments from fish ponds. *Disc. Inno.*, 5: 135.

30. Begley, T.H. et al. 1984. Determination of sodium chloride in meat by near-infrared diffuse reflectance. *J. Agric. Food Chem.*, 32: 984.

31. Gordon, D.T. 1978. Atomic absorption spectrometric and colorimetric determination of iron in seafoods. *J. AOAC Int.*, 61: 715.

32. Tee, E.S., A.C. Khor, and M.S. Siti. 1989. Determination of iron in foods by the atomic absorption spectrophotometric and colorimetric methods. *Pertanika*, 12: 313.

33. Hong, J. and K. Yasumoto. 1996. Near-infrared spectroscopic analysis of heme and nonheme iron in raw meats. *J. Food Compos. Anal.*, 9: 127.

34. Monsen, E.R. et al. 1978. Estimation of available dietary iron. *Am. J. Clin. Nutr.*, 31: 34.

35. Carter, P. 1971. Spectrophotometric determination of serum iron at the submicrogram level with a new reagent (ferrozine). *Anal. Biochem.*, 40: 450.

36. Buchowski, M.S. et al. 1998. Heating and the distribution of total and heme iron between meat and broth. *J. Food Sci.*, 53: 43.

37. Ahn, D.U., F.H. Wolfe, and J.S. Sim. 1993. Three methods for determining nonheme iron in turkey meat. *J. Food Sci.*, 58: 288.

38. Kalpalathika, P.V.M., E.M. Clark, and A.W. Mahoney. 1991. Heme iron content in selected ready-to-serve beef products. *J. Agric. Food Chem.*, 39: 1091.

39. Harrington C.F. et al. 2001. A method for the quantitative analysis of iron speciation in meat by using a combination of spectrophometric methods and high-performance liquid chromatography coupled to store field inductively coupled plasma mass spectrometry. *Anal. Chem.*, 73: 4422.

40. Demirbas, A. 1999. Proximate and heavy metal composition in chicken meat and tissues. *Food Chem.*, 67: 27.

41. Kadim, I.T. et al. 2005. Prediction of crude protein, extractable fat, calcium, and phosphorus contents of broiler chicken carcasses using near-infrared reflectance spectroscopy. *Asian-Aust. J. Anim. Sci.*, 18: 1036.

42. Christians, D.K. et al. 1991. Sulfuric-acid hydrogen-peroxide digestion and colorimetric determination of phosphorus in meat and meat-products-collaborative study. *J AOAC Int.*, 74: 22.

43. Tanner, D.K., E.N. Leonard, and J.C. Brazner. 1999. Microwave digestion method for phosphorus determination in fish tissue. *Limnol. Oceanogr.*, 44: 708.

44. Gökoölu, N. and P. Yerlikaya. 2003. Determination of proximate composition and mineral contents of blue crab (*Callinetes sapidus*) and swin crab (*Portunus pelagicus*) caught off Gulf of Antalya. *Food Chem.*, 80: 495.

45. Izquierdo Córser, P. et al. 2000. Análisis proximal, perfil de ácidos grasos, aminoácidos esenciales y contenido de minerals en doce especies de pescado de importancia commercial en Venezuela. *Arch. Latinoam. Nutr.*, 50: 187.

46. Izquierdo, P. et al. 2001. Análisis proximal, contenido de aminoácidos esenciales y relación calcio/fósforo en algunas especies de pescado. *Rev. Ci. FCV-Luz*, 11: 95.

47. Dusek, M. et al. 2003. Isotachophoretic determination of added phosphate in meat products. *Meat Sci.*, 65: 765.

48. Siong, T.E., K.S. Choo, and S.M. Shahid. 1989. Determination of iron in foods by the atomic absorption spectrophotometric and colorimetric methods. *Pertanika*, 12: 313.
49. Garwin, J.L., N.S. Rosenholtz, and A. Abdollahi. 1994. Two colorimetric assays for iodine in foods. *J. Food Sci.*, 59: 1135.
50. Tinggi, U., C. Reilly, and C.M. Patterson. 1992. Determination of selenium in foodstuffs using spectrofluorometry and hydride generation atomic absorption spectrometry. *J. Food Compos. Anal.*, 5: 269.
51. Bobcek, B. et al. 2004. Effects of dietary organic selenium supplementation on selenium content, antioxidative status of muscles and meat quality of pig. *Czech J. Anim. Sci.*, 49: 411.
52. Ragos, G.C., M.A. Demertxis, and P.B. Issopoulos. 1998. A high-sensitive spectrofluorimetric methods for the determination of micromolar concentration of iron (III) in bovine liver with 4-hydroxyquinoline. *Fármaco*, 53: 611.
53. Song, Z. and Q.Y. Changna Wang. 2006. Flow injection chemiluminiscence determination of femtogram-level cobalt in egg yolk, fish tissues, and human serum. *Food Chem.*, 94: 457.
54. Contis, E.T. 2000. Use of nuclear techniques for the measurement of trace element in food. *J. Radioanal. Nucl. Ch.*, 243: 53.
55. Alamin, M.B. et al. 2006. Determination of mercury and selenium in consumed food items in Libya using instrumental and radiochemical NAA. *J. Radioanal. Nucl. Ch.*, 270: 143.
56. Zikovsky, L. and K. Soliman. 2001. Determination of daily dietary intakes of Br, Ca, Cl, Co, and K in food in Montreal, Canada, by neutron activation analysis. *J. Radioanal. Food*, 247: 171.
57. Zikovsky, L. and K. Soliman. 1999. Epithermal neutron activation analysis of food. *J. Radioanal. Food Chem.*, 240: 681.
58. Anderson, D.L. 2000. Neutron capture prompt gamma-ray activation of meat homogenates. *J. Radioanal. Nucl. Ch.*, 244: 225.
59. Anderson, D.L. and E.A. Mackey. 2005. Improvements in food analysis by thermal neutron capture prompt gamma-ray spectrometry. *J. Radioanal. Nucl. Ch.*, 263: 683.
60. Helmke, P.A. and D.M. Ney. 1992. Relationships between concentrations of sodium, potassium, and chlorine in unsalted foods. *J. Agric. Food Chem.*, 40: 1547.
61. Cunningham, W.C., S.G. Capar, and D.L. Anderson. 1997. Determination of sodium in biological materials by instrumental neutron activation analysis. *J AOAC Int.*, 80: 871.
62. Chatt, A. et al. 1988. Determination of trace elements in foods by neutron activation analysis. *J. Radioanal. Nucl. Chem.*, 124: 65.
63. Ibrahim, N. and I. Mat. 1995. Trace element content in relation to the body weight of the marine bivalve, anadara granosa with especial reference to the application of INAA and ICP-AES as analytical techniques. *J. Radioanal. Nucl. Ch.*, 195: 203.
64. Sanchez, M., S. Landsberger, and J. Braisted. 2006. Evaluation of ^{40}K in foods by determining total potassium using neutron activation analysis. *J. Radioanal. Nucl. Chem.*, 269: 487.
65. McDowell, L.S., P.R. Giffen, and A. Chatt. 1987. Determination of selenium in individual food items using the short-lived nuclide ^{77m}Se. *J. Radioanal. Nucl. Chem.*, 110: 519.
66. Sarmani, S. et al. 1993. Analysis of toxic trace elements in seafood samples by neutron activation. *J. Radioanal. Nucl. Chem.*, 169: 255.
67. Khan, A.H. et al. 1989. The status of trace and minor elements in some Bangladeshi foodstuffs. *J. Radioanal. Nucl. Ch.*, 134: 367.
68. Nielson, K.K. et al. 1991. Mineral concentrations and variations in fast-food samples analyzed by X-ray fluorescence. *J. Agric. Food Chem.*, 39: 887.
69. Vives, A.E.S. et al. 2006. Analysis of fish samples for environmental monitoring and food safety assessment by synchrotron radiation total reflection X-ray fluorescence. *J. Radioanal. Nucl. Chem.*, 270: 231.
70. Carvalho, M.L., S. Santiago, and M.L. Nunes. 2005. Assessment of the essential element and heavy metal content of edible fish muscle. *Anal. Bioanal. Chem.*, 382: 426.
71. Tarafdar, S.A. et al. 1991. Level of some minor and trace elements in Bangladeshi meat products. *J. Radioanal. Nucl. Chem.*, 152: 3.

72. Carvalho, M.L., R.A. Pereira, and J. Brito. 2002. Heavy metals in soft tissues of *Tursiops truncatus* and *Delphinus delphis* from West Atlantic Ocean by X-ray spectrometry. *Sci. Total Environ.*, 292: 247.

73. Sandor, Z. et al. 2001. Trace metal levels in freshwater fish, sediment, and water. *Environ. Sci. Pollut. Res.*, 4: 265.

74. González-Martín, I. et al. 2002. Mineral analysis (Fe, Zn, Ca, Na, K) of fresh Iberian pork loin by near-infrared reflectance spectrometry. Determination of Fe, Na, and K with remote fibre-optic reflectance probe. *Anal. Chim. Acta*, 468: 293.

75. Locatelli, C. and G. Torsi. 2001. Heavy metal determination in aquatic species for food purposes. *Ann. Chim-Rome*, 91: 65.

76. Labar, Ch. and L. Lamberts. 1994. Determination of metals in animal tissues by potentiometric stripping analysis without chemical destruction of organic matter. *Electrochim. Acta*, 39: 317.

77. Sherigara, B.S. et al. 2007. Simultaneous determination of lead, copper, and cadmium onto mercury film supported on wax impregnated carbon paste electrode—assessment of quantification procedures by anodic stripping voltammetry. *Electrochim. Acta,* 52: 3137.

78. Locatelli, C. 2000. Proposal of new analytical procedures for heavy metal determinations in mussels, clams, and fishes. *Food Addit. Contam.*, 17: 769.

79. Stryjewska, E., S. Rubel, and A. Skowron. 1994. Wet digestion using microwave irradiation for mineralization of organicsamples before heavy-metal determination by differential pulse anodic stripping voltammetry 2. Food and animal tissue. *Chem. Analityczna*, 39: 609.

80. Filichkina, O.G., E.A. Zakharova, and G.B. Slepchenko. 2004. Determination of selenium in foodstuffs by cathodic stripping voltammetry at a mercury-graphite electrode. *J. Anal. Chem.*, 59: 481.

81. Lan, W.G., M.K. Wong, and Y.M. Sin. 1994. Microwave digestion of fish tissue for selenium determination by differential pulse polarography. *Talanta*, 41: 53.

82. Koplik, R. and S. Nemcova. 1993. Determination of cadmium, lead, copper, and zinc in foodstuffs. *Potravinárskévedy*, 11: 371.

83. Yebra-Biurrun, M.C. and A. García-Garrido. 2001. Continuous flow systems for the determination of trace elements and metals in seafoods. *Food Chem.*, 72: 279.

84. Miller-Ihli, N.J. 1996. Trace element determinations in foods and biological samples using inductively coupled plasma atomic emission spectrometry and flame atomic absorption spectrometry. *J. Agric. Food Chem.*, 44: 2675.

85. Cubadda, F. et al. 2002. Multielemental analysis of food and agricultural matrixes by inductively coupled plasma-mass apectrometry. *J. AOAC Int.*, 85: 113.

86. Kelly, S., K. Heaton, and J. Hoogewerff. 2005. Tracing the geographical origin of food: the application of multielement and multiisotope analysis. *Trends Food Sci. Technol.*, 16: 555.

87. Maichin, B., P. Kettisch, and G. Knapp. 2000. Investigation of microwave assisted drying of samples and evaporation of aqueous solutions in trace element analysis. *Fresenius J. Anal. Chem.*, 366: 26.

88. Jorhem L. et al. 1989. The levels of zinc, copper, manganese, selenium, chromium, nickel, cobalt, and aluminium in the meat, liver, and kidney of Swedish pigs and cattle. *Z. Lebensm. Unters. Forsch*, 188: 39.

89. Kuznetsova, A.I., O.V. Zarubina, and G.A. Leonova. 2002. Comparison of Zn, Cu, Pb, Ni, Cr, Sn, Mo concentrations in tissues of fish (Roach and perch) from Lake Baikal and Bratsk reservoir. *Russia Environ Geochem. Hlth.*, 24: 205.

90. Tinggi, U., C. Reilly, and C.M. Patterson. 1997. Determination of manganese and chromium in foods by atomic absorption spectrometry after wet digestion. *Food Chem.*, 60: 123.

91. Adeyeye, E.I. et al. 1996. Determination of some metals in *Clarias gariepinus* (Cuvier and Vallenciennes), *Cyprinus carpio* (L.), and *Oreochromis niloticus* (L.) fishes in a polyculture fresh water pond and their environments. *Aquaculture*, 147: 205.

92. Dias, V.M., S. Cadore, and N. Baccan. 2003. Determination of selenium in some food matrices by electrothermal atomic absorption spectrometry alters preconcentration with diethyldithiophosphate. *J. Anal. Atom Spectrom.*, 18: 783.

93. Chou, C.L. et al. 2000. Copper contamination and cadmium, silver, and zinc concentrations in the digestive glands of American lobster (*Homarus Americanus*) from the inner bay of Fundy, Atlantic Canada. *Bull. Environ. Contam. Toxicol.*, 65: 470.

94. Moreno-Cid, A. and M.C. Yebra. 2002. Flow injection determination of copper in mussels by flame atomic absorption spectrometry after on-line continuous ultrasound-assisted extraction. *Spectrochim. Acta B*, 57: 967.

95. Yebra, M.C. and A. Moreno-Cid. 2003. On-line determination of manganese in solid seafood samples by flame atomic absorption spectrometry. *Anal. Chim. Acta*, 477: 149.

96. Díaz-Alarcón, J.P. et al. 1996. Determination of selenium in meat products by hydride generation atomic absorption spectrometry-selenium levels in meat, organ meats, and sausages in Spain. *J. Agric. Food Chem.*, 44: 1494.

97. Plessi, M., D. Bertelli, and A. Monzani. 2001. Mercury and selenium content in selected seafood. *J. Food Compos. Anal.*, 14: 461.

98. Demirezen, D. and K. Uruc. 2006. Comparative study of trace elements in certain fish, meat, and meat products. *Meat Sci.*, 74: 255.

99. Haldimann, M., A. Eastgate, and B. Zimmerli. 2000. Improved measurement of iodine in food samples using inductively coupled plasma isotope dilution mass spectrometry. *Analyst*, 125: 1977.

100. Brumbaugh, W.G., C.J. Schmitt, and T.W. May. 2005. Concentrations of cadmium, lead, and zinc in fish from mining-influenced waters of Northeastern Oklahoma: sampling of blood, carcass, and liver for aquatic monitoring. *Arch. Environ. Contam. Toxicol.*, 49: 76.

101. Rhee, K.S. and Y.A. Ziprin. 1978. Modification of Schricker nonheme iron method to minimize pigment effects for red meats. *J. Food Sci.*, 52: 1174.

102. Hendricks, D.G. et al. 1987. Validity of assumptions used in estimating heme iron for determining available dietary iron. *Fed. Proc.*, 46: 1160.

103. Schricker, B.R., D.D. Miller, and J.R. Stauffer. 1982. Measurement and content of nonheme and total iron in muscle. *J. Food Sci.*, 47: 740.

104. Harrington C.F. et al. 2004. Quantitative analysis of iron-containing protein myoglobin in different foodstuffs by liquid chromatography coupled to high-resolution inductively coupled plasma mass spectrometry. *J. AOAC Int.*, 87: 253.

105. Szpunar, J. 2000. Bio-inorganic speciation analysis by hyphenated techniques. *Analyst*, 125: 963.

106. Önning, G. 2000. Separation of soluble selenium compounds in different fish species. *Food Chem.*, 68: 133.

107. Nischwitz, V., B. Michalke, and A. Kettrup. 2003. Extraction and characterisation of trace element species from porcine liver samples using on-line HPLC-ICP-MS and off-line HPLC-ESI-MS. *J. Anal. Atom Spectrom.*, 18: 444.

108. Jackson, B.P. et al. 2002. Trace element speciation in largemouth bass (*Micropterus salmoides*) from a fly ash settling basin by liquid chromatography- ICP-MS. *Anal. Bioanal. Chem.*, 374: 203.

109. Moreno, P. et al. 2001. Fractionation studies of selenium compounds from oysters, and their determination by high-performance liquid chromatography coupled to inductively coupled plasma mass spectrometry. *J. Anal. Atom. Spectrom.*, 16: 1044.

110. Diaz-Huerta, V., M.L. Fernandez -Sanchez, and A. Sanz-Medel. 2004. Quantitative selenium speciation in cod muscle by isotope dilution ICP-MS with a reaction cell: comparison of different reported extraction procedures. *J. Anal. Atom. Spectrom.*, 19: 644.

SENSORY QUALITY

Chapter 26

Color Measurements on Muscle-Based Foods

José Angel Pérez-Alvarez and Juana Fernández-López

Contents

26.1 General Aspects of Color

The first impression that a consumer receives concerning a food product is established visually, and among the properties observed are color, form, and surface characteristics. The power of color for food is not in doubt. Color appearance, color contrast, and color difference in muscle-based foods can have a significant effect on an individual's moods and feelings, and food technologists can exploit these effects.

26.1.1 Color Attributes

Color is the main aspect that defines a food's quality, and a product may be rejected simply because of its color, even before other properties, such as aroma, texture, and taste, can be evaluated [1].

All muscle-based foods possess a number of visually perceived attributes, all of which contribute to their color and appearance as well as to their overall quality. According to Lozano [2], the appearance, can be divided into three different categories: color, cesia, and spatial properties or spatiality. Color is related to optical power spectral properties of the stimulus detected by observers. Cesia includes transparency, translucence, gloss, luster, haze, lightness, opacity, matte and is related to the properties of reflecting, transmitting, or diffusing light by foods as evaluated by human observation. Spatial properties are divided into two main groups: (i) modes of appearance in which color is modified depending on the angle of observation relative to the angle of light incidence, such as metallic, pearlescent, or iridescent materials, and (ii) modes of appearance related to optical properties of surfaces or objects in which effects of ordered patterns (textures) or finishing characteristics of food (roughness, polish, etc.) may be apparent.

26.1.1.1 Appearance

The overall appearance of an object consists of visual structure, surface texture, and distributions of color, gloss, and translucency. It comprises the visual images within the observer. These images are controlled by viewer-dependent variables and scene-dependent variables. The first consists of the viewer's individual visual characteristics, upbringing and preferences, and immediate environment. The second consists of the physics of the constituent materials and their temporal properties combined with the way these are put together, as well as the scene illumination, that is, the light and shade as they define the volume and texture of the scene. The model considers the buildup of the appearance image [3]. As regards the specific characteristics that contribute to the physical appearance of meat, color is the quality that most influences consumer choice [4]. This is why the appearance (optical properties, physical form, and presentation) of meat products at the sales point is of such importance for the meat industry [5].

26.1.1.2 Color as Quality Parameter

The relation between meat color and quality has been the subject of study since the 1950s—since, indeed, Urbain [6] described how consumers had learned through experience that the color of fresh meat is bright red and any deviation from this color (nonuniform or anomalous coloring) is unacceptable [7]. The color of fresh meat and associated adipose tissue is, then, of great importance for its commercial acceptability, especially in the case of beef and lamb [8] and in certain countries such as the United States and Canada; there have been many studies attempting to identify the factors controlling its stability. Adams and Huffman [9] affirmed that consumers relate a meat's color with its freshness. In poultry, the consumers of many countries also associate the meat color with the way in which the animal was raised (intensive or extensive) and fed (cereals, animal feed, etc.) [10,11].

The color of foods greatly influences consumers' preferences [12,13]. Color as a quality factor in meat can be appreciated in different ways in different countries. For example, in Denmark, pork meat color has the fifth place among consumers' purchase decision criteria [14]. In the U.S. and Mexico, chicken skin color plays a significant role in the acceptance of chicken [15]. In the Balkans, visual impression when choosing young beef in retail stores was most important for the average consumer. In these countries the survey found color to be the deciding factor (red pink color was desirable) [16]. Sensorial quality, especially color and appearance [17], of meat can be affected by internal and external factors. For example, in poultry, if carbohydrate supplements are used prior to slaughter, producers should notify processing plant officials so that inspectors do not interpret light livers as an abnormal physiological state [18].

Food technologists, especially those concerned with the meat industry, have a special interest in the color of food for several reasons: first, because of the need to maintain a uniform color throughout processing; second, to prevent any external or internal agent from acting on the product during its processing, storage, and display; third, to improve or optimize a product's color and appearance; and, last, to bring the product's color into line with what the consumer expects [19].

In simple words, the color of meat is determined by the pigments present in the meat. These can be classified into four types: biological (carotenes and hemopigments), which are accumulated or synthesized in the organism antemortem [5]; pigments produced as a result of damage during manipulation or inadequate processing conditions; pigments produced postmortem (through enzymatic or nonenzymatic reactions) [20]; and, finally, those resulting from the addition of natural or artificial colorants [21].

The color also provides information about raw materials [22,23], processing technologies [24], storage conditions [25], shelf life [26], and defects [27]. As a quality parameter, color has been widely studied at slaughter [28] and in fresh meat [29–31] and cooked products [32–34]. Dry-cured meat products have received less attention [25,35,36] because in this type of product color formation takes place during the different processing stages [24,37]; recently, new hem pigment has been identified in this type of product [38–40].

From a practical point of view, color plays a fundamental role in the animal production sector, especially in meat production (beef and poultry, primarily) [41–44]; thus, supplementation of swine diets with 80 mg/kg of manganese may improve pork color and retard discoloration during retail display [45]. Ponsano et al. [46] found that *Rhodocyclus gelatinosus* supplementation resulted in more yellow breast skin and increased darkening and color purity of breast and thigh skins. Also, the consumer can evaluate beef fat as undesirable if it has yellow or dark color, or is excessive glossy or lustrous [44].

In many countries of the European Union (e.g., Spain and Holland), paleness receives a wholesale premium. The poultry sector is also affected by color characteristics, according to Zhang et al. [45]; skin color is used as indicator for chicken sex.

Color can also be used as an important tool by slaughter plant food inspectors. Trampel et al. [18] determined that livers from full-fed birds were lighter in color than normal, and consequently a significant number of chicken carcasses can be condemned for human consumption because they are associated with higher hepatic lipid concentrations.

There are some studies in which technological parameters and color coordinates are related. Pale, soft, exudative (PSE) meat is a growing problem in the poultry industry (characterized by rapid postmortem pH decline). The low pH condition while the body temperature has not yet chilled leads to protein denaturation, causing pale color and reduced water-holding properties. The water loss and protein damage from the PSE condition may impact the visible muscle properties [46]. Thus, Fraqueza et al. [47] stated that PSE and dark turkey meat quality can be defined by lightness (color coordinate: L^*) and pH, thus, dark turkey meat showed $L^* \leq 44$ and pH_{24} >5.8, while PSE turkey meat showed $L^* \geq 50$ and $pH_{24} <5.8$. Fasone and Priolo [48] stated that in ostrich meat, meat lightness (L^*) was strongly affected by stress; in fact, the color of meat from stressed ostriches was darker than meat from nonstressed birds (34.30 versus 38.10, respectively).

26.2   Color Measurement

The color of foods can be defined as the interaction of a light, an object, an observer, and the surroundings of the food. Recently, the International Commission on Illumination [49] described how background can influence the appreciation of color.

26.2.1 Objective Methods

Optical methods have the advantage of being nondestructive, fast, inexpensive, and are considered suitable for online measurement. The color of foods can be studied in two main ways: chemically by analyzing the pigments present or physically by measuring the interaction of light. Color necessarily requires a light source that illuminates an object, which, in turn, modifies the light and reflects (or transmits) it to an observer. The observer senses the reflected light, and the combined factors provide the stimulus that the brain converts into our perception of color, a property that has three quantitatively definable dimensions: hue, chroma, and lightness [50].

Several methods are available for objectively measuring the color of foods, some of which depend on the extraction of pigments from food products followed by spectrophotometric determination of pigment concentration [51,52]. However, since such pigment extraction methods are time-consuming and tedious, some researchers have sought simpler methods of color measurement. For example, several methods measure the light reflected from the surface of foods. There are also tabulated coefficients of various objective values that are correlated with panel scores [53]. These objective values consist of numerous combinations of percentage reflectance values and tristimulus values such as Hunter Lab [54], CIE XYZ, Munsell hue, chroma, and value or CIELAB [19]. The use of this color space was adopted as an internal standard. Thus, in many countries the use of this color space is related to quality control, using D_{65} as illuminant and $10°$ as standard observer [55]. L^* is a measure of lightness in which 0 equals black and 100 equals white. High, positive values of a^* indicate redness; large, negative values indicate greenness; b^* values indicate yellowness to blueness. In Table 26.1, CIELAB color coordinates of different raw materials used for meat products elaboration process are shown.

The univariate correlation between a^* and myoglobin concentration was 0.92. However, Lindahl et al. [56] in a recent study reported that a^* values explained more than 90% of the variation in pigment content and form of pork muscles. Another study demonstrated that L^* value was the instrumental value most highly correlated with subjectively assessed color when this was based on the Japanese Color Standards [57].

Table 26.1 Color Coordinates (L^* [Lightness], a^* [Redness], and b^* [Yellowness]) of Different Raw Materials Used for Meat Products Elaboration Process

Raw Material	L^*	a^*	b^*
Pork lard	65.04	2.91	5.28
Pork streaky bacon	62.18	3.65	4.68
Pork dewlap	63.12	3.88	3.60
Chicken breast	62.11	0.99	3.55
Chicken thigh	59.95	1.11	3.98
MDCM	43.42	18.79	13.68
Pork lean meat	42.67	8.76	3.96
Beef stifle	32.42	15.59	5.11
Pork heart	37.27	13.09	7.37
Pork lung	48.25	22.99	14.17

Note: MDCM: mechanically deboned poultry meat.

26.2.1.1 Reflectance Measurements

Diffuse reflectance measures photons that have survived absorption, been scattered diffusely in meat, and eventually escaped from the meat surface. Hence, the conventional absorbance is the combined result of the absorbing and scattering effects and is different from the derived absorption coefficient, which is independent of scattering. Absorbance cannot provide an accurate absorption spectrum because the scattering effect is not excluded. Similarly, the scattering coefficients are independent of sample chemical compositions and are solely determined by sample ultrastructure properties [58]. Color characteristics of dry-cured meat products are measured by reflectance values, which consist of indices and differences of reflectance at different wavelengths [59].

Myoglobin absorbs light in the ultraviolet region and through practically the complete visible region of light [60]. Metmyoglobin (MMb), oxymyoglobin (OMb), nitrosomyoglobin (NOMb), and deoxymyoglobin (DMb) have maximum absorbances (>400 nm) at about 410, 418, 419, and 434 nm, respectively [61]. The absorbance is typically much weaker at higher wavelengths (500–600 nm). Above 500 nm, myoglobin (OMb and NOMb) have absorption maxima at around 545 and 585 nm [61]. The NOMb complex maintains myoglobin in the ferrous state, but this is somewhat unstable and can be displaced and oxidized if stored with excess oxygen and light [62].

Objective color measurements may refer to several properties or various ratios or color difference indices [50]. By summarizing all the reflected colors (wavelengths) and expressing them as one color [63], the color a consumer sees can generally be described in one or two words, which indicate the main color and its shade. However, color measurements, whether descriptive or specific, must be made as carefully as other measurements [50].

Correlations between visual assessment and instrumental color measurement of muscle-based foods are not very high, generally due to both the technique and measurement conditions [25].

Rapid screening techniques to determine quality characteristics of meat are of great interest for both industry and consumers. Reflectance measurements closely relate to what the eye and brain see. This is a good method for examining the amount and chemical state of myoglobin in meat *in situ*. This method is also able to provide a procedure for estimating the percentages of myoglobin forms on the surface of meat. With this method, repeated measurements over time can be made on the same sample. In addition, the procedure is rapid and relatively easy. Reflectance measurements are affected by muscle structure, surface moisture, fat content, additives, and pigment concentrations [64]. Also, tissue structures are associated with the light-scattering properties of meat. In beef, light scattering could potentially be used as an indicator of beef tenderness [65].

Thus, the increase in reflectance values could be related to such factors as water-holding capacity (WHC). Its diminution, caused by falling pH, might cause the meat structure to close up, driving out the intracellular water, thus hindering light penetration into the myofibrils and increasing light scattering. Feldhusen [66] reported that this effect on reflectance values could be related to the sarcoplasmic protein denaturation and precipitation on the myofibrils, resulting in increased light scattering and less light penetration [25,64].

Reflectance spectra also were used to evaluate growth rates [67], detect poultry feces and ingesta [68] or fecal contamination of chicken carcasses [69], determine the use of nitrite in cooked meat products such as bologna type, etc. Swatland [70] reported that the use of nitrite was associated with lower reflectance at 400 and 410 nm, and with higher reflectance at 430 and 440 nm and from 600 to 700 nm.

Reflectance values of different myoglobin states (DMb, MMb, and OMb) can be equal at several wavelengths (isobestic points). Thus, myoglobin forms can be quantified by this method. Swatland [70] also found isobestic points at 580 nm in samples with and without use of nitrites.

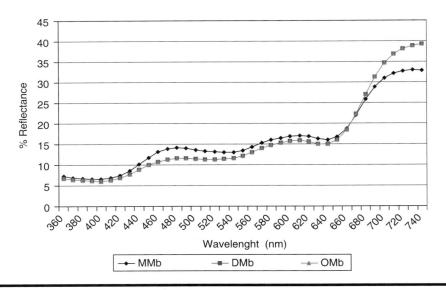

Figure 26.1   Reflectance spectra (360–740 nm) of different myoglobin states (OMb, oxymyoglobin; DMb, deoxymyoglobin; MMb, metmyoglobin) of tuna dark muscle (thynnus thynnus).

According to Snyder [71], several isobestic points are found in beef at 474, 525, 572, and 610 nm. All myoglobin states can be stated at 525 nm in beef. This behavior was also found in pork by Fernández-López et al. [72] and by Navarro [73]. This author also found more isobestic wavelengths for chicken meat and mechanically deboned poultry recovered meat (430, 440, 450, 460, 510, 560, 570, 610, 690 nm) than for beef and pork. In Figures 26.1 and 26.2, reflectance spectra of the different myoglobin states of tuna fish dark muscle and pork shoulder meat, respectively, can be observed. In tuna dark muscle, there are no differences in several wavelengths between OMb and DMb myoglobin states reflectance spectra.

Liu et al. [74] found that intensities of two visible bands at 445 and 560 nm increase with the storage temperature, possibly indicating a color change due to frozen storage. The reduction of spectral intensities probably indicate water loss and compositional alterations during the freeze-thaw process as well as the tenderization development in muscle storage. Chao et al. [75] determined that reflectance spectra in the range of 400–867 nm can be used for veterinarians to select wholesome and unwholesome carcasses.

Irie [44] reported that the main factors affecting bovine fat appearance are carotene concentration and hemoglobin concentration, which affect yellowness and redness, respectively. Also, the chemical state of hemoglobin and the translucency of fat affected the color and percentage reflectance or darkness of beef fat.

26.2.1.2 Near-Infrared Analysis

Near-infrared (NIR) technology by spectral analysis provides complete information about the molecular bonds and chemical constituents in a scanned sample. Optical devices coupled to computers offer potentially very fast data acquisition that may permit decision-making on meat eating quality, albeit from a selected small surface area only [76].

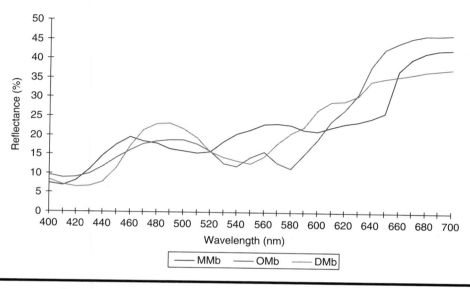

Figure 26.2 **Reflectance spectra (400–700 nm) of different myoglobin states (OMb, oxymyoglobin; DMb, deoxymyoglobin; MMb, metmyoglobin) of pork meat.**

When near-infrared reflectance (NIR) spectroscopy is used, it can be useful for quality control. This technique can predict the chemical composition of poultry meat and possibly some dietary treatments applied to the chickens [77]. Recent studies have therefore concentrated on looking for applications for this technique [78,79]. There is substantial interest in using NIR online to predict chemical parameters in the meat industry and to augment existing video image scanning and analysis (VISA) technology systems [80]. New VISA systems [81,82] provide a noninvasive method operating at normal abattoir chain speeds and enable automatic acquisition of data on carcasses using the side and back view. However, these systems cannot classify on the basis of meat quality and therefore need to be augmented with other suitable systems to measure traits related to meat eating quality. Post-rigor drip loss and WHC during storage can be predicted adequately by NIRS, which can determine WHC specifically, rather than the occurrence of PSE, using techniques such as near-infrared spectroscopy [83]. Also, this technique can be used to predict beef tenderness [84–88].

The USDA Agricultural Research Service has developed a method using a hyperspectral imaging system to detect feces (from duodenum, ceca, and colon) and ingesta on poultry carcasses. The method involves first the use of multivariate data analysis of visible and near-infrared (Vis/NIR) reflectance spectra of fecal and uncontaminated skin samples for classification of contaminants [68].

Myoglobin content varies with species and with fiber type within a particular species. For example, a type I muscle such as beef masseter contains more myoglobin, while a large type IIB pork muscle such as glutens medius contains little myoglobin. Beef latissimus dorsi is an intermediate muscle in myoglobin content.

However, some authors [89,90] have found evidence for the possibility of estimating myoglobin in meat samples using the visible region (400–800 nm) of the spectrum. This would be feasible due to the different forms of myoglobin in the meat samples (OMb, DMb, and MMb), thus giving rise to different colors (bright red, purple, and brown, respectively).

References

1. Pérez-Alvarez, J.A., Fernández-López, J., and Sayas-Barberá, M.E. Fundamentos físicos, químicos, ultraestructurales y tecnológicos en el color de la carne. In *Nuevas tendencias en la tecnología e higiene de la industria cárnica*, eds. M.R. Rosmini, J.A. Pérez-Alvarez, and J. Fernández-López, 51–71. Elche: Universidad Miguel Hernández. 2000.

2. Lozano, D. A new approach to appearance characterization. *Color Res Appl* 31(3):164–167. 2006.

3. Hutchings, J.B. The perception and sensory assessment of colour. In *Colour in food, improving quality*, ed. D.B. MacDougall, 9–32. Cambridge: Woodhead Publishing. 1999, 2002.

4. Krammer, A. Use of color measurements in quality control of food. *Food Technol* 48(10):62–71. 1994.

5. Lanari, M.C. et al. Pasture and grain finishing affect the color stability of beef. *J Food Sci* 67: 2467–2473. 2002.

6. Urbain, M.W. Oxygen is key to the color of meat. *Nat Prov* 127:140–141. 1952.

7. Diestre, A. Principales problemas de la calidad de la carne en el porcino. *Alimentación Equiposy Tecnología* 98:73–78. 1992.

8. Conforth, D. Colour—its basis and importance. In *Advances in meat research*, eds. A.M. Pearson, and T.R. Dutson, 34–78. London: Chapman Hall. 1994.

9. Adams, D.C. and Huffman, R.T. Effect of controlled gas atmospheres and temperature on quality of packaged pork. *J Food Sci* 37:869–872. 1972.

10. De Marchi, M. et al. Carcass characteristics and qualitative meat traits of the Padovana breed of chicken. *Int J Poult Sci* 4:233–238. 2005.

11. Pelicano, E.R.L. et al. Carcass and cut yields and meat qualitative traits of broilers fed diets containing probiotics and prebiotics. *Rev Bras Ciencia Avicola* 7(3):169–175. 2005.

12. Calvo, C., Salvador, A., and Fiszman, S.M. Influence of colour intensity on the perception of colour and sweetness in various fruit flavoured yoghurts. *Eur Food Res Technol* 213:99–103. 2001.

13. Hutchings, J.B. *Food colour and appearance* (2nd ed). Gaithersburg, MD: Aspen Publishers. 1999.

14. Bryhni, E.A. et al. Consumer perceptions of pork in Denmark, Norway and Sweden. *Food Qual Pref* 13:257–266. 2002.

15. Castaneda, M.P., Hirschler, E.M., and Sams, A.R. Skin pigmentation evaluation in broilers fed natural and synthetic pigments. *Poult Sci* 84(1):143–147. 2005.

16. Ostojic, D. et al. Criteria of consumers when purchasing beef in retail stores: position of beef compared to other meat types, reasons and frequency of its use in everyday nutrition and preparation methods. *Biotechnol Anim Husbandry* 22(3/4):45–53. 2006.

17. Brewer, M.S. and Mckeith, F.K. Consumer-rated quality characteristics as related to purchase intent of fresh pork. *J Food Sci* 64:171–174. 2006.

18. Trampel, D.W. et al. Preharvest feed withdrawal affects liver lipid and liver color in broiler chickens. *Poult Sci* 84(1):137–142. 2005.

19. Pérez-Alvarez, J.A. Color. In *Ciencia y tecnología de carnes*, eds. Y.H. Hui, I. Guerrero and M.R. Rosmini, 161–198. México: Limusa Noriega Editores. 2006.

20. Montero, P., Ávalos, A., and Pérez-Mateos, M. Characterization of polyphenoloxidase of prawns (*Penaeus japonicus*). Alternatives to inhibition: additives and high pressure treatment. *Food Chem* 75:317–324. 2001.

21. Fernández-López, J. et al. Effect of paprika (*Capsicum annum*) on color of Spanish-type sausages during the resting stage. *J Food Sci* 67:2410–2414. 2002.

22. Pérez-Alvarez, J.A. et al. Utilización de vísceras como materias primas en la elaboración de productos cárnicos: güeña. *Alimentaria* 291:63–70. 1998a.

23. Pérez-Alvarez, J.A. et al. Caracterización de los parámetros de color de diferentes materias primas usadas en la industria cárnica. *Eurocarne* 63:115–122. 1998b.

24. Pérez-Alvarez, J.A. et al. Chemical and color characteristics of "Lomo embuchado" during salting seasoning. *J Muscle Food* 8(4):395–411. 1997.

25. Pérez-Alvarez, J.A. Contribución al estudio objetivo del color en productos cárnicos crudo-curados. [PhD Thesis]. Valencia, Spain: Universidad Politécnica de Valencia. 1996.
26. Fernández-López, J. et al. Physical, chemical and sensory properties of bologna sausage made with ostrich meat. *J Food Sci* 68:1511–1515. 2003.
27. Ferrer, O.J., Otwell, W.S., and Marshall, M.R. Effect of bisulfite on lobster shell phenoloxidase. *J Food Sci* 54(2):478–480. 1989.
28. Bozek, R. and Juzl, M. Myopathy of slaughter chickens. *Acta Univ Agric Silv* 54:49–56. 2006.
29. MacDougall, D.B. Changes in the colour and opacity of meat. *Food Chem* 9(1/2):75–88. 1982.
30. Cassens, R.G. et al. Recommendation of reference method for assessment of meat color. In *Proceedings of 41st International Congress of Meat Science and Technology*, San Antonio, TX. C86: 410–411. 1995.
31. Faustman, C. et al. Strategies for increasing oxidative stability of (fresh) meat color. In *Proceedings 49th Annual Reciprocal Meat Conference*, Provo, UT. 73–78. 1996.
32. Anderson, H.J., Bertelsen, G., and Skibsled, L.H. Colour and colour stability of hot processed frozen minced beef. Result from chemical model experiments tested under storage conditions. *Meat Sci* 28(2):87–97. 1990.
33. Fernández-Ginés, J.M. et al. Effect of storage conditions on quality characteristics of bologna sausages made with citrus fiber. *J Food Sci* 68:710–715. 2003.
34. Fernández-López, J. et al. Evaluation of antioxidant potential of hyssop (*Hyssopus officinalis* L.) and rosemary (*Rosmarinus officinalis* L.) extract in cooked pork meat. *J Food Sci* 68:660–664. 2003.
35. Pagán-Moreno, M.J. et al. The evolution of colour parameters during "chorizo" processing. Fleischwirtschaft 78(9):987–989. 1998.
36. Aleson, L. et al. Utilization of lemon albedo in dry-cured sausages. *J Food Sci* 68:1826–1830. 2003.
37. Fernández-López, J., Pérez-Alvarez, J.A., and Aranda-Catalá, V. Effect on mincing degree on color properties in pork meat. *Color Res Appl* 25:376–380. 2000a.
38. Parolari, G., Gabba, L., and Saccani, G. Extraction properties and absorption spectra of dry cured hams made with and without nitrate. *Meat Sci* 64(4):483–490. 2003.
39. Wakamatsu, J., Nishimura, T., and Hattori, A. A Zn-porphyrin complex contributes to bright red colour in Parma ham. *Meat Sci* 67(1):95–100. 2004a.
40. Ponsano, E.H.G. et al. Performance and color of broilers fed diets containing *Rhodocyclus gelatinosus* biomass. *Revista Brasileira de Ciencia Avicola* 6(4):237–242. 2004.
40. Wakamatsu, J. et al. Establishment of a model experiment system to elucidate the mechanism by which Zn-protoporphyrin IX is formed in nitrite-free dry-cured ham. *Meat Sci* 68(2):313–317. 2004b.
41. Zhou, G.H., Yang, A., and Tume, R.K. A relationship between bovine fat colour and fatty acid composition. *Meat Sci* 35(2):205–212. 1993.
42. Esteve, E. Alimentación animal y calidad de la carne. *Eurocarne* 31:71–77. 1994.
43. Verdoes, J.C. et al. Isolation and functional characterisation of a novel type of carotenoid biosynthetic gene from *Xanthophyllomyces dendrorhous*. *Mol Gen Genet* 262(3):453–461. 1999.
44. Irie, M. Optical evaluation of factors affecting appearance of bovine fat. *Meat Sci* 57:19–22. 2001.
45. Apple, J.K. et al. Influence of dietary inclusion level of manganese on pork quality during retail display. *Meat Sci* 75:640–647. 2007.
45. Zhang, X.Y. et al. Observation on inheritance of skin colour in green-shell-egg fowl. *J Yunnan Agric Univ* 17(1):39–44. 2002.
46. Woelfel, R.L. and Sams, A.R. Marination performance of pale broiler breast meat. *Poult Sci* 80(10):1519–1522. 2001.
47. Fraqueza, M.J. et al. Incidence of pectoralis major turkey muscles with light and dark color in a Portuguese slaughterhouse. *Poult Sci* 85(11):1992–2000. 2006.
48. Fasone, V. and Priolo, A. Effect of stress on ostrich meat quality. In *Proceedings of the 3rd International Ratite Science Symposium of the World*, ed. E. Carbajo, Madrid, Spain: Poultry Science Association (WPSA). 393–396. 2005.

49. CIE. A colour appearance model for colour management systems: CIECAM02. *CIE Pub*, 159. 2004.

50. AMSA. *Guidelines for meat color evaluation*. American Meat Science Association. Chicago, IL: National Live Stock and Meat Board. 1991.

51. Hornsey, H.C. The colour of cooked cured pork. I. Estimation of the nitric oxide-haem pigments. *J Sci Food Agric* 7:534–540. 1956.

52. Agulló, E. et al. Determination of total pigments in red meats. *J Food Sci* 55:250–251. 1990.

53. Hunt, M.C. Meat color measurements. In *Proceedings of 33rd Reciprocal Meat Conference*, Chicago, IL: American Meat Science Association & National Live Stock and Meat Board. 41–46. 1980.

54. Zhu, S., Ramaswamy, H.S., and Simpson, B.K. Effect of high-pressure versus conventional thawing on color, drip loss and texture of Atlantic salmon frozen by different methods. *Lebensm Wiss u Technol* 37:291–299. 2004.

55. Sayas, M.E. Contribuciones al proceso tecnológico de elaboración del jamón curado: aspectos físicos, fisicoquímicos y ultraestructurales en los procesos de curado tradicional y rápido. [PhD Thesis]. Valencia, Spain: Universidad Politécnica de Valencia, 1997.

56. Lindahl, G., Lundström, K., and Tornberg, E. Contribution of pigment content, myoglobin forms and internal reflectance to the colour of pork loin and ham from pure breed pigs. *Meat Sci* 59(2): 141–151. 2001.

57. Brewer, M.S. et al. Measuring pork colour: effects of bloom time, muscle, pH and relationship to instrumental parameters. *Meat Sci* 57:169–176. 2001.

58. Xia, J.J. et al. Characterizing beef muscles with optical scattering and absorption coefficients in VIS-NIR region. *Meat Sci* 75:78–83. 2007.

59. Pagán-Moreno, M.J. et al. Entstehung von farbparametern während der Herstellung von "chorizo." *Fleischwirtschaft* 77:664–667. 1997.

60. Fox, J.B. The chemistry of meat pigments. *J Agric Food Chem* 14:20–27. 1966.

61. Millar, S.J., Moss, B.W., and Stevenson, M.H. Some observations on the absorption spectra of various myoglobin derivatives found in meat. *Meat Sci* 42(3):277–288. 1996.

62. Kanner, J. Oxidative processes in meat products: quality implications. *Meat Sci* 36:169–189. 1994.

63. Barbut, S. Effect of illumination source on the appearance of fresh meat cuts. *Meat Sci* 59:187–191. 2001.

64. Fernández-López, J. et al. Effect of sodium chloride, sodium tripolyphosphate and pH on color properties of pork meat. *Color Res Appl* 29:67–74. 2004.

65. Hildrum, K.I. et al. Near-infrared reflectance spectroscopy in the prediction of sensory properties of beef. *J Near Infrared Spectrosc* 3:81–87. 1995.

66. Feldhusen, F. Einflüsse auf die postmortale Farbveränderung der Oberfläche von Schweinemuskulatur. *Fleischwirtschaft* 74:989–991. 1994.

67. Correa, J.A. et al. Effects of slaughter weight on carcass composition and meat quality in pigs of two different growth rates. *Meat Sci* 72:91–99. 2006.

68. Windham, W.R. et al. Visible/NIR spectroscopy for characterizing fecal contamination of chicken carcasses. *Trans ASAE* 46(3):747–751. 2003a.

69. Windham, W.R. et al. Algorithm development with visible/near-infrared spectra for detection of poultry feces and ingesta. *Trans ASAE* 46(6):1733–1738. 2003b.

70. Swatland, H.J. A brief study of the effect of nitrite on bologna coloration measured with a Colormet fiber-optic spectrophotometer. *Can Inst Food Sci Technol J* 21(5):560–562. 1988.

71. Snyder, H.E. Analysis of pigments at the surface of fresh beef with reflectance spectrophotometry. *J Food Sci* 30:457–459. 1965.

72. Fernández-López, J. et al. Characterization of the different states of myoglobin in pork meat using colour parameters and reflectance ratios. *J Muscle Foods* 11:57–68. 2000b.

73. Navarro, C. Optimización del proceso de obtención de geles cárnicos a partir de carne de ave mecánicamente recuperada. [PhD Thesis]. Universidad Miguel Hernández. Elche, Alicante, Spain. 2005.

74. Liu, Y.L. et al. Two-dimensional correlation analysis of visible/near-infrared spectral intensity variations of chicken breasts with various chilled and frozen storages. *J Agric Food Chem* 52(3): 505–510. 2004.

75. Chao, K. et al. Characterizing wholesome and unwholesome chickens by CIELUV color difference. *Appl Eng Agric* 21(4):653–659. 2005.

76. Andrés, S.I. et al. Prediction of sensory characteristics of lamb meat samples by near infrared reflectance spectroscopy. *Meat Sci* 76:509–516. 2007.

77. Berzaghi, P. et al. Near-infrared reflectance spectroscopy as a method to predict chemical composition of breast meat and discriminate between different n-3 feeding sources. *Poult Sci* 84(1):128–136. 2005.

78. Brøndum, J. et al. Prediction of water-holding capacity and composition of porcine meat by comparative spectroscopy. *Meat Sci* 55:177–185. 2000.

79. Geesink, G.H. et al. Prediction of pork quality attributes from near infrared reflectance spectra. *Meat Sci* 65:661–668. 2003.

80. Schwarze, H. Continuous fat analysis in the meat industry. In *Third European symposium on near infrared (NIR) spectroscopy*, 43–49, Report no. 996-10-1. 1996.

81. Stanford, K. et al. Video image analysis for online classification of lamb carcasses. *Anim Sci* 67:311–316. 1998.

82. Hopkins, D.L. et al. Video image analysis in the Australian meat industry—precision and accuracy of predicting lean meat yield in lamb carcasses. *Meat Sci* 67:269–274. 2004.

83. Forrest, J.C. et al. Development of technology for the early postmortem prediction of water holding capacity and drip loss in fresh pork. *Meat Sci* 55:115–122. 2000.

84. Mitsumoto, M. et al. Near-infrared spectroscopy determination of physical and chemical characteristics in beef cuts. *J Food Sci* 56:1493–1496. 1991.

85. Hildrum, K.I. et al. Prediction of sensory characteristics of beef by near-infrared spectroscopy. *Meat Sci* 38:67–80. 1994.

86. Naes, T. and Hildrum, K.I. Comparison of multivariate calibration and discriminant analysis in evaluating NIR spectroscopy for determination of meat tenderness. *Appl Spectrosc* 51:350–357. 1997.

87. Byrne, C.E. et al. Non-destructive prediction of selected quality attributes of beef by near-infrared reflectance spectroscopy between 750 and 1098 nm. *Meat Sci* 49:399–409. 1998.

88. Park, B. et al. Near-infrared reflectance analysis for predicting beef longissimus tenderness. *J Anim Sci* 76:2115–2120. 1998.

89. Cozzolino, D. et al. The use of visible and near-infrared reflectance spectroscopy to predict colour on both intact and homogenised pork muscle. *Leb Wiss Technol Food Sci* 36:195–202. 2003.

90. Cozzolino, D. and Murray, I. Effect of sample presentation and animal muscle species on the analysis of meat by near infrared reflectance spectroscopy. *J Near Infrared Spectrosc* 10:37–44. 2002.

Chapter 27

Measuring Meat Texture*

Morse B. Solomon, Janet S. Eastridge, Ernie W. Paroczay, and Brian C. Bowker

Contents

---

* Mention of brand or firm names does not constitute an endorsement by the United States Department of Agriculture over others of a similar nature not mentioned.

27.1   Introduction

For consumers, tenderness is a critical sensory attribute of meat palatability. The perception of tenderness is influenced by many tactile senses including texture. The term "tenderness," which refers to hardness, is often used interchangeably with texture. Meat texture is a complex phenomenon that encompasses characteristics such as hardness, springiness, chewiness, cohesiveness, and even juiciness. Variations in meat texture originate from inherent differences within the structure of raw meat/muscle tissue relating to contractile protein structures, connective tissue framework, lipid, and carbohydrate components,[1,2] as well as external factors like cooking and sample handling. It has been argued that tenderness inadequately describes meat texture because it does not specify how much of the toughness sensation is due to the force to bite through meat compared with the cohesive forces that resist compression or deformation prior to rupture.[3] This chapter will not cover factors affecting texture as they are too numerous to present and discuss in this chapter and have been widely presented in many publications over the years.

The perception of texture is associated with mechanical failure properties that are related to the muscle structure.[4] Accurate and reliable measurement of texture is fundamental to the study of its variability and control.[5] Although texture is a trait that by definition can best be measured by sensory perception, researchers have continually sought to assess texture instrumentally. Instruments used for texture analysis generally measure muscle tissue's resistance to shearing, compression, and/or penetration. Although several older in-depth reviews and engineering assessments of specific meat texture devices exist,[6–8] this chapter provides an overview of the broad array of technologies developed for measuring meat texture and tenderness.

27.2   History of Meat Texture Measurements

27.2.1   Warner–Bratzler Shear

27.2.1.1   Development of the Warner–Bratzler Shear Device

The first and most widely accepted instrumental measure of meat texture over the past 80 years is the Warner–Bratzler shear (WBS) instrument. In 1928, a USDA scientist, Warner[9] presented a brief report of a mouse trap type machine equipped with a shear device that showed results strongly correlated to cooked meat tenderness and superior repeatability on raw meat measurements. In a more complete report,[10] the details of using this instrument to measure tenderness of beef were given. The instrument (Figure 27.1) consisted of a thin steel blade with a square hole in the middle slightly larger than the sample to be tested. A hand driven screw was used to pull the blade through a narrow slit in a wooden miter box, and the maximum force required to pull

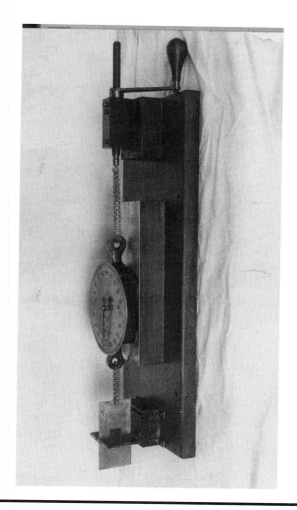

Figure 27.1   Original Warner shear device.

the blade through the sample was recorded by a spring-type dynamometer. The specifics of the instrument configuration were refined by Bratzler, a graduate student at Kansas State University, and the instrument became known as the Warner–Bratzler shear (WBS) device[11,12] (Figure 27.2). The thickness of the steel blade was standardized at 1.016 mm, the shape of the hole in the steel blade was changed to be triangular, and the points of the triangular hole were rounded. The triangular hole was selected because, as Bratzler demonstrated using a wooden dowel, without a platform to rest the sample on this would automatically align the cylindrical core of meat perpendicular to the blade as the slot moved downward past the blade.

Operation of the WBS consisted of excising 15 mm diameter cylindrical specimens from cooked meat and placing the sample in the cutout of the blade. As the slot moves past the blade, the meat specimen is compressed and the cross-sectional shape changes to conform to the restriction imposed by the triangular opening of the blade until it is eventually sheared into two pieces. The speed for shearing the cylindrical meat cores was set at 22.86 cm/min. After determining that

Figure 27.2 Modified Warner–Bratzler shear machine.

a machine powered by a constant speed motor was the easiest to use, Bratzler proceeded to have the G-R Electric Manufacturing Company (Manhattan, Kansas) build several copies based on his design. Some excellent discussions on the development of WBS were recounted by the inventors in early Reciprocal Meat Conferences.[13,14]

Studying the deformation and stresses developed in the Warner–Bratzler test and using a WBS test cell on other instruments, Voisey[8] concluded that despite the simplicity of the WBS devices, data curves are deceptively complex to interpret. The maximum force recorded during the WBS meat shear test (WBSF) is considered to be a measure of meat tenderness. Although WBS is primarily used to obtain peak force values, some researchers[15–21] have also charted the time–force curves during WBS operation to profile the forces measured. Based on the shape of these WBS time–force deformation curves, various mechanical parameters including peak force have been used to further characterize tenderness differences. The conventional interpretation, as implied in the name, is that shear forces are primarily responsible for cutting the meat specimen during WBSF measurement. Other researchers, however, argued that tensile strength[7,8,22–24] is the primary contributor in WBS shearing process and that WBS fails to account for meat's viscoelastic[25] nature.

Addressing the difficulties in standardizing WBS texture measurements, Bourne[26] concluded that frictional forces of the blade moving through the slot could be a substantial source of error and could vary considerably among instruments. The ideal texture test, according to Bourne, would be frictionless so that instruments could be calibrated in fundamental units of force, distance, or time.

Despite many years of research and innovation, the basic WBS technique or its variations remain the most widely accepted and utilized method for objectively measuring meat tenderness. In numerous studies relating WBSF to sensory panel tenderness scores, correlation coefficients have been found to range from −.92 to .7, depending on the tenderness range of the meat samples included in the study. Szczesniak[27] has an excellent discussion of correlating sensory with instrumental texture measurements. Table 27.1 lists some correlation coefficients (*R*) of various

Table 27.1   Correlation Coefficients (R) of Various Measures of Texture Reported in Literature

Comparison Factor	Correlated To	Correlation Coefficient[a]	Reference(s)
Sensory tenderness	WBSF	−.77 to −.92	42
	WBSF	−.63*	45
	WBSF	−.80[NR]	47
	WBSF	.68–.70**	123
	WBSF	−.63	90
	Sharp needle	−.77	75
	Sharp blade	−.52	75
	Plumb bob	−.53	75
	Plumb bob	−.71	100
	NIR raw	.42–.7	123
Sensory hardness	Needle probe	−.74	100
	Force at first yield	−.84***	66
	WBSF	.64	87
	BMORS	.54	87
Sensory toughness	BMORS cooked	.73	87
	WBSF	.8	87
	WBSF	.92***	171
WBSF raw	WBSF cooked	.82**	89, 99
WBSF, day 2 postmortem	WBSF, day 7	.66	116
	WBSF, day 14	.92	116
Resistance raw, day 2 postmortem	Resistance raw, day 8	.5	172
	Resistance raw, day 14	.8	172
BMORS raw, day 0	BMORS cooked, day 7	.56	87
NIR raw	WBSF cooked	.73–.82	123
Armour tenderometer raw	WBSF cooked	.69	89
	WBSF cooked	.48**	173, 174
	WBSF cooked	−.5	90
KT device	WBSF	.4**	45
Punch and die test	Instrumental	.92–.98**	50
BITE Master II	Instrumental	.86–.89[NR]	http://130.184.125.114/bite

[a] Significance of correlation: *$p < .05$, **$p < .01$, and ***$p < .001$. NR indicates not reported.

measures of meat texture/tenderness reported in various literature citations. Even though the WBS technique is the *de facto* standard to which other devices or techniques for measuring meat texture are compared, WBSF fails to comprehensively describe meat tenderness.

27.2.1.2   Standard Protocol for Measuring Warner–Bratzler Shear Force

The American Meat Science Association[28,29] (AMSA) first published standard protocol for determining WBS in 1978, which was updated in 1995. Since then, other scientists[30,31] have published variations of the AMSA guidelines on the World Wide Web. Standardized protocols also have been developed and finalized for the Organisation for Economic Co-operation and Development (OECD)[32–34] and for the Cooperative Research Center[35] (CRC) in Australia. The final protocol[34] for the OECD was presented at the International Congress on Meat Science and Technology in 1997–1998.

27.2.2 Other Texture Measurement Systems

Over the years, a number of devices have been developed for objectively measuring meat texture, and many have been reviewed[6,8,23,36–42] extensively. Szczesniak and Torgeson[6] reviewed about 20 texture devices covering shearing, penetration, biting, compression, and grinding. Many texture devices never gained the popularity of WBS or were deemed not acceptable due to low correlation to WBS measurements. Among these are the Christel Texturometer,[43,44] KT device,[45] Dassow's shear-jaw device,[41] the Slice Tenderness Evaluator (STE),[46,47] NIP tenderometer,[48] a dynamometer instrument that forced food through one or more perforated plates,[49] and a punch and die[50] test cell. Some instruments such as a hand-operated forceps-type biting instrument,[51] the L&W Texturemeter,[52] and tensiometers[53] were developed for measuring raw meat; however, most have failed to adequately predict cooked meat texture.

The Volodkevich[54] bite tenderometer (VBT), introduced in 1938 and modified by Winkler[55] to have rounded or blunt wedges that pinch the sample during compression, was not further developed as an independent apparatus. However, the VBT is widely available as an attachment to modern texture analyzers. Volodkevich-style bite wedges were incorporated into the Macfarlane and Marer[56] apparatus that was modified[57] (New Zealand patent #190945) to become the MIRINZ tenderometer (AgResearch MIRINZ, Hamilton, New Zealand). The wedges do not slide past each other as would incisor teeth[58] but meet end to end. Bouton and Harris[59] confirmed that the MIRINZ tenderometer and the WBS device had very significant linear relationships ($R = .94$) signifying that both instruments were measuring the same property. The MIRINZ tenderometer is no longer manufactured.[60]

Some specialized tenderometers and texturometers were not commercialized but have been important in texture research. The MIT Denture tenderometer (Massachusetts Institute of Technology, Cambridge, Massachusetts) was meant to simulate, but not duplicate, the masticatory action of the human mouth.[61,62] The MIT tenderometer, introduced by Proctor et al.[63] and later refined by Brody,[64] was more widely accepted for dental and biomedical applications rather than for food. It became the basis of the General Foods Texturometer (General Foods Corp., White Plains, New York) that was commercialized by Zenken Co, Ltd. (Tokyo, Japan). Later, the Bi-cyclical Instrument for Texture Evaluation (BITE) Master artificial panelist was built at the University of Georgia, followed by the BITE Master II (University of Arkansas, Fayetteville, Arkansas; http://130.184.125.114/bitemaster/about.htm). The BITE Master II is an electrognathograph used to track the three-dimensional jaw movements during biting and records corresponding forces exerted on the food. There is limited research[65] using the BITE Master II for muscle foods.

27.3 Developments in Meat Texture Measurement

27.3.1 Digital Texture Analyzers

All modern texture measurement instruments have common components and features such as a digital stress–strain gauge, controllable crosshead speed, data acquisition, and operation in compression or tensile mode. All feature a variety of interchangeable test cells, probes, or attachments. In the

United States, there are a number of manufacturers and distributors of texture analyzers (Table 27.2), and all offer test cells, jigs, or attachments equivalent to the Warner–Bratzler meat shear.

With the emergence of sophisticated data acquisition and software to analyze output, the potential number of variables for assessing texture has greatly expanded. From the time–force graphs generated with various instruments, it is possible to calculate a number of parameters, for example, total energy, work, force at first yield, slope at first yield, or other user-defined measurements. Huang et al.[66] applied step-wise regression analysis to compression test variables to calculate prediction models for tenderness and structural integrity of mutton. Among the four variables analyzed, the one most highly correlated to sensory panel evaluation of tenderness was force at first yield. When structural integrity, which they defined by a second-order mathematical model using force at first yield, was included in the model, the correlation to sensory data increased to $R^2 = .76$. Alternative texture parameters calculated from recorded data have not been utilized for elucidating the texture of meat as extensively as they have been for other food products.

27.3.2 Test Cells, Probes, and Attachments

Over the years many different test cells, probes, and attachments have been developed to fit on various instruments for the measurement of meat texture. Voisey[8] criticized equipment manufacturers because in their efforts to make attachments (test cells or jigs) interchangeable, test cells have not been standardized or verified. The various attachments described in the following sections are chiefly based on the principle of measuring the force necessary to shear, compress, or penetrate the tissue to assess meat texture. Following Bourne's[26] suggestion of using a frictionless test, tensile measurements[67] were performed as a measure of meat texture. With tensile testing, however, there were problems of attaching or gripping the meat sample. Use of a dumbbell shape[21,68] sample helped overcome gripping the sample and is the standard sample shape as outlined by Honikel.[34] Other examples of alternative tests are the texture profile analysis (TPA)[61,69] introduced by Friedman et al. and compression tests[33,34] adopted by the OECD.

27.3.2.1 Blades

27.3.2.1.1 Sharp and Blunt Blades

Bourne[7] suggested that using a flat-edge blade would simplify interpretation of time–force curves from WBS determinations. Several researchers[70–74] have investigated the possibility of using a flat-edge blade instead of the traditional V-notched Warner–Bratzler blade. Overall, it was determined that the V-shaped blade cutout resulted in lower shear values[74] compared to the flat blade for beef longissimus and semitendinosus muscles. Sharp and blunt blades and needles have been evaluated[73,75] for predicting tenderness of cooked longissimus beef using measurements on uncooked samples. Results of those studies indicated that the sharp versions of blade or needles were better than their blunt versions for predicting cooked meat tenderness and prediction results were of similar magnitude.

Shackelford et al.[70,71] presented a simplified technique for measuring longissimus muscle shear force, which they referred to as slice shear force (SSF). They developed this technique while working on a method to reduce variability in the WBS procedure and at the same time producing an online method[70] for assessment of meat tenderness. The SSF method involves cooking a 2.5 cm thick steak and then removing a 1 cm wide by 5 cm long slice parallel to the muscle fibers. The slices are then placed flat on a universal testing machine and sheared once using a blunt-end blade

Table 27.2 Sources of Some Available Texture Analyzers

Manufacturer	Texture Instrument/Features	Contact Information
GR Manufacturing Company Manhattan, Kansas	Original analog Warner–Bratzler Shear Machine with a Chattilon force dial; the WBS 2000D digital model has a Chattilon strain gauge	Richard Lundquist 1317 Collins Lane Manhattan, Kansas 66502; Tel.: +1-785-537-727
Ametek TCI Division Largo, Florida and Lloyd Instruments Ltd. Hampshire, UK	Single and dual column stands in analog and digital versions that employ NEXYGEN software. Models TCM2301 and the digital TCD200 have Chattilon test stands and force gauges	www.ametek.com; www.lloyd-instruments.com
BITE Master II, University of Arkansas Fayetteville, Arkansas	A three-dimensional movement electrognathograph used to track jaw movements during biting; records corresponding forces exerted on food. This artificial panelists was the concept of Dr Jean Meullenet based on the BITE Master	http://130.184.125.114/bitemaster/about.htm
Brookfield Engineering Laboratories Middleboro, Massachusetts	Viscometers, rheology, and texture analysis equipment	www.BrookfieldEngineering.com
Food Technology Corporation Sterling, Virginia	Range of tenderometers and texture measurement systems as stand-alone or running Windows-based software	www.foodtechcorp.com
Imada, Inc. Northbrook, Illinois	Many configurations for manual and motorized test stands, mechanical and digital force gauges and data acquisition systems for compression, tension, and torsion tests	www.imada.com
Instron Corporation Canton, Massachusetts	Universal testing machines for food texture analysis capable of measuring compression and tension. The working parts of most texture-measuring devices can be used on this universal testing machine, available in single and dual column models ranging from large table top to field portable units	www.instron.com

Table 27.2   (Continued)

Manufacturer	Texture Instrument/Features	Contact Information
Itin Scales Company Brooklyn, New York	Supplier of grips and fixtures for food testing and force gauges; products include Chattilon test stands, grips and accessories, Imada force gauges, and Dillon Quantrol force testing equipment	www.itinscales.com
Shimadzu Scientific Instruments Columbia, Maryland	Several universal testing instruments for materials testing (single and dual beam) with analysis using Trapezium software	www.shimadzu.com
Texture Technologies, Inc. Scarsdale, New York and Stable Micro Systems Ltd. Surry, UK	TA-XTplus, TA-XT2i, and TA.XT2iHR texture analyzers with dedicated software	www.texturetechnologies.com; www.stablemicrosystems.com
Tinius Olsen Horsham, Pennsylvania	Single column food testing machines. Warner–Bratzler shear cell available with 30°, 50°, 60°, and 70° V-angles, flat, and square cutout blades	www.TiniusOlsen.com
Zenken Company, Ltd. Tokyo, Japan	This company commercialized the General Foods Texturometer[61]	
Zwick/Roell, Ltd. Kennesaw, GA and Zwick GmbH & Co. KG Ulm, Germany	Materials testing systems for texture analysis, viscosity measurement, and packaging testing; has WBS fixture with flat and V-notch blade	www.zwick.com

at a crosshead speed of 500 mm/min. SSF values were reported[71] to correlate well with sensory panel tenderness values ($R = -.76$) and with WBSF values ($R = .80$).

27.3.2.1.2   Razor Blade

To obtain shear force of individual muscle fibers, Henrickson et al.[76] explored a microsensitive instrument that was fitted with a blunt-edge razor blade. Shear force and shear energy results obtained from this configuration were noted to be sensitive to differences in fiber diameter and muscle fiber type as well as to changes in chilling and deboning treatments.

A more recent variation in determining shear force is by use of a razor blade to penetrate 20 mm into cooked meat. Dubbed the Meullenet–Owen (MO) razor shear force[77–84] (RB, RBS, MOR, MORBS, or MORS), it was first used to gauge the texture of chicken breasts. MORS was reported to have high correlation to consumer sensory attributes of hardness and cohesiveness[77,79,81,85] and was correlated to the Allo-Kramer multiblade shear force of cooked broiler breasts. This novel test measures the force in Newtons (N) to penetrate an 8 mm wide × 20 mm high razor blade

20 mm into intact breast fillets[77] at crosshead speed of 60 mm/min. Improved correlation to sensory attributes was detected when a measure of shear energy (MORSE), defined as the area under the time–force curve, was used. Razor blade shear energy (MORSE = N mm) produced higher correlations ($R = .84–.87$) to sensory attributes than did the Allo-Kramer shear ($R = .68–.71$).

The effect of blade penetration depth, investigated using MORS,[79] revealed that a penetration depth of at least 10 mm was required to obtain consistent correlations (.82–.85) to expert and consumer panel sensory evaluations. As penetration depth increased to 20 mm, the correlation with the trained sensory panel hardness and cohesiveness scores increased, whereas consumer panel acceptance of texture and tenderness decreased. Xiong et al.[79] also showed strong correlation of MORSE values at various penetration depths when they were converted by a multiplication factor to standardize MORSE values equal to a 20 mm penetration depth.

Prediction of broiler breast tenderness by MORS was evaluated[85] using two instruments, a common texture measurement system and a portable texture analyzer. MORS and MORSE values from both instruments were correlated. For both texture analyzers evaluated, the shear energy MORSE values were more highly correlated to sensory and with a lower error rate than to peak shear MORS values. This indicates potential for online application by using a portable texture analyzer for predicting breast tenderness.

Changes in breast fillets occurring during long-term freezing for up to 8 months were detected[86] using MORSE values. A razor blade (8.9 mm wide × 0.5 mm thick × 30 mm high) was used to shear muscle fibers to a depth of 20 mm perpendicular to the fiber orientation. Crosshead speed was not given. Tenderness MORSE values were modeled[85] using a modified Goempertz equation to fit a sigmoid shape response that enabled establishing tenderness classifications based on MORSE values that correlated to sensory tenderness values.

Alpers, Priesmeyer, and Meullenet evaluated a blunt MO razor blade[87] (BMORS) for its effectiveness in predicting tenderness of beef using raw or cooked samples. Although not statistically different, the correlation of sensory hardness ratings to shear force was higher for WBS ($R^2 = .64$) than for BMORS ($R^2 = .54$) values for cooked meat. The relationship between BMORS shear values on raw meat to sensory ratings after 7 day aging was not as strong ($R^2 = .31$) as in cooked meat. It was demonstrated in this experiment that no-roll beef carcasses could be segregated into tenderness classifications within 48 h based on in-plant BMORS[87] values of the exposed ribeye muscle.

27.3.2.2 Needle Arrays

Researchers have also attempted to measure meat texture using needle or probe devices. The general idea is that the degree of penetration or compression and the associated force required relate to meat tenderness. These tests vary widely in the number of needles or probes, the dimensions, arrangement, and sharpness/bluntness of the individual needles or probes.

To measure tenderness in chicken breast samples, Peterson and Lilyblade[44] utilized a modified version of an earlier device, the Christel Texturometer,[43] attached to a Lee–Kramer apparatus. The modified attachment consisted of a circular pattern of 25 solid steel cylinder rods that were forced simultaneously through chicken breast sample in a modified Kramer shear attachment. A few years later, Hansen[88] developed the Armour Meat Tenderometer that recorded force to press 10 sharp tip 3.18 mm diameter needles into raw meat. In the design and testing, Hansen used the needle array attached to a universal testing (Instron) apparatus and demonstrated high correlation (.69) to sensory evaluation of tenderness and to WBS (.82). The final design of the portable instrument[89] for use in the packing plant measured the maximum force to press a 10-needle probe

5 cm into the ribeye (at the 12th rib location) of the beef carcass. Research using the Armour tenderometer yielded conflicting results[48,90–95] of efficacy because the force detector was highly dependent on operator technique,[8] and further use was abandoned.

Another needle penetrometer[96] was developed by the U.S. Army Food Engineering Laboratory, Natick, Massachusetts. The lack of additional reference to this penetrometer indicates it likewise, was not widely adopted for use. Morrow and Mohsenin[97] evaluated the mechanics of a multiple needle tenderometer for raw meat and concluded it would be possible to use a conical indenter to measure yield pressure independent of probe geometry and penetration depth.

A MIRINZ tenderness probe[98] consisting of two concentric sets of pins, one set static and one rotating, was evaluated in a series of experiments to predict beef tenderness based on measurements on raw samples. The design was considered nondestructive since researchers could obtain an objective estimate of cooked meat tenderness without introducing additional variation due to obtaining core samples needed to perform WBS. The amount of torque measured at 50° of rotation was most highly related to both trained panel and consumer sensory characteristics. On continued testing using both tension and shear heads on raw and cooked strip loin steaks, results confirmed that the probe values were measuring essentially the same characteristics. However, it was determined that the probe values from raw samples did not account for a sufficient amount of variation in either sensory or consumer judged traits to be useful. This MIRINZ tenderness probe outperformed WBS for predicting consumer sensory traits.

A recently evaluated[99,100] multineedle probe having two rows of three needles, 0.32 cm diameter with 10° tapered point, similar to those of the Armour tenderometer, was used to predict tenderness of strip loin steaks using mechanical measurements made on raw beef. The correlations[100] presented were of the same magnitude as those reported by Hansen[88,89] when the Armour tenderometer was used on a universal testing apparatus.

Some research on texture measurement devices has suggested that the shape and sharpness of the meat contact point is important. Hinnergardt and Tuomy[101] evaluated needle or probe ends including flat, pointed, dull, and tapered flat (blunt) to evaluate pork chops. They determined the optimal configuration would be an array of five flat-end (semiblunt) needles (3.175 mm diameter tapered to 0.18 cm diameter at the end) penetrating 1.27 cm into the meat. Using various mechanical probes[73,75] to predict tenderness of cooked beef longissimus based on measurements of uncooked samples that compared to sharp to blunt devices, sharp needles and sharp blades were more successful in predicting sensory panel scores with an overall ability to explain 47–50% of the variability.

27.3.2.3 Star and Hollow Probes

A star probe[102] (round punch with six tapered points) was employed for evaluating texture of pork loins, and was noted to have good correlation to sensory evaluation. The star probe is a cylindrical punch (solid or tubular) with four, five, or six tapered points and, most likely, is an adaptation of a cherry, date, or olive pitter. Huff-Lonergan et al.[102] described the probe as a five-pointed cherry pitter, 9 mm diameter and with 6 mm between each point. It was employed to determine the amount of force needed to punch and compress a pork chop to 80% of the sample height. The investigators noted highly significant correlation ($R = -.54$) to sensory tenderness scores. A star probe was also mentioned by researchers[103] at Lacombe in Canada; however, similar to the former report,[102] no additional information was presented on the background or why this probe was selected.

A novel probe called the "tensipresser"[104] consists of a round, 5.5 mm diameter hollow cylinder pressed repeatedly into meat to mimic chewing through a 1.0 cm thick pork sample. The first compression cycle is 0.5 mm deep and each successive chew cycle increases the depth by 0.04 mm. By increasing the compression depth in this manner, it takes 250 cycles to bite through the sample. Values representing peak stress at maximum penetration, severing threshold value, and a pliability measure were calculated from the tensipressor data.

27.3.2.4 Plumb Bob

The use of a plumb bob among other probes was mentioned in investigations of mechanical probes to predict tenderness of meat.[73,99] It is not clear whether the plumb bob was meant to represent a "conical indenter" mentioned by Morrow and Mohsenin[97] in their assessment of needle designs; although, the plumb bob would apply both tensile and compression forces as it penetrates muscle. The peak penetration force as the plumb bob (9.6 cm long, 20° taper, diameter maximum of 3.5 cm to 0 cm at the point) was pressed 6.9 cm into meat, perpendicular to the meat surface, at crosshead speed of 250 mm/min was measured on raw and cooked samples. Plumb bob prediction equations for tenderness were comparable to those derived using WBS.

27.3.2.5 Nondestructive Deformation Test

Nondestructive deformation tests[80] of raw broiler breast meat were performed using cylindrical probes 2, 5, 6, or 8 mm diameter for classifying breasts as tender or tough. A significant correlation coefficient, approximately $R = .6$, between the deformation test and razor blade shear energy was detected only when using the 8 mm probe. Tests demonstrated that 81% of all breast samples could be correctly classified as tender or tough in online testing and warrant further evaluation.

27.4 Innovations in Meat Texture Measurement

27.4.1 Isometric Tension

Isometric tension development during onset of rigor was measured in bovine,[105,106] porcine[107] and rabbit muscle,[53] poultry,[108] and salmon.[109] The RigoTech Muscle Texture Analyzer® (Rheologica Instruments, Inc., Lund, Sweden) was offered by ATS RheoSystems (Bordentown, New Jersey) as an analytical instrument for measuring muscle shortening and isometric tension in relation to rigor development and meat tenderness. Its primary use was for determining optimal cooling conditions to optimize meat tenderness. The consensus of two primary studies[53,108] employing isometric tension development was that postmortem isometric tension changes are probably not directly related to cooked meat texture due to relatively large differences in tension pattern that are required to cause small changes in tenderness.

27.4.2 Spectroscopy

Fluorescence[110–114] and other optical spectra probes have been evaluated for usefulness to segregate meat into tenderness classifications with varying degrees of success. An attempt to assess pork tenderness using WBS and online methods that included pH, light scattering, conductivity, and double density light transmission[115] was considered not successful. Egelandsdal et al.[116] attempted to determine tenderness of beef longissimus muscles at 2 days postmortem using fluorescence emission spectra; however, they concluded that wide variation in chilling rates between muscles and carcasses combined with variations in equipment wavelengths used and light sources restricted the application of this autofluorescence methodology.

27.4.2.1 Near-Infrared Reflectance Spectroscopy

Near-infrared reflectance (NIR) spectroscopy has been the subject of research[117–120] to evaluate its potential for predicting meat tenderness based on near-line or online measurements in raw meat. For many years, NIR spectrography was utilized for assessing chemical composition of foods; however, in recent years the technology has been investigated for meat texture measurements[121] with varying success. Correlation values (R^2) to predict shear force measurements in beef[118–120,122,123] ranged from .22 to .83 and are influenced by wavelength, environment, and instrumental variations such as lamp source. Initial NIR work sought to predict tenderness[118,119] measured by WBS. Later, NIR was related to SSF at 14 days postmortem and resulted in development of the QualitySpec® BT equipment (Analytical Spectral Devices, Inc., Boulder, Colorado) now in commercial production to perform online sorting of tough and tender beef.

NIR spectra were also compared to razor blade shear (MORS) and shear energy (MORSE)[117] for assessing and sorting poultry breasts for toughness. Calibration and validation R^2 values ranged from .9 to .95 and .84 to .89, respectively; although, regressions did not yield satisfactory model statistics. The investigators[117] suggested that more samples with greater range of tenderness would be needed to develop better predictions to make NIR a viable technology for estimating tenderness.

27.4.2.2 Raman Spectroscopy

Beattie et al.[124] confirmed the potential for using Raman spectroscopy for predicting sensory quality of beef. Light scattered from a molecule typically is elastic, that is, of the same energy (frequency) and wavelength as the incident light. In Raman spectroscopy, however, light is scattered with a shift in optical frequency. Raman spectroscopy is based on light scattering of monochromatic light, usually from a laser in the near-infrared range, and yields similar information to NIR spectroscopy.

27.4.3 Sonography

27.4.3.1 Ultrasonics

Ultrasonic properties such as velocity, attenuation, and backscatter intensity have been used to assess physicochemical properties of many foods. Sonography has been investigated for monitoring food cooling and freezing,[125–127] composition,[128–130] and has potential for predicting food texture.[129,131–135] Ultrasonic technology applied by Park and Chen[136] determined that shear wave velocity decreased at a rate of 2.97 m/s for every unit percentage increase in fat concentration in beef longissimus muscle. Based on the ability to quantify ultrasonic shear velocity, these researchers envisioned ultrasonic techniques could be used to evaluate tenderness of meat; although, earlier attempts to correlate ultrasonic spectral analysis[137] indicated accuracy of prediction models was not adequate. Connective tissue amount and shear force measurement correlated more with ultrasonic parameters at low frequency, however, and could be useful as a noninvasive tool to measure meat texture.[137]

27.4.3.2 Sonoelasticity

Nondestructive methods that may be useful for determining meat texture include sound wave velocity measurements made by ultrasonics,[136–139] sonoelasticity,[140] and nondestructive deformation tests.[80] Abouelkaram et al.[138] analyzed beef muscle textural characteristics using ultrasonic methods. Ultrasonic acoustic velocity, attenuation, and backscattering intensity revealed components of muscle structure that would influence texture. The changes in meat during rigor onset and aging were tracked by Ayadi et al.[140] using sonoelasticity. In sonoelasticity, low-frequency vibrations are applied to muscle and wave propagation characteristics are detected by simultaneously transmitted ultrasonic waves. Stress at constant 20% of strain was greatest at onset of rigor and correlated to muscle pH measurements. The velocity curve had the same qualitative changes as stress, whereas attenuation had an inverse shape. During the aging period, stress values decreased by a factor of eight to nine times representing a decrease in mechanical resistance of the meat. This decrease in mechanical resistance was accompanied by a decrease in the velocity of the mechanical wave and, inversely, an increase in attenuation. Ayadi et al.[140] interpreted this to indicate changes in mechanical resistance and wave velocity could be used as a nondestructive test to follow rigor development and aging. More research to refine the methodology and correlate findings to sensory textural evaluations are needed for this technology.

27.4.4 Image Analysis

Computer-assisted image capture, processing, and image analysis can reveal texture variations that can be related to sensory[141–147] and quality traits.[148–150] Color features have long been useful for quality classification of intact pork due to the relationship of color and pH for identifying pale, soft, and exudative pork[144] and for predicting beef texture.[151] Sorting beef carcasses using the Beef Cam[152,153] relies on grading (marbling) and color assessments, whereas other systems[142,143,154–158] identify morphological features such as the amount and thickness of connective tissue, marbling amount and distribution, and prominence of meat grain. Correlations between observed and predicted tenderness scores, however, indicated that image analysis has only marginal accuracy for predicting tenderness precisely. Use of wavelet, Gabor, and neural network modeling[147,159] has improved predictive value of image analysis to assess tenderness. Continued advances in image analysis may prove useful for assessing meat texture in the future.

Dual energy x-ray absorptiometry (DEXA) imaging showed some promise as a noninvasive measurement of meat tenderness properties.[160] For measurements on whole steaks, DEXA-predicted shear force was correlated ($R^2 = .69$) to mean peak force measured using a MIRINZ tenderometer. Small sample size and sample handling limited fully investigating this technique; although, the results are encouraging to continue further research into this novel technique for measuring meat texture.

A nondestructive shape profiling system[78] consisting of conveyor belt, laser displacement sensors, and portable computer was used to measure shape profiles of deboned breast fillets online. The shape profiles were related to shear energy (MORSE). Shear energy increased as breast thickness increased and decreased when breast length and width increased. Using shape analysis, 83% of breast samples could be correctly classified as tender or tough when assessed online.

27.4.5 Electromyography and Electrognathography

Electromyography (EMG) and electrognathography (EGN) are considered noninvasive techniques that make it possible to study the dynamics of mastication[65,161] as it relates to meat texture. For these evaluations, a sensory panelist has electrodes placed over facial muscles and the mandibular joint to measure muscle activity and forces during chewing. In the case of EGN, a magnet is also positioned on the lower incisor teeth for tracking jaw movements. These devices have revealed differences in individuals in number of chews, speed of chewing, and first bite force generated among other parameters. Lee et al.[65] noted that all subjects perceived texture of breast meat differently even though they were highly trained in descriptive attribute testing of meat. The EGN results showed that subjects' chewing behavior generally was not affected by meat tenderness and few correlations to tenderness were detected. EMG results for meat tenderness showed variation in 13 of 19 parameters measured; however, elucidation of how parameters sensed on the first bite relate to later chews is still needed.

27.4.6 Elastography

Elastography is quantitative imaging of strain and elastic modulus distributions in soft tissues[162–164] by measuring the internal displacement of small tissue elements in response to externally applied stress using ultrasonic pulses. The method is based on external tissue compression with subsequent computation of the strain profile along the transducer axis using standard ultrasound equipment. In conventional elastography, axial strain elastograms are generated by cross-correlating pre- and post-compression digitized radio frequency echo frames acquired from the tissue before and after small uniaxial compression, respectively.[165] The resulting two-dimensional elastogram shows softer, more elastic tissue as light-colored regions. Applied to meat science, this method[145] has potential to detect differences in elasticity of muscle bundles, connective tissue amounts, and quantity of intramuscular fat. Correlation of elastographic data to meat texture lags behind its application in the medical field where it is used extensively in cardiology and other fields such as physical therapy. Huang et al.[66,166] introduced ultrasonic elastography wavelet texture features[166,167] to predict meat quality. They reported high correlation of wavelet features to meat texture variables indicating potential for developing this technology for application in meat technology. Berg et al.[164] used elastography to calculate textural hardness and intramuscular fat in pork semimembranosus muscle. They determined that elastographic measurements were significantly correlated to shear force, but concluded that more research was necessary before being applied as a screening tool to identify tender meat.

A more recent development in elastography is magnetic resonance elastography (MRE),[168] which is capable of measuring *in vivo* elasticity and viscosity of living skeletal muscle tissue. In MRE, shear waves are induced into muscles using pneumatic and mechanical drivers. A magnetic resonance–compatible load cell records the force during contraction that is related to stiffness. The wavelength is sensitive to the morphology (unipennate or longitudinal) and fiber composition (type I or II) of muscles.[169] The viscoelastic property of muscle is detected by varying the vibration frequency applied to the muscles.

27.5 Conclusions

Owing to the complex and highly structured nature of muscle tissue, meat has inherently variable texture. Unfortunately, the multifaceted characteristic of meat texture has made it difficult to assess instrumentally. Some have argued that the absence of consistent terminology and lack of standardization of the test apparatus has hindered the interpretation of data and understanding of meat texture.[8] To better predict and control meat tenderness issues, accurate measures of meat texture are needed. Consequently, many technologies that measure the physical properties of meat as a measure of meat texture have been developed with varying degrees of success.

In the United States, there are at least 11 manufacturers or distributors of texture analyzers, and all offer test cells for the WBS test. Regardless of manufacturer, these texture measurement systems are remarkably similar in design: crosshead speed control, similar digital stress–strain gauges, interchangeable test cell attachments, advanced data acquisition, and software to manage and calculate data. The increased number of manufacturers for texture analysis benefits the researcher with competitive prices and choices for research. Although much has been learned over the years, instrumentally measuring and understanding characteristics of meat that correlate highly with the complex, sensory phenomenon of meat texture has proven difficult. The difficulty in predicting meat tenderness using instrumental measures also stems from the lack of understanding of the consumers' perception of texture and tenderness. It is a misconception that we can rely on a single WBS force value to adequately describe the complex nature of meat texture. Progress in measuring biomechanical characteristics[170] of raw meat and establishing correlation to overall sensory scores continues to improve, and there are some promising technologies that may allow noninvasive or nondestructive testing to achieve this goal.

References

1. Bailey, A.J. The basis of meat texture. *J. Sci. Food Agric.* 23, 995, 1972.
2. Dransfield, E., Béchet, D., and Ouali, A. Origins of variability in meat texture: an introduction to the workshop "Proteolysis and meat quality". *Sci. Alim.* 14, 369, 1994.
3. Hamann, D.D., Calkins, C.R., and Hollingsworth, C.A. Instrumental texture measurements for processed meat products. In *Proceedings of the 40th Annual Reciprocal Meat Conference*, National Live Stock and Meat Board, Chicago, IL, 1987, 19.
4. Dransfield, E. Intramuscular composition and texture of beef muscles. *J. Sci. Food Agric.* 28, 833, 1977.
5. Dransfield, E. and MacFie, H.J.H. Precision in the measurement of meat texture. *J. Sci. Food Agric.* 31, 62, 1980.
6. Szczesniak, A.S. and Torgeson, K.W. Methods of meat texture measurement viewed from the background of factors affecting tenderness. *Adv. Food Res.* 14, 33, 1965.
7. Bourne, M.C. Interpretation of force curves from instrumental texture measurements. In DeMan, J.M., Voisey, P.W., Rasper, V.F., and Stanley, D.W. (Eds.) *Rheology and Texture in Food Quality.* AVI Publishing Co., Inc., Westport, CT, 1976, 244.
8. Voisey, P.W. Engineering assessment and the critique of instruments used for meat tenderness evaluation. *J. Text. Stud.* 7, 11, 1976.
9. Warner, K.F. Progress report of the mechanical test for tenderness of meat. *Proc. Am. Soc. Anim. Prod.* 21, 114, 1928.
10. Black, W.H., Warner, K.F., and Wilson, C.V. *Beef Production and Quality as Affected by Grade of Steer and Feeding Grain Supplement on Grass.* USDA, Washington, DC, 1931, Vol. 217, 1.
11. Bratzler, L.J. *Measuring the Tenderness of Meat by Mechanical Shear.* MSc thesis, Kansas State College, Manhattan, KS, 1932.
12. Bratzler, L.J. *Physical Research in Meat (Final Report).* USDA and Kansas State Experiment Station, Manhattan, KS, 1933.
13. Bratzler, L.J. Determining the tenderness of meat by use of the Warner–Bratzler method. In *Proceedings of the 2nd Annual Reciprocal Meat Conference*, National Live Stock and Meat Board, Chicago, IL, 1949, 117.

14. Warner, K.F. Adventures in testing meat for tenderness. In *Proceedings of the 5th Annual Reciprocal Meat Conference*, National Live Stock and Meat Board, Chicago, IL, 1952, 156.
15. Purslow, P.P. The fracture properties of cooked beef muscle. In *30th International Congress of Meat Science and Technology*, Bristol, UK, 1984, 172.
16. Purslow, P.P., Donnelly, S.M., and Savage, A.W.J. Variations in the tensile adhesive strength of meat-myosin junctions due to test configurations. *Meat Sci.* 19, 227, 1987.
17. Purslow, P.P. Measuring meat texture and understanding its structural basis. In Vincent, J.F.V. and Lillford, P.J. (Eds.), *Feeding and the Texture of Food*. Cambridge University Press, New York, NY, 1991, 35.
18. Moller, A.J. Analysis of Warner–Bratzler shear pattern with regard to myofibrillar and connective tissue components of tenderness. *Meat Sci.* 5, 247, 1981.
19. King, N.L. and Jones, P.N. Analysing Warner–Bratzler curves. *J. Text. Stud.* 14, 283, 1983.
20. Voisey, P.W. Interpretation of force–deformation curves from the shear-compression cell. *J. Text. Stud.* 8, 19, 1977.
21. Bouton, P.E., Harris, P.V., and Shorthose, W.R. Possible relationships between shear, tensile, and adhesion properties of meat and meat structure. *J. Text. Stud.* 6, 297, 1975.
22. Pool, M.F. and Klose, A.A. The relation of force to sample dimensions in objective measurement of tenderness of poultry meat. *J. Food Sci.* 34, 524, 1969.
23. Voisey, P.W. and Larmond, E. Examination of factors affecting performance of the Warner–Bratzler meat shear test. *J. Inst. Can. Sci. Technol. Alim.* 74, 243, 1974.
24. Kapsalis, J.G. and Szczesniak, A.S. Instrumental testing of meat texture—comments on the past, present and future. *J. Text. Stud.* 7, 109, 1976.
25. Spadaro, V., Allen, D.H., Keeton, J.T., Moreira, R., and Boleman, R.M. Biomechanical properties of meat and their correlation to tenderness. *J. Text. Stud.* 33, 59, 2002.
26. Bourne, M.C. Standardization of texture measuring instruments. *J. Text. Stud.* 3, 379, 1972.
27. Szczesniak, A.S. Correlating sensory with instrumental texture measurements—an overview of recent developments. *J. Text. Stud.* 18, 1, 1986.
28. AMSA. *Research Guidelines for Cookery, Sensory Evaluation and Instrumental Tenderness Measurements of Fresh Meat*. American Meat Science Association and National Live Stock and Meat Board, Chicago, IL, 1995.
29. AMSA. *Guidelines for Cookery and Sensory Evaluation of Meat*. American Meat Science Association and National Live Stock and Meat Board, Chicago, IL, 1978.
30. Savell, J., Miller, R., Wheeler, T., Koohmaraie, M., Shackelford, S., Morgan, B., Calkins, C., Miller, M., Dikeman, M., McKeith, F., Dolezal, G., Henning, B., Busboom, J., West, R., Parrish, F., and Williams, S. http://savell-j.tamu.edu/shearstand.html, Accessed on January 8, 2000.
31. Wheeler, T.L., Koohmaraie, M., Shackelford, S.D. http://192.133.74.2...w.protocol/wbs.html, Accessed on 8/1/2000.
32. Chrystall, B.B., Culioli, J., Demeyer, D., Honikel, K.O., Moller, A.J., Purslow, P., Schwagele, F., Shorthose, R., and Uytterhaegen, L. Recommendation of reference methods for assessment of meat tenderness. In *40th International Congress of Meat Science and Technology*, The Hague, The Netherlands, 1994, 1, S-V.06.
33. Honikel, K.O. Reference methods supported by OECD and their use in Mediterranean meat products. *Food Chem.* 59, 573, 1997.
34. Honikel, K.O. Reference methods for the assessment of physical characteristics of meat. *Meat Sci.* 49, 447, 1998.
35. Boccard, R., Buchter, L., Casteels, E., Cosentino, E., Dransfield, E., Hood, D.E., Joseph, R.L., MacDougall, D.B., Rhodes, D.N., Schon, I., Tinbergen, B.J., and Touraille, C. Procedures for measuring meat quality characteristics in beef production experiments. Report of a working group in the Commission of the European Communities (CEC) beef production research programme. *Livest. Prod. Sci.* 8, 385, 1981.

36. Bourne, M.C., Moyer, J.C., and Hand, D.B. Measurement of food texture by a universal testing machine. *Food Technol.* 20, 522, 1966.

37. Voisey, P.W. Modernization of texture instrumentation. *J. Text. Stud.* 2, 129, 1971.

38. Voisey, P.W. and Kloek, M. Effect of cell size on the performance of the shear-compression texture test cell. *J. Text. Stud.* 12, 133, 1981.

39. Voisey, P.W. and Hansen, H. A shear apparatus for meat tenderness evaluation. *Food Technol.* 21, 37A, 1967.

40. Szczesniak, A.S. Correlating sensory with instrumental texture measurements—an overview of recent developments. *J. Text. Stud.* 18, 1, 1987.

41. Szczesniak, A.S. Instrumental methods of texture measurements. In Kramer, A. and Szczesniak, A.S. (Eds.), *Texture Measurements of Foods: Psychophysical Fundamentals; Sensory, Mechanical, and Chemical Procedures and Their Interrelationships.* D. Reidel Publishing, Co., Boston, MA, 1973, 71.

42. Pearson, A.M. Objective and subjective measurements for meat tenderness. In *Proceedings of the Meat Tenderness Symposium,* Campbell Soup Co., Camden, NJ, 1963.

43. Miyada, D.S. and Tappel, A.L. Meat tenderization. I. Two mechanical devices for measuring texture. *Food Technol.* 10, 142, 1956.

44. Peterson, D.W. and Lilyblade, A.L. Relative differences in tenderness of breast muscle of normal and two dystrophic mutant strains of chickens. *J. Food Sci.* 34, 142, 1969.

45. Kelly, R.F., Taylor, J.C., and Graham, P.P. Preliminary comparisons of a new tenderness measuring device with objective and subjective evaluations of beef. *J. Anim. Sci.* 19, 645, 1960.

46. Alsmeyer, R.H., Kulwich, R., and Hiner, R.L. Loin-eye tenderness variations measured by the STE. *J. Anim. Sci.* 21, 977, 1962.

47. Kulwich, R., Decker, R.W., and Alsmeyer, R.H. Use of a slice-tenderness evaluation device with pork. *Food Technol.* 17, 83, 1963.

48. Smith, G.C. and Carpenter, Z.L. Mechanical measurements of meat tenderness using the NIP tenderometer. *J. Text. Stud.* 4, 196, 1973.

49. Voisey, P.W., Kamel, B.S., Evans, G., and DeMan, J.M. A dynamometer instrument for measurement of textural characteristics of foods. *Can. Inst. Food Sci. Technol. J.* 7, 250, 1974.

50. Segars, R.A., Hamel, R.G., Kapsalis, J.G., and Kluter, R.A. A punch and die test cell for determining the textural qualities of meat. *J. Text. Stud.* 6, 211, 1975.

51. Purchas, R.W. Some aspects of raw meat tenderness. A study of some factors affecting its change with cooking and a new means of measurement. *J. Food Sci.* 38, 556, 1973.

52. Locker, R.H. and Wild, D.J.C. A machine for measuring yield point in raw meat. *J. Text. Stud.* 13, 71, 1982.

53. Busch, W.A., Goll, D.E., and Parrish Jr., F.C. Molecular properties of postmortem muscle. Isometric tension development and decline in bovine, porcine and rabbit muscle. *J. Food Sci.* 37, 289, 1972.

54. Volodkevich, N.N. Apparatus for measurements of chewing resistance or tenderness of foodstuffs. *J. Food Sci.* 3, 221, 1938.

55. Winkler, C.A. Tenderness of meat: I. A recording apparatus for its estimation, and relation between pH and tenderness. *Can. J. Res.* 17, 8, 1939.

56. Macfarlane, P.G. and Marer, J.M. An apparatus for determining the tenderness of meat. *Food Technol.* 20, 838, 1966.

57. Macfarlane, P.G. and Frazerhurst, L.F. Device for measuring tenderness of meat. New Zealand Patent #190945, 1980.

58. Phillips, D.M. A new technique for measuring texture and tenderness. In Valin, C. (Ed.), *38th International Congress of Meat Science and Technology.* Elsevier Applied Science, Barking, UK, 1992, 958.

59. Bouton, P.E. and Harris, P.V. The effects of cooking temperature and time on some mechanical properties of meat. *J. Food Sci.* 37, 140, 1972.

60. Rosenvold, K. Personal communication, July 10, 2007.
61. Friedman, H.H., Whitney, J.E., and Szczesniak, A.S. The texturometer—a new instrument for objective texture measurement. *J. Food Sci.* 28, 390, 1963.
62. Szczesniak, A.S. and Hall, B.J. Application of the General Foods Texturometer to specific food products. *J. Text. Stud.* 6, 117, 1975.
63. Proctor, B.E., Davison, S., Malecki, G.J., and Welch, M. A recording strain-gauge denture tenderometer for foods. I. Instrument evaluation and initial tests. *Food Technol.* 9, 471, 1955.
64. Brody, A.L. Masticatory properties of foods by the strain gage denture tenderometer. PhD dissertation, Massachusetts Institute of Technology, Boston, MA, 1957.
65. Lee, Y.-S., Saha, A., Owens, C.M., and Meullenet, J.-F. Prediction of tenderness of broiler breast fillets by electromyography (EMG) and electrognathography (EGN). In *IFT Annual Meeting Book of Abstracts*, IFT Press, Chicago, IL, 2007, 280.
66. Huang, F., Dethmers, A.E., and Robertson, J.W. A rheological method for evaluating the textural qualities of cooked mutton. *J. Food Sci.* 42, 721, 1977.
67. Stanley, D.W., McKnight, L.M., Hines, W.G.S., Usborne, W.R., and DeMan, J.M. Predicting meat tenderness from muscle tensile properties. *J. Text. Stud.* 3, 51, 1972.
68. Bouton, P.E. and Harris, P.V. A comparison of some objective methods used to assess meat tenderness. *J. Food Sci.* 37, 218, 1972.
69. Szczesniak, A.S., Brandt, M.A., and Friedman, H.H. Development of standard rating scales for mechanical parameters of texture and correlation between the objective and the sensory methods of texture evaluation. *J. Food Sci.* 28, 397, 1963.
70. Shackelford, S.D., Wheeler, T.L., and Koohmaraie, M. Tenderness classification of beef: II. Design and analysis of a system to measure beef longissimus shear force under commercial processing conditions. *J. Anim. Sci.* 77, 1474, 1999.
71. Shackelford, S.D., Wheeler, T.L., and Koohmaraie, M. Evaluation of slice shear force as an objective method of assessing beef longissimus tenderness. *J. Anim. Sci.* 77, 2693, 1999.
72. Shackelford, S.D., Wheeler, T.L., and Koohmaraie, M. Technical note: Use of belt grill cookery and slice shear force for assessment of pork longissimus tenderness. *J. Anim. Sci.* 82, 238, 2004.
73. Stephens, J.W., Unruh, J.A., Dikeman, M.E., Hunt, M.C., Lawrence, T.E., and Loughin, T.M. Mechanical probes can predict tenderness of cooked beef longissimus using uncooked measurements. *J. Anim. Sci.* 82, 2077, 2004.
74. Otremba, M.M., Dikeman, M.E., Milliken, G.A., Stroda, S.L., Unruh, J.A., and Chambers, E. Interrelationships among evaluations of beef longissimus and semitendinosus muscle tenderness by Warner–Bratzler shear force, a descriptive-texture profile sensory panel, and a descriptive attribute sensory panel. *J. Anim. Sci.* 77, 865, 1999.
75. Stephens, J.W., Unruh, J.A., Dikeman, M.C., Hunt, M.C., Lawrence, T.E., and Loughin, T.M. Mechanical probes used on uncooked strip loins can predict cooked longissimus beef tenderness. In *Proceedings of the 56th Annual Reciprocal Meat Conference*, American Meat Science Association and National Cattlemen's Beef Association, Savoy, IL, 2003, 124.
76. Henrickson, R.L., Marsden, J.L., and Morrison, R.D. An evaluation of a method for measuring shear force for an individual muscle fiber. *J. Food Sci.* 39, 15, 1974.
77. Cavitt, L.C., Owens, C.M., Meullenet, J.F., Gandhapuneni, R.K., and Youm, G.Y. Rigor development and meat quality of large and small broilers and the use of Allo-Kramer shear, needle puncture, and razor blade shear to measure texture. *Poult. Sci.* 80(Suppl. 1), 138, 2001.
78. Meullenet, J.-F.C., Xiong, R., Saha, A., and Owens, C. Novel shape profiling method for classifying tender and tough broiler breast meat. In *IFT Annual Meeting Book of Abstracts*. IFT Press, Chicago, IL, 2005.
79. Xiong, R., Meullenet, J.-F.C., Cavitt, L.C., and Owens, C. Effect of razor blade penetration depth on correlation of razor blade shear values and sensory texture of broiler major pectoralis muscles. In *IFT Annual Meeting Book of Abstracts*. IFT Press, Chicago, IL, 2005.

80. Xiong, R., Meullenet, J.-F.C., and Owens, C. Classification of tender and tough broiler breast meat by a non-destructive deformation test. In *IFT Annual Meeting Book of Abstracts*. IFT Press, Chicago, IL, 2005.

81. Cavitt, L.C., Youm, G.W., Meullenet, J.F., Owens, C.M., and Xiong, R. Prediction of poultry meat tenderness using razor blade shear, Allo-Kramer shear, and sarcomere length. *J. Food Sci.* 69, SNQ11, 2004.

82. Cavitt, L.C., Meullenet, J.F., Gandhapuneni, R.K., Youm, G.W., and Owens, C.M. Rigor development and meat quality of large and small broilers and the use of Allo-Kramer shear, needle puncture, and razor blade shear to measure texture. *Poult. Sci.* 84, 113, 2005.

83. Cavitt, L.C., Meullenet, J.-F.C., Xiong, R., and Owens, C.M. The relationship of razor blade shear, Allo-Kramer shear, Warner–Bratzler shear and sensory tests to changes in tenderness of broiler breast fillets. *J. Muscle Foods* 16, 223, 2005.

84. Xiong, R., Cavitt, L.C., Meullenet, J.F., and Owens, C.M. Comparison of Allo-Kramer, Warner–Bratzler and razor blade shears for predicting sensory tenderness of broiler breast meat. *J. Text. Stud.* 37, 179, 2006.

85. Lee, Y.S., Owens, C.M., and Meullenet, J.F. The application of the Meullenet-Owens-Razor-Shear (MORS) for the prediction of broiler *pectoralis major* muscle tenderness to the Instron InSpec. In *IFT Annual Meeting Book of Abstracts*, 2006.

86. Lee, Y.S., Xiong, R., Saha, A., Owens, C.M., and Meullenet, J.F. Tenderness of broiler breast fillets during long-term freezing. In *IFT Annual Meeting Book of Abstracts*, 2006.

87. Alpers, T.K. Effectiveness of blunt Meullenet-Owens razor shear (BMORS) for predicting tenderness in beef. In *IFT Annual Meeting Book of Abstracts*, 2006.

88. Alpers, T.K., Priesmeyer, T.S., and Meullenet, J.F. Meat tenderness testing. US Patent #3 602 038, 1971.

89. Hansen, L.J. Development of the Armour tenderometer for tenderness evaluation of beef carcasses. *J. Text. Stud.* 3, 146, 1972.

90. Campion, D.R., Crouse, J.D., and Dikeman, M.E. The Armour tenderometer as a predictor of cooked meat tenderness. *J. Food Sci.* 40, 886, 1975.

91. Dikeman, M.E., Tuma, H.J., Glimp, H.A., Gregory, K.E., and Allen, D.M. Evaluation of the tenderometer for predicting bovine muscle tenderness. *J. Anim. Sci.* 34, 960, 1972.

92. Carpenter, Z.L., Smith, G.C., and Butler, O.D. Assessment of beef tenderness with the Armour tenderometer. *J. Food Sci.* 37, 126, 1972.

93. Henrickson, R.L., Marsden, J.L., and Morrison, R.D. An evaluation of the Armour tenderometer for an estimation of beef tenderness. *J. Food Sci.* 37, 857, 1972.

94. Parrish Jr., F.C., Olson, D.G., Miner, B.E., Yound, R.B., and Snell, R.L. Relationship of tenderness measurements made by the Armour tenderometer to certain objective, subjective and organoleptic properties of bovine muscle. *J. Food Sci.* 38, 1214, 1973.

95. Huffman, D.L. An evaluation of the tenderometer for measuring beef tenderness. *J. Anim. Sci.* 38, 287, 1974.

96. Tuomy, J.M. Definition and measurement of meat texture in military development and procurement. *J. Text. Stud.* 7, 5, 1976.

97. Morrow, C.T. and Mohsenin, N.N. Mechanics of a multiple needle tenderometer for raw meat. *J. Text. Stud.* 7, 115, 1976.

98. Jeremiah, L.E. and Phillips, D.M. Evaluation of a probe for predicting beef tenderness. *Meat Sci.* 55, 493, 2000.

99. Timm, R.R., Unruh, J.A., Dikeman, M.E., Hunt, M.C., Lawrence, T.E., Boyer Jr., J.E., and Marsden, J.L. Mechanical measures of uncooked beef longissimus muscle can predict sensory panel tenderness and Warner–Bratzler shear force of cooked steaks. *J. Anim. Sci.* 81, 1721, 2003.

100. Timm, R.R., Unruh, J.A., Dikeman, M.E., Hunt, M.C., Lawrence, T.E., Boyer Jr., J.E., and Marsden, J.L. Mechanical measurements of uncooked beef longissimus muscle can predict tenderness of strip loin steaks. In *Proceedings of the 55th Annual Reciprocal Meat Conference*, American Meat Science Association and National Cattlemen's Beef Association, Savoy, IL, 2002, 124.

101. Hinnergardt, L.C. and Tuomy, J.M. A penetrometer test to measure meat tenderness. *J. Food Sci.* 35, 312, 1970.

102. Huff-Lonergan, E., Baas, T.J., Malek, M., Dekkers, J.C.M., Prusa, K., and Rothschild, M.F. Correlations among selected pork quality traits. *J. Anim. Sci.* 80, 617, 2002.

103. Anonymous. *Measuring Beef Tenderness Objectively*. Lancombe Research Center, Lancombe, Alberta, Bulletin, 2nd edition, 2000, 1.

104. Nakai, H., Tanabe, R., Ando, S., Ikeda, T., and Nishizawa, M. Development of a technique for measuring tenderness in meat using a "tensipresser". In Valin, C. (Ed.), *38th International Congress of Meat Science and Technology*. Elsevier Applied Science, Barking, UK, 1992, 947.

105. Devine, C.E., Wahlgren, N.M., and Tornberg, E. Effect of rigor temperature on muscle shortening and tenderisation of restrained and unrestrained beef m. longissimus thoracicus et lumborum. *Meat Sci.* 51, 61, 1999.

106. Tornberg, E., Wahlgren, M., Brondum, J., and Engelsen, S.B. Pre-rigor conditions in beef under varying temperature- and pH-falls studied with rigometer, NMR and NIR. *Food Chem.* 69, 407, 2000.

107. Josell, A., Martinsson, L., and Tornberg, E. Possible mechanism for the effect of the RN– allele on pork tenderness. *Meat Sci.* 64, 341, 2003.

108. Wood, D.F. and Richards, J.F. Isometric tension studies on chicken pectoralis major muscle. *J. Food Sci.* 39, 525, 1974.

109. Kiessling, A., Helge Stien, L., Torslett, O., Suontama, J., and Slinde, E. Effect of pre- and post-mortem temperature on rigor in Atlantic salmon muscle as measured by four different techniques. *Aquaculture* 259, 390, 2006.

110. Swatland, H.J. Feasibility of measuring meat texture and exudate using paired hypodermic needles for rheology, spectrophotometry and electrical impedance. *J. Text. Stud.* 30, 217, 1999.

111. Swatland, H.J. Stratification of connective tissue toughness in beef roasts as assessed by simultaneous fluorometry and penetrometry. *Food Res. Int.* 39, 1106, 2006.

112. Swatland, H.J. and Findlay, C.J. On-line probe prediction of beef toughness, correlating sensory evaluation with fluorescence detection of connective tissue and dynamic analysis of overall toughness. *Food Qual. Pref.* 8, 233, 1997.

113. Swatland, H.J., Brooks, J.C., and Miller, M.F. Possibilities for predicting taste and tenderness of broiled beef steaks using an optical-electromechanical probe. *Meat Sci.* 50, 1, 1998.

114. Swatland, H.J. Stratification of toughness in beef roasts. *Meat Sci.* 77, 2, 2007.

115. Van Oeckel, M.J., Warnants, N., and Boucque, C.H. Pork tenderness estimation by taste panel, Warner–Bratzler shear force and on-line methods. *Meat Sci.* 53, 259, 1999.

116. Egelandsdal, B., Wold, J.P., Sponnich, A., Neegard, S., and Hildrum, K.I. On attempts to measure the tenderness of *longissimus dorsi* muscles using fluorescence emission spectra. *Meat Sci.* 60, 187, 2002.

117. Meullenet, J.F., Jonville, E., Grezes, D., and Owens, C.M. Prediction of the texture of cooked poultry pectoralis major muscles by near-infrared reflectance analysis of raw meat. *J. Text. Stud.* 35, 573, 2004.

118. Park, B., Chen, Y.-R., Hruschka, W.R., Shackelford, S.D., and Koohmaraie, M. Near-infrared reflectance analysis for predicting beef longissimus tenderness. *J. Anim. Sci.* 76, 2115, 1998.

119. Park, B., Chen, Y.-R., Hruschka, W.R., Koohmaraie, M., and Shackelford, S.D. Assessment of beef tenderness using near-infrared spectroscopy. In *3rd Am. Soc. Agric. Eng.*, ASAE Annual International Meeting Technical Papers, Vol. 3, St. Joseph, MI, 1997.

120. Park, B., Chen, Y.-R., Hruschka, W.R., Shackelford, S.D., and Koohmaraie, M. Principal component regression of near-infrared reflectance spectra for beef tenderness prediction. *Trans. Am. Soc. Agric. Eng.* 44, 609, 2001.

121. Andres, S., Murray, I., Navajas, E.A., Fisher, A.V., Lambe, N.R., and Bunger, L. Prediction of sensory characteristics of lamb meat samples by near infrared reflectance spectroscopy. *Meat Sci.* 76, 509, 2007.

122. Rodbotten, R., Nilsen, B.N., and Hildrum, K.I. Prediction of beef quality attributes from early post mortem near infrared reflectance spectra. *Food Chem.* 69, 427, 2000.

123. Byrne, C.E., Downey, G., Troy, D.J., and Buckley, D.J. Non-destructive prediction of selected quality attributes of beef by near-infrared reflectance spectroscopy between 750 and 1098 nm. *Meat Sci.* 49, 399, 1998.

124. Beattie, R.J., Bell, S.J., Farmer, L.J., Moss, B.W., and Patterson, D. Preliminary investigation of the application of Raman spectroscopy to the prediction of the sensory quality of beef silverside. *Meat Sci.* 66, 903, 2004.

125. Miles, C.A. and Cutting, C.L. Changes in the velocity of ultrasound in meat during freezing. *J. Food Technol.* 9, 119, 1974.

126. Sigfusson, H., Ziegler, G.R., and Coupland, J.N. Ultrasonic monitoring of unsteady-state cooling of food products. *Trans. Am. Soc. Agric. Eng.* 44, 1235, 2001.

127. Sigfusson, H., Ziegler, G.R., and Coupland, J.N. Ultrasonic monitoring of food freezing. *J. Food Eng.* 62, 263, 2004.

128. Benedito, J., Carcel, J.A., Rossello, C., and Mulet, A. Composition assessment of raw meat mixtures using ultrasonics. *Meat Sci.* 57, 365, 2001.

129. Mulet, A., Benedito, J., Golas, Y., and Carcel, J.A. Noninvasive ultrasonic measurements in the food industry. *Food Rev. Int.* 18, 123, 2002.

130. Simal, S., Benedito, J., Clemente, G., Femenia, A., and Rossello, C. Ultrasonic determination of the composition of a meat-based product. *J. Food Eng.* 58, 253, 2003.

131. Benedito, J., Carcel, J.A., Gonzalez, R., and Sanjuan, N. Prediction of instrumental and sensory textural characteristics of mahon cheese from ultrasonic measurements. *J. Text. Stud.* 31, 631, 2000.

132. Llull, P., Simal, S., Femenia, A., Benedito, J., and Rossello, C. The use of ultrasound velocity measurement to evaluate the textural properties of sobrassada from Mallorca. *J. Food Eng.* 52, 323, 2002.

133. Llull, P., Simal, S., Benedito, J., Rossello, C., and Femenia, A. Erratum to: "Evaluation of textural properties of a meat-based product (sobrassada) using ultrasonic techniques" [*Journal of Food Engineering* 53 (2002) 279–285]. *J. Food Eng.* 55, 373, 2002.

134. Llull, P., Simal, S., Benedito, J., and Rossello, C. Evaluation of textural properties of a meat-based product (sobrassada) using ultrasonic techniques. *J. Food Eng.* 53, 279, 2002.

135. Bond, L.J. Ultrasonic measurements in food processing: Introduction and fundamentals. In *IFT Annual Meeting Book of Abstracts.* IFT Press, Chicago, IL, 2002.

136. Park, B. and Chen, Y.-R. Ultrasonic shear wave characterization in beef longissimus muscle. *Trans. Am. Soc. Agric. Eng.* 40, 229, 1997.

137. Park, B., Whittaker, A.D., Miller, R.K., and Hale, D.S. Ultrasonic spectral analysis for beef sensory attributes. *J. Food Sci.* 59, 697, 1994.

138. Abouelkaram, S., Suchorski, K., Buquet, B., Berge, P., Culioli, J., Delachartre, P., and Basset, O. Effects of muscle texture on ultrasonic measurements. *Food Chem.* 69, 447, 2000.

139. Park, B., Thane, B.R., and Whittaker, A.D. Ultrasonic image analysis for beef tenderness. In *1836th Proc. SPIE Int. Soc. Opt. Eng.*, SPIE, Boston, MA, 1993, 120.

140. Ayadi, A., Culioli, J., and Abouelkaram, S. Sonoelasticity to monitor mechanical changes during rigor and ageing. *Meat Sci.* 76, 321, 2007.

141. Jeyamkondan, S., Kranzler, G., Anand, L., Morgan, J.B., and Brooks, J.C. Predicting beef tenderness from textural features. In *Am. Soc. Agric. Eng. Ann. Int. Meet.*, ASAE, Chicago, IL, 2002.

142. Basset, O., Dupont, F., Hernandez, A., Odet, C., Abouelkaram, S., and Culioli, J. Texture image analysis: application to the classification of bovine muscles from meat slice images. *Opt. Eng.* 38, 1950, 1999.

143. Basset, O., Buquet, B., Abouelkaram, S., Delachartre, P., and Culioli, J. Application of texture image analysis for the classification of bovine meat. *Food Chem.* 69, 437, 2000.

144. Xing, J., Ngadi, M., Gunenc, A., Prasher, S., and Gariepy, C. Use of visible spectroscopy for quality classification of intact pork meat. *J. Food Eng.* 82, 135, 2007.

145. Chandraratne, M.R., Samarasinghe, S., Kulasiri, D., and Bickerstaffe, R. Prediction of lamb tenderness using image surface texture features. *J. Food Eng.* 77, 492, 2006.

146. Chandraratne, M.R., Kulasiri, D., and Samarasinghe, S. Classification of lamb carcass using machine vision: comparison of statistical and neural network analyses. *J. Food Eng.* 82, 26, 2007.

147. Zheng, C., Sun, D.W., and Zheng, L. Classification of tenderness of large cooked beef joints using wavelet and Gabor textural features. *Trans. ASABE* 49, 1447, 2006.

148. Brosnan, T. and Sun, D.W. Improving quality inspection of food products by computer vision—a review. *J. Food Eng.* 61, 3, 2004.

149. Zheng, C., Sun, D.W., and Zheng, L. Correlating colour to moisture content of large cooked beef joints by computer vision. *J. Food Eng.* 77, 858, 2006.

150. Tan, J. Meat quality evaluation by computer vision. *J. Food Eng.* 61, 27, 2004.

151. Goni, M.V., Beriain, M.J., Indurain, G., and Insausti, K. Predicting longissimus dorsi texture characteristics in beef based on early post-mortem colour measurements. *Meat Sci.* 76, 38, 2007.

152. Belk, K.E., Scanga, J.A., Wyle, A.M., and Smith, G.C. Prediction of beef palatability using instruments. In *Proceedings of the Beef Improvement Federation Convention*, Beef Improvement Federation, 2000, 1.

153. Wheeler, T.L., Vote, D., Leheska, J.M., Shackelford, S.D., Belk, K.E., Wulf, D.M., Gwartney, B.L., and Koohmaraie, M. The efficacy of three objective systems for identifying beef cuts that can be guaranteed tender. *J. Anim. Sci.* 80, 3315, 2002.

154. Li, J., Tan, J., and Martz, F.A. Predicting beef tenderness from image texture features. In *11th Am. Soc. Agric. Eng.*, ASAE Annual International Meeting Technical Papers, Vol. 1, St. Joseph, MI, 1997.

155. Li, J., Tan, J., and Shatadal, P. Classification of tough and tender beef by image texture analysis. *Meat Sci.* 57, 341, 2001.

156. Del Moral, F.G., O'Valle, F., Masseroli, M., and Del Moral, R.G. Image analysis application for automatic quantification of intramuscular connective tissue in meat. *J. Food Eng.* 81, 33, 2007.

157. Tian, Y.Q., McCall, D.G., Dripps, W., Yu, Q., and Gong, P. Using computer vision technology to evaluate the meat tenderness of grazing beef. *Food Aust.* 57, 322, 2005.

158. Tian, Y.Q., Tan, J., McCall, D.G., Gong, P. www.cnr.berkeley.edu/~ytian/tian_personal/Yong_Tian.files/meatpaper.pdf, Accessed on September 2, 2006.

159. Chandraratne, M.R., Kulasiri, D., and Samarasinghe, S. Classification of lamb carcass using machine vision: comparison of statistical and neural network analyses. *J. Food Eng.* 82, 26, 2007.

160. Kroger, C., Bartle, C.M., West, J.G., Purchas, R.W., and Devine, C.E. Meat tenderness evaluation using dual energy X-ray absorptiometry (DEXA). *Comp. Elect. Agric.* 54, 93, 2006.

161. Gonzalez, R., Montoya, I., and Carcel, J. Review: The use of electromyography on food texture assessment. *Food Sci. Technol. Int.* 7, 461, 2001.

162. Ophir, J., Miller, R.K., Ponnekanti, H., Cespedes, I., and Whittaker, A.D. Elastography of beef muscle. In Valin, C. (Ed.), *38th International Congress of Meat Science and Technology.* Elsevier Applied Science, Barking, UK, 1992, 153.

163. Ophir, J., Miller, R.K., Ponnekanti, H., Cespedes, I., and Whittaker, A.D. Elastography of beef muscle. *Meat Sci.* 36, 239, 1994.

164. Berg, E.P., Kallel, F., Hussain, F., Miller, R.K., Ophir, J., and Kehtarnavaz, N. The use of elastography to measure quality characteristics of pork semimembranosus muscle. *Meat Sci.* 53, 31, 1999.

165. Chandrasekhar, R., Ophir, J., Krouskop, T., and Ophir, K. Elastographic image quality vs. tissue motion in vivo. *Ultrasound Med. Biol.* 32, 847, 2006.

166. Huang, Y., Lacey, R.E., and Whittaker, A.D. Neural network prediction modeling based on elastographic textural features for meat quality evaluation. *Trans. Am. Soc. Agric. Eng.* 41, 1173, 1998.

167. Huang, Y., Lacey, R.E., Moore, L.L., Miller, R.K., Whittaker, A.D., and Ophir, J. Wavelet textural features from ultrasonic elastograms for meat quality prediction. *Trans. Am. Soc. Agric. Eng.* 40, 1741, 1997.

168. Kang, Y.B., Oida, T., Jung, D.Y., Fukuma, A., Azuma, T., Okamoto, J., Takizawa, O., Matsuda, T., and Tsutsumi, S. Non-invasive measurement of in-vivo elasticity of skeletal muscles with MR-elastography. *Key Eng. Mater.* 901, 342, 2007.

169. Bensamoun, S.F., Ringleb, S.I., Littrell, L., Chen, Q., Brennan, M., Ehman, R.L., and An, K.N. Determination of thigh muscle stiffness using magnetic resonance elastography. *J. Magn. Reson. Imaging* 23, 242, 2006.

170. Marburger, R.M., Keeton, J.T., Maddock, R.J., and Moreira, R.G. Biomechanical characterization of meat tenderness. In *IFT Annual Meeting Book of Abstracts*. IFT Press, Chicago, IL, 2000.

171. Campo, M.M., Santolaria, P., Sanudo, C., Lepetit, J., Olleta, J.L., and Panea, B. Assessment of breed type and ageing time effects on beef meat quality using two different texture devices. *Meat Sci.* 55, 371, 2000.

172. Lepetit, J. and Hamel, C. Correlations between successive measurements of myofibrillar resistance of raw longissimus dorsi muscle during ageing. *Meat Sci.* 49, 249, 1998.

173. Luckett, R.L., Bidner, T.D., and Turner, J.W. The tenderometer as a measure of beef tenderness. *J. Anim. Sci.* 34, 347, 1972.

174. Luckett, R.L., Bidner, T.D., Icaza, E.A., and Turner, J.W. Tenderness studies in straightbred and crossbred steers. *J. Anim. Sci.* 40, 468, 1975.

Chapter 28

Techniques for Sampling and Identification of Volatile Compounds Contributing to Sensory Perception

Saskia M. van Ruth

Contents

28.1 Introduction

Taste, aroma, and texture contribute to the palatability of meat. Flavor is an important sensory aspect of the overall acceptability of meat products [1]. There is no evidence that the acceptability of particular meat products is related to the ability to perceive particular odor or flavor qualities. For instance, although consumers in Asian markets show a poor acceptability of sheep meat odor, they are more sensitive to specific qualities. It is likely that in countries whose diet include a particular type of meat, a long history of consumption promotes a preference for the same flavor qualities that are disliked by those who do not regularly consume that type of meat [2]. These qualities, however, can be related to certain odor-active compounds [3]. Variation in beef quality is large and is due to many factors, such as genetic differences, as well as sex, age, management, and nutrition. These factors interact with one another [4]. The consumer's decision to purchase beef is based on the perception of health and a variety of sensory traits including color, tenderness, juiciness, and odor/flavor [5]. It is therefore worthwhile considering variations in meat quality at the consumer level, with respect to both sensory traits and health aspects.

As flavor is an important component of the quality of meat, there has been much research aimed at understanding the chemistry of meat flavor, and at determining those factors during the production and processing of meat that influence flavor quality. Meat flavor is thermally derived, since uncooked meat has little or no aroma and only a blood-like taste. During cooking, a complex series of thermally induced reactions occur between nonvolatile components of lean and fatty tissues, resulting in a large number of reaction products. Although the flavor of cooked meat is influenced by compounds contributing to the sense of taste, it is the volatile compounds, formed during cooking, that determine the aroma attributes, contributing to the characteristic flavors of meat [6]. Over 1000 volatile compounds have been reported in the literature [7]. Volatile compounds in cooked meat are derived from both lipid- and water-soluble precursors. They provide roast, boiled, fatty, and species-related flavors, as well as the characteristic meaty aromas associated with all cooked meats. Thermal degradation of lipid provides compounds that give fatty aromas to cooked meat and compounds that determine the flavors of the different species. The Maillard reaction is primarily responsible for the large number of heterocyclic compounds that have been found in the volatiles of cooked meat and are responsible for savory, roast, and boiled flavors. Pentoses, in particular ribose from meat ribonucleotides, and the sulfur-containing amino acid, cysteine, are important precursors for these reactions in meat. Furanthiols and furan sulfides and disulfides are very important flavor compounds, which have exceptionally low odor threshold values, and are responsible for characteristic meaty aromas. Compounds formed during the Maillard reaction may also react with other components of meat, adding new compounds to the complex profile of aroma compounds. For example, aldehydes and other carbonyls formed during lipid oxidation react readily with Maillard intermediates. Such interactions give rise to additional aroma compounds, but they also modify the overall profile of compounds contributing to meat flavor [6].

The presence of volatile compounds in meat is undoubtedly affected by the animal feeding system. The presence of certain classes of compounds, which are enhanced by a type of diet, can be identified. The duration of the treatment plays an important role in the accumulation of fat volatiles in ruminant meat. Concentrate-based diets produce high amounts of BCFA and lactones in ruminant meat, whereas aldehydes are markedly enhanced by fat-enriched diets.

Sesquiterpenes, 2,3-octanedione, and skatole are evidently associated with grass feeding systems, although some controversy exists among various studies. Moreover, the presence of secondary compounds in the diet could affect the biosynthesis of some volatiles. This is the case of dietary condensed tannins on skatole production [8].

Although many volatile compounds have been identified in cooked meat, it is well accepted that only a limited number of volatile compounds actually contribute to the overall aroma. In addition, it is known that some powerful odorants found in meat systems exist at concentrations too low to allow their identification by the usual gas chromatography–mass spectrometry (GC-MS) procedures [9]. Therefore, alternative approaches are required to understand meat flavor/aroma and to identify its odor-active compounds. Gas chromatography–olfactometry (GC-O) is a useful and powerful tool in aroma research. Various techniques are available to identify and rank odorants [10].

Flavor is an interaction between the food and the consumer, and not a property of the food alone. So, no study on flavor is complete unless the consumer is considered as well as the chemistry and physics of the food. The perception of flavor is not a single event but a dynamic process, involving a series of events, and every step must be considered if we are to truly understand flavor. When food is taken into the mouth, volatile and nonvolatile compounds are released from the food and must transfer to the receptors before there can be any sensation. Therefore, there will be a typical delay before anything happens, then a sharp rise in the concentrations of the stimulating molecules at the receptors, followed by a slower decline in concentrations. After swallowing, the decline will continue, possibly very slowly, allowing for a long aftertaste, until the stimulating molecules have all diffused away from the receptors [11]. Simultaneously, there are short-term fluctuations in the concentrations carried to the olfactory receptors, caused by breathing [12]. On the next bite, the sequence is repeated. Overlaid on this physical process of mass transfer, there is a process of continual sensory adaptation and recovery. Thus the apparently simple process of tasting a food or beverage is in fact composed of a sequence of complex processes, any of which can affect the sensation. For a real understanding of food flavor, such a dynamic process must be matched with dynamic research methods. Conventional sensory analysis gives a kind of integral or time-average of the total sensation. Chemical methods such as analysis of distilled or headspace volatiles provide measure of the compounds available as potential stimulants [11]. The truly dynamic methods provide some degree of time resolution of the changes in the stimulating molecules, or of the sensation [13].

28.2 Sampling of Volatiles Available for Sensory Perception

Distillation and extraction are common procedures to sample volatile compounds. They normally aim to analyze the totality of the volatiles present in a food and may not represent closely the volatiles available for perception when a food is consumed. Headspace analysis methods, whether static or dynamic, provide results that should more accurately represent the real consumption of food. To obtain data that better reflect the pattern of volatiles present at the olfactory receptors during consumption, a number of devices have been proposed that simulate in large or small measure the process of eating. These are variants of dynamic headspace analysis. The systems are often combined with some form of time-resolved sampling. Furthermore, systems have been developed to allow sampling of the headspace from the nose or mouth, on the assumption that this better reflects the volatiles that are available for perception [11]. In the following sections, headspace, *in vitro* (mouth analogs), and *in vivo* sampling techniques are described in greater detail.

28.2.1 Headspace Sampling

An important characteristic of volatile flavor compounds is that they exhibit sufficient vapor pressure to be present in the gas phase at a concentration detectable by the olfactory system. This basis of aroma isolation appears most reasonable. It is understandable that many aroma isolation techniques are based on volatility. Many of these techniques are focused on the volatility of the compounds in the system to be studied, and not on the concentration of the compounds in the food. The amounts of aroma compounds in the headspace do not follow in order of volatility (vapor pressure) of the pure compounds but, instead, depends on their vapor pressure over the food system.

28.2.1.1 Static Headspace Sampling

Many compounds exist as gases at the temperature at which they are being sampled or have sufficiently high vapor pressure to evaporate and produce a gas-phase solution. In these cases, the gas itself may be injected into a gas chromatograph, either by syringe or by transferring a known volume of vapor from a sample loop attached to a valve. The amount of gas that can be injected is limited by the capacity of the injection port and the analytical column. In practical terms, injections are almost always in the low milliliter range, with sizes of 0.1–2.0 mL being typical [14].

If a complex material, such as a food, is placed into a sealed vial and allowed to stand, the volatile compounds in the sample matrix will leave the sample and distribute over the headspace around it. The concentration of the compounds in the headspace depends on several factors, such as the concentration in the original sample, the volatility of the compound, the solubility of that compound in the sample matrix, the temperature of the sample, and combination of the size of the vial and the time the sample has been inside in the vial [14]. At equilibrium, the relationship between the concentration of the volatile compound in the product phase and in the vapor phase can be expressed by Henry's law. This law states that the mass of vapor dissolved in a certain volume of solvent is directly proportional to the partial pressure of the vapor that is in equilibrium with the solution [15]. Static headspace isolation normally involves taking a sample of the equilibrium headspace surrounding a sample. This can be directly injected onto a gas chromatographic column. In classical static headspace analysis, volatiles are removed without any attempt to simulate the conditions in the mouth during eating. Therefore, amounts determined do not necessarily represent the compounds and the quantities available for perception during eating. Classical headspace analysis is probably more closely related to the odor perceived as food approaches the mouth before eating [16].

More recently, solid-phase micro-extraction (SPME) has been developed for headspace sampling. It involves extraction of volatile compounds out of the headspace onto a fused-silica fiber coated with a polymeric phase. After equilibration, the fiber containing the adsorbed or absorbed analytes is thermally desorbed and subjected to analysis [17]. It should be kept in mind that a second equilibrium between headspace and fiber is involved that may affect how well the sample reflects the actual headspace composition.

28.2.1.2 Dynamic Headspace Sampling

In dynamic headspace analysis, a flow of gas is passed over the food to strip off volatiles, which results in a greater yield of material for analysis than in static headspace isolation. In a separate part of the apparatus, volatiles are trapped from the gas stream. The headspace of the sample is continually renewed. Often the sample is stirred or otherwise agitated to increase mass transfer from the sample into the headspace [18]. Normally, a solid adsorbent is used to trap the volatiles, such as charcoal, Porapak Q, Chromosorb 101–105, or Tenax, and they are usually thermally desorbed before gas chromatographic analysis. Dynamic headspace methods permit analyses with minimal introduction of artifacts developed or introduced during sampling [19]. The thermal desorption of the compounds from the absorbent, although convenient and rapid, has the disadvantage of causing molecular change in some important unstable aroma compounds, such as (Z)-3-hexenal, alkadienals, and certain sulfur compounds. This may be due to the elevated temperatures applied during desorption and the metal parts frequently used in thermal desorption units. In the actual eating process, equilibrium is not likely, if ever attained in the mouth. The dynamic mode seems, therefore, more closely aligned with what happens in the mouth during eating than static headspace analysis.

28.2.2   In Vitro *Sampling (Mouth Analogs)*

Mouth analogs have been developed to mimic aroma release in the mouth more precisely and to consider changes in aroma release during eating. In actual eating situations, aroma concentrations are determined kinetically rather than thermodynamically [20]. Only a few instrumental methods of aroma release have incorporated the crushing, mixing, dilution, and temperature conditions required to simulate aroma release in the mouth from solid foods. Lee III [21] reported an instrumental technique for measuring dynamic flavor release. A mass spectrometer was coupled with a dynamic headspace system, that is, a vial with several small metal balls that simulated mastication. van Ruth et al. [22] reported a model mouth system that consisted of a thermostated glass flask with a volume of 70 mL, with the option of (artificial) saliva addition. A plunger making screwing movements simulated mastication. This is one of the few mouth analogs that has been validated by comparison with the *in vivo* volatile flavor concentrations [23]. Naβl et al. [24] described a "mouth imitation chamber," which consisted of a large volume thermostated vessel with a stirrer, while artificial saliva was added to the system. Roberts and Acree [20] reported a "retronasal aroma simulator"—a purge-and-trap device made from a blender. It simulated mouth conditions by regulating the temperature to 37°C, adding artificial saliva, and using mechanical forces. Withers et al. [25] reported a simple mouth analog for drinks, which consisted of a glass flask filled with liquid sample and artificial saliva with glass beads, which was situated in a shaking water bath.

28.2.3   In Vivo *Sampling*

In the period 1985–2000, techniques were developed to measure the release of volatile compounds *in vivo*, that is, in the mouth and nose of subjects. The advantage of these techniques over model systems is that the data reflect volatile concentrations as they are during perception. However, one has to deal with the fact that the results are highly variable, because large inter-individual variations in volatile flavor release exist.

A simple way of measuring volatile release from foods is to analyze compounds in the expired air drawn from the nose of subjects consuming (model) foods. In the beginning of the nineties, attention turned to analyses that concentrated the volatiles over short time periods from the sample before analysis. Volatiles from expired breath were collected for certain time intervals (e.g., 15 s) either onto lengths of capillary tubes, by means of cryotrapping or by trapping on absorption materials, such as Tenax [26], and were subsequently analyzed.

Real-time analysis requires a detector capable of continuously monitoring compounds in air with a high temporal resolution. Several types of systems have been developed to allow real-time gas-phase analysis. A membrane separator method using direct MS was developed by Soeting and Heidema [27] for determining aroma profiles in the expired air of assessors as a function of time. Linforth and Taylor modified an atmospheric pressure chemical ionization source of a mass spectrometer to allow the introduction of gas-phase samples [28]. It is a soft ionization technique, which adds a proton to the compound of interest and does not normally induce fragmentation. Consequently, compounds present in the breath are monitored in selected ion mode, further enhancing sensitivity. Lindinger et al. [29] reported an online monitoring technique for measurement of volatile organic compounds based on proton transfer reaction mass spectrometry (PTR-MS). PTR-MS links the idea of chemical ionization introduced by Munson and Field in 1966 [30] with the swarm technique of flow-drift-tube type, invented by Ferguson and coworkers in the early 1970s [31]. Proton transfer reactions are used to induce chemical ionization of the vapors to be analyzed. The sample gas is continuously introduced into a drift tube, where it is mixed with H_3O^+ ions formed in a hollow cathode ion source. Volatile compounds that have proton affinities higher than water (>166.5 kcal/mol) are ionized by proton transfer from H_3O^+, mass analyzed in a quadrupole mass spectrometer, and eventually detected as ion counts/s (cps) by a secondary electron multiplier [32]. The outcome is either a mass-resolved fingerprint of the total volatile profile of a sample or a time-resolved profile for a selection of masses. PTR-MS is interesting for time-resolved analysis because it (i) requires no pretreatment of the sample, (ii) allows rapid measurements (typically 0.2 s/mass), and (iii) is extremely sensitive.

28.3 Identification of the Volatile Compounds Contributing to Sensory Perception: GC-O

The distinction between odor-active compounds and the whole range of volatiles present in a particular food product is an important task in flavor analysis. An interesting approach is sniffing the gas chromatographic effluent of a representative isolate of volatile compounds of a food to associate odor activity with the eluting compounds. Many of the "chemical" detectors are not as sensitive as the human nose for many odor-active compounds. Combining human perception of odor and chromatographic separation of compounds, that is, GC-O, offers great possibilities. The chromatographic separation of compounds can be a difficult task, which depends primarily on the complexity of the flavor. However, the registration and quantification of the sensory perception are also two of the important challenges of the technique.

Most methods aim to rank the volatile flavor compounds detected in order of sensory importance. The three important types of methods used are the dilution to detection threshold method, the detection frequency method, and the perceived intensity method. In the dilution analysis, an extract is diluted, and each dilution is sniffed until any odor is no longer detected. The last dilution at which a compound is detected is a measure for its odor potency [33,34]. The detection frequency method [35,36] is based on the number of assessors in a group of 6–12 assessors detecting an odor in the GC effluent simultaneously (detection frequency). The detection frequency is a measure for the sensory importance of a compound. The third group of GC-O techniques involves intensity methods. These methods measure the odor intensity of a compound in the GC effluent. They include the posterior intensity method [37,38], the cross-modality matching finger span technique [39], and the time-intensity methods [40].

28.4 Applications to Meat Flavor: Sampling and Identification of Volatiles Contributing to Sensory Perception

28.4.1 Extraction—GC-O

The odor-active compounds of cooked meats have been widely investigated using GC-O, but mostly using (solvent) extraction techniques for sampling. Examples of studies in which extractions were carried out followed by dilution analysis are those on cooked beef [41,42] as well as on pork and chicken [43,44]. Specht and Baltes [45] identified odor-active compounds in a similar manner in shallow-fried beef. Generally, compounds with higher boiling temperatures are sampled with this type of extraction.

28.4.2 Static Headspace Sampling—GC-O

An example of a study in which headspace sampling was used is the one of Carrapiso et al. [10]. They characterized the odor-active compounds of dry-cured ham using the detection frequency technique for GC-O evaluation. Compounds included aldehydes, ketones, esters, sulfur-containing compounds, nitrogen-containing compounds, and an alcohol. Moon et al. [46] reported a study on the comparison of different types of cooked beef. In this study, headspace sampling was applied using SPME. The odor-active compounds were identified by GC-O using the detection frequency analysis.

28.4.3 Dynamic Headspace Sampling—GC-O Analysis

Dynamic headspace sampling in combination with GC-O dilution analysis has also been applied to fermented sausages to evaluate various sampling procedures, such as molecular distillation, high vacuum distillation, and vacuum steam distillation [47]. Similarly, dynamic headspace sampling was compared to vacuum simultaneous distillation-solvent extraction in combination with GC-O dilution analysis for the determination of the odor-active compounds of alligator meat [48].

28.4.4 Mouth Analog Sampling—GC-O Analysis

Very few studies have been reported using a mouth analog for sampling. Machiels et al. determined the odor-active compounds in two commercial Irish beef meats [49] as well as in meat of differently fed bulls [50] using a mouth analog for sampling and the detection frequency technique for GC-O evaluation.

28.4.5 In Vivo Sampling

No studies on the analysis of odor-active compounds of meat using *in vivo* sampling were reported in the literature according to the author's knowledge. Neither was any meat study found using *in vivo* real-time analysis of meat volatiles or a form of intensity rating in GC-O analysis.

To summarize, most studies in which odor-active compounds have been analyzed in meat have reported sampling in the form of extractions or dynamic headspace analysis usually in combination with GC-O dilution analysis. Static headspace sampling and the use of mouth analogs have been applied in combination with GC-O analysis, using the detection frequency methodology.

28.5 Conclusions

For sampling of volatile compounds that contribute to sensory perception various techniques are available, such as static headspace sampling, dynamic headspace sampling, the use of mouth analogs, and *in vivo* samplings. GC-O analysis is an interesting technique for identification of the odor-active volatile compounds of food samples. Although most odor-active compounds in meat have been assessed by GC-O after a type of (solvent or vacuum) extraction, it is especially interesting to evaluate these characteristics of compounds at concentrations and in ratios that represent consumption conditions. Some studies report static or dynamic headspace sampling or the application of mouth analogs. The use of *in vivo* measurements for meat analysis with or without the combination of GC-O has not received attention yet. Therefore, this type of sampling may be an interesting way forward in meat flavor analysis.

References

1. Matsuishi, M., Igeta, M., Takeda, S., and Okitani, A. Sensory factors contributing to the identification of the animal species of meat. *J. Food Sci.* 69, 218–220. 2004.
2. Crandall, C. S. The liking of foods as a result of exposure: eating doughnuts in Alaska. *J. Soc. Psychol.* 125, 187–194. 1985.
3. Prescott, J., Young, O., and O'Neill, L. The impact of variations in flavour compounds on meat acceptability: a comparison of Japanese and New Zealand consumers. *Food Qual. Pref.* 12, 257–264. 2001.
4. Raes, K., Balcaen, A., Dirinck, P., De Winne, A., Claeys, E., Demeyer, D., and De Smet, S. Meat quality, fatty acid composition and flavour analysis in Belgian retail beef. *Meat Sci.* 65, 1237–1246. 2003.
5. Verbeke, W. and Viaene, J. Beliefs, attitude and behaviour towards fresh meat consumption in Belgium: empirical evidence from a consumer survey. *Food Qual. Pref.* 10, 437–445. 1999.
6. Mottram, D. S. Flavour formation in meat and meat products: a review. *Food Chem.* 62, 415–424. 1998.
7. Maarse, H. and Visscher, C. A. *Volatile Compounds in Food—Qualitative and Quantitative Data.* 7th edn. Zeist: TNO-CIVO. 1996.
8. Vasta, V. and Priolo, A. Ruminant fat volatiles as affect by diet. A review. *Meat Sci.* 73, 218–228. 2006.
9. Chevance, F. F. V. and Farmer, L. J. Identification of major volatile odor compounds in frankfurters. *J. Agric. Food Chem.* 47, 5151–5160. 1999.
10. Carrapiso, A. I., Ventanas, J., and García, C. Characterization of the most odor-active compounds of Iberian ham headspace. *J. Agric. Food Chem.* 50, 1996–2000. 2002.
11. Piggott, J. R. Dynamism in flavour science and sensory methodology. *Food Res. Int.* 33, 191–197. 2000.
12. Baek, I., Linforth, R.S.T., Blake, A., and Taylor, A. J. Sensory perception is related to the rate of change of volatile concentration in-nose during eating of model gels. *Chem. Senses* 24, 155–160. 1999.
13. Dijksterhuis, G. B. Dynamic sensory methods: some applications and developments. In *Cost 96 Symposium: Food and Flavour*, Udine, Italy, 23–24 September 1999. 1999.
14. Wampler, T. P. Analysis of food volatiles using headspace-gas chromatographic techniques. In *Techniques for Analyzing Food Aroma*, ed. R. Marsili, 27–58. New York: Marcel Dekker. 1997.
15. Morris, J. G. *A Biologist's Physical Chemistry*, London: Edward Arnold. 1968.
16. Taylor, A. J. Volatile flavor release from foods during eating. *Crit. Rev. Food Sci. Nutr.* 36, 765–784. 1996.

17. Machiels, D. and Istasse, L. Evaluation of two commercial solid-phase microextraction fibres for the analysis of target aroma compounds in cooked beef meat. *Talanta* 61, 529–537. 2003.

18. Sucan, M. K., Fritz-Jung, C., and Ballam, J. Evaluation of purge-and-trap parameters optimization using a statistical design. In *Flavor Analysis. Developments in Isolation and Characterisation*, eds C. J. Mussinan and M. J. Morello, 22–37. Washington: American Chemical Society. 1998.

19. Teranishi, R. Challenges in flavor chemistry: an overview. In *Flavor Analysis. Developments in Isolation and Characterization*, eds C. J. Mussinan and M. J. Morello, 1–6. Washington: American Chemical Society. 1998.

20. Roberts, D. D. and Acree, T. E. Simulation of retronasal aroma using a modified headspace technique: investigating the effects of saliva, temperature, shearing, and oil on flavor release. *J. Agric. Food Chem.* 43, 2179–2186. 1995.

21. Lee III, W. E. A suggested instrumental technique for studying dynamic flavor release from food products. *J. Food Sci.* 51, 249–250. 1986.

22. van Ruth, S. M., Roozen, J. P., and Cozijnsen, J. L. Comparison of dynamic headspace mouth model systems for flavour release from rehydrated bell pepper cuttings. In *Trends in Flavour Research*, eds H. Maarse and D. G. van der Heij, 59–64. Amsterdam: Elsevier. 1994.

23. van Ruth, S. M. and Roozen, J. P. Influence of mastication and artificial saliva on aroma release in a model mouth system. *Food Chem.* 71, 339–345. 2000.

24. Naßl, K., Kropf, F., and Klostermeyer, H. A method to mimic and to study the release of flavour compounds from chewed food. *Z. Lebensm. Unters. –Forsch.* 201, 62–68. 1995.

25. Withers, S. J., Conner, J. M., Piggott, J. R., and Paterson, A. A simulated mouth to study flavour release from alcoholic beverages. In *Food Science and Technology Cost 96, Interaction of Food Matrix with Small Ligands Influencing Flavour and Texture*, volume 3, ed. P. Schieberle, 13–18. Luxembourg: Office for Official Publications of the European Communities. 1998.

26. Roozen, J. P. and Legger-Huysman, A. Sensory analysis and oral vapour gas chromatography of chocolate flakes. In *Aroma. Perception, Formation, Evaluation*, eds M. Rothe and H.-P. Kruse, 627–632. Potsdam-Rehbrücke: Eigenverlag Deutsches Institut für Ernährungsforschung. 1995.

27. Soeting, W. J. and Heidema, J. A mass spectrometric method for measuring flavour concentration/time profiles in human breath. *Chem. Senses* 13, 607–617. 1988.

28. Linforth, R. S. T. and Taylor, A. J. Measurement of volatile release in the mouth. *Food Chem.* 48, 115–120. 1993.

29. Lindinger, W., Hansel, A., and Jordan, A. On-line monitoring of volatile organic compounds at pptv levels by means of Proton Transfer Reaction Mass Spectrometry (PTR-MS). Medical applications, food control and environmental research. *Int. J. Mass Spectrom. Ion Proc.* 173, 191–241. 1998.

30. Munson, M. S. B. and Field, F. H. Chemical ionization mass spectrometry. I. General introduction. *J. Am. Chem. Soc.* 88, 2621–2630. 1966.

31. McFarland, M., Albritton, D. L., Fehsenfeld, F. C., Ferguson, E. E., and Schmeltekopf, A. L. Flow-drift technique for ion mobility and ion-molecule reaction rate constant measurements. II. Positive ion reactions of N+, O+, and H with O2 and O+ with N2 from thermal to [inverted lazy s]2 eV. *J. Chem. Phys.* 59, 6620–6628. 1973.

32. Aprea, E., Biasioli, F., Gasperi, F., Märk, T. D., and van Ruth, S. In vivo monitoring of strawberry flavor release from model custards: effect of texture and oral processing. *Flav. Fragr. J.* 21, 53–58. 2006.

33. Acree, T. E., Barnard, J., and Cunningham, D. A procedure for the sensory analysis of gas chromatographic effluents. *Food Chem.* 14, 273–286. 1984.

34. Ullrich, F. and Grosch, W. Identification of most intense volatile flavour compounds formed during autoxidation of linoleic acid. *Z. Lebensm. Unters. –Forsch.* 184, 277–282. 1987.

35. Linssen, J. P. H., Janssens, J. L. G. M., Roozen, J. P., and Posthumus, M. A. Combined gas chromatography and sniffing port analysis of volatile compounds of mineral water packed in polyethylene laminated packages. *Food Chem.* 46, 367–371. 1993.

36. Pollien, P., Ott, A., Montigon, F., Baumgartner, M., Rafael Muňoz-Box, R., and Chaintreau, A. Hyphenated headspace-gas chromatography-sniffing techniques: screening of impact odorants and quantitative aromagram comparisons. *J. Agric. Food Chem.* 45, 2630–2637. 1997.

37. Berdagué, J. L., Tournayre, P., and Cambou, S. Novel multi-gas chromatography-olfactometry device and software for the identification of odour active compounds. *J. Chromatogr. A* 1146, 85–92. 2007.

38. Le Guen, S., Prost, C., and Demaimay, M. Critical comparison of three olfactometric methods for the identification of the most potent odorants in cooked mussels (*Mytilus edulis*). *J. Agric. Food Chem.* 48, 1307–1314. 2000.

39. Etievant, P. X., Callement, G., Langlois, D., Issanchou, S., and Coquibus, N. Odor intensity evaluation in gas chromatograpy-olfactometry by finger span method. *J. Agric. Food Chem.* 47, 1673–1680. 1999.

40. Miranda-Lopez, R., Libbey, L. M., Watson, B. T., and McDaniel, M. R. Odor analysis of Pinot noir wines from grapes of different maturities by a gas chromatography-olfactometry technique (Osme). *J. Food Sci.* 57, 985–993, 1019. 1992.

41. Gasser, U. and Grosch, W. Identification of volatile flavour compounds with high aroma values from cooked beef. *Z. Lebensm. Unters. –Forsch.* 186, 489–494. 1988.

42. Cerny, C. and Grosch, W. Evaluation of potent odorants in roasted beef by aroma extract dilution analysis. *Z. Lebensm. Unters. –Forsch.* 194, 322–325. 1992.

43. Werkhoff, P., Bruening, J., Emberger, R., Guentert, M., and Hopp, R. Flavour chemistry of meat volatiles: new results on flavor components from beef, pork and chicken. In *Recent Development in Flavor and Fragrance Chemistry: Proceedings of the 3rd International Haarmann & Reimer Symposium*, 183–213. Holzminden: VCH. 1993.

44. Gasser, U. and Grosch, W. Primary odourants of chicken broth. *Z. Lebensm. Unters. –Forsch.* 190, 3–8. 1990.

45. Specht, K. and Baltes, W. Identification of volatile flavour compounds with high aroma values from shallow-fried beef. *J. Agric. Food Chem.* 42, 2246–2253. 1994.

46. Moon, S.-Y., Cliff, M. A., and Li-Chan, E. C. Y. Odour active components of simulated beef flavour analysed by solid phase microextraction and gas chromatography-mass spectrometry and –olfactometry. *Food Res. Int.* 39, 294–308. 2006.

47. Schmidt, S. and Berger, R. G. Aroma compounds in fermented sausages of different origins. *Lebensm. –Wiss. Technol.* 31, 559–567. 1998.

48. Cadwallader, K. R., Baek, H. H., Chung, H. Y., and Moody, M. W. Contribution of lipid-derived components to the flavour of alligator meat. In *Lipids in Food Flavors, ACS Symposium Series* 558, 186–195. 1994.

49. Machiels, D., van Ruth, S. M., Posthumus, M. A., and Istasse, L. Gas chromatography-olfactometry analysis of the volatile compounds of two commercial Irish beef meats. *Talanta* 60, 755–764. 2003.

50. Machiels, D., Istasse, L., and van Ruth, S. M. Gas chromatography-olfactometry analysis of beef meat originating from differently fed Belgian Blue, Limousin and Aberdeen Angus bulls. *Food Chem.* 86, 377–383. 2004.

Chapter 29

Sensory Descriptors

Geoffrey R. Nute

Contents

29.1   Introduction

The science of sensory evaluation is still evolving as compared to the more established sciences of chemistry and physics. Indeed, the first textbook was written in Polish ca. 1957 by Tilgner[1] and the first textbook in English was by Amerine et al.[2] in 1965. The latter had its origins based on a review of the literature since 1940 and formed part of a course majoring in food science at the University of California, Davis, that started in 1957.

Stone and Sidel[3] defined sensory evaluation as *a scientific method used to evoke, measure, analyze and interpret those responses to products as perceived through the senses of sight, smell, touch, taste and hearing.*

An essential aspect of measuring these responses rely on the assessors who are able to accurately describe the characteristics of the stimuli that they are receiving. This leads to the need to understand and convey the characteristics of these responses to other assessors so that they understand and rate these characteristics in a similar way. This then forms the basis of the construction of either a category or an intensity scale or a present or absent rating.

An early work by Jones and Thurstone[4] and Jones et al.[5] used a list of 51 words and phrases that could be used as anchors for hedonic scales. These words or phrases were each rated on a 9-point category scale ranging from greatest dislike (−4) to greatest like (+4) and scale values and standard deviations produced for each word or phrase.

This work showed that words with an estimated value close to the assigned physical scale and with a small standard deviation tended to be the least ambiguous and were the preferred words to be included in a scale.

A systematic approach used to select texture descriptors[6] involved the tedious job of searching the *Concise Oxford Dictionary* for texture terms and their definitions; a short list by way of illustration are "Chalky, crisp, doughy, firm, flaky, fleshy, floury, flabby, greasy, hard, juicy, lean, limp, mushy, oily, powdery, ripe, rotten, rubbery, sandy, short, sleepy, slushy, soft, springy, sticky, syrupy, tender, thick, thin, tough, treacly, viscous, watery, waxy, woody." In all, approximately 60 descriptors were listed. This highlights the vast vocabulary than can arise from trying to describe just one facet of the eating experience. A further study involved the classification of odors; Harper et al.[7] listed some 300 terms of which about 69 were then presented to consumers in a questionnaire where respondents were required to judge the usefulness of the terms to describe odor. This list was then reduced to 44, which was subsequently published as a glossary of odor qualities.[8]

Later in the mid-1970s, Szczesniak et al.[9] developed the texture profile method that was used with a fixed set of force-related attributes to describe and characterize the rheological and tactile properties of food. Assessors were trained to recognize specific points on an intensity scale and were provided with effective standards for each intensity.

A further work tracked the changes in the perception of sensory attributes during the mastication process. A work by Brown et al.[10] concluded that eating roasted meat samples and recording changes in the perception of tenderness, using time intensity methods, revealed significant correlations between the amount of masticatory muscle activity during chewing as measured by electromyography and tenderness. They ascribed the differences among assessors to the difference in the way chewing broke down the structure of the samples. Therefore, the perceived stimulus and hence their rating of a descriptor was affected by inherent differences in chewing and masticatory behavior.

A work by Nute et al.[11] on ham used 22 descriptors covering appearance, texture, and flavor, and derived a consensus configuration using generalized Procrustes analysis (GPA) and then applied principal components to integrate instrumental and chemical determinations into the consensus space. This showed that traditional hams (containing less water than in the starting material) were linked with increased tensile parameters of stiffness, shear strength, ultimate tensile stress, and the sensory texture descriptors of flakiness, cohesiveness, firmness, and with the flavor descriptor showing increased ham flavor.

A recent work by Giboreau et al.[12] centered on how sensory descriptors are defined and suggested a set of guidelines that could aid the formulation of more accurate definitions of sensory descriptors. They analyzed 100 sensory descriptors that included olfactory, tactile, gustative, and visual by applying linguistic criteria: syntactics, for example, nouns, verbs, and adjectives; and semantics, for example, metaphors, analogy, and synonymy. They suggested that the accuracy of definitions could be improved if they were focused on three areas: the product under study is related to the project where differences are related to formulations, benchmarking; the perceiving subject is involved in a consumer study where new products are being tested and preference mapping techniques applied to identify the underlying dimensions of acceptability;[13] and the assessors are involved in the wider aspects and have expert knowledge of the individual products. Interestingly, sensory analysts usually insist that while using detailed descriptive profiles the training covers the range of samples that are in the proposed experiment. The argument is that if assessors are presented with samples that are outside their range of experience, they will not know how to rate the given intensity scales or may feel that they need a new descriptor.

This need to define sensory descriptors has been a topic of discussion among sensory analysts around the world and an attempt has been made to produce a vocabulary of terms (ISO 5492)[14] and terminology (ISO 1087)[15] used in sensory analysis. This approach has raised issues of translation because many of the definitions do not translate directly into other languages and with increasing globalization there is a need to standardize descriptive terms.

This chapter provides examples of the use of sensory descriptors in aspects of meat quality.

29.2   Category Scales

These scales are commonly used in animal production studies where basic sensory information is required; they have the advantage of being easy to use and understand from the assessors' viewpoint. Each category on the scale has a sensory descriptor that relates to a perceived intensity.

The concept of category scales has its origin based on the work of Jones et al.[5] This work describes a 9-point hedonic scale for liking that was originally developed at the U.S. Army Food and Container Institute (Quartermaster Corps) (Table 29.1).

These scales were subsequently adapted for and reduced to 8-point scales by removing the middle point (neither like nor dislike).

The uses and abuses of category scales in sensory measurement have been discussed by Riskey[16] who states that category ratings are affected by stimulus range, number of categories used, and the number of stimuli presented. He also makes the point that the same stimulus presented to an assessor on one day may not be rated the same on another day. His analogy is that a "warm day" in June may be quite different in an absolute sense to that in February. This raises the issue of contextual influences within a panel and should always be considered when organizing sensory panels, for example, a very tough sample in a panel followed by a tender sample will often result in the tender sample being rated higher than would normally be the case for those assessors that received the tough sample first; the opposite effect will be observed in those assessors that received the tender sample first. This potential bias can be accounted for by employing methods outlined by MacFie et al.[17] who published designs to balance the effect of order of presentation and first order carryover effects.

In a study on the effects of final endpoint cooked temperature on the eating quality of pork, Wood et al.[18] used 8-point category scales to evaluate tenderness, juiciness, pork flavor, and abnormal flavor. The study used material, in a highly structured way, from the same individual

Table 29.1   Quartermaster Corps

9-Point Hedonic Scale	Values Added after Assessment
Like extremely	9
Like very much	8
Like moderately	7
Like slightly	6
Neither like nor dislike	5
Dislike slightly	4
Dislike moderately	3
Dislike very much	2
Dislike extremely	1

Table 29.2 Category Scales Used in the Study of Endpoint Temperature Effects on the Eating Quality of Pork Loin Steaks

	Tenderness	Juiciness	Flavor/Abnormal Flavor
8	Extremely tender	Extremely Juicy	Extremely strong
7	Very tender	Very juicy	Very strong
6	Moderately tender	Moderately juicy	Moderately strong
5	Slightly tender	Slightly juicy	Slightly strong
4	Slightly tough	Slightly dry	Slightly weak
3	Moderately tough	Moderately dry	Moderately weak
2	Very tough	Very dry	Very weak
1	Extremely tough	Extremely dry	Extremely weak

Table 29.3 Influence of Endpoint Cooking Temperature on the Tenderness of Pork Loin Steaks from Different Sources

	Final Internal Endpoint Temperature (°C)			
	65	72.5	80	Significance
Commercial gilts	5.1	4.5	4.2	***
Stotfold gilts	5.6	5.2	4.6	***
Commercial entires	5.5	4.9	4.3	***
Stotfold entires	5.7	5.2	4.8	***

pig to avoid contextual influences that could occur when using pork from different pigs. It was shown that using category scales with individual scale descriptors enabled assessors to differentiate samples in a uniform way. Scales with individual category descriptors are given in Table 29.2.

In terms of tenderness, there was a consistent result for different types of pigs as shown in Table 29.3. The results showed that as endpoint temperature increases, tenderness decreases and this is evident in commercial gilts, commercial entires and Stotfold (experimental farm) entires and gilts.

The other attributes of juiciness, flavor, and abnormal flavor also showed significant trends for each attribute. To summarize, as the endpoint temperature increases, tenderness decreases, juiciness decreases, flavor increases, and abnormal flavor decreases. At low endpoint temperatures, tenderness increases, juiciness increases, flavor decreases, and abnormal flavor increases.

The category scale approach using a standard texture description of varying degrees of toughness/tenderness and in addition local scales for texture, juiciness, and flavor were used to compare beef quality across five European countries, given sets of steaks from the same animals.[19] Eight countries supplied steaks to all participating sensory panels. The results showed a good agreement among countries for the descriptor tenderness, but flavor and juiciness were poorly related, indicating that these descriptors were not rated in the same way. However, since the range of texture was large, it tended to dominate the other sensory attributes. A later study[20] produced steaks with a smaller range of tenderness so that juiciness and flavor would not be dominated by ranges in toughness. Their results showed that tenderness and juiciness were assessed in a consistent way, but flavor was the least consistent, indicating that the descriptor beef flavor has either a different meaning or that what constitutes beef flavor has different perceptions in different countries.

A later study on lamb[21] compared the eating quality of lamb by British and Spanish taste panels, using the same basic descriptors, but treating descriptors as categories in British panels and

Table 29.4   A Comparison of Interpretation of Eating Quality Attributes of Grilled Lamb Loins between Spanish and British Taste Panels

Breed	Spanish Merino (SM)	Rasa Aragonesa (RA)	British Export (BE)
Lamb odor intensity			
British panel	2.83[b]	2.69[b]	4.03[a]
Spanish panel	57.95[b]	58.47[b]	67.34[a]
Tenderness			
British panel	6.21[a]	5.30[b]	6.45[a]
Spanish panel	72.52[a]	58.46[b]	69.32[a]
Juiciness			
British panel	5.18[a]	5.13[a,b]	5.05[b]
Spanish panel	64.05[a]	61.69[a]	45.11[b]

Note: Different superscripts within a row indicate significant differences between the samples.

Table 29.5   A Comparison of Hedonic Ratings of Eating Quality of Grilled Lamb Loins between Spanish and British Taste Panels

Breed	SM	RA	BE
Overall appraisal			
British panel	3.49[b,c]	3.43[c]	4.74[a]
Spanish panel	59.48[a]	57.82[a]	32.01[b]

Note: Different superscripts within a row indicate significant differences between the samples.

using unstructured 100 mm line intensity scales with anchor points at each end in Spain. This was intended to explore two areas; first, whether the two panels tasting lamb from the same animals (left-side loins to Spain, right-side loins to the United Kingdom) achieved the same conclusion, indicating that the descriptors were used in the same way and second, whether the inclusion of a hedonic scale for overall appraisal was viewed in a similar way. Results showed that lamb odor intensity, tenderness, juiciness, and flavor intensity were assessed in the same way by assessors in different countries and produced similar conclusions (see Table 29.4).

However, the inclusion of the hedonic scale produced significant differences. The Spanish panel preferring Spanish lamb and the U.K. panel preferring British lamb are indicated in Table 29.5.

The categories with descriptive steps are used to establish major differences in eating quality; however, for further information and relationships between odor, texture, and flavor as perceived with information derived from chemical and instrumental measurements, more elaborate descriptors are employed.

29.3   Sensory Descriptive Profiles

Earlier, the use of category scales gave indications of the differences between lambs fed different diets; correlations between the descriptors and fatty acid composition[22] showed that lamb odor and lamb flavor were correlated with different fatty acids as shown in Table 29.6.

Lamb odor and flavor were positively correlated with oleic, EPA, and linolenic acids and negatively correlated with linoleic and arachidonic fatty acids.

Table 29.6 Sensory Analysis Descriptors Correlated with Fatty Acid Composition (mg/100 g of Muscle)

	Lamb Odor		Lamb Flavor	
	U.K. Lamb	Spanish Lamb	U.K. Lamb	Spanish Lamb
Oleic	0.59*	0.30	0.61*	0.48*
Linoleic	−0.63	−0.65	−0.69*	−0.67*
Linolenic	0.64	0.42	0.75	0.57
EPA	0.58	0.44	0.76	0.45
Arachidonic	−0.54	−0.49	−0.55	−0.54

Table 29.7 Flavor Descriptors Used to Distinguish Different Lamb Types/ Production Systems

Feeding/Finishing	Welsh Mountain Upland Flora	Soay Grass	Suffolk Grass	Suffolk Concentrates
Attribute				
Lamb flavor intensity	20.5[b]	13.5[a]	27.2[c]	14.9[a]
Abnormal lamb flavor	28.8[a]	41.8[b]	28.2[a]	45.0[b]
Fatty/greasy	21.6[b]	19.0[a]	18.2[a]	19.5[a]
Sweet	12.1[c]	7.2[a]	11.2[b,c]	9.1[b]
Metallic	10.6[a]	14.3[b]	9.2[a]	11.6[a,b]
Bitter	10.9[a]	15.9[b]	10.5[a]	14.0[b]
Stale	8.9[a]	12.3[b]	8.1[a]	11.9[b]
Rancid	6.9[a]	11.0[b]	6.8[a]	10.3[b]
Livery	16.2[a]	20.5[b]	14.9[a]	16.7[a]
Fishy	2.6[a]	4.8[b]	1.4[a]	2.0[a]
Ammonia	3.1[a]	5.9[b]	2.5[a]	5.4[b]

Note: Different superscripts within a row indicate significant differences, $p < .05$, between the lamb types.

On the basis of these correlations, a descriptive profile was developed to investigate variations in eating quality of lamb types from diverse production systems.[23]

This profile concentrated primarily on flavor and included the descriptors: lamb flavor intensity, abnormal flavor intensity, fatty/greasy, sweet, acidic, metallic, bitter, stale, rancid, livery, vegetable, grassy, fishy, and ammonia. Significant terms are given in Table 29.7.

The highest ratings for lamb flavor intensity came from Welsh Mountain and Suffolk lambs reared on grass and the lowest from Suffolk lambs reared on concentrates. Although the Soay was finished on grass their flavor profile was very similar to that of the Suffolk reared on concentrates. The Soay was also very high in n-6 and n-3 PUFA that are prone to peroxidation and were probably responsible for livery flavor rated as significantly higher than in other breeds.

A later study by Nute et al.[24] on the effects of dietary oil source on lamb flavor defined each of the flavor terms in both muscle and lamb fat (see Table 29.8).

Exploring the relationship between sensory attributes and fatty acid composition showed that lamb flavor was positively correlated with conjugated linoleic acid (CLA), 18:3n-3 (α-linolenic acid) and negatively correlated with DHA 22:6n-3 (docosahexaenoic acid).

Table 29.8 Flavor Descriptors and Definitions Used in the Assessment of Cooked Lamb Muscle

Term	Description
Lamb flavor	Intensity of cooked lamb flavor
Abnormal lamb flavor	Intensity of abnormal flavor
Fatty/greasy	The taste associated with oil and fat
Sweet	The taste associated with sugars
Livery	The taste associated with liver
Acidic	The taste associated with acids
Metallic	A tangy metal taste
Bitter	The taste associated with caffeine/quinine
Rancid	The taste associated with rancid, stale fat
Posset	The taste associated with the smell of warm sour milk
Fishy	The taste associated with fish
Grassy	The taste associated with vegetables/grass
Ammonia	Pungent, stale urine
Soapy	The taste associated with soap
Toughness	End points 0 (extremely tender) to 100 (extremely tough)
Juiciness	End points 0 (extremely dry) to 100 (extremely juicy) overall

Note: All terms were evaluated on 0–100 mm line intensity scales where 0 = nil intensity and 100 = extreme intensity.

Abnormal flavor was correlated with bitter, fishy, and rancid. The former two descriptors were correlated with 18:0 (stearic acid) and the latter with CLA.

The use of these descriptors have been effective in demonstrating that when formulating diets that are perceived to improve nutritional ratios in lamb meat, there can be adverse effects on flavor and the reason for these adverse flavors can now be related to the fatty acid composition of the meat. A review by Melton[25] provided information on the role of sensory analysis in red meat flavor and a later review by Wood et al.[26] concentrated on the effects of fatty acids on meat quality and included relationships between sensory descriptions and fatty acid concentrations in meat.

Lipid oxidation is of major importance in meat quality, especially with the demands of increasing shelf life required at the point of sale. Rancidity, as a measure of oxidation, is usually measured by the concentration of thiobarbituric acid-reactive substances (TBARS) and expressed as mg of malonaldehyde per kg of lean muscle. Values ranging between 0.6 and 2.0[27] have been quoted to be responsible for oxidized beef flavor thresholds.

In a study by Campo et al.,[28] meat from 73 Angus and Charolais cross steers reared on 10 different diets including grass silage, cereal concentrate, three diets with 3% added fat and a further three diets that contained protected fish oil, a constant amount of unprotected fish oil and a control unprotected fish oil diet were used to study the relationship between TBARS and sensory attributes.

Meat from these animals were aged for 10–13 days and then displayed under retail conditions for 0, 4, or 9 days.

A set of sensory descriptors for beef flavor was developed that included the terms, beef flavor intensity, abnormal flavor intensity, rancid, greasy/oily, bloody, metallic, livery, bitter, sweet, acidic, which are defined in Table 29.9. The mean values for each of these terms were correlated with TBARS.

Table 29.9   Definition of Beef Sensory Descriptors Used in the Assessment of Flavor Oxidation in Beef

Term	Description
Beef flavor intensity	Flavor associated with cooked beef
Abnormal flavor intensity	Abnormal flavor not found in cooked beef
Rancid	Rancid flavor found in meat
Greasy	Flavor associated with oil
Bloody	Flavor associated with blood or raw beef
Metallic	Flavor associated with meat taste
Livery	Flavor associated with liver
Bitter	Bitter taste
Sweet	Sweet taste
Fishy	Flavor associated with fish
Acidic	Sour taste
Cardboard	Flavor associated with wet cardboard and stale
Vegetable	Flavor associated with vegetables
Grassy	Flavor associated with fresh grass
Dairy	Flavor associated with dairy products

Beef flavor intensity decreased as display time increased and abnormal beef flavor increased. Rancid flavor increased with display time and followed a similar pattern to abnormal flavor intensity. Rancid flavor was also highly correlated with greasy flavor.

These relationships suggested that a TBAR value of approximately 2 could be considered as the limit for acceptability of beef.

29.4   Model Systems

The use of model systems, defining how compounds are perceived, is a useful route for characterizing odors and flavors that are associated with taints. A particular problem that has plagued meat science for many years has been the elucidation of what constitutes "boar taint." The compounds responsible have been identified as skatole and androstenone. Skatole is primarily produced in the hindgut by degradation of tryptophan by microbial action. Androstenone is produced in the Leydig cells of the testes and comes under the control of gonadotropin-releasing hormones. Once synthesized, the circulating androstenone is absorbed and stored in the salivary glands or in adipose tissue where it is concentrated as a result of its lipophilic nature.

As far as consumers are concerned, these two compounds can result in an objectionable taint that often leads to the rejection of the meat. The problem for sensory analysts is to produce a descriptive profile that will enable identification and distinction of these two compounds.

The approach used was described by Annor-Frempong et al.[29] They used spiked vegetable fat as a model carrier with varying concentrations of androstenone and skatole, initially as a single entity and then in combination.

Androstenone odor was characterized, summary descriptions in brackets, as acrid (bitter, pungent, caustic), ammonia (pungent, stale urine), sweaty (stale sweat), parsnip (smell of cooked parsnip), silage, (fresh smell, sweet, and sickly), dirty (soiled, unclean). Skatole was defined by mothball (naphthalene), musty (stale, old fabric).

Table 29.10 Mean Odor Panel Ratings of Relevant Descriptors of "Boar Taint" in Pork Fat Classified According to Convention

Descriptor	LA-LS	HA-LS	HA-HS
Acrid	9.92[a]	14.80[b]	21.03[c]
Mothball	2.50[a]	3.80[b]	23.50[c]

Note: LA-LS = Low androstenone and low skatole; HA-LS = High androstenone and low skatole; HA-HS = High androstenone and high skatole. Values with different superscripts, within a row, are significantly different at $p < .05$.

Subsequently, combinations of androstenone and skatole were presented to assessors. This showed that the perception of acrid and ammonia notes of androstenone were enhanced in the presence of skatole, whereas sweaty and parsnip notes were suppressed. The conclusion from this work on models shows that when skatole and androstenone are combined,[30] the rate of change based on skatole ratings is increased by a factor of 3, but when considering androstenone, the rate of change in intensity remains at a fairly constant level.

A later work by Annor-Frempong et al.[31] looked at the descriptions of androstenone and skatole in pork fat. Pork fats were classified according to the convention based on androstenone and skatole, that is, low androstenone ($<0.5\ \mu g\ g^{-1}$), low skatole ($<0.2\ \mu g\ g^{-1}$), termed LA-LS; high androstenone ($>0.5\ \mu g\ g^{-1}$), low skatole ($<0.2\ \mu g\ g^{-1}$), termed HA-LS; and high androstenone ($>0.5\ \mu g\ g^{-1}$), high skatole ($>0.2\ \mu g\ g^{-1}$) termed HA-HS.

Using the descriptive terms described earlier, assessors were able to differentiate the different classes of androstenone and skatole as shown in Table 29.10.

29.5   Conclusion

The use of sensory descriptors, by trained sensory assessors, provides an important tool in studying the components of eating quality. It is important to note that the statistical approaches used to study the ways in which assessors rate descriptors are crucial. It is vital that checks are made to measure the performance of the panel, including the methodology to check the understanding and meaning of descriptors and reproducibility of ratings. The works that have been described demonstrate the use of sensory descriptors in meat quality work.

They provide a synopsis of the different areas that can be studied using sensory descriptors; the examples are by no means exhaustive but cover the current areas of interest.

References

1. Tilgner, D.J. Analiza organoleptyczna zywnosci. Warszawa: Wydawnictwo przemyslu Lekkiego I Spozywczego. 1957.
2. Amerine, M.A., Pangborn, R.M., and Roessler, E.B. *Principles of Sensory Evaluation of Food*. Academic Press, New York, 1965.
3. Stone, H. and Sidel, J.L. *Sensory Evaluation Practices*, 2nd ed. Academic Press, San Diego, USA, 1993.
4. Jones, L.V. and Thurstone, L.L. The psychophysics of semantics: an experimental investigation. *J. Appl. Psychol.*, 39: 31–36, 1955.

5. Jones, L.V., Peryam, D.R., and Thurstone, L.L. Development of a scale for measuring soldier's food preferences. *Food Res.*, 20: 512–520, 1955.

6. Harper, R. Texture and consistency from the standpoint of perception: some major issues. In: *Rheology and Texture in Foodstuffs*. Monograph No. 27, Society of Chemical Industry, London, 11–28, 1968.

7. Harper, R., Bate-Smith, E.C., and Land, D.G. *Odour Description and Odour Classification*. Churchill, London, 1968.

8. Harper, R., Land, D.G., Griffiths, N.M., and Bate-Smith, E.C. Odour qualities: a glossary of usage. *Br. J. Psychol.*, 59: 231–252, 1968.

9. Szczesniak, A.S., Loew, B.J., and Skinner, E.Z. Consumer texture profile technique. *J. Food Sci.*, 40: 1253–1257, 1975.

10. Brown, W.E., Langley, K.R., Mioche, L., Marie, S., Gerault, S., and Braxton, B. Individuality of understanding and assessment of sensory attributes of foods, in particular, tenderness of meat. *Food Qual. Pref.*, 3/4: 205–216, 1996.

11. Nute, G.R., Jones, R.C.D., Dransfield, E., and Whelehan, O.P. Sensory characteristics of ham and their relationships with composition, visco-elasticity and strength. *Int. J. Food Sci. Technol.*, 22: 461–476, 1987.

12. Giboreau, A., Dacremont, C., Egoroff, C., Guerrand, S., Urdapilleta, I., Candel, D., and Dubois, D. Defining sensory descriptors: towards writing guidelines based on terminology. *Food Qual. Pref.*, 18: 265–274, 2007.

13. Nute, G.R., MacFie, H.J.H., and Greenhoff, K. Practical application of preference mapping. In: *Food Acceptability*, eds. Thomson, D.M.H. Elsevier Applied Science, London, UK, 377–386, 1988.

14. ISO (1992). Sensory analysis—Vocabulary. ISO 5492. International Organisation for Standardisation, Geneva.

15. ISO (1990). Terminology—Vocabulary. ISO 1087. International Organisation for Standardisation, Geneva.

16. Riskey, D.R. Use and abuses of category scales in sensory measurement. *J. Sensory Studies*, 3/4: 217–236, 1986.

17. MacFie, H.J., Bratchell, N., Greenhoff, K., and Vallis, L.V. Designs to balance the effect of order of presentation and first-order carry-over effects in hall tests. *J. Sensory Studies*, 4: 129–148, 1989.

18. Wood, J.D., Nute, G.R., Fursey. G.A.J., and Cuthbertson, A. The effect of cooking conditions on the eating quality of pork. *Meat Sci.*, 40: 127–135, 1995.

19. Dransfield, E., Rhodes, D.N., Nute, G.R., Roberts, T.A., Boccard, R., Touraille, C., Butcher, L., Hood, D.E., Joseph, R.L., Schon, I., Casteels, M., Cosentino, E., and Tinbergen, B.J. Eating quality of European beef assessed at five research institutes. *Meat Sci.*, 6: 163–184, 1982.

20. Dransfield, E., Nute, G.R., Roberts, T.A., Boccard, R., Touraille, C., Butcher, L., Casteels, M., Cosentino, E. Hood, D.E., Joseph, R.L., Schon, I., and Paardekooper, E.J.C. Beef quality assessed at European research centres. *Meat Sci.*, 10: 1–20, 1984.

21. Sanudo, C., Nute, G.R., Campo, M.M., Maria, G., Baker, A., Sierra, I., Enser, M., and Wood, J.D. Assessment of commercial lamb meat quality by British and Spanish taste panels. *Meat Sci.*, 1/2: 91–100, 1998.

22. Sanudo, C., Enser, M.E., Campo, M.M., Nute, G.R., Maria, G., Sierra, I., and Wood, J.D. Fatty acid composition and sensory characteristics of lamb carcasses from Britain and Spain. *Meat Sci.*, 54: 339–346, 2000.

23. Fisher, A.V., Enser, M.E., Richardson, R.I., Wood, J.D., Nute, G.R., Kurt, E., Sinclair, L.A., and Wilkinson, R.G. Fatty acid composition and eating quality of lamb types derived from four diverse breed x production systems. *Meat Sci.*, 55: 141–147, 2000.

24. Nute, G.R., Richardson, R.I., Wood, J.D., Hughes, S.I., Wilkinson, R.G., Cooper, S.L., and Sinclair, L.A. Effect of dietary oil source on the flavour and the colour and lipid stability of lamb meat. *Meat Sci.*, 77: 547–555, 2007.

25. Melton, S.L. Effects of feeds on flavour of red meat: a review. *J. Anim. Sci.*, 68: 4421–4435, 1990.

26. Wood, J.D., Richardson, R.I., Nute, G.R., Fisher, A.V., Campo, M.M., Kasapidou, E., Sheard, P.R., and Enser, M. Effects of fatty acids on meat quality: a review. *Meat Sci.,* 66: 21–32, 2003.
27. Greene, B.E. and Cumuze, T.H. Relationship between TBA numbers and inexperienced panellists' assessment of oxidised flavour in cooked beef. *J. Food Sci.,* 47: 52–54, 58, 1994.
28. Campo, M.M., Nute, G.R., Hughes, S.I., Enser, M., Wood, J.D., and Richardson, R.I. Flavour perception of oxidation in beef. *Meat Sci.,* 72: 303–311, 2006.
29. Annor-Frempong, I.E., Nute, G.R., Whittington, F.W., and Wood, J.D. The problem of taint in pork: 1. Detection thresholds and odour profiles of androstenone and skatole in a model system. *Meat Sci.,* 46: 45–55, 1997.
30. Annor-Frempong, I.E., Nute, G.R., Whittington, F.W., and Wood, J.D. The problem of taint in pork: 2. The influence of skatole, androstenone and indole, presented individually and in combination in a model lipid base, on odour perception. *Meat Sci.,* 47: 49–61, 1997.
31. Annor-Frempong, I.E., Nute, G.R., Whittington, F.W., and Wood, J.D. The problem of taint in pork:3. Odour profile of pork fat and the interrelationships between androstenone, skatole and indole concentrations. *Meat Sci.,* 47: 63–76, 1997.

Chapter 30

Sensory Perception

Rosires Deliza and M. Beatriz A. Gloria

Contents

30.1 Introduction

The sensory quality of meat and meat products has been considered an important factor since the beginning of the food industrialization process due to its influence on the overall quality of the product. Quality, in terms of sensory properties, is related to the adequate levels of sensory attributes considering the appearance, aroma, flavor, and texture. The sensory quality of a product has to be considered under several perspectives, mainly if it takes into account the consumer point of view. It is known that consumers create subjective impressions about the quality of a product based on various psychological processes. These processes are influenced by many factors such as the previous knowledge and cognitive competencies of each individual consumer [1]. Thus, from a consumer perspective, quality refers to the perceived quality and not to quality in an objective way.

Several approaches have been used to study the perception process of food quality by consumers. Within the widely accepted multiattribute approach, quality is a multidimensional phenomenon, described by a set of characteristics (attributes) that are subjectively perceived by consumers [2,3]. Perceived quality is generally considered as an overall concept and can be defined and analyzed, according to Oude Ophuis and Van Trijp [4], as the four "Ps" of the quality quadrant (Figure 30.1).

The first P in the perceived quality refers to the *Perception* process, which is related to the overall judgment formed on the basis of visible and invisible product characteristics that may have been experienced, or are believed to be associated with the evaluated product. Perceived quality may differ depending on the *Product* or product category under investigation (second P). Fat content, for example, may be a quality attribute (positive or negative) for meats, but it can have no relevance for fruits and vegetables. This leads to the consideration that perceived quality is based on consumer's judgments, the *Person* factor (third P), suggesting that perceived quality would vary among people, as they differ in their perceptual abilities, personal preferences, lifestyles, and experiences [5]. The fourth P of the quadrant is related to the context, and referred to as *Place*. There are many situations or circumstances that affect perceived quality, such as availability, price, social facilitation, and appropriateness of the eating environment (e.g., pizza may be judged as an excellent food

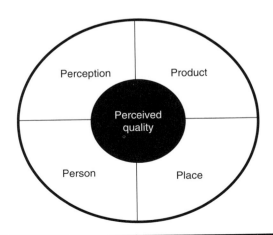

Figure 30.1 The quality quadrant. (From Oude Ophuis, P.A.M. and Van Trijp, H.C.M., *Food Qual. Prefer.*, 6, 177, 1995. With permission.)

to be consumed in evening meals, but inadequate for breakfast). Bell and Meilselman [6] have taken into account these factors when investigating the role of eating environments in determining food choice.

In this chapter, the perceived quality of meat and meat products will be considered, taking into account intrinsic and extrinsic quality cues, and their effects on consumer judgment regarding the products and consumer's food choice. The perceived quality of the product is multidimensional and includes sensory quality, healthiness, convenience, and for some consumers, product characteristics like animal welfare and organic production. The difference between intrinsic and extrinsic quality cues are presented and discussed in a wide scenario.

30.2 Factors Affecting Quality Perception

Food products are developed, produced, and marketed to appeal to the consumer, who is becoming more and more demanding about quality. The perception of food quality, particularly meat, is changing rapidly [7], which highlights the need for food producers to be innovative as a way to survive [5]. In other words, the success of a product depends on its acceptance by consumers, because they are the ultimate users of the product and thus, the ones who will be willing to purchase the product. Therefore, professionals in various industries are eager to understand consumer perceptions and attitudes toward a new product, a formulation change, or a new process. If a product is not liked by consumers, the research or manufacturing project is considered to be a failure [8].

The sensory attributes of food products can be either intrinsic or extrinsic (Table 30.1). Intrinsic attributes are concrete product characteristics that can be perceived by a consumer and, in many situations, can serve as a quality cue that can be observed, without actual consumption or use. It is related to the appearance, color, shape, size, and structure, all of them extremely important for meat. Intrinsic attributes are always related to the physical aspects of the product.

Extrinsic quality cues refer to product characteristics that are used to evaluate a product but are not physically part of it, such as price, brand, production and nutritional information, packaging design, country of origin, store, and convenience (Table 30.1). Extrinsic cues become more important when products are very similar in appearance [4]. The intrinsic and extrinsic cues are categorized and integrated by consumers to establish the quality attributes of meat.

Table 30.1 Intrinsic and Extrinsic Sensory Attributes of Food Products

Intrinsic	Extrinsic
Appearance	Price
Color	Brand name and familiarity
Shape	Label (packaging design)
Size	Advertisement
Structure	Nutritional information
Aroma	Production information (environment, organic)
	Origin (country)
	Store name
	Convenience

According to Steenkamp [9], quality attributes can also be *experience* or *credence cues*. The experience originates from the actual experience with or consumption of the product (e.g., aroma, taste, tenderness, leanness, etc.), whereas credence cannot be ascertained even after normal use (buying and consuming the product). Examples of credence cues are hormones, bovine spongiform encephalopathy (BSE), animal welfare, animal feeding, among others. These cues are gaining importance due to the increased consumer's concerns on safety, health, convenience, and ethical factors, in particular for meat and meat products.

In a study carried out by Bernués et al. [2], approximately 80% of the participants considered animal feeding and origin of meat, followed by animal welfare and environmentally friendly production, as the most appreciated extrinsic attributes of beef and lamb. Under the multiattribute approach, *experience* and *credence* quality are integrated by the consumer into an overall perceived quality, making the process very dynamic.

30.2.1 Consumer Expectations

Although sensory properties are by far the most important quality dimensions driving consumer's food choices, extrinsic product factors such as brand, price, and information can also affect food choice. Information given through the label of the package can be used to create positive expectation aimed to modify consumer perception and enhance purchase intention. During the decision-making process, previous experience and all the information available are processed in the consumer's mind.

Expectation is defined as the "action of mentally looking for something" in the food product to come. It plays an important role because it may improve or degrade the perception of a product, even before it is tasted. The higher the expectation levels about the product, the greater chances it has of being purchased. However, low expectation can cause the product to be ignored [10–13]. After making a choice, the consumer tastes the product, appreciating its sensory properties and other features that have created the expectation. Once confirmed, the consumer is pleased and possibly purchases the product again in a future experience. It means that the effects of expectation are likely to be important variables in determining satisfaction with the product [14]. A common hypothesis is that satisfaction is achieved when the product meets the consumer's expectation [15,16]. The ability to determine such expectation about a particular product becomes a vital strategy in promoting consumer's sensory satisfaction.

A model summarizing the role of expectations in product selection and evaluation is shown in Figure 30.2. As it can be seen, if disconfirmation of the expectation occurs, a mismatch between the expected and the actual product evaluation will occur. If the expectation is low and the sensory quality of the product is high, there will be a positive disconfirmation, and hence, consumer satisfaction is achieved. Conversely, high expectation on a poor sensory quality product will lead to a negative disconfirmation. In both cases, depending on the individual's behavior, the evaluation can be driven either toward or opposite to the initial expectation level [10,17].

There are two main models to explain the influence of information on food acceptance: assimilation and contrast [10]. Assimilation is the most described model and seems to come up mostly in cases of higher expectation and lower sensory quality. The individual tastes the product and receives information about it. Both dimensions—sensory and information—are integrated in his mind. If the product is perceived as worse than expected, cognitive dissonance takes place and to overcome it, the judgment moves toward expectation [13,17,18]. The second model—contrast—can be observed when sensory quality plays the most important role. The individual has high expectation; hence he evaluates the product and dislikes it. Therefore, the product is perceived as

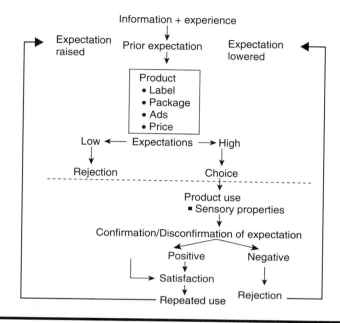

Figure 30.2 Model illustrating the influence of expectation on product selection and evaluation. (From Deliza, R. and MacFie, H.J.H., *J. Sens. Stud.*, 11, 103, 1996. With permission.)

worse than expected. This is called a negative contrast. However, a product perceived as good but with a low expectation level tends to be well accepted—a positive contrast [12].

Two other models are cited in the literature: generalized negativity and assimilation–contrast. In the first case, when the consumer has a different perception from the expected, it will be rated worse than if there had been no expectation. Assimilation–contrast, however, is related to the size of the discrepancy between expected and actual product performance. When the perceived discrepancy is small, the person tends to ignore it and there will be assimilation. Large discrepancies are not acceptable and contrast will occur [19].

To develop expectations about the products, consumers use intrinsic and extrinsic cues. Several factors affect the perception of meat quality such as the increased consumer health concerns, change in demographics, the need for convenience, changes in the distribution of meat, and price. Table 30.2 summarizes this issue considering the European consumer, emphasizing the meat industry's need for understanding the consumers and the measurement methods used to assess their attitudes and preferences [20].

In a study carried out on beef by Grunert et al. [21] in four European countries, the major cues used by consumers to elaborate quality expectations and the major dimension of meat quality were identified. Two factors formed the basis of quality expectations: perceived fat and place of purchase. The consumers had difficulty in evaluating meat quality, resulting in uncertainty and dissatisfaction. However, several possibilities were envisioned in the development of differentiated products, including improvement of eating quality, positive health effects, added convenience, and desirable process characteristics. The authors pointed out that product development is difficult and risky, and emphasized the need of consumer-led product development for the successful development of new products.

Table 30.2 European Consumer Expectations about Food of Animal Origin

Hygiene
Safety
Freshness
Nutritional value
Clear labeling
Ingredients
Price
Packaging
Brand reputation
Convenience
Product consistence
Suitability for specific situation
Origin
Ethical aspects
Environmental considerations
Appearance and sensory characteristics (flavor, color, aroma, etc.)

Source: ECA, www.esn-network.com/683.htwml, 1999; Ressurrec-cion, A.V.A., *Meat Sci.*, 66, 11, 2004. With permission.

In a recent study [5], three main trends were identified on the role of meat in the life of consumers, using the food-related lifestyle model as a conceptual framework. First, the increasing influence of extrinsic cues on quality perception of meat poses new requirements for the organization of the meat value chain, which has to fulfill the functions of delivering both meat and information. The second trend relates to the distinction of fast and efficient shopping in supermarkets compared to buying information-intensive specialized products from specific retail outlets. Although the bulk of meat will still be bought in supermarkets, in the future, there may also be room for other retail channels for specialized products. And finally, there is an increasing role of processed products, due to both convenience and trends toward meat avoidance in some consumer groups. Convenience is one of the major trends in food, whereas meat avoidance is a trend restricted to certain consumer groups; but both can lead to similar implications: making available products with a higher degree of processing, which enables more built-in convenience and less visibility of the meat ingredient, leads to a movement from bulk to differentiated, value-added products, which is probably the greatest trend of all.

Traditionally, quality perception of meat has been largely based on intrinsic cues such as color of the meat, the visible fat, and the cut. This is not mainly because consumers have been very competent in inferring quality from these cues, but because fresh meat is an unbranded product and only few extrinsic cues are available [5]. Focusing on intrinsic product characteristics, a huge diversity of analyses can be carried out to evaluate, control, and estimate product quality, either at a specific time, or during the product shelf life. A product can be evaluated in terms of its microbiological, chemical, or sensory characteristics; sensory cues fall under the domain of sensory evaluation, which has been defined as a scientific discipline used to evoke, measure, analyze, and interpret reactions to those characteristics of food as they are perceived by the senses of sight, smell, taste, touch, and hearing [22]. Owing to its relevance to the product evaluation, it is considered separately.

30.3 Sensory Evaluation and Consumer Studies

As mentioned earlier, the factors that affect food choice and intake are multidimensional and complex, but there is no doubt that the physical and sensory properties of a food must be among the main determinants. Therefore, it is very important that the methods used to measure the properties are accurate, not biased, and take into account the special requirements involved when humans are used as scientific measuring instruments [23]. During sensory evaluation, besides the sensory properties of the food, the responses from the sensory professional (individual providing the connection between the internal world of technology and product development and the external world of marketplace) are also analyzed and interpreted. This is done within the constraints of a product marketing brief, in a way that specialists can anticipate the impact of product changes in the marketplace [24].

The importance of sensory evaluation in the food area and, consequently, in the meat sector, is very well recognized and considered a cost-effective tool. It has its own challenges and should be viewed in broad terms. Its contribution far exceeds questions such as which flavor is best or whether ingredient X can be replaced with ingredient Z. This concept is especially important when looking at the impact of consumer response behavior as developed by marketing research.

Several studies on the physiological and psychological approaches used in the measurement of consumer behavior have been presented in the literature. Although research on sensory evaluation has only improved in recent years, there has been much information available on the physiology of the senses and the behavioral aspects of the perceptual process [24]. Comprehension of how sensory information is processed and integrated is important in understanding the evaluation process [25].

Sensory evaluation can be used for several purposes in a company. In Table 30.3, there is a list of activities to which sensory evaluation can contribute, directly or indirectly [24]. Every activity is important to the company, and the involvement of these activities will depend on the kind of company, as well as on the purpose of the study. For some companies the emphasis may be on the marketplace, new products, cost reduction and reformulation, line extension, and so on. For others, quality control is the primary focus.

Table 30.3 Sensory Evaluation Activities Within a Company

Product development
Product reformulation and cost reduction
Monitoring competition
Quality control
Quality assurance
Product sensory specification
Raw material specification
Storage stability
Process/ingredient/analytical/sensory relationship
Advertising claims

Source: Stone, H. and Sidel, J.L., in *Sensory Evaluation Practices*, Academic Press, San Diego, CA, 2004, 377. With permission.

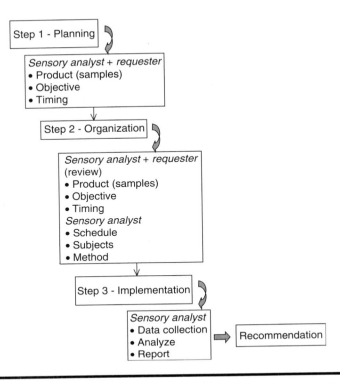

Figure 30.3 The representation of the steps involved in the sensory evaluation process. (Adapted from Stone, H. and Sidel, J.L., *Sensory Evaluation Practices*, Academic Press, San Diego, CA, 2004.)

In most companies sensory resources are located within the research and development area; however, it is sometimes part of the marketing research group. It depends on the company and the relationship with the involved areas. There is no rule stipulating where sensory evaluation has to be placed, but since most sensory professionals have some technical training and most often provide services to technical groups, sensory resources are frequently placed within the technical division. Regardless of the company goals and the location of the sensory resources, it is important to consider where the test will take place, who will serve as subjects, and which methods will be used. The accurate establishment of all mentioned factors is crucial to the achievement of the desired results [24]. The steps involved in the sensory evaluation process are presented in Figure 30.3.

An adequate planning involving the sensory analyst and the requester/client is essential. The aim of the study should be clear to the sensory analyst, that is, what is the purpose of the test, how many products will be evaluated, how the products will be prepared, and when the results are needed. In the next step, the test should be organized to guarantee that everything that is needed is available at the right time. Finally, the test is implemented, the data are collected and analyzed, and a written report is prepared, which is given to the requester in the form of recommended actions.

Physical and chemical properties are measured by instruments, whereas sensory properties are measured by a sensory panel. The panel consists of individuals selected on the basis of sensory acuity and ability to articulate the sensations experienced while viewing, smelling, and eating or drinking the food. Individuals are trained to describe products in terms of sensory characteristics and perceived product attributes' intensity. A trained panel is used for discriminative and descriptive tests. Preference and acceptability are determined by consumers, who are the users or potential users of the product.

Before presenting the methods available in sensory evaluation, it is important to make some comments about the place where the tests are carried out—the Sensory Laboratory—including the sensory booth, and other facilities. The test area should be of easy access for the people involved. A place in which most panel members pass on their way to lunch or to a break is a good location. If the panel members are coming from outside, the laboratory should be near the building entrance. Test rooms should not have noise and sources of odor, as they can affect the senses of the panel. The size of the booth, as well as the illumination, ventilation system, sinks, and temperature, should be adequate. The number of booths depends on the number and type of tests, which can be estimated before the planning of the laboratory. The booths should be individual and adjacent to the product preparation area so that samples can be served efficiently. A preparation area must contain at least refrigeration and storage facilities for ingredients and samples, cooking facility (electric or gas burners; conventional, convection, or microwave ovens), hoods with charcoal filters or venting to the outside, dishwasher, garbage disposers, waste basket, sink, and water sources. Other devices are specially used considering meat and meat products, such as heating trays, meat cutters, scales, and mixers. Another important area in the laboratory is the training area, designed to carry on training sessions with the panel. The desired characteristics of this area are as follows:

1. It should be close to the preparation area.
2. It should have a round or rectangular table (big enough for 6–12 panelists).
3. It should have controlled illumination and ventilation (similar to booth area).
4. It should be furnished with neutral color.
5. It should be nonodorous, and easy to clean.

More details about the facilities can be found in Stone and Sidel [24], American Society for Testing and Material (ASTM) [26], and Meilgaard et al. [27].

The preparation and presentation of the samples also have to be taken into account. Preliminary work is necessary before testing to determine the method of sample preparation, thawing and preparation times, equipment, and utensil. The sample should be representative or typical of the product to be tested. Proper sampling of the meat muscle is of critical importance. Every sample has to be tested at a given time and series. The experiments should be planned and performed using similar procedures, except for the factor under study.

Sample preparation guidelines for meat testing are available, wherein aspects such as thickness of steak and chops, size and thickness of roasts (beef, pork, and lamb), and meat cooking methods (roasting, broiling, and braising) are considered [28]. The order of presentation of samples has to follow a design [29], and a warmed-up sample should be used when required [30]. The serving temperature can be a cause of concern. Therefore, it has to be well controlled throughout the entire test. Generally, it is best to use the temperature at which it is usually consumed. However, in difference or descriptive tests, the temperature may be modified because taste acuity or perception is highest at temperatures between 20 and 40°C. It is recommended to provide a liquid (mineral water at room temperature) to rinse the mouth between samples. When tasting fatty foods, it is better to use warm water, dilute warm tea, lemon water, or a slice of apple. Crackers are used to remove residual flavors. The time of the test should also be relevant. If it is too early, it is difficult to evaluate hot spicy foods; however, if it is too late there is lack of motivation.

Motivation is an issue of concern to all sensory professionals, who need to develop a variety of practices to maintain the panel's interest in every study. Several recommendations were given by Stone and Sidel [24] as useful guidelines on motivation: subjects should be rewarded for participating, not for making correct scores; subjects' participation should be acknowledged on a regular basis, directly and indirectly; and management should visibly recognize sensory evaluation as a contributor to the company's growth and an indirect source of motivation.

30.3.1 Methods Used in Sensory Evaluation

30.3.1.1 Discriminative Tests

Focusing first on discriminative methods (a class of tests used to investigate whether there is a sensory difference between samples), a variety of specific methods can be found. The most common are the triangle test, duo-trio test, and paired comparison test. Triangle and duo-trio tests are used when the objective is to determine whether a sensory difference exists between two products. Paired comparison test is used to determine in which way a particular sensory characteristic differs between two samples (e.g., more or less tender). The importance of individuals' selection for a discriminative test has been known for a long time. Using unqualified people results in substantial variability that will mask a difference, leading to a wrong decision [24].

30.3.1.2 Descriptive Tests

Descriptive sensory tests are among the most sophisticated tools available to the sensory professional and involve the detection and description of both qualitative and quantitative sensory components of a product by trained panels. A relevant point of a descriptive test is its ability to allow the establishment of relationships between descriptive sensory and instrumental or consumer preference measurements. They can also be used to investigate product changes during its shelf life, and to investigate the effects of ingredients or processing variables on the final sensory properties of a product [31].

There are several different methods of descriptive analysis, such as Flavor Profile®, Texture Profile®, Quantitative Descriptive Analysis (QDA™), the Spectrum™, and Free-choice Profiling [32]. All of them require a panel with some degree of training and, in most of them (except for Free-choice Profiling), panelists are also required to have a reasonable level of sensory acuity. The selection and training of the panel and the monitoring of the performance of individuals are discussed in several publications [24,27,31]. The tests begin with the development of the descriptive terms, which describe the attributes of the product in a comprehensive and accurate way. It is achieved by exposing individuals to a wide range of products in the category under test. A consensus list of attributes is generated by the panelists (except for Free-choice Profiling, in which each panelist has his/her own list of attributes), which is used to evaluate samples. More details on descriptive tests can be gathered from Murray et al. [31].

Descriptive analysis has been extensively used in the meat industry and there are many studies published in the area [33–35]. They vary from studies evaluating a limited number of attributes to others with a well-trained panel, following careful preparation and presentation of the samples and developing and using a technically sound lexicon to describe the sensory characteristics of meats. Table 30.4 shows examples of flavor and texture lexicon for some types of meat.

The importance of establishing correlation between descriptive sensory tests and instrumental data or consumer preference measurements was illustrated by Boleman et al. [36] and Miller et al. [37], during analysis of meat and meat products for tenderness, which is considered one of the most important qualitative characteristic of meat. According to their studies, the consumers

Table 30.4 Flavor and Texture Lexicons for Some Types of Meat

	Beef	*Chicken*
Flavor		
Aromatics	Cooked beef/brothy	Chickeny
	Cooked beef fat	White chicken meat
	Browned	Dark chicken meat
	Liver/organy	Fat/skin
	Serum/bloody	Browned
	Grainy	Liver/organy
	Cardboard	Serum/bloody
	Painty	Cardboard
		Painty
Basic tastes	Sweetness	Sweetness
	Saltiness	Saltiness
	Sourness	Sourness
	Bitterness	Bitterness
Feeling factors	Metallic	Metallic
Texture		
Surface properties	Oiliness/wetness	Oiliness/wetness
	Roughness	Roughness
Partial compression	Springiness	Springiness
First bite	Firmness/hardness	Firmness/hardness
	Cohesiveness	Cohesiveness
	Juiciness/moisture release	Juiciness/moisture release
	Uniformity of bite	
Chew down	Cohesiveness of the mass	Cohesiveness of the mass
	Uniformity of the mass	Uniformity of the mass
	Juiciness	Juiciness
	Gristle	Gristle
	Connective tissue	
Residual	Toothpack	Toothpack
	Number of particles	Number of particles
	Oiliness/greasiness	Oiliness/greasiness

Source: Muñoz, A.M., *Meat Sci.*, 1, S287, 1998. With permission.

are willing to pay a higher price for beef in the marketplace as long as it is guaranteed to be tender. However, tenderness was a highly variable characteristic, depending on many intrinsic and extrinsic factors of the animal and on their interaction. Therefore, an instrumental means of evaluating tenderness that correlates with sensory perception would be desirable to overcome the practical problem of sensory evaluation. Destefanis et al. [38] investigated the ability of consumers to discern different levels of beef tenderness established by Warner–Bratzler shear force. The results indicated that beef with Warner–Bratzler shear force higher than 52.68 N and lower than 42.87 N were perceived by most consumers as tough and tender, respectively. Therefore, these values can represent reliable thresholds to classify beef tenderness. These values were within those obtained when a trained panel was used—42.28–58.76 N, corresponding to 4.31–5.99 kg [37].

30.3.1.3 Affective Tests

The last type of sensory test described is the affective test, which has the primary objective to assess the personal response (preference and acceptance) from users or potential users of a product or product idea. It is a valuable and important component of every sensory program and is referred to as acceptance, preference, or consumer testing. Subjects taking part in a sensory acceptance test are recruited based on demographic and usage criteria, and the number of participants in the test is an important issue in the design of the study. A large number of individuals are required; however, the number recommended in the literature varies, and about 100 consumers are usually considered adequate. In a recent study, Hough et al. [39] indicated that 112 consumers were necessary to perform a test with confidence.

Acceptance testing means the measurement of liking or preference for a product. Preference can be measured directly by comparing two or more products with each other, that is, which one of the two, three, or more products is preferred. It can also be measured indirectly, by determining which product is scored higher (more liked) than another product in a multiproduct test. By using appropriate scales it is possible to directly measure the degree of liking and to compare preference from these data. The two most used methods to measure preference and acceptance are the paired comparison and the 9-point hedonic scale tests. The paired comparison test is similar to the one described in the discrimination test, but in the present situation, the consumer will indicate which one of the two coded samples is preferred. The 9-point hedonic scale is probably the most useful. Despite some criticism about such scale, it is still the most used one for preference [24].

Similar to what was mentioned for descriptive and discriminative tests, attention has to be paid concerning sample preparation and sample presentation order when acceptance test is carried out. More details about the affective testing are provided by Stone and Sidel [24]. After collecting the data, it is necessary to analyze them to get the results, and enable the researcher to prepare a report, with recommendations and further actions. A classical statistical method to analyze preference data is analysis of variance. However, a mean of sample preference is obtained, which does not take into account the individual's preference. Advanced statistical methods have been successfully used to sort this problem out, such as preference mapping [40] and cluster analysis. Preference mapping has the advantage of identifying the sensory attributes that drive consumer preference, and cluster analysis identifies segments of consumers with similar preference, allowing the elaboration of marketing strategies. As an example of the usefulness of preference mapping, there is a study examining the sources of variation in restructured steaks, which indicated that the variations in fat and salt content were the most important determinants of consumer acceptability [23,41].

Consumer research and descriptive analysis play, by themselves, important roles in meat evaluation. Consumer research gives information on consumer perception and acceptance that only consumers can provide. However, consumers, because of their lack of training and limited vocabulary, cannot provide extensive product information. A descriptive panel, however, provides accurate and detailed product description. Therefore, to obtain the complete product information, both consumer and descriptive analysis should be used [40].

The main applications of consumer-descriptive studies are to achieve more thorough interpretation and understanding of consumer responses, to provide more specific product guidance, to enable the prediction of consumer responses based on descriptive and instrumental data, and to study different consumer segments [8]. This application was demonstrated on a hot dog study by Muñoz and Chambers [34]. A product highly accepted by consumers needed to have high intensity of smoke, cured meat, sweet and fat aroma, sweetness, saltiness, moisture release, cohesiveness of

Table 30.5   Examples of Sensory Issues Investigated in Qualitative and Quantitative Studies of Meats

Sensory issues in qualitative studies

What did you specifically like and dislike about the product?

How did you cook this meat product?

Please explain why this color of meat is more appealing than the other color

Please explain why this steak looks greasy to you

Please tell me more about why this amount of visible fat is unacceptable to you

Why in your opinion meat product 783 is tenderer than product 245?

Why your family preferred this meat product over the others?

Sensory issues in quantitative studies

The potential user of the meat product participates

A relatively large group of consumers participates (depending on the statistical power required)

Careful consideration to the meat product amount, consistency, and preparation for the large number of participants

Written questionnaire with attributes and scales to collect the perceived levels of sensory attributes by consumers

Sensory questions may include overall liking, liking and perceived intensity of attributes and preference

The selection of attributes in the questionnaire is critical

Data are statistically analyzed

Source: Muñoz, A.M., *Meat Sci.*, 1, S287, 1998. With permission.

the mass, residual oiliness, and also low intensity of skin awareness, firmness of the skin, grain aromatic, poultry, green herbs, pepper, onion, garlic, and speckles. Such a detailed guidance on how to formulate or reformulate this product would not have been obtained from consumers, in which the only attributes that could be asked for were color, size, hot dog flavor, spiciness, firmness, skin awareness, greasy/oily, spicy, smoky, and salty.

Consumer studies can be carried out through qualitative and quantitative studies. Qualitative studies involve the study of factors that motivate consumer opinions and behaviors. Consumer attitudes, perceptions, and beliefs that may explain the reasons for people's choices are explored. Qualitative research has an exceptional value, since the consumer can be probed to obtain information not easily obtained in quantitative studies. Consumers either participate in a group discussion (6–10 consumers) or are interviewed on an individual basis. Because of the nature of qualitative research, there are several sensory issues that may be addressed in meat studies (Table 30.5). The qualitative information can be the most important data that a researcher needs to collect and cannot be easily measured through a written questionnaire. The qualitative studies do not replace quantitative ones, but complement them [8].

Quantitative studies, however, are geared to collect data that can be summarized and analyzed statistically. The main characteristics of quantitative studies are also indicated in Table 30.5. Ultimately, researchers need this type of information and the power of statistics to make product decisions. However, if the quantitative research is conducted in conjunction with qualitative studies, more complete information on consumer responses is obtained [8].

Another challenge is the questionnaire design and the attributes selected to be evaluated. The attributes have to be carefully selected so that the consumers can provide meaningful information.

The terms used should be consumer terms and not a technical product lexicon. The attributes also depend on the type of meat product and the variables being studied. They have to be sufficient to provide all the information needed on the meat product [42].

The consumers included in the study must be selected according to the purpose of the research because there are considerable differences between young and elderly individuals on sensory perception and pleasantness of food flavors. Furthermore, there are changes with age in functions relating concentration with perceived intensity (psychophysical), concentration with pleasantness, and perceived intensity with pleasantness, which were specific for the different flavor qualities. These observations were confirmed in a study designed to determine the perceived intensity and pleasantness of bouillon flavor in water by a group of 32 young (mean age: 22, range 20–25) and 23 elderly individuals (mean age: 76, range 72–82). On average, the elderly subjects had lower perceived intensities for the highest concentrations of the series of bouillon, whereas the average responses to the lowest concentrations were almost equal. Optimal preferred concentrations were higher for the elderly than for the young subjects [43].

30.4   Factors Affecting the Sensory Quality of Meat

Several factors can affect the sensory properties of meat, such as type of meat, genetic factors, production and rearing conditions, effect of postmortem practices, product preparation, and ingredients and warmed-over flavor (WOF). Some of them are detailed in the following sections.

30.4.1   Type of Meat

There has been a strong trend in the last decades for most industries to expand their horizons and compete at a global level. Therefore, research must be conducted with a global perspective, which means that the quality of meat must be assessed by consumers of different countries or cultures. In fact, the acceptance of meat is unique to different countries or cultures. In conducting cross-cultural consumer research, attention must be paid to the cultural nuances and language of the population being tested. Therefore, these studies involve the close collaboration of several researchers, who know the country being studied [8]. Committee E18 on Sensory Evaluation of the ASTM is working on a manual that will cover these issues extensively for different cultures across Europe, North and South America, and Asia. Furthermore, the rapid change in consumers' attitudes must be taken into consideration. The changes in consumer and market orientation have resulted from several factors, including increased health concerns, change in demographics, the need for convenience, changes in distribution and price of meat, among others [20,44].

30.4.1.1   Chicken Meat

Two consumer studies in Spain (3100 consumers) and Germany (2000 consumers) were organized in the year 2000 with the objective of validating in meat the food quality attributes proposed by the European Community Association (ECA) as well as ranking them in order of importance. Safety, hygiene, and sensory characteristics were the most important meat quality characteristics as perceived by the consumers. Among the sensory characteristics, skin color of the chicken was considered very important [20,45].

Several consumer studies were conducted on chicken skin color. There was a significant preference for yellow-colored skin chickens sold in the open market of Guangdong, China (1993); the consumers were willing to pay even higher prices for the desired quality. In Barcelona, Spain (1999), a consumer survey in a supermarket indicated that 50% of the consumers associated yellow skin color with meat quality, 22% with fresh meat, and 11% with healthy birds. Additional consumer surveys have been carried out and published in South Africa and France where, once again, consumers showed their preferences for yellow broilers as synonymous of chicken meat quality [46]. Based on history and tradition, consumers consider color as an important quality attribute in poultry meat products. When given the choice, surveyed consumers from Europe and Africa always preferred the most intense golden yellow colors for chicken skin [45]. In France, 48% of the consumers preferred the yellow chicken skin, and the reasons were habit (32%) and quality/origin (29%). In Africa, 61% of the consumers preferred yellow chicken skin. They considered the color of the skin a very important quality aspect and were willing to pay extra for the preferred product (even 62% of the lower income group).

30.4.1.2 Pork Meat

In 2001, a book containing images of pork chops, based on a selection of 256 computer-modified images of pork chops differing in color, fat cover, marbling, and drip using digital photographs was published by Dransfield et al. [47] and Ngapo et al. [48]. Cho et al. [44] developed a novel methodology to conduct research for surveying consumer's preference of pork chops using the image book. Meat color was the most important characteristic; however, similar numbers of consumers chose the dark and light red colored pork. The presence of marbling and the absence of drip were also an important selection factor among characteristics for Korean consumers in choosing pork chops, whereas fat cover was not an important selection attribute. In a similar study undertaken by Ngapo et al. [48], French consumers also showed that color was the most important appearance attribute. However, fat cover was the second most important factor for the French consumer, with preference for lean meat.

30.4.1.3 Beef Meat

The role of the profile of fatty acids on the sensory perception of three muscles (longissimus dorsi, triceps brachii, and semimembranosus) from Korean Hanwoo and Australian Angus beef was investigated [49]. Cluster analysis showed that there was a significant difference in fatty acids (C16:0, C16:10, C18:0, C18:2n6, C18:3, C20:3n6, C20:4n6, C22:4n6, and C22:5n3) for tenderness, juiciness, flavor, and overall likeness of the beef from both origins. C14:0 had a significant effect on sensory perception only for Korean Hanwoo beef.

Lipid oxidation is one of the major causes of quality deterioration of processed meat, imposing an adverse effect on flavor, color, and texture. Furthermore, it causes a decrease in perception of meat flavor/odor and a concomitant increase in the off-flavor/odor described as linseed and rancid [50]. To relate humans' perception of lipid oxidation in beef, determined by a trained panel, to a chemical measurement of oxidation, Campo et al. [51] studied meat from animals with a wide range of potential oxidation through differences in their polyunsaturated fatty acid (PUFA) composition. The meat was obtained from Angus and Charolais cross steers from different trials that had been raised on 10 different diets. Thiobarbituric acid reactive substances (TBARS) and sensory analysis were performed in longissimus dorsi muscle displayed under simulated retail conditions. Positive attributes, such as beef flavor or overall liking, decreased throughout display, whereas negative attributes, such as abnormal and rancid flavors, increased. The correlations between sensory and analytical attributes were high. TBARS were a good predictor of the perception of rancidity, and a TBARS value of 2 could be considered the limiting threshold for the acceptability of oxidized beef.

30.4.1.4 Goat Meat

The major demand for goat meat in the United States seems to come from various ethnic groups, influenced by their cultural traditions or religious beliefs [52]. In sensory studies conducted with consumers, goat meat generally was rated less desirable than other red meats. When oven-baked rib or loin samples were evaluated by five sensory panels (24–94 untrained panelists), goat meat was not markedly different in flavor desirability than beef or lamb, but tended to be less desirable in flavor than pork [53]. Overall satisfaction scores were lower for goat meat than pork, beef, or lamb.

In another study [54], goat and sheep meats (broiled loin chops and leg steaks) were evaluated by U.S. and foreign panelists (natives of China, India, Mexico, Taiwan, Saudi Arabia, Venezuela, or Vietnam). All the participants had eaten lamb and goat earlier and had expressed no dislike for either meat when asked before the sensory test session. Goat meat was rated lower than lamb and mutton in overall desirability, by both domestic and foreign panels. Flavor desirability ratings were similar for goat meat, lamb, and mutton, regardless of the panel makeup.

When pan-fried ground meat patties containing various levels of goat meat and beef were evaluated by 45 consumers (predominantly nongoat meat eaters), flavor desirability scores were highest for 20%-goat/80%-beef patties, followed by all (100%)-beef patties and 80%-goat/20%-beef patties, which received similar scores. All goat patties were given lower flavor desirability scores when compared to all-beef patties [55].

Plain meat loaves and chili were prepared with ground goat meat or beef containing 15% fat. Acceptability scores by a consumer sensory panel were similar for plain goat meat and beef loaves when the goat meat product was served before the meat product, but were lower for the goat meat with reverse serving order. In triangle tests, goat meat was differentiated from beef, whether plain or seasoned [52].

30.4.2 Genetic Factors

Sensory perception and technological quality of meat are highly influenced by both genetic and environmental factors. Crouse et al. [56] investigated the tenderness of meat traits of 422 steers differing in the ratio of Brahman, Sahiwal, or Pinzgauer to Hereford or Angus inheritance. *Bos indicus* breed crosses were less ($p < 0.01$) tender and more variable in tenderness than *Bos taurus* breed crosses. As the percentage of *B. indicus* inheritance increased, shear values increased and sensory panel tenderness scores decreased. Decreases in tenderness were associated with less desirable sensory panel ease-of-fragmentation scores and, to a lesser extent, with a sensory panel perception of more abundant connective tissue content of meat samples. Flavor characteristics were similar among all breed groups of cattle.

The effects of RN genotype on the sensory perception of cured-smoked loins were investigated in 30 female pigs crossbred with Hampshire. RN-carriers (RN^-/rn^+) were more tender, juicy, and acidic than noncarriers (rn^+/rn^+). They were also patty-like and less salty. Despite the superior eating quality, in terms of higher tenderness and juiciness, most consumers preferred cured-smoked loins from noncarriers [57].

Table 30.6   Sensory Descriptive Terms with Definitions Developed for the Evaluation of Warmed-Over Flavor Pork Meat

Sensory Attribute	Definition with Reference Material
Aroma	*Odor associated with:*
Cardboard-like	Shredded wet cardboard
Linseed oil–like	Warmed linseed oil
Rubber-like/sulfur-like	Warmed rubber or the white of a warm boiled and peeled egg
Flavor	*Aromatic taste sensation and taste sensation associated with:*
Cooked pork meat–like	Cooked pork meat
Rancid	Oxidized vegetable oil
Bread-like	The surface of French style bread
Vegetable oil–like	Fresh vegetable oil
Fish-like	Cod liver oil
Nut-like	Whole hazel nuts
Monosodium glutamate/umami	Monosodium glutamate, 0.05% in water
Metallic	Ferrous sulfate, 0.01% in water
Bitter taste	Quinine chloride, 0.005% in water
Sweet taste	Sucrose, 0.1% in water
Salt taste	NaCl, 0.05% in water
Sour taste	Citric acid, 0.03% in water
Aftertaste	*Chemical feeling factor on skin surfaces of oral cavity described as dry associated with:*
Astringent	Aluminum sulfate 0.002% solution in water

Source: Byrne, D. V., Bak, L. S., Bredie, W. L. P., Bertelsen, G., and Martens, M., *J. Sens. Stud.*, 14, 47, 1999. With permission.

30.4.3   Warmed-Over Flavor

WOF is recognized as a major quality concern by the food industry in the marketing of pre-cooked, ready to heat and serve products. It is characterized by a rapid oxidative deterioration that occurs in a variety of cooked foods during refrigerated storage. Byrne et al. [58] developed a list of 16 terms as a vocabulary to describe the aroma, flavor, and taste character notes of porcine meat samples (Table 30.6). This list resulted from an initial one containing 45 descriptive terms. Selection criteria were that the terms should have relevance to the product, discriminate clearly between samples, be not redundant, and have cognitive clarity to assessors. According to Byrne et al. [59], WOF development in cooked, chill-stored, and reheated pork patties made with musculus semimembranosus involved lipid oxidation, which caused off-flavor and odor notes, for example, rancid-like flavor and linseed oil – like odor, in association with a concurrent decrease in cooked pork meat-like flavor.

30.4.4   Production and Rearing Conditions

The production and rearing conditions of the animal can significantly affect the sensory perception of the meat. Rearing entire instead of castrated male pigs for meat production has a number of advantages including lower production costs, leaner carcass, lower output of nitrogen in the environment, and reduction of suffering for the animal. However, in most countries, male pigs are still castrated to avoid the potential problem of boar taint at a young age. Boar taint is an unpleasant odor/flavor that can be perceived when cooking/eating the meat from some entire male pigs. Consumer dissatisfaction for the odor of entire male pig was mostly associated with skatole (tryptophan metabolite) levels, with little influence of androstenone (testicular steroid androstenone). However, androstenone and skatole had similar contributions to the level of dissatisfaction for flavor [60–62]. Therefore, chemical measurements for skatole and androstenone were highly predictive of specific sensory descriptors of boar taint in cooked pork meat [63].

Trained analytical sensory panels in seven European countries, assessing pork meat with known levels of skatole and androstenone, were able to differentiate between the two compounds and among different levels of the compounds. Androstenone was found to relate mostly to urine attribute, whereas skatole related to manure and, to a lesser extent, to naphthalene [64]. Sorting carcasses on the basis of androstenone/skatole would reduce, but not eliminate differences in consumer dissatisfaction between entire male and gilt pork [65]. Chicory root (*Cichorium intybus* L.) feeding reduced the boar taint off-flavor in longissimus dorsi and psoas major muscles of intact male, and, therefore, can be considered to have the potential for utilization [63].

The influence of castration of entire male pigs of two different crossbreeds on the eating quality of dry-cured ham was evaluated by Banon et al. [66]. Sensory analysis was carried out by a trained panel, and preference and acceptability paired test was performed by consumers. Significant differences were found between entire and castrated animals for the mean values given in all investigated sensory attributes. The dry-cured ham from castrates was scored as more flavored, more marbled, and softer. It was also perceived as less grainy, less salty, and having less boar odor and flavor. Dry-cured ham from castrated males was also more accepted and preferred by consumers, especially women and habitual consumers. Castration of male pigs contributed to an improvement in the quality of dry-cured ham. The rejection caused by boar odor and flavor was reduced, improving the overall flavor, texture, and juiciness. In addition, saltiness was less pronounced in ham from castrates.

Sales of organically produced food have increased significantly in value throughout the world. Organic and conventional lamb loin chops from three major supermarket chains of the United Kingdom were analyzed for fatty acid composition, and eating quality was assessed by a trained sensory panel [67]. Organic lamb had a better eating quality than conventional lamb in terms of juiciness, flavor, and overall liking. The differences in juiciness were attributed to the higher level of linolenic acid (18:3) and total n-3 PUFA in organic chops. Conventional chops had a higher percentage of linoleic acid (18:2).

Bryhni [68] investigated the influence of feed composition on pork meat sensory traits analyzed by a trained panel and by consumers. The quality of the pork influenced consumer satisfaction, and flavor was ranked by consumers as the most important sensory attribute. The presence of off-flavors indicated reduced liking for pork. The use of food wastes in the pig's diet did not affect sensory quality, but the back fat was darker and more susceptible to lipid oxidation. Pig dietary fatty acid composition affected fatty acid of the back fat and the trained panel sensory evaluation of both pork loin and sausage. Feeding a diet with low levels of fish oil or with high PUFA increased rancidity of the sausages.

Colditz et al. [69] investigated the impact of regrouped British breed steers 1, 2, and 4 weeks before dispatch for slaughter. Regrouping cattle less than 2 weeks before slaughter reduced meat quality. Also, the results confirmed the impact of flight time on growth rate during feedlot finishing.

30.4.5 Effect of Postmortem Practices, Product Preparation, and Ingredients on Meat Quality

Bryhni et al. [70] investigated the sensory quality and liking for pork as affected by pH, 24 h after slaughter, cooking temperature, and WOF among consumers ($n = 288$) in Scandinavia. The consumers preferred meat with higher pH (6.0), cooked to the lowest temperature (65 vs. 80°C). Samples with WOF were the least preferred and described as metallic, acidic, and off-flavor by a trained panel. Elevated-pH meat cooked to 65°C resulted in a sweeter and tender meat. Juiciness, tenderness, and the absence of off-flavor were the most important characteristics that governed the consumers' liking for pork. Consumption frequency and liking for pork were positively related. The most satisfied consumers reported highest consumption frequency. Elderly people and males expressed the highest liking score and consumption frequency, respectively.

Pork samples (longissimus dorsi) were cooked in an oven to a core temperature of 62 or 75°C and subsequently evaluated by a sensory panel. Juiciness and tenderness decreased with increasing temperature, while hardness, crumbliness, and chewing time increased. Such changes were associated with changes in the size of pores, confining the myofibrillar water together with an expulsion of water, determined by low-field nuclear magnetic resonance (NMR) T_2 relaxation [71].

A radio frequency cooking protocol was developed and its effect on the quality of pork-based white pudding was examined [72]. Results from a sensory similarity test involving 60 panelists indicated that they were not able to detect differences between radio frequency and conventional methods (water bath and steam oven heated products).

The irradiation of mechanically deboned chicken meat is of crucial importance to the safety of the industrial productive chain as well as to guarantee pathogen-free raw material. Gomes et al. [73] investigated the effect of gamma-radiation on the sensory characteristics of refrigerated mechanically deboned chicken meat. It was observed that the volatile compounds associated with the odor of irradiation procedures were dissipated from the samples during storage, and that the oxidation odor perceived in the samples irradiated with doses of 3.0 and 4.0 kGy was more pronounced than in the nonirradiated samples, as from the 8th and 12th day of refrigeration, respectively. Al-Bachir [74] investigated the influence of gamma-radiation in spices and packaging materials (10 kGy) and packaged luncheon meat (2 kGy) on the microbiological, chemical, and sensory characteristics of luncheon meat. Irradiation decreased the microbial counts and increased shelf life without affecting moisture, protein, fat, pH, total acidity, lipid peroxide, and volatile basic nitrogen. Sensory evaluation showed that all the combinations of treated luncheon meats were acceptable; however, the taste, odor, appearance, and texture scores of irradiated packaged products were significantly lower than those of nonirradiated samples.

A time-intensity evaluation was performed to investigate the effect of texture and two different chili products on the intensity of oral burn and meat flavor of pork patties. The pork patties spiced with chili powder were perceived significantly hotter and with a less pronounced meat flavor than the minced chili patties. A multivariate model showed an agreement with univariate analysis that both chili products masked the meat flavor. No effect of texture was found on the perceived oral burn or meat flavor. Chili burn and meat flavor were perceived as less intense to regular eaters of chili compared to noneaters of chili [75].

30.5 Influence of the Improvement of Nutritional Value on Sensory Quality

Today there is a tendency of adding functional ingredients to meat products. Shiitake mushroom powder is one of a growing number of ingredients that have shown functional properties even in small amounts. It is a source of potassium, amino acids, and vitamins; furthermore, it has medicinal properties (used for cholesterol, tumors, and diabetes). Chun et al. [76] investigated consumer liking and perception of pork patties prepared with 0–6% shiitake mushroom powder and 0 or

0.5% sodium tripolyphosphate (STP—used to improve water-binding capacity). The patties made with both ingredients were acceptable by Korean consumers, but increases in mushroom powder in patties with STP decreased acceptability in the case of U.S. consumers. However, addition of mushroom powder to patties without STP increased acceptability for U.S. consumers, as it tended to increase texture acceptance and juiciness.

Textured soybean proteins (TSP) and other soy ingredients have been used in ground beef products to improve their functional properties and aid in reducing fat and cholesterol content, providing consumers with healthier products. Besides, there are potential cost savings in the production of ground beef, which can help bridging the high price barrier of animal products. However, the incorporation of high concentrations of TSP into meat can adversely affect flavor, color, and texture. Ground beef patties were processed replacing meat with hydrated and colored TSP (0, 15, and 30%), and the resulting cooked patties were evaluated by eight trained judges for tenderness, juiciness, number of chews, beef flavor, and overall flavor quality. The results indicated that ground beef patties with 15% TSP had sensory attributes similar to those of the control. However, patties with 30% TSP were tenderer than the control and had less beef flavor and overall flavor quality [77].

30.6 Concluding Remarks

Sensory properties (appearance, aroma, flavor, and texture) are the most important quality attributes driving consumers' choices for meat and meat products. However, cultural, religious, and food-related lifestyle can affect meat acceptance and expectation. Furthermore, there are several other factors such as brand, price, nutritional composition and information, production information, advertisement, convenience, safety, hygiene, and so on, which can create and affect the expectations and decisions toward meat purchase. Several of these factors and a flow chart describing the role of expectations at the point of choice and during meat perception at the time of consumption are presented followed by scientific studies developed recently in the area.

Different models available to explain consumer expectations and demands are also provided along with their applications and advantages, as well as methods to investigate product sensory properties and acceptance. The demand for meat has changed over the years. Today, the nutritional quality (associated with lean products), convenience, and the movement from bulk to value-added products with a longer shelf life are probably the greatest trend of all. Several studies related to genetic improvement of meat quality, production and rearing conditions, postmortem practices, product preparation and incorporation of ingredients with functional properties, and their influence on meat acceptance have been conducted. Furthermore, their influence on shelf life and quality (WOF) have been discussed. The impact of product development with increased nutritional or functional properties on meat acceptance has also been illustrated. This has led to the development, production, and marketing of new meat products that appeal to the consumer, who is becoming more and more demanding about quality. Professionals in the industries must understand consumers' demands, perceptions, and attitudes to succeed and market the new product.

Abbreviations

ASTM	American Society for Testing and Materials
ECA	European Community Association
kGy	kilo gray

NMR	nuclear magnetic resonance
PUFA	polyunsaturated fatty acids
STP	sodium tripolyphosphate
TBARS	thiobarbituric acid reactive substances
TSP	textured soybean proteins
WOF	warmed-over flavor

References

1. Bredahl, L., Cue utilization and quality perception with regard to branded beef, *Food Qual. Prefer.*, 15, 65, 2003.
2. Bernués, A., Olaizola, A. and Corcoran, K., Extrinsic attributes of red meat as indicators of quality in Europe: an application for market segmentation, *Food Qual. Prefer.*, 14, 265, 2003.
3. Grunert, K.G., What's in a steak? A cross-cultural study on the quality perception of beef, *Food Qual. Prefer.*, 8(3), 157, 1997.
4. Oude Ophuis, P.A.M. and Van Trijp, H.C.M., Perceived quality: a market driven oriented approach, *Food Qual. Prefer.*, 6, 177, 1995.
5. Grunert, K.G., Future trends and consumer lifestyle with regard to meat consumption, *Meat Sci.*, 74, 149, 2006.
6. Bell, R. and Meilselman, H., *The role of eating environments in determining food choice*, in Marshall, D.W., Ed., *Food Choice and the Consumer*, Blackie Academic & Professional, Cambridge, UK, 292, 1995.
7. Grunert, K.G. and Valli, C., Designer-made meat and dairy products: consumer-led product development, *Livest. Prod. Sci.*, 72, 83, 2001.
8. Muñoz, A.M., Consumer perceptions of meat. Understanding these results through descriptive analysis, *Meat Sci.*, 1, S287, 1998.
9. Steenkamp, J.B.E.M., Conceptual model of the quality perception process, *J. Bus. Res.*, 21, 309, 1990.
10. Cardello, A.V. and Sawyer, F.M., Effects of disconfirmed consumer expectations on food acceptability, *J. Sens. Stud.*, 7, 253, 1992.
11. Deliza, R., *The Effects of Expectation on Sensory Perception and Acceptance*, PhD thesis, The University of Reading, UK, 1996, 198.
12. Deliza, R., MacFie, H.J.H. and Hedderley, D., Evaluation of consumer expectation, in Almeida, T.C., Hough, G., Damasio, M.H. and Silva, M.A.A.P., Eds., *Avanços em análise sensorial*, Varela, São Paulo, Brazil, 1999, 111.
13. Behrens, J.H., Villanueva, N.D.M. and Silva, M.A.A.P., Effect of nutrition and health claims on the acceptability of soyamilk beverages, *Int. J. Food Sci. Technol.*, 42, 50, 2007.
14. Deliza, R. and MacFie, H.J.H., Product packaging and branding, in Frewer, L.J., Risvik, E. and Schifferstein, H., Eds., *Food, People and Society: A European Perspective of Consumers' Food Choices*, Springer-Verlag, Berlin, 2001, 55.
15. Martin, D., The impact of branding and marketing on perception of sensory qualities, *Food Sci. Technol. Today: Proceed.*, 4(1), 44, 1990.
16. Kopalle, P.K. and Lehmann, D.R., The effects of advertised and observed quality on expectations about new product quality, *J. Marketing Res.*, 32, 280, 1995.
17. Deliza, R. and MacFie, H.J.H., The generation of sensory expectation by external cues and its effect on sensory perception and hedonic ratings: a review, *J. Sens. Stud.*, 11(2), 103, 1996.
18. Issanchou, S., Consumer expectations and perceptions of meat and meat product quality, *Meat Sci.*, 43S, 5, 1996.

19. Schifferstein, H.N.J., Effects of products beliefs on product perception and liking, in Frewer, L.J., Risvik, E. and Schifferstein, H., Eds., *Food, People and Society: A European Perspective of Consumers' Food Choices*, Springer-Verlag, Berlin, 2001, 73.

20. Ressurreccion, A.V.A., Sensory aspects of consumer choices for meat and meat products, *Meat Sci.*, 66, 11, 2004.

21. Grunert, K.G., Bredahl, L. and Brunsø, K., Consumer perception of meat quality and implications for product development in the meat sector—a review, *Meat Sci.*, 66(2), 259, 2004.

22. Institute of Food Technology (IFT), *Minutes of Sensory Evaluation*, Division Business Meeting, 35th Annual IFT Meeting, Chicago, 1975.

23. MacFie, H.J.H., Assessment of the sensory properties of food, *Nutr. Rev.*, 48(2), 87, 1990.

24. Stone, H. and Sidel, J.L., *Sensory Evaluation Practices*, Elsevier Academic, San Diego, CA, 2004, 377.

25. McBride, R.L. and MacFie, H.J.H., *Psychological Basis of Sensory Evaluation*, Elsevier Science, New York, NY, 1990.

26. American Society for Testing and Material (ASTM), in Eggert, J. and Zook, K., Eds., *Physical Requirement Guidelines for Sensory Evaluation Laboratories,* ASTM Special Technical Publication 913, Philadelphia, 1986.

27. Meilgaard, M., Civille, G.V. and Carr, B.T., *Sensory Evaluation Techniques*, 3rd ed., Boca Raton, New York, NY, 1999, 354.

28. American Meat Science Association (AMSA), *Guidelines for Cookery and Sensory Evaluation of Meat*, Academic Press Inc., Chicago, IL, 1978.

29. MacFie, H.J., Bratchell, N., Greenhoff, K. and Vallis, L.V., Designs to balance the effect of order of presentation and first-order carry-over effects in hall tests, *J. Sens. Stud.*, 4, 129, 1989.

30. Plemmons, L.E. and Resurreccion, A.V.A., A warm-up sample improves reliability of responses in descriptive analysis, *J. Sens. Stud.*, 13, 359, 1998.

31. Murray, J.M., Delahunty, C.M. and Baxter, I.A., Descriptive sensory analysis: past, present and future, *Food Res. Int.* 36, 6, 461, 2001.

32. Langron, S.P., The application of Procrustes statistics to sensory profiling, in William, A.A. and Atkin, R.K., Eds., *Sensory Quality in Food and Beverages: Definitions, Measurement and Control*, Ellis Horwood Ltd., Chichester, UK, 1983, 89.

33. Horsfield, S. and Taylor, L.J., Exploring the relationship between sensory data and acceptability of meat, *J. Sci. Food Agric.*, 27, 1044, 1976.

34. Muñoz, A.M. and Chambers, E., Relating sensory measurements to consumer acceptance of meat products, *Food Technol.*, 47, 11, 128, 1993.

35. Delores, H., Chambers IV, E. and Bowers, J.R., Consumer acceptance of commercially available frankfurters, *J. Sens. Stud.*, 11, 85, 1996.

36. Boleman, S.J., Boleman, S.L., Savell, J.W., Miller, S.K., Cross, R.H., Wheller, T.L., Koohmaraie, M., Shackelford, S.D., Miller, M.F., West, R.L. and Johnson, D.D., Consumer evaluation of beef of known tenderness levels, in *Proceedings of 41th International Congress of Meat Science and Technology*, Santo Antonio, TX, 1995, 494.

37. Miller, M.F., Carr, M.A., Ramsey, C.B., Crockett, K.L. and Hoover, L.C., Consumer threshold for establishing the value of beef tenderness, *J. Anim. Sci.*, 79, 3062, 2001.

38. Destefanis, G., Brugiapaglia, A., Barge, M.T. and Dal Molin, E., Relationship between beef consumer tenderness perception and Warner–Bratzler shear force, *Meat Sci.*, 78, 153, 2007.

39. Hough, G., Wakeling, I., Mucci, A., Chambers IV, E., Gallardo, I.M. and Alves, L.R., Number of consumers necessary for sensory acceptability tests, *Food Qual. Prefer.*, 17, 522, 2006.

40. Greenhoff, K. and MacFie, H.J.L., Preference mapping in practice, in MacFie, H.J.H. and Thomsom, D.M.H., Eds., *Measurements of Food Preferences*, Blackie Academic & Professional, London, UK, 1994, 137.

41. Berry, B.W. and Cicille, G.V., Development of a texture profile panel for evaluating restructured beef steaks varying in meat particle size, *J. Sens. Stud.*, 1(1), 15, 1986.

42. Deliza, R., *How to Understand and Interpret Consumer's Needs*, Short Course Notes, Embrapa Food Technology, Rio de Janeiro, 2007, 53.

43. De Graaf, C., Polet, P. and van Staveren, W.A., Sensory perception and pleasantness of food flavors in elderly subjects, *J. Gerontol.*, 49(3), 93, 1994.

44. Cho, S., Park, B., Ngapo, T., Kim, J., Dransfield, E., Hwang, I. and Lee, J., Effects of meat appearance on south Korean consumer's choice of pork chops determined by image methodology, *J. Sens. Stud.*, 22, 99, 2007.

45. European Consumer Association (ECA), http://www.esn-network.com/683.html, 1999.

46. Hernández Gimeno, J.M., Sensory perception of quality of products across Europe: a case study on poultry quality, http://www.esn-network.com/683.html, 2005.

47. Dransfield, E., Martin, J.F., Miramonte, J. and Ngapo, T.M., *Meat Appearance: Pork Chops. A Tool for Surveying Consumers' Preference*, INRA, Paris, France, 2001.

48. Ngapo, T.M., Martin, J.F. and Dransfield, E., Consumer choices of pork chops, *Food Qual. Prefer.*, 15(4), 349, 2004.

49. Cho, S.H., Park, B.Y., Kim, J.H., Hwang, I.H., Kim, J.H. and Lee, J.M., Fatty acids profiles and sensory properties of longissimus dorsi, triceps brachii, and semimenbranosus muscles from Korean Hanwoo and Australian Angus beef, *Asian-Aust. J. Anim. Sci.*, 18(12), 1786, 2005.

50. Nissen, L.R., Byrne, D.V., Bertelsen, G. and Skibsted, L.H., The antioxidant activity of plant extracts in cooked pork patties as evaluated by descriptive sensory profiling and chemical analysis, *Meat Sci.*, 68, 485, 2004.

51. Campo, M.M., Nute, G.R., Hughes, S.I., Enser, M., Wood, J.D. and Richardson, R.I., Flavor perception of oxidation in beef, *Meat Sci.*, 72(2), 303, 2006.

52. Rhee, K.S., Myers, C.E. and Waldron, D.F., Consumer sensory evaluation of plain and seasoned goat meat and beef products, *Meat Sci.*, 65, 785, 2003.

53. Smith, G.C., Pike, M.I. and Carpenter, Z.L., Comparison of the palatability of goat meat and meat from four other animal species, *J. Food Sci.*, 39, 1145, 1974.

54. Griffith, C.L., Orcutt, M.W., Riley, R.R., Smith, G.C., Savell, J.W. and Shelton, M., Evaluation of palatability of lamb, mutton and chevon by sensory panels of various cultural backgrounds, *Small Ruminant Res.*, 8, 67, 1982.

55. James, N.A. and Berry, B.W., Use of chevon in the development of low-fat meat products, *J. Anim. Sci.*, 75, 571, 1997.

56. Crouse, J.D., Cundiff, L.V., Koch, R.M., Koohmaraie, M. and Seideman, S.C., Comparisons of *Bos indicus* and *Bos taurus* inheritance for carcass beef characteristics and meat palatability, *J. Anim. Sci.*, 67, 2661, 1989.

57. Hullberg, A., Sohansson, L. and Lundstrom, K., Sensory perception of cured-smoked loin from carriers and noncarriers of the RN– allele and its relationship with technological meat quality, *J. Muscle Foods*, 16, 54, 2005.

58. Byrne, D.V., Bak, L.S., Bredie, W.L.P., Bertelsen, G. and Martens, M., Development of a sensory vocabulary for warmed-over flavor: part I. In porcine meat, *J. Sens. Stud.*, 14, 47, 1999.

59. Byrne, D.V., Bredie, W.L.P., Bak, L.S., Bertelsen, G., Martens, H. and Martens, M., Sensory and chemical analysis of cooked porcine meat patties in relation to warmed over flavor and pre-slaughter stress, *Meat Sci.*, 59, 229, 2001.

60. Bonneau, M., Kempster, A.J., Claus, R., Claudi-Magnussen, C., Diestre, A., Tornberg, E., Walstra, P., Chevillon, P., Weiler, U. and Cook, G.L., An international study on the importance of androstenone and skatole for boar taint: I. Presentation of the program and measurement of boar taint compounds with different analytical procedures, *Meat Sci.*, 54, 251, 2000.

61. Matthews, K.R., Homer, D.B., Punter, P., Beague, M.P., Gispert, M., Kempster, A.J., Agerhem, H., Claudi-Magnussen, C., Fischer, K., Siret, F., Leask, H., Font i Furnols, M. and Bonneau, M., An international study on the importance of androstenone and skatole for boar taint: III. Consumer survey in seven European countries, *Meat Sci.*, 54, 271, 2000.

62. Weiler, U., Font i Furnols, M., Fischer, K., Kemmer, H., Oliver, M.A., Gispert, M., Dobrowolski, A. and Claus, R., Influence of differences in sensitivity of Spanish and German consumers to perceive skatole and androstenone concentrations, *Meat Sci.*, 54, 297, 2000.

63. Byrne, D.V., Thamsborg, S.M. and Hansen, L.L., A sensory description of boar taint and the effects of crude and dried chicory roots (*Cichorium intybus* L.) and inulin feeding in male and female pork, *Meat Sci.,* 79, 252, 2008.

64. Dijksterhuis, G.B., Engel, B., Walstra, P., Font i Furnols, M., Agerhem, H., Fischer, K., Oliver, M.A., Claudi-Magnussen, C., Siret, F., Béague, M.P., Homer, D.B. and Bonneau, M., An international study on the importance of androstenone and skatole for boar taint: II. Sensory evaluation by trained panels in seven European countries, *Meat Sci.*, 54, 261, 2000.

65. Bonneau, M., Walstra, P., Claudi-Magnussen, C., Kempster, A.J., Tornberg, E., Fischer, K., Diestre, A., Siret, F., Chevillon, P., Claus, R., Dijksterhuis, G., Punter, P., Matthews, K.R., Agerhem, H., Béague, M.P., Oliver, M.A., Gispert, M., Weiler, U., von Seth, G., Leask, H., Font i Furnols, M., Homer, D.B. and Cook, G.L., An international study on the importance of androstenone and skatole for boar taint: IV. Simulation studies on consumer dissatisfaction with entire male pork and the effect of sorting carcasses on the slaughter line, main conclusion and recommendations, *Meat Sci.*, 54, 285, 2000.

66. Banon, S., Gil, M.D. and Garrido, M.D., The effects of castration on the eating quality of dry-cured ham, *Meat Sci.*, 65(3), 1031, 2003.

67. Angood, K.M., Wood, J.D., Nute, G.R., Whittington, F.M., Hughes, S.I. and Sheard, P.R., A comparison of organic and conventionally-produced lamb purchased from three major UK supermarkets: price, eating quality and fatty acid composition. *Meat Sci.*, 78, 176, 2008.

68. Bryhni, E.A., *Consumer Perception and Sensory Analysis of Pork Flavor: Effect of Fatty Acid Composition and Processing*, DS thesis, Norway, 2002, 39.

69. Colditz, I.G., Fergunson, D.M., Greenwood, P.L., Doogan, V.J., Petherick, J.C. and Kilgour, R.J., Regrouping unfamiliar animals in the weeks prior to slaughter has few effects on physiology and meat quality in *Bos taurus* feedlot steers, *Aust. J. Exp. Agric.*, 47(7), 763, 2007.

70. Bryhni, E.A., Byrne, D.V., Rodbotten, M., Moller, S., Claudi-Magnussen, C., Karlson, A., Agerhem, H., Johansson, M. and Martens, M., Consumer and sensory investigations in relation to physical/ chemical aspects of cooked pork in Scandinavia, *Meat Sci.*, 65(2), 737, 2003.

71. Bertram, H.C., Aaslyng, M.D. and Andersen, H.J., Elucidation of the relationship between cooking temperature, water distribution and sensory attributes of pork—a combined NMR and sensory study, *Meat Sci.*, 70(1), 75, 2005.

72. Brunton, N.P., Lyng, J.G., Cronin, D.A., Morgan, D. and Mckenna, B., Effect of radio frequency heating on the texture, color and sensory properties of a comminuted meat product, *Food Res. Int.*, 38(3), 337, 2005.

73. Gomes, H.A., Silva, E.N., Cardello, H.M.A.B. and Cipolli, K.M.V.A.B., Effect of gamma radiation on refrigerated mechanically deboned chicken meat quality, *Meat Sci.*, 65, 919, 2003.

74. Al-Bachir, M., The irradiation of spices, packaging materials and luncheon meat to improve the storage life of the end products, *Int. J. Food Sci. Technol.*, 40(2), 197, 2005.

75. Reinbach, H.C., Meinert, L., Ballabio, D., Aaslyng, M.D., Bredie, W.L.P., Olsen, K. and Moller, P., Interactions between oral burn, meat flavor and texture in chili spiced pork patties evaluated by time-intensity, *Food Qual. Prefer.*, 18(6), 909, 2007.

76. Chun, S., Chambers IV, E. and Chambers, D., Perception of pork patties with shiitake (*Lentinus edode* P.) mushroom powder and sodium tripolyphosphate as measured by Korean and United States consumers, *J. Sens. Stud.*, 20, 156, 2005.

77. Deliza, R., Saldivar, S., Germani, R., Benassi, V.T. and Cabral, L.C., The effects of colored textured soybean protein (TSP) on sensory and physical attributes of ground beef patties, *J. Sens. Stud.*, 17, 121, 2002.

Chapter 31

Sensory Aspects of Cooked Meats

Nelcindo Nascimento Terra, Rogério Manoel Lemes de Campos, and Paulo Cézar B. Campagnol

Contents

31.1 Introduction

Organoleptic characteristics of meat determine the appeal it will have for consumers. The organoleptic quality of meat may be evaluated in a more objective way through instrumental or sensory methods used by scientists as measurement tools.

In cooked meat, flavor is one of the most important attributes. Many studies that have been conducted aimed at identifying volatile compounds responsible for its characteristic smell; there is lot of evidence substantiating the significant contribution of certain branched-chain fatty acids to it.

In regular storage conditions, color is the main appeal in meat. The color reflects the amount and chemical state of its main pigment, myoglobin. During cooking of meat, myoglobin is denatured exposing the heme element and breaking the bond between the heme and the protein. However, the heme tends to form new bonds with the denatured myoglobin and other denaturated myoglobin.

The formation of the bond between the heme and the denatured proteins, different from myoglobin, may occur during the first stages of myoglobin denaturation when the heme is partially exposed, since the myoglobin precipitation on the meat happens at lower temperatures than those of pure solutions. The hemychrom pigment with the denatured protein, Fe^{2+} as Fe^{3+}, which is formed during cooking, gives the cooked meat a brownish color. The general precipitation of proteins increases the light diffusion on the surface and is responsible for the light color appearance of the cooked meat.

Meat texture is a sensory parameter that has the primary attributes such as softness, cohesivity, viscosity, and elasticity; secondary ones such as thickness, masticability, juiciness, fracturability, and adhesivity; as well as the residual ones such as velocity of break, humidity absorption, and cold sensation in the mouth. The most important attributes for meat texture are softness, juiciness, and masticability. Some scientists use the term "tenderness" for softness, when dealing with physical measurements of the cooked meat resistance to compression or bond strength, and "sensory tenderness" to describe the resistance to mastication detected by the tasters.

31.2   Color

The raw meat presents red color because of the myoglobin, a chromoprotein responsible for the fixation of oxygen in the muscles of live animals. Since there is no meat without blood, the hemoglobin present in it also contributes, in a little amount, to the red color of the meat.

Myoglobin consists of a protein portion (globin) attached to the heme group that contains, in its central portion, an atom of iron whose electronic lability facilitates the creation of other pigments with colors different from red, characteristic of meat. Myoglobin is found in meat in three main forms: deoxymyoglobin, oxymyoglobin, and metmyoglobin. Both deoxymyoglobin and oxymyoglobin are physiologically active, due to the action of the metmyoglobin reductase enzyme. This action declines in the postmortem.[1] Deoxymyoglobin is found in the central portion of the meat cuts, where oxygen is not present. When iron changes from ferrous to ferric, the metmyoglobin, which is a brown-colored pigment, is created. The oximyoglobin, resulting from the oxygenation of myoglobin, is responsible for the bright red color of raw meat.[2]

The behavior of these pigments under the action of heat determines the color of cooked meat. The heating determines the denaturation of the globin that precipitates with other meat proteins; in other words, this denaturation is a function of temperature.[3] The treatment of meat is a technological need, either for eliminating the pathogenic flora or for decreasing the saprophyte flora, as well as to produce meat products. In this condition, while globin is denatured the metmyoglobin creates the hemichromogen globine (brown color) also known as ferrihemochrome.

The meat color can be measured visually, chemically, or instrumentally. In the first case, sensory panels are used, whereas the typical instrumental technique is based on chroma meter searching for the values of L (brightness), a (red tendency), and b (yellow tendency), known as Hunter values. Technological procedures usually need the quantification of pigments in meat because these are directly related to the final color of cured meat products.

In procedures described by Terra and Brum,[4] the meat pigments are extracted in acidified acetone using the following technique: weighing 10 g of the sample in a 100 mL beaker, adding the solution consisting of 40 mL of acetone and 3 mL of distilled water, mixing it for 5 min without light, and filtering it through Whatman number 1 filter paper. The optic density (OD) can be quantified in spectrophotometer at 540 nm, applying a mixture with 80% of acetone and 20% of distilled water as "a blank." By multiplying the OD by 290, the nitrous pigment in ppm is obtained. To quantify the gross pigments, 10 g of the sample is measured in a beaker. To this sample 40 mL of acetone, 2 mL of distilled water, and 1 mL of concentrated chloride acid are added. This is then mixed using a glass stick, covered with a clock glass, and kept in dark for 1 h, followed by filtration and measurement of the OD at 640 nm. A mixture containing 80% of acetone, 2% of concentrated chloride acid, and 18% of distilled water is used as white. The gross pigments are obtained by multiplying the optic densities by 680. The heat induces changes in the meat because its components are dependent on the temperature and time of cooking to have a higher or lower effect on softening.[5] Cooking the meat induces structural changes that lessen the water retention capacity reflecting mainly on the juiciness.[6]

31.3 Texture

Many techniques are used to cook the meat because the variation in the time of cooking and the palatability does not allow the use of one single technique. The use of microwaves provides fast heating rates,[7] meat tenderness, with fewer flavors and inferior cooking performance, when compared to conventional techniques.[8] The use of ultrasound in the cooking of meat has been shown to be a fast and efficient method to improve the attributes of meat texture.[9] The mechanical properties of meat affect the proteins of the conjunctive tissue and collagen. The texture of meat is influenced not only by the quantity of collagen, but also by its solubility in the heating.

The heat induces changes in the muscle, and its components are dependent on the temperature and the time of cooking. The effect on hardness or softness is influenced by the process.[5] The loss of structure of the perimysium in the cooked meat is due to the shrinking and the denaturation of the collagen fibers.[10] Granule deposits are observed in the openings between the endomysium and the myofibrillar mass, which are from the plasmolemma, which in turn are noncolloidal particles. The presence of granules indicates the passive fat of the fibrillar muscle.

These extensive deposits are due to the rupture (distortions) of the endomysium and perimysium that occurs inside the cooked meat at high temperatures.[5] This loss of myofibrillar structure is due to the progressive distortion of the endomysium and perimysium during cooking, which also causes distortions in the myofibrillar mass, conducting this way the loss of the structural integrity. This happens due to the softening of the collagen from the endomysium, which causes the rupture of the fiber and the progressive weakening of the adhesion between them. All these qualitative alterations indicate progressive changes of denaturation inside the collagen, which leads to the softening of the meat.

Discontinuity and the marked distortions of the endomysium and perimysium with loss of structural integrity were observed in meat cooked under pressure. Palka[11] observed this in the semimembranosus muscle of bovines, cooked at 80–90°C. The texture of the cooked meat may also be altered by an increase in humidity.[12] The presence of an amount of intramuscular fat increases the juiciness and the flavor of cooked meat.[13] The thermal conditions (internal temperature) during the cooking of the meat have a significant effect on all parameters of the meat texture profile (cohesivity, elasticity, adhesivity, hardness, and masticability), reaching its best level in the scale between 70 and 80°C.[6]

Researches by Palka and Daun[14] reported that increasing the temperature of cooking to 100°C makes the structure of the meat turn into something like a compact case due to a significant decrease in the diameter of the fiber. Vasanthi et al.[5] cooked buffalo meat at 100°C for 45 min in water and achieved a softness, if compared with other techniques, similar to that obtained with cooking under pressure. Straadt et al.[15] used the combination of confocal laser microscopy with the nuclear magnetic resonance to visualize the changes in the meat during the aging and heat treatment, characterizing the water distribution among the muscular fibers. These authors concluded that the combination of these methods is ideal for acquiring basic information about the microstructure and the structural changes of fresh and cooked meat.

Sensory attributes of foods are of great importance to the preference, which is affected by endogenous factors (hereditary, sex, age, and activity) as well as exogenous ones (culture, society, and economy);[16] the instrumental tests and determinations are generally faster, simpler, and exact, while the sets of information about the quality of the food are not sufficient.[17] The relation between the sensory attributes and the instrumental techniques is another field under investigation;[18] thus, an analysis must help the other to ensure the quality of the elaborated products. Nicod[19] describes that many phases are necessary to obtain a sensory evaluation group. These include recruiting, preliminary selection, specific selection, and training and control of the group (confidence in the answers, individual and collective repetition, etc.). The consumer when buying food, such as meat or a meat product, looks for all the answers to his/her patterns of smell, taste, color, appearance, and texture; and for the description of these attributes, he/she may use a trained panel of tasters.

The texture of a food is a multidimensional property that is defined as the set of mechanical, geometric, and surface properties of a product, perceivable by the mechanic receptors, the tact receptors, and, in some way, the hearing receptors.[20] More recently, Szczesniak[21] defined texture as the sensory and functional manifestation of the structural, mechanic, and surface properties of foods, detected through the senses of vision, hearing, tact, and smelling. This affirmation leads to important concepts.[22] (1) Texture is a sensory characteristic; thus, only human beings and animals may perceive it. The measurement instruments can only detect and quantify certain physical parameters that soon shall be interpreted in terms of sensory concepts. (2) It is a multiparametric attribute. (3) It is derived from the food structure (molecular, microscopic, or macroscopic). (4) It is detected by many senses, the tact being the most important.

Texture is one of the attributes evaluated by a customer for accepting a certain type of meat; it is determined mainly by the presence of proteins of the connective tissue and of myofibrils.[23]

The increase in texture caused by the proteins of the conjunctive tissue is explained by the formation and stability of the crossed bridges of the collagen molecule established with the aging of the animal. There are many factors that influence the softness of the meat, such as genetic characteristics, physiologic factors, type of feeding, how the animals are raised, and, specially, the physical-chemical state of the meat, which is closely related to the previously mentioned factors. The texture promoted by the myofibrillar proteins is affected by the development of the rigor mortis and depends on the carcass formation that may promote the softening of the meat by proteases. The structural organization of the muscle proteins is decisive to the distribution of water inside the meat and this way it has a direct effect on the characteristics of the meat.[24,25]

Hardness depends mostly on the quantity of connective tissue of the meat and on the stability of unifying points among the many molecules of collagen, such as the level of shortening of the sarcomere.[26] The collagen is a major protein inside the conjunctive tissue that is present in the muscle, encircling every muscular fiber (endomysium) to every fiber beam (perimysium) and to the muscle set (epimysium). Collagen is formed by helicoidal protein molecules that appear in groups of three by three, coiling one over the other to form molecules of tropocollagen.

One of the possibilities to improve the uniformity of texture consists in homogenizing the raw material destined to a determined process.[27] Among the parameters of raw material that affect the texture, the important ones are: the content of intramuscular fat;[28] its composition in fatty acids;[29] the pH of the muscles and the proteolytic potential of the meat;[30] the residual enzyme activity of cathepsin B in raw material;[31] the residual enzyme activity of cathepsins B + L;[32] the genotype;[33] the weight and feeding of the animals; the conformity and type of cut; the postmortem time and the process of freezing the meat; the storage and defrosting of the meat;[27] the electric stimulation of the meat;[34] and the method of cooking,[35] among others.

It is also possible to use sensory tests to evaluate the texture of a food. In this technique, the evaluation of texture is done by means of a panel of trained tasters; this measure is going to be the closest to what the costumer will get when consuming the food and, therefore, the best prediction of customer acceptance. There is no doubt that having an analytical tool like this is not always possible, since it usually results in high cost, effort, time and money. Thus, it is not remarkable that some laboratories and companies prefer the instrumental measurers to be able to estimate the texture characteristics that will define their product quality.

Many instrumental methods are developed to determine the texture of food.[36] In general, the instrumental measure of the texture in food implies the use of mechanical tests that can replace the sensory panels imitating or trying to reproduce the way they measure.[37] From the instrumental methods of measure, the imitative methods are based on the concept of texture and its characteristics, described by Szczesniak.[38]

The objective of these methods is to determine the profile of the texture in a certain food, measuring instrumentally distinct parameters. There are many methods that may be classified as imitative; however, the most used among them is the instrumental texture profile (TPA, texture profile analysis), for its good correlations with the sensory quantification, and hardness is the most studied measure about meat and its derivatives.[22] This method imitates the circumstances under which the food is submitted during the chewing process.[39] In this test the mechanical characteristics correlate to the reaction of the food to the effort and obtained primary parameters such as hardness, cohesivity, viscosity, elasticity, and adhesivity, besides secondary parameters such as fragility, masticability, and thickness.[38]

This technique consists of compressing the food sample (meat samples, for example, cut in slices of approximately 1.5 cm of length and 2.5 cm of width) in a texturometer or between two plain surfaces, two times consecutively, with the equivalent strength used in a bite, with the objective of imitating the action of chewing. The parameters of TPA (hardness, masticability, elasticity, and cohesivity) according to the strength–time curves were detailed by Breene[40] and Bourne.[39] Once a 5 kg weight is applied, the same procedure is conducted with the compression probe making it reach down to the plate where the sample is put. In the compression test we apply two consecutive bites, obtaining a graphic where the time of the test (abscissa) represents the function of the applied strength (ordinate). The conditions in the analysis are the following: 2.0 mm/s for the velocity of the probe in compression; 10.0 mm/s for the velocity of the probe rising; and 50% for the degree of compression of the sample. When the test is realized, the following texture parameters are defined:

Hardness: It is defined as the necessary strength to compress a substance between the teeth or between the tongue and the palate, depending on the kind of the product. It is taken from the height of the higher peak in the first compression of the TPA curve. It is expressed in grams (g).

Cohesivity: It is equivalent to the unifying strength that keeps connected distinct parts of a food giving the structure of the sample, that is, the degree to which a substance is compressed between the teeth before it is torn apart. It is calculated as the relation between the two positive areas in a strength–time curve. It is an adimentional parameter.

Elasticity: It is the distance between the first and the second contact of the probe with the sample, that is, the degree in which the sample, once deformed, recovers its initial nondeformed condition when the deformation forces are suspended. It is expressed in nm.

Adhesivity: It is equivalent to the work necessary to overcome the attraction forces between the surface and the sample, when they get in contact. It is reflected in the area of the curve that may appear under the baseline after the first cycle of compression (first bite). It is measured in g × s (seconds).

Masticability: It is the energy required to chew the sample, reducing it to an adequate consistence for swallowing. It is calculated from the measures of hardness, cohesivity, and elasticity. It is expressed in g × mm.

The parameters of compression obtained with TPA are used by many authors to evaluate the meat and its derivatives, as a rate to determine the quality of the products or to select the best functional formula.[41–44] The compression test was used to study the mechanical properties of the muscle fibers. Many authors used the test of compression to monitor the mechanical properties of myofibers during the beginning of the rigor mortis and after it.[45]

More recently, Christensen et al.[46] applied this test to study the changes in the mechanical properties of fibers during the storage postmortem. Some authors employed tests of meat compression in fibers from the muscle and the conjunctive tissue, the perimysium, to determine the effect of the cooking temperature.[47,48]

In meat and in meat products, something similar occurs, even if there are clear preferences depending on the kind of meat and if it is consumed as fresh meat or as a meat product.[30] For pork meat the flavor and the texture seem to have similar importance when the product is not excessively dry and fibrous, situation in which the texture properties result determined. From the attributes of palatability of the meat, the softness is evaluated as the most important factor and a great amount of research has been done to improve this attribute.[5] The intramuscular fat influences the texture, juiciness, and flavor, thus playing an important role in the perception of meat quality.[13]

In irradiated pork meat, the characteristic smell is caused by the production of sulfur volatiles from sulfurous amino acids, such as methionine and cysteine.[49] The chemical changes in irradiated meat are initiated by sulfur free radicals or carbon monoxide resulting from reactions among meat components and radiolytic free radicals.[50]

Ruiz de Huidoro et al.[51] have tested two methods to determine the meat texture and concluded that in raw meat, TPA method is more useful than Warner–Braztler's (WB) test, and when such a method is applied to cooked meat the contrary is observed.

People are familiar to the basic five senses: sight (eyes), taste (tongue), smell (nose), hearing (ears), and touch (fingers and mouth sensation). The touch, responsible for mouth sensation, may be decomposed into three sensations: pressural, trigeminal, and kinesthetic. The pressural represents the sensation that is experimented when a force is applied to the food surface; trigeminal refers to pain sensation; and kinesthetic means retrofeeding from muscles of mastication during chewing process.[52]

31.4 Flavor

While purchasing, the consumer first checks food appearance and color, later on the smell. Finally, during the process of mastication he/she checks its texture, taste, and smell, which all together determine the final impression on flavor.[53] Flavor impression is caused first by food nonvolatile compounds, interacting with the tongue surface, palate mucosa, and tongue areas. Smell impression is caused by volatiles that evaporate from the food during mastication process and dislocate to the nasal cavity, where they react with olfactory sensors producing an electric signal, which is transmitted to the olfactory bulb in the frontal brain cortex.[53] Volatile concentrations with low molecular weight may sensitively affect the smell.[54]

Buck and Axel, who were awarded Nobel Prize in Physiology/Medicine, have shown how the signal, detected in the nose, is transmitted to the brain. First, the signal is sent to the olfactory bulb in the brain, which works as an initial piece of information organization center. Even though neurons are randomly distributed in the nose, the ones that carry the same type of receptive protein connect at the same fixed point in the olfactory bulb; therefore, there is a topographic map in the olfactory bulb where each one of the thousands reception proteins is represented and in which the smell activates the only combination between these points. Afterwards, this set of information is sent to the "superior" parts of the brain, where the conscious smell perception is generated, or to the "primitive" brain structures, which rule the inborn behavior, emotions, and memory. It is estimated that human beings are sensitive to 5,000–10,000 aromatic compounds.[55]

Up to now, more than 2600 chemical substances are known as smell volatile components. They are organic compounds of low molecular weight (less than 300), which are partially lipid soluble. Many of these compounds were identified after the invention of the gas chromatograph (GC). It is feasible to only identify and quantify the volatile compounds of a food without setting the direct contribution of individual compounds to flavor.[52] Different chemical groups add smell to the meat, such as aldehydes, alcohols, ketones, acids, esters, lactones, furans, phenols, terpenes, sulfur compounds, and so on.

The meats are cooked, desiccated, or even smoked, aiming at developing the flavors. The application of heat produces complex reactions among the amino acids (frequently those with sulfur) and sugars (which contain carbonyl) producing the Maillard reaction.[52] The time in which the meat is cooked, either using a dry method (roasting) or, a humid one, and the temperature reached during the cooking process may change the formed compounds and drastically change the *flavors*.

Taking into consideration that cooking methods give place to different flavor reactions, in each animal there is a singular relation of amino acids, fatty acids, and sugars, generating, this way, its own flavor. In bovines, sheep, and pig, the lipids have mainly saturated fatty acids, which do not degrade as fast as the unsaturated fatty acids. The poultry and fish have a great amount of unsaturated lipids, which generate flavor and small chemical molecules that interact with the products from the reaction of amino acids with sugars, forming more complex flavors. The taste of rancidity is developed more easily in fish and poultry due to such unsaturated lipids.

During water heating, many flavors are modified as well as new ones appear from nonvolatiles precursors. This way, these changed flavors, caused by the cooking process, are part of a different category. The thermal decomposition of the amino acids and peptides require higher temperatures than the ones normally applied while cooking the food. In roasted food only the superficial areas, where dehydration occurs, allow the temperature to significantly increase, over the boiling point of water, leading to decarboxylation and desamination of the amino acids, with the consequent formation of aldehydes, hydrocarbons, nitriles, and amino compounds.

The Maillard reaction is one of the most important means for the formation of flavor compounds in meat and meat products, constituting the basis for numerous patents for the industrial production of meat flavors.

Louis Maillard, a French chemist, discovered in 1912 the reaction between sugars and amino acids, while working on color compounds (melanoidins), formed by heating a solution of glucose and glycine. The reaction basically involves a protein compound (protein, peptide, amino, or amino acid) and a sugar.[56] Although apparently simple, the Maillard reactions are very complex because they comprise condensation, dehydration, reordering, and degradation to form furans, furfural and derivatives, aldehydes, dicarboniles, cetons, and so on. These compounds may undergo new reactions among themselves or with some other lipid carbonylic compounds to generate many volatile compounds (secondary reactions).

The Maillard reaction does not demand high temperatures as the ones for caramerization and pyrolysis of proteins. Frozen sugar and amino acid mixtures show, within the time, Maillard darkening. No doubt that the reactions of Maillard occur at higher speed and temperature, mainly temperatures that correspond to cooking, thermal processing, and desiccation.[52] The speed of the reactions also increases at low humidity. Therefore, flavor compounds produced by the Maillard reaction tend to associate with the surface of the foods that were dehydrated by heat.

The first step of the Maillard reaction consists of the addition of the carbonyl group of an open-chain reductive sugar to the amino group of an amino acid or peptide (a primary amino group). From the subsequent elimination of distilled water results a Schiff base. Water is the limiting factor, and over this phase, the reaction is irreversible; the set of oxidation and Maillard reactions generate a large number of volatile compounds that may group in the following categories: carbonyls (aldehydes and ketones), furans, fatty acids, pirazines, and sulfur compounds (sulfurs, thiazoles, thiols, and thiophenols).[57] The final flavor is determined by the set of all volatile compounds, considering their concentrations, minimal detection number, and so on.[56]

Strecker reactions characterize another set of volatile compounds such as 2-methyl propanal, 2-methyl butanal, and 3-methyl butanal, which derive from valine, isoleucine, and leucine, respectively.[57] Sulfur volatile compounds are also formed from amino acids rich in sulfur such as methionine, cysteine, and cystine.

Rancid flavor of the deteriorated fat is due to the lipid autoxidation of the unsaturated fatty acids. The course of the reaction is the same if the acids are esterified or free and occurs in three steps. Initiation step produces a small number of highly reactive free radicals (molecules with unpaired electrons).

In the propagation step, the oxygen reacts with the free radicals produced in the initiation step, forming hydroperoxides, which degrade to generate more free radicals that can attack more fatty acids and react with more oxygen to result in a chain of reactions. When the concentrations of free radicals are formed, they tend to react among themselves to form stable final products, a characteristic of the rancified fat: aldehydes, alcohols, and cetons. The oxidation reactions with the formation of free radicals that are produced while cooking, as well as rancidity, follow the same basic paths previously described and generate similar types of volatile products.

There are differences in the precise mechanisms of oxidation in refrigeration or heating conditions. In cooking, the reactions contribute to the agreeable taste, while in rancid products, the radical reactions produce undesirable flavors. Hydroperoxides are extremely thermolabile. At lower temperatures they are more stable and are formed in a higher amount before their decomposition. Different proportions of the many intermediate radicals are formed, resulting in different volatile products.

Aldehydes and ketones produced by lipid oxidation may also undergo secondary reactions in the presence of amino acids, generating other flavor compounds. Lipid oxidation in cooking plays an important role in the development of the complex flavor profile of the meat volatiles; in slightly cooked or roasted meats, especially in poultry, during processing and storage of the food, polyunsaturated fatty acids (PUFAs) tend to be oxidized.[58] The lipid autooxidation also occurs during long-term meat storage, producing strange and undesirable flavors. The heating flavor becomes evident when cooked meats are kept at refrigeration temperature, for a few days, and after that are reheated, a phenomenon that is attributed to lipid oxidation products. It is believed that it is due to the autoxidation of the phospholipids, which are more unsaturated than the triglycerides. During the initial heating, the rupture and disintegration of the membranes of the muscular cells leave the phospholipids more susceptible to oxidation. This reaction is catalyzed by metallic ion traces, iron in particular, proceeding from the degradation of the heme pigments during cooking.[52]

Cooking the meat results in a product with different appearance due to the alteration of characteristics such as texture, taste, and smell. The appearance of cooked meat may be influenced by the pH, origin of the meat (animal species, animal age, and anatomic position of the muscle), quality of processing, fat rate, additives, packaging conditions, and conservation treatments (heating, irradiation, and pressure).[3] Cooked meat is more susceptible to lipid oxidation than raw meat, during its storage, considering that the heating operations alter the muscle endogenous prooxidative/antioxidative balance, mixing the oxidation catalyzers with the lipid, thus promoting the myoglobin oxidation, liberating iron to the limit, inactivating the antioxidizing enzymes.[59] Meat cooking promotes the formation of smell compounds from lipids during the process of oxidation, Maillard reaction, and interactions among the reaction products.[60]

The heat releases the iron ion from the heme pigment, which is known to be an important catalyzer to the oxidation of lipids in cooked meat.[61] The oxidation induced by the heat produces degradation products such as aldehydes, ketones, and aliphatic alcohols, which lead to undesired tastes.[62] The potential losses of n-3 PUFA during the processing and cooking of the meat could be significant. The cooking of the meat may lead to a loss or to an alteration of fatty acids, specially the PUFAs; however, data concerning the effect of cooking chicken meat in relation to the quantity and type of fatty acid left after cooking are inconsistent. The same authors suggest that the more aggressive thermal processes may cause significant losses of PUFAs. Elmore et al.[62] report that this lower concentration of fatty acids reduces the oxidizing stability, resulting in alterations in the volatile composition of the aroma produced during cooking.

The flavor of the cooked meat decreases sharply during the storage, due to formation of free radicals by oxidation of intramuscular phospholipids of the membranes, which contain relatively high levels of polysaturated fatty acids. The retention of the flavor in meat is a little problematic, and shows that the high concentrations of free phospholipids in turkey chest are more susceptible to oxidative deterioration than other meats. Studies carried out by Insausti et al.[13] revealed differences in the volatile compounds in beef of autochthonous Spanish breeds, which may contribute to the taste and smell in the cooking of the meat.

Baek and Cadwallader,[63] using the technique of headspace gas chromatography/mass spectrometry, found in alligator cooked meat 56 volatile compounds, including 23 aldehydes, 10 alcohols, 11 ketones, 9 terpenes, and 3 varied compounds. The aliphatic saturated aldehydes derive from the oxidative degradation of the PUFAs and normally have an unpleasant smell. Based on the results obtained, it can be concluded that some oxidation occurred after the meat was cooked. Thus, the lipid oxidation must be avoided to improve the aroma of the cooked meat.

References

1. Warriss P. *Meat Science: An Introductory Text*. Oyon: Cabi Publishing; 2000.
2. Varnam A, Sutherland I. *Meat and Meat Products*. London: Chapman & Hall; 1995.
3. King N, White R. Does it look cooked? A review of factors that influence cooked meat color. *Journal of Food Science* 71, 31–40; 2006.
4. Terra NN, Brum MAR. *Carne e seus derivados*. São Paulo: Nobel; 1988.
5. Vasanthi C, Venkataramanujam V, Dushyanthan K. Efect of cooking temperature and time on the physico-chemical, histological and sensory properties of female carabeef (búfalo) meat. *Meat Science* 76, 274–280; 2007.
6. Barbera S, Tassone S. Meat cooking shrinkage: Measurement of a new meat quality parameter. *Meat Science* 73, 467–474; 2006.

7. Bakanowski SM, Zoller JM. End point temperature distribuitionss in microwave and conventionally cooked pork. *Food Technology* 38, 45–51; 1984.

8. El Shimi NM. Influence of microwave and conventional cooking and reheating on sensory and chemical characteristics of roast beef. *Food Chemistry* 45, 11–14; 1992.

9. Pohlman FW, Dikeman ME, Zayas JF, Unruh JA. Efects of ultrasound and convection cooking to diferent end point temperatures on cooking characteristics, shear force and sensory properties, composition and microscopic morphology of beef Longissimus and Pectoralis muscles. *Journal of Animal Science* 75, 386–401; 1997.

10. Lewis GJ, Purslow PP. The strength and stiffness of Perimysial connective tissue isolated from cooked beef muscle. *Meat Science* 26, 255–269; 1989.

11. Palka K. The influence of post-mortem ageing and roasting on the microsctructure, texture and collagen solubility of bovine *semitendinosus* muscle. *Meat Science* 64, 191–198; 2003.

12. Farouk MM, Swan JE. Boning and storage temperature effects on the attributes of soft jerky and frozen cooked free-flow mince. *Journal of Food Science* 64, 465–468; 1999.

13. Insausti K, Gon V, Petri E, Gorraiz C, Beriain MJ. Effect of weight at slaughter on the volatile compounds of coged beef from Spanish cattle breeds. *Meat Science* 70, 83–90; 2005.

14. Palka K, Daun H. Changes in texture, cooking losses, and miofibrillar structure of bovine M. Semitendinosus during heating. *Meat Science* 51, 237–243; 1999.

15. Straadt IK, Rasmussen M, Andersen J, Bertram HC. Aging-induced changes in microstruture and water distribution in fresh and cooked pork in relation to water-holding capacity and cooking loss—A combined confocal laser scanning microscopy (CLSM) and low-field nuclear magnetic resonance relaxation study. *Meat Science* 75, 687–695; 2007.

16. Risvik E. Sensory properties and preferences. *Meat Science* 34, 67–77; 1994.

17. Soriano AP, Quiles RZ, Garcia R. Selección, entrenamiento y control de un panel de análisis sensory especializado en jamón curado. *Eurocarne* 99, 25–34; 2001.

18. Elortondo FGP, Eguía PB, Aguado MA. Análisis sensory en quesos con denominación de origen. *Alimentaria* 309, 165–167; 2000.

19. Nicod H. L'organisation practique de la mesure sensorielle. La formation du groupe, en Evaluation sensorielle. *Manuel mèthodologique, SSHA et ISHA*. Paris: Technique et Documentation-Lavoisier; 1990.

20. ISO. Sensory Analysis. *Vocabulary*. ISO 5492, 1–36; 1992.

21. Szczesniak AS. Texture is a sensory property. *Food Quality and Preference* 13, 215–225; 2002.

22. Ruiz-Ramírez J, Gou P, Arnau J. Textura en productos cárnicos crudos curados: medidas instrumentales y sensoryes. *Eurocarne* 116, 95–106; 2003.

23. Shimokomaki M, Ida EL, Kriese PR, Soares AL. Calpaínas e Calpastatinas. In: Shimokomaki M, Olivo R, Terra NN. *Atualidades em Ciência e Tecnología de Carnes*. Sao Paulo: Livraria Varela; 2006. pp. 185–194.

24. Melody JL, Lonergan SM, Rowe LJ, Huiatt TW, Mayes MS, Huff-Lonergan E. Early post mortem biochemical factors influence tenderness and water-holding capacity of three porcine muscles. *Journal of Animal Science* 82, 1195–1205; 2004.

25. Huff-Lonergan E, Lonergan SM. Mechanisms of waterholding capacity of meat: the role post mortem biochemical and structural changes. *Meat Science* 71, 194–204; 2005.

26. López-Bote C. Calidad de la carne. In: Martín Bejarano S. *Manual práctico de la carne*. Madrid: M & M Ediciones; 1992. pp. 143–180.

27. Arnau J, Gou P, Comaposada J. Control del secado para uniformizar la textura del jamón curado. *Eurocarne* 115, 51–58; 2003.

28. Virgili R, Porta C, Schovazappa C. Effect of raw material on the end-product characteristics. *44th ICoMST*. Barcelona; 1998. pp. 26–38.

29. Ruiz-Carrascal J, Ventanas J, Cava R, Andrés AI, García C. Texture and appearance of dry-cured ham as affected by fat content and fatty acid composition. *Food Research International* 33, 91–95; 2000.

30. Guerrero L, Guàrdia MD. La medida de las propiedades mecánicas en la carne y en los derivados cárnicos. *Eurocarne* 77, 41–49; 1999.

31. Virgili R, Parolari G, Schovazappa C, Soresi Bordini C, Borri M. Sensory and texture quality of dry-cured ham as affected by endogenous cathepsin B activity and muscle composition. *Journal Food Science* 60, 1183–1186; 1995.
32. García-Garrido JA, Quiles-Zafra R, Tapiador J, Luque de Castro MD. Activity of cathepsins B, D, H and L Spanish dry-cured hamo f normal and defective textura. *Meat Science* 56, 1–6; 2000.
33. Fernández X, Gilbert S, Vendeuvre JL. Effects of halothane genotype and pre-slaughter on pig meat quality. Part 2. Physico-chemical traits of cured-cooked ham and sensory traits of cured-cooked and dry-cured hams. *Meat Science* 62, 439–446; 2002.
34. Aalhus JL, Jones SDM, Tong AKW, Jeremiah LE. The combined effects of time on feed, electrical stimulation and aging on beef quality. *Canadian Journal of Animal Science* 72, 525–535; 1992.
35. Panea B, Mons'n A, Olleta JL, Martínez-Cerezo S, Pardos JJ, Sañudo C. Estudio textural de la carne de vacuno. II. Análisis sensory. Textura study of bovine meta. II. Sensory analysis. *Información Técnico-Económica Agraria* 24, 31–33; 2003.
36. Kilcast D. Force/deformation techniques for measuring texture. In: Kilcast D. *Texture in Food*. Cambridge: Woodhead Publishing Ltd; 2004, 109–145.
37. Sastre CM, Domínguez BM, Fernández CG. Tipificación de la Cecina de León: características físico-químicas, nutricionales y sensoryes. Instituto Tecnológico Agrario de Castilla y León; 2006, 110.
38. Szczesniak AS. Classification of textural characteristics. *Journal of Food Science* 28, 385–389; 1963.
39. Bourne MC. Textura profile analysis. *Food Technology* 32, 62–72; 1978.
40. Breene WM. Applications of texture profile analysis to instrumental food texture evaluation. *Journal Texture Study* 6, 53–82; 1975.
41. Hoz L, D'Arrigo M, Cambero I, Ordóñez JA. Development of a n-3 fatty acid and α-tocopherol enriched dry fermented saudage. *Meat Science* 67, 485–495; 2004.
42. Visesssanguan W, Soottawat B, Riebroy S, Thepkasikul P. Changes in composition and functional properties of proteins and their contributions to Nham characteristics. *Meat Science* 66, 579–588; 2004.
43. Houben JH, Hooft BJ. Variations in product-related parameters during standardised manufacture of a semi-dry fermented sausage. *Meat Science* 69, 283–287; 2005.
44. Campos RML, Hierro E, Ordóñez JA, Bertol TM, Terra NN, Hoz L. Fatty acid and volatile compounds from salami manufactured with yerba mate (*Ilex paraguariensis*) extract and pork back fat and meat pigs fedo n diets with partial replacement of maize with rice bran. *Food Chemistry* 103, 1159–1167; 2007.
45. Willems MET, Purslow PP. Effect of postrigor sarcomere length on mechanical and structural characteristics of raw and heat-denature single porcine musclu fibers. *Journal of Texture Studies* 27, 217–233; 1996.
46. Christensen M, Young RD, Lawson MA, Larsen LM, Purslow PP. Effect of added μ-calpain and post-mortem storage on the mechanical properties of bovine single muscle fibers extended to fracture. *Meat Science* 66, 105–112; 2003.
47. Lepetit J, Culioli J. Mechanical properties of meat. *Meat Science* 36, 203–237; 1994.
48. Christensen M, Purslow PP, Larsen LM. The effect of cooking temperature on mechanical properties of whole meat, single muscle fibers and perimysial connective tissue. *Meat Science* 55, 301–307; 2000.
49. Ahn DU. Production of volatiles from acid homopolymers by irradiation. *Journal of Food Science* 67, 2565–2570; 2002.
50. Nam KC, Ahn DU. Mechanisms of pink color formation in irradiated precooked turkey breast. *Journal of Food Science* 67, 600–607; 2002.
51. Ruiz de Huidoro F, Miguel E, Blázquez B, Onega, E. A comparison between two methods (Warner-Bratzler and texture profile anlysis) for testing either raw meta or cooked meat. *Meat Science* 69, 527–536; 2005.
52. Fisher C, Scott TR. *Flavores de los alimentos, biología e química*. Barcelona: Acribia; 2000.
53. Röthe M. *Introduction to Aroma Research*. Dordrecht: Kluwer Academic Publishing; 1988.
54. Dalton P, Doolittle N, Nagata H, Breslin PAS. The merging of the senses: integration of sbthreshold taste and smell. *Nature Neuroscience* 3, 431–432; 2000.

55. Malnic B. O código dos aromas. *Ciência Hoje* 211, 14–15; 2004.
56. Arnau J, Monfort JM. El jamón curado: Tecnología y análisis de consumo. *Simposio Especial—44th International Congress of Meat Science and Technology*. Barcelona; 1998, 205.
57. Flores M, Grimm CC, Toldrá F, Spanier AM. Correlations of sensory and volatile compounds of Spanish "Serrano" dry-cured ham as a function of two processing times. *Journal of Agricultural and Food Chemistry* 45, 2178–2186; 1997.
58. Du M, Nam KC, Ahn DU. Cholesterol and lipid oxidation products in cooked meat as affected by raw-meat packaging and storage time. *Journal of Food Science* 66, 1396–1401; 2001.
59. Decker EA, Mei L. Antioxidant mechanisms and applications in muscle. *49th Proceedings of the Reciprocal Meat Conference* 49, 64–72; 1996.
60. Whitfield FB. Volatiles from interactions of Maillard reactions and lipids. *Critical Reviews in Food Science Nutrition* 31, 1–58; 1992.
61. Han D, McMillin KW, Golber JS, Bindner TD, Younathan MT, Marshall DL, Hart LT. Iron distribution in heated beef and chicken muscles. *Journal Food Science* 58, 697–700; 1993.
62. Stephen Elmore J, Mottram DS, Enser M, Wood JD. Effect of the polyunsaturated fatty acid composition of beed muscle on the profile of aroma volatiles. *Journal of Agricultural Food Chemistry* 47, 1619–1625; 1999.
63. Baek HH, Cadwallader KR. Aroma volatiles in cooked alligator meat. *Journal of Food Science* 62, 321–325; 1997.

SAFETY

Chapter 32

Methods for Evaluating Microbial Flora in Muscle Foods

Daniel Y.C. Fung

Contents

32.1   Introduction of Microbiological Testings in Muscle Foods

Meat microbiology and safety are of great importance to food science and technology nationally and internationally. Fung et al. [1] made a comprehensive review on the following topics related to meat safety: history of meat industry safety, microbiological hazards associated with meats, chemical hazards associated with meats, physical hazards associated with meats, identification and control, current regulatory policies and inspection, and meat safety in the future. This chapter presents methods for evaluating microbial flora in muscle foods especially: (1) enumeration and estimation of microbial populations in muscle foods and (2) detection and identification of normal and pathogenic microbes in muscle foods.

In the past 100 years many "conventional methods" have been developed throughout the world and are presented in detail in classic food microbiology books such as *Modern Food Microbiology* [2]; *Quality Attributes and Their Measurements in Meat, Poultry, and Fish Products* [3]; *Compendium of Methods for the Microbiological Examination of Foods* [4]; *Bacteriological Analytical Manual* [5]; *Microbiological Control for Foods and Agricultural Products* [6]; *Laboratory Methods in Food Microbiology* [7]; *Official Methods of Analysis* [8]; and many others. This chapter presents more recent developments in this area especially concerning rapid methods and automation in microbiology related to food microbiology and muscle foods. Excellent reference materials on this topic can be found in *Handbook for Rapid Methods and Automation in Microbiology Workshop* [9] and *Encyclopedia of Rapid Microbiological Methods Vols. 1, 2, and 3* [10].

Rapid methods and automation is a dynamic area in applied microbiology dealing with the study of improved methods in the isolation, early detection, characterization, and enumeration of microorganisms and their products in clinical, food, industrial, and environmental samples. In the past 15 years this field has emerged into an important subdivision of the general field of applied microbiology and is gaining momentum nationally and internationally as an area of research and application to monitor the numbers, kinds, and metabolites of microorganisms related to food spoilage, food preservation, food fermentation, food safety, and foodborne pathogens. Fung [11] in a comprehensive review of the entire field of rapid methods presented the following subjects: history and key developments, advances in sample preparations and treatments, advances in total viable cell count methodologies, advances in miniaturization and diagnostics kits, advances in immunological testing, advances in instrumentation and biomass measurements, advances in genetic testings, advances in biosensors, US and world market and testing trends and prediction of the future. For further information, see Ref. 11.

32.2 Sample Preparations and Treatments of Muscle Foods

One of the most important steps for successful microbiological analysis of any material is sample preparation. Without proper sampling procedures the data obtained will have limited meaning and usefulness. With the advancement of microbiological techniques and miniaturization of kits and test systems to even smaller sizes, proper sample preparation becomes critical. Statistic sampling plans for various foods for microbiological analysis is beyond the scope of this review. Sample preparation methods are discussed for muscle foods in general and also the environments in which muscle foods are prepared and processed.

Microbiological samples can be grouped as solid samples, liquid samples, surface samples, and air samples. Each type of sample has its unique properties and concerns in sample preparation and analysis. This section discusses the improvement of methods for solid, liquid, surface, and air sampling procedures. These procedures are important for both conventional microbiological techniques as well as new and sophisticated rapid methods directly and indirectly involved with muscle foods.

32.2.1 Solid Samples

Common laboratory procedures for solid samples include aseptic techniques to collect sample, rapid transport (less than 24 h) to laboratory site in frozen state for frozen foods, and chilled state for most other foods. The purpose is to minimize growth or death of the microorganisms in the muscle food to be analyzed. The next step is to aseptically remove a subsample such as 5, 10, 25 g, or more for testing. Sometime samples are obtained from different lots and composited for analysis. In food microbiology, almost always the food is diluted to 1:10 dilution (i.e., 1 part of food in 9 parts of sterile diluent) and then homogenized by a variety of methods. It should be noted that 1 g of food sample is equivalent to 1 mL of diluent (based on the specific gravity of water) for ease of calculation of dilution factors in microbiological manipulations. To make a 1:10 dilution, the procedure is simple, but when an analyst has to make 10 or more samples this becomes laborious and time-consuming. An instrument called Gravimetric Diluter marketed by Spiral Biotech (Bethesda, MD) can automatically perform this function. The analyst simply puts an amount of food (e.g., 10.5 g) into a sterile Stomacher bag and sets the desired dilution (1:10), the instrument will then deliver the appropriate amount of sterile diluent (e.g., 94.5 g). Thus, the dilution operation can be done automatically and efficiently. The dilution factor can be programmed to deliver other factors, such as 1:25, 1:50, etc. Manninen and Fung [12] found this system to be efficient and accurate over a wide range of dilutions. A product named Diluflo has been in use satisfactorily in this author's laboratory for about 10 years. A similar system called Dilumacher is marketed by PBI of Milano, Italy for dispersing diluents to samples automatically. After dilution, the sample needs to be homogenized. Traditionally, a sterile blender or osterizer is used to homogenize the food suspension for 1–2 min before further diluting the sample for microbiological analysis. The disadvantages of using a blender include the following: (1) the blender must be cleansed and resterilized between each use; (2) aerosols may be generated, which contaminate the environment; and (3) heat may be generated mechanically, which may kill some bacteria. In the past 25 years, the Stomacher invented by Anthony Sharpe has become standard equipment in food analysis laboratories. About 40,000 Stomacher units are in use worldwide. The sample is placed in a sterile plastic bag and an appropriate amount of sterile diluent is added. The sample in the bag is then "massaged" by two paddles of the instrument for 1–2 min and then the content can be analyzed with or without further dilution. The advantage of the Stomacher include the following: (1) no need to resterilize the instrument between samples because the sample (housed in a sterile plastic bag) does not come in contact with the instrument, (2) disposable bags allow analysis of large number of samples efficiently, (3) no heat or aerosols will be generated, and (4) the bag with the sample can serve as a container for time-course studies. A similar instrument called Masticator is marketed by IUL Instruments (Erlanger, KY). Recently Anthony Sharpe invented the Pulsifier for dislodging microorganisms from foods without excessively breaking the food structure. The Pulsifier has an oval ring that can house a plastic bag with sample and diluent. When the instrument is activated the ring will vibrate vigorously for a predetermined time (30–60 s). During this time microorganisms on the food surface or in the food will be dislodged into the diluent with minimum destruction of the food. Fung et al. [13] evaluated the Pulsifier against the Stomacher with 96 food items (including beef, pork, veal, fish, shrimp, cheese, peas, a variety of vegetables, cereal, and fruits) and found that the systems gave essentially the same viable cell count in the food but the "Pulsified" samples were much clearer than the "Stomached" samples. A more recent report by Kang et al. [14] found that the Pulsifier and Stomacher had a correlation coefficient of .971 and .959 for total aerobic count and coliform count, respectively, with 50 samples of lean meat tissues. The Pulsified samples, however, contained much less meat debris than Stomached samples. In the case of Stomached samples, many meat debris occurred which interfered with plating samples on agar. The superior quality of microbial suspensions with minimum food particles from the Pulsifier has positive implications for general analysis as well as for techniques such as ATP bioluminescence tests, DNA/RNA hybridization, PCR amplifications, enzymatic assays, etc. These instruments have been tested by this author for many years. One of the concerns is the level of noise generated by the Stomacher and the Pulsifier. Recently, AEC company developed a robust instrument called the "Smasher" (AES, Bruz, France) with almost no noise during the operation of mixing samples. Currently (year 2007) this author is evaluating all three instruments for the effectiveness of recovery of microorganisms from muscle foods and other foods as well as the noise level in a laboratory setting.

32.2.2 Liquid Samples

Although muscle foods are not considered liquid foods, they can be found in liquid suspensions such as clam chowder, meat soups, broths, etc. So, a discussion of liquid or semiliquid foods is included in this chapter. Liquid samples are easier to manipulate than solid samples. After appropriate mixing (by vigorous hand shaking or by instrument), one only needs to aseptically introduce a known volume of liquid sample into a container and then add a desired volume of sterile diluent to obtain the desired dilution ratio (1:10, 1:100, etc.). Further dilutions can be made as required. There are now many automated pipetting instruments available for sample dilutions such as the Rapid Plate 96 Pipetting Workstation marketed by Zymark Corp., Hopkinton, MA. Viscous and semisolid samples need special considerations such as the use of large mouth pipettes during operation. Regardless of the consistency of the semisolid sample, 1 mL of sample is considered as 1 mL of liquid for ease of making dilution calculations. It should also be noted that in a dilution series there are dilution errors involved; thus, the more dilutions one makes the more errors one will introduce.

32.2.3 Surface Samples

Sampling of surfaces of food or the environment presents a different set of concerns. The analyst needs to decide on the proper unit to report the findings, such as the number of bacteria per inch square, per centimeter square, or other units. One can analyze different shapes of the surface such as square, rectangle, triangle, circle, etc. A sterile template will be useful for this purpose. Occasionally, one has to analyze unusual shapes such as the surface of an egg, apple, the entire surface of a chicken, etc. The calculation of these areas becomes quite complex. For intact meat or other soft tissues, one can excise an area of the food by using a sterile knife assuming that all the organisms are on the surface and that the meat itself is sterile. Often a sterile moist cotton swab is used to obtain microbes from the surface of a known area and then the swab is placed into a diluent of known volume (e.g., 5 mL), shaken, and then plated on a general purpose agar or a selective agar. Instead of a cotton swab, one can use contact materials to sample surfaces. This includes selective and nonselective agar in Rodac plate, adhesive tape, sterile gauge, sterile sponge, etc. Under swabbing methods one can include tests such as ATP measurement, residue protein, lipid, carbohydrate, catalase enzymes, etc. The nature and characteristics of the surfaces are also very important. Methods of obtaining microbiological samples from dry surfaces, wet surfaces, oily surfaces, slimy surfaces, meat, chicken skin, orange skin, stainless steel, concrete, rocks, hairnets, etc. are very different. A lot of microorganisms will remain on the surface even after repeatedly sampling the same area. Biofilms are very difficult to remove completely from any surface. This, however, should not be a deterrent to use surface sampling techniques if one can relate the numbers obtained to another parameter such as cleanliness of the surface or quality of a food product. Lee and Fung [15] made a comprehensive review of methods for sampling meat surfaces. The first group of sampling is for nondestructive microbial sample methods that include swab methods, rinses method, direct agar contact method, director surface agar plate method, scraping method, vacuum method, etc. The second group of sampling is for destructive microbial sample methods, which include obtaining a proper weight of muscle food and making dilutions as stated before.

Fung et al. [16] made a comprehensive evaluation of using adhesive tape method for estimation of microbial load on meat surfaces and found that the surface counts from the adhesive tape correlated exceeding well with the conventional swab method with a correlation coefficient of .95 for psychrotrophs and .90 for mesophiles from 60 meat surfaces. The adhesive method is far more convenient to use and economical. More recently Fung et al. [17] developed a convenient method to obtain surface samples called "hands-free, 'pop-up' adhesive tape method" for microbial sampling of meat surfaces. In this procedure the 3M "pop-up" tape unit is placed on the wrist of an analyst, while both hands can be free to manipulate experimental materials such as obtaining the meat sample, arranging agar plates, labeling samples, etc. When the time is ready, the analyst simply pulls one tape out of the unit from the wrist and uses the tape to obtain microbial sample from the meat surface (15 s) and then transfers the tape to an agar surface (15 s) and finally incubates the plate for viable cell count of the meat surface. The correlation coefficient of the pop-up tape method and the more cumbersome conventional swab/rinse method for obtaining viable cell counts was .91. Thus the simple pop-up tape method is a viable alternative to other methods for estimating microbial surface contamination.

32.2.4 Air Samples

Air sampling in food microbiology received much less attention compared with other sample techniques already discussed. Because muscle foods are processed in factories and meat plants, the air quality in these environments are of great importance to the microbial safety of the meat products. Owing to recent concerns of environmental air pollution, in-door air quality, public health, and the threat of bio-terrorism, there is a renewed interest in rapid techniques to monitor microbes and their toxins in the air. The most common way to estimate air quality is the use of "air plates" where the lid of an agar plate is removed and the agar surface exposed to environmental air for a determined time such as 10 min, 30 min, or a couple of hours. The plate is then covered and incubated and later colonies are counted. If the colony numbers exceed a certain value, for example, 15 per plate, the air quality may be considered as unacceptable. However this simple method is "passive" and the information is not too quantitative. A much better way is to "actively" pass a known volume of air, through an instrument used to measure biological particles, over an agar surface (impaction) to obtain viable cell numbers, after incubation of the agar; or trap microorganisms in a liquid sample (impingement) and then analyze the liquid for various viable cells. There are a variety of commercially available air samplers. Some of them are quite sophisticated such as the Anderson air sampler, which can separate particle sizes from the environment in six stages from large particles (more than 5 μm in diameter) to small particles (0.2 μm). The SAS sampler (PBI, Milan, Italy) has been used for many years with good results. With this instrument, a Rodac plate or an ordinary plate with a suitable agar is clipped in place. A cover with precision pattern of holes (to direct air flow precisely) is then screwed on. After activating the instrument, a known volume of air is sucked through the holes and the particles hit and lodged onto the surface of the agar. After operation (e.g., 60 L of air in 20 s), the air sampler cover is removed, the lid of the agar plate replaced, and the plate incubated. The number of colonies developed on the agar can be converted into colony forming units (CFU) per cubic meter. A similar system named MAS 100 air sampler is marketed by EM Science, Darmstadt, Germany. Applied food microbiologists are constantly searching for better sample preparation methods to improve recovery of microbes from foods and the environment. This section only dealt with improvements related to solid and liquid foods, surfaces of food and food contact areas, and air samples. A great variety of physical, chemical, physicochemical, and biological sampling methods used in clinical sampling, industrial sampling, meat samples, and environmental sampling can also be explored by food microbiologists to make sampling of microorganisms in foods more precise and accurate.

32.3 Total Viable Cell Counts for Muscle Foods

One of the most important information concerning muscle food quality, general food quality, food spoilage, food safety, and potential implication of foodborne pathogens is the total viable cell count of food, water, food contact surfaces, and air of the food plants. Table 32.1a is a summary of the "Fung scale" concerning spoilage potential of total viable cell counts in solid, liquid, and surface samples [18]. Table 32.1b provides Fung scales for air samples in food plants [19] and food contact

Table 32.1a Fung Scale for Liquid, Solid, and Surface Samples

Total Counts for Spoilage Considerations	Ranges (CFU/mL, g, cm²)
Low count	10^{0-2}
Intermediate count	10^{3-4}
High count	10^{5-6}
Index of spoilage	10^{7}
Odor development	10^{8}
Slime development	10^{9}
Unacceptable, too high	10^{10}
No pathogens allowed in cooked ready to eat foods	

Source: Fung et al. (1980a).

Table 32.1b Fung Scale for Air Samples and Food Contact Surfaces

Total Counts for Air Samples	Ranges (CFU/m³)	A. Total Counts for Food Surfaces (Knives, Dishes, Chopping Blocks, etc.)	Ranges (CFU/cm²)
Acceptable	0–100	Acceptable	0–10
Intermediate	100–300	Intermediate	10–100
Too high, needs corrective action	>300	Not acceptable	>100

Notes: For air samples, see Al-Dagal, M.M., and Fung, D.Y.C., *J. Environ. Health*, 56, 7, 1993; for food contact surfaces, see Fung, D.Y.C. et al., *J. Rapid Methods Autom. Microbiol.*, 8, 171, 1995.

surfaces [20]. These scales were developed after more than 30 years of research and practical experiences of this author in muscle foods, food microbiology, and general microbiology. These scales are for general microbial counts related to contamination level and spoilage potential. Foodborne pathogens are not allowed in cooked food, ready-to-eat food, fermented foods and drinks, canned foods, etc. For raw muscle foods sometimes the presence of pathogens may be unavoidable, such as *Salmonella* in raw poultry meat; thus, all precautions must be taken to prepare food properly before consumption and avoid cross contamination with cooked foods.

The conventional "standard plate count" method has been in use for the past 100 years in applied microbiology. The method involves preparing the sample, diluting the sample, plating the sample with a general nonselective agar, incubating the plates at 35°C (or other temperature as needed), and counting the colonies after 48 h (or other time frames). There is a great variety of combinations of volumes to be plated, the use of nonselective and selective agars, incubation times, incubation temperatures, incubation gaseous environments, etc. The conventional "standard plate count" method, although simple, is time-consuming both in terms of operation and data collection. Also this method utilizes a large number of test tubes, pipettes, dilution bottles, dilution buffer, sterile plates, incubation space, and the related disposable and cleanup of reusable materials, and resterilizing them for further use.

Several methods have been developed, tested, and used effectively in the past 20 years as alternative methods for viable cell count. Most of these methods were first designed to perform viable cell counts and relate the counts to "standard plate counts." Later coliform count, fecal coliform count, yeast and mold counts were introduced into these systems. Further developments in these systems included differential counts, pathogen counts, and even pathogen detection after further manipulations. Many of these methods have been extensively tested in many laboratories throughout the world and have gone through AOAC International collaborative study approvals. The aim of these methods was to provide reliable viable cell counts of food and water in more convenient, rapid, simple, and cost-effective alternative formats compared to the cumbersome "standard plate count" method.

The spiral plating method is an automated system to obtain viable cell count (Spiral Biotech, Bethesda, MD). By use of a stylus, this instrument can spread a liquid sample on the surface of a prepoured agar plate (selective or nonselective) in a spiral shape (the Archimedes spiral) with a concentration gradient starting from the center and decreasing as the spiral progresses outward on the rotating plate. The volume of the liquid deposited at any segment of the agar plate is known. After the liquid containing microorganisms is spread, the agar plate is incubated overnight at an appropriate temperature for the colonies to develop. The colonies appearing along the spiral pathway can be counted either manually or electronically. The time for plating a sample is only several seconds compared to minutes used in the conventional method. Also using a laser counter an analyst can obtain an accurate count in a few seconds as compared with a few minutes, in the tiring procedure, of counting colonies by the naked eye. The system has been used extensively in the past 20 years with satisfactory microbiological results from meat, poultry, seafood, vegetable, fruits, diary products, spices, etc. Manninen et al. [21] evaluated the spiral plating system against the conventional pour-plate method using both manual and laser count and found that the counts were essentially the same for bacteria and yeast. Newer versions of the spiral plater are introduced as "Autoplater" (Spiral Biotech, Bethesda, MD) and Whitley Automatic Spiral Plater (Microbiology International, Rockville, MD). With these automatic instruments an analyst needs only to present the liquid sample and the instrument completely and automatically processes the sample, including resterilizing the unit for the next sample.

The ISOGRID system (Neogen, Lansing, MI) consists of a square filter with hydrophobic grids printed on the filter to form 1600 squares for each filter. A food sample is first weighted, homogenized, diluted, and enzyme treated and then passed through the filter assisted by vacuum. Microbes are trapped on the filter and into the squares. The filter is then placed on prepoured non-selective or selective agar and then incubated for a specific time and temperature. Because a growing microbial colony cannot migrate over the hydrophobic material, all colonies are confounded into a square shape. The analyst can then count the squares as individual colonies. Since there is a chance that more than one bacterium is trapped in one square, the system has a most probable number (MPN) conversion table to provide statistically accurate viable cell counts. Automatic instruments are also available to count these square colonies in seconds. Again this method has been used to test a great variety of foods in the past 20 years.

Petrifilm (3M Co., St. Paul, MN) is an ingenious system with appropriate rehydratable nutrients embedded in a series of films in the unit. The unit is little larger than the size of a credit card. To obtain viable cell count the protective top layer is lifted and 1 mL of liquid sample is introduced in the center of the unit and then the cover is replaced. A plastic template is placed on the cover to make a round mold. The rehydrated medium will support the growth of microorganisms after suitable incubation time and temperature. The colonies are directly counted in the unit. This system has a shelf life of over 1 year in cold storage. The advantages of this system are (1) simple to use, (2) smaller in size, (3) has long shelf life, (4) no need to prepare agar, and (5) easy to read results. Recently the company also introduced a petrifilm counter so that an analyst only needs to place the petrifilm with colonies into the unit and the unit will automatically count and record the viable cell count in the computer. The manual form of the petrifilm has been used for many food systems and is gaining international acceptance as an alterative method for viable cell count.

Redigel system (3M Co., St. Paul, MN) consists of tubes of sterile nutrient with a pectin gel in the tube but no conventional agar. This liquid system is ready for use and no heat is needed to "melt" the system since there is no agar in the liquid. After an analyst mixes 1 mL of liquid sample with the desirable volume of liquid in the tube, the resultant contents are poured into a special petri dish coated with calcium. The pectin and calcium will react and form a gel, which will solidify in about 30 min. The plate is then incubated at the proper time and temperature and the colonies will be counted the same way as the conventional standard plate count method. The new name of the Redigel system is "Micrology" system and the company is located in Goshen, IN.

The four methods described in the preceding text have been in use for almost 20 years. Chain and Fung [22] made a comprehensive evaluation of all four methods against the conventional standard plate count method on 7 different foods, 20 samples each, and found that the alternative systems and the conventional method were highly comparable at an agreement of $r = .95$. In the same study they also found that the alternative systems cost less than the conventional system for making viable cell counts.

A latest alternative method, the SimPlate system (BioControl, Bellevue, WA) has 84 wells imprinted in a round plastic plate. After the lid is removed, a diluted food sample (1 mL) is dispensed onto the center-landing pad and 10 mL of rehydrated nutrient liquid provided by the manufacturer is poured onto the landing pad. The mixture (food and nutrient liquid) is distributed evenly into the wells by swirling the SimPlate in a gentle, circular motion. Excessive liquid is absorbed by a pad housed in the unit. After 24 h of incubation at 35°C, the plate is placed under UV light. Positive fluorescent wells are counted and the number is converted according to the MPN table to determine the number of bacteria present in the SimPlate. The method is simple to use with minimum amount of preparation. A 198-well unit is also available for samples with high counts. Using different media, the unit can also make counts of total coliforms and *E. coli* counts, as well as yeast and mold counts.

The 3- and 5-tube MPN systems have been in use for more than 100 years in meat, food, and water microbiology. These methods are used widely in Public Health laboratories around the world. However these methods are very time-consuming, are labor intensive, and use large quantities of glasswares and expensive culture media. The author miniaturized the 3-tube MPN system in the microtiter plate in 1969 [23] in an effort to speed up operation of this tedious method and reduce the number of tubes and media to be used in the conventional MPN method. Following the concept of miniaturization of MPN system, a truly innovative new MPN system called TEMPO is being marketed by bioMerieux (Hazelwood, MO), which utilizes a 16-tube, 3 dilution series miniaturized MPN hands-off system. This system has three components. First, the heart of the system is a plastic card (4 in. × 3 in. × 1/8 in.) that has three series of 16 wells each. The first 16 wells are very small, the second series of 16 wells is 10 times larger than the first series, and the final series of 16 wells is 10 times again larger than the second series. The entire 48 wells will house exactly 1 mL of the sample to be tested in three, 16 MPN well series. To start the test an analyst will make a 1:10 dilution of a solid or liquid food sample by the Stomacher, Pulsifier, or the Smasher. One milliliter of the 1:10 diluted sample is added into a nutrient medium tube (commercially available from bioMerieux) containing a special nutrient for special application such as total count, coliform count, yeast and mold count, etc. After application of the 1:10 diluted sample, 3 mL of sterile water is added into the test tube to facilitate interaction of the sample with the specific medium to make a 1:40 dilution solution. A small plastic delivery tube is then placed in the diluted sample. The delivery tube is attached to the sterile 16 wells, 3 dilution MPN plastic card. The unit is then placed in a holder that can house five such units at one time. The holder with loaded plastic cards is placed into a "pressurization chamber" (supplied by bioMerieux). When the chamber is closed and activated, exactly 1 mL of the diluted sample is pressurized into each of the 48 wells (16 wells in three series MPN format) plastic card. After the application the delivery tube is automatically cut from the plastic card.

The loaded plastic cards are then placed in another holder and all the cards are incubated in an incubator at a specific temperature (depending on the test to be performed such as total count, coliform count, etc.) and incubated overnight. After incubation the cards are placed in a dedicated reading and recording instrument (supplied by bioMerieux), which senses the presence or absence of fluorescence in each of the 48 MPN wells and automatically calculates the MPN of the food sample with correct dilution factors. The entire procedure is highly automated with no need for a technician to transfer samples into large numbers of tubes, and after incubation read the reaction of each tube to determine MPN of the original sample. The time of operation and materials saved in TEMPO method, compared with the conventional MPN method, are truly impressive. Preliminary results by several laboratories in the United States indicated that the TEMPO MPN data are highly correlated with the conventional MPN methods. The TEMPO system should receive excellent reception by meat, food, and water microbiologists in years to come.

The above methods are designed to count aerobic microorganisms in muscle foods. To count anaerobic microorganisms, one has to introduce the sample into the melted agar and after solidification the plates need to be incubated in an enclosed anaerobic jar. In the anaerobic jar, oxygen is removed by the hydrogen generated by the "gas pack" in the jar to create an anaerobic environment. After incubation, the colonies can be counted and reported as anaerobic count of the food. The method is simple but requires expensive anaerobic jars and disposable "gas packs." It also takes about 1 h for the interior of the jar to become anaerobic. Some strict anaerobic microorganisms may die during this 1 h period of reduction of oxygen. The author developed a simple anaerobic double tube system (Fung Double Tube system or FDT) that is easy to use and provides instant anaerobic condition for the cultivation of anaerobes from foods and water [24]. In FDT system the desired agar (e.g., 23 mL of Shahidi Ferguson Perfringens [SFP] agar for enumeration of *Clostridium perfringens*) is first autoclaved in a large test tube (OD 25 × 150 mm). When needed, the agar is melted and tempered at 48°C (Figure 32.1, left set). A liquid or solid food sample (1 mL or 1 g) is added into the melted agar. A smaller sterile test tube (OD 16 × 150 mm) is inserted into the large tube with the food or water sample and the melted agar (Figure 32.1, right set). By so doing, a thin film is formed between the two test tubes (Figure 32.1, right set). The unit is tightly closed by a screw cap (Figure 32.1, right set). The entire unit is placed in an incubator (37°C, 42°C, or other temperatures) for the colonies to develop. No anaerobic jar is needed for this simple anaerobic system. After incubation, the colonies developing in the agar film can be counted and provide an anaerobic count of the food being tested (Figure 32.2). Tubes A, B, C, and D has decreasing number of anaerobes.

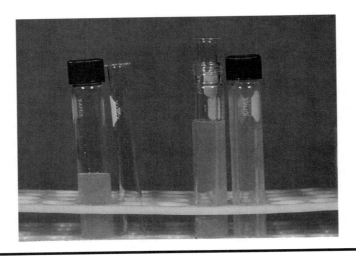

Figure 32.1   Fung double tube system (FDT).

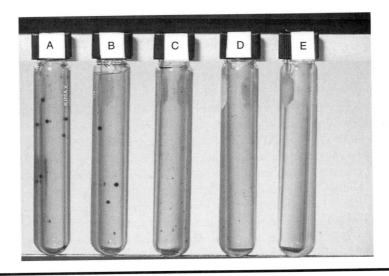

Figure 32.2   Anaerobic colony development in FDT.

Tube E is a negative control. This simple method has been used extensively for ground beef, dairy products, and water microbiology for about 20 years [25,26]. This FDT system been used extensively for enumeration of *C. perfringens* in ground beef and *Clostridium sporogenes* in meat products with great efficiency. *C. perfringens* will develop tiny colonies in 4 h of incubation in SFP agar 42°C. In 6 h distinctly countable colonies are formed. This simple system provides rapid and convenient enumeration of anaerobic bacteria from muscle foods as well as other food samples and water. To confirm the identity of the colonies one only needs to remove the inner tube and by using a sterile needle pick the desired colonies from the agar and purify the colonies for further identification. The FDT system has recently been evaluated as a 6 h test for enumeration of *C. perfringens* in recreational water of the State of Hawaii with great efficiency [27].

A few "real-time" viable cell count methods have been developed and tested in recent years. Many of these methods are applicable to muscle foods. These methods rely on using "vital" stains to stain "live" cells or ATP detection of live cells. All these methods need careful sample preparation, filtration, careful selection of dyes and reagents, and instrumentation. Usually the entire system is quite costly. However, they can provide one-shift results (<8 h) and can handle large number of samples.

The direct epifluorescent filter techniques (DEFT) method has been tested for many years and is in use in the United Kingdom for raw milk quality assurance programs and in Nordic countries for quality assurance in ground beef. In this method, the microorganisms are first trapped on a filter and then the filter is stained with acridine orange dye. The slide is observed under UV microscopy. "Live" cells usually fluoresce orange-red, orange-yellow, or orange-brown whereas "dead" cells fluoresce green. The slide can be read by the eye or by a semiautomated counting system marketed by Bio-Foss. A viable cell count can be made in less than an hour.

The Chemunex Scan RDI system (Monmouth Junction, NJ) involves filtering cells on a membrane and staining them with vital dyes (Fluorassure), and after about 90 min incubation (for bacteria), the membrane with stained cell is read in a scanning chamber that can scan and count fluorescing viable cells. This system has been used to test disinfecting solutions against organisms such as *Pseudomonas aeruginosa*, *Serratia marcescens*, *Escherichia coli*, and *Staphylococcus aureus* with satisfactory results.

The MicroStar System developed by Millipore Corporation utilizes ATP bioluminescence technology by trapping bacteria in a specialized membrane (Milliflex). Individual live cells are trapped in the matrix of the filter and grown into microcolonies. The filter is then sprayed with permeablizing reagent in a reaction chamber to release ATP. The bioluminescence reagent is then sprayed onto the filter. Live cells will give off light due to the presence of ATP and the light is measured by a CCD camera and fluorescent particles (live cells) are counted.

These are new developments in staining technology, ATP technology, and instrumentation for viable cell counts. The application of these methods for the food industry is still in the evaluation stage. The future looks promising.

32.4 Identification of Microbes by Miniaturized Methods

Identification of normal flora, spoilage organisms, foodborne pathogens, starter cultures, etc. in food microbiology and muscle foods is an important part of microbiological manipulations. Conventional methods, dating back to more than 100 years ago, utilize large volumes of medium (10 mL or more) to test for a particular characteristic of a bacterium (e.g., lactose broth for lactose fermentation by *E. coli*). Inoculating a test culture into these individual tubes one at a time is also very cumbersome. Through the years many microbiologists have devised vessels and smaller tubes to reduce the volumes used for these tests [28]. This author has systematically developed many miniaturized methods to reduce the volume of reagents and media (from 5 to 10 mL to about 0.2 mL) for microbiological testing in a convenient microtiter plate which has 96 wells arranged in an 8 × 12 format. The basic components of the miniaturized system are the commercially sterilized microtiter plates for housing the test cultures, a multiple inoculation device, and containers to house solid media (large Petri dishes) and liquid media (in another series of microtiter plates with 0.2 mL of liquid per well). The procedure involves placing liquid cultures (pure cultures) to be studied into sterile wells of a microtiter plate (ca. 0.2 mL for each well) to form a master plate. Each microtiter plate can hold up to 96 different cultures, 48 duplicate cultures, or various combinations as desired. The cultures are then transferred by a sterile multipoint inoculator (96 pins protruding from a template) to solid or liquid media. The inoculator is sterilized by alcohol flaming. Each transfer represents 96 separate inoculations in the conventional method. After incubation at an appropriate temperature, the growth of cultures on solid media or liquid media can be observed and recorded, and the data can be analyzed. These methods are ideal for studying large numbers of isolates or for research involving challenging large numbers of microbes against a host of test compounds. Through the years using the miniaturized systems this author has characterized thousands of bacterial cultures isolated from meat and other foods, studied the effect of organic dyes against bacteria and yeasts, and performed challenge studies of various compounds against microbes with excellent results. Many useful microbiological media were discovered through this line of research. For example, an aniline blue *Candida albicans* medium was developed and marketed by DIFCO under the name of Candida isolation agar. The sensitivity and specificity were 98.0 and 99.5%, respectively, with a predictive value of 99.1% [29].

Other scientists also have miniaturized many systems and developed them into diagnostic kits around late 1960s to 1970s. Currently, API systems, Enterotube, Minitek, Crystal ID system, MicroID, RapID systems, Biolog, and Vitek systems are available. Most of these systems were first developed for identification of enterics (*Salmonella, Shigella, Proteus, Enterobacter*, etc.). Later, many of the companies expanded the capacity to identify nonfermentors, anaerobes, gram-positive organisms, and even yeast and molds. Most of the early comparative analyses centered around evaluation of these kits for clinical specimens. Comparative analysis of diagnostic kits and selection criteria for miniaturized systems were made by Cox et al. [30] and Fung et al. [31]. They concluded that miniaturized systems are accurate, efficient, labor saving, space saving, and cheaper than the conventional methods. Originally, an analyst needs to read the color reaction of each well in the diagnostic kit and then use a manual identification code to "key" out the organisms. Recently, diagnostic companies have developed automatic readers phasing in with computer to provide rapid and accurate identification of the unknown cultures.

The most successful and sophisticated miniaturized automated identification system is the Vitek system (bioMerieux, Hazelwood, MO), which utilizes a plastic card that contains 30 tiny wells in which each has a different reagent. The unknown culture in a liquid form is "pressurized" into the wells in a vacuum chamber and then the cards are placed in an incubator for a period of time ranging from 4 to 12 h. The instrument periodically scans each card and compares the color changes or gas production of each tiny well with the database of known cultures. Vitek can identify a typical *E. coli* culture in 2–4 h. Each Vitek unit can handle 120 cards or more automatically. There are a few thousand Vitek units in use currently in the world. The database is especially good for clinical isolates. Recently bioMerieux introduced Vitek 2 system in which each identification card has 64 chambers to perform 64 different tests. This greatly increases identification of many more pathogens with more confidence.

Biolog system (Hayward, CA) is also a miniaturized system using the microtiter format for growth and reaction information similar to the system developed by this author. Pure cultures are first isolated on agar and then suspended in a liquid to the appropriate density (ca. 6 log cell/mL). The culture is then dispensed into a microtiter plate containing different carbon sources in 95 wells and one nutrient control well. The plate with the pure cultures is then incubated overnight after which the microtiter plate is removed and the color pattern of the wells with carbon utilization is observed and compared with profiles of typical patterns of microbes. This manual evaluation is too tedious to perform and the company developed a software system for the users to enter the data in a computer and then receive the identification. A more convenient mode is to put the microtiter plate in an instrument that can scan the pattern of the positive wells and conduct a match with known cultures to make an identification. This system is easy to operate and with the use of the automatic data analysis the instrument is a useful tool to characterize and identify unknown cultures. This system is very ambitious and tries to identify more than 1400 genera and species of environment, meat, food, and medical isolates from major groups of gram-positive, gram-negative, and other organisms. The database of many cultures is still limited and it needs further development to identify cultures from food and the environment. Nonetheless this system provides a simple operational format with good identification for typical isolates.

There is no question that miniaturization of microbiological methods has saved much materials and operational time and has provided needed efficiency and convenience in diagnostic microbiology for muscle foods and other food samples. The flexible systems developed by this author and others can be used in many research and development laboratories for studying large number of cultures. The commercial systems have played key roles in diagnostic microbiology and have saved many lives due to rapid and accurate characterization of pathogenic bacteria. These miniaturized systems and diagnostic kits will continue to be very useful and important in the medical and food microbiology arenas.

Another area in improving the viable cell count procedure is miniaturization. This is possible in two areas. The first area is to actually miniaturize the conventional viable cell count procedure that involves growing bacteria on agar after dilution of the sample. The second area is miniaturization of the entire 3- or 5-tube MPN procedure used extensively for water testing for almost 100 years in public health laboratories.

In Section 32.3, the discussion on viable cell count involving conventional and alternative methods to manipulate the standard plate count method did not describe miniaturization of the procedures. More than 30 years ago, this author and colleagues [32,33] miniaturized the viable cell count procedure by diluting the samples in the microtiter plate using 0.025 mL size calibrated loops in 1:10 dilution series. One can simultaneously dilute 12 samples to 8 series of 1:10 dilutions in a matter of minutes. After dilution, the samples can be transported by a calibrated pipette and spot plating 0.025 mL on agar. One conventional agar plate can accommodate four to eight spots. After incubation, colonies in the spots can be counted and the number of viable cells in the original sample can be calculated, since all the dilution factors are known. The accepted range of colonies to be counted in one spot is 10–100. The conventional agar plate standard is from 25 to 250 colonies per plate. This procedure actually went through an AOAC International collaborative study with satisfactory results [34]. However, the method has not received much attention and is waiting to be "rediscovered" in the future.

In a similar vein, this author also miniaturized the MPN method in the microtiter plate by diluting a sample in a 3-tube miniaturized series [23]. In one microtiter plate one can dilute four samples each in triplicate (3-tube MPN) to eight series of 1:10 dilution. After incubation, the turbidity of the wells is recorded and a modified 3-tube MPN table can be used to calculate the MPN of the original sample. This procedure recently received renewed interests in the scientific community. Walser [35] in Switzerland reported the use of an automated system for microtiter plate assay to perform classical MPN of drinking water. He used a pipetting robot equipped with sterile pipetting tips for automatic dilution of the samples and after incubation placed the plate in a microtiter plate reader and obtained MPN results with the use of a computer. The system can cope with low or high bacterial load from 0 to 20,000 colonies per milliliter. The system takes out the tedious and personnel influence of routine microbiological works and can be applied to determine MPN of fecal organisms in water as well as other microorganisms of interest in food microbiology. Irwin et al. [36] in the United States also worked on a similar system by using a modified Gauss–Newton algorithm and 96-well microtechnique for calculating MPN using Microsoft EXCEL spreadsheets.

These improvements are possible in 2000s compared with the original work of this author in 1969 because (1) automated instruments are now available in many laboratories to dispense liquid into the microtiter plate. Automated dilution instruments are also available to facilitate rapid and aseptic dilutions of samples; (2) automated readers of microtiter wells are now common place to efficiently read turbidity, color, and fluorescence of the liquid in the wells for calculation of MPN; and (3) elegant mathematic models, computer interpretations and analysis, and printout of data are now available which could not have been envisioned back in 1969.

As mentioned in Section 32.3, the TEMPO system is in fact a miniaturized MPN system using 16 wells, in three dilution series with very little manual labor.

The future is very bright for miniaturized viable cell count procedures in muscle foods, food and water microbiology.

32.5 Detection of Microbes by Immunological Method

Antigen and antibody reaction has been used for decades for detecting and characterizing micro-organisms and their components obtained from muscle foods, foods in general and medical and diagnostic microbiology. Antibodies are produced in animal systems when a foreign particle (antigen) is injected into the system. By collecting the antibodies and purifying the antibodies one can use these antibodies to detect the corresponding antigens. Thus when a *Salmonella* or a component of *Salmonella* is injected into a rabbit the animal will produce antibodies against *Salmonella* or the component (e.g., somatic antigen). By collecting and purifying these antibodies, one can use these antisera to react with a culture of suspected *Salmonella*. When a positive reaction commences, agglutination of antigens (*Salmonella*) and antibodies (antibodies against *Salmonella*) will occur and can be observed on a slide by a trained technician. This is the basis for serotyping bacteria such as *Salmonella*, *E. coli* O157:H7, *Listeria monocytogenes*, etc. These antibodies can be polyclonal (a mixture of several antibodies in the antisera that can react with different sites of the antigens) or monoclonal (there is only one pure antibody in the antiserum that will react with only one epitope of the antigens). Both polyclonal and monoclonal antibodies have been used extensively in applied food microbiology. There are many ways to perform antigen–antibody reactions but the most popular format in recent years is the "Sandwiched" enzyme linked immunosorbent assay or popularly known as the ELISA test.

Briefly, antibodies (e.g., anti-*Salmonella* antibody) are fixed on a solid support (e.g., wells of a microtiter plate). A solution containing a suspect target antigen (e.g., *Salmonella*) is introduced into the microtiter well. If the solution has *Salmonella*, the antibodies will capture the *Salmonella*. After washing away food debris and excess materials another anti-*Salmonella* antibody complex is added into the solution. The second anti-*Salmonella* antibody will react with another part of the trapped *Salmonella*. This second antibody is linked with an enzyme such as horseradish peroxidase. After another washing to remove debris, a chromagen complex such as tetramethyl-benzidine and hydrogen peroxide is added. The enzyme will react with the chromagen and will produce a color compound that will indicate that the first antibody has captured *Salmonella*. If all the reaction procedures are done properly and the liquid in a microtiter well exhibits a color reaction, then the sample is considered positive for *Salmonella*.

This procedure is simple to operate and has been used for decades with excellent results. It should be mentioned that these ELISA tests need about 1 million cells to be reactive and therefore before performing the ELISA tests the food sample has to go through an overnight incubation so that the target organism reaches a detectable level. The total time to detect a pathogen by these systems should include the enrichment time of the target pathogens.

Many diagnostic companies (such as BioControl, Organon Teknika, Tecra, etc.) have marketed ELISA test kits for foodborne pathogens and toxins such as *Salmonella*, *E. coli*, staphylococcal enterotoxins, etc. However, the time involved in samples addition, incubation, washing and discarding of liquids, addition of another antibody complex, washing, and finally addition of reagents for color reaction all contribute to inconvenience of the manual operation of the ELISA test. Recently some companies have completely automated the entire ELISA procedure.

The Vidas system (bioMerieux, Hazelwood, MO) is an automated system that can perform the entire ELISA procedure automatically and can complete an assay from 45 min to 2 h depending on the test kit. Vidas utilizes a more sensitive fluorescent immunoassay, named ELFA for reporting the results. All the analyst needs to do to present to the reagent strip a liquid sample of an overnight-enriched sample. The reagent strip contains all the necessary reagents in a ready-to-use format. The instrument will automatically transfer the sample into a plastic tube called the solid phase receptacle (SPR), which contains antibodies to capture the target pathogen or toxin. The SPR will be automatically transferred to a series of wells in succession to perform the ELFA test. After the final reaction, the result can be read and interpretation of positive or negative test will be automatically determined by the instrument. Presently, Vidas can detect *Listeria*, *L. monocytogenes*, *Salmonella*, *E. coli* O157, staphylococcal enterotoxin, and *Campylobacter*. They also market an immuno-concentration kit for *Salmonella* and *E. coli* O157. In 2007, more than 10,000 units of Vidas units were in use.

BioControl (Bellevue, WA) markets an Assurance EIA system that can be adapted to automation for high-volume testing. Assurance EIA is available for *Salmonella*, *Listeria*, *E. coli* O157:H7, and *Campylobacter*. The message of this above discussion is that many ELISA test kits are now highly standardized and can be performed automatically to increase efficiency and reduce human errors.

Another exciting development in immunology is the use of lateral flow technology to perform antigen–antibody tests. In this system, the unit has three reaction regions. The first well contains antibodies to react with target antigens. These antibodies have color particles attached to them. A liquid sample (after overnight enrichment) is added to this well and if the target organism (e.g., *E. coli* O157:H7) is present it will react with the antibodies. The complex will migrate laterally by capillary action to the second region, which contains a second antibody designed to capture the target organism. If the target organism is present, the complex will be captured and a blue line will be formed due to the color particles attached to the first antibody. Excess antibodies will continue to migrate to the third region, which contains another antibody that can react with the first antibody (which has now become an antigen to the third antibody) and form a blue color band. This is a "control" band indicating that the system is functioning properly. The entire procedure takes only about 10 min. This is truly a rapid test!

Reveal system (Neogen, Lansing, MI) and VIP system (BioControl, Bellevue, WA) are the two main companies marketing this type of system for *E. coli* O157, *Salmonella,* and *Listeria*. The newest entry to this field is Eichrom Technologies that markets a similar lateral migration system called Eclipse for the detection of *E. coli* O157:H7. Merck KgaA (Darmstadt, Germany) is also working on a similar lateral migration system for many common foodborne pathogens using a more sensitive gold particle system to report the reactions.

A number of interesting methods utilizing growth of the target pathogen are also available to detect antigen–antibody reactions.

The BioControl 1-2 test (BioControl, Bellevue, WA) is designed to detect motile *Salmonella* from foods. In this system, the muscle food or other food sample is first preenriched for 24 h in a broth and then 0.1 mL is inoculated into one of the chambers in an L-shaped system. The chamber contains selective enrichment liquid medium for *Salmonella*. There is a small hole connecting the liquid chamber with a soft agar chamber through which *Salmonella* can migrate. An opening on the top of the soft agar chamber allows the analyst to deposit a drop of polyvalent anti-H antibodies against flagella of *Salmonella*. The antibodies move downward in the soft agar due to gravity and diffusion. If *Salmonella* is present, it will migrate throughout the soft agar. As the *Salmonella* and the anti-H antibodies meet, they will react and form a visible V-shaped "immunoband." The presence of the immunoband indicates the presumptive positive for *Salmonella* in the food sample. This reaction occurs after overnight incubation of the unit. This system is easy to use and interpret, and has gained popularity because of its simplicity.

Tecra (Roseville, Australia) developed a unique *Salmonella* detection system that combines immuno-capturing and growth of the target pathogen and ELISA test in a simple to use self-contained unit. The food is first preenriched in a liquid medium overnight and an aliquot is added into the first tube of the unit. Into this tube a dipstick coated with *Salmonella* antibodies is introduced and left in place for 20 min in which time the antibodies will capture the *Salmonella*, if present. The dipstick with *Salmonella* attached will then be washed and placed into a tube containing growth medium. The dipstick is left in this tube for 4 h. During this time if *Salmonella* is present, it will start to replicate and the newly produced *Salmonella* will automatically be trapped by the coated antibodies. Thus, after 4 h of replication, the dipstick will be saturated with trapped *Salmonella*. The dipstick will be transferred to another tube containing a second antibody conjugated to enzyme and allowed to react for 20 min. After this second antigen–antibody reaction the dipstick is washed in the fifth tube and then placed into the last tube for color development similar to other ELISA tests. A purple color developed on the dipstick indicates the presence of *Salmonella* in the food. The entire process from incubation of food sample to reading of the test results is about 22 h, making it an attractive system for detection of *Salmonella*. The system can now also detect *Listeria*.

The BioControl 1-2 test and the unique *Salmonella* test are designed for laboratories with a low volume of tests. Thus, both the automatic systems and the hands-on unit systems have their place in different food testing laboratory situations.

A truly innovative development in applied microbiology is the immuno-magnetic separation system. Vicam (Somerville, MA) pioneered this concept by coating antibodies against *Listeria* on metallic particles. Large numbers of these particles (in the millions) are added into a liquid suspected to contain *Listeria* cells. The antibodies on the particles will capture the *Listeria* cells after rotating the mixture for about an hour. After the reaction, the tube is placed next to a powerful magnet that will immobilize all the metallic particles to the side of the glass test tube regardless of whether the particles have or do not have captured the *Listeria* cells. The rest of the liquid will be decanted. By removing the magnet from the tube, the metallic particles can again be suspended in a liquid. At this point, the only cells in the solution will be the captured *Listeria*. By introducing a smaller volume of liquid (e.g., 10% of the original volume), the cells are now concentrated by a factor of 10. Cells from this liquid can be detected by direct plating on selective agar, ELISA tests, PCR reaction, or other microbiological procedures in almost pure culture state. Immuno-magnetic capture can save at least one day in the total protocol of preenrichment and enrichment steps of pathogen detection in food.

Dynal (Oslo, Norway) developed this concept further by use of very homogenized paramagnetic beads that can carry a variety of molecules such as antibodies, antigens, DNA, etc. Dynal has developed beads to capture *E. coli* O157, *Listeria*, *Cryptosporidium*, *Giardia*, etc. Furthermore, the beads can be supplied without any coating materials and scientists can tailor according to their own needs by coating the necessary antibodies or other capturing molecules for detection of target organisms. Currently, many diagnostic systems (ELISA test, PCR, etc.) are combining immunomagnetic capture step to reduce incubation and increase sensitivity of the entire protocol. Fluorescent antibody techniques have been used for decades for the detection of *Salmonella* and other pathogens. Similar to the DEFT test designed for viable cell count, fluorescent antibodies can be used to detect a great variety of target microorganisms. Tortorello and Gendel [37] used this technique to detect *E. coli* O157:H7 in milk and juice.

Recently a powerful system named Pathatrix (Matrix MicroScience, Golden, CO) has been introduced as a circulating magnetic capturing system for pathogens in food microbiology. In this system the entire 250 mL of food sample (25 mL of sample in 225 mL of nutrient broth or selective broth) after brief enrichment time of ca. 4.5 h is circulated for 0.5 h in this unit. Inside the unit there is a magnetized plate with paramagnetic beads containing specific antibodies (e.g., antibodies against *E. coli* O157:H7; Figure 32.3). When the broth passes over the plate, *E. coli* O157:H7 will be captured by the antibodies (Figure 32.4). Those targeted cells not captured during the first circulation will be captured in subsequent circulations. Theoretically all the *E. coli* O157:H7 will be captured at the completion of a 0.5 h capture cycle (Figure 32.5). The beads with captured target organisms will be released into 0.1 mL liquid. From this liquid 0.05 mL of the sample can be plated on selective agar for isolation of target pathogen and the other 0.05 mL can be tested by ELISA test, PCR test, or other subsequent detection methods. This author's laboratory at Kansas State University published the first paper on this system and was able to detect *E. coli*

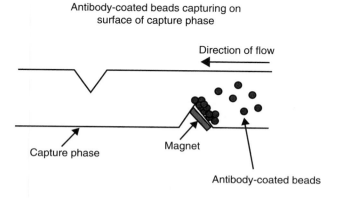

Figure 32.3   **Pathatrix system.**

Capture of target in food

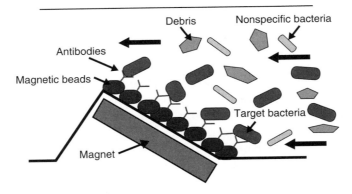

The sample is recirculated repeatedly across the capture phase with the whole 250 mL sample passing over the phase approximately twice every minute.

Figure 32.4 Pathatrix capturing targets in food.

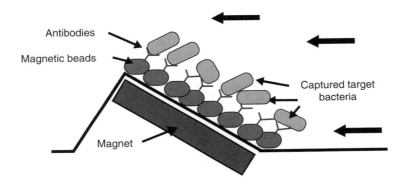

When the recirculation is complete the captured bacteria (bound to the magnetic particles) can be washed extensively.

Figure 32.5 Pathatrix with captured bacteria.

O157:H7 from raw ground beef in 5.25 h. Raw ground beef was placed in a modified EC broth for 4 h at 37°C, then circulated in the Pathatrix instrument for 0.5 h, and later the 0.1 sample eluded from Pathathrix was subjected to ELISA test for 0.25 h [38]. The procedure was able to detect inoculated 0.7–2.1 log CFU of *E. coli* O157:H7/25 g of raw beef. There are more than 30 publications on Pathathrix around the world reporting the superior recovery of pathogens by this system, including tracking of *E. coli* O157:H7 in the notorious spinach leaf outbreak in United States a year or so ago.

Antigen–antibody reactions is a powerful system for rapid detection of all kinds of pathogens and molecules. This section describes some of the useful methods developed for applied food microbiology. Some systems are highly automated and some systems are exceedingly simple to operate. It should be emphasized that many of the immunological tests described in this section provide presumptive positive or presumptive negative screening test results. For negative screening results, the food in question is allowed to be shipped for commerce. For presumptive positive test results, the food will not be allowed for shipment until confirmation of the positive is done by the conventional microbiological methods. This field of immunological testing will continue to evolve as detection methodologies are being explored.

32.6 Instrumentation and Biomass Measurements of Microbes

As the field of rapid methods and automation develops, the boundaries between instrumentation and diagnostic tests will merge. As mentioned in the miniaturization and diagnostic kit section, instrumentation is now playing an important function in diagnostic kit systems and the trend will continue. The following discussions are mainly on instrumentation related to signal measurements of microbial growth in muscle foods and other materials.

Instruments can be used to monitor changes in a population such as ATP levels, specific enzymes, pH, electrical impedance, conductance and capacitance, generation of heat, radioactive carbon dioxide, etc. It is important to note that for the information to be useful, these parameters must be related to viable cell count of the same sample series. In general, the larger the number of viable cells in the sample, the shorter the detection time of these systems. A scattergram is then plotted and used for further comparison of unknown samples. The assumption is that as the number of microorganisms increases in the sample, these physical, biophysical, and biochemical events will also increase accordingly. When a sample has 5 log or 6 log organisms/mL, detection time can be achieved in about 4 h.

All living things utilize ATP. In the presence of a firefly enzyme system (luciferase and luciferin system), oxygen, and magnesium ions, ATP will facilitate the reaction to generate light. The amount of light generated by this reaction is proportional to the amount of ATP in the sample. Thus, the light units can be used to estimate the biomass of cells in a sample. The light emitted by this process can be monitored by a sensitive and automated fluorimeter. Some of the instruments can detect as little as 100–1000 fg (1 fg is −15 log in gram). The amount of ATP in 1 CFU has been reported as 0.47 fg with a range of 0.22–1.03 fg. Using this principle, many researchers have used ATP to estimate microbial cells in solid and liquid foods.

Earlier, scientists attempted to use ATP to estimate the total number of viable cell counts in foods. The results are inconsistent because (1) different microorganisms have different amount of ATP per cell. For example, a yeast cell can have 100 times more ATP than a bacterial cell, (2) even for the same organism the amount of ATP per cell is different at different growth stages, and (3) background ATP from other biomass such as blood and biological fluids in the foods interferes with the target bacterial ATP. Only after much research and development scientists can separate nonmicrobial ATP from microbial ATP and obtain reasonable accuracy in relating ATP to viable cell counts in foods. Since obtaining an ATP reading takes only a few minutes, the potential of exploring this method further exists. To date, not much routine work has been applied using ATP to estimate viable cell counts in food microbiology laboratories.

From another viewpoint, the presence of ATP in certain food such as wine is undesirable regardless of the source. Thus, monitoring ATP can be a useful tool for quality assurance in the winery.

There is a paradigm shift in the field of ATP detection in recent years. Instead of detecting ATP of microorganisms, the systems are now designed to detect ATP from any source for hygiene monitoring. The idea is that a dirty food processing environment will have a high ATP level and a properly cleansed environment will have low ATP level regardless of what contributed to the ATP in these environments. Once this concept is accepted by the food industry, there will be an explosion of ATP systems on the market. In all of these systems, the key is to be able to obtain an ATP reading in the form of relative light units (RLUs) and relate these units to cleanliness of the food processing surfaces. Most systems design an acceptable RLU, an unacceptable RLU, and a marginal RLU for different surfaces in food plants. Because there is no standard in what constitutes an absolute acceptable ATP level on any given environment, these RLUs are quite arbitrary. In general, a dirty environment will have high RLUs and after proper cleaning the RLUs will decrease. Besides the sensitivity of the instruments, for an analyst to select a particular system the following attributes are considered: simplicity of operation, compactness of the unit, computer adaptability, cost of the units, support from the company, and documentation of usefulness of the system.

Currently the following ATP instruments are available: Lumac (Landgraaf, the Netherlands), BioTrace (Plainsboro, NJ), Lightning (BioControl, Bellevue, WA), Hy-Lite (EM Science, Darmstadt, Germany), Charm 4000 (Charm Sciences, Malden, MA), Celsis system SURE (Cambridge, U.K.), Zylux (Maryville, TN), Profile 1 (New Horizon, Columbia, MD), and others.

As microorganisms grow and metabolize nutrients, large molecules change to smaller molecules in a liquid system and cause a change in electrical conductivity and resistance in the liquid as well as at the interphase of electrodes. These changes can be expressed as impedance, conductance, and capacitance changes. When a population of cells reach to about 5 log/mL it will cause a change of these parameters. Thus, when a food has a large initial population the time to make this change will be shorter than a food that has a smaller initial population. The time for the curve change from the baseline and accelerates upward is the detection time of the test sample, which is inversely proportional to the initial concentration of microorganisms in the food. To use these methods a series of standard curves must be constructed by making viable cell counts of a series of food with different initial concentration of cells and then measuring the resultant detection time. A scattergram can then be plotted. Thereafter, in the same food system, the number of the initial population of the food can be estimated by the detection time on the scattergram.

The Bactometer (bioMerieux, Hazelwood, MO) has been in use for many years to measure impedance changes in foods, water, cosmetics, etc. by microorganisms. Samples are placed in the wells of a 16-well module which is then plugged into the incubator to start the monitoring sequence. As the cells reach the critical number (5–6 log/mL), the change in impedance increases sharply and the monitor screen shows a slope similar to the log phase of a growth curve. The detection time can then be obtained to determine the initial population of the sample. If one sets a cutoff point of 6 log organisms/g of food for acceptance or rejection of the product and the detection time is 4 h ± 15, min then one can use the detection time as a criterion for quality assurance of the product. Food that exhibits no change of impedance curve more than 4 h 15 min in the instrument is acceptable whereas food that exhibits a change of impedance curve before 3 h 45 min will not be acceptable. For convenience the instrument is designed such that the sample bar for a food on the screen will flash red for unacceptable sample, green for acceptable sample, and yellow for marginally acceptable sample. A similar system called rapid automated bacterial impedance technique (RABIT), marketed by Bioscience International (Bethesda, MD) is available for monitoring microbial activities in food and beverages. Instead of a 16-well module used in the Bactometer, individual tubes containing electrodes are used to house the food samples.

The Malthus system (Crawley, UK) uses conductance changes of the fluid to indicate microbial growth. It generates conductance curves similar to impedance curves used in the Bactometer. It uses individual tubes for food samples. Water heated to desirable temperature (e.g., 35°C) is used as the temperature control instead of heated air in the previous two systems. All these systems have been evaluated by various scientists in the past 10–15 years with satisfactory results. Depending on the type of food being analyzed all have their advantages and disadvantages. These systems can also be used to monitor target organisms such as coliform, yeast, and mold by specially designed culture media. In fact, the Malthus system has a *Salmonella* detection protocol that has AOAC International approval.

BacT/Alert Microbial Detection System (Organon Tecknika, Durham, NC) utilizes colorimetric detection of carbon dioxide production by microorganisms in a liquid system using sophisticated computer algorithms and instrumentation. Food samples are diluted and placed in special bottles with appropriate nutrients for growth of microorganisms and production of carbon dioxide. At the bottom of the bottle there is a sensor, which is responsive to the amount of carbon dioxide in the liquid. When a critical amount of the gas is produced, the sensor changes from dark green to yellow and this change is detected by reflectance colorimetry automatically. The units can accommodate 120 or 240 culture bottles. Detection time of a typical culture of *E. coli* is about 6–8 h.

BioSys (Neogen, Lansing, MI) utilizes color changes of media during the growth of cultures to detect and estimate organisms in foods and liquid systems. The uniqueness of the system is that the color compounds developed during microbial growth are diffused into an agar column in the unit and the changes are measured automatically without the interference of food particles. Depending on the initial microbial load in the food, same shift microbial information can be obtained. The system is easy to use and can accommodate 32 samples for 1 incubation temperature or 128 samples for 4 independent incubation temperatures in different models. The system is designed for bioburden testing and Hazard Analysis Critical Control Points (HACCP) control and can test for indirect total viable cell, coliform, *E. coli*, yeast, mold, lactic acid bacteria counts, swab samples, and environmental samples.

Basically, any type of instrument that can continuously and automatically monitor turbidity and color changes of a liquid in the presence of microbial growth can be used for rapid detection of the presence of microorganisms. There will definitely be more systems of this nature on the market in years to come.

32.7 Genetic Testings for Microbes

Up to this point, all the rapid tests discussed for detection and characterization of microorganisms were based on phenotypic expressions of genotypic characteristics of microorganisms. Phenotypic expression of cells is subject to growth conditions such as temperature, pH, nutrient availability, oxidation–reduction potentials, environmental and chemical stresses, toxins, water activities, etc. Even immunological tests depend on phenotypic expression of cells to produce the target antigens to be detected by the available antibodies or vice verse. The conventional "gold standards" of diagnostic microbiology rely on phenotypic expression of cells and are inherently subject to variation.

Genotypic characteristics of a cell is far more stable. Natural mutation rate of a bacterial culture is about 1 in 100 million cells. Thus, there is a push in recent years to make genetic test results as the confirmative and definitive identification step in diagnostic microbiology. The debate is still continuing and the final decision has not been reached by governmental and regulatory bodies for microbiological testing. Genetic-based diagnostic and identification systems are discussed in this section. Hybridization of the DNA sequence of an unknown bacteria by a known DNA probe is the first stage of genetic testings. Genetrak system (Framingham, MA) is a sensitive method and convenient system to detect pathogens such as *Salmonella, Listeria, Campylobacter,* and *E. coli* O157 in foods. Earlier, the system utilized radioactive compounds bound to DNA probes to detect DNA of unknown cultures. The drawbacks of the first generation of this type of probes are (1) most food laboratories are not eager to work with radioactive materials in routine analysis and (2) there are limited copies of DNA in a cell. The second generation of probes uses enzymatic reactions to detect the presence of the pathogens and they use RNA as the target molecule. In a cell, there is only one complete copy of DNA; however, there may be 1,000–10,000 copies of ribosomal RNA. Thus, the new generation of probes is designed to probe target RNA using color reactions. After enrichment of cells (e.g., *Salmonella*) in a food sample for about 18 h, the cells (target cells as well as other microbes) are lysed by a detergent to release cellular materials (DNA, RNA, and other molecules) into the enrichment solution. Two RNA probes (designed to react with one piece of target *Salmonella* RNA) are added into the solution. The capture probe with a long tail of a nucleotide (e.g., adenine, AAAAA) is designed to capture the RNA onto a dipstick with a long tail of thymine (TTTTT). The reporter probe with an enzyme attached will react with another part of the RNA fragment. If *Salmonella* RNA molecules are present, the capture probes will attach to one end of the RNA and the report probes will attach to the other end. A dipstick coated with many copies of a chain of complementary nucleotide (e.g., Thymine, TTTTT) will be placed into the solution. Because adenine (A) will hybridize with thymine (T), the chain (TTTTT) on the dipstick will react with the AAAAA and thus capture the target RNA complex onto the stick. After washing away debris and other molecules in the liquid, a chromagen is added. If the target RNA is captured, then the enzyme present in the second probe will react with the chromagen and will produce a color reaction indicating the presence of the pathogen in the food. In this case, the food is positive for *Salmonella.* The Genetrak has been evaluated and tested for many years and has AOAC International approval for many food types. More recently Genetrak has adapted a microtiter format for more efficient and automated operation of the system.

PCR is now an accepted method to detect pathogens by amplification of the target DNA, detecting the target PCR products. Basically, a DNA molecule (double helix) of a target pathogen (e.g., *Salmonella*) is first denatured at about 95°C to form single strands, then the temperature is lowered to about 55°C for two primers (small oligonucleotides specific for *Salmonella*) to anneal to specific regions of the single-stranded DNA. The temperature is increased to about 70°C for a special heat stable polymerase, the TAQ enzyme from *Thermus aquaticus,* to add complementary bases (A, T, G, or C) to the single-stranded DNA and complete the extension to form a new double strand of DNA. This is called a thermal cycle. After this cycle, the tube will be heated to 95°C again for the next cycle. After one thermal cycle one copy of DNA will become two couples. After about 21 and 31 cycles, one million and one billion copies of the DNA will be formed, respectively. This entire process can be accomplished in less than an hour in an automatic thermal cycler. Theoretically, if a food contains one copy of *Salmonella* DNA, the PCR method can detect the presence of this pathogen in a very short time. After PCR reactions, one still needs to detect the presence of the PCR products to indicate the presence of the pathogen. The following are brief discussions of four commercial kits for PCR reactions and detection of PCR products.

The BAX® for screening family of PCR assays for foodborne pathogens (Qualicon, Inc., Wilmington, DE) combines DNA amplification and automated homogeneous detection to determine the presence or absence of a specific target. All primers, polymerase, and deoxynucleotides necessary for PCR as well as a positive control and an intercalating dye are incorporated into a single tablet. The system works directly from an overnight enrichment of the target organisms. No DNA extraction is required. Assays are available for *Salmonella* [39], *E. coli* 0157:H7 [40–42], *Listeria* genus, and *L. monocytogenes* [43–45]. The system uses an array of 96 blue LEDs as the excitation source and a photomultiplier tube to detect the emitted fluorescent signal. This integrated system improves the ease-of-use of the assay. In addition to simplifying the detection process, the new method converts the system into a homogeneous PCR test. The homogenous detection process monitors the decrease in fluorescence of a double-stranded DNA (dsDNA) intercalating dye in solution with dsDNA as a function of temperature. Following amplification, melting curves are generated by slowly ramping the temperature of the sample to a denaturing level (95°C). As the dsDNA denatures, the dye becomes unbound from the DNA duplex, and the fluorescent signal decreases. This change in fluorescence can be plotted against temperature to yield a melting curve waveform. This assay thus eliminates the need for gel-based detection and yields data amenable to storage and retrieval in an electronic database. In addition, this method reduces the hands-on time of the assay and reduces the subjectivity of the reported results. Further, melting curve analysis makes possible the ability to detect multiple PCR products in a single tube. The inclusivity and exclusivity of the BAX system assays reach almost 100% meaning that false-positive and false-negative rates are almost zero. Additionally, the automated BAX system can now be used with assays for the detection of *Cryptosporidium parvum* and *Campylobacter jejuni/coli* and for the quantitative and qualitative detection of genetically modified organisms in soy and corn. Also the BAX system can detect yeast and mold after proper enrichment of sample for these higher organisms. The new BAX system is far more convenient than the old system in which a gel electrophoresis step was required to detect PCR products after thermal cycling.

The following two methods have been developed also to by-pass the electrophoresis step to detect PCR products.

TaqMan system of Applied Biosystems (Foster City, CA) also amplifies DNA by PCR protocol. However, during the amplification step a special molecule is annealed to the single-stranded DNA to report the linear amplification. The molecule has the appropriate sequence for the target DNA. It also has two attached particles. One is a fluorescent particle and another is a quencher particle. When the two particles are close to each other, no fluoresce occurs. However, when the TAQ polymerase is adding bases to the linear single strand of DNA, it will break this molecule away from the strand (like the PacMan in computer games). As this occurs, the two particles will separate from each other and fluorescence will occur. By measuring fluoresce in the tube, a successful PCR reaction can be determined. Note that the reaction and reporting of a successful PCR protocol occur in the same tube.

A new system called molecular beacon technology (Stratagene, La Jolla, CA) is developed and can be used for food microbiology in the future [46]. In this technology, all reactions are again in the same tube. A molecular beacon is a tailor-made hairpin-shaped hybridization probe. The probe is used to attach to the target PCR products. On one end of the probe there is attached a fluorophore and on the other end a quencher of the fluorophore. In the absence of the target PCR products the beacon is in a hairpin shape and there is no fluoresce. However, during PCR reactions and the generation of target PCR products, the beacons will attach to the PCR products and cause the hairpin molecule to unfold. As the quencher moves away from the fluorophore, fluorescence will occur, and this can be measured. The measurement can be done, as the PCR reaction is progressing thus allowing real-time detection of target PCR products and thus the presence of the target pathogen in the sample. This system has the same efficiency as the TaqMan system but the difference is that the beacons detect the PCR products themselves, whereas in the TaqMan system, it only reports the occurrence of a linear PCR reaction and not the presence of the PCR product directly. By using molecular beacons containing different fluorophores, one can detect different PCR products in the same reaction tubes, thus can be able to perform "multiplex" tests of several target pathogens or molecules. The use of this technology is new and not well known in food microbiology areas.

Theoretically, PCR system can detect one copy of target pathogen from a food sample (e.g., *Salmonella* DNA). In practice, about 200 cells are needed to be detected by current PCR methods. Thus, even in a PCR protocol the food must be enriched for a period of time, for example, overnight or at least 8 h incubation of food in a suitable enrichment liquid, so that there are enough cells for the PCR process to be reliable. Besides the technical manipulations of the systems that can be complicated for many food microbiology laboratories, two major problems need to be addressed: inhibitors of PCR reactions and the question of live and dead cells. In food, there are many enzymes, proteins, and other compounds that can interfere with the PCR reaction and result in false negatives. These inhibitors must be removed or diluted. Since PCR reaction amplifies target DNA molecules, even DNA from dead cells can be amplified and thus food with dead *Salmonella* can be declared as *Salmonella* positive by PCR results. Thus, food properly cooked but containing DNA of dead cells may be unnecessarily destroyed because of a positive PCR test.

PCR can be a powerful tool for food microbiology once all the problems are solved and analysts are convinced of the applicability in routine analysis of foods.

The aforementioned genetic methods are for detection of target pathogens in foods and other samples. They do not provide identification of the cultures to the species and subspecies level, which is so critical in epidemiological investigations of outbreaks or routine monitoring of occurrence of microorganisms in the environment. The following discussions will center around the developments in genetic characterization of bacterial cultures.

The RiboPrinter Microbial Characterization System (Du Pont Qualicon, Wilmington, DE) characterizes and identifies organisms to genus, species, and subspecies levels automatically. To obtain a RiboPrint of an organism, the following steps are followed:

1. A pure colony of bacteria suspected to be the target organism (e.g., *Salmonella*) from an agar plate is picked by a sterile plastic stick.
2. Cells from the stick are suspended in a buffer solution by mechanical agitation.
3. An aliquot of the cell suspension is loaded into the sample carrier to be placed in the instrument. Each sample carrier has space for eight individual colony picks.
4. The instrument will automatically prepare the DNA for analysis by restriction enzyme and lysis buffer to open the bacteria, release, and cut DNA molecules. The DNA fragments will go through an electrophoresis gel to separate DNA fragments into discrete bands. Finally, the DNA probes, conjugate, and substrate will react with the separated DNA fragments and light emission from the hybridized fragments is then photographed. The data are stored and compared with known patterns of the particular organism.

The entire process takes 8 h for eight samples. However, at 2 h intervals, another eight samples can be loaded for analysis.

Different bacteria will exhibit different patterns (e.g., *Salmonella* versus *E. coli*) and even the same species can exhibit different patterns (e.g., *L. monocytogenes* has 49 distinct patterns). Some examples of numbers of RiboPrint patterns for some important food pathogens are *Salmonella*, 145; *Listeria*, 89; *E. coli*, 134; *Staphylococcus*, 406; and *Vibrio*, 63. Additionally, the database includes 300 *Lactobacillus*, 43 *Lactococcus*, 11 *Leuconostoc*, and 34 *Pediococcus*. The current identification database provides 3267 RiboPrint patterns representing 98 genera and 695 species.

One of the values of this information is that in the case of a foodborne outbreak, scientists not only can identify the etiological agent (e.g., *L. monocytogenes*) but also can pinpoint the source of the responsible subspecies. For example, in the investigation concerning an outbreak of *L. monocytogenes*, cultures were isolated from the slicer of the product and also from the drains of the plant. The question is which source is responsible for the outbreak. By matching RiboPrint patterns of the two sources of *L. monocytogenes* against the foodborne outbreak culture, it was found that the isolate from the slicer matched the outbreak culture, thus determining the true source of the problem. The RiboPrinter system is a very powerful tool for electronic data sharing worldwide. These links can monitor the occurrence of foodborne pathogens and other important organisms as long as different laboratories utilize the same system for obtaining the RiboPrint patterns.

Another important system is the pulsed-field gel electrophoresis patterns of pathogens. In this system, pure cultures of pathogens are isolated and digested with restriction enzymes and the DNA fragments are subjected to a system known as pulsed-field gel electrophoresis that effectively separates DNA fragments on the gel (DNA fingerprinting). For example, in a foodborne outbreak of *E. coli* O157:H7, biochemically identical *E. coli* O157:H7 cultures can exhibit different patterns. By comparing the gel patterns from different sources one can trace the origin of the infection or search for the spread of the disease and thereby control the problem.

To compare data from various laboratories across the country the pulse net system is established under the National Molecular Subtyping Network for Foodborne Disease Surveillance at the Centers for Disease Control and Prevention (CDC). An extensive training program has been established so that all the collaborating laboratories use the same protocol and are electronically linked to share DNA fingerprinting patterns of major pathogens. As soon as a suspect culture is noted as a possible source of an outbreak, all the collaborating laboratories are alerted to search for the occurrence of the same pattern to determine the scope of the problem and share information in real time.

There are many other genetic base methods but they are not directly related to food microbiology and are beyond the scope of this review. It is safe to say that many genetic base methods are slowly but surely finding their ways into food microbiology laboratories and they will provide valuable information for quality assurance, quality control, and food safety programs in the future.

32.8 Biosensors for Microbes

Biosensor is an exciting field in applied microbiology. The basic idea is simple but the actual operation is quite complex and involves much instrumentation. Basically, a biosensor is a molecule or a group of molecules of biological origin attached to a signal recognition material.

When an analyte comes in contact with the biosensor, the interaction will initiate a recognition signal that can be reported in an instrument.

Many types of biosensors have been developed such as enzymes (a great variety of enzymes have been used), antibodies (polyclonal and monoclonal), nucleic acids, cellular materials, etc. Sometime whole cells can also be used as biosensors. Analytes detected include toxin (staphylococcal enterotoxins, tetrodotoxins, saxitoxin, botulinum toxin, etc.), specific pathogens (e.g., *Salmonella*, *Staphylococcus*, *E. coli* O157:H7, etc.), carbohydrates (e.g., fructose, lactose, galactose, etc.), insecticides and herbicides, ATP, antibiotics (e.g., penicillins), and others. The recognition signals used include electrochemical (e.g., potentiometry, voltage changes, conductance and impedance, light addressable, etc.), optical (such as UV, bioluminescence and chemiluminescence, fluorescence, laser scattering, reflection and refraction of light, surface plasmon resonance, polarized light, etc.), and miscellaneous transducers (such as piezoelectric crystals, thermistor, acoustic waves, quartz crystal, etc.).

An example of a simple enzyme biosensor is sensor for glucose. The reaction involves the oxidation of glucose (the analyte) by glucose oxidase (the biosensor) with the end-products being gluconic acid and hydrogen peroxide. The reaction was reported by a Clark oxygen electrode that monitors the decrease in oxygen concentration amperometrically. The range of measurement is from 1 to 30 mM, and response time of 1–1.5 min and the recovery time of 30 s. Lifetime of the unit is several months. Some of the advantages of enzyme biosensors are binding to the subject, highly selective, and rapid acting. Some of the disadvantages are expensiveness, loss of activity when they are immobilized on a transducer, and loss of activities due to deactivation. Other enzymes used include galactosidase, glucoamylase, acetylcholinesterase, invertase, lactate oxidase, etc. Excellent review articles and books on biosensors are presented by Eggins [47], Cunningham [48], Goldschmidt [49], and others. Recently, much attention has been directed to the field of "biochips" and "microchips" developments to detect a great variety of molecules including food-borne pathogens.

Owing to the advancement in miniaturization technology as many as 50,000 individual spots (e.g., DNA microarrays) with each spot containing millions of copies of a specific DNA probe can be immobilized on a specialized microscope slide. Fluorescent-labeled targets can be hybridized to these spots and be detected. An excellent article by Deyholos et al. [50] described the application of microarrays to discover genes associated with a particular biological process such as the response of the plant (*Arabidopsis*) to NaCl stress and detailed analysis of a specific biological pathway such as one-carbon metabolism in maize.

Biochips can also be designed to detect all kinds of foodborne pathogens by imprinting a variety of antibodies or DNA molecules against specific pathogens on the chip for the simultaneous detection of pathogens such as *Salmonella, Listeria, E. coli, S. aureus*, etc. on the same chip. According to Elaine Heron of Applied Biosystems (Foster City, CA) [51], biochips are an exceedingly important technology in life sciences and the market value is estimated to be as high as $5 billion by the middle of this decade. This technology is especially important in the rapidly developing field of proteomics, which requires massive amount of data that generate valuable information.

Certainly, the developments of these biochips and microarray chips are impressive for obtaining a large amount of information for biological sciences. As for foodborne pathogen detection, there are several important issues to consider. These biochips are designed to detect minute quantities of target molecule. The target molecules must be free from contaminants before being applied to the biochips. In food microbiology, the minimum requirement for pathogen detection is 1 viable target cell in 25 g of a food such as ground beef. A biochip will not be able to seek out such a cell from the food matrix without extensive cell amplification (either by growth or PCR) or sample preparation by filtration, separation, absorption, centrifugation, etc. as described in this chapter. Any food particle in the sample will easily clot the channels used in biochips. These preparations will not allow the biochips to provide "real-time" detection of pathogens in foods.

Another concern is viability of the pathogens to be detected by biochips. Monitoring the presence of some target molecule will only provide the presence or absence of the target pathogen and will not provide viability of the pathogen in question. Some form of culture enrichment to ensure growth is still needed to obtain meaningful results. It is conceivable that biomass of microbes can be monitored by biochips but instantaneous detection of specific pathogens such as *Salmonella, Listeria, Campylobacter*, etc. in food matrix during food processing operation is still not possible. The potential of biochip and microarrays for food pathogen detection is great but at this moment much more research is needed to make this technology a reality in applied food microbiology.

32.9 Conclusion

In conclusion, this chapter presented many novel techniques in applied microbiology developed in recent years. These methods can be used to rapidly and efficiently monitor total or selected viable cell counts of muscle foods. Also many modern techniques are presented that can effectively detect, identify, and characterize harmful as well as beneficial microorganisms in muscle foods. The field of rapid methods and automation in microbiology for muscle foods, food in general, environmental, industrial, and medical microbiology is very bright indeed.

References

1. Fung, D. Y. C., M. N. Hajmeer, C. L. Kastner, J. J. Kastner, J. L. Marsden, K. P. Penner, R. K. Phebus, J. S. Smith, and M. A. Vanier. 2001. Meat Safety. In *Meat Science Applications*. R. Rogers, Ed. Marcel Dekker, New York, NY. pp. 171–205.
2. Jay, J. M. 2005. *Modern Food Microbiology*, 7th Ed. Chapman and Hall, New York, NY.
3. Pearson, A. M. and T. R. Dutson. 1994. *Quality Attributes and Their Measurements in Meat, Poultry, and Fish Products*. Blackie Academic and Professional, New York.
4. Downes, F. P. and K. Ito. 2001. *Compendium of Methods for the Microbiological Examination of Foods*, 4th Ed. American Public Health Association, Washington DC.
5. Food and Drug Administration. 2005. *Bacteriological Analytical Manual*, 8th Ed. AOAC International, Gaithersburg, MD.
6. Bourgeois, C. M., J. Y. Leveau, and D. Y. C. Fung (eds). 1995. *Microbiological Control for Foods and Agricultural Products*. VCH, New York.
7. Harrigan, W. F. 1998. *Laboratory Methods in Food Microbiology*. Academic Press, New York.
8. AOAC International. 2005. *Official Methods of Analysis*. AOAC International, Gaithersburg, MD.
9. Fung, D. Y. C. 2007. *Handbook for Rapid Methods and Automation in Microbiology Workshop*. Kansas State University, Manhattan, KS. pp. 750.
10. Miller, J. M. 2005. *Encyclopedia of Rapid Microbiological Methods*. Vols 1, 2, and 3. Parental Drug Association, Bethesda, MD.
11. Fung, D. Y. C. 2002. Rapid methods and automation in microbiology. *Compr. Rev. Food Sci. Food Safety* 1(1):3–22. <IFT.org> Electronic Journal.
12. Manninen, M. T. and D. Y. C. Fung. 1992a. Use of the gravimetric diluter in microbiological work. *J. Food Prot.* 55:59–61.
13. Fung, D. Y. C., A. N. Sharpe, B. C. Hart, and Y. Liu. 1998. The Pulsifier: A new instrument for pre paring food suspensions for microbiological analysis. *J. Rapid Methods Autom. Microbiol.* 6:43–49.
14. Kang, D. H., R. H. Dougherty, and D. Y. C. Fung. 2001. Comparison of Pulsifier and Stomacher to detach microorganisms from lean meat tissues. *J. Rapid Methods Autom. Microbiol.* 9(1):27–32.
15. Lee, J. Y. and D. Y. C. Fung. 1986. Surface sampling technique for bacteriology. *J. Environ. Health* 48:200–205.
16. Fung, D. Y. C., C. Y. Lee, and C. L. Kastner. 1980b. Adhesive tape method of estimation of microbial load on meat surfaces. *J. Food Prot.* 43:295–297.
17. Fung, D. Y. C., L. K. Thompson, B. A. Crozier-Dodson, and C. L. Kastner. 2000. Hands-free "Pop-up" adhesive type method for microbial sampling of meat surfaces. *J. Rapid Method Autom. Microbiol.* 8(3):209–217.
18. Fung, D. Y. C., C. L. Kastner, M. C. Hunt, M. E. Dikeman, and D. Kropf. 1980a. Mesophilic and psychrotrophic population on hot-boned and conventionally processed beef. *J. Food Prot.* 43:547–550.
19. Al-Dagal, M. M. and D. Y. C. Fung. 1993. Aeromicrobiology: An assessment of a new meat research complex. *J. Environ. Health* 56(1):7–14.

20. Fung, D. Y. C., R. Phebus, D. H. Kang, and C. L. Kastner. 1995. Effect of alcohol-flaming on meat cutting knives. *J. Rapid Methods Autom. Microbiol.* 3(27):237–243.
21. Manninen, M. T., D. Y. C. Fung, and R. A. Hart. 1991. Spiral system and laser counter for enumeration of microorganisms. *J. Food Prot.* 11:177–187.
22. Chain, V. S. and D. Y. C. Fung. 1991. Comparison of Redigel, Petrifilm, Spiral Plate System, Isogrid, and aerobic plate count for determining the numbers of aerobic bacteria in selected food. *J. Food Prot.* 54:208–211.
23. Fung, D. Y. C. and A. A. Kraft. 1969. Rapid evaluation of viable cell counts using the microtiter system and MPN technique. *J. Milk Food Technol.* 322:408–409.
24. Fung, D. Y. C. and C. M. Lee. 1981. Double-tube anaerobic bacteria cultivation system. *Food Sci.* 7:209–213.
25. Ali, M. S. and D. Y. C. Fung. 1991. Occurrence of *Clostridium perfringens* in ground beef and turkey evaluated by three methods. *J. Food Prot.* 11:197–203.
26. Schmidt, K. A., R. H. Thakur, G. Jiang, and D. Y. C. Fung. 2000. Application of a double tube system for the enervation of *Clostridium tyrobutyricum*. *J. Rapid Methods Autom. Microbiol.* 8(1):21–30.
27. Fung, D. Y. C., R. Fujioka, K. Vijayavel, D. Sato, and D. Bishop. 2007. Evaluation of Fung double tube test for *Clostridium perfringens* and EasyPhage test for F-specific RNA coliphages as rapid screening tests for fecal contamination in recreational waters of Hawaii. *J. Rapid Methods Autom. Microbiol.* 15(3):217–229.
28. Hartman, P. A. 1968. *Miniaturized Microbiological Methods*. Academic Press, New York.
29. Goldschmidt, M. C., D. Y. C. Fung, R. Grant, J. White, and T. Brown. 1991. New aniline blue dyes medium for rapid identification and isolation of *Candida albicans*. *J. Clin. Microbiol.* 29(6):1095–1099.
30. Cox, N. A., D. Y. C. Fung, M. C. Goldschmidt, and J. S. Bailey. 1984. Selecting a miniaturized system for identification of Enterobacteriaceae. *J. Food Prot.* 47:74–77.
31. Fung, D. Y. C., N. A. Cox, M. C. Goldschmidt, and J. S. Bailey. 1989. Rapid methods and automation in microbiology: A survey of professional microbiologists. *J. Food Prot.* 52:65–68.
32. Fung, D. Y. C. and A. A. Kraft. 1968. Microtiter method for the evaluation of viable cells in bacterial cultures. *Appl. Microbiol.* 16:1036–1039.
33. Fung, D. Y. C. and W. S. LaGrange. 1969. Microtiter method for bacterial evaluation of milk. *J. Milk Food Technol.* 32:144–146.
34. Fung, D. Y. C., R. Donahue, J. P. Jensen, W. W. Ulmann, W. J. Hausler, Jr., and W. S. LaGrange. 1976. A collaborative study of the microtiter count method and standard plate count method on viable cell count of raw milk. *J. Milk Food Technol.* 39:24–26.
35. Walser, P. E. 2000. Using conventional microtiter plate technology for the automation of microbiology testing of drinking water. *J. Rapid Methods Autom. Microbiol.* 8(3):193–208.
36. Irwin, P., S. Tu, W. Damert, and J. Phillips. 2000. A modified Gauss-Newton algorithm and ninety-six well micro-technique for calculating MPN using EXCEL spread sheets. *J. Rapid Methods Autom.* 8(3):171–192.
37. Tortorello, M. and S. M. Gendel. 1993. Fluorescent antibodies applied to direct epifluorescent filter techniques for microscopic enumeration of *Escherichia coli* O157:H7 in milk and juice. *J. Food Prot.* 56:672.
38. Wu, V. C. H., V. Gill, R. Oberst, R. P. Phebus, and D. Y. C. Fung. 2004. Rapid Protocol (5.25 hr) for the detection of *Escherichia coli* O157:H7 in raw ground beef by an immuno-capture system (Pathatrix) in combination with Colortrix and CT-SMAC. *J. Rapid Methods Autom. Microbiol.* 12(1): 57–67.
39. Mrozinski, P. M., R. P. Betts, and S. Coates. 1998. Performance tested methods: Certification process for BAX for screening/*Salmonella*: A case study. *J. AOAC Int.* 81:1147–1154.
40. Johnson, J. L., C. L. Brooke, and S. J. Fritschel. 1998. Comparison of BAX for screening/*E. coli* O157:H7 vs. conventional methods for detection of extremely low levels of *Escherichia coli* O157:H7 in ground beef. *Appl. Environ. Microbiol.* 64:4390–4395.

41. Hochberg, A. M., P. N. Gerhardt, T. K. Cao, W. Ocasio, W. M. Barbour, and P. M. Morinski. 2000. Sensitivity and specificity of the test kit BAX for screening/*E. coli* 0157:H7 in ground beef: Independent laboratory study. *J. AOAC Int.* 83(6):1349–1356.

42. Hochberg, A. M., A. Roering, V. Gangar, M. Curiale, W. M. Barbour, and P. M. Mrozinski. 2000. Sensitivity and specificity for the BAX for screening/E-Coli 0157:H7 in ground beef independent laboratory study. *J. AOAC Int.* 83(6):1349–1356.

43. Steward, D. and S. M. Gendel. 1998. Specificity of the BAX polymerase chain reaction system for detection of the foodborne pathogen *Listeria monocytogenes*. *J. AOAC Int.* 81:817–822.

44. Norton, D. M., M. McCamey, K. J. Boor, and M. Wiedmann. 2000. Application of the BAX for screening/genus *Listeria* polymerase chain reaction system for monitoring *Listeria* species in cold-smoked fish and in the smoked fish processing environment. *J. Food Prot.* 63:343–346.

45. Norton, D. M., M. McCamey, K. L. Gall, J. M. Scarlett, K. J. Boor, and M. Wiedmann. 2001. Molecular studies on the ecology of *Listeria monocytogenes* in the smoked fish processing industry. *Appl. Environ. Microbiol.* 67:198–205.

46. Robinson, J. K., R. Mueller, and L. Filippone. 2000. New molecular beacon technology. *Am. Lab.* 32(24):30–34.

47. Eggins, B. 1997. *Biosensors: An Introduction*. Wiley, New York.

48. Cunningham, A. J. 1998. *Bioanalytical Sensors*. Wiley, New York.

49. Goldschmidt, M. C. 1999. Biosensors: Scope in Microbiological Analysis. In *Encyclopedia of Food Microbiology*. Robinson, R., C. Batt, and P. Patel, eds. Academic Press, New York. pp. 268–278.

50. Deyholos, M., H. Wang, and D. Galbraith. 2001. Microarrays for gene discovery and metabolic pathway analysis in plants. *Life Sci.* 2(1):2–4.

51. Heron, E. 2000. Applied biosystem: Innovative technology for the life sciences. *Am. Lab.* 32(24):35–38.

Chapter 33

Methods to Predict Spoilage of Muscle Foods

Geraldine Duffy, Anthony Dolan, and Catherine M. Burgess

Contents

33.1 Introduction

All animals, birds, fish, etc. contain a host of microorganisms in their intestinal tract and on their exposed outer skins, membranes, etc. During the slaughter and processing of the live organism into food, the muscle surface can become contaminated with microorganisms. Microbial contamination on the food and its composition/diversity is dependent on both the microbial load of the host organism and the hygiene practices employed during slaughter, processing, and distribution [1]. For example, during beef slaughter cross-contamination of microbial flora from the bovine hide, feces, and gut contents are recognized as the main cause of microflora on the beef carcass [2]. Among the principal genera of bacteria that are present on postslaughter muscle surfaces are *Pseudomonas* spp., *Acinetobacter* spp., *Aeromonas* spp., *Brochothrix thermosphacta*, members of the lactic acid bacteria (LAB) such as *Lactobacillus* and *Leuconostoc*, as well as many members of the Enterobacteriaceae including *Enterobacter* and *Serratia* spp. [3–7].

From a microbiological standpoint, muscle foods have a particularly unique nutritional profile, with intrinsic factors such as a neutral pH, high water content, a high protein content, and fat providing an excellent platform for microbial growth. Thus, during storage of muscle foods, favorable environmental conditions (temperature, pH, a_w, etc.) will allow the microflora to grow. As the microorganisms grow they metabolize the food components into smaller biochemical constituents, many of which emit unacceptable flavors, odors, colors, or appearance [6,8]. Spoilage may be defined as the time when the microorganisms reach a critical level, usually at around $\log_{10}$ 7–8 colony forming units (CFU) g^{-1}, to induce sufficient organoleptic changes to render the food unacceptable to the consumer. The particular species of bacteria that contaminate the muscle, along with the environmental conditions, will determine the spoilage profile of the stored muscle food [5]. Under aerobic storage conditions, certain species of the genus *Pseudomonas* are generally considered to significantly contribute to spoilage. This is due to the organisms' ability to utilize amino acids and grow well at refrigeration temperatures. Although it is a facultative anaerobe, under anaerobic conditions, the bacterium *B. thermosphacta* is considered a dominant member of the spoilage flora of meat products, producing lactic acid and ethanol as by-products of glucose utilization [9]. Recently, the use of modified atmosphere packaging (MAP) has gained popularity as a method of preservation. Gas mixtures containing variable O_2 and CO_2 concentrations are used to inhibit the growth of different spoilage-related bacteria. Under certain MAP conditions, lactic acid bacteria dominate and are prolific spoilers [10].

The storage period under a particular set of environmental conditions until the spoilage microflora reaches a threshold level is known as shelf life. To extend the shelf life of muscle foods, a range of procedures to prevent or retard microbial growth are deployed. When storing fresh muscle foods, where only chill storage temperatures (<5°C) are employed to retard microbial growth, the shelf life can be measured in days. Modified atmosphere or vacuum packaging can extend shelf life to several weeks or months. Extension of shelf life beyond this period requires the use of more robust and invasive preservation techniques such as freezing, mild or severe heat treatment (canning), reducing water activity (a_w), altering pH (acidic or alkaline), or the use of chemical or biological preservatives. However, all of above preservation processes generally have an unwanted deleterious influence on the organoleptic quality of the food. Therefore, there is an ever increasing move away from heavily preserved food to fresh and minimally preserved foods with a limited shelf life, imposing a greater need for industry to be able to accurately predict when spoilage of the food will occur. As there is a direct correlation between microbial load and spoilage, food hygiene regulators and industry set microbiological guidelines and criteria for specific foodstuffs, which are used to predict spoilage and determine shelf life. A number of direct and indirect techniques are available to assess the microbial load or its metabolites in food at the point of food production, which will give a predicted shelf life under a defined set of storage conditions. This chapter will review a selection of commonly used and emerging technologies that are used to directly or indirectly enumerate the total microbial load and predict spoilage.

33.2 Culture-Based Methods

Microbial cultural assays are generally dependent on the growth of a microbial population to form colonies on an agar plate, which are visible to the analyst. Specific conditions such as temperature, moisture content, atmosphere, and nutrient availability on solid media (agar) are used to induce this growth.

33.2.1 Agar Plate Count Methods

The gold standard method to assess microbial numbers remains the aerobic standard plate count (SPC). This cultural method has been widely and successfully used for many years in the food, pharmaceutical, and medical sectors. Serial dilutions of the sample material are prepared, plated onto agar (plate count agar), and incubated under specific conditions. When visible colonies appear, the number of CFU per gram of food can be readily calculated. The Association of Official Analytical Chemists (AOAC) Official Method 966.23 [11] and the International Organization for Standardization (ISO) (No. 4833:2003) [12] have standardized the test protocol. All alternative methods must generally be correlated or validated against these methods.

Although "gold standard" indicates the method is perfect, there are in fact some drawbacks to the method. The SPC result is often referred to as "total viable count" implying that "all" viable microorganisms will be incorporated in results of the assay. This is not so, as certain microorganisms, referred to as viable but nonculturable (VBNC) [13], may have growth requirements not met by the incubation conditions. The failure of the assay to account for these organisms may lead to an underestimation of the *true* microbial load. From a practical perspective the method is also very slow and labor intensive, requiring 3 days for the colonies to form and thus, a result to be obtained. For products with a short shelf life this delay is very impractical and a product may be in retail distribution before microbial counts are obtained.

33.2.2 Alternative Culture Methods

There are alternative agar-based methods, such as Petrifilm® (3M Microbiology Products, USA), that are AOAC accredited (Method 990.12) [14] and show comparable counts to SPC for a wide variety of meat samples [15]; although some problems have been noted [16]. Another product, SimPlate® (IDEXX Labs Inc., USA), has also been applied to meat muscle with relative success and is an approved AOAC method [17,18].

There is also an automated method based on a liquid media–based most probable number (MPN) technique (TEMPO®, bioMériuex, France). The system is based on wells containing a traditional culture media formula with a fluorescent indicator. Each well corresponds to an MPN dilution tube. Once the sample is distributed in the wells, the microorganisms metabolize the culture media producing a fluorescent signal. The system uses an MPN calculation to assess the number of microorganisms in the original sample. Apart from the obvious advantage that this type of automated instrumentation offers, the TEMPO system has a reduced incubation time ($\leq$48 h) compared with the ISO SPC method that takes 3 days. When applied to meat samples, the technology shows a high correlation with the SPC ($r = .99$) [19].

33.3 Direct Epifluorescent Filtration Technique

An alternative approach to culture is to directly extract the microorganisms from the muscle food by membrane filtration. When concentrated onto the membrane surface, the microorganisms can be stained using a fluorescent dye and the cells then detected and enumerated using epifluorescent microscopy.

The first step in this direct epifluorescent filtration technique (DEFT) is the use of membrane filtration to recover the bacteria from the food and this step poses some challenges in relation to muscle foods. When membrane filtration is used to recover microorganisms from muscle foods, they must be first placed in a liquid media and homogenized, stomached, or pulsified (Microgen Bioproducts, UK) [20] to remove the bacteria from the food surface or matrix into the liquid diluent. A problem encountered is that food particles in the liquid have a tendency to clog the pores of the membrane during filtration. This may mean that the required volume cannot be filtered and that any food debris on the membrane can interfere with the enumeration of bacterial cells. Some approaches to improve filterability of muscle foods have been employed to physically or chemically remove as much of the food suspension as possible before filtration. These have included the use of low-speed centrifugation, appropriate surfactants such as Tween 80 and sodium dodecyl sulfate (SDS), and the proteolytic enzyme, Alcalase [21].

Once the microorganisms are concentrated onto the membrane surface, the membrane is overlaid with a fluorescent dye such as acridine orange and mounted on a glass slide. The microorganisms are viewed using fluorescent microscopy, and the total numbers of organisms in a defined number of fields of view are counted. The microscopic count is used to predict the "gold standard" plate count using a calibration curve relating the DEFT count to the aerobic plate count.

DEFT (Figure 33.1) has been applied to the estimation of microbial numbers in a range of muscle foods (Table 33.1). Although acridine orange is the most commonly used fluorescent dye, it

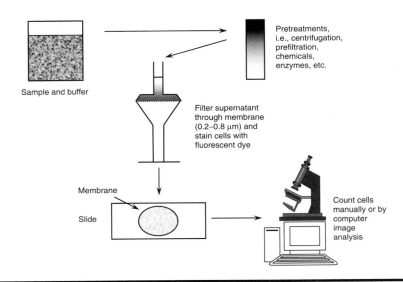

Figure 33.1   Flow diagram of a direct epifluorescent filtration technique (DEFT) for enumeration of microorganisms from muscle foods.

Table 33.1   Correlation of Direct Epifluorescent Filtration Technique (DEFT) with the Standard Aerobic Plate Count (SPC) for Enumeration of Microorganisms in a Range of Muscle Foods

Muscle Food	Treatment of Sample before Filtration through Membrane (0.4–0.8 µm)	Fluorescent Dye	Correlation with SPC	Reference
Fresh meat	Stomached 2 min	Acridine Orange	$r = .91$	[97]
Canned hams	Stomached 2 min, prefiltration through glass microfiber filter	Acridine Orange	poor	[98]
Raw ground beef	Stomached 2 min, prefiltered through nylon filter, Triton X, and bactotrypsin	Acridine Orange	$r = .79$	[99]
Raw beef pieces	Stomached 30 s, prefiltered through glass microfiber filter, Triton X	Acridine Orange	$r^2 = .91$	[100]
Raw pork mince	Stomached 30 s, Tween 80, SDS, Alcalase 0.6 L	Acridine Orange	$r = .97$	[101]
Raw beef mince	Stomached 30 s, low-speed centrifugation, Tween 80, SDS, Alcalase 0.6 L	Acridine Orange	$r = .97$	[102]
Lamb carcasses	Stomached 30 s, low-speed centrifugation, Tween 80, SDS, Alcalase 2.5 L	Acridine Orange	$r^2 = .87$	[103]
Minced beef	Stomached 30 s, low-speed centrifugation, Tween 80, SDS, Alcalase 2.5 L	Acridine Orange	$r^2 = .97$	[21]
Processed meat (minced beef, cooked ham, bacon rashers, frozen burgers)	Stomached 30 s, low-speed centrifugation, Tween 80, SDS, Alcalase 2.5 L	BacLight	$r^2 = (.90, .87, .82, .80)$	[22]

does not distinguish between live and dead cells and so may overestimate the bacterial load in processed meat samples containing large numbers of dead cells. To overcome this problem, a viability stain BacLight® (Molecular Probes Inc., The Netherlands) was reported to successfully distinguish between live and dead cells and in a DEFT gave a good correlation with the SPC for microorganisms in processed meat ($r^2 = .87–.93$) [22]. The DEFT takes approximately 15–20 min to analyze one sample and so at most 20 samples can be analyzed manually in a working day.

The DEFT has been successfully automated for high throughput enumeration of microorganisms in milk samples [23]. Commercial systems for analysis of milk include the Bactoscan® (Foss, Denmark) and Cobra® systems (Biocom, France). However, the DEFT has not been automated for muscle foods. This has hugely impacted its uptake commercially by this industry sector as, apart from the small number of samples that can be analyzed manually daily, the approach is labor intensive and requires significant operator skills. Manual enumeration is particularly difficult when there are very high or low numbers of microorganisms on the slide or when there is particulate debris on the slide. Future developments to make this approach commercially suitable for muscle foods may incorporate the initial membrane filtration approach to extract the microorganisms with an automated detection system. A solid-phase cytometry method has been proposed by D'Haese and Nelis [24] and could use a laser beam to detect microorganisms recovered onto a membrane filter. This method would potentially be very rapid and automated but a potential problem could arise from any food debris remaining on the membrane surface.

33.4   ATP Bioluminescence Methods

Enzyme-mediated light production, bioluminescence, is a widespread phenomenon in nature [25]. Bioluminescent organisms are widely distributed throughout the oceans and include bacteria, sea anemones, worms, crustaceans, and fish. Fireflies and glow worms are the best-known terrestrial organisms producing light. The principles of firefly bioluminescence were discovered over 40 years ago [26]. In the firefly bioluminescence reaction, adenosine 5′-triphosphate (ATP) reacts with the enzyme luciferase and the substrate luciferin producing a photon of light. ATP is a high-energy substance found only in living cells. It takes part in all metabolic pathways and therefore its concentration in all cells including bacterial cells is strictly regulated. When luciferin and luciferase are added to a cell suspension the amount of light emitted is proportional to the amount of ATP present. The amount of light can be measured using a photometer to give an indirect indication of the microbial population density.

$$\text{Luciferin} + \text{ATP} + O_2 \xrightarrow[\text{Mg}^{2+}]{\text{Luciferase}} \text{AMP} + CO_2 + \text{pyrophosphate} + \text{oxyluciferin} + \text{photon}$$

The firefly bioluminescence reaction has been exploited as a rapid and sensitive method for measuring cell numbers, including microbial cells.

The ATP bioluminescent assay has been widely applied to assess hygiene, based on detection of all ATP present [27,28]. However, a major problem in the use of bioluminescence to predict the microbial SPC of foods is interference from nonmicrobial ATP. If an accurate estimation of the microbial load is to be obtained, nonmicrobial somatic ATP must be destroyed before the bioluminescence test is carried out. The most common approach is the enzymatic destruction of nonmicrobial ATP, followed by release and estimation of residual ATP from the microbial cells [29,30]. Another approach is to separate the microorganisms from the rest of the material and estimate the ATP in the microbial fraction. Stannard and Wood [31] used this approach to estimate bacterial numbers in minced beef. The results show a linear relationship ($r = .94$) between colony counts and microbial ATP content in raw beef. An ATP bioluminescence test was shown by Siragusa et al. [32] to be an adequate means to assess the microbial load of poultry carcasses. This assay utilized differential extraction and filtration to separate somatic ATP from microbial ATP in a very rapid time frame. The assay required approximately 5 min to complete: approximately 3.5 min to sample and 90 s analytical time. The correlation coefficient (r) between aerobic colony counts and the ATP test was .82. Ellerbroek and Lox [33] used an ATP bioluminescence approach to investigate the total bacterial counts on poultry neck and carcasses. The correlation between the bioluminescence method and the total viable counts of neck skin samples was $r = .85$, whereas a lower correlation was reported between the bioluminescence count and the total viable counts on the carcass ($r = .66$).

Commercially available bioluminescent systems include Celsis® (Celsis International plc, UK) and Bactofoss® (Foss) but their application to date has been aimed at hygiene testing and liquid foods rather than muscle foods.

33.5   Electrical Methods

Electrical methods for assessing bacterial numbers include impedance and conductance. Impedance is the opposition to flow of an alternating electrical current in a conducting material [34]. The conductance of a solution is the charge carrying capacity of its components and capacitance is the ability to hold a charge [34].

When monitoring the growth of microorganisms, the conducting material is a microbiological medium. As microorganisms grow they utilize nutrients in the medium, converting them into smaller more highly charged molecules, for example, fatty acids, amino acids, and various organic acids [35]. If electrodes are immersed in the medium and an alternating current is applied, the metabolic activity of the microorganisms results in detectable changes in the flow of current. Typically, impedance decreases while conductivity and capacitance increase [35]. When the microbial population reaches a threshold of 10^6–10^7 CFU mL^{-1} an exponential change in impedance can be observed [34]. The elapsed time until this exponential change occurs is defined as impedance detection time and is inversely proportional to the initial microbial numbers in the sample.

The most commonly used application of impedance is shelf-life testing. This test determines whether a sample contains above or below a predetermined concentration of microorganisms. Impedance testing has been used in conjunction with a calibration curve with the SPC for a number of products including raw milk ($r = -.96$) [36], frozen vegetables (92.6% agreement between methods) [37] and meat [38]. A conductance method was used to predict microbial counts on fish [39] with a correlation of $r = -.92$ to $-.97$ using brain heart infusion.

Of all developed alternative methods to predict microbial load, the impedance technique has been most widely accepted within the food industry. Commercially available automated systems include the Malthus® (Malthus Instruments Ltd., UK) system, which measures conductance, and the Bactometer® (Bactomatic Inc., USA) system, which can measure impedance, conductance, and capacitance. Both systems can measure several hundred samples simultaneously and have detection limits of ≥1.0 CFU mL^{-1}. Using these systems to predict the SPC count on meat, correlations of $r = -.83$ and $r = -.80$ were reported for the Bactometer and Malthus machines, respectively [40]. In the muscle food sector, uptake has been in the processing sector rather than for raw foods.

33.6   *Limulus* Amoebocyte Lysate Assay

Gram-negative bacteria are important food spoilage organisms in muscle foods [41]. They differ from gram-positive bacteria in that their cell wall contains lipopolysaccharides (LPS). Based on this difference a *Limulus* amoebocyte lysate (LAL) assay method that targets LPS has been developed. LPS contains an endotoxin that activates a proteolytic enzyme found in the blood cells (amoebocytes) of the horseshoe crab (*Limulus polyphemus*). The enzyme activates a clotting reaction, which results in gel formation. The concentration of LPS is determined by making serial dilutions of the sample and noting the greatest dilution at which a gel is formed within a given time [41]. The reaction has been used to develop a colorometric assay.

LAL has been applied to the evaluation of microbial contamination on pork carcasses [42]. Although the test correlated well with coliform numbers, it did not correlate well with total numbers of organisms, indicating its limited usefulness as a spoilage indicator. However, more recently a chromogenic LAL was reported by Siragusa et al. [43] to rapidly predict microbial contamination on beef carcasses. A high correlation ($r^2 = .90$) was reported with the standard aerobic plate count.

33.7 Spectroscopic Methods

Various spectroscopic methods have been proposed as rapid, noninvasive methods for the detection of microbial spoilage in muscle foods. Such methods are based on the measurement of biochemical changes that occur in the meat as a result of the decomposition and formation of metabolites caused by the growth and enzymatic activity of microorganisms, which eventually results in food spoilage.

Fourier transform infrared (FT-IR) spectroscopy involves the observation of vibrations in molecules when excited by an infrared beam. An infrared absorbance spectrum gives a fingerprint-like spectral signature, which is characteristic of any chemical or biochemical substance [44]. Such a method is therefore potentially useful to measure biochemical changes in muscle foods due to microbial growth and could be used as an indicator for spoilage. FT-IR spectroscopy has been successfully employed to discriminate, classify, and identify microorganisms. Some examples include discrimination between *Alicyclobacillus* strains associated with spoilage in apple juice [45], the discrimination of *Staphylococcus aureus* strains from different staphylococci [46], and the setting up of a spectral database for the identification of coryneform strains [47]. Mariey et al. [48] provide a review of many other characterization methods using FT-IR. This also gives an overview of the statistical methods used to interpret spectroscopic data.

In addition to these discriminatory uses, FT-IR has shown promise for use as a spoilage detection method. FT-IR has been used to predict spoilage of chicken breasts in a rapid, reagentless, noninvasive manner [49]. The metabolic snapshot correlated well with the microbial load. Ellis et al. [50] applied FT-IR to predict microbial spoilage of beef; although the correlation with the microbial load was less accurate than for poultry.

Another spectroscopic method that has been used in recent times for detection of microbial spoilage is short-wavelength-near-infrared (SW-NIR) diffuse reflectance spectroscopy (600–1100 nm). It has the advantage over FT-IR in that it is useable through food packaging and can be used to examine bulk properties of a food due to its greater pathlength [51]. This technique was applied to predict spoilage of chicken breast muscle and the results showed that SW-NIR could be used in a partial least squares model to predict microbial load [51]. Lin et al. [52] have used this technique with success in predicting spoilage of rainbow trout.

33.8 Developmental Methods

There are a number of emerging methods and technologies that are being shown to be suitable for the rapid and specific identification or enumeration of microorganisms from clinical or liquid samples. Although most have not yet been applied to predict the total microbial flora or spoilage of muscle foods, they have the potential with further development to be applied in the future. Some of these technologies are summarized in the following sections.

33.8.1 Flow Cell Cytometry

Flow cell cytometry is a technique that can be used to detect and enumerate cells as they are passed on an individual cell basis, suspended in a stream of fluid, past a laser beam. A flow cytometer typically has several key components including a light or excitation source, a laser that emits light at a particular wavelength, and a liquid flow that moves liquid-suspended cells through the instrument past the laser and a detector, which is able to measure the brief flashes of light emitted as cells flow past the laser beam. Thus, individual cells can be detected and counted by the system.

The technique has been successfully applied to the enumeration of microorganisms in raw milk [53] and milk powder [54], but it has not yet been applied to muscle foods. As described in Section 33.3, a solid-phase cytometry method could potentially be applied to muscle foods, based on the assumption that the microorganisms could be successfully extracted from the food onto a filter and the filter then scanned by a laser beam [24].

33.8.2   Molecular Methods

Major advances in biotechnology have rapidly progressed the use of genetic tools for microbial detection. In particular, developments in the level of genomic information available for foodborne pathogens have been widely exploited in methods to detect and genetically characterize microorganisms. Genetic tools are now commonly used to detect specific pathogens or groups of spoilage microorganisms. However, to date the use of molecular technology to detect and enumerate, in a single assay, all microorganisms in a food sample is limited by the huge diversity of microorganisms likely to be present and identification of a common gene target present in all the foodborne microorganisms. The use of the 16S ribosomal RNA (rRNA) gene has been reported for this purpose [55]. If technological complexity can be overcome, this approach has enormous potential as a very rapid and specific test to predict spoilage in muscle foods.

33.8.2.1   Polymerase Chain Reaction

Nucleic acid methods that include an amplification step for the target DNA/RNA are now routinely employed in molecular biology. These amplification methods can increase the target nucleic acid material more than a billion fold and are particularly important in the arena of food microbiology where one of the major hurdles is the recovery and detection of very low numbers of a particular species. The most popular method of amplification is the polymerase chain reaction (PCR) technique (Figure 33.2). In PCR, a nucleic acid target (DNA) is extracted from the cell

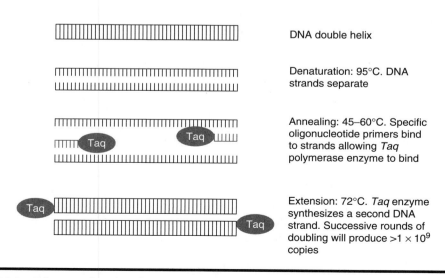

Figure 33.2   Diagram showing the main events in a typical polymerase chain reaction (PCR).

and denatured into single-stranded nucleic acid. An oligonucleotide primer pair specific for the selected gene target, along with an enzyme (usually *Taq* polymerase, a thermostable and thermoactive enzyme originally derived from *Thermus aquaticus*) in the presence of free deoxynucleoside triphosphates (dNTPs), is used to amplify the gene target exponentially, resulting in a double replication of the starting target material. This reaction is carried out in an automated, programmable block heater called a thermocycler, which provides the necessary thermal conditions needed to achieve amplification. Following amplification, the PCR products are separated by gel electrophoresis, stained with ethidium bromide, and visualized using ultraviolet light. This type of PCR, sometimes referred to as conventional PCR, can be used for the identification of specific groups of spoilage bacteria in meat including lactic acid bacteria [56,57].

A quantitative method using electrochemiluminescence to measure the PCR product was applied to predict the spoilage bacterial load on aerobically stored meat [58]. The correlation of this method with the SPC was $r = .94$. Gutierrez et al. [55] combined conventional PCR with an enzyme-linked immunosorbent assays (ELISA) to allow enumeration of microorganisms. On applying this technique to the detection of the microbial load in meat samples, a good correlation was achieved between the SPC counts and the PCR-ELISA ($r = .95$). The authors did express concerns about the complexity of the assay and about its suitability as a routine assay.

Recently, a more advanced quantitative PCR technology, in the form of real-time PCR (RT-PCR), has entered and revolutionized the area of molecular biology [59]. RT-PCR allows continuous monitoring of the amplification process through the use of fluorescent double-stranded DNA intercalating dyes or sequence-specific probes [60]. The amount of fluorescence after each amplification cycle can be measured and visualized in real time on a computer monitor attached to the RT-PCR machine. A number of dye chemistries have been reported for use in RT-PCR, from DNA-binding dyes such as SYBR® Green (Molecular Probes Inc.) to more complex fluorescent probe technologies such as TaqMan® (Roche Molecular Systems, UK), molecular beacons, and HybProbes® (Roche Molecular Systems) [61]. Whatever signal chemistry is used, RT-PCR not only allows quick determination of the presence/absence of a particular target, but can also be used for the quantification of a target that may then be related to microbial counts.

To quantify microorganisms by RT-PCR, a set of standards of known concentration must first be analyzed. The standards may be of known CFU per milliliter or known gene copy number and then related to the C_T (threshold cycle) of the reaction to generate a standard curve, which can be used to quantify unknown samples. An important factor to be considered when quantifying bacteria is that relying on a DNA-based RT-PCR will lead to a count that comprises of live, dead, and VBNC bacteria, which could potentially lead to an overestimation of the numbers present. A way to overcome this is by coupling RT-PCR with reverse transcription. This technique transcribes RNA (present only in viable cells) into complementary DNA (cDNA), which can then be employed in a RT-PCR reaction. Because the cDNA originates from RNA the quantification will be based on viable cells only, leading to a more accurate determination of the number of metabolically active bacteria.

To date, RT-PCR has been mainly used for the sensitive and rapid detection of a wide range of pathogens, such as *Salmonella* spp., *Escherichia coli*, and *Listeria* on meat [62–65], and a multiplex assay that has the ability to detect more than one pathogen at the same time has been described [66]. The quantitative feature of RT-PCR has been examined for the enumeration of the spoilage organism *Lactobacillus sakei* in meat products, and the application of a live staining method in combination with real-time technology for quantitative analysis has been reported using model organisms [67,68]. There is huge potential for this technology to quantify total microorganisms using an RNA gene target common to all microflora likely to be present.

33.8.2.2 *Fluorescent* In Situ *Hybridization*

In situ hybridization (ISH) using radiolabeled DNA was first reported by Pardue and Gall [69] and John et al. [70] for direct examination of cells. It was applied to bacteria for the first time in 1988 [71], and with the advent of fluorescent labels the technique became more widely used [72]. Fluorescent *in situ* hybridization (FISH) is a technique that specifically detects nucleic acid sequences in a cell using a fluorescently labeled probe that hybridizes specifically to its complementary target gene within the intact cell. The target gene is the intercellular rRNA in the microorganism as these genes are relatively stable, occur in high copy numbers, and have variable and conserved sequence domains, which allows for the design of discriminatory probes either specific to an individual species or to particular genera [73]. FISH generally involves four steps, fixation of the sample, permeabilization of cells to release the nucleic acids, hybridization with the fluorescent labeled probe, and detection by fluorescent microscopy. Traditionally, FISH methods have been implemented using DNA oligonucleotide probes. A typical oligonucleotide probe is between 15 and 30 base pairs in length. Short probes have easier access to the target but also may have fewer labels [74]. There are a number of ways in which probes can be labeled. Direct labeling is most commonly used where the fluorescent molecule is directly bound to the oligonucleotide either chemically during synthesis or enzymatically using terminal transferase at the 3′-end. This method is considered to be the fastest, cheapest, and most convenient [75]. Sensitivity of FISH assays can be increased using indirect labeling, where the probe is linked to a reporter molecule that is detected by a fluorescent antibody [76] or where the probe is linked to an enzyme and a fluorescent substrate can be added [77]. A more recent development in probe technology is the development of peptide nucleic acid (PNA) probes. PNAs are uncharged DNA analogs in which the negatively charged sugar–phosphate backbone is replaced by an achiral sugar–phosphate backbone formed by repetitive units of *N*-(2-aminoethyl) glycine [78]. PNA probes can hybridize to target nucleic acids more rapidly and with higher affinity and specificity than DNA probes [79].

FISH technology has been applied to the detection of bacterial pathogens in clinical and food samples [80,81] and can allow direct identification and quantification of microbial species. A FISH assay has been developed for the *Pseudomonas* genus, which is important in milk spoilage, allowing for the specific detection and enumeration of this group of organisms in milk much more rapidly than a cultural method [82]. In the wine industry, lactic acid bacteria can be detrimental or beneficial depending on the species, and when they develop in the process. A FISH technique has been described as the one that utilizes probes to differentiate between different LAB genera so that it is possible to identify potential spoilage strains from the species responsible for successful fermentation [83]. FISH technology has potential as a method to enumerate all microorganisms in a food sample using a common gene target but could be limited in its uptake by its current reliance on microscope-based detection.

33.9 Electronic Nose

It is well known that microorganisms produce a range of volatiles as they grow on food and can be used to identify particular species of microorganisms that have a unique volatile fingerprint or potentially, to determine the total level of microbial contamination on a food and predict spoilage. Gardener and Bartlett [84] defined an electronic nose as "an instrument which comprises an array of electronic, chemical sensors with partial specificity and an appropriate pattern recognition system, capable of recognising simple or complex odours." An electronic nose normally consists of a vapor-phase flow over the sensor, interaction with the sensor, and analyses of the interaction using computer software. The field of sensor development is highly active and includes a range of sensor types based on metal oxide, metal oxide silicon, piezoelectric, surface acoustic waves, optical, and electrochemical premises [85].

Blixt and Borch [86] investigated the use of an electronic nose to predict the spoilage of vacuum-packaged beef. The volatile compounds were analyzed using an electronic nose containing a sensory array composed of 10 metal oxide semiconductor field-effect transistors, four Tagushi type sensors, and one CO_2-sensitive sensor. Two of the Tagushi sensors performed best and correlated well with evaluation of spoilage by a sensory panel. They did not attempt to correlate the results with microbial counts.

Du et al. [87] used an electronic nose (AromaScan) to predict spoilage of yellowfin tuna fish. The change in fish quality as determined by AromaScan (AromaScan plc., UK) followed increases in microbiological counts in tuna fillets, indicating that electronic nose devices can be used in conjunction with microbial counts and sensory panels to evaluate the degree of decomposition in tuna during storage.

33.10   Time–Temperature Integrators

One of the key contributors to the spoilage of fresh muscle foods is a breakdown in the chill chain during distribution. The prediction of spoilage and the application of an optimized quality and safety assurance scheme for chilled storage and distribution of fresh meat and meat products would be greatly aided by the continuous monitoring of temperature during distribution and storage.

A time–temperature integrator (TTI) is defined as a small, inexpensive device that can be incorporated into a food package to show a visible change according to the time and temperature history of the stored food [88]. TTIs are devices that contain a thermally labile substance, which can be biological (microbiological or enzymatic), chemical, or physical. Of these groups, biological TTIs are the best studied. TTIs can be used to determine both whether a heat treatment has worked effectively and whether temperature abuse has occurred during storage. Different types of TTIs have been used for determining the effectiveness of heat treatment. α-Amylase from *Bacillus* species has been evaluated in a number studies [88–91] and a recent report describes the use of amylase from the hyperthermophile *Pyrococcus furiosus* as a sterilization TTI [92].

TTIs can be used to monitor temperature abuse during storage, transportation, and handling and thus to monitor such abuses that may lead to a shortened shelf life and spoilage. This type of TTI must (1) be easily activated and sensitive; (2) provide a high degree of precision; (3) have tamper-evident characteristics; (4) have a response that is irreversible, reproducible, and correlated with food quality changes; (5) have determined physical and chemical characteristics; and (6) have an easily readable response [93].

Giannakourou et al. [94] demonstrated that TTI readings using a commercially available enzyme could be adequately correlated to the remaining shelf life of the product (in this case marine-cultured gilt-head seabream) at any point of its distribution. The same TTI has also shown positive results for fresh chicken storage [95]; although the TTI predictions would be inaccurate following an extreme instance of temperature abuse, with the TTI indicator changing color before the product had actually spoiled [96].

Numerous time–temperature indicators are commercially available for all types of food stuffs. However, it is necessary to validate the TTI of choice with the product and process of choice before correlations can be made between the TTI and the potential for spoilage of the product.

33.11 Conclusion

Muscle foods pose considerably more challenges than other foods for development and successful application of spoilage detection methods. These include a highly complex food tissue matrix in which the microorganisms may be embedded and strongly attached, and from which they must be detached to detect and enumerate the microorganisms. This means the food must generally be placed in a liquid diluent and then physically manipulated to release the microorganisms into the liquid. This dilution effect obviously creates a need for a more sensitive detection method than a sample to which the method could be applied directly, such as a liquid food (i.e., milk). In addition, the microflora is generally quite diverse and the dominant flora is very much dependent on the storage environment. At the early stage of the food process the levels of microorganisms on the raw muscle food may be as low as log 2.0 CFU g^{-1}, thus posing additional challenges for the sensitivity of the detection method.

The gold standard method to predict spoilage remains the aerobic SPC, but it is still required by the industry that alternative methods are validated against this method. However, as previously described in this chapter, the SPC is far from perfect and more rapid methods to predict spoilage are urgently needed by the muscle foods sector (fish, meat, and poultry industries). These alternative spoilage methods must generally give results that are comparable and validated against "gold standard" cultural microbial methods. The methods must be sensitive, rapid, suited to online use, and at least semiautomated. They must be suited to routine use, without the need for highly skilled operators, as high staff turnover is often a major issue in the muscle food industry, and it is neither practical nor possible to keep retraining staff to carry out a test that is highly complex.

The muscle food sector has undoubtedly been the slowest sector in the food industry to take on board alternative technologies for spoilage prediction. However, they are now being compelled by their customers, regulatory authorities, and consumers to more accurately predict shelf life. This is even more pertinent with the continued market move toward chilled prepared foods with minimal preservatives and a short shelf life. Although there has been much research and development in the area of rapid spoilage detection methods, recent research on rapid microbial methods has tended to focus more on methods for identification of specific species of microorganisms rather than on the total microbial load. However, some of the emerging technologies developed, albeit for other applications, have enormous potential to be further developed for enumeration of the total microbial load and to predict spoilage. More research efforts should now be refocused in this direction using the newer technologies to overcome the hurdles that have to date prevented the widespread uptake of rapid methods to predict spoilage by the muscle food sector.

References

1. Bell, R.G., Distribution and sources of microbial contamination on beef carcasses. *J Appl Microbiol* 82, 292, 1997.
2. Elder, R.O., Keen, J.E., Siragusa, G.R., Barkocy-Gallagher, G.A., Koohmaraie, M. and Laegreid, W.W., Correlation of enterohemorrhagic *Escherichia coli* O157 prevalence in feces, hides, and carcasses of beef cattle during processing. *Proc Natl Acad Sci USA* 97, 2999, 2000.
3. Ercolini, D., Russo, F., Torrieri, E., Masi, P. and Villani, F., Changes in the spoilage-related microbiota of beef during refrigerated storage under different packaging conditions. *Appl Environ Microbiol* 72, 4663, 2006.
4. Hinton, A., Jr., Cason, J.A. and Ingram, K.D., Tracking spoilage bacteria in commercial poultry processing and refrigerated storage of poultry carcasses. *Int J Food Microbiol* 91, 155, 2004.

5. Borch, E., Kant-Muermans, M.L. and Blixt, Y., Bacterial spoilage of meat and cured meat products. *Int J Food Microbiol* 33, 103, 1996.

6. Huis in't Veld, J.H.J., Microbial and biochemical spoilage of foods: an overview. *Int J Food Microbiol* 33, 1, 1996.

7. Gustavsson, P. and Borch, E., Contamination of beef carcasses by psychrotrophic *Pseudomonas* and *Enterobacteriaceae* at different stages along the processing line. *Int J Food Microbiol* 20, 2, 1993.

8. Dainty, R.H., Edwards, R.A. and Hibbard, C.M., Time course of volatile compound formation during refrigerated storage of naturally contaminated beef in air. *J Appl Bacteriol* 59, 303, 1985.

9. Pin, C., Garcia de Fernando, G.D. and Ordonez, J.A., Effect of modified atmosphere composition on the metabolism of glucose by *Brochothrix thermosphacta*. *Appl Environ Microbiol* 68, 4441, 2002.

10. Chenoll, E., Macian, M.C., Elizaquivel, P. and Aznar, R., Lactic acid bacteria associated with vacuum-packed cooked meat product spoilage: population analysis by rDNA-based methods. *J Appl Microbiol* 102, 498, 2007.

11. AOAC, *Official Method of Analysis*, 15th ed., AOAC, Washington, DC, 1990.

12. International Organization for Standardization, *EN ISO 4833:2003*, Microbiology of food and animal feeding stuffs—horizontal method for the enumeration of microorganisms—colony count technique at 30°C, International Organization for Standardization, Geneva, Switzerland, 2003.

13. Besnard, V., Federighi, M., Declerq, E., Jugiau, F. and Cappelier, J.M., Environmental and physicochemical factors induce VBNC state in *Listeria monocytogenes*. *Vet Res* 33, 359, 2002.

14. AOAC, *Official Method of Analysis*, 16th ed., AOAC, Washington, DC, 1995.

15. Park, Y.H., Seo, K.S., Ahn, J.S., Yoo, H.S. and Kim, S.P., Evaluation of the Petrifilm plate method for the enumeration of aerobic microorganisms and coliforms in retailed meat samples. *J Food Prot* 64, 1841, 2001.

16. Dawkins, G.S., Hollingsworth, J.B. and Hamilton, M.A., Incidences of problematic organisms on Petrifilm aerobic count plates used to enumerate selected meat and dairy products. *J Food Prot* 68, 1506, 2005.

17. Beuchat, L.R., Copeland, F., Curiale, M.S., Danisavich, T., Gangar, V., King, B.W., Lawlis, T.L., Likin, R.O., Okwusoa, J., Smith, C.F. and Townsend, D.E., Comparison of the SimPlate total plate count method with Petrifilm, Redigel, conventional pour-plate methods for enumerating aerobic microorganisms in foods. *J Food Prot* 61, 14, 1998.

18. Feldsine, P.T., Leung, S.C., Lienau, A.H., Mui, L.A. and Townsend, D.E., Enumeration of total aerobic microorganisms in foods by SimPlate Total Plate Count-Color Indicator methods and conventional culture methods: collaborative study. *J AOAC Int* 86, 257, 2003.

19. Paulsen, P., Schopf, E. and Smulders, F.J., Enumeration of total aerobic bacteria and *Escherichia coli* in minced meat and on carcass surface samples with an automated most-probable-number method compared with colony count protocols. *J Food Prot* 69, 2500, 2006.

20. Kang, D.-H., Dougherty, R.H. and Fung, D.Y.C., Comparison of Pulsifier and Stomacher to detach microorganisms from lean meat tissues. *J Rapid Methods Auto Microbiol* 9, 27, 2001.

21. Duffy, G., Sheridan, J.J., McDowell, D.A., Blair, I. and Harrington, D., The use of Alcalase 2.5L in the acridine orange direct count technique for the rapid enumeration of bacteria in beef mince. *Lett Appl Microbiol* 13, 198, 1991.

22. Duffy, G. and Sheridan, J.J., Viability staining in a direct count rapid method for the determination of total viable counts on processed meats. *J Microbiol Methods* 31, 167, 1998.

23. Hermida, M., Taboada, M., Menendez, S. and Rodriguez-Otero, J.L., Semi-automated direct epifluorescent filter technique for total bacterial count in raw milk. *J AOAC Int* 83, 1345, 2000.

24. D'Haese, E. and Nelis, H.J., Rapid detection of single cell bacteria as a novel approach in food microbiology. *J AOAC Int* 85, 979, 2002.

25. Thore, A., Technical aspects of the bioluminescent firefly luciferase assay of ATP. *Sci. Tools* 26, 30, 1979.

26. McElroy, W.D., The energy source for bioluminescence in an isolated system. *Proc Natl Acad Sci USA* 33, 342, 1947.

27. Davidson, C.A., Griffith, C.J., Peters, A.C. and Fielding, L.M., Evaluation of two methods for monitoring surface cleanliness-ATP bioluminescence and traditional hygiene swabbing. *Luminescence* 14, 33, 1999.
28. Aycicek, H., Oguz, U. and Karci, K., Comparison of results of ATP bioluminescence and traditional hygiene swabbing methods for the determination of surface cleanliness at a hospital kitchen. *Int J Hyg Environ Health* 209, 203, 2006.
29. Bossuyt, R., Determination of bacteriological quality of raw milk by an ATP assay technique. *Milchwissenschaft* 36, 257, 1981.
30. Bossuyt, R., A 5 minute ATP platform test for judging the bacteriological quality of raw milk. *Neth Milk Dairy J* 36, 355, 1982.
31. Stannard, C.J. and Wood, J.M., The rapid estimation of microbial contamination of raw meat by measurement of adenosine triphosphate (ATP). *J Appl Bacteriol* 55, 429, 1983.
32. Siragusa, G.R., Dorsa, W.J., Cutter, C.N., Perino, L.J. and Koohmaraie, M., Use of a newly developed rapid microbial ATP bioluminescence assay to detect microbial contamination on poultry carcasses. *J Biolumin Chemilumin* 11, 297, 1996.
33. Ellerbroek, L. and Lox, C., The use of neck skin for microbial process control of fresh poultry meat using the bioluminescence method. *Dtsch Tierarztl Wochenschr* 111, 181, 2004.
34. Firstenberg-Eden, R., Electrical impedance method for determining microbial quality of foods. In: *Foodborne Microorganisms and Their Toxins: Developing Methodology*, Pierson, M.D. and Stern, N.J., Eds., Marcel Dekker, New York, NY, 679, 1985.
35. Dziezak, J.D., Rapid methods for microbiological analysis of food. *Food Technol* 41, 56, 1987.
36. Firstenberg-Eden, R. and Tricarico, M.K., Impedimetric determination of total, mesophilic and psychrotrophic counts in raw milk. *J Food Sci* 48, 1750, 1983.
37. Hardy, D., Kraeger, S.J., Dufour, S.W. and Cady, P., Rapid detection of microbial contamination in frozen vegetables by automated impedance measurements. *Appl Environ Microbiol* 34, 14, 1977.
38. Firstenberg-Eden, R., Rapid estimation of the number of microorganisms in raw meat by impedance measurements. *Food Technol* 37, 64, 1983.
39. Ogden, I.D., Use of conductance methods to predict bacterial counts in fish. *J Appl Bacteriol* 61, 263, 1986.
40. Bollinger, S., Casella, M. and Teuber, M., Comparative impedance evaluation of the microbial load of different foodstuffs. *Lebensm Wiss Technol* 27, 177, 1994.
41. Heeschen, W., Sudi, J. and Suhren, G., Application of the *Limulus* test for detection of gram negative micro-organisms in milk and dairy products. In: *Rapid Methods and Automation in Microbiology and Immunology*, Habermahl, K.O., Ed., Springer-Verlag, Berlin, 638, 1985.
42. Misawa, N., Kumamoto, K., Nyuta, S. and Tuneyoshi, M., Application of the *Limulus* amoebocyte lysate test as an indicator of microbial contamination in pork carcasses. *J Vet Med Sci* 57, 351, 1995.
43. Siragusa, G.R., Kang, D.H. and Cutter, C.N., Monitoring the microbial contamination of beef carcass tissue with a rapid chromogenic *Limulus* amoebocyte lysate endpoint assay. *Lett Appl Microbiol* 31, 178, 2000.
44. Gillie, J.K., Hochlowski, J. and Arbuckle-Keil, G.A., Infrared spectroscopy. *Anal Chem* 72, 71R, 2000.
45. Lin, M., Al-Holy, M., Chang, S.S., Huang, Y., Cavinato, A.G., Kang, D.H. and Rasco, B.A., Rapid discrimination of *Alicyclobacillus* strains in apple juice by Fourier transform infrared spectroscopy. *Int J Food Microbiol* 105, 369, 2005.
46. Lamprell, H., Mazerolles, G., Kodjo, A., Chamba, J.F., Noel, Y. and Beuvier, E., Discrimination of *Staphylococcus aureus* strains from different species of *Staphylococcus* using Fourier transform infrared (FT-IR) spectroscopy. *Int J Food Microbiol* 108, 125, 2006.
47. Oberreuter, H., Seiler, H. and Scherer, S., Identification of coryneform bacteria and related taxa by Fourier-transform infrared (FT-IR) spectroscopy. *Int J Syst Evol Microbiol* 52, 91, 2002.
48. Mariey, L., Signolle, J.P., Amiel, C. and Travert, J., Discrimination, classification, identification of microorganisms using FT-IR spectroscopy and chemometrics. *Vib Spectrosc* 26, 151, 2001.

49. Ellis, D.I., Broadhurst, D., Kell, D.B., Rowland, J.J. and Goodacre, R., Rapid and quantitative detection of the microbial spoilage of meat by Fourier transform infrared spectroscopy and machine learning. *Appl Environ Microbiol* 68, 2822, 2002.

50. Ellis, D.I., Broadhurst, D. and Goodacre, R., Rapid and quantitative detection of the microbial spoilage of beef by Fourier transform infrared spectroscopy and machine learning. *Anal Chim Acta* 514, 193, 2004.

51. Lin, M., Al-Holy, M., Mousavi-Hesary, M., Al-Qadiri, H., Cavinato, A.G. and Rasco, B.A., Rapid and quantitative detection of the microbial spoilage in chicken meat by diffuse reflectance spectroscopy (600–1100 nm). *Lett Appl Microbiol* 39, 148, 2004.

52. Lin, M., Mousavi, M., Al-Holy, M., Cavinato, A.G. and Rasco, B.A., Rapid near infrared spectroscopic method for the detection of spoilage in rainbow trout (*Oncorhynchus mykiss*) fillet. *J Food Sci* 71, S18, 2006.

53. Holm, C., Mathiasen, T. and Jespersen, L., A flow cytometric technique for quantification and differentiation of bacteria in bulk tank milk. *J Appl Microbiol* 97, 935, 2004.

54. Flint, S., Walker, K., Waters, B. and Crawford, R., Description and validation of a rapid (1 h) flow cytometry test for enumerating thermophilic bacteria in milk powders. *J Appl Microbiol* 102, 909, 2007.

55. Gutierrez, R., Garcia, T., Gonzalez, I., Sanz, B., Hernandez, P.E. and Martin, R., Quantitative detection of meat spoilage bacteria by using the polymerase chain reaction (PCR) and an enzyme linked immunosorbent assay (ELISA). *Lett Appl Microbiol* 26, 372, 1998.

56. Yost, C.K. and Nattress, F.M., The use of multiplex PCR reactions to characterize populations of lactic acid bacteria associated with meat spoilage. *Lett Appl Microbiol* 31, 129, 2000.

57. Goto, S., Takahashi, H., Kawasaki, S., Kimura, B., Fujii, T., Nakatsuji, M. and Watanabe, I., Detection of *Leuconostoc* strains at a meat processing plant using polymerase chain reaction. *Shokuhin Eiseigaku Zasshi* 45, 25, 2004.

58. Venkitanarayanan, K.S., Faustman, C., Crivello, J.F., Khan, M.I., Hoagland, T.A. and Berry, B.W., Rapid estimation of spoilage bacterial load in aerobically stored meat by a quantitative polymerase chain reaction. *J Appl Microbiol* 82, 359, 1997.

59. Bellin, T., Pulz, M., Matussek, A., Hempen, H.G. and Gunzer, F., Rapid detection of enterohemorrhagic *Escherichia coli* by real-time PCR with fluorescent hybridization probes. *J Clin Microbiol* 39, 370, 2001.

60. Wittwer, C.T., Herrmann, M.G., Moss, A.A. and Rasmussen, R.P., Continuous fluorescence monitoring of rapid cycle DNA amplification. *Biotechniques* 22, 130–131, 134, 1997.

61. Mackay, I.M., Real-time PCR in the microbiology laboratory. *Clin Microbiol Infect* 10, 190, 2004.

62. Josefsen, M.H., Krause, M., Hansen, F. and Hoorfar, J., Optimization of a 12-hour TaqMan PCR-based method for detection of *Salmonella* bacteria in meat. *Appl Environ Microbiol* 73, 3040, 2007.

63. Navas, J., Ortiz, S., Lopez, P., Jantzen, M.M., Lopez, V. and Martinez-Suarez, J.V., Evaluation of effects of primary and secondary enrichment for the detection of *Listeria monocytogenes* by real-time PCR in retail ground chicken meat. *Foodborne Pathog Dis* 3, 347, 2006.

64. Perelle, S., Dilasser, F., Grout, J. and Fach, P., Screening food raw materials for the presence of the world's most frequent clinical cases of Shiga toxin-encoding *Escherichia coli* O26, O103, O111, O145 and O157. *Int J Food Microbiol* 113, 284, 2007.

65. Holicka, J., Guy, R.A., Kapoor, A., Shepherd, D. and Horgen, P.A., A rapid (one day), sensitive real-time polymerase chain reaction assay for detecting *Escherichia coli* O157:H7 in ground beef. *Can J Microbiol* 52, 992, 2006.

66. Nguyen, L.T., Gillespie, B.E., Nam, H.M., Murinda, S.E. and Oliver, S.P., Detection of *Escherichia coli* O157:H7 and *Listeria monocytogenes* in beef products by real-time polymerase chain reaction. *Foodborne Pathog Dis* 1, 231, 2004.

67. Rudi, K., Moen, B., Dromtorp, S.M. and Holck, A.L., Use of ethidium monoazide and PCR in combination for quantification of viable and dead cells in complex samples. *Appl Environ Microbiol* 71, 1018, 2005.

68. Guy, R.A., Kapoor, A., Holicka, J., Shepherd, D. and Horgen, P.A., A rapid molecular-based assay for direct quantification of viable bacteria in slaughterhouses. *J Food Prot* 69, 1265, 2006.

69. Pardue, M.L. and Gall, J.G., Molecular hybridization of radioactive DNA to the DNA of cytological preparations. *Proc Natl Acad Sci USA* 64, 600, 1969.

70. John, H.A., Birnstiel, M.L. and Jones, K.W., RNA–DNA hybrids at the cytological level. *Nature* 223, 582, 1969.

71. Giovannoni, S.J., DeLong, E.F., Olsen, G.J. and Pace, N.R., Phylogenetic group-specific oligodeoxy-nucleotide probes for identification of single microbial cells. *J Bacteriol* 170, 720, 1988.

72. Amann, R.I., Krumholz, L. and Stahl, D.A., Fluorescent-oligonucleotide probing of whole cells for determinative, phylogenetic, and environmental studies in microbiology. *J Bacteriol* 172, 762, 1990.

73. Amann, R., Fuchs, B.M. and Behrens, S., The identification of microorganisms by fluorescence *in situ* hybridisation. *Curr Opin Biotechnol* 12, 231, 2001.

74. Bottari, B., Ercolini, D., Gatti, M. and Neviani, E., Application of FISH technology for microbiological analysis: current state and prospects. *Appl Microbiol Biotechnol* 73, 485, 2006.

75. Moter, A. and Gobel, U.B., Fluorescence in situ hybridization (FISH) for direct visualization of microorganisms. *J Microbiol Methods* 41, 85, 2000.

76. Zarda, B., Amann, R., Wallner, G. and Schleifer, K.H., Identification of single bacterial cells using digoxigenin-labelled, rRNA-targeted oligonucleotides. *J Gen Microbiol* 137, 2823, 1991.

77. Schonhuber, W., Fuchs, B., Juretschko, S. and Amann, R., Improved sensitivity of whole-cell hybridization by the combination of horseradish peroxidase-labeled oligonucleotides and tyramide signal amplification. *Appl Environ Microbiol* 63, 3268, 1997.

78. Lehtola, M.J., Loades, C.J. and Keevil, C.W., Advantages of peptide nucleic acid oligonucleotides for sensitive site directed 16S rRNA fluorescence in situ hybridization (FISH) detection of *Campylobacter jejuni, Campylobacter coli* and *Campylobacter lari. J Microbiol Methods* 62, 211, 2005.

79. Jain, K.K., Current status of fluorescent in-situ hybridisation. *Med Device Technol* 15, 14, 2004.

80. Stender, H., Lund, K., Petersen, K.H., Rasmussen, O.F., Hongmanee, P., Miorner, H. and Godtfredsen, S.E., Fluorescence *in situ* hybridization assay using peptide nucleic acid probes for differentiation between tuberculous and nontuberculous *Mycobacterium* species in smears of *Mycobacterium* cultures. *J Clin Microbiol* 37, 2760, 1999.

81. Fang, Q., Brockmann, S., Botzenhart, K. and Wiedenmann, A., Improved detection of *Salmonella* spp. in foods by fluorescent in situ hybridization with 23S rRNA probes: a comparison with conventional culture methods. *J Food Prot* 66, 723, 2003.

82. Gunasekera, T.S., Dorsch, M.R., Slade, M.B. and Veal, D.A., Specific detection of *Pseudomonas* spp. in milk by fluorescence in situ hybridization using ribosomal RNA directed probes. *J Appl Microbiol* 94, 936, 2003.

83. Blasco, L., Ferrer, S. and Pardo, I., Development of specific fluorescent oligonucleotide probes for in situ identification of wine lactic acid bacteria. *FEMS Microbiol Lett* 225, 115, 2003.

84. Gardener, J.W. and Bartlett, P.N., *Electronic Noses—Principles and Applications*, Oxford University Press, Oxford, 1999.

85. Magan, N. and Sahgal, N., Electronic nose for quality and safety control. In: *Advances in Food Diagnostics*, Nollet, L.M.L., Toldrá, F. and Hui, Y.H., Eds., Blackwell Publishing, Oxford, 119, 2007.

86. Blixt, Y. and Borch, E., Using an electronic nose for determining the spoilage of vacuum-packaged beef. *Int J Food Microbiol* 46, 123, 1999.

87. Du, W.X., Kim, J., Cornell, J.A., Huang, T., Marshall, M.R. and Wei, C.I., Microbiological, sensory, and electronic nose evaluation of yellowfin tuna under various storage conditions. *J Food Prot* 64, 2027, 2001.

88. Van Loey, A., Hendrickx, M., Ludikhuyze, L., Weemaes, C., Haentjens, T., De Cordt, S. and Tobback, P., Potential *Bacillus subtilis* alpha-amylase-based time–temperature integrators to evaluate pasteurization processes. *J Food Prot* 59, 261, 1996.

89. Mehauden, K., Cox, P.W., Bakalis, S., Simmons, M.J.H., Tucker, G.S. and Fryer, P.J., A novel method to evaluate the applicability of time temperature integrators to different temperature profiles. *Innov Food Sci Emerg Technol* 8, 507–514, 2007.

90. Guiavarch, Y., Zuber, F., van Loey, A. and Hendrickx, M., Combined use of two single-component enzymatic time–temperature integrators: application to industrial continuous rotary processing of canned ravioli. *J Food Prot* 68, 375, 2005.

91. Guiavarc'h, Y., Van Loey, A., Zuber, F. and Hendrickx, M., Development characterization and use of a high-performance enzymatic time–temperature integrator for the control of sterilization process' impacts. *Biotechnol Bioeng* 88, 15, 2004.

92. Tucker, G.S., Brown, H.M., Fryer, P.J., Cox, P.W., Poole II, F.L., Lee, H.S. and Adams, M.W.W., A sterilisation time–temperature integrator based on amylase from the hyperthermophilic organism *Pyrococcus furiosus*. *Innov Food Sci Emerg Technol* 8, 63, 2007.

93. Ozdemir, M. and Floros, J.D., Active food packaging technologies. *Crit Rev Food Sci Nutr* 44, 185, 2004.

94. Giannakourou, M.C., Koutsoumanis, K., Nychas, G.J. and Taoukis, P.S., Field evaluation of the application of time temperature integrators for monitoring fish quality in the chill chain. *Int J Food Microbiol* 102, 323, 2005.

95. Moore, C.M. and Sheldon, B.W., Use of time–temperature integrators and predictive modeling to evaluate microbiological quality loss in poultry products. *J Food Prot* 66, 280, 2003.

96. Moore, C.M. and Sheldon, B.W., Evaluation of time–temperature integrators for tracking poultry product quality throughout the chill chain. *J Food Prot* 66, 287, 2003.

97. Pettipher, G.L. and Rodrigues, U.M., Rapid enumeration of microorganisms in foods by the direct epifluorescent filter technique. *Appl Environ Microbiol* 44, 809, 1982.

98. Liberski, D.J.A., Bacteriological examinations of chilled, cured canned pork hams and shoulders using a conventional microbiological technique and the DEFT method. *Int J Food Microbiol* 10, 19, 1990.

99. Qvist, S.H. and Jakobsen, M., Application of the direct epifluorescent filter technique as a rapid method in microbiological quality assurance in the meat industry. *Int J Food Microbiol* 2, 139, 1985.

100. Walls, I., Sheridan, J.J. and Levett, P.N., A rapid method of enumerating microorganisms from beef, using an acridine orange direct-count technique. *Irish J Food Sci Technol* 13, 23, 1989.

101. Sheridan, J.J., Walls, I. and Levett, P.N., Development of a rapid method for enumeration of bacteria in pork mince. *Irish J Food Sci Technol* 14, 1, 1990.

102. Walls, I., Sheridan, J.J., Welch, R.W. and McDowell, D.A., Separation of microorganisms from meat and their rapid enumeration using a membrane filtration-epifluorescent microscopy technique. *Lett Appl Microbiol* 10, 23, 1990.

103. Sierra, M.L., Sheridan, J.J. and McGuire, L., Microbial quality of lamb carcasses during processing and the acridine orange direct count technique (a modified DEFT) for rapid enumeration of total viable counts. *Int J Food Microbiol* 36, 61, 1997.

Chapter 34

Microbial Foodborne Pathogens

Marios Mataragas, Dafni Kagli, and George-John E. Nychas

Contents

34.1 Introduction

Pathogenic bacteria are responsible for a large proportion of all foodborne illnesses. The presence of these bacteria on raw meat (beef, lamb, and pork) and poultry is the result of contamination from live animals, equipments, employees, and environment. Pathogens such as *Salmonella, Listeria monocytogenes, Staphylococcus aureus, Yersinia enterocolitica, Escherichia coli* (mainly *E. coli* O157:H7), *Campylobacter jejuni*, and *Clostridium perfringens* have been implicated in foodborne outbreaks associated with the consumption of meat and poultry. *C. jejuni* frequently occurs on poultry meat whereas *E. coli* is rarely found on poultry surfaces. However, beef has been implicated in many foodborne outbreaks associated with *E. coli, Salmonella*, and *L. monocytogenes* may be found on all types of meat, including beef, lamb, pork, and poultry but *Y. enterocolitica* is usually present on pork meat surface.[1,2] Psychrotrophic pathogens such as *L. monocytogenes* and *Y. enterocolitica* are of great concern because they are able to multiply and attain large numbers even at refrigeration temperatures, especially when the products are kept at abused temperatures (>7–$8°C$) for extended periods of time.[3] *St. aureus* and *Cl. perfringens* are of great concern because their growth results in the production of toxins in food. For more detailed information on the protocols and the culture media (including their preparation) mentioned in the following paragraphs for both conventional and rapid microbiological methods, interested readers should consult *Compendium of Methods for the Microbiological Examination of Foods, Handbook of Microbiological Media for the Examination of Food*, and *The Compendium of Analytical Methods*.[1,4,5]

34.2 Conventional (Traditional) Microbiological Methods

Conventional or traditional methods are simple and relatively inexpensive but they are time-consuming. Food sample (usually 25 g) is homogenized in a stomacher bag with 225 mL of diluent, using a stomacher machine for 1:10 dilution. Diluent must be correctly prepared in terms of buffer capacity and osmotic pressure (0.1%, w/v, saline peptone water—0.1% peptone and 0.85% NaCl), otherwise the microbial cells of the target microorganism may suffer stress, influencing the results. The sample, withdrawn for microbiological analysis, should be representative and randomly selected from different parts of the food. This is to ensure, to some degree, the detection of the target microorganisms, which are often irregularly distributed in the solid foods. Further 1:10 dilutions from the first one may be required depending on the population level of the target microorganism present in the food. An adequate volume of sample from the appropriate dilution is spread (0.1 mL), poured (1 mL), or streaked on selective agars to differentiate or enumerate the target microorganism. Non-selective agars may also be used to perform confirmatory biochemical and serological tests. In some cases, an enrichment and, if it is necessary, a preenrichment step may precede the first 1:10 dilution to suppress the growth of other microorganisms and allow the recovery of the injured cells.

34.2.1 Enumeration Methods

In general, two enumeration methods are used more often: the plate count and the most probable number. The second method is used for certain microorganisms, such as coliforms and *E. coli*.[6,7]

34.2.1.1 Plate Count

Plate count is the most popular conventional enumeration method. The procedure involves homogenization of the food sample, dilution, plating on various media, and incubation at selected temperatures. After incubation for a sufficient period of time, counting of the specific colonies of the target microorganism is performed. If confirmation of the target microorganism is required, and a number of randomly selected colonies are obtained for confirmatory purposes, then the ratio of the colonies confirmed as the target microorganism to total colonies tested should be multiplied by the average number of colonies and the dilution factor to obtain the number of viable cells per gram of food sample. For instance, if the mean number of *Cl. perfringens* colonies from two pour agar plates is 20 at the second dilution (10^{-2}) and the confirmed colonies of 10 randomly selected (5 per plate) colonies are 8, then the number of viable *Cl. perfringens* per gram of food sample will be as follows: $20 \times 10^2 \times (8/10) = 1.6 \times 10^3$. Laboratory media, used to subculture the microorganisms present in the food sample, are divided into three categories: elective, selective, and differential.[8] Elective media contain agents (e.g., microelements) that support the growth of the target microorganism but do not inhibit the growth of the accompanied microflora. This is achieved by the use of selective media containing some inhibitory agents, such as antibiotics. These agents inhibit the growth of the nontarget microorganisms as well as, in some cases, the growth of the microorganisms under examination, but to a lesser degree. Differential media contain agents that allow the differentiation of the microorganisms (e.g., chromogenic media). These agents react with the colonies, changing the color of the media. Usually, the nutrient substrates contain all the above-mentioned agents to ensure proper identification of the target microorganism.

34.2.1.2 Most Probable Number

The number of viable cells in a food sample is assessed on the basis of probability tables. The food sample is diluted (10-fold dilutions), and then the samples from each dilution are transferred to three tubes containing a growth medium (broth). After incubation of the tubes, the turbidity is measured and the tubes showing turbidity (growth) are compared to probability tables to find the population level of the target microorganisms present in the food.[1]

34.2.2 Detection Methods

Detection methods are used when the presence or absence of a specific pathogen is of concern. These methods include additional steps, such as enrichment and preenrichment, to allow injured cells to recover or increase to a detectable population because the target microorganism may be present at very low levels compared with the population levels of the dominant microflora. Usually, 25 g of food sample is aseptically weighted in a stomacher bag, homogenized in an enrichment broth (225 mL), and incubated for a certain period of time at a known temperature. After incubation, a sample from the broth is streaked on a selective agar plate, using a bacteriological loop. If the examined microorganism is present, then it is identified by its characteristic colonies formed on the agar. To confirm or identify further the microorganism at strain level, some additional biochemical or serological tests may be needed. These tests are performed on a pure culture; therefore, colonies from the selective agar plates are purified (streaking) on nonselective agar plates, for example, nutrient agar.

34.3 Injured or Stressed Cells

During processing of foods, the microbial cells may suffer, to some degree, a damage resulting in an inability to form visible colonies on plate count agars. These cells may remain undetected on selective agars, but are still viable and, under conditions that favor their growth, may recover and become active. This is of great importance for foodborne pathogens that may lead to a food poisoning outbreak. Therefore, additional steps such as the previously mentioned enrichment steps are included in the analytical procedures to allow injured cells to resuscitate and repair. There are many factors that influence the resuscitation of the injured cells such as composition and characteristics of the medium and environmental parameters.[9] Therefore, the analytical methods for the detection of the microorganisms are constructed in such a way as to allow maximum performance (recovery of stressed cells).

34.4 Rapid Microbiological Methods

Pathogen detection is of utmost importance, primarily for health and safety reasons.[10] In contrast to conventional methods, rapid microbiological methods are much faster but have the disadvantage that they are expensive. The most popular methods are, by far, those based on culture and colony counting methods,[11] as well as the polymerase chain reaction.[12] A careful look at the requirements of a laboratory or a food industry is required before the adaptation of a method. These methods also include an enrichment step, termed as concentration step aimed to separate and concentrate the target microorganism or toxin. In this way, the detection time is shorter and the specificity is improved. Such methods that concentrate the target microorganism or toxin are as follows:[4,13–15] (1) the immunomagnetic separation (IMS), in which antibodies linked to paramagnetic particles are added and the target microorganism is trapped because of antigen–antibody reaction. Commercial kits are available for immunomagnetic separation of various foodborne pathogens such as *L. monocytogenes*, *Salmonella* spp., and *E. coli* O157:H7 (Dynabeads™, Dynal Botech, Oslo, Norway). The immunomagnetic separation for *Salmonella* (10 min) has been proved to successfully replace the enrichment step (overnight incubation) of the standard procedure for the detection of *Salmonella*, shortening the time needed to obtain the results. (2) An alternative of the previous method is the technique of metal hydroxide-based bacterial concentration. Metal (hafnium, titanium, or zirconium) hydroxide suspensions react with the opposing charge of the bacterial cells. Then the cells are separated by centrifugation, resuspended, and plated. (3) The hydrophobic grid membrane filter is a filtration method similar to the method used for water. First the food sample is filtered to remove the large particles (>5 μm) and is then filtered through a grid membrane on which the microorganisms remain. The membrane is placed on a selective agar, and after an appropriate time for incubation, the colony counts estimated. (4) The direct epifluorescent technique (DEFT) is used for enumerating viable bacteria in milk and milk products. The cells of microorganisms are concentrated through filtration on a membrane and then retained microorganisms are colored, usually with acridine orange (fluorescent dyes), and counted. Viable cells are red (acridine orange fluoresces red with RNA) and nonviable green (acridine orange fluoresces green with DNA).

Some of the widely used methods for the identification and detection of the desired microorganisms are the following: (1) improved culture media containing chromogenic ingredients that produce a specific color or reaction because of bacterial metabolism. For instance, the chromogenic media ALOA (Agar Listeria Ottaviani and Agosti) and RAPID' L.Mono use the following properties to differentiate *Listeria* spp. and *L. monocytogenes* from the other *Listeria* species. ALOA contains a chromogenic compound that colors the listeriae colonies due to its degradation by the enzyme β-glucosidase. This enzyme is produced by all *Listeria* species. The differentiation of pathogenic *Listeria* from the nonpathogenic species is based on the formation of phosphatidylinositol phospholipase C (PI-PLC). This compound hydrolyzes a specific substrate added to the growth medium, resulting in a turbid halo (ALOA) or a specific color of colonies (RAPID' L.mono).[16] Petrifilm method (3M, Minneapolis, MN), is another method that uses a plastic film together with the appropriate medium in dried form. It is used mainly for coliforms (red colonies with gas bubbles) and *E. coli* (blue colonies with gas bubbles). One milliliter of the sample is added directly to the plates to rehydrate the medium. Plates are then incubated and counted. Validation and collaborative studies have found the Petrifilm method to be not significantly different from the traditional methods.[4,15,17] (2) Next are the PCR-based methods coupled to other techniques: most probable number counting method (MPN-PCR),[18] surface plasmon resonance and PCR acoustic wave sensors,[19] LightCycler real-time PCR (LC-PCR), PCR enzyme-linked immunosorbent assay (PCR-ELISA),[20] sandwich hybridization assays (SHAs),[21] and fluorescent *in situ* hybridization (FISH) detection test.[22] Of these methods, ELISA has been widely used for pathogen detection and identification, especially for *Salmonella* spp. and *L. monocytogenes*. The detection limit is 10^4 cfu/g; therefore, a conventional enrichment step before testing is required. Specific antibodies for the target microorganism, contained in microtiter plates, react with the antigen, which is detected using a second antibody conjugated to an enzyme (horse radish peroxidase or alkaline phosphatase) to give a colorimetric reaction after the addition of substrate.[4,14,15] (3) Adenosine triphosphate (ATP) bioluminescence can be used as an indicator of microbial contamination in foods and processing plants. The method detects the presence of bacterial ATP. In a buffer containing magnesium, luciferase is added to a sample along with luciferin. The latter is oxidized (oxyluciferin) and the photons of light produced are measured by a luminometer. A standard curve is used to calculate the contamination level, and the sensitivity of the method is 10^4 cfu/mL. (4) Reversed passive latex agglutination is used for the detection of toxins such as Shiga toxins from *E. coli*. Latex beads containing antibodies (rabbit antiserum) specific for the target microorganism react with the target antigen, if present. The particles agglutinate, and a V-shaped microtiter well has a diffuse appearance. If the antigen is not present, then a dot will appear.[4,14]

A rapid microbiological method frequently used for enumeration is the impedance or conductance technique. The method rapidly detects the growth of a specific microorganism, based on the production of charged metabolites (direct method) or carbon dioxide liberation (indirect method). In the first method, detection is achieved by the change in the conductivity of the culture medium because of the accumulation of various products produced by the microorganism, such as organic acids. These changes are recorded at constant time intervals. "Time to detection" is the time needed for the conductance value to be changed. Because the "time to detection" is dependent on the inoculum size, a calibration curve is made for a known wide range of population level of the desired microorganism. This curve facilitates the calculation of the population level of an unknown sample after the automatic determination of the "time to detection" by the equipment. In the other method, the sample is distinguished from the potassium hydroxide bridge by a head space in the test tube. The carbon dioxide produced during the microbial growth in the head space reacts with potassium hydroxide, forming potassium carbonate, which is less conductive. The conductance decrease is the recorded parameter.[14,15]

Genotypic, molecular methods are useful to identify bacteria as complement or alternative to phenotypic methods; besides enhancing the sensitivity and specificity of the detection process, they reduce much of the subjectivity inherent in interpreting the results. DNA is invariant throughout microbial life cycle and after short-term environmental stress factors. This is the reason why molecular methods targeting genomic DNA (gDNA) are generally applicable.[23] Restriction fragment length polymorphism (RFLP) of total gDNA represents a technique belonging to the first-generation molecular methods[24] widely used in microbial differentiation. Southern blot hybridization tests, which enhance the result of agarose gel electrophoresis by marking specific DNA sequences, have also been used. Second-generation molecular techniques (known as PCR-based technologies) such as PCR-RFLP and randomly amplified polymorphic DNA-PCR (RAPD-PCR) have been used for differentiation and identification of microbial isolates.[23] Recent advances in PCR technology, namely real-time PCR,[25] enable obtaining results within a few hours.[10] Quantification of microorganisms is of major importance, especially in the case of toxigenic bacteria, on the concentration of which depends the toxin production.[23]

Biosensors technology comes with promises of equally reliable results in a much shorter time, and is currently gaining profound interest. Many biosensors rely on either specific antibodies or DNA probes to provide specific results.[10]

The current trend is toward culture-independent PCR-based methods, which unlike the previously mentioned ones, are believed to overcome problems associated with selective cultivation and isolation of microorganisms from natural samples. The most commonly used method, among the culture-independent fingerprinting techniques, is PCR followed by denaturing gradient gel electrophoresis (DGGE). PCR-DGGE provides information about the variation of the PCR products, of the same length, but with different sequences upon differential mobility in an acrylamide gel matrix of increasing denaturant concentration.[23,26]

34.5 *Listeria monocytogenes*

L. monocytogenes is widely distributed in the environment and can be found in many food commodities.[3,27] It is a very persistent microorganism that survives on the equipment in case of inefficient cleaning.[28–32] Postprocessing contamination from the plant environment (equipment, personnel, floors, etc.) is the most frequent reason for its presence on the meat surface. Cross-contamination may also occur at the retail as well as at home, especially when the products have been mishandled and improper hygiene practice has been followed.[32,33] Various foods such as milk and dairy products (e.g., cheese), meat (including poultry) and meat products, vegetables, and fish and fish products have been implicated in outbreaks of foodborne *L. monocytogenes*.[34] It is usually killed during pasteurization but it is capable of growing in foods stored at refrigeration temperatures (psychrotrophic microorganism).[35,36] High salt concentrations and acid conditions do not allow *L. monocytogenes* growth.[36] However, it may survive even under these stressful environmental conditions.[37] Therefore, consumption of unpasteurized products or manufacture of products without a killing step, for example, pasteurization, products with little preservation factors, such as high initial pH, low salt content or high water activity, and products supporting growth of the pathogen stored at refrigeration temperatures for long periods of time are factors that increase the potential of acquiring listeriosis after the consumption of food contaminated with *L. monocytogenes*.[36,38] *L. monocytogenes* is a major hazard, particularly for elderly, immunocompromised people, infants, and pregnant women.

34.5.1   *Detection of* Listeria monocytogenes

The scheme to be followed for the conventional detection of *L. monocytogenes* in raw meat and poultry is shown in Figure 34.1 (based on Ref. 39). Two enrichment steps are employed by the method to detect the presence of *Listeria*. These steps are followed because enrichment makes it feasible to detect low number of *Listeria*, such as one cell per 25 g of food, because the microorganism is allowed to grow to a level of *ca.* 10^4–10^5 cfu/g. The first enrichment step is made with 1/2 Fraser broth (half concentrated Fraser broth) containing only half concentration of the inhibitory agents (antibiotics) because these agents may also have a negative effect on stressed or injured *Listeria* cells.[40,41] Antibiotics (acriflavin and nalidixic acid) are used to suppress the growth

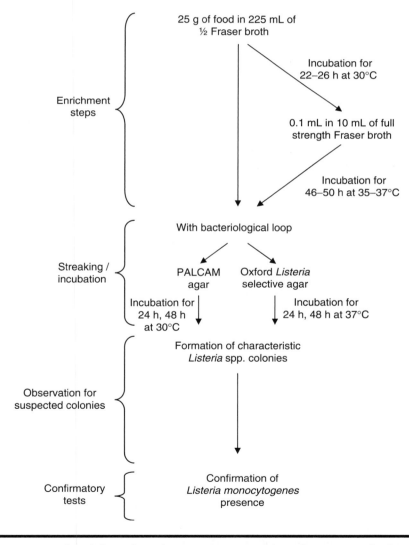

Figure 34.1   Conventional detection of *L. monocytogenes* based on ISO method.

of the accompanying microflora, which may outgrow *Listeria* due to its slow growth. The presence of *Listeria* on the selective agar plates is observed by the formation of characteristic colonies. They are grey-green with a black center surrounded by a black zone. Aesculin accompanied with ferrous iron is also added to Fraser broth, in conjunction with antibiotics, to allow detection of β-D-glycosidase activity by *Listeria*-induced blackening of the medium.[14,41]

Alternative molecular methods that monitor the incidence of *Listeria* spp. in foods are also applied. Suggested techniques include fluorescent antibody assay, enzyme immunoassay (EIA), flow cytometry (FCM), and DNA hybridization.[42]

DNA hybridization is the simplest molecular method used for the detection of *Listeria* spp. and *L. monocytogenes* in foods. The presence of a target sequence is detected using an oligonucleotide probe of a sequence complementary to the target DNA sequence, which contains a label for detection. Radioactive isotopes, biotinylated probes, and probes incorporating digoxygenin or fluorescent markers allow detection of target sequences.[41] PCR combined with DNA hybridization in a microtiter plate is a convenient, highly sensitive, and specific approach for detection of *Listeria* spp. in a high-throughput 96-well format.[43] Commercially available DNA hybridization tests are routinely used for food testing and have been proven to be extremely sensitive and accurate.

In contrast to DNA hybridization, where large amounts of DNA or RNA are necessary for detection, PCR provides amplification results starting from very little amounts of target DNA.[41] Detection using PCR is carried out after selectively enriching samples for 24–48 h. Multiplex PCR (MPCR) allows the simultaneous detection of more than one pathogen in the same sample, such as *L. monocytogenes* and *Salmonella*[44,45] or *L. monocytogenes* and other *Listeria* species.[46,47] This approach is the most attractive for food analysis, where testing time, reagents, and labor costs are reduced. To detect only living pathogens, RNA can be used instead of DNA. The presence of specific RNA sequences is an indication of live cells. When an organism dies, its RNA is quickly eliminated whereas DNA can last for years, depending on the storage conditions. Klein and Juneja[48] used reverse transcription-PCR (RT-PCR) to detect live *L. monocytogenes* in pure culture and artificially contaminated cooked ground beef.

The use of DNA microarrays is a recent technique that has found applicability in the detection of *L. monocytogenes*. Call et al.[49] used probes specific for unique portions of the 16S rRNA gene in *Listeria* spp. to demonstrate how each *Listeria* species can be differentiated by this method. In this procedure, first PCR is performed using universal primers to amplify all the 16S rRNA genes present in a sample. The various amplified DNA fragments bind only to the probes for which they have a complementary sequence. Because one of the oligonucleotides used in the PCR contains a fluorescent label, the spots where the amplified DNA has bound fluoresce. Pathogens are identified by the pattern of fluorescing spots in the array.[50] Lampel et al.[51] and Sergeev et al.[52] claim that in pure culture, the detection limit of the array is 200 cfu of *L. monocytogenes*. It is also claimed that the array is appropriate for the detection of pathogens in food and environmental samples.[52] Microarrays are able to identify a number of pathogens or serotypes at once, but they still require culture enrichment and PCR steps to improve sensitivity and specificity of detection.[50]

34.5.2 Enumeration of Listeria monocytogenes

Figure 34.2 shows the conventional enumeration method of *L. monocytogenes* (based on Ref. 53). The method has a detection limit ≥100 cfu/g. If less than 100 cfu/g of *Listeria* are expected, then the following procedure might be applied which allows detection equal to or above 10 cfu/g.

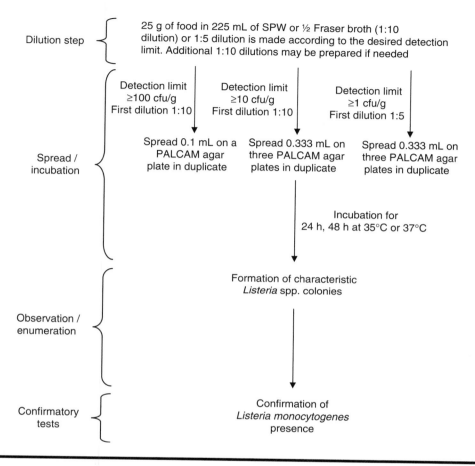

Figure 34.2 Conventional enumeration of *L. monocytogenes* based on ISO method.

One milliliter of sample from the first 1:10 dilution is spread on three PALCAM agar plates (0.333 mL on each agar plate), and after incubation, the colonies on all three plates are measured as a single plate. However, if even lower *Listeria* concentration is expected (1 cfu/g), then the previously mentioned procedure is followed. The difference is that the first dilution is one part of food sample and four parts of diluent (1:5) (saline peptone water [SPW] or 1/2 Fraser broth). SPW (0.1% peptone and 0.85% NaCl) or 1/2 Fraser broth has a large buffer capacity, which favors the cell growth and repair of stressed or injured cells.

Traditional PCR methods are able to detect the presence of a pathogen, but are not able to quantify the level of the pathogen contamination. One way to approach this problem is the use of competitive PCR. In this method, a competitor fragment of DNA, which matches the gene to be amplified, is introduced into the sample. In general, the competitor fragment is synthesized as a deletion mutant that can be amplified by the same primers being used to amplify the target DNA. The competitor fragment is distinguished from the pathogen gene fragment by its smaller size.[50] To determine the level of pathogen contamination, DNA purified from the food sample is serially diluted and added to a constant amount of competitor DNA. PCR is performed and the intensity of the pathogen's gene signal is compared to that of the competitor DNA on an agarose gel.

The number of cells in the original sample can be estimated by comparing the intensity of the two DNA fragments (target vs. competitor),[54] using a standard curve.

Choi and Hong[55] used a variation of competitive PCR, based on the presence of a restriction endonuclease site in the amplified gene for *L. monocytogenes* detection. The method was completed within 5 h without enrichment and was able to detect 10^3 cfu/0.5 mL milk, using the *hlyA* gene as target. The detection limit could be reduced to 1 cfu if culture enrichment for 15 h was conducted first.[55]

34.5.3 Confirmation of Listeria monocytogenes

L. monocytogenes presence can be confirmed by various biochemical tests. The tests are performed on purified cultures. From the PALCAM or Oxford agars, five suspected and randomly selected colonies are isolated and streaked on tryptone soya agar, containing 0.6% yeast extract (TSYEA). *Listeria* species are easily identified by Gram staining, motility, catalase, and oxidase reaction. *Listeria* spp. is gram-positive, small rod-shaped, motile, catalase-positive, and oxidase-negative. Motility test should be performed in a semisolid TSYEA tube (TSYE broth or TSYEB, supplemented with 0.5% agar) incubated at 25°C because at incubation temperatures above 30°C, the motility test is negative (nonmotile). The tube is inoculated by stabbing and observed for growth around the stab (appearance of turbidity with a characteristic umbrella-like shape).[56] Sugar fermentation, hemolysis, and CAMP (Christie, Atkins and Munch-Petersen) test may be used to differentiate the *Listeria* species (Figure 34.3). *L. monocytogenes*, *L. ivanovii*, and *L. seeligeri* are β-hemolytic species on horse or sheep blood agar. The CAMP test distinguishes the three species of *Listeria* and should be done on sheep blood agar. An enhanced β-hemolysis zone is observed close to *St. aureus* NCTC 1803 when either *L. monocytogenes* or *L. seeligeri* are streaked on blood agar. *L. seeligeri* shows a lesser-enhanced β-hemolysis zone than *L. monocytogenes*, *L. ivanovii* shows a wide enhanced β-hemolysis zone with *Rhodococcus equi* NCTC 1621. The plates are incubated at 37°C for no longer than 12–18 h. The *Listeria* isolates streaked on blood agar for the CAMP test are derived from the hemolysis plates used to examine β-hemolysis. The *Listeria* streaks should not touch the streaks of the *St. aureus* and *R. equi* control strains. The control strains are streaked parallel to each other and the suspected *Listeria* isolated in between the two streaks.[41,56] Alternatively, various commercial identification kits such as API 10 Listeria (BioMerieux, Marcy Etoile, France) might be used instead of traditional biochemical tests, which are time-consuming. Finally, the previous selective agars, for example, PALCAM and Oxford agars, may be substituted by other selective chromogenic media such as ALOA and RAPID' L.Mono, which allow the direct differentiation between *Listeria* species by specific reactions on the agar plates.[16] In this way, the direct detection or enumeration of a specific *Listeria* species is feasible from the dilutions of the original sample.

34.6 Escherichia coli O157:H7

Pathogenic *E. coli* include a variety of types having different pathogenicity based on the virulence genes involved. The different types of pathogenic *E. coli* are the enteropathogenic *E. coli* (EPEC), the enteroinvasive *E. coli* (EIEC), the enterotoxigenic *E. coli* (ETEC), the enteroaggregative *E. coli* (EAEC), and the enterohemorrhagic *E. coli* (EHEC). The latter belongs to Verocytotoxigenic *E. coli* (VTEC), producing Verocytotoxins or Shiga toxins. VTEC *E. coli* are of great concern because they include the most predominant foodborne pathogen, *E. coli* O157:H7. The letters and

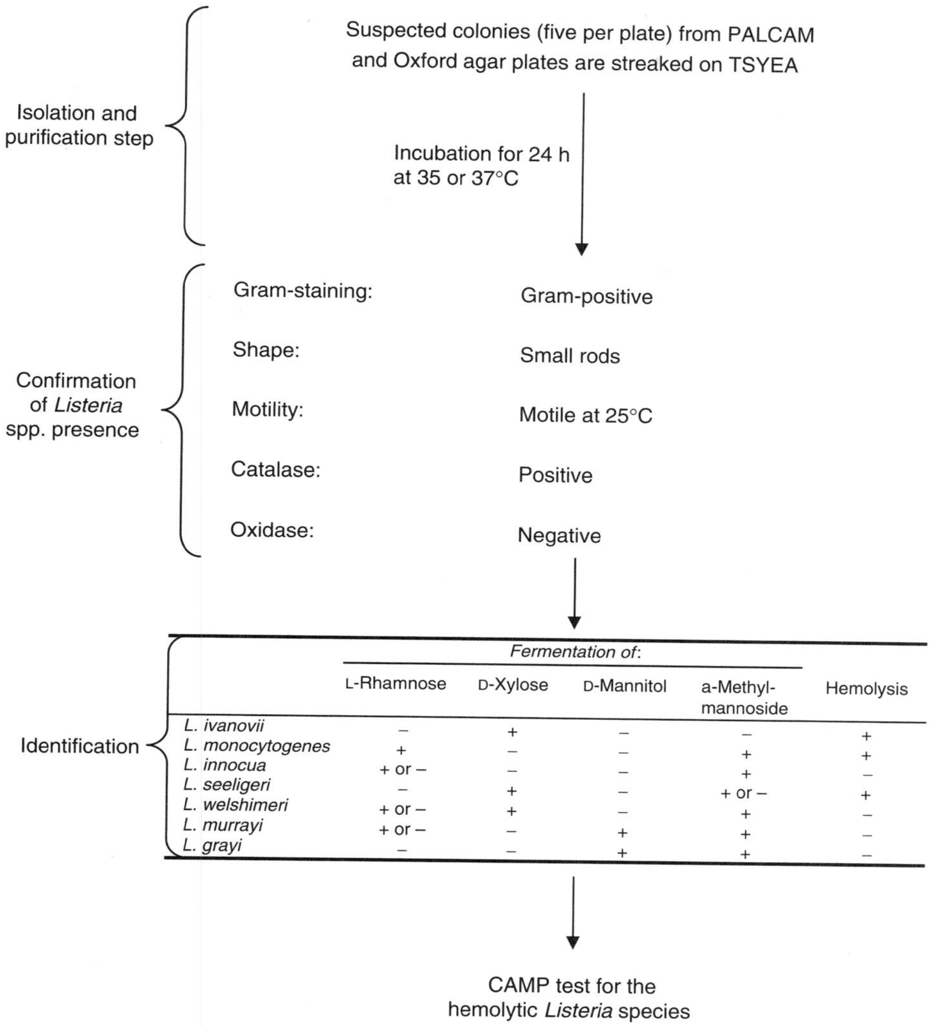
Isolation and purification step
Suspected colonies (five per plate) from PALCAM and Oxford agar plates are streaked on TSYEA
Incubation for 24 h at 35 or 37°C
Confirmation of *Listeria* spp. presence
Gram-staining:
Gram-positive
Shape:
Small rods
Motility:
Motile at 25°C
Catalase:
Positive
Oxidase:
Negative
Identification
CAMP test for the hemolytic *Listeria* species

	Fermentation of:				
	L-Rhamnose	D-Xylose	D-Mannitol	a-Methyl-mannoside	Hemolysis
L. ivanovii	–	+	–	–	+
L. monocytogenes	+	–	–	+	+
L. innocua	+ or –	–	–	+	–
L. seeligeri	–	+	–	+ or –	+
L. welshimeri	+ or –	+	–	+	–
L. murrayi	+ or –	–	+	+	–
L. grayi	–	–	+	+	–

Figure 34.3 ***L. monocytogenes*** **confirmation by the use of various biochemical tests.**

numbers, for example, O157:H7, are referred to the microorganism serogroup. The somatic antigens are stated with the letter "O" and the flagella antigens with the letter "H".[57] *E. coli* O157:H7 can be found on raw and processed meat.[58–61] Most often, it has been isolated from beef, which is believed to be the main vehicle for outbreaks associated with pathogenic *E. coli* O157:H7. The source of contamination of meat is usually the bovine feces or the intestinal tube during slaughtering. Their conduct with the muscle tissue results in meat contamination.[57] Heat treatment and the fermentation process are usually sufficient for producing a safe finished product. However, if these processes are not adequate, then *E. coli* O157:H7 may survive during manufacturing, if the microorganism is present in the raw material.[62–64] Also, factors other than process play a significant role in producing safe products, such as the implementation of good manufacturing practices (GMP) or good hygiene practices (GHP) to avoid postprocess contamination.[32,33,64] For the detection of

EPEC, EIEC, ETEC, and EAEC, there is no standard sensitive procedure and usually the food sample is diluted in brain heart infusion (BHI) broth, incubated at 35°C for 3 h to allow microbial cells to resuscitate. Then, an enrichment step (at 44°C for 20 h) in tryptone phosphate broth and plating on Levine eosin methylene blue agar and MacConkey agar are followed. Lactose-positive (typical) and lactose-negative (nontypical) colonies are collected for characterization using various biochemical, serological, or PCR-based tests.[17]

34.6.1 Detection of Escherichia coli O157:H7

Figure 34.4 shows the conventional method for detecting and identifying *E. coli* O157:H7 (based on Ref. 65). Pathogenic *E. coli* O157:H7 does not ferment sorbitol and does not possess β-glucuronidase, produced almost by all *E. coli* strains. The selective media exploit these attributes to distinguish the pathogenic *E. coli* O157:H7 from the other nonpathogenic *E. coli* strains. The method includes an enrichment step using a selective enrichment broth (tryptone soya broth [TSB] supplemented with novobiocin) to resuscitate the stressed cells and suppress the growth of the background flora. Before plating onto agar plates, intermediate steps may be comprehended. The cell antigen O157:H7 is revelatory of the microorganism pathogenicity, and therefore the immunomagnetic separation method (manufacturer instructions are followed to implement the technique) increases the detection of *E. coli* O157:H7.[14] *E. coli* O157:H7 is captured onto immunomagnetic particles, washed with sterile wash buffer, and resuspended in sterile wash buffer; a sample of the washed and resuspended magnetic particles is inoculated on selective medium to obtain isolated colonies. The selective agars used to subculture the sample are the modified MacConkey agar containing sorbitol instead of lactose, and selective agents such as tellurite and cefixine (CT-SMAC), and the tryptone bile glucuronic medium (TBX).[14,57] Because sorbitol-negative microorganisms other than *E. coli* O157:H7, such as *Proteus* spp. and some other *E. coli* strains, may grow on the agar plates, the addition of cefixine (inhibits *Proteus* spp. and not *E. coli*) and tellurite (inhibits *E. coli* strains other than *E. coli* O157:H7) substantially improves the selectivity of the medium.[17] CT-SMAC agar medium has been found to be most effective for the detection of Shiga toxin–producing *E. coli* O157:H7.[66] Typical *E. coli* O157:H7 colonies have 1 mm diameter, and are colorless (sorbitol-negative) or pellucid with a very slight yellow-brown color. However, because sometimes *E. coli* O157:H7 forms colonies similar to other *E. coli* strains (pink to red surrounded by a zone), further purification (streaking) on nutrient agar and confirmation of the typical and nontypical colonies is required.

Biochemical methods require time; hence, PCR-based protocols, including MPCR, have been developed. Detection of STEC strains by MPCR had first been described by Osek.[67] A protocol was developed using primers specific for genes that are involved in the biosynthesis of the O157 *E. coli* antigen (*rfb* O157), and primers that identify the sequences of Shiga toxins 1 and 2 (*stx*1 and *stx*2), and the intimin protein (*eae*A) involved in the attachment of bacteria to enterocytes.[23] The different strains were identified by the presence of one to four amplicons.[67] More protocols have been developed and applied for the detection and identification of *E. coli* in feces and meat (pork, beef, and chicken) samples.[68,69] Later, Kadhum et al.[70] designed and introduced an MPCR to determine the prevalence of cytotoxic necrotizing factors (CFNs) and cytolethal distending toxin (CDT)–producing *E. coli* on animal carcasses and meat products from Northern Ireland into a preliminary investigation of whether they could be a source of human infection.

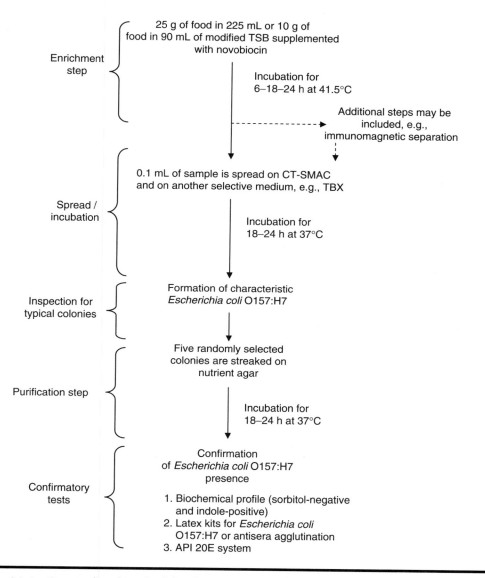

Enrichment step
{
25 g of food in 225 mL or 10 g of food in 90 mL of modified TSB supplemented with novobiocin

Incubation for 6–18–24 h at 41.5°C

Additional steps may be included, e.g., immunomagnetic separation
}

Spread / incubation
{
0.1 mL of sample is spread on CT-SMAC and on another selective medium, e.g., TBX

Incubation for 18–24 h at 37°C
}

Inspection for typical colonies
{
Formation of characteristic *Escherichia coli* O157:H7
}

Purification step
{
Five randomly selected colonies are streaked on nutrient agar

Incubation for 18–24 h at 37°C
}

Confirmatory tests
{
Confirmation of *Escherichia coli* O157:H7 presence

1. Biochemical profile (sorbitol-negative and indole-positive)
2. Latex kits for *Escherichia coli* O157:H7 or antisera agglutination
3. API 20E system
}

Figure 34.4   Conventional method for detecting and identifying *E. coli* O157:H7 based on ISO standard.

34.6.2   *Enumeration of* Escherichia coli *O157:H7*

Figure 34.5 shows the conventional enumeration method of *E. coli* O157:H7 (based on Ref. 71). The key step is the additional incubation period required (at 37°C for 4 h) before the incubation at 44°C for 18–24 h, in case of stressed cells. Typical *E. coli* O157:H7 colonies have a blue color, and plates with colonies (blue) less than 150 and less than 300 in total (typical and nontypical) are measured. Detection limit of the method is the cell concentration of 10 cfu/g.

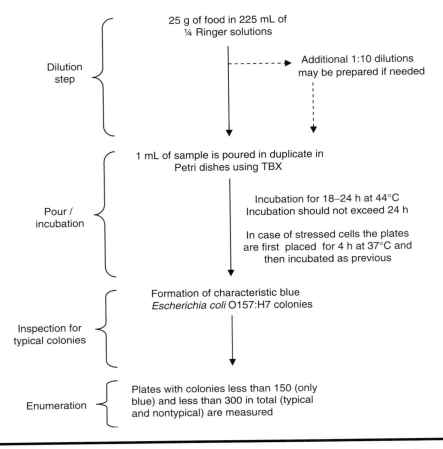

Figure 34.5 Conventional enumeration of *E. coli* O157:H7 based on ISO standard.

34.6.3 Confirmation of Escherichia coli *O157:H7*

To confirm *E. coli* O157:H7 presence, the following tests should be carried out. *E. coli* O157:H7 is negative to sorbitol, unlike most nonpathogenic *E. coli* strains, and indole positive. After defining the biochemical profile of the suspected colonies, latex kits for *E. coli* O157:H7 or antisera agglutination can be used to confirm *E. coli* O157:H7. Commercial kits such as API 20E (BioMerieux) comprise an alternative for *E. coli* O157:H7 confirmation. *E. coli* O157:H7 toxins can be detected, using reversed passive latex agglutination and cultured Vero cells. Polymyxin B may be used in the culture to facilitate Shiga toxins release.[14]

34.7 Salmonella spp.

Salmonella spp. has been isolated from all types of raw meat such as poultry, pork, beef, and lamb. All these products have been implicated in *Salmonella* spp. outbreaks.

Most often *Salmonella* spp. occurs in poultry and pork meat. The main source of contamination of the raw meat is the transfer of the microorganism from the feces to the meat tissue during slaughtering and the processing that follows.[72] Also, postprocess contamination may occur, and therefore, the GHP regarding the equipment and the personnel are essential.

34.7.1 Detection and Confirmation of Salmonella spp.

Figure 34.6 shows the conventional method for detecting and identifying *Salmonella* spp. (based on Ref. 73). The microbiological criterion for *Salmonella* spp. is "absence in 25 g". The method includes two enrichment steps: the preenrichment step to allow injured cells to resuscitate and the selective enrichment step to favor the growth of *Salmonella* cells. For this reason, a nonselective but nutritious medium is used (buffered peptone water [BPW]) in the first step, whereas in the second step the selective medium contains selective agents to suppress the growth of the accompaning microflora. Also, in the second step, two different selective media are used because the culture media have different selective ability against the numerous *Salmonella* serovars.[14] Time and temperature of incubation, during the preenrichment and selective enrichment steps play a significant role for the selectivity of the media. One of the selective media used in the second enrichment step is the selenite cystine (SC) broth, which however contains a very toxic substance (sodium biselenite); for this reason its use has been replaced by other media such as Muller–Kauffmann tetrathionate/novobiocin (MKTTn) broth. Rappaport-Vassiliadis Soya peptone (RVS) broth is the standard Rappaport-Vassiliadis (RV) broth with tryptone substituted by soya peptone because it has a better performance than RV broth.[17] The next step is the plating of the samples on selective agars that also contain selective agents such as bile salts and brilliant green, and use various diagnostic characteristics (e.g., lactose fermentation, H_2S production, and motility) to differentiate *Salmonella* spp. from the accompanying microflora such as *Proteus* spp., *Citrobacter* spp., and *E. coli*. The Oxoid Biochemical Identification System (O.B.I.S) *Salmonella* test (Oxoid, Basingstoke, UK) is also a rapid test to differentiate *Salmonella* spp. from *Citrobacter* spp. and *Proteus* spp. The principle of the test is based on the determination of pyroglutamyl aminopeptidase (PYRase) and nitrophenylalanine deaminase (NPA) activity to which *Salmonella* spp. is negative, *Citrobacter* spp. PYRase-positive and NPA-negative, and *Proteus* spp. NPA-positive and PYRase-negative. Also, more than one medium is used (xylose lysine desoxycholate [XLD] or xylose lysine tergitol 4 [XLT4], and phenol red/brilliant green agar) in parallel because selective agars also differ in their selectivity. The last steps include biochemical and serological confirmation of the suspected *Salmonella* colonies to confirm the identity and identify the serotype of the isolates.[14,17] *Salmonella* spp. is lactose-negative, H_2S-positive, and motile. However, lactose-positive strains have been isolated from human infections, and therefore an additional selective medium agar may be needed. Bismuth sulfite agar is considered as the most suitable medium for these strains.[17,74,75]

Salmonella is reported as one of the major pathogens of foodborne disease outbreaks throughout the world, causing salmonelliosis in human and animals.[76] The most frequently isolated serovars from foodborne outbreaks are *S. enterica* serovar *Typhimurium* and *S. enterica* serovar *Enteritidis*. Traditional phenotypic methods such as biotyping, serotyping, and phage typing of isolates, as well as antimicrobial susceptibility testing, provide sufficient information for epidemiological purposes. Molecular genetic methods have revolutionized the fingerprinting of microbial strains. However, not all of them have been internationally standardized, and problems in interpreting the results of different laboratories might occur. Nevertheless, the accuracy and speed by which results are obtained have rendered them more and more applicable.

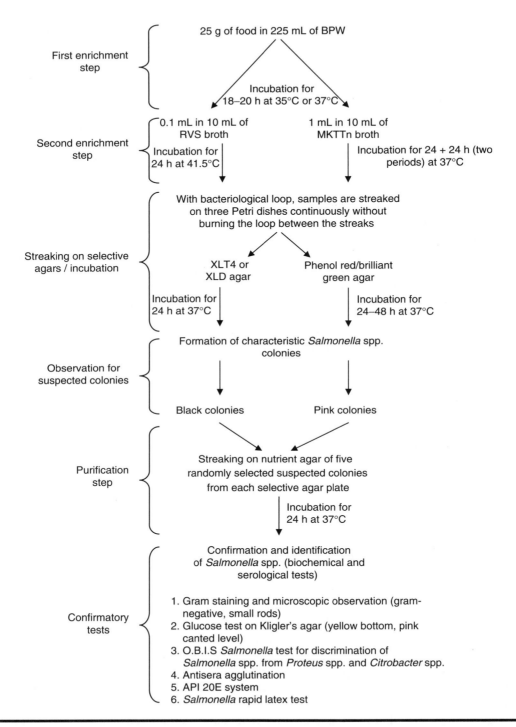

Figure 34.6   Conventional method for detecting and identifying *Salmonella* spp. based on ISO standard.

The assay generally used to identify *Salmonella* serovars is represented by a serological method, which needs the preparation of specific antibodies for each serovar, thus being extremely complex and time-consuming.[23] Plasmids are characteristic of *Salmonella* and therefore plasmid analysis can often be used to differentiate these strains.[77] A rapid alternative is given by PCR approaches. Based on primers, designed for detecting O4, H:I, and H:1,2 antigen genes from the antigen specific genes *rfbJ, fliC*, and *fljB* (coding for phase 2 flagellin), respectively, Lim et al.[78] described a MPCR for the identification of *S. enterica* serovar Typhimurium, whose presence was associated with the appearance of three amplification products. MPCR targeted to the *tyv* (CDP-tyvelose-2-epimerase), *prt* (paratose synthase), and *invA* (invasion) genes were designed to identify *S. enterica* serovar Typhi and *S. enterica* serovar Paratyphi A be the production of three or two bands, respectively.[79]

PCR amplifications of the 16S-to-23S spacer region of bacterial rRNA as well as specific monoclonal antibodies to the lipopolysaccharide of *S. enterica Typhimurium* DT104 have been used.[80]

34.8 *Staphylococcus aureus*

Origin of the *St. aureus* is the animal in which it is a part of the normal microflora. Food contamination with *St. aureus* may happen by humans who also carry staphylococci. Food poisoning by *St. aureus* is the result of enterotoxin ingestion produced by the microorganism. Enterotoxin is a heat-stable substance and high cell numbers are required to produce sufficient amounts of the toxin. Temperatures usually above 15°C favor the rapid growth of the microorganism and the production of enterotoxin. The minimum temperatures for microorganism growth and enterotoxin production are 7 and 10°C, respectively. Because staphylococci are a part of the natural microflora of humans and animals, these microorganisms will always be present in raw materials. Therefore, attention is required in the implementation of GMP and GHP to minimize the contamination of raw materials with *St. aureus* and to avoid postprocess contamination of processed meat products.[81]

34.8.1 *Enumeration and Confirmation of* Staphylococcus aureus

Figure 34.7 shows the conventional enumeration method of *Staphylococcus* spp. (based on Ref. 82). The method has a detection limit ≥100 cfu/g. If staphylococci lesser than 100 cfu/g are expected, then the procedure followed for *L. monocytogenes* enumeration may be applied. Low numbers of *St. aureus* are of little significance because extensive growth is needed to produce sufficient amounts of enterotoxin to cause illness; therefore, an enrichment step is not required for the isolation of the microorganism. The most widely used and accepted medium for *St. aureus* is the Baird–Parker (BP) agar (egg yolk–glykine–tellurite–pyruvate). Sodium pyruvate assists the resuscitation of stressed cells whereas tellurite, glykine, and lithium chloride enhance the medium selectivity. *Staphylococcus* spp. forms black colonies (tellurite reduction) whereas *St. aureus* colonies are also surrounded by a halo (clearance of egg yolk due to lipase activity). Plates having 15–300 colonies in total (*Staphylococcus* spp. and *St. aureus* if present) are measured. Coagulase test, reversed passive latex agglutination, or ELISA kits for enterotoxin detection may be used as confirmatory tests for *St. aureus* presence. Coagulase test is considered positive for enterotoxin presence only in case of strong positive reaction. API Staph (BioMerieux) may be also used to identify the isolated colonies from the agar plates.[14]

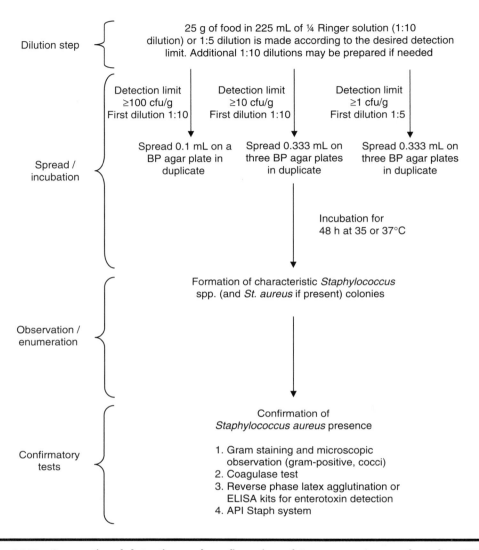

Figure 34.7 Conventional detection and confirmation of *St. aureus* presence based on ISO standard.

Molecular techniques have been applied in the case of *St. aureus* to quickly determine its presence and identification. Occasionally, isolates of *St. aureus* give equivocal results in biochemical and coagulase tests.[83] Most *St. aureus* molecular identification methods have been PCR-based. Primers targeted to the nuclease (*nuc*), coagulase (*coa*), protein A (*spa*), *femA* and *femB, Sa442*, 16S-rRNA, and surface-associated fibrinogen-binding genes have been developed.[84,85]

St. aureus food poisoning is caused by the ingestion of preformed toxins (staphylococcal enterotoxins [SEs]) produced in foods. It has been reported that nearly all SEs are superantigens and are encoded by mobile genetic elements, including phages, plasmids, and pathogenicity islands.[86,87] Several methods for SEs detection from isolated strains and foods have been described in the recent years; these include biological, immunological, chromatographical, and molecular assays.[88,89] The four SEs originally described can be detected with commercial antisera or by PCR reactions.[90,91]

Detection and identification of methicillin-resistant *St. aureus* (MRSA) has received great attention because in immunocompromised patients it can cause serious infections, which may finally lead to septicemia. Because MRSA strains mainly appear in nosocomial environments, most of the techniques developed for their detection are focused on clinical or blood isolates.[92] Such techniques include DNA probes,[93,94] peptide nucleic acid probes,[95] multiplex-PCR,[85] real-time PCR,[96–98] LightCycler PCR,[97,98] and combination of fluorescence *in situ* hybridization and FCM.[99] Recent advances include the development of segment-based DNA microarrays.[92] Even though MRSA strains are mainly encountered in nosocomial environments, food can be considered an excellent environment for introducing pathogenic microorganisms in the general population, especially in immunocompromised people; and in the intestinal tract, transfer of resistant genes between nonpathogenic and pathogenic or opportunistic pathogens might occur.[100] A community-acquired case was reported in 2001 in which a family was involved in an outbreak after ingesting MRSA with baked port meat, contaminated by the handler.[101] Therefore, even the techniques applied in different samples might have applicability in food products.

34.9   *Yersinia enterocolitica*

Infections with *Y. enterocolitica* involve meat and meat products. Especially, pork meat has been implicated with *Y. enterocolitica* outbreaks (yersiniosis). Not all *Y. enterocolitica* strains cause illness. The most common serotypes causing yersiniosis are the serotypes O:3, O:9, O:5,27, and O:8. Because contamination of meat with high numbers of *Y. enterocolitica* may occur during preprocess (e.g., slaughtering), precautionary measures such as GHP are essential.[102] Contamination with *Y. enterocolitica* is important because of its capability to grow even at refrigerated temperatures (4°C). However, contamination with *Y. enterocolitica* is considered of little importance because illness caused by other members of *Enterobacteriaceae*, such as *Salmonella* spp. or *E. coli*, is considered more severe.[17,81]

34.9.1   *Detection and Confirmation of* Yersinia enterocolitica

Figure 34.8 shows the conventional method for detecting *Y. enterocolitica* (based on Ref. 103). If specific serotypes are considered (e.g., O:3), then two isolation procedures are proposed to run in parallel.[17] The procedure involving enrichment with irgasan ticarcillin chlorate (ITC) broth is selective for serotype O:3 and possibly O:9, although some workers found poor recovery of the serotype O:9 from ground pork, using ITC.[104] After enrichment with ITC, plating of the samples should be done on *Salmonella–Shigella* deoxycholate calcium chloride (SSDC) instead of cefsulodin irgasan novobiocin (CIN) because the second medium is inhibitory for the serotype O:3. Furthermore, the isolation and identification of *Y. enterocolitica* from ground meat on CIN medium agar has proved to cause problems because many typical *Yersinia*-like colonies may grow.[105] After enrichment (primary) with TSB or peptone sorbitol bile salts (PBS) broth (peptone buffered saline with 1% sorbitol and 0.15% bile salts), an alkali treatment (potassium hydroxide [KOH]) may be used to increase recovery rates of *Yersinia* strains instead of secondary enrichment with bile oxalate sorbose (BOS).[106] This method should not be used with the procedure involving the ITC broth as selective enrichment step.[107] On SSDC agar, the *Yersinia* colonies are 1 mm in diameter, round, and colorless or opaque. On CIN agar, the colonies have a transparent border

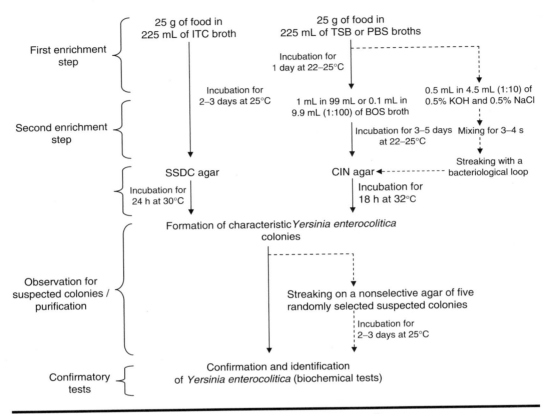

Figure 34.8 Conventional detection of *Y. enterocolitica* based on ISO standard and its identification.

with a red circle in the center (bull's eye). *Yersinia* strains and *Y. enterocolitica* serotypes may be distinguished, using some biochemical tests.[107] *Y. enterocolitica* may be identified using urease and citrate utilization tests as well as sucrose, raffinose, rhamnose, α-methyl-D-glucoside, and melibiose fermentation tests. *Y. enterocolitica* is urease and sucrose positive and negative in the other tests. The most used tests to identify pathogenic *Y. enterocolitica* strains are the following: calcium-dependent growth at 37°C, Congo red binding on Congo red magnesium oxalate (CR-MOX) agar, or low-calcium Congo red BHI agarose agar (CR-BHO), which determines the Congo red dye uptake, pyrazinamidase activity, and salicin–esculin fermentation.[107–112] Because the last two tests are not plasmid dependent as the other tests, the pyrazinamidase, salicin, and esculin tests are considered as the most reliable biochemical screening tests for pathogenicity because plasmid may be lost during subculture. Before testing, suspected colonies may be subcultured on a nonselective medium incubated at 25°C to reduce the risk of plasmid loss.[107] Pathogenic strains are negative to these three tests. Esculin fermentation and pyrazinamidase activity tests should be conducted at 25°C whereas salicin fermentation at 35°C or 37°C. Commercial kits for *Y. enterocolitica* identification such as API 20E (BioMerieux) may also be used as an alternative, which have been proved to be suitable for routine laboratory diagnostics.[105]

From a food hygiene point of view, *Y. enterocolitica* is of major importance and is a very heterogeneous species. Nonpathogenic strains may contaminate food products to the same extent as pathogenic *Y. enterocolitica*, and the main goal for nucleic acid-based methods has been to separate this group of pathogenic bacteria. Both polynucleotide and oligonucleotide probes, as well as PCR-based methods have been applied for its detection and quantification in meat and meat products.[113,114] Nested-PCR has also been developed for its detection in meat food products and can satisfactorily detect pathogenic *Y. enterocolitica* even in the presence of high background of microflora.[115] Comparative gDNA microarray analysis has been recently developed to differentiate between nonpathogenic and pathogenic biotypes.[116]

34.10   *Bacillus cereus*

B. cereus can be found in meat and especially in dishes containing meat. Also, outbreaks attributed to *B. cereus* infections have been associated with cooked meats. Its presence in food is not considered significant because high numbers ($>10^5-10^6$ cfu/g) are needed to cause diarrheal or emetic syndrome. The two types of illnesses are caused by an enterotoxin (diarrheagenic or emetic) produced by the microorganism. Because other *Bacillus* species are closely related physiologically to *B. cereus* such as *B. mycoides*, *B. thuringiensis*, and *B. anthracis*, further confirmatory tests are required to differentiate a typical (egg yolk reaction, inability to ferment mannitol) *B. cereus* from the other species.[117]

34.10.1   *Enumeration and Confirmation of* **Bacillus cereus**

The presence of low numbers of *B. cereus* is not considered significant unless *B. cereus* growth may occur; hence, an enrichment step is not needed (Figure 34.9). However, if enrichment must be applied, this can be done using BHI broth supplemented with polymyxin B and sodium chloride.[118] To enhance the selection of *B. cereus*, the following attributes of the microorganism are employed: its resistance to the antibiotic polymyxin, the production of phospholipase C, causing turbidity around colonies grown on agar containing egg yolk, and its inability to ferment mannitol. The media used for selection are usually the mannitol–egg yolk polymyxin (MYP)[119] and the Kim-Goepfert (KG)[120] agars. Because of the similarity in composition and functionality of the KG medium with the polymyxin pyruvate egg yolk mannitol bromothymol blue agar (PEMBA) (Holbrook and Anderson, 1980, in Ref. 115), the latter medium may be used instead of KG.[117]

Colonies on MYP agar have a surrounding precipitate zone (turbidity) and both colonies and zone are pink (not fermentation of mannitol). On PEMBA agar, the colonies are peacock-blue with a blue egg yolk precipitate zone. Finally, on KG agar, the colonies are translucent or white cream. Because turbidity zones may overlap each other and a clear measurement of the colonies with a precipitate zone may be not possible, plates having 10–100 colonies per plate are measured instead of 30–300 colonies per plate. For low numbers (<1000 cfu/g) of *B. cereus* in the food sample, the MPN technique may be used. A suitable medium for this purpose is the trypticase soy polymyxin broth. Each of the three tubes of 1:10, 1:100, and 1:1000 is inoculated with 1 mL of sample and the tubes are incubated at 30°C for 48 h and examined for dense turbidity. Confirmation of *B. cereus* presence is required before determining the MPN.[117,118] If only spores have to be counted, then the vegetative cells should be destroyed first by heat (the initial 1:10 dilution is heated for 15 min at 70°C) or by alcohol treatment (1:1 initial dilution in 95% ethyl alcohol for 30 min at room temperature).[14] Afterwards, the same procedure described in Figure 34.9 is followed.

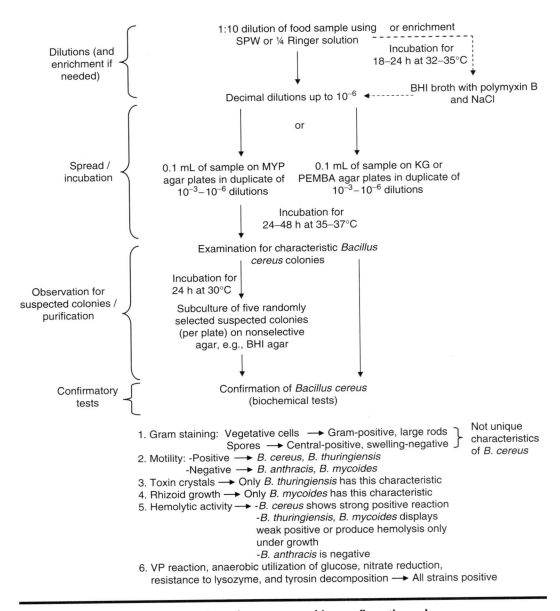

Figure 34.9 Conventional detection of *B. cereus* and its confirmation scheme.

Before testing isolated colonies for *B. cereus* identity, the culture should be purified on a non-selective agar (e.g., BHI agar) to promote sporulation. However, if KG was used as selective agar, then the isolated colonies may be tested directly because KG medium favors sporulation.

ELISA and RPLA tests are commercially available for *Bacillus* diarrheal enterotoxin. No tests have been developed for emetic enterotoxin due to purification problems. However, tissue culture assay using HEp-2 cells, may be useful for the detection and purification of the emetic toxin.[14,117]

Several molecular techniques have also been developed for the detection and characterization of *B. cereus,* derived from food products. Immunological methods for semiquantitative identification of enterotoxins are available (ELISA, RPLA), which demand at least 2 days to obtain a result because enterotoxin expression during growth is necessary.[121] Even though genetic probes are applied as well for the detection of *B. cereus,* the information provided would involve the presence of the gene and not the level of the enterotoxin production. It seems that the production of enterotoxins from enterotoxin-positive strains is too low to cause food poisoning.[122] A good choice for the detection of *B. cereus* would be the use of probes directed to the phospholipase C genes, which are present in the majority of the strains.

Different confirmatory tests exist for *B. cereus.* For enterotoxic *B. cereus,* molecular diagnostic assays (PCR-based),[123,124] biochemical assays, and immunological assays[123,125,126] are commercially available. Three methods for detection of the emetic toxin have been described in the past years: a cytotoxicity assay, LC-MS analysis, and a sperm-based bioassay.[127,128] They have however been proved difficult to use for routine applications and are not specific enough. Recently, a novel PCR-based detection system has been developed, based on the emetic toxin cereulide gene.[129]

The latest trend is toward the development of molecular tools that would be able to characterize virulence mechanisms of bacterial isolates within minutes.[130] The next-generation assays, such as biosensors and DNA chips have already been developed.[131] They can be classified into high-density DNA arrays[132] and low-density DNA sensors.[133] An automated electrochemical detection system, which allows simultaneous detection of presently described toxin-encoding genes of pathogenic *B. cereus,*[130] and a nanowire labeled direct-charge transfer (DCT) biosensor, which is capable of detecting *Bacillus* species, have also been developed.[134]

34.11   *Clostridium perfringens*

Cl. perfringens type A is considered one of the most common human foodborne pathogens in the United States and is acquired through the consumption of contaminated meat and poultry.[135] Foods usually associated with *Cl. perfringens* infections are cooked meat and poultry. Its presence in raw meats and poultry is not unusual. The illness (diarrhea) is caused by a heat-sensitive enterotoxin, produced only by sporulating cells. Usually, large numbers of the microorganism are required to cause illness. As a consequence, the microorganism is enumerated using direct plating without enrichment. Also, *Cl. perfringens* does not sporulate in the food and therefore there is no need to heat the sample before enumerating the microorganism.[136]

34.11.1   *Enumeration and Confirmation of* Clostridium perfringens

The selective media used for enumeration of *Cl. perfringens* contain antibiotics to inhibit other anaerobic microorganisms, iron, and sulfite because clostridia reduce the latter to sulfide, which reacts with iron forming a black precipitate (black colonies) characteristic for clostridia. The most commonly used and useful medium to recover *Cl. perfringens* is the egg yolk free tryptose sulfite cycloserine (EY-free TSC) agar (Figure 34.10).[137] EY-free TSC agar is used in pour plates. Cycloserine is added to inhibit the growth of enterococci. Because other sulfite-reducing clostridia that produce black colonies may grow on EY-free TSC agar, further confirmatory tests are needed to identify the presence of *Cl. perfringens* (Figure 34.11). If low numbers are expected, the MPN technique or enrichment using buffered trypticase peptone glucose yeast extract (TPGY) broth

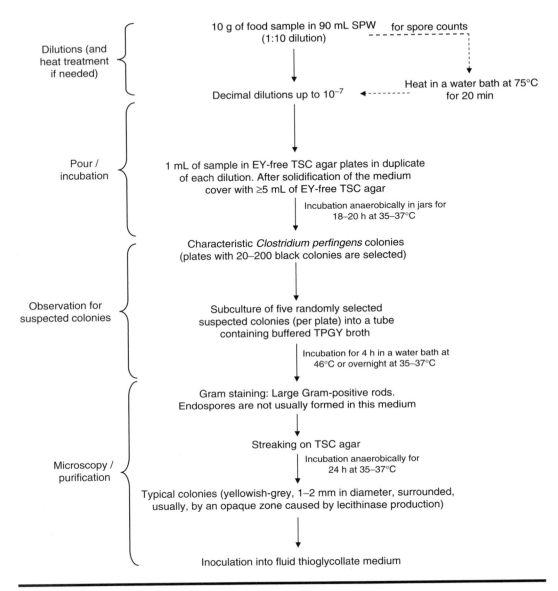

Figure 34.10 Conventional detection scheme of *Cl. perfringens*.

may be used. Two grams of food sample is inoculated into 15–20 mL of medium in a tube. The tube is incubated for 20–24 h at 35–37°C. With a bacteriological loop, a sample from the positive tubes (turbidity and gas production) is streaked on EY-free TSC agar plates.[136] Enterotoxin of *Cl. perfringens* can be detected using commercial kits such as ELISA and RPLA.

Nonisotopic colony hybridization technique has been developed for the detection and enumeration of *Cl. perfringens*, which proved to be more sensitive than the conventional culture methods.[138] It provides quantitative assessment of the presence of potentially enterotoxigenic strains of *Cl. perfringens*, as determined by the presence of the enterotoxin A gene and the results are

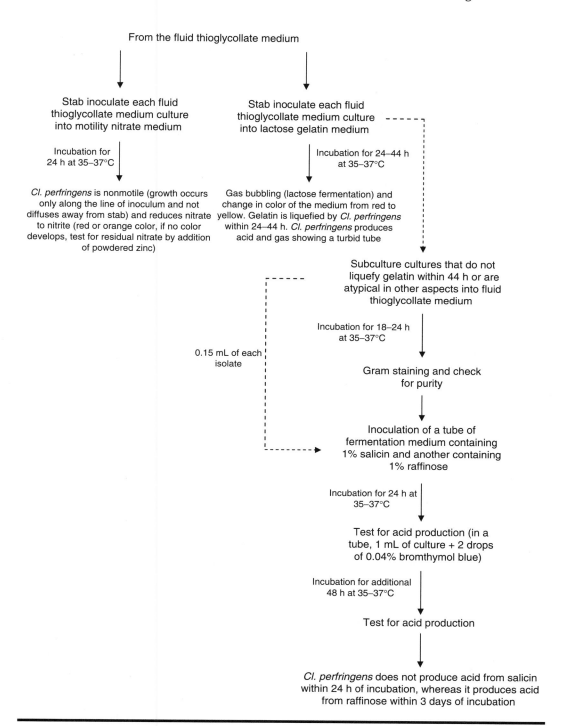

Figure 34.11   Identification scheme of *Cl. perfringens*.

acquired within 48 h. A MPCR assay has also been developed for the detection of *Cl. perfringens* type A[139] and has been evaluated in relation with American retail food by Wen et al.[140] Similar methods as the ones described previously for *B. cereus*[121] have been applied as well.

34.12 *Campylobacter jejuni*

Campylobacter species are part of intestinal tract microflora of animals and thus may contaminate foods such as meat, poultry, and their products. The most frequent *Campylobacter* species causing illness is *C. jejuni*. The microorganism is gram-negative, motile, and oxidase-positive, forming curved rods. Poultry is considered as the most important vehicle of *Campylobacter* illness. Several outbreaks have been associated with poultry.[141,142] *C. coli* and *C. lari* have also been isolated from poultry and recognized as potential hazards to human health, causing illness but less frequently than *C. jejuni*.[143]

34.12.1 *Detection and Confirmation of* Campylobacter jejuni

In general, *Campylobacter* species are sensitive microorganisms stressed during processing, and therefore an enrichment step is needed to resuscitate injured cells. Also, the microorganism fails to grow under normal atmospheric conditions because *Campylobacter* is microaerophilic; hence, gas jars should be used to provide the right oxygen conditions (5% oxygen, 10% carbon dioxide, and 85% nitrogen). Because of its sensitivity to oxygen, the food samples should be kept, before the analyses, in an environment without oxygen (100% nitrogen) with 0.01% sodium bisulfite and under refrigeration. For this purpose the Wang's medium may be used.[144]

Figure 34.12 shows the conventional detection of *Campylobacter* spp. (based on Ref. 145). Usually, 10 g of food sample (ground beef) are added to 90 mL of enrichment broth. For sampling poultry carcasses and moderately large pieces of foods, the surface rinse technique may be used.[144] Briefly, the sample is placed in a sterile stomacher bag with 250 mL of Brucella broth and the surface is rinsed by shaking and massaging. The broth (rinse/suspension) is passed through a filter and is then centrifuged at $16,000 \times g$ for 10 min at 4°C. The supernatant fluid is discarded, and the pellet is suspended in 2–5 mL of enrichment broth. After enrichment, or during the direct plating without enrichment, two selective agars are used: Karmali agar and one of the following agars: Butzler agar, Campy-BAP or Blaser agar, *Campylobacter* charcoal differential agar (CCDA)-Preston blood-free agar, or Skirrow agar. It has been found that CCDA-Preston blood-free medium has an excellent selectivity and is good for quantitative recovery of *C. jejuni* (Compendium book,[144] Chapter 29).[13] The oxygen tolerance of *Campylobacter* may be enhanced adding to growth media 0.025% of each of the following: ferrous sulfate, sodium metabisulfite, and sodium pyruvate.

Purification of the culture is performed as follows for conducting confirmatory tests: colonies from the selective agar plates are transferred to a heart infusion agar with 5% difibrinated rabbit blood (HIA-RB) and plates are incubated at 42°C for 24 h under microaerophilic conditions. The culture is transferred to 5 mL of heart infusion broth (HIB) and the density of the cells is adjusted to meet the McFarland No. 1 turbidity standard (BioMerieux). This cell suspension is used further for biochemical testing in tubes or on agar plates.[144] Finally, the commercial kit API Campy (BioMerieux) may be used as an alternative for differentiation of *Campylobacter* spp.

Polynucleotide and oligonucleotide probes have been used for the detection of *C. jejuni* and are reviewed by Olsen et al.[121] A rapid and sensitive method, based on PCR, for the detection of

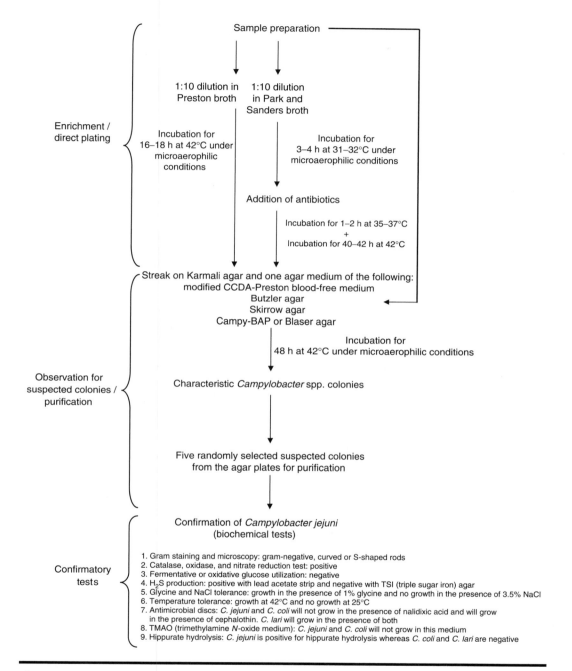

Figure 34.12 **Conventional detection and confirmation of *C. jejuni* based on ISO method.**

Campylobacter spp. from chicken products has been described by Giesendorf et al.,[146] which provided results within 48 h with the same sensitivity as the conventional method. Konkel et al.[147] developed a detection and identification method based on the presence of the *cad*F virulence gene, an adhesin to fibronectin, which aids the binding of *C. jejuni* to the intestinal epithelial cells.

This method may be useful for the detection of the microorganism in food products because it does not require bacterial cultivation before its application. Since then, further techniques have been developed with the incorporation of an enrichment step before the PCR and real-time PCR amplification, respectively.[148,149] A more recent evaluation of a PCR assay for the detection and identification of *C. jejuni* in poultry products reduced the time of analysis to 24 h or less depending on the necessity of the enrichment step.[150] Nevertheless, this method did not seem to be appropriate for ready-to-eat products but was proven to be useful in naturally contaminated poultry samples. Further improvements and trends include MPCRs, reviewed by Settanni and Corsetti[23] as well as real-time nucleic acid sequence-based amplifications (NASBA) with molecular beacons.[151]

References

1. Downes, F. P. and Ito, K. 2001. *Compendium of Methods for the Microbiological Examination of Foods*, fourth edition. American Public Health Association, Washington, DC.
2. van Nierop, W., Duse, A. G., and Marais, E. et al. 2005. Contamination of chicken carcasses in Gauteng, South Africa, by *Salmonella*, *Listeria monocytogenes* and *Campylobacter*. *International Journal of Food Microbiology* 99:1–6.
3. National Advisory Committee on Microbiological Criteria for Foods. 2005. Considerations for establishing safety-based consume-by date labels for refrigerated ready-to-eat foods. *Journal of Food Protection* 68:1761–1775.
4. Anonymous. 2007. *The Compendium of Analytical Methods*. Ottawa, Canada.http://www.hc-sc.gc.ca/fn-an/res-rech/analy-meth/microbio/index_e.html. Accessed on January 10, 2007.
5. Atlas, R. M. 2006. *Handbook of Microbiological Media for the Examination of Food*, second edition. CRC Press, Boca Raton, FL.
6. ISO. 2006. *International Standard, ISO 4831: Microbiology of Food and Animal Feeding Stuffs—Horizontal Method for the Detection and Enumeration of Coliforms—Most Probable Number Technique*. International Organization for Standardization, Geneva.
7. ISO. 2005. *International Standard, ISO 7251: Microbiology of Food and Animal Feeding Stuffs—Horizontal Method for the Detection and Enumeration of Presumptive Escherichia coli—Most Probable Number Technique*. International Organization for Standardization, Geneva.
8. Betts, R. and Blackburn, C. 2002. Detecting pathogens in food. In C. de W. Blackburn and P. J. McClure (eds), *Foodborne Pathogens: Hazards, Risk Analysis and Control*. CRC Press, Boca Raton, FL, 13–52.
9. Besse, N. G. 2002. Influence of various environmental parameters and of detection procedures on the recovery of stressed *L. monocytogenes*: a review. *Food Microbiology* 19:221–234.
10. Lazcka, O., Del Campo, F. J., and Xavier Munoz, F. 2007. Pathogen detection: A perspective of traditional methods and biosensors. *Biosensors and Bioelectronics* 22:1205–1217.
11. Leoni, E. and Legnani, P. P. 2001. Comparison of selective procedures for isolation and enumeration of *Legionella* species from hot water systems. *Journal of Applied Microbiology* 90:27–33.
12. Bej, A. K., Mahbubani, M. H., Dicesare, J. L., and Atlas, R. M. 1991. Polymerase chain reaction-gene probe detection of microorganisms by using filter-concentrated samples. *Applied and Environmental Microbiology* 57:3529–3534.
13. Otero, A., Garcia-Lopez, M. L., and Moreno, B. 1998. Rapid microbiological methods in meat and meat products. *Meat Science* 49:S179–S189.
14. Forsythe, Stephen J. 2000. *The Microbiology of Safe Food*. Blackwell Science, Malden, MA.
15. Gracias, K. S. and McKillip, J. L. 2004. A review of conventional detection and enumeration methods for pathogenic bacteria in food. *Canadian Journal of Microbiology* 50:883–890.
16. Becker, B., Schuler, S., Lohneis, M., Sabrowski, A., Curtis, G. D. W., and Holzapfel, W. H. 2006. Comparison of two chromogenic media for the detection of *Listeria monocytogenes* with the plating media recommended by EN/DIN 11290-1. *International Journal of Food Microbiology* 109:127–131.

17. de Boer, E. 1998. Update on media for isolation of Enterobacteriaceae from foods. *International Journal of Food Microbiology* 45:43–53.
18. Blais, B. W., Leggate, J., Bosley, J., and Martinez-Perez, A. 2004. Comparison of fluorogenic and chromogenic assay systems in the detection of *Escherichia coli* O157 by a novel polymyxin-based ELISA. *Letters in Applied Microbiology* 39:516–522.
19. Deisingh, A. K. and Thomson, M. 2004. Strategies for the detection of *Escherichia coli* O157:H7 in foods. *Journal of Applied Microbiology* 96:419–429.
20. Perelle, S., Dilasser, F., Malorny, B., Grout, J., Hoorfar, J., and Fach, P. 2004. Comparison of PCR-ELISA and LightCycler real-time PCR assays for detecting *Salmonella* spp. in milk and meat samples. *Molecular and Cellular Probes* 18:409–420.
21. Leskela, T., Tilsala-Timisjarvi, A., Kusnetsov, J., Neubauer, P., and Breitenstein, A. 2005. Sensitive genus-specific detection of *Legionella* by a 16S rRNA based sandwich hybridization assay. *Journal of Microbiological Methods* 62:167–179.
22. Lehtola, M. J., Loades, C. J., and Keevil, C. W. 2005. Advantages of peptide nucleic acid oligonucleotides for sensitive site directed 16S rRNA fluorescence in situ hybridization (FISH) detection of *Campylobacter jejuni, Campylobacter coli* and *Campylobacter lari. Journal of Microbiological Methods* 62:211–219.
23. Settanni, L. and Corsetti, A. 2007. The use of multiplex PCR to detect and differentiate food- and beverage-associated microorganisms: a review. *Journal of Microbiological Methods* 69:1–22.
24. Rossello-Mora, R. and Amann, R. 2001. The species concept for prokaryotes. *FEMS Microbiology Reviews* 25:39–67.
25. Levi, K., Smedley, J., and Towner, K. J. 2003. Evaluation of a real-time PCR hybridization assay for rapid detection of *Legionella pneumophila* in hospital and environmental water samples. *Clinical Microbiology and Infection* 9:754–758.
26. Muyzer, G., de Waal, E. C., and Uitterlinden, A. G. 1993. Profiling of complex microbial populations by denaturing gradient gel electrophoresis analysis of polymerase chain reaction-amplified genes coding for 16S rRNA. *Applied and environmental Microbiology* 59:695–700.
27. National Advisory Committee on Microbiological Criteria for Foods. 1991. *Listeria monocytogenes. International Journal of Food Microbiology* 14:185–246.
28. Samelis, J. and Metaxopoulos, J. 1999. Incidence and principal sources of *Listeria* spp. and *Listeria monocytogenes* contamination in processed meats and a meat processing plant. *Food Microbiology* 16:465–477.
29. Tompkin, R. B. 2002. Control of *Listeria monocytogenes* in the food processing environment. *Journal of Food Protection* 65:709–725.
30. Barbalho, T. C. F., Almeida, P. F., Almeida, R. C. C., and Hofer, E. 2005. Prevalence of *Listeria* spp. at a poultry processing plant in Brazil and a phage test for rapid confirmation of suspect colonies. *Food Control* 16:211–216.
31. Wilks, S. A., Michels, H. T., and Keevil, C. W. 2006. Survival of *Listeria monocytogenes* Scott A on metal surfaces: Implications for cross-contamination. *International Journal of Food Microbiology* 111:93–98.
32. Gibbons, I., Adesiyun, A., Seepersadsingh, N., and Rahaman, S. 2006. Investigation for possible source(s) of contamination of ready-to-eat meat products with *Listeria* spp. and other pathogens in a meat processing plant in Trinidad. *Food Microbiology* 23:359–366.
33. Reij, M. W., den Aantrekker, E. D., and ILSI Europe Risk Analysis in Microbiology Task Force. 2004. Recontamination as a source of pathogens in processed foods. *International Journal of Food Microbiology* 91:1–11.
34. Jay, J. M. 1996. Prevalence of *Listeria* spp. in meat and poultry products. *Food Control* 7:209–214.
35. Mataragas, M., Drosinos, E. H., and Metaxopoulos, J. 2003. Antagonistic activity of lactic acid bacteria against *Listeria monocytogenes* in sliced cooked cured pork shoulder stored under vacuum or modified atmosphere at 4 ± 2°C. *Food Microbiology* 20:259–265.
36. Mataragas, M., Drosinos, E. H., Siana, P., Skandamis, P., and Metaxopoulos, I. 2006. Determination of the growth limits and kinetic behavior of *Listeria monocytogenes* in a sliced cooked cured meat product: validation of the predictive growth model under constant and dynamic temperature storage conditions. *Journal of Food protection* 69:1312–1321.

37. Glass, K. A. and Doyle, M. P. 1989. Fate of *Listeria monocytogenes* in processed meat during refrigerated storage. *Applied and Environmental Microbiology* 55:1565–1569.

38. Drosinos, E. H., Mataragas, M., Veskovic-Moracanin, S., Gasparik-Reichardt, J., Hadziosmanovic, M., and Alagic, D. 2006. Quantifying nonthermal inactivation of *Listeria monocytogenes* in European fermented sausages using bacteriocinogenic lactic acid bacteria or their bacteriocins: a case study for risk assessment. *Journal of Food Protection* 69:2648–2663.

39. ISO. 1996. *International Standard, ISO 11290-1: Microbiology of Food and Animal Feeding Stuffs— Horizontal Method for the Detection and Enumeration of Listeria monocytogenes—Part 1: Detection Method*. International Organization for Standardization, Geneva.

40. Patel, J. R. and Beuchat, L. R. 1995. Enrichment in Fraser broth supplemented with catalase or Oxyrase*, combined with the microcolony immunoblot technique, for detecting heat-injured *Listeria monocytogenes* in foods. *International Journal of Food Microbiology* 26:165–176.

41. Gasanov, U., Hughes, D., and Hansbro, P. M. 2005. Methods for the isolation and identification of *Listeria* spp. and *Listeria monocytogenes*: a review. *FEMS Microbiology Reviews* 29:851–875.

42. Klinger, J. D., Johnson, A., and Croan, D. et al. 1988. Comparative studies of nucleic acid hybridization assay for *Listeria* in foods. *Journal of the Association of Official Analytical Chemists* 71:669–673.

43. Cocolin, L., Manzano, M., Cantoni, C., and Comi, G. 1997. A PCR-microplate capture hybridization method to detect *Listeria monocytogenes* in blood. *Molecular and Cellular Probes* 11:453–455.

44. Li, X., Boudjellab, N., and Zhao, X. 2000. Combined PCR and slot blot assay for detection of *Salmonella* and *Listeria monocytogenes*. *International Journal of Food Microbiology* 56:167–177.

45. Hsih, H. Y. and Tsen, H. Y. 2001. Combination of immunomagnetic separation and polymerase chain reaction for the simultaneous detection of *Listeria monocytogenes* and *Salmonella* spp. in food samples. *Journal of Food Protection* 64:1744–1750.

46. Bubert, A., Kohler, S., and Goebel, W. 1992. The homologous and heterologous regions within the *iap* gene allow genus- and species-specific identification of *Listeria* spp. by polymerase chain reaction. *Applied and Environmental Microbiology* 58:2625–2632.

47. Wesley, I. V., Harmon, K. M., Dickson, J. S., and Schwartz, A. R. 2002. Application of a multiplex polymerase chain reaction assay for the simultaneous confirmation of *Listeria monocytogenes* and other *Listeria* species in turkey sample surveillance. *Journal of Food protection* 65:780–785.

48. Klein, P. G. and Juneja, V. K. 1997. Sensitive detection of viable *Listeria monocytogenes* by reverse transcription-PCR. *Applied and Environmental Microbiology* 63:4441–4448.

49. Call, D. R., Borucki, M. K., and Loge, F. J. 2003. Detection of bacterial pathogens in environmental samples using DNA microarrays. *Journal of Microbiological Methods* 53:235–243.

50. Churchill, R. L. T., Lee, H., and Hall, C. J. 2006. Detection of *Listeria monocytogenes* and the toxin listeriolysin O in food. *Journal of Microbiological Methods* 64:141–170.

51. Lampel, K. A., Orlandi, P. A., and Kornegay, L. 2000. Improved template preparation for PCR-based assays for detection of food-borne bacterial pathogens. *Applied and Environmental Microbiology* 66:4539–4542.

52. Sergeev, N., Distler, M., and Courtney, S. et al. 2004. Multipathogen oligonucleotide microarray for environmental and biodefense applications. *Biosensors and Bioelectronics* 20:684–698.

53. ISO. 1998. *International Standard, ISO 11290-2: Microbiology of Food and Animal Feeding Stuffs— Horizontal Method for the Detection and Enumeration of Listeria monocytogenes—Part 2: Enumeration Method*. International Organization for Standardization, Geneva.

54. Schleiss, M. R., Bourne, N., Bravo, F. J., Jensen, N. J., and Bernstein, D. I. 2003. Quantitative–competitive PCR monitoring of viral load following experimental guinea pig cytomegalovirus infection. *Journal of Virological Methods* 108:103–110.

55. Choi, W. S. and Hong, C. H. 2003. Rapid enumeration of *Listeria monocytogenes* in milk using competitive PCR. *International Journal of Food Microbiology* 84:79–85.

56. Prentice, G. A. and Neaves, P. 1992. The identification of *Listeria* species. In R. G. Board, D. Jones, and F. A. Skinner (eds), *Applied Bacterial Symposium*. Blackwell Science Ltd, Oxford, 283–296.

57. Bell, C. and Kyriakides, A. 2002. Pathogenic *Escherichia coli*. In C. de W. Blackburn, and P. J. McClure (eds), *Foodborne Pathogens: Hazards, Risk Analysis and Control*. CRC Press, Boca Raton, FL, 280–306.

58. Smith, H. R., Cheasty, T., and Roberts, D. et al. 1991. Examination of retail chickens and sausages in Britain for Vero cytotoxin-producing *Escherichia coli*. *Applied and Environmental Microbiology* 57:2091–2093.

59. Heuvelink, A. E., Wernars, K., and de Boer, E. 1996. Occurrence of *Escherichia coli* O157 and other Verocytotoxin-producing *E. coli* in retail raw meats in the Netherlands. *Journal of Food Protection* 59:1267–1272.

60. Bolton, F. J., Crozier, L., and Williamson, J. K. 1996. Isolation of *Escherichia coli* O157 from raw meat products. *Letters in Applied Microbiology* 23:317–321.

61. Chapman, P. A., Siddons, C. A., and Cerdan Malo, A. T. et al. 2000. A one year study of *Escherichia coli* O157 in raw beef and lamb products. *Epidemiology and Infection* 124:207–213.

62. Getty, K. J. K., Phebus, R. K., and Marsden, J. L. et al. 2000. *Escherichia coli* O157:H7 and fermented sausages: A Review. *Journal of Rapid Methods and Automation in Microbiology*, 8:141–170.

63. Pond, T. J., Wood, D. S., Mumin, I. M., Barbut, S., and Griffiths, M. W. 2001. Modeling the survival of *Escherichia coli* O157:H7 in uncooked, semidry, fermented sausage. *Journal of Food protection* 64:759–766.

64. Adams, M. and Mitchell, R. 2002. Fermentation and pathogen control: a risk assessment approach. *International Journal of Food Microbiology* 79:75–83.

65. ISO. 2001. International Standard, ISO 16654: *Microbiology of food and animal feeding stuffs-Horizontal method for the detection of Escherichia coli O157*, International Organization for Standardization, Geneva.

66. Heuvelink, A. E., Zwartkruis-Nahuis, J. T. M., and de Boer, E. 1997. Evaluation of media and test kits for the detection and isolation of *Escherichia coli* O157 from minced beef. *Journal of Food Protection* 60:817–824.

67. Osek, J. 2002. Rapid and specific identification of Shiga toxin producing *Escherichia coli* in faeces by multiplex PCR. *Letters in Applied Microbiology* 34:304–310.

68. Kim, J. Y., Kim, S. Y., and Kwon, N. H. et al. 2005. Isolation and identification of *Escherichia coli* O157:H7 using different detection methods and molecular determination by multiplex PCR and RAPD. *Journal of Veterinary Science* 6:7–19.

69. Muller, D., Hagedorn, P., and Brast, S. et al. 2006. Rapid identification and differentiation of clinical isolates of enteropathogenic *Escherichia coli* (EPEC), atypical EPEC, and Shiga toxin-producing *Escherichia coli* by a one-step multiplex PCR method. *Journal of Clinical Microbiology* 44:2626–2629.

70. Kadhum, H. J., Ball, H. J., Oswald, E., and Rowe, M. T. 2006. Characteristics of cytotoxic necrotizing factor and cytolethal distending toxin producing *Escherichia coli* strains isolated from meat samples in Northern Ireland. *Food Microbiology* 23:491–497.

71. ISO. 2001. *International Standard, ISO 16649-2: Microbiology of Food and Animal Feeding Stuffs—Horizontal Method for the Enumeration of Beta-Glucuronidase-Positive Escherichia coli—Part 2: Colony-Count Technique at 44 degrees C Using 5-Bromo-4-chloro-3-indolyl Beta-D-glucuronide*. International Organization for Standardization, Geneva.

72. Bell, C. and Kyriakides, A. 2002. *Salmonella*. In C. de W. Blackburn, and P. J. McClure (eds) *Foodborne Pathogens: Hazards, Risk Analysis and Control*. CRC Press, Boca Raton, FL, 307–335.

73. ISO. 1993. *International Standard, ISO 6579: Microbiology of Food and Animal Feeding Stuffs—Horizontal Method for the Detection of Salmonella spp.* International Organization for Standardization, Geneva.

74. D'Aoust, J. Y., Sewell, A. M., and Warburton, D. W. 1992. A comparison of standard cultural methods for the detection of foodborne *Salmonella*. *International Journal of Food Microbiology* 16:41–50.

75. Ruiz, J., Nunez, J., Diaz, J., Sempere, M. A., Gomez, J., and Usera, M. A. 1996. Note: Comparison of media for the isolation of lactose-positive *Salmonella*. *Journal of Applied Bacteriology* 81:571–574.

76. Herikstad, H., Motarjemi, Y., and Tauxe, R. V. 2002. *Salmonella* surveillance: a global survey of public health serotyping. *Epidimiology and Infection* 129:1–8.

77. Lukinmaa, S., Nakari, U. M., Eklund, M., and Siitonen, A. 2004. Application of molecular genetic methods in diagnostics and epidemiology of food-borne bacterial pathogens. *Acta Pathologica, Microbiologica et Immunologica Scandinavica* 112:908–929.

78. Lim, Y. H., Hirose, K., and Izumiya, H. et al. 2003. Multiplex polymerase chain reaction assay for selective detection of *Salmonella enterica* serovar *typhimurium*. *Japanese Journal of Infectious diseases* 56:151–155.

79. Ali, K., Zeynab, A., Zahra, S., and Saeid, M. 2006. Development of an ultra rapid and simple multiplex polymerase chain reaction technique for detection of *Salmonella typhi*. *Saudi Medical Journal* 27:1134–1138.

80. Pritchett, L. C., Konkel, M. E., Gay, J. M., and Besser, T. E. 2000. Identification of DT104 and U302 phage types among *Salmonella enterica* serotype *typhimurium* isolates by PCR. *Journal of Clinical Microbiology* 38:3484–3488.

81. Sutherland, J. and Varnam, A. 2002. Enterotoxin-producing *Staphylococcus, Shigella, Yersinia, Vibrio, Aeromonas* and *Plesiomonas*. In C. de W. Blackburn, and P. J. McClure (eds), *Foodborne Pathogens: Hazards, Risk Analysis and Control*. CRC Press, Boca Raton, FL, 385–415.

82. ISO. 1999. *International Standard, ISO 6888-1: Microbiology of Food and Animal Feeding Stuffs—Horizontal Method for the Enumeration of Coagulase-Positive Staphylococci (Staphylococcus aureus and other Species)—Part 1: Technique using Baird–Parker Agar Medium*. International Organization for Standardization, Geneva.

83. Brown, D. F. J., Edwards, D. I., and Hawkey, P. M. et al. 2005. Guidelines for the laboratory diagnosis and susceptibility testing of methicillin-resistant *Staphylococcus aureus* (MRSA). *Journal Antimicrobial Chemotherapy* 56:1000–1018.

84. Geha, D. J., Uhl, J. R., Gustaferro, C. A., and Persing, D. H. 1994. Multiplex PCR for identification of methicillin-resistant staphylococci in the clinical laboratory. *Journal of Clinical Microbiology* 32:1768–1772.

85. Mason, W. J. Blevins, J. S., Beenken, K., Wibowo, N., Ojho, N., and Smeltzer, M. S. 2001. Multiplex PCR protocol for the diagnosis of staphylococcal infection. *Journal of Clinical Microbiology* 39:3332–3338.

86. Alouf, J. E. and Muller-Alouf, H. 2003. Staphylococcal and streptococcal superantigens: molecular, biological and clinical aspects. *International Journal of Medical Microbiology* 292:429–440.

87. Orwin, P. M., Fitzgerald, J. R., Leung, D. Y., Gutierrez, J. A., Bohach, G. A., and Schlievert, P. M. 2003. Characterization of *Staphylococcus aureus* enterotoxin L. *Infection and Immunity* 71:2916–2919.

88. Martin, M. C., Fueyo, J. M., Gonzalez-Hevia, M. A., and Mendoza, M. C. 2004. Genetic procedures for identification of enterotoxigenic strains of *Staphylococcus aureus* from three food poisoning outbreaks. *International Journal of Food Microbiology* 94 :279–286.

89. Nakano, S., Kobayashi, T., Funabiki, K., Matsumura, A., Nagao, Y., and Yamada, T. 2004. PCR detection of *Bacillus* and *Staphylococcus* in various foods. *Journal of Food Protection* 67:1271–1277.

90. McLauchlin, J., Narayanan, G. L., Mithani, V., and O'Neill, G. 2000. The detection of enterotoxins and toxic shock syndrome toxin genes in *Staphylococcus aureus* by polymerase chain reaction. *Journal of Food Protection* 63:479–488.

91. Martin, G. S., Mannino, D. M., Eaton, S., and Moss, M. 2003. The epidemiology of sepsis in the United States from 1979 through 2000. *The New England Journal of Medicine* 348:1546–1554.

92. Palka-Santini, M., Pützfeld, S., Cleven, B. E., Krönke, M., and Krut, O. 2007. Rapid identification, virulence analysis and resistance profiling of *Staphylococcus aureus* by gene segment-based DNA microarrays: application to blood culture post-processing. *Journal of Microbiological Methods* 68:468–477.

93. Levi, K. and Towner, K. J. 2003. Detection of methicillin-resistant *Staphylococcus aureus* (MRSA) in blood with the EVIGENE MRSA detection kit. *Journal of Clinical Microbiology* 41:3890–3892.

94. Poulsen, A. B., Skov, R., and Pallesen, L. V. 2003. Detection of methicillin resistance in coagulase-negative staphylococci and in staphylococci directly from simulated blood cultures using the EVIGENE MRSA detection kit. *Journal of Antimicrobial Chemotherapy* 51: 419–421.

95. Oliveira, K., Brecher, S. M., and Durbin, A. et al. 2003. Direct identification of *Staphylococcus aureus* from positive blood culture bottles. *Journal of Clinical Microbiology* 41:889–891.

96. Tan, T. Y., Corden, S., Barnes, R., and Cookson, B. 2001. Rapid identification of methicillin-resistant *Staphylococcus aureus* from positive blood cultures by real-time fluorescence PCR. *Journal of Clinical Microbiology* 39:4529–4531.

97. Wellinghausen, N., Wirths, B., Essig, A., and Wassill, L. 2004. Evaluation of the Hyplex Blood-Screen Multiplex PCR-Enzyme-linked immunosorbent assay system for direct identification of gram-positive cocci and gram-negative bacilli from positive blood cultures. *Journal of Clinical Microbiology* 42:3147–3152.

98. Wellinghausen, N., Wirths, B., Franz, A. R., Karolyi, L., Marre, R., and Reischl, U. 2004. Algorithm for the identification of bacterial pathogens in positive blood cultures by real-time LightCycler polymerase chain reaction (PCR) with sequence-specific probes. *Diagnostic Microbiology and Infectious Disease* 48:229–241.

99. Kempf, V. A., Mandle, T., Schumacher, U., Schafer, A., and Autenrieth, I. B. 2005. Rapid detection and identification of pathogens in blood cultures by fluorescence in situ hybridization and flow cytometry. *International Journal of Medical Microbiology* 295:47–55.

100. Sorum, H. and L'Abee-Lund, T. M. 2002. Antibiotic resistance in food-related bacteria-a result of interfering with the global web of bacterial genetics. *International Journal of Food Microbiology* 78:43–56.

101. Jones, T. F., Kellum, M. E., Porter, S. S., Bell, M., and Schaffner, W. 2002. An outbreak of community-acquired foodborne illness caused by methicillin-resistant *Staphylococcus aureus*. *Emerging Infectious Diseases* 8:82–84.

102. Nesbakken, T., Eckner, K., Hoidal, H. K., and Rotterud, O. J. 2003. Occurrence of *Yersinia enterocolitica* and *Campylobacter* spp. in slaughter pigs and consequences for meat inspection, slaughtering, and dressing procedures. *International Journal of Food Microbiology* 80:231–240.

103. ISO. 1994. *International Standard, ISO 10273: Microbiology of Food and Animal Feeding Stuffs—Horizontal Method for the Detection of Presumptive Pathogenic Yersinia enterocolitica*. International Organization for Standardization, Geneva.

104. de Zutter, L., Le Mort, L., Janssens, M., and Wauters, G. 1994. Short-comings of irgasan ticarcillin chlorate broth for the enrichment of *Yersinia enterocolitica* biotype 2, serotype 9 from meat. *International Journal of Food Microbiology* 23:231–237.

105. Arnold, T., Neubauer, H., Nikolaou, K., Roesler, U., and Hensel, A. 2004. Identification of *Yersinia enterocolitica* in minced meat: a comparative analysis of API 20E, *Yersinia* identification kit and a 16S rRNA-based PCR method. *Journal of Veterinary Medicine B* 51:23–27.

106. Logue, C. M., Sheridan, J. J., Wauters, G., Mc Dowell, D. A., and Blair, I. S. 1996. *Yersinia* spp. and numbers, with particular reference to *Y. enterocolitica* bio/serotypes, occurring on Irish meat and meat products, and the influence of alkali treatment on their isolation. *International Journal of Food Microbiology* 33:257–274.

107. Weagant, S. D. and Feng, P. 2001. In F. P. Downes, and K. Ito (eds), *Compendium of Methods for the Microbiological Examination of Foods*. American Public Health Association. Washington, DC, 421–428.

108. Gemski, P., Lazere, J. R., and Casey, T. 1980. Plasmid associated with pathogenicity and calcium dependency of *Yersinia enterocolitica*. *Infection and Immunity* 27:682–685.

109. Shiemann, D. A. 1982. Development of a two-step enrichment procedure for recovery of *Yersinia enterocolitica* from food. *Applied and Environmental Microbiology* 43:14–27.

110. Kandolo, K. and Wauters, G. 1985. Pyrazinamidase activity in *Yersinia enterocolitica* and related organisms. *Journal of Clinical Microbiology* 21:980–982.

111. Farmer, J. J., Carter, G. P., Miller, V. L., Falkow, S., and Wachsmuth, I. K. 1992. Pyrazinamidase, CR–MOX agar, salicin fermentation-esculin hydrolysis, and D-xylose fermentation for identifying pathogenic serotypes of *Yersinia enterocolitica*. *Journal of Clinical Microbiology* 30:2589–2594.

112. Bhaduri, S. and Cottrell, B. 1997. Direct detection and isolation of plasmid-bearing virulent serotypes of *Yersinia enterocolitica* from various foods. *Applied and Environmental Microbiology* 63:4952–4955.

113. Johannessen, G. S., Kapperud, G., and Kruse, H. 2000. Occurrence of pathogenic *Yersinia enterocolitica* in Norwegian pork products determined by a PCR method and a traditional culturing method. *International Journal of Food Microbiology* 54:75–80.

114. Lambertz, S. T., Granath, K., Fredriksson-Ahomaa, M., Johansson, K. E., and Danielsson-Tham, M. L. 2007. Evaluation of a combined culture and PCR method (NMKL-163A) for detection of presumptive pathogenic *Yersinia enterocolitica* in pork products. *Journal of Food Protection* 70:335–340.

115. Lucero-Estrada, C. S. M., del Carmen Velazquez, L., Di Genaro, S., and de Guzman, A. M. S. 2007. Comparison of DNA extraction methods for pathogenic *Yersinia enterocolitica* detection from meat food by nested PCR. *Food Research International* 40:637–642.

116. Howard, S. L., Gaunt, M. W., Hinds, J., Witney, A. A., Stabler, R., and Wren, B. W. 2006. Application of comparative phylogenomics to study the evolution of *Yersinia enterocolitica* and to identify genetic differences relating to pathogenicity. *Journal of Bacteriology* 188:3645–3653.

117. Bennett, R. W. and Belay, N. 2001. *Bacillus cereus*. In F. P. Downes, and K. Ito (eds), *Compendium of Methods for the Microbiological Examination of Foods*. American Public Health Association. Washington, DC, 311–316.

118. Shinagawa, K. 1990. Analytical methods for *Bacillus cereus* and other *Bacillus* species. *International Journal of Food Microbiology* 10:125–142.

119. Mossel, D. A. A., Koopman, M. J., and Jongerius, E. 1967. Enumeration of *Bacillus cereus* in foods. *Applied Microbiology* 15:650–653.

120. Kim, H. U. and Goepfert, J. M. 1971. Enumeration and identification of *Bacillus cereus* in foods. I. 24-hour presumptive test medium. *Applied Microbiology* 22:581–587.

121. Olsen, J. E., Aabo, S., and Hill, W. et al. 1995. Probes and polymerase chain reaction for detection of food-borne bacterial pathogens. *International Journal of Food Microbiology* 28:1–78.

122. Granum, P. E., Tomas, J. M., and Alouf, J. E. 1995. A survey of bacterial toxins involved in food poisoning: a suggestion for bacterial food poisoning toxin nomenclature. *International Journal of Food Microbiology* 28:129–144.

123. Hansen, B. M., Leser, T. D., and Hendriksen, N. B. 2001. Polymerase chain reaction assay for the detection of *Bacillus cereus* group cells. *FEMS Microbiology Letters* 202:209–213.

124. Manzano, M., Cocolin, L., Cantoni, C., and Comi, G. 2003. *Bacillus cereus*, *Bacillus thuringiensis*, and *Bacillus mycoides* differentiation using a PCR-RE technique. *International Journal of Food microbiology* 81:249–254.

125. Pruss, B. M., Dietrich, R., Nibler, B., Martlbauner, E., and Scherer, S. 1999. The hemolytic enterotoxin HBL is broadly distributed among species of the *Bacillus cereus* group. *Applied and Environmental Microbiology* 65:5436–5442.

126. Stenfors, L. P., Mayr, R., Scherer, S., and Granum, P. E. 2002. Pathogenic potential of fifty *Bacillus weihenstephanensis* strains. *FEMS Microbiology Letters* 215:47–51.

127. Haggblom, M. M., Apetroaie, C., Andersson, M. A., and Salkinoja-Salonen, M. S. 2002. Quantitative analysis of cereulide, the emetic toxin of *Bacillus cereus*, produced under various conditions. *Applied and Environmental Microbiology* 68:2479–2483.

128. Anderson, M. A., Jaaskelainen, E. L., Shaheen, R., Pirhonen, T., Wijnands, L. M., and Salkinoja-Salonen, M. S. 2004. Sperm bioassay for rapid detection of cereulide-producing *Bacillus cereus* in food and related environments. *International Journal of Food Microbiology* 94:175–183.

129. Horwood, P. F., Burgess, G. W., and Oakey, H. J. 2004. Evidence for non-ribosomal peptide synthetase production of cereulide (the emetic toxin) in *Bacillus cereus*. *FEMS Microbiology Letters* 236:319–324.

130. Liu, Y., Elsholz, B., Enfors, S. O., and Gabig-Ciminska, M. 2007. Confirmative electric DNA array based test for food poisoning *Bacillus cereus*. *Journal of Microbiological Methods* 70:55–64.

131. Homs, M. C. I. 2002. DNA sensors. *Analytical Letters* 35:1875.

132. Chee, M., Yang, R., and Hubbell, E. et al. 1996. Accessing genetic information with high-density DNA arrays. *Science* 274:610–614.

133. Albers, J., Grunwald, T., Nebling, E., Piechotta, G., and Hintsche, R. 2003. Electrical biochip technology—a tool for microarrays and continous monitoring. *Analytical and Bioanalytical Chemistry* 377:521–527.

134. Pal, S., Alocilja, E. C., and Downes, F. P. 2007. Nanowire labelled direct-charge transfer biosensor for detecting *Bacillus* species. *Biosensors and Bioelectronics* 22:2329–2336.

135. McClane, B. A. 1992. *Clostridium perfringens* enterotoxin: structure, action and detection. *Journal of Food Safety* 12:237–252.

136. Labbe, R. G. 2001. *Clostridium perfringens*. In . F. P. Downes, and K. Ito (eds), *Compendium of Methods for the Microbiological Examination of Foods*. American Public Health Association, Washington, DC, 325–330.

137. Hauschild, A. H. W. and Hilsheimer, R. H. 1974. Enumeration of foodborne *Clostridium perfringens* in egg yolk-free tryptose-sulfite cycloserine agar. *Applied and Environmental Microbiology* 27:521.

138. Baez, L. A. and Juneja, V. K. 1995. Nonradioactive colony hybridization assay for detection and enumeration of enterotoxigenic *Clostridium perfringens* in raw beef. *Applied and Environmental Microbiology* 61:807–810.

139. Garmory, H. S. Chanter, N., French, N. P., Bueschel, D., Songer, J. G., and Titball, R. W. 2000. Occurrence of *Clostridium perfringens* beta2-toxin amongst animals, determined using genotyping and subtyping PCR assays. *Epidemiology and Infection* 124:61–67.

140. Wen, Q., Miyamoto, K., and McClane, B. A. 2004. Development of a duplex PCR genotyping assay for distinguishing *Clostridium perfringens* type A isolates carrying chromosomal enterotoxin (*cpe*) genes from those carrying plasmid borne enterotoxin (*cpe*) genes. *Journal of Clinical Microbiology* 41:1494–1498.

141. Solomon, E. B. and Hoover, D. G. 1999. *Campylobacter jejuni*: a bacterial paradox. *Journal of Food Safety* 19:121–136.

142. Rautelin, H. and Hanninen, M. L. 2000. Campylobacters: the most common bacterial enteropathogens in the Nordic countries. *Annals of Medicine* 32:440–445.

143. Fernandez, H. and Pison, V. 1996. Isolation of thermotolerant species of *Campylobacter* from commercial chicken livers. *International Journal of Food Microbiology* 29:75–80.

144. Stern, N. J., Line, J. E., and Chen, H. C. 2001. *Campylobacter*. In F. P. Downes and K. Ito (eds), *Compendium of Methods for the Microbiological Examination of Foods*. American Public Health Association, Washington, DC, 301–310.

145. ISO. 1995. *International Standard, ISO 10272: Microbiology of Food and Animal Feeding Stuffs—Horizontal Method for Detection and Enumeration of Campylobacter spp.* International Organization for Standardization, Geneva.

146. Giesendorf, B. A. J., Quint, W. G., Henkens, M. H., Stegeman, H., Huf, F. A., and Niesters, H. G. 1992. Rapid and sensitive detection of *Campylobacter* spp. in chicken products by using polymerase chain reaction. *Applied and Environmental Microbiology* 58:3804–3808.

147. Konkel, M. E., Gray, S. A., Kim, B. J., Garvis, S. G., and Yoon, J. 1999. Identification of enteropathogens *Campylobacter jejuni* and *Campylobacter coli* based on the *cad*F virulence gene and its product. *Journal of Clinical Microbiology* 37:510–517.

148. Denis, M., Refregier-Petton, J., Laisney, M. J., Ermel, G., and Salvat, G. 2001. *Campylobacter* contamination in French chicken production from farm to consumer, use of a PCR assay for detection and identification of *Campylobacter jejuni* and *Campylobacter coli*. *Journal of Applied Microbiology* 91:255–267.

149. Sails, A. D., Fox, J. A., Bolton, F. J., Wareing, D. R. A., and Greenway, D. L. A. 2003. A real-time PCR assay for the detection of *Campylobacter jejuni* in foods after enrichment culture. *Applied and Environmental Microbiology* 69:1383–1390.

150. Mateo, E., Carcamo, J., Urquijo, M., Perales, I., and Fernandez-Astorga, A. 2005. Evaluation of a PCR assay for the detection and identification of *Campylobacter jejuni* and *Campylobacter coli* in retail poultry products. *Research in Microbiology* 156:568–574.

151. Churruca, E., Girbau, C., Martínez, I., Mateo, E., Alonso R., and Fernandez-Astorga, A. 2007. Detection of *Campylobacter jejuni* and *Campylobacter coli* in chicken meat samples by real-time nucleic acid sequence-based amplification with molecular beacons. *International Journal of Food Microbiology* 117:85–90.

Chapter 35

Parasites

Anu Näreaho

Contents

In the industrialized countries, people are relatively seldom bothered by parasites in meat, because of the high standards of meat hygiene and the advanced parasite control in animal production. However, when a primary toxoplasmosis is diagnosed in a pregnant woman, or a spot epidemic of trichinellosis is found in a village, it surely is devastating for the people involved. Also, the hospitalization costs for even one patient with a severe parasitic infection are considerable. Thus, meat parasites cannot be ignored.

To prevent meat-borne parasitic infections, it is good advice to cook the meat thoroughly. This is not, however, always the most desirable procedure from a culinary standpoint or the most tasteful way to prepare a meal. Tartar steak, dried ham, raw-marinated meat, and several kinds of smoked meat products are traditional and very much appreciated foods, but those treatments as such certainly are not adequate for destroying possible parasites. Also, in home cooking it is not always possible to follow the temperature of the meat very accurately. Color change of the meat is not a reliable indicator of sufficient temperature. In microwave cooking, in which the temperature is elevated unevenly, the risk is high even though the meal seems to be thoroughly cooked. Therefore, constant control work and research has to be done to maintain a high level of meat hygiene and safety.

In mammal muscles, both protozoan and helminth parasites can be found. Protozoans are microscopic organisms that can live intracellularly in a host. *Toxoplasma* and *Sarcocystis* are examples of zoonotic protozoans, which are infective to humans. Helminths are larger in size, and many of the species can be seen by naked eye. Of helminths, *Taenia* and *Trichinella* species can be transmitted to humans via the consumption of raw mammal meat or meat products [1].

Several parasite species live in the alimentary tract and may also be found in meat inspection, though not from the muscles. These parasites can affect the general welfare of the animal, cause production losses, and lower the quality of the meat. They can also have hygienic implications in food if parts of the alimentary tract are used as foodstuffs or if the meat is contaminated with the contents of the gut during the slaughtering process. Fecal examination for parasites or parasite eggs is not of importance in slaughterhouses, but could be a valuable diagnostic tool in animal husbandry. Intestinal parasites or external parasites either are not, however, discussed here. The focus in this chapter is on the parasites whose life cycle directly involves mammal muscles; some parasites affecting the liver are also briefly described.

In the meat inspection protocols at slaughterhouses, several checking, palpation, and incision steps are followed to verify parasites. Visual inspection and microscopic analysis of parasite morphology is useful and suitable for many species. Closer parasite species or strain differentiation, however, often requires molecular biological methods, such as polymerase chain reaction (PCR). Indirect methods that measure the immunological reaction of the host animal against parasite, including ELISA and immunoblot, are also useful in diagnostics of parasite infections, but also have their disadvantages, as described later.

Official meat inspection is regulated by law, and the analyses necessary are strictly stated. This chapter gives an overview of the diagnostic laboratory methods that are used or could be used to detect certain parasites in meat, and also introduces some methods that are designed for research purposes. The methods are introduced along with descriptions of the most important meat parasites in the industrialized countries.

35.1 *Trichinella* spp.

Trichinella nematodes are found worldwide, and can infect mammals, birds, and reptiles. They are well-known for their ability to cause illness to humans who eat undercooked infective meat; even deaths have been reported. Pork, horse, wild boar, and certain game meats are common sources for human infection. *Trichinella* larvae are freed from the muscle tissue in the stomach; molting and reproduction take place in the small intestine. The newly hatched larvae migrate through the circulatory system to the host's striated muscle, penetrate muscle cells, transform them into so-called "nurse cells," and settle in. Most species induce formation of a connective tissue capsule

Table 35.1 Summary of *Trichinella* Species and Genotypes

Isolates	Species/Genotype	Capsule	Geographical Distribution	Freeze Resistance	Host Examples
T1	*T. spiralis*	Yes	Cosmopolitan	−	Pig, wild boar, rat
T2	*T. nativa*	Yes	Arctic, subarctic	+++	Bear, wolf, fox
T3	*T. britovi*	Yes	Europe, Asia	+	Wild boar, horse
T4	*T. pseudospiralis*	No	Cosmopolitan	−	Birds, marsupials
T5	*T. murrelli*	Yes	North America	−	Bear, raccoon
T6		Yes	North America	++	Bear, wolf
T7	*T. nelsoni*	Yes	Africa	−	Hyena, lion
T8		Yes	South Africa	−	Hyena, lion
T9		Yes	Japan	−	Bear, raccoon dog
T10	*T. papuae*	No	Papua New Guinea	−	Reptiles
T11	*T. zimbabwensis*	No	Africa	−	Reptiles

Sources: Dupoy-Camet, J. et al., in *Trichinellosis. Proceedings of the Eighth International Conference on Trichinellosis 1993*, F. Istituto Superiore de Sanità Press, Rome, Italy, 1994, 83; Pozio, E., *Vet. Parasitol.*, 93, 241, 2000; and Pozio, E. et al., *Parasitology*, 128, 333, 2004.

around them (Table 35.1). The larvae can survive inside the muscle cells in a dormant state for years, until the muscle is ingested by a new host. The life cycle is straightforward and involves only one host animal; *Trichinella* does not have separate definitive and intermediate hosts for different development stages [2–4].

In Europe, four species, *T. spiralis*, *T. nativa*, *T. britovi*, and *T. pseudospiralis*, are found. Owing to the risk of human infection, they are actively searched for with laboratory analysis in the meat inspection in the EU [5]. Swine and other potential *Trichinella* hosts (horse, wild boar, and certain game) are examined. An approved freezing protocol is an alternative for the *Trichinella* inspection of pork, because in pork, the infective species usually is *T. spiralis*, which is not resistant to below-zero temperatures. Some *Trichinella* species can, however, tolerate freezing. *T. britovi* has a moderate tolerance for low temperatures, and are infective to swine. Freezing alone is not recommended without *Trichinella* testing in the areas where *T. britovi* is endemic.

In horse and game meat, freeze-resistant *Trichinella* species can be more predominant than in pork. Freezing is never an alternative to testing in these animals. Large recent epidemics of human trichinellosis have unexpectedly been caused by the consumption of horse meat. These herbivore infections can be a consequence of rodents accidentally getting crushed in the feed or of feeding the horses on purpose with animal protein. Also, pig infections result from similar causes, but as omnivorous species, pigs are naturally more willing to ingest meat or even kill rodents by themselves.

35.1.1 Direct Detection Methods

In official meat inspection, only direct methods of detecting *Trichinella* larvae are used. Samples are taken as described in legislation (Commission regulation [EC] No. 2075/2005 [5], in EU countries). Sample sites are set according to the vulnerability of muscles of each animal species. If the predilection muscles of the animal are not known, diaphragm or tongue should be used. Sample size also varies depending on the animal species. Samples are taken at the slaughter line with a knife or with *Trichinella* forceps and arranged in a special tray in such a way that traceability is guaranteed. Pure muscle tissue, without fat or tendons, should be taken as a sample.

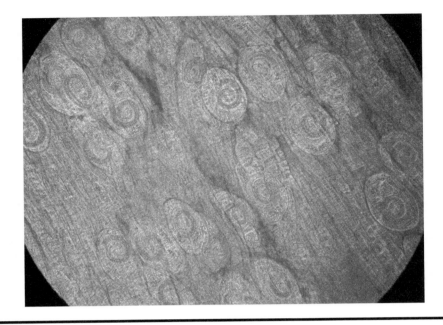

Figure 35.1 *Trichinella britovi* encapsulated in experimentally infected mouse muscle. Light microscopic picture of compressorium plates. (Photo by Dr. J. Bien and Dr. K. Pastusiak, Witold Stefanski Institute of Parasitology, PAS.)

The purpose of *Trichinella* inspection is to prevent clinical trichinellosis in humans [6]. This means that absolutely *Trichinella*-free meat is not guaranteed with the methods used for testing at the moment, but the possible infection level is so low that people do not get sick. If more precise results are needed, the sensitivity of current methods can easily be improved by increasing the sample size.

35.1.1.1 Trichinoscopic Examination

Trichinoscopy, or compression technique, is the classic method of *Trichinella* inspection. Several (the number depending on the animal species) small, oat-kernel-size pieces of meat are pressed tightly between two glass plates, and paper-thin slices are then carefully scanned through with a microscope with 30–40 times magnification (Figure 35.1). The method is labor-intensive, slow to perform, and sensitivity is poor. Early infections, nonencapsulated species of *Trichinella* (*T. pseudospiralis, T. papuae, T. zimbabwensis*), or low infection levels are not easily recognized in trichinoscopy. It is currently used only in exceptional conditions, and is normally replaced by digestion methods.

35.1.1.2 Methods for Digestion

Artificial digestion is the method most commonly used in *Trichinella* examination. Digestion methods mimic the conditions in the stomach, with hydrochloride acid (HCl) and pepsin enzyme. The treatment enzymatically dissolves the muscle and the connective tissue of the capsule, enabling the count of the released, sedimented larvae (Figure 35.2). The method consists of three basic steps: digestion, sieving, and microscopic detection of the larvae.

Figure 35.2 *Trichinella spiralis* **muscle larvae after digestion. Light microscopic picture with 480× magnification. (Photo by Dr. J. Bien and Dr. K. Pastusiak, Witold Stefanski Institute of Parasitology, PAS.)**

In meat inspection, 1 g of muscle tissue from the diaphragm of 100 fattening pigs is pooled as one *Trichinella* digestion sample. For species other than pig, and also for sows and boars, larger samples per animal are required. In the event of a positive finding in the pooled sample, the potentially infected individuals are searched by repeating the examination with smaller subsets of the samples pooled together, until single sample digestions finally reveal the infected animal(s). In the EU, infected carcasses are condemned and removed from the food chain.

The theoretical sensitivity of 1 g muscle sample digestion is naturally one larva per gram (1 lpg). In practice, it is lower; a true sensitivity of 3–5 lpg is evaluated [7]. The sensitivity of the method can be improved by increasing the size of the sample. For example, 5 g samples are estimated to give a true sensitivity of 1 lpg [6].

All the digestion methods operate on the same principle, but the magnetic stirrer method (Table 35.2) is considered the gold standard [8]. Some digestion methods involve specialized laboratory apparatus for homogenization and warming of the samples. Certain filters with a pump can be used as well. The results of the different digestion methods may vary because of the different ways of handling the digestion fluids. The magnetic stirrer method is performed in a glass container, in which the freed larvae do not easily stick to the surfaces. In the Stomacher˚ apparatus (Seward Ltd., Worthing, U.K.) and similar methods, the plastic bags used in the liquid handling may offer tight corners or form capillary forces when the digestion fluid is poured out, and some of the larvae may get trapped. This could result in lower larvae per gram values or even false negative results if the infection level is low. These machines, however, enable several digestions at the same time.

Other steps in digestion may affect the results as well. Care must be taken not to inactivate the pepsin enzyme with concentrated HCl before digestion. The correct order to measure and add the regents for digestion is: water, HCl, and, when those are well mixed together, pepsin. The water

Table 35.2   Magnetic Stirrer Digestion for *Trichinella* (100 g Pooled Sample)

- Add 16 mL of 25% hydrochloric acid into a beaker containing 2 L of water (46–48°C).
- Start magnetic stirring on a preheated plate. The digestion fluid must rotate at high speed but without splashing.
- Add 10 g of pepsin (1:10,000 U.S. National Formulary).
- Grind 100 g pool of samples with a kitchen blender, meat mincer, or scissors, and add to the beaker.
- Rinse all the equipment used for mincing with the digestion fluid to ensure that all the meat is included to the examination.
- Cover the beaker with aluminum foil to balance the temperature. Constant temperature of 44–46°C throughout the digestion must be maintained. Overheating will inactivate the pepsin and interrupt the digestion.
- Continue stirring for 30 min, until the meat particles have disappeared. Longer digestion times may be necessary (not exceeding 60 min) for tongue, game meat, etc.
- Pour the digestion fluid through 180 µm mesh sieve into the sedimentation funnel. Not more than 5% of the starting sample weight should remain on the sieve. You may cool the sample with ice before pouring it into the funnel.
- Let the fluid stand in the funnel for 30 min to sediment the larvae. The sedimentation may be aided with periodic mechanical vibration.
- From the bottom tap of the funnel, quickly run 40 mL sediment sample of digestion fluid into a measuring cylinder or large centrifuge tube.
- Allow the 40 mL sample to stand for 10 min. Carefully remove with suction 30 mL of supernatant and leave a volume of not more than 10 mL.
- Pour the remaining 10 mL sample of sediment into a petri dish marked with 10 × 10 mm grid to ease the examination.
- Rinse the cylinder or centrifuge tube with not more than 10 mL of water, which has to be added to the sample for larval count.
- Examine the sample by trichinoscope or stereomicroscope at 15–40 times magnification. For suspect areas or parasite-like shapes, use higher magnifications of 60–100 times. Count the larvae.

Sources: EC (European Commission) Regulation No. 2075/2005, of 5 December 2005, laying down specific rules on official controls for *Trichinella* in meat. *Official Journal of the European Union* 22.12.2005; ICT recommendations: Gamble, H.R. et al., *Vet. Parasitol.*, 93, 393, 2000; and OIE standards: World Organisation for Animal Health. Health standards. Manual of diagnostic tests and vaccines of terrestrial animals. Trichinellosis (updated 21.11.2005). http://www.oie.int/eng/normes/mmanual/A_00048.htm.

temperature must be controlled carefully before adding pepsin. At temperatures over 50°C, the enzyme is inactivated. Below that temperature, the inactivation happens so slowly that it does not have an effect on the digestion.

After digestion, the mesh size used for sieving the fluid to remove excess undigested material is 180 µm [5]. Higher sensitivity (better larval recovery) with a larger mesh size, 355 µm, has been reported [9], however. No more than 5% of the original pig sample weight should be retained in the sieve—the digestion should be considered inadequate if more material is found.

The meat sample itself could affect the digestion outcome as well. The structure of the muscle can slow the digestion. For certain muscle types (e.g., tongue) and animal species (horse, game), the time of digestion often has to be prolonged. Fat in the sample, besides decreasing the muscle mass and thus affecting the larvae per gram value, makes the digestion fluid hazy, and detecting the larvae with a microscope thus becomes more difficult. The separation of fat to the top of the digestion fluid, thereby diminishing the remaining lipids in the sample, can be made faster by cooling the fluid with ice before pouring it into the sedimentation funnel. Cooling also induces coiling and faster sedimentation of the larvae.

In meat inspection, the presence of bone pieces in the *Trichinella* sample is not a concern, but in research work various kinds of samples set limitations for the methods. In plastic bag digestion, sharp bones may break the plastic; the magnetic stirrer method is recommended. Bone, as well as fat and connective tissue, in the sample lowers the larvae per gram value—only muscle tissue should be used if infection intensity is analyzed.

The microscopic examination of the digestion fluid should be performed immediately after the digestion. If the examination is delayed, the fluid must be clarified [5]. Microscopic examination should not be postponed until the next day; digestion fluid containing acid and pepsin can cause degradation of the larvae.

All the containers, materials, and fluids should await the result of *Trichinella* examination before washing or disposing them; otherwise, they should be handled as would be done in the case of a positive sample. Digestion fluid and the instruments which have been in direct contact with a positive sample, should be sterilized to kill the *Trichinella* larvae. A few minutes in well-boiling water are enough. If potentially infective waste exists, it should be autoclaved or decontaminated in some other acceptable way. Of disinfectants, 1:1 mixtures of xylol and ethanol (95%), or xylol and phenol are lethal for infective larvae [10].

35.1.1.3   Histology

Trichinella can be found with a microscope in histological samples using common dyeing techniques for muscle and connective tissue—hematoxylin and eosin (HE), for example. If the section is not sufficiently representative, immunohistochemical confirmation can be done. Immunofluorescence techniques are also suitable. Histological methods are not used in slaughterhouses or in routine work, but in research, and sometimes in diagnostics, they can be useful. Biopsy is, however, an insensitive method because the size of the sample is so small. In the histological samples, the presence or absence of the capsule or differences in capsule formation and shape, as well as the cellular reaction around the parasite, can give hints about the infecting *Trichinella* species [11], but definitive differentiation is made using molecular biological methods.

35.1.1.3.1   PCR Methods

Identification of infecting *Trichinella* species is required in the EU when infection is found in meat inspection [5]. This analysis is performed in a national reference laboratory, or the examination may be ordered from some other qualified laboratory. Community Reference Laboratory for Parasites (Istituto Superiore di Sanita, Rome, Italy) offers services and detailed procedure descriptions for species identification [12].

Many methods for *Trichinella* species differentiation are published, but it is currently usually done with multiplex-PCR. In this method, several primer pairs are added to a single PCR reaction mixture, and thereby several DNA fragments can be amplified at the same time. *Trichinella* primers are generated according to sequence data from internal transcribed spacers ITS1 and ITS2 and from expansion segment V (ESV) of the ribosomal DNA repeat [13,14]. All the *Trichinella* species can be screened from one sample. Even one larva is enough for species analysis, but all the larvae found in the original digestion, plus some of the infected meat, should be sent as a sample to the laboratory. The laboratory analyzing for species gives detailed instructions for sending the samples, but usually digested, washed larvae in 90% ethanol are requested. Freeze-thaw cycles of the sample should be avoided before species analysis. Even more detailed diagnostics of the *Trichinella* isolates for epidemiological research purposes can be made with techniques based on sequence analysis.

Several methods are available to extract DNA from *Trichinella* larvae. When using commercial kits, usually the protocol application for tissue samples is the most suitable. A good yield of genomic DNA from even a single muscle larva can be achieved with a simple protocol in a small volume of buffer (10–20 µl) [15]. There are several modifications of this extraction method, and the yield can be enhanced with a longer incubation with proteinase K (overnight), by increasing the proteinase K concentration, adding detergents, such as sodium dodecyl sulfate (SDS), and combining the commercial Gene Releaser® (BioVentures Inc., Murfreesboro, TN) protocol after the extraction.

35.1.2 Indirect Detection Methods

In addition to the direct detection methods, *Trichinella* infection can be diagnosed by searching *Trichinella*-specific antibodies in serum, plasma, and whole blood or tissue fluid samples of the host. These methods are used in surveys and monitoring, not in the individual carcass testing. They are called indirect, since instead of detecting the parasite itself, they detect the immunological reaction of the host animal against the invader—antibodies show that the immune system has faced *Trichinella* antigens at some point. In addition to exposure to the parasite, the reaction is dependent on the host's immunological status and capability to produce antibodies. Production of antibodies to a detectable level takes time; early infections may be falsely diagnosed as negative reactions using these methods. Indirect methods are not used in meat inspection.

The antigen used in the test and the antibody type searched strongly affect the results. Validation of the antigens, dilutions, and cut-off levels to separate the positive samples from the negative ones should be done with a significant number of samples before accepting any method for routine use; otherwise, misdiagnosis may occur. Positive and negative control samples should always be included in the test, to ensure the proper performance of the test. There are several ways to count the cut-off level, which affects the interpretation of the results, that is, the sensitivity and specificity of the test. Serological methods can be very sensitive, with a detection level of one larva/100 g meat [16], but due to the disadvantages related to the indirect detection, they are not recommended for use in meat inspection to replace the conventional direct methods [17]. For surveillance studies and *Trichinella* diagnostics at herd or population levels, indirect methods are, however, very useful.

The immunological reaction of the host species should be studied before using any immunological test for diagnostics. In the horse, for example, the diagnostic value of serological *Trichinella* tests is questionable. False results are common. The circulating specific IgG antibodies are undetectable in horse already 4–5 months after the infection, even though the capsulated *Trichinella* larvae are alive and well, and ready to infect a new host [18]. Therefore, serological diagnostics of *Trichinella* infections should not be used in horses. Also, insufficient information about game animal and other wildlife testing is available; without decent validation of the method, serology should not be used even in surveillance studies.

Of serological laboratory techniques, bentonite flocculation, indirect immunofluorescence microscopy, latex agglutination, and enzyme-linked immunosorbent assay (EIA, ELISA) are commonly used for *Trichinella* diagnostics [19], ELISA being the most common method. Immunoblot is also often utilized.

In ELISA, antibodies from the sample are bound to antigens that are coated onto a microwell plate. The bound antibodies are detected with specific antibodies that are linked to an enzyme. The enzyme reacts with a substrate to form a color reaction, which is then spectrophotometrically measured. ELISA can be used for detection of antigens as well.

In immunoblot, antigens are first separated by their molecular weight in SDS polyacrylamide gel electrophoresis (PAGE) and transferred onto a nitrocellulose membrane. The membrane is exposed to the antibodies in the sample. Antigen–antibody binding is visualized, and the molecular weight of immunoreactive antigens can also be analyzed.

Noticeable effort in several laboratories has been put toward developing an optimal serological test for *Trichinella* infection. The test should recognize the infection as early as possible, with as low infection level as possible, but without cross-reactions with other parasites. Several antigens, crude, excretory–secretory (ES), and synthetic, have been used for detection. The replacement of crude parasite antigens with more specific antigens has lowered the risk of cross-reaction, and crude antigens are not recommended in immunological testing any more. To obtain more reliable results in immunological tests, several methods can be used for analysis of the samples.

Fast "patient-side" tests are developed for diagnostics, especially as preliminary tests, or when laboratory facilities are not available. A "dipstick assay" [20,21] has been reported to show useful sensitivity and specificity in analysis of human and swine *Trichinella* cases. In this method, the antigen is dotted onto a nitrocellulose membrane. The membrane is then dipped into a serum sample, and the antigen–antibody reaction is visualized. A commercial, so-called "lateral flow card" has been tested with blood, serum, and tissue fluids with satisfactory results [22]. It is recommended for use as a farm screening test or for preliminary screening of suspected pigs during slaughter, especially in countries with high *Trichinella* prevalence, to improve the food hygienic quality. In the muscle fluid there is about 10 times lower concentration of antibodies than in serum, but if serodiagnostics are not possible, muscle or body fluid samples can offer a good alternative [23,24].

35.2   *Taenia* spp.

Taenia tapeworms form larval cysts, cysticerci, in intermediate hosts. These are the infective forms for definitive hosts. There are three human *Taenia* species (*T. saginata, T. solium,* and *T. asiatica*), and several other species that can be found in the postmortem inspection of the intermediate hosts that are used for food.

Bovine cysticercosis is caused by *Cysticercus bovis*, the larval stage of *T. saginata*. The parasites spread through the bloodstream to skeletal muscles and the heart, forming 5–10 mm diameter, thick-walled, pearl-like infective cysts. The cyst remains viable for about 6 months, after which it starts to calcify. However, in light infections viable cysts can be found years after the onset of infection. The sites of predilection in cattle are the masticatory muscles, tongue, heart, and diaphragm. Incision of these muscles in the postmortem inspection reveal the possible infection. Infections can be classified as light, local, heavy, or generalized, and the judgment during meat inspection is made according to this analysis. In the case of a light or local infection, condemnation of the carcass is not necessary, though the meat should at minimum be heated or frozen before consumption.

T. solium is the infective agent in porcine cysticercosis, *Cysticercus cellulosae*. The species is significant, because humans can act as both definitive and intermediate hosts for the parasite. Thus, the infection can be manifested in humans not only as intestinal taeniosis, but also as cysticercosis or neurocysticercosis. Human cases are generally linked to inadequate sanitation, free-ranging swine, ineffective meat inspection, or ingestion of inadequately cooked pork [25]. Because pigs are slaughtered young, all cysts found in meat inspection in the industrialized countries should be considered as viable.

There are limitations with diagnosing cysticercosis by visual inspection only. Cases of low infection level, or early infections, are inevitably not diagnosed. On the other hand, overestimation of the disease may also be made in visual inspection: In a study of cysts diagnosed as cysticercosis in abattoirs, only 52.4% were confirmed positive in further studies with PCR [26]. In another study, 97% of viable *T. saginata* cysts were confirmed by PCR-restriction fragment polymorphism, while the percentage for dead cysts was approximately 73% [27].

The use of serological methods, such as ELISA and immunoblot, together with the visual inspection would improve the efficacy of meat inspection for the detection of cysticercosis [28]. Many serological methods are available for differentiation between viable and degenerated cysts [29–32].

35.3 *Toxoplasma gondii*

T. gondii is a common, worldwide, mammal- and bird-infecting, zoonotic, protozoan parasite. Its sexual reproduction takes place only in feline hosts, but asexual multiplication is possible in all host species. Infection can be acquired by ingestion of oocysts (for example, from soil or water contaminated with cat feces) or through tissue cysts ingested with infected meat. Transplacental infection from mother to fetus may also occur, and contact infection through mucous membranes has been reported. The infection is relatively common among humans but does not cause any severe symptoms for healthy adults; most of them are unaware that they have had *Toxoplasma* infection. The most severe threat in *Toxoplasma* infections is when a pregnant woman gets infected for the first time with no previous immunological protection. Damage caused to the fetus can be fatal. Also, immunocompromised people, such as AIDS patients or transplant recipients taking immunosuppressive medication, are at risk. In a study defining predisponating factors for pregnant women's *Toxoplasma* infections, eating undercooked meat, contact with soil, and traveling outside Europe and North America were risk factors, but contact with cats was not [33].

The majority of *Toxoplasma* isolates can be classified in three genetic lineages: type I, II, and III. There are differences in geographic distribution of the types. Their virulence, in addition to host genetic factors, has been shown to have an influence on the severity of the disease. Most human cases, especially in Europe and North America, are associated with type II, whereas lesions in the eye are often reported to be caused by type I [34,35].

Noticeable economic losses occur among farmed sheep due to *Toxoplasma* abortions and stillbirths. Goats and pigs are also susceptible; cattle and horses are more resistant to the disease. Several studies of prevalence among domestic animals over the world show that *Toxoplasma* infection is common. For example, in a German study, 19% of sows (*n* = 2041) were seropositive [36].

Toxoplasma is not searched during meat inspection. Instead, consumers at risk are advised not to eat undercooked meat. Nonspecific lesions in postmortem inspection, such as necrotic foci in organs, might be suggestive of *Toxoplasma* infection, but the tissue cysts are invisible to naked eye; thus, the diagnostics requires microscopic, immunological, or molecular biological analysis. Microscopic *Toxoplasma*-like findings (predilection in brain and placenta) can be confirmed with immunohistochemical or immunofluorescence methods. PCR identification can be performed from DNA isolated from tissue samples. For isolation and culturing the organism for research, infection of mice with placental or brain homogenate can be used. *In vitro* cultivation in cell cultures is commonly done to maintain the strains for research purposes.

In addition to conventional serology, analyzing immunoglobulin avidity has been successfully done to determine the onset of infection, which is of great importance in regard to infection in pregnant women [37]. The method measures the strength of the antigen–antibody binding: If the infection is new, the antibody is not yet mature and does not bind as strongly to the antigen as an older antibody would.

Genotyping of the isolates is done with multiplex-PCR combined with restriction fragment length polymorphism (RFLP) of the amplified loci. Closer analysis of the strains can be done with sequencing methods. Serotyping, based on the infected serum reaction in ELISA with polymorphic peptides derived from certain *Toxoplasma* antigens, is also used in human toxoplasmosis [38].

35.4 *Sarcocystis* spp.

Carnivores are the definitive hosts for *Sarcocystis* protozoan, whereas herbivores act as intermediate hosts. Birds and reptiles can be infected as well. Human infections occur, but they are usually asymptomatic. This parasite has worldwide distribution. There are several species of *Sarcocystis* with differing host preferences.

In meat inspection, light-colored *Sarcocystis* bradyzoite cysts between the muscle fibers of ruminants and pigs can be found. The size of the cysts varies according to the host species, and may be visible to the naked eye. Mild infections are asymptomatic, but in heavy infections various kinds of clinical signs may appear, depending on the location of the cysts. Death of the intermediate host has even been reported.

Diagnosis can be confirmed in histological samples; cysts stain basofilic in HE. *Sarcocystis* species can be ultrastructurally identified according to their cyst wall with transmission electron microscopy (TEM), but PCR with DNA isolated from the muscle with cycts reveals the species as well. This analysis can be completed with sequencing techniques. Common immunological methods, described earlier with other parasites, in addition to direct methods, can be used in diagnostics.

35.5 Some Other Parasites of Importance in Meat Inspection

35.5.1 Ascaris suum

Ascaris suum roundworms are macroscopic parasites of the pig's small intestine, but their life cycle also involves lungs and can affect abdominal organs as well. *Ascaris* eggs are secreted in the feces, and infection is feco-oral. Microscopic analysis of feces to find the eggs can be used in diagnostics. In meat inspection, common findings are so-called "milk spots" in the liver. These are light-colored fibrotic tissue areas in the tunnels made by the *Ascaris* larvae when migrating through the liver. They can easily be recognized by visual inspection of the liver surface. Economic losses result because the affected parts are condemned. *Ascaris* control should be done at the farm level. Besides pigs, other animals and humans might suffer from a condition called "visceral larva migrans" if *Ascaris* eggs are hatched in their intestines and the larvae begin to migrate, though the infection is not permanent in these coincidental hosts.

35.5.2 Fasciola hepatica *and Other Liver Flukes*

Liver trematodes are large and rather easily visible in meat inspection. Several species exist. *Fasciola hepatica* is 2–3 cm long and about 1 cm wide. It is predominantly a parasite of ruminants, but can also be found in horses, pigs, and humans. In heavy infections, bleeding and damage of the parenchymal liver tissue may even cause the death of the animal. Sheep are especially sensitive to acute fascioliasis. In milder infections, fairly small amounts of metacercaria forms can predisponate to secondary bacterial infections. The life cycle includes a development period in an intermediate host snail; wet grazing conditions are favorable for the infection, and seasonal variation occurs.

In meat inspection, in addition to the actual presence of the flukes, thickening or calcification of the bile ducts, darkish parasitic material in bile, and color changes on the liver surface or in the carcass (anemia, icterus) can be suggestive for fascioliasis (Figure 35.3). In addition, the lungs might be affected.

Dicrocoelium dendriticum, the lancet fluke, is smaller in size (5–10 mm long) than *Fasciola* but still visible in opened bile ducts. Cattle, sheep, and swine can be affected. The damages caused are small because this worm does not migrate in the liver parenchyme—usually no clinical signs or notable alterations are noticed in postmortem examination.

Liver fluke diagnosis is made in postmortem inspection of liver or by observing eggs in fecal samples. Immunological methods are available as well, and milk can also be conveniently used as a sample.

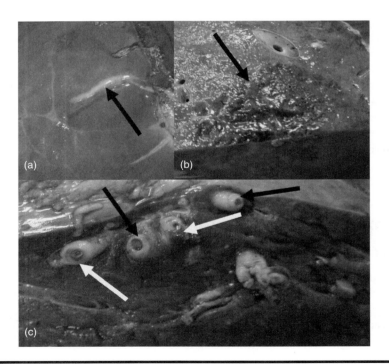

Figure 35.3 *Fasciola hepatica* lesions in bovine liver. (a) Thickened bile duct on the surface of the liver (arrow). (b) Hemorrhage area at the incision surface of the liver. (c) Thickened bile ducts are prominent at the incision surface of the liver (black arrows); the parasites are indicated with the white arrows. (Photo by Dr. M. Kozak, Witold Stefanski Institute of Parasitology, PAS.)

35.5.3   Echinococcus *spp.*

Echinococcus worms, belonging to the cestodes, cause lesions (cysts) mainly in the liver or lungs of livestock. These parasites are briefly discussed here because their cysts can be found in meat inspection, and the parasites are notable for their ability to cause the severe disease, hydatidosis, in humans.

To date, eight species (*E. multilocularis*, *E. shiquicus*, *E. vogeli*, *E. oligarthrus*, *E. granulosus*, *E. equinus*, *E. ortleppi*, and *E. canadensis*) are recognized, but the taxonomy is still incomplete. The distribution of *Echinococcus* is worldwide, with climate preferences varying among the species. Adult forms of these parasites reproduce in carnivores, but cattle, pigs, and sheep, for example, as well as humans, can act as intermediate hosts after ingestion of *Echinococcus* eggs, which are spread through feces of the definitive hosts. All the species except *E. equinus* are infective to humans. *E. multilocularis* is the most harmful because of its alveolar, cancer-like appearance in the liver. The main intermediate hosts for *E. multilocularis* are rodents. *E. granulosus* is important for its global distribution and common infections in humans. The main intermediate host for *E. granulosus* is sheep. Tissue cysts are not a source of infection for the intermediate hosts, such as humans, but it is necessary to control the *Echinococcus* life cycle at the meat inspection stage to prevent further carnivore infections.

Echinococcus worms are some millimeters long depending on species. The cysts in the intermediate hosts' organs can grow up to tens of centimeters in diameter, and are thus often easily noticed in meat inspection. The cysts are fluid-filled with concentrically calcified particles (hydatide sand) and protoscolices, preliminary heads of the worms.

In addition to the typical cyst findings in postmortem inspection, diagnosis can be confirmed histologically in uncertain cases. Formalin-fixed samples can be dyed with periodic acid Schiff (PAS) to observe the *Echinococcus* metacestode characteristic PAS-positive, acellular laminated layer.

Although PCR or coproantigen ELISA from fecal samples might be used to detect parasite eggs in the definitive host, molecular methods are not of great importance in intermediate host diagnostics of *Echinococcus*. For small and calcified lesions of *E. multilocularis*, PCR can be used for identification and confirmation of the diagnosis [39]. The analysis of the species of *Echinococcus* is also done with PCR-based methods. Immunological tests, which are successfully used in human diagnostics, are not sensitive or specific enough to replace careful conventional postmortem inspection of livestock. Furthermore, immunological methods do not distinguish current and past infections, and cross-reactions between *Taenia* species can occur as well. In a study of sheep echinococcosis, however, macroscopic diagnosis at the time of slaughter was found to have limitations, and histology or immunoblot were used with success [40]. In surveillance studies, immunological methods, mainly serology, are useful; suitable antigens are numerous [41]. Even ultrasonography has been used in mass screening for ovine hydatid cysts [42].

35.5.4   Parafilaria bovicola

The filarial nematode, *Parafilaria bovicola*, which is transmitted by a fly vector (*Musca*), can cause severe economic losses in the meat industry in areas where it is endemic. Filaria gets into a fly while it is feeding on damaged skin. After a 3-week development, the parasite is secreted in the fly's saliva and transmitted to a new host when the fly is again having a meal at a wound. A parafilaria female penetrates the skin of the host and deposits the new filaria into the surrounding tissue fluids. The length of the adult nematodes can be 3–6 cm. The lesions resemble a bruise and have a greenish color due to the presence of large numbers of eosinophilic granulocytes. The condition is therefore sometimes called "green meat." The subcutaneous tissue may appear swollen, and bleeding in the muscle might be found during inspection. Localized lesions can be trimmed, but in heavy, general infections the condemnation of the whole carcass should be considered. The infection can be diagnosed, in addition to a direct indication of the nematodes, with immunological methods such as ELISA.

35.6   Future Visions

Parasitological safety of muscle foods can be further improved in the future. Continuous research efforts will develop diagnostic tools, but legislation and administrative decisions can also have a powerful influence on safety: Control can be directed to the problematic areas based on risk analysis. If, for example, *Trichinella* epidemics due to private consumption of uninspected meat are a concern, inspection could be encouraged by removing inspection fees and by making the meat inspection easily achievable. Or if the meat is meant to be used in products that are consumed raw, the inspection could be stricter; *Toxoplasma* inspection or serological confirmation of *Trichinella* inspection, for example, could be demanded in addition. Even now, lower demands for *Trichinella* inspection are made in the areas that have officially recognized negligible risk of infection [5]. Careful monitoring of the infection pressure, however, must be performed with epidemiological studies on suitable indicator animals [43].

Modern molecular biological techniques are routinely used in current parasitological research. Meat production could benefit from these research tools as well, if supplemental parasite diagnostics is desired in addition to the official inspection. Serological laboratory procedures, for example, are easy, cheap, and can be almost fully automated. An antigen microarray for serodiagnostics [44,45], which detects several infections, including parasite infections, from a single sample at the same time, could be applied to slaughterhouse samples. Serology could be used together with conventional methods to complement the meat inspection; methods that actually replace parts of the parasite inspection should, however, be carefully evaluated—no elevated risk for the consumer can be allowed. On the other hand, consumers may considerably benefit from serological testing of animals, because parasites that might not be found in the official inspection can be diagnosed (*Toxoplasma*), or lower infection levels might be noticed (*Trichinella, Cysticercus*). Also, more precise diagnostics could be done in the future from serum samples, using, for example, a serological test differentiating the *Trichinella* species.

Modernization of meat inspection may change the practices at slaughterhouses. In the future, parasite control at the farm level may replace control at meat inspection, at least in the low prevalence areas. The tendency in animal production is to stress the overall welfare of the animals with health care programs and preventive actions. Part of the parasite inspection could be done at the farm as a normal part of the health care program. When farm testing is coordinated with serological testing at the time of slaughter, paired serum samples could be achieved, and the diagnostic value of the serological testing enhanced. In the endemic areas for severe parasitic diseases, however, only direct detection methods should be used for individual carcasses to prevent parasites in meat.

In the future, education on different levels of food production, the combination of good animal production practices with frequent control visits by authorities, and serological testing on the farms as well as in the slaughterhouses, together with risk-based, directed use of conventional methods for parasitic examination, could guarantee parasite-free meat.

References

1. Pozio, E. Foodborne and waterborne parasites. *Acta Microbiol. Pol.,* 52, 83, 2003.
2. Dupoy-Camet, J. et al. Genetic analysis of *Trichinella* isolates with random amplified polymorphic DNA markers, in: *Trichinellosis. Proceedings of the Eighth International Conference on Trichinellosis 1993.* Campbell, W.C., Pozio, E., and Bruschi, F. Istituto Superiore de Sanita Press, Rome, Italy. 1994, 83.
3. Pozio, E. Factors affecting the flow among domestic, synanthropic and sylvatic cycles of *Trichinella. Vet. Parasitol.,* 93, 241, 2000.
4. Pozio, E. et al. *Trichinella papuae* and *Trichinella zimbabwensis* induce infection in experimentally infected varans, caimans, pythons and turtles. *Parasitology,* 128, 333, 2004.
5. EC (European Commission) Regulation No. 2075/2005, of 5 December 2005, laying down specific rules on official controls for *Trichinella* in meat. *Official Journal of the European Union* 22.12.2005.
6. Gamble, H.R. et al. International Commission on Trichinellosis: recommendations on methods for the control of *Trichinella* in domestic and wild animals intended for human consumption. Review. *Vet. Parasitol.* 93, 393, 2000.
7. Webster, P. et al. Meat inspection for *Trichinella* in pork, horse meat and game within the EU: available technology and its implementation. *Eurosurveillance,* 11, 50, 2006.
8. World Organisation for Animal Health. Health standards. Manual of diagnostic tests and vaccines of terrestrial animals. Trichinellosis (updated 21.11.2005). http://www.oie.int/eng/normes/mmanual/A_00048.htm.
9. Gamble, H.R. Factors affecting the efficiency of pooled sample digestion for the recovery of *Trichinella spiralis* from muscle tissue. *Int. J. Food Microbiol.,* 48, 73, 1999.
10. Public Health Agency of Canada, Office of Laboratory Security. Material safety data sheet—infectious substances. *Trichinella* spp. http://www.phac-aspc.gc.ca/msds-ftss/index.html.
11. Sukura, A. et al. *Trichinella nativa* and *T. spiralis* induce distinguishable histopathologic and humoral responses in the raccoon dog (*Nyctereutes procyonoides*). *Vet. Pathol.,* 39, 257, 2002.
12. Community Reference Laboratory for Parasites. Istituto Superiore di Sanità. Identification of *Trichinella* muscle stage larvae at the species level by multiplex PCR. http://www.iss.it/binary/crlp/cont/PCR%20method%20WEB%20SITE.1177083731.pdf.
13. Zarlenga, D.S. et al. A multiplex PCR for unequivocal differentiation of all encapsulated and non-encapsulated genotypes of *Trichinella. Int. J. Parasitol.,* 29, 1859, 1999.
14. Zarlenga, D.S. et al. A single, multiplex PCR for differentiating all species of *Trichinella. Parasite,* 8, 24, 2001.
15. Rombout, Y.B., Bosch, S., and Van Der Giessen, J.W.B. Detection and identification of eight *Trichinella* genotypes by reverse line blot hybridization. *J. Clin. Microbiol.,* 39, 642, 2001.
16. Gamble, H.R. et al. Serodiagnosis of swine trichinosis using an excretory-secretory antigen. *Vet. Parasitol.,* 13, 349, 1983.
17. Gamble, H.R. et al. International Commission on Trichinellosis: Recommendations on the use of serological tests for the detection of *Trichinella* infection in animals and man. *Parasite,* 11, 3, 2004.
18. Pozio, E. et al. Evaluation of ELISA and Western blot analysis using three antigens to detect anti-*Trichinella* IgG in horses. *Vet. Parasitol.,* 108, 163, 2002.
19. Bruschi, F. and Murrel, K.D. New aspects of human trichinellosis: the impact of new *Trichinella* species. *Postgrad. Med. J.,* 78, 15, 2002.
20. Zhang, G.P. et al. Development and evaluation of an immunochromatographic strip for trichinellosis detection. *Vet. Parasitol.,* 137, 286, 2006.
21. Al-Sherbiny, M.M. et al. Application and assessment of a dipstick assay in the diagnosis of hydatidosis and trichinosis. *Parasitol. Res.,* 93, 87, 2004.
22. Patrascu, I. et al. The lateral flow card test: an alternative method for the detection of *Trichinella* infection in swine. *Parasite,* 8, 240, 2001.
23. Moller, L.N. et al. Comparison of two antigens for demonstration of *Trichinella* spp. antibodies in blood and muscle fluid of foxes, pigs and wild boars. *Vet. Parasitol.,* 132, 81, 2005.

24. Tryland, M. et al. Persistence of antibodies in blood and body fluids in decaying fox carcasses, as exemplified by antibodies against *Microsporum canis. Acta Vet. Scand.,* 48, 10, 2006.

25. Hoberg, E. *Taenia* tapeworms: their biology, evolution and socioeconomic significance. *Microbes Infect.,* 4, 859, 2002.

26. Abuseir, S. et al. Visual diagnosis of *Taenia saginata* cysticercosis during meat inspection: is it unequivocal? *Parasitol. Res.,* 99, 405, 2006.

27. Geysen, D. et al. Validation of meat inspection results for *Taenia saginata* cysticercosis by PCR-restriction fragment length polymorphism. *J. Food Prot.,* 70, 236, 2007.

28. European Commission, Health and Consumer Protection Directorate-General. Opinion of the scientific committee on veterinary measures relating to public health on: the control of taeniosis/cysticercosis in man and animals. (adopted on 27–28 September 2000).

29. Harrison, L.J. et al. Specific detection of circulating surface/secreted glycoproteins of viable cysticerci in *Taenia saginata* cysticercosis. *Parasite Immunol.,* 11, 351, 1989.

30. Brandt, J.R. et al. A monoclonal antibody-based ELISA for the detection of circulating excretory-secretory antigens in *Taenia saginata* cysticercosis. *Int. J. Parasitol.,* 22, 471, 1992.

31. Onyango-Abuje, J.A. et al. Diagnosis of *Taenia saginata* cysticercosis in Kenyan cattle by antibody and antigen ELISA. *Vet. Parasitol.,* 61, 221, 1996.

32. Wanzala, W. et al. Serodiagnosis of bovine cysticercosis by detecting live *Taenia saginata* cysts using a monoclonal antibody-based antigen-ELISA. *J. S. Afr. Vet. Assoc.,* 73, 201, 2002.

33. Cook, A.J. et al. Sources of *Toxoplasma* infection in pregnant women: European multicentre case-control study. European Research Network on Congenital Toxoplasmosis. *Br. Med. J.,* 15, 142, 2000.

34. Vallochi, A.L. et al. The genotype of *Toxoplasma gondii* strains causing ocular toxoplasmosis in humans in Brazil. *Am. J. Ophthalmol.,* 139, 350, 2005.

35. Switaj, K. et al. Association of ocular toxoplasmosis with type I *Toxoplasma gondii* strains: direct genotyping from peripheral blood samples. *J. Clin. Microbiol.,* 44, 4262, 2006.

36. Damriyasa, I.M. et al. Cross-sectional survey in pig breeding farms in Hesse, Germany: seroprevalence and risk factors of infections with *Toxoplasma gondii, Sarcocystis* spp. and *Neospora caninum* in sows. *Vet. Parasitol.,* 126, 271, 2004.

37. Hedman, K. et al. Recent primary *Toxoplasma* infection indicated by a low avidity of specific IgG. *J. Infect. Dis.,* 159, 736, 1989.

38. Kong, J.T. et al. Serotyping of *Toxoplasma gondii* infections in humans using synthetic peptides. *J. Infect. Dis.,* 187, 1484, 2003.

39. Deplazes, P., Dinkel, A., and Mathis, A. Molecular tools for studies on the transmission biology of *Echinococcus multilocularis.* Review. *Parasitology,* 127, 53, 2003.

40. Gatti, A. et al. Ovine echinococcosis I. Immunological diagnosis by enzyme immunoassay. *Vet. Parasitol.,* 143, 112, 2007.

41. Lorenzo, C. et al. Comparative analysis of the diagnostic performance of six major *Echinococcus granulosus* antigens assessed in a double-blind, randomized multicenter study. *J. Clin. Microbiol.,* 43, 2764, 2005.

42. Lahmar, S. et al. Ultrasonographic screening for cystic echinococcosis in sheep in Tunisia. *Vet. Parasitol.,* 143, 42, 2007.

43. Community Reference Laboratory for Parasites. Istituto Superiore di Sanità. Identification and development of sampling methods and design of suitable protocols for monitoring of *Trichinella* infection in indicator species. http://www.iss.it/binary/crlp/cont/Guideline%20sampling%20indicator%20species.1166800931.pdf.

44. Mezzasoma, L. et al. Antigen microarrays for serodiagnosis of infectious diseases. *Clin. Chem.,* 48, 121, 2002.

45. Bacarese-Hamilton, T. et al. Serodiagnosis of infectious diseases with antigen microarrays. *J. Appl. Microbiol.,* 96, 10, 2004.

Chapter 36

Mycotoxin Analysis in Muscles

Jean-Denis Bailly and Philippe Guerre

Contents

36.1 Introduction

Mycotoxins are toxic substances produced by fungi. They constitute a heterogeneous group of secondary metabolites with diverse potent pharmacological and toxic effects in humans and animals. More than 300 secondary metabolites have been identified but around 30 are of real concern for human and animal health (for review, see Ref. 1). Mycotoxins are produced during mould development on plants in the field or during storage period. They can be found as natural contaminants of not only many vegetal foods or feeds, primarily cereals, but also fruits, nuts, grains, forage, as well as compound foods intended for human or animal consumption. Most important mycotoxins are produced by moulds belonging to *Aspergillus*, *Penicillium*, and *Fusarium* genus

Table 36.1 Mycotoxins and Producing Fungal Species Associated with Human and Animal Nutrition

Mycotoxins	Main Producing Fungal Species
Aflatoxins B1, B2, G1, G2	*Aspergillus flavus, A. parasiticus, A. nomius*
Ochratoxin A	*Penicillium verrucosum, Aspergillus ochraceus, Aspergillus carbonarius*
Fumonisins B1, B2, B3	*Fusarium verticillioides, F. proliferatum*
Trichothecenes	*Fusarium graminearum, F. culmorum, F. sporotrichioides, F. poae, F. tricinctum, F. acuminatum*
Zearalenone	*Fusarium graminearum, F. culmorum, F. crookwellense*
Patulin	*Penicillium expansum, Aspergillus clavatus, Byssochlamys nivea*
Ergot alkaloids	*Claviceps purpurea, C. paspali, C. africana*
Citrinin	*Aspergillus terreus, A. carneus, A. niveus, Penicillium verrucosum, P. citrinum, P. expansum*
Cyclopiazonic acid	*Aspergillus flavus, A. versicolor, A. tamarii, Penicillium camemberti*
Sterigmatocystin	*Aspergillus flavus, A. versicolor, A. nidulans*
Sporidesmins	*Pythomyces chartarum*
Stachybotryotoxins	*Stachybotrys chartarum, S. atra*
Endophyte toxins	*Neotyphodium coenophialum, N. nolii*
Tremorgenic toxins	*Penicillium roqueforti, P. crustosum, P. puberrelum, Aspergillus clavatus, A. fumigatus*

(Table 36.1) [2–4]. Mycotoxins are usually classified depending on the fungal species that produce them (Table 36.1).

Even if some toxins display an acute toxicity (after unique exposure to one high dose), chronic effects (observed after repeated exposure to weak doses) are probably more important in humans. Mycotoxins are suspected to be responsible for several pathological syndromes in human: ochratoxin A (OTA) and Balkan endemic nephropathy (BEN), esophageal cancer and fumonisin B1, etc. Mycotoxin exposure of humans is usually directly linked with alimentary habits.

Mycotoxin toxicity is variable (Table 36.2). Some have a hepatotoxicity (aflatoxins), others have an estrogenic potential (zearalenone—ZEA) or are immunotoxic (trichothecenes, fumonisins) [1]. Some mycotoxins are considered as carcinogenic or suspected to have carcinogenic properties [5]. Moreover, several mycotoxins can contaminate the same product or diet.

Humans are exposed to mycotoxins mainly by cereals and cereal-based products [6–8]. However, humans may be also exposed to these toxic compounds indirectly due to the presence of residues in foods prepared from animals that have been fed with contaminated feeds. Animal-derived products such as meat, eggs, and milk may also represent a vector of mycotoxins. Moreover, the stability of these compounds allows them to resist to classical process of cooking and sterilization [9,10].

Table 36.2 Toxic Effects of Main Mycotoxins; Cellular and Molecular Mechanisms of Action

Toxin	Toxic Effect	Mechanism of Action
Aflatoxin B1 and M1	Hepatotoxic Genotoxic Carcinogenic Immunomodulation	Bioactivation by P450 cytochromes Lipids peroxidation Formation of DAN adducts
Ochratoxin A	Nephrotoxic Genotoxic Immunomodulation	Effect on protein synthesis Inhibition of ATP production
Trichothecenes (T-2 toxin, DON, etc.)	Hematotoxicity Immunomodulation skin toxicity	Impact on protein synthesis Apoptosis of hematopoietic stem cells and on immune cells Alteration of immunoglobulin
Zearalenone	Fertility and reproduction troubles	Bioactivation by reductases Link to estrogenic receptors
Fumonisin B1	Lesion of central nervous system Hematotoxicity Genotoxicity Immunomodulation	Inhibition of ceramide synthesis Modification of sphinganine/ sphingosine ratio Alteration of cell cycle

Many studies have aimed to evaluate the role of animal-derived products as a source of mycotoxin. For all mycotoxin analysis procedures, sampling is a key step especially for vegetal raw material in which contamination is always heterogeneous. Therefore, for foods and feeds of vegetal origin, several norms for sampling have been validated [11,12]. For milk, contamination is usually more homogenous and sampling step easier. Nevertheless, norms have also been set up: three sub-samples of the batch to control must be taken and mixed up to obtain the final sample for analysis [12]. For animal-derived products, due to mycotoxin circulation within the body, contamination is more homogenous and sampling less complicated.

Depending on the mycotoxins, the residues may correspond to the native toxin or to metabolites that keep all or part of the toxic properties of the parental molecule. Therefore, the passage through an "animal filter" may correspond to a detoxification process or, on the contrary, lead to the appearance of toxic compounds for humans. Owing to the important structural diversity of mycotoxins and to the variations in their metabolism, it is impossible to form general rules, and therefore each toxin and each product must be investigated as a particular case.

Currently, only aflatoxins and, to a lesser extent, OTA are regulated in foods from animal origin. For other toxins, the risk management is based on the control of the contamination of food from vegetal origin intended for both human and animal consumption. Regulatory values or recommendations are mainly built on available knowledge on toxicity of these molecules in animal. Therefore, by limiting animal exposure through feed ingestion, one can guarantee against the presence of residues of mycotoxins in animal-derived products. However, accidentally a high level of contamination may lead to a sporadic contamination of products coming from exposed animals. Moreover, only few toxins are regulated [13] and others, mainly toxins from *Penicillium* species, may also contaminate products from animal origin, particularly during ripening of dry-cured meat products.

Although many methods have been developed and validated for vegetal matrix, due to the absence of regulation, only few data are available on techniques that may be used for animal-derived foods. With the exception of the detection of aflatoxin M1 (AFM1) in milk and milk products [14], no official method is available for animal products.

Owing to the great structural differences between mycotoxins and their metabolism, no multi-detection method can be carried out and methods have to be developed specifically for each toxin or metabolite.

This work presents methodology for mycotoxin quantification in muscle foods. Therefore, data on most important toxins are presented toxin by toxin, providing successively the origin and toxicological features of each molecule. Analytical methods for quantification in animal tissues are also reported.

36.2   Aflatoxins

36.2.1   *Introduction*

Aflatoxins are probably the most studied and documented mycotoxins. They were discovered following a toxic accident in Turkeys fed with groundnut oilcake–supplemented diet (Turkey X disease) [15–17]. After their discovery, the toxicity of aflatoxins was studied rapidly in many animal species. Moreover, it has been demonstrated that aflatoxin B1 (AFB1) ingested by dairy cows is partially metabolized into a molecule called "milk aflatoxin 1" (AFM1) [18]. These studies demonstrated that mycotoxins could enter human food not only through the direct vegetal–human way but also through a more complex progress through food chain: vegetal → animal feed → animal tissues and derived products → human consumer.

36.2.2   *Origin and Toxicological Properties*

36.2.2.1   *Synthesis*

The four natural aflatoxins (B1, B2, G1, and G2) can be produced by strains of fungal species belonging to *Aspergillus* genus, mainly *Aspergillus flavus* and *Aspergillus parasiticus* [19,20]. These are worldwide common contaminants of a wide variety of commodities and therefore aflatoxins may be found in many vegetal products, from cereals to groundnuts, cotton seeds, dry fruits, spices, etc. [21–26]. These fungal species can grow and produce toxins in the field or during storage; climatic conditions required for their development are similar to tropical areas (high humidity and temperature ranging from 25 to 40°C) [27–31]. However, following extreme climatic conditions (abnormally hot summer period), aflatoxins could be found in other parts of the world. For example, in 2003, controls on maize harvested in Europe were found contaminated by unusual AFB1 concentrations [32,33] whereas European crops were usually considered as aflatoxin free. Also, a survey done in the United States in 1988, after a hot and dry period, showed that an abnormally high proportion of maize (8% of samples) was contaminated with mycotoxins [34].

When present in vegetal matrix, aflatoxins are stable and less sensitive to thermal treatments (sterilization or freezing) or drying step [9,35]. Many studies have focused on the treatments that may be applied to foods and feeds to decrease aflatoxin contamination level. Among these treatments, detoxification procedures based on ammoniac treatment can be performed to decrease aflatoxin concentration in oilcakes intended for animal feed [9,36–38].

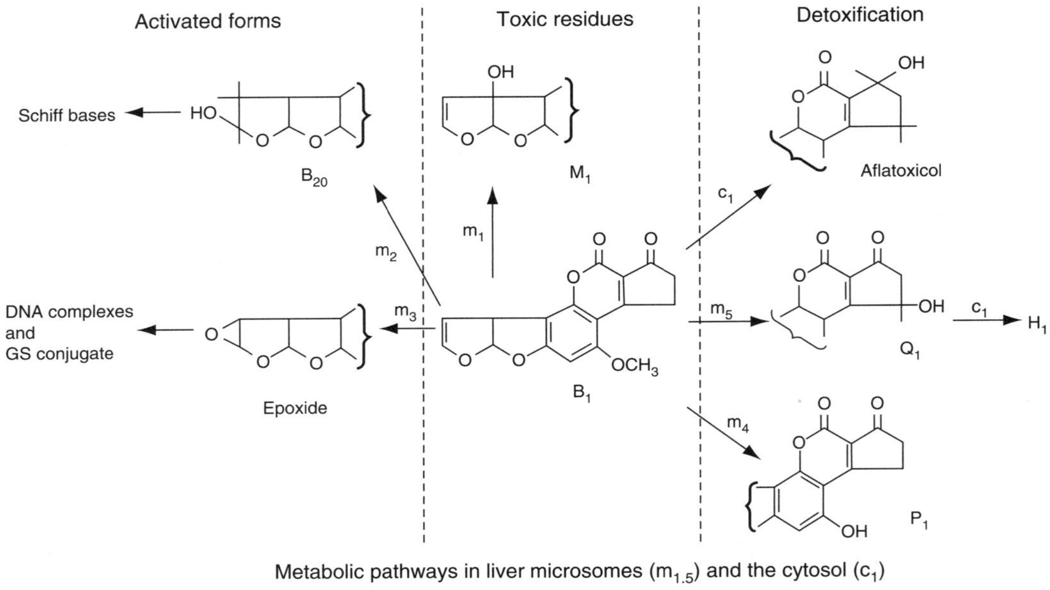

Activated forms | Toxic residues | Detoxification

Schiff bases

B_{20}

M_1

Aflatoxicol

DNA complexes and GS conjugate

Epoxide

B_1

Q_1

H_1

P_1

Metabolic pathways in liver microsomes (m_{1-5}) and the cytosol (c_1)

Figure 36.1   Phase I metabolism of aflatoxin B1. (From Paterson, D.S.P., *Pure Appl. Chem.*, 49, 1723, 1977. With permission.)

36.2.2.2   Metabolism (for Review, See Ref. 39)

Absorption of AFB1 administrated by oral route is rapid and nearly complete [40]. Absorption takes place in the small intestine, mainly in jejunum [41]. In plasma, AFB1 is strongly linked to albumin, part of this fixation being covalent. AFB1 then goes through the liver and most part of the toxin is metabolized, with 1–10% of the AFB1 staying fixed to macromolecules. AFB1 is a lipophilic molecule that goes through classical phase I and II biotransformation processes. Main phase I metabolites are epoxide; hydroxylated compounds AFM1, AFQ1, AFP1; a reduced compound—aflatoxicol (AFL); and a hydrogenated and hydroxylated molecule—AFB2a (Figure 36.1) [42]. The most important phase II metabolites of the epoxide derivative are conjugated with GSH and glucorono- and sulfoconjugated forms of AFM1, AFQ1, and AFP1.

Epoxide formation can be considered as a bioactivation process caused by the high reactivity of epoxide with liver macromolecules. Other metabolic pathways may be considered as detoxification processes leading to compounds without any toxicity (AFQ1, AFP1, AFB2a) or that keep residual toxicity (AFM1 and aflatoxicol).

Excretion of aflatoxins is quite slow (70–80% of a single dose in 4 days), biliary excretion being the most important route (50%) and AFM1 the major excreted metabolite.

Intense metabolism of AFB1 in the liver explains that only a very small part of the native molecule can be detected in animal tissues (Table 36.3). Whatever the study and the method used for quantification, liver and kidney always contain more toxin and metabolites than muscles [44,46,49–52]. In muscles of animals, even after their exposure to high doses of AFB1, always low levels are found, often below detection limits of the methods used [44,50,53]. Many studies have evaluated the transfer of aflatoxin in the milk of lactating cows [54–61], whereas no complete data are available on the transfer in the muscle of cattle. However, for other species, residues can be found in liver and kidney that are edible parts of these animals [62,63].

Table 36.3 Residues of Aflatoxin in Animal Tissues (Only the Most Demonstrative Studies Are Reported)

Animal Species	Dose and Duration of Exposure	Tissues	Residues (µg/kg)	Metabolites	Reference
Poultry					
Turkey	50 and 150 µg/kg feed for 11 weeks	Liver	0.02–0.009 and 0.11–0.23	AFB1 + AFM1	43
		Kidney	0.02–0.04 and 0.11–0.21	AFB1 + AFM1	
		Gizzard	0.04–0.16 and 0.01–0.12	AFB1 (AFM1 < 0.01)	
Turkey	50 and 150 µg/kg feed for 11 weeks and 1 week with toxin free feed	Liver	<0.01	AFB1 + AFM1	
		Kidney	<0.01	AFB1 + AFM1	
		Gizzard	0.04–1.9 and 0.09–0.24	AFB1 (AFM1 < 0.01)	
Quail	3000 µg/kg feed for 8 days	Liver	7.83 ± 0.49 and 5.31 ± 0.22	Free and conjugated AFB1	44
		Muscle	22.34 ± 2.4 and 10.54 ± 0.42	Free and conjugated metabolites	
			0.38 ± 0.03 and <0.03	Free and conjugated AFB1	
			0.82 ± 0.05 and 0.32 ± 0.08	Free and conjugated metabolites	
Duck	3000 µg/kg feed for 8 days	Liver	0.52 ± 0.04 and 0.44 ± 0.16	Free and conjugated AFB1	
		Muscle	2.74 ± 0.15 and 3.81 ± 0.25	Free and conjugated metabolites	
			<0.03 and <0.03	Free and conjugated AFB1	
			0.21 ± 0.09 and 0.14 ± 0.05	Free and conjugated metabolites	
Chicken	3000 µg/kg feed for 8 days	Liver	0.15 ± 0.09 and 0.10 ± 0.01	Free and conjugated AFB1	
		Muscle	1.54 ± 0.36 and 0.93 ± 0.04	Free and conjugated metabolites	
			<0.03 and <0.03	Free and conjugated AFB1	
			0.11 ± 0.02 and 0.08 ± 0.05	Free and conjugated metabolites	
Hen	3000 µg/kg feed for 8 days	Liver	0.34 ± 0.03 and 0.23 ± 0.08	Free and conjugated AFB1	
		Muscle	2.38 ± 0.36 and 4.04 ± 0.1	Free and conjugated metabolites	
			<0.03 and <0.03	Free and conjugated AFB1	
			0.14 ± 0.04 and 0.11 ± 0.04	Free and conjugated metabolites	
Laying hen	10,000 µg/kg feed for 7 days	Eggs	0.28 ± 0.1 and 0.38 ± 0.11	AFB1 and total aflatoxicol	45
Laying hen	8000 µg/kg feed for 7 days	Liver	0.49 ± 0.28 and 0.2 ± 0.09	AFB1 and total aflatoxicol	46
		Kidney	0.32 ± 0.18 and 0.1 ± 0.04	AFB1 and total aflatoxicol	
		Muscle	0.08 ± 0.03	Aflatoxicol	
		Eggs	0.24 ± 0.07 and 0.25 ± 0.09	AFB1 and total aflatoxicol	

(Continued)

Table 36.3 (Continued)

Animal Species	Dose and Duration of Exposure	Tissues	Residues (µg/kg)	Metabolites	Reference
Laying hen	2500 µg/kg feed for 4 weeks	Liver	4.13 ± 1.95	AFB1	47
		Eggs	<0.5 and <0.01	AFB1 and AFM1	
Chicken	55 µg/kg feed for 9 days	liver	0.26 and 0.02	AFB1 and AFM1	48
Chicken	4448 µg/kg feed for 9 days	Liver	1.52 and <0.1	AFB1 and AFM1	
Swine					
Adult pigs	395 µg/kg feed for 14 days	Liver	1.242 ± 0.44	AFB1, AFM1 and aflatoxicol	49
		Kidney	1.606 ± 0.63	AFB1, AFM1 and aflatoxicol	
		Muscle	0.16 ± 0.22	AFB1, AFM1 and aflatoxicol	
Growing mixed breed swine	400 µg/kg feed for 10 weeks	Liver	1.43	Total aflatoxins + AFM1	50
		Kidney	0.83	Total aflatoxins + AFM1	
		Muscle	ND	Total aflatoxins + AFM1	
Growing mixed breed swine	800 µg/kg feed for 10 weeks	Liver	2.81	Total aflatoxins + AFM1	50
		Kidney	1.21	Total aflatoxins + AFM1	
		Muscle	0.64	Total aflatoxins + AFM1	
Male miniature swine	590 µg/kg feed for 15 days	Liver	3.17	AFB1 + AFB2 + AFM1	51
		Kidney	6.47	AFB1 + AFB2 + AFM1	
		Muscle	0.52	AFB1 + AFB2 + AFM1	
Crossbreed pigs	524 µg/kg feed for 35 days	Liver	2.01	AFB1 + AFB2 + AFM1	51
		Kidney	3.951	AFB1 + AFB2 + AFM1	
		Muscle	0.443	AFB1 + AFB2 + AFM1	

36.2.2.3 Toxicity

AFB1 is a highly carcinogenic agent leading to primary hepatocarcinoma [64–67]. This property is directly linked to the metabolism of aflatoxin and to the appearance of the highly reactive epoxide derivative. Formation of DNA adducts of AFB1-epoxide is well characterized [68]. The primary site of adduct formation is the N7 position of the guanine nucleotide. Differences in AFB1 metabolism within animal species could explain the variability of the response in terms of carcinogenic potential of the mycotoxin [69,70].

AFM1 can be considered as a genotoxic agent but its carcinogenic potential is weaker than that of AFB1 [71]. Taking into account the toxicity of these molecules, IARC classified AFB1 in the group I of carcinogenic agents, AFM1 in the 2B group of molecules that are carcinogenic in animals and possibly carcinogenic in human, and AFG1 in the group III of noncarcinogenic compounds [5].

36.2.3 Regulation

Owing to their carcinogenic potential, JECFA did not define maximal tolerable daily intake for aflatoxins. Indeed, these molecules being cancer initiators, the most realistic way to protect consumers against these contaminants is to reduce human exposure to the "as low as reasonably achievable" level [72].

In 2003, among the 99 countries that had implemented regulation for mycotoxins content in foods, all had at least regulatory value for AFB1 or total aflatoxins content [13]. Most of these regulations are related to vegetal raw material intended for human or animal consumption and milk, as illustrated by European Union regulation (Table 36.4). However, in some particular cases such as Ukraine or Serbia, regulation is specific for meat and meat products and in many other countries, regulatory limits are applied to all foods intended for human consumption [13].

36.2.4 Aflatoxin Analysis

36.2.4.1 Physicochemical Properties

The structures of the main aflatoxins are shown in Figure 36.2. Molecular weights of aflatoxins range from 312 to 320 g/mol. These toxins are weakly soluble in water, insoluble in nonpolar solvents, and highly soluble in mildly polar organic solvents (i.e., chloroform and methanol). They are fluorescent under UV light (blue fluorescence for aflatoxin "B" and green for aflatoxin "G") [73].

36.2.4.2 Methods of Analysis

Most common solvent systems used for extraction of aflatoxins are chloroform–water [74–76] and methanol–water [77–81] mixtures. The latter is mainly used for multiextraction of mycotoxins and not specific for aflatoxin extraction [82]. Whatever the solvent system used, the obtained extract still contains various impurities and requires further cleanup steps. The most commonly used extraction technique is solid-phase extraction (SPE) that replaced the traditional liquid–liquid partition for cleanup [76]. Stationary phase SPE columns used may be silica gel, C18 bonded phase, and magnesium silicate (commercially available as Florisil) [74,83]. Antibody-affinity SPE columns are also widely used.

Immunoaffinity chromatography using antitoxin antibodies allowed the improvement of both specificity and sensitivity [84,85]. These techniques have been validated for matrixes that are the most frequently contaminated with aflatoxins and controlled by specific regulation. Indeed, these

Table 36.4 EU Regulation for Aflatoxins Contamination (µg/kg)

Destination	Toxin	Matrix	Maximal Concentration (µg/kg)
Human food	Aflatoxin B1	Groundnuts + grains + dry fruits	2, 5, or 8 depending on the product and the processing step
		Cereals	2 or 5 depending on the product and the processing step
		Spices	5
		Cereal-based foods for young children	0.1
	Aflatoxins B1 + B2 + G1 + G2	Groundnuts + grains + dry fruits	4, 10, or 15 depending on the product and the processing step
		Cereals	4 or 10 depending on the product and the processing step
		Spices	10
	Aflatoxin M1	Milk	0.05
		Preparation for young children	0.025
Animal feed	Aflatoxin B1	Raw material for animal feeds	20
		Compound feeds	5 to 20 depending on animal species

validated methods were developed for grains [86], cattle feed [87,88], maize, groundnuts and groundnut butter [89], pistachio, figs, paprika [90], or baby food [91]. Analytical methods of the same kind were validated for quantification of AM1 in milk [92] and in powder milk [93,94]. These methods show limits of quantification below the regulatory limit of 0.05 µg/L.

Aflatoxins are usually quantified by thin layer chromatography (TLC), high pressure liquid chromatography (HPLC), or enzyme-linked immunosorbent assay (ELISA).

TLC was the first to be developed in the early 1980s. Using strong fluorescence of the molecules, the characterization of signals with naked eyes or densitometric analysis could give semiquantitative to quantitative results (AOAC methods 980.20 and 993.17) [95]. Therefore, AFB1 could be measured in concentrations ranging from 5 to 10 µg/kg. A TLC method for quantification of AFM1 in milk was also validated by AOAC (980.21) [96] and normalized (ISO 14675:2005) [97]. A method for semiquantitative analysis of AFB1 in cattle feeds was also published (ISO 6651:2001) [98]. Confirmation of identity of AFB1 and AFM1 in foods and feeds is still classically done by TLC after bidimensional migration and trifluoric acid–hexane (1:4) spraying of plates.

The replacement of TLC by HPLC allowed the reduction of detection limits together with an improvement of the specificity of the dosage [99]. Therefore, new methods were validated for aflatoxin quantification in grains (AOAC 990.33), cattle feed (ISO 14718:1998), and AFM1 in milk (ISO/FDIS 14501) [100–102]. These methods are based on the use of a fluorescence detector allowing the quantification of low levels of aflatoxins (e.g., AFM1 concentration as low as 0.015 µg/L).

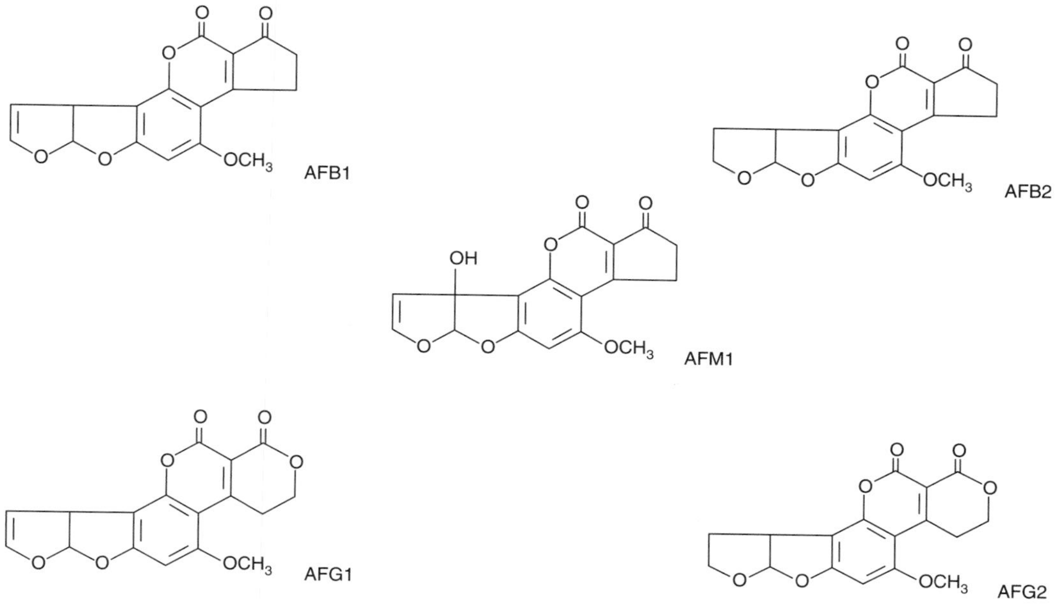

Figure 36.2   Structures of aflatoxins.

The sensitivity can be increased by the treatment of extracts with trifluoric acid to catalyze the hydratation of AFM1, AFB1, and AFG1 into their highly fluorescent M2a, B2a, and G2a derivatives.

ELISA has been developed for both total aflatoxins [103,104] and AFB1 detection in feeds and grains [105–108] and for AFM1 detection in milk [109]. These methods have limits of quantification in accordance with international regulations. Therefore, some commercially available kits have been validated by AOAC, such as AOAC 989.86 devoted to AFB1 dosage in animal feed. However, in spite of the development of ELISA methods for AFM1 detection [110], no ELISA kit has been validated following harmonized protocol ISO/AOACC/IUPAC for AFM1 quantification in milk. AOAC has edited rules for characterization of antibodies used in immunochemical methods [111].

Detection limits in the low ppt range can be achieved by these classical LC-fluorescence methods. Therefore, methods like LC-MS may represent only a minor alternative or confirmation technique for already well-established methodologies [112]. It may although be useful to confirm positive results of TLC- or ELISA-based screening analysis [113]. Today, few quantitative methods are available for aflatoxin determination in food and milk [114–117].

Table 36.5   Methods for Aflatoxin B1 Determination in Animal Tissues

Separation Detection	Purification	Tissues and Animal Species	Recovery (%)	LOQ	Reference
2D TLC fluorodensitometry	Column chromatography	Liver, kidney, heart, muscle (chicken, swine, beef)	92.9 ± 5.03% for AFB1	≤0.1 ng/g	118
			85.4 ± 5.22% for AFM1		
2D TLC fluorodensitometry	Solvent partition of methanol–water–chloroform Silica-gel chromatography Delipidation with hexane	Liver, kidney, muscle, heart (swine)	43–123%	≤0.1 ng/g	119, 120
TLC fluorodensitometry	Partition against hexane	Liver, kidney, muscle (swine)	ND	0.03 ng/g for AFB1	49
	Silica-gel column chromatography			0.05 ng/g for AFM1	
HPLC fluorescence after TFA derivatization of samples	Column chromatography	Liver, ground beef (beef, swine)	58 ± 9% to 128 ± 12%	0.05–0.1 ng/g	121
HPLC fluorescence	Column chromatography	Liver, kidney, muscle (swine)	85–115%	≤0.1 ng/g	50
ELISA optical density	Immunoaffinity columns	Liver (chicken)	50%	1 ng/g	122
HPLC fluorescence	Immunoaffinity columns	Liver (swine, chicken, beef, turkey)	0	0.008 ng/g	123

36.2.4.3   Aflatoxin Analysis in Animal Tissues and Muscles

Techniques described for aflatoxin analysis in animal tissues mainly use native fluorescence of aflatoxins after purification and separation of extract with chromatographic methods (TLC, HPLC) (Table 36.5). Since the 1980s, only few studies and surveys have been carried out to characterize aflatoxin presence in meat products [122,123]. Indeed, risk management is based on the control of animal feed quality that may guarantee the absence of toxin residues in animal-derived products. These surveys demonstrated that muscle foods were not an important source of aflatoxin exposure in humans. It is however likely that recent alerts for unusual aflatoxin contamination of cereals produced under temperate climates and the possible consequent animal exposure may strengthen the interest of aflatoxin testing in animal-derived foods. That is why some authors investigated the possible use of ELISA determination of aflatoxins residues in animal organs.

36.3   Ochratoxin A

36.3.1   *Introduction*

Ochratoxins A, B, and C are secondary metabolites produced by several *Aspergillus* and *Penicillium* species. Owing to its prevalence and toxicity, only OTA will be treated in this section. This toxin has been found as a contaminant of many foodstuffs, primarily cereals but also wine and coffee. Meat and meat products also may contain OTA if animals are exposed to contaminated feed. Therefore, many studies have tried to characterize human exposure related to ingestion of this contaminant and many methods have been developed to determine OTA content in foods.

36.3.2   *Origin and Toxicological Properties*

36.3.2.1   *Synthesis*

OTA was first isolated from *Aspergillus ochraceus* in 1965 [124]. It has also been demonstrated that OTA could be produced by other *Aspergillus* species such as *A. carbonarius* [125,126], *A. alliaceus* [127], and *A. niger* [128], although the frequency of toxigenic strains in this species appears moderate [129–131]. OTA can also be synthesized by *Penicillium* species, mainly *P. verrucosum* (previously named *P. virridicatum*) [132,133]. The ability of both *Aspergillus* and *Penicillium* species to produce OTA makes it a worldwide contaminant of numerous foodstuffs. Indeed, *Aspergillus* is usually found in tropical or subtropical regions whereas *Penicillium* is common in temperate and cold climate areas [134–137]. Many surveys revealed the contamination of a large variety of vegetal products such as cereals [138,139], grapes [140,141], and coffee [139,142]. For cereals, OTA contamination generally occurs during storage of raw materials, especially when moisture and temperature are abnormally high, whereas for coffee and wine, contamination occurs in the field or during the drying step [137,143–145]. When ingested by animals, OTA can be found at residue level in several edible organs (see Section 36.3.2.2). Therefore, the consumption of meat contaminated with OTA is also a source of exposure for humans [146]. Recent surveys done in European countries have demonstrated that the role of meat products in human exposure to OTA can be considered as low [6,147]. The presence of toxigenic fungal strains on dry-cured meat products [148,149] raises the possibility of the direct contamination of these foods. However, in this case, the production of OTA on meat products after contamination with toxigenic strains seems to remain quite low, even if this mycotoxin appears to be stable in meat products [148,150].

36.3.2.2   *Metabolism (for Review, See Ref. 151)*

OTA is partly absorbed at the stomach level but main absorption takes place in the small intestine [152]. OTA is hydrolyzed to a nontoxic derivative of OTA not only by carbopeptidase and chymotripsin but also by microbial flora of the digestive tract [153–155]. In ruminants, this degradation takes place before the absorption of the toxin. This considerably limits the risk of having OTA contamination of meat and milk but metabolites can be detected in these tissues [156,157].

At the hepatic level, OTA is detoxified in several minor metabolites such as to 4-hydroxy-ochratoxin.

Table 36.6 Occurrence of OTA in Meat Products from Different Countries

Sample (Country)	Number of Sample Analyzed	% Positive (Level ng/g)	% of Samples with Level >1 ng/g	Maximum Observed Level (ng/g)	Reference
Pork meat (DK)	76	84.2 (>0.02)	2.6	1.3	159
Pork kidney (DK)	300	94.6 (>0.02)	8	15	160
Pork kidney (F)	300	1 (>0.4)		1.4	161
	710	7.6 (>0.5)		5	
Pork meat (DK)	300	76 (>0.02)	3	2.9	160
Ham (I)	42	100 (>0.04)	35.7	2.3	162
Pork meat products (I)	106	23 (>0.03)		28.4	163
Pork edible offals (D)	102	69 (0.01)	0.03	9.33	164
Sausage (D)	201	52 (>0.01)	0.01	4.56	164
Ham (D)	57	72 (>0.01)	0	0.17	164

The very strong affinity of the toxin for plasmatic proteins may slow down the elimination [153]. Depending on animal species, big differences have been reported for OTA half-life. Humans appear to display the longest half-life, more than 30 days [158].

Tissue distribution of the toxin revealed that the toxin concentration was, in decreasing order, as follows: kidney > liver > muscle > fat (Table 36.6).

The re-absorption of OTA in kidney tubules via anionic transporters favors its renal accumulation. Meat and meat products represent only 3% of the OTA source in human diet [139]. The prevalence of contamination appears to be more important in northern Europe [165].

36.3.2.3 Toxicity

Kidney is the primary target of OTA. This molecule is nephrotoxic in all animal species studied. For example, OTA is considered as responsible for a porcine nephropathy that has been studied intensively in the Scandinavian countries [166,167]. This disease is endemic in Denmark where rates of porcine nephropathy and ochratoxin contamination of pig feed are highly correlated [168]. OTA is suspected to play a role in this human syndrome, because the renal lesions observed in pig kidneys after exposure to OTA are quite similar to that observed in kidneys of patients suffering from BEN [169–171]. BEN is a progressive chronic nephropathy that occurs in populations living in areas bordering Danube River in Romania, Bulgaria, Serbia, and Croatia [172,173].

OTA disturbs cellular physiology in multiple ways. The primary effect could be associated with the enzymes involved in the phenylalanine metabolism, mostly by inhibiting the synthesis of the phenylalanine tRNA complex [174]. Moreover, OTA inhibits mitochondrial ATP production [175] and stimulates lipid peroxidation [176].

In addition to its nephrotoxic effect, OTA appears to be a potent teratogen and a carcinogenic agent in animals [177] and therefore has been classified by the IARC in the 2B group of molecules that are carcinogenic in animals and potentially carcinogenic in humans [5]. This property could

Table 36.7   Maximum Level of OTA in Some Food and Feed as Regulated in Different Countries

Commodity	Level (ng/g)	Country
Raw cereal grains	5	EU
All cereal-derived products	3	EU
	50	Israel
	2	Switzerland
Dried vine fruits	10	EU
Children foods	6	Czech Republic
Infant foods	1	
Pork meat and derived products	1[a]	Italy
Pig kidneys	10[b]	Denmark
	25[c]	
Coffee	5[a]	Finland
Roasted and instant coffee	4[a]	Italy
Raw coffee beans	20	Greece
Cocoa-derived products	0.5[a]	Italy
Grains for feed	300	Romania
Foods (all)	5	
Foodstuffs for poultry	200	Sweden
Foodstuffs for pigs	100	
Rice, barley, beans, coffee, maize	50	Uruguay

[a] Guideline values.
[b] Viscera condemned.
[c] Whole carcass condemned.

Source: FAO, worldwide regulations for mycotoxins in food and feed in 2003. FAO food and nutrition papers, Rome, Italy, 81, 2004.

be related to the effect of OTA on DNA, leading to the appearance of DNA breakage [178] and adducts [179–181].

36.3.3   Regulation

Several maximal tolerable doses were determined for OTA. The first one, proposed by JECFA [182], corresponds to a daily dose of 16 ng/kg bw. It has been calculated taking into account the renal toxicity of OTA in pigs after a subchronic toxicity study.

By contrast, European Scientific Committee on Human Nutrition and the French High Committee of Public Hygiene (CSHPF) proposed a tolerable daily dose of 5 ng/kg bw, taking into account the carcinogenic effects observed in rats [183].

These doses were used to build regulatory values in different foods and feeds (Table 36.7). In most cases, these values are of few ng/g. Some countries included specific regulation for meat and meat products. For example, Denmark has a specific regulation for OTA content in pig kidneys and Italy has one for pig meat and derived products.

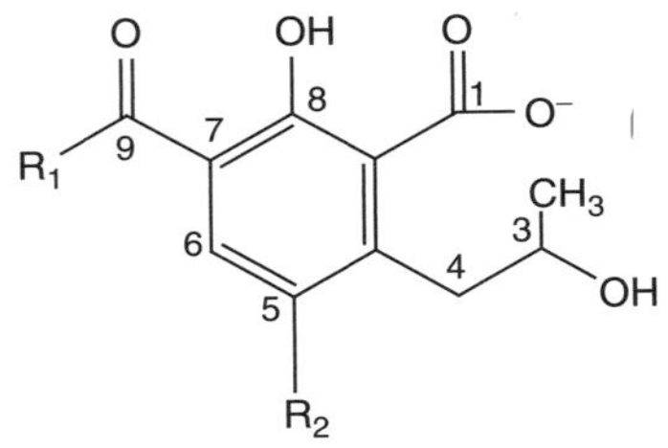

Figure 36.3 OTA structure.

36.3.4 Ochratoxin A Analysis

36.3.4.1 Physicochemical Properties

The structure of ochratoxins is presented in Figure 36.3. OTA has a molecular weight of 403.8 g/mol. It is a weak organic acid with a pK_a of 7.1. At an acidic or neutral pH, it is soluble in polar organic solvents and weakly soluble in water. At a basic pH, it is soluble and stable in aqueous solution of sodium bicarbonate (0.1 M; pH 7.4) as well as in alkaline aqueous solutions in general.

OTA is fluorescent after excitation at 340 nm and emits at 428 nm when nonionized and at 467 nm when ionized.

36.3.4.2 Methods of Analysis

Extraction of OTA is often achieved by using a mixture of acidified water and organic solvents. An IUPAC/AOC method validated for OTA determination in barley uses chloroform–phosphoric acid mixture [184]. For coffee or wine, chloroform is successfully used [185,186]. Mixtures of methanol–water or acetonitrile–water have also been reported [187,188]. *tert*-Butylmethylether has been used for OTA extraction from baby food and may be used as an alternative to chlorinated solvents [189].

Several efficient cleanup procedures based on immunoaffinity columns (IACs) and SPE using C8, C18, and C-N stationary phases have been developed to replace, when possible, conventional liquid–liquid extraction [165]. Stationary phases based on the principle of molecular printing (MIP) have emerged [190,191]. The specificity of such methods is comparable to that of IAC. Although their applicability in real matrixes has not been established, they may represent alternatives to IAC and SPE methods in the future.

For separation and detection of OTA, many methods were developed, including TLC methods [192–194]. However, both specificity and sensitivity of TLC are limited and interferences with the sample matrix often occur [195]. These drawbacks may be overcome by two-dimensional TLC [196]. However, HPLC is the most commonly used method for determination of OTA [165,197].

Most described HPLC methods use reverse-phase C18 column and an acidic mobile phase composed of acetonitrile or methanol with acetic, formic, or phosphoric acid [188,198–200]. The property of OTA to form an ion-pair on addition of a counterion to the mobile phase has been used [162]. This led to a shift in OTA fluorescence from 330 to 380 nm and allowed an improvement of the signal. Ion-pair chromatography was also used for detection of OTA in plasma and human and cows' milk, with detection level of 0.02 and 10 ng/mL for plasma and milk, respectively

[201,202]. The major limit of the method is that small changes in composition of mobile phase may change retention time of OTA.

HPLC methods using fluorescence detection are applicable to OTA detection in barley, wheat, and rye at concentrations of about 10 μg/kg [203]. For baby foods, a quantification limit of 8 ng/kg has been reached by postcolumn derivatization with ammoniac [186,189].

Today, several validated methods are available for OTA detection in cereals and derived products [204], in barley and coffee [205–207], and in wine and beer [208].

Immunoassays such as ELISA and radioimmunoassays have been developed [157,209–211] and may be regarded as qualitative or semiquantitative methods, useful for rapid screening.

Owing to its toxicity and regulatory values, OTA analysis has to be performed down to the ppb range in foods and feeds. In addition, plasma and urine samples are analyzed to monitor OTA exposure in humans and animals. In this context, methods using LC-MS may be used to confirm OTA positive results obtained by ELISA or HPLC-FL. It may be also a powerful tool to elucidate structure of *in vivo* metabolites and OTA adducts in biological fluids. Many studies have described LC-MS methods for OTA determination (for review, see Ref. 112).

36.3.4.3 OTA Analysis in Animal Tissues and Muscles

All previously described methodologies were used to analyze OTA content of animal tissues and animal-derived products. The aim of such studies was to characterize the potential carryover of the mycotoxin in animal tissues and assess the consecutive human exposure. Many methods have been devoted to the OTA analysis in animal tissues and the most recent ones are summarized in Table 36.8. They have essentially been set up in pigs and pig tissues because this species appears to be the most sensitive and exposed to OTA. For meat samples, the solvent extraction step cannot be avoided and precedes purification step. Typical procedures include extraction with acidic chloroform or acidic ethyl acetate, followed by back extraction into $NaHCO_3$ before cleanup on IAC or C18 columns.

It appears that detection limits exhibited by HPLC-FL are sufficient to control meat products according to existing regulations. The use of IAC for cleanup reduces the limit of quantification (LOQ) below 1 ng/g [213]. However, a 10-fold OTA fluorescence enhancement obtained by using the alkaline eluent in HPLC permitted the determination of very low level of OTA in muscle without any column purification or concentration step [214].

By contrast, the use of HPLC-MS does not strongly increase the sensitivity of detection but may be used as a confirmatory method in case of positive results.

ELISA tests usually display LOQ higher than other methods. Nevertheless, due to their simplicity and rapidity, these tests could be useful as screening methods in slaughterhouses.

36.4 Zearalenone

36.4.1 Introduction

ZEA is a mycotoxin with estrogenic effect that is produced by *Fusarium* species. Recently, endocrine disrupters received a lot of public attention since they are suspected to reduce male fertility in human and wildlife populations and possibly involved in several cancer development [218]. This molecule is well known by farmers, often responsible for reproduction perturbation, especially in pigs. Therefore, ZEA content is regulated in many foodstuffs and the carryover of this molecule in animal tissues has been investigated.

Table 36.8   Recent Methods for OTA Determination in Animal Tissues and Muscles

Quantification	Tissue	Extraction	Cleanup	LOQ (ng/g)	Reference
Fluorimetry	Ham	Methanol–1% sodium bicarbonate (70:30)	IAC	0.7[a]	212
HPLC-FL				0.04[a]	
HPLC-FL	Ham	Chloroform–orthophsphoric acid centrifugation	Back extraction with NaHCO₃, pH 7.5	0.03	163
			IAC		
HPLC-FL	Salami	Ethyl acetate (0.5 mol NaCl)–phosphoric acid	Back extraction with NaHCO₃, pH 8.0	0.2	205, 213
			IAC		
HPLC-FL	Pig tissues (kidney, muscle)	Ethyl acetate–phosphoric acid centrifugation	Back extraction with NaHCO₃, pH 8.4	0.52 (kidney)	165
			Acidification to pH 2.5 with phosphoric acid Back extraction with ethyl acetate	0.67 (muscle)	
HPLC-FL	Pig liver–derived pate	Acidified acetonitrile–water	C8 columns	0.84	215
HPLC-FL[b]	Dry-cured pork meat	Chloroform–phosphoric acid	Back extraction with Tris-HCl, pH 8.5, addition of chloroform until 90:10 ratio	0.06	214
HPLC-MS	Pig tissues (kidney, liver, and muscle)	Ethyl acetate–phosphoric acid centrifugation	Back extraction with NaHCO₃, pH 8.4	1.5	216
			Acidification to pH 2.5 with phosphoric acid Back extraction in ethyl acetate		
RIA	Pig kidney	Chloroform–phosphoric acid	Back extraction with NaHCO₃ (1 M) Acidification to pH 4.5 with HCl C18 column	0.2	210
	Pig kidney	Ethyl acetate–phosphoric acid centrifugation		7.8	217

[a] Limits of detection.
[b] Mobile phase: $NH_3/NH_4Cl:CH_3$-CN (85:15), pH 9.8 instead of acetonitrile–water–acetic acid (99:99:2) in others.

36.4.2 Origin and Toxicological Properties

36.4.2.1 Synthesis

ZEA has been isolated for the first time from maize contaminated with *Gibberella zeae*, the anamorph of *Fusarium graminearum* [219]. It has been demonstrated that ZEA can be synthesized by several *Fusarium* species such as *F. graminearum*, *F. proliferatum*, *F. culmorum*, and *F. oxysporum* [220,221]. These are fungal species that usually develop on living plants, and ZEA contamination occurs in the field, at harvest or early storage, when the drying step was not sufficient. Indeed, *Fusarium* growth and mycotoxin production usually occur at high water activity (>0.90) [222,223]. Temperature of ZEA production is lower than optimal temperature for mycelium development and is about 20–25°C [224–226]. Moreover, ZEA production is favored in substrates with high glucid/protein ratio. Owing to these factors, ZEA is a frequent contaminant of cereals and cereal-derived products in European and other countries with temperate climate [8,227].

36.4.2.2 Metabolism

ZEA is quickly absorbed after oral ingestion [228,229]. Although no quantification has been reported, urinary excretion of ZEA and its metabolites suggest that the absorption rate is high [230,231]. For example, the uptake in a pig after a single oral dose of 10 mg/kg bw was estimated to be 80–85% [232]. ZEA can be metabolized in digestive tracts by both microflora and intestinal mucosa [233,234]. This metabolism results in the appearance of α- and β-zearalenol and α- and β-zearalanol. The proportion of these two metabolites may change depending on the animal species [227,234,235]. Because among ZEA metabolites, α-zearalenol has a higher affinity for estrogenic receptors, its appearance during metabolic pathways can be considered as a bio-activation. Therefore, metabolism is a key factor of ZEA toxicity and differences in toxin transformation within organism can explain differences in toxicity observed in several animal species [236].

After absorption, two major hepatic biotransformation pathways have been suggested for ZEA in animals [237,238]:

- hydroxylation resulting in the formation of α-zearalenol and β-zearalenol
- conjugation of ZEA and reduced metabolites with glucuronic acid

Differences between species in hepatic biotransformation have been demonstrated; pigs seem to convert ZEA predominantly into α-zearalenol, whereas β-zearalenol is the main metabolite in cattle [239]. In human, as in pigs, ZEA was found mainly as glucuronide conjugates of ZEA and α-zearalenol in urine. All of the metabolites found in humans during the 24 h of sampling were glucuronides [231].

ZEA and its metabolites are excreted in urine or bile [240]. In ruminants, they are detected in bile at respective rates of 68% β-zearalenol, 24% α-zearalenol, and 8% ZEA [241]. In this study, neither ZEA nor its metabolites were detected in muscles, kidney, liver, or dorsal fat of bovine receiving 0.1 mg ZEA per day per kg feed. Only few studies are available on the potential carryover of this mycotoxin in animal edible organs of other species. It appears that, at least in pigs, meat and other edible parts may not be contaminated, even after exposure of the animals to high concentrations of the toxin [242–244]. Few studies conducted on poultry exposed to very high doses of ZEA detected the toxin at detectable level in muscles [245]. More recently, an experiment performed on laying hens exposed to 1.58 mg ZEA/kg feed for 16 weeks did not detect any residues in muscles, fat, or eggs [246].

36.4.2.3 Toxicity

Acute toxicity of ZEA is usually considered as weak with LD50 after oral ingestion ranging from 2,000 to more than 20,000 mg/kg bw [230,240]. Subacute and chronic toxicity of the mycotoxin is more frequent and may be observed at the natural contamination levels of feeds. The effects are directly related to the fixation of ZEA and its metabolites on estrogenic receptors [239]. Affinity with estrogenic receptors is, in decreasing order, as follows: α-zearalanol > α-zearalenol > β-zearalanol > ZEA > β-zearalenol. Pig and sheep appear more sensitive than other animal species: in multiple exposure experiments, the NOEL in pigs was 40 μg/kg bw whereas it was 100 μg/kg bw in rats [230,240].

ZEA induces alteration in the reproductive tracts of both laboratory and farm animals. Variable estrogenic effects have been described such as a decrease in the fertility, a decrease in the litter size, an increase in embryo-lethal resorptions, change in adrenal, thyroid, and pituitary gland weights. In male pigs, ZEA can depress testosterone, weight of testes, and spermatogenesis while inducing feminization and suppressing libido [227,230,240]. No teratogenic effect was observed in laboratory animals.

Studies reported that several alterations in immunological parameters could be observed *in vitro* after ZEA exposure of mice or human cells [247–249].

Long-term exposure studies did not demonstrate any carcinogenic potential of this mycotoxin. Therefore, ZEA has been classified by IARC as an estrogenic molecule in the group III of noncarcinogenic molecules [5].

36.4.3 Regulation

In 1999, JECFA established a temporary maximal daily tolerable dose of 0.5 μg/kg bw. It is based on the hormonal effects observed in the most sensitive species (pigs) and the NOEL of 50 μg/kg bw per day with a security factor of 100 [240]. In France, the CSPHF proposed a daily tolerable dose of 0.1 μg/kg bw per day calculated on the basis of effects observed on monkey's reproduction [183].

In 2003, ZEA was regulated in foods and feeds by 16 countries and in 2005 and 2006, the European Union adopted regulation and recommendation for ZEA in human foods and animal feeds [250,251] (Table 36.9).

36.4.4 Zearalenone Analysis

36.4.4.1 Physicochemical Properties

The structure of ZEA is shown in Figure 36.4. α- and β-Zearalenol, the natural metabolites of the native toxin, correspond to the reduction of the ketone function in C_6.

ZEA has a molecular weight of 318 g/mol. This compound is weakly soluble in water and in hexane. Its solubility increases with polarity of solvents: benzene, chloroform, ethyl acetate, acetonitrile, acetone, methanol, and ethanol [252]. The molecule has three maximal absorption wavelengths in UV: 236, 274, and 314 nm. The 274 nm peak is the most characteristic and commonly used for UV detection of the toxin.

ZEA emits a blue fluorescence with maximal emission at 450 nm after excitation between 230 and 340 nm [253].

36.4.4.2 Methods of Analysis

Owing to regulatory limits, methods for analysis of ZEA content in foods and feeds may allow the detection of several ng/g. Reviews have been published to detail analytical methods available

[112,254,255]. It has to be noted that ZEA is sensitive to light exposure, especially when in solution. Therefore, preventive measures must be taken to avoid this photodegradation.

Solvents used for liquid extraction of ZEA and its metabolites are mainly ethylacetate, methanol, acetonitirle, and chloroform, alone or mixed. The mixture acetonitrile–water is the most commonly used. For solid matrixes, more sophisticated and efficient methods may be applied: for example, ultrasounds or microwaves [256,257].

In biological matrixes (plasma, urine, feces, etc.), a step of hydrolysis of phase II metabolites is necessary before the purification procedure. It can be achieved by enzymatic or chemical protocol [258]. In vegetal materials, the demonstrated presence of sulfate [259] or glucoside conjugates [260] is rarely taken into account in routine methods. Since these conjugates have a biological activity comparable to that of the native molecule, it may lead to an underestimation of the real contamination of grains with ZEA.

Table 36.9 EU Regulation Concerning Zearalenone in Foods and Feeds

	Nature	*Maximal Value (ng/g)*	*Reference*
Food	Cereals (except maize)	100	250
	Maize	200	
	Cereal flour (except maize flour)	75	
	Maize flour	200	
	Bread, biscuits, pastries	50	
	Breakfast cereals	50	
	Maize-based baby food	20	
	Cereal-based baby food	20	
Feed[a]	Cereals and cereal-based products (except maize)	2000	251
	Maize by-products	3000	
	Complementary feed for		
	Piglets and gilts	100	
	Sows and bacon pigs	250	
	Calves, lactating cattle, ovines, caprines	500	

[a] Recommendations.

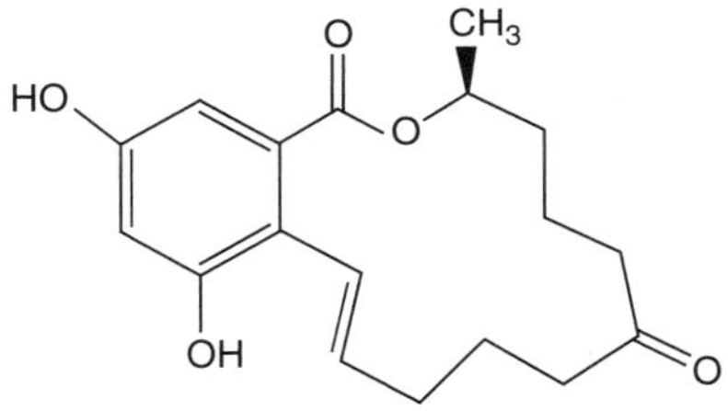

Figure 36.4 Structure of zearalenone.

Purification may be achieved using liquid–liquid extraction (LLE), SPE, or IAC procedures. For SPE, most stationary phases may be used: inverse phase (C18, C8, or C4), normal phase (florisil, SiOH, NH_2), or anion exchange (SAX) [261]. Another approach using ready-to-use column Mycosep (Romer Labs Inc, MT, USA) may be applied. These columns are adsorbants (charcoal, celite, ion-exchange resin) mixed in plastic tube. It allows a rapid purification of sample without rinsing and with a selective retention of impurities [262].

IAC columns have also been developed for ZEA and are very popular [263–269]. Although purification is very selective and extraction yields usually high, several points have to be highlighted:

■ antibody may not have the same affinity for all metabolites, some being not accurately extracted
■ fixation capacity of columns are limited; a great number of interfering substances may perturb the purification by saturation of the fixation sites [261]
■ these columns may be re-used, raising the risk of cross contamination of samples

For quantification of ZEA and metabolites in cereals and other matrixes, several immunological methods have been set up, including RIA and ELISA [270–274]. The LOQ of these methods is about several tens ng/g. ELISA kits show a cross reactivity with α- and β-zearalenol [275].

Physicochemical methods are also widely used. These mainly include HPLC and gas chromatography (GC), TLC being nearly withdrawn [276,277]. Many methods using C18 as stationary phase and CH_3CN/H_2O as mobile phase have been described. More specific stationary phases have also been proposed such as MIP [278]. Detectors are often fluorimeters [263–266,268,279] or UV detector [261,265]. Sensitivity of these methods varies depending on the metabolites and is less important for reduced metabolites (α- and β-zearalenol).

ZEA and its metabolites can also be detected by GC. However, the usefulness of such method is limited due to the time-consuming need of derivatization of phenolic hydroxy groups. Consequently, only GC-MS has been applied for confirmation of positive results [280,281].

Many LC-MS methods have also been proposed for ZEA and its metabolites detection (for review, see Ref. 112). The method of chemical ionization at atmospheric pressure is the most commonly used followed by electrospray [269,282–285]. These methods allow the detection of ZEA and its metabolites at levels below 1 ng/g [112].

Results from an international interlaboratory study revealed that important variations were observed between results from the participants' laboratory, probably related to differences in sample preparation (LLE, SPE, or IAC) and quantification (HPLC, GC, TLC, and ELISA) [272,286].

36.4.4.3 ZEA Analysis in Animal Tissues and Muscles

Owing to metabolism of the native molecule and the very weak carryover of ZEA in edible parts of farm animals (see Section 36.4.2.2), only few "classical" physicochemical or immunological methods were developed for ZEA detection in edible parts of farm animals [244,246,287–289]. HPLC-UV or HPLC-FL was used for quantification. The recent developments of LC-MS and GC-MS have improved sensitivity in ZEA and its metabolites detection (Table 36.10). It also makes possible the discrimination between the use of estrogen as growth promoters during breeding (Zeranol) and the exposure of animals to a contaminated feed [292,293].

Table 36.10 Overview of LC-MS and GC-MS Methods for Zearalenone Analysis in Animal Tissues

Analytes	Matrix	Sample Cleanup	Column	Mobile Phase	Ionization	Scan Mode	LOQ (ng/g)	Reference
			LC		*MS*			
ZEA, α-zearalenol; β-zearalenol, α-zearalanol, β-zearalanol	Pig tissues	SPE with RP-18 columns	RP-18	H_2O/MeOH/ ACN (45:45:10) with 15 mM NH_4AOc	APCI	SRM	0.5–1	285
ZEA, α-zearalenol; β-zearalenol, α-zearalanol, β-zearalanol, zeranol	Liver, urine, and muscle of pigs	SPE with RP-18 columns	RP-18	H_2O/MeOH/ ACN (45:45:10) with 15 mM NH_4AOc	ACPI	SRM	0.5–1 (muscle) 0.1–1 (liver)	258

Analytes	Matrix	Sample Cleanup	Column	Carrier Gas	Ionization	Scan Mode		
			GC		*MS*			
ZEA, α-zearalenol, zeranol	Bovine muscle	IAC	Hp-5 ms capillary	Helium	ACPI	SIM	1 ng/g	290, 291

36.5 Trichothecenes

36.5.1 Introduction

Trichothecenes constitute a large group of secondary metabolites produced by numerous species of *Fusarium*, such as *F. graminearum*, *F. culmorum*, *F. poae*, *F. sporotrichoides*. More than 160 trichothecenes have been identified, notably deoxynivalenol (DON), nivalenol (NIV), T-2 toxin, HT-2 toxin, diacetoxyscirpenol (DAS), fusarenon X. DON is the most frequently found trichothecene. These mycotoxins are important for public health concern; however, due to their metabolism pathways in animals, they do not represent a significant hazard as residual contaminant of muscle foods.

36.5.2 Origin and Toxicological Properties

36.5.2.1 Synthesis

Trichothecenes can be produced by a large variety of fungus (Table 36.11). They mainly belong to the *Fusarium* genus; however, other fungal species such as *Trichoderma viridae* or *Myrothecium roridum* can also produce some trichothecenes [294,295]. One fungal species may be able to produce several trichothecenes.

Table 36.11   Main Fungal Species Able to Produce Trichothecenes

Fungal Genus	Fungal Species	Toxins	Type
Fusarium	Fusarium tricinctum	DAS, T-2 toxin, HT-2 toxin	A
	Fusarium sporotrichoides	DAS, T-2 toxin, HT-2 toxin	
	Fusarium poae	DAS, T-2 toxin, HT-2 toxin, acetyl T-2 toxin	
	Fusarium solani	DAS, T-2 toxin, HT-2 toxin	
	Fusarium roseum	Scirpenol, monoacetoxyscirpenol	
	Fusarium nivale	DON, NIV, Fusarenon X	B
	Fusarium crookwellense	DON, Fusarenon X	
	Fusarium oxysporum	DON, diacetyl-NIV, Fusarenon X	
	Fusarium avanaceum	DON, Fusarenon X	
	Fusarium graminearum	DON, Fusarenon X	
	Fusarium solani	DAS, T-2 toxin, HT-2 toxin, DON, Fusarenon X	A and B
	Fusarium culmorum	DAS, T-2 toxin, HT-2 toxin, Diacetyl-NIV	
	Fusarium equiseti	DAS, T-2 toxin, HT-2 toxin, Diacetyl-NIV, Fusarenon X	
Trichoderma	Trichoderma viridae	Trichodermin	A
	Trichoderma polysporum	Trichodermin, roridin C	
	Trichoderma lignorum	Trichodermin, T-2 toxin	
Myrothecium	Myrothecium roridum	Trichodermadienediol, roridin C	A
Trichothecium	Trichothecium roseum	Trichothecin	B

These fungal species mainly belong to the field mycoflora, developing on living plants or during the early postharvest period. Indeed, *Fusarium* species are hygrophilic fungus and the drying step will block their development [296]. These fungal contaminants are well known for being responsible of *Fusarium* head blight of small grain cereals and ear rot of maize. These fungal pathologies reduce yields, decrease milling and malting qualities of grains, and may lead to mycotoxin contamination of infected grains [297,298]. As for all mycotoxin production, climatic conditions directly influence the trichothecene synthesis. Moreover, fungal development and subsequent trichothecene production may be related to agricultural practices such as crop rotation [299]. For example, it is generally accepted that wheat that follows an alternative host for *Fusarium* pathogen (i.e., maize) is at greater risk for DON contamination of grain [300].

Owing to their synthesis conditions, trichothecenes are worldwide common contaminants of cereals, mainly wheat and maize, and cereal-based products [8,301–304]. The conditions during malting of grains may allow *Fusarium* development and trichothecenes production. DON has been found as a frequent contaminant of beers, even if contamination levels are usually low [305–307].

All studies on the carryover of trichothecenes in edible parts of exposed animals revealed that meat and meat products cannot be considered as a source of exposure to these toxins (see Section 36.5.2.2).

36.5.2.2 Metabolism

Absorption kinetics have not been reported for all trichothecenes; however, they seem to be rapidly and efficiently absorbed in gastrointestinal tract, whatever the animal species [308]. DON, NiV, or T-2 toxin can be detected in blood less than 30 min after ingestion [309]. Absorption rates range from 10 to 55%. Trichothecenes can be metabolized within the digestive tracts by ruminal flora. It may partially explain the ruminant resistance to these toxic compounds [310]. In other species, intestinal microflora may also metabolize trichothecene [311]. DON transformation in digestive tract is well documented [312,313]. Few traces of de-epoxy-DON are found in pig stomach and small intestine whereas its quantity increases in large intestine to reach 80% of the toxin in rectum [314].

After absorption, metabolism of trichothecenes mainly consists in hydrolysis, hydroxylation, and de-epoxydation. This metabolism differs according to the animal species but it generally leads to the reduction of the epoxy group of the molecules and glucuronidation [308,310]. For example, in pigs, more than 95% of the ingested dose of DON is excreted without any transformation and less than 5% is found as its glucuronide metabolite [315].

When absorbed, trichothecenes are rapidly eliminated without any accumulation in the organism [309]. Only traces of compounds can be still detected 24 h after ingestion [289,314,315]. Plasmatic half-lives of trichothecenes are of several hours [308].

These data explain that meat and meat products are not considered as a potential source of trichothecenes for human consumers. Indeed, the starvation diet that always precedes animal slaughtering is usually long enough to allow trichothecene elimination from edible parts of the organism [315,316]. In case of animal exposure till slaughtering, trichothecene are only found in traces, even after repeated exposure in pigs [317,318] as well as in poultry (Table 36.12).

36.5.2.3 Toxicity

Because trichothecenes are a large family grouping many compounds of variable structure and properties, their toxicity can be very different depending on the molecule, the animal species, the dose, and the exposure period. There are many reviews available on trichothecenes toxicity [323–326] and only the main features are presented in this section.

Trichothecenes are potent inhibitors of eukaryotic protein synthesis, interfering with initiation, elongation, or termination stages.

Concerning their toxicity in animals, DAS, DON, and T-2 toxin are the most studied molecules. The symptoms include effects on almost all major systems of organisms; many of them being secondary processes initiated by poorly understood metabolic process in relation with protein synthesis inhibition.

Among naturally occurring trichothecenes, DAS and T-2 toxin seem to be the most potent in animal. They have an immunosuppressive effect, decreasing resistance to microbial infections [325]. They also cause a wide range of gastrointestinal, dermatological, and neurological symptoms [327]. In human, these molecules have been suspected to be associated with alimentary toxic aleukia. The disease, often reported in Russia during the nineteenth century, is characterized by inflammation of the skin, vomiting, damage to hematopoietic tissues [328,329].

When ingested at high concentrations, DON causes nausea, vomiting, and diarrhea. At lower doses, pigs and other farm animal display weight loss and food refusal [325]. For this reason, DON is often called vomitoxin or food refusal factor.

Table 36.12 Residues of Trichothecenes in Animal Tissues (Expressed as Equivalent-Toxin)

After a Single Administration

Toxin	Species	Route	Dose (mg/kg bw)	Tissues	Residues (µg/kg)					Half-life	Reference
					6 h	12 h	24 h	2 j	4 j		
DON	Hen	VO	1.3–1.7	Muscle	8.46	6.6	4.3	2.1	ND		319
				Liver	74	56	30	13	ND	15.7 h	
				Kidney	165	123	44	19	2	8.2 h	
T-2	Chicken	VO	0.126/1.895	Muscle			17/220				320
				Liver			32/416				
				Kidney			24/327				
	Chicken/duck	VO	5	Muscle	30	30	<10	<10			321
				Liver	130/90	30/40	10/<10	<10			
				Kidney	30	20	<10	<10			

After a Repeated Administration

Toxin	Species	Route	Length	Dose	Tissues	Residues (µg/kg)						Reference
						2 j	4 j	6 j	8 j	10 j	12 j	
DON	Hen	VO	6 j	1.3–1.7 mg/kg bw	Muscle	16	17	10	11	7	3	319
					Liver	37	41	39	25	15	9	
					Kidney	60	51	55	21	15	9	
	Chicken	VO	28–190 j	5 mg/kg feed	Muscle	<10						322
					Liver							
					Kidney							

Note: ND, not detectable.

36.5.3   Regulation

In 1993, IARC classified trichothecenes (T-2 toxin, DON, and NIV) in the group III of compounds with inadequate data in human and animals to rule on their carcinogenicity [5].

However, several daily tolerable doses were fixed by JECFA, according to toxic effects observed in pigs or rodents. These doses are 0.06, 1, and 0.7 µg/kg bw per day for T-2, DON, and NIV, respectively.

Several countries have implemented a regulation for trichothecenes content in food and feed. These regulatory values or recommendations mainly concern DON and are active in countries where climatic conditions are suitable for *Fusarium* development and toxinogenesis (Europe and northern countries).

36.5.4   Trichothecene Analysis

36.5.4.1   Structure

Trichothecenes belong to the sesquiterpenoid group. They all contain a 12,13-epoxytrichothene skeleton and an olefinic bond with various side chain substitutions. Trichothecenes are classified as macrocyclic or nonmacrocyclic, depending on the presence of a macrocyclic ester or an ester–ester bridge between C-4 and C-15 [330]. The nonmacrocyclic trichothecene can be classified into two groups: type A that do not have a ketone group on C-8 (T-2 toxin, HT-2 toxin, DAS) and the type B with ketone group on C-8 (DON, NIV, fusarenon X) (Figure 36.5) [331].

36.5.4.2   Physicochemical Properties

Trichothecenes have a molecular weight ranging from 154 to 697 Da but is often between 300 and 600 Da. They do not absorb UV or visible radiations, with the exception of the type D trichothecenes that absorb UV light at 260 nm. They are neutral compounds, usually soluble in mildly polar solvents such as alcohols, chlorinated solvents, ethyl acetate, or ethyl ether. They are sometimes weakly soluble in water [331].

These molecules are very stable, even if stored for long time at room temperature. They are not degraded by cooking or sterilization process (15 min at 118°C) [332].

36.5.4.3   Methods of Analysis

Methods reported mainly concern the most frequently found toxins in cereals, such as DON, NIV, T-2 toxin, and HT-2 toxin [333]. If validated methods are now available for DON [334], it is not the case for type A trichothecenes and reference material and interlaboratory studies are still required [335].

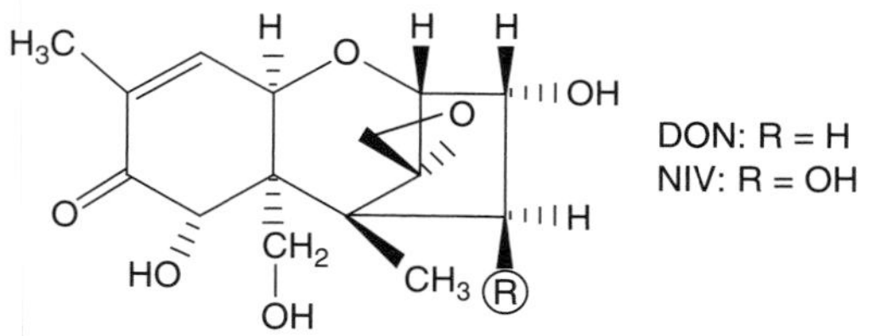

DON: R = H
NIV: R = OH

Figure 36.5   Structure of type B trichothecenes.

36.5.4.3.1 Type A Trichothecenes

Extraction from solid matrixes is usually done with binary mixtures such as water and acetonitrile, water and methanol, chloroform and methanol, or methanol alone.

Purification is done with SPE columns working with normal phase (silica, florisil) or inverse phase (C18). Multifunctional columns (Mycosep) are more popular.

Immunoassays are the main methods used in routine for T-2 and HT-2 determination in cereals. Detection limits are in accordance with contamination level that are observed for these contaminants and range from 0.2 to 50 ng/g for T-2 toxin [336].

Other methods have also been described but type A trichothecenes cannot be analyzed by HPLC-UV due to the absence of ketone group in C-8 position. That is why, GC is the most popular approach for this family of compounds. The derivatization of the native compounds by silylation or fluoroacylation is necessary to increase sensitivity of the measure. Detection can be performed with electron capture detector or by mass spectrometry. The limits of detection of these methods are about few tens ng/g [337]. Another method was reported using HPLC with fluorescence detection after IAC purification of extract and derivatization of T-2 toxin with 1-anthroylnitrile. This procedure allowed a limit of detection of 5 ng/g [338].

36.5.4.3.2 Type B Trichothecenes

Extraction of type B trichothecenes is done with mixtures such as acetonitrile–water or chloroform–methanol [339].

Many purification procedures were reported for type B trichothecenes such as LLE, SPE, or IAC [340]. However, the use of mixed columns (charcoal/alumina/celite) is still widespread [341]. Once again, the Mycosep column is of increasing use for DON analysis.

Thin layer chromatographic methods are still used for screening, particularly in countries where GC or HPLC are not easily available [342]. Since trichothecenes are not fluorescent, the detection of the molecules requires the use of revelators such as sulfuric acid, *p*-anisaldehyde, or aluminum chloride. Detection limits of TLC methods are ranging from 20 to 300 ng/g.

ELISA can also be of interest to get rapid and semiquantitative results with only minor purification of the extract. Many kits are commercially available for DON analysis in cereals [343,344].

GC coupled to an electron capture detector, a mass spectrometer, or in tandem (MS/MS) is regularly used after derivatization of the analyte [345–348]. Derivatization reactions are trimethylsilylation or perfluoroacylation. The fluoroacylation with anhydride perfluorated acid improves detection limits with electron capture detector or mass spectrometry. However, a European interlaboratory assay concerning AOAC official method for DON measurement revealed that the coefficient of variation between laboratories was very important (around 50%) despite the relatively high level of contamination of the material used (between 350 and 750 µg/kg). These observations increased the interest for HPLC-MS methodology in trichothecene determination. This progressively becomes the choice method [112].

36.5.4.4 Trichothecenes Analysis in Animal Tissues and Muscles

Only few methods were developed for trichothecenes analysis in animal tissues. Indeed, first experiments on the pharmacokinetics and distribution of these mycotoxins were performed using radiolabeled toxins [319–321,349–351]. As these experiments revealed that trichothecenes were rapidly excreted and carryover of the toxins in edible part of animals was minimal, only few studies were

Table 36.13   Methods for Trichothecene Analysis in Animal Muscles and Tissues

Toxin	Organ (Species)	Extraction and Cleanup	Derivatization–Quantification	LOD[a] (ng/g)	Reference
T-2	Liver, kidney, heart (chicken)	Acetonitrile Amberlite XAD-2 resin column	TFAA—tri-Sil TBT GC-MS	—	352
DON	Liver, kidney, muscle (pigs)	Ethyl acetate– sodium acetate IAC	HPLC-UV	4	244
	Liver, kidney, muscle (pigs)	Acetonitrile–water Alumina-charcoal column	HPLC-MS (SIM mode)	2	317, 353
	Liver, kidney, muscle (hens)	Acetonitrile–water Alumina-charcoal column	Heptafluorobutyryl imidazole Gas–liquid chromatography	10	322, 354

[a] Detection limit.

carried out to evaluate trichothecene presence in muscle and other tissues of animal after exposure to unlabeled toxins. The methods used in these works are summarized in Table 36.13.

36.6   Fumonisins

36.6.1   Introduction

Even if their effects on animal health, especially equines, had been known for long time, fumonisins were identified in 1988. Since then, this new family of mycotoxins has been extensively studied and has revealed several particular characteristics concerning physicochemical properties, metabolism, and mechanism of action. If these toxins appear to be of great importance for human and animal health concern, it has also been demonstrated that, due to their absorption and kinetic properties, meat and meat products may not be an important source of exposure for human consumers. Therefore, as for trichothecenes, only few methods were reported for fumonisins analysis in animal tissues. This section focuses on fumonisin B1 (FB1), the most abundant and toxic compound of this family.

36.6.2   Origin and Toxicological Properties

36.6.2.1   Synthesis

Fumonisins were first described and characterized in 1988 from *Fusarium verticillioides* (formerly *F. moniliforme*) culture material [355,356]. The most abundant and toxic member of the family is FB1. These molecules can be produced by few species of *Fusarium* fungi: *F. verticillioides*, *F. proliferatum*, and *F. nygamai* [357–359]. These fungal species are worldwide contaminants of maize that represent the main source of fumonisins [183]. *F. verticillioides* grows as an endophyte in corn, and even if it can cause plant pathology such as seedling blight, stalk rot, and ear rot [360], fumonisin contamination of grains can occur without any visible alteration.

Fumonisin production mainly occurs during periharvest period, at temperatures about 20–25°C and high moisture content of grains [359,361,362]. Many surveys have revealed that fumonisins are major contaminants of maize and maize products worldwide [363–370].

36.6.2.2 Metabolism

After oral ingestion, FB1 is only weakly absorbed (1–5%) in all studied animal species [371–374]. More than 95% of the toxin is found in native form in feces of exposed animals. Although FB1 is distributed in all tissues, most part of the absorbed toxin is found in liver and kidney [375,376]. The toxin is not excreted in milk [377–379]. Both *in vitro* and *in vivo* experiments failed to demonstrate any metabolism of fumonisins [380,381].

FB1 is excreted in feces, bile, and urine as native molecule or in partially hydrolyzed form [375,382,383]. Taking into account the toxicokinetic parameters observed for FB1, it appears that edible parts of animals, and especially muscles, should not represent a source for human exposure. This was confirmed in swine after exposure of the animal to high dose of FB1 [384]. However, in some particular species such as ducks, taking into account the high proportion of maize in their diet (up to 99%), it may be important to evaluate the possibility of having FB1 residues in liver and muscles in case of exposure during forced feeding period.

36.6.2.3 Toxicity

One major characteristic of fumonisins is that they induce different syndromes depending on the animal species. FB1 is responsible for equine leukoencephalomalacia characterized by necrosis and liquefaction of cerebral tissues [385,386]. Horses appear to be the most sensitive species since clinical signs may appear after exposure to doses as low as 5 mg FB1/kg feed during few weeks. Pigs are also sensitive to FB1 toxicity. In this species, fumonisins induce pulmonary edema after exposure to high doses (higher than 20 mg FB1/kg feed) of mycotoxins and hepatotoxic and immunotoxic at lower doses [387–389]. By contrast, poultry and ruminants are more resistant to this mycotoxin and clinical signs only appear after exposure to doses higher than 100 mg FB1/kg, which may be encountered in natural condition but is quite rare [390–395]. In rodents, FB1 is hepatotoxic and carcinogenic, leading to appearance of hepatocarcinoma in long-term feeding studies [396,397]. In human, FB1 exposure has been correlated with high prevalence of esophageal cancer in some part of the world, mainly South Africa, China, and Italy [398]. Finally, fumonisins can cause neural tube defects in experimental animals, and thus may also have a role in human cases [399–402]. At the cellular level, FB1 interacts with sphingolipid metabolism by inhibiting ceramide synthase [403]. This leads to the accumulation of free sphinganine (Sa) and, to a lesser extent, of free sphingosine (So). Therefore, the determination of the Sa/So ratio has been proposed as a biomarker of fumonisin exposure in all species that have been studied [404–407]. The accumulation of these active second messengers may perturb many cell function and lead to apoptosis, cell proliferation, membrane integrity disturbance, etc. [408–410].

Owing to its carcinogenic properties in laboratory animals, FB1 has been classified by IARC in the group 2B of molecules carcinogenic to animals and possibly carcinogenic in human [5].

36.6.3 Regulation

The European regulation set up maximal concentrations for FB1 and FB2 in maize and maize-derived products for human food [12]. Maximal limits are of 200–2000 µg/kg depending on the food. The European Union also recommends the respect of maximal values in animal feeds. These values may change depending on the animal species and their sensitivity to the toxins. They range from 5 mg/kg for horses and pigs to 50 mg/kg for adult ruminants [251]. Only few other countries such as the United States or Iran have regulatory limits for fumonisins in maize and maize products [13].

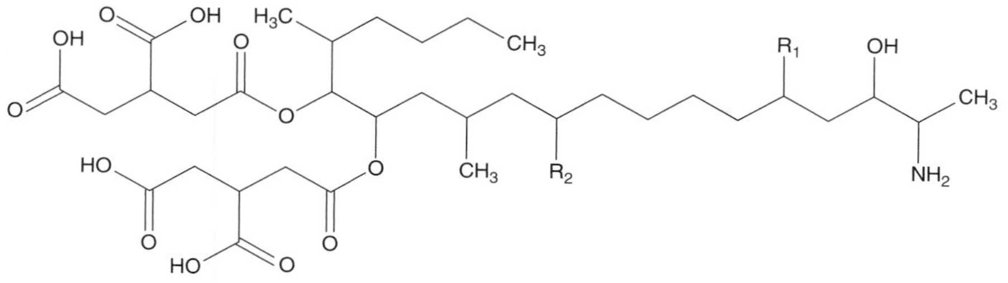

	R1	R2	Formula	CAS number	Weight
Fumonisin B1	OH	OH	$C_{34}H_{59}NO_{15}$	116355-83-0	721,838
Fumonisin B2	OH	H	$C_{34}H_{59}NO_{14}$	116355-84-1	705,839
Fumonisin B3	H	OH	$C_{34}H_{59}NO_{14}$	136379-59-4	705,839
Fumonisin B4	H	H	$C_{34}H_{59}NO_{13}$	136379-60-7	689,840

Figure 36.6 Structure of fumonisins B.

36.6.4 Fumonisin B1 Analysis

36.6.4.1 Physicochemical Properties

The structure of FB1 and related compounds is shown in Figure 36.6. FB1 has a molecular weight of 722 g/mol. It is a polar compound, soluble in water and not soluble in apolar solvents. FB1 do not absorb UV light nor is fluorescent. Fumonisins are thermostable [411]. However, extrusion cooking may reduce fumonisin content of maize products [412].

36.6.4.2 Methods of Analysis

Because of their relatively recent discovery, analytical methodology for fumonisin analysis is still undergoing development. In most described methods, food or foodstuff is corn. An HPLC method has been adopted by AOAC and European Committee for standardization (CEN) as reference methodology for FB1 and FB2 in maize [413–415].

An efficient extraction of fumonisins in solid matrix can be obtained with acetonitrile–water or methanol–water mixtures [416,417]. This was assessed by interlaboratory assay [418]. Increased contact time and solvent/sample ratio also increase yield of extraction step.

Purification of extracts is usually based on SPE with anion exchange (SAX), inverse phase (C18), or immunoaffinity [419,420].

Quantification of FB1 can be done by TLC, HPLC, or GC-MS. However, derivatization of the fumonisins is usually required. For TLC, this is usually done by spraying *p*-anisaldehyde on the plates after development in a chloroform–methanol–acetic acid mixture. It leads to the appearance of blue-violet spots that can be quantified by densitometry [411,421]. Quantification limits obtained with TLC methods often range from 0.1 to 3 mg/kg, which may be sufficient for rapid and costless screening of raw materials [383,422].

For HPLC analysis, fluorescent derivatives are formed with *o*-phthaldialdehyde (OPA), naphthalene-2,3-dicarboxaldehyde (NDA), or 4-fluoro-2,1,3-benzoxadiazole (DBD-F) [423]. OPA derivatization offers the best response and has been generally adopted, but the derivatization product is very unstable and analysis of samples has to be quickly performed after derivatization [424]. HPLC-FL methods have detection limits usually ranging from 10 to 100 µg/kg [420,423,424].

GC has also been proposed for FB1 determination. It is based on partial hydrolysis of fumonisins before re-esterification and GC-MS analysis. However, this structural change does not allow anymore the distinction of different fumonisins molecules [425]. Another GC-MS method was described, developing a derivatization step with trimethylsilylation coupled to detection with flame ionization [426].

The introduction of LC-MS with atmospheric pressure ionization has increased specificity and sensitivity of the detection. The majority of published fumonisin analysis with LC-MS was performed to the low ppb level in grains and maize-derived products. Furthermore, this methodology also appeared powerful in investigating for new fumonisin molecules, elucidating structures and biosynthetic pathways and behavior during food processing (for review, see Ref. 112).

ELISA kits are also commercially available for fumonisin quantification in vegetal matrix [427–429]. They usually offer detection limits of around 500 µg/kg. However, the comparison with HPLC-FL shows that ELISA often overestimates fumonisin content of samples. This may be due to cross reactions between antibody and coextracted impurities [430]. This drawback could be overcome by purification of extracts before ELISA realization. This method can nevertheless be useful for rapid screening of maize and maize products. One ELISA kit has been validated by AOAC for total fumonisins determination in corn [431].

36.6.4.3 FB1 Analysis in Animal Tissues and Muscles

Measurement of FB1 in tissues is poorly documented. Moreover, most of the data concerns its toxicokinetic in animals and were obtained by using labeled molecules [374,432]. Finally, only three methods were described concerning the determination of nonradiolabeled FB1 in tissues. The first one employs an extraction with strong anion exchange (SAX) cartridges and quantification with HPLC connected to a quadrupole mass spectrometer with electrospray ionization [384]. The second method is based on an SPE extraction with HPLC and fluorescence detection [433]. The last one is based on the use of IACs for the extraction of mycotoxin and quantification of derivatized FB1 by fluorescence detection after its separation by HPLC [434]. These methods are summarized on Table 36.14.

Table 36.14 Fumonisin B1 Analysis in Animal Tissues

Detection	Extraction	Cleanup	Tissues (Species)	LOQ (ng/g)	Reference
LC-MS	Methanol– water (3:1)	Washing with n-hexane SAX column	Kidney, liver, muscle (swine)	5–10	384
HPLC-FL (OPA derivatization)	Methanol– water (80:20)	Washing with n-hexane SAX column	Liver (swine)	50	433
HPLC-FL (OPA derivatization)	Water– methanol– acetonitrile (2:1:1), NaCl	Washing with n-hexane IAC	Liver, kidney, muscle (duck)	25	434

36.7   Conclusion

Mycotoxins are widely found contaminants of cereals and other vegetal products. When contaminated feeds are consumed by farm animals, mycotoxin may be found as residues in the edible parts of the animals. For most important toxins, the available data on absorption, distribution within animal organism, and metabolism revealed that mainly aflatoxins and OTA may be found at significant levels in muscles and muscle foods. For these molecules, sensitive and specific methods are required to allow safety control of meat products because levels of contamination are usually of the low ppb range. Most of the commonly used methodologies are based on HPLC-FL detection of molecules. In the future, the increasing development of mass spectrometry may strengthen the role of such method for both confirmation and screening. Moreover, because of frequent multi-contamination of foods and feeds by several mycotoxins, many efforts will focus on the development of multidetection methods for vegetal matrix and animal tissues [435].

References

1. Bennett, J.W. and Klich, M., Mycotoxins, *Clin. Microbiol. Rev.*, 16, 497, 2003.
2. Bhatnagar, D., Yu, J., and Ehrlich, K.C., Toxins of filamentous fungi, *Chem. Immunol.*, 81, 167, 2002.
3. Pitt, J.I., Biology and ecology of toxigenic *Penicillium* species, *Adv. Exp. Med. Biol.*, 504, 29, 2002.
4. Conkova, E. et al., Fusarial toxins and their role in animal diseases, *Vet. J.*, 165, 214, 2003.
5. IARC, Some naturally occurring substances, food items and constituents, heterocyclic aromatic amines and mycotoxins, in *Monographs on the Evaluation of Carcinogenic Risks to Humans*, World Health Organization, Lyon, 1993, 56, 245.
6. Leblanc, J.C., Etude de l'alimentation totale en France: mycotoxines, mineraux et aliments traces, *INRA (ed.)*, 2004.
7. SCOOP reports on Tasks 3.2.7. Assessment of dietary intake of Ochratoxin A by the population of EU members states, 2000.
8. Schothorst, R.C. and Van Egmond, H.P., Report from SCOOP task 3.2.10 "collection of *Fusarium* toxins in food and assessment of dietary intake by the population of EU member states"; subtask: trichothecenes, *Toxicol. Lett.*, 153, 133, 2004.
9. Park, D.L., Effect of processing on aflatoxin, *Adv. Exp. Med. Biol.*, 54, 173, 2002.
10. Ryu, D., Jackson, L.S., and Bullerman, L.B., Effects of processing on zearalenone, *Adv. Exp. Med. Biol.*, 504, 205, 2002.
11. ISO, 6497. Animal feeding stuffs: sampling, 2002.
12. European Union, Commission Regulation (EC) No 466/2001 setting maximum levels for certain contaminants in foodstuffs, L77, 1, 2001.
13. FAO, worldwide regulations for mycotoxins in food and feed in 2003. FAO food and nutrition papers, Rome, Italy, 81, 2004.
14. AOAC 2000.08. *Determination of Aflatoxin M1 in Liquid Milk*, 2000.
15. Nesbitt, B.F. et al., *Aspergillus flavus* and turkey X disease. Toxic metabolites of *Aspergillus flavus*, *Nature*, 195, 1062, 1962.
16. De Iongh Berthuis, R.K. et al., Investigation of the factor in groundnut meal responsible for "turkey X disease", 65, *Biochim. Biophys. Acta*, 548, 1962.
17. Asao, T. et al., Aflatoxins B and G, *J. Am. Chem. Soc.*, 85, 1706, 1963.
18. Allcroft, R. and Carnaghan, R.B.A., Groundnut toxicity: an examination for toxin in human food products from animals fed toxic groundnut meal, *Vet. Res.*, 75, 259, 1963.
19. Rapper, K.B. and Fennel, D.I., *The genus Aspergillus*, Williams & Wilkins, Baltimore, MD, 1965.
20. Klich, M.A. and Pitt, J.I., Differentiation of *Aspergillus flavus* from *Aspergillus parasiticus* and other closely related species, *Trans. Br. Mycolog. Soc.*, 91, 99, 1968.

21. Detroy, R.W., Lillehoj, E.B., and Ciegler, A., Aflatoxin and related compounds, in *Microbial Toxins Vol VI: Fungal Toxins*, Ciegler, A., Kadis, S., and Ajl, S.J., Eds., Academic Press, New York, 3, 1971.

22. Diener, U.L. et al., Epidemiology of aflatoxin formation by *Aspergillus flavus*, *Annu. Rev. Phytopathol.*, 25, 249, 1987.

23. Senyuva, H.Z., Gilbert, J., and Ulken, U., Aflatoxins in Turkish dried figs intended for export to the European Union, *J. Food Prot.*, 70, 1029, 2007.

24. Zinedine, A. et al., Limited survey for the occurrence of aflatoxins in cereal and poultry feeds from Rabat, Morocco, *Int. J. Food Microbiol.*, 115, 124, 2007.

25. Toteja, G.S. et al., Aflatoxin B1 contamination in wheat grain samples collected from different geographical regions of India: a multicenter study, *J. Food Prot.*, 69, 1463, 2006.

26. Fazekas, B., Tar, A., and Kovacs, M., Aflatoxin and ochratoxin A content of spices in Hungary, *Food Addit. Contam.*, 22, 856, 2005.

27. Kaaya, A.N. and Kyamuhangire, W., The effect of storage time and agroecological zone on mould incidence and aflatoxin contamination of maize from traders in Uganda, *Int. J. Food Microbiol.*, 110, 217, 2006.

28. Thompson, C. and Henke, S., Effect of climate and type of storage container on aflatoxin production in corn and its associated risks to wildlife species, *J. Wild. Dis.*, 36, 172, 2000.

29. Trenk, H.L. and Hartman, P.A., Effect of moisture content and temperature on aflatoxin production in corn, *Appl. Microbiol.*, 19, 781, 1970.

30. Northolt, M.D. and van Egmond, H.P., Limits of water activity and temperature for the production of some mycotoxins, *4th Meeting Mycotoxins in Animal Disease*, 106, 1981.

31. Sanchis, V. and Magan, N., Environmental conditions affecting mycotoxins, in *Mycotoxins in Food: Detection and Control*, Magan, N. and Olsen, M., Eds., Woodhead Publishing Ltd, Oxford, 174, 2004.

32. Giorni, P. et al., Studies on *Aspergillus* section flavi isolated from maize in northern Ital, *Int. J. Food Microbiol.*, 113, 330, 2007.

33. Battilani, P. et al., Monitoraggio della contaminazione da micotossine in mais, *Infor. Agro.*, 61, 47, 2005.

34. Ezzel, C., Aflatoxin contamination of US corn, *Nature*, 335, 757, 1988.

35. Hawkins, L.K., Windham, G.L., and Williams, W.P., Effect of different postharvest drying temperature on *Aspergillus flavus* survival and aflatoxin content in five maize hybrids, *J. Food Prot.*, 68, 1521, 2005.

36. Bailey, G.S. et al., Effect of ammoniation of aflatoxin B1 contaminated cottonseed feedstock on the aflatoxin M1 content of cows milk and hepatocarcinogenicity in the trout assay, *Food Chem. Toxicol.*, 32, 707, 1994.

37. Martinez, A.J., Weng, C.Y., and Park, D.L., Distribution of ammonia/aflatoxin reaction products in corn following exposure to ammonia decontamination procedure, *Food Addit. Contam.*, 11, 659, 1994.

38. Weng, C.Y., Martinez, A.J., and Park, D.L., Efficacy and permanency of ammonia treatment in reducing aflatoxin levels in corn, *Food Addit. Contam.*, 11, 649, 1994.

39. Guengerich, F.P. et al., Activation and detoxification of aflatoxin B1, *Mutat. Res.*, 402, 121, 1998.

40. Gregory, J.F., Goldstein, S.L., and Edds, G.T., Metabolite distribution and rate of residue clearance in turkeys fed a diet containing aflatoxin B1, *Food Chem. Toxicol.*, 21, 463, 1983

41. Kumagai, S., Intestinal absorption and excretion of aflatoxin in rats, *Toxicol. Appl. Pharmacol.*, 97, 88, 1989.

42. Paterson, D.S.P., Metabolism of aflatoxin and other mycotoxins in relation to their toxicity and the accumulation of residues in animal tissues, *Pure Appl. Chem.*, 49, 1723, 1977.

43. Richard, J.L. et al., Distribution and clearance of aflatoxins B1 and M1 in turkeys fed diets containing 50 or 150 ppb aflatoxin from naturally contaminated corn, *Avian Dis.*, 30, 788, 1986.

44. Bintvihok, A. et al., Residues of aflatoxins in the liver, muscle and eggs of domestic fowls, *J. Vet. Med. Sci.*, 64, 1037, 2002.

45. Qureshi, M.A. et al., Dietary exposure of broiler breeders to aflatoxin results in immune dysfunction in progeny chicks, *Poult. Sci.*, 77, 812, 1998.

46. Trucksess, M.W. et al., Aflatoxicol and aflatoxins B1 and M1 in eggs and tissues of laying hens consuming aflatoxin contamined feed, *Poult. Sci.*, 62, 2176, 1983.

47. Zaghini, A. et al., Mannanoligosillicate sorbent, B1 in feed for laying hens: effects on eggs quality, aflatoxins B1 and M1 residues in eggs, and aflatoxin B1 levels in liver, *Poult. Sci.*, 84, 825, 2005.

48. Madden, U.A. and Stahr, H.M., Effect of soil on aflatoxin tissue retention in chicks added to afla-toxin-contaminated poultry rations, *Vet. Hum. Toxicol.*, 34, 521, 1992.

49. Truckess, M.W. et al., Aflatoxicol and aflatoxins B1 and M1 in the tissues of pigs receiving aflatoxin, *J. Assoc. Off. Anal. Chem.*, 65, 884, 1982.

50. Miller, D.M. et al., High performance liquid chromatographic determination and clearance time of aflatoxin residues in swine tissues, *J. Assoc. Off. Anal. Chem.*, 65, 1, 1982.

51. Beaver, R.W. et al., Distribution of aflatoxins in tissues of growing pigs fed an aflatoxin-contaminated diet amended with a high affinity aluminosillicate sorbent, *Vet. Hum. Toxicol.*, 32, 16, 1990.

52. Stubblefield, R.D., Honstead, J.P., and Shotwell, O.L., An analytical survey of aflatoxins in tissues from swine grown in regions reporting 1988 aflatoxin contaminated corn, *J. Assoc. Off. Anal. Chem.*, 74, 897, 1991.

53. Hirano, K. et al., An improved method for extraction and cleanup of aflatoxin B1 from liver, *J. Vet. Med. Sci.*, 54, 567, 1992.

54. Kiermeier, F., Aflatoxin residues in fluid milk, *Pure Appl. Chem.*, 35, 271, 1973.

55. Viroben, G., Fremy, J.M., and Delort-Laval, J., Traitement à froid des tourteaux d'arachide par une solution aqueuse d'ammoniaque: conséquence sur la réduction de la teneur en Aflatoxine M1 du lait, *Le Lait*, 63, 171, 1983.

56. Fremy, J.M. and Quillardet, P., The "carry-over" of Aflatoxin into milk of cows fed ammoniated rations: use of an HPLC method and a genotoxicity test for determining milk safety, *Food Addit. Contam.*, 2(3), 201, 1985.

57. Frobish, R.A. et al., Aflatoxin residues in milk of dairy cows after ingestion of naturally contaminated grain, *J. Food Protect.*, 49, 781, 1986.

58. Veldman, A. et al., Carry-over of aflatoxin from cows' food to milk, *Animal-Production*, 55, 163, 1992.

59. Whitlow, L.W. et al., Mycotoxins and milk safety: the potential to block transfer to milk. Biotech-nology in the feed industry. *Proceedings of Alltech's 16th annual Symposium: the Future of Food*, 391, 2000.

60. Battacone, G. et al., Excretion of aflatoxin M1 in milk of dairy ewes treated with different doses of aflatoxin B1, *J. Dairy Sci.*, 86, 2667, 2003.

61. Battacone, G. et al., Transfer of aflatoxin B1 from feed to milk and from milk to curd and whey in dairy sheep fed artificially contaminated concentrates, *J. Dairy Sci.*, 88, 3063, 2005.

62. Stubblefield, R.D. et al., Fate of aflatoxins in tissues, fluids and excrements from cows dosed orally with aflatoxin B1, *Am. J. Vet. Res.*, 44, 1750, 1983.

63. Shreeve, B.J., Patterson, D.S.P., and Roberts, B.A., The carry over of aflatoxin, ochratoxin and zeara-lenone from naturally contaminated feed to tissues, urine, and milk of dairy cows, *Food Cosmet. Toxicol.*, 17, 151, 1979.

64. Newberne, P.M. and Butler, W.H., Acute and chronic effect of aflatoxin B1 on the liver of domestic animals: a review, *Cancer Res.*, 29, 236, 1969.

65. Peers, F.G. and Linsell, M.P., Dietary aflatoxins and human liver cancer—a population study based in Kenya, *Br. J. Cancer*, 27, 473, 1973.

66. Shank, R.C. et al., Dietary aflatoxin and human liver cancer IV. Incidence of primary liver cancer in two municipal population in Thailand, *Food Cosmetol. Toxicol.*, 10, 171, 1982.

67. JECFA Evaluation of certain food additives and contaminants. Forty-nine report. WHO Technical Report Series, Geneva, 40, 1999.

68. Cullen, J.M. and Newberne, P.M., Acute hepatotoxicity of aflatoxins, in *The Toxicology of Aflatoxins*, Eaton, D.L. and Groopman, J.D., Eds., Academic Press, San Diego, CA, 3, 1994.

69. Eaton, D.L. and Ramsdel, H.S., Species and related differences in aflatoxin biotransformation, in *Handbook of Applied Mycology, Vol 5: Mycotoxins in Ecological Systems*, Bhatnagar, D., Lillehoj, E.B., and Arora, D.K., Eds., Marcel Dekker, New York, 1992, 157.

70. Gallagher, E.P. and Eaton, D.L., In vitro biotransformation of aflatoxin B1 (AFB1) in channel catfish liver, *Toxicol. Appl. Pharmacol.*, 132, 82, 1995.

71. JECFA, Aflatoxin M1. Fifty-six report. WHO Technical Report Series, Geneva, 47, 2001.

72. Trischer, A.M., Human health risk assessment of processing-related compounds in food, *Toxicol. Lett.*, 149, 177, 2004.

73. Asao, T. et al., Structures of aflatoxins B$_1$ and G$_1$, *J. Am. Chem. Soc.*, 87, 822, 1965.

74. Van Egmond, H.P., Heisterkamp, S.H., and Paulsch, W.E., EC collaborative study on the determination of aflatoxin B1 in animal feeding stuffs, *Food Addit. Contam.*, 8, 17, 1991.

75. Otta, K.H. et al., Determination of aflatoxins in corn by use the personal OPLC basic system, *J. Planar Chromatogr.*, 11, 370, 1998.

76. Papp, E. et al., Liquid chromatographic determination of aflatoxins, *Microchem. J.*, 73, 39, 2002.

77. Vega, V.A., Rapid extraction of aflatoxin from creamy and crunchy peanut butter, *J. Assoc. Anal. Chem.*, 88, 1383, 2005.

78. Truckess, M.W., Brumley, W.C., and Nesheim, S., Rapid quantification and confirmation of aflatoxins in corn and peanut butter, using a disposable silica gel column, thin layer chromatography and gas chromatography/mass spectrometry, *J. Assoc. Off. Anal. Chem.*, 67, 973, 1984.

79. AOAC *Official Methods of Analysis*, 17th edition, 49, 20, 1995.

80. Truckess, M.W. et al., Multifunctional column coupled with liquid chromatography for determination of aflatoxins B1, B2, G1 and G2 in corn, almonds, brazil nuts, peanuts and pistachio nuts: collaborative study, *J. Assoc. Off. Anal. Chem.*, 77, 1512, 1994.

81. Otta, K.H., Papp, E., and Bagocsi, B., Determination of aflatoxins in foods by overpressured-layer chromatography, *J. Chromatogr. A.*, 882, 11, 2000.

82. Takeda, Y. et al., Simultaneous extraction and fractionation and thin layer chromatographic determination of 14 mycotoxins, *J. Assoc. Off. Anal. Chem.*, 62, 573, 1979.

83. Van Egmond, H.P., Paulsch, W.E., and Sizoo, E.A., Comparison of six methods of analysis for the determination of aflatoxin B1 in feeding stuffs containing citrus pulp, *Food Addit. Contam.*, 5, 321, 1988.

84. Truckess, M.W. et al., Immunoaffinity column coupled with solution fluorimetry or liquid chromatography postcolumn derivatization for determination of aflatoxins in corn, peanuts and peanut butter: collaborative study, *J. Assoc. Anal. Chem.*, 74, 81, 1991.

85. Grosso, F. et al., Joint IDF-IUPAC-IAEA (FAO) interlaboratory validation for determining aflatoxin M1 in milk by using immunoaffinity clean up before thin layer chromatography, *Food Addit. Contam.*, 21, 348, 2004.

86. AOAC. 990.33. *Aflatoxin Determination in Corn and Peanut Butter*, 1995.

87. ISO. 14718. *Aflatoxin B1 Determination in Compound Feeds with HPLC*, 1998.

88. AOAC. 2003.2. *Determination of Aflatoxin in Animal Feed by IAC Clean Up and HPLC with Post Column Derivatization*, 2003.

89. AOAC. 991.31. *Aflatoxins Determination in Corn, Raw Peanut and Peanut Butter*, 1994.

90. AOAC. 999.07. *Aflatoxin B1 and Total Aflatoxins in Peanut Butter, Pistachio Paste, Fig Paste and Paprika Powder*, 1999.

91. AOAC. 2000.16. *Aflatoxin B1 in Baby Food*, 2000.

92. AOAC. 2000.08. *Aflatoxin M1 in Milk*, 2000.

93. ISO. 14501. *Milk and Powder Milk. Determination of Aflatoxin M1. IAC Purification and HPLC Determination*, 1998.

94. ISO. 14674. *Milk and Milk Powder. Aflatoxin M1 Determination. Clean Up with Immunoaffinity Chromatography and Determination by Thin Layer Chromatography*, 2005.

95. Park, D.L. et al., Solvent efficient thin layer chromatographic method for the determination of aflatoxins B1, B2, G1 and G2 in corn and peanut products: collaborative study, *J. Assoc. Anal. Chem.*, 77, 637, 1994.

96. AOAC. 980.21. *Aflatoxin M1 in Milk and Cheese*, 1990.

97. International Organisation for Standardization. ISO 14675:2005 Norm. *Milk and Milk Powder: Determination of Aflatoxin M1 Content: Clean-up by Immunoaffinity Chromatography and Determination by Thin Layer Chromatography*, 2005.

98. International Organisation for Standardization. ISO 6651:2001 Norm. *Animal Feeding Stuffs: Semiquantitative Determination of Aflatoxin B1: Thin Layer Chromatographic Methods*, 2001.

99. Park, D.L. et al., Liquid chromatographic method for the determination of aflatoxins B1, B2, G1 and G2 in corn and peanut products: collaborative study, *J. Assoc. Anal. Chem.*, 73, 260, 1990.

100. AOAC. 990.33. *Aflatoxins Determination in Corn and Peanut Butter*, 1995.

101. International Organisation for Standardization. ISO 14718:1998 norm. *Animal Feeding Stuffs: Determination of Aflatoxin B1 Content of Mixed Feeding Stuffs: Method using High Performance Liquid Chromatography*, 1998.

102. International Organisation for Standardization. ISO/FDIS 14501 norm. *Milk and Milk Powder: Determination of Aflatoxin M1 Content: Clean Up by Immunoaffinity Chromatography and Determination by High Performance Liquid Chromatography*, 2007.

103. Truckess, M.W. et al., Enzyme linked immunosorbent assay of aflatoxins B1, B2, G1 and G2 in corn, cottonseed, peanuts, peanut butter and poultry feed: collaborative study, *J. Assoc. Anal. Chem.*, 72, 957, 1989.

104. Zheng, Z. et al., Validation of an ELISA test kit for the detection of total aflatoxins in grain and grain products by comparison with HPLC, *Mycopathologia*, 159, 255, 2005.

105. Chu, F.S. et al., Improved enzyme linked immunosorbent assay for aflatoxin B1 in agricultural commodities, *J. Assoc. Anal. Chem.*, 70, 854, 1987.

106. Chu, F.S. et al., Evaluation of enzyme linked immunosorbent assay of clean up for thin layer chromatography of aflatoxin B1 in corn, peanuts and peanut butter, *J. Assoc. Anal. Chem.*, 71, 953, 1988.

107. Park, D.L. et al., Enzyme linked immunosorbent assay for screening aflatoxin B1 in cottonseed products and mixed feed: collaborative study, *J. Assoc. Anal. Chem.*, 72, 326, 1989.

108. Kolosova, A.Y. et al., Direct competitive ELISA based on monoclonal antibody for detection of aflatoxin B1; stabilisation of ELISA kit component and application to grain samples, *Anal. Bioanal. Chem.*, 384, 286, 2006.

109. Thirumala-Devi, T. et al., Development and application of an indirect competitive enzyme-linked immunoassay for aflatoxin m(1) in milk and milk-based confectionery, *J. Agric. Food Chem.*, 50, 933, 2002.

110. Fremy, J.M. and Chu, F.S., Immunochemical methods of analysis for aflatoxin M1, in *Mycotoxin in Dairy Products*, Van Egmond, E., Ed., Elsevier Science, London, 1989, 97.

111. Fremy, J.M. and Usleber, E., Policy on the characterization of antibodies used in immunochemical methods of analysis for mycotoxins and phycotoxins, *J. AOAC Int.*, 86(4), 868, 2003.

112. Zölner, P. and Mayer-Helm, B., Trace mycotoxin analysis in complex biological and food matrices by liquid chromatography–atmospheric pressure ionisation mass spectrometry, *J. Chromatogr. A*, 1136, 123, 2006.

113. Blesa, J. et al., Determination of aflatoxins in peanuts by matrix solid hase dispersion and liquid chromatography, *J. Chromatogr. A*, 1011, 49, 2003.

114. Sorensen, L.K. and Elbaek, T.H., Determination of mycotoxins in bovine milk by liquid chromatography tandem mass spectrometry, *J. Chromatogr. B*, 820, 183, 2005.

115. Kokkonen, M., Jestoi, M., and Rizzo, A., Determination of selected mycotoxins in mould cheeses with liquid chromatography coupled tandem with mass spectrometry, *Food Addit. Contam.*, 22, 449, 2005.

116. Takino, M. et al., Atmospheric pressure photoionisation liquid chromatography/mass spectrometric determination of aflatoxins in food, *Food Addit. Contam.*, 21, 76, 2004.

117. Cavaliere, C. et al., Liquid chromatography/mass spectrometric confirmatory method for determining aflatoxin M1 in cow milk: comparison between electrospray and atmospheric pressure photoionization sources, *J. Chromatogr. A*, 1101, 69, 2006.

118. Stubblefield, R.D. and Shotwell, O.L., Determination of aflatoxins in animal tissues, *J. Assoc. Off. Anal. Chem.*, 64, 964, 1981.

119. Brown, S. et al., Method for the determination of aflatoxin in animal tissue, *J. AOAC Int.*, 56, 1437, 1973.

120. Murthy, T.R.K. et al., Aflatoxin residues in tissues of growing swine: effect of separate and mixed feeding of protein and protein free portion of the diet, *J. Anim. Sci.*, 41, 1339, 1975.

121. Gregory, J.F. and Manley, D.B., High performance liquid chromatographic determination of aflatoxins in animal tissues and products, *J. Assoc. Off. Anal. Chem.*, 64, 144, 1981.

122. Gathumbi, J.K. et al., Application of immunoaffinity chromatography and enzyme immunoassay in rapid detection of aflatoxin B1 in chicken liver tissues, *Poult. Sci.*, 82, 585, 2003.

123. Tavca-Kalcher, G. et al., Validation of the procedure for the determination of aflatoxin B1 in animal liver using immunoaffinity columns and liquid chromatography with postcolumn derivatisation and fluorescence detection, *Food Control*, 18, 333, 2007.

124. Van der Merwe, K.J. et al., Ochratoxin A, a toxic metabolite produced by *Aspergillus ochraceus* Wilh, *Nature*, 205, 1112, 1965.

125. Belli, N. et al., *Aspergillus carbonarius* growth and ochratoxin A production on a synthetic grape medium in relation to environmental factors, *J. Appl. Microbiol.*, 98, 839, 2005.

126. Abarca, M.L. et al., *Aspergillus carbonarius* as the main source of ochratoxin A contamination in dried vine fruits from their Spanish market, *J. Food Prot.*, 66, 504, 2003.

127. Bayman, P. et al., Ochratoxin production by the *Aspergillus ochraceus* group and *Aspergillus alliaceus*, *Appl. Environ. Microbiol.*, 68, 2326, 2002.

128. Abarca, M.L. et al., Ochratoxin A production by strains of *Aspergillus niger* var niger, *Appl. Environ. Microbiol.*, 60, 2650, 1994.

129. Teren, J. et al., Immunochemical detection of ochratoxin A in black *Aspergillus* strains, *Mycopathologia*, 134, 171, 1996.

130. Romero, S.M. et al., Toxigenic fungi isolated from dried vine fruits in Argentina, *Int. J. Food Microbiol.*, 104, 43, 2005.

131. Hajjaji, A. et al., Occurrence of mycotoxins (ochratoxin A, deoxynivalenol) and toxigenic fungi in Moroccan wheat grains: impact of ecological factors on the growth and ochratoxin A production, *Mol. Nutr. Food Res.*, 50, 494, 2006.

132. Pitt, J.I., *Penicillium viridicatum*, *Penicillium verrucosum*, and production of ochratoxin A, *Appl. Environ. Microbiol.*, 53, 266, 1987.

133. Pardo, E. et al., Ecophysiology of ochratoxigenic *Aspergillus ochraceus* and *Penicillium verrucosum* isolates; predictive models for fungal spoilage prevention: a review, *Food Addit. Contam.*, 23, 398, 2006.

134. Pitt, J.I. and Hocking, A.D., Influence of solute and hydrogen ion concentration on the water relations of some xerophilic fungi, *J. Gen. Microbiol.*, 101, 35, 1977.

135. Pardo, E. et al., Prediction of fungal growth and ochratoxin A production by *Aspergillus ochraceus* on irradiated barley grain as influenced by temperature and water activity, *Int. J. Food Microbiol.*, 95, 79, 2004.

136. Pardo, E. et al., Effect of water activity and temperature on mycelial growth and ochratoxin A production by isolates of *Aspergillus ochraceus* on irradiated green coffee beans, *J. Food Prot.*, 68, 133, 2005.

137. Magan, N. and Aldred, D., Conditions of formation of ochratoxin A in drying, transport, and in different commodities, *Food Addit. Contam.*, 22, 10, 2005.

138. Sangare-Tigori, B. et al., Preliminary survey of ochratoxin A in millet maize, rice and peanuts in Cote d'Ivoire from 1998 to 2002, *Hum. Exp. Toxicol.*, 25, 211, 2006.

139. Jorgensen, K., Occurrence of ochratoxin A in commodities and processed food: a review of EU occurrence data, *Food Addit. Contam.*, 22, 26, 2005.

140. Battilani, P. et al., Black Aspergilli and ochratoxin A in grapes in Italy, *Int. J. Food Microbiol.*, 111, S53, 2006.

141. Battilani, P., Magan, N., and Logrieco, A., European research on ochratoxin A in grapes and wine, *Int. J. Food Microbiol.*, 111, S2, 2006.

142. Taniwaki, M.H., An update on ochratoxigenic fungi and ochratoxin A in coffee, *Adv. Exp. Med. Biol.*, 571, 189, 2006.

143. Cairns-Fuller, V., Aldred, D., and Magan, N., Water, temperature and gas composition interaction affect growth and ochratoxin A production isolates of *Penicillium verrucusum* on wheat grain, *J. Appl. Microbiol.*, 99, 1215, 2005.

144. MacDonald, S. et al., Survey of ochratoxin A and deoxynivalenol in stored grains from the 1999 harvest in UK, *Food Addit. Contam.*, 21, 172, 2004.

145. Bucheli, P. and Taniwaki, M.H., Research on the origin and on the impact of post harvest handling and manufacturing on the presence of ochratoxin A in coffee, *Food Addit. Contam.*, 19, 655, 2002.

146. JECFA. Safety evaluation of certain mycotoxins in food. Fifty-six report. WHO Technical Report Series, Geneva, 47, 2001.

147. Miraglia, M. and Brera, C., Assessment of dietary intake of Ochratoxin A by the population of EU Member States, 2002.

148. Escher, F.E., Koehler, P.E., and Ayers, J.C., Production of ochratoxins A and B on country cured ham, *Appl. Microbiol.*, 26, 27, 1973.

149. Tabuc, C. et al., Toxigenic potential of fungal mycoflora isolated from dry cured meat products: preliminary study, *Rev. Med. Vet.*, 156, 287, 2004.

150. Bailly, J.D. et al., Production and stability of patulin, ochratoxin A citrinin, and cyclopiazonic acid on dry cured ham, *J. Food Protect.*, 68, 1516, 2005.

151. Ringot, D. et al., Toxicokinetics and toxicodynamics of ochratoxin A an update, *Chem. Biol. Interact.*, 159, 18, 2006.

152. Kumagai, S. and Aibara, K., Intestinal absorption and secretion of ochratoxin A in the rat, *Toxicol. Appl. Pharmacol.*, 64, 94, 1982.

153. Galtier, P. and Alvinerie, M., In vitro transformation of ochratoxin A by animal microbial floras, *Annu. Rech. Vet.*, 7, 91, 1976.

154. Hult, K., Teiling, A., and Gatenberg, S., Degradation of ochratoxin A by a ruminant, *Appl. Environ. Microbiol.*, 32, 443, 1976.

155. Höhler, D. et al., Metabolism and excretion of ochratoxin A fed to sheep, *J. Anim. Sci.*, 77, 1217, 1999.

156. Boudra, H. et al., Aflatoxin M1 and ochratoxin A in raw bulk milk from French dairy herds, *J. Dairy Sci.*, 90, 3197, 2007.

157. Valenta, H. and Goll, M., Determination of ochratoxin A in regional samples of cow milk from Germany, *Food Addit. Contam.*, 13, 669, 1996.

158. Creppy, E.E., Human ochratoxicosis, *J. Toxicol. Toxin Rev.*, 18, 277, 1999.

159. Hult, K. et al., Ochratoxin A in pig blood: method of analysis and use as a tool for feed studies, *Appl. Environ. Microbiol.*, 38, 772, 1979.

160. Jorgensen, K., Survey of pork, poultry, coffee, beer and pulses for ochratoxin A, 1998.

161. Dragacci, S. et al., A french monitoring programme for determining ochratoxin A occurrence in pig kidneys, *Nat. Toxins*, 7, 167, 1999.

162. Terada, H. et al., Liquid chromatographic determination of ochratoxin A in coffee beans and coffee products, *J. Assoc. Off. Anal. Biochem.*, 69, 960, 1986.

163. Pietri, A. et al., Occurrence of ochratoxin A in raw ham muscle and in pork products from northern Italy, *J. Food Sci.*, 18, 1, 2006.

164. SCOOP reports on Tasks 3.2.7. Assessment of dietary intake of Ochratoxin A by the population of EU members states, 2000.

165. Monaci, L., Tantillo, G., and Palmisano, F., Determination of ochratoxin A in pig tissues by liquid extraction and clean up and high performance liquid chromatography, *Anal. Bioanal. Chem.*, 378, 1777, 2004.

166. Krogh, P., Ochratoxin A residues in tissues of slaughter pigs with nephropathy, *Nord Vet. Med.*, 29, 402, 1977.

167. Elling, F., Feeding experiments with ochratoxin A contaminated barley for bacon pigs. 4. Renal lesions, *Acta Agric. Scand.*, 33, 153, 1983.

168. Krogh, P., Porcine nephropathy associated with ochratoxin A, in *Mycotoxins and animal foods*, Smith, J.E. and Anderson, R.S., Eds., CRC Press, Boca Raton, FL, 1991, 627.

169. Plestina, R. et al., Human exposure to ochratoxin A in areas of Yugoslavia with endemic nephropathy, *Environ. Pathol. Toxicol. Oncol.*, 10, 145, 1982.

170. Castegnaro, M. et al., Balkan endemic nephropathy: role of ochratoxin A through biomarkers, *Mol. Nutr. Food Res.*, 50, 519, 2006.

171. Fuchs, R. and Peraica, M., Ochratoxin A in human diseases, *Food Addit. Contam.*, 22, 53, 2005.

172. Abouzied, M.M. et al., Ochratoxin A concentration in food and feed from a region with Balkan endemic nephropathy, *Food Addit. Contam.*, 19, 755, 2002.

173. Vrabcheva, T. et al., Analysis of ochratoxin A in foods consumed by inhabitants from an area with Balkan endemic nephropathy: a 1 month follow up study, *J. Agric. Food Chem.*, 52, 2404, 2004.

174. Marquadt, R.R. and Frohlich, A.A., A review of recent advances in understanding ochratoxicosis, *J. Anim. Sci.*, 70, 3968, 1992.

175. Meisner, H. and Meisner, P., Ochratoxin A, an inhibitor of renal phosphoenolpyruvate carboxylase, *Arch. Biochem. Biophys.*, 208, 146, 1991.

176. Rahimtula, A.D. et al., Lipid peroxidation as a possible cause of ochratoxin A toxicity, *Biochem. Pharmacol.*, 37, 4469, 1988.

177. Bendele, A.M. et al., Ochratoxin A carcinogenesis in the (C57BL/6J X C3H)F1 mouse, *J. Natl. Cancer Inst.*, 75, 733, 1985.

178. Creppy, E.E. et al., Genotoxicity of ochratoxin A in mice: DNA single strand break evaluation in spleen, live rand kidney, *Toxicol. Lett.*, 28, 29, 1985.

179. Pfohl-Leszkowicz, A. et al., DNA adduct formation in mice treated with ochratoxin A, *IARC Sci. Publ.*, 115, 245, 1991.

180. Pfohl-Leszkowicz, A. et al., Differential DNA adduct formation and disappearance in three mouse tissues after treatment with the mycotoxin ochratoxin A, *Mutat. Res.*, 289, 265, 1993.

181. Obrecht-Pflumio, S. and Dirheimer, G., Horseradish peroxidase mediates DNA and deoxyguanosine 3-monophosphate adduct formation in the presence of ochratoxin A, *Arch. Toxicol.*, 75, 583, 2001.

182. JECFA. Ochratoxin A. WHO food additive, Geneva, 35, 1990.

183. Conseil Superieur d'Hygiène Publique de France (CSHPF), Les mycotoxines dans l'alimentation: évaluation et gestion du risque, éds, Tec&Doc, 1999.

184. Battaglia, R. et al., Fate of ochratoxin A during breadmaking, *Food Addit. Contam.*, 13, 25, 1996.

185. AOAC. 973.37. *Official Methods of Analysis, Ochratoxins in Barley*. Part 2, 1207, 1990.

186. Zimmerli, B. and Dick, R., Determination of ochratoxin A at the ppt level in human blood, serum, milk and some foodstuffs by high-performance liquid chromatography with enhanced fluorescence detection and immunoaffinity column cleanup: methodology and Swiss data, *J. Chromatogr. B*, 666, 85, 1995.

187. Entwisle, A.C. et al., Liquid chromatographic method with immunoaffinity column cleanup for determination of ochratoxin A in barley: collaborative study, *J. AOAC Int.*, 83, 1377, 2000.

188. Sharman, M., MacDonald, S., and Gilbert, J., Automated liquid chromatographic determination of ochratoxin A in cereals and in animal products using immunoaffinity column clean up, *J. Chromatogr.*, 603, 285, 1992.

189. Burdaspal, P., Determination of Ochratoxin A in Baby Food, CEN/TC 275/WG5, N 219, 1977.

190. Baggiani, C., Giraudi, G., and Vanni, A., A molecular imprinted polymer with recognition properties towards the carcinogenic mycotoxin ochratoxin A, *Bioseparation*, 10, 389, 2001.

191. Jodibauer, J., Maier, N.M., and Lindner, W., Towards ochratoxin A selective molecularly imprinted polymers for solid phase extraction, *J. Chromatogr. A*, 945, 45, 2002.

192. Le Tutour, B., Tantaoui-Elaraki, A., and Aboussalim, A., Simultaneous thin layer chromatographic determination of aflatoxin B1 and ochratoxin A in black olives, *J. Assoc. Anal. Chem.*, 67, 611, 1984.
193. Pittet, A. and Royer, D., Rapid low cost thin layer chromatographic screening method for the detection of ochratoxin A in green coffee at a control level of 10 μg/kg, *J. Agric. Food Chem.*, 50, 243, 2002.
194. Santos, E.A. and Vargas, E.A., Immunoaffinity column clean up and thin layer chromatography for determination of ochratoxin A in green coffee, *Food Addit. Contam.*, 19, 447, 2002.
195. Betina, V., *Chromatography of Mycotoxins*, Elsevier, Amsterdam, 1993.
196. Ventura, M. et al., Two dimensional thin layer chromatographic method for the analysis of ochratoxin A in green coffee, *J. Food Prot.*, 68, 1920, 2005.
197. Scott, P.M., Methods of analysis for ochratoxin A, *Adv. Exp. Med. Biol.*, 504, 117, 2002.
198. Levi, C.P., Collaborative study of a method for the determination of ochratoxin A in green coffee, *J. Assoc. Off. Anal. Chem.*, 58, 258, 1975.
199. Tangni, E.K. et al., Ochratoxin A in domestic and imported beers in Belgium: occurrence and exposure assessment, *Food Addit. Contam.*, 19, 1169, 2002.
200. Markaki, P. et al., Determination of ochratoxin A in red wine and vinegar by immunoaffinity high pressure liquid chromatography, *J. Food Prot.*, 64, 533, 2001.
201. Breitholtz, A. et al., Plasma ochratoxin A levels in three Swedish populations surveyed using an ion-paired HPLC, *Food Addit. Contam.*, 8, 183, 1991.
202. Breitholtz-Emanuelsson, A. et al., Ochratoxin A in cow's milk and in human milk with corresponding human blood samples, *J. AOAC Int.*, 76, 842, 1993.
203. Larsson, K. and Möller, T., Liquid chromatographic determination of ochratoxin A in Barley, wheat bran and rye by the AOAC/IUPC/NMKL method: NMKL—collaborative study, *J. AOAC Int.*, 79(5), 1102, 1996.
204. EN-ISO 15141-1, *Dosage of Ochratoxin A in Cereals and Derived Products: HPLC Method*, 1998.
205. EN 14132, *Dosage of Ochratoxin A in Barley and Roasted Coffee: Method with IAC Purification and HPLC Analysis*, 2003.
206. AOAC. 2000.09. *Ochratoxin A Determination in Roasted Coffee*, 2000.
207. AOAC, 2000.03. *Ochratoxin A Determination in Barley*, 2000.
208. EN 14133, *Dosage of Ochratoxin A in Wine and Beer. HPLC Method after IAC Purification*, 2004.
209. Kuhn, I., Valenta, H., and Rohr, K., Determination of ochratoxin A in bile of swine by high performance liquid chromatography, *J. Chromatogr. B*, 668, 333, 1995.
210. Rousseau, D.M. et al., Detection of ochratoxin A in porcine kidneys by a monoclonal antibody based radioimmunoassay, *Appl. Environ. Microbiol.*, 53, 514, 1987.
211. Solti, L. et al., Ochratoxin A content of human sera determined by sensitive ELISA, *J. Anal. Toxicol.*, 21, 44, 1997.
212. Chiavaro, E. et al., Ochratoxin A determination in ham by immunoaffinity clean up and a quick fluorometric method, *Food Addit. Contam.*, 19, 575, 2002.
213. Monaci, L. et al., Determination of ochratoxin A at part per trillion level in Italian salami by immunoaffinity clean up and high performance liquid chromatography with fluorescence detection, *J. Chromatogr. A.*, 1090, 184, 2005.
214. Toscani, T. et al., Determination of ochratoxin A in dry cured meat products by a HPLC-FLD quantitative method, *J. Chromatogr. B*, 855, 242, 2007.
215. Jimenez, A.M. et al., Determination of ochratoxin A in pig derived pates by high performance liquid chromatography, *Food Addit. Contam.*, 18, 559, 2001.
216. Losito, I. et al., Determination of ochratoxin A in meat products by high performance liquid chromatography coupled to electrospray ionisation sequential mass spectrometry, *Rapid Commun. Mass Spectrom.*, 18, 1965, 2004.
217. Clarke, J.R. et al., Quantification of ochratoxin A in swine kidneys by enzyme linked immunosorbent assay using a simplified sample preparation procedure, *J. Food Prot.*, 57, 991, 1994.

218. Stopper, H., Schmitt, E., and Kobras, K., Genotoxicity of phytoestrogens, *Mutat. Res.*, 574, 139, 2005.
219. Stob, M. et al., Isolation of an anabolic, uterotrophic compound from corn infected with *Gibberella zeae*, *Nature*, 29, 196, 1962.
220. Molto, G.A. et al., Production of trichothecenes and zearalenone by isolates of *Fusarium* spp. From Argentinian maize, *Food Addit. Contam.*, 14, 263, 1997.
221. Sydenham, E.W. et al., Production of mycotoxins by selected *Fusarium graminearum* and *F. crookwellense* isolates, *Food Addit. Contam.*, 8, 31, 1991.
222. Montani, M.L. et al., Influence of water activity and temperature on the accumulation of zearalenone in corn, *Int. J. Food Microbiol.*, 6, 1, 1988.
223. Jimenez, M., Manez, M., and Hernandez, E., Influence of water activity and temperature on the production of zearalenone in corn by three *Fusarium* species, *Int. J. Food Microbiol.*, 29, 417, 1996.
224. Llorens, A., Influence of the interactions among ecological variables in the characterization of zearalenone producing isolates of *Fusarium* spp., *Syst. Appl. Microbiol.*, 27, 253, 2004.
225. Ryu, D. and Bullerman, L.B., Effect of cycling temperatures on the population of deoxynivalenol and zearalenone by *Fusarium graminearum* NRRL 5883, *J. Food Prot.*, 62, 1451, 1999.
226. Milano, G.D. and Lopez, T.A., Influence of temperature on zearalenone production by regional strains of *Fusarium graminearum* and *Fusarium oxysporum* in culture, *Int. J. Food Microbiol.*, 13, 329, 1991.
227. Zinedine, A. et al., Review on the toxicity, occurrence, metabolism, detoxification, regulations and intake of zearalenone: an oestrogenic mycotoxin, *Food Chem. Toxicol.*, 45, 1, 2007.
228. Dailey, R.E., Reese, R.E., and Brouwer, E.A., Metabolism of (14C)zearalenone in laying hens, *J. Agric. Food Chem.*, 28, 286, 1980.
229. Olsen, M. et al., Plasma and urinary levels of zearalenone and alpha-zearalenol in a prepubertal gilt fed zearalenone, *Acta. Pharmacol. Toxicol. (Copenh.)*, 56, 239, 1985.
230. Kuiper-Goodman, T., Scott, P.M., and Watanabe, H., Risk assessment of the mycotoxin zearalenone, *Regul. Toxicol. Pharmacol.*, 7, 253, 1987.
231. Mirocha, C.J., Pathre, S.V., and Robinson, T.S., Comparative metabolism of zearalenone and transmission into bovine milk, *Food Cosmet. Toxicol.*, 19, 25, 1981.
232. Biehl, M.L. et al., Biliary excretion and enterohepatic cycling of zearalenone in immature pigs, *Toxicol. Appl. Pharmacol.*, 121, 152, 1993.
233. Kollarczik, B., Garels, M., and Hanelt, M., In vitro transformation of the Fusarium mycotoxins deoxynivalenol and zearalenone by the normal gut microflora of pigs, *Nat. Toxins*, 2, 105, 1994.
234. Olsen, M. et al., Metabolism of zearalenone by sow intestinal mucosa in vitro, *Food Chem. Toxicol.*, 25, 681, 1987.
235. Kallela, K. and Vasenius, L., The effects of ruman fluid on the content of zearalenone in animal fodder, *Nord. Vet. Med.*, 34, 336, 1982.
236. Gaumy, J.L. et al., Zearalénone: origine et effets chez les animaux d'élevage, *Revue Med. Vét.*, 152, 123, 2001.
237. Olsen, M., Petterson, H., and Kiessling, K.H., Reduction of zearalenone in female rat liver by 3 alpha-hydroxysteroid dehydrogenase, *Acta Pharmacol. Toxicol. (Copenh.)*, 48, 157, 1981.
238. Kiessling, K.H. and Petterson, H., Metabolism of zearalenone in rat liver, *Acta. Pharmacol. Toxicol. (Copenh.)*, 43, 285, 1978.
239. Malekinejad, H., Colenbrander, B., and Fink-Gremmels, J., Hydroxysteroid dehydrogenases in bovine and porcine granulosa cells convert zearalenone into its hydroxylated metabolites alpha-zearalenol and beta-zearalenol, *Vet. Res. Commun.*, 30, 445, 2006.
240. JECFA, 53rd Report. Safety evaluation of certain food additives. *WHO Food Additives Series* 44, 2000.
241. Dänicke, S. et al., Effects of Fusarium toxin contaminated wheat and of a detoxifying agent on performance of growing bulls, on nutrient digestibility in wethers and on the carry over of zearalenone, *Arch. Tierernahr.*, 56, 245, 2002.

242. Sundlof, S.F. and Strickland, C., Zearalenone and zeranol: potential residue problems in livestock, *Vet. Hum. Toxicol.*, 28, 242, 1986.

243. Baldwin, R.S., Williams, R.D., and Terry, M.K., Zeranol: a review of the metabolism, toxicology, and analytical methods for detection of tissue residues, *Regul. Toxicol. Pharmacol.*, 3, 9, 1983.

244. Goyarts, T. et al., Carry-over of *Fusarium* toxins (deoxynivalenol and zearalenone) from naturally contaminated wheat to pigs, *Food Addit. Contam.*, 24, 369, 2007.

245. Mirocha, C.J. et al., Distribution and residue determination of (3H)zearalenone in broilers, *Toxicol. Appl. Pharmacol.*, 66, 77, 1982.

246. Dänicke, S. et al., Effect of addition of a detoxifying agent to laying hen diets containing uncontaminated or *Fusarium* toxin-contaminated maize on performance of hens and on carryover of zearalenone, *Poult. Sci.*, 81, 1671, 2002.

247. Takemura, H. et al., Characterization of the estrogenic activities of zearalenone and zeranol in vivo and in vitro, *J. Steroid Biochem. Mol. Biol.*, 103, 170, 2007.

248. Marin, M.L. et al., Effects of mycotoxins on cytokine production and proliferation in EL-4 thymona cells, *J. Toxicol. Environ. Health*, 48, 379, 1996.

249. Berek, L. et al., Effects of mycotoxins on human immune functions in vitro, *Toxicol. In Vitro*, 15, 25, 2001.

250. European Union, règlement no 856/2005, Les toxines du Fusarium, *Off. J. Eur. Union*, L 143, 3, 2005.

251. European Union, Recommendation (576/2005), La présence de déoxynivalénol, de zéaralénone, d'ochratoxine A, des toxines T-2 et HT-2 et des fumonisines dans les produits destinés à l'alimentation animale, *Off. J. Eur. Union*, L229, 7, 2006.

252. Hidy, P.H. et al., Zearalenone and some derivates: production and biological activities, *Adv. Appl. Microbiol.*, 22, 59, 1977.

253. Gaumy, J.L. et al., Zéaralénone: propriétés et toxicité expérimentale, *Revue Med. Vét.*, 152, 219, 2001.

254. Krska, R., Performance of modern sample preparation techniques in the analysis of *Fusarium* mycotoxins in cereals, *J. Chromatogr. A*, 815, 49, 1998.

255. Krska, R. and Josephs, R., The state-of-the-art in the analysis of estrogenic mycotoxins in cereals, *Fresenius J. Anal Chem.*, 369, 469, 2001.

256. Pallaroni, L. et al., Microwave-assisted extraction of zearalenone from wheat and corn, *Anal. Bioanal. Chem.*, 374, 161, 2002.

257. Pallaroni, L. and Von Holst, C., Comparison of alternative and conventional extraction techniques for the determination of zearalenone in corn, *Anal. Bioanal. Chem.*, 376, 908, 2003.

258. Zöllner, P. et al., Concentration levels of zearalenone and its metabolites in urine, muscle tissue, and liver samples of pigs fed with mycotoxin-contaminated oats, *J. Agric. Food Chem.*, 50, 2494, 2002.

259. Plasencia, J. and Mirocha, C.J., Isolation and characterization of zearalenone sulfate produced by Fusarium spp., *Appl. Environ. Microbiol.*, 57, 146, 1991.

260. Garels, M. et al., Cleavage of zearalenone-glycoside, a "masked" mycotoxin, during digestion in swine, *Zentralbl Veterinarmed B*, 37, 236, 1990.

261. Llorens, A. et al., Comparison of extraction and clean-up procedures for analysis of zearalenone in corn, rice and wheat grains by high-performance liquid chromatography with photodiode array and fluorescence detection, *Food Addit. Contam.*, 19, 272, 2002.

262. Silva, C.M. and Vargas, E.A., A survey of zearalenone in corn using Romer Mycosep 224 column and high performance liquid chromatography, *Food Addit. Contam.*, 18, 39, 2001.

263. De Saeger, S., Sibanda, L., and Van Peteghem, C., Analysis of zearalenone and α-zearalenol in animal feed using high-performance liquid chromatography, *Anal. Chim. Acta*, 487, 137, 2003.

264. Eskola, M., Kokkonen, M., and Rizzo A., Application of manual and automated systems for purification of ochratoxin A and zearalenone in cereals with immunoaffinity columns, *J. Agric. Food Chem.*, 50, 41, 2002.

265. Fazekas, B. and Tar, A., Determination of zearalenone content in cereals and feedstuffs by immunoaffinity column coupled with liquid chromatography, *J. AOAC Int.*, 5, 1453, 2001.

266. Kruger, S.C. et al., Rapid immunoaffinity-based method for determination of zearalenone in corn by fluorometry and liquid chromatography, *J. AOAC Int.*, 82, 1364, 1999.

267. Zöllner, P., Jodlbauer, J., and Lindner, W., Determination of zearalenone in grains by high-performance liquid chromatography-tandem mass spectrometry after solid-phase extraction with RP-18 columns or immunoaffinity columns, *J. Chromatogr.*, 858, 167, 1999.

268. Visconti, A. and Pascale, M., Determination of zearalenone in corn by means of immunoaffinity clean-up and high-performance liquid chromatography with fluorescence detection, *J. Chromatogr.*, 815, 133, 1998.

269. Rosenberg, E. et al., High-performance liquid chromatography-atmospheric-pressure chemical ionization mass spectrometry as a new tool for the determination of the mycotoxin zearalenone in food and feed, *J. Chromatogr.*, 819, 277, 1998.

270. Meyer, K. et al., Zearalenone metabolites in bovine bile, *Arch. Für Lebensmittelhygiene*, 53, 115, 2002.

271. Pichler, H. et al., An enzyme-immunoassay for the detection of the mycotoxin zearalenone by use of yolk antibodies, *Fresenius' J. Anal. Chem.*, 362, 176, 1998.

272. Josephs, R.D., Schuhmacher, R., and Krska, R., International interlaboratory study fort the *Fusarium* mycotoxins zearalenone and deoxynivalenol in agricultural commodities, *Food Addit. Contam.*, 18, 417, 2001.

273. Lee, M.G. et al., Enzyme-linked immunosorbent assays of zearalenone using polyclonal, monoclonal and recombinant antibodies, *Methods Mol. Biol.*, 157, 159, 2001.

274. Bennet, G.A., Nelsen, T.C., and Miller, B.M., Enzyme-linked immunosorbent assay for detection of zearalenone in corn, wheat, and pig feed: collaborative study, *J. AOAC Int.*, 77, 1500, 1994.

275. Maragos, C.M. and Kim, E.K., Detection of zearalenone and related metabolites by fluorescence polarization immunoassay, *J. Food Protect.*, 67, 1039, 2004.

276. Dawlatana, M. et al., An HPTLC method for the quantitative determination of zearalenone in maize, *Chromatographia*, 47, 217, 1998.

277. De Oliveira Santos Cazenave, S. and Flavio Midio, A., A simplified method for the determination of zearalenone in corn-flour, *Alimentaria*, 298, 27, 1998.

278. Weiss, R. et al., Improving methods of analysis for mycotoxins: molecularly imprinted polymers for deoxynivalenol and zearalenone, *Food Addit. Contam.*, 20, 386, 2003.

279. Ware, G.M. et al., Preparative method for isolating α-zearalenol and zearalenone using extracting disk, *J. AOAC Int.*, 82, 90, 1999.

280. Tanaka, T. et al., Simultaneous determination of trichothecene mycotoxins and zearalenone in cereals by gas chromatography-mass spectrometry, *J. Chromatogr.*, 882, 23, 2000.

281. Ryu, J.C. et al., Survey of natural occurrence of trichothecene mycotoxins and zearalenone in Korean cereals harvested in 1992 using GC-MS, *Food Addit. Contam.*, 13, 333, 1996.

282. Pallaroni, L., Björklund, E., and Von Holst, C., Optimization of atmospheric pressure chemical ionization interface parameters for the simultaneous determination of deoxynivalenol and zearalenone using HPLC/MS, *J. Liquid Chromatogr. Relat. Technol.*, 25, 913, 2002.

283. Jodlbauer, J., Zöllner, P., and Lindner, W., Determination of zeranol, taleranol, zearalenone, α- and β-zearalenol in urine and tissue by high-performance liquid chromatography-tandem mass spectrometry, *Chromatographia*, 51, 681, 2000.

284. Zöllner, P. et al., Determination of zearalenone and its metabolites a- and b-zearalenol in beer samples by high-performance liquid chromatography-tandem mass spectrometry, *J. Chromatogr. B Biomed. Sci. Appl.*, 738, 233, 2000.

285. Kleinova, M. et al., Metabolic profiles of the zearalenone and of the growth promoter zeranol in urine, liver, and muscle of heifers, *J. Agric. Food Chem.*, 50, 4769, 2002.

286. Schuhmacher, R. et al., Interlaboratory comparison study for the determination of the *Fusarium* mycotoxins deoxynivalenol in wheat and zearalenone in maize using different methods, *Fresenius' J. Anal. Chem.*, 359, 510, 1997.

287. Medina, M.B. and Sherman, J.T., High performance liquid chromatography separation of anabolic oestrogens and ultraviolet detection of 17 beta-oestradiol, zeranol, diethylstilboestrol or zearalenone in avian muscle tissue extracts, *Food Addit. Contam.*, 3, 263, 1986.

288. Curtui, V.G. et al., Survey of Romanian slaughtered pigs for the occurrence of mycotoxins ochratoxins A and B, and zearalenone, *Food Addit. Contam.*, 18, 730, 2001.

289. Dänicke, S. et al., On the interactions between *Fusarium* toxin-contaminated wheat and non-starch-polysaccharide hydrolysing enzymes in turkey diets on performance, health and carry-over of deoxynivalenol and zearalenone, *Br. Poult. Sci.*, 48, 39, 2007.

290. Zhang, W. et al., Multiresidue determination of zeranol and related compounds in bovine muscle by gas chromatography/mass spectrometry with immunoaffinity cleanup, *J. AOAC Int.*, 89, 1677, 2006.

291. Roybal, J.E. et al., Determination of zeranol/zearalenone and their metabolites in edible animal tissue by liquid chromatography with electrochemical detection and confirmation by gas chromatography/ mass spectrometry, *J. Assoc. Off. Anal. Chem.*, 71, 263, 1988.

292. Launay, F.M. et al., Confirmatory assay for zeranol, taleranol and the *Fusarium* spp. Toxins in bovine urine using liquid chromatography-tandem mass spectrometry, *Food Addit. Contam.*, 21, 52, 2004.

293. Blokland, M.H. et al., Determination of resorcyclic acid lactones in biological samples by GC-MS. Discrimination between illegal use and contamination with *Fusarium* toxins, *Anal. Bioanal. Chem.*, 384, 1221, 2006.

294. Bean, G.A., Jarvis, B.B., and Aboul-Nasr, M.B., A biological assay for the detection of *Myrothecium* spp. Produced macrocyclic trichothecenes, *Mycopathology*, 119, 175, 1992.

295. Wilkins, K., Nielsen, K.F., and Din, S.U., Patterns of volatile metabolites and non-volatile trichothecenes produced by isolates of *Stachybotrys*, *Fusarium*, *Trichoderma*, *Trichothecium* and *Memnoniella*, *Environ. Sci. Pollut. Res. Int.*, 10, 162, 2003.

296. Schrödter, R., Influence of harvest and storage conditions on trichothecenes levels in various cereals, *Toxicol. Lett.*, 153, 47, 2004.

297. Parry, D.W., Jenkinson, P., and McLeod, L., Fusarium ear blight (scab) is small grain cereals-a review, *Plant Pathol.*, 44, 207, 1995.

298. Logrieco, A. et al., Toxigenic *Fusarium* species and mycotoxins associated with maize ear rot in Europe, *Eur. J. Plant Pathol.*, 108, 597, 2002.

299. Edwards, S.G., Influence of agricultural practices on *Fusarium* infection of cereals and subsequent contamination of grain by trichothecene mycotoxins, *Toxicol. Lett.*, 153, 29, 2004.

300. Obst, A. et al., The risk of toxins by *Fusarium graminearum* in wheat—interactions between weather and agronomic factors, *Mycotoxin Res.*, 16, 16, 2000.

301. Pan, D. et al., Deoxynivalenol in barley samples from Uruguay, *Int. J. Food Microbiol.*, 114, 149, 2007.

302. Trucksess, M.W. et al., Determination and survey of deoxynivalenol in white flour, whole wheat flour, and bran, *J. AOAC Int.*, 79, 883, 1996.

303. Li, F.Q. et al., *Fusarium* toxins in wheat from an area in Henan Province, PR China, with a previous human red mould intoxication episode, *Food Addit. Contam.*, 19, 163, 2002.

304. Tanaka, T. et al., Worldwide contamination of cereals by the *Fusarium* mycotoxins nivalenol, deoxynivalenol, and zearalenone. I. Survey of 19 countries, *J. Agric. Food Chem.*, 36, 979, 1988.

305. Molto, G. et al., Occurrence of trichothecenes in Argentinean beer: a preliminary exposure assessment, *Food Addit. Contam.*, 17, 809, 2000.

306. Wolf-Hall, C.E. and Schwarz, P.B., Mycotoxins and fermentation-beer production, *Adv. Exp. Med. Biol.*, 504, 217, 2002.

307. Papadopoulou-Bouraoui, A. et al., Screening survey of deoxynivalenol in beer from the European market by an enzyme-linked immunosorbent assay, *Food Addit. Contam.*, 21, 607, 2004.

308. Cavret, S. and Lecoeur, S., Fusariotoxin transfer in animal, *Food Chem. Toxicol.*, 44, 444, 2005.

309. Eriksen, G.S. and Petterson, H., Toxicological evaluation of trichothecenes in animal feed, *Anim. Feed Sci. Technol.*, 114, 205, 2004.

310. Yiannikouris, A. and Jouany, J.P., Mycotoxins in feeds and their fate in animals: a review, *Anim. Res.*, 51, 81, 2002.

311. Young, J.C. et al., Degradation of trichothecene mycotoxins by chicken intestinal microbes, *Food Chem. Toxicol.*, 45, 136, 2007.

312. He, P., Young, L.G., and Forsberg, C., Microbial transformation of deoxynivalenol (vomitoxin), *Appl. Environ. Microbiol.*, 58, 3857, 1992.

313. Swanson, S.P. et al., The role of intestinal microflora in the metabolism of trichothecene mycotoxins, *Food Chem. Toxicol.*, 26, 823, 1988.

314. Dänicke, S., Valenta, H., and Döll, S., On the toxicokinetics and the metabolism of deoxynivalenol (DON) in the pig, *Arch. Anim. Nutr.*, 58, 169, 2004.

315. Prelusky, D.B. et al., Pharmacokinetic fate of 14C-labeled deoxynivalenol in swine, *Fundam. Appl. Toxicol.*, 10, 276, 1988.

316. Coppock, R.W. et al., Preliminary study of the pharmacokinetics and toxicopathy of deoxynivalenol (vomitoxin) in swine, *Am. J. Vet. Res.*, 46, 169, 1985.

317. Prelusky, D.B. and Locksley Trenholm, H., Nonaccumulation of residues in swine tissue following extended consumption of deoxynivalenol-contaminated diets, *J. Food Sci.*, 57, 801, 1992.

318. Pollmann, D.S. et al., Deoxynivalenol-contaminated wheat in swine diets, *J. Anim. Sci.*, 60, 239, 1985.

319. Prelusky, D.B. et al., Tissue distribution and excretion of radioactivity following administration of 14C-labeled deoxynivalenol to white leghom hens, *Fundam. Appl. Toxicol.*, 7, 635, 1986.

320. Chi, M.S. et al., Excretion and tissue distribution of radioactivity from tritium-labeled T-2 toxin in chicks, *Toxicol. Appl. Pharmacol.*, 45, 391, 1978.

321. Giroir, L.E., Ivie, G.W., and Huff, W.E., Comparative fate of the tritiated trichothecene mycotoxin, T-2 toxin, in chickens and ducks, *Poult. Sci.*, 70, 1138, 1991.

322. El-Banna, A.A. et al., Nontransmission of deoxynivalenol (vomitoxin) to eggs and meat in chickens fed deoxynivalenol-contaminated diets, *J. Agric. Food Chem.*, 31, 1381, 1983.

323. Rocha, O., Ansari, K., and Doohan, F.M., Effects of trichothecene mycotoxins on eukaryotic cells: a review, *Food Add. Contam.*, 22, 369, 2005.

324. Pestka, J.J. and Smolinski, A.T., Deoxynivalenol: toxicology and potential effects on humans, *J. Toxicol. Environ. Health B Crit. Rev.*, 8, 39, 2005.

325. Rotter, B.A., Prelusky, D.B., and Pestka, J.J., Toxicology of deoxynivalenol (vomitoxin), *J. Toxicol. Environ. Health*, 48, 1, 1996.

326. Pieters, M.N. et al., Risk assessment of deoxynivalenol in food: concentration limits, exposure and effects, *Adv. Exp. Med. Biol.*, 504, 235, 2002.

327. Trenholm, H.L. et al., Lethal toxicity and nonspecific effects, in *Trichothecene mycotoxicosis: Pathophysiologic Effects*, Beasley, V.L., Ed., Vol. 1, CRC Press, Boca Raton, FL, 1989, 107.

328. Joffe, A.Z., *Fusarium poae* and *Fusarium sporotrichioides* as a principal causal agents of alimentary toxic aleukia, in *Mycotoxic Fungi, Mycotoxins, Mycotoxicoses*, Wyllie, T.D. and Morehouse, L.G., Eds., Vol. 3, Marcel Dekker, New York, 1978, 21.

329. Lutsky, I.N. et al., The role of T-2 toxin in experimental alimentary toxic aleukia: a toxicity study in cats, *Toxicol. Appl. Pharmacol.*, 43, 111, 1978.

330. Benett, J.W. and Klich, M., Mycotoxins, *Clin. Microbiol. Rev.*, 16, 497, 2003.

331. Balzer, A. et al., Les trichothécènes: nature des toxines, présence dans les aliments et moyens de lutte, *Revue Med. Vet.*, 155, 299, 2004.

332. Hazel, C.M. and Patel, S., Influence of processing on trichothecene levels, *Toxicol. Lett.*, 153, 51, 2004.

333. Krska, R., Baumgartner, S., and Josephs, R., The state-of-the-art in the analysis of type-A and -B trichothecene mycotoxins in cereals, *Fresenius J. Anal. Chem.*, 371, 285, 2001.

334. AOAC. 986-18. *Deoxynivalenol in Wheat*, 1995.

335. Josephs, R.D. et al., Trichothecenes: reference materials and method validation, *Toxicol. Lett.*, 153, 123, 2004.

336. Yoshizawa, T. et al., A practical method for measuring deoxynivalenol, nivalenol, and T-2 + HT-2 toxin in foods by an enzyme-linked immunosorbent assay using monoclonal antibodies, *Biosci. Biotechnol. Biochem.*, 68, 2076, 2004.

337. Koch, P., State of the art of trichothecenes analysis, *Toxicol. Lett.*, 153, 109, 2004.

338. Pascale, M., Haidukowski, M., and Visconti, A., Determination of T-2 toxin in cereal grains by liquid chromatography with fluorescence detection after immunoaffinity column clean-up and derivatization with 1-anthroylnitrile, *J. Chromatogr. A*, 989, 257, 2003.

339. Trenholm, H.L., Warner, R.M., and Prelusky, D.B., Assessment of extraction procedures in the analysis of naturally contaminated grain products deoxynivalenol (vomitoxin), *J. Assoc. Off. Anal. Chem.*, 68, 645, 1985.

340. Scott, P.M. and Kanhere, S.R., Comparison of column phases for separation of derivatised trichothecenes by capillary gas chromatography, *J. Chromatogr.*, 368, 374, 1986.

341. Romer, T.R., Use of small charcoal/alumina cleanup columns in determination of trichothecene mycotoxins in foods and feeds, *J. Assoc. Off. Anal. Chem.*, 69, 699, 1986.

342. Betina, V., Thin layer chromatography of mycotoxins, *J. Chromatogr.*, 334, 211, 1985.

343. Morgan, M.R.A., Mycotoxin immunoassays: with special reference to ELISAs, *Tetrahedron*, 45, 2237, 1989.

344. Schneider, E. et al., Rapid methods for deoxynivalenol and other trichothecenes, *Toxicol. Lett.*, 153, 113, 2004.

345. Park, J.C., Zong, M.S., and Chang, I.M., Survey of the presence of the *Fusarium* mycotoxins nivalenol, deoxynivalenol and T-2 toxin in Korean cereals of the 1989 harvest, *Food Addit. Contam.*, 8, 447, 1991.

346. Ryu, J.C. et al., Survey of natural occurrence of trichothecene mycotoxins and zearalenone in Korean cereals harvested in 1992 using gas chromatography mass spectrometry, *Food Addit. Contam.*, 13, 333, 1996.

347. Langseth, W. and Rundberget, T., Instrumental methods for determination of non-macrocyclic trichothecenes in cereals, foodstuffs and cultures, *J. Chromatogr. A*, 815, 103, 1998.

348. Klötzel, M. et al., Determination of 12 type A and B trichothecenes in cereals by liquid chromatography-electrospray ionization tandem mass spectrometry, *J. Agric. Food Chem.*, 53, 8904, 2005.

349. Yoshizawa, T., Swanson, S.P., and Mirocha, C.J., T-2 metabolites in the excreta of broiler chickens administered 3H-labeled T-2 toxin, *Appl. Environ. Microbiol.*, 39, 1172, 1980.

350. Prelusky, D.B. et al., Pharmacokinetic fate of ^{14}C-labeled deoxynivalenol in swine, *Fundam. Appl. Toxicol.*, 10, 276, 1988.

351. Prelusky, D.B. et al., Transmission of (14)deoxynivalenol to eggs following oral administration to laying hens, *J. Agric. Food Chem.*, 35, 182, 1987.

352. Visconti, A. and Mirocha, C.J., Identification of various T-2 toxin metabolites in chicken excreta and tissues, *Appl. Environ. Microbiol.*, 49, 1246, 1985.

353. Prelusky, D.B. and Trenholm, H.L., Tissue distribution of deoxynivalenol in swine dosed intravenously, *J. Agric. Food Chem.*, 39, 748, 1991.

354. Lun, A.K. et al., Effects of feeding hens a high level of vomitoxin-contaminated corn on performance and tissue residues, *Poult. Sci.*, 65, 1095, 1986.

355. Bezuidenhout, S.C. et al., Structure elucidation of the fumonisins, mycotoxins from *Fusarium moniliforme*, *J. Chem. Commun.*, 1988, 743, 1988.

356. Gelderblom, W.C. et al., Fumonisin-novel mycotoxins with cancer-promoting activity produced by *Fusarium moniliforme*, *Appl. Environ. Microbiol.*, 54, 1806, 1988.

357. Marasas, W.F.O. et al., Fumonisins-occurrence, toxicology, metabolism and risk assessment, in *Fusarium*, Summerell, B.A. et al., Eds., APS Press, St. Paul, MN, 2001, 332.

358. Rheeder, J.P., Prasanna, W.F., and Vismer, H.F., Production of fumonisin analogs by *Fusarium* species, *Appl. Environ. Microbiol.*, 68, 2102, 2002.

359. Marin, S. et al., Fumonisin-producing strains of *Fusarium*: a review of their ecophysiology, *J. Food Prot.*, 67, 1792, 2004.

360. Nelson, P.E., Desjardins, A.E., and Plattner, R.D., Fumonisins, mycotoxins produced by Fusarium species: biology, chemistry and significance, *Annu. Rev. Phytopathol.*, 31, 233, 1993.

361. Le Bars, J. et al., Biotic and abiotic factors in fumonisin B1 production and stability, *J. AOAC Int.*, 77, 517, 1994.

362. Le Bars, P. and Le Bars, J., Ecotoxinogenesis of *Fusarium moniliforme*: appearance of risks of fumonisins, *Cryptogamie Mycol.*, 16, 59, 1995.

363. Silva, L.J. et al., Occurrence of fumonisins B1 and B2 in Portuguese maize and maize-based foods intended for human consumption, *Food Addit. Contam.*, 24, 381, 2007.

364. Leblanc, J.C. et al., Estimated dietary exposure to principal food mycotoxins from the first French total diet study, *Food Addit. Contam.*, 22, 652, 2005.

365. SCOOP reports on tasks 3.2.10, Assessment of dietary intake of *Fusariums* by the population of EU members states, 2003.

366. Afolabi, C.G. et al., Effect of sorting on incidence and occurrence of fumonisins and *Fusarium verticilloides* on maize from Nigeria, *J. Food Prot.*, 69, 2019, 2006.

367. Sugita-Konishi, Y. et al., Occurrence of aflatoxins, ochratoxin A, and fumonisins in retail food in Japan, *J. Food Prot.*, 69, 1365, 2006.

368. Curtui, V. et al., A survey on the occurrence of mycotoxins in wheat and maize from western Romania, *Mycopathology*, 143, 97, 1998.

369. Kim, E.K. et al., Survey for fumonisin B1 in Korean corn-based food products, *Food Addit. Contam.*, 19, 459, 2002.

370. Shephard, G.S. et al., Fumonisin B1 in maize harvested in Iran during 1999, *Food Addit. Contam.*, 19, 676, 2002.

371. Prelusky, D.B., Trenholm, H.L., and Savard, M.E., Pharmacokinetic fate of 14C-labelled fumonisin B1 in swine, *Nat. Toxins* 2, 73, 1994.

372. Prelusky, D.B., Savard, M.E., and Trenholm, H.L., Pilot study on the plasma pharmacokinetics of fumonisin B1 in cows following a single dose by oral gavage or intravenous administration, *Nat. Toxins*, 3, 389, 1995.

373. Prelusky, D.B. et al., Biological fate of fumonisin B1 in food-producing animals, *Adv. Exp. Med. Biol.*, 392, 265, 1996.

374. Vudathala, D.K. et al., Pharmacokinetics fate and pathological effect of 14C-fumonisin B1 in laying Hens, *Nat. Toxins*, 2, 81, 1994.

375. Norred, W.P., Plattner, R.D., and Chamberlain, W.J., Distribution and excretion of (14C)fumonisin B1 in male Sprague-Dawley rats, *Nat. Toxins*, 1, 341, 1993.

376. Martinez-Larranaga, M.R. et al., Toxicokinetics and oral bioavailability of fumonisin B1, *Vet. Hum. Toxicol.*, 41, 357, 1999.

377. Becker, B.A. et al., Effects of feeding fumonisin B1 in lactating sows and their suckling pigs, *Am. J. Vet. Res.*, 56, 1253, 1995.

378. Richard, J.L. et al., Absence of detectable fumonisins in the milk of cows fed *Fusarium proliferatum* (Matsushima) *Nirenberg* culture material, *Mycopathology*, 133, 123, 1996.

379. Spotti, M. et al., Fumonisin B1 carry-over into milk in the isolated perfused bovine udder, *Vet. Hum. Toxicol.*, 43, 109, 2001.

380. Spotti, M., Pompa, G., and Caloni, F., Fumonisin B1 metabolism by bovine liver microsomes, *Vet. Res. Commun.*, 25, 511, 2001.

381. Cawood, M.E. et al., Interaction of 14C-labelled fumonisin B mycotoxins with primary rat hepatocyte cultures, *Food Chem. Toxicol.*, 32, 627, 1994.

382. Shephard, G.S. et al., Determination of the mycotoxin fumonisin B1 and identification of its partially hydrolysed metabolites in the faeces of non-human primates, *Food Chem. Toxicol.*, 32, 23, 1994.

383. Shephard, G.S. and Sewram, V., Determination of the mycotoxin fumonisin B1 in maize by reversed-phase thin layer chromatography: a collaborative study, *Food Addit. Contam.*, 21, 498, 2004.

384. Meyer, K. et al., Residue formation of fumonisin B1 in porcine tissues, *Food Addit. Contam.*, 20, 639, 2003.

385. Bailly, J.D. et al., Leuco-encéphalomalacie des tudes; cas rapportés au CNITV, *Rev. Méd. Vét.*, 147, 787, 1996.
386. Marasas, W. et al., Leucoencephalomalacia in a horse induced by fumonisine B1 isolated from *Fusarium moniliforme*, *Onderstepoort J. Vet. Res.*, 55, 197, 1988.
387. Harrison, L.R. et al., Pulmonary edema and hydrothorax in swine produced by fumonisin B1 a toxic metabolite of *Fusarium moniliforme*, *J. Vet. Diagnosis Investig.*, 2, 217, 1990.
388. Harvey, R.B. et al., Effects of dietary fumonisin B1-containing culture material, deoxynivalenol-contaminated wheat, or their combination on growing barrows, *Am. J. Vet. Res.*, 57, 1790, 1996.
389. Oswald, I.P. et al., Mycotoxin fumonisin B1 increases intestinal colonization by pathogenic *Escherichia coli* in pigs, *Appl. Environ. Microbiol.*, 69, 5870, 2003.
390. Diaz, D.E. et al., Effect of fumonisin on lactating dairy cattle, *J. Dairy Sci.*, 83, 1171, 2000.
391. Osweiller, G.D. et al., Effects of fumonisin-contaminated corn screenings on growth and health of feeder calves, *J. Anim. Sci.*, 71, 459, 1993.
392. Bermudez, A.J., Ledoux, D.R., and Rottinghaus, G.E., Effects of *Fusarium moniliforme* culture material containing known levels of fumonisin B1 in ducklings, *Avian Dis.*, 39, 879, 1995.
393. Brown, T.P., Rottinghaus, G.E., and Williams, M.E., Fumonisin mycotoxicosis in broilers: performance and pathology, *Avian Dis.*, 36, 450, 1992.
394. Bailly, J.D. et al., Toxicity of *Fusarium moniliforme* culture material containing known levels of fumonisin B1 in ducks, *Toxicology*, 11, 22, 2001.
395. Ledoux, D.R. et al., Fumonisin toxicity in broiler chicks, *J. Vet. Diagn. Invest.*, 4, 330, 1992.
396. Gelderblom, W.C.A. et al., The cancer initiating potential of the fumonisin B1 mycotoxins, *Carcinogenesis*, 13, 433, 1992.
397. Gelderblom, W.C. et al., Fumonisin-induced hepatocarcinogenesis: mechanisms related to cancer initiation and promotion, *Environ. Health Perspect.*, 109, 291, 2001.
398. Marasas, W.F., Fumonisins: their implications for human and animal health, *Nat. Toxins*, 3, 193, 1995.
399. Hendricks, K., Fumonisins and neural tube defects in south Texas, *Epidemiology*, 10, 198, 1999.
400. Hendricks, K.A., Simpson, J.C., and Larsen, R.D., Neural tube defects along the Texas-Mexico border, 1993–1995, *Am. J. Epidemiol.*, 149L, 1119, 1999.
401. Missmer, S. et al., Fumonisins and neural tube defects, *Epidemiol.*, 11, 183, 2000.
402. Marasas, W.F. et al., Fumonisins disrupt sphingolipid metabolism, folate transport, and neural tube development in embryo culture and in vivo: a potential risk factor for human neural tube defects among populations consuming fumonisin-contaminated maize, *J. Nutr.*, 134, 711, 2004.
403. Merrill, A.H. Jr. et al., Spingholipid metabolism: roles in signal tranduction and disruption by fumonisins, *Environ. Health Perspect.*, 109, 283, 2001.
404. Tran, S.T. et al., Sphinganine to sphingosine ratio and radioactive biochemical markers of fumonisin B1 exposure in ducks, *Chem. Biol. Interact.*, 2003, 61, 2003.
405. Goel, S. et al., Effects of *Fusarium moniliforme* isolates on tissue and serum sphingolipid concentrations in horses, *Vet. Hum. Toxicol.*, 38, 265, 1996.
406. Garren, L. et al., The induction and persistence of altered sphingolipid biosynthesis in rats treated with fumonisin B1, *Food Addit. Contam.*, 18, 850, 2001.
407. Van der Westhuizen, L., Shephard, G.S., and Van Schalkwyk, D.J., The effect of repeated gavage doses of fumonisine B1 on the sphinganine and sphingosine levels in vervet monkeys, *Toxicon*, 39, 969, 2001.
408. Riley, R.T. et al., Evidence for disruption of sphingolipid metabolism as a contributing factor in the toxicity and carcinogenicity of fumonisins, *Nat. Toxins*, 4, 3, 1996.
409. Strum, J.C., Ghosh, S., and Bell, R.M., Lipid second messengers: a role in cell growth regulation and cell cycle progression, *Adv. Exp. Med. Biol.*, 407, 421, 1997.
410. Desai, K. et al., Fumonisins and fumonisin analogs as inhibitors of ceramide synthase and inducers of apoptosis, *Biochim. Biophys. Acta*, 1585, 188, 2002.
411. Dupuy, J. et al., Thermostability of fumonisin B1: a mycotoxin from *Fusarium moniliforme*, in corn, *Appl. Environ. Microbiol.*, 59, 2864, 1993.

412. Castells, M. et al., Fate of mycotoxins in cereals during extrusion cooking: a review, *Food Addit. Contam.*, 22, 150, 2005.

413. AOAC. 995.15. *Fumonisins B1, B2 and B3 in Corn*, 2000.

414. Sydenham, E.W. et al., Liquid chromatographic determination of fumonisins B1, B2 and B3 in corn: AOAC-IUPAC collaborative study, *J. AOAC Int.*, 79, 688, 1996.

415. AFNOR, NF. EN. 13585, Produits alimentaires: dosage des fumonisines B1 et B2 dans le maïs. Méthode CHLP avec purification par extraction en phase solide, 2002.

416. Rice, L.G. et al., Evaluation of a liquid chromatographic method for the determination of fumonisins in corn, poultry feed, and *Fusarium* culture material, *J. AOAC Int.*, 78, 1002, 1995.

417. De Girolamo, A. et al., Comparison of different extraction and clean-up for the determination of fumonisins in maize-based food products, *Food Addit. Contam.*, 18, 59, 2001.

418. Visconti, A. et al., European intercomparison study for the determination of the fumonisins content in two maize materials, *Food Addit. Contam.*, 13, 909, 1996.

419. Shephard, G.S., Chromatographic determination of the fumonisin mycotoxins, *J. Chromatogr. A*, 815, 31, 1998.

420. Dilkin, P. et al., Robotic automated clean-up for detection of fumonisins B1 and B2 in corn-based feed by high-performance liquid chromatography, *J. Chromatogr. A*, 925, 151, 2001.

421. Bailly, J.D. et al., Production and purification of fumonisins from a highly toxigenic *Fusarium verticilloides* strain, *Rev. Med. Vet.*, 156, 547, 2005.

422. Preis, R.A. and Vargas, E.A., A method for determining fumonisin B1 in corn using immunoaffinity column clean-up and thin layer chromatography/densitometry, *Food Addit. Contam.*, 17, 463, 2000.

423. Dorner, J.W., Residues and other component analysis, in *Mycotoxins in Food: Methods of Analysis*, Vol. 2, Nollet L.M.L, New York, USA, 1996, 1089.

424. Shephard, G.S. et al., Quantitative determination of fumonisins B1 and B2 by high-performance liquid chromatography with fluorescence detection, *J. Liquid Chromatogr.*, 13, 2077, 1990.

425. Plattner, R. et al., Analysis of corn and cultured corn for fumonisin B1 by HPLC and GC/MS by four laboratories, *J. Vet. Diagn. Invest.*, 3, 357, 1991.

426. Jackson, M.A. and Bennett, G.A., Production of fumonisin B1 by *Fusarium moniliforme* NRRL 13616 in submerged culture, *Appl. Environ. Microbiol.*, 56, 2296, 1990.

427. Wang, S. et al., Rapid determination of fumonisin B1 in food samples by enzyme-linked immunosorbent assay and colloidal gold immunoassay, *J. Agric. Food Chem.*, 54, 2491, 2006.

428. Barna-Vetro, I. et al., Development of a sensitive ELISA for the determination of fumonisine B1 in cereals, *J. Agric. Food Chem.*, 48, 2821, 2000.

429. Bird, C.B. et al., Determination of total fumonisins in corn by competitive direct enzyme-linked immunosorbent assay: collaborative study, *J. AOAC Int.*, 85, 404, 2002.

430. Kulisek, E.S. and Hazebroek, J.P., Comparison of extraction buffers for the detection of fumonisin B1 in corn by immunoassay and high-performance liquid chromatography, *J. Agric. Food Chem.*, 48, 65, 2000.

431. AOAC. 2001.06. *Determination of Total Fumonisins in Corn*, 2001.

432. Shephard, G.S. et al., Fate of a single dose of 14C-labelled fumonisin B1 in vervet monkeys, *Nat. Toxins*, 3, 145, 1995.

433. Pagliuca, G. et al., Simple method for the simultaneous isolation and determination of fumonisin B1 and its metabolite aminopentol-1 in swine liver by liquid chromatography-fluorescence detection, *J. Chromatogr. B Anal. Technol. Biomed. Life Sci.*, 819, 97, 2005.

434. Tardieu, D. et al., Determination of fumonisin B1 in animal tissues with immunoaffinity purification, submitted, 2007.

435. Sulvok, M. et al., Development and validation of a liquid chromatography/tandem mass spectrometric method for the determination of 39 mycotoxins in wheat and maize, *Rapid Commun. Mass Spectrom.*, 20, 2649, 2006.

Chapter 37

Detection of Genetically Modified Organisms

Andrea Germini, Alessandro Tonelli, and Stefano Rossi

Contents

37.1 Introduction

37.1.1 GMO Production for Food and Feed

The advancement of biotechnologies applied to the agro-food industry has resulted, during the last few years, in an increasing number of genetically modified organisms (GMOs) being introduced into the food chain at various levels. Although regulatory approaches to this matter differ depending on the attitudes of different legislative bodies, the traceability of genetically modified products or ingredients coming from genetically modified products must be guaranteed to correctly inform final consumers and to be able to guarantee the safety of food production chains.

GMOs can be defined as organisms in which the genetic material has been altered by recombinant DNA technologies in a way that does not occur naturally by mating or natural recombination. Recombinant DNA techniques allow the direct transfer of one or a few genes between either closely or distantly related organisms; in this way, only the desired characteristic should be safely transferred from one organism to another, speeding up the process of improving the characteristics of interesting organisms and facilitating the tracking of the genetic changes and of their effects.

The first transgenic plants obtained by recombinant DNA technologies were produced in 1984, and since then more than 100 plant varieties, many of which are economically important crop species, have been genetically modified. The majority of these GMOs have been approved, albeit with differences according to the various legislations worldwide, for use in livestock feed and human nutrition.[1]

While only a few crops have been modified so far to improve their nutritional value, most of the first generation of GM crops (i.e., those currently in, or close to, commercialization) aim to increase yields and to facilitate crop management. This is achieved through the introduction of resistance to viral, fungal, or bacterial diseases or insect pests, or through herbicide tolerance. So far the majority of GM crops can be clustered according to the three main characteristics introduced:

- *Insect-protected plants*: The majority of the commercialized products belonging to this category are engineered to express a gene deriving from the soil bacterium *Bacillus thuringiensis* (Bt) that encodes for the production of a protein, the delta endotoxin, with insecticidal activity. Other genes used in developing this category of crops encode inhibitors of digestive enzymes of pest organisms, such as insect-specific proteinases and amylases, or direct chemically mediated plant defense by plant secondary metabolites.
- *Herbicide-tolerant plants*: A variety of products have been genetically engineered to create crops in which the synthesis of essential amino acids is not inhibited by the action of broad-spectrum herbicides such as glufosinate, as happens in conventional plants.
- *Disease-resistant plants*: Using GM technology, specific disease resistance genes can be transferred from other plants that would not interbreed with the crops of interest, or from other organisms; this allows the transformed crops to express proteins or enzymes that interfere with bacterial or fungal growth. GM virus-resistant crops have also been developed using "pathogen derived resistance" in which plants expressing genes for particular viral proteins are "immunized" to resist subsequent infection.

Other phenotype characteristics, less common than those mentioned above, include: modified fatty acid composition, fertility restoration, male sterility, modified color, and delayed ripening.

According to the latest statistics available, GMO crop cultivation has been continuously growing since their introduction into agricultural practice, in both industrialized and developing countries.

Although the first commercial GM crop (a tomato) was planted in 1994, it is only in the last few years that a dramatic increase in planting has been observed, bringing the estimated global area of GM crops to around 102 million hectares, involving 10.3 million farmers in 22 countries worldwide and with a global market value for biotech crops estimated to be around $6.15 billion.[2] As for the kinds of cultivated crops, four GM crops represented almost 100% of the market in 2005: GM soybeans accounted for the largest share (62%), followed by corn (22%), cotton (11%), and canola (5%). As for the traits inserted, GM herbicide-tolerant soybeans alone accounted for 58% of the total in 2005, followed by insect-resistant (largely Bt) corn and cotton with respective shares of 16% and 8%. In total, herbicide-tolerant crops accounted for around 76%, and insect-resistant crops accounted for almost 24% of global GM plantings.

These figures confirm how globally widespread GM cultivation is and how important the numbers are becoming compared to traditional crops: in particular, in 2005, GM soybeans accounted for 59% of total soybean plantings worldwide, while corn, cotton, and canola represented 13, 27, and 18% of their respective global plantings.

A new wave of genetically modified products, the second generation of GM-derived food and feed, is now at the end of its development stage or already under evaluation by the competent authorities for approval. These products mainly respond with similar approaches to the same issues addressed by the first generation (herbicide resistance, pest protection, and disease resistance). However, an increasing number of products are trying to respond to various new problems, including removing detrimental substances, enhancing health-promoting substances, enhancing vitamin and micronutrient content, and altering fatty acids and starch composition, reducing susceptibility to adverse environmental conditions and improving carbon and nitrogen utilization. The second generation of GMOs will constitute a new class of products which will try to respond to the needs of the consumers and of industries in the near future.

37.1.2 Legislative Framework

The need to monitor the presence of GM plants in a wide variety of food and feed matrices has become an important issue both for countries with specific regulations on mandatory labeling of food products containing GM ingredients or products derived from GMOs, and for countries without mandatory labeling of food products but which are required to test for the presence of unapproved GM varieties in food products.

Among the countries with mandatory labeling, the European Union has devised an articulated regulatory framework for GMOs to guarantee efficient control of food safety-related issues and to ensure correct information to European consumers: the use and commercialization of GM products and their derivatives has been strictly regulated both in food and feedstuffs, and compulsory labeling applies to all products containing more than 0.9% of genetically modified product (an adventitious presence threshold of 0.5% applies for GMOs which have already received a favorable risk evaluation but have not yet been approved). Other mandatory schemes for labeling exist worldwide in various countries, including Australia and New Zealand, Brazil, Cameroon, Chile, China, Costa Rica, Ecuador, India, Japan, Malaysia, Mali, Mauritius, Mexico, Norway, the Philippines, Russia, Saudi Arabia, South Africa, South Korea, Switzerland, Taiwan, Thailand, and Vietnam. Most of these countries have established mandatory labeling thresholds ranging from 0 to 5% GMO content.[3,4] In other countries in which labeling is voluntary, among the most important of which are the United States, Canada, and Argentina, being able to detect GM varieties is nevertheless of great importance to prevent unauthorized transgenes from entering their food production chains.

37.1.3 Analytical Methods for GMO Traceability

One of the main challenges related to the use of GMOs is their traceability along the food chain. In general, to be able to correctly identify the presence of transgenic material, a three-stage approach is needed:[5]

- *Detection.* A preliminary screening is performed to detect characteristic transgenic constructs used to develop GMOs (e.g., promoter and terminator sequences in the case of DNA analysis) and to gain a first insight into the composition of the sample being analyzed.
- *Identification.* This stage allows researchers to gain information on the presence of specific transgenic events in the sample being analyzed. Depending on the specific regulation framework in which the analysis is performed the presence of authorized GMOs should be then quantified, while the presence of unauthorized GMOs should be reported to competent authorities and the product prevented from entering the food chain.
- *Quantitation.* Transgene-specific quantification methods should be used at this stage to determine the amount of one or more authorized GMOs in the sample and to assess compliance with the labeling thresholds set in the context of the applicable regulatory framework.

All along this analytical scheme for the detection of GMOs, particular attention should be paid to the evaluation of (i) the degradation of the target DNA/protein during sampling and processing and (ii) the robustness of the analytical methods. A thorough knowledge and understanding of the problems associated with both the sample to be analyzed and the method for the analysis are fundamental prerequisites for obtaining reliable results.

While the first two stages of this scheme of analysis can essentially be accomplished by qualitative methods, semiquantitative or quantitative methods need to be used to accomplish the third stage of analysis.

At present, the two most important approaches for the detection of GMOs are (i) immunological assays based on the use of antibodies that bind to the novel proteins expressed, and (ii) polymerase chain reaction (PCR) based methods using primer oligonucleotides that selectively recognize DNA sequences unique to the transgene.

The two most common immunological assays are enzyme-linked immunosorbent assay (ELISA) based methods and immunochromatographic assays (e.g., lateral flow strip tests). While the former can produce qualitative, semiquantitative, and quantitative results according to the method employed, the latter, although fast and easy to perform, produces only qualitative results. However, both techniques require a sufficient protein concentration to be detected by specific antibodies, and thus their efficiency is strictly related to the plant environment, tissue-specific protein expression, and, not least, protein degradation during sampling and processing.

The most powerful and versatile methods for tracking transgenes are, however, based on the detection of specific DNA sequences by means of PCR methods. These methods are reported to be highly specific and have detection limits of only a few copies of the target DNA sequence. Qualitative and semiquantitative detection of GMOs can be easily achieved via end point PCR combined with gel electrophoresis, while quantitative detection can only be obtained by applying specific real-time PCR protocols, which rely on the quantification of fluorescent reporter molecules, which increase during the analysis with the amount of PCR product.

In addition to the above mentioned methods, other detection methods based on chromatography, mass spectrometry, and near-infrared (NIR) spectroscopy have been developed[5] and found to be suitable for specific applications, in particular when the genetic modifications create significant changes in the chemical composition of the host organism.

37.1.4   The Fate of Transgenic Material in Livestock Production

The significant increase of GM production since the commercialization of the first genetically modified crop has generated interest and concern regarding the fate of transgenic material along the food chain. Questions have been posed at both the public and scientific levels about the potential appearance of novel proteins and recombinant DNA in products for human consumption as a result of its presence in animal products containing GMOs. Because livestock consume large amounts of plant-derived material, it has become necessary to evaluate the effects of GMOs in the animals' diet and the possible consequences for human health. From a legislative point of view, however, countries that have implemented labeling regulation concerning GM feed have at present no mandatory regulations for products derived from livestock fed transgenic feed.

In the last few years, several attempts to investigate the fate of transgenic proteins and DNA within the gastrointestinal tract of livestock and the incorporation of transgenic material into tissues have been reported.[3] Since each novel protein expressed in a GM plant could be a potential source of allergy risk, the attempt to understand the fate of GM-derived proteins in livestock has become a major public issue. Despite numerous studies that have been reported in the literature to date, no transgenic proteins have been detected in animal tissues. Cry1Ab protein has been detected in cattle ruminal solids and gastrointestinal fluids but no positive signal was observed in any analyzed epithelial tissue.[6] A similar result was observed for Cry1Ab and CP4 EPSPS proteins in pigs.[7] The presence of Cry1Ab and Cry9C proteins has been also investigated in chickens and no positive samples were obtained from breast muscle, blood, liver, or muscle.[8,9] According to the reported experimental data, it appears quite difficult to detect GM proteins in livestock products using current methods of analysis. This could depend on multiple factors that may influence plant protein content and stability: (i) the kind of genetic modification and the type of plant tissue in which protein is expressed; (ii) different environmental conditions during plant growing; (iii) post harvest feed processing; (iv) the fact that proteins undergo rapid degradation in the gastrointestinal tract, reducing the absorption of GM proteins or protein fragments across the epithelial tissues to undetectable levels.

Several studies investigating the fate of GMO-derived nucleic acids have also been reported recently. The potential transfer of genetic material from GMOs to other organisms has given rise to concern regarding GMOs' food safety and effects on human health. Several studies have attempted to detect transgenic DNA in livestock in various organs and tissues.[10–14] However, only very common plant endogenous genes such as Rubisco and chloroplast were detected in blood, liver, spleen, kidney, lymph node, and muscle samples. The difficulty in detecting GMOs' nucleic acids in livestock tissues may be due to several factors: (i) feed processing such as ensiling, steeping, wet-milling, and heating often degrade DNA to undetectable levels; (ii) the high degradation rate in the gastrointestinal tract dramatically decreases the chances of finding DNA fragments long enough to be amplifiable by PCR; (iii) although the passage of dietary DNA fragments has been suggested by several researchers, it appears to be dependent on the concentration of DNA in the feed. Thus, the passage of exogenous DNA through tissues is clearly established only with "high copy number genes" such as chloroplast genes, which can be present at rates between 500 and 50,000 copies per genome. In contrast, transgenes are in most cases a single insertion event, making detection in tissues a difficult challenge with current PCR-based techniques.

According to the published literature, the fate of GM plant-derived DNA[7–18] and protein[7–9,15,19] in livestock productions seems to be mainly dependent upon two factors: (i) the initial amount and nature of the ingested plant derived materials and (ii) the anatomy and physiology of the gastrointestinal apparatus of the different animals. Since the detectable presence of GM-derived materials in livestock outside the gastrointestinal tract appears to be an extremely rare event, the principal route for GM material access could be an external event such as (i) contamination during slaughtering phases, when the gastrointestinal tract could come in contact with other animal parts or (ii) addition of foreign proteins or other components to enhance meat product properties (e.g., isolated soy proteins, starch, etc.).

37.2 Detection of GMOs

Approved transgenes and detection methods are continually updated, and official detection methods are validated and reported by the different national control agencies.[20] Online databases of protein- and DNA-based methods which have been validated by various research agencies are available for consultation.[21]

37.2.1 DNA-Based Methods

GMOs currently available are the result of transformation events that allow the stable insertion of an exogenous DNA fragment in a host's genome by means of DNA recombinant technology. The insert contains at least three elements: the gene coding for a specific desired feature and the transcriptional regulatory elements, typically a promoter and a terminator. Several additional elements could be present, depending on the transformation system employed, selection marker such as antibiotic resistance; introns or sequences coding for signaling peptides are commonly used.[22]

A wide spectrum of analytical methods based on PCR have been developed during the last decade, and PCR-based assays are generally considered the method of choice for regulatory compliance purposes. The general procedure for performing PCR analysis includes four sequential phases: sample collection, DNA isolation, DNA amplification, and detection of products. The latter two steps may occur simultaneously in certain PCR applications, such as real-time PCR.

Sampling, DNA extraction, and purification seem to constitute a crucial step in GMO detection. Sampling plans must be carefully designed to meet important statistical requirements involving the level of heterogeneity, the type of material (raw material, ingredients, or processed food), and the threshold limit for acceptance.[23,24] DNA quality and purity are also parameters that dramatically affect PCR efficiency.[25] DNA quality is strictly dependent on degradation caused by temperature, the presence of nucleases, and low pH, all of which determines the minimum length of DNA-amplifiable fragments. Moreover, the presence of contaminants from the food matrix or chemicals from the method used for DNA isolation can severely affect DNA purity and could cause inhibition of PCR reactions.

The PCR scheme involves subsequent steps at different temperatures during which: (i) the DNA is heated to separate the two complementary strands of the DNA template (denaturation, 95°C); (ii) the oligonucleotide primers anneal to their complementary sequences on the single-strand target DNA (annealing step, 50–60°C); (iii) the double-strand DNA region formed by the annealing is extended by the enzymatic activity of a thermostable DNA polymerase (extension step, 72°C). All these cycles are automatically repeated in a thermal cycler for a certain number of cycles, and at the end of the process the original target sequence will show an exponential increase in the number of copies.

Several authors have classified PCR-based GMO assays according to a "level of specificity" criterion.[5,26] (i) *Methods for screening purposes* are usually focused on target sequences commonly present in several GMOs. The most commonly targeted sequences pursued by this strategy are two genetic control elements, the cauliflower mosaic virus (CaMV) 35S promoter (P-35S) and the nopaline synthase gene terminator (T-NOS) from *Agrobacterium tumefaciens*. (ii) *Gene-specific methods* target a portion of the DNA sequence of the inserted gene. These methods amplify a gene tract directly involved in the genetic modification event, typically structural genes such as CrylA(b), coding for endotoxin B_1 from *B. thuringiensis*, or the *EPSPS* gene, coding for a herbicide-specific enzyme.

Both the screening and the gene-specific approach are useful for investigating the presence of GMOs but fail to reveal the GMO's identity. Moreover, these methods are based on the detection of sequences naturally occurring in the environment. This fact could lead to a significant level of false positive results.

(iii) Junction regions between two artificial construct elements such as the promoter and the functional gene are targeted by *construct-specific methods*; these reduce the risks of false positive appearances and increase the chances of identifying the GMO source of the DNA. However, more than one GMO could share the same gene construct, thus preventing unambiguous identification. (iv) The highest level of specificity is obtained using *event-specific methods,* which target the integration locus at the junction between the inserted DNA and the recipient genome.

An overview of validated PCR methods for the different strategies of GMO detection is reported in Table 37.2.

PCR assays can be combined with confirmation methods suitable for discriminating specific from nonspecific amplicons. Gel electrophoresis is the simplest method for confirming the expected size of PCR products but fails to identify the presence of nonspecific amplicons having the same size as the expected PCR product. Sequencing the amplicons is the most reliable method of confirming the identity of PCR products, but it is an expensive approach and requires specific instrumentation not often available in control laboratories. Nested-PCR is commonly used both in optimization steps and routine analyses; it is based on a second PCR reaction in which a PCR product is reamplified using primers specifically designed for an inner region of the original target sequence. Since nested-PCR consists of two PCR reactions in tandem, increased sensitivity is obtained. At the same time, however, it increases the risk of false positives by carry over or cross contamination. Southern blot assays are another reliable confirmation method; after gel electrophoresis, DNA samples are fixed onto nitrocellulose or nylon membranes and hybridized to a specific DNA probe. Southern blot is time consuming and quite labor-intensive and its implementation in routine analysis is limited.

37.2.1.1 DNA Extraction Methods

Isolation of nucleic acids is one of the most crucial steps in genetic studies. The existence of a great variety of extraction and purification methods arises from the different parameters that researchers must take into account. The most suitable isolation technique has to be selected according to the target (DNA, RNA, etc.), the source organism (mammalian, plant, etc.), the starting material (whole organ, tissue, blood, bone marrow, etc.), the desired results (yield, quality, purity, time-to-result, etc.), and the downstream application (PCR, real-time PCR, complementary DNA [cDNA] synthesis, etc.). Regardless of the technique chosen, the overall aim of this part of the detection process is to obtain an adequate yield of recovered DNA of high quality and purity to be used in the subsequent steps of the PCR analysis. DNA quality essentially refers to the degree of degradation of the nucleic acids recovered: the presence of DNA fragments long enough to be amplifiable by PCR is a key factor to be taken into account when designing and performing a PCR test. DNA purity mainly refers to the possible presence of PCR inhibitors in the extracted solution: the presence of proteins, polyphenols, polysaccharides, bivalent cations, and other secondary metabolites can interfere with enzyme activity and dramatically reduce the efficiency of PCR amplification.

The extraction of nucleic acids from biological material essentially requires three steps: cell lysis, inactivation of cellular nucleases, and separation of the desired nucleic acid from other cellular components.[27]

Common lysis procedures include mechanical disruption, chemical treatment, and enzymatic digestion. Several commonly adopted methods combine membrane disruption and inactivation of intracellular nucleases in a single step. The extraction is then completed by a purification step in which nucleic acid is selectively recovered from the sample matrix and purified. Effective DNA purification is absolutely essential for performing reliable PCR tests. Since food matrices in general, and meat samples in particular, can vary greatly in their physical and chemical properties, it is difficult to devise an all-purposes extraction procedure suitable for the different matrices and meeting all the necessary criteria. For this reason, customized DNA extraction methods need to be developed or adapted from more general methods to respond to the specific matrix to be analyzed and to optimize extraction efficiency. Common extraction and purification methods for the recovery of nucleic acids reported in the literature are fundamentally based in one of the following:

- combination of phenol and chloroform for protein removal followed by selective precipitation of nucleic acids with isopropanol or ethanol;
- use of the ionic detergent cetyltrimethylammonium bromide (CTAB) to lysate cells and selectively insolubilize nucleic acids in a low-salt environment, followed by solubilization and precipitation with isopropanol or ethanol;
- use of detergents and chaotropic agents followed by DNA binding on silica supports (e.g., spin column or magnetic silica particles) and elution in a low salt buffer.

Several commercial methods are currently available that employ combinations of the above mentioned detection strategies to perform fast and reliable extractions for specific food and feed matrices. An overview of customized DNA extraction procedures, grouped according to the different meat samples to be analyzed, is reported in Table 37.1 together with the corresponding bibliographic references.

37.2.1.2 PCR-Based Assay Formats

37.2.1.2.1 Qualitative PCR-Based Methods

Conventional end point PCR has been extensively used as a qualitative method to detect the presence of transgenic plants in raw materials and processed foods. PCR products are usually separated and visualized using agarose gel electrophoresis in combination with DNA staining.

The main advantages of this technique are cost effectiveness and simplicity. Conventional PCR is carried out using instrumentation commonly available in control laboratories. The amplification and detection steps, performed separately, extend the analysis time, increase the risk of contamination, and reduce the possibilities for automation. Despite these potential limitations, several qualitative methods have been developed for sensitive detection of GM crops.

Conventional PCR assays have been improved by performing simultaneous amplification of several GMOs in the same reaction, using more than one primer pair. The multiplex PCR format often requires longer optimization procedures but results in more rapid and inexpensive assays. Several multiplex PCR methods have been developed allowing simultaneous screening of different GM events in the same reaction tube.[29–32]

37.2.1.2.2 Quantitative PCR-Based Methods

The threshold for compulsory labeling of products containing GMOs set by many countries greatly accelerated the development of quantitative PCR-based GMO assays to comply with legislative requirements. Usually, the efficiency of quantitative methods is described using at least two fundamental parameters: the limit of detection (LOD) and the limit of quantification (LOQ). A principal drawback is that these values are usually determined using standard reference material with high-quality DNA, and their value dramatically decreases when faced with complex matrices or processed products. The availability of reference material containing known amounts of GMOs is another problematic factor in calibrating and standardizing quantitative assays, since certified reference materials (CRMs) are commercially available for only a limited number of GMOs (e.g., JRC-IRMM in Europe[33]). To overcome problems related to CRMs, alternative strategies have been proposed, such as the use of plasmid constructs carrying the sequence to be quantified, which seems to represent a valuable alternative strategy.[34,35]

37.2.1.2.2.1 Quantitative Competitive PCR (QC-PCR)

In QC-PCR, the target amplification is coupled with co-amplification of quantified internal controls that compete with target DNA for the same primers. The assay is carried out by amplifying samples with varying amounts of a previously calibrated competitor and finding the point that gives the same quantity of amplification products—the equivalence point. The end-point quantitation is then usually performed by agarose gel electrophoresis. QC-PCR methods for Roundup Ready soybeans and Maximizer maize have been developed[36] and tested in an interlaboratory trial at the EU level.[37] A screening method targeting the 35S promoter and the NOS terminator has been reported.[38] Even though the QC-PCR method potentially allows GMO detection with low limits of quantification, some drawbacks have limited the diffusion of this technique. The use of pipetting on a large scale increases the risk of cross-contamination and makes automation of procedures difficult. Moreover, QC-PCR is time consuming and often needs long optimization procedures.

37.2.1.2.2.2 Real-Time PCR

Real-time PCR-based methods have become more and more often recognized in the last few years as methods of choice for GMO quantitation. The most distinctive feature of this technique is that the amplicon can be monitored and quantified during each cycle of the PCR reaction. The increasing amount of amplicon is indirectly measured as a fluorescence signal variation during amplification. Quantitation by real-time PCR relies on the setting of two parameters: (i) the threshold fluorescence signal, defined as the value statistically significant above the noise, and (ii) the threshold cycle (C_t), which is the cycle number related to a fluorescence value above the threshold. Quantitation can be calculated directly by comparing C_t values of a GM-specific targeted gene with a reference gene. To obtain reliable measures it is essential to perform the reactions starting with the same concentration of DNA template. Moreover, this quantitation method relies

Table 37.1 Customized DNA Extraction Procedures for Different Meat Samples

Sample Type	Sample Pretreatment/Extraction	Purification	References
Pig (muscle, liver spleen, lymph nodes, blood) Poultry (turkey: breast muscle, liver; chicken: leg muscle, breast muscle, wings)	Stored –20°C (until DNA extraction) Samples cut out from the middle of the organs with sterile individual blade	Commercial kits using silica-columns (Roche, Mannheim, Germany)	14
Broiler chicks (blood, liver, muscles)	Heparinized blood Tissues frozen in liquid nitrogen and finely ground before DNA extraction	Commercial kits 1. Blood: Dr. GenTLE (Takara Shuzo Co. Ltd, Kyoto, Japan) 2. Tissues: Isotissue (Nippon Gene Co. Ltd, Toyama, Japan)	9
Broiler (tissue: heart, liver, kidney, bursa spleen, breast, blood gizzard; digesta samples)	Mechanical disruption using a tissue homogenizer in the presence of CLB lysis buffer (10 mM Tris, pH 8.0, 1% w/v SDS, and 50 mM EDTA); CLB:tissue (v/w) of 4:1	Ammonium acetate (proteins/cell debris elimination) Isopropanol/ethanol (DNA extraction)	16
Pig (blood, spleen, liver, kidney, muscle)	Stored –20°C (until DNA extraction) Samples cut out from the middle of the organs	Commercial kits 1. Wizard_Genomic DNA 2. Wizard_Magnetic DNA Purification System for Food (used only for muscle tissues) (Promega, Madison, USA)	18
Broiler (breast muscle)	Samples cut out from the middle of the organ and stored at –80°C until analyzed Mechanical disruption in the presence of lysis buffer (10 mM Tris, pH 8.0, 1% w/v SDS, and 50 mM EDTA); CLB:tissue (v/w) of 4:1	Ammonium acetate (proteins/cell debris elimination) Isopropanol/ethanol (DNA extraction)	8

Sample	Procedure	Methods	Ref.
Pork (loin muscle)	Frozen muscle samples (1 g circa) thoroughly homogenized with mortar and pestle	Method 1 (chloroform): 2/3 vol. chloroform (extraction); 0.8 vol. 2-propanol + glycogen (precipitation); 75 % (v/v) ethanol (washing); 0.2× TE (recovering buffer)	11
Beef (brisket muscle)	60°C under agitation with CTAB buffer (1.4 M NaCl, 2% (w/v) CTAB, 100 mM Tris, 15 mM EDTA, pH 8.0) + proteinase K + RNase A	Method 2 (CTAB): 2 vol. of CTAB precipitation buffer (40 mM NaCl, 0.5% (w/v); 1.2 M NaCl [pellet recovery]; 1 vol. chloroform (extraction); 1 vol. of 2-propanol + glycogen (precipitation); 75% (v/v) ethanol (washing); 0.2 × TE (recovery buffer)	
Chicken (breast muscle)		Method 3 (silica gel purification): DNeasy Plant Mini Kit (Qiagen, Valencia, CA) used according to manufacturer's instructions with slight modifications	
Ground beef	Stored at −20°C until subsampled. Divided into subsamples (1 g circa, wet weight); grounded in liquid nitrogen with a mortar and pestle. CTAB extraction buffer (1.4 M NaCl, 2% CTAB, 100 mM Tris, 20 mM EDTA, pH 8.0. 1% polyvinylpyrrolidone-40), 2% of 2-mercaptoethanol, and proteinase K for 2 h at 55°C with shaking	Method 1: phenol/chloroform/isoamyl alcohol (25:24:1, PCI). Samples extracted twice with equal volumes of PCI; 2/3 vol. cold 2-propanol (precipitation); TE buffer (10 mM Tris HCl, pH 8.0, 1 mM EDTA [recovery buffer] Method 2: see Nemeth et al.[11] Method 3: DNeasy Tissue Kit (Qiagen, Valencia, CA) silica gel purification (QIA column) used according to the manufacturer's instructions for the purification of DNA from animal tissues protocol	28

on the assumption that both amplicons are amplified with the same efficiency. As an alternative to overcome this limitation, quantitation can be done by building a standard curve with a series of PCR reactions using different known initial amounts of reference material. This method allows only C_t values of the same amplicons to be compared, reducing errors in measurements.

Several chemical strategies are currently available for real-time PCR analysis. Nonspecific methods use intercalating agents such as SYBR Green and others.[39] These assays have good sensitivity but often require postanalysis confirmation methods to distinguish the amplicons' identity and avoid false positives. This objective is achieved by some commercial instruments, which offers the possibility of analyzing the thermal denaturation curve to define the amplicons' identity.[40]

Specific methods, on the other hand, allow simultaneous detection and confirmation using specific probes or primers labeled with fluorescent dyes. The most widely adopted technology in real-time PCR analysis of GMOs is the TaqMan approach: a DNA oligonucleotide containing both a fluorophore and a quencher conjugated at each side of the molecule. During the extension step the probe is degraded by the 5'–3' exonuclease activity of the DNA polymerase, the quenching molecule results physically separated from the fluorophore reporter, and the intensity of the fluorescence signal increases. A further improvement compared to TaqMan assays was achieved through the use of minor groove binding (MGB) probes, in which a minor groove binder group increases the melting temperature of the duplex, improving the probe's selectivity and sensitivity. Alternatives based on the same principle of physical separation between fluorophore and quencher have been developed in scorpion primers and in molecular beacons. In these approaches a conformational change induced by the specific annealing, instead of a degradation event, drives the mechanism of fluorescence emission (a passage from a hairpin-shaped structure in solution to an unfolded conformation upon target hybridization). Other alternative technologies such as fluorescence resonance energy transfer (FRET) probes and light up probes could also be promising tools for the detection of GMOs.[39]

Compared to the other PCR-based methods, real-time PCR offers several advantages: (i) by performing both reaction and detection in a closed tube format, the risk of cross-contamination is greatly reduced; (ii) the high degree of automation makes real-time PCR less labor-intensive and time consuming; (iii) due to the possibility of setting multiplex assays and simultaneously performing several tests, the sample throughput result is increased compared to other PCR quantitation methods.

Real-time PCR has been successfully used for quantitative analysis of genetically modified maize, soybeans, rapeseed, cotton, potato, rice, tomatos, and sugarbeets (see Table 37.2). Several composite feed diets such as silage, commercial feed, and pellet mixed diet have also been investigated for their possible GMO content using real-time PCR.[41,42] Comparison of the different chemistries currently available for GMO detection has recently been reported.[43,44]

37.2.1.2.2.3 PCR-ELISA

An alternative method to perform end-point quantitation is coupling a conventional PCR with an enzymatic assay. In PCR-ELISA, a capture probe specific for the PCR amplicon is used to capture the amplicon in a well plate. PCR products, labeled during amplification, are then quantified by a conventional ELISA assay targeting the labeled amplicon. The main advantages of PCR-ELISA are that it offers a cheaper alternative to real-time PCR assays and requires less expensive instruments. Some PCR-ELISA applications have been developed for GMO detection and quantitation.[45,46] However, this technique does not seem to be widely adopted for accurate GMO quantitation.

Table 37.2 Validated PCR Methods for the Different Strategies of GMO Detection

P	Target	Primer Sequences	TaqMan Probe (If Real Time) (5'-FAM-3'-TAMRA)	References
1	Animal mtDNA 16S rRNA gene	5'-GGTTTACGACCTCGATGTT-3' 5' CCCGTCTGAACTCAGATCAC-3'		47
1	Myostatin gene of mammals and poultry species	5'-TTGTGCAAATCCTGAGACTCAT-3'	5'-CCCATGAAAGACGGTACAAGGTATACTG-3'	48
1	Cattle	5'-ATACCAGTGCCTGGGTTCAT-3' 5'-ACTCCTACCCATCATGCAGAT-3'	5'-AACATCAGGATTTTTGCTGCATTTGC-3'	49
1	Chicken	5'-TTTTAAATATTTCAGCTAAGAAAAAAG-3' 5'-TGTTACCTGCGGAGAAGTGGTTACT-3'	5'-TGAAGAAAGAAACTGAAGATGACACTGAAATTAAAG-3'	49
1	Lamb	5'-TTTTCGATATTTGAATAGCAGTTACAA-3' 5'-ACCCGTCAAGCAGACTCTAACG-3'	5'-CAGGATTTTTGCCGCATTCGCTT-3'	49
1	Pig	5'-TAAATATTTCAGCTAAGGAAAAAAAGAAG-3' 5'-CCCCACCTCAAGTGCCT-3'	5'-CACAGCAAGCCCCTTAGCCC-3'	49
1	Turkey	5'-CACAGACTTTATTTCTCCACTGCC-3' 5'-TGTATTTCAGTAGCACTGCTTATGACTACT-3' 5'-TTTATTAATGCTGGAAGAATTTCCAA-3'	5'-TTATGGAGCATCGCTATCACCAGAAAA-3'	49
2	Chloroplast gene for vegetal species	5'-CGAAATCGGTAGACGCTACG-3' 5'-GGGGATAGAGGGACTTGAAC-3'		50
2	Canola	5'-GGCCAGGGTTTCCGTGAT-3' 5'-CCGTCGTTGTAGAACCATTGG-3'	5'-AGTCCTTATGTGCTCCACTTTCTGGTGCA-3' (5'-VIC)	51
2	Cotton	5'-AGTTTGTAGGTTTGATGTTACATTGAG-3' 5'-GCATCTTTGAACCGCCTACTG-3'	5'-AAACATAAAATAATGGAACAACCATGACACATGT-3'	52
2	Maize	5'-CTCCCAAATCCTTTGACATCTGC-3' 5'-TCGATTTCTCTTCTTGGTGACAGG-3'	5'-AGCAAAGTCAGAGCGGTGCAATGCA-3'	53
2	Potato	5'-GGACATGTGAAGAGACGGAGC-3' 5'-CCTACCTCTACCCTCCGC-3'	5'-CTACCACCATTACCTGCACCTCCTCA-3'	54
2	Rice	5'-TGGTGAGCGTTTGCAGTCT-3' 5'-CTGATCCACTAGCAGGAGGTCC-3'	5'-TGTTGTGCTGCCAATGTGGCCCTG-3'	55

(Continued)

Table 37.2 (Continued)

P	Target	Primer Sequences	TaqMan Probe (If Real Time) (5'-FAM–3'-TAMRA)	References
2	Soybean	5'-TCCACCCCCATCCACATTT-3' 5'-GGCATAGAAGGTGAAGTTGAAGGA-3'	5'-AACCGGTAGCGTTGCCAGCTTCG-3'	56
2	Sugarbeet	5'-GACCTCCATATTACTGAAAGGAAG-3' 5'-GAGTAATTGCTCCATCCTGTTCA-3'	5'-CTACGAAGTTTAAAGTATGTGCCGCTC-3'	57
2	Tomato	5'-GGATCCTTAGAAGCATCTAGT-3' 5'-CGTTGGTGCATCCCTGCATGG-3'		58
3	CaMV 35S promoter	5'-CCACGTCTTCAAAGCAAGTGG-3' 5'-TCCTCTCCAAATGAAATGAACTTCC-3'		59
3	Coat protein gene from potato potyvirus Y (PVY)	5'-GAATCAAGGCTATCACGTCC-3' 5'-CATCCCGCACTGCCTCATACC-3'		60
3	CP4 EPSPS	5'-GCGTCGCCGATGAAGGTGCTGTC-3' 5'-CGGTCCTTCATGTTCGGCGGTCTC-3'		7
3	CryIA(b)	5'-CCCGCACCCTGAGCAGCAC-3' 5'-GGTGGCACGTTGTTGTTCTGA-3'		56
3	Figwort mosaic virus (P-FMV) promoter	5'-GCCAAAAGCTACAGGAGATCAATG-3' 5'-GCTGCTCGATGTTGACAAGATTAC-3'		61
3	Hygromycin phosphotransferase (hph) gene	5'-CGCCGATGGTTTCTACAA-3' 5'-GGCCGTCGGTTTCCACTAT-3'		62
3	Neomycin phosphotransferase II (nptII) gene	5'-GGATCTCCTGTCATCT-3' 5'-GATCATCCTGATCGAC-3'		63
3	Nopaline synthase (NOS) terminator	5'-GCATGACGTTATTTATGAGATGGG-3' 5'-GACACCGCGCGCGATAATTTATCC-3'		59
4	Canola GT73	5'-CCATATTGACCATCATACTCATTGCT-3' 5'-GCTTATACGAAGGCAAGAAAAGA-3'	5'-TTCCCGGACATGAAGATCATCCTCCTT-3'	64
4	Canola Ms8	5'-GTTAGAAAAAGTAAACAATTAATATAGCCGG-3'	5'-AATATAATCGACGGATCCCCGGAATTC-3'	65

4	Canola Rf3	5'-GGAGGGTGTTTTGGTTATC-3' 5'-AGCATTTAGCATGTACCATCAGACA-3'	5'-CGCACGCTTATCGACCATAAGCCCA-3'	66
4	Canola T45 (HCN28)	3'-CATAAAGGAAGATGGAGACTTGAG-3' 5'-CAATGGACACATGAATTATGC-3' 5'-GACTCTGTATGAACTGTTCGC-3'	5'-TAGAGGACCTAACAGAACTCGCCGT-3'	51
4	Cotton MON 1445	5'-GGAGTAAGACGATTCAGATCAAACAC-3' 5'-ATCGACCTGCAGCCCAAGCT-3'	5'-ATCAGATTGTCGTTTCCCGCCTTCAGTTT-3'	67
4	Cotton 281-24-236	5'-CTCATTGCTGATCCATGTAGATTTC-3' 5'-GGACAATGCTGGCCTTTGTG-3'	5'-TTGGGTTAATAAAGTCAGATTAGAGGGAGACAA-3'	52
4	Cotton 3006-210-23	5'-AAATATTAACAATGCATTGCAGTATGATG-3' 5'-ACTCTTTCTTTTCTCCATATTGACC-3'	5'-TACTCATTGCTGATCCATGTAGATTTCCCG-3'	52
4	Cotton MON 531	5'-TCCCATTCGAGTTTCTCACGT-3' 5'-AACCAATGCCACCCCACTGA-3'	5'-TTGTCCCTCCACTTCTTCTC-3'	68
4	Cotton LLCotton25	5'-CAGATTTTTGTGGGATTGGAATTC-3' 5'-CAAGGAACTATTCAACTGAG-3'	5'-CTTAACAGTACTCGGCCGTCGACCGC-3'	69
4	Maize Bt10	5'-CACACAGGAGATTATTATAGGGC-3' 5'-GGGAATAAGGGCGACACGG-3'		70
4	Maize Bt11	5'-AAAAGACACAACAACAGCCGC-3' 5'-CAATGCGTTCTCCACCAAGTACT-3'	5'-CGACCATGGACAACAACCCAAACATCA-3'	53
4	Maize CBH-351	5'-CCTTCGCAAGACCCTTCCTCTATA-3' 5'-GTAGCTGTCGGTGTAGTCCTCGT-3'		29
4	Maize DAS-59122-7	5'-GGGATAAGCAAGTAAAAGCGCTC-3' 5'-CCTTAATTCTCCGCTCATGATCAG-3'	5'-TTTAAACTGAAGGCGGGAAACGACAA-3'	71
4	Maize Bt176	5'-TGTTCACCCAGCAGCAACCAG-3' 5'-ACTCCACTTTGTCCAGAACAGATCT-3'	5'-CCGACGGTGACCGACTACCACATCGA-3'	53
4	Maize GA21	5'-GAAGCCTCGGCAACGTCA-3' 5'-ATCCGGTTGGAAAGGCGACTT-3'	5'-AAGGATCCGGTGCATGGCCG-3'	53
4	Maize MIR604	5'-GCGCACGCAATTCAACAG-3' 5'-GGTCATAACGTGACTCCCTTAATTCT-3'	5'-AGGCGGGAAACGACAATCTGATCATG-3'	72
4	Maize MON810	5'-GATGCCTTCTTCCCTAGTGTTGA-3' 5'-GGATGCACTCGTTGATGTTTG-3'	5'-AGATACCAAGCGGCCATGGACAACAA-3'	53
4	Maize MON863	5'-GTAGGATCGGAAAGCTTGGTAC-3' 5'-TGTTACGGCCTAAATGCTGAACT-3'	5'-TGAACACCCATCCGAACAAGTAGGGTCA-3'	73

(Continued)

Table 37.2 (Continued)

P	Target	Primer Sequences	TaqMan Probe (If Real Time) (5'-FAM–3'-TAMRA)	References
4	Maize NK603	5'-ATGAATGACCTCGAGTAAGCTTGTTAA-3' 5'-AGAGATAAACAGGATCCACTCAAACACT-3'	5'-TGGTACCA CGCGACACACTTCCACTC-3'	74
4	Maize T25	5'-GCCAGTTAGGCCAGTTACCCA-3' 5'-TGAGCGAAACCCTATAAGAACCCT-3'	5'-TGCAGGCATGCCCGCTGAAATC-3'	53
4	Maize TC1507	5'-TAGTCTTCGGCCAGAATGG-3' 5'-CTTTGCCAAGATCAAGCG-3'	5'-TAACTCAAGGCCCTCACTCCG-3'	75
4	Potato EH92-527-1	5'-GTGTCAAAACACAATTTACAGCA-3' 5'-TCCCTTAATTCTCCGCTCATGA-3'	5'-AGATTGTCGTTTCCGCGCCTTCAGTT-3'	54
4	Rice LLRICE601	5'-TCTAGGATCCGAAGCAGATCGT-3' 5'-GGAGGGCGCGGAGTGT-3'	5'-CCACCTCCCAACAATAAAAGCGCCTG-3'	76
4	Rice LLRICE62	5'-AGCTGGCGTAATAGGCGAAGAGG-3' 5'-TGCTAACGGGTGCATCGTCTA-3'	5'-CGCACCGATTATTATACTTTTAGTCCACCT-3'	55
4	Soybean A2704-12	5'-GCAAAAAAGCGGTTAGCTCCT-3' 5'-ATTCAGGCTGCGCAACTGTT-3'	5'-CGGTCCTCCGATCGCCCTTCC-3'	77
4	Soybean GTS 40-3-2	5'-CCGGAAAGGCCAGAGGAT-3' 5'-GGATTTCAGCATCAGTGGCTACA-3'	5'-CCGGCTGCTTGCACCGTGAAG-3'	78
4	Sugarbeet H7-1	5'-TGGGATCTGGGTGGCTCTAACT-3' 5'-AATGCTGCTAAATCCTGAG-3'	5'-AAGGCGGGAAACGACAATCT-3'	57
4	Tomato Nema 282F	5'-GGATCCTTAGAAGCATCTAGT-3' 5'-CATCGCCAAGACCGGCAACAG-3'		58

Notes: P, purpose of the analysis (1, presence of animal amplifiable material/identification of animal species; 2, presence of vegetal amplifiable material/identification of vegetal species; 3, identification of transgenic constructs; 4, identification of transgenic events).

37.2.1.3 Applications in Meat Analysis

With the recent interest in the fate of transgenic DNA after consumption by human and animals, several studies have attempted to detect plant DNA fragments in livestock fed GMOs using PCR-based technologies.

The fate of chloroplast specific gene fragments of different lengths (199 and 532 bp) and a Bt176 specific fragment have been evaluated in cattle and chickens fed a diet containing either conventional or GM maize.[10] Only short DNA fragments (<200 bp) from chloroplasts were detected in blood lymphocytes of cows, but no plant DNA was detectable in muscle, liver, spleen, or kidney. In contrast, in all chicken tissues (muscle, liver, spleen, kidney) the short maize chloroplast gene fragment was amplified. However, the Cry1A(b) sequence was not detectable in any analyzed sample.

An optimized DNA extraction protocol combined with PCR has been used to detect feed-derived plant DNA in muscle meat from chickens, swine, and beef steers fed MON 810 maize.[11] Short fragments (173 bp) amplified from the high copy number chloroplast-encoded maize Rubisco gene (*rbcL*) were detected in 5, 15, and 53% of the muscle samples from beef steers, broiler chickens, and swine, respectively. Interestingly, 1 pork sample out of 118 tested positive for the screening of P-35S; however, further analysis performed with a specific MON 810 PCR method generated indeterminate results, suggesting that the number of target copies in the sample, where present, was below the detection limit of the method.

Other authors employed PCR to investigate the fate of maize intrinsic and recombinant genes in calves fed genetically modified maize Bt11.[12] Rubisco and chloroplast gene fragments (231 and 196 bp) were detected in liver, spleen, kidney, mesenteric lymph nodes, and in longissimus muscle of calves; in contrast, the Cry1Ab gene, specific for maize Bt11, was never detected in tissues.

PCR has also been used to investigate the fate of feed-ingested foreign DNA in pigs fed Bt maize.[13] Fragments of transgenic DNA (211 bp) were detected in the gastrointestinal tract of pigs up to 48 h after the last feeding with transgenic maize. Chloroplast DNA was detected in blood, liver, spleen, kidney, lymphatic glands, ovary, musculus longissimus dorsi, musculus trapezius, and gluteus maximus. The single copy Cry1Ab gene was never detected in tissue samples.

An attempt to detect the presence of CryIa, chloroplast, and maize specific (zein) gene fragments in pigs fed diets containing Bt-maize and in supermarket poultry samples has been investigated.[14] Chloroplast was successfully amplified from the intestinal juices of pigs up to 12 h after the last feeding, while it was not found in blood, muscle, liver, spleen, or lymph nodes of examined pigs. Specific gene fragments from transgenic maize were never detected in any pig sample. The analysis of supermarket poultry samples (leg, breast and wing muscle, stomach) led to frequent detections of the short chloroplast DNA fragments; faint signals for the maize specific zein gene fragment were also detected.

37.2.2 Protein-Based Methods

Apart from transformation events bearing an antisense sequence, GM plants usually undergo the insertion of transgenes coding for novel proteins. In most cases, these proteins represent suitable targets for GMO detection. A wide spectrum of immunoassay-based technologies has been developed in the last decades, covering an enormous range of purposes and scientific disciplines.

37.2.2.1 Antibody-Based Assay Formats

37.2.2.1.1 ELISA

ELISA is the most commonly employed technique among immunoassay strategies. ELISA assays allow the detection, and often the quantitation, of several classes of molecules such as proteins, peptides, antibodies, hormones, and other small molecules able to elicit immune response (haptens). A standard 96-well (or 384-well) polystyrene plate is the most common format used to perform ELISAs. The first step of the assay usually involves the target protein (antigen) absorption to a solid surface (direct ELISA) or the bonding of the antigen to a specific antibody, fixed at the bottom of a plate well (sandwich ELISA). The antigen is then bonded by an antibody coupled with an enzyme (typically the horseradish peroxidase [HRP] or the alkaline phosphatase [AP]). After the complex formation, a substrate that produces a detectable product is added. Several substrates and instruments (luminometers, spectrophotometers, fluorometers) are available to meet different technical needs.

Variants of the ELISA assay with improved sensitivity have been developed using signal amplification strategies. The most common approach is based on the addition of a secondary enzyme-labeled antibody which bonds a primary antibody specifically linked to the antigen. The bonding of several secondary antibodies to a single primary immunoglobulin results in a strong signal enhancement. Another strategy consists of forming a biotine/streptavidine derived complex linking more copy numbers of the enzyme to the same antibody.

Competing ELISA formats have also been developed. These assays are particularly suitable for molecules which have only one epitope or when only one specific antibody is available. Different applications of this format are available. One of the most common uses of an enzyme-conjugated antigen as the standard; unlabeled antigen (from sample) competes with known amounts of labeled antigens for a limited number of specific binding sites of a capture antibody fixed on the well plate.

The main advantages of ELISA assays are related with the potential to derive quantitative information using an economical, high-throughput, and not especially labor-intensive approach.

37.2.2.1.2 Lateral Flow Strips

Lateral flow strip technology consists of a nitrocellulose strip containing specific antibodies conjugated to a color reactant. One end of the strip is placed in a tube containing plant extract, which then starts to flow to the other end of the strip. When the target protein is present, a complex with color reagent-conjugated antibodies is formed and passes through two capture zones containing, respectively, a second antigen-specific antibody (test line) and an antibody for the labeled immunoglobulin excess (control line). When both lines give a positive signal, the test indicates a positive sample. When only the control line is positive, the test shows a negative sample. Lateral flow strip tests are very inexpensive, take a short time to analysis, and do not require a high degree of technical skills. All these reasons make this assay particularly suitable for "field tests."

Several drawbacks have limited the application of antibody-based assay formats in GMO detection: (i) The presence of other substances in complex matrices (other proteins, phenolic compounds, surfactants, fatty acids) can interfere with the assay; (ii) GM protein can be expressed in very low amounts, and the amount of the target protein expressed could be highly variable in different plant tissues or developmental stages;[24] (iii) matrices that undergo industrial processing, such as heating, could change the conformational structure of active epitopes, resulting in nonreactive proteins. These problems should be carefully evaluated for each sample when choosing the appropriate assay format.

Although protein-based methods have not found wide application in GMO detection compared to PCR, several applications have been developed and tested.[23,79,80]

37.2.2.2 Applications in Meat Analysis

The potential allergenic risk of novel proteins as a consequence of GMO diffusion has become, in the last few years, a relevant issue at national and international policy levels, and has aroused concern among citizens as well. On account of this, several attempts to investigate the fate of transgenic proteins have been performed on livestock productions.

The possible transfer of the Cry9C protein to blood, liver, and muscle in broiler chicks fed with StarLink corn has been investigated.[9] The determination of Cry9C protein in analytical materials was performed using a commercial GMO Bt9 maize test kit, and no positive samples were detected in examined tissues.

A study was conducted to determine the extent of GM protein from Roundup Ready soybeans in tissues and eggs of laying hens.[19] A commercial double antibody sandwich incorporated in a lateral flow strip format, specific for the CP4 EPSPS protein, has been used. Whole egg, egg albumen, liver, and feces were all negative for GM protein.

The attempt to detect the Cry1Ab protein in chicken breast muscle samples from animals fed YieldGard Corn Borer Corn event MON 810 has been published.[8] Analyses were performed using a competitive ELISA developed in-house with a LOD of approximately 60 ng of protein per gram of chicken muscle. Neither the Cry1Ab protein nor immunoreactive peptide fragments were detectable in the breast.

Using a similar strategy, the same author also investigated the presence of CP4 EPSPS protein in muscle of pigs fed a diet containing Roundup Ready soybeans.[7] A competitive immunoassay with a LOD of approximately 94 ng of CP4 EPSPS protein per gram of pork muscle was developed by the authors and used to test samples; neither the CP4 EPSPS protein nor immunoreactive peptide fragments were detected in any samples.

Three different assays to detect Cry1Ab protein in the gastrointestinal contents of pigs fed genetically modified corn Bt11 have been employed.[15] Two commercial kits (a conventional microplate-format ELISA and a test strip format immunochromatography assay) and an immunoblotting procedure were used to test pig samples. The Cry1Ab protein was only detected in the contents of the stomach, duodenum, ileum, cecum, and rectum.

37.2.3 Alternative Techniques for GMO Detection

With the number of GMOs developed by biotech companies constantly increasing and expected to have an even greater impact on worldwide cultivation and markets in the coming years,[2] new technologies and instruments will be needed to face the challenges of high throughput and affordable detection of an increasing number of transgenes. For both qualitative and quantitative analysis, routine procedures such as PCR and immunodetection methods appear to be inadequate when confronted with the future demand for screening for very large numbers of different GMOs. Several analytical approaches have been used to develop new detection systems able to implement the currently available methodologies in terms of sensitivity, specificity, robustness, and sample throughput.

Although most of the work on the development of new detection methods cited in the literature mainly focus on analytical systems for the detection of GMOs in grains or plant products, several approaches also seem to be suitable for performing analysis on more complex matrices such as meat products.

NIR spectroscopy, usually employed for the nondestructive analysis of grains for the prediction of moisture, protein, oil, fiber, and starch, has been described as a tool to discriminate between sample sets of Roundup Ready soybeans and nontransgenic soybeans.[81] More recently, visible (vis)/NIR spectroscopy combined with multivariate analysis was used to analyze tomato leaves and successfully discriminated between genetically modified and conventional tomatoes.[82] Although NIR techniques combine rapidity, ease of use, and cost effectiveness, their ability to resolve small quantities of GM varieties is assumed to be low: in fact, the technique discriminates according to structural changes which are larger than those produced by single gene modifications. Further advancement in the development of the technique is still needed before it could be evaluated for use in complex matrices.

Some authors have proposed chromatographic techniques for the detection of GMOs. Conventional chromatographic methods combined with efficient detection systems such as mass spectrometry could be applicable when significant changes occur in the composition of GM plants or derived products. This approach has been used to investigate the triglyceride patterns of oil derived from GM canola, showing that an increased content of triacylglycerols characterizes the transgenic canola variety.[83] Matrix-assisted laser desorption/ionization time-of-flight (MALDI-TOF) and nanoelectrospray ionization quadrupole time-of-flight (nano ESI-QTOF) were successfully applied to the detection of the transgenic protein CP4 EPSPS in 0.9% GM soybean after fractionation by gel filtration, anion-exchange chromatography, and sodium dodecyl sulfate/polyacrylamide gel electrophoresis (SDS-PAGE).[84]

Again, these methodologies, although very sensitive, appear at present to be suitable only for differentiating between GM and conventional varieties, but they lack the specificity needed for detection in composite food matrices.

A recent application has been described which uses anion exchange liquid chromatography coupled with a fluorescent detector in combination with peptide nucleic acid (PNA) probes to detect and univocally identify PCR amplicons of Roundup Ready soybeans or Bt-176 maize, both on CRM and commercial samples.[85]

37.2.3.1 DNA Microarray Technology

With the number of genetic targets to be monitored constantly increasing, detection of GMOs in the near future appears to be moving toward the need for higher throughput analysis that can simultaneously detect a high number of targets of interest and lower the cost of detecting an increased variety of genetic targets. In this context, one of the more promising technologies available appears to be microarray systems. In their general form, microarray systems are oligonucleotide probe-based platforms on which a high number of nucleic acid targets can be simultaneously detected with high specificity. This would imply, in the case of GMO detection, the potential for rapid and efficient screening of a large number of controls, gene-, and transgene-specific nucleic acid targets.

The principal advantages of DNA microarray technology are miniaturization, high sensitivity, and screening throughput. Its main limitation is at present the strict dependence on PCR or other amplification techniques to amplify and label DNA or mRNA target sequences before performing the microarray analysis of a sample. The necessity of this PCR step, at present still not likely to be overcome, imposes on this technology all the limitations discussed in the earlier section on

PCR. Moreover, the possibility of quantifying GMO content in the sample is lost because amplification and labeling are performed using end-point PCR, which is strictly qualitative. Different DNA microarray approaches, at both the research and commercial stages, have been described for the detection of GMOs in food and feed systems, and these approaches could be valuable for the specific analysis of meat products.

A recent paper describes the development of a method for screening GMOs using multiplex-PCR coupled with an oligonucleotide microarray.[86] The authors developed a 20 oligonucleotide probe array for the detection of a majority of the genetic constructs, covering 95% of commercially available transgenes (soybean, maize, cotton, and canola) with a detection limit of 0.5 and 1.0% for transgenic soybeans and maize, respectively.

A multiplex DNA microarray chip was developed for simultaneous identification of nine GMOs, five plant species, and three GMO screening elements.[87] The targets were labeled with biotin during amplification, and the arrays could be detected using a colorimetric analysis with a detection limit below 0.3%.

A commercial microarray system for the qualitative detection of EU-approved GMOs has recently been commercialized in Europe.[88] The system combines the identification of GMOs by characterization of their genetic elements with a colorimetric detection based on silver.

A multiplex quantitative DNA array-based PCR method (MQDA-PCR) has been described for the quantification of seven different transgenic maize types in food and feed samples.[89] The authors were able to correctly characterize the presence of transgenic maize in the range of 0.1–2.0% using a two-step PCR, which used opportunely labeled primers, and a DNA array spotted on nylon membrane.

Ligation detection reaction (LDR) in combination with multiplex-PCR and a universal array has been described as a sensitive tool for GMO detection. The authors were able to detect trace amounts of five transgenic events (maize and soybean) in heterogeneous samples both in reference materials and in commercial samples.[90]

A class of synthetic oligonucleotide analogs with increased hybridization sensitivity and specificity has been described in a recent paper, in which the authors used PNAs as capture probes for the detection of five transgenic maize and soybeans amplified by a multiplex PCR with a LOD of 0.25%.[91]

37.2.3.2 Biosensors

Although only in the research stage, several biosensor-based methods have so far been developed and tested for the detection of GMOs. Their main advantage is the fact that detection is based on physical principles, resulting in the possibility of performing the analysis in a faster and more economical way than by conventional techniques. Their major drawback is that, as with the previously described techniques, they rely on PCR since their sensitivity is not high enough for stand-alone analysis. As research on biosensors has continually improved over the last few years, innovative techniques and detection systems are likely to be developed that could in the near future adequately fulfill the requirements of GMO detection.

A biosensor based on quartz crystal microbalance (QCM) has been described for the detection of sequences of the 35S promoter and NOS terminator.[92] PCR products obtained from CRM and real samples were correctly identified in a label-free hybridization reaction, showing how this approach could be a sensitive and specific method for the detection of GMOs in food samples.

An electrochemical biosensor based on disposable screen-printed gold electrodes has recently been described for detecting characteristic sequences of soybeans and the 35S promoter.[93]

The applied detection scheme, based on the enzymatic amplification of hybridization signals by a streptavidin–alkaline phosphate conjugate, led to a highly sensitive detection of the target sequences without the need for chemical or physical treatment of the electrodic surfaces.

A biosensor based on surface plasmon resonance (SPR) has been reported to allow for discrimination between samples containing 0.5 and 2.0% Bt-176 maize reference material.[94] The PCR products amplified by multiplex-PCR were immobilized on the surface of the sensor and oligonucleotide probes were flowed through the cell and hybridized to their specific target, generating a quantifiable signal.

References

1. AGBIOS, http://www.agbios.com/, 2007.
2. Brookes, G. and Barfoot, P., *GM crops: the first ten years—global socio-economic and environmental impacts, ISAAA Briefs* 36, 2006. ISAAA, Ithaca, NY.
3. Alexander, T. W., Reuter, T., Aulrich, K., Sharma, R., Okine, E. K., Dixon, W. T., and McAllister, T. A., A review of the detection and fate of novel plant molecules derived from biotechnology in livestock production, *Animal Feed Science and Technology* 133 (1–2), 31–62, 2007.
4. The Center for Food Safety, http://www.centerforfoodsafety.org/, 2007.
5. Anklam, E., Gadani, F., Heinze, P., Pijnenburg, H., and Van den Eede, G., Analytical methods for detection and determination of genetically modified organisms in agricultural crops and plant-derived food products, *European Food Research and Technology* 214 (1), 3–26, 2002.
6. Einspanier, R., Lutz, B., Rief, S., Berezina, O., Zverlov, V., Schwarz, W., and Mayer, J., Tracing residual recombinant feed molecules during digestion and rumen bacterial diversity in cattle fed transgene maize, *European Food Research and Technology* 218 (3), 269–273, 2004.
7. Jennings, J. C., Kolwyck, D. C., Kays, S. B., Whetsell, A. J., Surber, J. B., Cromwell, G. L., Lirette, R. P., and Glenn, K. C., Determining whether transgenic and endogenous plant DNA and transgenic protein are detectable in muscle from swine fed Roundup Ready soybean meal, *Journal of Animal Science* 81 (6), 1447–1455, 2003.
8. Jennings, J., Albee, L. D., Kolwyck, D. C., Surber, J. B., Taylor, M. L., Hartnell, G. F., Lirette, R. P., and Glenn, K. C., Attempts to detect transgenic and endogenous plant DNA and transgenic protein in muscle from broilers fed YieldGard1 Corn Borer Corn, *Poultry Science* 82, 371–380, 2003.
9. Yonemochi, C., Fujisaki, H., Harada, C., Kusama, T., and Hanazumi, M., Evaluation of transgenic event CBH 351 (StarLink) corn in broiler chicks, *Animal Science Journal* 73 (3), 221–228, 2002.
10. Einspanier, R., Klotz, A., Kraft, J., Aulrich, K., Poser, R., Schwagele, F., Jahreis, G., and Flachowsky, G., The fate of forage plant DNA in farm animals: a collaborative case-study investigating cattle and chicken fed recombinant plant material, *European Food Research and Technology* 212 (2), 129–134, 2001.
11. Nemeth, A., Wurz, A., Artim, L., Charlton, S., Dana, G., Glenn, K., Hunst, P., Jennings, J., Shilito, R., and Song, P., Sensitive PCR analysis of animal tissue samples for fragments of endogenous and transgenic plant DNA, *Journal of Agricultural and Food Chemistry* 52 (20), 6129–6135, 2004.
12. Chowdhury, E. H., Mikami, O., Murata, H., Sultana, P., Shimada, N., Yoshioka, M., Guruge, K. S., Yamamoto, S., Miyazaki, S., Yamanaka, N., and Nakajima, Y., Fate of maize intrinsic and recombinant genes in calves fed genetically modified maize Bt11, *Journal of Food Protection* 67, 365–370, 2004.
13. Reuter, T. and Aulrich, K., Investigations on genetically modified maize (Bt-maize) in pig nutrition: fate of feed-ingested foreign DNA in pig bodies, *European Food Research and Technology* 216 (3), 185–192, 2003.
14. Klotz, A., Mayer, J., and Einspanier, R., Degradation and possible carry over of feed DNA monitored in pigs and poultry, *European Food Research and Technology* 214 (4), 271–275, 2002.

15. Chowdhury, E. H., Kuribara, H., Hino, A., Sultana, P., Mikami, O., Shimada, N., Guruge, K. S., Saito, M., and Nakajima, Y., Detection of corn intrinsic and recombinant DNA fragments and Cry1Ab protein in the gastrointestinal contents of pigs fed genetically modified corn Bt11, *Journal of Animal Science* 81 (10), 2546–2551, 2003.

16. Deaville, E. R. and Maddison, B. C., Detection of transgenic and endogenous plant DNA fragments in the blood, tissues, and digesta of broilers, *Journal of Agricultural and Food Chemistry* 53 (26), 10268–10275, 2005.

17. Aeschbacher, K., Messikommer, R., Meile, L., and Wenk, C., Bt176 corn in poultry nutrition: physiological characteristics and fate of recombinant plant DNA in chickens, *Poultry Science* 84 (3), 385–394, 2005.

18. Mazza, R., Soave, M., Morlacchini, M., Piva, G., and Marocco, A., Assessing the transfer of genetically modified DNA from feed to animal tissues, *Transgenic Research* 14 (5), 775–784, 2005.

19. Ash, J., Novak, C., and Scheideler, S. E., The fate of genetically modified protein from Roundup Ready soybeans in laying hens, *The Journal of Applied Poultry Research* 12 (2), 242–245, 2003.

20. JRC-IHCP, http://gmo-crl.jrc.it/statusofdoss.htm.

21. JRC-IHCP, http://biotech.jrc.it/home/ict/methodsdatabase.htm.

22. Garcia-Canas, V., Cifuentes, A., and Gonzalez, R., Detection of genetically modified organisms in foods by DNA amplification techniques, *Critical Reviews in Food Science and Nutrition* 44 (6), 425–436, 2004.

23. Emslie, K. R., Whaites, L., Griffiths, K. R., and Murby, E. J., Sampling plan and test protocol for the semiquantitative detection of genetically modified canola (*Brassica napus*) seed in bulk canola seed, *Journal of Agricultural and Food Chemistry* 55 (11), 4414–4421, 2007.

24. Miraglia, M., Berdal, K. G., Brera, C., Corbisier, P., Holst-Jensen, A., Kok, E. J., Marvin, H. J. P., Schimmel, H., Rentsch, J., van Rie, J., and Zagon, J., Detection and traceability of genetically modified organisms in the food production chain, *Food and Chemical Toxicology* 42 (7), 1157–1180, 2004.

25. Cankar, K., Štebih, D., Dreo, T., Žel, J., and Gruden, K., Critical points of DNA quantification by real-time PCR—effects of DNA extraction method and sample matrix on quantification of genetically modified organisms, *BMC Biotechnology* 6, 37, 2006.

26. Holst-Jensen, A., Ronning, S. B., Lovseth, A., and Berdal, K. G., PCR technology for screening and quantification of genetically modified organisms (GMOs), *Analytical and Bioanalytical Chemistry* 375 (8), 985–993, 2003.

27. Somma, M., The analysis of food samples for the presence of genetically modified organisms. Session 4: extraction and purification of DNA, http://gmotraining.jrc.it/, 2007.

28. He, X., Brandon, D. L., Chen, G. Q., McKeon, T. A., and Carter, J. M., Detection of castor contamination by real-time polymerase chain reaction, *Journal of Agricultural and Food Chemistry* 55 (2), 545–550, 2007.

29. Matsuoka, T., Kuribara, H., Akiyama, H., Miura, H., Goda, Y., Kusakabe, Y., Isshiki, K., Toyoda, M., and Hino, A., A multiplex PCR method of detecting recombinant DNAs from five lines of genetically modified maize, *Journal of the Food Hygienic Society of Japan* 42 (1), 24–32, 2001.

30. Hernandez, M., Rodriguez-Lazaro, D., Zhang, D., Esteve, T., Pla, M., and Prat, S., Interlaboratory transfer of a PCR multiplex method for simultaneous detection of four genetically modified maize lines: Bt11, MON810, T25, and GA21, *Journal of Agricultural and Food Chemistry* 53 (9), 3333–3337, 2005.

31. Germini, A., Zanetti, A., Salati, C., Rossi, S., Forre, C., Schmid, S., and Marchelli, R., Development of a seven-target multiplex PCR for the simultaneous detection of transgenic soybean and maize in feeds and foods, *Journal of Agricultural and Food Chemistry* 52 (11), 3275–3280, 2004.

32. James, D., Schmidt, A. M., Wall, E., Green, M., and Masri, S., Reliable detection and identification of genetically modified maize, soybean, and canola by multiplex PCR analysis, *Journal of Agricultural and Food Chemistry* 51 (20), 5829–5834, 2003.

33. JRC-IRMM, http://www.irmm.jrc.be/.

34. Lee, S. H., Kim, J. K., and Yi, B. Y., Detection methods for biotech cotton MON 15985 and MON 88913 by PCR, *Journal of Agricultural and Food Chemistry* 55 (9), 3351–3357, 2007.

35. Weighardt, F., Barbati, C., Paoletti, C., Querci, M., Kay, S., De Beuckeleer, M., and Van den Eede, G., Real-time polymerase chain reaction-based approach for quantification of the pat gene in the T25 *Zea mays* event, *Journal of AOAC International* 87 (6), 1342–1355, 2004.

36. Studer, E., Rhyner, C., Luthy, J., and Hubner, P., Quantitative competitive PCR for the detection of genetically modified soybean and maize, *Zeitschrift für Lebensmittel-Untersuchung und -Forschung A-Food Research and Technology* 207 (3), 207–213, 1998.

37. Hubner, P., Studer, E., Hafliger, D., Stadler, M., Wolf, C., and Looser, M., Detection of genetically modified organisms in food: critical points for duality assurance, *Accreditation and Quality Assurance* 4 (7), 292–298, 1999.

38. Hardegger, M., Brodmann, P., and Herrmann, A., Quantitative detection of the 35S promoter and the NOS terminator using quantitative competitive PCR, *European Food Research and Technology* 209 (2), 83–87, 1999.

39. Kubista, M., Andrade, J. M., Bengtsson, M., Forootan, A., Jonak, J., Lind, K., Sindelka, R., Sjoback, R., Sjogreen, B., Strombom, L., Stahlberg, A., and Zoric, N., The real-time polymerase chain reaction, *Molecular Aspects of Medicine* 27 (2–3), 95–125, 2006.

40. Hernandez, M., Rodriguez-Lazaro, D., Esteve, T., Prat, S., and Pla, M., Development of melting temperature-based SYBR Green I polymerase chain reaction methods for multiplex genetically modified organism detection, *Analytical Biochemistry* 323 (2), 164–170, 2003.

41. Alexander, T. W., Sharma, R., Deng, M. Y., Whetsell, A. J., Jennings, J. C., Wang, Y. X., Okine, E., Damgaard, D., and McAllister, T. A., Use of quantitative real-time and conventional PCR to assess the stability of the cp4 epsps transgene from Roundup Ready (R) canola in the intestinal, ruminal, and fecal contents of sheep, *Journal of Biotechnology* 112 (3), 255–266, 2004.

42. Novelli, E., Balzan, S., Segato, S., De Rigo, L., and Ferioli, M., Detection of genetically modified organisms (GMOs) in food and feedstuff, *Veterinary Research Communications* 27, 699–701, 2003.

43. LaPaz, J. L., Esteve, T., and Pla, M., Comparison of real-time PCR detection chemistries and cycling modes using Mon810 event-specific assays as model, *Journal of Agricultural and Food Chemistry* 55 (11), 4312–4318, 2007.

44. Andersen, C. B., Holst-Jensen, A., Berdal, K. G., Thorstensen, T., and Tengs, T., Equal performance of TaqMan, MGB, molecular beacon, and SYBR green-based detection assays in detection and quantification of Roundup Ready soybean, *Journal of Agricultural and Food Chemistry* 54 (26), 9658–9663, 2006.

45. Petit, L., Baraige, F., Balois, A.-M., Bertheau, Y., and Fach, P., Screening of genetically modified organisms and specific detection of Bt176 maize in flours and starches by PCR-enzyme linked immunosorbent assay, *European Food Research and Technology* 217 (1), 83–89, 2003.

46. Brunnert, H. J., Spener, F., and Borchers, T., PCR-ELISA for the CaMV-35S promoter as a screening method for genetically modified Roundup Ready soybeans, *European Food Research and Technology* 213 (4–5), 366–371, 2001.

47. Sawyer, J., Wood, C., Shanahan, D., Gout, S., and McDowell, D., Real-time PCR for quantitative meat species testing, *Food Control* 14 (8), 579–583, 2003.

48. Laube, I., Spiegelberg, A., Butschke, A., Zagon, J., Schauzu, M., Kroh, L., and Broll, H., Methods for the detection of beef and pork in foods using real-time polymerase chain reaction, *International Journal of Food Science and Technology* 38 (2), 111–118, 2003.

49. Laube, I., Zagon, J., Spiegelberg, A., Butschke, A., Kroh, L. W., and Broll, H., Development and design of a "ready-to-use" reaction plate for a PCR-based simultaneous detection of animal species used in foods, *International Journal of Food Science and Technology* 42 (1), 9–17, 2007.

50. Taberlet, P., Gielly, L., Pautou, G., and Bouvet, J., Universal primers for amplification of three noncoding regions of chloroplast DNA, *Plant Molecular Biology* 17 (5), 1105–1109, 1991.

51. JRC-IHCP, Event-specific method for the quantification of oilseed rape line T45 using real-time PCR, http://gmo-crl.jrc.it/statusofdoss.htm.

52. JRC-IHCP, Event specific methods for the quantification of the hybrid cotton line 281-24-236/3006-210-23 using real-time PCR, http://gmo-crl.jrc.it/statusofdoss.htm.
53. Shindo, Y., Kuribara, H., Matsuoka, T., Futo, S., Sawada, C., Shono, J., Akiyama, H., Goda, Y., Toyoda, M., Hino, A., Asano, T., Hiramoto, M., Iwaya, A., Jeong, S. I., Kajiyama, N., Kato, H., Katsumoto, H., Kim, Y. M., Kwak, H. S., Ogawa, M., Onozuka, Y., Takubo, K., Yamakawa, H., Yamazaki, F., Yoshida, A., and Yoshimura, T., Validation of real-time PCR analyses for line-specific quantitation of genetically modified maize and soybean using new reference molecules, *Journal of AOAC International* 85 (5), 1119–1126, 2002.
54. JRC-IHCP, Event-specific method for the quantification of event EH92-527-1 potato using real-time PCR, http://gmo-crl.jrc.it/statusofdoss.htm.
55. JRC-IHCP, Event-specific method for the quantification of rice line LLRICE62 using real-time PCR, http://gmo-crl.jrc.it/statusofdoss.htm.
56. Pauli, U., Liniger, M., Schrott, M., Schouwey, B., Hübner, P., Brodmann, P., and Eugster, A., Quantitative detection of genetically modified soybean and maize: method evaluation in a Swiss ring trial, *Mitt. Lebensm. Hygiene* 92, 145–158, 2001.
57. JRC-IHCP, Event-specific method for the quantitation of sugar beet line H7-1 using real-time PCR, http://gmo-crl.jrc.it/statusofdoss.htm.
58. LMBG, Food analysis, collection of official methods under Article 35 of the German Federal Foodstuffs Act. L 25.03.01, 1999.
59. Lipp, M., Bluth, A., Eyquem, F., Kruse, L., Schimmel, H., Van den Eede, G., and Anklam, E., Validation of a method based on polymerase chain reaction for the detection of genetically modified organisms in various processed foodstuffs, *European Food Research and Technology* 212 (4), 497–504, 2001.
60. Akiyama, H. et al., A detection method of recombinant DNA from genetically modified potato (NewLeaf Plus potato) and detection of NewLeaf Plus potato in snack, *Journal of the Food Hygienic Society of Japan* 43, 301–305, 2002.
61. Jaccaud, E., Hohne, M., and Meyer, R., Assessment of screening methods for the identification of genetically modified potatoes in raw materials and finished products, *Journal of Agricultural and Food Chemistry* 51 (3), 550–557, 2003.
62. LMBG, Food analysis, collection of official methods under Article 35 of the German Federal Foodstuffs Act. L 24.01.01, 1997.
63. Beck, E., Ludwig, G., Auerswald, E. A., Reiss, B., and Schaller, H., Nucleotide sequence and exact localization of the neomycin phosphotransferase gene from transposon Tn5, *Gene* 19 (3), 327–336, 1982.
64. JRC-IHCP, A recommended procedure for real-time quantitative TaqMan PCR for Roundup Ready canola RT73, http://gmo-crl.jrc.it/detectionmethods/MON-Art47-pcrGT73rapeseed.pdf.
65. JRC-IHCP, Event-specific method for the quantification of oilseed rape line Ms8 using real-time PCR, http://gmo-crl.jrc.it/summaries/Ms8_validated_Method.pdf.
66. JRC-IHCP, Event-specific method for the quantification of oilseed rape line Rf38 using real-time PCR, http://gmo-crl.jrc.it/summaries/Rf3_validated_Method.pdf.
67. JRC-IHCP, A recommended procedure for real-time quantitative TaqMan PCR for Roundup Ready cotton 1445, http://gmo-crl.jrc.it/detectionmethods/MON-Art47-pcr1445cotton.pdf.
68. JRC-IHCP, A recommended procedure for real-time quantitative TaqMan PCR for Bollgard cotton 531, http://gmo-crl.jrc.it/detectionmethods/MON-Art47-pcr531cotton.pdf.
69. JRC-IHCP, Event-specific method for the quantification of cotton line LLCotton25 using real-time PCR, http://gmo-crl.jrc.it/summaries/LLCotton25_validated_Method.pdf.
70. JRC-IHCP, PCR assay for the detection of maize transgenic event Bt10, http://gmo-crl.jrc.it/summaries/Bt10%20Detection%20Protocol.pdf.
71. JRC-IHCP, Event-specific method for the quantitation of maize line DAS-59122-7 using real-time PCR, http://gmo-crl.jrc.it/statusofdoss.htm.

72. JRC-IHCP, Event-specific method for the quantification of maize line MIR604 using real-time PCR, http://gmo-crl.jrc.it/summaries/MIR604_validated_Method.pdf.

73. JRC-IHCP, Event-specific method for the quantitation of maize line MON 863 using real-time PCR, http://gmo-crl.jrc.it/statusofdoss.htm.

74. JRC-IHCP, Event-specific method for the quantitation of maize line NK603 using real-time PCR, http: //gmo-crl.jrc.it/statusofdoss.htm.

75. JRC-IHCP, Event-specific method for the quantitation of maize line TC1507 using real-time PCR, http://gmo-crl.jrc.it/statusofdoss.htm.

76. JRC-IHCP, Report on the verification of an event-specific detection method for identification of rice GM-event LLRICE601 using a real-time PCR assay, http://gmo-crl.jrc.it/LLRice601update.htm.

77. JRC-IHCP, Event-specific method for the quantification of soybean line A2704-12 using real-time PCR, http://gmo-crl.jrc.it/summaries/A2704-12_soybean_validated_Method.pdf.

78. Hird, H., Powell, J., Johnson, M. L., and Oehlschlager, S., Determination of percentage of RoundUp Ready® soya in soya flour using real-time polymerase chain reaction: interlaboratory study, *Journal of AOAC International* 86 (1), 66–71, 2003.

79. Fantozzi, A., Ermolli, M., Marini, M., Scotti, D., Balla, B., Querci, M., Langrell, S. R. H., and Van den Eede, G., First application of a microsphere-based immunoassay to the detection of genetically modified organisms (GMOs): quantification of Cry1Ab protein in genetically modified maize, *Journal of Agricultural and Food Chemistry* 55 (4), 1071–1076, 2007.

80. Ermolli, M., Prospero, A., Balla, B., Querci, M., Mazzeo, A., and Van den Eede, G., Development of an innovative immunoassay for CP4EPSPS and Cry1AB genetically modified protein detection and quantification, *Food Additives and Contaminants* 23 (9), 876–882, 2006.

81. Roussel, S. A., Hardy, C. L., Hurburgh, C. R., and Rippke, G. R., Detection of Roundup Ready™ soybeans by near-infrared spectroscopy, *Applied Spectroscopy* 55 (10), 1425–1430, 2001.

82. Xie, L. I., Ying, Y., and Ying, T., Quantification of chlorophyll content and classification of nontransgenic and transgenic tomato leaves using visible/near-infrared diffuse reflectance spectroscopy, *Journal of Agricultural and Food Chemistry* 55(12), 4645–4650, 2007.

83. Byrdwell, W. C. and Neff, W. E., Analysis of genetically modified canola varieties by atmospheric pressure chemical ionization mass spectrometric and flame ionization detection, *Journal of Liquid Chromatography and Related Technologies* 19 (14), 2203–2225, 1996.

84. Ocana, M. F., Fraser, P. D., Patel, R. K. P., Halket, J. M., and Bramley, P. M., Mass spectrometric detection of CP4 EPSPS in genetically modified soya and maize, *Rapid Communications in Mass Spectrometry* 21 (3), 319–328, 2007.

85. Rossi, S., Lesignoli, F., Germini, A., Faccini, A., Sforza, S., Corradini, R., and Marchelli, R., Identification of PCR-amplified genetically modified organisms (GMOs) DNA by peptide nucleic acid (PNA) probes in anion-exchange chromatographic analysis, *Journal of Agricultural and Food Chemistry* 55 (7), 2509–2516, 2007.

86. Xu, J., Miao, H. Z., Wu, H. F., Huang, W. S., Tang, R., Qiu, M. Y., Wen, J. G., Zhu, S. F., and Li, Y., Screening genetically modified organisms using multiplex-PCR coupled with oligonucleotide microarray, *Biosensors & Bioelectronics* 22 (1), 71–77, 2006.

87. Leimanis, S., Hernández, M., Fernández, S., Boyer, F., Burns, M., Bruderer, S., Glouden, T., Harris, N., Kaeppeli, O., Philipp, P., Pla, M., Puigdomènech, P., Vaitilingom, M., Bertheau, Y., and Remacle, J., A microarray-based detection system for genetically modified (GM) food ingredients, *Plant Molecular Biology* 61 (1), 123–139, 2006.

88. Eppendorf, DualChip® GMO kit, http://www.eppendorf.com, 2007.

89. Rudi, K., Rud, I., and Holck, A., A novel multiplex quantitative DNA array based PCR (MQDA-PCR) for quantification of transgenic maize in food and feed, *Nucleic Acids Research* 31 (11), e62, 2003.

90. Bordoni, R., Germini, A., Mezzelani, A., Marchelli, R., and De Bellis, G., A microarray platform for parallel detection of five transgenic events in foods: a combined polymerase chain reaction—ligation detection reaction—universal array method, *Journal of Agricultural and Food Chemistry* 53 (4), 912–918, 2005.

91. Germini, A., Rossi, S., Zanetti, A., Corradini, R., Fogher, C., and Marchelli, R., Development of a peptide nucleic acid array platform for the detection of genetically modified organisms in food, *Journal of Agricultural and Food Chemistry* 53 (10), 3958–3962, 2005.

92. Mannelli, I., Minunni, M., Tombelli, S., and Mascini, M., Quartz crystal microbalance (QCM) affinity biosensor for genetically modified organisms (GMOs) detection, *Biosensors & Bioelectronics* 18 (2–3), 129–140, 2003.

93. Carpini, G., Lucarelli, F., Marrazza, G., and Mascini, M., Oligonucleotide-modified screen-printed gold electrodes for enzyme-amplified sensing of nucleic acids, *Biosensors & Bioelectronics* 20 (2), 167–175, 2004.

94. Feriotto, G., Gardenghi, S., Bianchi, N., and Gambari, R., Quantitation of Bt-176 maize genomic sequences by surface plasmon resonance-based biospecific interaction analysis of multiplex polymerase chain reaction (PCR), *Journal of Agricultural and Food Chemistry* 51 (16), 4640–4646, 2003.

Chapter 38

Detection of Adulterations: Addition of Foreign Proteins

Maria Concepción García López and Maria Luisa Marina Alegre

Contents

38.1 Introduction

The addition of foreign proteins in minced meats has been a common practice in the past decades. Moreover, new developments have also made possible the introduction of foreign proteins into large pieces of muscle tissue (poultry breast, etc.). The introduction of foreign proteins in minced meat and muscle tissue requires different procedures. In the first case, the minced meat is directly mixed with the foreign protein, whereas in the second case, a brine containing water, salts, and foreign proteins is pump-injected or massaged (tumble technology) into the muscle tissue. The tumble technology has been employed primarily in the poultry industry, and to a lesser extent, in the pork and beef industry where injection is the most usual way to introduce foreign proteins.[1,2]

However, the addition of these proteins is subjected to regulations that limit or forbid their addition, which make necessary the development of analytical methods to avoid their fraudulent presence in meats. This chapter revises the type of proteins added and the most usual analytical methods employed for their detection in raw meats.

38.2 Reasons for the Addition of Foreign Proteins to Raw Meats

38.2.1 Reduction of Meat Fat Content

The claim of modern societies for healthier and less fatty diets has resulted in an increasing demand for low-fat meats.[3] However, the development of low-fat meats is not as simple as removing fat because it plays a significant role in assuring suitable organoleptic and functional properties in meat.[4] Without fat, meat tends to be tough and lacks the richness of flavor expected in meats.

There are several strategies for fat reduction: substitution of fat with leaner meats, with water, with a fat replacer (proteins, carbohydrates, or alternative fats), or with a synthetic compound. The possibility of using a protein-based fat replacer has been commonly used. In these cases, the added proteins mimic some fat characteristics, but not all because their functionality as imitation fats is limited (e.g., they do not work like fat at high temperatures, such as frying).[4] The other option is the substitution of fat with water. However, the addition of water alone is detrimental to cook yield, juiciness, and tenderness, making meats too soft, with additional shrink during cooking, and excess accumulation of purge in packages.[5] To avoid detrimental effects due to the excess water, it is advisable to use water-binding compounds such as foreign proteins.[1]

38.2.2 Improvement of Meat Characteristics

Tenderness and juiciness are considered to be the most important quality attributes of meat. This has prompted the meat industry to methodically incorporate water in meats (*brine-enhanced* or *water-added meats*). As mentioned earlier, to avoid the effects of the addition of water, foreign proteins with water-binding capacity are added.[6]

38.2.3 Reduction of Meat Cost

The addition of cheap foreign proteins can also result in an economic benefit. In this case, foreign proteins are added as *meat extenders,* partially substituting meat proteins. To accomplish this, the foreign proteins should present functional properties similar to those of meat proteins. Soybean proteins are an important candidate for supplementing meat as they are a cheap source of proteins and present functional properties similar to those of meat proteins.[7] Another approach has been the development of *restructured meats*. Restructuration of meats uses less-valuable meats to produce palatable meat products at reduced cost. Cohesion among meat pieces of structured meat products is accomplished by the formation of a protein matrix after extraction of muscle proteins, which requires the addition of salts and tumbling.[8–10] However, due to the damages produced in the muscle texture during the tumbling and to the increasing concern of consumers over the sodium content of food, nonmeat proteins have been in demand as binders.[11–13]

Moreover, the commercialization of the *water-added meats* also results in an economic benefit by increasing the yield of the product to be sold. Therefore, the commercialization of these meats prepacked for case-ready merchandising is increasing considerably and is gradually replacing traditional, nontreated meats.[6]

38.2.4   *Health Benefits*

The consumption of some foreign proteins, such as soybean proteins, is also related with health benefits. New food-based recommendations issued by the American Heart Association with the objective of reducing the risk for cardiovascular disease promoted the inclusion in the diet of specific foods such as soybean with cardioprotective effects. The available evidence indicates that the daily consumption of 25 g/day of soybean proteins could decrease total and LDL-cholesterol levels in hypercholesterolemic individuals.[1,14]

38.2.5   *Fat-Binding Properties*

Foreign proteins can also be added for their fat-binding properties. Despite the limitations from a nutritional point of view, fat presents excellent properties that make meats containing fat having better texture and flavor. However, when lean meat is combined with fat in some meat products (especially in minced meats), it is necessary to use a protein to carry the fat. Soybean proteins have widely been employed to enhance and stabilize fat emulsions.[1]

38.3   Kinds of Foreign Proteins Added to Raw Meats

The most usual foreign proteins added to meats are soybean proteins, wheat gluten, milk proteins, and egg proteins. Moreover, blood proteins can also be added although its use is very limited.[5]

38.3.1   *Soybean Proteins*

Soybean proteins were introduced as *meat extenders* in the 1970s with the purpose of lowering meat prices. The first soybean products used for this purpose were soybean flour and grits and textured soybean (all having 50% soybean proteins). However, as a consequence of the misuse of soybean in meat industry and the beany flavor derived from its use, this practice became less popular.[15] In many cases, this led to the establishment of meat regulations restricting or prohibiting the addition of soybean to meat products, but nowadays, this poor image is being overcome by the use of high-quality soybean products: soybean protein concentrates (70% soybean proteins) and soybean protein isolates (90% soybean proteins). In these products, most compounds responsible for beany flavors and flatulence have been removed, improving the taste and functionality of the soybean product and becoming an excellent supplement from a nutritional point of view.[15–17]

Soybean protein concentrate is prepared from low heat undenatured defatted soybean flour by eluting soluble components (carbohydrates, ash, peptides, and phytic acid) using acidic (pH 4.5), aqueous ethanol (70%) or hot water leaching agents. Moreover, soybean protein isolate is produced from dehulled and defatted soybeans by removing most of the nonprotein components with water or mild alkali (pH 8–9), followed by centrifugation to remove the insoluble fibrous residue. Afterward, the protein is precipitated at pH 4.5 forming a curd that is centrifuged, washed several times to remove soluble oligosaccharides, and then spray-dried. Usually, the isolate is neutralized forming sodium and potassium proteinates, which are more soluble and functional.[1] Soybean protein isolate presents a greater emulsion capacity (sixfold) than soybean concentrate; it brings about lesser moisture loss than whey protein concentrate and sodium caseinate.[18]

Hydrolyzed soybean proteins, which have a high solubility in water, have also been used in the meat industry. Hydrolyzed vegetable proteins, in general, are used for their flavoring properties. In particular, in the case of meat products they are added to enhance the meaty flavor.[19] Moreover, hydrolyzed soybean proteins present interesting water-holding properties due to strong hydrophilicity of soybean peptides.[2]

38.3.2   Other Vegetable Proteins

Wheat gluten is another vegetable protein usually incorporated into meat products. The major problem with gluten is its solubilization. Therefore, soluble wheat gluten is manufactured by non-enzymatic acid deamidation. Deamidation generally improves the solubility and the emulsifying properties of proteins by imparting additional negative charges that decrease their isoelectric point. Since gluten has a considerable number of amino acids containing amide groups (glutamine and aspargine), which are turned into glutamic and aspartic acids by deamidation, its solubility is significantly improved by this procedure.[5,9] Wheat gluten enhances product sliceability, yield, flavor, and color. Wheat gluten can also act as a water binder and presents limited emulsifying capacity.

38.3.3   Milk Proteins

Various milk products (nonfat dry milk, whey proteins [delactose whey, lactalbumin, whey protein concentrate, dried sweet whey], sodium caseinate, etc.) have also been added to meat products. Milk proteins enhance emulsion stability, water immobilization, texture, color, and organoleptic qualities of meat.[5,20]

Various sodium caseinates have been used in meat industry as one of the best-known foreign protein sources. Sodium caseinate contains approximately 90% protein and is completely soluble in water and in solutions with pH lower than 9. Moreover, its solubility at these pHs is negligibly affected by the presence of salt. Sodium caseinate can emulsify up to 188 mL of oil/g of protein. The incorporation of sodium caseinate results in a high moisture loss compared with soybean protein isolate. Moreover, the swelling capacity of sodium caseinate disappears with heat treatment, observing the highest moisture loss. Potassium caseinate and soybean-sodium-proteinate are more effective as emulsifiers at higher pH (10.5) and tend to have the greatest emulsifying capacity at lower ionic strength.[20]

Several authors have studied the influence of the addition of milk preparations to meat products. Kumar et al. utilized milk coprecipitates (obtained by heating skim milk at 90°C for 2 min with continuous stirring and addition of calcium chloride) for the preparation of restructured buffalo meat, reporting a significant cooking yield, less percentage of shrinkage, and improved sensory attributes.[13] Ozimek and Poznánski studied the effect of the addition of textured milk proteins on the physicochemical properties (water absorption capacity, water-holding capacity, thermal shrinkage, consistency, viscosity, fat absorption capacity, emulsifying capacity, emulsion stability, and active acidity) in ground beef meats.[21]

One drawback of milk proteins in comparison with soybean proteins is its higher cost. In fact, the increased cost of meat products containing milk proteins has led to the partial replacement of these proteins with a cheaper one with similar quality such as soybean proteins. Owing to the sweet flavor of nonfat dried milk derived from its lactose content, its partial substitution with soybean proteins can be recognized. The addition of sugar or corn syrup to give to the finished product a sweet taste similar to the original product is a common practice in these cases.

38.3.4 Egg Proteins

Egg white proteins have been used as binders in meat industry. Lu and Chen compared the performance of egg white proteins to other binders (wheat gluten and soybean protein isolate) in meats from different species (bovine, porcine, and lamb). Results indicated that egg white proteins were superior to all other binders in binding muscle to muscle, muscle to fat, and fat to fat.[11]

38.4 Need for Analytical Methods to Control the Use of Foreign Proteins in the Manufacture of Meats

The increasing use of foreign proteins in the manufacture of meat products has paralleled the demands for regulations controlling this practice. Nowadays, all developed countries have elaborate regulations limiting, or even forbidding the addition of foreign proteins in the meat industry. These laws not only regulate the addition of these proteins but also control the adequate labeling of meat products.[22] The need for these regulations is reasonable. In addition to possible economic advantages from the substitution of meat proteins by cheaper ones, most of the foreign proteins used by meat industry are allergens and their undeclared addition can result in serious consequences for sensitive people.[23,24] For example, in the case of soybean proteins, different allergens (trypsin inhibitor, Gly m Bd 68 K, Gly m Bd 30 K, also called P34 or oil body-associated protein, Gly m Bd 28 K, etc.) have been detected.[25–29] Gluten is another protein used in the manufacture of meat whose uncontrolled addition leads to nutrition disorders in people suffering from the celiac disease.[25,30]

Therefore, these regulations require the existence of suitable analytical methods enabling their fulfillment. However, this has been a concern due to the absence of reliable, sensitive, and quantitative analytical methods. If the addition of foreign proteins is forbidden, then the method must be sensitive enough to detect small additions. For example, in the case of soybean proteins, concentrations well below 100 ppm (0.01%) can cause allergic reactions and the recommended detection limit is 1–2 ppm.[31] Moreover, when the addition is admitted to a certain extent, quantitative methods are necessary.

38.5 Methods Used for the Detection of Foreign Proteins in Raw Meats

The identification of foreign proteins in meat products poses several difficulties. One of the most important limitations is that the proportion of these proteins is very low in comparison with meat proteins that can elude their detection. Most efforts have been focused on the determination of soybean proteins in meats that, on occasions, also contain other added proteins. Several approaches have been investigated and reviewed.[32–34] The oldest methods were based on the observation of certain singular cells under the microscope, on the analysis of stains developed from carbohydrates or proteins (histological methods), or on the indirect determination of soybean proteins. Normally, these methods were applied to the detection of soybean flour or textured soybean and are of limited use when adding concentrates and isolates.[32–34]

The direct determination of the foreign proteins themselves is the most commonly used strategy. Electrophoretic and immunologic methods are the most significantly used. Moreover, chromatographic methods based on the analysis of proteins, certain peptides, or amino acids have also been applied for this purpose.

38.5.1 Electrophoretic Methods

Polyacrylamide gel electrophoresis (PAGE) in both slab and tube modes has been the electrophoretic method applied for the determination of foreign proteins in meat products. The extraction or solubilization of proteins from the meat requires the use of detergents or concentrated solutions of urea or guanidine hydrochloride in the presence of a reducing agent such as mercaptoethanol to disrupt hydrogen, hydrophobic, and disulfide bonds.[32,33,35] Moreover, most PAGE methods developed used sodium dodecyl sulfate (SDS), which changes the separation principle. In the absence of SDS, proteins move to the electrodes according to their net charge depending on the pH; a careful control of this movement is necessary. In the presence of SDS, proteins are surrounded by a shell of SDS molecules that neutralize the positive charges of proteins; hence, the motion of molecules is dependent on their molecular weights. Protein separations on the basis of their molecular weights provide more consistent and reproducible electrophoretic separations.[36]

The first application of PAGE to the detection of foreign proteins in meat products was reported in 1969 when Olsman applied this technique in the slab mode for the detection of casein and soybean proteins in heated meats.[37] Subsequent works applied to the analysis of foreign proteins in raw meats have been grouped in Table 38.1.

Parsons and Lawrie used PAGE in slabs for the first time for the determination of soybean proteins in fresh meats. Laser densitometry of electropherograms corresponding to proteins extracted with 10 M urea enabled the identification of two different bands, one characteristic of soybean proteins and the other of meat proteins. A linear relationship was achieved when the ratio of the areas of these two bands in different soybean–meat mixtures were plotted against composition. This calibration plot was used to estimate the quantity of soybean proteins in meat products.[38] Later, van Gils and Hidskes developed a more comprehensive electrophoretic method enabling the simultaneous detection of different foreign proteins in meats extracted with 8 M urea. They could separate casein, soybean proteins, and egg white proteins without interference from meat proteins. To achieve this, they employed cellulose acetate membranes instead of polyacrylamide gels, which made possible removal of many interfering bands from meat and the detection of up to 0.1% of casein and soybean proteins and 0.5% of egg white proteins.[39]

A comparative study of two different methods for the extraction of foreign proteins in raw meat (beef) using different concentrations of urea (8 M urea in method 1 and 10 M urea in method 2), different concentrations of 2-mercaptoethanol (1% in method 1 and 4% in method 2), different extraction temperatures (18–20°C in method 1 and 100°C in method 2), and different extraction times (16 h in method 1 and 30 min in method 2) was accomplished by Guy et al.[40] They concluded that both methods were suitable for the determination of soybean proteins in raw

Table 38.1 Electrophoretic Methods Used for the Analysis of Foreign Proteins in Raw Meats

Sample	Fat Extraction Medium	Protein Solubilization Medium	Technique and Detection	Running Buffer	Detection Limit	References
Model samples consisting of fresh beef, lamb, and pork meats mixed with soybean protein concentrate	HCl/EtOH and acetone	0.05 M Tris-citrate buffer (pH 8.6) containing 10 M urea and 2% 2-mercaptoethanol	Urea-PAGE (in slab) with pore gradient (3–8% acrylamide) and staining with Naphthalene Black	0.05 M Tris-citrate buffer (pH 8.6) containing 10 M urea and 2% 2-mercaptoethanol	—	38
Model samples consisting of meat mixed with casein, soybean proteins, and egg white proteins	Hot distilled water	Soybean proteins and caseins: 8 M urea and incubation of the extract with acetate buffer (pH 4.6) Egg white proteins: 8 M urea/0.15 M thioglycol	Electrophoresis (in slab) on cellulose acetate membranes and staining with Nigrosin	Casein: barbitone/urea buffer Soybean and egg-white proteins: formic acid/urea buffer	0.1% for casein and soybean proteins and 0.5% for egg white proteins	39
Beef burger elaborated in the laboratory and containing soybean proteins, milk powder, egg proteins, and gluten	Acetone	Method 1: 1% 2-mercaptoethanol and 8 M urea Method 2: 4% 2-mercaptoethanol and 10 M urea	Urea-PAGE (in tube) (6% acrylamide) with 6 M urea (method 1) or 8 M urea (method 2) and staining with 0.1% Naphthalene Black	0.06 M Tris-glycine buffer (pH 8.6)	—	40, 41
Model samples consisting of beef meat and soybean protein isolate or textured soybean	Acetone	0.0625 M Tris-HCl buffer (pH 6.8) with 3% SDS and 1% 2-mercaptoethanol	Stacking SDS-PAGE (in tube) (10% acrylamide) and detection by staining with Coomassie Blue	0.025 M Tris-HCl buffer containing 0.1% SDS and 0.192 M glycine	—	42

(Continued)

Table 38.1 (Continued)

Sample	Fat Extraction Medium	Protein Solubilization Medium	Technique and Detection	Running Buffer	Detection Limit	References
Model meat samples consisting of soybean proteins, cottonseed proteins, peanut proteins, casein, whey proteins, nonfat milk powder, and egg white proteins						43
Fresh meats containing soybean proteins	HCl/EtOH and acetone	Phenol:acetic acid: water (2:1:1) and urea	Urea-PAGE (in tube) (7.5% acrylamide) and staining with Aniline Blue	10% acetic acid and 0.1% mercaptoethanol	2%	45
Model samples consisting of pork and beef meats mixed with soybean, egg white, milk, egg yolk, and wheat proteins	—	Tris-boric acid buffer (pH 8.2) containing 1% SDS	SDS-PAGE (in slab) (8% acrylamide) and detection by staining with Coomassie Blue	Tris-boric acid buffer (pH 8.2) containing 1% SDS	—	36, 47, 48
Model samples consisting of beef, chicken, lamb or pork mixed with soybean proteins	—	Urea and mercaptoethanol	IEF in polyacrylamide gels (in slab)	Urea and carrier ampholytes (pH 3–11)	1%	49
Model samples consisting of pork or beef meat containing soybean protein isolate and casein	—	Soybean proteins: 8 M urea Casein: hot water	PAGE (in tube) (7.5% acrylamide) and staining with Amidoblack	Tris-HCl buffer (for casein) with 8 M urea (for soybean proteins)	1% for soybean proteins and casein	50

Model samples consisting of fresh beef and pork meat and containing soybean flour, textured soybean, or soybean protein isolate and casein, milk powder, whole blood, etc.	Acetone	3% SDS and 3% 2-mercaptoethanol in a 0.248 M Tris-borate buffer (pH 8.2)	SDS-PAGE (in slab) (12% acrylamide) and detection by staining with Coomassie Blue	2.1 M Tris-HCl buffer (pH 9.18)	1% for soybean proteins	51
Model samples consisting of soybean and beef meat	—	0.01 M phosphate buffer (pH 7) containing 1% SDS, 1% 2-mercaptoethanol, and 8 M urea	SDS-PAGE (in tube) (10% acrylamide) and staining with Coomassie Blue	0.01 M phosphate buffer (pH 7) containing 1% SDS	—	52

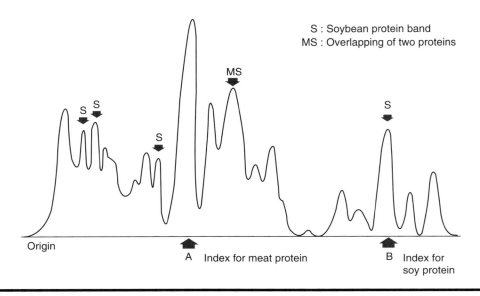

Figure 38.1 Densitometer scan of a gel containing a soybean–meat blend. The unlabeled bands were from meat proteins whereas S and MS bands were from soybean proteins. (From Lee, Y.B., Rickansrud, D.A, Hagberg, E.C., Briskey, E.J., and Greaser, M.L., *J. Food Sci.*, 40, 380, 1975. With permission.)

meat products although method 1 provided a better separation of proteins in cooked meats.[40] Later, the same group investigated the reproducibility and the presence of interferences from other nonmeat proteins (milk proteins, egg proteins, and gluten) in the application of method 1 for the quantitation of soybean proteins in raw beef. They observed that this method was reproducible and free from serious inteferences.[41]

Lee et al. tried to increase the sensitivity in the detection of soybean proteins in meats and applied, for the first time, stacking SDS-PAGE in tubes to the quantitative analysis of soybean proteins in raw beef.[42] This technique preconcentrates proteins into a very thin zone, enabling a higher resolution than other electrophoretic techniques that do not include this stacking gel feature. Figure 38.1 shows the densitometer scan of a gel containing both meat and soybean proteins. Soybean proteins from both soybean protein isolate and textured soybean were separated in five bands, one overlapping with meat proteins (MS band). Peak B was chosen for the quantitation of soybean proteins because it was completely separated from meat protein bands and had a large peak area. To avoid irreproducible results derived from the fact that the peak area was susceptible to the degree of destaining, a correction was made by plotting the area ratio peak B/peak A against the concentration ratio of soybean/meat proteins. They also applied this method to the qualitative detection of soybean, cottonseed, and peanut protein in model mixtures of beef-vegetable protein and to the detection of meat and soybean proteins in meat products also containing nonfat milk powder, casein, whey proteins, and egg white proteins. Vegetable proteins yielded characteristic patterns, making possible their clear identification. Moreover, meat and soybean proteins were also successfully identified and quantified in the presence of milk and egg white proteins.[43] This method was later modified by Persson and Appelqvist for its application in gel slabs. They could also discriminate the electrophoretic patterns obtained from different foreign proteins (egg, milk, whey, and soybean proteins) in meats and applied it to the determination of soybean proteins in raw meats.[44]

In 1977, Homayounfar adopted a different approach to the determination of soybean proteins in fresh meats.[45] He used urea-PAGE on glass tubes and performed protein separation in an acid environment instead of the neutral or slightly alkaline conditions used so far. Four bands pertaining to soybean proteins were observed when the concentration of soybean proteins in the meat was higher than 5%. When soybean proteins were present in a lower proportion, only two bands could be detected. This method was later used by Baylac et al. for the determination of different foreign proteins (soybean proteins, caseinate, whey proteins, egg white proteins, gluten, and blood plasma) in model raw sausages prepared in the laboratory.[46]

Hofmann also applied SDS-PAGE in gel slabs to the identification and determination of meat (pork and beef), soybean, egg white, milk, egg yolk, and wheat proteins in meat products. Every protein showed a characteristic pattern that enabled its identification in model mixtures of meat containing foreign proteins. Moreover, he found that the heat treatment did not affect the position of the protein bands but made them less intense. In addition, soybean protein pattern was independent of the source and the type of soybean product used. Egg yolk and wheat proteins could not be identified because the protein pattern was very complex in the case of egg yolk and the proteins of wheat were not stained properly.[36,47,48]

Another innovation in the detection of foreign proteins in meats was made by Flaherty who applied, for the first time, isoelectric focusing (IEF) for this purpose. A multitude of finely focused protein bands were obtained from soybean proteins that made IEF profiles of meat–soybean protein mixtures very complex. For analyzing such mixtures, it was necessary to coagulate and remove interfering meat proteins.[49]

Vállas-Gellei tried to reduce the complexity and difficulties of the methods previously developed by using disc electrophoresis.[50] To achieve this, he employed fresh beef and pork meats spiked with soybean protein isolate and sodium caseinate. He found that light heat treatment at 74°C for 150 min weakened the intensity of meat protein bands due to the high sensitivity of meat proteins to thermal denaturation. Soybean and casein bands remained unchanged or even stronger, facilitating the detection of these foreign proteins in fresh meats.[50]

Molander developed a SDS-PAGE method in gel slabs for the determination of soybean proteins in meats and compared standard graphs obtained with different soybean protein sources to study the dependence of the method on the soybean source. Figure 38.2 shows the densitometry scans obtained from the electropherograms corresponding to soybean proteins, meat proteins, and mixtures of soybean and meat proteins. Soybean proteins were separated in five major bands (S1–S5) isolated from meat bands (A—actin and M—myosin). Raw blends and slightly heated products showed almost identical standard curves using soybean protein isolate and soybean flour. He also observed that the presence of other ingredients (milk powder, potato flour, bread-crumbs, casein, and whole blood) did not seem to affect the determination of soybean proteins.[51]

The results obtained by SDS-PAGE in the determination of soybean proteins in soybean–beef blends were used as reference in the evaluation of the soybean protein content forecasted by applying statistical analysis to the data related to the contents of some major chemical constituents (protein, fat, fiber, ash, and total carbohydrates) of the soybean–meat blends, observing good correlations.[52]

38.5.2 Immunological Methods

Most of the limitations found in the application of immunoanalysis for the determination of foreign proteins in raw meats are related with the fact that the antigenic properties of the proteins should be retained during extraction from the sample to be analyzed. To avoid any risk, dilute buffer solutions are mostly used for their extraction. However, these mild extraction conditions result in low extraction

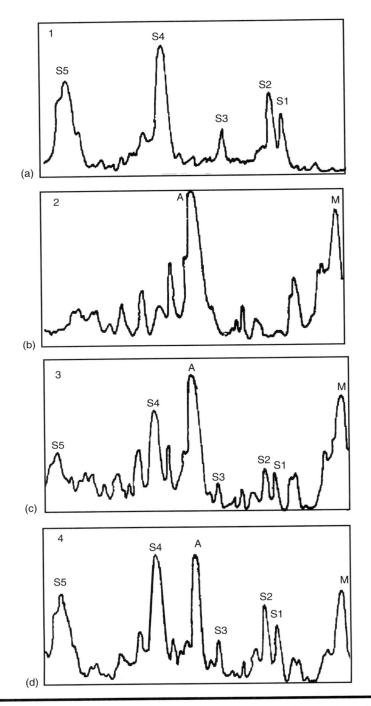

Figure 38.2 Densitometer scan of a gel containing soybean proteins (a), meat (b), and two soybean–meat blends containing 6% of soybean proteins (c) and 10% of soybean proteins (d). A: actin, M: myosin, S1–S5: soybean bands. (From Molander, E., *Z. Lebensm. Unters. Forsch.*, 174, 278, 1982. With permission.)

efficiency. Obviously, the use of reagents employed in electrophoresis (e.g., urea or SDS) could not only increase the extraction efficiency of the methods but also modify the proteins that could not form complexes with their antibodies. Moreover, in the case of processed meats, there is an additional limitation related with the structural changes occurring in foreign proteins due to the processing, which is not a problem while working with raw meats. All these problems have limited the establishment of a routine analytical method for the detection of soybean proteins in food analysis.[32,53]

Table 38.2 groups the immunological methods that have been developed and applied to the analysis of foreign proteins in raw meat products.

Rossebo and Nordal designed a serological method to detect the presence of soybean proteins by using a specific antiserum against soybean trypsin inhibitor. The method was valid only for raw meats because trypsin inhibitors were inactivated by heat treatment.[54]

Immunoelectrophoretic methods, in which proteins were first separated and subsequently diffused against an antiserum applied in a gel medium, appeared later, first, in the single version and second, in the double version (Ouchterlony technique). Immunoelectrophoresis was also applied for the first time in heated meat products (sausages) because it is not useful in the case of raw meats. Kamm applied immunoelectrophoresis for the quantitation of soybean proteins obtaining a high sensitivity at low cost. The method was useful with raw meats but failed with heated meats.[55]

In 1978, Koh published a different approach and developed a method enabling the identification and quantification of soybean proteins in both raw and heated beef mixtures. To achieve this, soybean proteins were extracted under denaturing conditions using urea and mercaptoethanol, which were removed later by dialysis. Thus, whatever be the initial protein conformation, all soybean proteins attained a common conformation. Antibodies were then raised against these renatured proteins, which were used to assay similarly renatured antigens by immunoelectrophoresis. Good agreements were obtained between the real and observed contents in both raw and cooked meats.[56]

Based on Koh's idea for the extraction of proteins, Hitchcock et al.[57] developed an enzyme-linked immunosorbent assay (ELISA) in which antibodies were raised against soybean proteins renatured from a hot urea solution by removing or diluting the denaturant. In this method, the sample was reduced with acetone-dried powder, which was solubilized in a hot concentrated solution of urea in aqueous buffer.[57] The solution was then cooled, diluted, and treated with a known excess of diluted soybean protein antiserum. The soybean protein in the sample (the antigen) interacted with the antibody whereas the unreacted antibody was trapped on an immunosorbent that contained an immobilized standard of soybean antigen. The captured antibody was a rabbit serum globulin that was determined after adding a second antibody (goat antirabbit) to which an enzyme had been covalently attached (conjugate). Each interaction was followed by a washing step to remove any nonimmobilized species. The captured enzyme (alkaline phosphatase) was determined by adding *p*-nitrophenyl phosphate as chromogenic substrate. Finally, the optical density after incubation was measured at 405–410 nm. This method was subjected to various collaborative studies in 1984 and 1985 for the determination of soybean proteins in uncooked sausages. The results showed that the designed procedure provided a reliable identification of soybean proteins in the meat being useful for (semi)quantitative determinations. In fact, reliable quantitative analysis could be obtained only if the source of soybean proteins was known and available for calibration. Other limitations are different analytical responses displayed by raw and sterilized products, high time consumption (several days were needed to prepare samples), and reagent requirements.[58–60] Despite these limitations, this method was recommended for its adoption as AOAC official first action with an applicability statement that the method is semiquantitative but may be quantitative when the nature of the added soybean protein is known, especially when a sample is available for calibration.

Table 38.2 Immunological Methods Employed in the Analysis of Foreign Proteins in Raw Meats

Sample	Fat Extraction Medium	Protein Solubilization Medium	Technique	Detection Limit	References
Model samples consisting of raw sausage and soybean protein isolate	—	Extraction with a solution of 2% NaCl	Ouchterlony's double immunodiffusion Sera against soybean trypsin inhibitor	—	54
Model samples consisting of raw meat and soybean proteins	—	Extraction with sodium carbonate (pH 8.5)	Immunoelectrophoresis in barbital buffer (pH 8.7) Sera against crude soybean globulin	0.1%	55
Model samples consisting of beef skeletal and cardiac muscles and soybean protein isolate	HCl/ethanol/acetone	0.05 M Tris-HCl buffer (pH 8.6) containing 10 M urea and 2% 2-mercaptoethanol	Immunoelectrophoresis in barbital buffer (pH 8.6) Sera against renatured soybean proteins	—	56
Model beef burgers containing soybean proteins	Cl₃CH/MeOH EtOH/HCl Acetone	1 M Tris-HCl buffer (pH 8.6) containing urea and mercaptoethanol	ELISA Sera against renatured soybean proteins	—	59
Bacon and minced beef meat containing soybean flour or textured soybean	—	13.3 M urea, 18.8 mM DTT, 0.05 M Tris-HCl buffer (pH 8.6)	ELISA Sera against renatured soybean proteins	—	61, 62
Raw beef meat spiked with soybean proteins	—	0.1 M Tris-HCl buffer (pH 8.1)	ELISA	0.22%	63

Note: DTT: dithiothreitol; SDS: sodium dodecyl sulfate; DTE: dithioerythritol.

Most of the analytical methods proposed so far did not satisfy the needs of a control laboratory for routine analysis of meats. To reduce the time consumed in these analyses, Rittenburg et al.[61] proposed an improved quantitative ELISA method in which standardized and commercial reagents were employed. Since these reagents were commercially available, the analysis, including sample preparation, could be completed within a working day, leading to a significant reduction of analysis time in comparison with the previous methods; recovery was higher in raw meats than in heated meats and the kind of soybean product added in the meat product influenced its quantitative determination. Moreover, the assay enabled the measurement of soybean protein additions ranging from 1 to 10%.[62] These works resulted in the commercialization of a soybean protein enzyme immunoassay kit. The performance of this kit was evaluated by Hall et al.,[62] who reported the results obtained in a collaboration trial in which 14 different laboratories determined soybean proteins in raw and heated meat products. They concluded that using a single arbitrary soybean standard as a reference, the level of soybean protein in a raw or pasteurized meat product of entirely unknown composition may be reliably estimated with suitable repeatability and reproducibility (RSD values of 1 and 2%, respectively) and with recoveries ranging from 80 to 100% of soybean proteins. These results were better than the ones previously obtained, especially for heated meat products.[62]

SDS-PAGE coupled with immunoblotting has also been applied to the detection of soybean proteins in raw meats.[63] Berkowitz and Webert tried to prepare a polyclonal antibody that could recognize a broad spectrum of soybean proteins. Despite the good results obtained and the low detection limits yielded by this method, its use was limited with textured products due to the structural changes caused by processing.

More recently, Koppelman has proposed a different extraction media to improve sensitivity in the determination of soybean proteins in processed foods. Detection limits obtained so far are around 0.1% (1000 ppm), which is enough for the detection of soybean added to food products as nutritional fortifier or functional ingredient, but not to detect trace additions, which can cause terrible consequences for allergic people. The use of an extremely high pH (12) for the extraction of soybean proteins yields higher recoveries than that observed while using native conditions (Tris buffer, pH 8.2) or while employing the commercial available test in which soybean proteins are extracted with urea and dithiothreitol. The use of these conditions in ELISA analysis of soybean proteins enabled the detection of up to 1 ppm of soybean proteins in meats.[31]

Finally, Meyer et al.[64] compared the performance of a commercial ELISA test (based on polyclonal antibodies against renatured soybean proteins) in the analysis of soybean proteins in meats with the results obtained by the analysis of the soybean DNA (corresponding to the Lectin gene *Le*1). The trial consisted of the analysis of 47 products and included different meats (beef, pork, turkey, chicken, and sheep) with and without different amounts of soybean proteins from different sources. The ELISA kit yielded higher recoveries and could quantify soybean proteins in meat products. However, sample preparation using a denaturation and renaturation step was found to be time-consuming. In contrast, the oligonucleotides used in PCR were synthesized rapidly and could be stored for several years. They concluded that PCR could be an interesting method to confirm ELISA results.[64]

Other immunological methods for the determination of foreign proteins, which are different from soybean proteins, in raw meats have not been found.

38.5.3 Chromatographic Methods

Chromatography has also been applied to the determination of foreign proteins through the analysis of amino acids, peptides, or the whole proteins. Table 38.3 groups the chromatographic methods developed for the determination of foreign proteins in raw meat products following some of these three strategies.

The chromatographic analysis of amino acids is based on the comparison of the amino acid pattern of the meat containing the foreign protein(s) with those corresponding to the foreign protein(s) and meat proteins. A computer program based on a regression method is used to determine the different types of proteins present in the sample.[32] Amino acid analysis is advantageous since they are least prone to changes on processing. The major difficulties arise from the fact that all proteins contain all the major 17 amino acids, although in varying amounts, and that soybean and muscle proteins present a similar amino acid composition.[35,65–69]

One of the first applications of amino acid analysis for the identification and quantitation of proteins was carried out by Lindqvist et al.[65] They used a stepwise multiregression analysis adapted to perform the comparison of the amino acid pattern from a composite sample with those of simple substances arranged in a data bank. This publication focused on describing the program performance that automatically selected from the bank those proteins whose amino acid patterns best matched with that of the sample and calculated the proportion of every protein in the mixture. The program was applied to two cases, one of them consisting of a protein mixture containing soybean proteins. Ten years later, Lindberg et al. developed methods consisting of acid hydrolysis of the sample, derivatization, and separation by high-performance liquid chromatography (HPLC) to obtain a pattern that was subjected to multivariate processing (partial least-squares calibration).[66,67] Proteins were first calibrated and the calibration coefficients were then used to predict the amount of these proteins in the model systems. They first applied this procedure to the prediction of meat proteins (muscle and collagen proteins) in model systems containing interference at different levels of soybean proteins.[66] Later on, they slightly modified the procedure and applied it to the simultaneous determination of muscle protein, collagen, soybean proteins, casein, and whey in raw and heated model meat pork systems. They observed that the accuracy for raw systems (3%) was higher than that obtained for heated systems.[67]

Zarkadas et al. also applied amino acid chromatography for quantitative purposes in hydrolysates from meat and soybean protein mixtures. They observed that the amino acid composition and the total collagen in composite meats could be used as indices for evaluating their protein quality and for the estimation of the nutritive value of these blends.[68]

More recently, Zhi-Ling et al. have applied a similar strategy to the determination of muscle, collagen, wheat, shrimp, and soybean proteins in model meat systems. Owing to the similar amino acid composition of soybean, shrimp, and muscle proteins, their determination was less accurate than in the case of wheat and collagen proteins.[69]

Another approach for the determination of foreign proteins in meat samples is the chromatographic analysis of soluble peptides obtained by enzymatic digestion. In fact, the existence of specific peptide characteristics of soybean proteins and meat proteins has been used for the identification and determination of these proteins. Denaturation of proteins by heating had a negative impact on the sensitivity and accuracy of methods due to the aggregation of individual protein molecules into large groups, which were then difficult to dissolve. These problems were overcome by hydrolyzing the proteins to more readily dissolved peptides. This idea, first applied to the determination of meat proteins, was translated by Bailey to the determination of soybean proteins.[70] Bailey isolated characteristic peptides from soybean proteins and meat obtained by trypsin digestion with a previous denaturation step to bring all proteins to the same random coil conformation, rendering the results independent of manufacturer processing conditions. The peptides were separated by isocratic elution using an ion-exchange resin, and a particular pentapeptide (SP$_1$: *Ser-Gln-Gln-Ala-Arg* from 11S globulin) characteristic of soybean enabled the identification of its presence in soybean–meat blends from raw, cooked, sterilized, or dried products containing soybean flour, soybean

Table 38.3 Chromatographic Methods Used for the Analysis of Foreign Proteins in Raw Meats

Sample	Sample Treatment	Separation Conditions	Detection	Detection Limit	References
Determination of amino acids					
Model samples consisting of ground beef meat (muscle and collagen) mixed with soybean (textured and isolate)	Hydrolysis with 6 M HCl and dissolved in 0.04 M lithium carbonate (pH 9.5)	Reversed-phase (Hypersyl ODS, 150 × 4 mm) Elution gradient 0–100% B in 15 min Mobile phases: A, ACN/water 20:80 with 0.03 M phosphate buffer (pH 6.5); B, ACN/water 40:60	Derivatization with dansyl chloride (250 nm)	—	66
Model samples consisting of muscle pork meat, rind, soybean, casein, and whey		Reversed-phase (Hypersil C18, 150 × 4 mm) Elution gradient 0–100% B in 25 min Mobile phases: A, ACN/water 15:85 with 0.03 M phosphate buffer (pH 7.5); B, ACN/water 35:65		—	67
Model samples consisting of muscle meat and soybean	Fat extraction with Cl₃CH/MeOH/water (1:2:0.8) and hydrolysis with 6 M HCl	—	—	—	68
Model samples consisting of muscle meat, collagen, shrimp, soybean, and wheat proteins	Hydrolysis with 6 M HCl	Reversed-phase (PICO TAG, 150 × 3.9 mm) Elution gradient 54–100% in 10 min Mobile phases: A, acetate/ TEA buffer (pH 6.4):ACN (47:3); B, ACN:water (60:40)	Derivatization with TEA-phenylisothiocyanide (254 nm)	—	69

(Continued)

Table 38.3 (Continued)

Sample	Sample Treatment	Separation Conditions	Detection	Detection Limit	References
Determination of Peptides					
Model samples consisting of meat–soybean mixtures	Fat extraction with Cl₃CH/MeOH, HCl/EtOH, and acetone Denaturation with a buffer (pH 8.1) and trypsin digestion	Cation-exchange (Aminex A5, 250 × 9 mm) Elution buffer: 0.2 N sodium citrate (pH 5.47)	Derivatization with ninhydrin (570 nm)	5–10%	70–72
Model samples constituted by fresh beef meat–soybean protein isolate mixtures	Fat extraction with Cl₃CH/MeOH and sample treatment with a 0.05 M Tris-HCl buffer containing 0.0086 M CaCl₂ and trypsin digestion	Cation-exchange (Aminex A5, 650 × 6 mm) Elution gradient 0.033–0.067 M sodium citrate Flow rate, 0.4 mL/min	Derivatization with ninhydrin (570 nm)	2%	74
Determination of Proteins					
Model samples consisting of beef, chicken, lamb, or pork meats mixed with soybean proteins	Protein extraction with acetate buffer (pH 5)	Anion-exchange (Al-Pellionex-wax) Elution with acetate buffer (pH 5)	UV at 254 nm	—	49
Model samples consisting of meats from different species (beef, pork, chicken, and turkey) containing soybean proteins, whey, and casein	Protein extraction with 0.05 M phosphate buffer (pH 7) containing 0.5% SDS and 0.1% 2-mercaptoethanol	Reversed-phase (Hi-Pore RP-304, 250 × 4.6 mm) Elution gradient 30–75% B in 70 min Flow rate, 1.5 mL/min Mobile phases: A, 0.1% TFA in water; B, 95:5:0.1 ACN:water:TFA	UV at 280 nm	1% for soybean proteins, casein, and whey proteins	75

Model samples consisting of beef meat spiked with soybean protein isolate	Protein extraction with 0.02 M Tris-HCl buffer (pH 7) containing 8 M urea, 0.01% EDTA, and 0.01% DTT	Anion-exchange (Aquapore AX-300, 220 × 4.6 mm and 30 × 4.6 mm) Elution gradient 0–0.5 M NaCl (for the column 200 × 4.6 mm) in 30 min and 0–2.0 M NaCl (for the column 30 × 4.6 mm) in 20 min Flow rate, 1 mL/min	UV at 280 nm	—	76
	Protein extraction with 0.05 M Tris-HCl buffer (pH 7.5) containing 0.01% EDTA and 0.1% DTT	Reversed-phase (Aquapore RP-300, 220 × 4.6 mm) Elution gradient 20–70% B in 50 min Flow rate, 1 mL/min Mobile phases: A, 0.05% TFA in water; B, 0.05% TFA in ACN	Derivatization with 4-vynylpyridine (280 nm)	2%	
Pork, turkey, chicken, and beef meats products containing soybean protein isolate and milk proteins	Fat extraction with acetone Protein extraction with a 50 mM Tris-HCl buffer containing 0.5% 2-mercaptoethanol	Reversed-phase POROS R2/H (50 × 4.6 mm) Elution gradient 5–25% B in 0.8 min; 25–42% B in 0.8 min; 42–50% B in 0.6 min Flow rate, 3 mL/min Mobile phases: A, 0.05% TFA in water; B, 0.05% TFA in ACN	UV at 280 nm	0.07% for soybean proteins	77–79

Note: TEA: triethylamine.

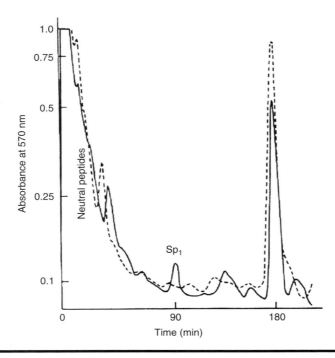

Figure 38.3 Chromatographic separation of the peptides obtained by trypsin digestion of a raw meat (- - - -) and a soybean protein isolate (-------). SP₁: soybean polypeptide. (From Bailey, F.J., *J. Sci. Food Agric.*, 27, 827, 1976. With permission.)

concentrate, soybean isolate, or texturized soybean.[31,70–72] Figure 38.3 shows the chromatographic pattern obtained for soybean proteins and for a raw meat, observing the possibility of detecting the polypeptide SP₁ pertaining to soybean proteins in soybean–meat blends. However, the peak of interest from soybean proteins was not totally resolved from a preceding peak from meat and the results obtained were subject to error. Moreover, the continuous use of the cation-exchange resin column proposed by Bailey rapidly deteriorated the column. Llewellyn et al. improved Bailey's method by the filtration of extracts before separation and the use of a larger column.[73] They applied this method to heated meat products observing that the previous SP₁ pentapeptide disappeared because it did not pass through the ultrafiltration membrane and another two peaks (designated as SP 2A and SP 2B) that did not appear in the meat were used for the determination of soybean proteins. Despite these efforts, the method was not accurate because these two soybean peptides overlapped with some minor peaks from meat.[73] Finally, Agater et al.[74] improved the method using an even longer column from which peptides were eluted with a mobile-phase gradient and applied the method to the determination of soybean and meat proteins in beef–soybean protein isolate mixtures. Despite these improvements, the method presented some limitations for quantitative purposes, especially if the soybean protein source was unknown. Moreover, its application for routine analysis was not possible since the total time required for a single analysis was 5–6 days.

Another approach for the analysis of soybean proteins in meats was based on the chromatographic profiling of proteins. Flaherty used different chromatographic modes for that purpose: molecular exclusion, cation exchange, and anion exchange.[49] Molecular exclusion and strong cation-exchange supports did not enable the separation of soybean and meat protein peaks. The use

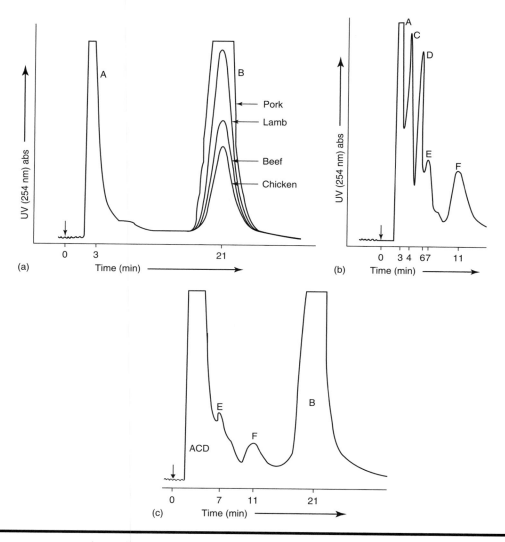

Figure 38.4 Proteic profiles obtained by RP-HPLC from different meat protein extracts (a), a soybean protein isolate (b), and a beef–soybean protein isolate (15%) blend (c). Peaks A and B correspond to meat proteins and peaks A, C–F correspond to soybean proteins. (From Flaherty, B., *Chem. Ind.,* **12, 495, 1975. With permission.)**

of a weak anion-exchange material for the separation of proteins before extraction with an acetone buffer (pH 5) yielded separations good enough to enable the identification of soybean proteins in meat–soybean mixtures. Figure 38.4 shows the chromatograms corresponding to different meats (pork, lamb, beef, and chicken), to a soybean protein isolate, and to a mixture of beef meat and soybean proteins. Meat proteins eluted with the dead time (band A) and in a band appearing at about 21 min (band B). Some soybean proteins were not retained in the column and eluted in band A, whereas the rest eluted in peaks C–F. Detection and determination of soybean proteins could be performed from peak F, which was totally isolated from meat peaks.[49]

Ashoor and Stiles used HPLC in the reversed-phase mode for the determination of soybean proteins, whey proteins, and casein in raw meats from different species (beef, pork, chicken, and turkey).[75] The method consisted of a linear gradient in 70 min with mobile phases containing trifluoroacetic acid as ion-pairing agent. To achieve this, they prepared model meat samples by spiking meats with the different foreign proteins. They observed three characteristic peaks for soybean proteins and one for casein that were well separated from meat proteins. In the case of whey, its main peak coeluted with meat proteins. They applied the method to the quantitation of these foreign proteins, obtaining detection limits of 1%.[75]

Parris and Guillespie used both anion-exchange and reversed-phase HPLC for the simultaneous separation of soybean and meat proteins.[76] Ion-exchange chromatography enabled the separation of soybean and meat proteins but the excess overlapping of soybean protein peaks did not enable its application for quantitative purposes. Reversed-phase chromatography also enabled the separation of soybean and meat proteins with a previous derivatization with 4-vinylpyridine. Detection limit in this case was 2%.[76]

Recently, Marina's research team published different articles related to the determination of soybean proteins in meats from different species (pork, turkey, chicken, and beef) that could also contain milk proteins by reversed-phase HPLC. These works were mostly devoted to heat-processed meat products although they also analyzed some nonheated products to study the effect of the heat treatment.[77–79] They optimized the method by studying the effect of different variables: fat extraction conditions (fat extraction solvent and number of extractions), protein extraction conditions (extraction time, extraction temperature, extraction media, and number of extractions), and separation conditions. Separation of soybean and meat proteins was performed, for the first time,

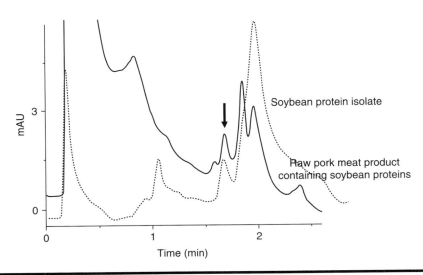

Figure 38.5 Chromatographic profiles obtained from a raw pork meat product containing soybean proteins and from a soybean protein isolate. Chromatographic conditions: column, POROS R2H (50 × 4.6 mm); temperature, 50°C; flow rate, 3 mL/min; gradient, 5–25% B in 0.8 min, 25–42% B in 0.8 min, 42–50% B in 0.6 min; mobile phases, water with 0.05% (v/v) trifluoroacetic acid (phase A) and acetonitrile with 0.05% (v/v) trifluoroacetic acid (phase B); injected volume, 20 μL; detection, 280 nm. Sample preparation: fat extraction with acetone followed by protein solubilization in 50 mM Tris-HCl buffer containing 0.5% 2-mercaptoethanol.

with perfusion columns that enabled the drastic reduction of analysis times because separation took place in only 2.5 min. Figure 38.5 shows the separation obtained from a raw pork meat product containing soybean proteins and a soybean protein isolate observing the existence of a peak at about 1.7 min, enabling the detection of soybean additions in meats. Moreover, the methods were validated following a standardized validation procedure for quantitative methods in food chemistry laboratories, evaluating the linearity of the calibration plot, detection limit, and quantitation limit, presence of matrix interferences, specificity, precision, robustness, and accuracy. Moreover, the results obtained were compared with those observed by applying the official ELISA method, concluding that the proposed method could be a serious alternative to this method leading to a significant reduction of analysis time, cost, and the complexity of the method itself.

Acknowledgments

The authors thank the Comunidad Autónoma de Madrid for project S-0505/AGR/0312.

References

1. Endres, J.G., *Soy Protein Products. Characteristics, Nutritional Aspects, and Utilization*, AOAC Press, Champaign, IL, 2001.
2. Xiong, Y.L., Role of myofibrillar proteins in water-binding in brine-enhanced meats, *Food Res. Int.*, 38, 281, 2005.
3. Keeton, J.T., Low-fat meat products—technological problems with processing, *Meat Sci.*, 36, 261, 1994.
4. Shand, P.J., Mimetic and synthetic fat replacers for the meat industry, in *Production and Processing of Healthy Meat, Poultry and Fish Products*, Pearson, A.M. and Dutson, T.R., Eds., Blackie Academic & Professional, London, UK, 1997.
5. Eilert, S.J. and Mandigo, R.W., Use of additives from plant and animal sources in production of low fat meat and poultry products, in *Production and Processing of Healthy Meat, Poultry and Fish Products*, Pearson, A.M. and Dutson, T.R., Eds., Blackie Academic & Professional, London, UK, 1997.
6. Siegel, D.G., Tuley, W.B., Norton, H.W., and Schmidt, G.R., Sensory, textural and yield properties of a combination ham extended with soybean protein isolate, *J. Food Sci.*, 44, 1049, 1979.
7. Abdel-Aziz, S.A., Esmail, S.A., Hussein, L., and Janssen, F., Chemical composition and levels of non-meat proteins in meat brands extended with soy protein concentrate, *Food Chem.*, 60, 389, 1997.
8. Pietrasik, Z. and Li-Chan, E.C.Y., Binding and textural properties of beef gels as affected by protein, κ-carrageenan and microbial transglutaminase addition, *Food Res. Int.*, 35, 91, 2002.
9. Comfort, S. and Howell, N.K., Gelation properties of salt soluble meat protein and soluble wheat protein mixtures, *Food Hydrocoll.*, 17, 149, 2003.
10. Means, W.J. and Schmidt, G.R., Restructuring fresh meat without the use of salt or phosphate, in *Advances in Meat Research*, AVI B, Person, A.M. and Dutson, T. R., Eds., Van Nostrand, New York, NY, 1987, Chap. 4.
11. Lu, G.H. and Chen, T.C., Application of egg white and plasma powders as muscle food binding agents, *J. Food Eng.*, 42, 147, 1999.
12. Tsai, S.J., Unklesbay, N., Unklesbay, K., and Clarke, A., Textural properties of restructured beef products with five binders at four isothermal temperatures, *J. Food Qual.*, 21, 397, 1998.
13. Kumar, S., Sharma, B.D., and Biswas, A.K., Influence of milk co-precipitates on the quality of restructured buffalo meat blocks, *Asian-Aust. J. Anim. Sci.*, 17, 564, 2004.
14. Kris-Etherton, P.M., Etherton, T.D., Carlson, J., and Gardner, C., Recent discoveries in inclusive-based approaches and dietary patterns for reduction in risk for cardiovascular disease, *Curr. Opin. Lipidol.*, 13, 397, 2002.

15. Rakosky, J., Jr., Soy grits, flour, concentrates, and isolates in meat products, *J. Am. Oil Chem. Soc.*, 51, 123A, 1974.

16. McMindes, M.K., Application of isolated soy protein in low-fat meat products, *Food Technol.*, 45, 61, 1991.

17. Anonymous, Soy protein enhances value of meat, fish and poultry products, *Food Product Dev.*, 15, 38, 1981.

18. Feng, J. and Xiong, Y.L. Interaction of myofibrillar and preheated soy proteins, *J. Food Sci.*, 67, 2851, 2002.

19. Rakosky, J., Soy products for the meat industry, *J. Agric. Food Chem.*, 18, 1005, 1970.

20. Mittal, G.S. and Usborne, W.R., Meat emulsion extenders, *Food Technol.*, 39, 121, 1985.

21. Ozimek, G. and Poznánski, S., Influence of an addition of textured milk proteins upon physico-chemical properties of meat mixtures, *J. Food Sci.*, 47, 234, 1981.

22. Mussman, H.C., Regulations governing the use of soy protein in meat and poultry products in the US, *J. Am. Oil Chem. Soc.*, 51, 104A, 1974.

23. Malmhesen Yman, I., Eriksson, A., Everitt, G., Yman, L., and Karlsson, T., Analysis of food proteins for verification of contamination or mislabelling, *Food Agric. Immunol.*, 6, 167, 1994.

24. Malmhesen Yman, I., Detection of inadequate labelling and contamination as causes of allergic reactions to food, *Acta Alim.*, 33, 347, 2004.

25. Tsuji, H., Kimoto, M., and Natori, Y., Allergens in major crops, *Nutr. Res.*, 21, 925, 2001.

26. Tsuji, H., Okada, N., Yamanishi, R., Bando, N., Kimoto, M., and Ogawa, T., Measurement of Gly mBd 30K, a major soybean allergen, in soybean products by a sandwich enzyme-linked immunosorbent assay, *Biosci. Biotechnol. Biochem.*, 59, 150, 1995.

27. Poms, R.E., Klein, C.L., and Anklam, A., Methods for allergen analysis in food: a review, *Food Addit. Contam.*, 21, 1, 2004.

28. Brandon, D.L. and Friedman, M., Immunoassays of soy proteins, *J. Agric. Food Chem.*, 50, 6635, 2002.

29. Borissowa, M.A., Lissizyn, A.B., Kalinowa, J.E., and Danilenko, A.N., Formation of meat myosin complexes with proteins of leguminous plants, *Fleischwirtschaft*, 4, 129, 2000.

30. Baudner, S., Analysis of plant proteins using immunological techniques based on the antigen–antibody precipitation, *Ann. Nutr. Alim.*, 31, 165, 1977.

31. Koppelman, S.J., Lakemond, C.M.M., Vlooswijk, R., and Hefle, S.L., Detection of soy proteins in processed foods: literature overview and new experimental work, *J. AOAC Int.*, 87, 1398, 2004.

32. Olsman, W.J. and Hitchcock, C., Detection and determination of vegetable proteins in meat products, in *Developments in Food Analysis Techniques-2*, King, R.D., Ed., Applied Science Publishers, London, UK, 1980.

33. Eldrigde, A.C., Determination of soya protein in processed foods, *J. Assoc. Off. Anal. Chem*, 58, 483, 1981.

34. Belloque, J., García, M.C., Torre, M., and Marina, M.L., Analysis of soyabean proteins in meat products: a review, *CRC Crit. Rev. Food Sci. Nutr.*, 42, 507, 2002.

35. Olsman, W.J., Methods for detection and determination of vegetable proteins in meat products, *J. Am. Oil Chem. Soc.*, 56, 285, 1979.

36. Hofmann, K., Identification and determination of meat and foreign proteins by means of dodecyl sulphate polyacrylamide gel electrophoresis, *Ann. Nutr. Alim.*, 31, 207, 1977.

37. Olsman, W.J., Detection of non-meat proteins in meat products by electrophoresis, *Z. Lebensm. Unters. Forsch.*, 141, 253, 1969.

38. Parsons, A.L. and Lawrie, R.A., Quantitative identification of soya protein in fresh and heated meat products, *J. Food Technol.*, 7, 455, 1972.

39. van Gils, W.F. and Hidskes, G.G., Detection of casein, soy proteins and coagulated egg-white in meat products by electrophoresis on cellulose acetate-membranes, *Z. Lebensm. Unters. Forsch.*, 151, 175, 1973.

40. Guy, R.C.E., Jayaram, R., and Willcox, C.J., Analysis of commercial soya additives in meat products, *J. Sci. Food Agric.*, 24, 1551, 1973.
41. Guy, R.C.E. and Willcox, C.J., Analysis of soya proteins in commercial meat products by polyacrylamide gel electrophoresis of the proteins extracted in 8 M urea and 1% 2-mercaptoethanol, *Ann. Nutr. Alim.*, 31, 193, 1977.
42. Lee, Y.B., Rickansrud, D.A., Hagberg, E.C., Briskey, E.J., and Greaser, M.L., Quantitative determination of soybean protein in fresh and cooked meat-soy blends, *J. Food Sci.*, 40, 380, 1975.
43. Lee, Y.B., Rickansrud, D.A., Hagberg, E.C., and Forsythe, R.H., Detection of various nonmeat extenders in meat products, *J. Food Sci.*, 41, 589, 1976.
44. Persson, B. and Appleqvist, L.A., The determination of non-meat proteins in meat products using polyacrylamide gel electrophoresis in dodecylsulphate buffer, *Ann. Nutr. Alim.*, 31, 225, 1977.
45. Homayounfar, H., Identification électrophorétique des proteins étrangéres et en particulier des proteins de soja dans les produits alimentaires carnés frais ou en conserve, *Ann. Nutr. Alim.*, 31, 187, 1977.
46. Baylac, P., Luigi, R., Lanteaume, M., Pailler, F.M., Lajon, A., Bergeron, M., and Durand, P., Vérification des limites de l'identification de différents liants protéiques dans un produit carné soumis a des valeurs cuisatrices connues, par utilisation de methods électrophorétiques immunologiques et histologiques, *Ann. Fals. Exp. Chim.*, 870, 333, 1988.
47. Hofmann, K., Identification and determination of meat and foreign protein by means of dodecylsulphate polyacrylamide gel electrophoresis, *Z. Anal. Chem.*, 267, 355, 1973.
48. Hofmann, K. and Penny, I.F., A method for the identification and quantitative determination of meat and foreign proteins using sodium-dodecylsulphate-polyacrylamide gel electrophoresis, *Fleischwirtschaft*, 53, 252, 1973.
49. Flaherty, B., Progress in the identification of non-meat food proteins, *Chem. Ind.*, 12, 495, 1975.
50. Vállas-Gellei, A., Detection of soya and milk proteins in minced meat, *Acta Alim.*, 6, 215, 1977.
51. Molander, E., Determination of soya protein in meat products by standard curves obtained from SDS gel electrophoresis, *Z. Lebensm. Unters. Forsch.*, 174, 278, 1982.
52. Abd Allah, M.A., Foda, Y.H., El-Dashlouty, S., El-Sanafiry, N.Y.A., and Abu Salem, F.M., Detection of soybean in soy-based meat substitutes, *Die Nahrung*, 5, 549, 1986.
53. Fukal, L., Modern immunoassays in meat-product analysis, *Die Nahrung*, 35, 431, 1991.
54. Rossebo, L. and Nordal, J., A serological method for the detection of trypsin inhibitor in commercial soy proteins and its use in detecting soy protein addition to raw meat products, *Z. Lebensm. Unters. Forsch.*, 148, 69, 1972.
55. Kamm, L., Immunochemical quantitation of soybean proteins in raw and cooked meat products, *J. Assoc. Off. Anal. Chem*, 53, 1248, 1970.
56. Koh, T.Y., Immunochemical method for the identification and quantitation of cooked or uncooked beef and soya proteins in mixtures, *J. Inst. Can. Sci. Technol. Alim.*, 11, 124, 1978.
57. Hitchcock, C.H.S., Bailey, F.J., Crimes, A.A., Dean, D.A.G., and Davis, P.J., Determination of soya protein in food using an enzyme-linked immunosorbent assay procedure, *J. Sci. Food Agric.*, 32, 157, 1981.
58. Crimes, A.A., Hitchcock, C.H.S., and Wood, R., Determination of soya protein in meat products by an enzyme-linked immunosorbent assay procedure: collaborative study, *J. Assoc. Publ. Analysts*, 22, 59, 1984.
59. Griffiths, N.M., Billington, M.J., Crimes, A.A., and Hitchcock, C.H.S., An assessment of commercially available reagents for an enzyme-linked immunosorbent assay (ELISA) of soya protein in meat products, *J. Sci. Food Agric.*, 35, 1255, 1984.
60. Mcneal, J.E., Semiquantitative enzyme-linked immunosorbent assay of soy protein in meat products: summary of collaborative study, *J. Assoc. Off. Anal. Chem.*, 71, 443, 1988.
61. Rittenburg, J.H., Adams, A., Palmer, J., and Allen, J., Improved enzyme-linked immunosorbent assay for determination of soy protein in meat products, *J. Assoc. Off. Anal. Chem.*, 70, 582, 1987.

62. Hall, C.C., Hitchcock, C.H.S., and Wood, R., Determination of soya protein in meat products by a commercial enzyme immunoassay procedure: collaborative trial, *J. Assoc. Publ. Analysts*, 25, 1, 1987.
63. Berkowitz, D.B. and Webert, D.W., Determination of soy in meat, *J. Assoc. Off. Anal. Chem.*, 70, 85, 1987.
64. Meyer, R., Chardonnens, F., Hübner, P., and Lüthy, J., Polymerase chain reaction (PCR) in the quality and safety assurance of food: detection of soya in processed meat products, *Z. Lebensm. Unters. Forsch.*, 203, 339, 1996.
65. Lindqvist, B., Ostgren, J., and Lindberg, I., A method for the identification and quantitative investigation of denatured proteins in mixtures based on computer comparison of amino-acid patterns, *Z. Lebensm. Unters. Forsch.* 159, 15, 1975.
66. Lindberg, W., Ohman, J., Wold, S., and Martens, H., Simultaneous determination of five different food proteins by high-performance liquid chromatography and partial least-squares multivariate calibration, *Anal. Quim. Acta*, 174, 41, 1985.
67. Lindberg, W., Ohman, J., Wold, S., and Martens, H., Determination of the proteins in mixtures of meat, soymeal and rind from their chromatographic amino-acid pattern by partial least-squares method, *Anal. Quim. Acta*, 171, 1, 1985.
68. Zarkadas, C.G., Karatzas, C.N., and Khanizadeh, S., Evaluating protein quality of model meat/soybean blends using amino acid compositional data, *J. Agric. Food Chem.*, 41, 624, 1993.
69. Zhi-Ling, M., Yan-Ping, W., Chun-Xu, W., and Fen-Zhi, M., HPLC determination of muscle, collagen, wheat, shrimp, and soy proteins in mixed food with the aid of chemometrics, *Am. Lab.*, 29, 27, 1997.
70. Bailey, F.J., A novel approach to the determination of soya proteins in meat products using peptide analysis, *J. Sci. Food Agric.*, 27, 827, 1976.
71. Bailey, F.J. and Hitchcock, C., A novel approach to the determination of soya protein in meat products using peptide analysis, *Ann. Nutr. Alim.*, 31, 259, 1977.
72. Bailey, F.J., Llewellyn, J.W., Hitchcock, C.H.S., and Dean, A.C., Determination of soya protein in meat products using peptide analysis and the characterisation of the specific soya peptide used in calculations, *Chem. Ind.*, 13, 477, 1978.
73. Llewellyn, J.W., Dean, A.C., Sawyer, R., Bailey, F.J., and Hitchcock, C.H.S., Technical note: the determination of meat and soya protein in meat products by peptide analysis, *J. Food Technol.*, 13, 249, 1978.
74. Agater, I.B., Brian, K.J., Llewellyn, J.W., Sawyer, R., Bailey, F.J., and Hitchcock, C.H.S., The determination of soya and meat protein in raw and processed meat products by specific peptide analysis. An evaluation, *J. Sci. Food Agric.*, 37, 317, 1986.
75. Ashoor, S.H. and Stiles, P.G., Determination of soy protein, whey proteins, and casein in unheated meats by high-performance liquid chromatography, *J. Chromatogr.*, 393, 321, 1987.
76. Parris, N. and Guillespie, P.J., HPLC separation of soy and beef protein isolates, in *Methods for Protein Analysis*, Chemy, J.P. and Barford, R.A., Eds., The American Oil Chemists' Society, Champaign, IL, 1988, p. 142.
77. Castro-Rubio, F., García, M.C., Rodríguez, R., and Marina, M.L., Simple and inexpensive method for the reliable determination of additions of soybean proteins in heat-processed meat products: an alternative to the AOAC Official method, *J. Agric. Food Chem.*, 53, 220, 2005.
78. Castro, F., Marina, M.L., Rodríguez, J., and García, M.C., Easy determination of the addition of soybean proteins to heat-processed meat products prepared with turkey meat or pork-turkey meat blends that could also contain milk proteins, *Food Addit. Contam.*, 22, 1209, 2005.
79. Castro, F., García, M.C., Rodríguez, R., Rodríguez, J., and Marina, M.L., Determination of soybean proteins in commercial heat-processed meat products prepared with chicken, beef or complex mixtures of meats from different species, *Food Chem.*, 100, 468, 2007.

Chapter 39

Stable Isotope Analysis for Meat Authenticity and Origin Check

Hanns-Ludwig Schmidt, Andreas Rossmann,
Susanne Rummel, and Nicole Tanz

Contents

39.1   Introduction: Stable Isotopes in Food Traceability and Consumer Protection

The stable isotope ratios of the main elements of organic compounds and biomass (bioelements H, C, N, O, and S) are predominantly determined by their individual primary compounds and the conditions of their (bio)synthesis. Therefore, they preserve the history of the compounds' origin and are analyzed for its reconstruction. In contrast, concentrations and isotope ratios of secondary or trace elements record the local geological particularities of the origin. The information thus obtained is of fundamental importance for the investigation of environmental, ecological, archeological, and nutritional questions and of forensic background and last but not the least of food authenticity and origin assessment.[1,2] The latter application has originally been used for the authenticity check of fruit juices, wine and spirits, flavors and honey, mainly on the basis of the carbon isotopes of sugars and organic acids and of oxygen isotopes of water from these products. Since 1995 quality check analyses on the basis of isotope ratio analyses have become common for origin and authenticity investigations on animal products such as milk,[3] butter,[4] cheese,[5] eggs,[1] and meat.[6]

The worldwide trade with meat and meat products together with events such as the bovine spongiform encephalopathy (BSE) scandal were the causes that the European Commission (EC)/ European Union (EU) issued several regulations concerning the quality demands of food and food commodities and the information of purchasers and consumers about their origin. The first regulation was the Council Regulation No. 2081/92 "on the protection of geographical indications and designations of origin for agricultural products and foodstuffs," also an issue of the EU protected food names schemes, introducing the quality indications protected denomination of origin (PDO), protected geographical indication (PGI), and certificate of specific character (CSI). The following Council Regulations No. 1760/2000, "providing for the general rules for a compulsory beef labeling system" and 178/2002, "defining traceability as a basic principle of consumer protection," and laying down the general principles and requirements of food laws (Council Regulation No. 178/2002, for details see Ref. 7). On this basis the origin indication of a product thus implies the guarantee for a typical ware, produced by a defined feeding regime and with a special quality and hence prize (e.g., Parma Ham, Orkney Beef, Camembert de Normandie). A corresponding regulation for products of aquaculture and fishery came into effect in 2000 (Council Regulation No. 104/2000, for details see Ref. 8). Corresponding regulations also exist for the definition of "products from organic farming": in 1991 (Council Regulation No. 2092/91) the community stated among others that products traded under this denomination must originate from animals grown with fodder from farm's own production, thus characterized by elements from the local geography. With the aim of a better consumer protection on an international level the EU has also initiated in 2005 the research program traceability of food commodities in Europe (TRACE; www.trace.eu.org), in which actually more than 60 participants in 12 European countries and associated partners from other continents (U.S., Australia, South America, China) elaborate standard methods and data banks for the origin and authenticity control of food, hence also meat. Characterizations of meat quality, as they have been developed, for example, for honey or maple syrup carbon isotope analysis by the Association of Official Analytical Chemists (AOAC), are, according to our knowledge, not yet issued in the United States.

Like the indications of defined origin, the attainable higher prizes for products labeled with indications for a certain feeding regime (organic beef, Atlantic wild salmon) may seduce some producers to adulterations. The resulting task for laboratories is to find objective analytical criteria for the assignment of meat and fish to defined geographical origins and diets. Recent reviews compile sophisticated methods for traceability basing not only on biomarkers such as defined organic compounds and their patterns[8,9] or trace elements and genetic characteristics[6,10,11] but also on stable isotope ratios.[2] Actually the latter are the most common indicators and the methodology of their analysis is fast and highly advanced. The experience showed that most information about an unknown sample could be attained by a multielement and sometimes multicompound isotope ratio analysis, as the individual elements of some easily available distinct fractions (fat, protein) contribute independently to the desired information. To understand the corresponding correlations, the general fundamentals of natural isotope abundances and discriminations and some special causes influencing meat isotope characteristics have to be discussed.

39.2 Isotope Discriminations of Bioelements and Information from Resulting Isotope Ratios in Biomass

As already indicated, the relative isotope abundances of organic material are primarily determined by the isotope characteristics of the starting material and isotope discriminations in the biological cycles of the elements and by isotope discriminations in the course of food chains. Isotope discriminations are changes of isotope abundance ratios under the conditions of metabolism.[12] Their physical basis are isotope effects (IEs), ratios of equilibrium constants in phase transitions or (bio)chemical reactions (thermodynamic IEs), or of rate constants of (bio)chemical reactions (kinetic IEs) of isotopolog and isotopomer molecules (molecules differing by any kind of isotopes in any molecular position or by the same isotopic atom in different molecular positions, respectively).

The bioelements occur in nature as mixtures of stable isotopes in "mean natural abundances." Usually the main ("light") isotopes dominate by far (Table 39.1), whereas the "heavy" ones are present in minor concentrations. The observed variations of isotope abundances are not expressed in atom-% but in δ-values, relative differences of the isotope ratio R of a given element (actual abundance ratio of a heavy isotope to the main isotope) to that of an international standard, for example, for carbon (standard V-PDB = Vienna PeeDee-Belemnite, $R = 0.0112372$):

$$\delta^{13}C\,[\%o]_{\text{V-PDB}} = \left[\left(\frac{R_{\text{Sample}} - R_{\text{V-PDB}}}{R_{\text{V-PDB}}}\right)\right] \times 1000 = \left[\left(\frac{R_{\text{Sample}}}{R_{\text{V-PDB}}}\right) - 1\right] \times 1000 \qquad (39.1)$$

Table 39.1 Characteristics and Standards of the Stable Isotopes of the Bioelements

Element	Stable Isotopes	Average Natural Abundance (Atom %)	Atomic Mass (AMU)	Spin-quantum Number I	Standard International Name	Standard Isotope Ratio R	Measuring Gas for IRMS
Hydrogen	^{1}H	99.985	1.00782522	1/2			
	^{2}H	0.015	2.0141022	1	V-SMOW	0.00015576	H_2
Carbon	^{12}C	98.90	12.0000000	0			
	^{13}C	1.10	13.0033543	1/2	V-PDB	0.011802	CO_2
Nitrogen	^{14}N	99.635	14.0030744	1			
	^{15}N	0.367	15.0001081	1/2	AIR	0.0036782	N_2
Oxygen	^{16}O	99.762	15.9949149	0			
	^{17}O	0.038	16.9991334	5/2		0.0003799	
	^{18}O	0.200	17.9991598	0	V-SMOW	0.0020052	CO_2, CO
Sulfur	^{32}S	95.03	31.972074	0			
	^{33}S	0.75	32.97146	3/2			
	^{34}S	4.22	33.967864	0	V-CDT	0.0441509	SO_2, SF_6
	^{36}S	0.02	35.96709	0			

Notes: AMU = atomic mass unit; V-SMOW = Vienna standard mean ocean water; V-PDB = Vienna Pee-Dee Belemnite; AIR = air-N2; V-CDT = Vienna canyon diablo troilite; isotope ratio R = [isotope]/[main isotope]; IRMS = isotope ratio mass spectrometry.

Source: Adapted from Schmidt, H.-L., et al. in *Flavourings, Production, Composition, Applications, Regulations*, WILEY-VCH Verlag GmbH & Co. KgaA, Weinheim, 2007, 602. With permission.

The international standards are available from the International Atomic Energy Agency (IAEA) in Vienna (Table 39.1); practical measurements of bioelement isotope ratios are performed by isotope ratio mass spectrometry (IRMS) or by quantitative positional ^{2}H-NMR (site-specific natural isotope fractionation [SNIF]-NMR®, see Ref. 14) relative to laboratory standards.

39.2.1 Contributions by Primary Materials and Elemental Cycles

39.2.1.1 Hydrogen and Oxygen and Their Isotopes

Meteoric water is the primary source of any physically bound water in biological material and of organically bound hydrogen and oxygen. As the natural water cycle starts from the ocean, the "Vienna standard mean ocean water" (V-SMOW, Table 39.1) is the isotopic reference for hydrogen and oxygen with—per definition—a δ^2H- and a δ^{18}O-value 0‰. By evaporation, the "light" water molecules ^{1}H^{1}H^{16}O are—due to their higher vapor pressure—relatively enriched in the vapor phase and depleted in condensates. The consequence is that any precipitate and ground water has a more negative δ^2H- and δ^{18}O-value, the farther it falls from the ocean. This "continental effect" is superimposed by influences of seasonal variations and of local altitude and climate conditions. Nonetheless, the isotopic characteristics of water are more or less typical for a given place; for display of the δ^{18}O-values of the global precipitation see Ref. 15. The δ^2H- and δ^{18}O-values are correlated by the meteoric water line (Equation 39.2, Ref. 16), and both are correlated to the average annual air surface temperature T_{ann} in °C (Equation 39.3, Ref. 17):

$$\delta^2 H \; [‰]_{\text{V-SMOW}} = 8 \times \delta^{18} O \; [‰]_{\text{V-SMOW}} + 10 \tag{39.2}$$

$$\delta^{18} O = (0.695 T_{ann} - 13.6) \; [‰]_{\text{V-SMOW}}; \; \delta^2 H = (5.6 T_{ann} - 100) \; [‰]_{\text{V-SMOW}} \tag{39.3}$$

In plants, the water is reenriched in heavy isotopes by IEs on the evapo-transpiration; the enrichment by C_4-plants (see Section 39.2.1.2) is less than that by C_3-plants. Therefore, the isotopic data of leaf water or of the water from plant products (vegetables, fruit and juices, wine) are indicative of their geographic origin. The isotopic characteristics of water in animal tissues are determined by the overlap of those of the drinking water, the water in the food, and that from its oxidation.[18]

The photosynthetic binding of hydrogen to carbon atoms is always accompanied by isotope discriminations. In principle, only carbon-bound hydrogen is indicative, whereas oxygen- and nitrogen-bound hydrogen atoms can exchange and equilibrate with protons of the surrounding water. Correspondingly, an exchange of oxygen atoms of carbonyl and carboxyl groups with those of the surrounding water is possible; in isotopic equilibrium these are enriched in ^{18}O by ~28 and ~19‰, respectively.[19]

39.2.1.2   Carbon and Isotopes

The $\delta^{13}C$-value of atmospheric CO_2 is worldwide constant near −8‰ versus V-PDB; HCO_3^- in water, in isotopic equilibrium with CO_2, is relatively enriched by 7‰. The kinetic IE on the photosynthetic CO_2-fixation by land plants with C_3-metabolism (the binding of CO_2 by the ribulose bisphosphate carboxylase reaction leads to the primary product 3-phosphoglycerate with three C-atoms), such as trees, bushes, vegetables, most grasses, grain plants, and sugar beets, results in a biomass with $\delta^{13}C$-values between −24 and −30‰, whereas the products of C_4-plants (primary assimilation product of the phosphoenolpyruvate carboxylase reaction is oxaloacetate with four C-atoms), such as corn, sugar cane, sorghum, and millet, show $\delta^{13}C$-values between −11 and −15‰.[20] The variations within both groups are due to climatic factors such as local humidity and temperature influencing the leaf stomata opening and hence the efficiency of the photosynthesis.[21]

The most common fodder plants except corn and some African grasses belong to the C_3-plant group, and therefore the $\delta^{13}C$-values of animal products (eggs, milk, and meat) reflect primarily the ration of corn in the animals' diet. However, local climatic effects additionally influence the ^{13}C-content of the fodder, and animal products are normally enriched by 1–3‰ relative to the diet.[22] Finally, in the case of products from fish or water fowl it has to be taken into account that their natural food chain starts, at least partially, from plants with aquatic origin; phytoplankton has, due to the above-mentioned ^{13}C-enrichment of HCO_3^- as primary source, a $\delta^{13}C$-value near −20 to −18‰.

39.2.1.3   Nitrogen and Isotopes

The most important primary source of nitrogen in plants is NO_3^-. When originating from industrial production or from stratospheric processes, the ion has a $\delta^{15}N$-value near 0‰ versus air. It is essentially enriched in ^{15}N (up to +25‰) when it is formed by bacterial nitrification of

[15]N-enriched precursors (urea and uric acid in natural manure) or when it is a remainder of nitrate, which had been submitted to partial denitrification.[23] NH_3 from any origin is enriched in ^{15}N by partial evaporation, preferring the "light" molecules with a higher vapor pressure. Although the uptake of nitrate by higher plants proceeds without isotope discrimination, the assimilatory nitrate reduction implies a remarkable IE leading to a ^{15}N-depletion of the organically bound N relative to the assimilated NO_3^-.[23]

Because natural fertilizers are normally enriched in ^{15}N, it is often assumed that a higher δ^{15}N-value of plant and animal products is indicative of "organic" production. However, as in organic plant production sometimes "green fertilization" by legumes (N_2-assimilation) is used, nitrogen with a δ^{15}N-value near 0‰, like mineral nitrate, is applied to the soil. The situation becomes even more complicated by the fact that any kind of added nitrogen fertilizer is partially mixed with the large reservoir of natural "soil nitrogen" except for plants cultivated in green houses with a limited soil reservoir. Nonetheless, one can assume that plant biomass with a δ^{15}N-value $>+7$‰ may probably be from organic production.[24]

39.2.1.4 Sulfur and Isotopes

The primary source of sulfur for plants is sulfate, which is obtained from either the soil, industrial fertilization, or sea spray (aerosol in coastal areas up to 100 km from the sea shore). In addition, SO_2 released from combustion processes (industry, private heating devices, cars) or natural sources (e.g., volcanoes) can be used by plants. The δ^{34}S-values of these sources range from -10 to $+24$‰, and they are locally quite different.[25] The assimilation of sulfur by higher plants proceeds without large isotope discrimination (1–3‰ depletion relative to the source[26]). Therefore, in the practice of origin assignments, the average δ^{34}S-value of plant biomass is assumed to be identical to that of the primary source. The S-content of higher plant dry matter is between 0.1 and 0.3%; 90% of this sulfur occurs in the reduced form and is predominantly bound to the amino acids cysteine and methionine, except for some secondary products in Brassicaceae.[26] In this form, the sulfur is also transferred into animal biomass, especially meat. Cartilage will, in addition, contain some sulfur in the form of sulfate, bound to chondroitin sulfate, and the origin of this sulfate is from plant biomass ([34]S-enriched) and drinking water.[27] The primary source of sulfur for wild fish is the sulfate of the water in which they live (sea water sulfate is ~+24‰; fresh water sulfate is lower), depleted in context with reduction processes in the food chain. In farmed fish this is overlapped by the influence of the individual diet.[28]

39.2.1.5 Minor and Trace Elements and Their Isotopes

Minor elements such as Na, K, Mg, Ca, Cl, and P are widespread in biological systems and their concentration or isotope ratio is not characteristic for a given origin; their isotope ratios are also not influenced by any bioactivity or metabolism. Some trace elements are essential in biological systems and are actively accumulated; some others are incorporated by chance and can even be toxic. An example of the latter group is lead, which has been successfully investigated in forensic and archeometric questions and may be a future candidate for food origin assignment.[29]

Trace elements as biomarkers for meat origin identification have been reviewed by Gremaud et al.[10] and Franke et al.,[6] the required instrumental equipment has been compiled by Hölzl et al.,[30] and some applications for origin assignment proof are demonstrated in Table 39.2. Franke et al.[6] discriminate between trace elements in meat caused by natural deposits and through feed supplementation (for details, see Ref. 31); they discuss the potential and limits of these elements as indicators for the geographic origin of meat. As a matter of fact, some essential trace elements such as Fe, Zn, Cu, Co, Se are found in most animal tissues and some nonessential ones are accumulated in special organs. A general application of type and its concentration as origin indicators would demand an enormous work for the elaboration of data banks and is so far not described.

Therefore, in the present context we shall concentrate on a nonessential trace element that is characterized by a very distinct isotope ratio typical for its geographical origin, has so far led to very promising results in the origin identification of other animal products, and is actually checked for meat origin assignment in the frame of the European program TRACE.[4,32] The ratio of the strontium isotopes $^{87}Sr/^{86}Sr$ is exclusively determined by local geological prerequisites, namely the primordial relative concentration of Rb and Sr in rocks, the isotope ^{87}Sr being the radiogenic product of the β-decay of ^{87}Rb ($t_{1/2} = 4.8$ Ga), and it is not modulated by IEs in context with physical or metabolic factors. We are therefore convinced that it will be a valuable additional and independent indicator in the origin identification of meat.

39.2.2 Influences of Animal Nutrition on Meat Isotope Characteristics

From correlations outlined in Section 39.2.1 one can derive that the isotope ratios of hydrogen and oxygen in biomass are predominantly indicative of the climate at the origin of a sample; its carbon isotope ratios indicate primarily the photosynthetic characteristics of the producing plants, and some climatic properties of their geographical origin; nitrogen isotope ratios are indicators of the local fertilizers applied; and sulfur isotopes and even more those of the heavy metals preserve soil and geological properties and human activities at the growth area. The isotope characteristics of animal products are based on those of the fodder plants but with a certain shift in the δ-values. This is expressed in the title of one of the first studies on the correlation between the isotope characteristic of diet and animal body mass by DeNiro and Epstein:[22] "You are what you eat (plus a few ‰)." These few ‰ of the isotopic shift between diet and animal tissue ($\Delta\delta$-values) are different for the individual elements; they not only depend on the diet but are also modulated by the actual nutritional state of the animals and the type and metabolic activity of the tissue in question.[33] The knowledge and understanding of these correlations is of importance for a better interpretation of experimental results on isotope measurements in respect to the origin assignment of unknown meat samples.

The aims and results of representative isotope ratio and trace element determinations for meat origin and authenticity investigations so far available are compiled in Table 39.2. Most of these data are from samples of known origin and treatment and elaborated under controlled experimental conditions. This is important because so far no international data banks for meat exist (however, see "TRACE" in Section 39.1.1), and any laboratory must actually provide its own reference data for a given problem. However, the synopsis of these data and their discussion on the basis of the knowledge about the causes for the isotopic shifts between nutrition and meat will contribute to their understanding and interpretation.

39.2.2.1 Particularities of Meat Tissue Water and Influences of Meat Processing and Conservation

The δ^2H- and $\delta^{18}O$-values of water can easily be determined by isotopic equilibration with H_2 and CO_2,[56,57] respectively. This may be one of the reasons that originally lot of importance had been

Table 39.2 Representative Examples of Isotope Investigations on Origin of Meat and Correlations Between Meat and Animal Diet

Muscle Food from	Investigated Elements and Isotopes	Method, Aim, and Result of Investigation	Reference
Beef	^{13}C	Fattening of steers with grass or maize silage, respectively, for 230 days; $\delta^{13}C$-value of maize group meat attains −16.9, fat −22.4‰.	34
	^{13}C	Feeding of young bulls with various mixed C_3-/C_4-plant diet until isotopic equilibrium. Muscle enriched, liver identical, fat depleted in ^{13}C relative to diet.	35
	^{13}C, ^{15}N	Shift from grass to maize silage results in diet correlated δ-values in lipid-free muscle meat and fat (the latter relatively depleted 4‰).	36
	^{2}H, ^{13}C, ^{15}N, ^{18}O, ^{34}S	$\delta^{2}H$- and $\delta^{18}O$-values of muscle water permit meat origin assignment to Germany or Argentina; additional discrimination possible by $\delta^{15}N$- and $\delta^{34}S$-values. $\delta^{13}C$- and $\delta^{15}N$-values are indicative for conventional and organic farming, respectively.	37
	^{13}C, ^{15}N, ^{34}S	$\delta^{13}C$- and $\delta^{15}N$-values permit discrimination between beef from the United States, Brasilia, and Europe; additional $\delta^{34}S$-values permit differentiation within Ireland. Combination of $\delta^{15}N$-/$\delta^{34}S$-values are different for organic and conventional farming.	38
	^{2}H, ^{13}C, ^{15}N, ^{18}O, trace elements	Combination of $\delta^{13}C$-/$\delta^{15}N$-values of fat free muscle with $\delta^{2}H$-/$\delta^{18}O$-values of lipid permit, together with data from trace elements and assisted by statistical methods, meat assignments to England, Ireland, Brazil.	39
	^{2}H, ^{13}C, ^{18}O (NMR)	^{18}O-IRMS of meat water and ^{2}H- and ^{13}C-NMR analysis of extracted fat permit origin and diet (maize) identification of Charolais steers from different sites in F.	40
	^{2}H, ^{18}O	Chances and problems of isotope analysis of meat water as sole criterion for beef geographical origin assignment (also for pork, lamb, fowl).	41–43
Lamb	^{13}C, ^{15}N	$\delta^{13}C$- and $\delta^{15}N$-values of defatted meat and $\delta^{13}C$-values of fat are correlated to geographical origin, feeding regime, and genotype of the animals. By discriminant analysis it is possible to differentiate between the individual influences of the parameters.	44
	^{13}C, ^{15}N, ^{1}H-NMR Ca, Mg, Na, K, Zn, Cu, Fe, Cr	Lamb meat from three Apulian areas and with different diet regimes is analyzed for stable isotopes, mean, and trace metals with ^{1}H-NMR. Multivariate statistical analysis of the data permits satisfactory origin differentiation.	45
	^{2}H, ^{13}C, ^{15}N, $H^{34}S$	Data of multielement isotope ratio analysis of defatted neck meat from 12 European regions correlate to precipitation (^{2}H), diet, climate (^{13}C, ^{15}N), and local geological characteristics. Statistical discriminant analysis permits origin assignment.	46

Pork	^{13}C	δ^{13}C-value of adipose tissue of Spanish swine is linearly correlated to the fraction of acorn in diet during fattening.	47
	^{13}C, ^{34}S	δ^{13}C- and δ^{34}S-values of Spanish swine liver is indicative for diet (acorn or other feed) and permits in addition differentiation of animal breeds.	48
	^{13}C, ^{15}N	Nondefatted muscle tissue of Brazilian pig was depleted in ^{13}C by 2.2–3.0‰ relative to diet.	49
Fish	^{2}H, ^{13}C, ^{2}H-NMR	Differentiation between wild and farmed Atlantic salmon by combination of IRMS of muscle, and by GC- and site-specific ^{2}H-NMR of fatty acids from extracted fat.	50
	^{34}S	Reconstruction of hatchery and nature grown salmon life history and diet by ^{34}S-micro isotope analysis of otolith layers.	28
	^{13}C, ^{15}N	Discrimination between wild and farmed New Foundland salmon by IRMS. The wild fish are significantly enriched in ^{15}N and depleted in ^{13}C in muscle tissue. The farmed fish are enriched to diet by 5‰ in ^{15}N.	51
	^{13}C, ^{15}N	IRMS of fillets of salmon combined with GC of fatty acids and statistical data analysis permits discrimination between wild, organically and conventionally farmed animals.	52
	^{13}C, ^{15}N	Carps fed with the same diet in different feeding rates show a decrease of the isotopic shift in ^{13}C and ^{15}N between diet and fish by ~1‰ with increasing feed.	53
	^{13}C, ^{15}N	The δ^{13}C-value of Gilthead sea bream permits a discrimination between wild and farmed animals, the δ^{15}N-value is indicative for geographical origin and feed.	54
Beef, pork, lamb, chicken, fish	^{2}H, ^{13}C, ^{15}N, ^{18}O, ^{34}S, ^{87}Sr	Dependence of δ^{2}H-, δ^{13}C-values of beef on feeding regime (stable or meadow keeping), of pork and chicken on geographical origin. Origin assignment of carps by δ^{15}N-, δ^{34}S-, and δ^{87}Sr-values and of lamb meat by multielement IRMS and discriminant analysis.	55

given to the measurement of these values of tissue water for the origin identification of meat. As a matter of fact, differences were found between meat samples from Argentina and Europe, and even between samples from Northern and Southern Germany. However, a more detailed study by Hegerding et al.[41] showed that the $\delta^{18}O$-values of samples from different parts of Germany overlapped and that discriminations between samples from Germany and England were not possible. Systematic investigations by Thiem et al.[42] confirmed this result and demonstrated that even influences of the meat's storage and processing could overlap geographical and seasonal differences. Also, in spite of constant δ^2H- and $\delta^{18}O$-values of the ground (drinking) water, seasonal differences of the tissue water were observed,[37] and Renou et al.[40] even found that the $\delta^{18}O$-value of meat water of samples from different areas of France was more dependent on the diet than on the growth area (=drinking water) of the cattle. Finally, a change of drinking water,[18] for example, due to transport of the animals before slaughter and water evaporation from or condensation on the meat surface, or salting and freezing may cause artifacts in the isotope ratios of the product.

Therefore, the isotope characteristics of tissue water can be a hint but not an absolute indication for the origin of meat, and the isotope determination of the organically bound (nonexchangeable) oxygen or hydrogen will be more reliable for meat origin assignment.[39]

39.2.2.2 Trophic Shift, Nutritional State of Animals, Half-Life and Kind of Tissues

The biochemical background of isotopic shifts between trophic levels has been studied in more detail by Schoeller[58] and has been the topic of recent studies for ecologists and (animal) physiologists.[59–61] From here it turns out that it is a quite complex question recommending some caveats with the discussion of experimental results.

In the present context, muscle protein from herbivores needs to be considered. At first, we have to keep in mind that cattle diet is in most cases derived from several sources; mixing models may help to take this fact into account.[62] In practice, empirical values of isotope abundance shifts between diet and muscle meat may be sufficient; they are 1–2.5‰ for carbon and 2–4‰ for nitrogen but may underlie variations.[63] For example, the nutritional state or metabolic stress of the subject interferes with the actual N-balance and from here the $\delta^{15}N$-value, as measured on hair keratin.[64–66] Gaye-Siessegger et al.[53] studied the influence of the actual ration of the diet: after feeding carps with different amounts of the same diet they found that the $\delta^{13}C$- and the $\delta^{15}N$-value shifts ($\Delta\delta$-values) between diet and muscle decreased by 1–2‰ with an increase in feeding rate. The same group also observed with Nile tilapia a decrease in the trophic shifts of C and N with increasing protein accretion.[61] Trueman et al.[67] found with salmons that the isotopic shift between diet and defatted body mass was inversely related to the growth rate; the difference in the $\Delta\delta^{15}N$-value between fast and slowly growing animals was 1‰, approximately 50% of the average normal shift between diet and muscle protein. The authors attribute their results to the difference of the N-use efficiency. In a recent systematic study on the influence of temperature, diet ration, body size, and age on $\Delta\delta$-values of European sea bass (*Dicentrarchus labrax*), the temperature rise caused a small negative effect on $\Delta\delta^{34}S$, an increase in $\Delta\delta^{13}C$, and a decrease in $\Delta\delta^{15}N$; $\Delta\delta^{34}S$ was positively related to body size and age of the animals.[68] An example for a quite dramatic change of an isotopic shift between diet and body protein under metabolic stress (protein deficiency) has been shown for the $\Delta\delta^{34}S$-value.[69] The authors observed, after feeding horses with hay of different $\delta^{34}S$-values, a ^{34}S-fractionation of −1‰ to hair keratin when the hay was "a protein-adequate C_3 plant diet," but a fractionation of +4‰ when the hay was "a possible low-protein C_4 diet." They believe that their finding is due to the integration of endogenous sulfur from the mobilization of body proteins under protein deficiency. Recently, even the shift of the δ^2H-value has become of practical importance.[70]

The prerequisite for a practical use of these observations is that the protein under investigation is in isotopic equilibrium with the diet. Experiments with quails (*Coturnix japonica*) on the animals' ^{13}C-adaptation after a diet switch gave a half-life time for collagen of 173.3 days, whereas that for liver was only 2.6 days.[60] Corresponding experiments with alpacas (*Lama pacos*) yielded 37 days for liver and 178.7 days for muscle tissue.[71] However, Gebbing et al.[34] found that the meat (nondefatted dorsal muscle) of steers, even 230 days after a switch from grazing feed to corn silage, did not attain the $\delta^{13}C$-value of the new diet, probably because a complete equilibration between diet and meat had not been attained under these conditions. However, this was obviously the case in a study with cattle receiving pure C_3- and moderately C_4-plant containing diet from the beginning: here an enrichment of (nondefatted) muscle meat (m. longissimus thoracis) relative to the diet of +2.5‰ and a depletion of the kidney fat by −3.5‰ were observed (Ref. 35; for model considerations on isotopes in diet changes, see Ref. 72).

To summarize, muscle tissue of animals under normal feeding conditions is enriched in ^{13}C relative to the mean of the diet by 1–2.5‰ but needs a long time until an isotopic equilibrium is attained. The isotopic half-life time of animal tissues depends on the general mean metabolic turnover rate of the animal (quite fast for small animals) and on that of the individual tissues (faster for "smooth" tissues). As the bulk of our meat food is muscle protein from quite large animals, short changes of their diet will therefore not largely interfere with its isotope characteristics. Nonetheless, longer periods of diet change influence the isotope characteristics, as can be seen on the tail hair of sheep, which shows alternating isotope data in accordance with summer and winter feeding.[73,74] The maximal differences reported for muscle meat from different parts of the same animal are <1‰ for $\delta^{13}C$ and <2‰ for $\delta^{15}N$;[75] this may be due to variable contents of proteins other than actin and myosin.

39.2.2.3 Diet Differences: Conventional and Organic Meat Production

The most fundamental principles of organic farming are that no pesticides and no artificial fertilizers are used with diet production and that any cattle feed is from (own) organic farming. The common opinion is that the organic products must have a relatively high $\delta^{15}N$-value due to the application of natural fertilizers for feed production, and a relatively negative $\delta^{13}C$-value due to small amounts of corn in the diet. However, as explained earlier, not any organic feed must be enriched in ^{15}N, and vice versa commercially available conventional cattle feed can be enriched in ^{15}N.[76] Correspondingly, organically produced corn silage also is enriched in ^{13}C. Nonetheless, Boner and Förstel[37] postulate that a $\delta^{13}C$-value of beef above −20‰ is not compatible with organic production, probably because a high percentage of (even organically produced) corn silage in the diet should be a proof that the cattle has not been kept appropriate to the species.

Schmidt et al.[38] observed that beef from conventional farming in Ireland was more enriched in ^{15}N and less depleted in ^{13}C than that from organic production; the authors conclude that the first group probably got more concentrated foodstuff and that, although in their study the two groups of animals could be distinguished from each other on the basis of combined C-, N-, and S-isotope characteristics but not absolutely assigned to organic or conventional diet regime, at present this absolute assignment is not possible. The correlation of the $\delta^{15}N$-value with the diet has also been the basis for a proposed proof of illegal meat and bone meal (MBM) feeding to ruminants in context with BSE-transfer.[77] The authors found that hair and milk products from cattle grown under conditions of intensive farming (use of protein concentrates, MBM not excluded) showed relatively more positive δ^{15}N-values ($>+6.4‰$) than those from organic farming and that all cases of BSE belonged to the first group. Correspondingly, Carrijo et al.[78] replaced up to 25% of soybean meal as protein source by MBM in the diet of chicken and found a decrease of the δ^{13}C-value of 1.6‰ and an increase of the δ^{15}N-value of 0.9‰.

However, the δ^{15}N-value of animal products by itself is not an absolute proof for an illegal feeding with MBM because in cases of unknown meat samples the isotope characteristic of the first link of the food chain, the plant diet, is not known, and, as discussed earlier, even plants from organic farming must not be enriched in ^{15}N. Correspondingly, we also believe that a high δ^{15}N-value of animal products is not indicative for their origin from organic farming, but that this finding must be combined with and discussed on the basis of further criteria, for example, the isotope data of additional elements (H, Sr) and on information from other biomarkers.[6,40]

39.2.3 Particularities of Individual Meat Ingredients and in Defined Molecular Positions

Additional information on origin and authenticity of meat is also to be expected from the analysis of meat fractions. The normal sample pretreatment (Section 39.3.1) provides meat water, fat, and defatted protein. The isotope analysis of the latter implies the information from C, N, and S; the potential of meat water isotope analysis has been discussed in Section 39.2.2.1. Whereas the O- and H-isotope data of protein and water must be treated with some caution, fat contains these elements in nonexchangeable positions and their δ-values are correlated to the climate of the sample's origin.[39] Also, δ^2H- and δ^{13}C-values and even positional δ^2H-values of fat and fatty acids by NMR have remarkably contributed to the discrimination between farmed and wild salmon and the identification of the feeding diet of beef,[50,40] respectively. NMR techniques have even been applied to whole lamb meat and contributed to origin assignments.[45]

Finally, it is to be mentioned that the δ^{18}O-value of the *p*-position of L-tyrosine depends on the pathway of the amino acid's biosynthesis and indicates whether it originates from plants or animals[19,79] This is the basis for an absolute method to prove the illegal feeding of MBM to ruminants.[80]

39.3 Meat Sample Preparation and Isotope Ratio Analysis

39.3.1 Light Elements: Pretreatment and Fractionation of Samples

According to the recommendations of several authors[39,46,55] about 50 g meat from defined muscles, freed from tendons and fasciae, are cut into small pieces (<1 cm^3); these are frozen and lyophilized for at least 24 h. The remainder is homogenized in a mortar or ball mill and the lipids are extracted with petroleum ether for 6 h in a Soxhlet device. The extract is discarded or evaporated to dryness for the isotope analysis of the lipid fraction, and the defatted raw protein is dried and homogenized again by grinding. A control of the completeness of the separation is possible by elemental analysis (in some cases automatically performed with the isotope ratio analysis): the product should have ~14% N and ~46–47% C (nonextracted sea fish muscle is reported to have a C:N ratio $<5:1$,[54]). If necessary, the extraction should be repeated. Fish muscle fillets are often analyzed without lipid extraction, Molkentin et al.[52] grounded and lyophilized the material; a fat extraction with methanol:chloroform:water in the ratio of 10:5:4 is recommended by Trueman et al.[67]

Concerning fish lipid extraction for NMR-analysis, see Ref. 50; for the preparation of fatty acid methyl esters (FAMEs) for GC-analysis, see Ref. 52 and cited references. The hydrogen atoms of the triglycerides do not exchange with water and are relatively depleted in deuterium. With some precautions and by standardization, even the δ^2H-value of the protein or the nondefatted meat is indicative of the origin of a sample, provided the material is equilibrated with a suitable standard and the δ^2H-values are correspondingly corrected.[81] All samples are analyzed in comparison to laboratory references (e.g., casein) or to the same material obtained by the identical procedure and standardized by several laboratories.

Meat tissue water is obtained for isotope analysis by heating 50–100 g samples for 15 min to 100°C in impermeable plastic bags or in vapor-tight brass pistons,[41] where the yield can even be enhanced by pressure.[43] The meat is put between two pieces of acrylamide glass and pressed by means of a bench vice.

39.3.2 Combustion and Multielement Isotope Ratio Measurement of Organic Material and of Tissue Water

Samples for the H-, C-, N-, and S-isotope analysis are weighed in tin capsules, and for the O-isotope analysis in silver capsules; these are sealed and applied to the autosampler of a combustion or pyrolysis unit of an elemental analyzer (EA). The amount of required material is determined by its elemental composition, the isotopes to be analyzed, and the properties of the mass spectrometer coupled to the EA. In the case of meat and for the simultaneous determination of the δ^{13}C- and δ^{15}N-values, 0.8–1.5 mg of dried and defatted material is necessary whereas for the integrated determination of the δ^{34}S- and δ^2H-values 3.0 mg is needed. The samples are converted in an EA (e.g., Carlo Erba Strumentazione, Rodano/Italy, NA 1500; Costech, Milano/Italy, ECS 4010; Elementar Analysensysteme GmbH, Hanau/Germany, Vario EL III; ThermoQuest, Milano/Italy, CE Flash EA 1112; Thermo Electron, Bremen/Germany, TC/EA 200) into suitable gases (see Table 39.1: N_2, masses 28 and 29 for ^{15}N-analysis; CO_2, masses 44 and 45 for ^{13}C-analysis; SO_2, masses 64 and 66 for ^{34}S-analysis). The pyrolytic preparation of the gases H_2 (masses 2 and 3 for ^{2}H-analysis) and CO (masses 28 and 30 for ^{18}O-analysis) for the hydrogen and oxygen isotope ratio determination (double isotope method) demands approximately 1.0–1.5 mg sample (corresponding to 80–120 µg H or 80–160 µg O); for the ^{2}H-analysis of organically bound hydrogen the meat must be equilibrated for 24 h with a laboratory (meat protein) standard (comparative equilibration technique[81]). In any case, three to four repeats of the samples have to be analyzed and compared with working standards of similar material (e.g., casein, collagen) or defined organic compounds.

The EA is coupled to an isotope ratio mass spectrometer (IRMS, e.g., GV Instruments, Wythenshawe/UK, IsoPrime; Thermo Finnigan, Delta Plus XP; SerCon Ltd, Crewe, UK, 20–20) via a suitable software. In most devices, after the combustion of the organic material and the reduction of NO_x, SO_2 and H_2O are eliminated and CO_2 and N_2 are separated by GC; therefore, only a simultaneous isotope ratio analysis of C and N is possible. The isotope ratio of S must normally be determined independently,[48] and that of H and O is obtained by high temperature (HT) pyrolysis (~1450°C) of a separate sample to H_2 and CO. A simultaneous isotope ratio and elemental analysis of H, N, C, and S in a single run is only possible with the modified EA of Vario EL III of Elementar Analysensysteme GmbH,[82] in which the combustion gases are separated by reversible adsorption. ^{2}H and ^{18}O in water are transferred to the measuring gases H_2 and CO_2, respectively, by isotopic equilibration,[56,57] following the recommendations of the equipment providers. In any case, the δ-values are referred to international standards (Table 39.2); error limits ±0.1‰ for ^{13}C, ±0.2‰ for ^{15}N, ±0.3‰ for ^{34}S, and ±3.0‰ for ^{2}H.

39.3.3 Sample Preparation and Isotope Analysis of Minor and Trace Elements

The (isotope ratio) analysis of trace elements demands special precautions for sample preparation to avoid any contamination; for example, the use of ceramic scissors is recommended for meat cutting. For this special field we recommend to refer to the corresponding literature.[39,45] In contrast, the sample preparation for isotope ratio determinations of minor or contaminant elements (Sr) is less sophisticated because losses of these elements will not interfere with their isotope ratios. Although losses of material do not falsify the isotope ratios of trace elements (e.g., Sr) any contaminations have to be avoided. Therefore, all chemical separation steps have to be performed in a class 100 clean laboratory and three times distilled reagents have to be used.

For the isotopic analysis of Sr in meat, about 400 mg of dried and defatted sample (see Section 39.3.3) are combusted at 800°C in a quartz crucible; the ashes are dissolved in conc. HNO_3. The Sr-isotope analysis demands a separation of the element from contaminants (Ca, Ba, and Rb; meat often contains relatively high concentrations of ^{87}Rb, which interferes with the analysis of ^{87}Sr) and an analyte concentration. This is performed by ion chromatography on Sr-Spec®, an Sr-specific crown ether resin, following a procedure adapted from Horwitz et al.[83] and Pin and Bassin.[84] The fraction with the Sr is transferred to the filament of the thermal ionization mass spectrometer (TIMS) and the isotope ratio measurement (50 repeats) is done in comparison to the international Sr-standard (SrCO$_3$, NIST SRM 987, $^{87}Sr/^{86}Sr = 0.710235$); in some laboratories a standard with $^{87}Sr/^{86}Sr = 0.7093$, originating from Baltic sea water, is in use. Mass fractionations implied in the ionization are corrected on the basis of the naturally invariant ratio $^{88}Sr/^{86}Sr = 8.37521$. The precision of the isotope ratio measurement is better than 0.005%.

39.3.4 Presentation and Statistical Analysis of Results

The isotopic properties of biomass from a given origin are never constant and absolute but underlie local, temporal, and biological alterations. So, the δ^2H- and $\delta^{18}O$-values of meteoric water of the same place show annual and seasonal variations; the $\delta^{15}N$- and $\delta^{13}C$-values of plant material are modulated by actual fertilization practices and local climatic influences, respectively. Therefore, in spite of the availability of general data and experience as compiled in Table 39.2 and further data to be expected in the frame of the "TRACE" program of the EU mentioned in Section 39.1, the assessment of unknown samples will always need the additional simultaneous analysis of corresponding reference material. This and the multielement multicompound isotope ratio analysis of unknown and reference material will provide data and information that demand a statistical treatment of the results for optimal interpretation.

Quite common and efficient, and in many cases valuable and sufficient, is the two-dimensional display of isotope data for two different δ-values representing origin or treatment characteristics as shown in Figure 39.1. The combination of two independent indications for cattle keeping (diet and drinking water, Figure 39.1a) or for diet production (fertilization and soil characteristics, Figure 39.1b) permit the assignment of beef to Northern and Southern Germany and simultaneously to fattening in stables or on meadows, and the discrimination between carps from Poland and Germany, respectively.

The more sophisticated statistical treatment of the data by multivariate data analysis, for example, in the form of the canonical discriminant analysis (CDA), permits to condense the information of more than two variables into a two-dimensional scale.[85] The different variables (δ-values,

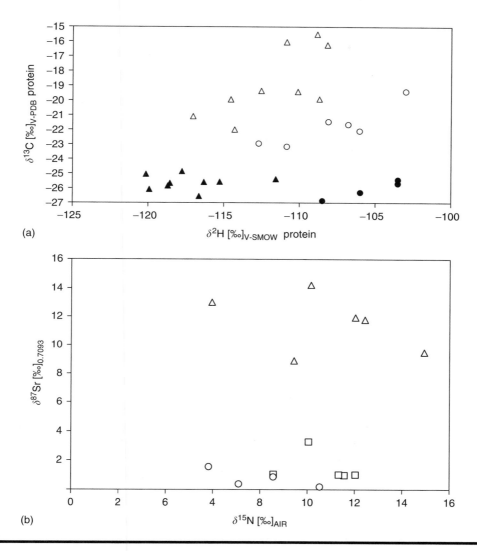

(a)

(b)

Figure 39.1   (a) δ^{13}C-values in [‰]$_{V\text{-PDB}}$ of fat-free dry matter and δ^{18}O-values in [‰]$_{V\text{-SMOW}}$ of water from German beef from different areas and diets. Cattle from Bavaria, kept in stables $\triangle$ and on meadows ▲, respectively, and from Northern Rhenania-Westfalia in stables $\bigcirc$ and meadows ●, respectively. The more positive δ^{13}C-values of the meat from stable kept cattle correspond to the corn content in their feed. The δ^{2}H-values of the meadow kept cattle meat correspond to their drinking water, whereas those of the stable kept animals indicate an overlap of the influences of geographical origin and C_4-plant diet. (b) δ^{15}N-values in [‰]$_{AIR}$ and δ^{87}Sr-values in [‰]$_{0,7093}$ (laboratory standard) of fat-free carp meat from hatcheries in Bavaria and neighboring regions. $\triangle$ = East Bavaria/Bohemia, $\square$ = North Bavaria, $\bigcirc$ = Saxonia/Poland. The δ^{87}Sr-values are indicative for the geological characteristics of the sample origin, which can additionally be differentiated by means of δ^{34}S-values (not shown); the δ^{15}N-values are dominated by the production of the diet. (After Rossmann, A. and Schlicht, C., *Fleischwirtschaft*, 2007. With permission.)

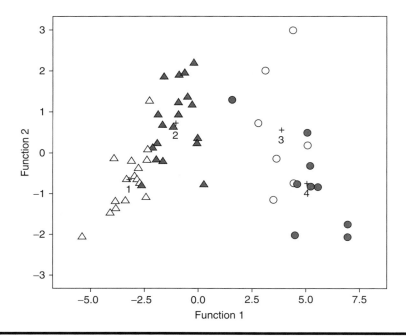

Figure 39.2 Numeric results of the first two discriminant functions after a linear discriminant analysis (statistic program SPSS 14 for windows) of the multielement isotope ratios (δ^2H-, δ^{13}C-, δ^{15}N-, δ^{34}S-values) of defatted lamb meat from two areas in Bavaria and Greece. Δ = West Bavaria (Allgäu, centroid 1), ▲ = North Bavaria (Franconia, centroid 2), ○ = South Greece (Lakonia, centroid 3), ● = North Greece (Chalkidiki, centroid 4). A classification by the discriminant functions permitted a 100% separation between Bavarian and Greek samples; within the countries 12% of the samples from West Bavaria were assigned to North Bavaria, 5% of the latter to West Bavaria, the samples from South Greece were classified 100% correctly, those from South Greece 89%. (After Rossmann, A. and Schlicht, C., *Fleischwirtschaft*, 2007. With permission.)

representing climate, fertilization, geological properties) are condensed by a linear combination to canonical discriminant functions. Earlier, the significance of each variable is tested by the analysis of variance (ANOVA) to obtain functions for an optimal separation of the individual data groups.[86] Each group should consist of many more samples than variables.

The CDA or other statistical principles have been applied by several authors;[39,44,45,52] in the present example CDA has been used for the assignment of lamb meat to different areas in Greece and Bavaria on the basis of its δ^2H-, δ^{13}C-, δ^{15}N-, and δ^{34}S-values (Figure 39.2). Although the number of samples available has not been very large for all groups, satisfactory differentiations are obtained.

39.4 Conclusion and Future Aspects

Multifunctional bioelement isotope ratio analysis of whole meat is a powerful tool for its origin and authenticity assignment. Additional information is obtained by the isotope analysis of fractions such as water, fat, and defatted protein. In some cases, this can even be supported by ^{2}H-IRMS and NMR measurements of the fat fraction or the determination of trace elements, although these analyses are laborious and demand additional expensive instrumentation. More promising is the integration of an additional determination of the Sr-isotopes. In any case, the selection and application of the different possibilities depend on the question to be solved.

The most efficient data evaluation is by multivariate data analysis. This demands the availability of laboratory or international data banks and the interlaboratory exchange of data from authentic material. However, the correct interpretation of the isotope data of unknown material depends on the researcher's individual experience and his knowledge about the parameters and factors influencing the isotopic shifts between nutrition and meat. This can be further supported by external information from animal labeling, nutrition state, and trophic level of the animals and about meat processing and conservation. Although, on the basis of isotope data, it will never be possible to assign an unknown meat sample worldwide to a geographical origin or to a given kind of animal breeding and feeding, it will always be possible to discriminate between defined limited alternatives of origins and treatments.

Summarizing the potential of isotope ratio analysis in meat origin and authenticity determination, one can state:

1. A worldwide absolute origin identification of an unknown sample will not be possible but will normally also not be requested. However, the method permits the control of origin and production indications, preferably on the basis of its comparison with authentic reference samples.

2. Multielement isotope ratio determination of fat-free muscle food discriminates between samples of beef and lamb meat from Europe, Argentina, the United States, and New Zealand with approximately 80% probability. Discriminations between samples from individual regions within these countries can also be performed but will often demand additional information from other analytical methods.

3. The isotope ratio determination permits the distinction between samples from defined production methods, for example, pork from special pig fattening rations and corresponding surrogates. Correspondingly, the differentiation between fish from wild and farm breeding will be possible. Discrimination between meat from conventional and organic production, respectively, can only be attained at the availability of additional information.

4. The method has so far not been applied to meat products (sausage, etc.)

Acknowledgment

The authors thank Roland A. Werner, ETH Zürich, for providing many references and Christine Lehn, LMU München, for assisting with the discriminant analysis.

Notes

Section 39.2.2   Very recently a paper by Nakashita et al. about beef isotopic data appeared. Its particularity is that it is the first paper reporting results from Japan and Australia and includes data on organically bound oxygen.

Nakashita, R., Suzuki, Y., Akamatsu, F., Iizumi, Y., Korenaga, T., Chikaraiski, Y. Stable carbon, nitrogen, and oxygen isotope analysis as a potential tool for verifying geographical origin of beef. *Anal. Chim. Acta.* 617, 148, 2008.

Section 39.2.2.1   In contrast to Thiem et al., 2004 (Ref. 42), who had observed an increase of the $\delta^{18}O$-value of meat water, when they kept small meat pieces in open vessels at room temperature for 2 h, M. Horacek and coworkers (private communication) stored meat under industrial cool house conditions over longer periods and found no change in the $\delta^{18}O$-value of the water.

Section 39.2.2.3   In a recent paper Bahar et al. (2008) studied the seasonal variation of the isotope characteristics of beef from conventional and organic breeding. They found that the $\delta^{13}C$-value of the meat from organic beef was more negative and less variable than that from conventional beef and, whilst the $\delta^{15}N$-value of conventional beef remained nearly 7‰ throughout the year, that of organic beef was more variable and significantly more negative.

Bahar, B., Schmidt, O., Moloney, A.P., Scrimgeour, C.M., Begley, I.S., Monahan, F.J. Seasonal variation in the C, N and S stable isotope composition of retail organic and conventional Irish beef. *Food Chem.* 106, 1299–1305, 2008.

References

1. Rossmann, A. Determination of stable isotope ratios in food analysis. *Food Rev. Intern.* 17, 347, 2001.
2. Kelly, S., Heaton, K., and Hoogewerff, J. Tracing the geographical origin of food: the application of multi-element and multi-isotope analysis. *Trends Food Sci. Technol.* 16, 555, 2005.
3. Kornexl, B.E. et al. Measurement of stable isotope abundances in milk and milk products: a possible tool for origin assignment and quality control. *Z. Lebensm. Unters. Forsch.* 205, 19, 1997.
4. Rossmann, A. et al. The potential of multielement stable isotope analysis for regional origin assignment of butter. *Eur. Food Res. Technol.* 211, 32, 2000.
5. Manca, G. et al. Correlation between multielement stable isotope ratio and geographical origin in Peretta cow's milk cheese. *J. Dairy Sci.* 89, 831, 2006.
6. Franke, B.M. et al. Geographic origin of meat: elements of an analytical approach to its authentication. *Eur. Food Res. Technol.* 221, 493, 2005.
7. Schwägele, F. Traceability from a European perspective. *Meat Sci.* 71, 164, 2005.
8. Moretti, V.M. et al. Traceability issues in fishery and aquaculture products. *Vet. Res. Commun.* 27, Suppl. 1, 497, 2003.
9. Prache, S. et al. Traceability of animal feeding diet in the meat and milk of small animals. *Small Ruminant Res.* 59, 157, 2005.
10. Gremaud, G., Carlen, S., and Hulliger, K. Analytical methods for the authentication of meat and meat products: recent developments. *Mitt. Lebensm. Hyg.* 93, 481, 2002.
11. Peres, B. et al. Review of the current methods of analytical traceability allowing determination of the origin of foodstuffs. *Food Control* 18, 228, 2007.
12. Schmidt, H.-L. Fundamentals and systematics of the non-statistical distributions of isotopes in natural compounds. *Naturwissenschaften* 90, 537, 2003; 91, 148, 2004.
13. Schmidt, H.-L. et al. Stable isotope ratio analysis in quality control of flavourings. In *Flavourings, Production, Composition, Applications, Regulations,* 2nd Ed., ed. H. Ziegler, 2007, 602. Weinheim, WILEY-VCH Verlag GmbH & Co. KgaA.
14. Martin, M.L. and Martin, G.J. Deuterium NMR in the study of site-specific natural isotope fractionation (SNIF-NMR), in *NMR Basic Principles and Progress,* Vol. 23, eds. P. Diehl, E. Fluck, H. Günther, R. Kosfeld, and J. Seelig, 1990, 1–61. Berlin, Springer Verlag.
15. Bowen, G.J. and Wilkinson, B. Spatial distribution of $\delta^{18}O$ in meteoric precipitation. *Geology* 30, 315, 2002.
16. Craig, H. Isotopic variations in meteoric water. *Science* 133, 1702, 1961.
17. Dansgaard, W. Stable isotopes in precipitation. *Tellus* 16, 436, 1964.
18. Tatner, P. A model of the natural abundance of oxygen-18 and deuterium in the body water of animals. *J. Theor. Biol.* 133, 267, 1988.

19. Schmidt, H.-L., Werner, R.A., and Rossmann, A. ^{18}O pattern and biosynthesis of natural plant products. *Phytochemistry* 58, 9, 2001.
20. O'Leary, M.H. Carbon isotopes in photosynthesis. *Bioscience* 38, 328, 1988.
21. Farquhar, G.D., Ehleringer, J.R., and Hubick, K.T. Carbon isotope discrimination and photosynthesis. *Ann. Rev. Plant Physiol. Plant Mol. Biol.* 40, 503, 1989.
22. DeNiro, M.J. and Epstein, S. You are what you eat (plus a few ‰): the carbon isotope cycle in food chains. *Geol. Soc. Am. Abstr. Prog.* 8, 834, 1976.
23. Werner, R.A. and Schmidt, H.-L. The in vivo nitrogen isotope discrimination among organic plant compounds. *Phytochemistry* 61, 465, 2002.
24. Schmidt, H.-L. et al. Isotope characteristics of vegetables and wheat from conventional and organic production. *Isot. Environ. Health Stud.* 41, 223, 2005.
25. Thode, H.G. Sulfur isotopes in nature and the environment: an overview. In *Stable Isotopes. Natural and Anthropogenic Sulphur in the Environment* (SCOPE 43), eds. H.R. Krouse and V.A. Grinenko, 1991, 1–26. New York, Wiley.
26. Krouse, H.R., Stewart, J.B.W., and Grinenko, V.A. Pedosphere and biosphere. In *Stable Isotopes. Natural and Anthropogenic Sulphur in the Environment* (SCOPE 43), eds. H.R. Krouse and V.A. Grinenko, 1991, 267–306. New York, Wiley.
27. Tanz, N. and Schmidt, H.-L., unpublished data.
28. Weber, P.K. et al. Otolith sulfur isotope method to reconstruct salmon (*Oncorhynchus tshawytscha*) life history. *Can. J. Fish. Aquat. Sci.* 59, 587, 2002.
29. Aberg, G., Fosse, G., and Stray, H. Man, nutrition and mobility: a comparison of teeth and bone from the medieval era and the present from Pb and Sr isotopes. *Sci. Total Environ.* 224, 109, 1998.
30. Hölzl, S. et al. Isotope-abundance of light (bio) and heavy (geo) elements in biogenic tissues: methods and applications. *Anal. Bioanal. Chem.* 378, 270, 2004.
31. Hintze, K.J. et al. Selenium accumulation in beef: effect of dietary selenium and geographical area of animal origin. *J. Agric. Food Chem.* 50, 3938, 2002.
32. Pillonel, L. et al. Stable isotope ratios, major, trace and radioactive elements in Emmental cheeses of different origins. *Lebensm. -Wiss. u. -Technol.* 36, 615, 2003.
33. McCutchan, Jr., J.H. et al. Variation in trophic shift for stable isotope ratios of carbon, nitrogen and sulfur. *OIKOS* 102, 378, 2003.
34. Gebbing, T., Schellberg, J., and Kühbauch, W. Switching from grass to maize diet changes the C isotope signature of meat and fat during fattening of steers. In *Proceedings of the 20th General Meeting of the European Grassland Federation*, eds. A. Lüscher, B. Jeangros, W. Kessler, O. Huguenin, M. Lobsiger, N. Miller et al., Zürich, Hochschulverlag an der ETH (*Grassland Sci. Eur.*) 9, 1130, 2004.
35. De Smet, S. et al. Stable carbon isotope analysis of different tissues of beef animal in relation to their diet. *Rapid Commun. Mass Spectrom.* 18, 1227, 2004.
36. Bahar, B. et al. Alteration of the carbon and nitrogen stable isotope composition of beef by substitution of grass silage with maize silage. *Rapid Commun. Mass Spectrom.* 19, 1937, 2005.
37. Boner, M. and Förstel, H. Stable isotope variation as a tool to trace the authenticity of beef. *Anal. Bioanal. Chem.* 378, 301, 2004.
38. Schmidt, O. et al. Inferring the origin and dietary history of beef from C, N, and S stable isotope ratio analysis. *Food Chem.* 91, 545, 2005.
39. Heaton, K. et al. Verifying the geographical origin of beef: the application of multi-element isotope and trace element analysis. *Food Chem.* 107, 506, 2008.
40. Renou, J.P. et al. Characterization of animal products according to geographic origin and feeding diet using nuclear magnetic resonance and isotope ratio mass spectrometry. Part II: beef meat. *Food Chem.* 86, 251, 2004.
41. Hegerding, L. et al. Sauerstoffisotopen-Verhältnis-Analyse von Rindfleisch. *Fleischwirtschaft* 82, 95, 2002.
42. Thiem, I., Lüpke, M., and Seifert, H. Factors influencing the ^{18}O/^{16}O-ratio in meat juices. *Isot. Environ. Health Stud.* 40, 191, 2004.

43. Thiem, I., Lüpke, M., and Seifert, H. Extraction of meat juices for isotopic analysis. *Meat Sci.* 71, 334, 2005.
44. Piasentier, E. et al. Stable isotope ratio analysis for authentication of lamb meat. *Meat Sci.* 64, 239, 2003.
45. Sacco, D. et al. Geographical origin of Apulian lamb meat samples by means of analytical and spectroscopic determinations. *Meat Sci.* 71, 542, 2005.
46. Camin, F. et al. Multielement (H, C, N, S) stable isotope characterisation of lamb meat from different European regions. *Anal. Bioanal. Chem.* 389, 309, 2008.
47. González-Martin, I. et al. Use of isotope analysis to characterize meat from Iberian-breed swine. *Meat Sci.* 52, 437, 1999.
48. González-Martin, I. et al. Differentiation of dietary regimen of Iberian swine by means of isotopic analysis of carbon and sulphur in hepatic tissue. *Meat Sci.* 58, 25, 2001.
49. Bielefeld Nardoto, G. et al. Stable carbon and nitrogen isotopic fractionation between diet and swine tissue. *Sci. Agric. (Piracicaba, Braz.)* 63, 579, 2006.
50. Aursand, M., Mabon, F., and Martin, G.J. Characterization of farmed and wild salmon (*Salmo salar*) by a combined use of compositional and isotopic analyses. *J. Am. Off. Chem. Soc.* 77, 659, 2000.
51. Dempson, J.B. and Power, M. Use of stable isotopes to distinguish farmed from wild Atlantic salmon, *Salmon salar. Ecol. Freshw. Fish* 12, 176, 2004.
52. Molkentin, J. et al. Identification of organically farmed Atlantic salmon by analysis of stable isotopes and fatty acids. *Eur. Food Res. Technol.* 224, 535, 2007.
53. Gaye-Siessegger, J. et al. Feeding level and individual metabolic rate affect δ^{13}C and δ^{15}N values in carp: implications for food web studies. *Oecologia* 138, 175, 2004.
54. Rojas, J.M.M. et al. The use of stable isotope ratio analyses to discriminate wild and farmed gilthead sea bream (*Sparus aurata*). *Rapid Commun. Mass Spectrom.* 21, 207, 2007.
55. Rossmann, A. and Schlicht, C. Feststellung der geografischen Herkunft von Fleisch durch massenspektrometrische Multielement-Stabilisotopenanalyse (H,C,N,S). *Fleischwirtschaft* 8, 104, 2008.
56. Brand, W.A. et al. New methods for fully automated isotope ratio determination from hydrogen at the natural abundance level. *Isot. Environ. Health Stud.* 32, 263, 1996.
57. Epstein, S. and Mayeda, T. Variation of the ^{18}O content of water from natural sources. *Geochim. Cosmochim. Acta* 4, 213, 1953.
58. Schoeller, D.A. Isotope fractionation: why aren't we what we eat? *J. Archaeol. Sci.* 26, 667, 1999.
59. Gannes, L.Z., O'Brien, D.M., and del Rio, C.M. Stable isotopes in animal ecology: assumptions, caveats, and a call for more laboratory experiments. *Ecology* 78, 1271, 1997.
60. Gannes, L.Z., Martinez del Rio, C., and Koch, P. Natural abundance variations in stable isotopes and their potential uses in animal physiological ecology. *Comp. Biochem. Physiol.* 119A, 725, 1998.
61. Gaye-Siessegger, J. et al. Individual protein balance strongly influences δ^{15}N and δ^{13}C values in Nile tilapia, *Oreochromis niloticus. Naturwissenschaften* 91, 90, 2004.
62. Phillips, D.L. Mixing models in analyses of diet using multiple stable isotopes: a critique. *Oecologia* 127, 166, 2001.
63. Koch, P.L., Fogel, M.L., and Tuross, N. Tracing the diets of fossil animals using stable isotopes. In *Stable Isotopes in Ecology and Environmental Sciences*, eds. K. Lajtha and R.H. Michener, 1994, 63. Oxford, Blackwell Scientific.
64. Fuller, B.T. et al. Nitrogen balance and δ^{15}N: why you're not what you eat during nutritional stress. *Rapid Commun. Mass Spectrom.* 19, 2497, 2005.
65. Hatch, K.A. et al. An objective means of diagnosing anorexia nervosa and bulimia nervosa using ^{15}N/^{14}N and ^{13}C/^{12}C ratios in hair. *Rapid Commun. Mass Spectrom.* 20, 3367, 2006.
66. Mekota, A.-M. et al. Serial analysis of stable nitrogen and carbon isotopes in hair: monitoring starvation and recovery phases of patients suffering from anorexia nervosa. *Rapid Commun. Mass Spectrom.* 20, 1604, 2006.

67. Trueman, C.N., McGill, R.A.R., and Guyard, P.H. The effect of growth rate on tissue-diet isotopic spacing in rapidly growing animals. An experimental study with Atlantic salmon (*Salmo salar*). *Rapid Commun. Mass Spectrom.* 19, 3239, 2005.

68. Barness, C. and Jennings, S. Effect of temperature, ration, body size and age on sulphur isotope fractionation in fish. *Rapid Commun. Mass Spectrom.* 21, 1461, 2007.

69. Richards, M.P. et al. Sulphur isotopes in palaeodietary studies: a review and results from a controlled experiment. *Int. J. Osteoarchaeol.* 13, 37, 2003.

70. Birchall, J. et al. Hydrogen isotope ratios in animal body reflect trophic level. *J. Anim. Ecol.* 74, 877, 2005.

71. Sponheimer, M. et al. Turnover of stable carbon isotopes in the muscle, liver, and breath CO_2 of alpacas (*Lama pacos*). *Rapid Commun. Mass Spectrom.* 20, 1395, 2006.

72. Phillips, D.L. and Eldridge, P.M. Estimating the timing of diet shifts using stable isotopes. *Oecologia* 147, 195, 2006.

73. Schwertl, M., Auerswald, K., and Schnyder, H. Reconstruction of the isotopic history of animal diets by hair segmental analysis. *Rapid Commun. Mass Spectrom.* 17, 1312, 2003.

74. Schwertl, M. et al. Carbon and nitrogen stable isotope composition of cattle hair: ecological fingerprints of production systems? *Agric. Ecosys. Environ.* 109, 153, 2005.

75. Boner, M. Überprüfung der Authentizität von Rindfleisch (Bio) mit Hilfe der stabilen Isotope der Bioelemente. *Thesis at the Rheinische Friedrich-Wilhelms-Universität Bonn*, 71, 2005.

76. Anonymous. Aktueller Stand der Möglichkeiten zur Unterscheidung von Nahrungsmitteln aus konventioneller bzw. ökologischer Produktion durch Isotopenverhältnisanalyse. *Lebensmittelchemie* 59, 50, 2005.

77. Delgado, A. and Garcia, N. $\delta^{15}N$ and $\delta^{13}C$ analysis to identify cattle fed on feed containing animal proteins. A safety/quality index in meat, milk and cheese. In *Proceedings of the 6th International Symposium on Food Authenticity and Safety*, Nantes, Eurofins, 2001.

78. Carrijo, A.S. et al. Traceability of bovine meat and bone meal in poultry by stable isotope analysis. *Brazil. J. Poultry Sci.* 8, 63, 2006.

79. Fronza, G. et al. The $\delta^{18}O$-value of the *p*-OH group of L-tyrosine permits the assignment of its origin to plant or animal sources. *Eur. Food Res. Technol.* 215, 55, 2002.

80. Tanz, N. and Schmidt, H.-L., unpublished date.

81. Wassenaar, L.I. and Hobson, K.A. Comparative equilibration and online technique for determination of non-exchangeable hydrogen of keratins for use in animal migration studies. *Isot. Environ. Health Stud.* 39, 211, 2003.

82. Sieper, H.-P. et al. A measuring system for the fast simultaneous isotope ratio and elemental analysis of carbon, hydrogen, nitrogen and sulfur in food commodities and other biological material. *Rapid Commun. Mass Spectrom.* 20, 2521, 2006.

83. Horwitz, E.P., Chiarizia, R., and Dietz, M.L. A novel strontium-selective extraction chromatographic resin. *Solv. Extr. Ion Exch.* 10, 313, 1992.

84. Pin, C. and Bassin, C. Evaluation of a strontium-specific extraction chromatographic method for isotopic analysis in geological material. *Anal. Chim. Acta* 269, 249, 1992.

85. Afifi, A.A. and Clark, V. *Computer-Aided Multivariate Analysis*, 2nd Ed. New York, Van Nostrand Reinhold Company, 1990.

86. Shaw, A.D. et al. Discrimination of the variety and region of origin of extra virgin olive oils using ^{13}C NMR and multivariate calibration with variable reduction. *Anal. Chim. Acta* 348, 357, 1997.

Chapter 40

Determination of Persistent Organic Pollutants in Meat

Anna Laura Iamiceli, Igor Fochi, Gianfranco Brambilla, and Alessandro di Domenico

Contents

40.1 Introduction

40.1.1 General Remarks

As defined by the United Nations Environment Programme (UNEP), persistent organic pollutants (POPs) are "chemical substances that persist in the environment, bioaccumulate through the food web, and pose a risk of causing adverse effects to human health and the environment" [1]. Although many POPs are already strictly regulated or are no longer in production, they are found in the environment and can enter the food chain mainly through the intake of animal fats (meat, fish, and milk) [2]. The measurement of POPs in food and, in particular, in products of animal origin is particularly important for the protection of human health. Maximum residue limits (MRLs) for some POPs (organochlorine pesticides) in a variety of food commodities were established by the European Union (EU), thus making necessary the development of sensitive methods to analyze these pollutants in food.

This chapter evaluates the methods in use for the determination of POPs in meat. The chapter is divided into three parts. The first part provides an overview of the POPs under the Stockholm Convention [3] and of those proposed for inclusion. The second part deals with the chemical methods available for the determination of some selected classes of POPs. The similarities among these methods often result in the simultaneous determination of several families of pollutants—for example, polychlorinated dibenzo-*p*-dioxins (PCDDs), polychlorinated dibenzofurans (PCDFs), and polychlorinated biphenyls (PCBs), and polybrominated diphenyl ethers (PBDEs)—after a common preparative procedure of the sample. Nevertheless, for practical reasons separate sections will be devoted to the methods used in the analysis of each class of pollutants: organochlorine pesticides (Section 40.2.1); PCDDs, PCDFs, and dioxin-like PCBs (DL-PCBs) (Section 40.2.2); non-dioxin-like PCBs (NDL-PCBs) and PBDEs (Section 40.2.3); and polyfluorinated alkylated substances (PFAS) (Section 40.2.4). The third part of the chapter provides some information about the use of bioassays for PCDDs, PCDFs, and DL-PCBs for screening.

40.1.2 Laboratory Safety

The POPs under investigation should be treated as a potential health hazard. A strict safety program for handling these substances and the chemicals used for their determination should be developed by the laboratory.

Following are suggested readings taken from the vast literature available: Organochlorine pesticides were evaluated for their risk to human health and the environment within the

International Programme on Chemical Safety (IPCS) [4–14]. Evaluation of carcinogenic risk for humans from PCDDs and PCDFs was performed by IARC in 1997 [15]. More recently, the Scientific Committee on Food (SCF) of the European Commission (EC) adopted an opinion on PCDDs, PCDFs, and DL-PCBs in food [16], updating its opinion of 2000 [17]. As regards PBDEs, a draft risk profile on pentabromodiphenyl ether (penta-BDE) (commercial mixture) was adopted by the POPs Review Committee [18]; a risk assessment report on octabromodiphenyl ether (octa-BDE) (commercial mixture) was prepared on behalf of the EU [19]. A draft risk profile on perfluorooctane sulfonate (PFOS) was adopted by the POPs Review Committee [20], while a risk assessment opinion is being recently finalized at the European Food Safety Authority (EFSA).

Waste handling and decontamination of glassware, towels, laboratory coats, etc., are described in Refs 21–25.

40.1.3 The Stockholm Convention on POPs

POPs are by definition persistent, thus representing a risk of a long-time exposure. Their persistence is generally correlated to their chemical stability, which makes these substances highly resistant to biological and chemical degradation. POPs can be found worldwide, even in areas where human activities are almost completely absent, that is, in the Antarctic and Arctic regions. POPs are also characterized by their ability to bioaccumulate. Bioaccumulation magnitude depends on several factors, one among which is the solubility of the substance in lipids [2]. Highly lipophilic substances are substantially insoluble in water, as commonly shown by the high values of n-octanol–water partition coefficient (K_{OW}). This strengthens the tendency of these substances to be concentrated in the fatty tissue of a living organism. As a result of bioaccumulation, several organic persistent substances are subjected to biomagnification process and are found at higher concentrations in animals at higher levels of the food chain [26].

Potential adverse effects on the environment and human health caused by exposure to POPs are of considerable concerns for governments, nongovernmental organizations, and the scientific community. The persistence of POPs in the environment and the capacity of covering long distances away from the point of their release have required that concerted international measures were adopted to efficiently control release. To this end, the global Stockholm Convention on POPs, opened for signatures in 2001 and entered into force in May 2004 [27], provides an international framework, based on the precautionary principle that seeks to guarantee the elimination of POPs or the reduction of their production and use. Since the beginning, the Convention has concerned 12 chlorinated chemicals, but every country for which the Convention is in force may submit a proposal for listing new POPs in Annex A (substances to be eliminated), Annex B (substances whose production and use is restricted), or Annex C (substances unintentionally produced whose releases have to be reduced and finally eliminated). The substances actually under the Convention are listed in Table 40.1 and include eight individual organochlorine pesticides, hexachlorobenzene, PCBs, PCDDs, and PCDFs. At present, a second group of chemicals (candidate POPs) is under consideration for inclusion in the Convention on the basis of their risk profile prepared by the POP Review Committee (Table 40.2); a third group (proposed POPs) has been proposed for risk evaluation to the Review Committee (Table 40.3). The inclusion of organic chemicals in the frame of the Convention presupposes that some requirements be met:

a. Persistence, measured as half-life, greater than 2 months in water, or 6 months in soil or sediments, or any evidence of sufficient persistence to justify the consideration of a substance within the Convention;

Table 40.1 Some Physical–Chemical Properties of the POPs under the Stockholm Convention

Compound	Molecular Structure	Molecular Weight	Water Solubility	logK$_{OW}$	Vapor Pressure (mmHg)	Half-life in Soil (years)	Reference
Aldrin (CAS 309-00-2)		365	27 µg/L (25°C)	5.17–7.4	2.3 × 10^{-5} (20°C)	<1.6	28
Chlordane (CAS 57-74-9)		410	56 µg/L (25°C)	4.58–5.57	0.98 × 10^{-5} (20°C)	4	28
pp′-DDT[a] (CAS 50-29-3)		355	1.2–5.5 µg/L (25°C)	6.19	0.2 × 10^{-6} (20°C)	15	28
Dieldrin (CAS 60-57-1)		381	140 µg/L (20°C)	3.69–6.2	1.78 × 10^{-7} (20°C)	3–4	28

Compound	Structure					
PCDDs[b,c]		322–460	19.3–0.074 ng/L (25°C)	6.80–8.20	1.5×10^{-9}–8.25×10^{-13} (25°C)	10–12[d]	28, 29
PCDFs[e,f]		306–444	419–1.16 ng/L (25°C)	6.53–8.7	1.5×10^{-8}–3.75×10^{-13} (25°C)	—	28, 29
Endrin (CAS 72-20-8)		381	220–260 µg/L (25°C)	3.21–5.34	2.7×10^{-7} (25°C)	12	28
Hexachlorobenzene (CAS 118-74-1)		285	50 µg/L (20°C)	3.93–6.42	1.09×10^{-5} (20°C)	2.7–5.7	28
Heptachlor (CAS 76-44-8)		373	180 ng/mL (25°C)	4.4–5.5	3×10^{-4} (20°C)	0.75–2	28

(Continued)

Table 40.1 (Continued)

Compound	Molecular Structure	Molecular Weight	Water Solubility	$\log K_{OW}$	Vapor Pressure (mmHg)	Half-life in Soil (years)	Reference
Mirex (CAS 2385-85-5)		546	0.07 µg/L (25°C)	5.28	3×10^{-7} (25°C)	10	28
PCBs[g]		189–499	0.0001–0.01 µg/L (25°C)	4.3–8.26	$0.003–1.6 \times 10^{-6}$ (25°C)	>6	28
Toxaphene (CAS 8001-35-2)		414	550 µg/mL (20°C)	—	0.2–0.4 (25°C)	0.3–12	28

[a] 1,1,1-trichloro-2,2-*bis*(4-chlorophenyl)ethane.
[b] Polychlorinated dibenzo-*p*-dioxins.
[c] Data refer to the seven 2,3,7,8-chlorosubstituted toxic congeners only.
[d] Data refer to 2,3,7,8-T4CDD.
[e] Polychlorinated dibenzofurans.
[f] Data refer to the ten 2,3,7,8-chlorosubstituted toxic congeners only.
[g] Polychlorinated biphenyls.

Table 40.2 Some Physical–Chemical Properties of POPs Candidate for Inclusion in Annex A, B, or C of the Stockholm Convention

Compound	Molecular Structure	Molecular Weight	Water Solubility	$logK_{OW}$	Vapor Pressure (mmHg)	Half-life in Soil (years)	Reference
Chlordecone (CAS 143-50-0)		490.6	1–3 mg/L	4.50–5.41	2.25×10^{-5}–3×10^{-5} (25°C)	1–2	28, 30
Hexabromobiphenyl[a] (CAS 6355-01-8)		627.58	3–11 µg/L	6.39	5.2×10^{-8}–5.6×10^{-6} (25°C)	—	31
Lindane (gamma-hexachlorocyclohexane) (CAS 58-89-9)		290.83	7 mg/L (20°C)	3.8	4.2×10^{-5} (20°C)	>1	28, 32

(Continued)

Table 40.2 (Continued)

Compound	Molecular Structure	Molecular Weight	Water Solubility	$\log K_{OW}$	Vapor Pressure (mmHg)	Half-life in Soil (years)	Reference
PFOS[b] (CAS 2795-39-3)		538	519–680 mg/L (20–25°C)	Not measurable	2.45×10^{-6}	—	20
Penta-BDE[c] (commercial mixture)		485.8–564.7	13 µg/L (25°C)	5.9–7.0	7.2×10^{-10} –3.5×10^{-7} (20–25°C)	—	33

[a] Only one isomeric structure is shown.
[b] Perfluorooctane sulfonate (the potassium salt is shown).
[c] Pentabromodiphenyl ether. The commercial mixture contains penta- through heptabromo-substituted homologs.

Table 40.3 Some Physical–Chemical Properties of the POPs Proposed for Risk Evaluation as Prescribed for Inclusion in the Stockholm Convention

Compound	Molecular Structure	Molecular Weight	Water Solubility	logK_{OW}	Vapor Pressure (mmHg)	Half-life in Soil (days)	Reference
Pentachlorobenzene (CAS 608-93-5)		250.32	0.56 mg/L (25°C)	4.8–5.18	1.65×10^{-2}	194–345	34
Octa-BDE[a] (commercial mixture) (CAS 32536-52-0)		801.38	0.0005 mg/L	6.29	4.94×10^{-8}	—	35
SCCPs[b] (CAS 85535-84-8)		320–500	0.0224–0.994 mg/L	4.39–8.69	2.1×10^{-9} –1.9 $\times 10^{-2}$	>365	36

(Continued)

Table 40.3 (Continued)

Compound	Molecular Structure	Molecular Weight	Water Solubility	$\log K_{OW}$	Vapor Pressure (mmHg)	Half-life in Soil (days)	Reference
alpha-HCH[c] (CAS 319-84-6)	(+)-alpha-HCH (-)-alpha-HCH	290.83	10 mg/L (28°C)	3.46–3.85	2×10^{-2} (20°C)	48–125	33, 37
beta-HCH[d] (CAS 319-85-7)		290.83	5 mg/L (20°C)	3.78–4.50	5×10^{-3} (20°C)	91–122	33, 38

[a] Octabromodiphenyl ether. The commercial mixture contains penta- through decabromo-substituted homologs.
[b] Short-chained chlorinated paraffins.
[c] alpha-Hexachlorocyclohexane.
[d] beta-Hexachlorocyclohexane.

b. Bioaccumulation, measured as bioconcentration factor (BCF) or bioaccumulation factor (BAF), greater than 5000, or as $\log K_{OW}$, greater than 5, or any evidence of bioaccumulation for consideration within the Convention;

c. Long-range environmental transport, evidenced by the measured levels of the chemical far from the source of release, or by modeled data demonstrating the potentiality for the substance to be transported through air, water, or migratory species;

d. Adverse effects to human health and the environment.

Owing to their physical–chemical properties, bioaccumulative behavior in lipid tissues, and possible toxicological effects, POPs represent a relevant and growing interest for human beings, with the food of animal origin representing the main source of exposure. For most of them (i.e., organochlorine pesticides), regulatory limits have been already set on meat commodities at European level and in non-European countries, with possible different maximum levels (MLs) of acceptance according to the animal species [39].

40.1.4 Organochlorine Pesticides

The term organochlorine pesticides refers to a wide range of organic chemicals containing chlorine atoms and used in agriculture and public health activity to effectively control pest. Although most of them have been banned during the 1970s and 1980s, they are still found in the environment [40–42] and in biological matrices [43,44]. In fact, the intrinsic characteristics of these substances (i.e., highly lipophilic, low chemical and biological degradation rate) have led to their accumulation in the biosphere where they magnify in concentrations progressing through the food web. Organochlorine pesticides under the Stockholm Convention and their principal chemical–physical properties are listed in Table 40.1.

Food is considered to represent a constant source of exposure. For this reason, regulations concerning pesticides in food have become more and more severe in the past decades. To harmonize registration of pesticides and tolerances throughout the community, the EU Directive 91/414/EEC [45] lays down the basic rules with respect to plant protection products. Regulation 396/2005/EC [46] indicates that temporary maximum residue limits (TMRLs) have to be established at the EU level for all the active substances for which harmonized MRLs are not yet set, that is, pesticides currently not covered by the EU MRL Directives 86/362/EEC [47], 86/363/EEC [39], and 90/642/EEC [48]. Current EU MRLs established for the organochlorine pesticides of interest in animal products are set between 0.02 and 1 mg/kg on fat basis. In the United States, legislation was enacted in 1996 with the Food Quality Protection Act, including stricter safety standards, especially for infants and children, and a complete reassessment of all existing pesticide tolerances. For the pesticides of our concern, U.S. residue limits are established between 0.1 and 7 mg/kg fat. At the international level, MRLs in meat and meat products recommended by Food and Agriculture Organization (FAO)/World Health Organization (WHO) vary from 0.05 to 3 mg/kg fat [49].

40.1.5 Polychlorodibenzo-p-Dioxins and Polychlorodibenzofurans

PCDDs and PCDFs (altogether also commonly known as "dioxins") are two groups of tricyclic aromatic compounds containing between one and eight chlorine atoms, thus resulting in 210 congeners (75 PCDDs and 135 PCDFs), different in the number and position of chlorine atoms. As shown in Table 40.1, PCDDs and PCDFs are insoluble in water, exhibit a strong lipophilic character, and are very persistent. Neither PCDDs nor PCDFs are produced intentionally. In fact, their formation and release into the environment occur primarily in thermal or combustion processes or as unwanted by-products of industrial processes involving chlorine. Of the 210 positional isomers, only the 17 congeners with chlorines at positions 2, 3, 7, and 8 are of toxicological interest. To facilitate risk assessment/management, a toxicity equivalency factor (TEF) relative to $2,3,7,8-T_4CDD$ was assigned to each of the toxic congeners: for food and feeding stuffs, the WHO-TEFs adopted in 1997 by the WHO [50] are presently used. An update of WHO-TEFs was carried out in 2005 [51]. The "international" system (I-TEFs) [52] is still used for environmental samples.

Humans are exposed to PCDDs and PCDFs through the diet. The contribution of meat and meat products and fish and fishery products together may be higher than 90% of the total exposure to PCDDs, PCDFs, and DL-PCBs [53–57]. To reduce human exposure and protect consumer health, the EU has progressively issued regulatory measures setting MLs for PCDDs, PCDFs, and DL-PCBs in food and feeding stuffs. For example, an ML of 3.0 pg WHO-TE/g fat was established for PCDDs and PCDFs in bovine and sheep meat corresponding to an ML of 4.5 pg WHO-TE/g fat when DL-PCBs are considered [58]; in pork meat, the corresponding ML values are 1.0 and 1.5 pg WHO-TE/g fat.

40.1.6 Polychlorobiphenyls

PCBs are a family of 209 chlorinated compounds produced commercially under various trade names by direct chlorination of biphenyl. As PCDDs and PCDFs, PCBs are substantially insoluble in water, strongly lipophilic, and very persistent (Table 40.1). For their chemical–physical stability and dielectric properties, they were used worldwide as or in transformer and capacitor oils, hydraulic and heat exchange fluids, and lubricating and cutting oils. PCBs are divided into two groups according to their toxicological mode of action: the DL-PCBs consist of 12 congeners with toxicological properties similar to PCDDs and PCDFs, and NDL-PCBs, with a different toxicological profile. $2,3,7,8-T_4CDD$ WHO-TEFs have also been assigned to DL-PCBs [50,51]. Both NDL-PCBs and DL-PCBs bioaccumulate in animals and humans and biomagnify in the food chain. No MLs for NDL-PCBs in food and feeding stuffs have been set at the community level as yet.

40.1.7 Polybrominated Diphenyl Ethers

PBDEs are a group of 210 congeners, differing in the number of bromine atoms and in their position on two phenyl rings linked by oxygen. Their nomenclature is identical to that of PCBs. These chemicals are persistent and lipophilic, which results in bioaccumulation in fatty tissues of organisms and enrichment through the food chain [59]. PBDEs were first introduced into the market in the 1960s and used as flame retardants to improve fire safety in various consumer products and in electronics. There are three types of commercial PBDE products—penta-BDE, octa-BDE, and decabromodiphenyl ether (deca-BDE)—each product being a mixture of various PBDE congeners [60]. The EU has prohibited the uses of penta- and octa-BDE [61], but these substances are still on the market in many regions of the world. In any case, a substantial reservoir of PBDEs exists in products that could release them to the environment.

Despite the fact that dietary intake is probably the main route of exposure to PBDEs for the general population [62,63], no MLs for PBDEs in food have yet been set by the EU.

Table 40.4 Examples of PFAS of Environmental Interest

Compound	Molecular Structure	Acronym	Molecular Weight
Perfluorobutyl sulfonate (CAS 29420-49-3)		PFBS	299.21
Perfluorooctanoic acid (CAS 335-67-1)		PFOA	414.07
6:2 Fluorotelomer sulfonate (CAS 29420-49-3)		6:2 FTS	427.16
Perfluorooctane sulfonamide (CAS 754-91-6)		PFOSA	499.14
Perfluorooctyl sulfonate (CAS 2795-39-3; 1763-23-1 (acid))		PFOS	499.23
N-Methyl perfluorooctane sulfonamidoethanol (CAS 24448-09-7)		N-MeFOSE	557.23
Perfluorotetradecanoic acid (CAS 376-06-7)		PFTeDA	714.12

40.1.8 Polyfluorinated Alkylated Substances

The PFAS are compounds consisting of a hydrophobic alkyl chain of variable length (typically C_4 to C_{16}) and a hydrophilic end group. The hydrophobic part may be fully or partially fluorinated: for instance, the "6:2" formula (Table 40.4) indicates that, in the C_8-chain, six carbons are fully fluorinated whereas the remaining two bear hydrogen atoms. When fully fluorinated, the molecules are called perfluorinated alkylated substances, whereas the partially fluorinated ones, because of the telomerization production process, are named telomers. The hydrophilic end group can be neutral or positively or negatively charged. The resulting compounds are nonionic, cationic, or anionic surface-active agents: due to their amphiphilic features, most of the perfluorinated compounds will not accumulate in fatty tissues as is usually the case with other persistent halogenated compounds.

PFAS can be widely found in the environment, primarily resulting from anthropogenic sources as a consequence of industrial and consumer applications, including stain-resistant coatings for fabrics and carpets, oil-resistant coatings for paper products, fire-fighting foams, mining and oil well surfactants, floor polishes, and insecticide formulations [64].

At present, PFOS and perfluorooctanoic acid (PFOA) are the most investigated molecules in the environment and humans, due to their widespread occurrence, bioaccumulation, and persistence. The latter is determined by the strong covalent C–F bond [65–69]. Many of the neutral PFAS—such as perfluorooctane sulfonamide (PFOSA) and *n*-ethyl perfluorooctane sulfonamidoethanol (*n*-EtFOSE)—are considered to be potential precursors of PFOS. In addition, PFOA could be generated from the 8:2 fluorotelomer alcohol (8:2 FTOH), PFOSA, and *n*-EtFOSE. Some selected PFAS of environmental interest are reported in Table 40.4.

40.2   Chemical Methods

40.2.1   Organochlorine Pesticides

40.2.1.1   Analytical Methods

The EU MRLs set for organochlorine pesticides in products of animal origin require the development of highly sensitive methods to analyze these pesticides in different sample matrices. The most suitable and efficient approach involves the use of multiresidue procedures whose properties were recently reported by Hercegovà et al. [70] and are summarized as follows: (a) possibility to determine a number of pesticides as high as possible in a single analysis, (b) high recoveries, (c) high selectivity obtained by means of effective removal of potential interferences from the sample, (d) high sensitivity, (e) high precision, (f) good ruggedness, (g) low cost, (h) high speed, and (i) use of less harmful solvents and in low amounts.

Multiresidue methods developed for organochlorine pesticides follow the general scheme shown in Figure 40.1. After a pretreatment step aimed at obtaining a homogenized and dried sample, the analytes and fat are extracted together from the test matrix and the extract is purified to obtain a suitable sample for instrumental determination. The lipids coextracted with the analytes are separated using different nondisruptive procedures, which include liquid–liquid partitioning and gel permeation chromatography (GPC). Pesticides are further cleaned up by adsorption chromatography with Florisil® (U.S. Silica Company, Berkeley Springs), alumina, or silica, used as adsorbent phases. Instrumental determination is performed by high-resolution gas chromatography (HRGC) coupled to electron capture detection (ECD) or mass spectrometry (MS). The principal procedures currently used for the analysis of organochlorine pesticides in meat samples (Table 40.5) are examined in the following sections. Most of them were reviewed by the Codex Committee on Pesticide Residues [71] and included in the Official Methods of Analysis of the Association of Official Analytical Chemists (AOAC International, 2005) [72] and in the Pesticide Analytical Manual of the Food and Drug Administration (FDA) [73].

40.2.1.1.1   Pretreatment

A good preparation of sample matrix is an essential step to enhance extraction efficiency. The ideal sample for extraction is a dry, finely divided solid. In fact, a high surface area of the test matrix is recommended to improve the contact of the solvent with test molecules and, finally, to obtain quantitative recoveries. As reported in the general procedure related to preparation of test samples for meat and meat products [80], samples of animal origin are minced and homogenized. Anhydrous sodium sulfate or Hydromatrix™ (Varian Associates, Inc.) are added until a friable mixture is obtained.

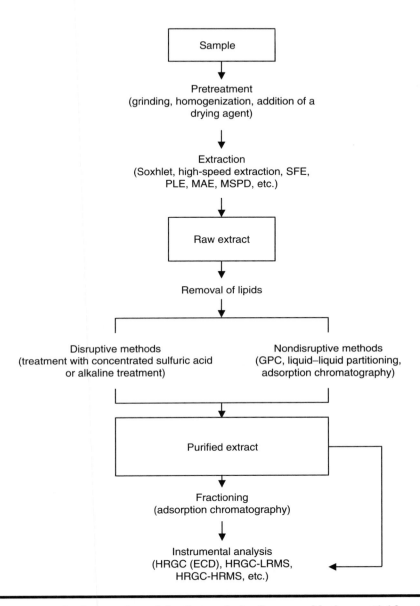

Figure 40.1   General scheme adopted for the analysis of organochlorine pesticides, PCDDs, PCDFs, PCBs, and PBDEs.

40.2.1.1.2   Extraction

Extraction techniques include classical procedures (i.e., Soxhlet extraction, column extraction, partitioning extraction, high-speed extraction, etc.) and innovative methods, such as supercritical fluid extraction (SFE), pressurized liquid extraction (PLE)—also known as accelerated solvent extraction (ASE)—microwave-assisted extraction (MAE), and matrix solid-phase dispersion (MSPD) extraction.

Table 40.5   Examples of Organochlorine Pesticide Determination in Samples of Animal Origin

Sample	Pretreatment	Extraction	Cleanup	Instrumental Analysis	Method Performance	Reference
Cattle fat, swine internal organ tissues 5 g	Homogenization, addition of Hydromatrix™	PLE, 1:1 dichloromethane–acetone, 1500 psi, two cycles™	Fat removal with GPC, SX-3 BioBeads column; fractioning into two fractions on silica gel SPE cartridges (PCBs, PBDEs, and nonpolar chlorinated pesticides in fraction I, polar chlorinated pesticides in fraction II)	HRGC-LRMS, DB-5 ms GC column; MS operating in NCI mode with methane as a reagent gas	Mean recoveries, 24–111%	74
Fatty samples 4–5 g	Addition of Hydromatrix	SFE, CO_2 modified with 3% acetonitrile at 27.58 MPa, 60°C; fat removal on a C_1-bonded phase at 95°C	Adsorption chromatography on Florisil column	HRGC coupled to an electrolytic conductivity detector; DB-1 GC column	Mean recoveries, 85–115%	75
Meat (chicken, pork, and lamb) 5 g	Homogenization, addition of anhydrous Na_2SO_4	Extraction at high speed with ethyl acetate; alternatively, Soxhlet extraction with ethyl acetate	GPC, SX-3 BioBeads column	HRGC-MS/MS (triple quadrupole)	Mean recoveries, 75–96% (extraction at high speed); mean recoveries, 67–86% (Soxhlet); LOQ, 0.8–2.7 µg/kg	76

Matrix	Sample preparation	Extraction	Cleanup	Detection	Results	Reference
Meat (chicken, pork, and lamb) 5 g	Homogenization, freeze-drying, addition of Hydromatrix™	PLE, ethyl acetate, 120°C, 1800 psi, static extraction time 5 min	GPC, SX-3 BioBeads column		Mean recoveries, 64–87%; LOQ, 0.8–2.7 µg/kg	76
Liver of chicken, pork, and lamb 5 g	Homogenization	Extraction at high speed with ethyl acetate	GPC, SX-3 BioBeads column	HRGC-MS/MS (triple quadrupole)	Mean recoveries, 65–111%	77
Liver of chicken, pork, and lamb 0.5 g	Homogenization, addition of C_{18}	MSPD, sample/C_{18} transferred in a cartridge containing 2 g Florisil; elution with ethyl acetate	No additional cleanup steps	HRGC-MS/MS (triple quadrupole)	Mean recoveries, 69–86%, except for lindane and endrin <30 % (MSPD); LOQ, 3.5–4.9 µg/kg	77
Pork fat 1.25 g		Blending with 1:1 ethyl acetate–cyclohexane	GPC, two Environsep-ABC columns, elution with 1:1 cyclohexane–ethyl acetate	HRGC-MS/MS (triple quadrupole); VF-5 ms GC column	Mean recoveries, 66–101%; LOD, 0.1–2 µg/kg	78
Meat 50–100 g		Extraction at high speed with 2:1 petroleum ether–acetone	GPC, Environgel™ column; adsorption chromatography on Florisil column	GC-ECD; DB-5 GC column	Mean recoveries, 101–103%[a]; LOQ, 0.002–0.05 µg/kg[a]	79

[a] Data refer to chlordane only.

Soxhlet extraction has been widely used in the organochlorine pesticide analysis [81,82]. The continuous contact of the sample with freshly distilled solvent ensures high extraction efficiency, usually higher than 70% for the pesticides of interest [73], so that it allows its employment as reference method.

Besides Soxhlet, other classical techniques are widely used for the extraction of meat sample. Among these, extraction with solvent at high speed is largely used, in which the sample is transferred into a blender cup or into a homogenizer and extracted with organic solvent [73,76,82]. The extract is decanted and separated from the matrix by filtration or centrifugation. Alternatively, extraction of meat sample (in the form of a friable product) is carried out directly in centrifuge in presence of organic solvent [82].

Fast extraction of organochlorine pesticides is also performed by the column extraction technique. This is carried out in a glass column where the sample, dried and homogenized, is transferred and eluted with organic solvent [82].

The need for determining a high number of residues in a time as short as possible had led to the development of innovative techniques for the extraction of pesticides from fatty food. Among these techniques, SFE with CO_2 and PLE with pressurized solvents appear to be equivalent to liquid-base techniques in the extraction of pesticides from fatty matrices. In the past decades, SFE has received wide attention for its low solvent consumption and its high degree of selectivity [72,83]. The use of supercritical fluids, characterized by densities close to those of the liquid solvents but with lower viscosity and higher diffusion capability, results in extraction agents that are more penetrative and with a higher solvating power. In addition, the combination with solid sorbent traps allows to obtain more purified extracts, eventually resulting in a single-step extraction and cleanup. Applications of SFE to the analysis of pesticides in fatty matrices are reported by Snyder et al. [84] and Hopper [72].

PLE is an innovative method for the rapid extraction of analytes. It is based on the use of solvents at temperatures (from 60 to 200°C) higher than their boiling points at atmospheric pressure, and at high pressure (from 5×10^5 to 2×10^7 Pa) to maintain the solvent in a liquid state. PLE has found wide applications, especially in the field of environmental samples (soil, sediments, sludges, dust, etc.). More recently, the extraction efficiency of PLE as well as its application to the analysis of pesticides in fatty matrices was investigated in the isolation of lipids from biological tissues [71,85]. This technique appears to be effective in the quantitative extraction of organochlorine compounds from tissue samples (liver, heart, kidney, and adipose tissue) and from muscle of chicken, pork, and lamb (recoveries in the range from 70 to 93%) [73].

In the MAE, a sample is suspended in a suitable solvent and the mixture irradiated in a microwave oven. The irradiation step is generally repeated until the maximum yield of extraction is obtained. In the field of organochlorine pesticides, this technique has been applied especially for the extraction of environmental samples, such as soil, sediments, and vegetables [86,87], whereas applications to the extraction of fatty samples are limited [88,89].

The MSPD extraction was introduced by Barker in 1989 [90] and has proven to be an efficient procedure for the extraction of a wide range of drugs, pesticides, and naturally occurring constituents in samples of vegetable and animal origin [91,92]. It involves the use of octadecyl-bonded silica (C_{18}), octyl-bonded silica (C_8), or other sorbents obtained by chemical modification of silica surface, blended with the sample by means of a mortar and pestle. The material is successively transferred to a syringe, compressed by a syringe plunger, and eluted with a suitable organic solvent. Applications of the MSPD technique to the extraction of organochlorine pesticides in animal fat and animal tissue are reported in the studies by Long et al. [93], Furusawa [94,95], and Frenich et al. [77].

40.2.1.1.3 Cleanup

Owing to the lipophilic properties of organochlorine pesticides and their tendency to accumulate in fat, their extraction from the matrix is always accompanied by coextraction of lipidic material, which makes the instrumental analysis difficult without a preliminary purification of the extract. In the determination of some chlorinated environmental contaminants (e.g., PCDDs, PCDFs, PCBs), fat is often efficiently removed by treatment with sulfuric acid. However, pesticides such as dieldrin, endrin, and DDT (different isomers and metabolites) are not sufficiently stable and are decomposed by this method [71,96]. Therefore, nondestructive procedures, such as GPC and liquid–liquid partitioning, are widely applied for the elimination of the lipidic fraction in the analysis of organochlorine residues in fatty samples.

GPC is an automated procedure that is highly effective in removing high-molecular-weight substances (i.e., lipids, proteins, and pigments) due to the difference in molecular size between interferences and the target analytes. In the analysis of organochlorine pesticides, the divinyl-benzene-linked polystyrene gel (BioBeads SX-3) is the most commonly used sorbent [97]; several solvent mixtures have been recommended as eluents [71]. Owing to separation principle, which is not selective with respect to interferences with low molecular weight, the application of additional cleanup steps is generally necessary. In exceptional cases when highly selective instrumental detectors are used (e.g., MS/MS detector), samples can be directly injected after GPC cleanup [73].

Liquid–liquid partitioning uses the differences in polarity between the analytes and the interfering species to separate pesticides from the lipidic fraction. Partitioning between petroleum ether (or light petroleum) and acetonitrile is one of the most traditional procedures to separate pesticides from fat [98,99]. Owing to the low solubility of lipidic compounds in acetonitrile, fat is retained in petroleum ether whereas partition of organochlorine compounds into acetonitrile is a function of their partitioning coefficients. In a subsequent step, residues in acetonitrile are partitioned back into petroleum ether when acetonitrile is diluted with excess water, which is added to reduce pesticide solubility in acetonitrile. As observed for GPC, liquid–liquid partitioning is not effective for the separation of organochlorine pesticides and other coextractive species. This is generally accomplished by various cleanup procedures, based on the use of different adsorbent phases (commonly Florisil, alumina, or silica) employed either in traditional column chromatography or in solid-phase extraction (SPE) cartridges [98–100]. Recently, this last approach has gained wide acceptance because of its simplicity and low time and solvent consumption. Today several commercial SPE cartridges are available for cleanup of organochlorine pesticides in fatty samples. The most widely used SPE cartridges include octadecyl (C_{18})-bonded porous silica, silica gel, Florisil, and alumina [101].

40.2.1.1.4 Instrumental Analysis

Common methods for the quantification of organochlorine pesticides involve HRGC (ECD) and HRGC-MS.

40.2.1.1.4.1 HRGC (ECD)

The high efficiency of capillary columns allows the separation of a large number of organochlorine pesticides. In this field, the most widely used phases are the nonpolar 100% dimethylpolysiloxane and (5% phenyl) methylpolysiloxane columns. Operative conditions and relative retention data obtained on these two GC columns can be found in Refs 81, 102, and 103. ECD is the most common detector used for the detection of organochlorine pesticides. It presents high sensitivity but, according to the guidelines proposed by EC DG Health and Consumer Protection [104] for pesticide residue analysis, does not provide enough selectivity. To overcome this problem, the use of two columns of different polarity is mandatory to obtain unambiguous identification when HRGC (ECD) is used as a confirmatory method in the determination of residues of organic contaminants in live animals and their products [105]. In two-dimensional GC, two columns of different selectivity are serially coupled via a modulation device that cuts small portions of the effluent from the first column and refocuses them onto the second column, thus obtaining an improvement of the overall resolution [106]. Applications of two-dimensional GC to pesticide analysis were reported by Focant et al. [107], Korytár et al. [108], Seemamahannop et al. [109], and Chen et al. [110].

40.2.1.1.4.2 HRGC-MS

As observed, the determination of organochlorine pesticides in samples with high fat content requires selective techniques to unambiguously confirm their presence, since interfering species often mask the analytical signal of the target compounds. With respect to ECD, MS detectors coupled to HRGC provide greater identification and confirmation power, thus generally avoiding false positive and false negative errors due to matrix interferences [111]. Nowadays, the most widely used MS technique in the field of organochlorine pesticides relies on low-resolution (LR) apparatuses such as the quadrupole analyzer, operating with an electron impact (EI) ion source, an electron energy of 70 eV, and in the selected ion monitoring (SIM) mode. It allows the determination of the target analytes at levels in compliance with the regulations established for a wide range of pesticides and food commodities, including those of animal origin. However, confidence in the confirmation of identity could be reduced if one or more of the selected ions are affected by matrix interferences [78]. A remarkable enhancement in terms of selectivity and sensitivity is observed with tandem MS (MS/MS, ion trap, or triple quadrupole analyzers) where a single ion is subjected to a second fragmentation to confirm the identity of the parent compound. Application of triple quadrupole MS in the analysis of organochlorine pesticides in samples of animal origin has recently been reported by Patel et al. [78] and Frenich et al. [76,77].

40.2.2 Polychlorodibenzo-p-Dioxins, Polychlorodibenzofurans, and Dioxin-Like Polychlorobiphenyls

40.2.2.1 Analytical Methods

The standard methods developed to determine these contaminants in food samples generally include the assessment of the 17 2,3,7,8-chlorosubstituted PCDDs and PCDFs, and the 12 DL-PCBs. The latter include four non-*ortho* congener PCBs (PCBs 77, 81, 126, and 169), and eight mono-*ortho* congener PCBs (PCBs 105, 114, 118, 123, 156, 157, 167, and 189). The analysis of these classes of substances in food is complicated by their low contamination levels (in the order of pg/g) and by the complexity of the matrix. This generally contains large amounts of interfering species (i.e., lipids) whose appropriate removal requires laborious cleanup procedures. To give an overall view of the analytical methods in use for the quantification of PCDDs, PCDFs, and DL-PCBs in meat samples, the most topical literature has been reviewed (Table 40.6). Recent advances in determination of PCDDs, PCDFs, and DL-PCBs are in particular reported by Reiner et al. [112]; the importance of matrix pretreatment, sample extraction, cleanup, and fractionation of PCBs from food matrices are exhaustively described by Ahmed [97]; and a critical review of the various methods used in the analysis of DL-PCBs is given by Hess et al. [113]. Reference is also

Table 40.6 Examples of PCDD/F, DL-PCB, NDL-PCB, and PBDE Determinations in Samples of Animal Origin

Analytes	Sample	Pretreatment	Extraction	Cleanup	Instrumental Analysis	Method Performances[a]	Found Levels[a,b]	Reference
PCDDs, PCDFs	Meat (beef, pork, chicken, lamb) 5–10 g	Homogenization, freeze-drying, addition of anhydrous Na_2SO_4	Soxhlet, toluene, 24 h	Fat removal with adsorption chromatography on a multilayer column; fractioning with an alumina column	HRGC-HRMS, DB-5 GC column; MS operating in the EI mode at resolution of 10,000	LOD, 0.02–0.2 ng/kg dw	Beef: 0.7 pg TEQ/g fat Pork: 0.3 pg TEQ/g fat Chicken: 0.8 pg TEQ/g fat Lamb: 0.7 pg TEQ/g fat	115
PCDDs, PCDFs, DL-PCB	Meat (beef, chicken) 25 g	Homogenization	PLE, toluene, 135°C, 1500 psi, static extraction time 10 min, three cycles	Fat removal with 20 g Extrelut impregnated with 40 g H_2SO_4; fractioning with disposable prepacked columns containing multilayer silica, alumina, and carbon	HRGC-HRMS, DB-5 GC column; MS operating in the EI mode at 35 eV and a resolution of 10,000		Beef: PCDD/Fs 2.95 pg TEQ/g fat; DL-PCBs 3.44 pg TEQ/g fat Chicken: PCDD/Fs 1.37 pg TEQ/g fat; DL-PCB 1.12 pg TEQ/g fat	116
PCDDs, PCDFs	Meat (beef, pork, poultry)	Homogenization, freeze-drying, crushing with blender, addition of anhydrous Na_2SO_4	Blending with 2:1 n-hexane–acetone	Adsorption chromatography on multilayer silica, charcoal, and Florisil columns	HRGC-HRMS, DB-5 and SP2331 GC column; MS operating in the EI mode and a resolution of 10,000		Beef: 0.72 pg I-TEQ/g fat Pork: 0.27 pg I-TEQ/g fat Poultry: 0.46 pg I-TEQ/g fat	117

(Continued)

Table 40.6 (Continued)

Analytes	Sample	Pretreatment	Extraction	Cleanup	Instrumental Analysis	Method Performances[a]	Found Levels[a,b]	Reference
PCDDs, PCDFs, DL-PCBs, PBDE	Fatty samples (aliquots tested corresponding to 10 g of fat)	Homogenization	Blending with *n*-hexane and acidified silica gel (1:1.5 H_2SO_4:silica); extraction through a multilayer column	Fractioning of the extract on a carbon column into two fractions (mono- to tetra-*ortho*-PCBs, and PBDEs in fraction I, non-*ortho*-PCBs, PCDDs, and PCDFs in fraction II); purification of fraction II by treatment with H_2SO_4 and elution through silica gel and alumina columns	HRGC-HRMS, (non-*ortho*-PCBs, PCDD/Fs), DB-5 GC column; MS operating in the EI mode and a resolution of 10,000; HRGC-MS, (mono- to tetra-*ortho*-PCBs), DB-5 GC column; MS operating in the EI mode	LOD, 0.01–0.05 ng/kg; precision, 5–11%; recoveries, >50%		118
PCDDs, PCDFs, DL-PCB	Meat (poultry)	Homogenization, freezing under liquid nitrogen, addition of anhydrous Na_2SO_4, freeze-drying, grounding	PLE, *n*-hexane, 1500 psi, static extraction time 5 min, two cycles	Fat removal with GPC (4 g fat loaded), SX-3 BioBeads column; fractioning with disposable prepacked columns containing multilayer silica, alumina, and carbon	HRGC-HRMS, RTX-5SIL-MS GC column; MS operating in the EI mode at 60 eV and a resolution of 10,000	Recoveries, >75 %	PCDD/Fs 1.6 pg I-TEQ/g fat; non-*ortho*-PCBs 2.2 pg I-TEQ/g fat	119

Analyte	Sample	Pretreatment	Extraction	Cleanup	Determination	Recovery/LOD	Results	Reference
NDL-PCBs	Animal fat (pork, poultry) 1–5 g	Homogenization, melting	Blending with n-hexane–acetone	Adsorption chromatography on acidified silica column (1:1 concentrated H_2SO_4-silica)	HRGC-ECD, HT-8 GC column; HRGC-LRMS, DB-5 ms GC column	Recoveries, 72–78%	Poultry: 171–3753 ng/g fat[c] Pork: 591–2855 ng/g fat[c]	120
PCDDs, PCDFs, DL-PCBs	Animal fat (pork, poultry) 1–5 g	Homogenization	Blending with n-hexane–acetone	Fat removal with concentrated H_2SO_4 adsorbed on and Florisil; fractioning on activated carbon	HRGC-HRMS	LOD, <0.2 pg/g (tetra- to hexa-PCDD/Fs)	Poultry: PCDD/Fs 3–118 pg TEQ/g fat; non-ortho-PCBs 3–6 pg TEQ/g fat[d]; mono-ortho-PCBs 8–187 pg TEQ/g fat Pork: PCDD/Fs 3–118 pg TEQ/g fat; non-ortho-PCBs 1–1.5 pg TEQ/g fat[d] mono-ortho-PCBs 13–63 pg TEQ/g fat	120
PCDDs, PCDFs, DL-PCB	Meat (chicken, pork)	Homogenization, addition of anhydrous Na_2SO_4 (Na_2SO_4/sample 1.5–2.0)	PLE with fat retainer, n-heptane, 100°C, static extraction time 5 min, two cycles	Fractioning of extract on activated carbon column AX-21 with Celite® (Celite Corporation) into three fractions (bulk PCBs in fraction I, mono-ortho-PCBs in fraction II, non-ortho-PCBs, PCDD/Fs in fraction III)	HRGC-HRMS, DB-5 GC column; MS operating in the EI mode at 65 eV and a resolution of 10,000	Recoveries, 74–92%		121

(Continued)

Table 40.6 (Continued)

Analytes	Sample	Pretreatment	Extraction	Cleanup	Instrumental Analysis	Method Performances[a]	Found Levels[a,b]	Reference
NDL-PCBs	Meat (beef, pork, poultry, horse)	Homogenization, freezing under liquid nitrogen, freeze-drying	PLE, *n*-hexane, 1500 psi, static extraction time 5 min, two cycles	GPC (4 g fat loaded), SX-3 BioBeads column; fractioning with disposable prepacked columns containing multilayer silica, alumina, and carbon	HRGC-ITMS/MS, RTX-5SIL-MS GC column; MS operating in the EI mode	Recoveries, 60–101%	Beef: 5910 pg/g fat[c] Pork: 8828 pg/g fat[c] Poultry: 4770 pg/g fat[c] Horse: 21,588 pg/g fat[c]	122
PCDDs, PCDFs, DL-PCBs	Fast food samples containing meat 10–15 g dw	Homogenization, freezing under liquid nitrogen, freeze-drying, addition of anhydrous Na_2SO_4	PLE, *n*-hexane, 1500 psi, static extraction time 5 min, two cycles	Fractioning (4 g fat) with disposable prepacked columns containing multilayer silica, alumina, and carbon	HRGC-HRMS (non-*ortho*-PCBs, PCDD/Fs), GC column RTX-5SIL-MS; MS operating in the EI mode at 60 eV and a resolution of 10,000; HRGC-ITMS/MS, (mono-*ortho*-PCBs), RTX-5SIL-MS GC column; MS operating in the EI mode		McDonald's Big Mac® PCDD/Fs ND–1.07 pg TEQ/g fat; DL-PCBs ND–2.31 pg TEQ/g fat	123

Analytes	Sample	Sample Preparation	Extraction	Cleanup	Instrumentation	Results	Reference
PBDEs	Meat 0.2–10 g	Homogenization, addition of anhydrous Na₂SO₄	Soxhlet with 3:1 n-hexane–acetone	Column chromatography on silica impregnated with concentrated H_2SO_4	GC-NCI/MS, HT-8 GC column (for tri- to hepta-BDE congeners) and AT-5 GC column (for 209 congener)	Beef steak: 31 pg/g ww[e] Minced meat: 110 pg/g ww[e] Hamburger: 120 pg/g ww[e] McDonald's Big Mac: 160 pg/g ww[e]	124
PCDDs, PCDFs, DL-PCBs, PBDEs	Meat (hamburger, fat of chicken, pork, and beef) 5 g	Homogenization in dichloromethane and drying with anhydrous Na₂SO₄ (fat samples); mixing with Celite (hamburger samples)	PLE, 35:30:35 2-propanol–n-hexane–dichloromethane, 125°C, 1500 psi (hamburger samples)	Cleanup and fractioning with disposable prepacked jumbo columns containing multilayer silica, alumina, and carbon	HRGC-HRMS, (non-*ortho*-PCBs, PCDD/Fs), GC column DB-5; MS operating in the EI mode at 35 eV and a resolution of 10,000; HRGC-MS, (PBDEs), DB-5 ms GC column; MS operating in EI mode at a resolution of 2500	Recoveries, 35–150% (except for PBDE 209 occasionally <20%); accuracy and precision better than 20%	Hamburger: PCDD/Fs 1.3 pg I-TEQ/g fat; non-*ortho*-PCBs 0.2 pg I-TEQ/g fat; PBDEs 2911 pg/g fat[f] Pork fat: PCDD/Fs 0.2 pg I-TEQ/g fat; non-*ortho*-PCBs 0.01 pg I-TEQ/g fat; PBDEs 2588 pg/g[f] Beef fat: PCDD/Fs 0.6 pg I-TEQ/g fat; non-*ortho*-PCBs 0.1 pg I-TEQ/g fat; PBDEs 244 pg/g fat[f] Chicken fat: PCDD/Fs 0.3 pg I-TEQ/g fat; non-*ortho*-PCBs 0.1 pg I-TEQ/g fat; PBDEs 648 pg/g fat[f]	125

(Continued)

Table 40.6 (Continued)

Analytes	Sample	Pretreatment	Extraction	Cleanup	Instrumental Analysis	Method Performances[a]	Found Levels[a,b]	Reference
PCDDs, PCDFs, DL-PCBs, PBDEs	Meat and eggs (processed meat products, beef, pork, poultry, eggs)	Freeze-drying	Soxhlet, toluene, 24 h	Fat removal by elution through a silica gel column containing acidic and neutral silica layers; fractioning on an activated carbon column into two fractions (PCBs and PBDEs in fraction I, and PCDD/Fs in fraction II); purification of the two fractions on an activated alumina column; further fractioning of fraction I into two subfractions on activated alumina column (sub-fraction IA containing non-*ortho*-PCBs, and sub-fraction IB containing other PCBs and PBDEs)	HRGC-HRMS, DB-Dioxin GC column; MS operating in the EI mode at a resolution of 10,000	Recoveries, >50%; LOQ, 0.0007–0.63 pg/g ww (PCDD/Fs); LOQ, 0.0007–0.13 pg/g ww (non-*ortho*-PCBs); LOQ, 0.048–3.2 pg/g ww (mono-*ortho*-PCBs and NDL-PCBs); LOQ, 0.035–13 pg/g ww (PBDEs)	Meat and eggs (market basket): PCDD/Fs 0.0082 pg I-TEQ/g ww; non-*ortho*-PCBs 0.59 pg I-TEQ/g fw[d]; mono-*ortho*-PCBs 52 pg I-TEQ/g ww; NDL-PCBs 410 pg/g ww[g]; PBDEs 13 pg/g ww[h]	126

Compound	Sample	Preparation	Extraction	Cleanup	Instrumentation	LOD/Recovery	Results	Reference
PBDEs	Chicken fat 1 g	Homogenization, filtration through anhydrous Na_2SO_4		Stirring with 10 g of 40% acid silica, purification with prepacked disposable columns containing multilayer silica, and alumina	HRGC-MS, DB-5 ms GC column; MS operating in EI mode at 70 eV	Recoveries, >75%	1.76–39.43 ng/g[i]	127
PBDEs	Meat (pork, beef, chicken) 20 g	Homogenization	Saponification with 1 M KOH/EtOH containing 10% H_2O and extraction with n-hexane	Purification with multilayer column chromatography	HRGC-LRMS, SPB-5 GC column; MS operating at 70 eV; HRGC-MRMS, SPB-5 GC column; MS operating at 38 eV and a resolution of 5000	LOD, 9.0–27 pg/g (HRGC-LRMS) LOD, 2.0–8.0 pg/g (HRGC-MRMS)	Pork: 63.6 pg/g ww[j] Beef: 16.2 pg/g fw[j] Chicken: 6.25 pg/g fw[j]	128
PBDEs	Meat (pork, beef, chicken) 5–200 g	Homogenization, addition of anhydrous Na_2SO_4	Column extraction, cyclohexane–dichloromethane	Fat removal with acid treatment; adsorption chromatography on activated silica gel and alumina columns	HRGC-HRMS, DB-5 GC column; MS operating at a resolution of 10,000		Pork: 41 pg/g ww[k] Ground beef: 78.3 pg/g ww[k] Chicken breast: 283 pg/g ww[k]	62

(Continued)

Table 40.6 (Continued)

Analytes	Sample	Pretreatment	Extraction	Cleanup	Instrumental Analysis	Method Performances[a]	Found Levels[a,b]	Reference
PBDEs	Cattle fat, swine internal organ tissues 5 g	Homogenization, addition of Hydromatrix	PLE, 1:1 dichloromethane–acetone, 1500 psi, two cycles	Fat removal with GPC, SX-3 BioBeads column; fractioning into two fractions on silica gel SPE cartridges (PCBs, PBDEs, and nonpolar chlorinated pesticides in fraction I, polar chlorinated pesticides in fraction II)	HRGC-LRMS, DB-5 ms GC column; MS operating in NCI mode with methane as a reagent gas	Recoveries, >68%		74
DL-PCBs, NDL-PCBs PBDEs	Animal fat (beef, chicken) 0.5 g	Homogenization, addition of anhydrous Na_2SO_4 and Florisil	MSPD, sample/Florisil transferred to a cartridge containing 5 g acidic silica; elution with n-hexane	Fractioning into two fractions on silica gel SPE cartridges (PCBs in fraction I, PBDEs in fraction II)	HRGC-ECD, HP-5 GC column; HRGC-MS/MS (ion trap), HP-5 GC column	Mean recoveries, 74–99 % LOQ, 0.4–3 ng/g		129

[a] dw, dry weight; ww, wet weight.
[b] Where not specified, TEQs were obtained by the WHO-TEFs system of 1997.
[c] Sum of NDL-PCBs 28, 52, 101, 138, 153, and 180.
[d] Sum of non-*ortho*-PCBs 77, 126, and 169.
[e] Sum of PBDEs 28, 47, 99, 100, 153, 154, 183, and 209.
[f] Sum of PBDEs 28, 47, 99, 100, 153, 154, and 183.
[g] Sum of NDL-PCBs 18, 28, 33, 49, 52, 60, 66, 74, 99, 101, 110, 122, 128, 138, 141, 153, 170, 180, 183, 187, 194, 206, and 209.
[h] Sum of PBDEs 47, 99, 100, 153, and 154.
[i] Sum of PBDEs 47, 99, 100, 153, 154, 183, and 209.
[j] Sum of PBDEs 28, 47, 99, 100, 153, and 154.
[k] Sum of PBDEs 17, 28, 47, 66, 77, 85, 99, 100, 138, 153, 154, 183, and 209.

made to the methods elaborated by the U.S. Environmental Protection Agency (EPA) for determination of the tetra- through octachloro-substituted PCDD and PCDF toxic congeners [23] and for PCB congeners [24] by HRGC-HRMS (HRMS—high-resolution mass spectrometry). Basic requirements for analytical methods used in the EU for official controls of PCDD, PCDF, and DL-PCB levels in foodstuffs are reported in EU Regulation 1883/2006 [114].

Many analytical methods follow the general scheme of Figure 40.1. As observed for organochlorine pesticide analysis, the test sample has to be homogenized and dehydrated with anhydrous sodium sulfate to obtain a friable product. Because some amounts of the chemicals under investigation may be lost during the complex preparative procedures, the internal standard (IS) technique is generally adopted to provide proper correction for analyte losses [23,24]. To this aim, known quantities of isotopically labeled analytes are added to the samples at the earliest possible stage of extraction. The analytes of interest are extracted with a suitable organic solvent and the extract is purified to remove interfering compounds and prepare the sample for instrumental determination. Many of the purification procedures are based on the use of sulfuric acid, generally adsorbed on an inert support such as Extrelut® (Merck KGaA, Darmstadt, Germany). The rationale is that all the analytes of interest are resistant to acid treatment and this property is exploited to selectively destroy most of the interfering species coextracted with the target compounds. Owing to the difference in concentrations between planar (PCDDs, PCDFs, and non-*ortho* DL-PCBs) and nonplanar analytes (mono-*ortho* DL-PCBs) and the presence of other coextractive compounds resistant to cleanup procedure (i.e., chlorinated pesticides), fractioning steps are generally included during purification before instrumental analysis by HRGC-HRMS.

40.2.2.1.1 Pretreatment

Tissue samples are dissected into small pieces and preserved by fast freezing in liquid nitrogen [97] or, else, normal deep-freeze. Before analysis, samples are grinded to rupture cell membranes and homogenized. Addition of anhydrous sodium sulfate in the ratio sodium sulfate-to-sample 1.5–2.0 (w/w) is carried out to dry sample [121]. Alternatively, a freeze-drying procedure is sometimes adopted.

40.2.2.1.2 Extraction

Extraction techniques for meat samples are generally based on the principle that lipophilic organic compounds such as PCDDs, PCDFs, and PCBs are predominantly associated with the fat fraction of the matrix. Therefore, the extraction methods used for removal of these compounds are based on general methods employed for the isolation of the lipidic fraction. As observed for organochlorine pesticides (Section 40.2.1.1.2), a number of well-established techniques, including Soxhlet extraction or sonication with solvent, are available for the extraction of PCDDs, PCDFs, and DL-PCBs in fatty samples. These procedures are shown to be highly efficient (Soxhlet is the extraction method indicated by the U.S. EPA Methods 1613 and 1668A in the case of tissue samples) and do not require expensive instrumentation. For these reasons they are still in use in several routine laboratories. However, the main disadvantages presented by these procedures, that is, the large solvent consumption and the long time required for extraction, are determining their gradual replacement with more sophisticated extraction techniques such as PLE, MAE, and SFE. As previously observed, the possibility of working at elevated temperatures and pressures drastically improves the speed of the extraction process. These innovative techniques were object of evaluation within the DIFFERENCE research project [130], requested by the EU as a result of the "Belgian dioxin crisis" to develop fast and cheap analytical procedures for determination of PCDDs, PCDFs, and PCBs. In this context, a promising new procedure is the inclusion of a fat retainer (sulfuric acid impregnated silica) in PLE extraction cells [131,132]. As demonstrated by Sporring and Björklund [133], the presence of a fat retainer efficiently removes lipidic substances by oxidizing them and hindering their coeluting compounds.

40.2.2.1.3 Cleanup and Fractionation

The nonselective nature of the exhaustive extraction procedures results in complex extracts that contain the analytes of interest together with lipidic material and other organohalogen compounds (e.g., organochlorine pesticides, polychlorinated naphthalenes, polychlorinated camphenes, toxaphene). Therefore, the purification methodology used for PCDD, PCDF, and PCB analysis requires first lipid elimination, then fractionation to separate the groups of analytes from other coextractive species.

For the removal of lipids, two approaches are generally employed: destructive and nondestructive methods. The nondestructive lipid removal principally includes the use of GPC with SX-3 BioBeads columns, and adsorption chromatography with alumina, silica, and Florisil. Destructive methods comprise oxidative dehydration by concentrated sulfuric acid mixed with the lipid extract [134] or adsorbed on solid support through which the extract is eluted [135].

Fractionation of the extract into groups of analytes is normally required before instrumental analysis. In fact, with the exception of DL-PCB 118, and to a minor extent DL-PCB 105, all mono-*ortho* and non-*ortho* DL-PCBs, PCDDs, and PCDFs are present at substantially lower concentrations with respect to the remaining NDL-PCBs [97,113]. Therefore, the range of concentrations of target compounds is normally too large to measure all congeners without additional dilution or concentration. The methods available for the isolation of the analytes of interest into separate fractions utilize spatial planarity of these molecules to selectively adsorb them on the surface of carbonaceous material such as activated or graphitized carbon. Recently, an automated cleanup system (Power-Prep™, Fluid Management Systems, Inc.) has been developed, which is capable of rapidly separating planar and nonplanar organochlorine molecules [119]. This system uses high-capacity disposable multilayer silica columns, basic alumina columns, and PX-21 carbon columns. Fractionation allows isolation of two fractions, one containing NDL-PCBs and the eight mono-*ortho* DL-PCBs, the other containing the 17 PCDDs and PCDFs and the four coplanar non-*ortho* DL-PCBs [123,136].

40.2.2.1.4 HRGC-HRMS Instrumental Analysis

In the determination of PCDDs, PCDFs, and DL-PCBs, NDL-PCB interferences can be eliminated by fractioning the extract into analyte groups or by analyzing the final extract on multiple column [24]. In the attempt to reduce the need for multicolumn analysis, a number of analyte-specific columns have been developed. The low-polarity 5% phenyl columns exhibit multiple coelution for PCDDs, PCDFs, and DL-PCBs. However, they are generally considered sufficiently selective for biological samples, containing a smaller number of congeners in comparison with environmental samples [137].

The HRMS based on magnetic sector instruments is the reference method for the determination of PCDDs, PCDFs, and DL-PCBs [23,24] at $10^{-12}–10^{-15}$ g/g levels in complex matrices. EI ion sources are normally used in the HRMS determination of these compounds, with conventional electron energies of 30–35 eV. SIM mode is canonically employed to improve specificity and sensitivity. MS/MS with triple quadrupole and ion trap detectors has also been investigated for the analysis of PCDDs, PCDFs, and dioxin-like compounds [138,139]. For PCDDs and PCDFs, the selectivity of MS/MS is usually higher, due to the specific loss of the COCl fragment, never observed in any other halogenated organic compounds [140]. In the case of DL-PCBs, this enhanced selectivity is not observed, because the loss of Cl_2 from the parent molecule is not uniquely related to PCB molecules [112]. The sensitivity of MS/MS instruments is generally lower than HRMS [141], but it can be compensated by adjustments to sample size and final extract volume.

40.2.3 Non-Dioxin-Like Polychlorobiphenyls and Polybrominated Diphenyl Ethers

40.2.3.1 Analytical Methods

Most analytical studies on NDL-PCBs and PBDEs are limited to the determination of a small number of congeners as indicators of the presence of NDL-PCBs and PBDEs, respectively. In the case of NDL-PCBs, data on their occurrence in food are generally reported as the sum of the six congeners—PCBs 28, 52, 101, 138, 153, and 180—often termed as "indicator PCBs" or "marker PCBs," that represent some 50% of the total NDL-PCBs in food [142]. For PBDEs, the EFSA Scientific Panel on Contaminants in the Food Chain has recently recommended the inclusion of the following congeners in a European monitoring program for feed and food: PBDEs 28, 47, 99, 100, 153, 154, 183, and 209 [143]. As for NDL-PCBs, this "core group" reflects the most frequently found PBDEs in food and biological samples.

The analytical procedures for NDL-PCBs and PBDEs are reviewed here (Table 40.7) on the basis of the recent literature. Particular attention is given to the articles by Ahmed [97] on PCB analysis in food and by Covaci et al. [144,145] on the advances in the analysis of brominated flame retardants.

The analytical methods developed for NDL-PCBs and PBDEs are based on the same protocols used for PCDDs, PCDFs, and DL-PCBs. Differences may be observed in the chromatographic and detection systems used for instrumental determination. The general scheme adopted for the analysis of these pollutants is reported in Figure 40.1. NDL-PCBs and PBDEs are extracted with an organic solvent, most frequently by Soxhlet or PLE. Lipids are removed by GPC or treatment with sulfuric acid and coextracted substances are eliminated by adsorption chromatography. The final determination is performed by HRGC (ECD) or, preferably, by HRGC-MS. As a function of the detection system used for the analysis, the IS technique is generally adopted in accord with the U.S. EPA Methods 1668 and 1614 (draft) [24,25].

40.2.3.1.1 Pretreatment

The pretreatment step for NDL-PCBs and PBDEs is similar to that adopted for PCDD, PCDF, and DL-PCB analysis. Specific information and references can be found in Section 40.2.2.1.1.

40.2.3.1.2 Extraction

Given the similarity with the extraction methods used for PCDDs, PCDFs, and DL-PCBs, more detailed information of the extractive procedures applied in the case of NDL-PCBs can be found in Section 40.2.2.1.2. For PBDEs, the use of Soxhlet [25,152], elution through multilayer column [121], extraction at high speed [153], MSPD [114], PLE [154], and MAE [155] are reported for the extraction of these contaminants in samples of animal tissue.

Table 40.7 Examples of PFAS Determinations in Samples of Animal Origin

Analytes	Sample	Pretreatment	Extraction	Cleanup	Instrumental Analysis	Method Performances[a]	Found Levels[a]	Reference
Acid compounds, PFOSA	Rabbit liver 0.20 g	Homogenization	IPE with tetrabutyl ammonium hydrogen sulfate; extraction with methyl *tert*-butyl ether (pH 10)	Centrifuging (speed not specified); filtration with 0.2 μm nylon mesh filter	HPLC-MS/MS, C$_{18}$ column 50 × 2 mm (5 μm), water (2 mM ammonium acetate)–methanol mobile phase	Recovery, 87% (PFOA), 100% (PFOS); LOD, 5 ng/g ww (PFOA), LOD, 8.5 ng/g ww (PFOS)		146
N-EtFOSA, N,N-Et$_2$FOSA, PFOSA	Hamburger 10 g	Homogenization	Extraction at high speed with 2:1 *n*-hexane–acetone	Centrifuging and adsorption chromatography on silica gel column impregnated with concentrated sulfuric acid	GC-PCI/MS, DB-1701 GC column, reagent gas, methane	Recovery, 74–101%; MDL 0.10–0.25 ng/g ww	0.23–0.70 ng/g ww	147
Acid compounds, PFOSA	Cod, gull liver 1 g	Homogenization	Mixing with Vortex; ultrasonic extraction with methanol–water (50:50; 2 mM ammonium acetate)	Filtration at high speed with YM-3 centrifugal filter	HPLC-HRMS (mass tolerance, 0.06 u), C$_{18}$ column 150 × 2.1 mm (3 μm); methanol and water (2 mM ammonium acetate each phase) mobile phase	Recovery, 83–84% (PFOA), 79–90% (PFOS); MDL, 1.25–1.28 ng/g ww (PFOA), MDL, 0.23–0.30 ng/g ww (PFOS)	Gull liver: PFOA <1.28 ng/g ww; PFOS 183 ng/g ww	148

Analyte	Sample	Homogenization	Extraction/cleanup	Instrumental analysis	Recovery/LOD	Results	Ref.
Acid compounds, N-EtFOSA, alcoholic telomer	Beaver liver 1 g	Homogenization	Blending with 0.01 N KOH methanolic solution	Fractioning on weak anionic exchange solid-phase column into two fractions (nonacid compounds in fraction I, acid compounds in fraction II)	HPLC-MS/MS, C_{18} column 50 × 2.1 mm (5 μm), water (2 mM ammonium acetate)/methanol mobile phase	Mean recovery, 85%; MDL, 0.03–3 ng/g ww	PFOA 0.29 ng/g ww; PFOS 133 ng/g ww	149
Acid compounds	Fish 0.01 g	Homogenization	IPE with tetrabutyl ammonium hydrogen sulfate; extraction solvent methyl *tert*-butyl ether (pH 10)	Centrifuging (speed not specified)	HPLC-ITD, C_{18} column 50 × 2.1 mm (5 μm), water (1 mM ammonium acetate)/methanol mobile phase	Recovery, 80–81% (PFOA), 99–102%, (PFOS) LOD, 10 ng/g ww (PFOA); LOD, 2.5 ng/g ww (PFOS)	PFOA 100 ng/g ww, PFOS 200 ng/g ww	150
Acid compounds	Meat (composite samples) 2 g	Homogenization	Blending with methanol	Centrifuging at high speed	HPLC-MS/MS, C_{18} column 50 × 2.1 mm, water (5 mM ammonium formate)/acetonitrile–methanol (2:1) mobile phase	Recovery 91–116% (PFOA), 85–108% (PFOS); LOD 0.5–1 ng/g ww	PFOA <0.5–2.6 ng/g ww, PFOS <0.6–2.7 ng/g ww	151

[a] ww, wet weight.

40.2.3.1.3   Cleanup and Fractionation

As observed for PCDDs, PCDFs, and DL-PCBs, the nonselective nature of exhaustive extraction procedures and the complexity of sample matrices result in a complex extract that requires efficient purification. Lipid elimination, performed by destructive or nondestructive methods (see Section 40.2.2.1.3), is generally followed by isolation of the target analytes from other organohalogenated compounds. Fractionation of NDL-PCBs and PBDEs from coextractive species with similar chemical–physical properties (i.e., organochlorine pesticides, DL-PCBs, PCDDs, and PCDFs) is based on the different polarity of NDL-PCBs and PBDEs in comparison with other chlorinated compounds, and the attitude of NDL-PCBs and PBDEs to be easily eluted from activated carbon with respect to other molecules with planar structure (i.e., PCDDs, PCDFs, and non-*ortho* DL-PCBs). With regard to this, the use of silica, alumina, Florisil, and activated carbon is widely described in the literature [63,156,157]. Recently, the PowerPrep automated cleanup procedure used for separation of PCDDs, PCDFs, and PCBs [122] has been extended also to include PBDEs, after optimization of the type and volume of the solvent necessary to isolate the different chemical families [158].

40.2.3.1.4   Instrumental Analysis

HRGC combined with ECD or MS detectors is the method of choice for the analysis of NDL-PCBs. A comprehensive review on developments in the HRGC of PCBs is given by Cochran and Frame [159], who evaluated a variety of stationary phases commonly used for PCB analysis. The 5%-phenyl type column has substantially become the standard for PCB analysis. Although alternative phases, such as phenyl carborane and that present in DB-XLB columns, have been an attempt to overcome the problem of coelution of the most significant congeners, no column phase can resolve all PCBs in a single injection. More complete separation can be achieved with a different column configuration based on the use of a single injection split coupled to two columns in parallel that end in two ECD detectors [97,160]. ECD is the most utilized detection method for PCBs for its high sensitivity, low cost, and ease in use and maintenance. As observed for organochlorine pesticides (Section 40.2.1.1.4.1), the main disadvantages are its poor selectivity and, as observed by Cochran and Frame [159], nonlinear response over a relative narrow concentration range. The application of low-resolution mass spectrometry (LRMS) operating either in the EI mode or with negative chemical ionization (NCI) [161] provides higher specificity than ECD and allows to obtain qualitative information for analyte identification along with HRGC retention time. Recently, the use of ion trap tandem MS systems has been evaluated for the analysis of PCBs in environmental samples and biota [162,163].

PBDEs are generally quantified by HRGC-MS. Given the degradation problems sometimes experienced for certain congeners (i.e., PBDE 209), the characteristics of the GC system have to be properly selected. In fact, as observed by Björklund et al. [164], the column brand, type of retention gap, press-fit connector, stationary phase, column length, and injection technique strongly influence the accuracy and precision of nona- and deca-PBDE analysis. Determination of PBDEs can be relatively easily performed on nonpolar or semipolar columns such as 100% methyl-polysiloxane (DB-1) and 5% phenyl-dimethyl-polysiloxane (DB-5, CP-Sil 8, or AT-5) [144]. A selection of the most suitable GC columns for PBDE congener-specific analysis can be easily done on the base of the work of Korytár et al. [165] who reported the elution order of 126 PBDEs on seven different GC stationary phases; a DB-XLB column was found to be the most efficient for the separation of PBDE congeners, with a DB-1 column as runner-up. The most commonly used detectors for PBDE analysis is MS operating in the NCI or EI mode [166]. Although NCI presents a higher sensitivity than EI, it is less selective, since only bromine can be monitored, and less accurate since IS method with ^{13}C-labeled PBDEs cannot be utilized. HRMS with EI ionization is preferred in principle over LRMS for its higher sensitivity and selectivity. Nevertheless, due to the complexity of the analysis and cost of HRMS, the LRMS is the most widely used. Recently the use of ion trap MS or quadrupole MS has been evaluated for PBDE analysis. Application of these analytical approaches to the determination of PBDE in abiotic, biotic, and food samples is reported by Wang et al. [167], Gómara et al. [168], Petinal et al. [169], and Yusà et al. [170].

40.2.4 Polyfluorinated Alkylated Substances

In the recent scientific literature, only few works deal with specific analytical methods for PFAS in meat (Table 40.7). More information can be drawn from analyses carried out on biota of environmental interest. In this section, when not further specified, the assay of PFOS and PFOA is mainly dealt with.

Tittlemier et al. [151] describe a liquid chromatography in tandem with mass spectrometry (LC-MS/MS) multiresidue method to analyze PFOS, PFOA, and related compounds in composite samples of several foods (e.g., chicken, lamb, beef, pork) with a limit of determination (LD) ranging from 0.5 to 1 ng/g fresh weight. Nevertheless, the following possible problems in the analysis of PFAS have been reported by Martin et al. [171]:

a. Ion suppression in electrospray interface (ESI), a widely reported phenomenon using such interface
b. Presence of teflonated materials in the analytical tools, which can release PFAS
c. Capability of glassware to sequestrate PFAS when in aqueous solutions
d. Ambiguous quantification of branched and linear isomers, possibly due to poor resolution in liquid chromatography or insufficient purity of the standards
e. Limited availability of ^{13}C-labeled PFAS to be used as ISs for quantification
f. Nonavailability of reference materials

40.2.4.1 Sampling and Sample Storage

Contamination may occur during sampling if the gloves, dresses, or tools that the operators use are made of material releasing PFAS (e.g., Gore-Tex®, W. L. Gore & Associates, Inc.; Teflon®, Dupont). Polypropylene sample bottles should be precleaned by rinsing with polar solvents, such as methanol. Sample storage at −20°C seems to be generally appropriate to preserve the analytes [172].

40.2.4.2 Extraction and Cleanup

Owing to their tensioactive properties, PFAS tend to interact with materials that are used in an undedicated analytical laboratory. Therefore, it is recommended to limit the extraction and cleanup procedures to the essential steps, capable of guaranteeing quantitative recoveries and a selectivity that can minimize the ion-suppression phenomenon during the instrumental acquisition of data [172,173]. The following examples describe some reference methods for PFAS analysis according to their evolution in time.

Ylinen et al. [174] proposed an ion-pair extraction (IPE) using tetra-*n*-butylammonium hydrogen sulfate as a counterion; the approach was subsequently modified by Hansen et al. [146]. This extraction technique was also applied to several biological and environmental matrices

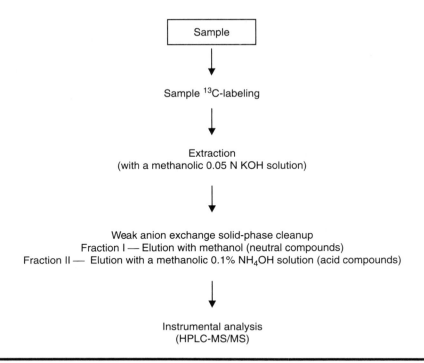

Figure 40.2 Analytical method employing solid-phase cleanup. (From Taniyasu, S. et al., *J. Chromatogr. A*, 1093, 89, 2005.)

[150,175,176]. More recently, a sample extraction with alkalinized methanolic solution followed by a weak anionic exchange solid-phase cleanup was applied by Taniyasu et al. [149] on fish samples (Figure 40.2), with an optimization of recovery rates. A selective elution from an SPE column of the analytes related to their polarity was achieved, thereby obtaining two fractions with neutral and ionic analytes, respectively, with a low matrix overload. Powley et al. [177,178] proposed a dispersive solid phase with graphitized carbon as a cleanup step. Carbon was directly added to the extract and mixed thoroughly by vortexing, thus allowing to sequestrate the hydrophobic substances that were further removed by centrifugation. This procedure yielded a decreased ion-suppression phenomenon and the possible release of PFAS from SPE column cartridges was avoided. Berger and Haukas [148] proposed to use a polar solvent extraction followed by a clarification of the extract via centrifugation and a selective filtration at 3000 nominal molecular weight limit cutoff, before instrumental analysis.

40.2.4.3 Instrumental Identification and Determination

The performance of analytical instruments may condition the choice of extraction and cleanup procedures, allowing the injection of extracts more or less diluted, which possibly did not undergo a cleanup. Moreover, according to the geometry of LC-MS interfaces, the ion-suppression phenomena can be reduced. LC coupled to MS/MS detectors with an ESI is the instrumental technique of choice to identify and determine PFAS. Data can be acquired in the selected reaction monitoring (SRM) mode as reported in Table 40.8.

Table 40.8   Principal MS/MS Transition of Some PFAS

Compound/Acronym	Precursor Ion $(M - H)^-$ m/z	Transition	Nature of Product Ion
PFBS	299	$299 \rightarrow 99$	FSO_3^-
PFOA	413	$413 \rightarrow 369$	$(M - COOH)^-$
6:2 FTS	427	$427 \rightarrow 81$	HSO_3^-
PFOSA	498	$498 \rightarrow 78$	SNO_2^-
PFOS	499	$499 \rightarrow 80$	SO_3^-
N-MeFOSE	556	$556 \rightarrow 526$	$(M - CH_2OH)^-$
PFTeDA	713	$713 \rightarrow 669$	$(M - COOH)^-$

A particular attention is required when using ion trap mass spectrometers. Owing to cutoff limitations, very wide transitions—such as SO_3^- produced by the molecular ion of PFOS—cannot be achieved in a quantitative way [150,179]. Berger and Haukas [148] analyzed carboxylic acid, sulfonate compounds, and PFOSA with LC coupled to HRMS such as a time of flight (TOF), as an alternative to the MS/MS technique. Chromatographic separation is generally achieved on reverse phase C_8 or C_{18} columns, using methanol and ammonium acetate or formic acid or acetic acid aqueous solutions as mobile phases. Possible background contributions, originating from teflonated parts in the LC system, should be carefully evaluated. LC-grade water should be decontaminated through Amberlite XAD-7 resin to remove any possible perfluorinated compound [151].

It is generally agreed to prepare calibration curves using the real matrix to account for ion-suppression phenomena as well, especially when the ^{13}C-labeled ISs are not available for all the analytes. A volumetric standard is also employed.

Nonpolar fluorinated compounds, such as PFOSA, can be directly determined with HRGC-MS with positive chemical ionization (PCI) [180]. An inventory of MS-based techniques is reported in Table 40.7.

40.2.4.4   Observations

As the first two international intercalibration studies yielded unsatisfactory results [181,182], the organizers of the third round (2007) decided to meet all the participating laboratories to define the "best" analytical method(s) to be adopted for the determination of PFAS. This concern provides an indication that the aforesaid analytical methods still need to be consolidated to be adequately reliable.

40.3   Bioassays to Screen in Meat Polychlorodibenzo-*p*-Dioxins, Polychlorodibenzofurans, and Dioxin-Like Polychlorobiphenyls

40.3.1   Introduction

The use of bioassay as a screening tool aims at dosing the biological activity of contaminants, by comparing their effects with those of a standard preparation or a reference material, on a culture of living cells [183]. Within this frame, rather than the amount of contaminant(s) bound to a biological macromolecule, as in the case of immunoassay determinations [184], bioassays allow to dose the response elicited as a result of the interaction between the analyte(s) and a specific receptor. As a consequence, the signal measured on the selected biological substrate results from the cumulative effects of the different substances present in the extract to be analyzed that share the same mode of action, according to their concentration and their relative potency (REP) [185]. This can be the case of PCDD, PCDF, and DL-PCB congeners [58], and of other categories of pesticides and contaminants, for which a cumulative assessment on a toxicological basis has been suggested by regulatory agencies [186].

40.3.2 Cell-Based Bioassays

In the literature, the most consolidated applications of bioassays on food samples are based on the use of chemically activated fluorescence or luminescence gene expression in engineered cell lines [187]. Briefly, the contaminants present in the extract interact with the specific aryl hydrocarbon receptor (AhR) expressed on cell membranes. The complex is transported to the nucleus where it activates the deoxyribonucleic acid (DNA) sequence for the synthesis of a specific enzyme (i.e., luciferase). After extraction and cleanup, and incubation of the extract on the cell culture, the addition of luciferine as substrate to the supernatant from cell lysis—containing the induced luciferase—produces a chemoluminescence or fluorescence signal whose intensity is related to both the amount(s) and REP(s) of the contaminant(s) present.

40.3.3 Bioassay Based on Polymerase Chain Reaction

Another bioassay, which has been only preliminarily applied to food matrices, is based on DNA real-time amplification and fluorescence detection [188]. This technique has the advantage that no cell lines and related laboratory facilities are needed to perform the test. The target compounds activate the AhR to a form that binds to DNA. The activated complex is then trapped onto a micro-well; the receptor-bound DNA is amplified through the polymerase chain reaction (PCR) and read in real-time mode.

40.3.4 Bioassay Reliability and Applicability

The EU legislation has recently established some specific requirements that should be fulfilled in the cell-based bioassay screening of PCDDs, PCDFs, and DL-PCBs [114]. A series of reference concentrations of $2,3,7,8\text{-T}_4\text{CDD}$ or a PCDD, PCDF, and DL-PCB mixture should be tested to obtain a significant full dose–response curve; it is recommended to use reference materials and build appropriate quality control charts to ensure that the relative standard deviation shall not be above 15% in a triplicate determination for each sample (repeatability) and not above 30% between three independent experiments (reproducibility). For quantitative calculations, the induction of the sample dilution used must fall within the linear portion of the response curve, with an LD sixfold the standard deviation of the solvent blank or the background. Information on the correspondence between bioassay and HRGC-HRMS results should also be provided. For official use, positive results from screening must always be confirmed; false negative rates must be below 1%.

The following critical points can be identified as the main causes of possible inconsistencies between bioassay screening and confirmatory (HRGC-HRMS) analysis outputs:

a. The samples should be appropriately processed, allowing an exhaustive fat extraction and the removal of other possible AhR ligands—such as polycyclic aromatic hydrocarbons—capable of eliciting a bioassay response, if present in large quantities.

b. The congener REPs may differ from the consensus-based TEFs used for conversion of HRGC-HRMS data into toxicology-based WHO-TEQs: this may cause deviations from HRGC-HRMS results (the magnitude of deviations is affected by the contamination profile) [189].

c. Possible deviations from simple additivity of the bioassay measured effects may be expected in the presence of PCDD, PCDF, and PCB mixtures (e.g., Aroclors 1242, 1254, 1260) [190].

40.4   Abbreviations

AhR	aryl hydrocarbon receptor
AOAC	Association of Official Analytical Chemists
ASE	accelerated solvent extraction
pp'-DDT	1,1,1-trichloro-2,2-*bis*(4-chlorophenyl)ethane
deca-BDE	decabromodiphenyl ether
DG	direction-general
DL-PCB	dioxin-like polychlorinated biphenyls
DNA	deoxyribonucleic acid
EC	European Commission
ECD	electron capture detector
EFSA	European Food Safety Authority
EI	electron impact
EPA	Environmental Protection Agency
ESI	electrospray interface
n-EtFOSE	*n*-ethyl perfluorooctane sulfonamidoethanol
EU	European Union
FAO	Food and Agriculture Organization
FDA	Food and Drug Administration
8:2 FTOH	8:2 fluorotelomer alcohol
6:2 FTS	6:2 fluorotelomer sulfonate
GPC	gel permeation chromatography
alpha-HCH	alpha-hexachlorocyclohexane
beta-HCH	beta-hexachlorocyclohexane
HPLC	high-performance liquid chromatography
HRGC	high-resolution gas chromatography
HRMS	high-resolution mass spectrometry
IPCS	International Programme on Chemical Safety
IPE	ion-pair extraction
I-TEF	international toxicity equivalency factor
IS	internal standard
LC	liquid chromatography
LD	limit of determination
LOD	limit of detection
LOQ	limit of quantification
LRMS	low-resolution mass spectrometry
MAE	microwave-assisted extraction
MDL	minimum detection level

N-MeFOSE	*N*-methyl perfluorooctane sulfonamidoethanol
ML	maximum level
MRL	maximum residue limit
MSPD	matrix solid-phase dispersion
NCI	negative chemical ionization
NDL-PCB	non-dioxin-like polychlorinated biphenyl
octa-BDE	octabromodiphenyl ether
PBDE	polybrominated diphenyl ether
PCDD	polychlorinated dibenzo-*p*-dioxin
PCDF	polychlorinated dibenzofuran
PCI	positive chemical ionization
penta-BDE	pentabromodiphenyl ether
PFAS	polyfluorinated alkylated substances
PFBS	perfluorobutyl sulfonate
PFOA	perfluorooctanoic acid
PFOS	perfluorooctane sulfonate
PFOSA	perfluorooctane sulfonamide
PFTeDA	perfluorotetradecanoic acid
PLE	pressurized liquid extraction
POP	persistent organic pollutant
REP	relative potency
SCCP	short-chained chlorinated paraffin
SCF	Scientific Committee on Food
SFE	supercritical fluid extraction
SIM	single (or selected) ion monitoring
SRM	selected reaction monitoring
SPE	solid-phase extraction
TMRL	temporary maximum residue limit
TOF	time of flight
UNEP	United Nations Environment Programme
WHO	World Health Organization
WHO-TEF	WHO toxicity equivalency factor

References

1. Available at http://www.chem.unep.ch/pops/default.html.
2. Bernes, C., Where do persistent pollutants come from? in *Persistent Organic Pollutants: A Swedish View of an International Problem*, Swedish Environmental Protection Agency, Stockholm, Sweden, 1998, chap. 2.
3. Available at http://www.pops.int/.
4. IPCS, Aldrin and Dieldrin, *Environmental Health Criteria* 91, WHO, Geneva, 1989.
5. IPCS, Camphechlor, *Environmental Health Criteria* 45, WHO, Geneva, 1984.
6. IPCS, Chlordane, *Environmental Health Criteria* 34, WHO, Geneva, 1984.
7. IPCS, Chlordecone, *Environmental Health Criteria* 43, WHO, Geneva, 1984.
8. IPCS, DDT and its derivatives, Environmental Health Criteria 9, WHO, Geneva, 1979.
9. IPCS, Endrin, *Environmental Health Criteria* 130, WHO, Geneva, 1992.
10. IPCS, Heptachlor, *Environmental Health Criteria* 38, WHO, Geneva, 1984.
11. IPCS, Hexachlorobenzene, *Environmental Health Criteria* 195, WHO, Geneva, 1997.

12. IPCS, *Alpha-* and *beta*-hexachlorocyclohexane, *Environmental Health Criteria* 123, WHO, Geneva, 1992.

13. IPCS, Lindane, *Environmental Health Criteria* 124, WHO, Geneva, 1991.

14. IPCS, Mirex, *Environmental Health Criteria* 44, WHO, Geneva, 1984.

15. IARC, Polychlorinated dibenzo-para-dioxins and dibenzofurans. *IARC Monographs on the Evaluation of Carcinogenic Risks to Humans.* Vol. 69, IARC, Lyon, France, 1997.

16. SCF, Opinion of the SCF on the Risk Assessment of Dioxins and Dioxin-like PCBs in food. Update based on new scientific information available since the adoption of the SCF opinion of 22nd November 2000 (adopted on 30 May 2001), http://ec.europa.eu/food/fs/sc/scf/out90_en.pdf.

17. SCF, Opinion of the SCF on the Risk Assessment of Dioxins and Dioxin-like PCBs in food (adopted on 22 November 2000), http://ec.europa.eu/food/fs/sc/scf/out78_en.pdf.

18. UNEP, Draft risk profile: pentabromodiphenyl ether, 2006, available at http://www.pops.int/documents/meetings/poprc_2/meeting_docs.htm.

19. European Commission, Diphenyl ether, octabromo derivative. Summary Risk Assessment Report, Joint Research Centre, 2003.

20. UNEP, Draft risk profile: perfluorooctane sulfonate (PFOS), 2006, available at http://www.pops.int/documents/meetings/poprc_2/meeting_docs.htm.

21. FDA, General Analytical Operations and Information, in *Pesticide Analytical Manual, Volume I,* FDA. 3rd Edition, 1994, chap. 2.

22. U.S. EPA, Method 1618A, *Organo-halide Pesticides, Organo-phosphorus Pesticides, and Phenoxy-acid Herbicides by Wide Bore Capillary Column Gas Chromatography with Selective Detectors,* Industrial Technology Division, Office of Water, US Environmental Protection Agency (Washington), 1989.

23. U.S. EPA, Method 1613, Tetra- through octachlorinated dioxins and furans by isotope dilution HRGC-HRMS, Engineering and Analysis Division (4303), Office of Water, US Environmental Protection Agency (Washington), 1994.

24. U.S. EPA Method 1668, Revision A. Chlorinated biphenyl congeners in water, soil, sediment, biosolids and tissue by HRGC/HRMS, Engineering and Analysis Division (4303), Office of Water, US Environmental Protection Agency (Washington), 1999.

25. U.S. EPA Method 1614 Draft, Brominated diphenyl ethers in water, soil, sediment, and tissue by HRGC/HRMS. Engineering and Analysis Division (4303), Office of Water, US Environmental Protection Agency (Washington), 2003.

26. Moriarty, F., Prediction of ecological effects, in *Ecotoxicology: The Study of Pollutants in Ecosystems,* 2nd Edition, Academic Press, San Diego, CA, 1988, chap. 6.

27. EU Council Decision 2006/507/EC of 14 October 2004 concerning the conclusion, on behalf of the European Community, of the Stockholm Convention on Persistent Organic Pollutants. OJ L 209, 31.7.2006, 1–2.

28. United Nation/UNEP, Global Report 2003. Regionally Based Assessment of Persistent Toxic Substances, 2003.

29. Iamiceli, A.L., Turrio-Baldassarri, L., and di Domenico, A., Determination of PCDDs and PCDFs in water, in *Handbook of Water Analysis*, Nollet, L.M.L., ed, Marcel Dekker, 2000, chap. 31.

30. UNEP, Draft risk profile: chlordecone, 2006, available at http://www.pops.int/documents/meetings/poprc_2/meeting_docs.htm.

31. UNEP, Draft risk profile: hexabromobiphenyl, 2006, available at http://www.pops.int/documents/meetings/poprc_2/meeting_docs.htm.

32. UNEP, Draft risk profile: lindane, 2006, available at http://www.pops.int/documents/meetings/poprc_2/meeting_docs.htm.

33. Joint WHO/Convention task force on the health aspects of air pollution, Health risks of persistent organic pollutants from long-range transboundary air pollution, WHO, 2003.

34. UNEP, Summary of pentachlorobenzene proposal, 2006, available at http://www.pops.int/documents/meetings/poprc_2/meeting_docs.htm.

35. UNEP, Summary of octabromodiphenyl ether proposal, 2006, available at http://www.pops.int/documents/meetings/poprc_2/meeting_docs.htm.

36. UNEP, Summary of short-chained chlorinated paraffins proposal, 2006, available at http://www. pops.int/documents/meetings/poprc_2/meeting_docs.htm.

37. UNEP, Summary of *alpha*-hexachlorocyclohexane proposal, 2006, available at http://www.pops.int/ documents/meetings/poprc_2/meeting_docs.htm.

38. UNEP, Summary of *beta*-hexachlorocyclohexane proposal, 2006, available at http://www.pops.int/ documents/meetings/poprc_2/meeting_docs.htm.

39. EU Council Directive 86/363/EEC of 24 July 1986 on the fixing of maximum levels for pesticide residues in and on foodstuffs of animal origin. *OJ L 221, 7.8.1986, 43–47.*

40. Kostantinou, I.K. et al., The status of pesticide pollution in surface waters (rivers and lakes) of Greece. Part I. Review on occurrence and levels, *Environ. Pollut.*, 141, 555, 2006.

41. Li, J. et al., Organochlorine pesticides in the atmosphere of Guangzhou and Hong Kong: regional sources and long-range atmospheric transport, *Atmos. Environ.*, 41, 3889, 2007.

42. Vagi, M.C. et al., Determination of organochlorine pesticides in marine sediments samples using ultrasonic solvent extraction followed by GC/ECD, *Desalination*, 210, 146, 2007.

43. Torres, M.J. et al., Organochlorine pesticides in serum and adipose tissue of pregnant women in Southern Spain giving birth by cesarean section, *Sci. Total Environ*, 372, 32, 2006.

44. Meeker, J.D., Altshul, L., and Hauser, R., Serum PCBs, *p,p'*-DDE and HCB predict thyroid hormone levels in men, *Environ. Res.*, 104, 296, 2007.

45. EU Council Directive 91/414/EEC of 15 July 1991 concerning the placing of plant protection products on the market. *OJ L 230, 19.8.1991, 1–32.*

46. EU Regulation 396/2005/EC of the European Parliament and of the Council of 23 February 2005 on maximum residue levels of pesticides in or on food and feed of plant and animal origin and amending Council Directive 91/414/EECText with EEA relevance. *OJ L 70, 16.3.2005, 1–16.*

47. EU Council Directive 86/362/EEC of 24 July 1986 on the fixing of maximum levels for pesticide residues in and on cereals. *OJ L 221, 7.8.1986, 37–42.*

48. EU Council Directive 90/642/EEC of 27 November 1990 on the fixing of maximum levels for pesticide residues in and on certain products of plant origin, including fruit and vegetables. *OJ L 350, 14.12.1990, 71–79.*

49. FAO/WHO Codex Alimentarius Commission. Maximum Residue Limits for pesticides, FAO/ WHO, Rome, Italy, 2001.

50. Van den Berg, M. et al., Toxic equivalency factors (TEFs) for PCBs, PCDDs, PCDFs for humans and wildlife, *Environ. Health Perspect.*, 106, 775, 1998.

51. Van den Berg, M. et al., The 2005 World Health Organization reevaluation of human and mammalian toxic equivalency factors for dioxins and dioxin-like compounds, *Toxicol. Sci.*, 93, 223, 2006.

52. NATO/CCMS, International toxicity equivalency factor (I-TEF) method of risk assessment for complex mixtures of dioxins and related compounds, Report No. 176, Committee on the Challenges of Modern Society, North Atlantic Treaty Organization, 1988.

53. Galliani, B. et al., Occurrence of ndl-PCBs in food and feed in Europe, *Organohalogen Compd.*, 66, 3561, 2004.

54. Fattore, E. et al., Current dietary exposure to polychlorodibenzo-*p*-dioxins, polychlorodibenzofurans, and dioxin-like polychlorobiphenyls in Italy, *Mol. Nutr. Food Res.*, 50, 915, 2006.

55. Fattore, E. et al., Assessment of the dietary exposure to non-dioxin-like PCBs of the Italian general population, *Chemosphere*, submitted.

56. Brambilla, G. et al., Persistent organic pollutants in meat: a growing concern, *Meat Sci.*, 78, 25–33, 2008.

57. Domingo, J.L., and Bocio, A., Levels of PCDD/PCDFs and PCBs in edible marine species and human intake: a literature review, *Environ. Int.*, 33, 397, 2007.

58. EU Commission Regulation 1881/2006/EC of 19 December 2006 setting maximum levels for certain contaminants in foodstuffs. OJ L 364, 20.12.2006.

59. Law, R.J. et al., Levels and trends of polybrominated diphenylethers and other brominated flame retardants in wildlife, *Environ. Int.*, 29, 757, 2003.

60. Alaee, M. et al., An overview of commercially used brominated flame retardants, their applications, their use patterns in different countries/regions and possible modes of release, *Environ. Int.*, 29, 683, 2003.

61. EU Directive 2002/95/EC of the European Parliament and of the Council of 27 January 2003 on the restriction of the use of certain hazardous substances in electrical and electronic equipment. OJ L 37, 13.2.2003, 19–23.

62. Schecter, A. et al., Polybrominated diphenyl ethers contamination of United States Food, *Environ. Sci. Technol.*, 5306, 2004.

63. Schuhmacher, M. et al., Concentrations of polychlorinated biphenyls (PCBs) and polybrominated diphenyl ethers (PBDEs) in milk of women from Catalonia, Spain, *Chemosphere*, 67, S295, 2007.

64. Renner, R., Growing concern over perfluorinated chemicals, *Environ. Sci. Technol.*, 35, 154A, 2001.

65. Hoff, P. et al., Perfluorooctane sulfonic acid in bib (trisopterus luscus) and plaice (pleuronectes platessa) from the Belgian North Sea: distribution and biochemical effects, *Environ. Toxicol. Chem.*, 22, 608, 2003.

66. Kannan, K. et al., Perfluorooctanesulfonate and related fluorochemicals in human blood from several countries, *Environ. Sci. Technol.*, 38, 4489, 2004.

67. Kannan, K. et al., Perfluorinated compounds in aquatic organisms at various trophic levels in a great lakes food chain, *Arch. Environ. Contam. Toxicol.*, 48, 559, 2005.

68. Yamashita, N. et al., A global survey of perfluorinated acids in oceans, *Mar. Pollut. Bull.*, 51, 658, 2005.

69. Haukas, M. et al., Bioaccumulation of per- and polyfluorinated alkyl substances (PFAS) in selected species from the Barents Sea food web, *Environ. Pollut.*, 148, 360, 2007.

70. Hercegová, A., Dömötörová, M., and Matisová, E., Sample preparation methods in the analysis of pesticide residues in baby food with subsequent chromatographic determination, *J. Chromatogr.*, doi:10.1016/j.chroma.2007.01.008.

71. Codex Alimentarius Commission, Analysis of Pesticide Residues: Recommended Methods, CODEX STAN 229-1993, REV.1-2003.

72. AOAC International, Official Methods of Analysis of AOAC International, 18th Edition, 2005.

73. FDA, Multiresidue methods, in *Pesticide Analytical Manual, Volume I*, FDA, 3rd Edition, 1994, chap. 3.

74. Saito, K. et al., Development of a accelerated solvent extraction and gel permeation chromatography analytical method for measuring persistent organohalogen compounds in adipose and organ tissue analysis, *Chemosphere*, 57, 373, 2004.

75. Hopper M.L., Automated one-step supercritical fluid extraction and clean-up system for the analysis of pesticide residues in fatty matrices, *J. Chromatogr. A*, 840, 93, 1999.

76. Frenich, A.G. et al., Multiresidue analysis of organochlorine and organophosphorus pesticides in muscle of chicken, pork and lamb by gas chromatography–triple quadrupole mass spectrometry, *Anal. Chim. Acta*, 558, 42, 2006.

77. Frenich, A.G., Bolaños, P.P., and Vidal, J.L.M., Multiresidue analysis of pesticides in animal liver by gas chromatography using triple quadrupole tandem mass spectrometry, *J. Chromatogr. A*, doi: 10.1016/j.chroma.2007.01.066.

78. Patel, K. et al., Evaluation of gas chromatography–tandem quadrupole mass spectrometry for the determination of organochlorine pesticides in fats and oils, *J. Chromatogr. A*, 1068, 289, 2005.

79. Janouskova, E. et al., Determination of chlordane in foods by gas chromatography, *Food Chem.*, 93, 161, 2005.

80. AOAC Official method 983.18, Meat and Meat Products. Preparation of Test Sample Procedure.

81. RIVM, Pesticides amenable to gas chromatography: multiresidue method 1. RIVM report No. 638817014, 1996.

82. UNI EN 1528-2, Fatty food. Determination of pesticides and polychlorinated biphenyl (PCBs). Part 2: extraction of fat, pesticides and PCBs, and determination of fat, 1996.

83. Valcarel, M., and Tena, M.T., Applications of supercritical fluid extraction in food analysis, *Fresenius J. Anal. Chem.*, 35, 561, 1997.

84. Snyder , J.M. et al., Supercritical fluid extraction of poultry tissues containing incurred pesticide residues, *J. AOAC Int.* 76, 888, 1993.

85. Gallina Toschi, T. et al., Pressurized solvent extraction of total lipids in poultry meat, *Food Chem.*, 83, 551, 2003.

86. Eskilsson, C.S., and Björklund, E., Analytical-scale microwave-assisted extraction, *J. Chromatogr. A*, 902, 227, 2000.

87. Pereira, M.B. et al., Comparison of pressurized liquid extraction and microwave assisted extraction for the determination of organochlorine pesticides in vegetables, *Talanta*, 71, 1345, 2007.

88. Hummert, K., Vetter, W., and Luckas, B., Fast and effective sample preparation for determination of organochlorine compounds in fatty tissue of marine mammals using microwave extraction, *Chromatographia*, 42, 300, 1996.

89. Hummert, K., Vetter, W., and Luckas, B., Combined microwave assisted extraction and gel permeation chromatography for the determination of organochlorine compounds in fatty tissue of marine mammals, *Organohalogen Compd.*, 27, 360, 1996.

90. Barker, S.A., Long, A.R., and Short, C.R., Isolation of drug residues from tissues by solid phase dispersion, *J. Chromatogr. A*, 475, 353, 1989.

91. Barker, S.A., Matrix solid phase dispersion (MSPD), *J. Biochem. Biophys. Methods*, 70, 151, 2007.

92. Bogialli, S., and Di Corcia, A., Matrix solid-phase dispersion as valuable tool for extracting contaminants from foodstuffs, *J. Biochem. Biophys. Methods*, 70, 163, 2007.

93. Long, A.R., Soliman, M.M., and Barker, S.A., Matrix solid phase dispersion (MSPD) extraction and gas chromatographic screening of nine chlorinated pesticides in beef fat, *J. AOAC Int.*, 74, 493, 1991.

94. Furusawa, N., A toxic reagent-free method for normal-phase matrix solid-phase dispersion extraction and reversed-phase liquid chromatographic determination of aldrin, dieldrin, and DDTs in animal fats, *Anal. Bioanal. Chem.*, 378, 2004, 2004.

95. Furusawa, N., Determination of DDT in animal fats after matrix solid-phase dispersion extraction using an activated carbon filter, *Chromatographia*, 62, 315, 2005.

96. van der Hoff, G.R., and van Zoonen, P., Trace analysis of pesticides by gas chromatography, *J. Chromatogr. A*, 843, 301, 1999.

97. Ahmed, F.E., Analysis of pesticides and their metabolites in foods and drinks, *Trends Anal. Chem.*, 20, 649, 2001.

98. AOAC Official Method 970.52, Organochlorine and Organophosphorus Pesticides Residues. General Multiresidue Method, in *Official Methods of Analysis of AOAC International*, 18th Edition, AOAC International, Gaithersburg, MD, 2005.

99. UNI EN 1528-3, Fatty food. Determination of pesticides and polychlorinated biphenyl (PCBs). Part 3: Clean-up methods, 1996.

100. EPA Method 3620C, Florisil clean-up. Revision 3. US Environmental Protection Agency, Washington, DC, November 2000.

101. Burke, E.R., Holden, A.J., and Shaw, I.C., A method to determine residue levels of persistent organochlorine pesticides in human milk from Indonesian women, *Chemosphere*, 50, 529, 2003.

102. UNI EN 1528-4, Fatty food. Determination of pesticides and polychlorinated biphenyl (PCBs). Part 4: Determination, confirmatory tests, miscellaneous, 1996.

103. U.S. EPA Method 8081A, Organochlorine pesticides by gas chromatography. Revision 1. US Environmental Protection Agency, Washington, DC, December 1996.

104. European Commission, Quality Control Procedures for Pesticide Residue Analyses, SANCO/10232/2006, 24 March 2006, available at http://ec.europa.eu/food/plant/protection/resources/qualcontrol_en.pdf.

105. EU Decision/2005/657/EC implementing Council Directive 96/23/EC concerning the performance of analytical methods and the interpretation of results, OJ L 221 of 17.8.2002.

106. Zrostlíková, J., Hajšlová, J., and Cajka, T., Evaluation of two-dimensional gas chromatography–time-of-flight mass spectrometry for the determination of multiple pesticide residues in fruit, *J. Chromatogr. A*, 1019, 173, 2003.

107. Focant, J.F., Sjödin, A., and Patterson, D.G., Qualitative evaluation of thermal desorption-programmable temperature vaporization-comprehensive two-dimensional gas chromatography–time-of-flight mass spectrometry for the analysis of selected halogenated contaminants, *J. Chromatogr. A*, 1019, 143, 2003.

108. Korytár, P. et al., Group separation of organohalogenated compounds by means of comprehensive two-dimensional gas chromatography, *J. Chromatogr. A*, 1086, 29, 2005.

109. Seemamahannop, R. et al., Uptake and enantioselective elimination of chlordane compounds by common carp (*Cyprinus carpio*, L.), *Chemosphere*, 59, 493, 2005.

110. Chen, S. et al., Determination of organochlorine pesticide residues in rice and human and fish fat by simplified two-dimensional gas chromatography, *Food Chem.*, doi: 10.10167j.foodchem.2006.10.032, 2006.

111. Reyes, J.F.G. et al., Determination of pesticide residues in olive oil and olives, *Trends Anal. Chem.*, 26, 239, 2007.

112. Reiner, E.J. et al., Advances in analytical techniques for polychlorinated dibenzo-*p*-dioxins, polychlorinated dibenzofurans and dioxin-like PCBs, *Anal. Bioanal. Chem.*, 386, 797, 2006.

113. Hess, P. et al., Critical review of the analysis of non- and mono-*ortho*-chlorobiphenyls, *J. Chromatogr. A*, 703, 417, 1995.

114. EU Commission Regulation 1883/2006/EC of 19 December 2006 laying down methods of sampling and analysis for the official control of levels of dioxins and dioxin-like PCBs in certain foodstuffs, OJ L 364, 20.12.2006, 32–43.

115. Bocio, A., and Domingo, J.L., Daily intake of polychlorinated dibenzo-*p*-dioxins/polychlorinated dibenzofurans (PCDD/PCDFs) in foodstuffs consumed in Tarragona, Spain: a review of recent studies (2001–2003) on human PCDD/PCDF exposure through the diet, *Environ. Res.*, 97, 1, 2005.

116. Loutfy, N. et al., Monitoring of polychlorinated dibenzo-*p*-dioxins and dibenzofurans, dioxin-like PCBs and polycyclic aromatic hydrocarbons in food and feed samples from Ismailia city, Egypt, *Chemosphere*, 66, 1962, 2007.

117. Mayer, R., PCDD/F levels in food and canteen meals from Southern Germany, *Chemosphere*, 43, 857, 2001.

118. Fernandes, A. et al., Simultaneous determination of PCDDs, PCDFs, PCBs and PBDEs in food, *Talanta*, 63, 1147, 2004.

119. Focant, J.F. et al., Fast clean-up for polychlorinated dibenzo-*p*-dioxins, dibenzofurans and coplanar polychlorinated biphenyls analysis of high-fat-content biological samples, *J. Chromatogr. A*, 925, 207, 2001.

120. Covaci, A., Ryan, J.J., and Schepens, P., Patterns of PCBs and PCDD/PCDFs in chicken and pork fat following a Belgian food contamination incident, *Chemosphere*, 47, 207, 2002.

121. Wiberg, K. et al., Selective pressurized liquid extraction of polychlorinated dibenzo-*p*-dioxins, dibenzofurans and dioxin-like polychlorinated biphenyls from food and feed samples, *J. Chromatogr. A*, 1138, 55, 2007.

122. Pirard, C., Focant, J.F., and De Pauw, E., An improved clean-up strategy for simultaneous analysis of polychlorinated dibenzo-*p*-dioxins (PCDD), polychlorinated dibenzofurans (PCDF), and polychlorinated biphenyls (PCB) in fatty food samples, *Anal. Bioanal. Chem.*, 372, 373, 2002.

123. Focant, J.F., Pirard, C., and De Pauw, E., Levels of PCDDs, PCDFs and PCBs in Belgian and international fast food samples, *Chemosphere*, 54, 137, 2004.

124. Voorspoels, S. et al., Dietary PBDE intake: a market-basket study in Belgium, *Environ. Int.*, 33, 93, 2007.

125. Huwe, J.K., and Larsen, G.D., Polychlorinated dioxins, furans, and biphenyls, and polybrominated diphenyl ethers in a U.S. meat market basket and estimates of dietary intake, *Environ. Sci. Technol.*, 39, 5606, 2005.

126. Kiviranta, H., Ovaskainen, M.J., and Vartiainen, T., Market basket study on dietary intake of PCDD/Fs, PCBs, and PBDEs in Finland, *Environ. Int.*, 30, 923, 2004.

127. Huwe, J.K. et al., Analysis of mono- to deca-brominated diphenyl ethers in chickens at the part per billion level, *Chemosphere*, 46(5), 635, 2002.

128. Ohta, S. et al., Comparison of polybrominated diphenyl ethers in fish, vegetables, and meats and levels in human milk of nursing women in Japan, *Chemosphere*, 46, 689, 2002.

129. Martínez, A. et al., Development of a matrix solid-phase dispersion method for the screening of polybrominated diphenyl ethers and polychlorinated biphenyls in biota samples using gas chromatography with electron-capture detection, *J. Chromatogr. A*, 1072, 83, 2005.

130. European Commission, DIFFERENCE Project G6RD-CT-2001-00623, available at www.dioxins.nl.
131. Sporring, S., Holst, C., and Björklund, E., Selective pressurized liquid extraction of PCBs from food and feed samples: effects of high lipid amounts and lipid type on fat retention, *Chromatographia*, 64, 553, 2006.
132. Björklund, E. et al., New strategies for extraction and clean-up of persistent organic pollutants from food and feed samples using selective pressurized liquid extraction, *Trends Anal. Chem.*, 25, 318, 2006.
133. Sporring, S., and Björklund, E., Selective pressurized liquid extraction of polychlorinated biphenyls from fat-containing food and feed samples. Influence of cell dimensions, solvent type, temperature and flush volume, *J. Chromatogr. A*, 1040, 155, 2004.
134. Harrad, S.J. et al., A method for the determination of PCB congeners 77, 126 and 169 in biotic and abiotic matrices, *Chemosphere*, 24, 1147, 1992.
135. Berdié, L., and Grimalt, J.O., Assessment of the sample handling procedures in a labor-saving method for the analysis of organochlorine compounds in a large number of fish samples, *J. Chromatogr. A*, 823, 373, 1998.
136. Pirard, C., Focant, J.F., and De Pauw, E., An improved clean-up strategy for simultaneous analysis of polychlorinated dibenzo-*p*-dioxin (PCDD), polychlorinated dibenzofurans (PCDF), and polychlorinated biphenyls (PCB) in fatty samples, *Anal. Bioanal. Chem.*, 372, 373, 2001.
137. Maier, E.A., Griepink, B., and Fortunati, U., Round table discussions. Outcome and recommendations, *Fresenius J. Anal. Chem.*, 348, 171, 1994.
138. March, R.E. et al., A comparison of three mass spectrometric methods for the determination of dioxins/furans, *Int. J. Mass Spectrom.*, 194, 235, 2000.
139. Lorán, S. et al., Evaluation of GC-ion trap-MS/MS methodology for monitoring PCDD/Fs in infant formula, *Chemosphere*, 67, 513, 2007.
140. Focant, J.F. et al., Recent advances in mass spectrometric measurement of dioxins, *J. Chromatogr. A*, 1067, 265, 2005.
141. Petrovic, M. et al., Recent advances in the mass spectrometric analysis related to endocrine disrupting compounds in aquatic environmental samples, *J. Chromatogr. A*, 974, 23, 2002.
142. EFSA, Opinion of the Scientific Panel on contaminants in the food chain on a requested from the Commission related to the presence of non dioxin-like polychlorinated biphenyls (PCB) in feed and food. Question N° EFSA-Q-2003-114. Adopted on 8 November 2005, *EFSA J.*, 284, 1, 2005.
143. EFSA, Advice of the scientific panel on contaminants in the food chain on a request from the commission related to relevant chemical compounds in the group of brominated flame retardants for monitoring in feed and food. Question N° EFSA-Q-2005-244. Adopted on 24 February 2006, *EFSA J.*, 328, 1, 2006.
144. Covaci, A., Voorspoels, S., and de Boer, J., Determination of brominated flame retardants, with emphasis on polybrominated diphenyl ethers (PBDEs) in environmental and human samples—a review, *Environ. Int.*, 29, 735, 2003.
145. Covaci, A. et al., Recent developments in the analysis of brominated flame retardants and brominated natural compounds, *J. Chromatogr. A*, 1153, 145, 2007.
146. Hansen, K.J. et al., Compound-specific, quantitative characterization of organic fluorochemicals in biological matrices, *Environ. Sci. Technol.*, 35, 766, 2001.
147. Tittlemier, S.A. et al., Development and characterization of a solvent extraction-gas chromatographic/mass spectrometric method for the analysis of perfluorooctane sulfonamide compounds in solid matrices, *J. Chromatogr. A*, 1066, 189, 2005.
148. Berger, U., and Haukas, M., Validation of a screening method based on liquid chromatography coupled to high-resolution mass spectrometry for analysis of perfluoroalkylated substances in biota, *J. Chromatogr. A*, 1081, 210, 2005.
149. Taniyasu, S. et al., Analysis of fluorotelomer alcohols, fluorotelomer acids, and short- and long-chain perfluorinated acids in water and biota, *J. Chromatogr. A*, 1093, 89, 2005.
150. Tseng, C.L. et al., Analysis of perfluorooctanesulfonate and related fluorochemicals in water and biological tissue samples by liquid chromatography-ion trap mass spectrometry, *J. Chromatogr. A*, 1105, 119, 2006.

151. Tittlemier, S.A. et al., Dietary exposure of Canadians to perfluorinated carboxylates and perfluorooctane sulfonate via consumption of meat, fish, fast foods, and food items prepared in their packaging, *J. Agric. Food Chem.*, 55, 3203, 2007.

152. Morris, S. et al., Determination of the brominated flame retardant, hexabromocyclodocane, in sediments and biota by liquid chromatography-electrospray ionisation mass spectrometry, *Trends Anal. Chem.*, 25, 343, 2006.

153. de Boer, J. et al., Method for the analysis of polybrominated diphenylethers in sediments and biota, *Trends Anal. Chem.*, 20, 591, 2001.

154. Eljarrat, E. et al., Occurrence and bioavailability of polybrominated diphenyl ethers and hexabromocyclododecane in sediment and fish from the Cinca River, a tributary of the Ebro River (Spain), *Environ. Sci. Technol.*, 38, 2603, 2004.

155. Bayen, S., Lee, H.K., and Obbard, J.P., Determination of polybrominated diphenyl ethers in marine biological tissues using microwave-assisted extraction, *J. Chromatogr. A*, 1035, 291, 2004.

156. Basu, N., Scheuhammer, A.M., and O'Brien, M., Polychlorinated biphenyls, organochlorinated pesticides, and polybrominated diphenyl ethers in the cerebral cortex of wild river otters (*Lontra canadensis*), *Environ. Pollut.*, 2007, doi:10.1016/j.envpol.2006.12.026.

157. Ingelido, A.M. et al., Polychlorinated biphenyls (PCBs) and polybrominated diphenyl ethers (PBDEs) in milk from Italian women living in Rome and Venice, *Chemosphere*, 67, S301, 2007.

158. Pirard, C., De Pauw, E., and Focant, J.F., New strategy for comprehensive analysis of polybrominated diphenyl ethers, polychlorinated dibenzo-*p*-dioxins, polychlorinated dibenzofurans and polychlorinated biphenyls by gas chromatography coupled with mass spectrometry, *J. Chromatogr. A*, 998, 169, 2003.

159. Cochran, J.W., and Frame, G.M., Recent developments in the high-resolution gas chromatography of polychlorinated biphenyls, *J. Chromatogr. A*, 843, 323, 1999.

160. Galceran, M.T., Santos, F.J., Barceló, D., and Sanchez, J., Improvements in the separation of polychlorinated biphenyl congeners by high-resolution gas chromatography: application to the analysis of two mineral oils and powdered milk, *J. Chromatogr. A*, 655, 275, 1993.

161. Chernetsova, E.S. et al., Determination of polychlorinated dibenzo-p-dioxins, dibenzofurans, and biphenyls by gas chromatography/mass spectrometry in the negative chemical ionization mode with different reagent gases, *Mass. Spectrom. Rev.*, 21, 373, 2002.

162. Verenitch, S.S. et al., Ion-trap tandem mass spectrometry-based analytical methodology for the determination of polychlorinated biphenyls in fish and shellfish: performance comparison against electron-capture detection and high-resolution mass spectrometry detection, *J. Chromatogr. A*, 1142, 199, 2007.

163. Gómara, B. et al., Feasibility of gas chromatography– ion trap tandem mass spectrometry for the determination of polychlorinated biphenyls in food, *J. Sep. Sci.*, 123, 2006.

164. Björklund, J. et al., Influence of the injection technique and the column system on gas chromatographic determination of polybrominated diphenyl ethers, *J. Chromatogr. A*, 1041, 201, 2004.

165. Korytár, P. et al., Retention-time database of 126 polybrominated diphenyl ether congeners and two Bromkal technical mixtures on seven capillary gas chromatographic columns, *J. Chromatogr. A*, 1065, 239, 2005.

166. Thomsen, C. et al., Comparing electron ionization high-resolution and electron capture low-resolution mass spectrometric determination of polybrominated diphenyl ethers in plasma, serum and milk, *Chemosphere*, 46, 641, 2002.

167. Wang, D. et al., Gas chromatography/ion trap mass spectrometry applied for the determination of polybrominated diphenyl ethers in soil, *Rapid. Commun. Mass Spectrom.*, 19, 83, 2005.

168. Gómara, B. et al., Quantitative analysis of polybrominated diphenyl ethers in adipose tissue, human serum and foodstuff samples by gas chromatography with ion trap tandem mass spectrometry and isotope dilution, *Rapid Commun. Mass Spectrom.*, 20, 69, 2006.

169. Petinal, C.S. et al., Headspace solid-phase microextraction gas chromatography tandem mass spectrometry for the determination of brominated flame retardants in environmental solid samples, *Anal. Bioanal. Chem.*, 385, 637, 2006.

170. Yusà, V. et al., Optimization of a microwave-assisted extraction large-volume injection and gas chromatography–ion trap mass spectrometry procedure for the determination of polybrominated diphenyl ethers, polybrominated biphenyls and polychlorinated naphthalenes in sediments, *Anal. Chim. Acta*, 557, 304, 2006.

171. Martin, J.W. et al., Analytical challenges hamper perfluoroalkyl research, *Environ. Sci. Technol.*, 38, 248A, 2004.

172. van Leeuwen, S.P.J., and de Boer, J., Extraction and clean-up strategies for the analysis of poly- and perfluoroalkyl substances in environmental and human matrices, *J. Chromatogr. A*, 1153, 172, 2007.

173. de Voogt, P., and Saez, M., Analytical chemistry of perfluoroalkylated substances, *Trends Anal. Chem.*, 25, 326, 2006.

174. Ylinen, M. et al., Quantitative gas chromatographic determination of perfluorooctanoic acid as the benzyl ester in plasma and urine, *Arch. Environ. Contam. Toxicol.*, 14, 713, 1985.

175. Giesy, J.P., and Kannan, K., Global distribution of perfluorooctane sulfonate in wildlife, *Environ. Sci. Technol.*, 35, 1339, 2001.

176. De Silva, A., and Mabury, S., Isolating isomers of perfluorocarboxylates in polar bears (ursus maritimus) from two geographical locations, *Environ. Sci. Technol.*, 38, 6538, 2004.

177. Powley, C.R., Ryan, T.W., and George, S.W., Matrix-effect free analytical methods for determination of perfluorinated carboxylic acids in environmental and biological samples, Proceeding in "Fourth SETAC World Congress, 25th Annual Meeting in North America", Portland, Oregon, USA, 2004.

178. Powley, C.R. et al., Matrix effect-free analytical methods for determination of perfluorinated carboxylic acids in environmental matrixes, *Anal. Chem.*, 77, 6353, 2005.

179. Langlois, I., and Oehme, M., Structural identification of isomers present in technical perfluorooctane sulfonate by tandem mass spectrometry, *Rapid Commun. Mass Spectrom.*, 20, 844, 2006.

180. Tittlemier, S.A. et al., Development and characterization of a solvent extraction-gas chromatographic/mass spectrometric method for the analysis of perfluorooctane sulfonamide compounds in solid matrices, *J. Chromatogr. A*, 1066, 189, 2005.

181. Van Leeuwen, S. et al., First worldwide interlaboratory study on perfluorinated compounds in human and environmental matrices, Joint report of the Netherlands Institute for Fisheries Research (ASG-RIVO) (The Netherlands), Man-Technology-Environment (MTM) Research Centre, Sweden, and Institute of Water Technology Laboratory, Malta, 2005.

182. Fluoros Report 2006, 2nd Worldwide Interlaboratory Study on PCFs. December 2006.

183. Behnisch, P.A., Hosoea, K., and Sakai, S., Bioanalytical screening methods for dioxins and dioxin-like compounds—a review of bioassay/biomarker technology, *Environ. Int.*, 27, 413, 2001.

184. Płaza, G., Ulfig, K., and Tien, A.J., Immunoassays and environmental studies, *Polish J. Environ. Studies*, 9, 231, 2000.

185. Scippo, M.L. et al., DR-CALUX® screening of food sample: evaluation of the quantitative approach to measure dioxin, furans and dioxin-like PCBs, *Talanta*, 63, 1193, 2004.

186. U.S. EPA, Guidance on Cumulative Risk Assessment of Pesticide Chemicals That Have a Common Mechanism of Toxicity, Office of Pesticide Programs U.S. Environmental Protection Agency, Washington, D.C. 20460 January 14, 2002, available at http://www.epa.gov/pesticides/trac/science/cumulative_guidance.pdf.

187. Hoogenboom, L. et al., The CALUX bioassay: current status of its application to screening food and feed, *Trends Anal. Chem.*, 25, 410, 2006.

188. U.S. EPA, Interim Report on the Evolution and Performance of the Eichrom Technologies Procept® Rapid Dioxin Assay for Soil and Sediment Samples, available at http://costperformance.org/monitoring/pdf/epa_eichrom_dioxin_assay.pdf.

189. Fochi, I. et al., Modeling of DR CALUX® bioassay response to screen PCDDs, PCDFs, and dioxin-like PCBs in milk from dairy herds, *Chemosphere*, submitted.

190. Schroijen, C. et al., Study of the interference problems of dioxin-like chemicals with the bio-analytical method CALUX, *Talanta*, 63, 1261, 2004.

Chapter 41

Growth Promoters

Milagro Reig and Fidel Toldrá

Contents

41.1 Introduction

Growth promoters include a wide range of substances that are generally used in farm animals for therapeutic and prophylactic purposes. These substances can be administered through the feed or the drinking water. In some cases, the residues may proceed from contaminated animal feedstuffs.[1] Anabolic promoters have been administered in the United States to meat-producing animals where estradiol, progesterone, and testosterone are some of the allowed substances. The regulations in 21 Code of Federal Regulations (CFR) Part 556 provide the acceptable concentrations of residues of approved new animal drugs that may remain in edible tissues of treated animals.[2] Other countries allowing the use of certain growth promoters are Canada, Mexico, Australia, and New Zealand. However, the use of growth promoters is officially banned in the European Union since 1988 due to concerns about harmful effects on consumers.[3]

Growth promoters increase growth rate and improve efficiency of feed utilization and thus, contribute to the increase in protein deposition that is usually linked to fat utilization, which means a reduction in the fat content in the carcass and an increase in meat leanness.[4] In addition, some fraudulent practices consist in the use of low amounts of several substances such as β-agonists (clenbuterol) and corticosteroids (dexamethasone) and anabolic steroids, mixtures known as "cocktails," that have a synergistic effect and exert growth promotion but make their analytical detection more difficult.

The presence of residues of growth promoters or their metabolites in meat and their associated harmful health effects on humans make necessary the continuous improvement of analytical methodologies to guarantee consumer protection. The use of veterinary drugs in food animal species is strictly regulated in the European Union and, in fact, only some of them can be permitted for specific therapeutic purposes under strict control and administration by a veterinarian.[5]

Sanitary authorities in different countries are concerned about the presence of residues of veterinary drugs or their metabolites in meat because they may exert some adverse toxic effects on consumers' health. The European Food Safety Authority has recently issued an opinion about substances with hormonal activity, specifically testosterone and progesterone, as well as trenbolone acetate, zeranol, and melengestrol acetate. The exposure to residues of the hormones used as growth promoters could not be quantified. Although epidemiological data in the literature provided evidence for an association between some forms of hormone-dependent cancers and red meat consumption, the contribution of residues of hormones in meat could not be assessed.[6] Other substances such as β-agonists have shown adverse effects on consumers. This was evident in the case of intoxications in Italy, with symptoms described as gross tremors of the extremities, tachycardia, nausea, headaches, and dizziness, after consumption of lamb and bovine meat containing residues of clenbuterol.[7]

Meat quality is also affected by the use of substances used as growth promoters.[4] The connective tissue production is increased and collagen cross links at a higher rate giving a tougher meat,[8–10] whereas muscle proteases responsible for protein breakdown in postmortem meat are inhibited.[8,11] The lipolysis rate and the breakdown of triacylglycerols are accelerated.[12,13] The use of anabolic steroids reduces marbling and tenderness and may have a negative effect on palatability.[14] The "aggressive" use of anabolic implants may compromise the quality grades of beef carcasses and increase the incidence of dark cutting carcasses.[14] When cocktails of clenbuterol and dexamethasone are used, the meat quality is also affected; but it has been reported to be less tough than when using clenbuterol alone.[15]

Meat must be monitored for the presence of residues of veterinary drugs. Control strategies also include sampling at farm, which helps in prevention before animals reach the slaughterhouse. The samples include hair and urine as well as feed and water. This chapter reports the important strategies for the control of growth promoters as part of the wide range of residues of veterinary drugs in meat. The analysis of antibiotic residues is discussed in Chapter 42.

Table 41.1   Lists of Substances Having Anabolic Effect Belonging to Groups A and B According to Council Directive 96/23/EC[16]

Group A: Substances having anabolic effect
   1. Stilbenes
   2. Antithyroid agents
   3. Steroids
      Androgens
      Gestagens
      Estrogens
   4. Resorcyclic acid lactones
   5. Beta-agonists
   6. Other compounds
Group B: Veterinary drugs
   1. Antibacterial substances
      Sulfonamides and quinolones
   2. Other veterinary drugs
      a. Antihelmintics
      b. Anticoccidials, including nitroimidazoles
      c. Carbamates and pyrethroids
      d. Sedatives
      e. Nonsteroidal antiinflammatory drugs
      f. Other pharmacologically active substances

41.2   Control of Growth Promoters

The monitoring of residues of substances having hormonal or thyreostatic action as well as β-agonists is regulated in the European Union through the Council Directive 96/23/EC[16] on measures to monitor certain substances and residues in live animals and animal products. The European Union Member States have set up national monitoring programs and sampling procedures following this directive.

The major veterinary drugs and substances with anabolic effect are listed in Table 41.1, where group A includes unauthorized substances that have anabolic effect and group B includes veterinary drugs some of which have established maximum residue limits (MRLs). Commission Decisions 93/256/EC[17] and 93/257/EC[18] gave criteria for the analytical methodology regarding the screening, identification, and confirmation of these residues. Council Directive 96/23/EC[16] was implemented by the Commission Decision 2002/657/EC,[19] which is in force since September 1, 2002. This directive provides rules for the analytical methods to be used in testing of official samples and specific criteria for the interpretation of analytical results of official control laboratories for such samples. This means that when using mass spectrometric detection, substances in group A would require four identification points whereas those in group B would only require a minimum of three. The relative retention of the analyte must correspond to that of the calibration solution at a tolerance of $\pm 0.5\%$ for gas chromatography (GC) and $\pm 2.5\%$ for liquid chromatography (LC). The guidelines given in this new directive also imply new concepts such as the decision limit (CCα) or the detection capability (CCβ) that are briefly defined in Table 41.2. Both limits permit the daily control of the performance of a specific method qualified when used with a specific

Table 41.2 Definitions of Main Performance Criteria and Other Requirements for Analytical Methods[19]

Term	Definition
Decision limit (CCα)	It is defined as the limit at and above which it can be concluded with an error probability of α that a sample is noncompliant
Detection capability (CCβ)	It is the smallest content of the substance that may be detected, identified, and quantified in a sample with an error probability of β
Minimum required performance limit (MRPL)	It means the minimum content of an analyte in a sample, which at least has to be detected and confirmed
Precision	The closeness of agreement between independent test results obtained under stipulated conditions
Recovery	The percentage of the true concentration of a substance recovered during the analytical procedure
Reproducibility	Conditions where test results are obtained within the same method on identical test items in different laboratories with different operators using different equipment
Specificity	Ability of a method to distinguish between the analyte being measured and other substances
Ruggedness	Susceptibility of an analytical method to changes in experimental conditions that can be expressed as a list of the sample materials, analytes, storage conditions, environmental, and sample preparation conditions under which the method can be applied as presented or with specified minor conditions
Interlaboratory study	Organization, performance, and evaluation of tests on the same sample by two or more laboratories in accordance with predetermined conditions to determine testing performance
Within-laboratory reproducibility	Precision obtained in the same laboratory under stipulated conditions

instrument and under specific laboratory conditions, and thus contribute to the determination of the level of confidence in the routine analytical result.

41.3 Sampling and Sample Preparation

41.3.1 Samples from Animal Farms

41.3.1.1 Water

It is the easiest sample because it requires no general treatment. It does not require homogenization, just mild centrifugation to remove any suspended particle before further analysis.

41.3.1.2 Feed

The sample must be representative, especially taking into account the heterogeneous nature of most feeds. Thus, feeds must be milled and well homogenized before sampling. Adequate liquid extraction and solid-phase extraction (SPE) are performed for sample cleanup and concentration.

41.3.1.3 Urine

As a fluid, it does not require homogenization and, after centrifugation, an aliquot is diluted with the buffer and pH adjusted to correct values. Many analytes form conjugates such as sulfates and glucuronides and must be hydrolyzed to release the free analyte. Enzymatic hydrolysis with the juice of *Helix pomatia*, which has sulfatase and β-glucuronidase, is a milder treatment that usually gives good results. Special caution must be taken if other types of hydrolysis are performed (i.e., an acidic or alkaline hydrolysis) because they might affect and degrade the analyte.

41.3.1.4 Hair

The possibility of using hair to detect the illegal addition of clenbuterol even after 3 weeks of withdrawal, which is undetectable in urine and tissues, has been reported.[20] The amount of clenbuterol increased up to 20 days in the washout period and then slightly decreased even though still detectable after 40 days.[21] It must be pointed out that black hairs accumulate more clenbuterol and steroids than colored hairs;[22,23] the residue has been detected at 23 weeks after treatment with clenbuterol.[24] The hair, previously cleaned with detergents (sodium dodecyl sulfate [SDS]), is extracted with methanol and evaporated to dryness. The residue is dissolved in phosphate buffer and immunoextracted by affinity chromatography using monoclonal antisalbutamol immunoglobulin (IgG) that displays cross-reactivity (75%) with clenbuterol.[25] The residue is purified with SPE and silyl derivatized for its analysis by GC-MS (mass spectrometry). Other residues such as 17β-estradiol-3-benzoate have also been detected in hair up to 2 weeks after administration. However, 17α-methyltestosterone and medroxyprogesterone acetate could not be detected in hair.[23] The confirmation was possible above 5 ng/g by using liquid chromatography with mass spectrometry (LC-MS/MS) detection.[26] The analysis of a wide range of steroid residues such as estrogens, resorcylic acid lactones, and stilbens has been recently reported. The hair was extracted with methanol before the acid hydrolysis followed by specific liquid–liquid extraction and SPE to get four different fractions that were analyzed separately.[27]

41.3.2 Meat Samples

Preparation procedures and handling of meat samples are very important to improve the sensitivity of the screening tests.[28] Typical procedures include cutting, blending, and homogenization of the meat in an appropriate buffer. Enzymatic digestion with proteases such as subtilisin may be alternatively performed. The homogenate is extracted with an organic solvent usually followed by an SPE for sample cleanup and concentration. Previously, the residues may be bound or conjugated (i.e., as sulfates or glucuronides) and need further cleavage by treatment with the juice of the snail *H. pomatia*, which has sulfatase and β-glucuronidase, to release the free analytes. Some authors prefer enzymatic digestion with subtilisin to release steroids as they state that using enzymatic hydrolysis with the juice of *H. pomatia* may not reflect the conjugated fraction of steroids.[29]

41.4   Methods for Cleanup of Growth Promoters and Their Residues

41.4.1   Extraction Procedures

Extraction is primarily performed to remove interfering substances while retaining most of the analyte. Extraction solvents must be carefully chosen for each analyte as determined by pH, polarity, and solubility in different solvents. For instance, polar extraction methods for the determination of anabolic steroids of beef are used because they avoid some cleanup problems when following nonpolar extraction, but they are insufficient. It has been reported that polar extraction followed by nonpolar extraction gives better results.[30] Supercritical fluid extraction of meat with unmodified supercritical CO_2 has also been used for certain residues such as steroids.[31]

Matrix solid-phase dispersion consists in the mechanical blending of the sample with a solid sorbent that progressively retains the analyte by hydrophobic and hydrophilic interactions. The solid matrix is then packed into a column and eluted with an adequate solvent.

SPE is extensively used for the isolation of a group or class of analytes. The type of extractant and cartridge depends on the target analyte.[32] Small cartridges (C18, C8, NH_2) are commercially available at reasonable prices and have low affinity and specificity but have high capacity. Furthermore, they can be performed in parallel and thus, they allow the simultaneous extraction of a large number of samples.

41.4.2   Immunoaffinity Chromatography

This type of chromatography is based on the antigen–antibody interaction, which is very specific for a particular residue. The columns are packaged with a specific antibody that is bound to the solid matrix, usually a gel. When the extract is injected, the analyte (antigen) is retained. These chromatographic columns are highly specific and are only limited by potential interferences (i.e., substances that may cross-react with the antibody) that must be checked. These columns are rather expensive and can only be reused a certain number of times. In any case, due to the nature of the specific antibody when preparing the immunosorbent material, an in-depth assessment is necessary before considering its use in a routine analytical method.[33]

41.4.3   Molecular Recognition

There are several methods based on molecular recognition mechanisms for cleanup. Molecular imprinted polymers (MIPs) have shown promising results for the isolation of low amounts of residues such as those found in meat. These are cross-linked polymers prepared in the presence of a template molecule like a β-agonist. When this template is removed, the polymer offers a binding site complementary to the template structure. MIPs have better stability than antibodies because they can support high temperatures, larger pH ranges, and a wide range of organic solvents. The choice of the appropriate molecule as template is the critical factor for a reliable analysis.[34] The extracted residues are then analyzed by LC-MS and have shown good quantitative results for cimaterol, ractopamine, clenproperol, clenbuterol, brombuterol, mabuterol, mapenterol, and isoxsurine but not for salbutamol and terbutaline.[35]

41.5 Screening Methods

The wide range of veterinary drugs and residues potentially present in a meat sample necessitates the use of screening procedures for routine monitoring. Screening methods are used to detect the presence of the suspect analyte in the sample at the level of interest. If the searched residue has a MRL, then the screening method can detect the residue below this limit. These controls are based on the screening of a large number of samples and thus must have a large throughput, low cost, and enough sensitivity to detect the analyte with a minimum of false negatives.[36,37] Compliant samples are accepted whereas the suspected noncompliant samples would be further analyzed using confirmatory methods. According to the Commission Decision 2002/657/EC,[19] the screening methods must be validated and have a detection capability (CCβ) with an error probability (β) less than 5%.

41.5.1 Immunological Techniques

Immunological techniques are specific for a given residue because they are based on the antigen–antibody interaction. The most well-known and extensively used technique is the enzyme-linked immunosorbent assay (ELISA). A wide range of assay kits with measurement based on color development are commercially available. The possibility of interferences by cross-reactions with other substances must be taken into account. Other immunological techniques are radioimmunoassay (RIA), based on the measurement of the radioactivity of the immunological complex;[38] dipsticks, based on membrane strips with the receptor ligands and measurement of the developed color;[39] or the use of luminiscence or fluorescence detectors.[40]

41.5.2 Biosensors

The need to screen a large number of meat samples in relatively short time has prompted the development of biosensors that are based on an immobilized antibody that interacts with the analyte in the sample and the optical or electronic detection of the resulting signal.[41,42] Biosensors can simultaneously detect residues of multiple veterinary drugs in a sample at a time[43,44] without the need for sample cleanup.[45] There are different types of biosensors, such as the surface plasmon resonance (SPR) that measures variations in the refractive index of the solution close to the sensor[46] and has been successfully applied to the detection of residues of different veterinary drugs,[47,48] or the biosensors based on the use of biochip arrays that are specific for a certain number of residues[49] and are also applied to the detection of residues.[50]

41.5.3 Chromatographic Techniques

High-performance thin-layer chromatography (HPTLC) has been successfully used for multiresidue screening purposes in meat. Samples are injected onto the plates and the residues eluted from the plate with the appropriate eluent. Once eluted, residues can be viewed under UV or fluorescent lights or visualized by spraying with a chromogenic reagent. HPTLC has been applied to meat to screen different residues in meat such as agonists,[51,52] nitroimidazol,[53] and thyreostatic drugs.[54,55]

GC and high-performance liquid chromatography (HPLC) are powerful separation techniques capable of separating the analyte from most of the interfering substances by varying the type of column and elution conditions.[37] In some cases, the analyte can be detected after appropriate derivatization.[56] In addition, these techniques can be used for multiresidue screening. The recent development of ultraperformance liquid chromatography systems and new types of columns with packagings of reduced size offer valuable improvements for residue detection such as a considerable reduction in elution times and the possibility of a larger number of samples per day.[37,57] This procedure has been applied to meat for detection of residues of a wide range of veterinary drugs,[58–62] anabolic steroids,[63,64] quinolone,[65] and corticosteroids.[66–69] Additional advantages of GC and HPLC are automation and the possibility to couple the chromatograph to mass spectrometry detectors for further confirmatory analysis. Recently, a rapid, specific, and highly sensitive multiresidue method has been reported for the determination of anabolic steroid residues in bovine, pork, and poultry meat.[29] The methodology involves enzymatic digestion, methanol extraction, and SPE for final purification. The detection is carried out with LC-MS/MS in both ESI$^+$ and ESI$^-$ with a CCα and CCβ below 0.5 ng/g, but the method shows good performance for qualitative screening but not for quantitation.[29]

41.6 Confirmatory Analytical Methods

Confirmatory methods are preferentially based on mass spectrometry because they provide direct information on the molecular structure of the suspect compound and thus an unambiguous identification and confirmation of the residue in meat. However, these methods are costly in terms of time, equipment, and chemicals. When the target analyte is clearly identified and quantified above the decision limit for a forbidden substance (i.e., substances of group A) or exceeding the MRL in the case of substances having an MRL, the sample is considered as noncompliant (unfit for human consumption). A suitable internal standard must be added to the test portion at the beginning of the extraction procedure. If no suitable internal standard is available, the identification of the analyte can be done by cochromatography. This consists in dividing the sample extract into two parts. The first part is injected into the chromatograph as such. The second part is mixed with the standard analyte to be detected and injected into the chromatograph. The amount of added standard analyte must be similar to the estimated amount of the analyte in the extract. Identification is easier for a limited number of target analytes and matrices of constant composition.[70]

GC with mass spectrometry detection has been used for many years even though derivatization (i.e., silyl or boronate derivatives) was required for some nonvolatile residues such as agonists. An example of some agonists such as boronate derivatives giving good identification ions is shown in Figure 41.1. But derivatization entails a serious limitation adding to time and cost of the analysis.

In recent years, the rapid development of mass spectrometry coupled to LC has expanded its applications in this field, especially for nonvolatile or thermolabile compounds. Tandem mass spectrometry (MS/MS) has shown high selectivity and sensitivity and thus allows the analysis of more complex matrices such as meat with easier sample preparation procedures. LC-MS/MS allows the selection of a precursor *m/z* that is performed first. This eliminates any

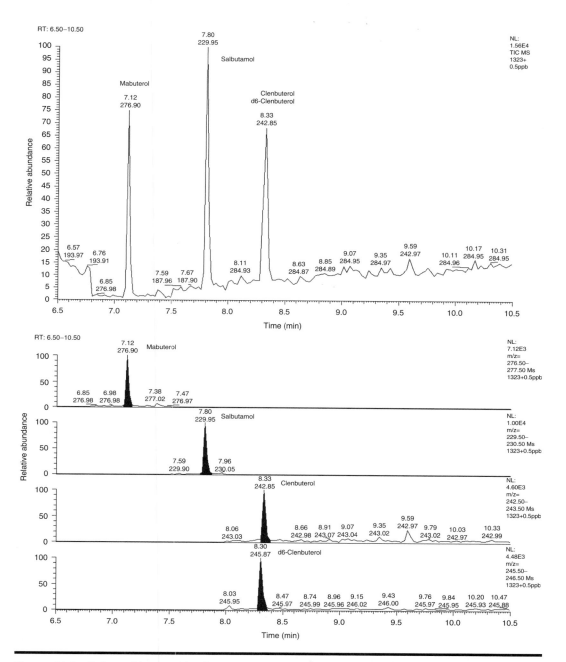

Figure 41.1 Selected ion monitoring (SIM) GC-MS chromatogram of bovine urine fortified with 0.5 ng/mL for clenbuterol, 0.5 ng/mL for mabuterol, 0.75 ng/mL for salbutamol, and 1 ng/mL for d6-clenbuterol as IS. (Reproduced from Reig, M. et al., *Anal. Chim. Acta*, 529, 293, 2005. With permission.)

uncertainty on the origin of the observed fragment ions, eliminates potential interferences from the meat sample or from the mobile phase, reduces the chemical noise, and increases the sensitivity.[71]

The interface technology has been rapidly developed. Electrospray ionization (ESI) and atmospheric pressure chemical ionization (APCI) interfaces are preferred depending on the polarity and molecular mass of analytes.[71] ESI ionization technique facilitates the analysis of small to relatively large and hydrophobic to hydrophilic molecules.[56,72,73] An important limitation of LC-MS/MS quantitative analysis is its susceptibility to matrix effect that is dependent on the ionization type, type of sample, and sample preparation. APCI ionization has been reported to be less sensitive than ESI to matrix effects.[74–77] ESI is preferred for the MS analysis of nonsteroidal antiinflammatory drugs (NSAIDs) due to their polar nature; however, some interfering substances of the matrix such as fat may lead to ion suppression problems.[78] The extraction of the analyte must be more selective and further purified and cleaned up.

A rapid qualitative method using online column-switching LC-MS/MS has been developed and validated for screening 13 target veterinary drugs in different animal muscles.[79] This system may reduce the cost and time for confirmatory analysis. A list of recent performance reports of the analysis of veterinary drug residues is shown in Table 41.3.

The ion suppression phenomenon in LC-MS must be taken into account because of matrix effect problems and the presence of interfering compounds that affect the analyte detection. A number of reviews about ion suppression phenomenon and its consequences for residue analysis has been recently published.[84] The major mechanism for ion suppression corresponds to the presence of matrix-interfering compounds that reduce the evaporation efficiency leading to reduced detection capability and repeatability. The ion ratios, linearity, and quantification are also affected. It could even lead to the lack of detection of an analyte or the underestimation of its concentration or the nonfulfilment of the identification criteria.[84] The prevention of this phenomenon includes an improved purification and cleanup of the sample as well as the use of an appropriate internal standard. Another strategy is to modify the elution conditions for the analytes to elute in an area nonaffected by ion suppression.[84]

According to the Commission Decision 2002/657/EC,[19] a system of identification points is used for confirmatory purposes with a minimum of 4 points required for the substances of group A and a minimum of 3 for group B substances. So, 1 identification point can be earned for the precursor ion with a triple quadrupole spectrometer and 1.5 points for each product ion. A high-resolution mass spectrometer acquires 2 identification points for the precursor ion and 2.5 for each product ion. Variable window ranges for MS peak abundances are also established in the new decision (EC 2002). So, the relative ion intensities must be >50, >20–50, >10–20, and ≤10%. In the case of electron impact-GC-MS (EI-GC-MS), the maximum permitted tolerances are ±10, ±15, ±20, and ±50%, respectively, whereas in the case of collision-induced GC-MS (CI-GC-MS), GC-MS^n, LC-MS, and LC-MS^n are ±20, ±25, ±30, and ±50%, respectively.

Other methods are allowed for group B substances.[19] So, liquid chromatography-full scan diode array detection (LC-DAD) can be used as a confirmatory method if specific requirements for absorption in UV spectrometry are met. This means that the absorption maxima of the spectrum of the analyte shall be at the same wavelengths of the calibration standard within a margin of ±2 nm for diode array detection. Furthermore, the spectrum of the analyte above 220 nm will not be visibly different (at no point greater than 10%) from the spectrum of the calibration standard.

Table 41.3 Performance of Some Recent Methods of Analysis of Growth Promoters in Antemortem (Farm Samples) and Postmortem (Meat) Samples

Analyte	Matrix	Extraction	Column	System/Detector	CCα (ng/g)	CCβ (ng/g)	Recovery (%)	Reference
17β-Estradiol-3-benzoate	Bovine hair	Methanol extraction, SPE NH₂	Nucleosil C18AB, 5 μm	LC-MS/MS, ESI⁺	LOD 4.1	LOI 5.0	—	23
Zeranol	Bovine hair	Methanol extraction, acid hydrolysis, SPE	OV-1, 0.25 μm	GC-MS/MS	LOD 2.66	LOI 4.48	—	27
17α-Trenbolone	Bovine hair	Methanol extraction, acid hydrolysis, SPE	OV-1, 0.25 μm	GC-MS/MS	LOD 0.76	LOI 1.99	—	27
	Bovine hair	Methanol extraction, acid hydrolysis, SPE	OV-1, 0.25 μm	GC-MS/MS	LOD 1.02	LOI 1.74	—	27
17-Estradiol	Bovine hair	Methanol extraction, acid hydrolysis, SPE	OV-1, 0.25 μm	GC-MS/MS	LOD 0.12	LOI 0.19	—	27
17α-Testosterone	Bovine hair	Methanol extraction, acid hydrolysis, SPE	OV-1, 0.25 μm	GC-MS/MS	LOD 0.29	LOI 0.85	—	27
Melengestrol	Bovine hair	Methanol extraction, acid hydrolysis, SPE	OV-1, 0.25 μm	GC-MS/MS	LOD 1.12	LOI 1.97	—	27
Hexestrol	Pork meat	Liquid extraction, SPE C18	DB-5, 30 m, 0.25 μm	GC-MS/MS	LOD 0.2	—	84.6	80
Diethylestilbestrol	Pork meat	Liquid extraction, SPE C18	DB-5, 30 m, 0.25 μm	GC-MS/MS	LOD 0.1	—	80.1	80
Androsterone	Pork meat	Liquid extraction, SPE C18	DB-5, 30 m, 0.25 μm	GC-MS/MS	LOD 0.2	—	91.0	80
Estradiol	Pork meat	Liquid extraction, SPE C18	DB-5, 30 m, 0.25 μm	GC-MS/MS	LOD 0.1	—	95.8	80

(Continued)

Table 41.3 (Continued)

Analyte	Matrix	Extraction	Column	Detector	CCα (ng/g)	CCβ (ng/g)	Recovery (%)	Reference
Zeranol	Pork meat	Liquid extraction, SPE C18	DB-5, 30 m, 0.25 μm	GC-MS/MS	LOD 0.1	—	95.3	80
α-Zearalenol	Pork meat	Liquid extraction, SPE C18	DB-5, 30 m, 0.25 μm	GC-MS/MS	LOD 0.1	—	94.6	80
17α-Hydroxyl-progesterone	Pork meat	Liquid extraction, SPE C18	DB-5, 30 m, 0.25 μm	GC-MS/MS	LOD 0.4	—	102.4	80
Diethylestilbestrol	Beef muscle	Solvent extraction, freezing-lipid filtration, SPE C8	DB-1MS, 30 m, 0.25 μm	GC-MS/MS	LOD 0.3	—	91.0	81
17β-Estradiol	Beef muscle	Solvent extraction, freezing-lipid filtration, SPE C8	DB-1MS, 30 m, 0.25 μm	GC-MS/MS	LOD 0.2	—	85.0	81
Testosterone	Beef muscle	Solvent extraction, freezing-lipid filtration, SPE C8	DB-1MS, 30 m, 0.25 μm	GC-MS/MS	LOD 0.1	—	100.0	81
Zeranol	Beef muscle	Solvent extraction, freezing-lipid filtration, SPE C8	DB-1MS, 30 m, 0.25 μm	GC-MS/MS	LOD 0.2	—	83.0	81
Progesterone	Beef muscle	Solvent extraction, freezing-lipid filtration, SPE C8	DB-1MS, 30 m, 0.25 μm	GC-MS/MS	LOD 0.3	—	80.0	81
Dexamethasone	Feed	Liquid extraction, SPE NH₂	Sinergy MAX-RP 80A, 4 μm	LC-DAD	190	217	108.9	69
Dexamethasone	Drinking water	Centrifugation			26	30	105.1	69
β-Boldenone glucuronide	Urine	Extraction, SPE	Nucleosil C18, 5 μm	LC-MS/MS, API⁻	0.40	0.55	75.0	82
β-Boldenone sulfates	Urine	Extraction, SPE	Nucleosil C18, 5 μm	LC-MS/MS, API⁻	0.75	0.99	72.0	82

β-Boldenone	Urine	Extraction, SPE	Nucleosil C18, 5 μm	LC-MS/MS, API−	0.52	0.70	76.0	82
α-Boldenone	Urine	Extraction, SPE	Nucleosil C18, 5 μm	LC-MS/MS, API−	0.70	0.93	71.0	82
5β-Androst-1en-17ol-3one	Urine	Extraction, SPE	Nucleosil C18, 5 μm	LC-MS/MS, API−	0.42	0.56	79.0	82
Trenbolone	Poultry muscle	Matrix solid-phase dispersion	Alltima C18, 5 μm	LC-MS/MS, APCI+		0.13	99	83
Testosterone	Poultry muscle	Matrix solid-phase dispersion	Alltima C18, 5 μm	LC-MS/MS, APCI+	0.03	0.21	97	83
Melengestrol acetate	Poultry muscle	Matrix solid-phase dispersion	Alltima C18, 5 μm	LC-MS/MS, APCI+	0.03	0.26	90	83
Progesterone	Poultry muscle	Matrix solid-phase dispersion	Alltima C18, 5 μm	LC-MS/MS, APCI+	0.21	0.16	96	83
α-Zeranol	Poultry muscle	Matrix solid-phase dispersion	Alltima C18, 5 μm	LC-MS/MS, TIS−	0.08	0.87	90	83
α-Estradiol	Poultry muscle	Matrix solid-phase dispersion	Alltima C18, 5 μm	LC-MS/MS, TIS−	0.11	0.85	100	83
Diethylestilbestrol	Poultry muscle	Matrix solid-phase dispersion	Alltima C18, 5 μm	LC-MS/MS, TIS−	0.04	0.33	80	83

Note: ESI, electrospray ionization; LOD, limit of detection; LOI, limit of identification; APCI, atmospheric pressure chemical ionization; TIS, turbo ion spray.

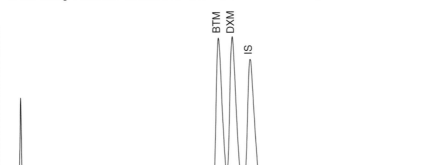

Figure 41.2 Example of detection of dexamethasone, a substance in group B 2f, and closely related substances in feed through LC-DAD. Retention times 10.14, 10.78, and 11.60 min corresponded to betamethasone (BTM), dexamethasone (DXM), and flumethasone (IS), respectively. (Reproduced from Reig, M. et al., *Meat Sci.* 74, 676, 2006. With permission.)

An example of identification of dexamethasone, a substance in group B 2f, in feed through LC-DAD is shown in Figure 41.2.

References

1. McEvoy, J.D.G. Contamination of animal feedstuffs as a cause of residues in food: a review of regulatory aspects, incidence and control. *Anal. Chim. Acta* 473: 3–26, 2002.
2. Brynes, S.D. Demystifying 21 CFR Part 556—Tolerances for residues of new animal drugs in food. *Regul. Toxicol. Pharmacol.* 42: 324–327, 2005.
3. Council Directive 88/146/EEC of March 7, 1988, prohibiting the use in livestock farming of certain substances having a hormonal action. *Off. J. Eur. Commun.* L070: 16, 1988.
4. Lone, K.P. Natural sex steroids and their xenobiotic analogs in animal production: growth, carcass quality, pharmacokinetics, metabolism, mode of action, residues, methods, and epidemiology. *Crit. Rev. Food Sci. Nutr.* 37: 93–209, 1997.
5. Van Peteguem, C. and Daeselaire, E. Residues of growth promoters. In: L.M.L. Nollet (Ed.), *Handbook of Food Analysis*, 2nd ed. New York, NY: Marcel Dekker, 1037–1063, 2004.
6. EFSA. Opinion of the scientific panel on contaminants in the food chain on a request from the European Commission related to hormone residues in bovine meat and meat products. *EFSA J.* 510: 1–62, 2007.
7. Barbosa, J. et al. Food poisoning by clenbuterol in Portugal. *Food Addit. Contam.* 22: 563–566, 2005.
8. Moloney, A. et al. Influence of beta-adrenergic agonists and similar compounds on growth. In: A.M. Pearson and T.R. Dutson (Eds.), *Growth Regulation in Farm Animals*. London, UK: Elsevier Applied Science, 455–513, 1991.
9. Miller, L.F. et al. Relationships among intramuscular collagen, serum hydroxyproline and serum testosterone in growing rams and wethers. *J. Anim. Sci.* 67: 698, 1989.

10. Miller, L.F., Judge, M.D., and Schanbacher, B.D. Intramuscular collagen and serum hydroxyproline as related to implanted testosterone and estradiol 17β in growing wethers. *J. Anim. Sci.* 68: 1044, 1990.

11. Fiems, L.O. et al. Effect of a β-agonist on meat quality and myofibrillar protein fragmentation in bulls. *Meat Sci.* 27: 29–35, 1990.

12. Duquette, P.F. and Muir, L.A. Effect of the β-adrenergic agonists isoproterenol, clenbuterol, L-640,033 and BRL 35135 on lipolysis and lipogenesis in rat adipose tissue *in vitro. J. Anim. Sci.* 61(Suppl. 1): 265, 1985.

13. Brockman, R.P. and Laarveld, R. Hormonal regulation of metabolism in ruminants. Review. *Livest. Prod. Sci.* 14: 313–317, 1986.

14. Dikeman, M.E. Effects of metabolic modifiers on carcass traits and meat quality. *Meat Sci.* 77: 121–135, 2007.

15. Monsón, F. et al. Carcass and meat quality of yearling bulls as affected by the use of clenbuterol and steroid hormones combined with dexamethasone. *J. Muscle Foods* 18: 173–185, 2007.

16. Council Directive 96/23/EC of April 29, 1996, on measures to monitor certain substances and residues thereof in live animals and animal products. *Off. J. Eur. Commun.* L125: 10, 1996.

17. Commission Decision 93/256/EC of May 14, 1993, laying down the methods to be used for detecting residues of substances having hormonal or a thyreostatic action. *Off. J. Eur. Commun.* L118: 64, 1993.

18. Commission Decision 93/257/EC of April 15, 1993, laying down the reference methods and the list of the national reference laboratories for detecting residues. *Off. J. Eur. Commun.* L118: 73, 1993.

19. Commission Decision 2002/657/EC of August 17, 2002, implementing Council Directive 96/23/EC concerning the performance of the analytical methods and the interpretation of results. *Off. J. Eur. Commun.* L221: 8, 2002.

20. Appelgren, L.-E. et al. Analysis of hair samples for clenbuterol in calves. *Fleischwirtsch.* 76: 398–399, 1996.

21. Panoyan, A. et al. Immunodetection of clenbuterol in the hair of calves. *J. Agric. Food Chem.* 43: 2716–2718, 1995.

22. Gleixner, A. and Meyer, H.H.D. Hair analysis for monitoring the use of growth promoters in meat production. *Fleichswirtsch.* 76: 637–638, 1996.

23. Rambaud, L. et al. Study of 17-estradiol-3-benzoate, 17β-methyltestosterone and medroxyprogesterone acetate fixation in bovine hair. *Anal. Chim. Acta* 532: 165–176, 2005.

24. Haasnoot, W. et al. A fast immunoassay for the screening of β-agonists in hair. *Analyst* 123: 2707–2710, 1998.

25. Adam, A. et al. Detection of clenbuterol residues in hair. *Analyst* 119: 2663–2666, 1994.

26. Hooijerink, H. et al. Liquid chromatography-electrospray ionisation-mass spectrometry based method for the determination of estradiol benzoate in hair of cattle. *Anal. Chim. Acta* 529: 167–172, 2005.

27. Rambaud, L. et al. Development and validation of a multi-residue method for the detection of a wide range of hormonal anabolic compounds in hair using gas chromatography–tandem mass spectrometry. *Anal. Chim. Acta* 586: 93–104, 2007.

28. McCracken, R.J., Spence, D.E., and Kennedy, D.G. Comparison of extraction techniques for the recovery of veterinary drug residues from animal tissues. *Food Addit. Contam.* 17: 907–914, 2000.

29. Blasco, C., Poucke, C.V., and Peteghem, C.V. Analysis of meat samples for anabolic steroids residues by liquid chromatography/tandem mass spectrometry. *J. Chromatogr. A* 1154: 230–239, 2007.

30. Schmidt, G. and Steinhart, H. Impact of extraction solvents on steroid contents determined in beef. *Food Chem.* 76: 83–88, 2002.

31. Stolker, A.A.M., Zoonties, P.W., and Van Ginkel, L.A. The use of supercritical fluid extraction for the determination of steroids in animal tissues. *Analyst* 123, 2671–2676, 1998.

32. Stubbings, G. et al. A multi-residue cation-exchange clean up procedure for basic drugs in produce of animal origin. *Anal. Chim. Acta* 547: 262–268, 2005.

33. Godfrey, M.A.J. Immunoaffinity extraction in veterinary residue analysis: a regulatory viewpoint. *Analyst* 123: 2501–2506, 1998.

34. Wistrand, C. et al. Evaluation of MISPE for the multi-residue extraction of β-agonists from calves urine. *J. Chromatogr. B* 804: 85–91, 2004.

35. Kootstra, P.R. et al. The analysis of beta-agonists in bovine muscle using molecular imprinted polymers with ion trap LCMS screening. *Anal. Chim. Acta* 529: 75–81, 2005.

36. Toldrá, F. and Reig, M. Methods for rapid detection of chemical and veterinary drug residues in animal foods. *Trends Food Sci. Technol.* 17: 482–489, 2006.

37. Reig, M. and Toldrá, F. Veterinary drug residues in meat: concerns and rapid methods for detection. *Meat Sci.* 78: 60–67, 2008.

38. Samarajeewa, U. et al. Application of immunoassay in the food industry. *Crit. Rev. Food Sci. Nutr.* 29: 403–434, 1991.

39. Link, N., Weber, W., and Fussenegger, M. A novel generic dipstick-based technology for rapid and precise detection of tetracycline, streptogramin and macrolide antibiotics in food samples. *J. Biotechnol.* 128: 668–680, 2007.

40. Roda, A. et al. A rapid and sensitive 384-well microtitre format chemiluminescent enzyme immunoassay for 19 nortestosterone. *Luminescence* 18: 72–78, 2003.

41. Patel, P.D. Biosensors for measurement of analytes implicated in food safety: a review. *Trends Food Sci. Technol.* 21: 96–115, 2002.

42. White, S. Biosensors for food analysis. In: L.M.L. Nollet (Ed.), *Handbook of Food Analysis*, 2nd ed. New York, NY: Marcel Dekker, 2133–2148, 2004.

43. Gründig, B. and Renneberg, R. Chemical and biochemical sensors. In: A. Katerkamp, B. Gründig, and R. Renneberg (Eds.), *Ullmann's Encyclopedia of Industrial Chemistry*. Verlag Wiley-VCH, Berlin, Germany, 87–98, 2002.

44. Franek, M. and Hruska, K. Antibody based methods for environmental and food analysis: a review. *Vet. Med. Czech.* 50: 1–10, 2005.

45. Elliott, C.T. et al. Use of biosensors for rapid drug residue analysis without sample deconjugation or clean-up: a possible way forward. *Analyst* 123: 2469–2473, 1998.

46. Gillis, E.H., Gosling, J.P., Sreenan, J.M., and Kane, M. Development and validation of a biosensor-based immunoassay for progesterone in bovine milk. *J. Immunol. Methods* 267: 131–138, 2002.

47. Bergweff, A.A. Rapid assays for detection of residues of veterinary drugs. In: A. van Amerongen, D. Barug, and M. Lauwars (Eds.), *Rapid Methods for Biological and Chemical Contaminants in Food and Feed*. Wageningen Academic Publishers, Wageningen, The Netherlands, 259–292, 2005.

48. Haughey, S.A. and Baxter, C.A. Biosensor screening for veterinary drug residues in foodstuffs. *J. AOAC Int.* 89: 862–867, 2006.

49. Johansson, M.A. and Hellenas, K.E. Sensor chip preparation and assay construction for immunobiosensor determination of beta-agonists and hormones. *Analyst* 126: 1721–1727, 2001.

50. Zuo, P. and Ye, B.C. Small molecule microarrays for drug residue detection in foodstuffs. *J. Agric. Food Chem.* 54: 6978–6983, 2006.

51. Degroodt, J.-M. et al. Clenbuterol residue analysis by HPLC–HPTLC in urine and animal tissues. *Z. Lebens. Unters. Forsch.* 189: 128–131, 1989.

52. Degroodt, J.-M. et al. Cimaterol and clenbuterol residue analysis by HPLC–HPTLC. *Z. Lebens. Unters. Forsch.* 192: 430–432, 1991.

53. Gaugain, M. and Abjean, J.P. High-performance thin-layer chromatographic method for the fluorescence detection of three nitroimidazole residues in pork and poultry tissue. *J. Chromatogr. A* 737: 343–346, 1996.

54. De Brabender, H.F., Batjoens, P., and Van Hoof, V. Determination of thyreostatic drugs by HPTLC with confirmation by GC-MS. *J. Planar Chromatogr.* 5: 124–130, 1992.

55. De Wasch, K. et al. Confirmation of residues of thyreostatic drugs in thyroid glands by multiple mass spectrometry after thin-layer chromatography. *J. Chromatogr. A* 819: 99–111, 1998.

56. Bergweff, A.A. and Schloesser, J. Residue determination. In: B. Caballero, L. Trugo, and P. Finglas (Eds.) *Encyclopedia of Food Sciences and Nutrition*, 2nd ed. London, UK: Elsevier, 254–261, 2003.

57. Aristoy, M.C., Reig, M., and Toldrá, F. Rapid liquid chromatography techniques for detection of key (bio)chemical markers. In: L.M.L. Nollet and F. Toldrá (Eds.), *Advances in Food Diagnostics.* Ames, IA: Blackwell Publishing, 229–251, 2007.

58. Cooper, A.D. et al. Development of multi-residue methodology for the HPLC determination of veterinary drugs in animal-tissues. *Food Addit. Contam.* 12: 167–176, 1995.

59. Aerts, M.M.L., Hogenboom, A.C., and Brinkman, U.A.T. Analytical strategies for the screening of veterinary drugs and their residues in edible products. *J. Chromatogr. B Biomed. Appl.* 667: 1–40, 1995.

60. Horie, M. et al. Rapid screening method for residual veterinary drugs in meat and fish by HPLC. *J. Food Hyg. Soc. Jpn.* 39: 383–389, 1998.

61. Kao, Y.M. et al. Multiresidue determination of veterinary drugs in chicken and swine muscles by high performance liquid chromatography. *J. Food Drug Anal.* 9: 84–95, 2001.

62. Reig, M. et al. Stability of β-agonist methyl boronic derivatives before GC-MS analysis. *Anal. Chim. Acta* 529: 293–297, 2005.

63. Gonzalo-Lumbrearas, R. and Izquierdo-Hornillos, R. High-performance liquid chromatography optimization study for the separation of natural and synthetic anabolic steroids. Application to urine and pharmaceutical samples. *J. Chromatogr. B* 742: 1–11, 2000.

64. De Cock, K.J.S. et al. Detection and determination of anabolic steroids in nutritional supplements. *J. Pharm. Biomed. Anal.* 25: 843–852, 2001.

65. Verdon, E., Hurtaud-Pessel, D., and Sanders, P. Evaluation of the limit of performance of an analytical method based on a statistical calculation of its critical concentrations according to ISO standard 11843: application to routine control of banned veterinary drug residues in food according to European Decision 657/2002/EC. *Accred. Qual. Assur.* 11: 58–62, 2006.

66. Shearan, P., O'Keefe, M., and Smyth, M. Reversed-phase high-performance liquid chromatographic determination of dexamethasone in bovine tissues. *Analyst* 116, 1365–1368, 1991.

67. Mallinson, E.T. et al. Determination of dexamethasone in liver and muscle by liquid chromatography and gas chromatography/mass spectrometry. *J. Agric. Food Chem.* 43: 140–145, 1995.

68. Stolker, A.A.M. et al. Comparison of different liquid chromatography methods for the determination of corticosteroids in biological matrices. *J. Chromatogr. A* 893, 55–67, 2000.

69. Reig, M. et al. A chromatography method for the screening and confirmatory detection of dexamethasone. *Meat Sci.* 74: 676–680, 2006.

70. Milman, B.L. Identification of chemical compounds. *Trends Anal. Chem.* 24: 493–508, 2005.

71. Gentili, A., Perret, D., and Marchese, S. Liquid chromatography-tandem mass spectrometry for performing confirmatory analysis of veterinary drugs in animal-food products. *Trends Anal. Chem.*, 24: 704–733, 2005.

72. Hewitt, S.A. et al. Screening and confirmatory strategies for the surveillance of anabolic steroid abuse within Northern Ireland. *Anal. Chim. Acta* 473: 99–109, 2002.

73. Thevis, M., Opfermann, G., and Schänzer, W. Liquid chromatography/electrospray ionization tandem mass spectrometric screening and confirmation methods for β_2-agonists in human or equine urine. *J. Mass Spectrom.* 38, 1197–1206, 2003.

74. Dams, R. et al. Matrix effects in bio-analysis of illicit drugs with LC-MS/MS: influence of ionization type, sample preparation and biofluid. *J. Am. Soc. Mass Spectrom.* 14: 1290–1294, 2003.

75. Puente, M.L. Highly sensitive and rapid normal-phase chiral screen using high-performance liquid chromatography-atmospheric pressure ionization tandem mass spectrometry (HPLC/MS). *J. Chromatogr.* 1055: 55–62, 2004.

76. Maurer, H.H. et al. Screening for library-assisted identification and fully validated quantification of 22 beta-blockers in blood plasma by liquid chromatography-mass spectrometry with atmospheric pressure chemical ionization. *J. Chromatogr.* 1058: 169–181, 2004.

77. Turnipseed, S.B. et al. Analysis of avermectin and moxidectin residues in milk by liquid chromatography-tandem mass spectrometry using an atmospheric pressure chemical ionization/atmospheric pressure photoionization source. *Anal. Chim. Acta* 529: 159–165, 2005.

78. Gentili, A. LC–MS methods for analyzing anti-inflammatory drugs in animal-food products. *Trends Anal. Chem.* 26: 595–608, 2007.

79. Tang, H.P., Ho, C., and Lai, S.S. High-throughput screening for multi-class veterinary drug residues in animal muscle using liquid chromatography/tandem mass spectrometry with on-line solid-phase extraction. *Rapid Commun. Mass Spectrom.* 20: 2565–2572, 2006.

80. Fuh, M.-R. et al. Determination of residual anabolic steroid in meat by gas chromatography–ion trap–mass spectrometer. *Talanta* 64: 408–414, 2004.

81. Seo, J. et al. Simultaneous determination of anabolic steroids and synthetic hormones in meat by freezing-lipid filtration, solid-phase extraction and gas chromatography–mass spectrometry. *J. Chromatogr. A* 1067: 303–309, 2005.

82. Buiarelli, F. et al. Detection of boldenone and its major metabolites by liquid chromatography—tandem mass spectrometry in urine samples. *Anal. Chim. Acta* 552: 116–126, 2005.

83. Gentili, A. et al. High- and low-resolution mass spectrometry coupled to liquid chromatography as confirmatory methods of anabolic residues in crude meat and infant foods. *Rapid Commun. Mass Spectrom.* 20: 1845–1854, 2006.

84. Antignac, J.P. et al. The ion suppression phenomenon in liquid chromatography-mass spectrometry and its consequences in the field of residue analysis. *Anal. Chim. Acta* 529: 129–136, 2005.

Chapter 42

Antibiotic Residues in Muscle Tissues of Edible Animal Products

Eric Verdon

Contents

42.1   Introduction

42.1.1   *Antibiotics and Antibacterials in Veterinary Practice*

Antibiotics are considered as the most important class of drugs. They play a key role in controlling bacterial infections in both human and animals. The need of food for human consumption is growing and expansion of intensive livestock farming is of major concern. Antibiotics contribute, for a large part, to this industrialization of the farming practice (treatment and prevention of animal diseases and growth-promoting feed additives, even though the latter is being reassessed in some countries such as in the European Union [EU]). Their use in animal husbandry requires them to be on the top of the veterinary drug production for the pharmaceutical industry. Both the rational usage of these substances and the monitoring of their residual concentrations in animal products for human consumption would contribute to prevent their excessive content in human food, reducing the risks for human health. Analytical methods dedicated to monitor these substances in food-producing animal products for human consumption are one of the tools for food safety control. However, it is the mutual concern of all the actors involved in the human food supply—farmers, veterinarians, feed manufacturers, food industry, and regulatory agencies. They should create the conditions congenial to human food safety.

For a better understanding of the terms "antibiotics" and "antibacterials," it is essential to clarify their meaning and usage. Today, the term "antibiotic" is often wrongly used in the place of "antibacterial" or "antimicrobial" because not only antibiotics possess an antibacterial activity. In fact, according to an internationally recognized classification, the term "antibiotic" should strictly apply to a range of compounds that are of biological origin; some are produced metabolically from molds of filamentous fungi such as from *Penicillium* species (i.e., benzylpenicillin, 6-aminopenicillenic acid [6-APA], cephalosporin C, 7-aminocephalosporanic acid [7-ACA]), others are extracted from cultures of specific bacteria such as from several *Streptomyces* species (i.e., streptomycin, gentamicin, tetracycline, spiramycin), and many others are semisynthetic substances that are additionally modified by chemical synthesis (i.e., florfenicol, amoxycillin, cephalexin). The terms "antibacterial" or "antimicrobial" apply to a broader set of compounds including not only the natural and semisynthetic antibiotics, but also several other classes of molecules having an antibacterial property, besides those that are produced by chemical synthesis; quinolones, nitrofurans, nitroimidazoles, sulfonamides belong to this category. The compounds covered by the term "antibiotics" fall into seven categories: aminoglycosides, amphenicols, cephalosporins, macrolides, penicillins, polypeptides, and tetracyclines. The penicillins and the cephalosporins are frequently merged into the wider family called beta-lactams. According to the recognized classification, the synthetic compounds from the four families of nitrofurans, nitroimidazoles, quinolones, and sulfonamides can only be considered as antibacterials or antimicrobials but do not belong to the antibiotic class. Finally, there are several other compounds that are included in this chapter dedicated to antibiotics. Some of them are considered as subfamilies or quasifamilies to those described in the preceding discussion: lincosamides and cephamycins fall into this category. Other compounds are from less important veterinary drug families that also present some antibacterial activities: carbadox, dapsone, malachite green, olaquindox, novobiocin, and virginiamycin. In chemical terms, the collections of substances that exhibit antibiotic properties feature diverse groups characterized by very different molecular structures and bearing widely divergent functionalities and mode of operation. This diversity poses a tremendous challenge to the analyst when subtle structural variations in closely related antibiotic compounds can lead to large variations in the chemical toxicity and biological activity of the antibiotic.

42.1.2 Veterinary Drug Residue Regulatory Control for Food Safety

The administration of licensed veterinary antimicrobials to food-producing animals may lead to the occurrence of residues in the food, primarily in the meat produced for human consumption. With the increasing concern for the safety of human food supply, monitoring for animal drug residues has become an important regulatory issue. To safeguard human health, safe tolerance levels or maximum residue limits (MRLs) in food products from animal origin have been established in various countries around the world. In the EU, the establishment of MRLs is governed by Council Regulation 2377/90/EEC [1]. These limits account as part of the regulation for controlling the safety of food with regard to residues of veterinary drugs in tissues and fluids of animals entering the human food chain. To ensure that human food is entirely free from potentially harmful concentrations of residues, MRLs are calculated from toxicological data and with a safety margin ranging from a factor of 10 to 100, depending on the drug considered. The Regulation 2377/90/EEC establishes the lists of compounds that have a fixed MRL (Annex 1) or that need no MRL (Annex 2). Provisional MRL can be supported for a limited period in certain cases (Annex 3); other substances including some antibiotics (chloramphenicol) and some antibacterials (nitrofurans, nitroimidazoles), which are excluded from Annexes 1, 2, or 3, are enlisted in the Annex 4 of the Regulation. This enlisting in Annex 4 has the consequence of prohibiting their use in livestock production. The Council Regulation 2377/90/EEC is amended continuously since 1990 for the implementation of new MRLs while authorizing new veterinary substances. Therefore, the surveillance of veterinary drug residues in food products is an issue for each country subjected to the EU legislation. In essence, there are two types of regulatory residue programs. One deals with the direct-targeted control where the animal or the product is under consignment pending the result of the analysis. The other is built through the implementation of a National Residue Monitoring Plan (NRMP) that is used to monitor the residue status of food from animal origin, without systematic rejection of the specific product from the food market. The regulatory NRMP is established under EU Council Directive 96/23/EC [2], and more recently, also included into Regulation 882/2004/EC [3]. This Directive describes the quantity of samples to be tested for each species of food-producing animals (i.e., bovine species, etc.) or of animal products (milk, etc.) and the different groups of residual compounds to be monitored (i.e., antimicrobials, anabolics, antiparasitics, etc.). In both cases, suspected samples should be efficiently separated from the bulk of negative samples.

The criteria establishing the performance expected from the analytical methods for the screening and for the confirmatory control of residues have been established in the EU by Commission Decision 2002/657/EC [4] replacing in 2002 the former EU Commission Decision 93/256/EEC [5]. Efforts have been made to develop analytical tools capable of supporting the surveillance of the residues in food products from animal origin according to this set of laws.

42.1.3 Strategies for Screening and Confirmation of Antimicrobial Residues in Meat

There is a need to develop rapid analytical methods for controlling drug residues in food products of animal origin, particularly antimicrobials. While developing or selecting analytical procedures for residue control programs, certain aspects have to be taken into account, some of which are governed by external economical/political factors or internal organizational constraints. These factors/constraints are important to build the strategy that must be, at the same time, in line with the national, the European community, and the international food safety legislations as it is described in a relevant paper dedicated to the control of chloramphenicol [6]. Various methodological options for the screening and confirmation of veterinary drug residues, particularly the antimicrobials, can be implemented. Traditionally, the microbiological assays involving bacterial inhibition of a probe microorganism on a medium containing the antibiotic are the basis of antimicrobial residue control in food. But these methods are time-consuming and labor-intensive. In addition, microbiological assays often cannot differentiate univocally among the various forms and derivatives of a given antibiotic family. The quantitative information offered by such an approach reflects a lack of selectivity with the total amount of all forms of a given antibiotic, rather than providing distinct information related to quantitation and identification on the different analogs. These drawbacks are counterbalanced by the cost-effective and high sample throughput implementation of these techniques. In contrast to microbiological methods, chemical chromatographic approaches such as gas chromatography (GC) and liquid chromatography (LC) with various detectors (gas chromatography–electron capture detector [GC-ECD], gas chromatography–flame ionization detector [GC-FID], gas chromatography–nitrogen-phosphorus detector [GC-NPD], liquid chromatography–ultraviolet detector [LC-UV], liquid chromatography–visible detector [LC-Vis], liquid chromatography–fluorescence detector [LC-FLD]) can provide a more selective response with both high sensitivity and good separation efficiencies for most of them. Thus, chromatographic methods hold a real potential to display many of the characteristics necessary for systematic screening of antimicrobial residues in food products. However, the extremely diverse chemical nature of antibiotic substances requires that a variety of separation modes, detection strategies, and sample preparation procedures be used to achieve the goals outlined previously as necessary for rapid and sensitive screening. Moreover, the changes in the regulations, enforced during the past 20 years, highlighted the need to monitor drug residues in food, starting from a single rapid cost-effective screening to now achieving a univocal confirmation of the residual substance(s) primarily suspected in the food products. The analytical strategies, applied then by the networks of control laboratories involved in the food safety legislation, now require at least a two-step analytical monitoring; sometimes, to adjust the cover of regulatory needs, a strategy involving three or more steps may be required. Fortunately, in the time legislation strengthened during the past 20 years, the introduction of versatile and highly sensitive detectors built from different modes of mass spectrometric analyzers (single-quadrupole, triple-quadrupole, also called tandem-quadrupole, ion-trap, time-of-flight [ToF], quadrupole ToF, and quadrupole ion-trap) improved and sometimes simplified the strategies in the veterinary residue control. These rather expensive instruments introduced into the residue control laboratories gave the opportunity to readily modify the strategies, improving the quality and enhancing the efficiency of the control.

42.1.4 Analytical Methods for Control of Antimicrobials in Meat

Trends in analytical method development for drug residue control, particularly for antimicrobial residue control in meat products, changed significantly during the 1990s and early 2000s, with the increasing reliability of high-performance liquid chromatographic (HPLC) instruments. During the past 20 years, there have been a number of reviews covering the analysis of residual antimicrobials in food [7–21], which indicate that comparatively few analytical methods capable of measuring residual concentrations of many antimicrobials, at or near their MRL, existed in the 1980s. For example, developing procedures to extract and concentrate their residues from biological matrices became difficult due to low solubility of some antimicrobials in organic solvents. The other antimicrobials are either insufficiently volatile or thermally unstable (or both) to permit their analysis using GC or GC coupled to mass spectrometry (GC-MS). As a consequence, many methods for measuring antimicrobial residues have been developed by using HPLC. Liquid chromatographic technologies received more and more attention in the late 1980s, and much innovations and reliability occurred in the 1990s [22–25]. However, HPLC with UV detection is not considered sufficiently specific for use as a reliable confirmatory technique, at least in the present 2000's EU legislation. A spectral recognition of the compound (by means of multiwavelength UV-visible detectors such as diode array detector [DAD] and photodiode array [PDA]) or a more specific signal such as with fluorescence detection is mandatory. The development of LC coupled to mass spectrometry (LC-MS) instruments has significantly increased the range of antimicrobials for which reliable and identificative assays based on molecular spectrometry can be developed.

Today, the strategy of surveillance for the presence of antimicrobial residues in meat can be divided into two main categories of analytical methods. The biological methods (inhibitory plate tests, receptor test kits, immunoenzymological kits, and immunochemical biosensors) are generally aimed at wide range of antimicrobial screening, sometimes proposing a reduced monitoring to only one antimicrobial family or even to a single substance. The chromatographic methods (GC, GC-MS, LC, and LC-MS) often bring a higher degree of selectivity, sensitivity, and chemical structure recognition. These physicochemical methods are dedicated to a more specialized control of single substances with monoresidue methods. However, they are also able to cover the control of a family or of a set of substances and hence considered as multiresidue methods. The trends with the brand-new versatile technologies provided by LC-tandem-MS and LC-hybrid-MS lead to even wider ranges of antimicrobials and families of antimicrobials potentially analyzed altogether, and in the near future, even possibly together with other veterinary drugs (antiparasitics, anticoccidials, and antiinflammatories).

42.2 Screening Analysis by Means of Biological Methods

42.2.1 Microbiological Methods for Antimicrobial Residues

Ideally, a screening method should allow to establish the presence or absence of veterinary drugs by detecting all suspect samples (avoiding false negatives), preferably using a simple, routinely applicable procedure. Microbiological methods able to control the inhibitory activity of a majority of antimicrobials are of premium interest in this regard. Yet, it may sometimes be rather complex in analyzing inhibitory data that result due to the variety of antimicrobials of interest or also due to the desired limit of detection. In certain cases of inhibitory tests, postscreening orientative information is needed before confirming with adequate identification/quantification. Most of the developed microbiological tests dedicated to the control of meat products focus on muscle or kidney as a target tissue. The obvious advantage of analyzing muscle tissue lies in the fact that this is the edible part of the animal for which MRLs have primarily been established. Another advantage is that false positives due to naturally inhibiting substances are not likely to occur or are at least considerably reduced compared to other potential target matrices (kidney, liver). A disadvantage is that a variety of microorganisms have to be used to meet the MRLs for the commonly used antimicrobials. The major advantage of using kidney is the highest factor of concentration of many antimicrobial veterinary drugs in that kind of tissue compared to muscle [26]. This is an advantage of first interest when the analytical technologies for residue testing seriously lack sensitivity. The major drawback is the false-positive samples that are likely to occur due, in part, to the inhomogeneity of this offal and also due to the lack of stability of this tissue and to the difficulties in extracting several of the protein-bound antimicrobial compounds.

Most of the microbiological methods for tissues detect inhibitory substances diffusing from a piece of tissue [27–37], or from a paper disk soaked with tissue fluid [26,38–40], into an agar layer seeded with a susceptible bacterial strain. These tests are multiresidue screening methods and use either only one plate [41,42] or different plates, different combinations of pH, media, and different test microorganisms to try to improve the detection of different families of drugs [29, 33–35,37,43]. Some of these methods have also been modified to perform a postscreening analysis [34,35,37] often proposed in antibiotic residue control strategy to orientate toward the appropriate antimicrobial family before more sophisticated chemical identification and quantification.

42.2.2 Other Biological Methods

Several other biologically derived analytical methods were recently or are still in use in some control laboratories: the radioimmunological Charm II Test[*] applied to muscle tissue and derived from milk control [44–46], the high voltage gel electrophoresis [47–50], or the TLC-bioautography [51–56].

Owing to the fact that the microbiological methods with wide-range antimicrobial residue screening are considered time-consuming incubation procedures with regard to the fast processing in agrifood industrial practice, several other microbiological or immunological receptor test technologies have been developed for residue control. The analytical strategy for meat control is very similar and can be compared to that built for antimicrobial residue control in milk. The most common tools used for time-saving strategies are rapid microbiological tube test assays such as the Charm Farm Test® [46,57,58], the more recently developed Premi Test® [59–61], rapid receptor tests such as the Tetrasensor® for tetracyclines [62], or solid-phase fluorescence immunoassays (SPFIA) for gentamicin, several antibiotics, and sulfonamides [63–65], or enzyme-linked immunosorbent assays (ELISA kits) based on monoclonal or polyclonal antibodies. These tools generally focus on an immunoenzymatic action for one or two specific antimicrobial families or, even in some cases, on only one very specific antimicrobial compound. For example, ELISA kits were developed not only for chloramphenicol, enrofloxacin, gentamicin, halofuginone, nicarbazin, specific nitrofuran metabolites, streptomycin, tylosin but also for beta-lactams, fluoroquinolones, macrolides, nitroimidazoles, and tetracyclines [6,36,66–84]. A new immunological technology developed in the 1990s, the surface plasmon resonance-based biosensor immunoassay (SPR-BIA), based on both immunological receptors and signal reading by a specific light-scattering property, has been of great interest in the 2000s. It has been applied first as screening/postscreening strategy in milk products for residues of sulfonamide compounds such as sulfamethazine and sulfadiazine [85–90] and then also extended in milk and in tissues to several other antimicrobial compounds or families of antimicrobials such as streptomycin, dihydrostreptomycin, nicarbazin metabolite dinitrocarbanilamide (DNC), and a range of penicillins and fluoroquinolones [83,91–96]. However, lack of wide-range screening for antimicrobials generally put these methodologies in the position to be prescribed for prescreening strategies or for very specific and selective control. In certain conditions, they might be useful when the regulatory residue control enforces a ban on antimicrobial substances such as chloramphenicol, nitrofurans, and nitroimidazoles. They are also available for some screening strategies when the microbiological wide-range screening methods lack sensitivity with regard to the requested MRL for specific families or substances (sulfonamides, aminoglocosides). A survey of screening methodologies implemented in the EU by the reference laboratories in the period 2000–2003 for monitoring authorized antimicrobials in meat is shown in Figure 42.1. The same survey is displayed in Figure 42.2 but for the screening in meat of the residues of a prohibited drug: chloramphenicol.

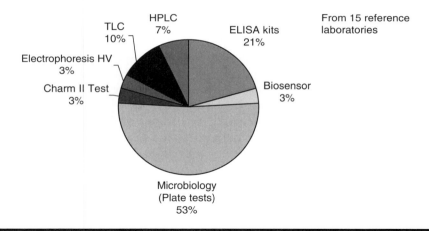

Figure 42.1 Screening strategy for meat control: methodologies used for screening in meat products considering 15 national reference laboratories from the European Union Member States (2003).

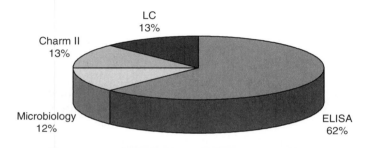

Figure 42.2 Screening strategy for meat control—chloramphenicol: methodologies used for screening in meat products for chloramphenicol residues considering 15 national reference laboratories from the European Union Member States (2003).

42.3 Confirmatory Analysis by Means of Chromatographic Methods

42.3.1 Chromatographic Methodologies for Antimicrobial Residues

The widely used methods in residue analysis are those based on chromatographic procedures. Their efficiency comes, for a large part, from their separative properties making them selective to the compounds to be analyzed with regard to the complex endogenous interfering substances extracted from the biological tissues. A part of their efficiency also derives from the sensitivity of their detectors, often enabling them to monitor traces of antimicrobials at low parts per billion equivalent to μg of residue/kg of tissue (ppb) level. Today they are the preferred methodologies for the confirmatory step in most of the analytical strategies. Ideally, a method set for drug residue confirmation should unequivocally establish the identity of the residue. During the regulatory control of nonprohibited drug residues, reliable quantification should be additionally carried out at an appropriate stage. Quantification is a mandatory procedure of the analytical residue method when the control has to reliably establish whether the residue concentration exceeds the MRL. On the contrary, for the regulatory control of residues of prohibited drugs, the unequivocal identification of the drug is necessary; its reliable quantification lies only at the second level even though, for chemical analytical methods, the quality of the identification is often correlated to the quality of the quantification. Among the chromatographic procedures, one can quote TLC as an efficient method when used for screening/identifying within a single assay several compounds from the same family of antimicrobials [97–104]. In its use of screening, this technique displays an acceptable resolutive property compared to microbiological methods, especially when these nonselective wide-range inhibitory methods significantly lack sensitivity as it is for the sulfonamides [103]. GC [105] has also been used for a long time since the 1970s for analyzing a large number of compounds in a variety of matrices but with complicated development and variable success for antimicrobial substances [106–110]. Antibiotics are, for most of them, nonvolatile, polar-to-very polar, and thermally unstable compounds precluding their correct analysis by GC. Chloramphenicol is the most often cited antibiotic to be analyzed by GC or GC-MS techniques [6,78,111–117].

HPLC is by far the most suitable and widely used technique for antimicrobial analysis [22–25]. It can be applied to the measurement of almost all the antimicrobial compounds. Considering its high resolutive properties, it is an appropriate technology for both the separation and the simultaneous quantification of closely related residues (parent drug and its possible metabolites). High precision in the measurement can easily be achieved with relative standard deviation (RSD) lower than any other technologies ever used for analyzing traces of antimicrobials in muscle tissue (1% < RSD < 15%). A wide choice in acidic–basic aqueous solvents combined with several organic ones (generally acetonitrile [ACN] or methanol [MeOH]) leads to numerous mobile phases enabling the control of the elution of antimicrobials on an extended library of column packings (normal phase, reverse phase, ion exchange phase, and mixed phases) and column geometries (conventional, narrow-bore, and micro-bore) [118]. Multiantimicrobial residue methods are then to be considered readily achievable in such a choice of analytical parameters. Still, some disadvantages need to be mentioned. The low sample throughput as compared to microbiological methods is probably the major drawback. The limited number of compounds to be separated within the same run should also be considered. Additionally, not all antimicrobials bear chromophore or fluorophore properties, making them undetectable by conventional HPLC detectors. Derivatization procedures before detection are generally employed at precolumn or postcolumn stage to recover the detectability by UV-visible or fluorescence detectors [119–122]. Another limitation derives from the implementation of mass spectrometers for which not all compositions of mobile phase can be applied to LC-MS instruments (phosphate buffers, etc.). Nevertheless, the LC-MS technology is readily becoming one of the most powerful LC techniques in the field of drug residue analysis. Research in new technologies in MS, notably LC-MS, with new quadrupolar MS detectors in the 1990s (LC-tandem MS, also called LC-triple quadrupolar MS), with ion-trap and hybrid-trap technologies (LC-ion trap MS and LC-quadrupolar ion trap MS), or with ToF and hybrid ToF technologies in the 2000s (LC-ToF-MS and LC-quadrupolar ToF-MS), has greatly benefited from the international need of protecting food quality. Monitoring nonvolatile and polar antimicrobials with a high degree of sensitivity and specificity is now possible, thanks to the large variety of interfaces developed for mass analyzers during the past 20 years: thermospray ionization (TSP), particle beam (PB), electrospray ionization (ESI), atmospheric pressure chemical ionization (APCI), and atmospheric pressure photochemical ionization (APPI). Several interesting review articles have been published recently, some of them dedicated to the advances in mass spectrometry analysis coupled to LC and GC separative systems [123–134], and others dealing with LC-MS analysis of drug residues in food [8,11,13,14,135–137].

42.3.2 Other Modes of Chemical Analysis

The other technologies potentially dedicated to polar nonvolatile compounds, such as many antimicrobials, were also investigated in the past 15 years. At the end of the 1990s, supercritical fluid chromatography (SFC) was reported to be a potentially innovative concept of separation evaluated on sulfonamide antimicrobials extracted from swine tissues [138–141]. Electrokinetic technologies with their major component, the capillary zone electrophoresis (CZE or CE), have focused on the high separation efficiencies possible with this separative mode [142–154]. Its three major variants are the micellar electrokinetic capillary chromatography (MEKC), the capillary isotachophoresis, and the electrochromatography (CEC) [155–161]. But none of these innovating analytical technologies reached a sufficient degree of reliability to trigger their implementation in the field of drug residue control in food. The capillary zone electrophoresis coupled to a mass spectrometer (CZE-MS or CE-MS) can become a potentially interesting technique as soon as the major technical problems related to the hyphenation between the electrophoretic nanocapillary tube and the atmospheric source of the mass spectrometer are reliably stabilized [162–169].

42.3.3 Sample Preparation for Liquid Chromatography

The sample preparation procedures developed for the antimicrobial residue analysis in muscle tissues by means of chromatographic instruments are all described with a similar process. The very first step always involves the extraction of the compounds of interest from the tissue by deproteinizing the sample using organic solvents such as ACN acidic or buffer aqueous solutions ethyl acetate (EtOAc), or through such as hydrochloric acid (HCl) or trichloroacetic acid (TCA) or phosphate buffer solutions (PBS) mixed with miscible organic solvents such as ACN or dichloromethane (DCM). The efficiency of the deproteinization depends on the degree and strength of the binding of the residues to the tissue proteins and on the adequateness of the deproteinizing solvent or mixture of solvents to be used. A centrifugation is often applied at this stage to separate the liquid phase containing the residues from the solid phase principally made of the precipitated proteins and other remaining substrates. The following step in sample preparation is the proper extraction from the biological liquid into a suitable solvent by partitioning process or liquid–liquid extraction (LLE). The choice of the solvent(s) depends on the polarity of the residue(s) of interest and must be adjusted (pH, volume, ionic strength, salt saturation, homogenization process, salting-out process, breaking emulsions) to give a maximum recovery of the residue(s). An acidic antimicrobial compound must be extracted by nonpolar solvents at low pH values where its acidic function is suppressed. A basic antimicrobial substance must be extracted by nonpolar solvents at high pH values where its basic function is neutralized. A neutral antimicrobial analyte is extracted whatever the pH of the sample/solvent is. An amphoteric antimicrobial molecule is extracted by nonpolar solvents at its specific pH of neutralization. Protein-bound antimicrobials can be released more efficiently by a stronger solvent at high acidic pH such as sulfuric acid (H_2SO_4), HCl, phosphoric acid (H_3PO_4) or at basic pH such as trisaminomethane (TRIS) and dithioerythreitol (DTE), enabling to cut the bind between the attached residue and the protein. Release of the residue is either as a neutral substance or more generally as an ionic species that can be neutralized by using an ion-pair extraction mode. The counterionic species neutralizes the ionicity of the residue and forms a neutral ion pair easily transferred into the organic phase. Further to this extraction step is carried out another possible centrifugation where separate liquid phases can be obtained leading to discard one phase and to retain of the other for further sample extract purification. Rather than neutralizing the antimicrobial residue by the ion pairing mode, it is also possible to make it react with a labeling substance to enhance the detectability of the residual antimicrobial by UV-visible or fluorescence detectors, in case it lacks chromophoric or fluorogenic properties. The last step in the sample preparation is generally the cleanup process. Purification is often necessary in biological matrices when too much endogenous substances still remain in the extract, leading to problems of chromatographic separation (column plugging or retention time variation) or interference in the detection. The liquid–liquid partition between immiscible solvents was the most common procedure. But now liquid–solid extraction with pouring the sample extract through sorbing material packed in a short column, also called the solid-phase extraction (SPE), is very often performed in residue analysis and the preferred cleanup for routine control. Most of the sorbents employed are derived from the different classes of chromatographic packings such as pure silica normal-phase sorbents, or alkyl-bonded silica such as 8-carbon-alkyl bonded silica stationary phase (C8), 18-carbon-alkyl bonded silica stationary phase (C18), and phenyl reverse-phase sorbents or ion-exchange material such as strong anionic or cationic exchange sorbents (SAX, SCX) or now mixed hydrophilic–lipophilic balanced cartridge (HLB), medium anion exchange cartridge (MAX), or medium cation exchange cartridge (MCX) mixed sorbents balanced with reverse-phase and ionic exchange properties. A number of reviews deal with this interesting SPE approach [170–175]. Further to the cleanup procedure is sometimes added another step to remove the last fat content from the extract by a liquid–liquid partition with apolar organic solvent (*n*-hexane or *iso*-octane). The very last sample preparation step is the transfer of the purified extract into the correct mixture of solvents, which should be closely similar to the mobile phase employed for the separative chromatography. Very often, this step requires a partial evaporation by rotary evaporator or a complete evaporation to dryness under gentle heating (40–50°C) combined with a gentle nitrogen flow. Direct extract is finally reconstituted in the mobile phase or in a closely related mixture of solvents before injection in the chromatograph.

42.3.4   Modes of Separation in Liquid Chromatography

Several modes of separation are applicable in liquid chromatography; the three major ones are the normal-phase mode (adsorption on silica stationary phases and elution driven by a mobile phase composed of mixtures of organic solvents), the reverse-phase mode (liquid partitioning between carbon-bonded stationary phases and a mobile phase mixing an aqueous buffer with an organic modifier), and the ion-exchange mode (cationic or anionic exchange with specific ion-bonded stationary phases). Because the heterogeneity in the chemical behavior of the antimicrobial substances is governed by their polarity or their weaker or stronger ionizability, different modes of separation can be considered depending on the compounds to be analyzed. Polar and highly ionizable antimicrobials can be successfully and specifically separated in the ion-exchange mode. Less polar antimicrobials can be separated in the normal-phase mode. Nonpolar antimicrobials or neutralized ionic antimicrobials are generally separated in the reverse-phase mode. Because it is easier to neutralize ionic compounds to chromatograph them together with the nonpolar ones, the reverse-phase mode has often been considered the preferred mode of separation for antimicrobial residues in food. Different techniques of neutralization have been investigated from the buffered pH displacement of the mobile phase for weak acidic and alkaline antimicrobials (penicillins, quinolones) to the use of ion-pairing agents for the more polar and ionic antimicrobials (aminoglycosides). All these strategies are generally employed to adjust the resolution and to improve the efficiency of the separation. It is convenient for chromatographing several compounds not only selected in the same family of antimicrobials, such as different neutral and amphoteric penicillins [176], but also for several families within the same runs, such as penicillins, cephalosporins, sulfonamides, and macrolides or aminoglycosides, quinolones, and tetracyclines [177]. The analytical parameters for the analyst to control the LC in the reverse-phase mode are numerous ranging from the chemistry of the mobile phase (pH, buffering, organic modifiers) to the chemistry of the stationary phase, including controlling the physical parameters of the pumping device by working in isocratic or gradient mode. The choice in the reverse stationary phase (C4, C8, C18, phenyl, etc.) can be relevant to resolve certain analytes even though C18-bonded silica stationary phase is actually the most common one. Polymeric reverse stationary phases (PLRP-S) are also sometimes useful when the ionic or highly polarizable antimicrobials may react with the silica of the column packing or with impurities contained in it. Some new brands of stationary phases appeared in the recent years such as the mixing of reverse-phase mode and weak ion-exchange mode or as the HLB stationary phase. A promising stationary phase for chromatographing highly polar antimicrobials is the hydrophilic liquid chromatography (HILIC) pseudonormal phase. Depending on the quality of detection mode, the separation can receive more or less attention. The science of chromatographic separation recently received less interest as soon as highly selective and sensitive detectors, such as new-generation mass spectrometers, have replaced the conventional UV-visible or diode array or fluorescence detectors.

42.3.5 Modes of Detection in Liquid Chromatography

The three commonly used detectors for analyzing antimicrobials are UV-visible detectors in a monowavelength version or now, since the 1990s, in a multiwavelength version, such as the DAD or the PDA detector; fluorescence detectors with a higher degree of specificity compared to the UV-visible ones while considering possible coextractive substances from the food matrix; and mass spectrometers with a large variety of combinations from single quadrupolar (SQ-MS) or triple quadrupolar (TQ-MS or tandem MS) instruments to ion-trap (IT-MS) devices and now to ToF-MS instruments, the most attractive and reliable instrument being the LC-TQ-MS with its high degree of reliability in identifying and quantifying drug residues in food matrix. Several hybrid instruments are now also commercially available, such as LC-Q/IT-MS or LC-Q/ToF-MS or LC-Q/Trap/Orbitrap-MS. To bring some antimicrobials to a more selective and sensitive analysis, derivatization by tagging the compounds of interest either with a chromogenic or with a fluorogenic labeling agent is also a useful alternative to extend the field of detection for some undetectable compounds such as for the aminoglycosides. Derivatization can also be proposed for some antimicrobials to enhance their signal in mass spectrometry by adjusting the antimicrobial(s) of interest to a correctly detectable range of mass and/or by improving its (their) capacity of ionization in the source of the mass spectrometer.

42.4   Applications of Chromatographic Methods to Antimicrobial Residues

Considering the wide range of antimicrobials potentially found as their residues in meat and particularly in muscle tissues of food-producing animals, a large number of methods have been developed during the last decades to monitor these compounds in accordance with the regulations enforced in various countries interested in food safety and food-producing animal husbandry.

The following paragraphs present some relevant methodological developments proposed in the field of antimicrobial residue analysis in the past 20 years. They are sorted according to the different families in an alphabetic order, although the more recent selective LC methods now open the field of multiresidue antimicrobial analysis, including more than 50 compounds and up to 100 compounds at once in some cases of LC-MS approaches.

42.4.1   Aminoglycosides

The aminoglycosides are broad-spectrum antibiotics produced by members of two types of bacterial genii, either *Streptomyces* sp. (streptomycin, neomycin) or *Micromonospora* sp. (gentamicin, amikacin). With the first streptomycin compound reported in 1944, this family of antibiotics is the second one discovered following the penicillins. Structurally, they belong to the chemical family of carbohydrates with two or more amino-sugars linked via a glycosidic bond to an aglycone moiety called the "aminocyclitol" ring (Figure 42.3). The aminoglycosides are divided into two subgroups: one small subgroup containing a streptamine moiety and the other larger one containing a 2-deoxystreptamine moiety. Streptomycin and dihydrostreptomycin belong to the streptamine subgroup. Neomycins, paromomycins, gentamicins, kanamycins, and apramycin belong to the deoxystreptamine subgroup. The deoxystreptamine subgroup is further structured in subclasses depending on the substituents attached to the deoxystreptamine moiety, leading to classes such as neomycins (neomycin A, neomycin B, neomycin C), or gentamicins (gentamicin C1, gentamicin C2, gentamicin C1a, gentamicin C2a). Useful reviews have been published on this family of antibiotics [18,19].

In food animal production the most commonly used aminoglycosides are gentamicin, neomycin, dihydrostreptomycin, and streptomycin. The aminoglycosides are not metabolized in the body but rather are bound to proteins (generally <30%) and excreted as the parent compound. Residues concentrate in the kidney (cortex) for more than 40% and in cochlear tissues, making these drugs both nephrotoxic and ototoxic. Muscle tissues are less subjected to aminoglycoside residue concentrations—more than 150 times lesser than in kidney tissues. Aminoglycosides are water-soluble, polar, and weakly basic compounds. They are stable at both high and low pH levels, and are heat resistant. A number of methods can be quoted for aminoglycoside residue analysis in tissues. For a screening strategy based on the antibacterial property of aminoglycosides, several bioassays have been developed such as microbiological inhibitory plate tests with different strains (*Bacillus subtilis*, *B. stearothermophilus*, *B. megaterium*, and *B. aureus*) [27–30, 32–35,37,39,40,42,43]. Several microbiological tube test kits such as the Charm Farm Test Kit

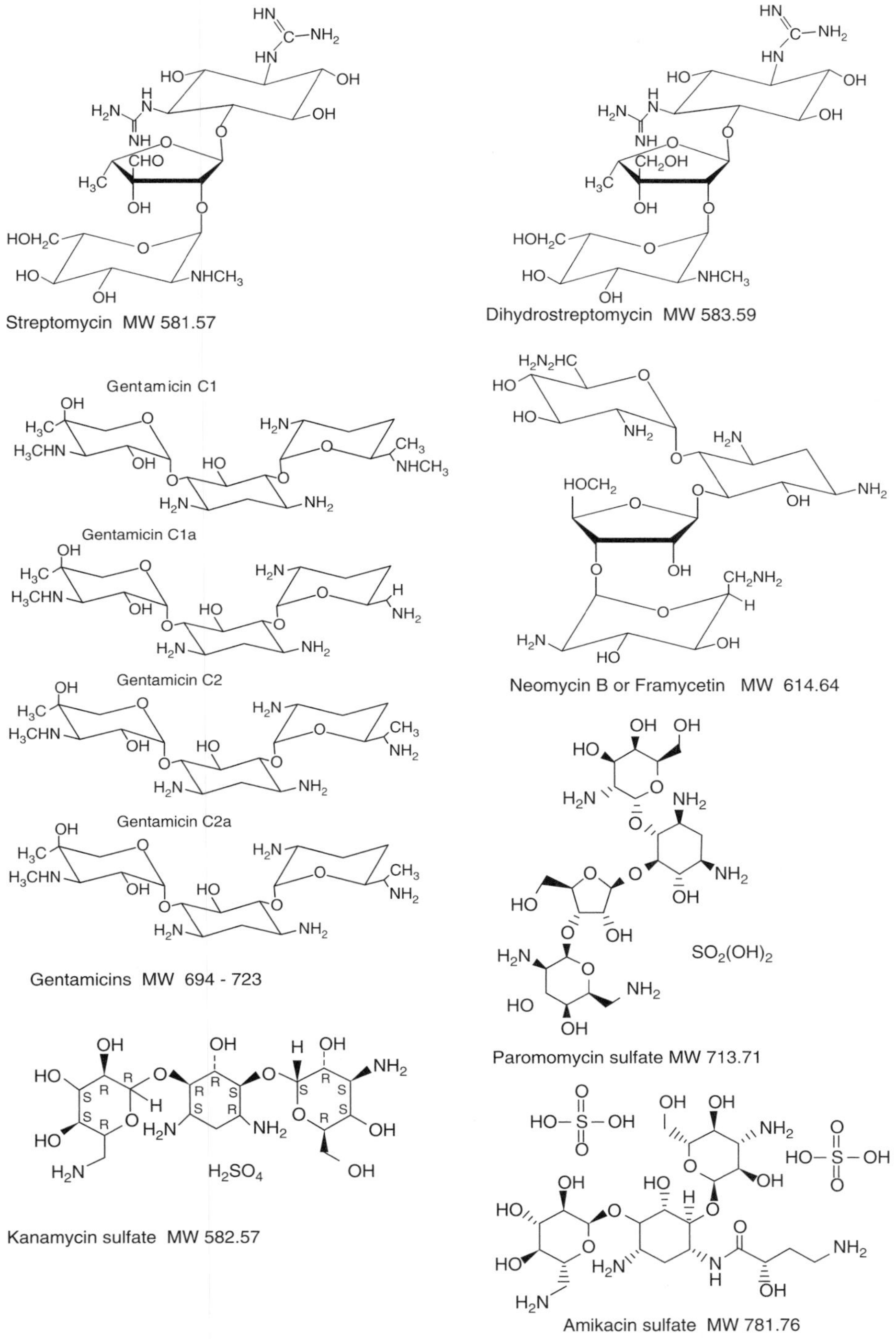

Figure 42.3 Structures of some aminoglycosides.

[57,58] or the Premi Test Kit [60,61] are also available. Some immunological bioassays are also dedicated to specific aminoglycosides: a fluorescence polarization immunoassay test (FPIA) for gentamicin [63]; and for streptomycin and dihydrostreptomycin, either the radioimmunological Charm II test [45–46], an ELISA test kit [71], or a SPR-BIA immunoassay [93]. TLC methods have also been proposed for some aminoglycosides in tissues and urines of swine and calves [178] or in the form of TLC-bioautography [46,55,56]. For their detection by analytical physicochemistry (Table 42.1), aminoglycosides lack chromophores and fluorophores; thus, derivatization is usually required. Because of their nonvolatility, HPLC procedures with fluorescent labeling are preferred; several of them were developed in the 1980s and 1990s [179–192]. Very few HPLC-UV methods have been proposed [193]. In addition, the ionicity of the aminoglycosides induces active use of the ion-pairing technique to achieve sufficient separation in the reverse-phase mode. Recently, in the 2000s, mass spectrometry (LC-MS techniques) has also become a convenient alternative to detecting and identifying these antibiotics [194–196]. But efficient and reliable extraction from animal tissue matrix still remains an issue in the analytical chemistry of aminoglycoside residues.

42.4.2 Amphenicols

The amphenicols are composed of three substances, namely chloramphenicol, thiamphenicol, and florfenicol. They are broad-spectrum bacteriostatic antibiotics. Chloramphenicol was the first of the family to be produced in 1947 from cultures of *Streptomyces venezuelae* and, due to its frequent use in veterinary and human medicine, was further produced synthetically starting from dichloroacetic acid. This natural compound is rather unique in that it contains a nitrobenzene moiety. But, following extensive reports of adverse reactions, primarily aplastic anemia, in humans and other side effects after chloramphenicol treatment, the drug was found toxic enough to be banned worldwide in the end of the 1980s and early 1990s in veterinary practice and also in most human medicines. Thiamphenicol and, more recently, florfenicol, which have chemical structures similar to that of chloramphenicol (Figure 42.4), have been permitted as substitutes but have been licensed with effective tolerance limits at the 50–500 ppb level such as for the EU-MRL ones [1]. Because they are chemically and thermally rather stable [74,197–199], they can be readily analyzed by GC techniques and residues can be found in frozen stored meat for several months. The chromatographic methods for the analysis of chloramphenicol in edible animal products were reviewed thoroughly by Allen in 1985 [200]. But the ban of chloramphenicol changed significantly the strategy for the analysis of this family of compounds. In the 1990s, chloramphenicol became an important issue in residue monitoring because of the particular public health concern due to considerable levels of drug residues that may occur in edible animal products from illegally treated animals. To replace the conventional screening methods inefficient in detecting chloramphenicol residues at very low ppb levels, such as microbiological inhibitory plate tests [201] and microbiological tube tests [58] or TLC analysis [66,103,202] or TLC/bioautography [55], several sensitive and rapid immunoenzymological test kits (ELISA) were developed specifically for the detection of chloramphenicol residues at <1.0 μg/kg in muscle tissues and other edible animal products [6,26,74,203,204].

At the confirmatory stage (Table 42.2) of the residue control strategy for chloramphenicol, it became mandatory in the 1990s to change from conventional GC [107,112,114,115] and HPLC-UV methods [6,77,197,199,205–208] to sensitive and unequivocally identifying techniques such as GC-MS(MS) [78,111,116] and LC-MS(MS) ones [78, 209–212]. Several mass spectrometric detection methods were also developed for the three amphenicols within a single method [113,195,213,214] and sometimes also including the major metabolite of florfenicol, the florfenicol amine [117,215].

Table 42.1 Summary of Literature Methods for Determination of Aminoglycoside Residues in Tissues Using Liquid Chromatography

Tissue	Analytes	Sample Treatment	Derivatization	LC Column Technique	Detection	Limit Range (ppm)	Year (Reference)
Kidney (cortex)	Gentamicin, kanamycin	Buffer extn, cell lysis, ion exchange SPE	Postcolumn: OPA	RPLC C18, HPSA ion pairing	Fluoresc. Ex:340 nm, Em:440 nm	—[a]	1983 (179)
Kidney, muscle	Neomycin, paromomycin	Buffer extn, heat deprot.	Postcolumn: OPA	RPLC C18, PSA ion pairing	Fluoresc. Ex:340 nm, Em:455 nm	0.5	1985 (180)
Kidney, muscle	Kanamycin	TCA deprot. ether defat, ion exchange SPE	Precolumn: OPA	RPLC C18, ion pairing	Fluoresc. Ex:335 nm, Em:440 nm	0.04	1986 (181)
Kidney, muscle	Streptomycin	PCA deprot., C8 SPE	Postcolumn: nihydrin	RPLC C18, OSA ion pairing	Fluoresc. Ex:400 nm, Em:495 nm	0.5	1988 (182)
Muscle	Gentamicin	Buffer extn, H$_2$SO$_4$ deprot., ion exchange, silica SPE	Precolumn: OPA	RPLC C18, HPSA ion pairing	Fluoresc. Ex:340 nm, Em:418 nm	0.2	1989 (183)
Muscle, liver, kidney, fat	Gentamicin	TCA deprot., ion exchange SPE	Postcolumn: OPA	RPLC C18, CPS ion pairing	Fluoresc. Ex:340 nm, Em:418 nm	0.5	1993 (184)
Muscle, liver, kidney	Streptomycin, DHS	PCA deprot., cation exchange SPE	Postcolumn: NQS	RPLC C18, HSA ion pairing	Fluoresc. Ex:347 nm, Em:418 nm	0.02	1994 (185)
Kidney, liver	Neomycin	TCA deprot., cation exchange SPE	Postcolumn: OPA	RPLC C18, CPS ion pairing	Fluoresc. Ex:340 nm, Em:440 nm	0.05	1995 (186)

(Continued)

Table 42.1 (Continued)

Tissue	Analytes	Sample Treatment	Derivatization	LC Column Technique	Detection	Limit Range (ppm)	Year (Reference)
Kidney, muscle	Streptomycin, DHS	TCA deprot., cation exchange SPE	Postcolumn: NQS	RPLC C18, OSA ion pairing	Fluoresc. Ex:375 nm, Em:420 nm	0.02	1997 (187)
Kidney, muscle	DHS	Acid deprot., ion exchange SPE, C8 SPE	Postcolumn: NQS	RPLC C18, HSA ion pairing	Fluoresc. Ex:375 nm, Em:420 nm	0.015	1998 (188)
Muscle, liver, kidney	Spectinomycin	TCA deprot., WCX SPE	Postcolumn: NQS	RPLC C18, ion pairing	Fluoresc. Ex:340 nm, Em:460 nm	0.05	1998 (189)
Feed	Amikacin, kanamycin, gentamicin, neomycin	HCl deprot.	Postcolumn: OPA	RPLC C18, ion pairing	Fluoresc. Ex:355 nm, Em:415 nm	0.2	1998 (190)
Muscle, kidney	Neomycin	PCA deprot., WCX SPE	Postcolumn: FMOCl	RPLC C4, ion pairing	Fluoresc. Ex:260 nm, Em:315 nm	0.12	1999 (191)
Muscle, liver, kidney	Gentamicin, neomycin	Buffer extn, TCA deprot., C18 SPE	Postcolumn: FMOCl	RPLC C18, HSA ion pairing	Fluoresc. Ex:260 nm, Em:315 nm	0.05, 0.10	2001 (192)
Muscle, liver, kidney, milk	Gentamicin netilmicin[IS]	Buffer extn, C18 SPE	Postcolumn: OPA	RPLC C18, HSA ion pairing	UV:330 nm	1.0	1995 (193)
Muscle, injection site	Gentamicin, neomycin, spectinomycin tobramycin[IS]	Buffer extn, TCA deprot., C18 SPE	—	RPLC C18	LC-MS[n] esi+	—[a]	2000 (194)

Muscle, liver, kidney, fat	Gentamicin, tobramycin (IS)	Deprot., WCX SPE	RPLC C18, PFPA ion pairing	LC-tandem MS esi+	—	0.025	2003 (195)
Muscle, liver	Spectinomycin, apramycin, streptomycin, DHS, neomycin, gentamicin, kanamycin, paromomycin, amikacin, tobramycin, sisomycin	TCA extn, SAX SPE, HLB SPE	RPLC C18, HFBA ion pairing	LC-tandem MS esi+	—	0.015–0.040	2005 (196)

[a] Not reported.
(IS)Internal standard.

Note: ppm: parts per million equivalent to mg of residue/kg of tissue; OPA: ortho-phthalaldehyde; RPLC: reverse-phase liquid chromatography; HPSA: 1-heptanesulfonic acid; Fluoresc.: fluorescence; Ex: excitation; Em: emission; PSA: 1-pentane sulfonic acid; TCA: trichloroacetic acid; SPE: solid-phase extraction; PCA: perchloric acid; OSA: octane sulfonic acid; CPS: dl-camphor-10-sulfonate; DHS: dihydrostreptomycin; extn: extraction; NQS: beta-naphtoquinone-4-sulfonate; HSA: hexane-1-sulfonic acid; WCX: weak cation exchange; FMOCl: 9-fluorenylmethylchloroformate; MS^n: ion trap; esi+: electrospray source in positive mode; PFPA: pentafluoropropionic acid; MSMS: triple quadrupole; SAX: strong anion exchange; HLB: hydrophilic lipophilic balance; HFBA: heptafluorobutyric acid.

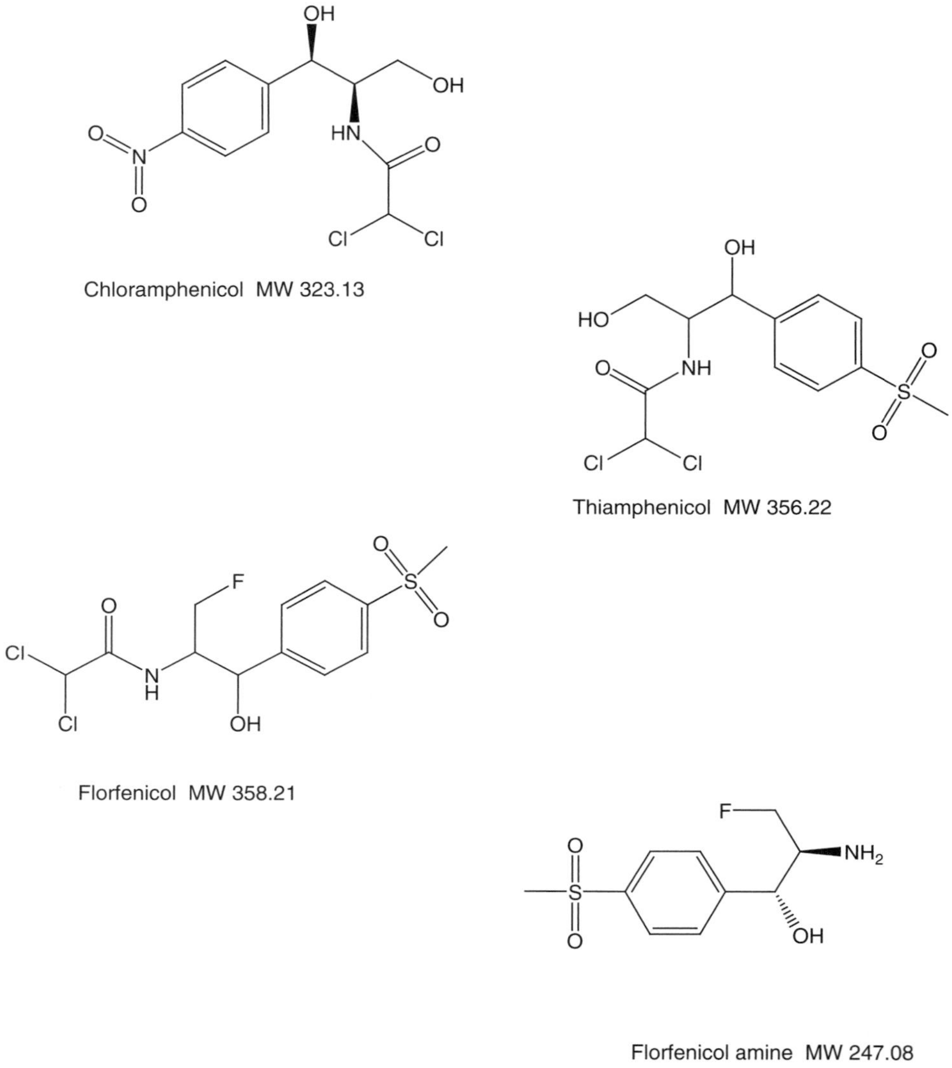

Chloramphenicol MW 323.13

Thiamphenicol MW 356.22

Florfenicol MW 358.21

Florfenicol amine MW 247.08

Figure 42.4 Structures of amphenicols.

42.4.3 Beta-Lactams: Penicillins and Cephalosporins

The beta-lactams are antibiotics active against gram-positive bacteria. They consist, basically, of two classes of thermally labile compounds: penicillins and cephalosporins. Most of the commonly used beta-lactam antibiotics are produced semisynthetically either from the 6-APA for semisynthetic penicillins or from the 7-ACA for semisynthetic cephalosporins. Only the benzylpenicillin (or penicillin G), the 6-APA, and the phenoxymethylpenicillin (or penicillin V) are three naturally occurring penicillins extracted directly from molds of *Penicillium chrysogenum* and *Penicillium notatum*. Cephalosporin C and 7-ACA are the naturally occurring cephalosporins extracted from

Table 42.2 Summary of Literature Methods for Determination of Amphenicol Residues in Tissues Using Gas or Liquid Chromatography

Tissue	Analytes	Sample Treatment	Derivatization	Separation Technique	Detection	Limit Range (ppb)	Year (Reference)
Muscle	CAP, TAP	TCA extn, ion exchange SPE	Pyr, TMSA, TMSI, TMCS	GC	FID	2.5–2.5	1992 (107)
Muscle, kidney, liver	CAP	ACN/NaCl extn, hexane defat., EtOAc purif, IAC cleanup	Pyr, TMCS, HDMDS	GC	ECD	0.2	1995 (112)
Muscle	CAP, metaCAP$^{(IS)}$	Water extn, hexane defat., silica SPE	Pyr, HDMDS, TMCS	GC	ECD	1.0	2002 (114)
Fish flesh, shrimp flesh	CAP	EtOAc extn, hexane defat.	TMSA,TMCS	GC	Micro-ECD	0.1	2005 (115)
Muscle	CAP	H_2O extn, silica SPE, toluene defat.	—	RPLC	UV-285 nm	1.5	1989 (205)
Muscle, kidney, liver	CAP	EtOAc extn, hexane/$CHCl_3$ defat.	—	RPLC	UV-278 nm	1.0	1991 (197)
Muscle	CAP PCB$^{(IS)}$	H_2O extn, silica SPE, toluene defat.	—	RPLC	DAD- 240–320 nm	1.0	1992 (6)
Muscle	CAP	MSPD, hexane defat.	—	RPLC	UV-290 nm	6.0	1997 (206)
Muscle, kidney, liver	CAP	Various extn tested (glucuronidase digestion for kidney), silica SPE, toluene defat.	—	RPLC	UV-285 nm	—[a]	1998 (207)
Muscle, kidney	CAP	EtOAc/ACN extn, acidic purif., heat, neutral, C$_{18}$ SPE	—	RPLC	UV-270 nm	0.5	2003 (77)
Chicken muscle	CAP	EtOAc/anh. Na_2SO_4 extn, silica SPE, DCM defat., ACN/EtOAc partition., hexane/$CHCl_3$ defat.	—	RPLC	UV-278 nm	2.0	2003 (199)
Fish feed	CAP	ACN/water extn, C$_{18}$ SPE	—	RPLC	UV-225 nm	200	2005 (208)
Muscle	CAP, metaCAP$^{(IS)}$	EtOAc extn, NaCl partition., C$_{18}$ SPE	Pyr, HDMDS, TMCS	GC	MS nci	0.5	1994 (111)
Shrimp flesh	CAP, CAP- d$_5$ $^{(IS)}$	EtOAc extn, hexane defat., C$_{18}$ SPE	MSTFA	GC	MSMS nci	0.1	2003 (78)
Shrimp flesh, crayfish flesh	CAP, metaCAP$^{(IS)}$, CAP-d$_5$$^{(IS)}$	ACN/NaCl extn, hexane defat., EtOAc purif, C$_{18}$ SPE	BSA, n-heptane	GC	MS nci	0.07	2006 (116)

(Continued)

Table 42.2 (Continued)

Tissue	Analytes	Sample Treatment	Derivatization	Separation Technique	Detection	Limit Range (ppb)	Year (Reference)
Muscle	CAP	ACN extn, CHCl$_3$ defat., C$_{18}$ SPE	—	RPLC	MS esi−	0.5	2001 (209)
Shrimp flesh	CAP, CAP-d$_5$ [IS]	EtOAc extn, hexane defat., C$_{18}$ SPE	—	RPLC	MS^n esi	0.1	2003 (78)
Muscle	CAP	EtOAc extn, NaCl partition., C$_{18}$ SPE	—	RPLC	MSMS apci-	0.02	2003 (210)
Crab meat	CAP	EtOAc extn, MeOH/NaCl partition., heptane defat., EtOAc purif.	—	RPLC	MSMS esi−	0.25	2005 (211)
Muscle, liver, kidney	CAP	ACN extn, hexane defat.	—	RPLC	MS esi−	0.2–0.6	2005 (212)
Fish flesh	CAP, TAP, FLF	EtOAc extn, hexane defat., EtOAc purif., silica SPE	BSA	GC	MS esi	5	1996 (113)
Muscle, injection sites	CAP, FLF	—^{a}	—	RPLC	MS^n esi+/esi−	—^{a}	2003 (195)
Kidney	CAP, TAP, FLF	—^{a}	—	RPLC	MSMS esi−	0.3	2006 (213)
Muscle	CAP, FLF	ACN/MeOH extn, hexane defat.	—	RPLC	MSMS esi+	3 and 2	2006 (214)
Shrimp flesh	CAP, TAP, FLF, FFA	EtOAc/ACN extn, hexane defat., C$_{18}$ SPE, cation exchange SPE	—	RPLC	MS^n esi+	0.5	2003 (215)
Fish flesh, shrimp flesh muscle	FLF, FFA	EtOAc extn, hexane defat., EtOAc purif., MCX SPE	—	GC	micro-ECD	0.5 and 1.0	2006 (117)

^{a} Not reported.

[IS] Internal standard.

Note: ppb: parts per billion equivalent to μg of residue/kg of tissue; CAP: chloramphenicol; TAP: thiamphenicol; Extn: extraction; SPE: solid-phase extraction; Pyr: pyridine; TMSA: *N*,*O*-bis(trimethylsilyl)acetamide; TMSI: *N*-trimethylsilylimidazole; TMCS: trimethylchlorosilane; GC: gas chromatography; FID: flamme ionization detector; ACN: acetonitrile; defat.: defattening; EtOAc: ethyl acetate; purif.: purification; IAC: immunoaffinity column; ECD: electron capture detector; HDMDS: hexadimethyldisilazane; RPLC: reverse-phase liquid chromatography; MSPD: matrix solid-phase dispersion; Neutral.: NaOH neutralization; DCM: dichloromethane; MS: single quadrupole; nci: negative ion chemical ionization; MSTFA: *N*-methyl-*N*-(trimethylsilyl) trifluoroacetamide; MSMS: triple quadrupole; BSA: *N*,*O*-Bis(trimethylsilyl)acetamide; esi−: electrospray source in negative mode;. MS^n: ion trap; MeOH: methanol; partition.: aqueous partitioning; FLF: florfenicol; esi+: electrospray source in positive mode; FFA: florfenicol amine; MCX: medium cation exchange; HLB: hydrophilic–lipophilic balance.

the molds of *Cephalosporium*. All other beta-lactam compounds are derived semisynthetically from these natural precursors. Penicillin G was the first among all antibiotics discovered accidentally in 1928 by Alexander Fleming, revealed by inhibiting the growth of *Staphylococcus aureus* strains. Both classes of beta-lactams contain bulky side chain attached, respectively, to 6-APA or 7-ACA nuclei (Figure 42.5). The penicillin nucleus is a reactive unstable beta-lactam (four-membered) ring coupled to a thiazolidine (five-membered) ring to form the penam ring system. The cephalosporin nucleus differs from the penicillin nucleus by having a dihydrothiazine (six-membered) ring coupled to the beta-lactam ring to form the 3-cephem ring system and conferring a better stability to the molecule compared to the penam ring.

Several other beta-lactam antibiotic subfamilies have been discovered and synthetically modified, such as the cephamycins (extracted from *Actinomycetes*) or the clavulanic acid (a beta-lactamase inhibitor). The beta-lactam antibiotics act on bacteria by binding to peptidoglycan transpeptidase and inhibiting its normal cross-linking role in completing the cell wall synthesis in bacteria, causing the bacterial cell to undergo lysis and death. Beta-lactams are used at therapeutic levels in veterinary practice primarily to treat disease and prevent infection. Used at subtherapeutic levels, they increase feed efficiency and promote growth of food-producing animals besides preventing spread of disease from the herd or flock of animals kept at a level of optimum productivity.

Penicillins are medium acidic polar drugs (p$K_{a\ (-COOH)}$ ranging from 2.4 to 2.7) and are relatively unstable in aqueous solutions. Their degradation is catalyzed by both acids and bases. They are also extremely susceptible to nucleophilic reactions in aqueous solutions. The optimum stability for the amphoteric penicillins (ampicillin, amoxicillin) with their second p$K_{a(NH_2)}$ ranging from 7.2 to 7.4, occurs at a pH that coincides with their respective isoelectric point; for the monobasic penicillins (penicillin G, cloxacillin, etc.), all of which have measured acid dissociation constants p$K_{a\ (-COOH)}$ less than 3, this generally occurs at a pH between 6 and 7. The presence of an unstable four-term ring in the beta-lactam moiety makes these compounds prone to degradation by heat and/or in the presence of alcohols. Penicillins are also readily isomerized in an acidic solution. The chemical instability of the beta-lactams, particularly of the penicillins, has important consequences for extraction and chromatography conditions. The relative stability of incurred penicillin G, amoxicillin and ampicillin residues in animal muscle tissues under various storage conditions has also been studied [216–223], demonstrating fairly good stability incurred in muscle tissues stored at –70 to –80°C (more than one year) compared to a lower stability at –20 to –30°C (less than 3 months [216,222]). Cephalosporins are somewhat more resistant to breakdown than many other beta-lactam compounds. But, as with the penicillins, they can be subjected to enzymatic degradation by specific beta-lactamases (cephalosporinases). New highly stable third generation of cephalosporins such as cefquinome and ceftiofur are available in veterinary practice since the 1990s.

Several useful reviews including methods for residue analysis in meat have been published on this family of antibiotics [9,11,13,14,16,20]. For screening the beta-lactams as residues in meat, many biological methods have been proposed. Most of them are based on the inhibitory properties of these antibiotics related to bacterial growth. They are readily detected in the presence of highly reactive specific strains such as *B. subtilis* [38,42,56,198], *B. stearothermophilus* [58,59,70,222,224], *B. megaterium* [39,40], or *Escherichia coli* [36]. The comparison of different screening methodologies has been published [32,42,60,61,225,226]. Often, these inhibitory screening methods, developed for beta-lactam at first, have been extended to large-scale multifamily methods, often called the "plate test methods." The first of these large screening tests, still officially in use in many countries in the 2000s, was the four-plate test method first described by Bogaerts and Wolf in 1980 [27] and revisited by Currie in 1998 [31]. Several other three-plate [227], five-plate [33,35], six-plate [34,37], seven-plate [29], and other multicombination/multistrain plate test [216,43] inhibitory

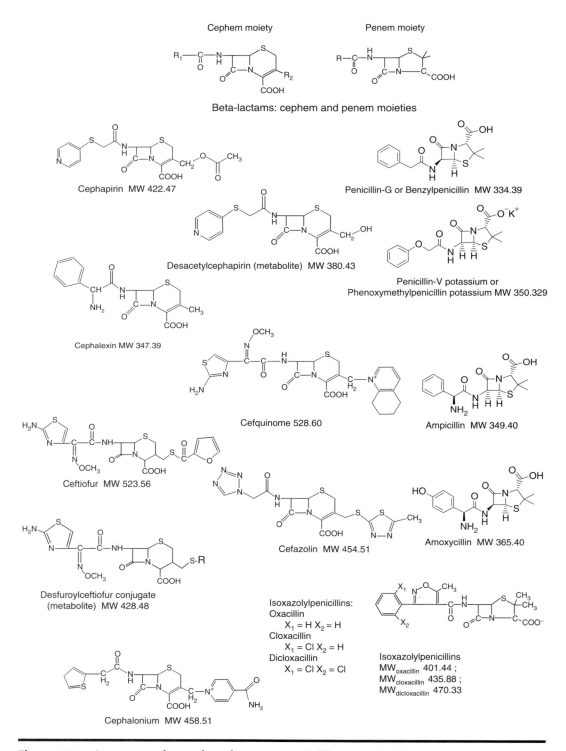

Figure 42.5 Structures of some beta-lactams—penicillins and cephalosporins.

methods have additionally been proposed either to extend the number of antimicrobials or to improve the sensitivity for some of them. TLC-bioautography was a method employed also for penicillins in the 1980s and 1990s [46,51,53–56]. A TLC method with fluorescence detection was published for ampicillin in milk and muscle tissue [104]. Beta-lactam antibiotics are also easily detected by means of rapid biological test kits such as the Charm Farm Test [46, 57,58] or the more recently developed European Premi Test [59–61]. Very few immunological ELISA test kits have been developed for beta-lactams [70,76,228]. The radioimmunological Charm II Test kit is also adapted to screen beta-lactam residues in animal tissues [46]. Finally, a SPR-BIA immuno-logical method for penicillins was proposed in 2001, but for detection in milk matrix [20, 229].

For the confirmation of the presence of beta-lactams in meat, several chromatographic meth-odologies have been investigated (Table 42.3). But, due to the unstable behavior of the beta-lactam ring and the thermal lability of these compounds, very few GC methods were developed in the 1990s [106,107], and fewer GC-MS methods [109]. Most of the developments required a liq-uid chromatographic separation. The UV detection was also quite problematic when considering extraction of the nonfluorescent and low chromophoric beta-lactams from biological matrices. Two key options were proposed: the first one was to improve the sample preparation by purifying as much as possible the biological extracts before detection at low UV wavelength (200–230 nm), where most of the beta-lactams display a good UV absorption [230–240]; the other option was to shift the UV detection by means of derivatizing reagents to higher UV wavelengths (>300 nm), where very few biological endogenous substances absorbed UV photons [176, 218,221,241–249], or, when possible, to move to fluorescent conditions by means of appropriate reagents as it was proposed for the two amphoteric penicillins: amoxycillin and ampicillin [236,250–252]. In the 2000s, most of the developments for beta-lactam residues in meat received the benefit of the mass spectrometry advances. From the methods with LC-MS thermospray (TSP) or PB sources and SQ detector employed in the 1990s [253–256], to those, in 2000s, with LC coupled to a tandem mass spectrometer (LC-MSMS) atmospheric pressure ionization (API) sources such as electrospray (ESI) or APCI and TQ detector (LC-tandem MS) [223,257–261] or ion-trap detector (LC-MS^n) [195,262], the confirmation of penicillin and cephalosporin residues became more easily achiev-able at the ppb levels requested by the regulations. However, because of the chemical behavior of beta-lactams, analytical chemistry poses a challenge for these substances present as residues in food from animal origin.

42.4.4 Macrolides and Lincosamides

The macrolide antibiotics are mostly produced by *Streptomyces* genera, characterized by a large macrocyclic lactone ring structure of 12–16-carbon-lactone ring to which several amino groups and neutral sugars are bound (Figure 42.6). Erythromycin was the first macrolide to be isolated in 1952 from *Saccharopolyspora erythraea*. It is active against gram-positive bacteria and mycoplasmas and is widely available in animal veterinary practice to treat respiratory diseases. Some macrolides (tylosin, spiramycin) have also been used as feed additives to promote growth. Other examples of macrolides include oleandomycin, tilmicosin, josamycin, and more recently, isovaleryltylosin. They are easily absorbed after oral administration and distribute extensively to tissues, especially the lungs, liver, and kidneys. They are weak bases slightly soluble in water but readily soluble in common organic solvents. Most of them are multicomponent systems containing lesser amounts of related compounds. For example, tylosin A, which is a commercialized compound, also contains small amounts of desmycosin (tylosin B), macrocin (tylosin C), and relomycin (tylosin D) [263].

Table 42.3 Summary of Literature Methods for Determination of Beta-Lactam Residues in Tissues Using Gas or Liquid Chromatography

Tissue	Analytes	Sample Treatment	Derivatization	GC and LC Columns Technique	Detection	Limit Range (ppb)	Year (Reference)
Muscle	PenG, PenV, MTH, OXA, CLX, DCX, NAF	PBS/ACN extn, water removal, ether/ PBS partition., buffer/DEE partition., buffer/ CH_2Cl_2 partition., H_3PO_4/ CH_2Cl_2 partition., SCX SPE, buffer/ CH_2Cl_2 partition., cyclohexane drying	Diazomethane	GC	NPD	3.0	1991 (106)
Muscle	PenG, PenV, CLX, DCX, AMP, AMOX	TCA extn, ion exchange SPE	Pyr, TMSA, TMSI, TMCS	GC	FID	5.0–10.0	1992 (107)
Muscle	PenG, $^{13}C_2$PenG$^{(IS)}$, PenV$^{(IS)}$	PBS/ACN extn, water removal, ether/ PBS partition., buffer/DEE partition., buffer/ CH_2Cl_2 partition., H_3PO_4/ CH_2Cl_2 partition., SCX SPE, buffer/ CH_2Cl_2 partition., cyclohexane drying	Diazomethane	GC	MS ei	3.0	1998 (109)
Muscle, liver, kidney	PenG	$NaWO_4$/H_2SO_4 extn-deprot., Al_2O_3 SPE, C18 SPE	—	RPLC-C18	UV, 210 nm	5.0	1985 (230)
Muscle	PenG, CEP AMP$^{(IS)}$	MSPD/MeOH extn, PBS extn	—	RPLC-C18	UV, 230 nm		1989 (231)
Muscle	PenG, PenV, CLX	ACN extn, H_3PO_4/CH_2Cl_2 partition., ACN extn, hexane defat., PBS extn	—	PLRP-S	UV, 210 nm	5.0	1992 (232)
Muscle	CEFT, DFCC	DTE extn, C18 SPE, SAX SPE, SCX SPE	Iodoacetamide	RPLC-C18	UV, 266 nm	100	1995 (233)
Muscle, kidney	PenG	Et_4NCl/ACN extn, LC fract. cleanup	—	RPLC-C18	UV, 215 nm	5.0	1998 (234)
Muscle, liver, kidney	DFCC	Et_4NCl/ACN extn, LC fract. cleanup	—	RPLC-C18, ion pairing DSF, DDSF	UV, 270 nm	—[a]	1998 (235)

Tissue	Analytes	Extraction/cleanup	Derivatization	LC	Detection	LOD	Year (Ref.)
Muscle, kidney, liver	AMOX, AMP, PenG, CLX, DFCC, DACEP	PBS extn, LC fract. cleanup	—	RPLC-C18	UV 210 nm, 270 nm	5.0	1998 (236)
Muscle, kidney, liver	PenG	PBS extn, ultrafiltration	—	RPLC-C4	DAD, 211 nm	40	2001 (237)
Muscle	DACEP, CEP, CEFQ, CEPX	Buffer extn, isooctane defat., C18 SPE	—	RPLC-C18	UV, 252 nm	12, 12, 9, 45	2002 (238)
Muscle, kidney	CEFT, DFCA	DTE extn, C18 SPE, SAX SPE, SCX SPE	Iodoacetamide	RPLC-C18	UV, 266 nm	100	2003 (239)
Muscle	PenG, PenV, CLX, DCX, OXA, NAF, AMP, AMOX, CEP, CEFT	ACN extn, hexane defat., C18 SPE	—	RPLC-C18	DAD, 21 nm and 228 nm and 270 nm	40	2004 (240)
Muscle, kidney, liver	PenG, PenV(IS)	PBS extn, C18 SPE	Triazole-HgCl$_2$ and acetic anhydride	RPLC-C18, ion pairing THS	UV, 325 nm	5.0	1991 (241)
Muscle, kidney, liver	PenG, PenV(IS)	PBS extn, C18 SPE	Triazole-HgCl$_2$ and acetic anhydride	RPLC-C18, ion pairing THS	UV, 325 nm	5.0	1992 (218)
Muscle	AMOX, CFDX	PBS extn, online dialysis, ion pair C18 SPE	Postcolumn: NaOH	RPLC-C18, ion pairing HTACl	UV, 260 nm	50, 200	1994 (242)
Muscle	AMP	PBS extn, hexane defat., C18 SPE	Triazole-HgCl$_2$ and acetic anhydride	RPLC-C8, ion pairing THS/TBA	UV, 325 nm	5.0	1996 (243)
Muscle, liver	AMOX, AMP	NaWO$_4$/H$_2$SO$_4$ extn-deprot., SCX SPE, PGC SPE	Triazole-HgCl$_2$ and acetic anhydride	RPLC-C8, ion pairing THS	UV, 325 nm	5.0	1997 (244)

(Continued)

Table 42.3 (Continued)

Tissue	Analytes	Sample Treatment	Derivatization	GC and LC Columns Technique	Detection	Limit Range (ppb)	Year (Reference)
Muscle, liver	PenG, PenV[(IS)]	PBS extn, C18 SPE	Triazole-HgCl$_2$ and acetic anhydride	RPLC-C18, ion pairing THS	UV, 325 nm	5.0	1997 (221)
Muscle	PenG, CLX, AMP, AMOX, PenV[(IS)]	PBS extn, t-C18 SPE	Triazole-HgCl$_2$ and acetic anhydride	RPLC-C18, ion pairing THS	UV, 325 nm, 340 nm	5.0	1998 (245)
Muscle	PenG, CLX, AMP, AMOX, PenV[(IS)]	PBS extn, t-C18 SPE	Triazole-HgCl$_2$ and acetic anhydride	RPLC-C18, ion pairing THS	UV, 325 nm, 340 nm	5.0	1998 (246)
Muscle	PenG, CLX, DCX, AMP, AMOX	MSPD extn, hexane defat., C18 SPE	Triazole-HgCl$_2$ and acetic anhydride	RPLC-C18	UV, 325 nm, 340 nm	20.0	1998 (247)
Muscle	PenG, NAF, CLX, DCX, OXA, AMP, AMOX	PBS extn, isooctane defat., HLB SPE	Triazole-HgCl$_2$ and benzoic anhydride	RPLC-C18	UV, 325 nm, 340 nm	8.0–11.0	1999 (248)
Muscle	PenG, PenV, CLX, DCX, OXA, NAF, AMP, AMOX	PBS extn, isooctane defat., C18 SPE	Triazole-HgCl$_2$ and benzoic anhydride	RPLC-C8, ion pairing THS/TBA	UV, 325 nm, 340 nm	3.0–10.0	1999 (176)
Muscle	AMP, AMOX, PenG, PenV, CLX, DCX, OXA, NAF	PBS extn, isooctane defat., C18 SPE	Triazole-HgCl$_2$ and benzoic anhydride	RPLC-C8, ion pairing THS/TBA	UV, 325 nm, 340 nm	3.0–10.0	2002 (249)
Muscle, kidney, liver	AMOX, AMP, PenG, CLX, DFCC, DACEP	PBS extn, LC fract. cleanup	Formaldehyde	RPLC-C18	FLD, Exc 358 nm, Em 440 nm	5.0	1998 (236)
Muscle	AMOX	PBS extn, C18 SPE	Formaldehyde	RPLC-C18	FLD, Exc 358 nm, Em 440 nm	5.0	2000 (250)

Matrix	Analytes	Extraction/cleanup	Derivatization	Column	Detection	LOD	Year (Ref.)
Muscle, kidney, liver	AMOX	TCA extn/deprot.	Salicylaldehyde	RPLC-C18	FLD, Exc 358 nm, Em 440 nm	6.0–16	2000 (251)
Feed	AMOX, AMP	Water/ACN extn,	Formaldehyde	RPLC-C18	FLD, Exc 358 nm, Em 440 nm	5000	2003 (252)
Muscle	PenG	—[a]	—	RPLC-C18	MS pb-nci	—[a]	1990 (253)
Muscle, kidney	PenG Nafcillin[(IS)]	Water/ACN extn, o-H_3PO_4 neutral., DCM purif.	—	RPLC-C18; ion pairing TBA	MS esi–	25	1994 (254)
Muscle	OXA, CLX, DCX	EtOAc acidic extn, C18 SPE	—	RPLC-C18	MS pb-nci	40–50	1994 (255)
Muscle, kidney, liver	PenG, PenV, CLX, DCX	TCA/Acetone extn, C18 SPE	—	RPLC-C18	MS esi–	15	1998 (256)
Muscle, kidney, liver	PenG, PenV, DCX, OXA, NAF	Extn, C18 SPE, QMA ion exchange SPE	—	RPLC-C18; ion pairing DBAA	MSMS esi–	20	2001 (257)
Muscle, kidney, liver	PenG, PenV, CLX, DCX, AMP Pheneticillin[(IS)]	Aqueous extn, C18 SPE	—	RPLC-C18	MSMS esi–	6.0–15.0	2003 (258)
Muscle, kidney, liver	AMOX	—[a]	—	RPLC-C18	MSMS esi–	—[a]	2003 (223)
Muscle, kidney, liver	PenG, PenV, CLX, DCX, OXA, NAF	Extn, C18 SPE, QMA ion exchange SPE	—	RPLC-C18	MSMS esi–	20	2003 (259)
Muscle, kidney, liver	PenG, PenV, CLX, DCX, OXA, NAF, PenG-d5[(IS)] Nafcillin-d6[(IS)]	NaCl aqueous extn, and $NaWO_4/H_2SO_4$ deprot. for liver and kidney	—	RPLC-C18	MSMS esi–	2.0–10.0	2004 (260)

(Continued)

Table 42.3 (Continued)

Tissue	Analytes	Sample Treatment	Derivatization	GC and LC Columns Technique	Detection	Limit Range (ppb)	Year (Reference)
Kidney	PenG, AMOX, CEP, DACEP, DCCD, AMP, CFZ, OXA, CLX, NAF, DCX, PenV[(IS)]	ACN extn, MSPD	—	RPLC-C18	MSMS esi+	1.0	2005 (261)
Muscle, injection sites	PenG	MeOH extn, evap., water dilution, C18 SPE, Evap., reconst.	—	RPLC-C18	MS^n esi+/esi−	—[a]	2003 (195)
Fish flesh	PenG, AMOX, CEPX, AMP, OXA, CLX, DCX	ACN Extn; hexane defat.; water/ACN; hexane defat.	—	RPLC-phenyl	MS^n esi+/esi−	100–1000	2005 (262)

[a] Not reported.
[(IS)] Internal standard.

Note: ppb: parts per billion equivalent to μg of residue/kg of tissue; PenG: benzylpenicillin or Penicillin G; PenV: phenoxymethylpenicillin or Penicillin V; MTH: methicillin; OXA: oxacillin; CLX: cloxacillin; DCX: dicloxacillin; NAF: nafcillin; PBS: phosphate buffer solution; ACN: acetonitrile; SCX: strong cation exchange; SPE: solid-phase extraction; NPD: nitrogen specific detector; TCA: trichloroacetic acid; TMSA: N,O-bis(trimethylsilyl) acetamide; TMSI: N-trimethylsilylimidazole; TMCS: trimethylchlorosilane; UV: ultraviolet detection; AMP: ampicillin; RPLC: reverse phase liquid chromatography; defat.: defattening; DSF: decanesulfonate; DDSF: dodecylsulfate; DFCC: desfuroylceftiofur cysteine; DACEP: desacetylcephapirin; CEP: cephapirin; CEFQ: cefquinome; CEPX: cephalexin; AMX: amoxycillin; DCCD: desfuroylceftiofur cysteine disulfide; DFCA: desfuroylceftiofuracetamide; CEFT: ceftiofur; CFZ: cefazolin; CFDX: cefadroxil; Extn: extraction; MeOH: methanol; Et_4NCl: tetraethylammonium chloride; DTE: dithioerythritol; deprot.: deproteinisation; SAX: strong anion exchange; PGC: porous graphitic carbon; HLB: hydrophilic Lipophilic Balance; MCX: medium cation exchange; LC Fract.: LC fractionation; MSPD: matrix solid phase dispersion; Neutral.: NaOH neutralization; DCM: dichloromethane; DBAA: di-n-butylamine acetate; Pyr.: pyridine; PLRP-S: copolymeric reverse phase column; HTACl: hexadecyltrimethylammonium chloride; THS: sodium thiosulfate; TBA: tetrabutylammonium hydrogenosulfate; FID: flamme ionization detector; FLD: fluorescence detection; DAD: UV diode array detection; MS:single quadrupole; MS^n: ion trap; nci: negative ei: electron impact source; pb-nci: particle beam source in negative ion chemical ionization; MSMS: triple quadrupole; MS[n]: ion trap; nci: negative ion chemical ionization; esi+: electrospray source in positive mode; esi−: electrospray source in negative mode.

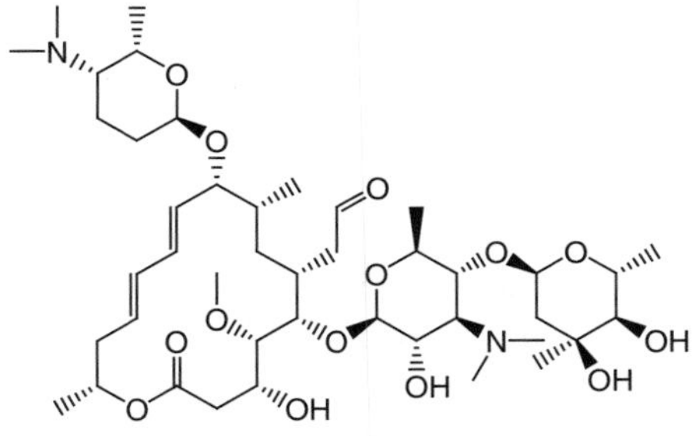

Spiramycin I, MW 843.06

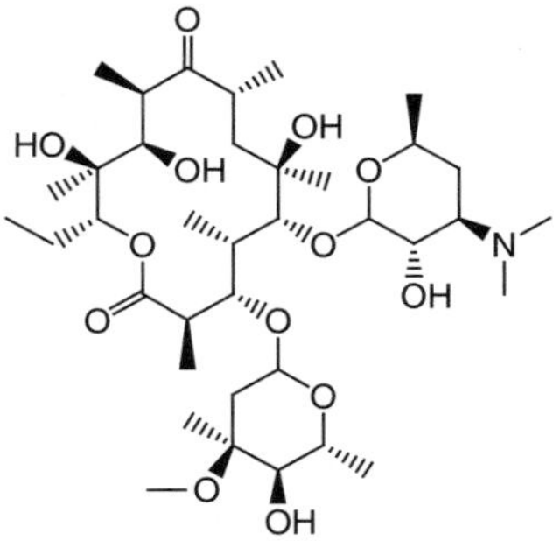

Erythromycin A, MW 733.93

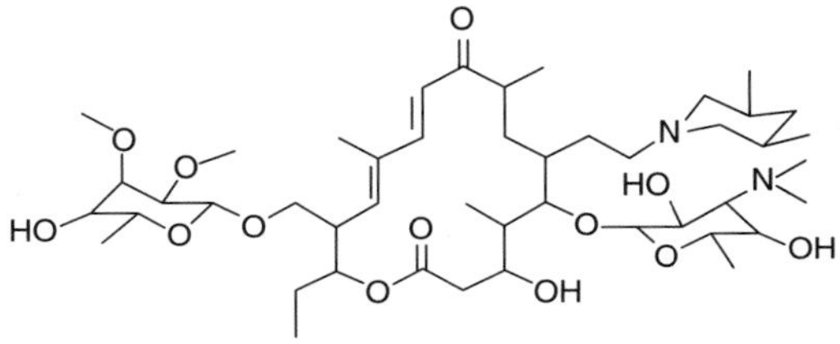

Tilmicosin, MW 865.15

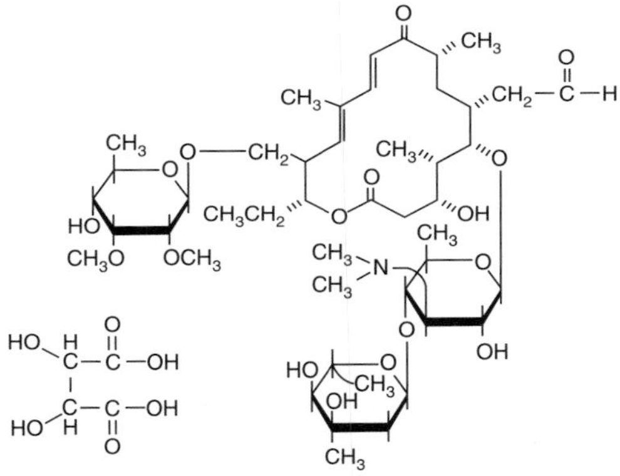

Tylosin tartrate
MW 1014.10

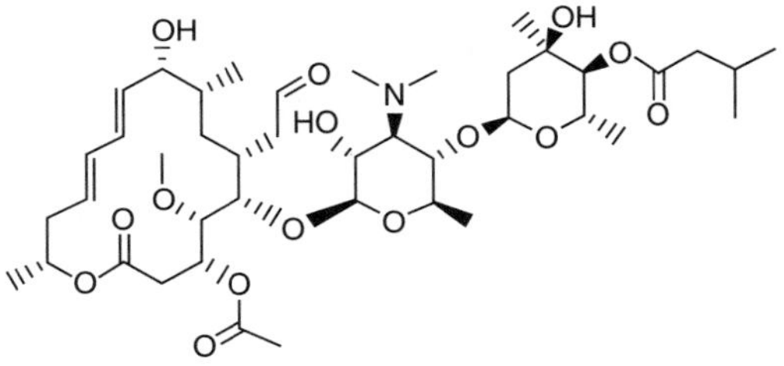

Josamycin, MW 827.99

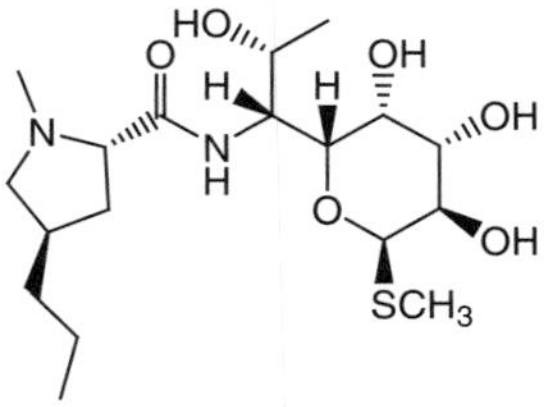

Lincomycin MW 406.54

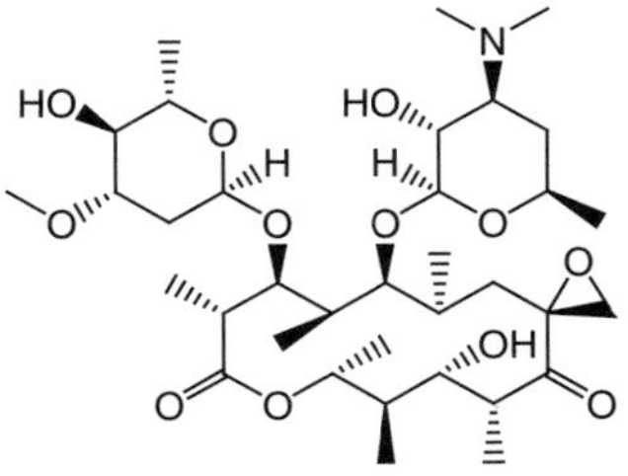

Oleandomycin MW 687.86

Figure 42.6   Structures of some macrolides and lincosamides.

In following the same principle, erythromycin A is commercialized with small but quantifiable amounts of erythromycin B and C [264]. Spiramycin I is generally found with spiramycins II and III [265]. Moreover, spiramycin can easily be degraded to neospiramycin under acidic conditions and the sum of neospiramycin and of spiramycin should be taken into account for spiramycin residue control in food as of regulation 1442/95/EC [266].

No comprehensive review specifically dedicated to macrolide residue analysis in food has been addressed in the recent years but several reviews dealing with antibiotic residue analysis attributed a part of their content to macrolide antibiotics [9,11,13–15,267].

For screening the macrolide antibiotics as their residues in meat, several biological methods have been proposed. Most of them are based on the inhibitory properties of these antibiotics related to bacterial growth. As for beta-lactam antibiotics, some of them (erythromycin, tylosin, tilmicosin, spiramycin, and lincomycin) are readily detected by microbiological inhibitory tube tests in the presence of specific strains such as *B. stearothermophilus*. It is the case of (the Charm Farm Test [58], the Premi Test [59]), or by microbiological inhibitory plate tests in multiplate configurations in the presence of specific strains such as *Micrococcus luteus* [29,37,43] or *B. megaterium* [40]. The comparison of different screening methodologies has been published to evaluate among several other antimicrobial families the response and detectability of macrolide residues in regard to regulated limits as for EU-MRLs [32,42,60,61]. TLC methods have also been proposed for some macrolides in the form of TLC-bioautography with *B. subtilis* as the revealing strain [46, 55]. An ELISA test kit for macrolide was also investigated by Draisci et al. [72]. Because of the complex structure and composition of macrolides and of their relatively weak UV absorption, the development of chromatographic methods for their determination in foods and muscle tissues has been rather limited in the 1980s [268]. Nevertheless, several HPLC-UV methods (Table 42.4), which were able to cover at least two, tylosin and tilmicosin, and often more macrolides within the same multiresidue run of analysis, were proposed in the past 15 years [240,269–272]. Few but potentially interesting methods proposed HPLC with fluorescence detection after fluorescent labeling derivatization of some macrolide compounds such as josamycin, erythromycin, and oleandomycin [273,274]. Considering the mass spectrometric detector, it opens widely the field of macrolide detection and identification. From the methods with LC-MS thermospray (TSP) or PB sources and SQ detector employed in the 1990s [275,276], to those in 2000s with LC-MS API sources such as electrospray (ESI), APCI, TQ detector (LC-tandemMS) [214,223,262,277–281], or LC-MS^n [195], the confirmation of macrolide residues became more easily achievable at the ppb levels requested by the regulations.

42.4.5 Nitrofurans

Nitrofurans are synthetic compounds adapted from the 5-nitrofuran nucleus, all displaying a broad-spectrum activity. They are bacteriostatic antimicrobials acting by inhibition of some microbial enzymes involved in carbohydrate metabolism. They were widely used in veterinary medicine against gastrointestinal infections in cattle, pigs, and poultry. The major members of this antibacterial family are furazolidone, furaltadone, nitrofurazone, and nitrofurantoin (Figure 42.7). Following evidence of mutagenicity and genotoxicity of furazolidone in the late 1980s and early 1990s, legislation changed regarding the 5-nitrofuran nucleus compounds, which were all prohibited for use in food-producing animals in many countries. Listed in the Annex IV of EU Council Regulation 2377/90/EC [1,266,282], a minimum required performance limit (MRPL) for analytical methods developed for the residue control of nitrofurans in food has been set in the

Table 42.4 Summary of Literature Methods for Determination of Macrolide and Lincosamide Residues in Tissues Using Gas or Liquid Chromatography

Tissue	Analytes	Sample Treatment	Derivatization	LC Column Technique	Detection	Limit Range (ppb)	Year (References)
Muscle, kidney	Tylo	ACN extn, CH_2Cl_2 partition., ACN/ether defat.	—	RPLC-C18	UV, 278 nm	200	1982 (268)
Muscle, kidney	Tylo, Tilmico	ACN extn, C18 SPE	—	RPLC-C18	UV, 287 nm	20, 10	1994 (269)
Muscle, kidney, liver	Spira, Tylo, Josa, Kitasa, Mirosa	MPA/MeOH deprot., SCX SPE	—	RPLC-C18	UV, 232 nm and 287 nm	50	1998 (270)
Muscle	Spira, Neospira, Tylo, Tilmico	ACN extn, C18 SPE	—	RPLC-C18	UV, 232 nm (spira, neospir) and 287nm (tylo, tilmico)	30, 25, 15, 15	1999 (271)
Muscle	Spira, Tilmico, Tylo, Josa, Kitasa, Erythro, Oleando	MPA/MeOH deprot., SCX SPE	—	RPLC-C18	UV, 232 nm and 287 nm	6 – 33 (S,T,J,K) and 400 (E,O)	2001 (271)
Muscle	Tylo, Spira, Neospira, Tilmico, Josa, Kitasa, Mirosa, Roxithro	ACN extn, hexane purif., HLB-SCX SPE	—	RPLC-C18	UV, 210 nm, 228 nm, and 287 nm	40	2004 (240)
Muscle, liver, kidney	Josa	PBS/ACN	Cyclohexane-1, 3-dione	RPLC-C18	FLD, Exc:375 nm, Em:450 nm	100	1994 (273)
Muscle, liver, kidney	Erythro, Oleando Roxithro[IS]	ACN extn, hexane purif., SCX SPE,	FMOC	RPLC-C18	FLD, Exc:260 nm, Em:305 nm	50–100	2002 (274)
Muscle	Tylo, Spira, Erythro	$CHCl_3$ extn, Diol SPE	—	RPLC-C18	MS pb-nci	20	1994 (275)
Muscle	Tylo, Spira, Erythro, Josa, Tilmico	$CHCl_3$ extn, Diol SPE	—	RPLC-C18	MS pb-pci and pb-nci	50	1996 (276)
Muscle	Erythro, Tylo, Tilmico	$CHCl_3$ extn, Diol SPE	—	RPLC-C18	MSMS esi+	25–40	2001 (277)

(Continued)

Table 42.4 (Continued)

Tissue	Analytes	Sample Treatment	Derivatization	LC Column Technique	Detection	Limit Range (ppb)	Year (References)
Muscle, liver, kidney	Spira, Tylo, Erythro, Timico, Josa Roxithro[(IS)]	Tris buffer extn, acetic acid/ NaWO$_4$ deprot., HLB SPE	—	RPLC-C18	MSMS esi+	25–35	2001 (278)
Muscle	Erythro, Roxithro, Tylo, Tiamul	—	—	RPLC-C18	MS esi+	1.0–10	2002 (279)
Muscle	Tylo A Spiramycin[(IS)]	—[a]	—	RPLC-C18	MSMS esi	—[a]	2003 (223)
Feed	Spira, Tylo	MeOH/H$_2$O extn, HLB SPE, Dilut ACN/H$_2$O,	—	RPLC-C18	MSMS esi+	<1000	2003 (280)
Fish flesh	Erythro, $^{13}C_2$-Erythro[(IS)]	ACN extn, hexane defat.	—	RPLC-C18	MSMS esi+	20	2005 (281)
Fish flesh	Tylo, Tilmico, Erythro, Linco	ACN extn, hexane defat., water/ACN,	—	RPLC-phenyl	MS^n esi+/ esi–	10	2005 (262)
Muscle	Erythro, Tylo, Tilmico, Josa, Kitasa, Linco, Clinda, Oleando	ACN/MeOH extn, hexane/ ACN defat.	—	RPLC-C18	MSMS esi+	0.2 – 2.0	2006 (214)
Muscle, injection site	Tilmico, Linco	MeOH extn, Evap, Water dilution, C18 SPE, Evap., Reconst.	—	RPLC-C18	MS^n esi+/esi–	—[a]	2003 (195)

[a] Not reported.
[(IS)] Internal standard.

Note: ppb: parts per billion equivalent to μg of residue/kg of tissue; ACN: acetonitrile; Extn: extraction; defat.: defattening; RPLC: reverse phase liquid chromatography; UV: ultraviolet detection; Tylo: tylosin; Tilmico: tilmicosin; MPA: metaphosphoric adic; MeOH: methanol; deprot.: deproteinisation; SCX: strong cation exchange; HLB: hydrophilic Lipophilic Balance; SPE: solid-phase extraction; Spira: spiramycin; Josa: josamycin; Kitasa: kitasamycin; Mirosa: mirosamycin; Neospira: neospiramycin; Roxithro: roxithromycin; Linco: lincomycin; Oleando: oleandomycin; Tiamul: tiamulin (not a macrolide but a pleuromutiline); Clinda: clindamycin; purif.: purification; partition.: aqueous partitioning; Neutral.: NaOH neutralization; Tris: tris(hydroxymethyl)aminomethane; DCM: dichloromethane; EtOAc: ethyl acetate; PBS: phosphate buffer solution; FMOC: 9-fluoromethylchloroformate; MSPD: matrix solid phase dispersion; PLRP-S: copolymeric reverse phase column; FLD: fluorescence detection; MS:single quadrupole; pci: positive ion chemical ionization; nci: negative ion chemical ionization; MSMS: triple quadrupole; MS^n: ion trap; esi$^+$: electrospray source in positive mode; esi$^-$: electrospray source in negative mode.

EU at 1.0 μg/kg [283]. Besides, the detection of nitrofurans is difficult in tissue matrices because of extremely rapid metabolization *in vivo* (<1 day). But it was found in the late 1980s that furazolidone was able to bind extensively to proteinaceous tissues, and that acidic treatment could be efficiently applied to release a metabolized compound directly related to furazolidone, that is, the 3-amino-2-oxazolidinone (AOZ) [284]. Following this finding, the corresponding metabolites for the other nitrofurans were also extracted from tissues, confirming that parent compound monitoring was actually ineffective in food-producing animal tissues. The 3-amino-5-morpholinomethyl-2-oxazolidinone (AMOZ) metabolite was extracted from protein-bound furaltadone, the 1-aminohydantoin (AHD) from nitrofurantoin, and the semicarbazide (SEM) from nitrofurazone (Figure 42.7).

Several reviews over the past 15 years covered the analysis of nitrofurans and their metabolites [11,13–15,267].

Owing to the aforesaid regulations and to the problematic issue in residue analysis of the rapidly metabolized nitrofurans, their monitoring in edible animal products has been, since the 2000,

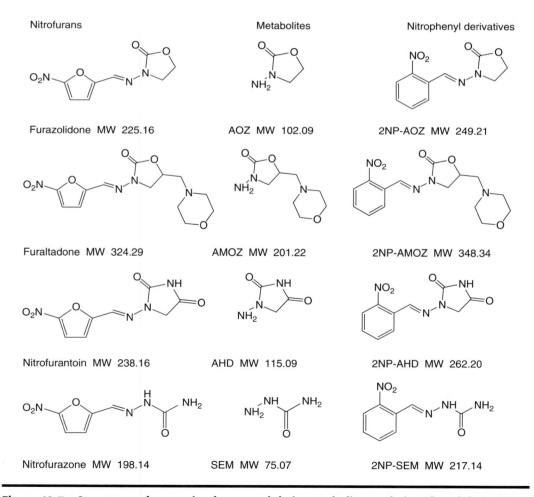

Figure 42.7 **Structures of some nitrofurans and their metabolites and nitrophenyl derivatives.**

focused essentially on the LC-MS technology, avoiding the screening by microbiological methods or TLC [103, 202], which were found unadapted for nitrofuran compounds, and also disregarding the conventional GC or LC techniques that had been developed in the 1980s and 1990s [107, 206, 285–288]. Yet, in relation to the screening step, it can be mentioned an immunological screening of nitrofurans by means of two recent ELISA kits able to detect AOZ and AMOZ at very low ppb level in different matrices including muscle tissues (<1 μg/kg) [79]. ELISA screening for SEM and AHD is still under development. HPLC-UV methods have also been investigated in the early 2000 (Table 42.5), considering the four major nitrofuran metabolites as target residues in muscle tissue [289,290]. But, most of the recent developments on nitrofuran residue control in meat has been extensively supported by the LC-MS technology, including seldom methods with SQ detectors [291] and frequently with tandem MS or ion-trap MS detectors [214,284,292–297]. Apart from the four major nitrofurans, a fifth one, the nifursol, was employed as a feed additive against histomoniosis poultry infections. As a consequence of its ban in 2002 under regulation 1756/2002/EC [298], LC-MSMS methods have been proposed to monitor the nifursol metabolite, the 3,5-dinitrosalicylic acid hydrazide (DNSH), either specifically as a single metabolite monitored in poultry muscle tissues [299,300] or as the fifth residue monitored in a multinitrofuran metabolite analysis [301]. In addition, the problematic analysis of semicarbazide, which can be found as a protein-bound, but also as unbound, compound in meat and in many other food products (e.g., baby food), should be cited. It is not generated by nitrofurazone metabolization after a veterinary treatment, but produced by several external contaminations of meat products, one of which being the flour coating of poultry meat. Cereal flours can be treated with legal concentrations of azodicarbonamide (ADC)—a chemical substance easily transformed to biurea and finally to free semicarbazide [302–303].

42.4.6 Nitroimidazoles

Nitroimidazoles are antiprotozoals and bactericidal antimicrobials active against gram-negative and also many gram-positive bacteria. They are obtained synthetically and their structure is based on a 5-nitroimidazole ring while two protonic positions in N1 and C2 can be substituted by several groups to give different members of the family. Two methyl substitutions lead to the dimetridazole compound. A methyl and an ethanolic substitution give the metronidazole (Figure 42.8). The destructive action of nitroimidazoles takes place in the bacterial cell when the 5-nitro group is reduced by nitro-reductase bacterial proteins of the anaerobic bacteria, leading to free radicals or intermediate products, most of them cytotoxic for the bacteria.

Four major nitroimidazoles were commonly used in veterinary medicine or employed as feed additives in poultry prophylactic treatments against histomoniosis and coccidiosis: the dimetridazole, the metronidazole, the ronidazole, and the ipronidazole. The metronidazole is known to be easily transformed *in vivo* into both its alcoholic metabolite, the hydroxymetronidazole that is even more active against anaerobic bacteria and into its acidic metabolite, the acetylmetronidazole that bears no bactericidal activity anymore.

Dimetridazole can also experience an *in vivo* metabolization to give the hydroxydimetridazole metabolite, that is, the 2-hydroxymethyl-1-methyl-5-nitroimidazole (HMMNI). Ronidazole is extensively and quite exclusively metabolized to compounds without the intact nitroimidazole ring structure, also generating small amounts of the HMMNI too. Ipronidazole is also metabolized *in vivo* to its specific hydroxylated counterpart. Owing to their mutagenic, carcinogenic, and toxic properties toward eukaryotic cells, the nitroimidazoles have been prohibited for use in food-producing

Table 42.5 Summary of Literature Methods for Determination of Nitrofuran Residues in Tissues Using Gas or Liquid Chromatography

Tissue	Analytes	Sample Treatment	Derivatization	LC Column Technique	Detection	Limit Range (ppb)	Year (Reference)
Muscle, liver	Nitrofurazone, furazolidone, nitromide, sulfanitran	CHCl₃/EtOAc/DMSO Extn, Alumina SPE	—	RPLC-C18	Electrochemical, reductive mode	2.0–6.0	1989 (285)
Muscle, liver	Furazolidone, nitrofurazone, furaltadone, nitrofurantoin	ACN/EtOAc/DCM Extn, hexane defat., PBS dilution	—	RPLC-CN	UV, 365 nm	1.0	1989 (286)
Muscle	Furazolidone, nitrofurazone	TCA extn, ion exchange SPE	Pyr, TMSA, TMSI, TMCS	GC	FID	100, 50	1992 (107)
Kidney	Furazolidone, nitrofurazone, furaltadone, nitrofurantoin	ACN extn, C18 SPE, Silica SPE	—	RPLC-C18	UV-DAD, 359 nm, 370 nm	2.0–3.0	1995 (287)
Muscle	Furazolidone, nitrofurazone[IS]	ACN extn, Partition., Reconst.	—	RPLC-C18	UV-DAD, 365 nm	3.0	1995 (288)
Muscle	Furazolidone	MSPD	—	RPLC-C18	UV-DAD, 365 nm	3.5	1997 (206)
Liver	AOZ	PED, HCl hydrol., 2-NBA deriv, neutral., MAX SPE, HLB SPE	2-NBA	RPLC-C18	UV, 275 nm	—ᵃ	2002 (289)
Liver	Protein-bound AOZ	H₂O,MeOH,EtOH, EtOAc washings, HCl hydrol. extn, 2-NBA deriv, Neutral/PBS dilution, EtOAc partition., Tris dilution; MAX SPE; HLB SPE	2-NBA	RPLC-C18	UV, 275 nm MSMS esi+	2.0, < 1.0	2003 (290)

(Continued)

Table 42.5 (Continued)

Tissue	Analytes	Sample Treatment	Derivati-zation	LC Column Technique	Detection	Limit Range (ppb)	Year (Reference)
Muscle	AOZ, AMOZ, AHD, SEM	TCA extn, hexane defat., 2-NBA deriv, neutral./PBS dilut., C18 SPE, CHCl₃ extn, H₂O dilut.	2-NBA	RPLC-C18	MS esi+	0.5	2004 (291)
Muscle, liver	Protein-bound AOZ	MeOH, EtOH, EtOAC washings, HCl hydrol. extn, 2-NBA deriv, neutral/PBS dilution, EtOAc partition., ACN/H₂O reconst.	2-NBA	RPLC-C18	MS TSP+	0.5	1997 (284)
Muscle	AOZ, AMOZ, AHD, SEM, 4NBA-SEM$^{(IS)}$	HCl hydrol. extn, 2-NBA deriv, neutral, PBS dilution, C18 SPE	2-NBA, 2-NBA-d4	RPLC-C18	MSMS esi+	0.5–5.0	2001 (292)
Muscle	AOZ, AMOZ, AHD, SEM	HCl hydrol. extn, 2-NBA deriv, neutral/ PBS dilution, hexane defat., HLB SPE	2-NBA	RPLC-C18	MSMS esi+	0.2–0.5	2003 (293)
Muscle	AOZ, AMOZ, AHD, SEM, AOZd4$^{(IS)}$, AMOZd5$^{(IS)}$	HCl hydrol. extn, 2-NBA deriv, Neutral/ PBS dilution, EtOAc partition., reconst. ACN/Acetic acid	2-NBA	RPLC-C18	MSMS esi+	0.1–0.5	2004 (294)
Muscle	AOZ, AMOZ, AHD, SEM, AOZd4$^{(IS)}$, AMOZd4$^{(IS)}$, AHDd4$^{(IS)}$, SEMd4$^{(IS)}$	HCl hydrol. extn, 2-NBA deriv, neutral/ PBS dilution, EtOAc partition., reconst. H₂O, hexane defat., C18 SPE, reconst. H₂O/ACN	2-NBA, 2-NBA-d4	RPLC-C18	MSMS esi+	0.2	2005 (295)

Matrix	Analytes	Method	Derivatization	Column	Detection	LOD	Year (Ref.)
Muscle, Egg	AOZ, AMOZ, AHD, SEM, AOZd4(IS), AMOZd5(IS), $^{13}C^{15}N_2SEM$(IS)	HCl hydrol. extn, 2-NBA deriv, Neutral/PBS dilution, EtOAc partition, reconst. H_2O/ACN	2-NBA	RPLC-C18	MSMS esi+	0.3	2005 (296)
Shrimp flesh	AOZ, AMOZ, AHD, SEM, AOZd4(IS), AMOZd5(IS)	HCl hydrol. extn, 2-NBA deriv, neutral/PBS dilution, EtOAc partition, reconst. acetic acid	2-NBA	RPLC-C18	MSMS esi+	< 0.5	2006 (297)
Muscle	Nifuroxazide	ACN/MeOH extn, Hexane/ACN defat.	—	RPLC-C18	MSMS esi+	0.2	2006 (214)
Muscle, liver	DNSH, SH(IS)	HCl hydrol. extn, 2-NBA deriv, ammonia neutral, reconst. ACN	2-NBA	RPLC-C18	MSMS esi–	0.05	2005(299)
Muscle, liver	DNSH, HBH(IS)	HCl hydrol. extn, 2-NBA deriv, neutral/PBS dilution, EtOAc partition, reconst. ACN/NH_4OH	2-NBA	RPLC-C18	MSMS esi+	0.10, 0.06	2005 (300)
Muscle	AOZ, AMOZ, AHD, SEM, DNSAH, AOZd4(IS), AMOZd5(IS), 13C3AHD(IS), $^{13}C^{15}N_2SEM$(IS), SH(IS)	HCl hydrol. extn, 2-NBA deriv, Neutral/PBS dilution, EtOAc partition, reconst. $MeOH/CH_3ONH_4$	2-NBA	RPLC-C18	MSMS esi+	0.08–0.20	2007 (301)
Flour-coated meat	AOZ, AMOZ, AHD, SEM, AOZd4(IS), AMOZd5(IS), ADC, Biurea	HCl hydrol. extn, 2-NBA deriv, EtOAc partition, reconst. ACN/H_2O/acetic acid	2-NBA	RPLC-C18	MSMS esi+	—[a]	2004 (302)

(Continued)

Table 42.5 (Continued)

Tissue	Analytes	Sample Treatment	Derivati-zation	LC Column Technique	Detection	Limit Range (ppb)	Year (Reference)
Muscle, liver, hen eye	Nitrofurazone, $^{13}C^{15}N_2NFZ^{(IS)}$, total SEM, bound SEM $^{13}C^{15}N_2SEM^{(IS)}$	For muscle, liver: Solvent washings, HCl hydrol. extn, 2-NBA deriv, Neutral/ PBS dilution, EtOAc partition., reconst. MeOH/H$_2$O For eye: EtOAc extn, reconst. ACN, hexane defat., reconst. MeOH/H$_2$O	For muscle: 2-NBA For eye tissue: –	RPLC-C18	MSMS esi+	< 0.5 –a	2005 (303)

a Not reported.
(IS) Internal standard.

Note: ppb: parts per billion equivalent to µg of residue/kg of tissue; EtOAc: ethyl acetate; DMSO: dimethyl sulfoxyde; Extn: extraction; RPLC: reverse phase liquid chromatography; SPE: solid-phase extraction; ACN: acetonitrile; DCM: dichloromethane; UV: ultraviolet detection; Pyr.: pyridine; TMSA: N,O-bis(trimethyl silyl)acetamide; TMSI: N-trimethylsilylimidazole; TMCS: trimethylchlorosilane; FID: flame ionization detection; MSPD: matrix solid phase dispersion; AOZ: 3-amino-2-oxazolidinone; PED: protease enzyme digestion; 2-NBA: 2-nitrobenzaldehyde; Neutral.: NaOH neutralization; MAX: medium anion exchange; HLB: hydrophilic Lipophilic Balance; MeOH: methanol; EtOH: ethanol; PBS: phosphate buffer solution; MSMS: triple quadrupole; esi+: electrospray source in positive mode; AMOZ: 3-amino-5-morpholinomethyl-2-oxazolidinone; AHD: 1-aminohydantoin; SEM: semicarbazide; AOZd4: 4 times deuterated AOZ; AMOZd5: 5 times deuterated AMOZ; $^{13}C^{15}N^2SEMd4^{(IS)}$; $^{13}C^{-15}N$ isotopic SEM; partition.: aqueous partitioning; Reconst.: aqueous solvent reconstitution; DNSH, DNSAH: 3,5-dinitrosalicylic acid hydrazine; esi–: electrospray source in negative mode; SH: salicylic acid; HBH: 4-hydroxy-3-5-dinitrobenzoic acid hydrazide; NFZ: nitrofurazone; ADC: azodicarbonamide; DAD: diode array detection; defat.: defattening; deprot.: deproteinisation; Hydrol.: hydrolytic extraction; MS: single quadrupole; MS^n: ion trap; nci: negative ion chemical ionization; pci: positive ion chemical ionization; purif.: purification; TSP: thermospray.

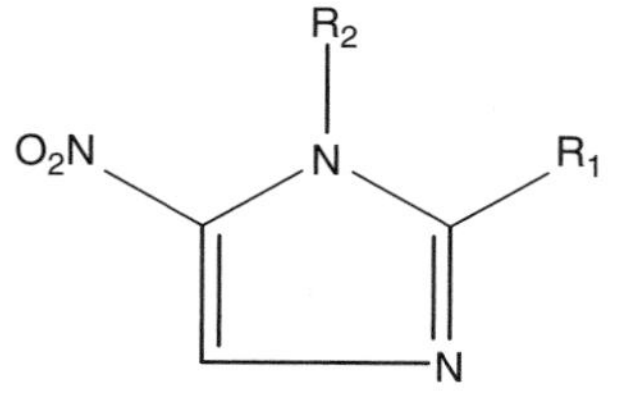

Compound	R1	R2	MW
Dimetridazole	CH_3	CH_3	141.13
Ronidazole	CH_2OOCNH_2	CH_3	200.15
Ipronidazole	$CH(CH_3)_2$	CH_3	169.18
Metronidazole	CH_3	C_2H_4OH	171.15
Hydrolylated metabolites			
HMMNI	CH_2OH	H	158.14
MNZOH	CH_2OH	C_2H_4OH	202.19
IPZOH	$C(CH_3)_2OH$	H	186.19

Figure 42.8 Structures of some nitroimidazoles and their hydroxylated metabolites.

animals in the mid-1990s as enforced in the EU by three regulations—3426/93/EC, 1798/95/EC, and 613/98/EC [304–306].

Several reviews have been published on nitroimidazole residues over the past 10 years [13–15]. Recently, an immunological method has been investigated with production of polyclonal antibodies against a range of nitroimidazoles: metronidazole, ronidazole, dimetridazole, and ipronidazole, and their hydroxy metabolites [75]. An ELISA test kit for the screening of several anticoccidials including the nitroimidazoles in chicken muscle and eggs was further developed with analytical limits ranging from 2 μg/kg (dimetridazole) to 40 μg/kg (ipronidazole) [84]. Not many HPLC methods were published during the late 1990s and early 2000s for nitroimidazole monitoring in animal food matrices (Table 42.6) [307–312] together with rare LC-MS methods [313–314]. More recently, several up-to-date GC-MS [110,315] and LC-MS(MS) methods [316–320] were proposed following the ban of nitroimidazoles and the request for higher level of sensitivity and for unequivocal identification for the confirmatory methods employed in the residue control for food safety. This request is legally stated in the EU Decision 657/2002/EC regarding the criteria for performance of the official analytical methods for residue monitoring in food from animal origin [4].

42.4.7 Quinolones

Quinolones are broad-spectrum synthetic antimicrobial compounds used in the treatment of livestock and in aquaculture. They act against bacteria by inhibiting the DNA gyrase—a key component in DNA replication. They are a relatively new family of antibacterials synthesized from 3-quinolonecarboxylic acid, the carboxylic group at position 3 providing them with acidic properties (Figure 42.9). Nalidixic acid, oxolinic acid, and flumequine represent the oldest subgroup of compounds from the first generation of acidic quinolones, generally called the pyridonecarboxylic acid (PCA) antibacterials. Oxolinic acid is more restricted to treating fish diseases such

Table 42.6 Summary of Literature Methods for Determination of Nitroimidazole Residues in Tissues Using Gas or Liquid Chromatography

Tissue	Analytes	Sample Treatment	Derivatization	LC Column/Technique	Detection	Limit Range (ppb)	Year (Reference)
Muscle, liver	DMZ, HMMNI, MNZ, MNZOH	EtOAc extn, evap., HCl/EtOAc dilution, hexane defat., aqueous neutral., C18 SPE, evap., reconst.	—	RPLC-C18	UV-DAD	—	1992 (307)
Muscle, liver	DMZ, MNZ	ACN extn, SPE cleanup	—	RPLC-C18	UV-DAD 450 nm	2.0, 5.0	1995 (308)
Muscle, liver	DMZ	ACN extn, SPE cleanup	—	RPLC-C18	UV-DAD 450 nm	2.0	1996 (309)
Muscle, liver	DMZ, RNZ, HMMNI	ACN extn, NaSO$_4$ deprot., acetic acid dilution, SCX SPE, reconst. PBS	—	RPLC-C18	UV 315 nm MS apci+	0.5 0.1–0.5	1998 (310)
Muscle	DMZ, HMMNI, RNZ	ACN extn, EtOAc/hexane partition., silica SPE	—	RPLC-C18	UV 315 nm	0.5	1999 (311)
Fish flesh	MNZ, MNZOH TNZ[IS]	ACN extn, C18 SPE	—	RPLC-C18	UV 325 nm	1.5–2.0	2000 (312)
Muscle, liver	DMZ, HMMNI, IPZ, IPZOH	EtOAc extn (DMZ-DMZOH), C18 SPE, reconst. benzene extn (IPZ-IPZOH), reconst.	—	RPLC-C18	MS TSP+	2.0	1992 (313)
Muscle, egg	DMZ, DMZ-d3[IS]	DCM extn, silica SPE, reconst. MeOH/H$_2$O, hexane defat.	—	RPLC-C18	MS TSP+	<5.0	1997 (314)
Muscle, liver	DMZ, MNZ, RNZ, HMMNI,	Protease+PBS hydrolysis extn, PBS partition., silica SPE, deriv	BSA-50	GC	MS nci	0.6–2.8, 5.2[IPPZOH]	2001 (110)

Analytes	Matrix	Sample prep		Separation	Detection	Range	Year (Ref.)
MNZOH, IPZOH, TNZ, IPZ-d3(IS), IPZOH-d3(IS), HMMNI-d3(IS), RNZ-d3(IS), DMZ-d3(IS)							
DMZ, MNZ, RNZ, HMMNI, MNZOH, IPZOH	Retina, plasma	Protease+PBS hydrolysis extn, PBS partition,, hexane defat., silica SPE, deriv	BSA-50	GC	MS nci	0.5–4.0	2002 (315)
DMZ, MNZ, RNZ, HMMNI	Muscle	EtOAc extn, Hexane/CCl_4/ Formic acid partition.,	—	RPLC-C18	MS esi+	—[a]	2000 (316)
DMZ, RNZ, MNZ	Muscle, liver	ACN extn, NaCl/DCM partition., H_2O dilut., hexane defat., silica SPE	—	RPLC-C18	MS esi+	2.0–4.0	2001 (317)
DMZ, RNZ, MNZ, IPZ, HMMNI, TNZ, RNZ-d3(IS), DMZ-d3(IS)	Muscle, liver	ACN extn, $NaSO_4$ deprot., acetic acid dilution, SCX SPE, reconst. PBS/ACN	—	RPLC-C18	MSMS esi+	2.0–5.0	2004 (318)

(Continued)

Table 42.6 (Continued)

Tissue	Analytes	Sample Treatment	Derivatization	LC Column/ Technique	Detection	Limit Range (ppb)	Year (Reference)
Muscle, liver	DMZ, RNZ, MNZ, IPZ, HMMNI, MNZOH, IPZOH, IPZ-d3[(IS)], IPZOH-d3[(IS)], HMMNI-d3[(IS)], RNZ-d3[(IS)], DMZ-d3[(IS)]	—[a]	—	RPLC-C18	MSMS apci+	—[a]	2004 (319)
Muscle, liver	RNZ, DMZ, MNZ, HMMNI	2xACN extn with NaCl partition., partial ACN evap., filtration	—	RPLC-C8	MSMS esi+	0.1–0.3	2006 (320)

[a] Not reported.
[(IS)]Internal standard.

Note: DMZ: dimetridazole; MNZ: metronidazole; EtOAc: ethyl acetate; Extn: extraction; RPLC: reverse phase liquid chromatography; MNZOH: hydroxymetronidazole; Reconst.: aqueous solvent reconstitution; Reconst: reconstitution prior to injection; ACN: acetonitrile; SPE: solid-phase extraction; RNZ: ronidazole; deprot.: deproteinisation; SCX: strong cation exchange; PBS: phosphate buffer solution; partition.: aqueous partitioning; TNZ: tinidazole; IPZ: ipronidazole; IPZOH: hydroxyipronidazole; BSA-50: N,O-bis(trimethylsilyl)acetamid; esi+: electrospray source in positive mode; MSMS: triple quadrupole; DAD: diode array detection; DCM: dichloromethane; defat.: defattening; DMSO: dimethyl sulfoxyde; DMZOH: hydroxydimetridazole; esi⁻: electrospray source in negative mode; HLB: hydrophilic Lipophilic Balance; MeOH: methanol; MS:single quadrupole; MSⁿ: ion trap; MSPD: matrix solid phase dispersion; nci: negative ion chemical ionization; Neutral.: NaOH neutralization; ppb: parts per billion equivalent to μg of residue/kg of tissue; purif.: purification; TSP: thermospray; UV: ultraviolet detection.

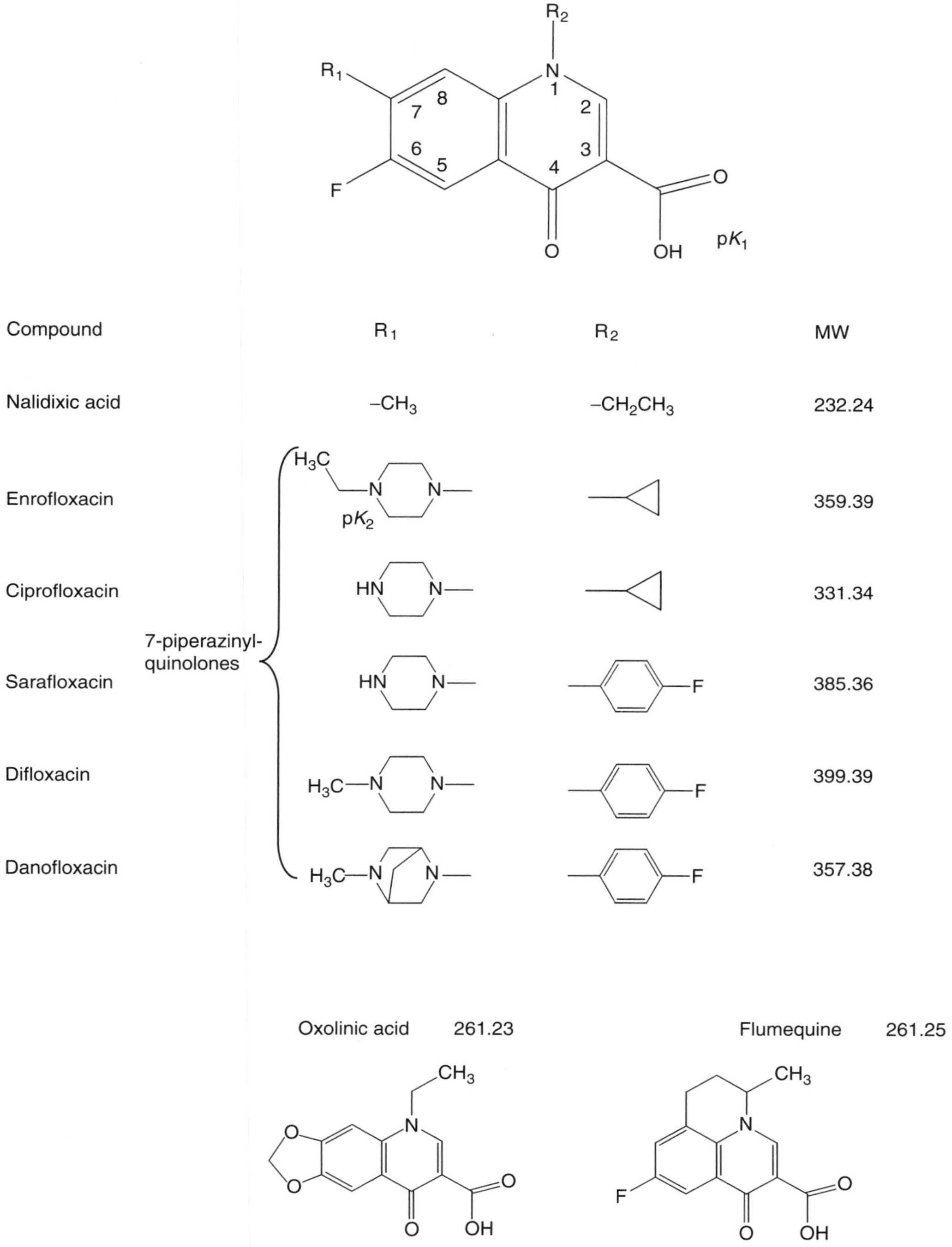

Compound	R₁	R₂	MW
Nalidixic acid	$-CH_3$	$-CH_2CH_3$	232.24
Enrofloxacin			359.39
Ciprofloxacin			331.34
Sarafloxacin			385.36
Difloxacin			399.39
Danofloxacin			357.38

Oxolinic acid   261.23

Flumequine   261.25

Figure 42.9   Structures of some quinolones and fluoroquinolones.

as killing the bacteria causing furunculosis in salmon. Nalidixic acid is not used in veterinary medicine. Representatives of the second generation are the fluoroquinolones, such as enrofloxacin, ciprofloxacin, or sarafloxacin, with higher potency in regard to the first generation. They bear a piperazinyl moiety in the C-7 position which gives these amino-quinolones some additional basic properties, and depending on the chemical environment leads to zwitterionic, cationic, or anionic behaviors in aqueous solution. At pH of 6–8, they hold poor water solubility due to their amphoteric characteristics. However, they are readily soluble in polar organic solvents and also in acidic or basic aqueous/organic solutions. Their extraction from biological matrices needs to be considered taking into account their incell intranuclear accumulation. They all display a native fluorescence, which is of particular interest when ppb residual quantities are detected from biological tissues and fluids of food-producing animals. Other fluorescent fluoroquinolones employed in veterinary medicine are the danofloxacin, the difloxacin, and the marbofloxacin. Several reviews on quinolone residue analysis have been published over the past 10 years, most of them dealing with LC methods [13–15]. Microbiological methods aimed at screening quinolones in meat have also been reported. Several papers present relevant inhibitory methods using strains of *B. subtilis* or *E. coli* [32,34,37,49,321]. Specific applications of microbiological inhibitory testing have also been reported for quinolones using particular strains of bacteria such as *Klebsellia pneumoniae* [322] or *Yersinia ruckeri* [323].

ELISA have also been recently developed [73,80,83]. A TLC method was proposed about 30 years ago for a very specific quinolonic substance, the decoquinate, for its monitoring in chicken muscle with a fluorescent detection [324], and was followed by a HPLC-FLD method [325]. In the 1990s, many papers were published on quinolone residue analysis (Table 42.7), especially for poultry meat monitoring, and a large part of them focused on RPLC-FLD technique [73,326–335]. In regard to potential multiresidue testing including quinolones from first and second generations, some of these papers present comparative studies of different LC methods, either HPLC-FLD to HPLC-UV/DAD [336] or HPLC-FLD to LC-MS methodologies [337–338] or even LC-UV to LC-MS [339].

Recently, a multifamily analysis of both quinolones and tetracyclines in chicken muscle has been proposed within the same HPLC-FLD method [340]. Few recent articles discuss the HPLC-UV analysis of quinolones in muscle tissues [341]. This is probably due to the lack of sensitivity of this type of detection with a reduced UV absorbance efficiency for most of the quinolones. Considering the chromatographic separation on reverse-phase mode, it is worth noting the ability of the ampholytic quinolones to interact with silanols and metal impurities of the stationary phases, causing peak tailing and reducing drastically the quality of the quantification in multiquinolone analysis. To cope with this problem, polymeric phases (PLRP-S) have sometimes been preferred to sustain reliable separative performance in LC methods [326,331]. Another alternative is to utilize the ion-pairing properties of sulfonic acids [333] or to use phenyl stationary phases instead of conventional C8 or C18 ones [337]. In the 2000s, investigations have been carried out for several quinolones in chicken muscle tissues with the capillary zone electrophoretic techniques (CZE or CE) with UV detector [342–343] or with laser-induced fluorescence detection (CZE-LIF) [344] or even more recently, with MS detection [344]. Following first articles published in the 1990s dealing with MS detection of quinolones in different tissues [346], the 2000s have been the period for exploring the use of LC-MSMS techniques in the analysis of very large multiquinolone residues in muscle and in other matrices (kidney, etc.) from animal origin [262,338,347–350]. Table 42.7 displays different techniques published for LC analysis of quinolones in muscle tissues of different food-producing animal species.

Table 42.7 Summary of Literature Methods for Determination of Quinolone Residues in Tissues Using Gas or Liquid Chromatography

Tissue	Analytes	Sample Treatment	Derivatization	LC Column Technique	Detection	Limit Range (ppb)	Year (Reference)
Muscle, liver, kidney	Decoquinate	MeOH/CHCl₃ Extn, Acid/ CHCl₃ Partition.	—	RPLC-Florisil	FLD: Ex290 nm Em370 nm	<100	1973 (325)
Muscle, milk	ENR, SAR[IS]	ACN/NH₄OH Extn, EtOAc/Hexane/NaCl partition., H₃PO₄ acidif.	—	PLRP-S	FLD: Ex278 nm Em440 nm	5	1994 (326)
Muscle, liver, egg, honey	OXA, NLA, FLU, DAN, ENR, CIP, SAR, MAR, NOR, ENO, LOM, OFL	Fluoroquinos: ACN/ Acetic acid/Na₂SO₄ extn, SCX SPE, Drying, Reconst. Acidic quinos: ACN/Na₂SO₄ extn	—	RPLC-C8	FLD: Ex278 nm Em445 nm UV (MAR); 302 nm	5; 10; 50	1998 (327)
Muscle	MAR, DAN, ENR, CIP, DIF, SAR, NOR	Drying, PBS dilution, SAX SPE, Evap., Reconst.	—	RPLC-C18	FLD: Ex278 nm Em440 nm	—ᵃ	2000 (328)
Muscle	FLU, OXA	DCM Extn, NaOH partition.	—	RPLC-C8	FLD: Ex328 nm Em365 nm	<1	2000 (329)
Fish muscle	7OH-NLA, NLA, OXA, CIN	A: CHCl₃ Extn, Evap., Reconst. B: NaOH Extn, CHCl₃ partition., ChloroAcetic acid partition., CHCl₃ extn, Na₂SO₄ drying, Evap., Reconst.	—	RPLC-C18	FLD: Ex260 nm Em360 nm FLD: Ex270 nm Em440 nm	<20	2000 (330)
Muscle	CIP,n Extn, Evap., Tris buffer Reconst., Hexane defat.		—	PLRP-S	FLD: Ex280 nm Em450 nm FLD: Ex294 nm Em514 nm FLD: Ex312 nm Em366 nm	0.5–35	2000 (331)

(Continued)

Table 42.7 (Continued)

Tissue	Analytes	Sample Treatment	Derivatization	LC Column Technique	Detection	Limit Range (ppb)	Year (Reference)
Fish muscle	OXA, FLU	ACN/NH$_4$OH Extn, EtOAc/Hexane/NaCl partition., H$_3$PO$_4$ acidif./Acetone defat, H$_2$O dilution	—	PLRP-S	FLD: Ex325 nm Em360 nm	20, 30	2001 (332)
Muscle, liver	ENR, CIP, SAR, DIF	TCA/ACN Extn, comparison: C8,C18,NH$_2$, BSA, SDB SPE cleanups	—	RPLC-C8; ion-pairing HSA	FLD: Ex278 nm Em440 nm	—[a]	2001 (333)
Muscle, liver	ENR	ACN Extn, Hexane defat., Evap., Reconst., Filtration	—	RPLC-C18	FLD: Ex290 nm Em455 nm	< 50	2002 (73)
Muscle	ENR, CIP, SAR, OXA, FLU	PBS Extn, C18 SPE	—	RPLC-C18	FLD: Ex280 nm Em450 nm; FLD: Ex312 nm Em366 nm	5.0, 10 (SAR)	2003 (334)
Muscle, liver, kidney, fish flesh, egg, milk	MAR, NOR, ENR, CIP, DAN, SAR, DIF, OXA, NLA, FLU	TCA extn, Filtration	—	RPLC-C18	FLD: Ex294 nm Em514 nm; FLD: Ex328 nm Em425 nm; FLD: Ex312 nm Em366 nm	4-36	2005 (335)
Feeds	CIP, ENR, DAN, OXA, NLA, FLU, DIF, NOR, OFL, ENO, RUF, PIP, CIN	ASE extn ACN/MPA, HLB SPE, Evap. Reconst.	—	RPLC-C5	DAD: 278 nm; FLD: Ex278 nm Em446 nm; FLD: Ex324 nm Em366 nm	500–1500	2003 (336)
Muscle, liver	DESCIP, NOR, CIP, DAN, ENR, ORB, SAR, DIF	PBS extn, ACN/NaOH extn, Hexane/EtOAc/NaCl cleanup, Evap., Reconst.	—	RPLC-Phenyl	FLD: Ex278 nm Em440 nm	0.1–0.5	2002 (337)

Sample	Analytes	Sample preparation		Separation	Detection	Range	Year (Ref)
Shrimp flesh	DESCIP, NOR, CIP, DAN, ENR, ORB, SAR, DIF	PBS extn, ACN/NH$_4$OH partition., Hexane/EtOAc/NaCl cleanup, Evap., Reconst.	—	RPLC-Phenyl	FLD: Ex278 nm Em465 nm	0.1–1.0	2005 (338)
Muscle	ENR, CIP, DIF, DAN, MAR, AR, OXA, FLU, NOR[IS]	ACN/H$_3$PO$_4$ Extn, ENV+isolute SPE	—	RPLC-C8	UV: 250 nm, 280 nm, 290 nm	7–13	2006 (339)
Muscle	DAN, CIP, ENR, DIF, SAR	ACN/Citrate buffer/MgCl$_2$ extn, Evap., Reconst. in Malonate/MgCl$_2$	—	RPLC-Phenyl	FLD: Ex275 nm Em425 nm	0.5–5.0	2007 (340)
Muscle	CIP, ENR, DAN, SAR, DIF, OXA, FLU	DCM extn, NaOH partition, comparison HLB-MAX-SDB SPEs	—	RPLC-C8	UV-DAD: 250 nm, 280 nm	16–30	2004 (341)
Muscle	ENR, CIP MAR[IS]	DCM extn, C18 SPE	—	CZE	UV-DAD: 270 nm	<25	2001 (342)
Muscle	CIP, ENR, DAN, DIF, MAR, OXA, FLU, PIR[IS]	DCM extn, NaOH partition., comparison C18-SCX-SAX-HLB-MAX-SDB SPEs	—	CZE	UV-DAD: 260 nm	7–30	2004 (343)
Muscle	ENR, CIP DIF[IS]	H$_2$O homogen., Buffer extn, DCM partition., H$_3$PO$_4$ Dilution, Evap., Hexane defat., Filtration	—	CZE	LIF: HeCd Exc 325 nm	5; 20	2002 (344)
Muscle, fish flesh	DAN, ENR, FLU, OFL, PIP	PBS extn, DCM partition., NaOH extn, H$_3$PO$_4$ neutral, Hexane defat., C18 SPE	—	CZE	MS^n esi+	20	2006 (345)
Muscle	ENR, CIP, DAN, MAR, SAR, DIF	PBS extn, C18 SPE	—	RPLC-C18	MS apci+	7.5	1998 (346)

(Continued)

Table 42.7 (Continued)

Tissue	Analytes	Sample Treatment	Derivatization	LC Column Technique	Detection	Limit Range (ppb)	Year (Reference)
Muscle, milk, prawn flesh, Eel flesh	ENR, CIP, DAN, SAR, LOM, ENO, OFL	ACN/Formic acid extn, C18 SPE, Dilution IPCC-MS3	—	RPLC-C18	MSMS esi+	1.0–2.0	2004 (347)
Muscle	ENR, CIP, DAN, SAR, DIF, OXA, FLU	ACN extn, Hexane defat., Evap. H$_2$O/ACN extn, Hexane defat.	—	RPLC-C18	MSMS esi+/esi−	10	2005 (262)
Kidney	NOR, MAR, ENR, CIP, DAN, OXA, NLA, FLU, CIN, OFL, ENO, LOM[IS], CIN[IS]	ACN extn, Evap., Reconct., SDB SPE, Dilution	—	RPLC-C8	MSMS esi+	0.3–2.0	2005 (348)
Kidney	NOR, MAR, ENR, CIP, DAN, OXA, NLA, FLU, CIN, OFL, ENO, LOM[IS], CIN[IS]	ACN extn, Evap., Reconct., SDB SPE, Dilution		RPLC-C8	MSMS esi+	0.3–2.0	2005 (349)
Muscle	DESCIP, NOR, CIP, DAN, ENR, ORB, SAR, DIF	ACN/NH$_4$OH Extn, EtOAc/Hexane/NaCl partition., Evap., PBS dilution	—	RPLC-Phenyl	MSMS apci+	0.1–1.0	2005 (338)
Muscle	ENR, CIP, DIF, NOR, OFL, ORB	ACN/MeOH extn, Hexane/ACN defat., Evap., Reconst.	—	RPLC-C18	MSMS esi+	0.3–3.0	2006 (214)

Muscle	ENR, CIP, DIF, DAN, MAR, SAR, OXA, FLU, NOR[IS]	ACN/ H$_3$PO$_4$ Extn, ENV+isolute SPE	RPLC-C8	MS esi+ MSMS esi+	—	0.3–1.8 < 0.2	2006 (339)
Muscle	ENR, CIP, DIF, DAN, DESCIP, SAR, OXA, NLA, FLU, OFL, PIR, NOR, ENO, CIN, LOM[IS], PIP[IS]	EtOH/ Acetic acid, HCl dilution, hexane defat, SCX SPE	RPLC-C18	MSMS esi+	—	0.1–0.4	2007 (350)

a Not reported.

[IS] Internal standard.

Note: ppb: parts per billion equivalent to µg of residue/kg of tissue; MeOH: methanol; partition.: aqueous partitioning; RPLC: reverse phase liquid chromatography; ENR: enrofloxacin; SAR: sarafloxacin; ACN: acetonitrile; EtOAc: ethyl acetate; PLRP-S: polymeric stationary phase; OXA: oxolinic acid; NLA: nalidixic acid; FLU: flumequine; DAN: danofloxacin; CIP: ciprofloxacin; NOR: norfloxacin; ENO: enoxacin; LOM: lomefloxacin; OFL: ofloxacin; SCX: strong cation exchange; SPE: solid-phase extraction; PBS: phosphate buffer solution; SAX: strong anion exchange; Evap.: evaporation before reconstitution; defat.: defattening; UV: ultraviolet detection; DIF: difloxacin; DCM: dichloromethane; Neutral.: neutralization; CIN: cincophen; TCA: trichloroacetic acid; BSA: benzene sulfonic acid; HSA: heptane sulfonic acid; SDB: styrenedivinylbenzene; RUF: rufoxacin; PIP: pipemidic acid; ASE: accelerated solvent extractor; MPA: metaphosphoric acid; HLB: hydrophilic Lipophilic Balance; DAD: diode array detection; DESCIP: desethylciprofloxacin; ORB: orbifloxacin; MAX: medium anion exchange; PIR: piromidic acid; LIF: laser-induced fluorescence; MSn: ion trap; esi+: electrospray source in positive mode; esi−: electrospray source in negative mode; MS: single quadrupole; MSMS: triple quadrupole; Extn: extraction; EtOH: ethanol; CIN: cinoxacin; PIR: piromidic acid; deprot.: deproteinisation; purif.: purification; Reconst. reconstitution prior to injection.

42.4.8 Sulfonamide Antiinfectives

Sulfonamides are an important antiinfective family of drugs with bacteriostatic properties. Owing to their broad-spectrum activity against a range of bacterial species, both gram-positive and gram-negative, they are widely used in veterinary medicine with more than 12 licensed compounds. Synthetically prepared from *para*-aminobenzenesulfonic acid, they act by competing with *para*-aminobenzoic acid in the enzymatic synthesis of dihydrofolic acid, leading to a decreased availability of the reduced folates that are essential molecules in the synthesis of nucleic acids. In practice, they are usually combined with synthetic diaminopyrimidine, trimethoprime, to enhance synergistic action against bacterial DNA synthesis even though the synergy was never really demonstrated. Most of the sulfonamide drugs are readily soluble in polar solvents such as ethanol, ACN, and chloroform but relatively insoluble in nonpolar ones. They are considered as weak acids but behave as amphoteric compounds due to the interaction between an acidic N–H link in the vicinity of a sulfonyl group (pK_a 4.6) and an alkaline character at the *para*-NH$_2$ group (pK_a 11.5), leading to a particular behavior in extraction and cleanup process in the 7–9 pH range. Sulfamethazine (also called sulfadimidine and sulfadimerazine) is probably the most widely used sulfa drug. But, several other sulfonamides are also employed in food-producing animal treatments such as sulfadiazine, sulfadoxine, sulfaquinoxaline, sulfapyridine, sulfapyridazine, sulfadimethoxine, sulfamerazine, sulfathiazole, sulfachloropyridazine, sulfamonomethoxine, and sulfamethoxazole (Figure 42.10). Extensive and updated reviews of analytical methods for sulfonamide analysis in food from animal origin have been published over the past 15 years [9,13–15,351].

Most of these reviews relate to chromatographic methods. Yet, several microbiological screening techniques have also been tentatively applied to these compounds even though sulfonamides do not react with much sensitivity to bacteria such as *B. subtilis* [29,37,42,43,60,66,198], *B. stearothermophilus* [32,58,59,61], or even *B. megaterium* [39–40]. It is of great concern that they are not easily detected in food products (muscle, milk, egg, honey, etc.) in the 10–100 ppb range where they are generally regulated even after enhanced sensitivity brought by the addition of trimethoprim. Therefore, TLC technique is one alternative for screening sulfa drugs, which is sometimes proposed in regulatory control achieving good sensitivity in the 100 ppb range [103,202,352]. TLC-bioautography was employed in the 1990s [46] and ELISA kits have also been proposed especially for sulfamethazine screening in urine and plasma with predictive concentration in porcine muscle tissues [67,68]. More recently, biosensor-based immunochemical screening assays for the detection at 10 ppb level of sulfamethazine and sulfadiazine in bile and in muscle extracts from pigs and in chicken serum were developed [86,87,89]. High cross-reactivities (50–150%) in chicken serum were found with several other sulfa drugs such as sulfamerazine, sulfathiazole, sulfachloropyrazine, sulfachloropyridazine, and sulfisoxazole [90]. On the confirmatory quantitative stage of sulfa drug strategy of control, many relevant chromatographic techniques have been developed (Table 42.8). From GC with flame ionization detection in the early 1990s [107] to GC-MS [108] for sulfamethazine-specific detection, it is the LC methods that took the leadership in an even wider scale and not only for sulfamethazine monitoring with multisulfa drug analysis by LC-UV or LC-PDA instruments [206,287,353–363], and by LC-Fluo detections after derivatization of the sulfonamides [206,354,364–366]. The investigation of LC-MS and LC-MSMS methods with a wide range of sulfonamides started in the mid-1990s [367–369]; they were still improved with enhanced mass detectors in the 2000s [195,214,262,370–375]. A representative overview of LC-UV, LC-Fluo, and LC-MS methods used to monitor sulfa drugs in the early 2000s is presented in a recent paper related to European proficiency testing studies on sulfonamide residue in muscle and milk [376].

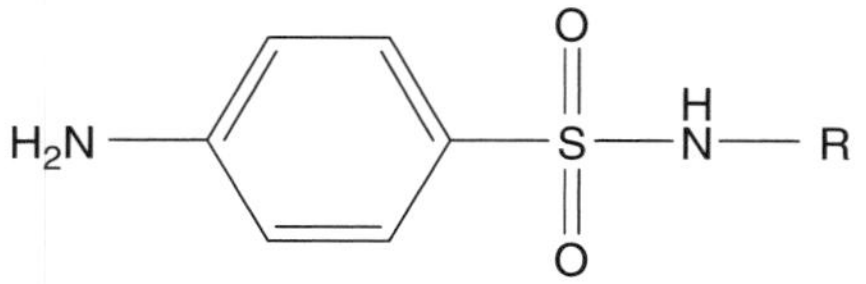

Sulfonamide	R	pKa	MW
Sulfadiazine	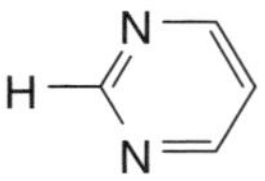	6.4	250.28
Sulfamethazine Sulfadimidine Suldimerazine	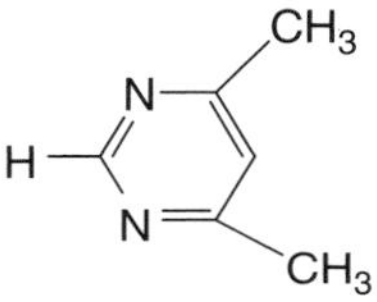	7.5	278.33
Sulfamethoxypyridazine	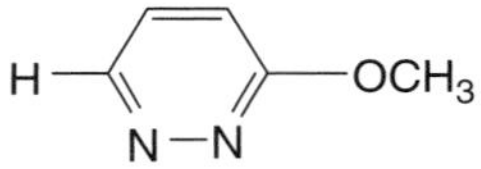	6.7– 7.2	280.30
Sulfadoxine	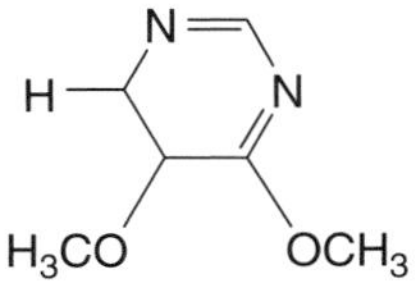	6.1	310.33
Sulfadimethoxine	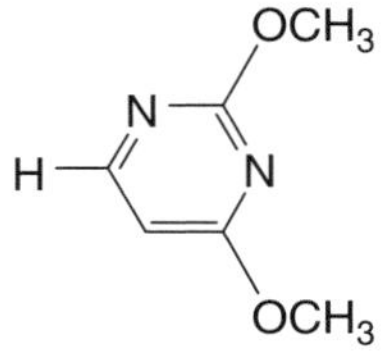	5.9	310.33
Sulfathiazole	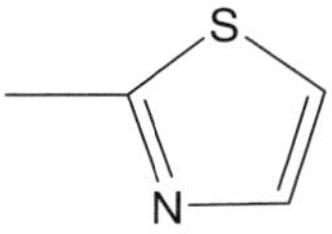	7.1	255.32

Figure 42.10 Structures of some sulfonamide drugs used in veterinary medicine.

Table 42.8 Summary of Literature Methods for Determination of Sulfonamide Residues in Tissues Using Gas or Liquid Chromatography

Tissue	Analytes	Sample Treatment	Derivatization	LC Column Technique	Detection	Limit Range (ppb)	Year (Reference)
Muscle	SMM, SDM, SMT, SMX, SQX	TCA extn, ion exchange SPE	Pyr, TMSA, TMSI, TMCS	GC	FID	100	1992 (107)
Muscle	SMT	CH_2Cl_2/acetone extn, silica SPE, SCX SPE, PBS/MTBE partition., evap., deriv.	methylation, silylation	GC	MS	10–20	1996 (108)
Muscle	SMT SEPDZ[IS]	$CHCl_3$ extn, alkaline NaCl partition., C18 SPE	—	RPLC-C18	UV: 265 nm	2	1994 (353)
Kidney	SMT,SQX, SDZ	ACN extn, drying evap., buffer dilution, C18 SPE, drying evap., DCM dilution, silica SPE, drying evap., buffer reconst.	—	RPLC-C8	UV-DAD: 220–400 nm; 246 nm (SQX), 251 nm (SDZ), 299 nm (SMT)	2–18	1995 (287)
Muscle, liver, kidney	SMT, N⁴-metabolites	EtOAc/acetic acid /Na_2WO_4, NH_2-SCX SPE, drying evap., HCl reconst.	—	RPLC-C18	UV: 270 nm	—[a]	1995 (354)
Muscle, liver, kidney	STZ, SMR, SCP, SMT, SMPDZ, SMX, SQX, SDM	MSPD extn, partition.n 1 with CH_2Cl_2 and partition.n 2 with EtOAc	—	RPLC-C18	UV: 270 nm	1–66	1997 (206)
Muscle	SMT	MAE	—	RPLC-C18	UV: 450 nm	—[a]	1998 (355)
Muscle	SMT, SDZ, SPD, SMR, SDX	—[a]	—	RPLC-C18	UV: 270 nm	—[a]	1998 (356)
Muscle, liver, kidney	SDZ, STZ, SPD, SMR, SMZ, SMT, SMPDZ	Saline extn, dialysis, C18 SPE	—	RPLC-C18	UV: 280 nm	40	1999 (357)
Muscle	SMM, SDM, SQX	MeOH/H_2O extn, IAC cleanup	—	RPLC-C18	UV: 370 nm	1–2	2000 (358)
Muscle, liver, kidney	SMT	EtOH/H_2O extn, ultrafiltration	—	RPLC-C4	UV-DAD: 263 nm	24–27	2001 (359)

Sample	Analytes	Sample preparation	Derivatization	Column	Detection	LOD	Year (Ref.)
Shrimp flesh	SDZ, STZ, SQX, SDM, SMR	EtOAc extn, Drying evap., SEC cleanup, Drying evap., ACN/acetic acid reconst.	—	RPLC-phenyl	UV: 270 nm	10	2003 (360)
Muscle, liver, kidney	SMT	PCA ultrasonic extn	—	RPLC-C4	UV: 267 nm	<90	2003 (361)
Muscle	SDZ, STZ, SPD, SMR, SMT, SMM, SCP, SMX, SQX, SDM	EtOAc/Na_2SO_4 extn, drying evap., EtOAc dilution, SCX SPE, drying evap., reconst. acetate buffer	—	RPLC-C8	UV-DAD: 270 nm	30–70	2004 (362)
Muscle	SDZ, STZ, SMT, SMR, SDX, SMM, SCP, SMX, SQX, SMZ	Acetone/$CHCl_3$ extn, SCX SPE, drying evap., MeOH reconst.	—	RPLC-C18	UV-DAD: 270 nm		2007 (363)
Muscle, liver, kidney, serum	SMT, SMM, SMX, SDM SDZ[IS]	ACNdeprot. Extn, Evap., H_2O/ACN dilution, evap., TCA dilution, hexane defat.	Fluoresc-amine	RPLC-C18	FLD: Ex390 nm Em475 nm	0.1	1995 (364)
Muscle, liver, kidney	SMT, N^4-metabolites	EtOAc/acetic acid /Na_2WO_4, NH_2-SCX SPE, drying evap., HCl reconst.	Fluoresc-amine	RPLC-C18	FLD: Ex405 nm Em495 nm	—[a]	1995 (354)
Muscle, liver, kidney	STZ, SMR, SCP, SMT, SMPDZ, SMX, SQX, SDM	MAE	DMABA	RPLC-C18	FLD: Ex405 nm Em495 nm	2.5	1997 (206)
Muscle	SCP, SDZ, SDM, SDX, SMT, SQX, STZ SPD[IS]	EtOAc extn, glycine/PBS/HCl purif., hexane defat., CH_2Cl_2/Na_2SO_4 purif., DEA dilution, drying evap., reconst. PBS/ACN	Fluoresc-amine	RPLC-C18	FLD: Ex405 nm Em495 nm	15	2004 (365)
Muscle	SDZ, SMR, SMT, SMPDZ, SMX, SDM	ACN extn, C18 MSPD, drying evap., reconst. acetate buffer	Fluoresc-amine	RPLC-C18	FLD: Ex405 nm Em495 nm	1–5	2005 (366)

(Continued)

Table 42.8 (Continued)

Tissue	Analytes	Sample Treatment	Derivatization	LC Column Technique	Detection	Limit Range (ppb)	Year (Reference)
Kidney	SMT, SMR, SDZ, SQX	Acidic EtOAc extn, NH₂ + SCX SPE, drying evap., acetone storage, drying evap., reconst.	—	RPLC-C18	MS esi+ / MSMS esi+	—[a]	1994 (367)
Muscle	SMT, SDM, SEPDZ[IS]	CHCl₃ extn, alkaline NaCl partition., C18 SPE, drying evap., reconst.	—	RPLC-C18	UV: 265 nm / MS TSP+	<10	1995 (368)
Muscle	TMP	CHCl₃/acetone extn, drying evap., MeOH/H₂O/acetic acid dilution, hexane defat.	—	RPLC-C18	MS TSP+	4	1997 (369)
Kidney Muscle	Sulfa drugs / SDX	On-line extn, sample cleanup / MeOH extn, evap., water dilution, C18 SPE, evap., reconst.	— / —	RPLC-C18 / RPLC-C18	MSMS esi+ / MSMS esi+	—[a] / —[a]	2000 (370) / 2003 (195)
Muscle	SDZ, STZ, SPD, SMR, SMT, SMZ, SMPDZ, SCP, SMX, SMM, SDM, SQX, SME[IS]	MSPD with 80°C water extn	—	RPLC-C18	MS esi+	3–15	2003 (371)
Muscle	SDZ, STZ, SMT, SMR, SDM	ACN/Na₂PO₄ extn, C18 SPE, drying evap., reconst.	—	RPLC-C8	MS apci+ + MS esi+	—[a]	2003 (372)
Raw meat, infant food	SIM, SDZ, SPD, SMR, SMO, SMT, SMTZ, SMPDZ, SCP, SMM, SMX, SQX, SDM	C18 ASE with 160°C/100 atm water extn	—	RPLC-C18	MSMS esi+	0.25	2004 (373)
Fish muscle	SDZ, SMT, SDM, TMP, OMP	ACN extn, hexane defat., evap. H₂O/ACN extn, hexane defat.	—	RPLC-C18	MSMS esi+/esi–	10	2005 (262)

Muscle	SMT, SCP, SBZ, SDZ, SMZ, SDX, SMR, SMX, SMM, SMPDZ, STZ, SPZ, SDM, SQX, SSZ, SOZ, TMP, SNT, SPD	ACN/MeOH extn, hexane/ACN defat., evap., reconst.	—	RPLC-C18	MSMS esi+	0.1–0.6	2006 (214)
Muscle	SMT, SDZ, SMM, SMX, SDM, SQX	Alumina MSPD extn, drying evap., reconst.	—	RPLC-C4	MS apci+	<50	2007 (374)
Muscle	SMT, SAA, SGN, SNL, SPD, SDZ, STZ, SMR, SMX, SMO, SOZ, SMPDZ, SMM, SDM, SQX, SCP	ACN/H$_2$O extn, hexane defat., CHCl$_3$ partition,, drying evap. reconst.	—	RPLC-C18	MSMS esi+	0.1 – 0.9	2007 (375)
Muscle	SMT, SDZ, SMM, SMX, SDM, SQX, STZ, SGN, SMPDZ	ACN extn, hexane defat., drying evap. reconst.	—	RPLC-C18	MSMS esi+	<2	2005 (376)

a Not reported.

(IS) Internal standard.

Note: ppb: parts per billion equivalent to μg of residue/kg of tissue; SMM: sulfamonomethoxine; SDM: sulfadimethoxine; SMT: sulfamethazine 'also called sulfadimidine or sulfadimerazine; SMX: sulfamethoxazole; SQX: sulfaquinoxaline; TCA: trichloroacetic acid; SPE: solid-phase extraction; Pyr.: pyridine; TMSA: N,O-bis(trimethylsilyl)acetamide; TMSI: N-trimethylsilylimidazole; TMCS: trimethylchlorosilane; SCX: strong cation exchange; PBS: phosphate buffer solution; MTBE: methyl-tert-butyl-ether; partition.: aqueous partitioning; Evap.: evaporation before reconstitution; derivat.: derivatization; MS: single quadrupole; SEPDZ: sulfaethoxypyridazine; Extn: extraction; SDZ: sulfadiazine; ACN: acetonitrile; DCM: dichloromethane; RPLC: reverse phase liquid chromatography; UV: ultraviolet detection; DAD: diode array detection; EtOAc: ethyl acetate; SMR: sulfamerazine; SMPDZ: sulfamethoxypyridazine; MSPD: matrix solid phase dispersion; MAE: microwave assisted extraction; SDX: sulfadoxine; SMZ: sulfamethizol; MeOH: methanol; SEC: size-exclusion chromatography; PCA: perchloric acid; defat. defattening; purif.: purification; DMABA: dimethylaminobenzaldehyde; DEA: diethylamine; esi+: electrospray source in positive mode; MSMS: triple quadrupole; TSP: thermospray; TMP: trimethoprim; SME: sulfameter; SIM: sulfisomidine; esi–: electrospray source in negative mode; ASE: accelerated solvent extraction; SMO: sulfamoxole; SMTZ: sulfamethizole; OMP: ormethoprim; SBZ: sulfabenzamide; SPZ: sulfaphenazole; SSZ: sulfasalazine; SOZ: sulfisoxazol; SNT: sulfanitran; SAA: sulfacetamide; SGN: sulfaguanidine; SNL: sulphanilamide; SCPZ: sulfachloropyridazine; Reconst: reconstitution prior to injection; PLRP-S: polymeric stationary phase; HSA: heptane sulfonic acid; MSn: ion trap.

42.4.9 Tetracyclines

Tetracyclines are broad-spectrum bacteriostatic antibiotics, some of which are produced by bacteria of the genus *Streptomyces* and others obtained as semisynthetic products. They act by inhibiting protein biosynthesis through their binding to the 30S ribosome. Owing to their high degree of activity against both gram-positive and gram-negative bacteria, they are commonly used in veterinary medicine to treat respiratory diseases in cattle, sheep, pig, and chicken. They may be employed prophylactically as additives in feed or in drinking water. Oxytetracycline, tetracycline, and chlortetracycline are the three major compounds licensed in veterinary medicine. Doxycycline is also, to some extent, a veterinary drug candidate to monitoring in tissues. Oxytetracycline can be found after treatment of various bacterial diseases in fish farming. The basic structure of tetracyclines is derived from the polycyclic naphthacenecarboxamide and contains four fused rings (Figure 42.11). They are polar compounds due to the different functional groups attached to the four fused rings. Particularly active are an acidic hydroxyl group in position 3 (pK_a 3.3), a dimethylamino group in position 4 (pK_a 7.5), and a basic hydroxyl group in position 12 (pK_a 9.4). Tetracyclines are photosensitive, nonvolatile compounds, existing as bipolar ions in aqueous solution in the pH range 4–7, able to lose their dimethylamino group in the pH range 8–9, and capable of reversible epimerization in the pH range 2–6. This chemical behavior leads to difficulties in extracting them from biological matrices where they can easily bind to proteins to form macromolecules. Acidic extraction is often utilized, but further purification by liquid–liquid partitioning and SPE cleanup through organic solvents remain a critical issue. Ion pairing and chelation process are also used to achieve acceptable recoveries. The analysis of tetracyclines by

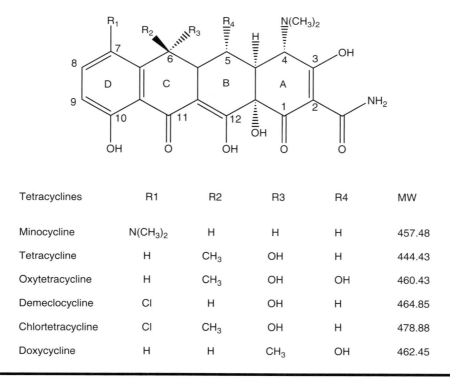

Tetracyclines	R1	R2	R3	R4	MW
Minocycline	N(CH$_3$)$_2$	H	H	H	457.48
Tetracycline	H	CH$_3$	OH	H	444.43
Oxytetracycline	H	CH$_3$	OH	OH	460.43
Demeclocycline	Cl	H	OH	H	464.85
Chlortetracycline	Cl	CH$_3$	OH	H	478.88
Doxycycline	H	H	CH$_3$	OH	462.45

Figure 42.11   Structures of tetracyclines used in veterinary medicine.

reverse-phase LC is also of concern. Silanol-encapped, metal-purified, alkyl-bonded silica stationary phases are required for their satisfactory separation and elution. Polymeric phases have also been successfully investigated. New mixed polymeric/alkyl-bonded silica stationary phases are now promising separative instruments.

An extensive development over the past years has been dedicated to analytical methods for monitoring tetracycline residues at the ppb level in meat products (MRL in muscle tissue is 100 µg/kg). Several reviews are reported on this subject [9,11,13–15]. On the part dealing with residue screening methods, several papers describe microbiological bioassays using the inhibitory properties regarding bacterial growth. Strains such as *B. subtilis*, *B. stearothermophilus*, *Bacillus cereus*, and *E. coli* have been employed to attempt developing inhibitory plate tests capable of detecting, with more or less success, the tetracyclines at the 100–500 ppb level in muscle or kidney tissues [29,32,34,37,40,42,43,58,69,198,216,321]. In the consideration to find the best strategy to screen in meat products antibiotic residues and thus tetracyclines as a part of it, some comparative studies of the performance between inhibitory plate tests and rapid test kits such as Tetrasensor have also been evaluated recently [59–62,377]. TLC with or without bioautography has also been an alternative to detect tetracycline residues in meat tissues but some 20 years ago [55,202].

Regarding the confirmatory methods (Table 42.9), GC was investigated in the previous years but with a limited extent [107], and due to the polar nonvolatile chemical properties of tetracyclines, HPLC was largely preferred to GC. First applied with UV or DAD detection [378–391], and also with fluorescence detection in regard to the high capacity of tetracycline to form fluorophoric metal complexes [340,370, 392–401], it was more recently coupled to different mass spectrometric detectors, with SQ detectors and now with TQ detectors (LC-MS/MS) or ion-trap mass spectrometric detectors (LC-IT/MS or LC-MS^n) or even time-of-flight mass spectrometric detectors (LC-TOF-MS) [195,223,262,402–405]. Capillary electrophoretic techniques have also been tested for tetracyclines [388]. One of the challenges in modern analysis of tetracycline residues in muscle or other food products is to separate and analyze simultaneously all the four tetracyclines along with the existing 4-epimers and some possible degradation compounds [223,382,384,387,390,396,402–405].

42.4.10 Polypeptidic Antibiotics

Polypeptide antibiotics include flavomycin (also named bambermycin or flavophospholipol), avoparcin, virginiamycin, and among polymyxin polypeptides, bacitracin and colistin (Figure 42.12). They are all derived from fungi or bacteria (*Streptomyces bambergiensis*, *Streptomyces candidus*, *Streptomyces virginiae*, *Streptomyces orientalis*, *B. subtilis*, *Bacillus polymyxa*) and exist as complexes of several related macromolecules. Avilamycin, a polysaccharide antibiotic, obtained from *Streptomyces viridochromogenes* can also be added to this group of substances. Most of these antibiotics were used as growth promoters and efficient feed converters except for colistin. Formerly regulated under the feed additive legislation in the EU by Directive 70/524/EC [406], they are now extensively proposed to be prohibited. The risk that resistance to antibiotics might be transferred through them to pathogenic bacteria was assessed at the end of the 1990s and beginning of the 2000s. It led to food safety recommendations from the antimicrobial resistance research program [407]. Avoparcin in 1997, and bacitracin along with virginiamycin in 1999 were immediately banned as feed additives in the EU by Directive 97/6/EC [408] and by Regulation 2821/98/EC [409]. Flavomycin is still on the market but on its way to be banned too. Under Regulation 2562/99/EC, a period of 5–10 years is granted from 2004 to 2014 for reevaluation of the drug by the supporting pharmaceutical stakeholders [410]. The polypeptidic antibiotics are macromolecular compounds. They often feature a mixture

Table 42.9 Summary of Literature Methods for Determination of Tetracycline Residues in Tissues Using Gas or Liquid Chromatography

Tissue	Analytes	Sample Treatment	Derivatization	LC Column Technique	Detection	Limit Range (ppb)	Year (Reference)
Muscle	TTC, OTC, CTC	TCA extn, ion exchange SPE	Pyr, TMSA, TMSI, TMCS	GC	FID	50	1992 (107)
Muscle	CTC, isoCTC	HCl/glycine extn, cyclohexyl SPE	pH 12 isoCTC conversion	PLRP-S	FLD: Ex:340 nm Em:420 nm	20–50	1994 (377)
Muscle	TTC, OTC, CTC	EDTAMIB extn, hexane/DCM purif., TCA deprot., C18 SPE	—	RPLC-C18	UV-DAD: 360 nm	50	1994 (379)
Muscle	TTC, OTC, CTC, DMCTC	SEPSA extn, C8 SPE or XAD2resin SPE, Cu^{2+} gel chelate purif.	—	MCAC + PLRP-S	UV: 350 nm	10–20	1996 (380)
Muscle	TTC, OTC, CTC	Liquid–Liquid extn, C18 SPE cleanup	—	RPLC-C18	UV: 360 nm	100	1996 (381)
Muscle	OTC, 4-epiOTC, alpha-apoOTC, beta-apoOTC	Oxalic acid extn, C18 SPE cleanup	—	RPLC-C18	UV-DAD	—[a]	1996 (382)
Kidney	OTC	Citrate buffer/EtOAc extn, Na_2SO_4 drying, filtration	—	MCAC + PLRP-S	UV: 350 nm	—[a]	1998 (383)
Muscle	OTC, 4-epiOTC	LiqLiq extn, C18 SPE cleanup	—	RPLC-C18	UV: 350 nm	5–10	1998 (384)
Muscle, kidney	TTC, OTC, CTC	Oxalic acid/ACN extn-dechelation-deprot., SDB SPE	—	PLRP-S	UV: 360 nm	10–40	1999 (385)

Matrix	Analytes	Extraction/Cleanup	Derivatization	Column	Detection	LOD	Year (Ref)
Muscle, kidney	TTC, OTC, CTC	ACN/H$_3$PO$_4$ extn, hexane/DCM defat., limited evap., filtration	—	RPLC-C18 + ion pairing DSA	UV: 370 nm	50–100	2000 (386)
Kidney	TTC, OTC, CTC, DC	LiqLiq extn, C18 SPE cleanup	—	RPLC-C18	UV: 350 nm	50–100	2001 (387)
Muscle, kidney, liver	OTC	EDTAMIB extn, C18 or HLB SPE	—	RPLC-C8 + CZE	LC-UV: 350 nm CZE-UV: 365 nm	80–160	2001 (388)
Muscle, milk	TTC, OTC, CTC, DC	TCA/EDTAMIB extn, HLB SPE	—	RPLC-C18	UV-DAD: 365 nm	10–30	2003 (389)
Kidney	CTC+4-epiCTC	Oxalic acid/TCA extn, SDB SPE	—	RPLC-C8	UV-DAD: 365 nm	70–90	2005 (390)
Plasma	OTC	MeOH/EDTAMIB extn, C18 SPE	—	RPLC-C18	UV: 360 nm	3.5–12	2006 (391)
Muscle	CTC, isoCTC	HCl/Glycine extn, cyclohexyl SPE	pH 12 isoCTC conversion	PLRP-S	FLD: Ex:340 nm Em:420 nm	20–50	1989 (392)
Muscle, liver, fish, milk, egg	OTC, TTC[IS]	ASTED dialysis, online enrichment SDB cartridge	NaOH + irradiation 366 nm	PLRP-S + HSA ion-pairing +	FLD: Ex:358 nm Em:460 nm	3–4	1992 (393)
Muscle, kidney, liver	TTC, OTC, CTC	HCl/glycine extn, cyclohexyl SPE	Al^{3+} postcol deriv	RPLC-C18	FLD: Ex:390 nm Em:490 nm	20–230	1995 (394)
Muscle, kidney	OTC	ACN/EDTAMgIB extn, hexane defat., ultrafiltration	Mg^{2+} deriv	RPLC-C18	FLD: Ex:380 nm Em:520 nm	40–50	1996 (395)
Muscle, kidney, liver fresh and lyophilized	CTC+4-epiCTC	HCl/glycine extn, cyclohexyl SPE	pH 12 isoCTC conversion	PLRP-S	FLD: Ex:340 nm Em:420 nm	20–50	1998 (396)
Muscle, liver	DC+4-epiDC DMCTC[IS]	Succinate buffer extn, MeOH dilut., MCAC cleanup, SDBRPS SPE cleanup	Postcol Zr^{2+}deriv	RPLC-C18	FLD: Ex:406 nm Em:515 nm	1.0	1998 (397)

(Continued)

Table 42.9 (Continued)

Tissue	Analytes	Sample Treatment	Derivatization	LC Column Technique	Detection	Limit Range (ppb)	Year (Reference)
Muscle, liver	DC+4-epiDC DMCTC[IS]	Succinate buffer extn, MeOH dilut., MCAC cleanup, SDBRPS SPE cleanup	Postcol Zr²⁺ deriv	RPLC-C18	FLD: Ex:406 nm Em:515 nm	1.0	2000 (398)
Fish muscle	TTC, OTC DMCTC[IS]	EDTAMIB extn, hexane defat, TCA deprot., HLB SPE	—	RPLC-C18	FLD: Ex:385 nm Em:500 nm	50	2003 (399)
Chicken muscle	TTC, OTC, CTC	EDTAMIB extn, HLB SPE	Tris/Eu³⁺ /CTAC deriv	—	TRL: Ex:388 nm Em:615 nm	3–20	2004 (400)
Fish muscle	OTC, 4-epiOTC, anhydroOTC, alpha-apoOTC, beta-apoOTC	EDTAMIB extn, C18 SPE, NH₂ SPE	Tris/Mg²⁺ deriv	RPLC-phenyl	FLD: Ex:378 nm Em:500 nm	100	2005 (401)
Muscle	TTC, OTC, CTC	ACN/citrate buffer/MgCl₂ extn, evap., reconst. in malonate/MgCl₂	—	RPLC-phenyl	FLD: Ex375 nm Em535 nm	1.0–2.0	2007 (340)
Muscle, kidney	TTC, OTC, CTC+isomers	HCl/glycine extn, cyclohexyl SPE	—	RPLC-C8 + ion pairing HFBA/EDTA	MS apci+ammoniac	10–20	1997 (402)
Kidney	TTC, OTC, CTC	On-line extn, sample cleanup	—	RPLC-C18	MSMS esi+	–ᵃ	2000 (370)
Kidney	OTC	MeOH extn, Evap., water dilution, C18 SPE, evap., reconst.	—	RPLC-C18	MSn esi–	–ᵃ	2003 (195)

Tissue	Analytes	Sample prep		Column	Detection	Range	Year (ref)
Muscle, liver, kidney	OTC+ 4-epiOTC DMCTC[(IS)]	Succinate buffer extn, TCA deprot., HLB SPE	—	PLRP-S	MSn esi+	1–48	2003 (403)
Muscle	OTC+ 4-epiOTC, selected[(IS)]	—[a]	—	PLRP-S	MSMS esi+/esi–	—[a]	2003 (223)
Fish muscle	TTC, OTC, CTC, DC	ACN extn, hexane defat., evap. H$_2$O/ ACN extn, hexane defat.	—	RPLC-C18	MSMS esi+/esi–	10–100	2005 (262)
Muscle	TTC, OTC, CTC, DC + 4-epimers, DMCTC[(IS)]	EDTAMIB extn, HLB SPE	—	RPLC-C18	MSMS esi+	10	2006 (404)
Muscle	TTC, OTC, CTC, DC, 4-epiOTC, 4-epiTTC, 4-epiCTC, DMCTC[(IS)]	EDTA/SW extn	—	RPLC-C8	MSMS esi+	1–10	2006 (405)

[a] Not reported.
[(IS)]Internal standard.

Note: TTC: tetracycline; OTC: oxytetracycline; CTC: chlortetracycline; TCA: trichloroacetic acid; SPE: solid-phase extraction; PLRP-S: polymeric stationary phase; EDTAMIB: Ethylenediaminetetraacetic acid and McIlvaine Buffer pH 4; DCM: dichloromethane; RPLC: reverse phase liquid chromatography; UV: ultraviolet detection; DAD: diode array detection; DMCTC: minocycline, demeclocycline or demethylchlortetracycline; SEPSA: succinate/EDTA/pentane sulfonic acid buffer; MCAC: metal-chelate affinity chromatographic precolumn; EtOAc: ethyl acetate; ACN: acetonitrile; SDB: styrene divylnilbenzene cartridge; DSA: decane sulfonic acid; DC: doxycycline; Extn: extraction; HLB: hydrophilic Lipophilic Balance; MeOH: methanol; HSA: heptane sulfonic acid; EDTAMgIB: Ethylenediaminetetraacetic acid and Mg²⁺ in imidazole buffer pH 7.2; 4-epiDC: 4-epimer of doxycycline; SDBRPS: polystyrene-divinylbenzene-reverse phase sulfonated cartridge; CTAC: cetyltrimethylammonium chloride; HFBA: heptafluorobutyric acid; EDTA: Ethylenediaminetetraacetic acid; MSMS: triple quadrupole; Evap.: evaporation before reconstitution; Reconst: reconstitution prior to injection; MSn: ion trap; esi+: electrospray source in positive mode; esi–: electrospray source in negative mode; SW: subcritical water; 4-epiOTC: 4-epimer of oxytetracycline; 4-epiCTC: 4-epimer of chlortetracycline; deprot.: deproteinisation; defat. defattening; purif.: purification; Dilut.: dilution; PBS: phosphate buffer solution; Tris: tris(hydroxymethyl)-aminomethane pH 9; PLRP-S: polymeric stationary phase; TRL: time resolved luminescene; MS: single quadrupole; ppb: parts per billion equivalent to µg of residue/kg of tissue.

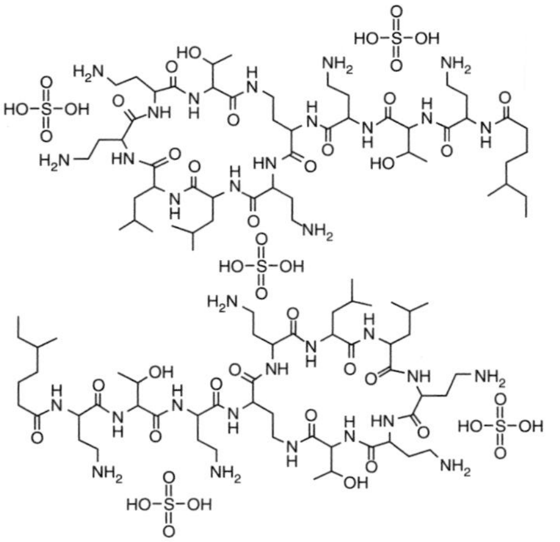

Colistin sulfate or polymyxin E

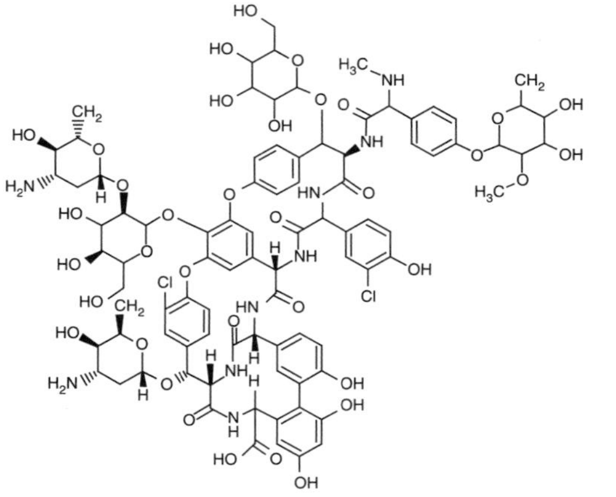

alpha-Avoparcin

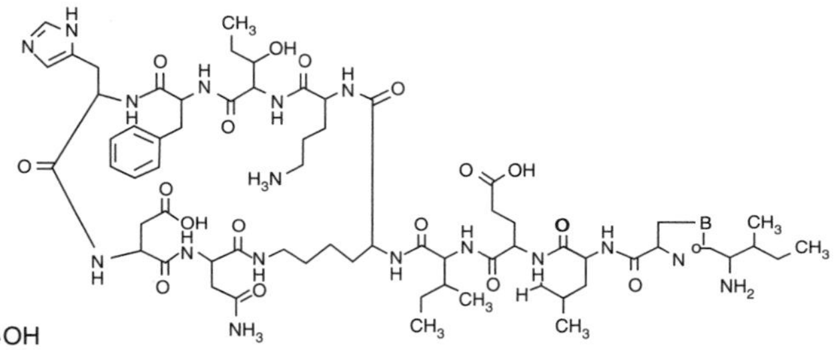

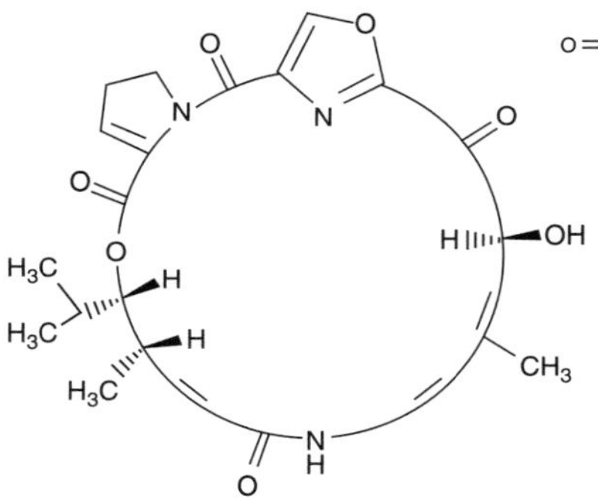

Virginiamycin M1 (streptomgramin A)

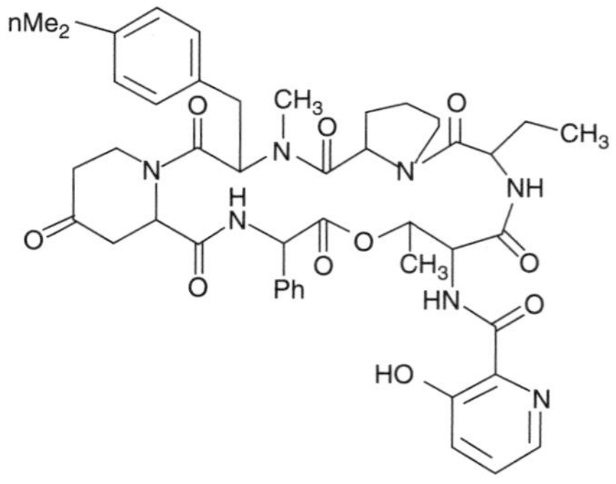

Bacitracin A

Virginiamycin S1 (streptogramin B)

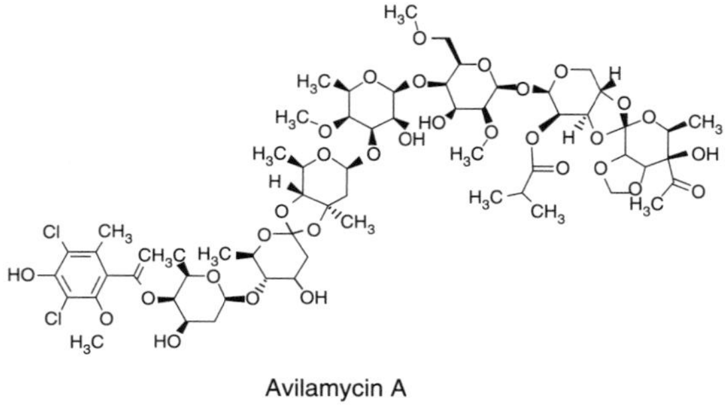

Avilamycin A

Figure 42.12 Structures of some polypeptidic antibiotics.

of several molecules, for example, factor M1 and factor S1 principal components for virginiamycin or compound A and compound F principal components for bacitracin or even alpha and beta major components for avoparcin. The macromolecular structure makes them difficult to selectively be extracted among and purified from the naturally occurring polypeptidic molecules found in food products from animal origin. Few attempts have been undertaken during the past 10 years for developing selective analytical methods aimed at monitoring polypeptides in meat (muscle and kidney tissues) in the ppb range in line with their illegal use. Table 42.10 displays some of these methods for the glycopeptidic antibiotic avoparcin [411], for the streptograminic antibiotic virginiamycin [206,214,412–415], for the polypeptidic antibiotic bacitracin, and for the cyclic polypeptidic antibiotic colistin [280,416]. In line with their use in feedingstuffs, most of the methods developed for their monitoring analyze polypeptides as additives in the feedingstuff matrices instead of residues in meat tissues. Two examples are presented in Table 42.10, one for colistin [190] and one for bacitracin and virginiamycin [280]. Polypeptidic antibiotics are still a challenge in drug residue analysis from biological matrix.

42.4.11 Polyether Antibiotics

The ionophores are polyether antibiotics obtained mostly by fermentation of several *Streptomyces*. They hold the specificity to be licensed essentially for use against protozoal coccidial infections in poultry instead of being directed against bacteria. They are, therefore, more generally considered as anticoccidials or coccidiostats even though formerly employed as feed additives for promoting growth in cattle and sheep. They are regulated as feed additives and growth-promoting agents under the feed additive legislation in the EU by Directive 70/524/EC [406]. As for the polypeptidic antibiotics, under regulation 2562/99/EC, a period of 5–10 years is granted from 2004 to 2014 for reevaluation of these drugs by the supporting pharmaceutical stakeholders [410]. Their principal compounds are lasalocid, maduramicin, monensin, narasin, salinomycin, and semduramicin. In terms of chemistry, the basis of their structure is a sequence of tetrahydrofuran and tetrahydropyran units linked together in the form of spiroketal moieties (Figure 42.13). In spite of hydroxylic and carboxylic functions at both ends of these macromolecules, they are rather poor soluble antibiotics in aqueous solutions due to their macrocyclic conformation with polar groups oriented inward and nonpolar groups oriented outward. As a consequence, organic solvent extraction is the preferred one. But, on the counterbalance, purification by liquid–liquid partitioning is difficult to achieve due to similar solubility properties of the ionophores in their free acid and salt forms and due to instability in acidic media. The term "ionophore" is attributed to these macromolecules in relation to their ability to stabilize by complexing with such alkaline cations as Ca^{2+} or Mg^{2+}. Four interesting reviews were published on ionophore polyethers: one by Weiss and MacDonald in 1985 essentially dedicated to their chemistry [417] and three other more recent ones by Asukabe and Harada in 1995 [418], by Botsoglou and Kufidis in 1996 [11], and by Elliott et al. in 1998 [419] and dedicated to their chemical analysis in food products from animal origin. Since then, the analysis of polyether ionophores in muscle tissues relied essentially on the confirmatory LC-MS technique as described in Table 42.11 [214,420–425]. Only few attempts are reported of the screening with fluoroimmunoassays [426].

42.4.12 Other Antibiotics (Novobiocin, Tiamulin)

Novobiocin is an antibiotic produced by *Streptomyces spheroides* and *Streptomyces niveus* with a narrow-spectrum activity against some gram-positive bacteria. It is soluble in polar organic solvents such as alcohols, acetone, and EtOAc but rather insoluble in aqueous solution below pH 7.5 and

Table 42.10 Summary of Literature Methods for Determination of Polypeptidic Antibiotic Residues in Tissues Using Liquid Chromatography

Tissue	Analytes	Sample Treatment	Derivati- zation	LC Column Technique	Detection	Limit (ppb)	Year (Reference)
Muscle	Virginiamycin M1 Virginiamycin S1	ACN extn, Evap., MeOH dilution, CHCl$_3$ partition., H$_2$O purif., drying evap.	—	RPLC-C18	FLD: Ex311 nm Em427 nm	100, 10	1987 (412)
Muscle, kidney, liver, serum	Virginiamycin M1	PBS/MeOH extn, PE defat.,	—	RPLC-C18	UV: 254 nm	10	1988 (413)
Muscle	Virginiamycin M1	MeOH/PTA extn, CHCl$_3$ partition., silica SPE	—	RPLC-C18	UV: 235 nm	50	1989 (414)
Muscle	Virginiamycin M1	C18 MSPD, EtOAc extn, evap.	—	RPLC-C18	UV-DAD: 254 nm	2–7	1997 (206)
Feed	Colistin A, Colistin B	HCl extn	Postcol. OPA	RPLC-C18	FLD: Ex355 nm Em415 nm	100	1998 (190)
Kidney	Avoparcin	Hot water/ EtOH ASE, XAD-7 MSPD, HILIC SPE	—	HILIC-LC	UV: 225 nm	500	2002 (411)
Feed	Virginiamycin M1, bacitracin A	MeOH/Water extn, HLB SPE	—	RPLC-C18	MSMS esi+	200–600	2003 (280)
Standards	Polymyxins, Bacitracin A	—	—	RPLC-C18	MS^n	—[a]	2003 (415)
Muscle	Virginiamycin	ACN/MeOH extn, hexane/ ACN defat., evap., reconst.	—	RPLC-C18	MSMS esi+	2–8	2006 (214)
Muscle	Bacitracin A, Colistin A, Colistin B	Acid extn, strata-X SPE	—	RPLC-C18	MSMS esi+	14–47	2006 (416)

[a] Not reported.
[IS]Internal standard.

Note: ACN: acetonitrile; Evap.: evaporation of volatile solvent; MeOH: methanol; RPLC: reverse phase liquid chromatography; PBS: phosphate buffer solution; PE: petroleum ether; UV: ultraviolet detection; PTA: phosphotungstic acid; SPE: solid-phase extraction; MSPD: matrix solid pagse dispersion; EtOAc: ethyl acetate; OPA: ortho-phthalaldehyde; EtOH: ethanol; ASE: accelerated solvent extractor; XAD-7 HP: acrylic polymer resin; HILIC: hydrophilic interaction chromatography; HLB: hydrophilic Lipophilic Balance; MSMS: triple quadrupole; esi+: electrospray source in positive mode; MS^n: ion trap; Extn: extraction; defat. defattening; purif.: purification; partition.: aqueous partitioning; DCM: dichloromethane; ppb: parts per billion equivalent to μg of residue/kg of tissue.

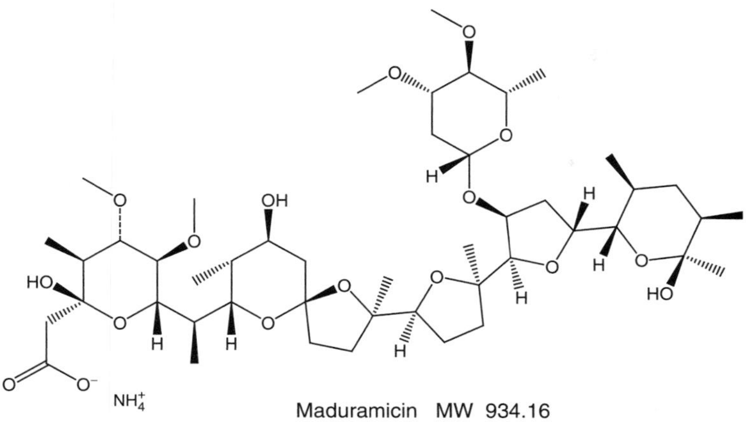

Maduramicin MW 934.16

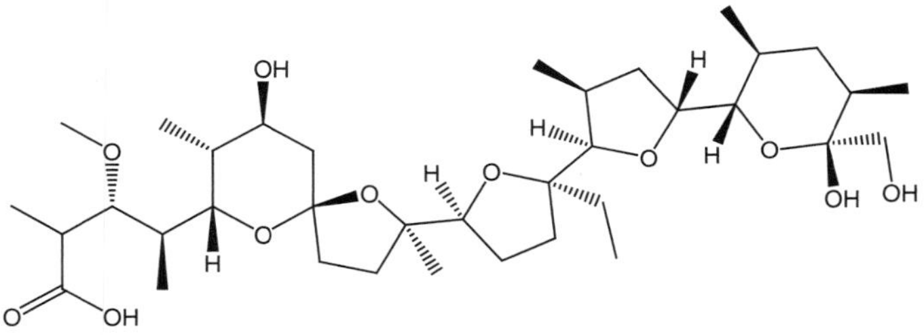

Monensin MW 670.88

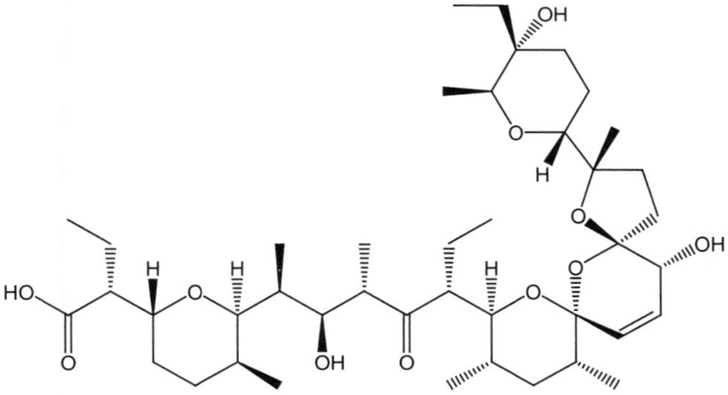

Salinomycin MW 751.01

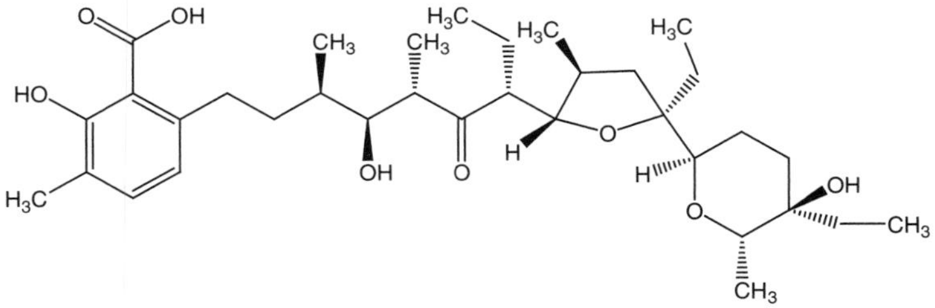

Lasalocid MW 590.79

Figure 42.13   Structures of ionophore polyethers used in food-producing animal feeding.

Table 42.11 Summary of Literature Methods for Determination of Polyether Antibiotic Residues in Tissues Using Liquid Chromatography

Tissue	Analytes	Sample Treatment	Derivati-zation	LC Column Technique	Detection	Limit (ppb)	Year (Reference)
Standards	Lasalocid, salinomycin, narasin, monensin	—	—	RPLC-C18	MS esi+	—[a]	1998 (420)
Standards	Salinomycin, narasin, lasalocid	—	—	RPLC-C18	MSMS esi+	—[a]	1999 (421)
Muscle	Lasalocid	—	—	RPLC-C18	MS esi+	—[a]	2002 (422)
Liver, eggs	Narasin, monensin, salinomycin, lasalocid	MeOH extn	—	RPLC-C18	MSMS esi+	0.5	2002 (423)
Muscle, eggs	Narasin, monensin, salinomycin, lasalocid, maduramycin, nigericin[(IS)]	ACN extn, silica SPE	—	RPLC-C18	MSMS esi+	0.1—0.5	2004 (424)
Eggs	Lasalocid, salinomycin, narasin, monensin, nigericin[(IS)]	Organic extn	—	RPLC-C18	MSMS esi+	1	2005 (425)
Muscle	Lasalocid, salinomycin, narasin, monensin	ACN/MeOH extn, hexane/ACN defat., evap., reconst.	—	RPLC-C18	MSMS esi+	0.6—2	2006 (214)

[a] Not reported.
[(IS)]Internal standard.

Note: RPLC: reverse phase liquid chromatography; MS: single quadrupole; esi+: electrospray source in positive mode; MSMS: triple quadrupole; MeOH: methanol; ACN: acetonitrile; SPE: solid-phase extraction; Extn: extraction; defat.: defattening; Evap.: evaporation before reconstitution; Reconst: reconstitution prior to injection; purif.: purification; EtOAc: ethyl acetate; SDB: styrene divinylbenzene; HLB: hydrophilic Lipophilic Balance; PLRP-S: polymeric stationary phase; UV: ultraviolet detection; DAD: diode array detection; LIF: laser-induced fluorescence; MS^n: ion trap; esi−: electrospray source in negative mode; ppb: parts per billion equivalent to μg of residue/kg of tissue.

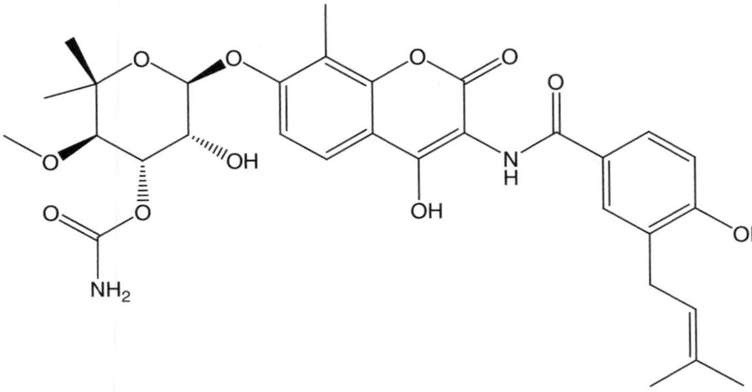

Novobiocin, MW 612.62

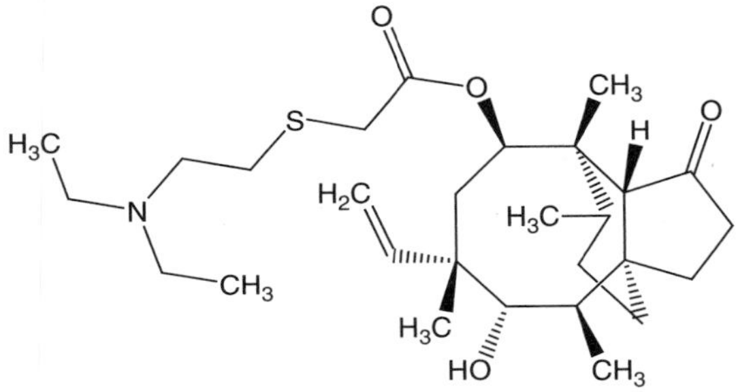

Tiamulin, MW 493.74

Figure 42.14   Structures of novobiocin and tiamulin.

in chloroform. It bears both an enolic (pK_a 4.3) and a phenolic (pK_a 9.2) character (Figure 42.14), leading to a weak dibasic acid behavior. As a consequence, liquid–liquid partitioning is not an efficient process for extraction-purification from biological matrices. Veterinary treatments for lactating cows may lead to novobiocin residues in milk, and feed additive practice in poultry may give novobiocin residues in chicken muscle. Microbiological methods have been tentatively applied to novobiocin residue detection in meat [29]. TLC-bioautography was a formerly screening method applied to novobiocin [55]. Although very few LC methods for monitoring novobiocin in meat are reported [427], recent articles relate to residues in egg [428] or in milk [429] as displayed in Table 42.12.

Tiamulin is a diterpene antibiotic with a pleuromutilin chemical structure similar to that of valnemulin with an eight-membered carbocyclic ring at the center of the structure. Pleuromutilines are biosynthetically produced by *Pleurotus mutilus*. Tiamulin's activity is largely confined to gram-positive microorganisms. This antibiotic acts by inhibiting bacterial protein synthesis at the ribosomal level. Its usage in veterinary medicine applies for treatment and prophylaxis of dysentery, pneumonia, and mycoplasmal infections in pigs and poultry. The principal residue to be monitored in muscle and in other meat tissues is the metabolite 8-alpha-hydroxymutilin. A bioscreening assay was reported in 2000 to monitor tiamulin activity [430]. Two LC methods (Table 42.12) are reported for tiamulin with one in meat by UV detection [431] and one in honey by MS detection [432].

Table 42.12 Summary of Literature Methods for Determination of Several Antibiotic Residues of Lower Interest in Tissues Using Gas or Liquid Chromatography

Tissue	Analytes	Sample Treatment	Derivatization	LC Column Technique	Detection	Limit (ppb)	Year (Reference)
Muscle, milk	Novobiocin	MeOH extn deprot., filtering, online SPE	—	RPLC-C18	UV: 340 nm	50	1988 (427)
Eggs	Novobiocin	MeOH extn, silica SPE, hexane defat.	—	RPLC-C18	MS^n esi+	3	2004 (428)
Milk	Novobiocin	Buffer dilution, MeOH deprot., filtering	—	RPLC-C18	UV: 340 nm	50	2005 (429)
Muscle	Tiamulin	ACN extn, Evap., hexane purif., C18 SPE	—	RPLC-C18	UV: 210 nm	25	2006 (431)
Honey	Tiamulin	Organic extn, SDB SPE	—	RPLC-C18	MS esi+	0.5—1.2	2006 (432)

[a] Not reported.
[IS] Internal standard.

Note: MeOH: methanol; SPE: solid-phase extraction; RPLC: reverse phase liquid chromatography; UV: ultraviolet detection; MS^n: ion trap; esi+: electrospray source in positive mode; ACN: acetonitrile; Extn: extraction; Purif.: purification; deprot.: deproteinisation; defat. defattening; DAD: diode array detection; MS: single quadrupole; MSMS: triple quadrupole; ppb: parts per billion equivalent to μg of residue/kg of tissue.

42.4.13 Other Antibacterials (Carbadox and Olaquindox)

Carbadox and olaquindox are two widely available antibacterial synthetic N,N'-di-N-oxide quinoxaline compounds used as growth promoters (Figure 42.15). Possible mutagenicity and carcinogenicity have been demonstrated for these active molecules, leading to their ban in many countries including EU under regulation 2788/98/EC [433]. As metabolism studies have shown the rapid conversion of these compounds into their monooxy- and desoxy-metabolites which are also possible mutagenic and carcinogenic entities, it was important to monitor either the intermediate desoxycarbadox or final metabolite quinoxaline-2-carboxylic acid (QCA) as residual targets in muscle or liver tissues and also the methyl-3-quinoxaline-2-carboxylic acid (MQCA) as the stable metabolite for olaquindox. As described in Table 42.13, several LC-UV and LC-MS methods were

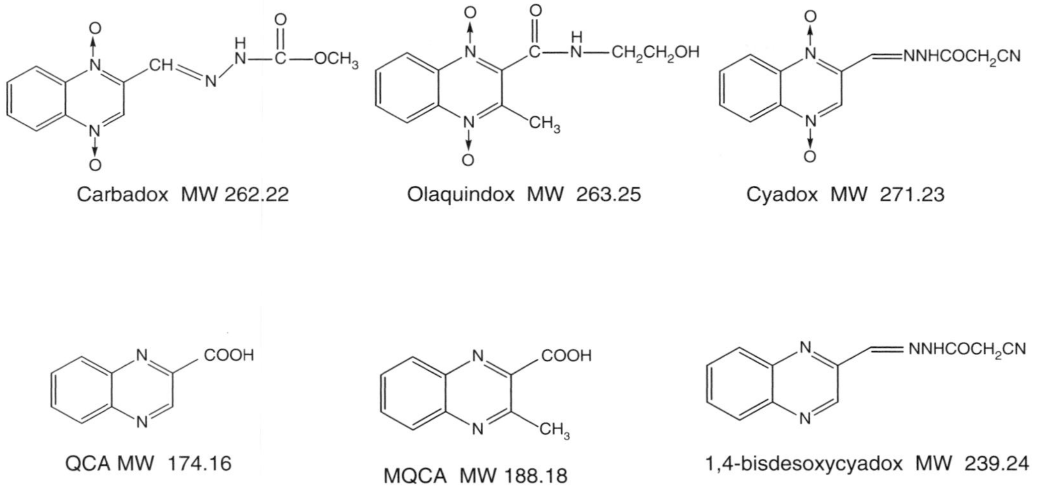

Figure 42.15 Structures of 3 *N,N'*-di-*N*-oxide quinoxalines and their major metabolites—QCA, MQCA, and 1,4-bisdesoxycyadox.

proposed recently to cover the control of the ban of these two quinoxaline compounds [434–441]. Another quinoxaline 1,4-dioxide with antimicrobial properties, the cyadox (CYX), might be of interest for monitoring as it is possibly used as a growth promoter. It metabolizes as carbadox and olaquindox in animal tissues to give, among other intermediate compounds, the 1,4-bisdesoxycyadox (BDCYX). A paper related to the analysis of its major metabolites by LC-UV in chicken muscle tissues is reported in the literature [442].

42.5 Conclusion

Antimicrobials are one of the largest families of pharmaceuticals used in veterinary medicine either to treat animal diseases or to prophylactically prevent their occurrence. Also used to promote the growth of food-producing animals, this practice is prone to drastic reduction in the near future. The control of veterinary drug residues in meat and other food from animal origin is one of the concerns for food safety regulation. It is important to prevent risks for human health. Farmers, veterinarians, feed manufacturers, food industry, and regulatory agencies together have to create the conditions of the food safety for the consumers. The control of antimicrobial resistance of certain bacteria is also another challenge for human health. Regulating antibacterials in animal husbandry is the first step to contain bacterial resistance and control human diseases and their potential cure.

Analytical methods developed for monitoring antimicrobial residues in meat and in other food products from animal origin can be ranged in two different stages. At first stage, there is the strategy of screening to be adapted for evaluating in a reduced period of time the presence or absence of antimicrobials. The screening should generally be as large as possible in terms of the residues tested. This is the concept applied in the microbiological inhibitory methods (plate tests, swab tests, or receptor tests). But, screening methods reduced to one family or even one compound

Table 42.13 Summary of Literature Methods for Determination of Carbadox and Olaquindox Residues in Tissues Using Gas or Liquid Chromatography

Tissue	Analytes	Sample Treatment	Derivatization	LC Column Technique	Detection	Limit (ppb)	Year (References)
Liver	QCA	Tris buffer/Subtilisin A enzymat. digestion heat extn; acetic acid neutral.; DEE LLE; EtOAc LLE; HCl dilution; EtOAc extn; evap.; H$_2$SO$_4$/propanol esterification; hexane defat.; evap. reconst.; HPLC fraction purif.; Evap.; hexane extn; evap.; EtOH reconst	H$_2$SO$_4$/propanol esterification	GC	MS	10	1990 (434)
Liver, muscle	QCA	Alkaline hydrolysis, neutral., EtOAc LLE, citric acid dilution, SCX SPE	Methylation for GC	RPLC-C18, GC	UV: 320 nm, MS nci	3–5	1996 (435)
Liver	QCA, QCA-d4[IS]	Alkaline hydrolysis, neutral., LLE	—	RPLC-C18	MSMS esi+	0.2	2002 (436)
Muscle, liver	QCA, MQCA, QCA-d4[IS], MQCA-d4[IS]	Alkaline hydrolysis, HCl acidif., SCX SPE, HCl acidif., EtOAc LLE, evap., reconst.	—	RPLC-C18	MSMS esi+	1–4	2005 (437)
Feeds	Carbadox, olaquindox	ACN/CHCl$_3$ extn, evap., reconst.	—	RPLC-C18	MSMS esi+	500	2005 (438)
Liver	QCA, MQCA	Protease digestion, LLE, SPE cleanup, LLE	—	RPLC-C18	MSMS esi+	1–3	2005 (439)
Liver	QCA, QCA-d4[IS]	MPA/MeOH deprot., PBS/EtOAc extn, MAX SPE	MTBSTFA, or TMSDM	GC	MS nci	0.7	2007 (440)
Liver, muscle	QCA, MQCA	Acid hydrolysis, LLE, MAX SPE	—	RPLC-C18	UV	1–5	2007 (441)
Muscle	QCA, BDCYX	QCA: alkaline hydrolysis, neutral., EtOAc LLE, citric acid dilution, IEC purif, HCl/ CHCl$_3$ partition., evap., MeOH BDCYX: ACN extn, evap., ACN/hexane defat., evap., MeOH reconst.	—	RPLC-C18	UV	20–25	2005 (442)

a Not reported.
[IS]Internal standard.

Note: QCA: quinoxaline-2-carboxylic acid; EtOAc: ethyl acetate; Evap.: evaporation before reconstitution; MS: single quadrupole; LLE: liquid–liquid extraction/ partitioning; RPLC: reverse phase liquid chromatography; UV: ultraviolet detection; SCX: strong cation exchange; SPE: solid-phase extraction; MSMS: triple quadrupole; esi+: electrospray source in positive mode; MQCA: methyl-3-quinoxaline-2-carboxylic acid; ACN: acetonitrile; Reconst: reconstitution prior to injection; MPA: metaphosphoric acid; MeOH: methanol; PBS: phosphate buffer solution; MAX: medium anion exchange; MTBSTFA: N-methyl-N-tert-butyldimethylsilyltrifluoroacetamide; TMSDM: , trimethylsilyldiazomethane; BDCYX: 1,4-bisdesoxycyadox; IEC: ion exchange cartridge; Extn: extraction; DEE: diethyl ether; HLB: hydrophilic Lipophilic Balance; PLRP-S: polymeric stationary phase; MS^n: ion trap; esi-: electrospray source in negative mode; nci: negative chemical ionization; ppb: parts per billion equivalent to μg of residue/kg of tissue.

(immunological tests) can also be proposed in some specific cases, for specific antimicrobial monitoring, or with particular food products. At the second stage, there is an increasing interest in the strategy of confirmation of the residues with unequivocal identification of the analyte(s). During the past 15 years, physicochemical technologies have been developed and implemented for that specific purpose. From TLC to HPLC, from HPLC to LC-MS, and from LC-MS to LC-MSMS or LC-MS^n systems, it is obvious that nowadays it is the innovative mass spectrometric technology that is in application at the confirmatory step. One methodology that has been increasingly disregarded because of the use of innovative mass spectrometric technologies is the chemical extraction/purification process. However, this is often a bad consideration because optimizing the extraction is still of great importance in antimicrobial residue testing. Following the same idea, a thorough purification before injecting into the analytical instruments, including the LC-MS ones, can be of great help to improve the reliability of the analysis. In fact, chemical diversity of antimicrobials always requires the setting up of different approaches for their extraction from food matrices. Extraction methods have considerably changed in the past 20 years. LLE has been miniaturized or largely replaced by solid-phase extraction. Matrix solid-phase extraction (MSPD) and accelerated solvent extraction (ASE) are also emerging techniques. Extracting and purifying residues of antimicrobials from meat and other food matrices still need to be satisfactorily undertaken for quality and reliability of the analyses. On the instrumental part, a chromatographic separation is always necessary to optimize the selective control of an analyte with regard to the others and to the interfering substances in the purified extract. At the detection level of the analytical instrument, the mass spectrometer is now considered the optimal detector for controlling reliable identification and sufficient quantification of antibiotic residues in the ppb (μg/kg) range of concentrations. Several food safety legislations including the EU one have now enforced this concept, in particular for the prohibited substances [4].

Abbreviations

6-APA	6-aminopenicillenic acid
7-ACA	7-aminocephalosporanic acid
ACN	acetonitrile
ADC	azodicarbonamide
AHD	1-aminohydantoin
AMOZ	3-amino-5-morpholinomethyl-2-oxazolidinone
AOZ	3-amino-2-oxazolidinone
APCI	atmospheric pressure chemical ionization source
API	atmospheric pressure ionization source
APPI	atmospheric pressure photochemical ionization source
ASE	accelerated solvent extraction
BDCYX	1,4-bisdesoxycyadox
C18	18-carbon alkyl-bonded silica stationary phase
C8	8-carbon alkyl-bonded silica stationary phase
CEC	electrochromatography
CE-MS	capillary zone electrophoresis coupled to mass spectrometry
CYX	cyadox
CZE or CE	capillary zone electrophoresis
DAD	diode array detector
DCM	dichloromethane

DNA	desoxyribonucleic acid
DNC	dinitrocarbanilamide
DNSH	3,5-dinitrosalicylic acid hydrazide
DTE	dithioerythreitol
ECD	electron capture detector
ELISA	enzyme-linked immunosorbent assay
ESI	electrospray source of ionization
EtOAc	ethyl acetate
EU	European Union
FID	flame ionization detector
FLD	fluorescence detector
Fluo	fluorescence detector
FPIA	fluorescence polarization immunoassay test
GC	gas chromatography
GC-MS	gas chromatography coupled to mass spectrometry
HCl	hydrochloric acid
HILIC	hydrophilic liquid chromatography
HLB	mixed hydrophilic–lipophilic balanced cartridge
HMMNI	2-hydroxymethyl-1-methyl-5-nitroimidazole
HPLC	high-performance liquid chromatography
HPLC-UV	high performance liquid chromatography connected to UV detector
IT	ion trap
LC	liquid chromatography
LC-MS	liquid chromatography coupled to mass spectrometry
LC-MSMS	liquid chromatography coupled to a tandem mass spectrometer also called triple quadrupolar mass spectrometer
LC-MS^n	liquid chromatography coupled to a ion trap mass spectrometer
LC-tandemMS	liquid chromatography coupled to a tandem mass spectrometer
LIF	laser-induced fluorescence detection
LLE	liquid–liquid extraction
MAX	medium anion exchange cartridge
MCX	medium cation exchange cartridge
MEKC	micellar electrokinetic capillary chromatography
MeOH	methanol
MQCA	methyl-3-quinoxaline-2-carboxylic acid
MRL	maximum residue limit
MRPL	minimum required performance limit
NPD	nitrogen-phosphorus detector
NRMP	national residue monitoring program (or plan)
MSPD	matrix solid-phase extraction
PB	particle beam source of ionization
PBS	phosphate buffer solution
PCA	pyridonecarboxylic acid
PDA	photodiode array
PLRP-S	polymeric reverse stationary phase
ppb	parts per billion equivalent to μg of residue/kg of tissue

ppm   parts per million equivalent to μg of residue/kg of tissue
QCA   quinoxaline-2-carboxylic acid
RPLC   reverse-phase liquid chromatography
RSD   relative standard deviation
SAX   strong anion exchange cartridge
SCX   strong cation exchange cartridge
SEM   semicarbazide
SFC   supercritical fluid chromatography
SPE   solid-phase extraction
SPFIA   solid-phase fluorescence immunoassay
SPR-BIA   surface plasmon resonance-based biosensor immunoassay
SQ   single quadrupole
TCA   trichloroacetic acid
TLC   thin layer chromatography
ToF   time-of-flight
TQ   triple quadrupole
TRIS   tris(hydroxymethyl)-aminomethane
TSP   thermospray source of ionization
UV   ultra-violet

References

1. Council Regulation (EEC) No. 2377/90 of 26th June 1990, *Off. J. Eur. Commun.*, L224 (1990) 1.
2. Council Directive 96/23/EC of 29th April 1996, *Off. J. Eur. Commun.*, L125 (1996) 10.
3. Council Regulation (EEC) No. 882/2004 of 29th April, 29th, 2004, *Off. J. Eur. Commun.*, L165 (2004) 1 and its Corrigendum of 28th May, 28th, 2004, *Off. J. Eur. Commun.*, L191 (2004) 1.
4. Commission Decision 2002/657/EC of 12th August 2002, *Off. J. Eur. Commun.*, L221 (2002) 8.
5. Commission Decision 93/256/EEC of 14th April 1993, *Off. J. Eur. Commun.*, L118 (1993) 64.
6. Keukens, H.J. et al., Analytical strategy for the regulatory control of residues of chloramphenicol in meat: preliminary studies in milk, *J. Assoc. Off. Anal. Chem.*, 75, 245, 1992.
7. Moats, W.A., Liquid chromatography approaches to antibiotic residue analysis, *J. Assoc. Off. Anal. Chem.*, 73, 343, 1990.
8. Bobbitt, D.R. and Ng, K.W., Chromatographic analysis of antibiotic materials in food, *J. Chromatogr.*, 624, 153, 1992.
9. Shaikh, B. and Moats, W.A., Liquid chromatographic analysis of antibacterial drug residues in food products of animal origin, *J. Chromatogr.*, 643, 369, 1993.
10. Aerts, M.M.L., Hogenboom, A.C., Brinkman, U.A.Th., Analytical strategies for the screening of veterinary drugs and their residues in edible products, *J. Chromatogr. B*, 667, 1, 1995.
11. Botsoglou, N.A. and Kufidis, D.C., Determination of antimicrobial residues in edible animal products by HPLC, in *Handbook of Food Analysis*, Vol. 2, Nollet, L.M.L., Ed., Marcel Dekker, New York, 1996.
12. Korsrud, G.O. et al., Bacterial inhibition tests used to screen for antimicrobial veterinary drug residues in slaughtered animals, *J. Assoc. Off. Anal. Chem. Int.*, 81, 21, 1998.
13. Di Corcia, A. and Nazzari, M., Liquid chromatographic-mass spectrometric methods for analyzing antibiotic and antibacterial agents in animal food products, *J. Chromatogr. A*, 974, 53, 2002.
14. Kennedy, D.G. et al., Use of liquid chromatography-mass spectrometry in the analysis of residues of antibiotics in meat and milk, *J. Chromatogr. A*, 812, 77, 1998.

15. Stolker, A.A.M. and Brinkman, U.A.Th., Analytical strategies for residue analysis of veterinary drugs and growth-promoting agents in food-producing animals—a review, *J. Chromatogr. A*, 1067, 15, 2005.

16. Moats, W.A., Liquid chromagraphic approaches to determination of beta-lactam antibiotic residues in milk and tissues, in *Analysis of Antibiotic Drug Residues in Food Products of Animal Origin*, Agarwal, V.K., Ed., Plenum Press, New York, NY, 1992, 133.

17. Boison, J.O., Review—Chromatographic methods of analysis for penicillins in food-animal tissues and their significance in regulatory programs for residue reduction and avoidance, *J. Chromatogr.*, 624, 171, 1992.

18. Shaikh, B. and Allen, E.H., Overview of physico-chemical methods for determining aminoglyco-side antibiotics in tissues and fluids of food-producing animals, *J. Assoc. Off. Anal. Chem.*, 68, 1007, 1985.

19. Salisbury, C.D.C., Chemical analysis of aminoglycoside antibiotics, in *Chemical Analysis for Antibiotics Used in Agriculture*, Oka, H. et al., Eds., AOAC International, Arlington, VA, 1995.

20. Boison, J.O., Chemical analysis of beta-lactam antibiotics, in *Chemical Analysis for Antibiotics Used in Agriculture*, Oka, H. et al., Eds., AOAC International, Arlington, VA, 1995.

21. Wang, S. et al., Analysis of sulphonamide residues in edible animal products: a review, *Food Addit. Contam.*, 23, 362, 2006.

22. Dorsey, J.G. et al., Liquid chromatography: theory and methodology, *Anal. Chem.*, 66, 1R, 1994.

23. Bruckner, C.A. et al., Column liquid chromatography: equipment and instrumentation, *Anal. Chem.*, 66, 1R, 1994.

24. La Course, W.R., Column liquid chromatography: equipment and instrumentation, *Anal. Chem.*, 72, 37R, 2000.

25. La Course, W.R., Column liquid chromatography: equipment and instrumentation, *Anal. Chem.*, 74, 2813, 2002.

26. Nouws, J.F.M. et al., The new Dutch kidney test, *Arch. Lebensmittelhygiene*, 39, 133, 1988.

27. Bogaerts, R. and Wolf, F., A standardized method for the detection of residues of antibacterial substances in fresh meat, *Die Fleischwirtschaft*, 60, 672, 1980.

28. Ellerbroek, L., The microbiological determination of the quinolone carbonic acid derivatives enrofloxacin, ciprofloxacin and flumequine, *Die Fleischwirtschaft*, 71, 187, 1991.

29. Calderon, V. et al., Evaluation of a multiple bioassay technique for determination of antibiotic residues in meat with standard solutions of antimicrobials, *Food Addit. Contam.*, 13, 13, 1996.

30. Korsrud, G.O. et al., Bacterial inhibition tests used to screen for antimicrobial veterinary drug residues in slaughtered animals, *J. Assoc. Off. Anal. Chem. Internat. Internat.*, 81, 21, 1998.

31. Currie, D., Evaluation of a modified EC four plate method to detect antimicrobial drugs, *Food Addit. Contam.*, 15, 651, 1998.

32. Okerman, L., De Wasch, K., and Van Hoof, J., Detection of antibiotics in muscle tissue with microbiological inhibition tests: effects of the matrix, *Analyst*, 123, 2361, 1998.

33. Fuselier, R., Cadieu, N., and Maris, P., S.T.A.R.: screening test for antibiotic residues in muscle—results of a European collaborative study, in *Proceedings Euroresidue IV Conference on Residues of Veterinary Drugs in Food*, Veldhoven, Netherlands, May 8–10, Van Ginkel, L.A. and Ruiter, A., Eds., Faculty of Medicine, Utrecht, Netherlands, 2000, 444.

34. Myllyniemi, A.-L. et al., A microbiological six-plate method for the identification of certain antibiotic groups in incurred kidney and muscle samples, *Analyst*, 126, 641, 2001.

35. Gaudin, V. et al., Validation of a microbiological method: the STAR protocol, a five-plate test, for the screening of antibiotic residues in milk, *Food Addit. Contam.*, 21, 422, 2004.

36. Pena, A. et al., Antibiotic residues in edible tissues and antibiotic resistance of faecal *Escherichia coli* in pigs from Portugal, *Food Addit. Contam.*, 21, 749, 2004.

37. Ferrini, A.-M., Mannoni, V., and Aureli, P., Combined plate microbial assay (CPMA): a 6-plate-method for simultaneous first and second level screening of antibacterial residues, *Food Addit. Contam.*, 23, 16, 2006.

38. Johnston, R.W., A new screening method for the detection of antibiotic residues in meat and poultry tissues, *J. Food Prot.*, 44, 828, 1981.
39. Dey, B.P. et al., Calf antibiotic and sulfonamide test (CAST) for screening antibiotic and sulfonamide residues in calf carcasses, *J. Assoc. Off. Anal. Chem. Internat. Internat.*, 88, 440, 2005.
40. Dey, B.P. et al., Fast antimicrobial screen test (FAST): improved screen test for detecting antimicrobial residues in meat tissue, *J. Assoc. Off. Anal. Chem. Internat. Internat.*, 88, 447, 2005.
41. Koenen-Dierick, K. et al., A one-plate microbiological screening test for antibiotic residue testing in kidney tissue and meat: an alternative to the EEC four-plate method ?, *Food Addit. Contam.*, 12, 77, 1995.
42. Cornet, V. et al., Interlaboratory study based on a one-plate screening method for the detection of antibiotic residues in bovine kidney tissue, *Food Addit. Contam.*, 22, 415, 2005.
43. Myllyniemi, A.-L. et al., Microbiological and chemical identification of antimicrobial drugs in kidney and muscle samples of bovine cattle and pigs, *Food Addit. Contam.*, 16, 339, 1999.
44. Charm, S.E. and Chi, R., Microbial receptor assay for rapid detection and identification of seven families of antimicrobial drugs in milk: collaborative study, *J. Assoc. Off. Anal. Chem.*, 71, 304, 1988.
45. Korsrud, G.O. et al., Evaluation and testing of Charm test II receptor assays for the detection of antimicrobial residues in meat, in *Analysis of Antibiotic Drug Residues in Food Products of Animal Origin*, Agarwal, V.K., Ed., Plenum Press, New York, NY, 1992, 75.
46. Korsrud, G.O. et al., Investigation of Charm test II receptor assays for the detection of antimicrobial residues in suspect meat samples, *Analyst*, 119, 2737, 1994.
47. Smither, R. and Vaughan, D.R., An improved electrophoretic method for identifying antibiotics with special reference to animal tissues and animal feeding stuffs, *J. Appl. Bacteriol.*, 44, 421, 1978.
48. Stadhouders, J., Hassing, F., and Galesloot, T.E., An electrophoretic method for the identification of antibiotic residues in small volumes of milk, *Neth. Milk Dairy J.*, 35, 23, 1981.
49. Lott, A.F., Smither, R., and Vaughan, D.R., Antibiotic identification by high voltage electrophoresis bioautography, *J. Assoc. Off. Anal. Chem.*, 68, 1018, 1985.
50. Tao, S.H. and Poumeyrol, M., Detection of antibiotic residues in animal tissues by electrophoresis, *Rec. Med. Vet.*, 161, 457, 1985.
51. Bossuyt, R., Van Renterghem, R., and Waes, G., Identification of antibiotic residues in milk by thin-layer chromatography, *J. Chromatogr.*, 124, 37, 1976.
52. Herbst, D.V., Applications of TLC and bioautography to detect contaminated antibiotic residues: tetracycline identification scheme, *J. Pharm. Sci.*, 69, 616, 1980.
53. Yoshimura, H., Itoh, O., and Yonezawa, S., Microbiological and thin-layer chromatographic identification of benzylpenicillin and ampicillin in animal body, *Jpn. J. Vet. Sci.*, 43, 833, 1981.
54. Kondo, F., A simple method for the characteristic differentiation of antibiotics by TLC-bioautography in graded concentration of ammonium chloride, *J. Food Prot.*, 51, 786, 1988.
55. Salisbury, C.D.C., Rigby, C.E., and Chan, W., Determination of antibiotic residues in Canadian slaughter animals by thin-layer chromatography-bioautography, *J. Agric. Food Chem.*, 37, 105, 1989.
56. Lin, S.-Y. and Kondo, F., Simple bacteriological and thin-layer chromatographic methods for determination of individual drug concentrations treated with penicillin-G in combination with one of the aminoglycosides, *Microbios*, 77, 223, 1994.
57. MacNeil, J.D., Current laboratory testing strategy for the identification and confirmation of antibiotic residues in fresh meat, in *Proceedings Euroresidue II Conference on Residues of Veterinary Drugs in Food*, Veldhoven, Netherlands, Haagsma, N., Ruiter, A., and Czedik-Eysenberg, P.B., Eds., Faculty of Medicine, Utrecht, Netherlands, 1993, 469.
58. Korsrud, G.O. et al., Laboratory evaluation of the Charm Farm test for antimicrobial residues in meat, *J. Food Prot.*, 58, 1129, 1995.
59. Stead, S. et al., Meeting maximum residue limits: an improved screening technique for the rapid detection of antimicrobial residues in animal food products, *Food Addit. Contam.*, 21, 216, 2004.
60. Fabre, J.-M. et al., Résidus d'antibiotiques dans la viande de porc et de volaille en France: situation actuelle et évaluation d'un nouveau test de détection, *Bull. des G.T.V.*, 23, 305, 2004.

61. Cantwell, H. and O'Keeffe, M., Evaluation of the Premi® test and comparison with the one plate test for the detection of antimicrobials in kidney, *Food Addit. Contam.*, 23, 120, 2006.

62. Okerman, L. et al., Evaluation and establishing the performance of different screening tests for tetracycline residues in animal tissues, *Food Addit. Contam.*, 21, 145, 2004.

63. Brown, S.A. et al., Extraction methods for quantitation of gentamicin residues from tissues using fluorescence polarization immunoassay, *J. Assoc. Off. Anal. Chem.*, 73, 479, 1990.

64. Okerman, L. et al., Simultaneous determination of different antibiotic residues in bovine and porcine kidneys by solid-phase fluorescence immunoassay, *J. Assoc. Off. Anal. Chem. Internat.*, 86, 236, 2003.

65. Korpimäki, T. et al., Generic lanthanide fluoroimmunoassay for the simultaneous screening of 18 sulfonamides using an engineered antibody, *Anal. Chem.*, 76, 3091, 2004.

66. Corrégé, I. et al., Comparaison des méthodes rapides de détection des résidus d'antibactériens dans la viande de porc, *Techni-Porc*, 17/04/94, 29, 1994.

67. Crooks, S.R.H. et al., The production of pig tissue sulphadimidine reference material, *Food Addit. Contam.*, 13, 211, 1996.

68. Haasnoot, W. et al., Application of an enzyme immunoassay for the determination of sulphamethazine (sulphadimidine) residues in swine urine and plasma and their use as predictors of the level in edible tissue, *Food Addit. Contam.*, 13, 811, 1996.

69. De Wasch, K. et al., Detection of residues of tetracycline antibiotics in pork and chicken meat: correlation between results of screening and confirmatory tests, *Analyst*, 123, 2737, 1998.

70. Medina, M.B., Poole, D.J., and Anderson, M.R., A screening method for beta-lactams in tissues hydrolyzed with penicillinase I and lactamase II, *J. Assoc. Off. Anal. Chem. Internat.*, 81, 963, 1998.

71. Edder, P., Cominoli, A., and Corvi, C., Determination of streptomycin residues in food by SPE and LC with post-column derivatization and fluorimetric detection, *J. Chromatogr. A*, 830, 345, 1999.

72. Draisci, R. et al., A new electrochemical enzyme-linked immunosorbent assay for the screening of macrolide antibiotic residues in bovine meat, *Analyst*, 126, 1942, 2001.

73. Watanabe, H. et al., Monoclonal-based enzyme-linked immunosorbent assay and immunochromatographic assay for enrofloxacin in biological matrices, *Analyst*, 127, 98, 2002.

74. Gaudin, V., Cadieu, N., and Maris, P., Inter-laboratory studies for the evaluation of ELISA kits for the detection of chloramphenicol residues in milk and muscle, *Food Agric. Immunol.*, 15, 143, 2003.

75. Fodey, T.L. et al., Production and characterisation of polyclonal antibodies to a range of nitroimidazoles, *Anal. Chim. Acta*, 483, 193, 2003.

76. Grunwald, L. and Petz, M., Food processing effects on residues: penicillins in milk and yoghurt, *Anal. Chim. Acta*, 483, 73, 2003.

77. Posyniak, A., Zmudski, J., and Niedzielska, J., Evaluation of sample preparation for control of chloramphenicol residues in porcine tissues by enzyme-linked immunosorbent assay and liquid chromatography, *Anal. Chim. Acta*, 483, 307, 2003.

78. Impens, S. et al., Screening and confirmation of chloramphenicol in shrimp tissue using ELISA in combination with GC-MSMS and LC-MSMS, *Anal. Chim. Acta*, 483, 153, 2003.

79. Gaudin, V., Cadieu, N., and Sanders, P., Validation of commercial ELISA kits for banned substances: chloramphenicol and nitrofurans, communication presented at *Workshop on Validation of Screening Methods*, Fougeres, France, June 3–4, 2005 (unpublished data).

80. Kuhlhoff, S. and Diehl, Y., Schneller Nachweis von Fluoroquinolonen bei Geflügelfleischproben aus dem Handel, *Deutsche Lebensmittel-Rundschau*, 101, 384, 2005.

81. Norgaard, A., Validation of a screening method for detection of chloramphenicol in muscle, casings and urine, sing previous obtained results, communication presented at *Workshop on Validation of Screening Methods*, Fougeres, France, June 3-4, 2005 (unpublished data).

82. Huet, A.-C. et al., Simultaneous determination of (fluoro)quinolone antibiotics in kidney, marine products, eggs and muscle by ELISA, communication presented at the *5th International Symposium on Hormone and Veterinary Drug Residue Analysis*, Antwerp, Belgium, May 16–19, 2006.

83. Huet, A.-C. et al., Simultaneous determination of (fluoro)quinolone antibiotics in kidney, marine products, eggs, and muscle by enzyme-linked immunosorbent assay (ELISA), *J. Agric. Food Chem.*, 54, 2822, 2006.

84. Huet, A.-C. et al., Development of an ELISA screening test for halofuginone, nicarbazin and nitro-imidazoles in egg and chicken muscle, communication presented at the *5th International Symposium on Hormone and Veterinary Drug Residue Analysis*, Antwerp, Belgium, May 16–19, 2006.

85. Sternesjö, A., Mellgren, C., and Björck, L., Determination of sulfamethazine residues in milk by a surface plasmon resonance-based biosensor assay, *Anal. Biochem.*, 226, 175, 1995.

86. Crooks, S.R.H. et al., Immunobiosensor—an alternative to enzyme immunoassay screening for residues of two sulfonamides in pigs, *Analyst*, 123, 2755, 1998.

87. Baxter, G.A. et al., Evaluation of an immunobiosensor for the on-site testing of veterinary drug residues at an abattoir. Screening for sulfamethazine in pigs, *Analyst*, 124, 1315, 1999.

88. Gaudin, V. and Pavy, M.L., Determination of sulfamethazine in milk by biosensor immunoassay, *J. Assoc. Off. Anal. Chem. Internat.*, 82, 1316, 1999.

89. Bjurling, P. et al., Biosensor assay of sulfadiazine and sulfamethazine residues in pork, *Analyst*, 125, 1771, 2000.

90. Haasnoot, W., Bienenmann-Ploum, M., and Kohen, F., Biosensor immunoassay for the detection of eight sulfonamides in chicken serum, *Anal. Chim. Acta*, 483, 171, 2003.

91. Gaudin, V., Fontaine, J., and Maris, P., Screening of penicillin residues in milk by a surface plasmon resonance based biosensor assay: comparison of chemical and enzymatic sample treatment, *Anal. Chim. Acta*, 436, 191, 2000.

92. Gaudin, V. and Maris, P., Development of a biosensor based immunoassay for screening of chloramphenicol residues in milk, *Food Agric. Immunol.*, 13, 77, 2000.

93. Ferguson, J.P. et al., Detection of streptomycin and dihydrostreptomycin residues in milk, honey and meat samples using an optical biosensor, *Analyst*, 127, 951, 2002.

94. McCarney, B. et al., Surface plasmon resonance biosensor screening of poultry liver and eggs for nicarbazin residues, *Anal. Chim. Acta*, 483, 165, 2003.

95. Gustavsson, E., SPR biosensor analysis of beta-lactam antibiotics in milk. Development and use of assays based on a beta-lactam receptor protein, Doctoral thesis, Swedish University of Agricultural Sciences, Uppsala, 2003.

96. Mellgren, C. and Sternesjö, A., Optical immunobiosensor assay for determining enrofloxacin and ciprofloxacin in bovine milk, *J. Assoc. Off. Anal. Chem. Internat.*, 81, 394, 1998.

97. Langner, H.J. and Teufel, U., Chemical and microbiological detection of antibiotics 1. Thin-layer chromatographic separation of various antibiotics, *Die Fleischwirtschaft*, 52, 1610, 1972.

98. Parks, O.W., Screening test for sulfamethazine and sulfathiazole in swine liver, *J. Assoc. Off. Anal. Chem.*, 65, 632, 1982.

99. Thomas, M.H., Soroka, K.E., and Thomas, S.H., Quantitative thin-layer chromatographic multi-sulfonamide screening procedure, *J. Assoc. Off. Anal. Chem.*, 66, 881, 1983.

100. Herbst, D.V., Identification and determination of four beta-lactam antibiotics in milk, *J. Food Prot.*, 45, 450, 1982.

101. Moats, W.A., Detection and semiquantitative estimation of penicillin-G and cloxacillin in milk by thin-layer chromatography, *J. Agric. Food Chem.*, 31, 1348, 1983.

102. Moats, W.A., Chromatographic methods for determination of macrolide antibiotic residues in tissues and milk of food-producing animals, *J. Assoc. Off. Anal. Chem.*, 68, 980, 1985.

103. Abjean, J.P., Planar chromatography for the multiclass, multiresidue screening of chloramphenicol, nitrofuran, and sulphonamide residues in pork and beef, *J. Assoc. Off. Anal. Chem. Internat.*, 80, 737, 1997.

104. Abjean, J.P. and Lahogue V., Planar chromatography for quantitative determination of ampicillin residues in milk and muscle, *J. Assoc. Off. Anal. Chem. Internat.*, 80, 1171, 1997.

105. Eiceman, G.A., Instrumentation of gas chromatography, in *Encyclopedia of Analytical Chemistry*, Meyers, R.A., Ed., Wiley, Chichester, 1994.

106. Meetschen, U. and Petz, M., Gas chromatographic method for the determination of residues of seven penicillins in foodstuffs of animal origin, *Z. Lebensm. Unters. Forsch.*, 193, 337, 1991.

107. Mineo, H. et al, An analytical study of antibacterial residues in meat: the simultaneous determination of 23 antibiotics and 13 drugs using gas chromatography, *Vet. Hum. Toxicol.*, 34, 393, 1992.

108. Kennedy, D.G. et al., Gas-chromatographic-mass spectrometric determination of sulfamethazine in animal tissues using a methyl/trimethylsilyl derivative, *Analyst*, 121, 1457, 1996.

109. Preu, M. and Petz, M., Isotope dilution GC-MS of benzylpenicillin residues in bovine muscle, *Analyst*, 123, 2785, 1998.

110. Polzer, J. and Gowik, P., Validation of a method for the detection and confirmation of nitroimidazoles and corresponding hydroxy metabolites in turkey and swine muscle by means of gas chromatography-negative ion chemical ionization mass spectrometry, *J. Chromatogr. B*, 761, 47, 2001.

111. Epstein, R.L. et al., International validation study for the determination of chloramphenicol in bovine muscle, *J. Assoc. Off. Anal. Chem. Internat.*, 77, 570, 1994.

112. Gude, Th., Preiss, A., and Rubach, K., Determination of chloramphenicol in muscle, liver, kidney and urine of pigs by means of immunoaffinity chromatography and gas chromatography with electron-capture detection, *J. Chromatogr. B*, 673, 197, 1995.

113. Nagata, T. and Oka, H., Detection of residual chloramphenicol, florfenicol, and thiamphenicol in yellowtail fish muscles by capillary gas chromatography-mass spectrometry, *J. Agric. Food Chem.*, 44, 1280, 1996.

114. Cerkvenik, V., Analysis and monitoring of chloramphenicol residues in food of animal origin in Slovenia from 1991 to 2000, *Food Addit. Contam.*, 19, 357, 2002.

115. Ding, S. et al., Determination of chloramphenicol residue in fish and shrimp tissues by gas chromatography with a microcell electron capture detector, *J. Assoc. Off. Anal. Chem. Internat.*, 88, 57, 2005.

116. Polzer, J. et al., Determination of chloramphenicol residues in crustaceans: preparation and evaluation of a proficiency test in Germany, *Food Addit. Contam.*, 23, 1132, 2006.

117. Zhang, S. et al., Simultaneous determination of florfenicol and florfenicol amine in fish, shrimp, and swine muscle by gas chromatography with a microcell electron capture detector, *J. Assoc. Off. Anal. Chem. Internat.*, 89, 1437, 2006.

118. Vissers, J.P.C., Claessens, H.A., and Cramers, C.A., Microcolumn liquid chromatography: instrumentation, detection and applications, *J. Chromatogr. A* 770, 1, 1997.

119. Krull, I.S. and Lankmayr, E.P., Derivatization reaction detector in HPLC, *Am. Lab.*, 14, 18, 1982.

120. Frei, R.W., Jansen, H., and Brinkman, U.A.Th., Postcolumn reaction detectors for HPLC, *Anal. Chem.*, 57, 1529, 1985.

121. Krull, I.S., Deyl, Z., and Lingeman, H., General strategies and selection of derivatization reactions for liquid chromatography and capillary electrophoresis, *J. Chromatogr. B Biomed. Appl.*, 659, 1, 1994.

122. Decolin, D. et al., Analyse de traces d'antibiotiques dans les tissus animaux par chromatographie en phase liquide couplée à la fluorescence, *Ann. Fals. Exp. Chim.*, 89, 11, 1996.

123. Yergey, A.L. et al., *Liquid Chromatography-Mass Spectrometry, Techniques and Applications*, Plenum Press, New York, NY, 1990, 316 pp.

124. Niessen, W.M.A. and Van der Greef, J., in *Liquid Chromatography-Mass Spectrometry: Principles and Applications, Chromatographic Science Series*, Vol. 58, Marcel Dekker, New York, 1992, 479 pp.

125. Kinter, M., Mass spectrometry, *Anal. Chem.*, 67, 493, 1994.

126. March, R.E., An introduction to quadrupole ion trap mass spectrometry, *J. Mass Spectrom.*, 32, 351,1997.

127. Cole, R.B., *Electrospray Ionization Mass Spectrometry—Fundamentals, Instrumentation and Applications*, Cole, R.B., Ed., Wiley, New York, NY, 1997.

128. Gaskell, S.J., Electrospray: principles and practice, *J. Mass Spectrom.*, 32, 677, 1997.

129. Burlingame, A.L., Boyd, R.K., and Gaskell, S.J., Mass spectrometry, *Anal. Chem.*, 70, 467, 1998.

130. Abian, J., The coupling of gas and liquid chromatography with mass spectrometry, *J. Mass Spectrom.*, 34, 157, 1999.

131. Niessen, W.M.A., Liquid chromatography-mass spectrometry, in *Chromatographic Science Series*, 2nd ed., Vol. 79, Niessen, W.M.A., Ed., Marcel Dekker, New York, NY, 1999.

132. Amad, M.H. et al., Importance of gas-phase proton affinities in determining the electrospray ionization response for analytes and solvents, *J. Mass Spectrom.*, 35, 784, 2000.

133. Shukla, A.K. and Futrell, J.H., Tandem mass spectrometry: dissociation of ions by collisional activation, *J. Mass Spectrom.*, 35, 1069, 2000.

134. Chernushevich, I.V., Loboda, A.V., and Thomson, B.A., An introduction to quadrupole-time-of-flight mass spectrometry, *J. Mass Spectrom.*, 36, 849, 2001.

135. Brown, M.A., Liquid chromatography-mass spectrometry, in *ACS Symposium Series*, Vol. 420, Brown, M.A., Ed., American Chemical Society, Washington, DC, 1990.

136. Strege, M.A., High-performance liquid chromatographic-electrospray ionization mass spectrometric analyses for the integration of natural products with modern high-throughput screening, *J. Chromatogr. B Biomed. Appl.*, 725, 67, 1999.

137. Careri, M., Bianchi, F., and Corradini, C., Recent advances in the application of mass spectrometry in food-related analysis, *J. Chromatogr. A*, 970, 3, 2002.

138. Perkins, J.R. et al., Analysis of sulphonamides using supercritical fluid chromatography and supercritical fluid chromatography-mass spectrometry, *J. Chromatogr.*, 540, 239, 1991.

139. Chester, T.L., and Pinkston, J.D., and Raynie, D.E., Supercritical fluid chromatography and extraction, *Anal. Chem.*, 66, 106, 1994.

140. Verillon, F., Dossier: supercritical-fluid chromatography, *Analusis*, 27, 671, 1999.

141. Chester, T.L. and Pinkston, J.D., Supercritical fluid and unified chromatography, *Anal. Chem.*, 72, 129, 2000.

142. Lloyd, D.K., Capillary electrophoretic analyses of drugs in body fluids: sample pretreatment and methods for direct injection of biofluids, *J. Chromatogr. A*, 735, 29, 1996.

143. Pesek, J.J. and Matyska, M.T., Column technology in capillary electrophoresis and capillary electro chromatography, *Electrophoresis*, 18, 2228, 1997.

144. El Rassi, Z., Capillary electrophoresis reviews, *Electrophoresis*, 18, 2121, 1997.

145. Robson, M.M. et al., Capillary electrochromatography: a review, *J. Microcol. Sep.*, 9, 357, 1997.

146. Thormann, W. and Caslavska, J., Capillary electrophoresis in drug analysis, *Electrophoresis*, 19, 2691, 1998.

147. Altria, K.D., Smith, N.W., and Turnbull, C.H., Analysis of acidic compounds using capillary electrochromatography, *J. Chromatogr. B Biomed. Appl.*, 717, 341, 1998.

148. Dermaux, A. and Sandra, P., Applications of capillary electrochromatography, *Electrophoresis*, 20, 3027, 1999.

149. Frazier, R.A., Ames, J.M., and Nursten, H.E., The development and application of capillary electrophoresis methods for food analysis, *Electrophoresis*, 20, 3156, 1999.

150. Colon, L.A. et al., Recent progress in capillary electrochromatography, *Electrophoresis*, 21, 3965, 2000.

151. Osbourn, D.M., Weiss, D.J., and Lunte, C.E., On-line preconcentration methods for capillary electrophoresis, *Electrophoresis*, 21, 2768, 2000.

152. Waterval, J.C.M. et al., Derivatization trends in capillary electrophoresis, *Electrophoresis*, 21, 4029, 2000.

153. Colon, L.A., Maloney, T.D., and Fermier, A.M., Packing columns for capillary electrochromatography, *J. Chromatogr. A*, 887, 43, 2000.

154. Pursch, M. and Sander, L.C., Stationary phases for capillary electrochromatography, *J. Chromatogr. A*, 887, 313, 2000.

155. Nishi, H. and Tsumagari, N., Effect of tetraalkylammonium salts on micellar electrokinetic chromato-graphy of ionic substances, *Anal. Chem.*, 61, 2434, 1989.

156. Nishi, H. et al., Separation of beta-lactam antibiotics by micellar electrokinetic chromatography, *J. Chromatogr.*, 477, 259, 1989.

157. Tsikas, D., Hofrichter, A., and Brunner, G., Capillary isotachophoretic analysis of beta-lactam antibiotics and their precursors, *Chromatographia*, 30, 657, 1990.

158. Zhu, Y.X. et al., Micellar electrokinetic capillary chromatography for the separation of phenoxymethylpenicillin and related substances, *J. Chromatogr. A*, 781, 417, 1997.

159. De Boer, T. et al., Selectivity in capillary electrokinetic separations, *Electrophoresis*, 20, 2989, 1999.

160. Smith, N.W. and Carterfinch, A.S., Electrochromatography, *J. Chromatogr. A*, 892, 219, 2000.

161. Legido-Quigley, C. et al., Advances in capillary electrochromatography and micro-high performance liquid chromatography monolithic columns for separation science, *Electrophoresis*, 24, 917, 2003.

162. Banks, J.F., Recent advances in capillary electrophoresis/electrospray/mass spectrometry, *Electrophoresis*, 18, 2255, 1997.

163. Bateman, K.P., Locke, S.J., and Volmer, D.A., Characterization of isomeric sulfonamides using capillary zone electrophoresis coupled with nano-electrospray quasi-MS/MS/MS, *J. Mass Spectrom.*, 32, 297, 1997.

164. Chen, Y.C. and Lin, X., Migration behavior and separation of tetracycline antibiotics by micellar electrokinetic chromatography, *J. Chromatogr. A*, 802, 95, 1998.

165. Berzas-Nevado, J.J., Castaneda-Penalvo, G., and Guzman-Bernardo, F.J., Micellar electrokinetic capillary chromatography as an alternative method for the determination of sulfonamides and their associated compounds, *J. Liq. Chromatogr Relat. Technol.*, 22, 975, 1999.

166. Spikmans, V. et al., Hyphenation of capillary electrochromatography with mass spectrometry: the technique and its application, *LC-GC Int.*, 13, 486, 2000.

167. Desiderio, C. and Fanali, S., Capillary electrochromatography and CE-electrospray MS for the separation of non-steroidal anti-inflammatory drugs, *J. Chromatogr. A*, 895, 123, 2000.

168. von Broke, A., Nicholson, G., and Bayer, E., Recent advances in capillary electrophoresis/electrospray-mass spectrometry, *Electrophoresis*, 22, 1251, 2001.

169. von Broke, A. et al., On-line coupling of packed capillary electrochromatography with coordination ion spray-mass spectrometry for the separation of enantiomers, *Electrophoresis*, 23, 2963, 2002.

170. Berrueta, L.A., Gallo, B., and Vicente, F., A review of solid-phase extraction: basic principles and new developments, *Chromatographia*, 40, 474, 1995.

171. Thurman, E.M. and Mills, M.S., *Solid Phase Extraction: Principles and Practice*, Wiley, New York, NY, 1998.

172. Franke, J.P. and De Zeeuw, R.A., Solid-phase extraction procedures in systematic toxicological analysis, *J. Chromatogr. B Biomed. Appl.*, 713, 51, 1998.

173. Hubert, P. et al., Préparation des échantillons d'origine biologique préalable à leur analyse chromatographique, *S.T.P. Pharma pratiques*, 9, 160, 1999.

174. Huck, C.W., and Bonn, G.K., Recent developments in polymer-based sorbents for solid-phase extraction, *J. Chromatogr. A*, 885, 51, 2000.

175. Buldini, P.L., Ricci, M.C., and Sharma, J.L., Recent applications of sample preparation techniques in food analysis, *J. Chromatogr. A*, 975, 47, 2002.

176. Verdon, E. and Couëdor, P., Multiresidue analytical method for the determination of eight penicillin antibiotics in muscle tissue by ion-pair reversed-phase HPLC after precolumn derivatization, *J. Assoc. Off. Anal. Chem. Internat.*, 82, 1083, 1999.

177. Delepine, B. and Hurtaud-Pessel D., Liquid chromatography/tandem mass spectrometry: screening method for the identification of residues of antibiotics in meat, communication presented at the *4th International Symposium on Hormone and Veterinary Drug Residue Analysis*, Antwerp, Belgium, June 3–7, 2002 (unpublished data).

178. Yoshimura, H., Itoh, O., and Yonezawa, S., Microbiological and TLC identification of aminoglycoside antibiotics in animal body, *Jpn. J. Vet. Sci.*, 44, 233, 1982.

179. Lachatre, G. et al., Séparation des aminosides et de leurs fractions par chromatographie en phase liquide—application à leur dosage dans le plasma, l'urine et les tissus, *Analusis*, 11, 168, 1983.

180. Shaikh, B., Allen, E.H., and Gridley, J.C., Determination of neomycin in animal tissues using ion-pair liquid chromatography with fluorometric detection, *J. Assoc. Off. Anal. Chem.*, 68, 29, 1985.

181. Nakaya, K.-I., Sugitani, A., and Yamada, F., Determination of kanamycin in beef and kidney of cattle by HPLC, *J. Food Hyg. Soc. Jpn*, 27, 258, 1986.

182. Okayama, A. et al., Fluorescence HPLC determination of streptomycin in meat using ninhydrin as a postcolumn labelling agent, *Bunseki Kagaku*, 37, 221, 1988.

183. Agarwal, V.K., HPLC determination of gentamicin in animal tissue, *J. Liq. Chromatogr.*, 12, 613, 1989.

184. Sar, F. et al., Development and optimization of a liquid chromatographic method for the determination of gentamicin in calf tissues, *Anal. Chim. Acta*, 275, 285, 1993.

185. Gerhardt, G.C., Salisbury, C.D.C., and MacNeil, J.D., Determination of streptomycin and dihydrostreptomycin in animal tissue by on-line sample enrichment liquid chromatography, *J. Assoc. Off. Anal. Chem. Internat.*, 77, 334, 1994.

186. Guggisberg, D. and Koch, H., Methode zur fluorimetrischen bestimmung von neomycin in der niere und in der leber mit HPLC und nachsaülenderivatisation, *Mitt. Gebiete Lebensmittel Hygiene*, 86, 449, 1995.

187. Hormazabal, V. and Yndestad, M., H.P.L.C. determination of dihydrostreptomycin sulfate in kidney and meat using post column derivatization, *J. Liq. Chromatogr. Relat. Technol.*, 20, 2259, 1997.

188. Abbasi, H. and Hellenäs, K.E., Modified determination of DHS in kidney, muscle and milk by HPLC, *Analyst*, 123, 2725, 1998.

189. Bergwerff, A.A., Scherpenisse, P., and Haagsma, N., HPLC determination of residues of spectinomycin in various tissue types from husbandry animals, *Analyst*, 123, 2139, 1998.

190. Morovjan, G., Csokan, P.P., and Nemeth-Konda, L., HPLC determination of colistin and aminoglycoside antibiotics in feeds by post-column derivatization and fluorescence detection, *Chromatographia*, 48, 32, 1998.

191. Reid, J.-A. and MacNeil, J.D., Determination of neomycin in animal tissue by liquid chromatography, *J. Assoc. Off. Anal. Chem. Internat.*, 82, 61, 1999.

192. Posyniak, A., Zmudski, J., and Niedzielska, J., Sample preparation for residue determination of gentamicin and neomycin by liquid chromatography, *J. Chromatogr. A*, 914, 59, 2001.

193. Fennell, M.A. et al., Gentamicin in tissue and whole milk: an improved method for extraction and cleanup of samples for quantitation on HPLC, *J. Agric. Food Chem.*, 43, 1849, 1995.

194. Cherlet, M., De Baere, S., and De Backer, P., Determination of gentamicin in swine and calf tissues by HPLC combined with electrospray ionization mass spectrometry, *J. Mass Spectrom.*, 35, 1342, 2000.

195. de Wasch, K. et al., Identification of "unknown analytes" in injection sites: a semi-quantitative interpretation, *Anal. Chim. Acta*, 483, 387, 2003.

196. Kaufmann, A. and Maden, K., Determination of 11 aminoglycosides in meat and liver by liquid chromatography with tandem mass spectrometry, *J. Assoc. Off. Anal. Chem. Internat.*, 88, 1118, 2005.

197. Sanders, P., LC determination of chloramphenicol in calf tissue: studies of stability in muscle, kidney and liver, *J. Assoc. Off. Anal. Chem.*, 74, 483, 1991.

198. O'Brien, J.J., Campbell, N., and Conaghan, T., Effect of cooking and cold storage on biologically active antibiotic residues in meat, *J. Hyg. Camb.*, 87, 511, 1981.

199. Ramos, M. et al., Chloramphenicol residues in food samples: their analysis and stability during storage, *J. Liq. Chromatogr. Relat. Technol.*, 26, 2535, 2003.

200. Allen, E.H., Review of chromatographic methods for chloramphenicol residues in milk, eggs, and tissues from food-producing animal, *J. Assoc. Off. Anal. Chem.*, 68, 5, 1985.

201. McCracken, A., O'Brien, J.J., and Campbell, N., Antibiotic residues and their recovery from animal tissues, *J. Appl. Bacteriol.*, 41, 129, 1976.

202. Kruzik, P. et al., Über den Nachweis und die Bestimmung antibiotsch wirksamer Substanzen in Lebensmitteln tierischer Herkunft: Sulfonamide, Nitrofurane, Nicarbazin, Tetracycline, Tylosin und Chloramphenicol, *Wiener Tierärztliche Monatsschirft*, Vol. 77, p. 141, 1989.

203. Cazemier, G., Haasnoot, W., and Stouten, P., Screening of chloramphenicol in urine, tissue, milk and eggs in consequence of the prohibitive regulation, in *Proceedings Euroresidue III Conference on Residues of Veterinary Drugs in Food*, Veldhoven, Netherlands, May 6–8, Haagsma, N. and Ruiter, A., Eds., Faculty of Medicine, Utrecht, Netherlands, 1996, 315.

204. Lynas, L. et al., Screening for chloramphenicol residues in the tissues and fluids of treated cattle by the four plate test, Charm II radioimmunoassay and Ridascreen CAP-glucuronid enzyme immunoassay, *Analyst*, 123, 2773, 1998.

205. Aerts, R.M.L., Keukens, H.J., and Werdmuller, G.A., Liquid chromatographic determination of chloramphenicol residues in meat: interlaboratory study, *J. Assoc. Off. Anal. Chem.*, 72, 570, 1989.

206. Le Boulaire, S., Bauduret, J.C., and André, F., Veterinary drug residues survey in meat: an HPLC method with a matrix solid phase dispersion extraction, *J. Agric. Food Chem.*, 45, 2134, 1997.

207. Cooper, A.D. et al., Aspects of extraction, spiking and distribution in the determination of incurred residues of chloramphenicol in animal tissues, *Food Addit. Contam.*, 15, 637, 1998.

208. Hayes, J.M., Determination of florfenicol in fish feed by liquid chromatography, *J. Assoc. Off. Anal. Chem. Internat.*, 88, 1777, 2005.

209. Hormazabal, V. and Yndestad, M., Simultaneous determination of chloramphenicol and ketoprofen in meat and milk and chloramphenicol in egg, honey and urne using LC-MS, *J. Liq. Chromatogr. Relat. Technol.*, 24, 2477, 2001.

210. Gantveg, A., Shishani, I., and Hoffman, M., Determination of chloramphenicol in animal tissues and urine. LC-tandem MS versus GC-MS, *Anal. Chim. Acta*, 483, 125, 2003.

211. Rupp, H.S., Stuart, J.S., and Hurlbut, J.A., Liquid chromatography/tandem mass spectrometry analysis of chloramphenicol in cooked crab meat, *J. Assoc. Off. Anal. Chem. Internat.*, 88, 1155, 2005.

212. Penney, L. et al., Determination of chloramphenicol residues in milk, eggs, and tissues by liquid chromatography/mass spectrometry, *J. Assoc. Off. Anal. Chem. Internat.*, 88, 645, 2005.

213. Batas, V. et al., Determination of fenicols (chloramphenicol, thiamphenicol and florfenicol) residues in live animals and primary animal products by liquid chromatography-tandem mass spectrometry, communication presented at the *5th International Symposium on Hormone and Veterinary Drug Residue Analysis*, Antwerp, Belgium, May 16–19, 2006 (unpublished data).

214. Yamada, R. et al., Simultaneous determination of residual veterinary drugs in bovine, porcine, and chicken muscle using liquid chromatography coupled with electrospray tandem mass spectrometry, *Biosci. Biotechnol. Biochem.*, 70, 54, 2006.

215. Turnipseed, S.B. et al., Use of ion-trap liquid chromatography-mass spectrometry to screen and confirm drug residues in aquacultured products, *Anal. Chim. Acta*, 483, 373, 2003.

216. Nouws, J.F.M. and Ziv, G., The effect of storage at 4°C on antibiotic residues in kidney and meat tissues of dairy cows, *Tijdschrift voor diergeneeskunde (Neth. J. Vet. Sci.)*, 101, 1145, 1976.

217. Wiese, B. and Martin, K., Determination of benzylpenicillin in milk at the pg/ml level by reversed-phase liquid chromatography in combination with digital subtraction chromatography technique part I, *J. Pharm. Biomed. Anal.*, 7, 95, 1989.

218. Boison, J.O., Effect of cold-temperature storage on stability of benzylpenicillin residues in plasma and tissues of food-producing animals, *J. Assoc. Off. Anal. Chem.*, 76, 974, 1992.

219. Boison, J.O., Keng, L., and MacNeil, J.D., Analysis of penicillin-G in milk by liquid chromatography, *J. Assoc. Off. Anal. Chem. Internat.*, 77, 565, 1994.

220. Gee, H.-E., Ho, K.-B., and Toothill, J., Liquid chromatographic determination of benzylpenicillin and cloxacillin in animal tissues and its application to a study of the stability at –20°C of spiked and incurred residues of benzylpenicillin in ovine liver, *J. Assoc. Off. Anal. Chem. Internat.*, 79, 640, 1995.

221. Rose, M.D. et al., The effect of cooking on veterinary drug residues in food. Part 8: benzylpenicillin, *Analyst*, 122, 1095, 1997.

222. Verdon, E. et al., Stability of penicillin antibiotic residues in meat during storage—Ampicillin, *J. Chromatogr. A*, 882, 135, 2000.

223. Croubels, S., De Baere, S., and De Backer, P., Practical approach for the stability testing of veterinary drugs in solutions and in biological matrices during storage, *Anal. Chim. Acta*, 483, 419, 2003.

224. Vilim, A.B. and Larocque, L., Determination of penicillin-G, ampicillin and cephapirin residues in tissues, *J. Assoc. Off. Anal. Chem.*, 66, 176, 1983.

225. McNeil, J.D. et al., Performance of five screening tests for the detection of penicillin-G residues in experimentally injected calves, *J. Food Prot.*, 54, 37, 1991.

226. Suhren, G. and Knappstein, K., Detection of cefquinome in milk by LC and screening methods, *Food Addit. Contam.*, 483, 363, 2003.

227. Ellerbroek, L. et al., Zur mikrobiologischen erfassung von rückstanden antimikrobiell wirksamer stoffe beim fisch, *Archiv für Lebensmittelhygiene*, 48, 1, 1997.

228. Fitzgerald, S.P. et al., Stable competitive enzyme-linked immunosorbent assay kit for rapid measurement of 11 active beta-lactams in milk, tissue, urine, and serum, *J. Assoc. Off. Anal. Chem. Internat.*, 90, 334, 2007.

229. Gustavsson, E. et al., Determination of beta-lactams in milk using a surface plasmon resonance-based biosensor, *J. Agric. Food Chem.*, 52, 2791, 2004.

230. Terada, H., Sakabe, Y., and Asanoma, M., Studies on residual antibacterials in foods. III. High performance liquid chromatographic determination of penicillin-G in animal tissues using an on-line pre-column concentration and purification system, *J. Chromatogr.*, 318, 299, 1985.

231. Barker, S.A., Long, A.R., and Short, C.R., Isolation of drug residues from tissues by solid phase dispersion, *J. Chromatogr.*, 475, 355, 1989.

232. Moats, W.A., High performance liquid chromatographic determination of penicillin-G, penicillin V and cloxacillin in beef and pork tissues, *J. Chromatogr.*, 593, 15, 1992.

233. Beconi-Barker, M.G. et al., Determination of ceftiofur and its desfuroylceftiofur-related metabolites in swine tissues by high-performance liquid chromatography, *J. Chromatogr. B*, 673, 231, 1995.

234. Moats, W.A., and Romanovski, R.D., Determination of penicillin-G in beef and pork tissues using an automated LC cleanup, *J. Agric. Food Chem.*, 46, 1410, 1998.

235. Moats, W.A. and Buckley, S.A., Determination of free metabolites of ceftiofur in animal tissues with an automated liquid chromatographic cleanup, *J. Assoc. Off. Anal. Chem. Internat.*, 81, 709, 1998.

236. Moats, W.A., Romanovski, R.D., and Medina, M.B., Identification of beta-lactam antibiotics in tissue samples containing unknown microbial inhibitors, *J. Assoc. Off. Anal. Chem. Internat.*, 81, 1135, 1998.

237. Furusawa, N., Liquid chromatographic determination/identification of residual penicillin G in food-producing animal tissues, *J. Liq. Chromatogr. Relat. Technol*, 24, 161, 2001.

238. Verdon, E. and Couëdor, P., HPLC/UV determination of residues of cephalosporins in pork muscle tissue, communication presented at the *116th AOAC International Meeting*, Los Angeles, CA, September 22–26, 2002 (unpublished data).

239. Hornish, R.E., Hamlow, P.J., and Brown, S.A., Multilaboratory trial for determination of ceftiofur residues in bovine and swine kidney and muscle, and bovine milk, *J. Assoc. Off. Anal. Chem. Internat.*, 86, 30, 2003.

240. Nagata, T., Ashizawa, E., and Hashimoto, H., Simultaneous determination of residual fourteen kinds of beta-lactam and macrolide antibiotics in bovine muscles by high-performance liquid chromatography with a diode array detector, *J. Food Hyg. Soc. Jpn*, 45, 161, 2004.

241. Boison, J.O. et al., Determination of penicillin-G residues in edible animal tissues by liquid chromatography, *J. Assoc. Off. Anal. Chem.*, 74, 497, 1991.

242. Snippe, N. et al., Automated column LC determination of amoxicillin and cefadroxil in bovine serum and muscle tissue using on-line dialysis for sample preparation, *J. Chromatogr. B*, 662, 61, 1994.

243. Verdon, E. and Couëdor, P., Determination of ampicillin residues in bovine muscle at the residue level (25 to 100 µg/kg) by HPLC with precolumn derivatization, in *Proceedings Euroresidue III Conference on Residues of Veterinary Drugs in Food*, Veldhoven, Netherlands, May 6–8, Vol. 2, Haagsma, N. and Ruiter, A., Eds., Faculty of Medicine, Utrecht, Netherlands, 1996, 963.

244. Rose, M.D. et al., Determination of penicillins in animal tissues at trace residue concentrations: II. Determination of amoxicillin and ampicillin in liver and muscle using cation exchange and porous graphitic carbon solid phase extraction and high-performance liquid chromatography, *Food Addit. Contam.*, 14, 127, 1997.

245. Boison, J.O. and Keng, J.-Y., Multiresidue liquid chromatographic method for determining residues of mono- and dibasic penicillins in bovine muscle tissues, *J. Assoc. Off. Anal. Chem. Internat.*, 81, 111, 1998.

246. Boison, J.O. and Keng, J.-Y., Improvement in the multiresidue LC analysis of residues of mono- and dibasic penicillins in bovine muscle tissues, *J. Assoc. Off. Anal. Chem. Internat.*, 81, 1257, 1998.

247. McGrane, M., O'Keeffe, M., and Smyth, M.R., Multi-residue analysis of penicillin residues in porcine tissue using matrix solid phase dispersion, *Analyst*, 123, 2779, 1998.

248. Sorensen, L.K. et al., Simultaneous determination of seven penicillins in muscle, liver and kidney tissues from cattle and pigs by a multiresidue HPLC method, *J. Chromatogr. B*, 734, 307, 1999.

249. Verdon, E. et al., Liquid chromatographic determination of ampicillin residues in porcine muscle tissue by a multipenicillin analytical method: European collaborative study, *J. Assoc. Off. Anal. Chem. Internat.*, 85, 889, 2002.

250. Luo, W. and Ang, C.W.Y., Determination of amoxicillin residues in animal tissues by solid-phase extraction and liquid chromatography with fluorescence detection, *J. Assoc. Off. Anal. Chem. Internat.*, 83, 20, 2000.

251. Csokan, P. and Bernath, S., A simple and fast HPLC assay for amoxicillin residues in swine and cattle tissues, in *Proceedings of the 8th International Congress of the European Association for Veterinary Pharmacology and Toxicology (EAVPT)*, Jerusalem, Israel, July 30–August 10, Soback, S. and McKellar, Q.A., Eds., *J. Vet. Pharmacol. Toxicol. Ther.*, 29, 1, 2000.

252. Gamba, V. and Dusi, G., LC with fluorescence detection of amoxicillin and ampicillin in feeds using pre-column derivatization, *Anal. Chim. Acta*, 483, 69, 2003.

253. Voyksner, R.D., Smith, C.S., and Knox, P.C., Optimization and application of particle beam high-performance liquid chromatography/mass spectrometry to compounds of pharmaceutical interest, *Biomed. Environ. Mass Spectrom.*, 19, 523, 1990.

254. Blanchflower, W.J., Hewitt, S.A., and Kennedy, D.G., Confirmatory assay for the simultaneous detection of five penicillins in muscle, kidney and milk using liquid chromatography-electrospray mass spectrometry, *Analyst*, 119, 2595, 1994.

255. Hurtaud, D., Delepine, B., and Sanders, P., Particle beam liquid chromatography-mass spectrometry method with negative ion chemical ionization for the confirmation of oxacillin, cloxacillin and dicloxacillin residues in bovine muscle, *Analyst*, 119, 2731, 1994.

256. Hormazabal, V. and Yndestad, M., Determination of benzylpenicillin and other beta-lactam antibiotics in plasma and tissues using liquid chromatography-mass spectrometry for residual and pharmacokinetic studies, *J. Liq. Chromatogr. Relat. Technol.*, 21, 3099, 1998.

257. Ito, Y. et al., Application of ion-exchange cartridge clean-up in food analysis. IV. Confirmatory assay of benzylpenicillin, phenoxymethylpenicillin, oxacillin, cloxacillin, nafcillin and dicloxacillin in bovine tissues by liquid chromatography-electrospray tandem mass spectrometry, *J. Chromatogr. A*, 911, 217, 2001.

258. Hatano, K., Simultaneous determination of five penicillins in muscle, liver and kidney from slaughtered animals using liquid chromatography coupled to electrospray ionization tandem mass spectrometry, *J. Food Hyg. Soc. Jpn*, 44, 1, 2004.

259. Ito, Y., Development of analytical methods for residual antibiotics and antibacterials in livestock products, *Yakugaku Zasshi*, 123, 19, 2003.

260. Ito, Y. et al., Application of ion-exchange cartridge clean-up in food analysis. VI. Determination of six penicillins in bovine tissues by liquid chromatography-electrospray ionization tandem mass spectrometry, *J. Chromatogr. A*, 1042, 107, 2004.

261. Fagerquist, C.K., Lightfield, A.R., and Lehotay, S.J., Confirmatory and quantitative analysis of beta-lactam antibiotics in bovine kidney tissue by dispersive solid-phase extraction and liquid chromatography-tandem mass spectrometry, *Anal. Chem.*, 77, 1473, 2005.

262. Smith, S. et al., Multiclass confirmation of veterinary drug residues in fish by LC-MS[n], communication presented at the *119th AOAC International Meeting*, Orlando, FL, USA, September 11–15, 2005 (unpublished data).

263. Zuzulova, M. et al., In vivo activity of tylosin and its derivatives against ureaplasma urealyticum, *Arzneimittel-Forschung*, 45, 1222, 1995.

264. Chepkwony, H.K., Liquid chromatographic determination of erythromycins in fermentation broth, *Chromatographia*, 53, 89, 2001.

265. Chepkwony, H.K., Development and validation of an reversed-phase liquid chromatographic method for analysis of spiramycin and related substances, *Chromatographia*, 54, 51, 2001.

266. Council Regulation (EEC) No. 2377/90 of 26th June 1990, *Off. J. Eur. Commun.*, L224 (1990) 1 as amended by regulation No. 1442/95 of 26th June 1995, *Off. J. Eur. Commun.*, L143 (1995) 26.

267. Woodward, K.N. and Shearer, G., Antibiotic use in animal production in the European Union—regulation and current methods for residue detection, in *Chemical Analysis for Antibiotics used in Agriculture*, Oka, H. et al., Eds., AOAC International, Arlington, VA, 1995.

268. Moats, W.A., Determination of tylosin in tissues, milk and blood serum by reversed phase high performance liquid chromatography, in *Instrumental Analysis of Foods*, Vol. 1, Charalambous, G. and Inglett, G., Eds., Academic Press, Orlando, 1983, 357.

269. Chan, W., Gerhardt, G.C., and Salisbury, C.D.C., Determination of tylosin and tilmicosin residues in animal tissues by reversed-phase liquid chromatography, *J. Assoc. Off. Anal. Chem. Internat.*, 77, 331, 1994.

270. Horie, M. et al., Simultaneous determination of five macrolide antibiotics in meat by high performance liquid chromatography, *J. Chromatogr. A*, 812, 295, 1988.

271. Gaugain-Juhel, M., Anger, B., and Laurentie, M., Multiresidue chromatographic method for the determination of macrolide residues in muscle by high performance liquid chromatography with UV detection, *J. Assoc. Off. Anal. Chem. Internat.*, 82, 1046, 1999.

272. Leal, C. et al., Determination of macrolide antibiotics by liquid chromatography, *J. Chromatogr. A*, 910, 285, 2001.

273. Leroy, P., Decolin, D., and Nicolas, A., Determination of josamycin residues in porcine tissues using high performance liquid chromatography with pre-column derivatization and spectrofluorimetric detection, *Analyst*, 119, 2743, 1994.

274. Edder, P. et al., Analysis of erythromycin and oleandomycin residues in food by high-performance liquid chromatography with fluorometric detection, *Food Addit. Contam.*, 19, 232, 2002.

275. Delepine, B., Hurtaud, D., and Sanders, P., Identification of tylosin in bovine muscle at the maximum residue limit level by liquid chromatography-mass spectrometry, using a particle beam interface, *Analyst*, 119, 2717, 1994.

276. Delepine, B., Hurtaud-Pessel, D., and Sanders, P., Multiresidue method for confirmation of macrolide antibiotics in bovine muscle by liquid chromatography/mass spectrometry, *J. Assoc. Off. Anal. Chem. Internat.*, 79, 397, 1996.

277. Draisci, R. et al., Confirmatory method for macrolide residues in bovine tissues by micro-liquid chromatography-tandem mass spectrometry, *J. Chromatogr. A*, 926, 97, 2001.

278. Dubois, M. et al., Identification and quantification of five macrolide antibiotics in several tissues, eggs and milk by liquid chromatography-electrospray tandem mass spectrometry, *J. Chromatogr. B*, 753, 189, 2001.

279. Hwang, Y.-H. et al., Simultaneous determination of various macrolides by liquid chromatography/mass spectrometry, *J. Vet. Sci.*, 3, 103, 2002.

280. Van Poucke, C. et al., Liquid chromatographic-tandem mass spectrometric detection of banned antibacterial growth promoters in animal feed, *Anal. Chim. Acta*, 483, 99, 2003.

281. Lucchetti, D. et al., Simple confirmatory method for the determination of erythromycin residues in trout: a fast liquid–liquid extraction followed by liquid chromatography-tandem mass spectrometry, *J. Agric. Food Chem.*, 53, 9689, 2005.

282. Council Regulation (EEC) No. 2377/90 of 26th June 1990, *Off. J. Eur. Commun.*, L224 (1990) 1 as amended by regulation No. 2901/93 of 18th October 1993, *Off. J. Eur. Commun.*, L264 (1993) 1.

283. Commission Decision No. 2003181/EC of 13th March 2003, *Off. J. Eur. Commun.*, L71 (2003) 17 and amending Decision 2002/657/EC.

284. McCracken, R.J. and Kennedy, D.G., Determination of the furazolidone metabolite, 3-amino-2-oxazolidinone, in porcine tissues using LC-thermospray MS and the occurrence of residues in pigs produced in Northern Ireland, *J. Chromatogr. B*, 691, 87, 1997.

285. Parks, O.W., Liquid chromatographic electrochemical detection screening procedure for six nitro-containing drugs in chicken tissues at low ppb level, *J. Assoc. Off. Anal. Chem.*, 72, 567, 1989.

286. Laurensen, J.J. and Nouws, J.F.M., Simultaneous determination of nitrofuran derivatives in various animal substrates by HPLC, *J. Chromatogr.*, 472, 321, 1989.

287. Cooper, A.D. et al., Development of multi-residue methodology for the high-performance liquid chromatography determination of veterinary drugs in animal tissues, *Food Addit. Contam.*, 12, 167, 1995.

288. Hormazabal, V. and Yndestad, M., Simple and rapid method of analysis for furazolidone in meat tissues by HPLC, *J. Liq. Chromatogr.*, 18, 1871, 1995.

289. Conneely, A., Nugent, A., and O'Keeffe, M., Use of solid phase extraction for the isolation and clean-up of a derivatised furazolidone metabolite from animal tissues, *Analyst*, 127, 705, 2002.

290. Conneely, A. et al., Isolation of bound residues of nitrofuran drugs from tissue by solid-phase extraction with determination by liquid chromatography with UV and tandem mass spectrometric detection, *Anal. Chim. Acta*, 483, 91, 2003.

291. Hormazabal, V. and Norman Asp, T., Determination of the metabolites of nitrofuran antibiotics in meat by liquid chromatography-mass spectrometry, *J. Liq. Chromatogr. Relat. Technol.*, 27, 2759, 2004.

292. Leitner, A., Zöllner, P., and Lindner, W., Determination of the metabolites of nitrofuran antibiotics in animal tissue by high-performance liquid chromatography-tandem mass spectrometry, *J. Chromatogr. A*, 939, 49, 2001.

293. Edder, P. et al., Analysis of nitrofuran metabolites in food by high-performance liquid chromatography with tandem mass spectrometry detection, *Clin. Chem. Lab. Med.*, 41, 1608, 2003.

294. O'Keeffe, M. et al., Nitrofuran antibiotic residues in pork - The FoodBRAND retail survey, *Anal. Chim. Acta*, 520, 125, 2004.

295. Mottier, P. et al., Quantitative determination of four nitrofuran metabolites in meat by isotope dilution liquid chromatography-electrospray ionisation-tandem mass spectrometry, *J. Chromatogr. A*, 1067, 85, 2005.

296. Finzi, J.K. et al., Determination of nitrofuran metabolites in poultry muscle and eggs by liquid chromatography-tandem mass spectrometry, *J. Chromatogr. B*, 824, 30, 2005.

297. Hurtaud-Pessel, D., Verdon, E., and Sanders, P., Proficiency study for the determination of nitrofuran metabolites in shrimps, *Food Addit. Contam.*, 23, 569, 2006.

298. Council Regulation (EEC) No. 2377/90 of 26th June 1990, *Off. J. Eur. Commun.*, L224 (1990) 1 as amended by regulation No. 1756/2002/EC of 23rd September 2002 *Off. J. Eur. Commun.*, L265 (2002) 1.

299. Vahl, M., Analysis of nifursol residues in turkey and chicken meat using liquid chromatography-tandem mass spectrometry, *Food Addit. Contam.*, 22, 120, 2005.

300. Mulder, P.P.J. et al., Determination of nifursol metabolites in poultry muscle and liver tissue. Development and validation of a confirmatory method, *Analyst*, 130, 763, 2005.

301. Verdon, E., Couëdor, P., and Sanders, P., Multi-residue monitoring for the simultaneous determination of five nitrofurans (furazolidone, furaltadone, nitrofurazone, nitrofurantoine, nifursol) in poultry muscle tissue through the detection of their five major metabolites (AOZ, AMOZ, SEM, AHD, DNSAH), by liquid chromatography coupled to electrospray tandem mass spectrometry—in-house validation in line with Commission Decision 657/2002/EC, *Anal. Chim. Acta*, 586, 336, 2007.

302. Pereira, A.S., Donato, J.L., and De Nucci, G., Implications of the use of semicarbazide as a metabolic target of nitrofurazone contamination in coated products, *Food Addit. Contam.*, 21, 63, 2004.

303. Cooper, A.D., McCracken, R.J., and Kennedy, D.G., Nitrofurazone accumulates in avian eyes—a replacement for semicarbazide as a marker of abuse, *Analyst*, 130, 824, 2005.

304. Council Regulation (EEC) No. 2377/90 of 26th June 1990, *Off. J. Eur. Commun.*, L224 (1990) 1 as amended by regulation No. 3426/93/EC, *Off. J. Eur. Commun.*, L312 (1993) 15.

305. Council Regulation (EEC) No. 2377/90 of 26th June 1990, *Off. J. Eur. Commun.*, L224 (1990) 1 as amended by regulation No. 1798/95/EC, *Off. J. Eur. Commun.*, L174 (1995) 20.

306. Council Regulation (EEC) No. 2377/90 of 26th June 1990, *Off. J. Eur. Commun.*, L224 (1990) 1 as amended by regulation No. 613/98/EC, *Off. J. Eur. Commun.*, L82 (1998) 14.

307. Mallinson, E.T. and Henry, A.C., Determination of nitroimidazole metabolites in swine and turkey muscle by liquid chromatography, *J. Assoc. Off. Anal. Chem.*, 75, 790, 1992.

308. Semeniuk, S. et al., Determination of nitroimidazole residues in poultry tissues, serum and eggs by high-performance liquid chromatography, *Biomed. Chromatogr.*, 9, 238, 1995.

309. Posyniak, A. et al., Tissue concentration of dimetridazole in laying hens, *Food Addit. Contam.*, 13, 871, 1996.

310. Sams, M.J. et al., Determination of dimetridazole, ronidazole and their common metabolite in poultry muscle and eggs by HPLC with UV detection and confirmatory analysis by apci-MS, *Analyst*, 123, 2545, 1998.

311. Rose, M.D., Bygrave, J., and Sharman, M., Effect of cooking on veterinary drug residues in food. Part 9. Nitroimidazoles, *Analyst*, 124, 289, 1999.

312. Sorensen, L.K. and Hansen, H., Determination of metronidazole and hydroxymetronidazole in trout by a high-performance liquid chromatography method, *Food Addit. Contam.*, 17, 197, 2000.

313. Matusik, J.E. et al., Identification of dimetridazole, ipronidazole, and their alcohol metabolites in turkey tissues by thermospray tandem mass spectrometry, *J. Agric. Food Chem.*, 40, 439, 1992.

314. Cannavan, A. et al., Gas chromatographic-mass spectrometric determination of sulfamethazine in animal tissues using a methyl/trimethylsilyl derivative, *Analyst*, 121, 1457, 1997.

315. Polzer, J. et al., Validation of a method for the determination of nitroimidazoles in plasma and retina of turkeys, communication presented at the *4th International Symposium on Hormone and Veterinary Drug Residue Analysis*, Antwerp, Belgium, June 4–7, 2002 (unpublished data).

316. Hurtaud-Pessel, D., Delepine, B., and Laurentie, M., Determination of four nitroimidazole residues in poultry meat by liquid chromatography-mass spectrometry, *J. Chromatogr. A*, 16, 89, 2000.

317. Hormazabal, V. and Yndestad, M., Determination of nitroimidazole residues in meat using mass spectrometry, *J. Liq. Chromatogr. Relat. Technol.*, 24, 2487, 2001.

318. Govaerts, Y., Degroodt, J.M., and Srebrnik, S., Nitroimidazole drug residues in poultry and pig tissues, in *Proceedings Euroresidue IV Conference on Residues of Veterinary Drugs in Food*, Veldhoven, Netherlands, Van Ginkel, L.A. and Ruiter, A., Eds., Faculty of Medicine, Utrecht, Netherlands, 2000, 470.

319. Radeck, W., Determination of nitroimidazoles using apci+ mass spectrometry, *Proceedings Euroresidue IV Conference on Residues of Veterinary Drugs in Food*, Veldhoven, Netherlands, May 8–10, Van Ginkel, L.A. and Ruiter, A., Eds., Faculty of Medicine, Utrecht, Netherlands, 2000.

320. Xia, X. et al., Determination of four nitroimidazoles in poultry and swine muscle and eggs by liquid chromatography/tandem mass spectrometry, *J. Assoc. Off. Anal. Chem. Internat.*, 89, 94, 2006.

321. Calderon, V. et al., Screening of antibiotic residues in Spain, communication presented at the *Workshop on Validation of Screening Methods*, Fougeres, France, June 3–4, 2005.

322. Schneider, M.J. and Donoghue, D.J., Comparison of a bioassay and a LC-Fluorescence-MS^n Method for the detection of incurred enrofloxacin residues in chicken tissues, *Poult. Sci.*, 83, 830, 2004.

323. Pikkemaat, M.G. et al., Improved microbial screening assay for the detection of quinolone residues in poultry and eggs, *Food Addit. Contam.*, 24, 842, 2007.

324. Laurent, M.R., Terlain, B.L., and Caude, M.C., Estimation of decoquinate residues in chicken tissues by TLC and measurement of fluorescence on plates, *J. Agric. Food Chem.*, 19, 55, 1971.

325. Stone, L.R., Fluorometric determination of decoquinate in chicken tissues, *J. Assoc. Off. Anal. Chem.*, 56, 71, 1973.

326. Hormazabal, V. and Yndestad, M., Rapid assay for monitoring residues of enrofloxacin in milk and meat tissues by HPLC, *J. Liq. Chromatogr. Relat. Technol.*, 17, 3775, 1994.

327. Rose, M.D., Bygrave, J., and Stubbings, G.W.F., Extension of multi-residue methodology to include the determination of quinolones in food, *Analyst*, 123, 2789, 1998.

328. Prat, M.D. et al., Liquid chromatographic separation of fluoroquinolone antibacterials used as veterinary drugs, *Chromatographia*, 52, 295, 2000.

329. Hernandez-Arteseros, J.A., Compano, R., and Prat, M.D., Analysis of flumequine and oxolinic acid in edible animal tissues by LC with fluorimetric detection, *Chromatographia*, 52, 58, 2000.

330. Duran-Meras, I. et al., Comparison of different methods for the determination of several quinolonic and cinolonic antibiotics in trout muscle tissue by HPLC with fluorescence detection, *Chromatographia*, 51, 163, 2000.

331. Yorke, J.C. and Froc, P., Quantitation of nine quinolones in chicken tissues by high-performance liquid chromatography with fluorescence detection, *J. Chromatogr. A*, 882, 63, 2000.

332. Hormazabal, V. and Yndestad, M., A simple assay for the determination of flumequine and oxolinic acid in fish muscle and skin by HPLC, *J. Liq. Chromatogr. Relat. Technol.*, 24, 109, 2001.

333. Posyniak, A., Zmudski, J., and Semeniuk, S., Effects of the matrix and sample preparation on the determination of fluoroquinolone residues in animal tissues, *J. Chromatogr. A*, 914, 89, 2001.

334. Ramos, M. et al., Simple and sensitive determination of five quinolones in food by liquid chromatography with fluorescence detection, *J. Chromatogr. B Analyt. Technol. Biomed. Life Sci.*, 789, 373, 2003.

335. Verdon, E. et al., Multiresidue method for simultaneous determination of ten quinolone antibacterial residues in multimatrix/multispecies animal tissues by liquid chromatography with fluorescence detection: single laboratory validation study, *J. Assoc. Off. Anal. Chem.*, 88, 1179, 2005.

336. Pecorelli, I. et al., Simultaneous determination of 13 quinolones from feeds using accelerated solvent extraction and liquid chromatography, *Anal. Chim. Acta*, 483, 81, 2003.

337. Schneider, M.J. and Donoghue, D.J., Multiresidue analysis of fluoroquinolone antibiotics in chicken tissue using liquid chromatography-fluorescence-multiple mass spectrometry, *J. Chromatogr. B*, 780, 83, 2002.

338. Schneider, M.J., Vazquez-Moreno, L., and Del Carmen Bermudez-Almada, M., Multiresidue determination of fluoroquinolones in shrimp by liquid chromatography-fluorescence-mass spectrometry, *J. Assoc. Off. Anal. Chem. Internat.*, 88, 1160, 2005.

339. Hermo, M.P., Barron, D., and Barbosa, J., Development of analytical methods for multiresidue determination of quinolones in pig muscle samples by liquid chromatography with ultraviolet detection, liquid chromatography-mass spectrometry and liquid chromatography-tandem mass spectrometry, *J. Chromatogr. A*, 1104, 132, 2006.

340. Schneider, M.J. et al., Simultaneous determination of fluoroquinolones and tetracyclines in chicken muscle using HPLC with fluorescence detection, *J. Chromatogr. B*, 846, 8, 2007.

341. Bailac, S. et al., Determination of quinolones in chicken tissues by liquid chromatography with ultraviolet absorbance detection, *J. Chromatogr. A*, 1129, 145, 2004.

342. Barron, D. et al., Determination of residues of enrofloxacin and its metabolite ciprofloxacin on biological materials by capillary electrophoresis, *J. Chromatogr. B*, 759, 73, 2001.

343. Jimenez-Lozano, E. et al., Effective sorbents for solid-phase extraction in the analysis of quinolones in animal tissues by capillary electrophoresis, *Electrophoresis*, 25, 65, 2004.

344. Horstkötter, C. et al., Determination of residues of enrofloxacin and its metabolite ciprofloxacin in chicken muscle by capillary electrophoresis using laser-induced fluorescence detection, *Electrophoresis*, 23, 3078, 2002.

345. Juan-Garcia, A., Font, G., and Pico, Y., Determination of quinolone residues in chicken and fish by capillary electrophoresis-mass spectrometry, *Electrophoresis*, 27, 2240, 2006.

346. Delepine, B., Hurtaud-Pessel, D., and Sanders, P., Simultaneous determination of six quinolones in pig muscle by liquid chromatography-atmospheric pressure chemical ionisation mass spectrometry, *Analyst*, 123, 2743, 1998.

347. Hatano, K., Simultaneous determination of quinolones in foods by LC/MS/MS, *J. Food Hyg. Soc. Jpn*, 45, 239, 2004.

348. Toussaint, B. et al., Determination of (fluoro)quinolone antibiotic residues in pig kidney using liquid chromatography-tandem mass spectrometry. Part I. Laboratory-validated method, *J. Chromatogr. A*, 1088, 32, 2005.

349. Toussaint, B. et al., Determination of (fluoro)quinolone antibiotic residues in pig kidney using liquid chromatography-tandem mass spectrometry. Part II: Intercomparison exercise, *J. Chromatogr. A*, 1088, 40, 2005.

350. Dufresne, G. and Fouquet, A., Multiresidue determination of quinolone and fluoroquinolone antibiotics in fish and shrimp by liquid chromatography-tandem mass spectrometry, *J. Assoc. Off. Anal. Chem. Internat.*, 90, 604, 2007.

351. Agarwal, V.K., High-performance liquid chromatographic methods for the determination of sulfonamides in tissue, milk and eggs, *J. Chromatogr. A*, 624, 411, 1992.

352. Diserens, J.-M., Renaud-Bezot, C., and Savoy-Perroud, M.-C., Simplified determination of sulfonamides residues in milk, meat and eggs, *Deutsche Lebensmittel Rundschau*, 87, 205, 1991.

353. Boison, J.O. and Keng, J.-Y., Determination of sulfamethazine in bovine and porcine tissues by HPLC, *J. Assoc. Off. Anal. Chem. Internat.*, 77, 558, 1994.

354. Rose, M.D., Farrington, W.H.H., and Shearer, G., The effect of cooking on veterinary drug residues in food: 3.sulfamethazine (sulphadimidine), *Food Addit. Contam.*, 12, 739, 1995.

355. Humayoun-Akhtar, M. et al., Extraction of incurred sulphamethazine in swine tissue by microwave assisted extraction and quantification without clean up by HPLC following derivatization with dimethylaminobenzaldehyde, *Food Addit. Contam.*, 15, 542, 1998.

356. Alfredsson, G. and Ohlsson, A., Stability of sulphonamide drugs in meat during storage, *Food Addit. Contam.*, 15, 302, 1998.

357. McGrane, M., O'Keeffe, M., and Smyth, M.R., The analysis of sulphonamide drug residues in pork muscle using automated dialysis, *Anal. Lett.*, 32, 481, 1999.

358. Li, J.S. et al., Determination of sulfonamides in swine meat by immunoaffinity chromatography, *J. Assoc. Off. Anal. Chem. Internat.*, 83, 830, 2000.

359. Furusawa, N., Determining the procedure for routine residue monitoring of sulfamethazine in edible animal tissues, *Biomed. Chromatogr.*, 15, 235, 2001.

360. Roybal, J.E. et al., Application of size-exclusion chromatography to the analysis of shrimp for sulphonamide residues, *Anal. Chim. Acta*, 483, 147, 2003.

361. Furusawa, N., A clean and rapid LC technique for sulfamethazine monitoring in pork tissues without using organic solvents, *J. Chromatogr. Sci.*, 41, 377, 2003.

362. Pecorelli, I. et al., Validation of a confirmatory method for the determination of sulphonamides in muscle according to the European Union regulation 2002/657/EC, *J. Chromatogr. A*, 1032, 23, 2004.

363. Di Sabatino, M. et al., Determination of ten sulphonamide residues in meat samples by liquid chromatography, *J. Assoc. Off. Anal. Chem. Internat.*, 90, 598, 2007.

364. Tsai, C.-E. and Kondo, F., A sensitive high-performance liquid chromatography method for detecting sulfonamide residues in swine serum and tissues after fluorescamine derivatization, *J. Liq. Chromatogr.*, 18, 965, 1995.

365. Salisbury, C.D.C., Sweet, J.C., and Munro, R., Determination of sulfonamide residues in the tissues of food animals using automated precolumn derivatization and liquid chromatography with fluorescence detection, *J. Assoc. Off. Anal. Chem. Internat.*, 87, 1264, 2004.

366. Posyniak, A., Zmudski, J., and Mitrowska, K., Dispersive solid-phase extraction for the determination of sulfonamides in chicken by muscle liquid chromatography, *J. Chromatogr. A*, 1087, 259, 2005.

367. Porter, S., Confirmation of sulphonamide residues in kidney tissue by liquid chromatography-mass spectrometry, *Analyst*, 119, 2753, 1994.

368. Boison, J.O. and Keng, L.J.-Y., Determination of sulfadimethoxine residues in animal tissues by liquid chromatography and thermospray mass spectrometry, *J. Assoc. Off. Anal. Chem. Internat.*, 78, 651, 1995.

369. Cannavan, A. et al., Determination of trimethoprim in tissues using liquid chromatography-thermospray mass spectrometry, *Analyst*, 122, 1379, 1997.

370. Van Eekhout, N. and Van Peteghem, C., The use of LC/MS/MS in the determination of residues of veterinary drugs and hormones in foodstuffs, in *Proceedings of the 8th International Congress of the European Association for Veterinary Pharmacology and Toxicology (EAVPT)*, Jerusalem, Israel, July 30–August 10, Soback, S. and McKellar, Q.A., Eds., 2000.

371. Bogialli, S. et al., A liquid chromatography-mass spectrometry assay for analyzing sulfonamide antibacterials in cattle and fish muscle tissues, *Anal. Chem.*, 75, 1798, 2003.

372. Kim, D.-H. and Lee, D.W., Comparison of separation conditions and ionization methods for the LC-MS determination of sulfonamides, *J. Chromatogr. A*, 984, 153, 2003.

373. Gentili, A. et al., Accelerated solvent extraction and confirmatory analysis of sulfonamide residues in raw meat and infant foods by liquid chromatography electrospray tandem mass spectrometry, *J. Agric. Food Chem.*, 52, 4614, 2004.

374. Kishida, K., Quantitative and confirmation of six sulphonamides in meat by liquid chromatography-mass spectrometry with photodiode array detection, *Food Control*, 18, 301, 2007.

375. Potter, R.A. et al., Simultaneous determination of 17 sulfonamides and the potentiators ormetoprim and trimethoprim in salmon muscle by liquid chromatography with tandem mass spectrometry detection, *J. Assoc. Off. Anal. Chem. Internat.*, 90, 343, 2007.

376. Juhel-Gaugain, M. et al., European proficiency testing of national reference laboratories for the confirmation of sulfonamide residues in muscle and milk, *Food Addit. Contam.*, 22, 221, 2005.

377. Stead, S.L. et al., New method for the rapid identification of tetracycline residues in foods of animal origin using Premi′test in combination with a metal ion chelation assay, *Food Addit. Contam.*, 24, 583, 2007.

378. Mc Evoy, J.D.G. et al., Origin of chlortetracycline in pig tissue, *Analyst*, 119, 2603, 1994.

379. Sokol, J. and Matisova, E., Determination of tetracycline antibiotics in animal tissues of food-producing animals by high-performance liquid chromatography using solid-phase extraction, *J. Chromatogr. A*, 669, 75, 1994.

380. Stubbings, G., Tarbin, J.A., and Shearer, G., On-line metal chelate affinity chromatography clean-up for the HPLC determination of tetracycline antibiotics in animal tissues, *J. Chromatogr. B Biomed. Appl.*, 679, 137, 1996.

381. MacNeil, J.D. et al., Chlortetracycline, oxytetracycline and tetracycline in edible animal tissues, liquid chromatographic method: collaborative study, *J. Assoc. Off. Anal. Chem. Internat.*, 79, 405, 1996.

382. Rose, M.D. et al., The effect of cooking on veterinary drug residues in food: 4. Oxytetracycline, *Food Addit. Contam.*, 13, 275, 1996.

383. Cooper, A.D., Effects of extraction and spiking procedures on the determination of incurred residues of tetracycline in cattle kidney, *Food Addit. Contam.*, 15, 645, 1998.

384. Juhel-Gaugain, M. et al., Results of a European interlaboratory study for the determination of oxytetracycline in pig muscle by HPLC, *Analyst*, 123, 2767, 1998.

385. Posyniak, A. et al., Validation study for the determination of tetracycline residues in animal tissues, *J. Assoc. Off. Anal. Chem. Internat.*, 82, 862, 1999.

386. Moats, W.A., Determination of tetracycline antibiotics in beef and pork tissues using ion-paired liquid chromatography, *J. Agric. Food Chem.*, 48, 2244, 2000.

387. Oka, H. et al., Survey of residual tetracyclines in kidneys of diseased animals in Aichi Prefecture, Japan (1985–1997), *J. Assoc. Off. Anal. Chem. Internat.*, 84, 350, 2001.

388. Hernandez, M., Borrull, F., and Calull, M., Capillary zone electrophoresis determination of oxytetracycline in pig tissue samples at maximum residue limits, *Chromatographia*, 54, 355, 2001.

389. Cinquina, A.L. et al., Validation of a HPLC method for the determination of oxytetracycline, tetracycline, chlortetracycline and doxycycline in bovine milk and muscle, *J. Chromatogr. A*, 987, 227, 2003.

390. Posyniak, A. et al., Analytical procedure for the determination of chlortetracycline and 4-epi-chlortetracycline in pig kidneys, *J. Chromatogr. A*, 1088, 169, 2005.

391. Kowalski, C., Pomorska, M., and Slavik, T., Development of HPLC with UV-VIS detection for the determination of the level of oxytetracycline in the biological matrix, *J. Liq. Chromatogr. Relat. Technol.*, 29, 2721, 2006.

392. Blanchflower, W.J., McCracken, R.J., and Rice, D.A., Determination of chlortetracycline residues in tissues using high-performance liquid chromatography with fluorescence detection, *Analyst*, 114, 421, 1989.

393. Agasoster, T., Automated determination of oxytetracycline residues in muscle, liver, milk and egg by on-line dialysis and post-column reaction detection HPLC, *Food Addit. Contam.*, 9, 615, 1992.

394. McCracken, R.J. et al., Simultaneous determination of oxytetracycline, tetracycline and chlortetracycline in animal tissues using liquid chromatography, post-column derivatization with aluminium and fluorescence detection, *Analyst*, 120, 1763, 1995.

395. Kawata, S. et al., LC determination of oxytetracycline in swine tissue, *J. Assoc. Off. Anal. Chem. Internat.*, 79, 1463, 1996.

396. McEvoy, J.D.G. et al., Production of CTC-containing porcine reference materials, *Analyst*, 123, 2535, 1998.

397. Croubels, S. et al., Liquid chromatography separation of doxycycline and 4-epidoxycycline in a tissue depletion study of doxycycline in turkeys, *J. Chromatogr. B*, 708, 145.

398. Croubels, S., De Baere, S., and De Backer, P., The proposed MRL for doxycycline: controversy about the inclusion of the 4-epimer as marker, in *Proceedings of the 8th International Congress of the European Association for Veterinary Pharmacology and Toxicology (EAVPT)*, Jerusalem, Israel, July 30–August 10, Soback, S. and McKellar, Q.A., Eds., 2000.

399. Pena, A.L., Lino, C.M., and Silveira, M.I.N., Determination of tetracycline antibiotics in salmon muscle by liquid chromatography using post-column derivatization with fluorescence detection, *J. Assoc. Off. Anal. Chem. Internat.*, 86, 925, 2003.

400. Schneider, M.J. and Chen, G., Time-resolved luminescence screening assay for tetracyclines in chicken muscle, *Anal. Lett.*, 37, 2067, 2004.

401. Rupp., H.S. and Anderson, C.R., Determination of oxytetracycline in salmon by liquid chromatography with metal-chelate fluorescence detection, *J. Assoc. Off. Anal. Chem. Internat.*, 88, 505, 2005.

402. Blanchflower, W.J. et al., Confirmatory assay for the determination of tetracycline, oxytetracycline, chlortetracycline and its isomers in muscle and kidney using liquid chromatography-mass spectrometry, *J. Chromatogr. B Biomed. Appl.*, 692, 351, 1997.

403. Cherlet, M., De Baere, S., and De Backer, P., Quantitative analysis of oxytetracycline and its 4-epimer in calf tissues by HPLC combined with positive electrospray ionization mass spectrometry, *Analyst*, 128, 871, 2003.

404. Berendsen, B.J.A. and Van Rhijn, J.A., Residue analysis of tetracyclines in poultry muscle: shortcomings revealed by a proficiency test, *Food Addit. Contam.*, 23, 1141, 2006.

405. Bogialli, S. et al., A rapid confirmatory method for analyzing tetracycline antibiotics in bovine, swine, and poultry muscle tissues: matrix solid-phase dispersion with heated water as extractant followed by liquid chromatography-tandem mass spectrometry, *J. Agric. Food Chem.*, 54, 1564, 2006.

406. Council Directive 70/524/EEC of 23rd November 1970, *Off. J. Eur. Commun.*, L270 (1970) 1.

407. Lönnroth, A., *Antimicrobial Resistance Research 1999–2002*, Revised and extended edition, European Commission-DG for Research, Directorate F: Life sciences, genomics and biotechnology for Health, Office for Official Publications of the European Communities, Luxembourg, 2003.

408. Council Directive 97/6/EC of 30th January 1997, *Off. J. Eur. Commun.*, L35 (1997) 11.

409. Council Regulation (EC) No. 2821/98 of 17th December 1998, *Off. J. Eur. Commun.*, L351 (1998) 4.

410. Council Regulation (EC) No. 2562/99 of 3rd December 1999, *Off. J. Eur. Commun.*, L310 (1999) 11.

411. Curren, M.S.S. and King, J.W., New sample preparation technique for the determination of avoparcin in pressurized hot water extracts from kidney samples, *J. Chromatogr. A*, 954, 41, 2002.

412. Nagasen, M. and Fukamachi, K., Determination of virginiamycin in swine, cattle and chicken muscles by HPLC with fluorescence detection, *Jpn Analyst*, 36, 297, 1987.
413. Moats, W.A. and Leskinen, L., Determination of virginiamycin residues in swine tissue using high performance liquid chromatography, *J. Agric. Food Chem.*, 36, 1297, 1988.
414. Saito, K. et al., Determination of virginiamycin in chicken or swine tissues by high performance liquid, *Eisei Kagaku*, 35, 63, 1989.
415. Govaerts, C. et al., Hyphenation of liquid chromatography to ion trap mass spectrometry to identify minor components in polypeptide antibiotics, *Anal. Bioanal. Chem.*, 377, 909, 2003.
416. Wan, E.C.-H. et al., Detection of residual bacitracin A, colistin A, and colistin B in milk and animal tissues by liquid chromatography tandem mass spectrometry, *Anal. Bioanal. Chem.*, 385, 181, 2006.
417. Weiss, G. and MacDonald, A., Methods for determination of ionophore-type antibiotic residues in animal tissues, *J. Assoc. Off. Anal. Chem.*, 68, 972, 1985.
418. Asukabe, H. and Harada, K.I., Chemical analysis of polyether antibiotics, in *Chemical Analysis for Antibiotics Used in Agriculture*, Oka, H. et al., Eds, AOAC International, Arlington, VA, 1995.
419. Elliott, C.T., Kennedy, D.G., and McCaughey, W.J., Methods for the detection of polyether ionophore residues in poultry, *Analyst*, 123, 45R, 1998.
420. Harris, J.A., Russell, C.A.L., and Wilkins, J.P.G., The characterisation of polyether ionophore veterinary drugs by HPLC-electrospray MS, *Analyst*, 123, 2625, 1998.
421. Davis, A.L. et al., Investigations by HPLC-electrospray mass spectrometry and NMR spectroscopy into the isomerisation of salinomycin, *Analyst*, 124, 251, 1999.
422. Lopes, N.P. et al., Fragmentation studies on lasalocid acid by accurate mass electrospray mass spectrometry, *Analyst*, 127, 1224, 2002.
423. Rosen, J., Efficient and sensitive screening and confirmation of residues of selected ionophore antibiotics in liver and eggs by liquid chromatography-electrospray mass spectrometry, *Analyst*, 126, 1990, 2002.
424. Dubois, M., Pierret, G., and Delahaut, Ph., Efficient and sensitive detection of residues of nine coccidiostats in egg and muscle by liquid chromatography-electrospray mass spectrometry, *J. Chromatogr. B*, 813, 181, 2004.
425. Mortier, L., Daeseleire, E., and Van Peteghem, C., Determination of the ionophoric coccidiostats narasin, monensin, lasalocid and salinomycin in eggs by liquid chromatography/tandem mass spectrometry, *Rapid Commun. Mass Spectrom.*, 19, 533, 2005.
426. Peippo, P. et al., Rapid time-resolved fluoroimmunoassay for the screening of narasin and salinomycin residues in poultry and eggs, *J. Agric. Food Chem.*, 52, 1828, 2004.
427. Moats, W.A. and Leskinen, L., Determination of novobiocin residues in milk, blood, and tissues by liquid chromatography, *J. Assoc. Off. Anal. Chem.*, 71, 776, 1988.
428. Heller, D.N. and Nochetto, C.R., Development of multiclass methods for drug residues in eggs: silica SPE cleanup and LC-MS/MS analysis of ionophore and macrolide residues, *J. Agric. Food Chem.*, 52, 6848, 2004.
429. Reeves, V.B., Liquid chromatographic procedure for the determination of novobiocin residues in bovine milk: interlaboratory study, *J. Assoc. Off. Anal. Chem. Internat.*, 78, 55, 2005.
430. Beechinor, J.G. and Bloomfield, J., Effect of cooking on residues of tiamulin in beef, in *Proceedings of the 8th International Congress of the European Association for Veterinary Pharmacology and Toxicology (EAVPT)*, Jerusalem, Israel, July 30–August 10, Soback, S. and McKellar, Q.A., Eds., 2000.
431. Chen, H.-C. et al., Determination of tiamulin residue in pork and chicken by solid phase extraction and HPLC, *J. Food Drug Anal.*, 14, 80, 2006.
432. Nozal, M.J. et al., Trace analysis of tiamulin in honey by liquid chromatography–diode array–electrospray ionization mass spectrometry detection, *J. Chromatogr. A*, 1116, 102, 2006.
433. Council Regulation 2788/98/EC of 22nd December 1998. *Off. J. Eur. Commun.*, L347 (1998) 31.
434. Van Ginkel, J.A. et al., The detection and identification of quinoxaline-2-carboxylic acid, a major metabolite of carbadox, in swine tissue, in *Proceedings Euroresidue Conference on Residues of Veterinary Drugs in Food*, Noordwijkerhout, Netherlands, Haagsma, N., Ruiter, A., and Czedik-Eysenberg, P.B., Eds., Faculty of Medicine, Utrecht, Netherlands, 1990, 189.

435. Rutalj, M. et al., Quinoxaline-2-carboxylic acid (QCA) in swine liver and muscle, *Food Addit. Contam.*, 13, 879, 1996.

436. Hutchinson, M.J. et al., Development and validation of an improved method for confirmation of the carbadox metabolite, quinoxaline-2-carboxylic acid, in porcine liver using LC-electrospray MS-MS according to revised EU criteria for veterinary drug residue analysis, *Analyst*, 127, 342, 2002.

437. Hurtaud-Pessel, D. et al., An LC/MS-MS method for the determination of QCA and MQCA, the metabolites of carbadox and olaquindox in porcine liver and muscle, communication presented at the *5th International Symposium on Hormone and Veterinary Drug Residue Analysis*, Antwerp, Belgium, May 16–19, 2006.

438. Hutchinson, M.J., Young, P.B., and Kennedy, D.G., Confirmatory method for the analysis of carbadox and olaquindox in porcine feedingstuffs using liquid chromatography–electrospray mass spectrometry, *Food Addit. Contam.*, 22, 113, 2005.

439. Hutchinson, M.J., Young, P.B., and Kennedy, D.G., Confirmation of carbadox and olaquindox metabolites in porcine liver using liquid chromatography–electrospray-tandem mass spectrometry, *J. Chromatogr. B*, 816, 15, 2005.

440. Sin, D.W.M. et al., Determination of quinoxaline-2-carboxylic acid, the major metabolite of carbadox, in porcine liver by isotope dilution gas chromatography–electron capture negative ionization mass spectrometry, *Anal. Chem. Acta*, 508, 147, 2004.

441. Wu, Y. et al., Development of a high-performance liquid chromatography method for the simultaneous quantification of quinoxaline-2-carboxylic acid and methyl-3-quinoxaline-2-carboxylic acid in animal tissues, *J. Chromatogr. A*, 1146, 1, 2007.

442. Zhang, Y. et al., Effects of cooking and storage on residues of cyadox in chicken muscle, *J. Agric. Food Chem.*, 53, 9737, 2005.

Index

N